THE FLORA OF NEW BRUNSWICK

A Manual for the Identification of all Vascular Plants, Including Ferns and Fern Allies and Flowering Plants Growing Without Cultivation in New Brunswick, Canada

Harold R. Hinds

Curator of the Connell Memorial Herbarium
and Senior Instructor
University of New Brunswick
Fredericton, N.B.

Available by mail from the author,
Biology Department
University of New Brunswick
Bag Service Number 45111
Fredericton, N.B. E3B 6E1

CANADIAN CATALOGUING IN PUBLICATION DATA

Hinds, Harold R.
Flora of New Brunswick

Bibliography: p.
Includes index.
ISBN 0-9692456-0-2

1. Botany - New Brunswick. 2. Plants - Identification.
I. Title.

QK203.N4H56 1986 583 C86-093999-5

Design and Photo Offset
by Graphic Services
University of New Brunswick
Fredericton, N.B.

CONTENTS

ABBREVIATIONS AND SIGNS

+/-	More or less; approximately
CAN	National Herbarium, Ottawa, Ont., Canada
c	Centimeter
f.	forma
GH	Gray Herbarium, Cambridge, MA, USA
Is.	Island
m	Meter
mm	Millimeters
mtns	Mountains
NBM	New Brunswick Museum, Saint John, N.B., Canada
nm	Nothomorph (see glossary)
NSAC	Nova Scotia Agricultural College, Truro, N.S., Canada
QK	Fowler Herbarium, Queens University, Kingston, Ont., Canada
R.	River
ssp.	Subspecies
var.	Varieties

FORWARD

The publication of the 'Flora of New Brunswick' marks an important event for botany in this Province. Until now we have had to depend on 'floras' of adjacent areas, such as Smith and Roland's 'Flora of Nova Scotia', and on wider regional floras such as 'Gray's Manual' or the 'New Britton and Brown'.

An identification manual restricted to plants of the New Brunswick area and unencumbered by references to northeastern North American plants of other habitats and areas will make identification of the species very much simpler. It will be of use not only to botanists, amateur and professional, but to agriculturalists, foresters, wildlife biologists, ecologists and naturalists.

Many years ago Dr. Patricia Roberts-Pichette began work on a New Brunswick flora. It has, however, fallen to Hal Hinds to bring this work to reality by his hard work, his dedication and perseverance. In the process he and his helpers have greatly extended our knowledge of where our vascular plants occur across the Province by the extensive collections he has made, studied and deposited in the expanding Connell Memorial Herbarium of the Biology Department of the University of New Brunswick in Fredericton.

It is fitting that this project has come to completion as part of our Bicentennial celebrations. I am sure that those early New Brunswick naturalists and scientists, Robb, Bailey, and Fowler would delight to see a New Brunswick Flora. We who have the opportunity to make use of it should be grateful for its publication. I am sure that it will help to stimulate interest in our native flora; it should also provide an incentive to further botanical exploration of the Province and a focus for study and revision as more becomes known. In short, we now have a baseline from which we can move forward with enthusiasm.

April, 1986

A.R.A. Taylor
Professor and Acting Chairman
Department of Biology
University of New Brunswick
Fredericton, N.B.

INTRODUCTION

The purpose of this book is to provide a means of identifying all the vascular plants reported as growing without cultivation in the Province of New Brunswick. There is also an attempt to update the nomenclature of scientific names based on recent monographs and the synonomies of Kartesz and Kartesz and the U.S. Department of Agriculture. The former publication has been most heavily relied upon for valid names. Common names have been taken principally from Canadian Flora, Terminology Bulletin No. 156., 1974. The order of the families departs only slightly from standard floras for our region which basically follow the Engler and Prantl system of classification beginning with the monocots and ending with the Asteraceae of the dicots. Within each family the genera and species are arranged alphabetically.

HOW TO USE THE BOOK

With this manual you should be able to identify any vascular plant growing wild in the Province, e.g. trees and shrubs, wildflowers, grasses and sedges, ferns, horsetails and clubmosses. It does not include the mosses, liverworts, lichens, algae and fungi.

To identify a plant specimen, first trace it through the Key to Families. Always read completely both choices in each couplet before deciding which description fits best. Continue as the key indicates until you arrive at a family. In some cases you may arrive directly at the identity of your specimen, but in most cases you will be directed to family keys elsewhere in the book. Follow a similar procedure to key the specimen to a genus within the family and finally to a species within that genus. Information for each species includes the scientific name, author or authors of the name, figure and map page numbers, common French and English names, frequency and distribution in the Province, habitat, general world-wide range, synonyms, subspecific taxa, e.g. forms and varieties, and special notes concerning rarity, hybridization and references.

The keys are so constructed that they can be read backwards, thus providing a brief descriptive sketch of each species. The illustrations and maps should further help the identification process. More complete descriptions can be found in Gray's Manual of Botany by M.L. Fernald, Flora of Canada by H. Scoggan, Manual of Vascular Plants by Gleason and Cronquist and Flora Laurentienne by Fr. Marie Victorin.

Distribution maps are based on records of New Brunswick plants housed in several herbaria and from published records. Where published records seemed especially atypical, they were carefully checked for positive identification. Those species that are widely distributed throughout the Province have not been mapped except in a few cases to illustrate the degree of collecting coverage in the Province. Also those plants that have been collected only a very few times are not mapped, but their distribution is described under the heading, Frequency.

There exists an extensive bibliography concerning the flora of New Brunswick, but only those references mentioned in the text have been provided under References Cited.

PHYSICAL BACKGROUND

New Brunswick lies on the Atlantic Coast of North America between 47 37' and 48 30' North Latitude and 63 46' and 69 30' West Longitude. Approximately square in outline, it measures about 270 km east-west and 320 km north-south with a total area of about 73,000 square km.

New Brunswick is bounded in general by the Bay of Fundy facing Nova Scotia on the south, the State of Maine (Washington and Aroostook Counties) on the West, the Gulf of St. Lawrence on the east and the Province of Quebec on the north. In the extreme southeast the Province joins Nova Scotia by means of the 30 km wide Isthmus of Chignecto. The Province of Prince Edward Island lies a few km off the east coast and the Gaspé Peninsula lies across the Baie of Chaleurs from the extreme northeast of the Province. To the southwest lie Machias Seal Island and the Grand Manan Archipelago.

The range in elevation is from sea level to 700 m at the top of Mt. Carleton in the northcentral uplands. About 85% of the Province is covered mainly by the heterogeneous Acadian Forest Region of red spruce, balsam fir, white and yellow birch and a complex mixture of many other tree species. There are however, a considerable number of breaks in this forest cover which allow a great variety of different plant species to survive. These breaks are provided mostly by the approximately 30,000 km of rivers that dissect the Province and by the +/- 2,000 lakes and ponds, rocky cliffs and talus slopes, marshlands and boglands, dunes and meadows and a seacoast of approximately 1,000 km. Man has also maintained a considerable area in an open condition, e.g. roadsides, railways, lawns, agricultural areas and similar marginal areas where the forest is not allowed to regenerate.

The drainage systems of the St. John, Miramichi, St. Croix and Restigouche Rivers are a dominant feature of the landscape and especially the St. John and Miramichi Rivers that drain approximately two-thirds of the land mass.

PHYSIOGRAPHY, GEOLOGY AND SOILS

New Brunswick is considered as part of the northern terminus of the Appalachian Mountain System which extends from southern U.S. to the Gaspé Peninsula and consists mainly of complex folded, more or less flat-topped, rolling uplands running approximately northeast-southwest. These flat summits are believed to have once formed a continuous surface but are now deeply dissected by the many rivers and scoured by glaciers.

The Province has been divided into three physiographic regions (Clayton et al. 1977). The Chaleur Uplands occupy much of the northwestern quarter of the Province. They are drained by both the St. John and Restigouche River drainage systems and rise to about 300 m before sloping southeast and merging with the next region.

The Central Highlands are divided by the St. John River into three sections; the Miramichi Highlands towards the northeast, the St. Croix Highlands towards the west and the Caledonian Uplands to the southeast of the river. In the northeastern and western sections of the Highlands large granite batholiths rise above slates, conglomerates and sandstones to elevations of 700 m on Mt. Carleton in the northern section and to about 300 m along the western shore of the Bay of Fundy.

The Eastern Lowlands or Maritime Plain occupy a basin east of the Central Highlands in a roughly triangular shape covering almost 1/2 the area of the Province. This large depositional basin, averaging less than 100 m in elevation, is largely swampy with glacially impeded drainage. Large peat deposits occupy considerable areas of the northeast and a sandy drift, probably of sedimentary origin covers large parts of the area.

GLACIAL HISTORY

With the possible exception of some of the Central Highlands, the Province was scoured by Pleistocene glaciations between 110,000 and 10,000 years B.P. which left moraines, kanes, eskers and drumlins as prominent surficial features. The glaciers significantly altered the surface of the Province denuding much of its soil mantle and depositing unconsolidated sand and gravel in many areas as the glaciers retreated.

The Provinces soils are mainly podzolic, characterized by a heavily leached, acidic surface horizon with formation in cool-temperate to temperate humid climates under coniferous or mixed forests. There is considerable variation, however, due to different parent materials, precipitation, glacial history and topography. Fine grained alluvial soils occur extensively along the major rivers.

CLIMATE

The division of the Province into 11 climatic regions (Van Groenewoud. 1984) is the latest analysis of the Province's climatic variability. However, Putnam's (1940) analysis provides a more general if simplistic view. He divided New Brunswick into three climatic regions; northern, southern and Bay of Fundy.

In general the northern region has short, warm summers and long, cold snowy winters. The mean temperatures for January and July are -12°C and 18°C respectively. The growing season is approximately 90 days long. The mean annual precipitation is about 90 cm even though the mean annual snowfall is 270 cm.

The southern region has less severe winters than the northern region with an average winter temperature of -10°C, and the growing season is longer with 113 frost free days. Precipitation is greatest in the fall and winter.

The Bay of Fundy region has a cool summer with a mean temperature of 15°C. The mean winter temperature is about -5°C. The growing season is similar to that of the southern region although the frost-free season is much longer with 160 days. Precipitation is highest in this region with 100 to 115 cm and frequent foggy conditions.

HISTORY OF PLANT COLLECTIONS*

Scholars were collecting plants and assigning them to categories three hundred and eighty years ago when explorers made the first detailed observations of the territory now known as New Brunswick. They are still collecting and analyzing. But the process over the centuries has not been merely one of adding increments from generation to generation to a growing stockpile of botanical information. Accumulation is only a small part of the story. The science of botany has undergone revolutionary changes. It has emerged from the medieval ideas of the European herbalists, through the rise of systematic botany and of biogeography, to our modern concepts of cytogenetics and biosystematics.

*contributed by Dr. C. Mary Young

In 1604 Champlain, while on Dochet's Island in the St. Croix river, attempted to discover the identity of the tree known to provide a cure for scurvy, while in 1606 Lescarbot described how a party including Champlain and the Parisian apothecary, Louis Hebert, investigated the lower reaches of the St. John River valley where they found oaks, beeches, butternuts, wild grapes, cedars and various herbs. An early French settler, Nicholas Denys, a fish merchant by trade, described the trees and plants of many coastal areas of Acadia, including the mouths of the Richibucto and Miramichi rivers in 1672. His observations of the uses of plants and plant products by the Indians are of particular interest.

Seeds and roots collected by these early visitors to Acadia were taken back to Paris where they were grown in the Jardin Royal and were investigated for their medicinal properties or "virtues". The first work of particular botanical significance, describing approximately 43 Canadian plants and illustrating 38, was the Canadensium Plantarum Historia (1635) by Jacques Philippe Cornuti. Cornuti never visited Canada but obtained the plants from Jean and Vespasian Robin, curators of the Jardin Royal in Paris. Associated with this garden in the late seventeenth century was the renowned French botanist and academician, Tournefort. Tournefort frequently commissioned surgeon-botanists to collect plants abroad, including the Sieur de Dièreville who visited the lower St. John river valley in 1699 and 1700. Among the specimens he took back to Paris was the turtlehead, **Chelone glabra** and the bush honeysuckle, **Diervilla lonicera**.

Between the period of French interest in the plants of Acadia and the revival of interest during the British colonial period in the nineteenth century, new systems of plant classification were devised and the binomial system of naming, the starting point of modern botanical nomenclature, advanced by Linnaeus in his Species Plantarum (1753), was universally adopted.

In the early years of the British colonial era interest in New Brunswick plants was largely limited to the economic potential of the forest trees but later some army and navy surveyors, notably Sir James Alexander in the 1840's made lists of plants they came across in their work. In his magisterial Flora Borealis Americana 1833-40, Sir William Hooker included references to a few New Brunswick plants obtained by E.N. Kendall, who had been on the 1825-27 Franklin expedition and who later resided at Fredericton for a time.

In the mid-nineteenth century intense interest in North American plants shifted from Europe to the American arena. The American, Asa Gray, after examining the type specimens of American plants in European herbaria, was actively encouraged by Sir William Hooker to become an authority on North American flora. Two works, The Flora of North America by John Torrey and Asa Gray 1838-40 and A Manual of the Botany of the Northern United States by Asa Gray 1848, marked the coming of age for North American botany. These books acted as catalysts for North American botanists and the eighth edition of Gray's Manual, greatly revised and expanded by M.J. Fernald, is still in use today.

A systematic collection of New Brunswick plants was initiated in 1838 by James Robb, the first professor of Natural Science and Chemistry at King's College, Fredericton, later the University of New Brunswick. Trained in Scotland and Paris where he was exposed to the teaching of the botanist Adrien de Jussieu, Robb collected many plants from around Fredericton and Carleton (Saint John) and also brought back specimens from an extensive field trip up the St. John, Tobique, and Grande Rivers and down the Restigouche River system to the Bay of Chaleur. The collection at the

University of New Brunswick continued to grow from 1861 to 1907 under the care of Loring Woart Bailey, James Robb's successor. In 1863 Bailey undertook a geological expedition up the Tobique River and down the Nepisiguit River, bringing back plants as well as geological information.

An extensive collection of New Brunswick plants was made by a contemporary of Dr. Bailey's, the Presbyterian minister, James Fowler, who lived at Bass River near Richibucto. His collections include many plants from Bass River, Kouchibouguac, the St. John River valley and the area around St. Andrews. Between 1878 and 1880 Fowler published a preliminary list of New Brunwick plants and in 1885 produced an extensive list. John Macoun of the Geological and Natural History Survey of Canada used Fowler's lists, incorporating the information into his Catalogue of Canadian Plants (1883-1902). In 1880 Fowler became Professor of Natural History at Queen's University, Kingston and took a large collection of New Brunswick plants to that university with him. Today there are a number of duplicate specimens from Fowler's collection in the Saint John Museum and at the University of New Brunswick.

In the latter part of the nineteenth century members of the Geological Survey of Canada collected information on plants they discovered on their surveys. D'Urban noted plants found along the Quebec side of the Patapedia and Restigouche Rivers in his report of 1865. In New Brunswick, Robert Chalmers, listed plants of western New Brunswick in his report of 1882-4 in which he included the Furbish's Lousewort, **Pedicularis furbishiae** and the snakeroot, **Polygala senega**. His reports of 1886 and 1894 demonstrate his interest in the New Brunswick forests.

Loring Bailey, the Reverend James Fowler and Robert Chalmers were all associated with the Natural History Society of New Brunswick, founded at Saint John in 1862, which had a remarkably active and talented group of members. Plants for a society herbarium were received from G.U. Hay and R. Chalmers and from J. Moser of York County, A.I. Trueman of Westmorland, and James Vroom of Charlotte County. One thousand specimens were contributed by James Fowler in 1895. One of the most active members of this society was G.U. Hay, who, accompanied by botanically interested friends, explored and collected along the major river systems of the province. The accounts of his exploits and the plants he discovered along the St. Francis, St. John, Restigouche, Nepisiguit and Tobique Rivers were published in the Bulletin of the New Brunswick Natural History Society. His large personal collection of New Brunswick plants now enriches the herbarium of Acadia University, N.S.

A very different approach to the botany of the region was displayed by W.F. Ganong. Educated in New Brunswick, the United States and Germany, Ganong wrote a classic paper on the ecology of the salt marshes at the head of the Bay of Fundy (1903).

During the same period, Philip Cox, an educator and active member of the Miramichi Natural History Association, was responsible for collecting many plants of that region between 1897 and 1899. He produced a preliminary catalogue of the herbarium at the Society's museum at Chatham in 1905. Cox followed Loring Bailey as professor of natural history and geology at the University of New Brunswick and added some plants to the university herbarium collection.

The early twentieth century was marked by an influx of trained botanists from outside the province investigating the local flora. A.B. Klugh, of Queen's University, for instance, listed plants of southern New Brunswick and of the St. Croix region. The most signifiant contributions to our knowledge of New Brunswick plants in this period, however, were made by American botanists, trained at Harvard University and who followed in the footsteps of Asa Gray in their interests in the geographical distribution of plant species. They returned to Harvard with many specimens thus swelling the gray Herbarium collection with New Brunswick plants.

Among the Americans who visited the province, the work of M.L. Fernald and his colleagues and students is of particular significance. Fernald made collections around the St. Croix, and Saint John regions, the Aroostook River and Bathurst and published his results in the journal of the New England Botanical Club, Rhodora. His discovery of relict species and plants with a discontinuous range led him to propose theories accounting for the isolation of certain plant species in this region. Among Fernald's students who made further investigations in New Brunswick, S.F. Blake worked in the Bathurst area in 1913, F. Tracey Hubbard around Shediac in 1914, and Norman C. Fassett 1927-28 made an extensive survey of the estuaries of the rivers which entered the Gulf of St. Lawrence. He too, noted the discontinuous distribution of many estuarine species. Duplicate specimens of his collection are housed at Acadia University.

Charles Weatherby, also of Harvard, assisted by his wife, Una, and by John Adams, studied the flora of Grand Manan between 1926 and 1944. More recently, 1960 to 1973 plants of the islands of the Grand Manan Archipelago and the Wolf Islands, have been investigated by other Americans, A.R. Hodgden and Radcliffe B. Pike from New Hampshire and of Kent Island by David Potter of Clark University.

Although Canada had no similarly trained corps of botanists in the early part of the twentieth century, the nucleus of a national collection had been brought together by members of the geological survey: this collection was later transferred to the National Museum. While John and James Macoun appear to have added little from New Brunswick to this collection, the chief botanist between 1921 and 1933, M.O. Malte, spent the summers of 1926, 1927 and 1929 in New Brunswick. He investigated the plants of the Campbellton region and with a student assistant, W.R. Watson made collections in the St. Leonard and Restigouche area and around St. Andrews returning to Ottawa with over 1400 species from New Brunswick.

The other national collection, that of the Department of Agriculture, started at the experimental farm at Ottawa by James Fletcher shortly after 1887, concentrated on plants of economic importance to farmers, including weeds, grasses, fodder crops and poisonous plants. Fletcher made plant collections at Dalhousie and at Youghall beach, Bathurst, in the early years of this century.

Both the national collections and those in the province suffered from neglect through the 1930's due, no doubt, to the economic crises of the time, but also due to the perceived lack of usefulness of plant collections in the biological community which became for a time so enamoured with experimental work that some basic aspects of the discipline were neglected.

The war years brought immediate practical problems but also a change of attitude towards plant collecting. R.P. Gorham of the Department of Agriculture, Fredericton, made a survey of the distribution of wild plum, **Prunus nigra** and buckthorn, **Rhamnus** spp., known to be overwintering sites of insect vectors of virus diseases of potatoes, while a national survey of the distribution of milkweeds, a possible source of a substitute for kapok and latex for rubber production was instituted by the Ottawa Experimental Farm. Cytogenetic and hybridization experiments at Ottawa with milkweeds and with the imported Russian dandelion, **Taraxacum kok-saghyz** were the initial steps towards the modern work now carried on at Ottawa on biosystematics.

W.G. Dore, H.A. Senn, and E. Gorham collected plants for the Department of Agriculture (Ottawa) in 1945 from Shippegan to Fredericton and Sackville. The Dominion

Forest Service also started a programe of plant collecting in the national parks and forest experimental stations which continued for many years. In New Brunswick G.C. Cunningham of the Acadia Forest Station was particularly active. Specimens from this collection are now integrated into the herbarium collection at the University of New Brunswick.

In the post war period at the National Museum, plants collected by M.O. Malte in 1929 were finally named by H.J. Scoggan in 1950. Scoggan, himself, came to the province during the summers of 1954-5 taking 1,461 specimens back to Ottawa with him. After planning originally to produce a Maritime flora, Scoggan wrote a four-volume flora of the whole of Canada published by the National Museum in 1978. J.A. Forsythe, also of the National Museum, in conjunction with D.S. Christie of the New Brunswick Museum, made plant collections in the somewhat neglected areas of central and northern New Brunswick and Albert County in 1978 and 1979. An extensive survey of plants of Kouchibouguac National Park was undertaken by Derek Munro of Agriculture Canada in 1977. A similar study in conjunction with Parks Canada was made in 1985 of the St. Croix watershed in New Brunswick by H.R. Hinds and M. Lulham.

At the provincial level the New Brunswick Museum plant collection which grew out of the collection of the New Brunswick Natural History Society, was supplemented by the many specimens collected from all parts of the Province by W. Austin Squires and David S. Christie, the natural history curators between 1940 and 1980.

The collection at the University of New Brunswick, rescued from oblivion by A.R.A. Taylor in the 1940's, was augmented by E.O. Hagmeier and in the 1960's coinciding with the wave of public interest in ecology, was revitalized by Patricia Roberts-Pichette and student assistants. Her collections from many parts of the Province were a major contribution to the herbarium. The collection of beautifully mounted plants from Carleton County of Dr. Katherine Connell was a welcome gift to the University herbarium and the herbarium was renamed in her honour. During the last fifteen years the Connell Memorial Herbarium has been under the able care of H.R. Hinds.

ACKNOWLEDGEMENTS

There have been many people who have helped to guide this volume along the way to publication. I am especially grateful to Dr. C. Mary Young for her many helpful suggestions, contributions, and assistance over the past few years. Her Furbish Lousewort and Marsh Violet drawings decorate the front and her glossary drawings, and those of Carol Bayley appear in the back of the book.

I would also like to thank Dr. Pierre Taschereau for his contribution on the genus **Atriplex**, and Dr. A.E. Roland for his helpful suggestions and permission to use several illustrations from the Flora of Nova Scotia.

Marilyn Bagley, Barbara Bell, Brian Birch, Leo Dionne, Dorothy Farmer, Marc Lulham, Christopher Pearce and Ross Wein have all been enormously helpful with various aspects of background and manuscript preparation.

The N.B. Department of Tourism deserves a note of thanks for their support by way of Provincial Park camping facilities during the several years of field work.

Financial help with manuscript preparation was provided by the N.B. Bicentennial Commission.

I gratefully acknowledge the numerous aspects of support provided by the University of New Brunswick and especially my colleagues of the Biology Department, Susan Montague of the Department of Public Relations and Development, Vaughn Bosma, William Ball and Martin Beaupre of Graphic Services and Julie Montgomery of the Science Library.

To Irene Johnston I owe a tremendous debt of gratitude for her persistance and excellence as principal manuscript typist along with the assistance of Cathy Dorcas.

The students of my taxonomy courses over the last few years have been used as guinea pigs to test many of the identification keys used in this work. To all of them my sincere thanks.

And to my many naturalist friends, thank you for your patience.

KEY TO FAMILIES

1a Plants reproducing by means of microscopic spores; producing neither flowers nor seeds. **PTERIDOPHYTA KEY A**, p. xiii

1b Plants normally reproducing by means of flowers and seeds. 2

2a(1b) Plants parasitic on branches of conifers (usually **Picea mariana**); leaves scale-like; stems to 2 cm long.
(**Arceuthobium pusillum**) **LORANTHACEAE**, p. 170

2b(1b) Plants not parasitic on conifers; leaves various; stems usually longer. 3

3a(2b) Plants very small, thalloid, without distinct stem and leaves; roots wanting or a few from lower surface; flowers inconspicuous; aquatics.
LEMNACEAE, p. 124

3b(2b) Plants otherwise. 4

4a(3b) Trees, shrubs or woody vines whose stems persist above ground throughout the winter. **KEY B**, p. xiv

4b(3b) Herbaceous plants which produce no woody tissues above the ground. 5

5a(4b) Leaves usually parallel-veined; parts of flower usually in threes or sixes; seed with a single cotyledon; vascular bundles scattered in stem.
CLASS LILIATAE (Moncots) **KEY C**, p. xvii

5b(4b) Leaves usually net-veined; parts of flower usually in fours or fives; seed with two cotyledons; vascular bundles arranged in a ring surrounding a central pith or cavity. **CLASS MAGNOLIATAE** (Dicots) 6

6a(5b) Stems and leaves normally under or floating in or on water. **KEY D**, p. **xix**

6b(5b) Stems and leaves not on or under water; base of stem may occur under water. 7

7a(6b) Plants without green leaves, either parasitic or saprophytic or without leaves at flowering time. **KEY E**, p. xxi

7b(6b) Plants with green leaves or other chlorophyllous green tissue. 8

8a(7b) Plants always without flowers or fruit when in leaf. 9

8b(7b) Plants with flowers and/or fruit present when in leaf. 10

9a(8a) Leaves cordate, angled or deeply palmately lobed, white tomentose beneath.
(**Petasites**, **Tussilago**) **ASTERACEAE**, p. 377

9b(8a) Leaves entire, cordate-ovate, to 6 dm broad, unpleasantly-scented when crushed. (**Symphoricarpus foetidus**) **ARACEAE** (actually a monocotyledon with confusing net-veined leaves), p. 123

10a(8b) Flowers wholly or partly unisexual. **KEY F**, p. xxi

10b(8b) Flowers usually perfect. 11

11a(10b) Plants fitting the complete description of one of the following common families. (A-I)

A. Stipular leaf bases or base of short petiole usually completely sheathing stem (ocrea); flowers relatively small, often aggregated; sepals 5-6, often petaloid; ovary single, superior; fruit achenous. **POLYGONACEAE**, p. 271

B. Flowers +/- irregular (nearly regular in **Mimulus**, **Veronica** and **Verbascum**), sympetalous, synsepalous; ovary 1, superior; fruit capsular, 2-celled, many seeded.
SCROPHULARIACEAE, p. 347

C. Leaves opposite; stem 4-angled; flowers irregular (nearly regular in **Lycopus** and **Mentha**), sympetalous, synsepalous; ovary deeply 4-parted, superior, surrounded by style base; fruit of 4 nutlets; plants often aromatic. **LAMIACEAE**, p. 337

D. Flowers borne in simple or compound umbels, regular, epigynous; leaves often compound or deeply dissected; fruit of 2 one-seeded mericarps. **APIACEAE**, p. 303

E. Flowers individually small, few to many in dense heads, each head surrounded by 4 to many bracts; ovary inferior; fruit achenous; corolla sympetalous; calyx wanting or often replaced by a ring of bristles (pappus) or scales; stamens 5, free, their anthers fused to form a tube. **ASTERACAEAE**, p. 377

F. Flowers regular, hypogynous; sepals 4; petals usually 4 and clawed at base; stamens 6, usually 2 short, 4 long; ovary normally 2-celled; fruit usually capsular with several seeds, mostly dehiscent by 2 valves exposing their thin, persistent partition (silicle or silique). **BRASSICACEAE**, p. 211

G. Plants commonly with fruit a 1-celled, several seeded pod (legume) splitting along both sides, occasionally divided into 1-seeded joints (**Desmodium**) or sometimes 1-few seeded and indehiscent; ovary superior; flowers irregular, 5-merous with large upper median petal (the standard), 2 lateral petals (the wings) and 2 keel petals usually partly fused and enclosing the 10 stamens and single pistil; leaves alternate, usually compound. **FABACEAE**, p. 350

H. Flowers epigynous, perfect, regular, mostly 4-merous (2 in **Circaea**); floral tube often prolonged beyond tip of ovary; fruit a many seeded capsule or small, indehiscent and bur-like. (**Circaea**) **ONAGRACEAE**, p. 295

I. Flowers usually perfect, mostly hypogynous, usually 5-merous, regular; sepals free or connate; petals distinct, sometimes wanting, white, pink or reddish, never yellow in our species; stamens usually twice as many as petals; ovary mostly 1-celled, with free-central or basal placentation; fruit a capsule or one-seeded utricle; nodes often swollen. **CARYOPHYLLACEAE**, p. 189

11b(10b) Plants not fitting the complete description of one of the above families. 12

12a(11b) Calyx present; corolla wanting. **KEY G**, p. xxiii
12b(11b) Calyx and corolla both present. 13
13a(12b) Petals separate. 14
13b(12b) Petals united. 15
14a(13a) Flowers irregular. **KEY H**, p. xiii
14b(13a) Flowers regular. 16
15a(13b) Flowers not in involucrate heads. 17
15b(13b) Flowers in involucrate heads. **KEY I**, p. xxiv
16a(14b) Pistil 1. 18
16b(14b) Pistil more than 1. **KEY J**, p. xxiv
17a(15a) Flowers irregular. **KEY K**, p. xxiv
17b(15a) Flowers regular. 19

18a(16a) Flowers irregular. **KEY L**, p. xxv

18b(16a) Ovary inferior. **KEY M**, p. xxv

19a(16b) Ovary superior. **KEY N**, p. xxvi

19b(17b) Ovary inferior. **KEY O**, p. xxvi

KEY A

1a Stems jointed and grooved; leaves whorled, scale-like, forming a sheath at each joint. **EQUISETACEAE**, p. 1

1b Stems not jointed and grooved (not as above). 2

2a(1b) Stems usually concealed, corm-like, at base of long linear awl-like leaves; sporangia on inner face of leaf bases; plants of shallow water or muddy shores. **ISOETACEAE**, p. 7

2b(1b) Stems erect or prostrate with scattered leaves; sporangia not as above. 3

3a(2b) Leaves scale-like, numerous along stems; sporangia in axils of leaves or bracts in cones (strobili). 4

3b(2b) Leaves not scale-like; growing from underground rootstock or on short prostrate or erect stems (ferns). 5

4a(3a) Plants moss-like, less than 5 cm long; spores of 2 sizes. **SELAGINELLACEAE**, p. 6

4b(3a) Plants much larger, erect or trailing; spores of 1 size. **LYCOPODIACEAE**, p. 3

5a(3b) Sterile leaf blade not divided, ovate-lanceolate; fertile portion erect, spike-like. (***Ophioglossum***) **OPHIOGLOSSACEAE**, p. 9

5b(3b) Sterile leaf blade variously divided or lobed. 6

6a(5b) Fertile frond or parts of the frond differing markedly from the sterile parts; sporangia not on lower surface of green fronds. 7

6b(5b) Fertile fronds similar to sterile ones; sporangia on lower surface of green fronds. **POLYPODIACEAE** , p. 12

7a(6a) Fronds 5-20 dm high, clumped from apex of conspicuous fibrous rooted rootstocks. 8

7b(6a) Fronds up to 5 dm high, usually single from inconspicuous fleshy roots; fertile portion of the frond forming a spike or panicle, erect or nearly so, and apparently subtended by the sterile portion. (***Botrychium***) **OPHIOGLOSSACEAE**, p. 9

8a(7a) Fertile portion of the frond either terminating the frond or borne near the middle of the frond with sterile pinnae above and below. **OSMUNDACEAE**, p. 11

8b(7a) Fertile and sterile fronds separate. 9

9a(8b) Sterile frond simple, pinnately lobed. (***Onoclea***) **POLYPODIACEAE**, p. 12

9b(8b) Sterile frond pinnately compound. 10

10a(9b) Fertile frond about as tall as sterile. (***Osmunda cinnamomea***) **OSMUNDACEAE**, p. 11

10b(9b) Fertile frond much shorter than sterile. (***Matteuccia struthiopteris***) **POLYPODIACEAE**, p. 12

++

KEY B

1a Trees or shrubs with narrow, needle-like or scale-like, mostly evergreen leaves; true flowers absent; ovules borne on surface of scales or embedded in fleshy disk, never enclosed in an ovary. **PINOPHYTA (GYMNOSPERMAE)** 2

1b Trees, shrubs or vines of various aspect, usually with broad, flat, deciduous leaves; true flowers produced, sometimes greatly reduced; ovules borne within a closed ovary. **MAGNOLIOPHYTA (ANGIOSPERMAE)** 3

2a(1a) Seeds borne in cones. **PINACEAE**, p. 24

2b(1a) Seeds not borne in cones, partially covered with a fleshy red covering; leaves alternate (spirally arranged), flat, sharp, narrowed to a subpetiolar base. (**Taxus canadensis**) **TAXACEAE**, p. 23

3a(1b) Leaves and leaf-scars opposite or whorled. 4

3b(1b) Leaves and leaf-scars alternate on the stem. 5

4a(3a) Leaves unexpanded at flowering. 6

4b(3a) Leaves partly or fully expanded at flowering. 7

5a(3b) Plants regularly or frequently dioecious; bearing either staminate or pistillate flowers but not both. 8

5b(3b) Plants not dioecious, flowers either perfect or unisexual; if unisexual, both female and male flowers borne on the same plant. 9

6a(4a) Perianth with both calyx and corolla. 10

6b(4a) Perianth lacking either calyx or corolla, or both. 11

7a(4b) Leaves compound. 12

7b(4b) Leaves simple. 13

8a(5a) Plants climbing or trailing. 14

8b(5a) Plants erect or nearly prostrate, but not climbing. 15

9a(5b) Flowers in catkins or catkin-like clusters, always unisexual and individually inconspicuous. 16

9b(5b) Flowers not in catkins; mostly perfect and conspicuous. 17

10a(6a) Petals separate; flowers mostly unisexual; stamens mostly eight; ovary superior, 2-lobed. **ACERACEAE**, p. 278

10b(6a) Petals united; flowers perfect; stamens 5; ovary inferior. (**Lonicera**) **CAPRIFOLIACEAE**, p. 367

11a(6b) Stamens usually 2; hypanthium none. 18

11b(6b) Stamens usually 8; hypanthium well developed, bearing sepals at the margin. (**Shepherdia**) **ELAEAGNACEAE**, p. 288

12a(7a) Plants climbing or trailing; corolla none or transitional into stamens, but sepals 4, petaloid; stamens numerous. (**Clematis**) **RANUNCULACEAE**, p. 199

12b(7a) Plants not climbing or trailing. 19

13a(7b) Perianth of a single series of parts (calyx). 20

13b(7b) Perianth of both calyx and corolla, the calyx often inconspicuous. 21

14a(8a) Plants producing tendrils; leaves palmately veined, with dentate lobes or compound; petals 4 or 5 and early deciduous. **VITACEAE**, p. 282

14b(8a) Plants twining; stamens 5; pistil 1; leaves pinnately veined. (**Celastrus**) **CELASTRACEAE**, p. 277

15a(8b) Flowers in catkins; corolla none; calyx small and inconspicuous or lacking. 22

15b(8b) Flowers not in catkins; either calyx or corolla or both present. 23

16a(9a) Staminate flowers in ellipsoid to elongate-cylindric catkins. 24

16b(9a) Staminate flowers in dense globose heads. (**Fagus**) **FAGACEAE**, p. 258

17a(9b) Perianth not differented into a distinct calyx and corolla; erect shrubs. 25

17b(9b) Perianth clearly differentiated into calyx and corolla. 26

18a(11a) Ovary not lobed. (**Fraxinus**) **OLEACEAE**, p. 324

18b(11a) Ovary prominently 2-lobed. **ACERACEAE**, p. 278

19a(12b) Petals well developed and conspicuous; corolla gamopetalous; stamens 5; leaves pinnately compound. (**Sambucus**) **CAPRIFOLIACEAE**, p. 367

19b(12b) Petals none. 27

20a(13a) Leaves palmately lobed. **ACERACEAE**, p. 278

20b(13a) Leaves entire, silvery, scurfy-lepidote. (**Shepherdia**) **ELAEAGNACEAE**, p. 288

21a(13b) Stamens more numerous than lobes of corolla. 28

21b(13b) Stamens as many as the petals or corolla lobes or fewer. 29

22a(15a) Ovary many-ovuled (capsules many-seeded); trees or shrubs; twigs and bracts not resinous dotted. **SALICACEAE**, p. 155

22b(15a) Ovary with 1 ovule (fruit with 1 seed); twigs and bracts resinous-dotted; shrubs. **MYRICACEAE**, p. 161

23a(15b) Leaves compound, present at flowering. **ANACARDIACEAE**, p. 276

23b(15b) Leaves simple or absent at flowering. 30

24a((16a) Pistillate flowers solitary or in small clusters. 31

24b(16a) Pistillate flowers in catkins, heads or cone-like structures. 32

25a(17a) Stamens more numerous than lobes or divisions of the perianth, or perianth nearly lacking. 33

25b(17a) Stamens as many as the lobes or divisions of perianth. 34

26a(17b) Ovaries 3 to many, distinct or nearly so; stamens more than 10; sepals 5, petals 5; leaves simple or compound. **ROSACEAE**, p. 231

26b(17b) Ovary 1 (styles or stigmas may be more than 1). 35

27a(19b) Stamens 2-4; ovary not 2-lobed. (**Fraxinus**) **OLEACEAE**, p. 324

27b(19b) Stamens mostly 8; ovary two-lobed. **ACERACEAE**, p. 278

28a(21a) Petals united. (**Kalmia**) **ERICACEAE**, p. 315

28b(21a) Petals distinct; stamens 10 or fewer; leaves palmately veined and lobed. **ACERACEAE**, p. 278

29a(21b) Corolla of separate petals. 36

29b(21b) Corolla gamopetalous. 37

30a(23b) Leaves up to 8 mm long, revolute, present at flowering; petals none. **EMPETRACEAE**, p. 272

30b(23b) Leaves much larger at maturity, or absent at flowering. 38

31a(24a) Leaves pinnately compound. **JUGLANDACEAE**, p. 161

31b(24a) Leaves simple, often lobed. **FAGACEAE**, p. 258

32a(24b) Pistillate flowers 2 or 3 behind each bract. **BETULACEAE**, p. 162

32b(24b) Pistillate flowers solitary behind each bract. **MYRICACEAE**, p. 161

33a(25a) Leaves glabrous beneath or still undeveloped at flowering. **THYMELAEACEAE**, p. 293

33b(25a) Leaves silvery-lepidote beneath. **ELAEAGNACEAE**, p. 288

34a(25b) Flowers in freely branched terminal cymes. (**Cornus**) **CORNACEAE**, p. 311

34b(25b) Flowers in lateral or axillary clusters. 39

35a(26b) Corolla irregular or petal one. 40

35b(26b) Corolla regular or essentially so. 41

36a(29a) Flowers in terminal heads or cymes. (Cornus) **CORNACEAE**, p.

36b(29a) Flowers axillary or in axillary clusters. (Rhamnus) **RHAMNACEAE**, p.

37a(29b) Flowers numerous in dense spherical heads; leaves entire.
(Cephalanthus) **RUBIACEAE**, p. 363

37b(29b) Flowers not in dense globose heads; leaves entire, toothed or lobed.
CAPRIFOLIACEAE, p. 367

38a(30b) Perianth not differentiated into calyx and corolla or lacking. 42

38b(30b) Perianth of calyx and corolla; calyx sometimes small and inconspicuous. 43

39a(34b) Vines, climbing by tendrils. **VITACEAE**, p. 282

39b(34b) Shrubs or small trees, not climbing, 44

40a(35a) Petal one, blue; flowers in dense spike-like racemes; leaves pinnately compound. (Amorpha) **FABACEAE**, p. 350

40b(35a) Petals or corolla-lobes 5, upper three petals separate or nearly so; lower two petals more or less united and concealing stamens. **FABACEAE**, p. 350

41a(35b) Corolla gamopetalous. 45

41b(35b) Corolla of separate petals. 46

42a(38a) Style and stigma one. **AQUIFOLIACEAE**, p. 276

42b(38a) Style divided above; stigmas 2-4. **RHAMNACEAE**, p.

43a(38b) Flowers in a loose or dense terminal panicle. **ANACARDIACEAE**, p. 276

43b(38b) Flowers axillary or in axillary clusters. 47

44a(39b) Style very short; stigma 1, almost sessile. **AQUIFOLIACEAE**, p. 276

44b(39b) Styles 2-4 lobed; stigmas 2-4. **RHAMNACEAE**, p. 281

45a(41a) Stamens as many as the corolla lobes. 48

45b(41a) Stamens more numerous than corolla lobes. **ERICACEAE**, p. 315

46a(41b) Ovary actually or apparently inferior. 49

46b(41b) Ovary superior. 50

47a(43b) Style very short, stigma almost sessile. **AQUIFOLIACEAE**, p. 276

47b(43b) Style divided above the middle. **RHAMNACEAE**, p. 281

48a(45a) Style very short; stigma nearly or quite sessile. **AQUIFOLIACEAE**, p. 276

48b(45a) Style well developed. **ERICACEAE**, p. 315

49a(46a) Stamens just as many as the petals. 51

49b(46a) Stamens twice as many as petals or more numerous. 52

50a(46b) Mature leaves cylindric, less than 1 cm long. **EMPETRACEAE**, p. 272

50b(46b) Mature leaves flat, with expanded blade more than 1 cm long. 53

51a(49a) Petals 4. 54

51b(49a) Petals 5, flowers in racemes or small corymbiform clusters.
(Ribes) **(GROSSULARIACEAE) SAXIFRAGACEAE** , p. 227

52a(49b) Style one (Vaccinium). **ERICACEAE**, p. 315

52b(49b) Styles 2-5. **ROSACEAE**, p. 231

53a(50b) Flowers produced in spring before the leaves have expanded or in autumn after the leaves have fallen. 55

53b(50b) Flowers and leaves (at least partially expanded) present together. 56

54a(51a) Flowers white, in terminal cymes (Cornus) **CORNACEAE**, p. 311

54b(51a) Flowers yellow, in small axillary clusters; opening in autumn.
(Hamamelis) **HAMAMELIDACEAE**, p. 230

55a(53a) Sepals, petals and stamens each 4; styles 2.
(Hamamelis) **HAMAMELIDACEAE**, p. 230

55b(53a) Sepals and petals each 5. 57

56a(53b) Stamens more than twice as many as petals. 58

56b(53b) Stamens twice as many as the petals <u>or</u> fewer. 59

57a(55b) Styles 3. **ANACARDIACEAE**, p. 276

57b(55b) Style 1. (<u>**Prunus**</u>) **ROSACEAE**, p. 231

58a(56a) Flowers yellow; style 1. (<u>**Hudsonia**</u>) **CISTACEAE**, p. 289

58b(56a) Flowers white to pink. 60

59a(56b) Leaves compound. 61

59b(56b) Leaves simple. 62

60a(58b) Flower clusters on a peduncle apparently arising from the middle of a large, narrowly oblong foliaceous bract. (<u>**Tilia**</u>) **TILIACEAE**, p. 283

60b(58b) Flower clusters not arising from a foliaceous bract. (<u>**Prunus**</u>) **ROSACEAE**, p. 231

61a(59a) Leaves evenly pinnate or evenly bipinnate; trees or tall shrubs; stems thorny. (<u>**Gleditsia**</u>) **CAESALPINACEAE** (Reported from Gagetown, Sunbury Co.)

61b(59a) Leaves odd-pinnate, trifoliate or palmate; trees, shrubs or vines; never thorny. 63

62a(59b) Stamens more numerous than petals. 64

62b(59b) Stamens as many as petals. 65

63a(61b) Flowers in terminal panicles. (<u>**Rhus**</u>) **ANACARDIACEAE**, p. 276

63b(61b) Flowers in lateral or axillary clusters. 66

64a(62a) Flowers yellow; low gray-green shrubs with small leaves. (<u>**Hudsonia**</u>) **CISTACEAE**, p. 289

64b(62a) Flowers white or pink. **ERICACEAE**, p. 315

65a(62b) Styles 2, separate to the base; flowers 4-merous, yellow. (<u>**Hamamelis**</u>) **HAMAMELIDACEAE**, p. 230

65b(62b) Style one. 67

66a(63b) Stamens alternate with petals; shrubs or vines; if climbing, the terminal leaflet of the 3-foliate leaves on a distinctly longer stalk than lateral ones. **ANACARDIACEAE**, p. 276

66b(63b) Stamens opposite the quickly deciduous petals; vine climbing by tendrils. **VITACEAE**, p. 282

67a(65b) Flowers yellow, 6-merous;.stems spiny. (<u>**Berberis**</u>) **BERBERIDACEAE**, p. 208

67b(65b) Flowers white or greenish; stems never spiny. 68

68a(67b) Flowers in slender elongate racemes; leaves evergreen, densely white or rusty tomentose beneath, strongly revolute. (<u>**Ledum**</u>) **ERICACEAE**, p. 315

68b(67b) Flowers axillary, or in axillary clusters <u>or</u> in terminal or lateral panicles. 69

69a(68b) Stamens opposite petals; style 3-cleft or 3-lobed. (<u>**Rhamnus**</u>) **RHAMNACEAE**, p. 281

69b(68b) Stamens alternate with petals; style short, stigma nearly sessile. **AQUIFOLIACEAE**, p. 276

KEY C

1a Flowers modified into bulblets or leaf tufts. 2

1b Flowers normally developed. 3

2a(1a) Leaves garlic or onion scented. (<u>**Allium**</u>) **LILIACEAE**, p. 134

++

2b(1a) Leaves without strong odor of garlic or onion. 4

3a(1b) Perianth wanting or scale-like, chaffy or bristle-like (not petal-like in color, texture or size). 5

3b(1b) Perianth present, petal-like in color, texture or size. 6

4a(2b) Leaf blades cross-septate; flowers regular; fruit a capsule. (Juncus) **JUNCACEAE**, p. 127

4b(2b) Leaf blades not cross-septate; sheath closed; stem usually 3-angled; fruit achenous. (Cyperus dentatus) **CYPERACEAE**, p. 78

5a(3a) Flowers embedded in a dense fleshy spike. **ARACEAE**, p. 123

5b(3a) Flowers not embedded in a fleshy spike. 7

6a(3b) Flowers +/- sessile in globose heads; corolla yellow, regular. (Xyris montana) **XYRIDACEAE**, p. 125

6b(3b) Flowers usually pedicelled, not in heads. 8

7a(5b) Perianth of 6 +/- similar segments, regular and chartaceous, greenish or brownish; fruit a capsule. **JUNCACEAE**, p. 127

7b(5b) Perianth absent, sac-like or reduced to bristles. 9

8a(6b) Ovary inferior. 10

8b(6b) Ovary superior. 11

9a(7b) Flowers not in axil of scales. 12

9b(7b) Flowers in axil of scales. 13

10a(8a) Flowers regular, borne from a spathe. **IRIDACEAE**, p. 142

10b(8a) Flowers irregular, not associated with a spathe. **ORCHIDACEAE**, p. 143

11a(8b) Perianth petaloid. 14

11b(8b) Perianth with green sepals; petals various. 15

12a(9a) Flowers in globose heads or spikes. 16

12b(9a) Flowers solitary or in small clusters in axils of sheathing leaves or subtended by a spathe, often inconspicuous or absent. 17

13a(9b) Leaf-sheaths split lengthwise opposite blades; leaves usually 2-ranked; stems rounded or flattened, never triangular in cross-section, usually hollow and jointed; anthers attached near the middle. **POACEAE**, p. 40

13b(9b) Leaf-sheaths continuous around stem or rupturing in age; leaves usually 3-ranked, or reduced to sheathing scales; stems often triangular in cross-section, usually with a pith, usually not jointed; anthers attached by the base. **CYPERACEAE**, p. 78

14a(11a) Carpels 6, mostly free; stamens 9. (Butomus umbellatus) **BUTOMACEAE**, p. 39

14b(11a) Carpels 3, united; stamens 3 or 6. 18

15a(11b) Carpels numerous in a ring or head; flowers trimerous, white or pinkish, pedicellate. **ALISMATACEAE**, p. 37

15b(11b) Carpels 1-6. 19

16a(12a) Flowers in globose heads; fruit large, glabrous. (Sparganium) **SPARGANIACEAE**, p. 28

16b(12a) Flowers in spikes, clusters or whorls. 20

17a(12b) Leaves opposite; flowers/fruit in axils of leaves. 21

17b(12b) Leaves alternate; flowers/fruit not in leaf axils; leaves to about 1.5 m long, 8 mm broad; marine habitat. (Zostera) **ZOSTERACEAE**, p. 29

18a(14b) Flowers irregular, in dense spikes, violet-blue with yellow spots. (Pontederia cordata) **PONTEDERIACEAE**, p. 126

18b(14b) Flowers regular, pedicilled, never blue (in our species).
LILIACEAE, p. 134

19a(15b) Flowers +/- sessile; perianth 4-merous, united; leaves basal, flattened at least along one side, not sheathing. **PLANTAGINACEAE** (not monocolyledonous, but often mistaken for such), p. 361

19b(15b) Flowers pedicelled, perianth trimerous, free; leaves round in cross-section with sheathing base. 22

20a(16b) Plants emergent, of marshy habitats; thick cylindrical spikes divided into basal pistillate and terminal staminate segments.
(**Typha**) **TYPHACEAE**, p. 27

20b(16b) Plants aquatic, +/- submerged; inflorescence of narrow short spikes or small clusters with bisexual flowers. 23

21a(17a) Stems bushy-branched; fruit solitary, fusiform; leaves 1-3 cm long.
(**Naja flexilis**) **NAJADACEAE**, p. 36

21b(17a) Stems long trailing; fruit +/- 4, curved and coarsely toothed on one side; leaves 3-10 cm long. (**Zannichellia palustris**) **ZANNICHELLIACEAE**, p. 36

22a(19b) Leaves all basal, without a terminal pore; mature fruit 2.5 - 9 mm long divided into 3-6 connivent 1-seeded carpels.
(**Triglochin**) **JUNCAGINACEAE**, p. 36

22b(19b) Leaves alternate and basal with terminal pore; mature fruit 7-10 mm long divided into 3 spreading, inflated 2-seeded follicles.
(**Scheuchzeria palustris**) **SCHEUCHZERIACEAE**, p. 37

23a(20b) Fruits +/- stalked in umbel-like clusters on long, usually coiled peduncles; leaves very narrow, to about 0.5 mm broad; plants of brackish or salt water. (**Ruppia maritima**) **RUPPIACEAE**, p. 35

23b(20b) Fruit sessile in short spikes or whorls on straight peduncles; leaves usually broader, often floating; plants of fresh to brackish waters.
(**Potamogeton**) **POTAMOGETONACEAE**, p. 30

KEY D

1a Leaves, at least those submerged, divided into linear or filiform segments. 2

1b Leaves not finely dissected, the uppermost often floating, lanceolate to orbicular. 3

2a(1a) Leaves or branches bearing bladder-traps; corolla yellow or purple, 2-lipped and spurred. (**Utricularia**) **LENTIBULARIACEAE**, p. 359

2b(1a) Leaves or branches without bladder-traps. 4

3a(1b) Leaves all basal. 5

3b(1b) Leaves cauline. 6

4a(2b) Leaves once divided from a central rachis; flowers axillary; immersed leaves usually not divided. (**Myriophyllum**) **HALORAGIDACEAE**, p. 300

4b(2b) Leaves more than once divided. 7

5a(3a) Flowers irregular, few to many in racemes. 8

5b(3a) Flowers regular, single; pedicels often very long. **NYMPHACEAE**, p. 198

6a(3b) Stem creeping, rooting at the nodes; flowers single; pistils several-many.
(**Ranunculus**) **RANUNCULACEAE**, p. 199

6b(3b) Stem not both creeping and rooting at nodes <u>or</u> flowers not single; pistil usually 1. 9

++

7a(4b) Plants attached to boulders in rapid currents by fleshy disks; leaves filiform, alternate, 2-ranked, rigid and algae-like. (**Podostemum ceratophyllum**) **PODOSTEMACEAE**, p. 226

7b(4b) Plants with normal root or free floating. 10

8a(5a) Corolla yellow; calyx free from ovary; leaves minute. (**Utricularia**) **LENTIBULARIACEAE**, p. 359

8b(5a) Corolla light blue; calyx fused to ovary; leaves linear, in a basal rosette. (**Lobelia**) **CAMPANULACEAE**, p. 374

9a(6b) Submerged leaves alternate or basal. 11

9b(6b) Submerged leaves opposite or whorled; flowers single, axillary. 12

10a(7b) Leaves alternate, dilated-sheathing at base; petals +/- 5, yellow or whitish. (**Ranunculus**) **RANUNCULACEAE**, p. 199

10b(7b) Leaves opposite or whorled, not dilated-sheathing at base. 13

11a(9a) Stipules sheathing stem. **POLYGONACEAE**, p. 171

11b(9a) Stipules not sheathing stem. 14

12a(9b) Leaves whorled, 6-12 at a node, linear or filiform; flowers sessile, perfect. (**Hippuris vulgaris**) **HIPPURIDACEAE**, p. 302

12b(9b) Leaves opposite; petals minute or wanting. 15

13a(10b) Upper leaves not usually densely overlapping, nearly black, entire; immersed leaves, if present, deeply to shallowly lobed; flowers in a closely bracted head; ligulate flowers yellow. (**Megalodonta beckii**) **ASTERACEAE**, p. 377

13b(10b) Upper leaves densely overlapping, dull green, often finely serrate; immersed leaves wanting, plants usually without roots, free floating; flowers inconspicuous in axils of leaves. (**Ceratophyllum**) **CERATOPHYLLACEAE**, p. 197

14a(11b) Leaves peltate, glutinous beneath, floating; flowers dull purple, long petioled. (**Brasenia schreberi**) **NYMPHAEACEAE**, p. 198

14b(11b) Leaves deeply lobed to serrate, the uppermost often emergent; flowers inconspicuous in leaf axils. (**Proserpinaca palustris**) **HALORAGIDACEAE**, p. 300

15a(12b) Ovary inferior; leaves +/- petioled, reddish-green. (**Ludwigia palustris**) **ONAGRACEAE**, p. 295

15b(12b) Ovary superior. 16

16a(15b) Fruit flattened, notched; floating leaves sometimes present; stem and leaves flaccid, collapsing out of water. (**Callitriche**) **CALLITRICHACEAE**, p. 272

16b(15b) Fruit not flattened nor notched; floating leaves wanting; stem and leaves turgid, not collapsing out of water. 17

17a(16b) Leaves linear, sessile, connate around the stem; pistils 3 or 4; fruit a follicle with few inconspicuous seeds.(**Tillaea aquatica**) **CRASSULACEAE**,p. 226

17b(16b) Leaves lanceolate to spatulate, petiolate to subsessile, not connate; pistil 1; fruit a capsule with many cylindrical, minutely areolate seeds. (**Elatine**) **ELATINACEAE**, p. 288

KEY E

1a Flowers individually small, numerous in bracteate heads. (Tussilago, Petasites) **ASTERACEAE**, p. 377
1b Flowers not in bracteate heads. 2
2a(1b) Flowers irregular; corolla bilabiate; ovary superior. 3
2b(1b) Flowers regular or petals wanting. 4
3a(2a) Corolla bright yellow, spurred; leaves inconspicuous, mostly subterranean; roots with minute bladders. (Utricularia cornuta) **LENTIBULARIACEAE**, p.359
3b(2a) Corolla white, violet or white with 2 purple brown lines, spurless. (Orobanche, Epifagus) **OROBANCHACEAE**, p. 358
4a(2b) Corolla sympetalous; leaves inconspicuous, reduced to minute scales. 5
4b(2b) Corolla gamopetalous or wanting; leaves bract-like or wanting. 6
5a(4a) Stems rooted in the ground, not viney but often twisted; inflorescence racemose or a loose panicle. (Bartonia) **GENTIANACEAE**, p. 326
5b(4a) Stems viney, yellowish, attached to various herbaceous plants. (Cuscuta) **CONVOLVULACEAE**, p. 329
6a(4b) Corolla gamopetalous; cespitose stems bearing bract-like, non-green leaves; woodland plants. (Monotropa) **PYROLACEAE**, p. 312
6b(4b) Corolla wanting; flowers embedded in brittle, fleshy, leafless stem; salt marsh plants. (Salicornia) **CHENOPODIACEAE**, p. 180

KEY F

1a Leaves dissected into linear or filiform segments; plants +/- aquatic. 2
1b Leaves otherwise. 3
2a(1a) Leaves pinnately dissected. (Myriophyllum) **HALORAGACEAE**, p. 300
2b(1a) Leaves palmately dissected. **CERATOPHYLLACEAE**, p. 197
3a(1b) Leaves scale-like or wanting. 4
3b(1b) Leaves neither scale-like nor dissected. 5
4a(3a) Plants parasitic on the roots of beech, without green color. (Epifagus virginiana) **OROBANCHACEAE**, p. 358
4b(3a) Plants not parasitic, with green color. 6
5a(3b) Leaves compound. 7
5b(3b) Leaves simple. 8
6a(4b) Plants of saltmarshes, stems fleshy jointed. (Salicornia europaea) **CHENOPODIACEAE**, p. 180
6b(4b) Plants with delicate tapering stems of fresh water or shores. (Myriophyllum tenellum) **HALORAGACEAE**, p. 300
7a(5a) Leaves trifoliate or palmately compound. 9
7b(5a) Leaves pinnately compound, or ternately or pinnately decompound. 10
8a(5b) Leaves all basal. 11
8b(5b) Leaves +/- cauline. 12
9a(7a) Flowers aggregated on a fleshy spadix subtended by a colored spathe. (Arisaema, a monocotyledon with net-veined leaves often confused for a dicotyledon). **ARACEAE**, p. 123
9b(7a) Flowers in umbels. 13
10a(7b) Inflorescence umbellate. **ARALIACEAE**, p. 302
10b(7b) Inflorescence spicate, paniculate or flowers solitary. 14
11a(8a) Flowers in spikes. **PLANTAGINACEAE**, p. 361

11b(8a) Flowers in open panicles. (**Rumex**) **POLYGONACEAE**, p. 171

12a(8b) Leaves opposite or whorled. 15

12b(8b) Leaves alternate. 16

13a(9b) Leaves alternate or basal. (**Sanicula**) **APIACEAE**, p. 303

13b(9b) Leaves cauline, in a single whorl. (**Panax**) **ARALIACEAE**, p. 302

14a(10b) Flowers in dense cylindric spikes. (**Sanguisorba canadensis**) **ROSACEAE**, p. 231

14b(10b) Flowers in panicles or solitary. 17

15a(12a) Flowers solitary; plants of mud, swamps or shallow water. 18

15b(12a) Flowers in axillary or terminal clusters. 19

16a(12b) Calyx and corolla both present, the latter usually white or colored. 20

16b(12b) Calyx present or absent; corolla always lacking; calyx sometimes petaloid. 21

17a(14b) Stem leaves opposite. 22

17b(14b) Stem leaves alternate; flowers with either calyx or corolla but not both, or perianth lacking. (**Thalictrum**) **RANUNCULACEAE**, p. 199

18a(15a) Leaves in whorls of 6-12. (**Hippuris vulgaris**) **HIPPURIDACEAE**, p. 300

18b(15a) Leaves opposite. (**Callitriche**) **CALLITRICHACEAE**, p. 272

19a(15b) Inflorescence axillary. 23

19b(15b) Inflorescence terminal. 24

20a(16a) Plants climbing by tendrils or trailing; stamens 3; pistil 1; ovary inferior. (**Echinocystis lobata**) **CUCURBITACEAE**, p. 374

20b(16a) Plants erect, spreading or creeping; stamens more than 6; pistils 4 or more, separate or united. 25

21a(16b) Flowers very small in small axillary clusters. 26

21b(16b) Flowers in spikes, racemes, or panicles. 27

22a(17a) Stamens numerous; pistils several to many. (**Clematis**) **RANUNCULACEAE**, p. 199

22b(17a) Stamens 3; pistil 1 with one style. (**Valeriana**) **VALERIANACEAE**, p. 373

23a(19a) Leaves +/- entire. **PLANTAGINACEAE**, p. 361

23b(19a) Leaves serrate. **URTICACEAE**, p. 168

24a(19b) Stamens 3; style 1. **VALERIANACEAE**, p. 373

24b(19b) Stamens 10; styles usually 5. (**Lychnis**) **CARYOPHYLLACEAE**, p. 189

25a(20b) Leaves +/- sessile, fleshy; flowers in panicles. (**Sedum**) **CRASSULACEAE**, p. 226

25b(20b) Leaves petioled; flowers single. (**Dalibarda**, **Rubus**) **ROSACEAE**, p. 231

26a(21a) Plants with pistillate flowers. 28

26b(21a) Plants with staminate flowers. 29

27a(21b) Perianth segment 6, (3 inner and 3 outer). (**Rumex**) **POLYGONACEAE**, p. 171

27b(21b) Perianth segments 5 or fewer or lacking. 30

28a(26a) Styles 3, irregularly branched. (**Acalypha**) **EUPHORBIACEAE**, p. 373

28b(26a) Styles 2 or 3, unbranched. 31

29a(26b) Bracts foliaceous, lobed or cleft. (**Acalypha**) **EUPHORBIACEAE**, p. 373

29b(26b) Bracts entire, scarious, acute. **AMARANTHACEAE**, p. 187

30a(27b) Sepals acute, scarious, subtended by scarious bracts; filaments often connate below. **AMARANTHACEAE**, p. 187

30b(27b) Sepals not scarious, often green; filaments usually free. **CHENOPODIACEAE**, p. 180

31a(28b) Sepals and bracts acute, scarious. **AMARANTHACEAE**, p. 187
31b(28b) Sepals (often lacking) and the bracts herbaceous. (Atriplex) **CHENOPODIACEAE**, p. 180

KEY G

1a Pistil 1. 2
1b Pistils more than 1. **RANUNCULACEAE**, p. 199
2a(1a) Ovary +/- superior. 3
2b(1a) Ovary inferior. 4
3a(2a) Fruit a loculicidal or septicidal capsule. 5
3b(2a) Fruit a pyxis or hard or leathery 1-seeded structures. 6
4a(2b) Fruit drupaceous (dry in Geocaulon). **SANTALACEAE**, p. 170
4b(2b) Fruit capsular or nut-like. 7
5a(3a) Capsules 3 or 5-lobed. 8
5b(3a) Capsules otherwise; plants decumbent with central root. (Mollugo verticillata) **AIZOACEAE**, p. 188
6a(3b) Leaves lobed, divided or compound, or plants vine-like and twining; fruit achenose. 9
6b(3b) Leaves otherwise. 10
7a(4b) Fruit capsular, several seeded. 11
7b(4b) Fruit nut-like one seeded; aquatics. 12
8a(5a) Capsules 3-lobed; seeds 1-2 per cell. **EUPHORBIACEAE**, p. 373
8b(5a) Capsule 5-lobed. (Penthorum sedoides) **SAXIFRAGACEAE**, p. 227
9a(6a) Flowers in stalked axillary clusters or spikes. **CANNABINACEAE**, p. 167
9b(6a) Flowers in sessile axillary clusters or terminal spikes. (Alchemilla, Sanguisorba) **ROSACEAE**, p. 231
10a(6b) Calyx petaloid; stipules sheathing; fruit an achene or small nut. **POLYGONACEAE**, p. 171
10b(6b) Calyx not petaloid. 13
11a(7a) Plant mat forming, of wet habitats. (Chrysosplenium americanum) **SAXIFRAGACEAE**, p. 227
11b(7a) Plants not mat forming, larger, of mesic habitats; leaves cordate-ovate; perianth trimerous. (Asarum canadense) **ARISTOLOCHIACEAE**, p. 170
12a(7b) Fruit 3-4 lobed; leaves +/- wanting, finely divided or sharply toothed. **HALORAGACEAE**, p. 300
12b(7b) Fruit unlobed; leaves entire, linear. **HIPPURIDACEAE**, p. 302
13a(10b) Inflorescence axillary. 14
13b(10b) Inflorescence terminal and axillary. 15
14a(13a) Leaves small, entire and 1-nerved. (Schleranthus annuus) **CARYOPHYLLACEAE**, p. 189
14b(13a) Leaves larger, usually toothed. **URTICACEAE**, p. 168
15a(13b) Flowers bractless, sepals not scarious. **CHENOPODIACEAE**, p. 180
15b(13b) Flowers bracted; sepals scarious. **AMARANTHACEAE**, p. 187

KEY H

1a Leaves simple and usually not deeply lobed. 2
1b Leaves compound, or deeply lobed or divided. 3
2a(1a) Flowers pendulous and spurred; capsule explosive when ripe. (Impatiens) **BALSAMINACEAE**, p. 280

2b(1a) Flower otherwise. 4

3a(1b) Plant viney, herbaceous. (**Adlumia fungosa**) **FUMARIACEAE**, p. 208

3b(1b) Plant not viney. 5

4a(2b) Sepals petaloid, unequal; petals variously united, upper two similar; stamens usually 8. **POLYGALACEAE**, p. 271

4b(2b) Sepals green, +/- similar; petals separate; stamens 5. (**Viola**) **VIOLACEAE**, p. 289

5a(3b) Pistil 1. 6

5b(3b) Pistils more than 1. **RANUNCULACEAE**, p. 199

6a(5a) Fruit a legumine; stamens usually 10. **FABACEAE**, p. 350

6b(5a) Fruit otherwise. 7

7a(6b) Ovary superior; plants delicate and +/- succulent. **FUMARIACEAE**, p. 208

7b(6b) Ovary inferior; fruit of two similar seed-like division; inflorescence umbellate. **APIACEAE**, p. 303

KEY I

1a Stamens separate. (**Knautia arvensis**) **DIPSACACEAE**, p. 374

1b Stamens united by their anthers. **ASTERACEAE**, p. 377

KEY J

1a Stamens borne on calyx tube or distinct hypanthium. **ROSACEAE**, p. 231

1b Stamens borne on receptacle or appearing so. 2

2a(1b) Stamens numerous. **RANUNCULACEAE**, p. 199

2b(1b) Stamens 5-10. 3

3a(2b) Plants +/- succulent; carpels as many as the petals, free or united only at base. **CRASSULACEAE**, p. 226

3b(2b) Plants not succulent; carpels 2-5, +/- united. **SAXIFRAGACEAE**, p. 227

KEY K

1a Petals 3; sepals 5, unequal; perianth variously united. **POLYGALACEAE**, p. 271

1b Petals 5. 2

2a(1b) Ovary superior. 3

2b(1b) Ovary inferior. 4

3a(2a) Fruit of 1-4 separate nutlets. 5

3b(2a) Fruit capsular. 6

4a(2b) Leaves opposite or connate; fruit a dry berry. (**Triosteum aurantiacum**) **CAPRIFOLIACEAE**, p. 367

4b(2b) Leaves alternate; fruit capsular. (**Lobelia**) **CAMPANULACEAE**, p. 374

5a(3a) Fruit of 4-nutlets; stem often 4-angled. **LAMIACEAE**, p. 337

5b(3a) Fruit of a single nutlet, reflexed and racemose in oppositely arranged pairs. (**Phryma leptostachya**) **PHRYMACEAE**, p. 361

6a(3b) Capsule 1-celled; leaves all basal. **LENTIBULARIACEAE**, p. 359

6b(3b) Capsule 2 or more celled; leaves all or mostly cauline. **SCROPHULARIACEAE**, p. 347

++

KEY L

1a Leaves hollow, pitcher-like or covered with glandular hairs. 2
1b Leaves otherwise. 3
2a(1a) Leaves hollow. **SARRACENIACEAE**, p. 225
2b(1a) Leaves covered with tentacle-like glandular hairs. **DROSERACEAE**, p. 225
3a(1b) Petals 4; stamens 6; flowers in racemes. **BRASSICACEAE**, p. 211
3b(1b) Plants otherwise. 4
4a(3b) Plants succulent; sepals 2 or 3. **PORTULACACEAE**, p. 188
4b(3b) Plants not succulent. 5
5a(4b) Leaves at least some of them, opposite or whorled. 6
5b(4b) Leaves alternate, basal or only 1 or 2. 7
6a(5a) Stem joints noticeably swollen; stamens less than twice the sepals. **CARYOPHYLLACEAE**, p. 189
6b(5a) Joints not especially swollen; stamens numerous. **HYPERICACEAE**, p. 286
7a(5b) Leaves with 3 equal obcordate subdivisions; sap sour. **OXALIDACEAE**, p. 269
7b(5b) Plants otherwise. 8
8a(7b) Stamens many. 9
8b(7b) Stamens not more than twice as many as petals. 10
9a(8a) Stamens united by their filaments and forming a sheath around the style. **MALVACEAE**, p. 283
9b(8a) Stamens separate. 11
10a(8b) Sepals 5 and very unequal or only 3. **CISTACEAE**, p. 289
10b(8b) Sepals otherwise. 12
11a(9b) Plants with milky or colored sap. **PAPAVERACEAE**, p. 208
11b(9b) Plants otherwise. 13
12a(10b) Capsules 5-lobed at base, 5-seeded, overtopped by long styler beak. **GERANIACEAE**, p. 270
12b(10b) Capsules otherwise. 14
13a(11b) Plants with sharply toothed compound leaves or palmately divided leaves. **RANUNCULACEAE**, p. 199
13b(11b) Leaves compound, each with 3 trifoliate divisions. **BERBERIDACEAE**, p. 208
14a(12b) Leaves all or chiefly basal or the cauline leaves few and small or paired. (**Mitella**, **Saxifraga**) **SAXIFRAGACEAE**, p. 227
14b(12b) Leaves chiefly cauline. 15
15a(14b) Plants less than 1 dm tall; leaves evergreen. (**Chimophila**, **Pyrola**) **PYROLACEAE**, p. 312
15b(14b) Plants taller than 1 dm; leaves not evergreen. 16
16a(15b) Flowers clustered in upper leaf or bract axils, purple or pink; stems angled. (**Lythrum salicaria**) **LYTHRACEAE**, p. 295
16b(15b) Flowers in open cymes, white or light blue; stems rounded. **LINACEAE**, p. 269

KEY M

1a Flowers in umbels or sometimes in head-like inflorescence. 2
1b Flowers otherwise. 3
2a(1a) Fruit drupaceous. **ARALIACEAE**, p. 302
2b(1a) Fruit dry, of 2 seed-like divisions. **APIACEAE**, p. 303
3a(1b) Fruit indehiscent, +/- nut-like. 4
3b(1b) Fruit capsular. **ONAGRACEAE**, p. 295

4a(3a) Terrestrial plants; fruit bristly. (Circaea) **ONAGRACEAE, p. 295**

4b(3a) Aquatic plants; fruit not bristly. **HALORAGACEAE, p. 300**

KEY N

1a Plants green. 2

1b Plants non-green, vine-like; flowers usually in dense clusters. (Cuscuta) **CONVOLVULACEAE, p. 329**

2a(1a) Stamens opposite lobes of corolla. 3

2b(1a) Stamens alternate lobes of corolla, or fewer. 4

3a(2a) Fruit capsular, many seeded. **PRIMULACEAE, p. 323**

3b(2a) Fruit utricular, one-seeded; leaves basal; flowers small, lavender. (Limonium carolinianum) **PLUMBAGINACEAE, p. 324**

4a(2b) Plants with milky juice; leaves opposite; fruit follicular. 5

4b(2b) Plants without milky juice; fruit not follicular. 6

5a(4a) Filaments distinct; corolla campanulate to tubular. (Apocynum) **APOCYNACEAE, p. 328**

5b(4a) Filaments and anthers united, surrounding stigma; corolla deeply 5-parted. **ASCLEPIADACEAE, p. 329**

6a(4b) Leaves basal. **PLANTAGINACEAE, p. 361**

6b(4b) Leaves cauline. 7

7a(6b) Fruit capsular. 8

7b(6b) Fruit either berry-like or 1-4 dry nutlets. 9

8a(7a) Leaves opposite or compound. 10

8b(7a) Leaves alternate, not compound. 11

9a(7b) Fruit dry. 12

9b(7b) Fruit berry-like or pod-like. **SOLANACEAE, p. 345**

10a(8a) Stigmas 3. **POLEMONIACEAE, p. 332**

10b(8a) Stigma 1, either capitate or 2-lobed. **GENTIANACEAE, p. 326**

11a(8b) Plants mostly twining, trailing or creeping. **CONVOLVULACEAE, p. 329**

11b(8b) Plants erect. **SOLANACEAE, p. 345**

12a(9a) Stamens as many as corolla lobes; inflorescence commonly circinate cymes. **BORAGINACEAE, p. 333**

12b(9a) Stamens fewer than corolla lobes; inflorescence otherwise. 13

13a(12b) Style arising apically from ovary. **VERBENACEAE, p. 337**

13b(12b) Style surrounded at base by 4 lobes of ovary; stems +/- square; leaves opposite, often aromatic. **LAMIACEAE, p. 337**

KEY O

1a Leaves opposite or whorled. 2

1b Leaves alternate. 3

2a(1a) Stamens 3. **VALERIANACEAE, p. 373**

2b(1a) Stamens 4 or 5. 4

3a(1b) Plants tendril bearing; leaves palmately lobed; fruit soft prickly. (Echinocystis lobata) **CUCURBITACEAE, p. 374**

3b(1b) Plants not tendril bearing; leaves entire or toothed; flowers bluish (rarely white in albino forms); fruit capsular. **CAMPANULACEAE, p. 374**

4a(2b) Leaves entire. **RUBIACEAE, p. 363**

4b(2a) Leaves toothed or pinnately lobed; outer corollas irregular. **DIPSACACEAE, p. 374**

Division I **PTERODOPHYTA** (Ferns and Fern Allies)

EQUISETACEAE (Horsetail Family)

EQUISETUM L. Horsetail. Prêle

1a Stems whitish or brownish, unbranched, appearing early in spring. 2
1b Stems soon green. 3
2a(1a) Teeth dark brown throughout. **E. arvense**
2b(1a) Teeth with white margins. **E. pratense**
3a(1b) Branches simple or absent. 4
3b(1b) Branches rebranched, regularly whorled. **E. sylvaticum**
4a(3a) Branches absent. 5
4b(3a) Branches present, mostly simple. 6
5a(4a) Stem cavity very large, walls thin and easily crushed. **E. fluviatile**
5b(4a) Stem stiffer, with central cavity and smaller surrounding cavities. 7
6a(4b) Lowermost internode on each branch longer than the adjacent sheath on the main stem. **E. arvense**
6b(4b) Lowermost internode on each branch as long as or shorter than the adjacent sheath on the main stem. 8
7a(5b) Plants small, tufted with wiry stems; sheaths with 3 teeth. **E. scirpoides**
7b(5b) Plants larger, not tufted; sheaths with many more than 3 teeth. 9
8a(6b) Branch sheaths with 3 teeth; branches very slender and lax; teeth light brown and white bordered. **E. pratense**
8b(6b) Branch sheaths with more than 3 teeth; branches usually thicker, often irregular; teeth darker. 10
9a(7b) Teeth persistant; stem to 4 mm thick; annual. **E. variegatum**
9b(7b) Teeth deciduous; stems much larger, biennial; sheath with 2 dark rings. **E. hyemale**
10a(8b) Stems soft and papery with very large central cavity; often reddish-orange at base. **E. fluviatile**
10b(8b) Stems firmer, central cavity smaller and with side cavities visible to the naked eye; rhizome blackish. 11
11a(10b) Side cavities +/- the same size as the central cavity; teeth with wide scarious margins. **E. palustre**
11b(10b) Side cavities much smaller than the central cavity; teeth with narrow scarious margins. **E.** X **litorale**

Equisetum arvense L. Fig. p. A1. Map p. B1.
COMMON NAME(S): Common Field-Horsetail. Queue de renard.
FREQUENCY: Common throughout.
HABITAT: Mostly sandy roadsides, railroad embankments, open poorly drained areas.
SUBSPECIFIC TAXA: A variable species with numerous named varieties and forms. Most of these are considered environmentally induced. Var. **boreale** (Bong.) Ledeb. has 3-angled branches.
GENERAL RANGE: Widely distributed in the Northern Hemisphere.

EQUISETACEAE

++

Equisetum fluviatile L. Fig. p. A1. Map p. B1.
COMMON NAME(S): Water-Horsetail. Pipes
FREQUENCY: Common throughout.
HABITAT: Quiet shallow waters.
SUBSPECIFIC TAXA: Several growth forms have been named based principally on the degree of branching. Forma **linnaeanum** (Döll) Broun has unbranched stems.

Equisetum hyemale L. Fig. p. A1. Map p. B1.
COMMON NAME(S): Scouring-rush. Prêle des tourneurs.
FREQUENCY: Apparently absent from the eastern lowlands and southwestern corner of the Province.
HABITAT: Open or shaded sandy or gravelly banks, occasionally in rich alluvial soil.
SUBSPECIFIC TAXA: The American plants belong to var. **affine** (Engelm.) A.A. Eaton.
GENERAL RANGE: Most of North America, south to Guatemala; Europe, Asia.

Equisetum x **litorale** Kuehl. Fig. p. A1. Map p. B1.
COMMON NAME(S): Shore-Horsetail. Prêle littorale.
FREQUENCY: Rare; mostly the northern part of the Province.
HABITAT: Moist rocky or gravelly shores.
SPECIAL NOTE: Evidently the hybrid between **E. arvense** and **E. fluviatile**.
GENERAL RANGE: Local; N.S. to B.C. south to Penn., Minn. and Ore.

Equisetum palustre L. Fig. p. A1. Map p. B1.
COMMON NAME(S): Marsh-Horsetail. Prêle des marais.
FREQUENCY: Rare in most of Province. Uncommon in Saint John River valley.
HABITAT: Marshes and seepy gravels often in shallow water.
SUBSPECIFIC TAXA: Numerous growth forms have been named.
GENERAL RANGE: Circumboreal.

Equisetum pratense Ehrh. Fig. p. A1. Map p. B1.
COMMON NAME(S): Meadow-Horsetail. Prêle des prés.
FREQUENCY: Locally common, mostly in the western portion of the Province.
HABITAT: Open woods and wet meadows, usually in calcareous areas.
SPECIAL NOTE: Hybridizes with **E. arvense** (see **E** x **litorale**).
GENERAL RANGE: Circumboreal.

Equisetum scirpoides Michx. Fig. p. A1. Map p. B1.
COMMON NAME(S): Dwarf Scouring-rush. Prêle faux-scirpe.
FREQUENCY: Uncommonly scattered throughout.
HABITAT: Rich moist woods, and mossy banks.
GENERAL RANGE: Nfld. to Alaska south to Mass., Wisc. and Wash.

Equisetum sylvaticum L. Fig. p. A1. Map p. B1.
COMMON NAME(S): Wood-Horsetail. Prêle des bois.
FREQUENCY: Common throughout.
HABITAT: Wet woods, banks, meadows.
GENERAL RANGE: Northern North America; Europe.

+++

Equisetum variegatum Schleich. Fig. p. A1. Map p. B1.
COMMON NAME(S): Variegated Scouring-rush. Prêle panachée.
FREQUENCY: Very rare. Occasionally found in quantity in calcareous areas such as Maplehurst Marl Mining Pit, Carleton Co. and shaley roadside of Trans-Canada Highway, Perth, Victoria Co.
HABITAT: Rocky banks and muddy shores in calcareous areas.
SPECIAL NOTE: For more information see The Rare Vascular Plants of New Brunswick. H.R. Hinds. Syllogeus No. 50. 1983.
GENERAL RANGE: Nfld. to Penn. and Wisc.; Alaska to Colo.; Eurasia.

LYCOPODIACEAE (Club-moss Family)

LYCOPODIUM L. Club-moss. Lycopode.

1a Spore producing structures (sporangia) in the axils of nearly normal leaves. 2
1b Sporangia aggregated in a spike. 3
2a(1a) Leaves broadest at the base, entire or minutely toothed, 3-7 mm long; uniformily distributed along the stem. **L. selago**
2b(1a) Leaves broadest above the middle, toothed, 7-12 mm long, alternating in zones of shorter and longer lengths along the stem. **L. lucidulum**
3a(1b) Sporophylls (bracts of the spike) green and leaf like; sterile branches prostrate. 4
3b(1b) Sporophylls yellowish, larger than leaves. 5
4a(3a) Leaves mostly entire, to 0.7 mm; sporophylls widely spreading; cones to 12 mm thick; sterile shoots elongating to 1.5 dm beyond the 1-3 fertile erect stems. **L. inundatum**
4b(3a) Leaves mostly ciliate-denticulate to 1.2 mm broad; sporophylls appressed to ascending; cones to 8 mm thick; sterile shoots elongating to 4 dm beyond the 2-10 erect fertile stems. **L. appressum**
5a(3b) Leaves linear to lanceolate, free for most of their length, up to 7 mm long. 6
5b(3b) Leaves small and scale-like. 7
6a(5a) Leaves with hair-like bristle-tips; spikes 1-5 on scaly bracted peduncles; stems creeping and forking dichotomously. **L. clavatum**
6b(5a) Leaves not bristle-tipped; cones sessile. 8
7a(5b) Leaves in 4-5 rows around the stem, attached less than half their length; sterile branches rounded; cones less than 3.5 cm long, sessile on peduncles to 3 cm long. **L. sitchense**
7b(5b) Leaves in 4 rows around the stem, attached for more than half their length. 9
8a(6b) Ascending branches bushy branched; prostrate stems deeply underground. 10
8b(6b) Ascending branches simple or few forked; prostrate stems trailing on or near surface of the ground. 11
9a(7b) Lower and upper leaves similar; cones to 4.5 cm long, on leafy peduncles to 8 cm long. **L. sabinaefolium**

++

9b(7b) Lower leaves much reduced; peduncles with widely separated bract-like leaves. 12

10a(8a) Branchlets mostly ascending or erect, congested; leaves narrowly linear, usually less than 1 mm broad, not in evident ranks, equally spreading, ascending or incurved, often yellowish-green. **L. dendroideum**

10b(8a) Branchlets more horizontally spreading; leaves usually in 4-5 evident ranks, about 1 mm or more wide, with at least 2 ranks widely spreading, upper and lower ranks often reduced so leafy branches appear flattened, dark green. **L. obscurum**

11a(8b) Leaves ascending, 2.5-6 mm long, entire, acuminate. **L. annotinum** var. **pungens**

11b(8b) Leaves spreading or reflexed, 5.5-11 mm long. 13

12a(9b) Lower leaves of ultimate branches not especially flattened, 1-1.5 mm, about as long as marginal ones, often with a waxy bloom; horizontal stems deeply underground, annual constrictions conspicuous. **L. tristachyum**

12b(9b) Lower leaves up to 1 mm, much smaller than lateral ones; horizontal stems near surface; ultimate branches definitely flattened. 14

13a(8b) Leaves distinctly finely toothed, especially near tip, lanceolate to lance oblong, 1 mm or more wide. **L. annotinum**

13b(8b) Leaves entire or slightly finely toothed, less than 1 mm wide. **L. annotinum** var. **acrifolium**

14a(12b) Branches spreading in a regular fan-like manner, annual constrictions inconspicuous; horizontal stems mostly on or near soil surface; cones often with sterile, finely pointed tips. **L. digitatum**

14(12b) Branches irregularly ascending, not definitely fan-like; annual constrictions readily evident on many branches; horizontal stems mostly underground; cones usually with rounded or blunt tips. **L. complanatum**

Lycopodium annotinum L. Fig. p. A1. Map p. B1.
COMMON NAME(S): Stiff Club-moss.
FREQUENCY: Common throughout
HABITAT: Most common in open, dry hardwoods.
SUBSPECIFIC TAXA: Three varieties are separated in the key. Var. **annotinum** and var. **acrifolium** Fern. overlap morphologically and ecologically. Var. **pungens** (Pylaie) Desv. occurs rarely in high upland or exposed coastal habitats.
GENERAL RANGE: Nfld. to Alaska south to Penn., Wisc. and Colo. Var. **acrifolium** ranges south to Va. and var. **pungens** south to mts. of New England.

Lycopodium appressum (Chapm.) Lloyd & Underw.
SPECIAL NOTE: Reported from the Province by Scoggan, 1978, based on map 198 in Hultèn, 1958. No specimens have been seen from N.B. but should be searched for along the Maine-N.B. border. Reported from adjacent Washington Co., Maine, Richards et al, 1983.
GENERAL RANGE: Fla. to Tex. north to N.H., Maine and N.S.: West Indies.
SYNONYMS: **L. inundatum** L. var. **bigelovii** Tuck.

++

Lycopodium clavatum L. Fig. p. A1. Map p. B1.
COMMON NAME(S): Common or Running Club-moss. Courants verts.
FREQUENCY: Common throughout.
HABITAT: Dry open woods, pastures, clearings.
SUBSPECIFIC TAXA: Var. **monostachyon** Grev. & Hook. with peduncles mostly bearing a solitary cone occurs sporadically throughout the range of the species.
GENERAL RANGE: Northern North America: Europe and Asia.

Lycopodium complanatum L. Map p. B1.
COMMON NAME(S): Ground-Cedar.
FREQUENCY: Frequent throughout the Province.
HABITAT: Dry open woods, thickets and clearings.
SUBSPECIFIC TAXA: Previously separated into var. **complanatum** and var. **flabelliforme** Fern. The latter now considered a separate species, **L. digitatum** A. Br.
GENERAL RANGE: Greenland to Alaska south to New England. Minn. and Wash.

Lycopodium dendroidium Michx. Fig. p. A1. Map p. B2.
COMMON NAME(S): Round-branch Ground-Pine. Petite-pins.
FREQUENCY: Uncommon, more frequent southward.
HABITAT: Dry open heathland, exposed headlands and uplands, often on old burned areas.
GENERAL RANGE: Nfld. to Alaska south to Ala.; northeastern Asia.
SYNONYMS: **L. obscurum** L. var. **dendroideum** (Michx.) D.C. Eaton.

Lycopodium digitatum A. Br. Fig. p.A2. Map p. B2.
COMMON NAME(S): Ground-Cedar.
FREQUENCY: Common in the southern half of the Province, uncommon northward.
HABITAT: Wet open woods, clearings and pastures.
GENERAL RANGE: Similar to **L. complanatum** but restricted in Canada from Ont. to Nfld., N.B., P.E.I. and N.S.
SYNONYMS: **L. complanatum** L. var. **flabelliforme** Fern.

Lycopodium inundatum L. Fig. p. A2. Map p. B2.
COMMON NAME(S): Bog Club-moss.
FREQUENCY: Locally common; mostly in the southern part of the Province.
HABITAT: Wet shores, old roadways and boggy areas.
SUBSPECIFIC TAXA: See **L. appressum**.
GENERAL RANGE: Nfld. to Alaska south to W. Va., Ind. and Ore.

Lycopodium lucidum Michx. Fig. p. A2. Map p. B2.
COMMON NAME(S): Shining Club-moss.
FREQUENCY: Common throughout.
HABITAT: Rich moist woods.
GENERAL RANGE: Eastern North America.

Lycopodium obscurum L. Fig. p. A1.
COMMON NAME(S): Flat-branch Ground-Pine.
FREQUENCY: Common throughout.

HABITAT: Damp open woods, pastures and clearings.
SUBSPECIFIC TAXA: See **L. dendroideum.** The two species intergrade.
GENERAL RANGE: Northern North America; Asia.

Lycopodium sabinifolium Willd. Fig. p. A2. Map p. B2.
COMMON NAME(S): Juniper Club-moss, Ground-Fir.
FREQUENCY: Uncommon.
HABITAT: Open mostly dry woods and clearings.
SUBSPECIFIC TAXA: See **L. sitchense.**
GENERAL RANGE: Northern North America; Europe, eastern Asia.

Lycopodium selago L. Fig. p. A2. Map p. B2.
COMMON NAME(S): Fir club-moss, Mountain Club-moss, Rat's Tail. Sélagine, Herbe aux porcs.
FREQUENCY: Rare in most of the Province; uncommon along the Fundy Coast.
HABITAT: Ledges, peaty barrens and exposed shores and headlands.
GENERAL RANGE: Northern North America; Eurasia.

Lycopodium sitchense Rupr. Map p. B2.
COMMON NAME(S): Sitka Club-moss.
FREQUENCY: Rare throughout the Province.
HABITAT: Open dry heathland.
SPECIAL NOTE: Sometimes considered a variety of **L. sabinifolium.**
GENERAL RANGE: Alaska to Nfld. south to Oreg., Mont., Mich. Pa., and New Eng.

Lycopodium tristachyum Pursh Fig. p. A2. Map p. B2.
COMMON NAME(S): Ground-Cedar. Lycopode à trois épis.
FREQUENCY: Scattered throughout the province.
HABITAT: Dry barrens, open hardwoods, clearings.
GENERAL RANGE: Northern Ont. to Nfld. south to Minn., Tenn., and N.C.; Europe.

SELAGINELLACEAE (Spikemoss Family)

SELAGINELLA Beauv. Spikemoss. Sélaginelle

1a Fertile erect stems unbranched, from base; fertile leaves of cone cylindrical, spreading to ascending, finely toothed; plants of calcareous wet rocks, banks, bogs. **S. selaginoides**

1b Fertile ascending stems profusely branched from base; fertile leaves of four-sided cone erect, compressed, finely ciliate; plants of dry, exposed rocky sites. **S. rupestris**

Selaginella rupestris (L.) Spring Fig. p. A2. Map p. B2.
FREQUENCY: Very rare throughout the Province. Only found in the southern half such as below the Mactaquac Dam, York Co., and the Bluff, Rockville, Kings Co.
HABITAT: Exposed dry ledges.

++

SPECIAL NOTE: For more information see The Rare Vascular Plants of New Brunswick. H.R. Hinds. Syllogeus No. 50. 1983.
GENERAL RANGE: Alta. to N.S. south to S. Dak., Tex. and Ga.; Greenland.

Selaginella selaginoides (L.) Link Fig. p. A2. Map p. B2.
FREQUENCY: Very rare; known in southern N.B. only from the Confluence on the Point Wolfe River, Fundy National Park, Albert Co. Scattered on calcareous ledges along the Restigouche River, Restigouche Co.
HABITAT: Wet calcareous ledges, mossy banks.
SPECIAL NOTE: For more information see The Rare Vascular Plants of New Brunswick. H.R. Hinds. Syllogeus No. 50. 1983.
GENERAL RANGE: Alaska to Nfld. south to Nev., Wyo., Minn. and Maine; Greenland, iceland; Eurasia.

ISOETACEAE (Quillwort Family)

ISOETES L. Quillwort. Isoète

This group is extremely difficult to identify especially since hybridization appears to occur frequently among the species. W. Carl Taylor (pers. comm.) has found intergrading, hybrid forms which produce abortive, polymorphic spores. These hybrids are not clearly understood and have been named **I. X heterospora** A.A. Eaton and **I. X harveyi** A.A. Eaton. The latter appears to be fairly frequent in the southern part of the Province.

1a Megaspores covered with sharp or blunt, sometimes forked or toothed spinules; microspores +/- smooth; plants mostly submerged. **I. echinospora**
1b Megaspores not spinulose. 2
2a(1b) Megaspores with low, rounded, smooth ridges forming a loose network; plants submerged. 3
2b(1b) Megaspores with a network of +/- sharp jagged ridges; some species immersed. 4
3a(2a) Megaspores averaging .635 mm in diameter; ridges very low and +/- moundlike. **I. hieroglyphica**
3b(2a) Megaspores averaging .519 mm in diameter; ridges higher, more distinct and continuous. **I. acadiensis**
4a(2b) Megaspores averaging over .600 mm; leaves usually stiff and dark green; plants submerged. **I. macrospora**
4b(2b) Megaspores averaging less than 600 m in diameter; leaves +/- soft and bright green or yellow-green.
5a(4b) Megaspore ridges honeycomb-like on rounded (distal) side; plants submerged. **T. tuckermanii**
5b(4b) Megaspore ridges close and +/- branched but not honeycomb-like, sometimes with isolated flared crested columns; plants often emerged on muddy shores. **I. riparia**

ISOETACEAE

++

Isoëtes acadiensis Kott

FREQUENCY: Very rare; only reported from Bald Mnt. Pond, Harvey Station, York Co., and Chamcook Lake, Charlotte Co.

HABITAT: Shallow water

SPECIAL NOTE: For more information see The Rare Vascular Plants of New Brunswick. H.R. Hinds. Syllogeus No. 50. 1983.

GENERAL RANGE: Northeastern U.S. and Canada from Mass. north to N.S.

Isoëtes echinospora Dur. Fig. p. A2. Map p. B2.

FREQUENCY: Probably common throughout the Province.

HABITAT: Shallow water and wet shores.

SUBSPECIFIC TAXA: Our material is ssp. **muricata** (Durieu) Love & Love.

GENERAL RANGE: Alaska to Nfld. south to Calif., Colo., Ind. Pa., and N.J.; Greenland, Iceland and Eurasia.

SYNONYMS: **Isoëtes muricata** Dur.

Isoëtes X harveyi A.A. Eaton Map p. B2.

FREQUENCY: Uncommon; reported from Chamcook Lake, Charlotte Co., Washademoak Lake, the Keyhole, Cumberland Pt., Queens Co. and the SW Miramichi River near Deersdale, York Co.

HABITAT: Shallow water with muddy bottom.

SPECIAL NOTE: Apparently a recurring hybrid of **I. echinospora** and **I. tuckermanii**.

Isoëtes hieroglyphica Eat. Fig. p. A2.

FREQUENCY: Very rare, probably overlooked. Collected in Tower Lake below Harvey Mtn., Harvey Station, York Co.

HABITAT: Clear, cold still waters, up to 3 m deep.

GENERAL RANGE: Reported from Maine, Ont., Que. and N.S.

SYNONYMS: **I. macrospora** Dur. forma **hieroglyphica** (Eat.) Pfeiff.

Isoëtes macrospora Dur. Fig. p. A2.

SPECIAL NOTE: Reported from the Province by Scoggan, 1978-79 and Boivin, 1966-67, but both included **I. tuckermanii** in their concept of the species. The report by Kott, 1980 may possibly be based on hybrid material. No specimens have been seen.

Isoëtes riparia Engelm. Fig. p. A2.

SPECIAL NOTE: Reported by Scoggan, 1978-79 and Boivin, 1966-67, but no specimens seen. Probably north of range.

Isoëtes tuckermanii A. Braun Fig. p. A3. Map p. B3.

FREQUENCY: Uncommon in southern half of Province.

HABITAT: Shallow pond margins and shores.

SUBSPECIFIC TAXA: Var. **borealis** A. A. Eat. with longer and stouter leaves, is reported from Tower Lake, Harvey Station, Charlotte Co.

GENERAL RANGE: Northeastern North America south to southern N. Eng. and N.Y.

SYNONYMS: Sometimes combined with **I. macrospora**.

+++

OPHIOGLOSSACEAE (Adder's-tongue Family)

1a Sterile blade undivided; veins reticulate; fertile spike undivided. **Ophioglossum vulgatum**

1b Sterile blade variously divided to simple and lobed; veins free; fertile portion in panicles (rarely simple spikes). **Botrychium**

BOTRYCHIUM Sw. Moonwort, Grape-fern. Botryche.

1a Sterile frond branching from common axis near or underground, evergreen. 2

1b Sterile frond branching from common axis some distance above ground, mostly withering in autumn. 3

2a(1a) Terminal segments of sterile frond more elongate than lower segments, lanceolate to lance-ovate, up to several times longer than broad. 4

2b(1a) Terminal segments of sterile fronds about the same size and shape as lower segments, not more than 3 times as long as broad; sterile blades remaining green over winter. 5

3a(1b) Sterile blade large, up to 20 cm long and 30 cm wide; twice or thrice compound and sessile from near middle of plant, 9-34 cm from ground. **B. virginianum**

3b(1b) Sterile blade smaller, less than 9 cm long and only divided once **or** entire or merely lobed. 6

4a(2a) Terminal segments of sterile frond elongate, broadly ovate, obtuse; roots averaging 2.5 mm alive, 2.0 mm dried, 1 cm from stem, light tan to ivory-gray with scarce circular ridges; plants of acid secondary woods or swamps; margins of pinnules nearly entire to shallowly denticulate; unfolding leaves limegreen, wintering leaves dark green. **B. oneidense**

4b(2a) Terminal segments of sterile frond oblong-ovate to oblong-lanceolate, acute; roots averaging 3.0 mm alive, 2.5 mm dried 1 cm from stem, dark gray-brown with well developed circular ridges; unfolding leaves reddish, overwintering leaves bronze where well exposed; margins of pinnules denticulate to very deeply and coarsely lacerate; plants of open fields, brushy meadows, dry woods, swamps. **B. dissectum**

5a(2b) Ultimate division of blade rather crowded and often overlapping, bi-tripinnate; blades less than 1 dm long with petioles 1-4 cm; pinnules usually ovate, obtuse; plants of sandy open areas. **B. multifidum**

5b(2b) Ultimate divisions usually not overlapping, less crowded, 3-4 pinnate; blades often more than 1 dm long with petioles 3-8 cm long, slightly pointed; plants of open fields and woods in acid soils. **B. multifidum** var. **intermedium**

6a(3b) Sterile frond sessile on main axis, triangular in outline, lowest pinnae elongate. **B. lanceolatum**

6b(3b) Sterile frond sessile or short stalked, usually narrower and oblong or ovate in outline. 7

7a(6b) Sterile blade once divided; leaf parts erect in bud. 8

7b(6b) Sterile blade twice divided; upper part of blade folded over and partly enclosing fertile part in bud. **B. matricariifolium**

8a(7a) Sterile blade mostly sessile from near the middle of the plant, with up to 7 or more pairs of fan-shaped pinnae; fertile blade paniculate. **B. minganense**

8b(7a) Sterile blade usually distinctly stalked, simple or with up to 5 pairs of obliquely ovate segments. 9

9a(8b) Sterile frond borne near the summit of the stem: often growing in shade. **B. simplex** var. **tenebrosum**

9b(8b) Sterile blade borne below middle of main axis; basal pinnae usually larger than second pair; plants usually growing in the open. **B. simplex**

Botrychium dissectum Spreng. Fig. p. A3. Map p. B3.
FREQUENCY: Scattered, mostly along the Saint John River valley.
HABITAT: Pastures, swampy woods and rich hardwoods.
SUBSPECIFIC TAXA: Forma **dissectum** has the divisions of the blade deeply incised or lacerate while forma **obliquum** (Muhl.) Fern. has the blade divisions nearly entire to minutely toothed.
GENERAL RANGE: Minn. and Ont. to N.S. south to Tex. and Fla.

Botrychium lanceolatum (Gmel.) Angstr. Fig. p. A3. Map. p. B3.
COMMON NAME(S): Lance-leaved Grape-fern.
FREQUENCY: Rare, especially in the northern part of the Province.
HABITAT: Rich hardwoods and clearings.
GENERAL RANGE: Alaska to Nfld. south to Ohio, Pa., and N.J.; Greenland, Iceland and Eurasia.
SUBSPECIFIC TAXA: Var. **angustisegmentum** Pease & Moore, with thinner less crowded pinnae usually more acute at apex and to 2.5 mm broad near base, is found rarely in the southern part of Province.

Botrychium matricariifolium A. Br. Fig. p. A3. Map p. B3.
FREQUENCY: Scattered throughout the Province.
HABITAT: Clearings and rich hardwoods.
GENERAL RANGE: B.C. to Nfld. south to Idaho, S. Dak., Ohio, and Va.; S. America; Eurasia.

Botrychium minganense Vict. Fig. p. A3. Map p. B3.
COMMON NAME(S): Moonwort.
FREQUENCY: Very rare; found only on Sagamook Mtn., Mt. Carleton Provincial Park, Restigouche Co.
HABITAT: Shaded mossy banks.
SUBSPECIFIC TAXA: Sometimes considered a variety of **B. lunaria**.
SPECIAL NOTE: For more information see The Rare Vascular Plants of New Brunswick. H. Hinds. Syllogeus No. 50. 1983.

Botrychium multifidum (Gmel.) Rupr. Fig. p. A3. Map p. B3.
COMMON NAME(s): Grape-fern.
FREQUENCY: Scattered throughout the Province.
HABITAT: Sandy thickets, stabilized dunes, clearings and rich hardwoods.
SUBSPECIFIC TAXA: Var. **intermedium** (D.C. Eat.) Farw., with blade usually slightly

broader than long and ultimate divisions not overlapping, has been collected sparingly in the southern part of the Province.

Botrychium oneidense (Gilbert) House
SPECIAL NOTE: Reported from the province by Scoggan, 1978-79 as **B. dissectum** forma **oneidense**, but no voucher specimens seen.

Botrychium simplex E. Hitch. Fig. p. A3. Map p. B3.
FREQUENCY: Very rare; scattered sparingly throughout the Province.
HABITAT: Rocky slopes, clearings, and rich hardwoods.
SUBSPECIFIC TAXA; Varieties separated as follows.

1a Blade borne near summit of stem, simple or with 1-5 pairs of well separated lobes. var. **tenebrosum** (A.A. Eat.) Clausen
1b Blade borne from near base to slightly above the middle of the stem, lateral broad-based lobes 1-3 pairs. var. **simplex**.

SPECIAL NOTE: Young specimens of **B. matricariifolium** are very similar to var. **tenebrosum**.
GENERAL RANGE: B.C. to Nfld. south to Calif. and N.J.; Eurasia.

Botrychium virginianum (L.) Sw. Fig. p. A3. Map p. B3.
COMMON NAME(s): Rattlesnake-fern. Botryche de Virginie.
FREQUENCY: Common throughout the Province.
HABITAT: Mostly found in rich hardwoods.
GENERAL RANGE: Alaska to Nfld. south to Oreg. and Fla.; Eurasia.

OPHIOGLOSSUM L. Adder's-tongue. Herbe sans couture

Ophioglossum vulgatum L. Fig. p. A3. Map p. B3.
COMMON NAME(S): Northern Adder's-tongue.
FREQUENCY: Rare; known only from a few scattered localities in the south of the Province such as on Grand Manan Island, Charlotte Co. and Waterford, Kings Co.
SUBSPECIFIC TAXA: Our plants are distinguished from others outside our range as var. **pseudopodium** (Blake) Farw.
SPECIAL NOTE: For more information see The Rare Vascular Plants of New Brunswick. H. Hinds. Syllogeus No. 50. 1983.

OSMUNDACEAE (Flowering Fern Family)

OSMUNDA L. Flowering Fern. Osmonde

1a Frond blade pinnate-pinnatifid; spore bearing segments not confined to the tip. 2
1b Frond blade completely bipinnate; spore bearing segments, if present, confined to the tip of the vegetative frond. **O. regalis**
2a(1a) Blades completely dimorphic, with the completely fertile fronds having reddish-brown sporangia; the vegetative frond with conspicuous, dense tufts of cinnamon hairs in axils of the pinnae; margins of vegetative segments

smooth; tips of pinnae obtuse; plants of mostly wet, open or partly shaded habitats. **O. cinnamomea**

2b(1a) Blades with dark brown spore bearing segments in the middle portion of the leaf; cinnamon tufts lacking or scant and inconspicuous on the vegetative fronds; margins of vegetative segments usually pubescent, tips of pinnae acuminate; plants of dryer shaded habitats. **O. claytoniana**

Osmunda cinnamomea L. Fig. p. A3. Map p. B3.
COMMON NAME(s): Cinnamon-fern. Osmonde cannelle.
FREQUENCY: Common throughout except rare in the extreme northwest of the Province.
HABITAT: Wet woods, pastures and swamps.
GENERAL RANGE: Ont. to Nfld. south to N. Mex. and Fla.; S. America; Asia.

Osmunda claytoniana L. Fig. p. A3. Map. p. B3.
COMMON NAME(S): Interrupted Fern. Osmonde de Clayton
FREQUENCY: Common throughout the Province.
HABITAT: Moist woods and thickets; bottomlands.
GENERAL RANGE: Man. to Nfld. south to Ark. and Ga; Asia.

Osmunda regalis L. Fig. p. A3. Map p. B3.
COMMON NAME(S): Royal Fern. Osmonde royale.
FREQUENCY: Common towards the southern part of the province, rarer northwards.
HABITAT: Wet, rocky woods, stream bands and shores.
GENERAL RANGE: Ont. to Nfld. south to Mexico and Fla.; Eurasia.
SUBSPECIFIC TAXA: North American material differs slightly from Eurasian plants and has been named var. **spectabilis** (Wills.) Gray.

POLYPODIACEAE (Fern Family)

Our ferns have been divided by fern specialists into five new families; **Adiantum** and **Cryptogramma** in the **ADIANTACEAE**, **Woodwardia** in the **BLECHNACEAE**, **Polypodium** in the **POLYPODIACEAE**, **Dennstaedtia** and **Pteridium** in the **DENNSTAEDTIACEAE**; and all other genera in the **ASPLENIACEAE**. For reasons of simplicity these are here all combined in the **POLYPODIACEAE**.

1a Fronds arising singly from a forking underground rhizome. 2
1b Fronds tufted or densely clumped from a more upright, mostly unbranched underground rhizome. 3
2a(1a) Frond once deeply pinnatifid; sori large, without indusia. **Polypodium virginianum**
2b(1a) Frond variously pinnate to pinnate-pinnatifid. 4
3a(1b) Frond undivided, strap-shaped, about 3 cm wide and to 30 cm long; veins free; sori paired, long and parallel. **Asplenium scolopendrium**
3b(1b) Frond variously pinnate to pinnate-pinnatifid. 5
4a(2b) Indusia peltate (umbrella-shaped). **Polystichum**
4b(2b) Indusia various, not peltate. 6
5a(3b) Sori elongate or curved; indusia attached along one side. 7

+++

5b(3b) Sori various, not elongate or curved. 8

6a(4b) Fronds ternately compound, divided into three more or less similar bipinnate segments. 9

6b(4b) Fronds not ternately compound. 10

7a(5a) Blade simple pinnate; small tufted mostly rock-inhabiting plants. **Asplenium**

7b(5a) Blade bi-tripinnate-pinnatifid; larger forest-inhabiting species. **Athyrium**

8a(5b) Frond tapering gradually to the base, the basal pinnae less than one-fifth the length of the largest pinnae; sterile fronds noticably different from the fertile, feather-like fronds with sori enclosed in inrolled pinnules. **Matteuccia struthiopteris**

8b(5b) Fronds not significantly narrowed at base; fertile and sterile fronds not significantly different. 11

9a(6a) Large, coarse plants with fronds to 150 cm; sori mostly continuous along margins, covered by revolute edge of frond. **Pteridium aquilinum**

9b(6a) Small delicate plants with fronds to 23 cm; sori near margins, separate, round without indusium. **Gymnocarpium**

10a(6b) Veins of sterile pinnae forming a network; sterile and fertile fronds very different, sori enclosed in inrolled pinnules. **Onoclea sensibilis**

10b(6b) Veins free (open branching towards margins) 12

11a(8b) Stipes eventually deciduous at a distinct joint near base, leaving stubble of uniform length; sori circular, the indusium attached below the sorus with the indusial hairs curving over it; small tufted plants of ledge crevices. **Woodsia**

11b(8b) Stipes not jointed; sori round, covered by a centrally attached kidney-shaped indusium; usually larger plants of various habitats. **Dryopteris**

12a(10b) Sori marginal, covered by recurved margin. 13

12b(10b) Sori various, not situated along margins. 14

13a(12a) Stipe shiny black or dark brown, divided at summit into two subequal rachises bearing pinnules along one side; sterile and fertile fronds similar. **Adiantum pedatum**

13b(12a) Stipe light green to brown, not divided at summit; sterile frond usually pinnate-pinnatifid with broad obtuse segments; very delicate; fertile frond more lanceolate, bipinnate with ovate or lanceolate pinnae, thicker and more erect than sterile. **Cryptogramma stelleri**

14a(12b) Blade mostly bipinnate-pinnatifid or bipinnate-bipinnatifid, minutely glandular pubescent especially beneath. **Dennstaedtia punctilobula**

14b(12b) Blade pinnate-pinnatifid to bipinnatifid with inconspicuous hairs especially along midribs of pinnules. **Thelypteris**

ADIANTUM L. Maidenhair. Adiante

Adiantum pedatum L. Fig. p. A4. Map p. B3.

COMMON NAME(S): Maidenhair Fern. Capillaire.

FREQUENCY: Uncommon; found occasionally in the Saint John River valley, in Victoria, Carleton, York and Kings Counties and also an island near the mouth of the Matapedia River, Restigouche Co.

++

HABITAT: Rich hardwoods in circumneutral soils.
GENERAL RANGE: Ont. to N.S. south to Calif. and Ga.

ASPLENIUM L. Spleenwort. Doradille

1a Frond undivided, strap-shaped, 3 cm wide; sori long, parallel, paired. **A. scolopendrium**

1b Frond once pinnate. 2

2a(1b) Rachis and stipe shiny purplish-brown; frond to 2.5 dm long. **A. trichomanes**

2b(1b) Rachis green; stipe brown at base, green above; fronds 1.5 dm long. **A. viride**

Asplenium scolopendrium L.
COMMON NAME(S): Hart's-tongue.
SPECIAL NOTE: First reported from the Meduxnekeag River, Carleton Co. by Lawson, 1880. Specimens supposedly collected from this site have been examined by B. Boivin and this author and conform closely with the European form of this species. It would thus seem that either a hoax was deliberately perpetrated or that greenhouse specimens of the European form of Hart's-tongue became locally established in a limestone ravine of the Meduxnekeag River.
GENERAL RANGE: Southern Ont., central N.Y. and Tenn.; Mexico; Europe; Japan; Africa.
SYNONYMS: **Phyllitis scolopendrium** (L.). Newm.

Asplenium trichomanes L.
COMMON NAME(S): Maidenhair-Spleenwort. Asplénie chevelue.
FREQUENCY: Very rare; only known from Jewetts Ck., York Co. and Mt. Zachy-Jonah, Albert Co.
HABITAT: Shaded, calcareous ledges.
SPECIAL NOTE: For more information see The Rare Vascular Plants of New Brunswick. H. Hinds. Syllogeus No. 50. 1983.
GENERAL RANGE: Alaska to Nfld. south to Oreg. and Ga.; Iceland; Eurasia.

Asplenium viride Huds. Fig. p. A4. Map. p. B3.
COMMON NAME(S): Green Spleenwort. Doradille verte.
FREQUENCY: Very rare; the range appears to be bipolar with a cluster of locations near the Bay of Fundy and another near or along the northern border of the Province.
HABITAT: Calcareous ledges.
SPECIAL NOTE: For more information see The Rare Vascular Plants of New Brunswick. H. Hinds. Syllogeus No. 50. 1983.
GENERAL RANGE: Nfld. to Wisc. and Colo. northwest to Alaska; Greenland; Eurasia.

ATHYRIUM Roth

1a Blade pinnate to deeply pinnatifid; pinnae segments rounded at apex; sori straight or only slightly curved. **A. thelypteroides**

1b Blade bi-tripinnate; pinnules narrowly rounded or acute at apex; sori usually curved or hooked at one end. **A. filix-femina**

++

Athyrium filix-femina (L.) Roth Fig. p. A4.
COMMON NAME(S): Lady-Fern. Fougère femelle.
FREQUENCY: Common throughout.
HABITAT: Moist woods.
SUBSPECIFIC TAXA: Our plants are var. **angustum** (Willd.) Lawson. Forma **rubellum** (Gilbert) Farw. has wine-red stipe and rachis. Var. **cyclosorum** Rupr., with kidney-shaped sori and glabrous indusia is found in eastern Canada from the Gaspé Pen. and Nfld. and should be searched for in northern N.B.
GENERAL RANGE: Man. to Nfld. south to Va. and Colo.
SYNONYMS: **Athyrium filix-femina** (L.) Roth var. **michauxii** (Spreng.) Farw.

Athyrium thelypteroides (Michx.) Desv. Fig. p. A4. Map p. B4.
COMMON NAME(S): Silvery Spleenwort. Athyrie argentée.
FREQUENCY: Locally common, except uncommon in the Eastern Lowlands.
HABITAT: Rich moist hardwoods.
GENERAL RANGE: N.S. and Gaspé; Pen. to Minn. south to Ga. and Ala.; Asia.
SYNONYMS: **Deparia acrostichoides** (Sw.) M. Kato ined.

CRYPTOGRAMMA R. Br. Rock-brake

Cryptogramma stelleri (Gmel.) Prentl Fig. p. A4. Map p. B4.
COMMON NAME(S): Slender Cliff-brake.
FREQUENCY: Rare; most commonly found along the Restigouche River, Restigouche Co.
HABITAT: Moist, shaded calcareous ledges.
GENERAL RANGE: Alaska to Nfld. south to Wash. and W.Va.; Asia.

CYSTOPTERIS Bernh. Bladder-Fern. Cystoptéride

1a Frond long attenuate, bearing abundant bulblets beneath, towards late summer; lowest pinnae longest; veins mostly ending at sinuses; basal part of pinnae not winged. **C. bulbifera**
1b Frond lanceolate but not attenuate, rarely bulblet-bearing; lowest pinnae not longest; veins mostly ending on a tooth or unnotched margin; basal part of pinnae winged. 2
2a(1b) Indusium glandular; frond 3.8-4.8 dm long. **C. X laurentiana**
2b(1b) Indusium glabrous, frond to 3 dm long. **C. fragilis**

Cystopteris bulbifera (L.) Bernh. Fig. p. A4. Map p. B4.
COMMON NAME(S): Bulblet Bladder-Fern. Cystoptéride bulbifère
FREQUENCY: Locally common in calcareous areas of the Province.
HABITAT: Wet limestone or other calcareous ledges or soils.
GENERAL RANGE: Ont. to Nfld. south to Ariz. and Ga.

Cystopteris fragilis (L.) Bernh. Fig. p. A4. Map p. B4.
COMMON NAME(S): Fragile Fern, Brittle Fern. Cystoptéride fragile

++

FREQUENCY: Scattered throughout the Province.
HABITAT: Shaded ledge crevices and wooded rocky slopes.
SUBSPECIFIC TAXA: Var. **mackayii** Lawson differs from var. **fragilis** in having smaller indusia (about 0.5 mm long) which are nearly entire or shallowly lobed instead of indusia to 1 mm long and deeply cleft at apex and base of pinnules wedge-shaped instead of rounded.
GENERAL RANGE: Alaska to Nfld. south to Calif. and Va.; S. America; Greenland; Iceland; Eurasia. Var. **mackayii** has been found from N.S. and southern Que. to Minn. south to N.C.

Cystopteris X laurentiana (Weath.) Blasdell Map p. B4.
SPECIAL NOTE: This is the hybrid between **C. bulbifera** and **C. fragilis**.
FREQUENCY: Very rare; found occasionally where the two species occur in the same general area such as at Limestone Jct., Victoria Co., near Walker Settl., Kings Co., and at Tetagouche Falls, Gloucester Co.
GENERAL RANGE: Ont. and Wisc. to eastern Que., to Nfld. and N.S.

DENNSTAEDTIA Bernh

Dennstaedtia punctilobula (Michx.) Moore Fig. p. A4. Map p. B4.
COMMON NAME(S): Hay-scented Fern. Fougère foin.
FREQUENCY: Common in the southern and rare in the northern half of the Province.
HABITAT: Rocky open woods, slopes, meadows and pastures.
GENERAL RANGE: Ont. to N.S. south to Ark. and Ga.

DRYOPTERIS Adans. Shield-Fern, Wood-Fern

1a Underside of blade with numerous scales; pinnae small, 4.8 mm broad; indusia large and often overlapping. **D. fragrans**
1b Underside of blade nearly smooth or glandular, but not scaley; pinnae 12-90 mm broad. 2
2a(1b) Sori positioned close to the margin; stipe very scaley especially towards base; pinnule margins usually entire or wavy, but occasionally toothed. **D. marginalis**
2b(1b) Sori medial or near the midvein. 3
3a(2b) Ultimate segments finely spinulose (teeth ending in mucronate or short bristle-like tips). 4
3b(2b) Ultimate segments shallow toothed, not spinulose. 5
4a(3a) Indusia glandular (especially when young) margin jagged; plants +/- ever-green. 6
4b(3a) Indusia glabrous, margin +/- entire; plants mostly yellow-brown in winter. 7
5a(3b) Blade linear to lance-oblong; basal pinnae broadly triangular. 8
5b(3b) Blade oblong-triangular; basal pinnae tapering gradually from middle to base; sori widely spaced, nearer midvein than margin. **D. goldiana**

++

6a(4a) Blade oval to narrowly triangular, bipinnate-pinnatifid to tripinnate; pinnae tapering gradually to tip. 9

6b(4a) Blade lance-oblong, pinnate-pinnatifid above, bipinnate below; fertile fronds longer and more erect than sterile; lower pinnae more broadly triangular, usually shorter than median pinnae. **D. X boottii**

7a(4b) Blade oval to triangular. 10

7b(4b) Blade lance-oblong; lower pinnules of basal pinnae about as long as opposing pinnae. **D. X uliginosa**

8a(5a) Frond linear-oblong to narrowly lance-oblong to 8 dm long, 15 cm wide; fertile fronds narrower and more erect than sterile; fertile pinnae often at right angles to rachis (ladder-like); scales concolorous. **D. cristata**

8b(5a) Frond lance-oblong to 13 dm long, 29-30 cm wide; fertile and sterile fronds more uniform and flat; scales near base of stipe with dark centre. **D. clintoniana**

9a(6a) Lower pinnules of basal pinnae as long or longer than the one next to it; pinnae oblique to the rachis; stipe-scales light brown. **D. X triploidea**

9b(6a) Lower pinnules of basal pinnae shorter than the one adjacent to it; pinnae mostly at right angles to the rachis and more lanceolate; stipe-scales with a dark centre. **D. intermedia**

10a(7a) Blades oval to oval-triangular; lower pinnules of basal pinnae 2-3 times as long as the opposing pinnules and closer to them than to second ones. **D. spinulosa**

10b(7a) Blades broadly triangular; lower pinnules of basal pinnae 3-5 times as long as opposing pinnules and closer to second upper pinnules (at least 5 mm distant from upper basal ones); the remaining pinnae gradually tapering to tip (upper and lower pinnules about the same length). **D. campyloptera**

Dryopteris X boottii (Tuck.) Underw. Fig. p. A4. Map p. B4.
COMMON NAME(S): Boott's Shield-Fern.
SPECIAL NOTE: This is the fertile hybrid between **D. cristata** and **D. intermedia**.
FREQUENCY: Occasionally found where the two parents occur in close proximity.

Dryoptris campyloptera (Kunze) Clarkson Fig. p. A4. Map p. B4.
COMMON NAME(S): Mountain Wood-Fern, Spreading Wood-Fern.
FREQUENCY: Frequent along the Bay of Fundy and in the Central Highlands, uncommon elsewhere.
HABITAT: Cool, moist deciduous or mixed woods.
GENERAL RANGE: Alaska to Nfld. south, to Wash. and N.C.
SYNONYMS: **D. spinulosa** (Muell.) Watt var. **americana** (Fisch.) Fern.; **D. austriaca** (Jacq.) Woynar; **D. dilatata** (Hoffm.) Gray ssp. **americana** (Fisch.) Hultén.

Dryopteris clintoniana (D.C. Eat.) Dowell
COMMON NAME(S): Clinton's Shield-Fern.
SPECIAL NOTE: Probably originally of hybrid origin between **D. goldiana** and **D. cristata**. It functions as a distinct species.
FREQUENCY: Very rare; found only once in the Province at Maplehurst Marl Pit, Carleton Co.
GENERAL RANGE: Northeastern North America north to Ont. and west central N.B.

++

SYNONYMS: **D. cristata** (L.) Gray var. **clintoniana** (D.C. Eat.) Underw.; **D. austriaca** (Jacq.) Woynar var. **clintoniana** (D.C. Eat.) Underw.

Dryopteris cristata (L.) Gray Fig. p. A4. Map p. B4.
COMMON NAME(S): Crested Wood-Fern. Dryoptéride accrêtée.
FREQUENCY: Common, becoming rarer northwards.
HABITAT: Swamps and wet woods.
GENERAL RANGE: B.C. to Nfld. south to Idaho and N.C.; Europe; Asia.

Dryopteris fragrans (L.) Schott Fig. p. A4. Map p. B4.
COMMON NAME(S): Fragrant Fern. Dryoptéride odorante.
FREQUENCY: Rare; occasionally found in quantity in widely separated localities such as along the Bay of Fundy in eastern St. John Co. and western Albert Co. and along the Restigouche River in Restigouche Co.
HABITAT: Dry, often exposed ledges.
SUBSPECIFIC TAXA: Our plants are all var. **remotiuscula** Komarov.
GENERAL RANGE: Yukon to Nfld. south to Wisc. and northern N. Eng.

Dryopteris intermedia (Muhl.) Gray Fig. p. A5. Map p. B4.
COMMON NAME(S): Glandular Wood-Fern, Fancy Fern, Evergreen Wood-Fern.
FREQUENCY: Common throughout the Province.
HABITAT: Moist shaded woods and rocky slopes.
GENERAL RANGE: Eastern North America.
SYNONYMS: **D. spinulosa** (Muhl.) Watt var. **intermedia** (Muhl.) Underw.; **D. austriaca** (Jacq.) Woynar var. **intermedia** (Muhl.) Morton.

Dryopteris marginalis (L.) Gray Fig. p. A5. Map. p. B5.
COMMON NAME(S): Marginal Shield-Fern. Dryoptèride marginale.
FREQUENCY: Common in most of the southern part of the Province; rare in the north and east.
HABITAT: Rocky slopes.
GENERAL RANGE: Ont. to N.S. south to Okla. and Ga.
SUBSPECIFIC TAXA: Forma **elegans** (Rob.) Gray, with the pinnae bipinnatifid is reported from St. Andrews, Charlotte Co. and Pokiok Falls, York Co.

Dryopteris spinulosa (Muell.) Watt Fig. p. A5. Map p. B5.
COMMON NAME(S): Spinulose Shield-Fern. Dryoptéride spinuleuse.
FREQUENCY: Common throughout the Province.
HABITAT: Swamps and wet woods.
GENERAL RANGE: Northern North America.
SYNONYMS: **D. austriaca** (Jacq.) Woynar var. **spinulosa** (Muell.) Fiori

Dryopteris X triploidea Wherry Fig. p. A5. Map p. B5.
SPECIAL NOTE: This is the hybrid between **D. spinulosa** and **D. intermedia**.
FREQUENCY: Uncommon, but usually found wherever the parent species occur together.
HABITAT: Rich swampy and/or rocky woods.

++

GENERAL RANGE: Minn. to Nfld. south to Ohio and Va.
SYNONYMS: **D. austriaca** (Jacq.) Woynar var. **fructuosa** (Gilbert) Morton; **D. spinulosa** (Muell.) Watt var. **fructuosa** (Gilbert) Trudell.

Dryopteris X uliginosa (A. Br.) Druce Map p. B5.
SPECIAL NOTE: This is the hybrid between **D. cristata** and **D. spinulosa**.
FREQUENCY: Very rare, occurring occasionally where the parents grow in close proximity such as on Heron Is., Restigouche Co., or near Cormier Village, Westmorland Co.

GYMNOCARPIUM Newm. Oak Fern

1a Rachis glabrous or rarely scattered glandular; normally only basal pinnae stalked, almost as large as remainder of blade; lower basal pinnule of basal pinnae of the same size as third pinna or slightly larger. **G. dryopteris**

1b Rachis, at least at the junction of second and third pinna pair, densely glandular; lowermost basal pinnule of basal pinnae of the same size as forth pinna. 2

2a(1b) Lowermost basal pinnule of basal pinnae usually much larger than corresponding upper pinnule; middle lobes or pinnules of the above pinnae nearly at right angles with the rachis; distance between glands usually shorter than the glands. **G. robertianum**

2b(1b) Lowermost basal pinnule of basal pinnae only slightly larger than corresponding upper pinnule; middle lobes or pinnules of basal pinnae oblique to rachis or curved; distance between glands usually longer than glands. **G. jessoense**

Gymnocarpium dryopteris (L.) Newm. Fig. p. A5.
COMMON NAME(S): Oak Fern. Dryoptéride disjointe.
FREQUENCY: Common throughout the Province.
HABITAT: Moist woodlands.
GENERAL RANGE: Alaska to Nfld. south to Oreg. and Va.; Greenland; Iceland; Eurasia.
SYNONYMS: **Dryopteris disjuncta** (Rupr.) Morton; **Phegopteris dryopteris** (L.) Fée; **Thelypteris dryopteris** (L.) Slosson.

Gymnocarpium jessoense (Koidz.) Koidz.
SPECIAL NOTE: Recently reported from the Province by Sarvela et al. (1981) from a collection by J. Brittain on the upper Restigouche River in 1888 (CAN, QK). This is the only N.B. record. For more information see The Rare Vascular Plants of New Brunswick. H. Hinds. Syllogeus No. 50. 1983.
GENERAL RANGE: Alas. to NWT: B.C. to N.B.; Minn, Mich. and Wis.
SUBSPECIFIC TAXA; The North American plants have been named ssp. **parvulum** Sarvela.

Gymnocarpium robertianum (Hoffm.) Newm. Fig. p. A5. Map p. B5.
COMMON NAME(S): Limestone Polypody, Northern Oak Fern. Dryoptéris calcicole.
FREQUENCY: Very rare; known only from the Restigouche River valley, Restigouche Co., and from Sisson Gorge, Victoria Co.

+++

HABITAT: Dry calcareous ledges.
SPECIAL NOTE: For more information see The Rare Vascular Plants of New Brunswick. H. Hinds. Syllogeus No. 50. 1983.
GENERAL RANGE: Alaska to Nfld. south to B.C. and Pa.; Eurasia.
SYNONYMS: **Dryopteris robertiana** (Hoffm.) Christens.; **Phegopteris robertianum** (Hoffm.) A.Br.

MATTEUCCIA Todaro

Matteuccia struthiopteris (L.) Todaro Fig. p. A5. Map p. B5.
COMMON NAME(S): Ostrich Fern, Fiddle-heads. Ptérétide noduleuse.
FREQUENCY: Locally common throughout.
HABITAT: Rich woods and meadows mostly in fine grained alluvial soils.
GENERAL RANGE: Alaska to Nfld. south to B.C. and Va.; Eurasia.
SYNONYMS: **Matteuccia pensylvanica** (Willd.) Raymond; **Onoclea struthiopteris** (L.) Hoffman; **Pteretis nodulosa** (Michx.) Nieuwl.
SUBSPECIFIC TAXA: North American material may be distinguished from Eurasian plants as var. **pensylvanica** (Willd.) Morton.

ONOCLEA L.

Onoclea sensibilis L. Fig. p. A5.
COMMON NAME(S): Sensitive Fern. Onoclée délicate.
FREQUENCY: Common throughout the Province.
HABITAT: Wet open woods and meadows.
SUBSPECIFIC TAXA: Forma **obtusiloba** (Schk.) Gilbert, with pinnules of the fertile frond sterile and leaf-like, occurs occasionally with the species.
GENERAL RANGE: Man. to Nfld. south to Tex. and Fla.

POLYPODIUM L. Polypody. Polypode

Polypodium virginianum L. Fig. p. A5. Map p. B5.
COMMON NAME(S): Rock Polypody, Polypody Fern. Polypode de Virginie.
FREQUENCY: Common throughout much of the Province.
HABITAT: Ledges and boulders.
SUBSPECIFIC TAXA: Forma **cambricoides** F.W. Gray, with segments deeply cleft into toothed lobes, is reported from Charlotte Co.
GENERAL RANGE: B.C. to Nfld. south to Ark. and Ga.; Asia.

POLYSTICHUM Roth

1a Fronds simply pinnate, each pinna with conspicuous basal lobe, evergreen; stipe and rachis moderately chaffy. **P. acrosticoides**

++

1b Frond bipinnate, the pinnae without ascending basal lobes, nonevergreen; stipe and rachis heavily chaffy. **P. braunii**

Polystichum acrosticoides (Michx.) Schott Fig. p. A5. Map p. B5.
COMMON NAME(S): Christmas Fern. Polystic faux-acrostic.
FREQUENCY: Abundant in its preferred habitat from Albert Co. west and north to Victoria Co. Rare or unknown elsewhere.
HABITAT: Rich woods and rocky slopes.
SUBSPECIFIC TAXA: Forma **incisum** (Gray) Gilbert, with pinnae +/- deeply incised, is reported from Albert Mines, Albert Co. (UNB).
GENERAL RANGE: Ont. to N.S. south Mexico and Fla.

Polystichum braunii (Spenner) Fée Fig. p. A5. Map p. B5.
COMMON NAME(S): Braun's Holly Fern. Polystic de Braun.
FREQUENCY: Rare except in the northern part of the Province where it is uncommon.
HABITAT: Rich woods, mostly in calcareous soils.
SUBSPECIFIC TAXA: Our plants are referable to var. **purshii** Fern.
GENERAL RANGE: Ont. to N.S. south to Wash. and Pa.; Eurasia.

PTERIDIUM Gled. Bracken. Ptéridium des aigles.

Pteridium aquilinum (L.) Kuhn Fig. p. A6.
COMMON NAME(S): Bracken, Brake. Fougére-aigle commune, Grande Fougére.
FREQUENCY: Common throughout the Province.
HABITAT: Open dry deciduous woods, clearings, heaths, and burns.
SUBSPECIFIC TAXA: Our plants are referable to var. **latiusculum** (Desv.) Underw.
GENERAL RANGE: Alaska to Nfld. south to Calif. and N.C.; Eurasia.

THELYPTERIS Schmidel

1a Pinnae gradually decreasing in size from about the middle of the blade to near the base. **T. noveboracensis**
1b Pinnae not decreasing in size to base. 2
2a(1b) Blade broadly triangular with longest pinnae near base; rachis mostly winged; indusium lacking. **T. phegopteris**
2b(1b) Blade lanceolate to elliptic-lanceolate with lowest pinnae as long as or shorter than middle ones; rachis not winged; indusia present. 3
3a(2b) Veins of the pinnules unbranched from midvein to margin; fertile pinnules not strongly inrolled. **T. simulata**
3b(2b) Veins of the pinnules branched between midvein and margin; fertile pinnules appearing more contracted due to inrolled margins. **T. palustris**

Thelypteris noveboracensis (L.) Nieuwl. Fig. p. A6. Map p. B5.
COMMON NAME(S): New York Fern. Dryoptéride de New York.
FREQUENCY: Common in the southern and rare or unknown in the northern half of the Province.

++

HABITAT: Moist mixed woods.
GENERAL RANGE: Minn. to Nfld. south to Ark. and Ga.
SYNONYMS: **Dryopteris noveboracensis** (L.) Gray.

Thelypteris phegopteris (L.) Slosson Fig. p. A6.
COMMON NAME(S): Long Beech-Fern. Dryoptéride du Hêtre, Phégopteris.
FREQUENCY: Common throughout the Province.
HABITAT: Moist woods.
GENERAL RANGE: Alaska to Nfld. south to Oreg. and N.C.; Greenland; Iceland; Eurasia.
SYNONYMS: **Dryopteris phegopteris** (L.) Slosson; **Phegopteris connectilis** (Michx.) Watt.

Thelypteris palustris (Salisb.) Schott Fig. p. A6. Map p. B5.
COMMON NAME(S): Marsh Fern. Dryoptéride thélyptéride.
FREQUENCY: Common. Scattered throughout the Province.
HABITAT: Wet meadows.
SUBSPECIFIC TAXA: Our material is referable to var. **pubescens** (Lawson) Fern.
GENERAL RANGE: Man. to Nfld. south to Tex. and Ga.; Eurasia.
SYNONYMS: **Dryopteris thelypteris** (L.) Sw. var. **pubescens** (Lawson) Prince.

Thelypteris simulata (Davenp.) Nieuwl. Fig. p. A6.
COMON NAME(S): Massachusetts Fern.
FREQUENCY: Very rare; only known from the southwest side of Grand Lake, Queens Co. (UNB).
HABITAT: Wet thickets and swampy woods.
GENERAL RANGE: Ont. to N.S. south to Wisc. and Va.
SPECIAL NOTE: For more information see The Rare Vascular Plants of New Brunswick. H. Hinds. Syllogeus No. 50. 1983.
SYNONYMS: **Dryopteris simulata** Davenp.

WOODSIA R.Br.

1a Stipes and leaves with dense covering of rusty scales; hairy on upper side. **W. ilvensis**

1b Stipes and leaves only sparingly scaley or smooth; smaller plants. 2

2a(1b) Blades sparingly hairy beneath; stipes and rachises brownish or blackish with scattered scales, about .05-1 mm in diameter; indusial hairs many, conspicuously overtopping the sporangia. **W. alpina**

2b(1b) Blade smooth, stipe smooth above the stipe joint, light green or stramineus; rachises green; very slender (nearly threadlike) and delicate about 0.4 mm in diameter; indusial hairs few and only slightly overtopping sporangia. **W. glabella**

Woodsia alpina (Bolton) S.F. Gray Fig. p. A6. Map p. 85.
COMMON NAME(S): Northern Woodsia, Alpine Woodsia.
FREQUENCY: Very rare; the most famous location is the Aroostook River Gorge, Victoria Co. It has also been found on Ministers Head, Long Island, Kings Co. and a few places along the Restigouche River in Restigouche Co.
HABITAT: Calcareous ledges.

SPECIAL NOTE: For more information see The Rare Vascular Plants of New Brunswick. H. Hinds. Syllogeus No. 50. 1983.
GENERAL RANGE: Alaska to Nfld. south to Minn. and Vt.; Greenland; Iceland; Eurasia.
SYNONYMS: **W. bellii** (Lawson) Porsild.

Woodsia glabella R. Br. Fig. p. A6. Map p. B6.
COMMON NAME(S): Smooth Woodsia.
FREQUENCY: Rare; scattered in a few southern and western parts of the Province.
HABITAT: Moist calcareous ledges.
GENERAL RANGE: Alaska to Nfld. south to B.C. and northern N. Eng.; Greenland; Iceland; Eurasia.

Woodsia X gracilis (Lawson) Butters
SPECIAL NOTE: This is the hybrid between **W. alpina** and **W. ilvensis**.
FREQUENCY: Very rare; only known in N.B. from Morrisey Rock near the railroad tunnel, Restigouche Co. (UNB).
HABITAT: Mossy calcareous ledges.
GENERAL RANGE: Western Ont., eastern Que. and northern N.B.

Woodsia ilvensis (L.) R. Br. Fig. p. A6. Map p. B6.
COMMON NAME(S): Rusty or Fragrant Woodsia. Woodsia de l'île d'Elbe.
FREQUENCY: Locally common except rare in the Eastern Lowlands.
HABITAT: Dry exposed ledges and talus, often noncalcareous.
GENERAL RANGE: Alaska to Nfld. south to B.C. and N.C.; Greenland; Iceland; Eurasia.

WOODWARDIA Sm. Chain-Fern. Woodwardie

Woodwardia virginica (L.) Sm. Fig. p. A6.
COMMON NAME(S): Virginia Chain-Fern. Woodwardie de Virginie.
FREQUENCY: Very rare; reported from Tower Lake, Harvey Station and Oromocto Lake, York Co.
HABITAT: Wet thickets usually in acid peat.
SPECIAL NOTE: For more information see The Rare Vascular Plants of New Brunswick. H. Hinds. Syllogeus No. 50. 1983.
GENERAL RANGE: Ill. to N.S. south to Tex. and Fla.; Bermuda.

Division II **PINOPHYTA** (Gymnosperms)

TAXACEAE (Yew Family)

TAXUS L. Yew. If

Taxus canadensis Nutt. Fig. p. A6.
COMMON NAME(S): Canada yew, Ground-Hemlock. Buis de sapin, If du Canada.
FREQUENCY: Common throughout the Province.

++

HABITAT: Rich woods.
GENERAL RANGE: Man. to Nfld. south to Iowa and Va.

PINACEAE (Pine Family)

1a Leaves linear, in bundles of 2-5, from dwarf branches. 2
1b Leaves short, linear or scale-like, alternate, solitary. 3
2a(1a) Leaves deciduous, numerous in each cluster and scattered on long twigs. **Larix laricina**
2b(1a) Leaves evergreen, 2-5 in a cluster. **Pinus**
3a(1b) Leaves all linear; seeds in a woody cone with many scales. 4
3b(1b) Leaves shorter, overlapping and often scale-like. 5
4a(3a) Leaves attached directly to twigs and leaving a smooth rounded scar when falling; cones erect, scales deciduous at maturity. **Abies balsamea**
4b(3a) Leaves attached to persistent elevated woody bases; cones pendulous, scales persistent. 6
5a(3b) Seeds in a cone with 4-6 leathery scales; trees. **Thuja occidentalis**
5b(3b) Seeds in a bluish berry-like rounded cone; low shrubs. **Juniperus**
6a(4b) Leaves squarish in cross-section, not two ranked. **Picea**
6b(4b) Leaves flattish in cross-section, two ranked. **Tsuga canadensis**

ABIES Mill. Fir. Sapin or Sapin baumier

Abies balsamea (L.) Mill Fig. p. A6.
COMMON NAME(S): Balsam-Fir. Sapin baumier, Sapin blanc.
FREQUENCY: Common throughout the Province.
HABITAT: Damp woods.
GENERAL RANGE: Alta. to Nfld. south to Iowa and Va.
SUBSPECIFIC TAXA: Var. **phanerolepis** Fern. is an uncommon segregate with the cone bracts conspicuously exserted beyond the cone-scales.

JUNIPERUS L. Juniper. Genévrier

1a Leaves on mature branches scale-like (needle-like on young growth), appressed, opposite; fruit terminal. **J. horizontalis**
1b Leaves on mature branches linear-subulate, mostly spreading in whorls of 3; fruit sessile in leaf axils. **J. communis**

Juniperus communis L. Fig. p. A6. Map. p. B6.
COMMON NAME(S): Common Juniper. Genévrier commun.
FREQUENCY: Locally common in the southern part of the Province and rare northward.
HABITAT: Dry exposed ledges, old pastures, rocky slopes and calcareous bogs.
SUBSPECIFIC TAXA: Several varieties based on erectness of the main trunk and the size of the fruit have been named. The delineation of these variations appears to be somewhat arbitrary and these varieties are not accepted in this work.
GENERAL RANGE: Alaska to Nfld. south to Calif. and Ga.; Greenland; Iceland; Eurasia.

++

Juniperus horizontalis Moench Fig. p. A7. Map p. B6.
COMMON NAME(S): Creeping Juniper. Savinier.
FREQUENCY: The distribution is bipolar being uncommon along the Bay of Fundy and more common along Baie des Chaleurs from Dalhousie, Restigouche Co. to Miscou Is., Gloucester Co.
HABITAT: Rocky, sandy or boggy open areas.
GENERAL RANGE: Alaska to Nfld. south to B.C. and N.Y.

LARIX Mill. Larch. Mélèze

Larix laricina (Du Roi) Koch Fig. p. A7. Map p. B6.
COMMON NAME(S): American Larch, Hackmatack, Tamarack. Mélèze d'Amerique, Epinette rouge, Tamarac.
FREQUENCY: Common throughout the Province.
HABITAT: Bogs and other poorly drained areas.
GENERAL RANGE: Alaska to Nfld. south to B.C. and N.J.

PICEA Dietr. Spruce. Epinette

1a Twigs and bud scales glabrous; pale brown cones 3.5-5 cm long with the scales thin and flexible at the entire margin; leaves blue-green and sharp. **P. glauca**

1b Twigs and lower bud scales pubescent; cones up to 4.5 cm long; scales with wavy or dentate firm margins. 2

2a(1b) Leaves mostly bluish-green, 6-18 mm long; cones dark purple before maturity 1.5-3.5 cm long, persistent; bark usually dark. **P. mariana**

2b(1b) Leaves yellowish-green, 10-30 mm long, blunter, less shiny; cones confined to top of tree, usually falling first autumn, bark reddish tinged. **P. rubens**

Picea glauca (Moench) Voss Fig. p. A7.
COMMON NAME(S): White Spruce, Cat Spruce. Epinette blanc, Prusse blanc.
FREQUENCY: Common throughout the Province.
HABITAT: Mostly dry, upland areas.
GENERAL RANGE: Alaska to Nfld. south to B.C. and Maine.

Picea mariana (Mill.) BSP. Fig. p. A7. Map p. B6.
COMMON NAME(S): Black Spruce. Epinette noire.
FREQUENCY: Common throughout the Province.
HABITAT: Swamps, bogs and similar poorly drained areas.
GENERAL RANGE: Alaska to Nfld. south to B.C. and Pa.

Picea rubens Sarg. Fig. p. A7. Map p. B6.
COMMON NAME(S): Red Spruce. Epinette rouge.

++

FREQUENCY: Common in the southern half of the Province, becoming rare northwards.
HABITAT: Well drained loamy areas.
GENERAL RANGE: Easternmost Ont. to N.S. south to Tenn. and N.C.

Picea rubens X P. mariana
SPECIAL NOTE: The hybrid and backcrosses to the parents are especially common in the eastern lowlands. Hybrid swarms occur frequently consisting of highly variable populations. For further information and methods of identifying **P. rubens**, **P. mariana** and their hybrids, see Manley, 1971.

PINUS L. Pine. Pin

1a Leaves in clusters of 5. **P. strobus**
1b Leaves in clusters of 2. 2
2a(1b) Leaves 9-16 cm long, stiff; large tree with reddish bark. **P. resinosa**
2b(1b) Leaves less than 8 cm long. 3
3a(2b) Larger branches orange-brown; leaves 4-6 cm long. **P. sylvestris**
3b(2b) Branches dark; leaves 1-4 cm long. **P. banksiana**

Pinus banksiana Lamb. Fig. p. A7. Map p. B6.
COMMON NAME(S): Jack or Scrub-Pine. Pin gris or Cyprès.
FREQUENCY: Common in the eastern part of the Province, rare westward.
HABITAT: Shallow gravelly soils, sand plains, on highly acid soils.
GENERAL RANGE: Dist Mack. to N.S. south to Alta. and New Engl.

Pinus resinosa Ait. Fig. p. A7. Map. p. B6.
COMMON NAME(S): Red Pine. Pin rouge.
FREQUENCY: Locally common; unknown from large areas of the Province and rarer towards the north.
HABITAT: Sandy or rocky soils, occasionally peaty heaths.
GENERAL RANGE: Man. to Nfld. south to Minn. and Mass.

Pinus strobus L. Fig. p. A7. Map p. B6.
COMMON NAME(S): White Pine. Pin blanc.
FREQUENCY: Common throughout most of the Province.
HABITAT: Well drained sandy or gravelly soils.
GENERAL RANGE: Man. to Nfld. south to Iowa and S.C.

Pinus sylvestris L.
COMMON NAME(S): Scotch Pine.
SPECIAL NOTE: This European species is frequently planted and has occasionally become naturalized as in St. Andrews, Charlotte Co.

THUJA L. Arbor Vitae. Thuier

Thuja occidentalis L. Fig. p. A7. Map p. B6.
COMMON NAME(S): White Cedar. Cédre or Balai.
FREQUENCY: Common throughout the Province.
HABITAT: Chiefly calcareous areas in swamps and ledge crevices.
GENERAL RANGE: Man. to N.S. south to Minn., Tenn. and N.C.

TSUGA (Endl.) Carr. Hemlock. Pruche

Tsuga canadensis (L.) Carr. Fig. p. A7. Map p. B6.
COMMON NAME(S): Eastern Hemlock. Pruche de l'Est.
FREQUENCY: Common in the southern half of the Province and becoming rare northwards.

Division III MAGNOLIOPHYTA (Flowering Plants)

Class I LILIOPSIDA
(Alternate name **MONOCOTYLEDONEAE**)

TYPHACEAE (Cat-tail Family)

Typha L. Cat-tail. Massette or Quenouille

1a Staminate and pistillate parts of spike separated by a gap of more than 5 mm; mature pistillate part of spike less than 2 cm thick, cinnamon brown; stigmas linear; leaves slightly rounded on back, less than 1 cm broad. **T. angustifolia**

1b Staminate and pistillate parts of spike +/- contiguous, occasionally separated less than 5 mm; mature pistillate part of spike up to 3.5 cm thick, dark brown with darker markings; stigmas lance-ovate; leaves flat, up to 2.5 cm broad. **T. latifolia**

Typha angustifolia L. Fig. p. A7. Map p. B7.
COMMON NAME(S): Narrow-leaved Cat-tail. Petite Massette.
FREQUENCY: Probably more common than collections seem to indicate, but truly rare in the north of the Province. Absent from the Gaspé.
HABITAT: Brackish swales, ditches, river backwaters and margins, and pond margins.
GENERAL RANGE: Man. to N.S. south to Calif. and S.C.; Eurasia.

Typha X glauca Godr.
SPECIAL NOTE: This is the hybrid between **T. angustifolia** and **T. latifolia**. The best way to identify the hybrid is to see it growing with the parents where its hybrid vigor and intermediate characteristics become obvious.
FREQUENCY: Locally common as in the Tantramar marshes, Westmorland Co.
GENERAL RANGE: Iowa, s Ont. to Me. south to Calif. and Ala.; Guatemala; Europe.

+++

Typha latifolia L. Fig. p. A7. Map p. B7.
COMMON NAME(S): Common Cat-tail, Broad-leaved Cat-tail. Grande Massette.
FREQUENCY: Scattered throughout the Province except in the extreme northern section.
HABITAT: Wet areas in fields, ditched, edges of streams and ponds, upper end of estuaries and in floating bog associations.
SUBSPECIFIC TAXA: Forma **ambigua** (Sonder) Kronf. mimics **T. angustifolia** in having the pistillate and staminate inflorescence separated. In all other aspects these plants resemble the species.
GENERAL RANGE: Alaska to Nfld. south to Calif. and Fla.; Eurasia; Africa.

SPARGANIACEAE (Bur-reed Family)

SPARGANIUM L. Bur-reed. Rubanier

1a Stigmas 2; achene sessile with rounded summit. **S. eurycarpum**

1b Stigma 1; achenes tapering into beak at summit. 2

2a(1b) Fruiting heads about 10-12 mm in diameter including beaks which are short and conical, .5-1.5 mm long; leaves mostly floating; staminate heads usually 1; pistillate heads all borne in axils of upper leaves or bracts. **S. minimum**

2b(1b) Fruiting heads 15-30 mm in diameter; staminate heads 2 or more before disintegrating. 3

3a(2b) Fruiting head 15-20 mm in diameter in a branched inflorescence; beak strongly curved from base, about 2(-3) mm long; sepals reaching to about middle of achene; leaves all floating; achenes dull, dark reddish brown. **S. fluctuans**

3b(2b) Fruiting heads 20-30 mm in diameter in a branched or unbranched inflorescence; beak +/- straight or slightly curved near summit. 4

4a(3b) Heads or branches of inflorescence all borne directly in axils of leaves or bracts 5

4b(3b) Heads (or at least one of them) borne above the leaf axils (supra-axillary). 6

5a(4a) Body of mature achene shiny, about 6 mm long with prominantly corrugated summit, the beak 5-7 mm long; anthers 1.1-1.4 mm long; branches of inflorescence often with only staminate heads. **S. androcladum**

5b(4a) Body of mature achene dull, about 4-5 mm long, slightly if at all corrugated at summit, the beak 2.5-4(-5) mm long; anthers 0.6-1.1 mm long; branches of inflorescence, if present, with both staminate and pistillate heads. **S. americanum**

6a(4b) Leaves erect, rarely floating; beak about as long as body of achene; achene shiny green, prominently corrugated between 3-angled summit. **S. chlorocarpum**

6b(4b) Leaves floating; beak shorter than body of achene; achene scarcely corrugated between 3-angled summit, 2-toned (dull greenish above, reddish brown below constriction). **S. emersum**

Sparganium americanum Nutt. Fig. p. A7. Map p. B7.
FREQUENCY: Scattered throughout the southern part of the Province.
HABITAT: Muddy shores of lakes and slow-moving streams.
GENERAL RANGE: B.C. to Nfld. south to Tex. and Fla.

Sparganium androcladum (Engelm.) Morong Fig. p. A8.
SPECIAL NOTE: Reported by Scoggan, 1978-79. Combined with **S. americanum** by Boivin. 1967-79.

Sparganium chlorocarpum Rydb. Fig. p. A8. Map p. B7.
FREQUENCY: Common throughout the Province.
HABITAT: Wet mucky shores or shallows.
GENERAL RANGE: Alta. to Nfld. south to Calif. and N.C.

Sparganium emersum Rehmann Fig. p. A8. Map p. B7.
FREQUENCY: Scattered throughout.
HABITAT: Fresh or brackish pond margins.
GENERAL RANGE: Alaska to Nfld. south to Calif. and N.J.; Greenland; Iceland; Europe; Kamchatka and Japan.
SYNONYMS: **S. angustifolium** Michx.; **S. multipedunculatum** (Morong) Reveal

Sparganium eurycarpum Engelm. Fig. p. A8. Map. p.B7.
FREQUENCY: Locally common as in the Jemseg River area. Rare northwards.
HABITAT: Rich marshy areas and shallow sloughs.
GENERAL RANGE: B.C. to Nfld. south to Calif. and Fla.

Sparganium fluctuans (Morong) Robins.
FREQUENCY: Uncommon; known only from the southeastern quarter of the Province.
HABITAT: Cold lakes and ponds.
GENERAL RANGE: B.C. to Nfld. south to Minn. and N. Eng.

Sparganium minimum (Hartm.) Fries Fig. p. A8. Map p. B7.
FREQUENCY: Rare; scattered widely through much of the Province.
HABITAT: Alkaline ponds, pond margins, lagoons.
GENERAL RANGE: Alaska to Nfld. south to Oreg. and N.J.; Iceland; Eurasia.

ZOSTERACEAE (Eelgrass Family)

The genus, **Zostera**, has been treated variously by different authors. Fernald (1950) included **Zostera** with **Potamogeton** and **Ruppia** in the Zosteraceae. Cronquist (1968) and Tomlinson (1982) and others consider **Zostera** distinct enough (along with the non-native genera **Phyllospadix** and **Heterozostera**) to warrant its own family.

ZOSTERA L. Eelgrass, Grass-wrack. Mousse de mer

Zostera marina L. Fig. p. A10. Map p. B9.
COMMON NAME(S): Eelgrass, Sea-wrack. Zostère marine, Herbe à bernaches.
FREQUENCY: Common along the Northumberland Straits and western shore of the Bay of Fundy.
HABITAT: Shallow coastal waters.
GENERAL RANGE: Alaska to Nfld. south to S.C.; Greenland; Iceland; Europe; Asia; Africa.
SUBSPECIFIC TAXA: Our material is referable to var. **stenophylla** Asch. & Graebn.

POTAMOGETONACEAE (Pondweed Family)

POTAMOGETON L. Pondweed. Potamot

1a Submersed leaves clasping the stem or sheath, abruptly narrowed at base, sessile; floating leaves none. 2
1b Submersed leaves not clasping stem or sheath; sheaths sometimes clasping stem. 3
2a(1a) Leaf margins usually finely serrate, +/- parallel except at ends; blades 3-9 mm broad to 11 cm long, 2-ranked, 20-60 nerved, fused to lower half of stipule. **P. robbinsii**
2b(1a) Leaf margins not serrate, nor parallel; blades 5-3 cm wide, not attached to sheath, with 3-7 heavy nerves and several light ones. 4
3a(1b) Submersed leaves non-linear, tapering to both ends; floating leaves often present, 10-75 mm wide. 5
3b(1b) Submersed leaves linear with margins parallel most of their length, 0.1-10.0 mm wide. 6
4a(2b) Stipules 3-10 mm long, whitish, persistant and conspicuous; blades 1-3 cm wide, 1-2.5 dm long, apex cucullate, splitting when pressed; stem 1.5-4 mm thick, rhizome rusty-red spotted. **P. praelongus**
4b(2b) Stipules 0.4-2 cm long, inconspicuous and rapidly disintegrating or persisting as whitish fibers; blades 1.5-10 cm long, 0.5-2 cm wide; stem to 2.3 mm wide near base. 7
5a(3a) Submersed leaves 27-37-nerved, 2.5-7.5 cm wide, ruffle-edged and arcuate. **P. amplifolius**
5b(3a) Submersed leaves less than 27-nerved, not ruffled nor arcuate. 8
6a(3b) Stipules completely free from leaf sheath. 9
6b(3b) Stipules adnate to leaf sheath. 10
7a(4b) Leaves ovate-lanceolate to narrowly-lanceolate, 3-10 cm long; stipules persisting as whitish fibers. **P. richardsonii**
7b(4b) Leaves suborbicular to ovate (more elongate-leaved plants occurring in softer water), 1-6 cm long, 0.5-2.0 cm wide; stipules filmy, evanescent. **P. perfoliatus**

8a(5b) Petiole of submersed leaves 2-13 cm long; leaf blades 1-3.5 cm wide, 7 to 15-nerved, lacunae in 2-5 rows each side of midrib; floating leaves to 11 cm long, acute; fruit 3.5-4.3 mm long. **P. nodosus**

8b(5b) Petiole absent, leaves sessile; fruit less than 3.5 mm long. 11

9a(6a) Submersed leaves with prominent lacunae on each side of midrib 1/3-1/2 as wide as blade; floating leaves tapering at base, with flattened petioles; fruit keel 0.2-1.2 mm high. **P. epihydrus**

9b(6a) Submersed leaves without prominent lacunae. 12

10a(6b) Leaf base fused to sheath for less than 5 mm; submersed leaves 0.5-2 mm wide, lacunae abundant; fruit compressed with evident spiral embryo; floating leaves often developed. **P. spirellus**

10b(6b) Leaf base fused to sheath for 10 mm or more (rarely for as little as 5-6 mm); submersed leaves 0.2-2 mm wide; floating leaves absent. 13

11a(8b) Floating blades delicate, translucent, on petioles 1-3 cm long, tapering gradually to base; stem mostly unbranched from base; submersed leaves with 7 major nerves and narrow band of lacunae often reddish tinged in upper parts when dry, acute or obtuse at apex; stipules blunt; fruit plump, pedicelled. **P. alpinus**

11b(8b) Floating blades leathery, opaque on petioles 2-10(15) cm long, rounded to short tapering at base; stem usually much branched; submersed leaves with 3-17 prominent veins, acute at apex or with awl-like tip; lacunae not obvious; stipules acute; fruit compressed, not pedicelled. **P. gramineus**

12a(9b) Floating leaves absent. 14

12b(9b) Floating leaves usually developed. 15

13a(10b) Leaves acute, apiculate on young plants and branches; fruit (2.5)3-4.5 mm long, not including short, but definite beak. **P. pectinata**

13b(10b) Leaves +/- blunt, obtuse or notched and often apiculate at tip; fruit 2-3 mm long with central wart-like beak. **P. filiformis**

14a(12a) Stipules +/- coarsely fibrous, whitish, disintegrating into fibers towards the base of stem; winter buds hardened towards base, strongly ribbed. 16

14b(12a) Stipules more delicate, membraneous, whitish to green or brownish, not disintegrating into fibers; winter buds, if present, not hardened nor strongly ribbed. 17

15a(12b) Floating leaves with well developed blades 1.2-4.7 cm wide (sometimes more narrowly elliptic and acute at both ends); submersed leaves phyllodial (stem-like) without expanded blade (often absent later in the season). 18

15b(12b) Floating leaves, if present, with blades less than 1-2 cm broad 5-9 veined (sometimes not developed); submersed leaves with definite flat blades 1-1.0 mm wide; winter buds common; fruit with distinct dorsal keel and recurved beak. **P. vaseyi**

16a(14a) Leaf apex obtuse, mucronate; leaves (3)5-7(9) nerved, 1.5-2.7(3) mm wide with narrow band of lacunae along midrib; nodal glands conspicuous; inner leaves of winter buds at right angles to outer leaves. **P. friesii**

16b(14a) Leaf apex usually tapering to a sharp, bristle-like tip; leaves 3(5) nerved, 0.8-1.3 mm wide, lacking lacunae; nodal glands inconspicuous or apparently lacking; inner and outer leaves of winter buds in same plane. **P. strictifolius**

++

17a(14b) Leaves thread-like, 0.1-0.5 mm wide, tapering to slender point; rhizome long, delicate and creeping; peduncles terminal, 15-24 cm long. **P. confervoides**

17b(14b) Leaves usually wider, 0.3-3.5 mm wide; rhizome short or lacking; peduncles 0.3-10 cm long, often curved, in axils of several leaves. 19

18a(15a) Blades of floating leaves cordate to subcordate at base, 2-4.7 cm wide, (3.2)3.7-9(10) cm long; fruit with beak (3.5)3.7-4.5 mm long, obscurely keeled. **P. natans**

18b(15a) Blades of floating leaves rounded to acute at base (0.5)1.2-2.2(2.8) cm wide; fruit 2.5-3.5 mm long, prominently keeled. **P. oakesianus**

19a(17b) Nodal glands present; fruit less than 3 mm long and not keeled. 20

19b(17b) Nodal glands usually absent. 21

20a(19a) Leaves rounded or apiculate at apex, 2.0-3.5 mm wide often tinted reddish brown; fruit keeled, 3-4 mm long including beak. **P. obtusifolius**

20b(19a) Leaves 0.2-2.5 mm wide usually green, apex obtuse, acute or apiculate; fruit not keeled, less than 3 mm long. **P. pusillus**

21a(19b) Leaves 2-5 mm wide with 15-20 close veins; stem wing-flattened; fruit 3.5-5.5 mm long. **P. zosteriformis**

21b(19b) Leaves 0.3-1.5(2.7) mm wide, acute; fruit 1.8-2.3(2.7) mm long with thin, often wavy keel. **P. foliosus**

Potamogeton alpinus Balbis Fig. p. A8. Map p. B7.
FREQUENCY: Uncommon.
HABITAT: Small ponds and streams.
SUBSPECIFIC TAXA: The American plants differ slightly from the typical Eurasian form and are referable to var. **subellipticus** (Fern.) Ogden.
GENERAL RANGE: Alaska to Nfld. south to Calif. and N. Eng.; Greenland; Iceland; Eurasia.
SYNONYMS: **P. alpinus** Balbis var. **tenuifolius** (Raf.) Ogden.

Potamogeton amplifolius Tuckerm. Fig. p. A8. Map p. B7.
FREQUENCY: Uncommon.
HABITAT: Lakes and streams.
GENERAL RANGE: B.C. to Nfld. south to Calif. and Ga.

Potamogeton confervoides Rchb. Fig. p. A8. Map p. B7.
FREQUENCY: Very rare; known only from Sterling Lake, York Co., Fox Creek, Westmorland Co., and Welsh Lake, Charlotte Co.
HABITAT: Shallow water of lakes and streams.
GENERAL RANGE: Wisc. to Nfld. south to Pa., N.Y. and N.J.
SPECIAL NOTE: For more information see H. Hinds, 1983. The Rare Vascular Plants of New Brunswick. Syllogeus No. 50.

Potamogeton epihydrus Raf. Fig. p. A8. Map p. B7.
FREQUENCY: Common throughout the Province.
HABITAT: Lakes and streams, backwaters of rivers.
GENERAL RANGE: Alaska to Nfld. south to Calif. and Ga.; Europe.

++

SYNONYMS: **P. epihydrus** Raf. var. **nuttallii** C. & S.; **P. epihydrus** Raf. var. **ramosus** (Peck) House.

Potamogeton filiformis Pers. Fig. p. A8. Map p. B8.
FREQUENCY: Rare; reported from Dolan Lk., St. John Co., Lac Baker, and Third Lk., Madawaska Co., and Maplehurst Marl Pit, Carleton Co.
HABITAT: Calcareous or brackish waters.
GENERAL RANGE: Alaska to Nfld. south to Calif., Pa., Vt. and Me.; Greenland; Iceland; Eurasia; Africa.
SUBSPECIFIC TAXA: Our plants are referable to var. **alpinus** (Blytt) Aschers & Graebn.
SYNONYMS: **P. filiformis** Pers. var. **borealis** (Raf.) St. John
SPECIAL NOTE: For more information see H. Hinds, 1983. The Rare Vascular Plants of New Brunswick. Syllogeus No. 50.

Potamogeton foliosus Raf. Fig. p. A9. Map p. B8.
FREQUENCY: Scattered throughout but especially common along the western border.
HABITAT: Mostly alkaline ponds and pools.
GENERAL RANGE: Alaska to N.S. south to Calif. and Fla.; Guatemala; W.I.; Hawaii.
SYNONYMS: **P. foliosus** Raf. var. **macellus** Fern.

Potamogeton friesii Rupr. Fig. p. A9.
FREQUENCY: Very rare; known only from Tidehead, Restigouche Co. and Westfield, Kings Co.
HABITAT: Calcareous or brackish waters.
GENERAL RANGE: Alaska to Nfld. south to Wash. and Va.; Eurasia.
SPECIAL NOTE: For more information see H. Hinds, 1983. The Rare Vascular Plants of New Brunswick, Syllogeus No. 50.

Potamogeton gramineus L. Fig. p. A9. Map p. B8.
FREQUENCY: Common except rare in the eastern lowlands.
HABITAT: Lakes, ponds and rivers, often in swiftly flowing waters.
GENERAL RANGE: Alaska to Nfld. south to Calif. and N.J.; Greenland; Iceland; Eurasia.
SYNONYMS: **P. gramineus L. var. maximus** Morong.
SUBSPECIFIC TAXA: Forma **terrestris** (Schlecht.) Carpenter, with emersed leaves crowded and submerged leaves absent, appears to result when the aquatic habitat suffers drought.

Potamogeton natans L. Fig. p. A9. Map p. B8.
FREQUENCY: Locally common but unknown from large areas of the Province.
HABITAT: Lakes and streams.
GENERAL RANGE: Alaska to Nfld. south to Calif. and N.J.; Greenland; Iceland; Eurasia.

Potamogeton nodosus Poir. Fig. p. A9.
FREQUENCY: Very rare; reported only from the South Branch of the Meduxnekeag River, Carleton Co. (UNB).
HABITAT: Swiftly moving water of relatively high alkalinity.
GENERAL RANGE: B.C. to s. Lab. south to Calif. and Va.; Eurasia; Africa.
SPECIAL NOTE: An apparent hybrid with **P. perfoliatus** is also reported at the above location in Carleton Co.

++

Potamogeton oakesianus Robbins Fig. p. A9. Map p. B8.
FREQUENCY: Rare; scattered in the southern half of the Province.
HABITAT: Peaty or sandy ponds.
GENERAL RANGE: Ont. to Nfld. south to Wisc. and N.J.

Potamogeton obtusifolius Mert. & Koch Fig. p. A9. Map p. B8.
FREQUENCY: Scattered throughout.
HABITAT: Still waters.
GENERAL RANGE: Yukon to Nfld. south to Kans. and N.J.; Eurasia.

Potamogeton pectinatus L. Fig. p. A9. Map p. B8.
FREQUENCY: Rare throughout.
HABITAT: Brackish and alkaline, mostly still, waters.
GENERAL RANGE: Alaska to Nfld. south to Calif. and Fla.; S. America; Eurasia; Africa.

Potamogeton perfoliatus L. Fig. p. A9. Map p. B8.
FREQUENCY: Common throughout the Province.
HABITAT: Brackish or alkaline waters.
GENERAL RANGE: Alaska to Nfld. south to Calif. and Fla.; Greenland; Iceland; Eurasia; Africa.
SYNONYMS: **P. perfoliatus** L. var. **bupleuroides** (Fern.) Hult.

Potamogeton praelongus Wulfen Fig. p. A9.
FREQUENCY: Very rare; known only from Dolan Lake, St. John Co.
HABITAT: Deep water of lakes.
GENERAL RANGE: Alaska to Nfld. south to Calif. and N.J.; Iceland; Eurasia.

Potamogeton pusillus L. Fig. p. A9. Map p. B8.
SUBSPECIFIC TAXA: Two varieties are recognized in our range.
1a Midrib of leaves paralleled by up to 2 rows of lacunae; peduncles mostly terminal, 1-3 per plant; inflorescence usually of 2-4 distinct, interrupted whorls. var. **pusillus**
1b Midrib of leaves paralleled by up to 5 rows of lacunae; peduncles axillary or terminal, more than 3 per plant; inflorescence crowded. var. **tenuissimus**
1. Var. **pusillus**
FREQUENCY: Mostly scattered in the western part of the Province.
HABITAT: Alkaline waters.
GENERAL RANGE: Alaska to Nfld. south to Mex. and Fla.; C. America.
SYNONYMS: **P. panormitanus** Biv. var. **minor** Biv., **P. pusillus** var. **minor** (Biv.) Fern. & Schub.
2. Var. **tenuissimus** Mert. & Koch
FREQUENCY: Scattered throughout much of the Province.
HABITAT: Acid, alkaline or brackish waters.
GENERAL RANGE: Alaska to Nfld. south to Calif., La. and Fla.
SYNONYMS: **P. berchtoldii** Fieber.

Potamogeton richardsonii (Benn.) Rydb. Fig. p. A9. Map p. B8.
FREQUENCY: Rare except in the extreme northwest of the Province.

HABITAT: Mostly alkaline waters.
GENERAL RANGE: Alaska to N.S. south to Calif. and N. Eng.

Potamogeton robbinsii Oakes Fig. p. A10. Map p. B8.
FREQUENCY: Uncommon.
HABITAT: Deep water and slow streams.
GENERAL RANGE: B.C. to N.S. south to Calif., Ala. and Del.
SUBSPECIFIC TAXA: Forma **cultellatus** Fassett with leaf margins entire instead of minutely serrulate has been reported from the Province.

Potamogeton spirellus Tuckerm. Fig. p. A10. Map p.8.
FREQUENCY: Scattered; more common towards the south of the Province.
HABITAT: Quiet waters.
GENERAL RANGE: S. Dak. to Ont. and Nfld. south to Iowa and Va.

Potamogeton strictifolius Benn. Fig. p. A10.
FREQUENCY: Very rare; tentatively identified by E.C. Ogden from the Portage Lakes, Restigouche Co. (UNB).
HABITAT: Calcareous waters.
GENERAL RANGE: Yukon to N.B. south to Utah, Pa. and Vt.
SYNONYMS: **P. strictifolius** Benn. var. **rutiloides** Fern.

Potamogeton vaseyi Robbins Fig. p. A10.
FREQUENCY: Very rare; only reported from Westfield, Kings Co.
HABITAT: Calcareous waters.
GENERAL RANGE: Ont. to N.B. south to Minn. and Mass.

Potamogeton zosterifolius Schum. Fig. p. A10. Map p. B9.
FREQUENCY: Uncommon; mostly along the St. John River drainage.
HABITAT: Quiet usually alkaline waters.
GENERAL RANGE: Alaska to N.S. south to Calif. and Va.; Eurasia.

RUPPIACEAE (Widgeon-grass Family)

RUPPIA L.

Ruppia maritima L. Fig. p. A10. Map p. B9.
COMMON NAME(S): Ditch-grass, Sea-grass. Ruppie maritime, Persil d'eau.
FREQUENCY: Common along the east coast.
HABITAT: Saline pools and ditches.
GENERAL RANGE: Alaska to Nfld. south to Calif. and Fla.; S. America; Eurasia.
SUBSPECIFIC TAXA; Fernald, 1950, recognized 7 varieties of this species. Richardson, 1980, found distinct clinal variation in vegetative morphology apparently in response to a diversity of habitats. This species needs further study to delimit the range of variation in the various suggested taxa. The present treatment combines the 7 varieties.

ZANNICHELLIACEAE (Horned Pondweed Family)

ZANNICHELLIA L.

Zannichellia palustris L. Fig. p. A10. Map p. B9.
COMMON NAME(S): Horned Pondweed. Zannichellie des marais.
FREQUENCY: Uncommon; reported mostly from the southern part of the Province. Unknown from the western or northern parts.
HABITAT: Mostly brackish pools.
GENERAL RANGE: B.C. to Nfld. south to Mexico and Fla.; S. America; Eurasia; Africa.
SYNONYMS: **Z. palustris** L. var. **major** (Boenn.) Koch.

NAJADACEAE (Naiad Family)

NAJAS L. Naiad

Najas flexilis (Willd.) Rostk. & Schmidt Fig. p. A10. Map p. B9.
COMMON NAME(S): Bushy Naiad, Slender Naiad. Naïas souple, Naïade flexible.
FREQUENCY: Common throughout except rare in the eastern half of the Province.
HABITAT: Shallow water.
GENERAL RANGE: Dist. Mackenzie to Nfld. south to Calif. and N.C.; Europe; Asia.

JUNCAGINACEAE (Arrow-grass Family)

TRIGLOCHIN L. Arrow-grass. Troscart

1a Fruit oblong or ovoid to 6 mm long, rounded at base and separating into 6 sections, each with recurved beak. 2
1b Fruit slender oblanceolate to 8 mm long, cuneate at base and separating into 3 blunt and beakless sections. **T. palustris**
2a(1a) Scapes robust, to over 50 cm tall; leaves 1.5-3 mm wide, erect or curving from sheath at angle of less than 30°, shorter than scapes; ligule 4-5 mm long. **T. maritima**
2b(1a) Scapes slender, 10-15 cm tall; leaves 0.5-1 mm wide, curving from sheath at an angle of 45-50°, as tall as or taller than scapes; ligule 0.5-1 mm long. **T. gaspense**

Triglochin gaspense Lieth & D. Löve Fig. p. A10. Map p. B9.
FREQUENCY: Rare and scattered; easily overlooked or confused with **T. maritima**.
HABITAT: Forming a "dwarf lawn" around small lagoons in salt marshes.
GENERAL RANGE: Quebec to Nfld. south to Maine.
SPECIAL NOTE: For more information see H. Hinds, 1983. The Rare Vascular Plants of New Brunswick. Syllogeus No. 50.

Triglochin maritima L. Fig. p. A10. Map p. B9.
FREQUENCY: Common in coastal areas.

HABITAT: Saline or brackish areas of salt marshes forming hummocks and also rarely in calcareous bogs inland.
GENERAL RANGE: Alaska to Nfld. south to Calif. and Pa.; Patagonia; Iceland; Eurasia; Africa.

Triglochin palustris L. Fig. p. A11. Map p. B9.
COMMON NAME(S): Marsh Arrow-grass. Troscart des marais.
FREQUENCY: Uncommon, with populations in the southeast and northeast and one station near Shediac, Westmorland Co.
HABITAT: Usually high on the landward side of salt marshes or in calcareous boggy areas.
GENERAL RANGE: Alaska to Nfld. south to Calif. and Pa.; S. America; Greenland; Iceland; Eurasia.

SCHEUCHZERIACEAE (Pod-grass Family)

SCHEUCHZERIA L.

Scheuchzeria palustris L. Fig. p. A10. Map p. B9.
COMMON NAME(S): Pod-grass, Scheuchzeria. Scheuchérie des marais.
FREQUENCY: Scattered in the southern half of the Province.
HABITAT: Bogs and boggy pond margins.
GENERAL RANGE: Alaska to Nfld. south to Calif. and N.J.; Eurasia.
SUBSPECIFIC TAXA: The North American material has been referred to var. **americana** Fern.

ALISMATACEAE (Water-plantain Family)

1a Pistils in a globose head; uppermost flowers usually staminate with numerous stamens; leaves unlobed or sagittate. **Sagittaria**

1b Pistils in a single flattened ring; flowers perfect, with 6-9 stamens; leaves unlobed. **Alisma**

ALISMA L. Water-plantain

Alisma plantago-aquatica L. Fig. p. A11. Map p. B9.
COMMON NAME(S): Alisma plantain-d'eau.
FREQUENCY: Common in the St. John River drainage area, rare elsewhere.
HABITAT: Shallow water and shores.
GENERAL RANGE: B.C. to Nfld. south to Calif. and Fla.; S. America; Eurasia; Africa.
SUBSPECIFIC TAXA: Our plants, with white petals, are var. **americana** Schultes & Schultes. Var. **plantago-aquatica** with pinkish or reddish petals, is the Eurasian-African taxon reported from a few areas of the Province such as near Saint John, St. John Co. and Moncton, Westmorland Co.

++

SYNONYMS: Var. **americana** is sometimes called **A. triviale** Pursh. Var. **plantago-aquatica** is also called var. **michaletii** Hendrics.

SAGITTARIA L. Arrowhead. Flèche d'eau

1a Flowering scape much shorter than leaves; pedicels thickened; sepals appressed to fruiting heads; leaves often reduced to spongy phyllodia with varying degrees of expanded blade; estuarine plants. **S. calycina**

1b Flowering scape as long as or longer than leaves; pedicels filiform; sepals reflexed; mostly fresh water plants. 2

2a(1b) Anther filaments glabrous, slender, longer than anther; leaves linear to ovate or sagittate. 3

2b(1b) Anther filaments obscurely scaly-pubescent, about as long as anther and dilated; leaves petioled with linear to lance-elliptic blades. **S. graminea**

3a(2a) Beak (to 2 mm) extending laterally from top of achene; fruiting heads to 3 cm thick; bracts to 1 cm. **S. latifolia**

3b(2a) Beak (less than .5 mm) erect on top of achene; fruiting heads less than 1.5 cm thick; bracts to 4 cm long; often with ribbon-like leaves when deeply submerged. **S. cuneata**

Sagittaria calycina Engelm. Fig. p. A11. Map p. B9.
FREQUENCY: Rare; only known from a few eastern localities.
HABITAT: Muddy estuarine shores.
GENERAL RANGE: Eastern Que. to Eastern N.B. south to Va.
SPECIAL NOTE: For more information see H. Hinds, 1983. The Rare Vascular Plants of New Brunswick. Syllogeus No. 50.
SUBSPECIFIC TAXA: Our plants are referable to var. **spongiosa** Engelm.
SYNONYMS: **S. montevidensis** Cham. & Schlecht.; **Lophotocarpus spathulatus** J.G. Sm.; **S. spathulata** (Sm.) Buch.

Sagittaria cuneata Sheldon Fig. p. A11. Map p. B9.
COMMON NAME(S): Arrowhead, Wapato. Sagittaire cunéaire.
FREQUENCY: Common throughout the Province.
HABITAT: Shallow water and muddy shores.
GENERAL RANGE: Alaska to Nfld. south to Calif. and N.Y.

Sagittaria graminea Michx. Fig. p. A11. Map p. B10.
COMMON NAME(S): Grass-like Arrowleaf. Sagittaire graminée
FREQUENCY: Uncommon in the southwestern part of the Province.
HABITAT: Muddy and sandy shores.
GENERAL RANGE: Ont. to Nfld. south to Tex. and Fla.

Sagittaria latifolia Willd. Fig. p. A11. Map p. B10.
COMMON NAME(S): Broad-leaved Arrowleaf. Wapato, Duck-potato. Sagittaire à grande feuilles.
FREQUENCY: Common throughout the Province.

HABITAT: Shallow water and muddy shores.
GENERAL RANGE: B.C. to N.S. south to Calif. and Tex.; S. America; W.I.; Hawaii.

BUTOMACEAE (Flowering Rush Family)

BUTOMUS L. Flowering Rush. Jonc fleuri

Butomus umbellatus L. Fig. p. A34.
FREQUENCY: Very rare; reported from the Eel River, Restigouche Co. (UNB).
HABITAT: Shallow water and shores.
GENERAL RANGE: Eurasian; introduced and spreading rapidly from Ont. to N.S. south to the Great Lakes and Idaho.
SPECIAL NOTE: This plant was first collected in the Province in 1982. It will be interesting to trace its spread.

HYDROCHARITACEAE (Frog's-bit Family)

1a Leaves long and ribbon-like in a basal rosette; anthers usually 2. **Vallisneria**

1b Leaves to 2 cm long, whorled; anthers usually 9. **Elodea**

ELODEA Michx. Waterweed

1a Leaves narrowly lanceolate 0.8-1.5 mm wide (averaging ca. 1 mm), (4)5-10 (13) times as long as wide; globose spathe body of staminate flower 2-3 mm long, sessile and released at maturity; pistillate sepals 1-1.5 mm long. **E. nuttallii**

1b Leaves broadly lanceolate to oblong or ovate 1.5-4(5) mm wide (averaging ca. 2 mm), 2-5 times as long as wide; elongate spathe of staminate flower 4-8.5(14) mm long at maturity and held by a delicate stalk; pistillate sepals 2-4.5 mm long. **E. canadensis**

Elodea canadensis Michx. Fig. p. A11. Map p. B10.
COMMON NAME(S): Canada Waterweed. Elodée du Canada.
FREQUENCY: Locally common especially in the lower St. John River valley.
HABITAT: Shallow ponds and streams, usually calcareous.
GENERAL RANGE: B.C. to N.S. south to Calif. and N.C.

Elodea nuttallii (Planch.) St. John Fig. p. A11. Map p. B10.
COMMON NAME(S): Nuttall's Waterweed. Elodée de Nuttall.
FREQUENCY: Rare; scattered on the St. John and Restigouche River systems.
HABITAT: Shallow quiet waters.
GENERAL RANGE: Idaho to N.B. south to Calif. and N.C.
SPECIAL NOTE: For more information see H. Hinds, 1983. The Rare Vascular Plants of New Brunswick. Syllogeus No. 50.

VALLISNERIA L. Tapegrass. Herbe à la barbotte

Vallisneria americana Michx. Fig. p. A11. Map p. B10.
COMMON NAME(S): Tape-grass, Water-celery, Wild Celery. Vallisnérie américaine.
FREQUENCY: Locally common in the lower St. John River valley and N.B./N.S. border area.
HABITAT: Shallow calm waters.
GENERAL RANGE: N. Dak. to N.S. south to Tex. and Fla.

POACEAE (Grass Family)

1a Spikelets unisexual, the staminate and pistillate conspicuously different, separate parts of one panicle; plant of shores or shallow water. **Zizania**

1b Spikelets bisexual or occasionally some sterile, rarely unisexual and then staminate and pistillate not conspicuously different nor widely separated in the inflorescence. 2

2a(1b) Spikelets all with a single fertile floret, with or without reduced florets below the perfect one. 3

2b(1b) Spikelets with 2 or more perfect florets, or occasionally with only 1 and then the fertile floret basal in the spikelet with 1 or more imperfect florets above it. 4

3a(2a) Spikelets in pairs, 1 sessile and 1 pedicellate; pedicellate spikelet usually rudimentary; lower glume longer than and tightly clasping or enclosing upper glume; lemma of fertile floret membranous, much thinner than lower glume; pedicels long hairy; racemes digitately arranged at end of culm or branches. **Schizachyrium**

3b(2a) Spikelets in pairs or not; lower glume shorter than or equal to upper glume, rarely absent, not clasping or enclosing it; lemma of fertile floret never much thinner than lower glume, herbaceous, cartilaginous or chartaceous and shiny. 5

4a(2b) Spikelets sessile or subsessile on the rachis of the main axis or of primary branches, forming 1 to several spikes, these occasionally very short. 6

4b(2b) Spikelets stalked, arranged in open or dense panicles. 7

5a(3b) One or more staminate or sterile florets present below the fertile one, sometimes reduced to small lemmas; occasionally only 1 glume present and then lemma of sterile floret appearing as upper glume; lemma of fertile floret often cartilaginous or ligneous and shiny, always so if glume single. 8

5b(3b) Staminate or sterile florets absent, or occasionally rudimentary structures present above the fertile floret; lemma never cartilaginous or ligneous and shiny. 9

6a(4a) Spikelets 2-3 at each joint, alternating on opposite sides of the rachis. 10

6b(4a) Spikelets solitary at each joint of the rachis. 11

7a(4b) Lemma of lowest floret usually shorter than or reaching to the tips of the glumes; lemma awned with a bent awn from the back or between the teeth, occasionally awnless or with a short straight awn. 12

7b(4b) Lemma of lowest floret extending beyond the tips of the glumes; lemma awnless or with a straight awn arising at or just below the tip or between the terminal teeth. 13

8a(5a) Spikelets disarticulating above the glumes; imperfect florets usually 2, if 1 then always staminate with lemma and palea similar to those of fertile one; lemma of fertile floret herbaceous, not enclosing membranous palea. 14

8b(5a) Spikelets disarticulating below the glumes (except **Setaria italica**); imperfect floret 1, usually sterile, rarely staminate, with lemma dissimilar to that of the fertile one and may be misinterpreted for a first glume (check for a reduced palea usually associated with the sterile lemma); lemma of fertile floret shiny, cartilaginous to ligneous, with inrolled edges enclosing palea of similar texture. 15

9a(5b) Spikelets sessile or subsessile on the rachis of the main axis or of primary branches, forming 1 or more spikes. 16

9b(5b) Spikelets stalked, sometimes shortly so, arranged in open or dense and spike-like panicles or racemes, rarely sessile and then remote and not in a spike. 17

10a(6a) Spikelets usually 2 at a joint, similar, 2- to 6-flowered. **Elymus**

10b(6a) Spikelets 3 at a joint, with the lateral ones sterile or reduced to awns, the central 1-flowered, fertile. **Hordeum**

11a(6b) Spikelets with the backs of one row of florets toward the rachis (back-on); glume next to the rachis wanting. **Lolium**

11b(6b) Spikelets with one side of both rows of florets toward the rachis (side-on); both glumes well developed. 18

12a(7a) Ligule a row of hairs; leaf sheaths generally hairy with tufts of longer hairs at collar. **Danthonia**

12b(7a) Ligule membranous; leaf sheaths glabrous or hairy but without any tuft of hairs at collar. 19

13a(7b) Rachilla bearing long silky hairs that at maturity exceed glumes, giving a plume-like panicle; plant 1.5-4 m high, reed-like with leaf blades 2-3 cm wide; ligule a row of matted hairs. **Phragmites**

13b(7b) Rachilla naked or short-pilose; plant less than 2 m high, not reed-like; leaf blades less than 2 cm wide. 20

14a(8a) Imperfect floret 1, staminate; lemmas 8-10 mm, the lower long-awned. **Arrhenatherum**

14b(8a) Imperfect florets 2, sterile, or if staminate, lemmas awnless; lemmas less than 6 mm long. 21

15a(8b) Spikelets not surrounded or subtended by bristles or prickles but sometimes stiffly hispid or bristly on the glumes and sterile lemmas; ligule of hairs or membranous or absent; inflorescence an open panicle or composed of digitate spikes. 22

15b(8b) Spikelets with one to several bristles or prickles below or surrounding; ligule a row of hairs. **Setaria**

16a(9a) Spikelets 3 at each joint, the lateral ones sterile or reduced to awns, the joints alternating at opposite sides of the rachis. **Hordeum**

++

16b(9a) Spikelets not in threes, all fertile, arranged on one side of the rachis.

17a(9b) Palea equaling the lemma and of similar chartaceous texture; glumes absent. **Leersia**

17b(9b) Palea smaller than the lemma and enclosed in it, often membranous, rarely absent; at least 1 glume present. 24

18a(11b) Plants perennial, rhizomatous or tufted; native or introduced. **Agropyron**

18b(11b) Plants annual; cultivated cereals. 25

19a(12b) Annual weeds or cereals. 26

19b(12b) Perennials. 27

20a(13b) Lemma conspicuously 3-nerved; spikelets in open panicles. 28

20b(13b) Lemma 5-to many-nerved, the nerves sometimes obscure; ligule membranous. 29

21a(14b) Lemmas of sterile florets much shorter than that of fertile floret; inflorescence greenish, pale or purple-tinged, appearing in mid or late summer; foliage not flavored or scented. **Phalaris**

21b(14b) Lemmas of sterile florets equaling or exceeding that of fertile floret; inflorescence golden brown, appearing in May or June; foliage bitter tasting when fresh, sweet-scented when dry. 30

22a(15a) Spikelets stiffly hispid, with coarsely awn-tipped glumes or sterile lemmas; ligule absent. **Echinochloa**

22b(15a) Spikelets often pubescent but not coarsely hispid, awnless. 31

23a(16b) Glumes broad, inflated, with abrupt short points, equal in length; soft, tufted plants; ligule long, membranous. **Beckmannia**

23b(16b) Glumes slender, not inflated, awned, the upper longer than the lower; coarse plants with heavy rhizomes; ligule a row of short hairs. **Spartina**

24a(17b) Lemmas firm, harder than the glumes, embracing the palea; florets dorsally compressed or terete; tufted grasses without rhizomes or stolons. 32

24b(17b) Lemmas as thin as or thinner than the glumes; florets generally compressed laterally. 33

25a(18a) Glumes ovate. **Triticum**

25b(18a) Glumes awn-shaped. **Secale**

26a(19a) Spikelets pendent, 2.2-2.5 cm long (excluding awns). **Avena**

26b(19a) Spikelets erect to erecto-patent, 1.0-1.5 cm long. **Ventenata**

27a(19b) Awns absent or rarely with a minute straight and nearly terminal awn. 34

27b(19b) Awns present, bent. 35

28a(20a) Ligule membranous; lemma hairless; soft creeping perennials, occurring rarely along northern coast. **Catabrosa**

28b(20a) Ligule a row of short hairs or absent; spikelets compressed 3- to many-flowered; paleas persistent; low annual weeds. **Eragrostis**

29a(20b) Spikelets crowded in one-sided clusters at ends of stiff naked panicle branches; sterile spikelets absent; leaf sheaths flattened. **Dactylis**

29b(20b) Spikelets not in dense one-sided clusters. 36

30a(21b) Glumes equal in size and shape, broad, but not wrapped around the florets; lower florets staminate, with broad acute lemmas; panicle open; basal leaf blades much longer than those of the culm and appearing later in the season, hairless. **Hierochloë**

30b(21b) Upper glumes much longer than the lower, wrapped around the florets; lower florets neuter with narrow lemmas 2-lobed at apex; panicle spike-like; basal and culm leaves similar, pubescent. **Anthoxanthum**

31a(22b) Lemma and palea of fertile floret cartilaginous, flexible, usually dark colored, with membranous margins; ligule membranous; lower glume minute or absent; spikelets in pairs, short stalked and appressed to one side of the digitate racemes. **Digitaria**

31b(22b) Lemma and palea of fertile floret ligneous, rigid, usually pale straw-colored, with thick inrolled margins; ligule a row of hairs or absent. 37

32a(24a) Spikelets dorsally compressed; lemma awnless, glabrous; spikelets 3-4 mm long; plants more than 50 cm high with broad soft leaf blades; glumes falling with grains. **Milium**

32b(24a) Spikelets +/- terete or slightly laterally compressed; lemma awned (awn sometimes caducous), glabrous or finely pubescent on back; plants less than 50 cm high, or if higher, spikelets 6-9 mm long or else leaf blades coarse and becoming narrowly involute; lemma with a delicate, generally deciduous awn, less than 2 cm long, lemma body elliptical, 2-9 mm long with a short blunt callus; glumes persisting after falling of grains. **Oryzopsis**

33a(24b) Spikelets, exclusive of awns, 8 mm or more long. 38

33b(24b) Spikelets, exclusive of awns, 1-7 mm long. 39

34a(27a) Spikelets falling away completely from naked pedicels at maturity; lower glume very narrow; upper glume obovate. **Sphenopholis**

34b(27a) Glumes persisting on pedicels at maturity, with individual florets breaking away; glumes somewhat unequal, but the upper not obovate; panicle somewhat contracted, branched; callus hairy. **Trisetum**

35a(27b) Foliage glabrous, sometimes scabrous; spikelets 2-flowered. 40

35b(27b) Foliage pubescent. 41

36a(29b) Lemmas laterally compressed, keeled on back, generally with fine webby hairs from callus. **Poa**

36b(29b) Lemmas rounded on back, at least near base, with no webby hairs from callus. 42

37a(31b) Soft-based annuals with shallow-rooting habit, seedling leaves, absence of stems of previous year, and panicle development in mid to late summer from every stem; spikelets glabrous; basal leaves not forming winter rosettes. **Panicum**

37b(31b) Perennials; firmly rooted with stems of previous year often present; spikelets hairy. **Dichanthelium**

38a(33a) Panicle densely flowered and spike-like; glumes as long as the lemma; lemma awnless or essentially so; coarse rhizomatous grass of sand beaches. **Ammophila**

38b(33a) Panicle slender with few spikelets; glumes short or obsolete; lemma tapering into a long awn; tufted grasses of woodland. **Brachyelytrum**

39a(33b) Inflorescence a dense, cylindrical, elliptical or ovate, uninterrupted, spike-like panicle with spikelets on very short pedicels; glumes united in lower part, fringed on keel. 43

39b(33b) Inflorescence a lax open or contracted panicle with spikelets on evident pedicles; if panicle contracted and spike-like, then more or less interrupted at base. 44

40a(35a) Both florets perfect with awned lemmas; spikelets 3-6 mm long. **Deschampsia**

40b(35a) Lower floret staminate, long-awned; upper floret perfect, awnless; spike-lets more than 7 mm long. **Arrhenatherum**

++

41a(35b) Spikelets 2-flowered; only the upper lemma awned with a short hook-like awn; lemmas 2-3 mm long; panicle contracted but not spike-like at maturity, usually pale purple or pink. **Holcus**

41b(35b) Spikelets usually more than 2-flowered; awns on each lemma, not hook-like; lemmas 4-10 mm long; panicles sometimes +/- spikelike, often green, greenish-brown or deep purple. **Trisetum**

42a(36b) Palea colorless; upper glume obovate. **Sphenopholis**

42b(36b) Palea green or brown, at least on nerves; upper glume usually broadest at or below the middle. 45

43a(39a) Lemma awnless. **Phleum**

43b(39a) Lemma awned from midback. **Alopecurus**

44a(39b) Ligule a row of hairs; lemma awnless; palea splitting down the centre; lemma hairless on the callus; annuals with us. **Sporobolus**

44b(39b) Ligule membranous, but sometimes short, firm and ciliate; lemma awned or awnless. 46

45a(24b) Nerves of lemma raised, equally prominent, +/- equally spaced, not converging at apex. 47

45b(24b) Nerves of lemma not prominent, or not equally so, and not equally spaced. 48

46a(44b) Callus of lemma with a fringe of fine hairs at least ¼ the length of the lemma. **Calamagrostis**

46b(44b) Callus of lemma hairless or with a few very short hairs; upper glume 1-nerved. 49

47a(45a) Lemma 7-nerved; upper glume 1-nerved; leaf sheaths closed, rupturing later; rhizomes present. **Glyceria**

47b(45a) Lemma 5-nerved; upper glume 3-nerved; leaf sheaths open; rhizomes lacking. **Torreyochloa**

48a(45b) Nerves of lemma +/- parallel, not converging at apex; lemma awnless, obtuse. **Puccinellia**

48b(45b) Nerves of lemma converging to apex; lemma often acute, awned or awnless. 50

49a(46b) Spikelet disarticulating below the glumes, leaving naked pedicels; rachilla prolonged as a slender bristle extending for more than one-half the length of the apparently 1-nerved palea; floret on a short recurved stalk, 0.2-0.4 mm long; stamens usually 1. **Cinna**

49b(46b) Spikelet disarticulating above the glumes, the empty glumes remaining on the branches for some time after maturity; rachilla not prolonged; palea 2-nerved; florets sessile; stamens 3; spikelets 1-3.5 mm long. 51

50a(48b) Lemma bifid at apex, usually awned from just below apex, between or in front of the teeth. 52

50b(48b) Lemma entire at apex, awnless or awned from tip; grain glabrous; glumes and lemmas lanceolate, +/- cuneate or truncate; perennials; glumes and lemmas firm; ligule short, almost obsolete. **Festuca**

51a(49b) Glumes awnless, equal to or longer than the lemma; lemma and palea loosely surrounding grain at maturity; lemma awnless or awned from back; rhizomes when present long, slender; leafy stolons sometimes present; culms smooth, glabrous. **Agrostis**

++

51b(49b) Glumes awned or not, the body shorter than the lemma; lemma and palea closely embracing the grain at maturity; lemma awnless or awned from the tip; rhizomes when present short, scaly; leafy stolons absent; culms smooth, hispidulous or nodulose. **Muhlenbergia**

52a(50a) Callus of lemma bearded with stiff straight hairs; grain free, glabrous. **Schizachne**

52b(50a) Callus of lemma not bearded; all florets similar, fertile; grain adherent to palea, pubescent at summit. **Bromus**

AGROPYRON Gaertn.

1a Anthers 1-2.2(2.4) mm long; rachilla disarticulating between the florets when mature; empty glumes often present; culms cespitose; rhizomes absent. **A. trachycaulum**

1b Anthers (2.7)3-5(6.1) mm long; spikelets tardily disarticulating +/- as a whole, not easily dislodged; empty glumes seldom if ever present; culms from elongate rhizomes. **A. repens**

Agropyron repens (L.) Beauv. Fig. p. A11. Map p. B10.
COMMON NAME(S): Witch-grass, Couch-grass. Chiendent rampant.
FREQUENCY: Common throughout the Province.
HABITAT: Sandy and gravelly shores and open mostly fertile soils.
CENTRAL RANGE: Alaska to Nfld. south throughout most of N. America.
SUBSPECIFIC TAXA: Numerous variations occur in almost all features of this species and have been given various infraspecific names. These appear to occur in all possible combinations and are not recognized in this work.

Agropyron trachycaulum (Link) Malte Fig. p. A11. Map p. B10.
COMMON NAME(S): Dog Couch-grass, Slender Wheat Grass. Chiendent à tiges rudes.
SUBSPECIFIC TAXA: Two well marked varieties occur with us.
1a Awns none or less than half as long as lemma. var. **novae-angliae**
1b Awns more than half as long as lemma. var. **glaucum**

1. Var. **novae-angliae** (Scribner) Fern.
FREQUENCY: Uncommon; reported scattered in Charlotte, Carleton, Westmorland, Gloucester, Northumberland and Madawaska Counties.
HABITAT: Calcareous gravel strands, ledges, peats and coastal beaches.
GENERAL RANGE: B.C. to Nfld. south to Nev., Neb., Mich. and N. Eng.

2. Var. **glaucum** (Pease & Moore) Malte
FREQUENCY: Common throughout much of the Province.
HABITAT: Rocky woods, thickets and shores.
GENERAL RANGE: B.C. to Nfld. south to Calif., Nev., Mich. and N. Eng.

AGROSTIS L. Bent-grass. Agrostide

Note: All inflorescences contracted when immature.

++

1a Ligule truncate, about (1.5) mm long; panicle branches usually contracted at first, remaining open when mature at about a 45° angle from main axis, flowerless near base; leaf blades 1-5 mm wide; spikelets usually bronze or purple tinged; palea about ½ as long as lemma. **A. capillaris**

1b Ligule rounded or acute, +/- lacerate, 2-6 mm long. 2

2a(1b) Culms tufted, without stolons or rhizomes. 3

2b(1b) Culms not tufted, arising from stolons or rhizomes (tardily so in **A. canina** and difficult to discern in mounted specimens). 4

3a(2a) Panicle branches abundantly scabrous, usually forking beyond the middle; spikelets reddish tinged, borne mostly near ends of branches; leaf blades mainly basal, +/- involute, less than 2(3) mm broad, the uppermost usually shorter than the lowest panicle branches; flowering in mid summer. **A. scabra**

3b(2a) Panicle branches smooth or only slightly scabrous, usually forking near or below middle; spikelets pale greenish (rarely reddish), not as confined to the ends of the branches; leaf blades distributed along stem, the uppermost about twice as long as lowest panicle branches; flowering in late summer. **A. perennans**

4a(2b) Lemma with a bent awn extending beyond glumes; palea minute or wanting. 5

4b(2b) Lemma usually awnless, if awned, awn straight, shorter than glumes; palea at least ½ as long as lemma. 6

5a(4a) Leaves mostly basal; anthers less than 1 mm long; spikelets averaging 2.8 mm long; panicle not contracting in fruit; rare native grass of open rocky habitats. **A. mertensii**

5b(4a) Leaves of stem as abundant as basal; anthers 1.2-1.8 mm long; spikelets averaging 2.1 mm long; greenish, yellowish or purplish panicle strongly contracted in fruit; introduced grass of meadows and fields. **A. canina**

6a(4b) Leaf blades usually more than 3 mm wide; panicle open during and after flowering; bases of middle branches meeting axis at an angle of about 30-40° when mature; spikelets +/- purple tinged; culms arising straight or curved at the very base, from underground scaly rhizomes and sometimes also stolons. **A. gigantea**

6b(4b) Leaf blades usually less than 3 mm wide; panicle open at anthesis, closed and +/- cylindrical after flowering; bases of middle panicle branches usually strongly ascending or appressed to axis, diverging about 15°; spikelets pale greenish (rarely reddish); culms arising from stolons, +/- decumbent at bases, lower nodes strongly bent and/or rooting; rhizomes wanting. **A. stolonifera**

Agrostis canina L. Fig. p. A11.

COMMON NAME(S): Velvet Bent-grass, Brown Bent. Agrostide des chiens

FREQUENCY: Very rare; reported from N.B. by Scoggan, 1978-79 and Boivin, 1966-67. No undoubted specimens have been seen.

HABITAT: Wet meadows and hillsides. Used extensively in golf courses.

GENERAL RANGE: European; introduced in B.C.; Minn. to Nfld. south to Tenn. and Del.

++

Agrostis capillaris L. Fig. p. A12. Map p. B10.
COMMON NAME(S): Fine Bent-grass. Agrostide capillaire.
FREQUENCY: Common throughout the Province.
HABITAT: Fields, lawns, roadsides.
GENERAL RANGE: Eurasian, introduced from southern Alaska to Nfld. south to N.C.
SYNONYMS: **A. tenuis** Sibth.

Agrostis gigantea Roth. Fig. p. A12 (see **A. stolonifera**). Map p. B10.
COMMON NAME(S): Redtop. Foin follette.
FREQUENCY: Common throughout the Province.
HABITAT: Old fields, pastures, ditch banks and wet beaches.
GENERAL RANGE: Introduced from Eurasia; Alaska to eastern Que. and N.S. south throughout N. America.
SYNONYMS: **A. stolonifera** L. var. **major** (Gaud.) Farw.

Agrostis mertensii Trin.
FREQUENCY: Very rare; collections are known from Pabineau Falls and Tremblay Gorge, Gloucester Co. (UNB).
HABITAT: Rock crevices and gravelly soil.
GENERAL RANGE: Alaska to Nfld. south to B.C. and Gaspé Pen.; also mtns. of N. Eng; Greenland; Eurasia.
SYNONYMS: **A. borealis** Hartm.

Agrostis perennans (Walt.) Tuckerm. Fig. p. A12. Map p. B10.
COMMON NAME(S): Upland Bent, Autumn Bent-grass.
FREQUENCY: Uncommon but probably frequently overlooked and under collected.
HABITAT: Moist woodlands and flooded shores.
GENERAL RANGE: Ont. to Nfld. south to Tex. and Fla.
SUBSPECIFIC TAXA: Var. **aestivalis** Vasey is considered a shade form and should probably have no formal taxonomic rank.

Agrostis stolonifera L. Fig. p. A12. Map p. 10.
COMMON NAME(S): Creeping Bent-grass, Carpet Bent.
FREQUENCY: Common throughout the Province.
HABITAT: Ditches, stream banks, wet pastures and shores.
GENERAL RANGE: Eurasian; introduced from Alaska to Nfld. south to Calif. and Va.; Greenland.
SUBSPECIFIC TAXA: This taxon varies considerably in the wild and has been divided into two varieties on the basis of stolons present or absent, panicle breadth in fruit and leaf breadth. Since these characteristics appear to combine randomly, varieties are not recognized in this work.
SYNONYMS: **A. alba** L. var. **palustris** (Hudson) Pers.; **A. stolonifera** L. var. **palustris** (Huds.) Farw.; **A. maritima** Lam.

Agrostis stolonifera L. x **A. capillaris** L.
SPECIAL NOTE: One collection from Edmundston, Madawaska Co., appears to be from this hybrid combination. It has been previously reported from P.E.I. and St. Paul Is., N.S. (UNB).

++

Agrostis scabra Willd. Map p. B10.
COMMON NAME(S): Fly-away Grass, Hair Grass, Ticklegrass. Agrostide scabre, Foin-des-fous.
FREQUENCY: Common throughout the Province.
HABITAT: Lakeshores, bog trails, roadsides, railroad yards, burnt-over areas.
SUBSPECIFIC TAXA: Forma **tuckermanii** Fern., with lemmas dorsally awned has been collected at Pabineau Falls, Gloucester Co.
GENERAL RANGE: Alaska to Nfld. south to Calif. and Fla.; Greenland; Asia.
SYNONYMS: **A. hiemalis** (Walt.) B.S.P. var. **tenuis** (Tuckerm.) Gleason.

ALOPECURUS L. Foxtail. Vulpin

1a Culms erect or ascending, usually over 6 dm tall; panicle about 10 mm thick; leaves to 10 mm wide; spikelets 4-6.5 mm long with awns exserted up to 7 mm beyond glumes; plants of dryish open areas. **A. pratensis**

1b Culms decumbent to tufted, erect, usually less than 6 dm tall; panicle 5-7 mm thick; leaves to 6 mm wide; spikelets not over 3 mm long with awns included or exserted 2-3 mm beyond glumes; plants of wet shores, ditches, and similar open areas. 2

2a(1b) Awn inconspicuous, included or rarely exsert to 2 mm beyond glumes, straight, attached near middle of lemma; anthers yellow or orange after anthesis. **A. aequalis**

2b(1b) Awn conspicuously exserted 2-3 mm beyond glumes, geniculate, attached near base of lemma; anthers purplish after anthesis. **A. geniculatus**

Alopecurus aequalis Sobol. Fig. p. A12. Map p. B11.
COMMON NAME(S): Short-awned Foxtail. Vulpin à courtes arêtes.
FREQUENCY: Mostly scattered throughout.
HABITAT: Marshy shores.
GENERAL RANGE: Alaska to Nfld. south to Calif. and Md.; Greenland; Iceland; Eurasia; Africa.

Alopecurus geniculatus L. Fig. p. A12.
COMMON NAME(S): Water Foxtail, Floating Foxtail. Vulpin genouillé.
FREQUENCY: Very rare; reported only from the city of Saint John, St. John Co., and Grand Manan Island, Charlotte Co.
HABITAT: Ditches and muddy flats.
GENERAL RANGE: Eurasian; scattered across Canada from B.C. to Nfld. south from Minn. to N.J.

Alopecurus pratensis L. Fig. p. A12. Map p. B11.
COMMON NAME(S): Meadow Foxtail. Vulpin des prés.
FREQUENCY: Locally common.
HABITAT: Moist meadows.
GENERAL RANGE: Eurasian; introduced from Alaska and B.C. to Nfld. south to Oreg. and Ga.

++

AMMOPHILA Host.

Ammophila breviligulata Fern. Fig. p. A12. Map p. B11.
COMMON NAME(S): Beachgrass. Gourbet, Oyat.
FREQUENCY: Scattered and locally common.
HABITAT: Coastal sandy shores and dunes.
GENERAL RANGE: Ont. to Nfld. south to N.C.

ANTHOXANTHUM L. Vernal Grass

Anthoxanthum odoratum L. Fig. p. A12. Map p. B11.
COMMON NAME(S): Sweet Vernal Grass. Flouve odorante, Foin d'odeur.
FREQUENCY: Common in the southern part of the Province.
HABITAT: Fields, pastures and similar open areas.
GENERAL RANGE: Eurasian; introduced over much of N. America.

ARRHENATHERUM Beauv. Oat-Grass

Arrhenatherum elatius (L.) Beauv. Fig. p. A12.
COMMON NAME(S): Tall Oat-Grass.
FREQUENCY: Very rare; only known from the Dalhousie area, Restigouche Co. (UNB).
HABITAT: Fields and similar open areas.
GENERAL RANGE: Eurasian; introduced in Canada in B.C. and Ont. to Nfld.
SUBSPECIFIC TAXA: Our material is forma **flavescens** (Niels.) Holmb., with yellowish rather than purplish panicles.

AVENA L. Oat. Avoine

1a Lemmas glabrous; awns straight or only slightly twisted at base or absent; florets not separating readily when mature. **A. sativa**
1b Lemmas +/- hirsute on back; awn twisted and bent near the middle; florets readily separating when mature. **A. fatua**

Avena fatua L. Fig. p. A12.
COMMON NAME(S): Wild Oat. Avoine sauvage, Folle Avoine.
FREQUENCY: Rare.
HABITAT: Fields, roadsides and similar waste areas.
GENERAL RANGE: Eurasian; found in most agricultural areas across Canada.

Avena sativa L. Fig. p. A12. Map p. B11.
COMMON NAME(S): Oat. Avoine
FREQUENCY: Scattered throughout but not persisting.
HABITAT: Fields, roadsides and railways.
GENERAL RANGE: Eurasian; introduced in agricultural areas across Canada.

POACEAE

++

BECKMANNIA Host.

Beckmannia syzigachne (Steud.) Fern.
COMMON NAME(S): Beckmann's-Grass, Slough Grass.
FREQUENCY: Very rare; reported only from Fredericton, York Co. (UNB).
HABITAT: Wet ground near grain elevators.
GENERAL RANGE: Alaska to N.S. south to Calif. and Pa.; Eurasia. Probably introduced from further west as a weed in grain seed.
SUBSPECIFIC TAXA: Our material is referable to ssp. **baicalensis** (Kusnez.) Koyama & Kawano.

BRACHYELYTRUM Beauv.

Brachyelytrum erectum (Schreb.) Beauv. Fig. p. A13. Map p. B11.
COMMON NAME(S): Bearded Short-husk.
FREQUENCY: Common throughout the Province.
HABITAT: Mixed woods or rich hardwoods.
GENERAL RANGE: Minn. to Ont. and Nfld. south to La. and Ga.
SUBSPECIFIC TAXA: Our material is referable to var. **glabratum** (Vasey) Koyama & Kawano.
SYNONYMS: **B. erectum** var. **septentrionale** Babel.

BROMUS L. Brome-Grass. Brome

1a Perennials (stubble of previous year's growth usually present). 2
1b Annuals (old growth absent, remains of grain often attached at base). 3
2a(1a) Plants rhizomatous; awns less than 1 mm long; sheaths usually glabrous; branches of panicle stiff, erect to spreading. **B. inermis**
2b(1a) Plants tufted, non-rhizomatous; awns about 3 mm long; sheaths glabrous or hairy; branches of panicle lax, drooping. **B. ciliatus**
3a(1b) Awns more than 10 mm long; lemmas lanceolate, terminal teeth 2-5 mm long. **B. tectorum**
3b(1b) Awns up to 10 mm long; lemmas oblong, terminal teeth less than 1 mm long. 4
4a(3b) Panicle small and +/- dense with short, erect or ascending branches; peduncles mostly shorter than spikelets. 5
4b(3b) Panicle large and open with long, spreading branches; peduncles mostly longer than spikelets. **B. secalinus**
5a(4a) Panicle branches mostly bearing more than 2 spikelets; lemmas pubescent with prominent nerves. **B. hordeaceus**
5b(4a) Panicle branches mostly bearing 1 or 2 spikelets; lemmas glabrous with obscure nerves. **B. racemosus**

Bromus ciliatus L. Fig. p. A13. Map p. B11.
COMMON NAME(S): Fringed Brome-Grass
FREQUENCY: Common throughout the Province.

++

HABITAT: Moist open woods and banks of streams.
GENERAL RANGE: Alaska to Nfld. south to Calif. and N.C.
SYNONYMS: **B. canadensis** Michx.: **B. dudleyi** Fern.

Bromus inermis Leyss. Fig. p. A13. Map p. 11.
COMMON NAME(S): Smooth or Hungarian Brome.
SPECIAL NOTE: European; introduced to fields and similar open areas throughout the Province. Known elsewhere in Canada from all provinces.

Bromus hordeaceus L. Fig. p. A13.
COMMON NAME(S): Soft Chess. Brome mou.
SPECIAL NOTE: Eurasian; uncommonly introduced to fields and similar open areas. Known elsewhere in Canada from B.C. to Que. and N.S.
SYNONYMS: **B. mollis** L.

Bromus racemosus L. Fig. p. A13.
COMMON NAME(S): Hairy Brome-Grass. Brome en grappe.
SPECIAL NOTE: Eurasian; introduced to fields and similar open areas as in Fredericton, York Co. Known elsewhere in Canada from Yukon and B.C. to Nfld., P.E.I. and N.S.

Bromus secalinus L. Fig. p. A13.
COMMON NAME(S): Cheat, Chess. Brome des seigles.
SPECIAL NOTE: Eurasian; uncommonly introduced to fields and similar open areas. Known elsewhere in Canada from B.C., Alta., Ont. to Nfld. and N.S.

Bromus tectorum L. Fig. p. A13.
COMMON NAME(S): Downy Chess. Brome des toits.
SPECIAL NOTE: Eurasian; introduced to roadsides and similar open areas as in St. Andrews, Charlotte Co. and Fredericton, York Co. Known elsewhere in Canada from Yukon and B.C. to Que.

CALAMAGROSTIS Adans. Reed-Bentgrass

1a Panicle widely open with loosely spreading branches at flowering, slightly contracted in fruit; lemma nearly smooth; callus hairs as long or longer than lemma. **C. canadensis**

1b Panicle contracted and narrow with strongly ascending branches at flowering; lemma prominently scabrous; callus hairs usually shorter than lemma. 2

2a(1b) Lemma awn +/- geniculate, often protruding from side of spikelet; callus hairs less than 1 mm long, up to ¼ as long as lemma, in two tufts; panicle mostly purplish and glaucous. **C. pickeringii**

2b(1b) Lemma awn +/- straight or slightly twisted, not exserted; callus hairs 1.7-2.2 mm long; panicle greenish, bronzed or rarely purplish, not glaucous. **C. stricta**

Calamagrostis canadensis (Michx.) Nutt. Fig. p. A13. Map p. B11.
COMMON NAME(S): Blue-joint. Foin bleu.
FREQUENCY: Common throughout the Province.
HABITAT: Wet meadows, open woods and bogs.
GENERAL RANGE: Alaska to Nfld. south to Calif. and Del.; Greenland; Eurasia.
SUBSPECIFIC TAXA: Var. **macouniana** (Vasey) Stebbins, with panicles densely flowered and spikelets 2.5-2.8 long, has been collected at Acadia Forest Station, Sunbury Co., and near Blue Mtn., Victoria Co. Several specimens with spikelets about 4 mm long have been collected in the Province but in other respects do not conform to the description of var. **robusta** Vasey.

Calamagrostis pickeringii Gray Fig. p. A13.
COMMON NAME(S): Pickering's Blue-joint.
FREQUENCY: Very rare; collected only near the Canaan River in the northwest corner of Westmorland Co. (UNB).
HABITAT: Boggy heath and wet sandy open woods.
GENERAL RANGE: Ont. and Nfld. south to Vt. and Mass.

Calamagrostis pickeringii Gray X **C. canadensis** (Michx.) Nutt.
SPECIAL NOTE: Plants intermediate between these two species were collected at the Canaan River site where both species occur and probably represent hybrids. (UNB).

Calamagrostis stricta (Timm) Koel. Fig. p. A13. Map p. B11.
FREQUENCY: Locally common along the eastern side of the Province.
HABITAT: Wet meadows and landward edge of saltmarshes.
GENERAL RANGE: Alaska to Nfld. south to Idaho and Minn.
SYNONYMS: **C. neglecta** (Ehrh.) Gaertn., Mey. & Scherb.

CATABROSA Beauv. Water Hairgrass

Catabrosa aquatica (L.) Beauv.
COMMON NAME(S): Brook-Grass, Water Whorl-Grass.
FREQUENCY: Very rare; known only from Dalhousie, Restigouche Co. Not seen in the field since 1922.
HABITAT: Fresh to brackish marshy ground.
GENERAL RANGE: Ont. to Nfld. south to Oreg., Ariz. and Mich.; Greenland; Eurasia.
SPECIAL NOTE: For more information see H. Hinds. 1983. The Rare Vascular Plants of New Brunswick. Syllogeus No. 50.

CINNA L. Wood Reedgrass

1a Spikelets 2.5-3.8(4) mm long (excluding awn); glumes +/- equal, the lower with lateral nerves obscure; anthers 0.4-0.9 mm long; ligules +/- hyaline; panicle typically lax and drooping. **C. latifolia**

++

1b Spikelets (4.2)4.5-6.5 mm long; glumes differing in size by 1-2 mm, the lower with well developed lateral nerves; anthers 0.8-1.8 mm long; ligules tinged reddish brown; panicle stiffer and usually more condensed. **C. arundinacea**

Cinna arundinacea L. Fig. p. A13.
COMMON NAME(S): Reed Cinna. Cinna roseau.
FREQUENCY: Very rare; collected only along the St. John River west of Jemseg, Queens Co. (UNB).
HABITAT: Swampy alluvial woods.
GENERAL RANGE: Minn. and Ont. to N.B. south to Tex. and Ga.
SPECIAL NOTE: For more information see H. Hinds, 1983. The Rare Vascular Plants of New Bunswick. Syllogeus No. 50.

Cinna latifolia (Trev.) Griseb. Fig. p. A13. Map. B11.
COMMON NAME(S): Drooping Wood Reed, Wide-leaved Cinna. Cinna à larges feuilles.
FREQUENCY: Common throughout the Province.
HABITAT: Woods and clearings.
GENERAL RANGE: Alaska to Nfld. south to Calif. and N.C.; Eurasia.

DACTYLIS L.

Dactylis glomerata L. Fig. p. A13. Map p. B11.
COMMON NAME(S): Orchard-Grass. Dactyle.
SPECIAL NOTE: Eurasian; introduced in fields and similar open areas scattered throughout the Province. Known elsewhere in Canada from B.C. to Nfld., P.E.I. and N.S.

DANTHONIA DC. Wild Oat-Grass

Danthonia spicata (L.) Beauv. Fig. p. A14. Map. p. B12.
COMMON NAME(S): Common Wild-Oat-Grass, Poverty-Grass, Junegrass. Avoine du pauvre, Herbe de pauvreté.
FREQUENCY: Common throughout the Province.
HABITAT: Poor soil of old pastures and open woods.
GENERAL RANGE: Alaska to Nfld. south to Oreg., N. Mex. and Fla.
SYNONYMS: **D. spicata** var. **pinetorum** Piper.

DESCHAMPSIA Beauv. Hair Grass. Canche

1a Basal leaves tufted, involute, mostly setaceous, about 1.5 dm long; panicle branches +/- glabrous; lemma awns geniculate, well exserted. **D. flexuosa**

1b Basal leaves flat or tardily involute, 1.5-5 mm wide, to 6 dm long; panicle branches +/- scabrous; lemma awns mostly straight, included or only slightly exserted. **D. caespitosa**

++

Deschampsia caespitosa (L.) Beauv. Fig. p. A14. Map p. B12.
COMMON NAME(S): Tufted Hair Grass.
SUBSPECIFIC TAXA: Two well marked varieties occur in the Province.

1a Culms to 170 cm high; basal leaf blades to 50 cm long; panicle 25 - 45 cm long; spikelets 2-3 mm long. var. **parviflora**

1b Culms to 75 cm high basal leaf blades to 30 cm long; panicle to 20 cm long; spikelets 3 - 4.5 mm long. var. **caespitosa**

1. Var. **caespitosa**
FREQUENCY: Scattered throughout except rare in the eastern part of the Province.
HABITAT: Flood shores of rivers and lakes.
GENERAL RANGE: Alaska to Nfld. south to Calif., Mexico and Va.; Greenland; Iceland; Eurasia.

2. Var. **parviflora** (Thuill.) Cosson & Germ.
FREQUENCY: Common in the city of Saint John, Saint John Co.
HABITAT: Roadsides, lawns, fields and similar open areas.
GENERAL RANGE: European; introduced from N.B. to N.S. south to N. Eng.

Deschampsia flexuosa (L.) Trin. Fig. p. A14. Map p. B12.
COMMON NAME(S): Wavy Hair Grass.
FREQUENCY: Scattered; most common along the Bay of Fundy.
HABITAT: Exposed rocky shores and similar exposed sites.
GENERAL RANGE: Ont. to Nfld. south to Okla., Tenn. and Ga.; Greenland; Eurasia.

DICHANTHELIUM (Hitchc. & Chase) Gould Panic-Grass. Panic

1a Leaves very long and narrow, 10-20 times as long as wide, 2-5 mm wide, mostly confined to near base; nodes +/- hidden within leaf sheaths, plants not forming rosettes of short, stubby leaves. 2

1b Leaves shorter and broader, not more than 10 times as long as wide, well distributed along the culm; nodes conspicuous; plants usually forming basal rosettes of short, stubby leaves. 3

2a(1a) Spikelets 3-4 mm long, +/- glabrous; glumes and lemma extending as short, pointed beak. **D. depauperatum**

2b(1a) Spikelets blunt, 2.2-2.8 mm long, sparsely pilose or glabrous; glumes and lemma not prolonged. **D. linearifolium**

3a(1b) Spikelets more than 3 mm long; leaf blades erect, rounded or cordate at base, glabrous, 10-15 mm broad. **D. xanthophysum**

3b(1b) Spikelets less than 3 mm long; leaf blades spreading to ascending, narrowed at base, glabrous or pilose, 2-12 mm broad. 4

4a(3b) Ligule hairs 2-5 mm long; spikelets usually less than 2 mm long (rarely 2.1 mm long); leaves generally less than 1 cm wide. **D. acuminatum**

4b(3b) Ligule hairs less than .5 mm long; spikelets 2-2.2 mm long; leaves generally greater than 1 cm wide. **D. boreale**

+++

Dichanthelium acuminatum (Sw.) Gould & Clark Fig. p. A14. Map p. B12.
GENERAL RANGE: B.C. to Nfld. south to Tex. and Fla.
SPECIAL NOTE: Three varieties of this species are recognized in the Province. These are separated as follows.

1a Spikelets 1.3-1.5(1.8) mm long; culms and sheaths glabrous or pilose. 2
1b Spikelets 1.6-2.1 mm long; culms and sheaths pilose. var. **acuminatum**
2a(1a) Culms and panicle axes glabrous; middle and upper leaf sheaths glabrous (except for ciliate margins); often purple tinged. var. **lindheimeri**
2b(1a) Culms and panicle axes pilose; middle and upper leaf sheaths pilose; pale green and lower growing with straggly stems and inflorescence. var. **implicatum**

1. Var. **acuminatum**
FREQUENCY: Scattered in the southwestern part of the Province.
HABITAT: Open sandy or gravelly areas.
SYNONYMS: **Panicum lanuginosum** Ell. var. **fasciculatum** (Torr.) Fern.; **P. lanuginosum** var. **tennesseense** (Ashe) Gleason; **P. lanuginosum** var. **septentrionale** (Fern.) Fern.; **P. subvillosum** Ashe.

2. Var. **implicatum** (Schribn.) Gould & Clark
FREQUENCY: Locally common.
HABITAT: Rocky outcrops, eroding slopes, clearings, old pastureland.
SYNONYMS: **P. lanuginosum** var. **implicatum** (Schribn.) Fern.; **P. meridionale** (Ashe) Freckmann.

3. Var. **lindheimeri** (Nash) Gould & Clark
FREQUENCY: Scattered mostly in the St. John River valley.
HABITAT: Flood shores.
SYNONYMS: **P. lanuginosum** var. **lindheimeri** (Nash) Fern.; **P. lindheimeri** Nash.

Dichanthelium boreale (Nash) Freckmann Fig. p. A14. Map. p. B12.
FREQUENCY: Probably common throughout the Province.
HABITAT: Banks of rivers and streams, moist hillsides, clearings.
GENERAL RANGE: Ont. to Nfld. south to Ind. and N.J.
SYNONYMS: **Panicum boreale** Nash.

Dichanthelium depauperatum (Muhl.) Gould Fig. p. A14. Map p. B12.
FREQUENCY: Very rare; known only from Pabineau Falls, Gloucester Co., Blissville, Sunbury Co., and White's Cove, Grand Lake, Queens Co.
HABITAT: Sandy or gravelly shores and open woods.
GENERAL RANGE: Man. to N.S. south to Tex. and S.C.
SUBSPECIFIC TAXA: Our material is referred to var. **psilophyllum** Fern., with +/- glabrous leaf-sheaths instead of leaf-sheaths copiously pilose in var. **depauperatum**.
SYNONYMS: **Panicum depauperatum** Muhl.
SPECIAL NOTE: For more information see H. Hinds. 1983. The Rare Vascular Plants of New Brunswick. Syllogeus No. 50.

++

Dichanthelium linearifolium (Scribn.) Gould Fig. p. A14.
FREQUENCY: Very rare; known only from near Porter Bk. on S.W. Miramichi River, near Zealand Station, York Co., and near jct. of Renous and S.W. Miramichi Rivers, Northumberland Co.
HABITAT: Sandy softwoods, gravel banks and roadsides.
GENERAL RANGE: Man. to N.S. south to Tex. and Ga.
SYNONYMS: **Panicum linearifolium** Scribn.

Dichanthelium xanthophysum (Gray) Freckmann Fig. p. A14.
FREQUENCY: Very rare; only known from Pabineau Falls, Gloucester Co.
HABITAT: Sandy or rocky ground or open woods.
GENERAL RANGE: Sask. to N.S. south to Minn. and N. Eng.
SYNONYMS: **Panicum xanthophysum** Gray
SPECIAL NOTE: For more information see H. Hinds, 1983. The Rare Vascular Plants of New Brunswick. Syllogeus No. 50.

DIGITARIA Heist. Finger-Grass, Crab-Grass

Digitaria ischaemum (Schreb.) Muhl. Fig. p. A14. Map p. B12.
COMMON NAME(S): Small Crab-Grass. Digitarie.
SPECIAL NOTE: Eurasian; introduced to roadsides, fields and similar open areas in the southern half of the Province. Known elsewhere in Canada from all Provinces except Nfld.

DISTICHLIS Raf. Spike-Grass, Alkali-Grass

Distichlis spicata (L.) Greene Fig. p. A14. Map p. B12.
FREQUENCY: Rare along the eastern shore and inland to salt springs at Plumweseep, Kings Co.
HABITAT: Saltmarshes, brackish sands and salt springs.
GENERAL RANGE: B.C. to N.S. south to Tex. and Fla.; S. America.
SPECIAL NOTE: For more information see H. Hinds, 1983. The Rare Vascular Plants of New Brunswick. Syllogeus No. 50.

ECHINOCHLOA Beauv.

Echinochloa crusgalli (L.) Beauv. Fig. p. A15. Map p. B12.
COMMON NAME(S): Barnyard-Grass. Pied-de-coq.
SPECIAL NOTE: Eurasian; introduced in rich soil of gardens and similar open areas scattered throughout the Province. Known elsewhere in Canada from all provinces.

++

ELYMUS L. Wild Rye, Lyme Grass

1a Lemmas awnless; culms finely pubescent below inflorescence; plants with extensively creeping rhizomes growing on sandy beaches. **E. mollis**

1b Lemmas definitely awned; culms +/- glabrous at summit; tufted plants without extensively creeping rhizomes growing in various habitats. 2

2a(1b) Lemma awns straight; glumes clearly expanded and flattened above the bowed, indurated base, expanded portion longer than expanded portion of lemma; paleas 5-8(9) mm long; spikes erect. **E. virginicus**

2b(1b) Lemma awns +/- flexuous; glumes straight, linear, the expanded portion shorter than the expanded portion of the lemma; paleas 8.6-12.7(14) mm long; spikes nodding. 3

3a(2b) Leaves flat, 10-12 per culm, 1.5-2.4 cm wide, pilose above. **E. wiegandii**

3b(2b) Leaves +/- involute, especially at tip, 5-8 per culm, 5-15 mm wide, glabrous above. **E. canadensis**

Elymus canadensis L. Fig. p. A15. Map p. B12.
COMMON NAME(S): Canada Wild Rye. Élyme du Canada.
FREQUENCY: Rare; known only from a few sites along the St. John River in Carleton and York Counties.
HABITAT: Sandy or gravelly banks, strands or nearby roadways.
GENERAL RANGE: B.C. to N.S. south to Calif., Tex. and N.C.

Elymus mollis Trin. Fig. p. A15. Map p. B13
COMMON NAME(S): Sea Lyme-Grass, Strand Wheat. Élyme des sables, Seigle de mer.
FREQUENCY: Common especially along the eastern coast.
HABITAT: Sea beaches and sands. Occasionally introduced along sandy roadsides inland.
GENERAL RANGE: Alaska to Nfld. south along the coasts to Calif. and Mass.; Greenland; Iceland; Asia.
SYNONYMS: **E. arenarius** L. var. **mollis** (Trin.) Koidzumi.

Elymus virginicus L. Fig. p. A15. Map p. B13.
COMMON NAME(S): Lyme Grass, Terrell Grass.
FREQUENCY: Most common in the southern half of the Province.
HABITAT: Open wet woods and gravel strands.
GENERAL RANGE: B.C. to Nfld. south to Wash., Ariz. and Fla.

Elymus wiegandii Fern. Fig. p. A15. Map p. B13.
COMMON NAME(S): Wiegand's Wild Rye.
FREQUENCY: Mostly scattered in the western and northern part of the Province.
HABITAT: Rich floodplain forest and meadows.
GENERAL RANGE: Sask. to N.S. south to N. Eng. and Pa.
SYNONYMS: **E. canadensis** L. var. **wiegandii** (Fern.) Bowden.

+++

ERAGROSTIS Beauv. Love-Grass

1a Spikelets longer than their pedicels, with 10 or more florets; panicle branches, pedicels, leaf edges and sometimes other parts with glandular depressions (often obscure). **E. minor**

1b Spikelets usually shorter than their pedicels, with less than 10 florets; plants not glandular. 2

2a(1b) Spikelets with less than 6 florets; glumes glabrous on keel; panicle slenderly ellipsoid. **E. frankii**

2b(1b) Spikelets with 6 or more florets; at least the upper glume scabrous on the keel; panicle more broadly ellipsoid with branches more spreading. **E. pectinacea**

Eragrostis frankii Meyer Fig. p. A15.
COMMON NAME(S): Frank's Love-Grass, Sandbar Love-Grass.
FREQUENCY: Very rare; collected only in Fredericton, York Co. (UNB).
HABITAT: Dry sandy or cindery roadsides or railways.
GENERAL RANGE: Native in eastern U.S.; introduced from Ont. to N.B.

Eragrostis minor Host. Fig. p. A15.
COMMON NAME(S): Little Love-Grass.
FREQUENCY: Rare; reported from Fredericton, York Co. and from St. Stephen, Charlotte Co.
HABITAT: Sandy or cindery roadsides, railways and similar open areas.
GENERAL RANGE: Eurasian; introduced in Canada from Ont. to NS.
SYNONYMS: **E. poaeoides** Beauv.; **E. eragrostis** (L.) Beauv.

Eragrostis pectinacea (Michx.) Nees Fig. p. A15. Map p. B13.
COMMON NAME(S): Tufted Love-Grass. Éragrostide pectiné.
FREQUENCY: Uncommon; scattered mostly in the southern counties.
HABITAT: Sandy shores and similar open areas.
GENERAL RANGE: B.C. to N.S. south to Calif., Mexico and Fla.

FESTUCA L. Fescue-Grass. Fétuque

1a Leaf blades flat, +/- inrolled when dry, (2.5)3-8 mm broad; lemmas awnless or awn rarely to 0.8 mm long. 2

1b Leaf blades (at least the basal) tightly folded, less than 3 mm wide, awned or awnless. 3

2a(1a) Spikelets borne beyond middle of ascending-spreading panicle branches, 2-5 flowered; lemmas 2.5-4.5 mm long, nerveless; anthers 0.8-1.4 mm long; rare native species of rich woodlands. **F. obtusa**

2b(1a) Spikelets borne uniformily throughout panicle, 5-10 flowered; lemmas 5.5-8 mm long, distinctly nerved on back; anthers 2.2-3.5(3.8) mm long; introduced species of roadsides, meadows and waste areas. 4

3a(1b) Leaf sheaths of youngest shoots closed to near summit (often very obscure), reddish and hirsute, older sheaths open, brown, disintegrating to pale fibers near base; +/- rhizomatous with shoots bursting laterally through the lowest sheaths (extravaginal); culm leaves sometimes expanded. **F. rubra**

3b(1b) All leaf sheaths open to near base with mostly overlapping edges, tan to dark brown or sometimes reddish, smooth, not disintegrating into fibers; plants +/- densely tufted; all leaves closely folded. 5

4a(2b) Branches of panicle closely ascending after flowering, each node with 1 branch or if 2, the second with a single spikelet; spikelets 6-8 flowered; leaf auricles not ciliate. **F. pratensis**

4b(2b) Branches of panicle widely spreading after flowering, each node with mostly 2 branches each with several spikelets; spikelets usually 4-8 flowered; leaf auricles ciliate. **F. arundinacea**

5a(3b) Lemmas 2.5-3.5 mm long, awnless or rarely with awn to 0.3 mm long, brownish; leaves capillary, 0.2-0.3 mm broad, flexuous; anthers 1.5-1.8 mm long. **F. tenuifolia**

5b(3b) Lemmas 3.8-5 mm long, awned, green; leaves 0.5-1.2 mm broad; anthers 2.7-3 mm long. **F. longifolia**

Festuca arundinacea Schreb. fig. p. A15.
COMMON NAME(S): Tall Fescue. Fétuque élevée.
FREQUENCY: Locally common in the Fredericton area, York Co. and apparently rapidly spreading.
HABITAT: Roadsides, meadows and similar open areas.
GENERAL RANGE: Eurasian; introduced across Canada.
SYNONYMS: **F. elatior** L. var. **arundinacea** (Schreb.) Wimm.

Festuca longifolia Thuill.
COMMON NAME(S): Hard Fescue.
FREQUENCY: Apparently rare and locally common as near Belleisle Creek, Springfield, Queens Co. and near Petitcodiac, Westmorland Co.
HABITAT: Dry sandy roadsides and similar open areas.
GENERAL RANGE: European; introduced widely in lawn and pasture seed mixtures across Canada.
SYNONYMS: **F. ovina** L. var. **duriuscula** (L.) Koch.; **F. trachyphylla** (Hackel) Krajina.

Festuca obtusa Biehler Fig. p. A15.
COMMN NAME(S): Nodding Fescue.
FREQUENCY: Very rare; known only from the valley of the Restigouche River, Restigouche Co. (UNB).
HABITAT: Rich deciduous woods.
GENERAL RANGE: Man. to n.S. south to Tex. and Fla.
SYNONYMS: **F. nutans** Willd.
SPECIAL NOTE: For more information see H. Hinds, 1983. The Rare Vascular Plants of New Brunswick. Syllogeus No. 50.

Festuca pratensis Hudson Fig. p. A15. Map p. B13.
COMMON NAME(S): Meadow Fescue.
FREQUENCY: Scattered and locally common.

HABITAT: pastures, meadows and similar open areas.
GENERAL RANGE: Eurasian; introduced in Canada from Yukon and B.C. to Nfld.
SYNONYMS: **F. elatior** L.

Festuca rubra L. Fig. p. A16. Map p. 13.
COMMON NAME(S): Red Fescue. Fétuque rouge.
SPECIAL NOTE: This species is extremely variable and consists of both native and introduced varieties. These are not well understood but the following key will help separate some of the principal variations.

1a Leaf blades slender, firm and +/- glaucous; culms much longer than leaves; spikelets usually heavily glaucous over purplish base; native plants of sandy open habitats such as headlands, dunes and beaches. var. **juncea** (Hackel) Richter
1b Leaves more lax, usually not strongly whitened. 2
2a Basal offshoots erect, arising within sheath producing tufted appearance; leaves 1.5-6 dm long; culms 2.5-9 dm tall; introduced plants of pastures, meadows and similar open areas. var. **commuta** Gaudin
2b(1b) Basal offshoots +/- spreading or decumbent (plants appearing loosely matted). 3
3a(2b) Spikelets 6-8 flowered, 1-1.7 cm long; panicle with widely spreading, short branches; plants mostly of seashore marshes. var. **megastachys** (Gaudin) Hegi
3b(2b) Spikelets 3-7 flowered, 7-10 mm long; introduced plants mostly of fresh, moist and fertile soils. var. **rubra**

Festuca tenuifolia Sibthorp map. p. B13.
COMMON NAME(S): Hair Fescue.
FREQUENCY: Common in the southern part of the Province.
HABITAT: Open sandy loams.
GENERAL RANGE: European; introduced in Canada in B.C. and from Ont. to Nfld.
SYNONYMS: **F. capillata** Lam.; **F. ovina** L. var. **capillata** (Lam.) Alef.; **F. filiformis** Pourret

GLYCERIA R. Br. Manna-Grass

1a Spikelets 10-18 mm long; pedicels appressed to main axis, shorter than spikelets; joints of rachilla 1-1.3 mm long; panicle contracted and slender; ligule prolonged 6-16 mm long; leaves often floating. **G. borealis**
1b Spikelets linear to ovate, 2-6.5 (7) mm long; pedicels erect to widely spreading, longer than spikelets; joints of rachilla less than 1 mm long; ligule mostly less than 6 mm long (in some cases less than 1 mm long); panicle open or contracted; leaves not floating. 2
2a(1b) Lemmas with corrugated appearance from conspicuously raised scabrous nerves; spikelets to 2.6 mm wide at maturity. 3
2b(1b) Lemmas +/- smoothly rounded on back or with glabrous nerves only slightly raised near base; spikelets to 4 mm wide at maturity. 4

++

3a(2a) Panicle narrow, with appressed branches, +/- drooping at summit; spikelets 3-4 flowered; lemmas 1.9-2.8 mm long. G. **melicaria**

3b(2a) Panicle branches spreading, often drooping at tip. 5

4a(2b) Panicle strict, dense, ovoid or oblong with short ascending branches; palea completely covered by lemma. G. **obtusa**

4b(2b) Panicle branches prolonged, widely spreading and drooping at maturity; palea extending slightly beyond edge (not tip) of lemma. 6

5a(3b). Spikelets 5-6 mm long; glumes 1 mm or longer; lemma ovate, usually purple tinged; anthers 0.7-1 mm long; more robust in nearly all aspects than the following species. G. **grandis**

5b(3b) Spikelets 2-4.5 mm long; lower glumes usually less than 1 mm long; lemma elliptic to obovate, usually not as conspicuously purple tinged; anthers 0.3-0.5 mm long. G. **striata**

6a(4b) Lower lemmas 3-4 mm long, 0.5-1 mm longer than paleas; secondary branches of panicle in 1s to 3s; spikelets 5-10 flowered, 4.5-7 mm long; flowers fertile; cauline leaves 3-5. G. **canadensis**

6b(4b) Lower lemmas 2-2.5 mm long, about as long as paleas; secondary branches of panicle in 3s to 5s; spikelets 3-6 flowered, 3-5 mm long; flowers sterile; cauline leaves 6-8. G. X **laxa**

Glyceria borealis (Nash) Batch. Fig. p. A16. Map p. B13.
COMMON NAME(S): Small Floating Manna-Grass.
FREQUENCY: Common throughout the Province.
HABITAT: Shallow water.
GENERAL RANGE: Alaska to Nfld. south to Calif. and N. Eng.

Glyceria canadensis (Michx.) Trin. Fig. p. A16. Map p. B13.
COMMON NAME(S): Rattlesnake-Grass.
FREQUENCY: Common throughout the Province.
HABITAT: Shallow water and damp ground.
GENERAL RANGE: Sask. to Nfld. south to Ill. and N.C.

Glyceria grandis Wats. Fig. p. A16. Map p. B14.
COMMON NAME(S): Reed-meadow Grass.
FREQUENCY: Scattered throughout the Province.
HABITAT: Wet ditches and similar wet areas.
GENERAL RNAGE: Alaska to Nfld. south to Oreg. and N. Mex. to Va.
SUBSPECIFIC TAXA: Forma **pallescens** Fern., with yellowish instead of purplish spikelets, is reported from Fredericton, York Co.

Glyceria X laxa (Scribn.) Scribn. Map p. B14.
COMMON NAME(S): Loose-panicled Manna-Grass.
SPECIAL NOTE: This species is now considered to be a sterile hybrid between G. **canadensis** and some other species, probably G. **striata**.
FREQUENCY: Occasionally found where the parent species occur commonly together.
HABITAT: Wet softwoods, bog edges, alder thickets.
GENERAL RANGE: Ont. to N.S. south to W. Va.
SYNONYMS: G. **canadensis** (Michx.) Trin. var. **laxa** (Scribn.) Hitchc.

++

Glyceria melicaria (Michx.) Hubbard Fig. p. A16. Map p. B14.
COMMON NAME: Slender Manna-Grass.
FREQUENCY: Common throughout except in the extreme northern sector.
HABITAT: Wet open woods.
GENERAL RANGE: Ont. to N.S. south to Tenn. and N.C.

Glyceria obtusa (Muhl.) Trin. Fig. p. A16. Map p. B14.
FREQUENCY: Rare; reported from a few sites in Charlotte Co.
HABITAT: Bog edges.
GENERAL RANGE: N.B. and N.S. south to Pa. and N.C.
SPECIAL NOTE: For more information see H. Hinds, 1983. The Rare Vascular Plants of New Brunswick. Syllogeus No. 50.

Glyceria striata (Lam.) Hitchc. Fig. p. A16. Map p. B14.
COMMON NAME(S): Fowl-manna Grass.
FREQUENCY: Common throughout the Province.
HABITAT: Damp ground in woods, pastures, ditches, stream margins and bogs.
GENERAL RANGE: Aggregate species Alaska to Nfld. south to Calif. and Fla.
SUBSPECIFIC TAXA: Two varieties occur in the Province.

1a Spikelets usually purple-tinged, to 4.5 mm; lower lemmas 2-2.2 mm long; panicle to about 1.5 dm long; leaves usually folded, about 5 mm wide; culms to about 0.5 m high. var. **stricta** (Scribn.) Fern.

1a Spikelets usually green, to 4.0 mm long; lower lemmas 1.5-1.8 mm long; panicle to 3 dm long; leaves flat, about 7 mm wide; culms to about 1 m high. var. **striata**

HIEROCHLOE R. Br. Holy Grass

Hierochloë odorata (L.) Beauv. Fig. p. A16. Map p. B14.
COMMON NAME(S): Indian Grass, Vanilla Grass, Sweet Grass. Foin d'odeur, Herbe sainte.
FREQUENCY: Scattered throughout the Province.
HABITAT: Wet meadows, shores, landward edge of saltmarshes.
GENERAL RANGE: Alaska to Nfld. south to Oreg. and N.J.; Greenland; Iceland; Eurasia.

HOLCUS L. Velvet-Grass

Holcus lanatus L. Fig. p. A16.
COMMON NAME(S): Common Velvet-Grass, Yorkshire Fog.
FREQUENCY: Very rare; reported only from Fredericton, York Co. Probably not persisting.
HABITAT: Fields and similar open areas.
GENERAL RANGE: Eurasian; introduced in Canada in B.C. and Ont. to N.S.

++

HORDEUM L. Barley. Orge

1a Leaves conspicuously auricled, glabrous; spike rachis not disarticulating at maturity; awns of lemmas conspicuously larger than those of the glumes. **H. vulgare**

1b Leaves without auricles, pubescent (at least on sheaths); spike rachis disarticulating at maturity; awns of glumes and lemmas similarly subulate. **H. jubatum**

Hordeum jubatum L. Fig. p. A16. Map p. B14.
COMMON NAME(S): Squirrel-tail Grass. Queue d'écureuil.
FREQUENCY: Common along the coast.
HABITAT: Coastal beaches and saltmarshes. Occasionally inland on roadsides and similar open areas.
GENERAL RANGE: Alaska to Nfld. south to Mexico; S. America; Eurasia.

Hordeum vulgare L. Fig. p. A16. Map p. B14.
COMMON NAME(S): Barley. Orge vulgaire.
FREQUENCY: Reported from scattered locations in the southern part of the Province. Not persisting.
HABITAT: Roadsides, railways near grain elevators and similar open areas.
GENERAL RANGE: Eurasian; introduced but not established across Canada.

HYSTRIX Moench.

Hystrix patula Moench Fig. p. A16. Map p. B14.
COMMON NAME(S): Bottle-brush Grass, Porcupine-Grass. Asprelle Hérisson.
FREQUENCY: Very rare; reported from a few locations along the St. John River in Carleton Co. and on Grog Brook near the Upsalquitch River, Restigouche Co.
HABITAT: Rich hardwoods and clearings.
GENERAL RANGE: Ont. to N.B. south to Okla. and Ga.
SPECIAL NOTE: For more information see H. Hinds, 1983. The Rare Vascular Plants of New Brunswick. Syllogeus No. 50.
SUBSPECIFIC TAXA: Our material is referable to var. **bigeloviana** (Fern.) Bowden with pubescent instead of glabrous lemmas.
SYNONYMS: **Elymus hystrix** L. var. **bigeloviana** (Fern.) Bowden.

LEERSIA Sw. Cutgrass, Whitegrass

1a Panicle branches single at each node; spikelets 3-4 mm long; stamens 2; rhizomes slender, elongate. **L. virginica**

1b Panicle branches (at least the lowest) more than 1 at the nodes; spikelets 4-6 mm long; stamens 3; scaly rhizomes short and thick. **L. oryzoides**

++

Leersia oryzoides (L.) Sw. Fig. p. A17. Map p. 14.
COMMON NAME(S): Rice-Cutgrass.
FREQUENCY: Scattered but locally common in the southern part of the Province.
HABITAT: Muddy shores, ditches, marshy areas.
GENERAL RANGE: B.C. to N.S. south to Calif. and Fla.; S. America; W.I.; Europe; Asia.

Leersia virginica Willd. Fig. p. A17.
COMMON NAME(S): Whitegrass.
FREQUENCY: Very rare; reported from near the Mouth of Keswick and the floodplain of the St. John River, Maugerville, York Co.
HABITAT: Shaded banks and alluvial woods.
GENERAL RANGE: Nebr., Minn. and Ont. to N.B.
SPECIAL NOTE: For more information see H. Hinds, 1983. The Rare Vascular Plants of New Brunswick. Syllogeus No. 50.
SYNONYMS: **L. virginica** var. **ovata** (Poir.) Fern.

LOLIUM L. Darnel. Ivraie

1a Glume as long as or longer than spikelet; lemma typically long awned. **L. temulentum**
1b Glume shorter than spikelet; lemma awned or not. **L. perenne**

Lolium perenne L. Fig. p. A17. Map p. B14.
COMMON NAME(S): Common Darnel, Perennial Rye-Grass. Ivraie vivace, Ray-grass anglais.
FREQUENCY: Scattered in the south of the Province.
HABITAT: Fields, roadsides and similar open areas.
GENERAL RANGE: Eurasian; introduced across Canada.

Lolium temulentum L. Fig. p. A17.
COMMON NAME(S): Bearded or Poison Darnel. Ivraie enivrante.
FREQUENCY; Rare; reported from Carleton and York Cos.
HABITAT: Grainfields, railways near grain elevators and similar open areas.
GENERAL RANGE: Eurasian; introduced sporadically across Canada. Not persisting.

MILIUM L.

Milium effusum L. Fig. p. A17. Map p. B14.
COMMON NAME(S): Spreading Millet Grass. Millet sauvage.
FREQUENCY: Uncommon but widely distributed in the Province.
HABITAT: Mostly deciduous upland or rich alluvial lowland woods.
GENERAL RANGE: Man. to Nfld. south to S. Dak. and Md.; Iceland; Eurasia.
SUBSPECIFIC TAXA: Our plants are referable to var. **cisatlanticum** Fern. to distinguish them from slightly different Eurasian material.

++

MUHLENBERGIA Schreb.

1a Panicle diffuse, with long peduncles and pedicels; spikelets less than 2 mm long. **M. uniflora**

1b Panicle dense; spikelets more than 2 mm long. 2

2a(1b) Panicle very strict, less than 18 mm wide; lemmas not pilose at base; leaves 1-2 mm wide, becoming involute; scaley rhizomes absent. **M. richardsonis**

2b(1b) Panicle wider, usually greater than 10 mm wide; lemmas pilose at base; leaves usually wider than 2 mm wide, flat to inrolled; culms from elongate scaley rhizomes. 3

3a(2b) Glumes (including awns) (3)3.5-6.5(7.5) mm long, distinctly longer than the body of the lemma; anthers 0.5-1.3 mm long. **M. glomerata**

3b(2b) Glumes (including awns) less than 3.6 mm long, equalling or shorter than body of lemma; anthers not over 0.5 mm long. 4

4a(3b) Culm internodes glabrous; base of lateral inflorescences often enclosed in upper leaf sheath; culms mostly decumbent at base. **M. frondosa**

4b(3b) Culm internodes puberulent, especially below the conspicuously swollen nodes; inflorescences +/- exserted; culm mostly erect at base. **M. mexicana**

Muhlenbergia frondosa (Poir.) Fern. Fig. p. A17.
COMMON NAME(S): Leafy Muhlenbergia.
FREQUENCY: Common along the lower St. John and St. Croix Rivers.
HABITAT: Alluvial woods, clearings and similar open areas.
GENERAL RANGE: Ont. to N.B. south to Tex. and Ga.
SUBSPECIFIC TAXA: Forma **commutata** (Scribn.) Fern. with lemmas short awned rather than awnless occurs occasionally with the species.

Muhlenbergia glomerata (Willd.) Trin. Fig. p. A17. Map p. B15.
FREQUENCY: Scattered throughout the Province.
HABITAT: Rocky shores, meadows and bogs.
GENERAL RANGE: Yukon and B.C. to Nfld. south to Oreg., Nev. and Conn.
SYNONYMS: **M. glomerata** var. **cinnoides** (Link) Herm.

Muhlenbergia mexicana (L.) Trin. Fig. p. A17. Map p. B15
FREQUENCY: Scattered throughout except along the east coast.
HABITAT: Shores, moist rocky scree slopes, rocky banks.
GENERAL RANGE: B.C. to N.S. south to Calif. and N.C.
SUBSPECIFIC TAXA: Forma **ambigua** (Torr.) Fern. with awned lemmas occurs with the awnless species extending north to Restigouche Co.
SYNONYMS: Forma **ambigua** is also called var. **filiformis** (Willd.) Scribn.

Muhlenbergia richardsonis (Trin.) Rydb. Fig. p. A17. Map p. B15.
COMMON NAME(S): Mat Muhly, Richardson's Muhlenbergia.
FREQUENCY: Rare; reported from a few localities along the St. John River in York, Carleton and Madawaska Counties. and on the Restigouche River, Restigouche Co.
HABITAT: Gravelly river strands, rocky shores.

GENERAL RANGE: Yukon and B.C. to N.B. south to Baja, Calif., Ohio and Maine.
SPECIAL NOTE: For more information see H. Hinds, 1983. The Rare Vascular Plants of New Brunswick, Syllogeus No. 50.

Muhlenbergia uniflora (Muhl.) Fern. Fig. p. A17. Map p. B15.
COMMON NAME(S): One-flowered Dropseed.
FREQUENCY: Scattered mostly throughout the southwestern part of the Province.
HABITAT: peaty or sandy clearings and shores.
GENERAL RANGE: Ont. to Nfld. south to Wisc. and N.J.

ORYZOPSIS Michx. Mountain-Rice

1a	Lemma awn conspicuous, 6-15 mm long, +/- persisting after flowering. 2
1b	Lemma awn inconspicuous, less than 2 mm long, usually falling after flowering; glumes minutely scabrous near apex. **O. pungens**
2a(1a)	Leaf blades usually flattened, glaucous, becoming slightly involute in drying, more than 4 mm broad; lemma 6-9 mm long (excluding awn); glumes distinctly nerved. **O. asperifolia**
2b(1a)	Leaf blades involute, less than 3 mm in diameter; lemma 2-4 mm long (excluding awn); glumes (except midnerve) inconspicuously nerved. **O. canadensis**

Oryzopsis asperifolia Michx. Fig. p. A17. Map p. B15.
COMMON NAME(S): Rough Mountain-Rice, White-grained Mountain Ricegrass.
FREQUENCY: Scattered throughout the Province.
HABITAT: Open dry woods.
GENERAL RANGE: Dist. Mack. and B.C. to Nfld. south to Utah, N. Mex. and Va.

Oryzopsis canadensis (Poir.) Torr. Fig. p. A17. Map p. B15.
COMMON NAME(S): Canadian Mountain-Rice.
FREQUENCY: Uncommon and scattered from Pabineau Falls, Gloucester Co. south to Petitcodiac, Westmorland Co. and west to Brockway, York Co.
HABITAT: Sandy barrens and rocky clearings.
GENERAL RANGE: B.C. to Nfld. south to Calif., N. Mex. and N.H.

Oryzopsis pungens (Torr.) Hitchc. Fig. p. A18. Map p. B15.
COMMON NAME(S): Slender Mountain-Rice.
FREQUENCY: Rare; reported from Youghall Beach and Pabineau Falls, Gloucester Co., Big Bald Mtn., Northumberland Co., Richibucto Harbour vicinity and on the Broad River, Albert Co.
HABITAT: Rocky or sandy open woods and clearings.
GENERAL RANGE: Yukon to N.S. south to Colo. and N.J.
SPECIAL NOTE: For more information see H. Hinds, 1983. The Rare Vascular Plants of New Brunswick, Syllogeus No. 50.

++

PANICUM L. Panic-Grass. Panic

1a Leaf sheaths, nodes and usually leaves glabrous; lower glume not over length of spikelet, truncate or triangular tipped. **P. dichotomiflorum**

1b Leaf sheaths and leaves papillose-hispid; nodes usually bearded; lower glume at least 1/3 length of spikelet. 2

2a(1b) Spikelets large, 4.5-5 mm long; grain 2 mm thick; panicle heavy and often drooping at tip. **P. miliaceum**

2b(1b) Spikelets smaller, less than 4 mm long; grain less than 1 mm thick; panicle erect. 3

3a(2b) Spikelets lance-acuminate, usually over 2 mm long; panicle well exserted at maturity, comprising more than ½ entire height of the plant; basal pulvini copiously long-hispid; grain stramineus. **P. capillare**

3b(2b) Spikelets ellipsoid to ovoid, acute to short-acuminate, usually less than 2 mm long; panicle included in upper sheath or short exserted at maturity, comprising less than ½ the entire height of the plant; basal pulvini glabrous or rarely slightly hispid; grain blackish. **P. tuckermanii**

Panicum capillare L. Fig. p. A18. Map p. B15.
COMMON NAME(S): Common Witch Grass. Mousseline.
FREQUENCY: Locally common in the southwest of the Province especially in the St. John River valley.
HABITAT: Beaches, banks, roadsides, gardens and similar open areas.
GENERAL RANGE: B.C. to N.S. south to Calif. and Fla.

Panicum dichotomiflorum Michx. Fig. p. A18.
COMMON NAME(S): Fall Panic-Grass.
FREQUENCY: Rare; reported from Fredericton, York Co. and Sheffield, Sunbury Co. (UNB).
HABITAT: Railways, roadsides and fields.
GENERAL RANGE: Minn. to Ont. and N.S. south to Mexico and Fla.
SPECIAL NOTE: Apparently recently introduced in the Province and slowly spreading.
SYNONYMS: **P. dichotomiflorum** var. **geniculatum** (Wood) Fern.

Panicum miliaceum L.
COMMON NAME(S): Millet, Broom-corn Millet. Panic Millet
FREQUENCY: Rare; collected in a few scattered locations in the south of the Province such as at Fredericton, York Co. and St. John, St. John Co. Not persisting.
HABITAT: Railways near grain elevators, waste areas and similar open areas.
GENERAL RANGE: Eurasian; introduced across Canada.

Panicum tuckermanii Fern. Fig. p. A18. Map p. B15.
COMMON NAME(S): Tuckerman's Panic-Grass, Tuckerman Witch Grass.
FREQUENCY: Scattered from the S.E. Miramichi River southwest to the St. Croix River valley and Grand Manan.
HABITAT: Flood shores of streams and lakes.
GENERAL RANGE: Ont. to N.S. south to Tex. and Ga.
SYNONYMS: **P. philadelphicum** Bernh. var. **tuckermanii** (Fern.) Stey. & Schmoll.

++

PHALARIS L. Canary-Grass. Phalaride

1a Inflorescence large, 5-20(30) cm long, branched, becoming closely contracted after flowering; glumes lanceolate, often purple-tinged, the keel inconspicuously winged. **P. arundinacea**

1b Inflorescence smaller, less than 4 cm long, ovoid, dense and spike-like; glumes white with green nerves, the keel broadly winged. 2

2a(1b) Sterile lemma 1, +/- 1 mm long; fertile lemma 3 mm long; glumes about 5 mm long. **P. minor**

2b(1b) Sterile lemmas 2, about ½ as long as fertile lemmas; fertile lemmas about 6 mm long; glumes about 7 mm long. **P. canariensis**

Phalaris arundinacea L. Fig. p. A18. Map p. B15.
COMMON NAME(S): Reed Canary-Grass. Phalaride vivace.
FREQUENCY: Common throughout except rare along the eastern coast.
HABITAT: Shores and wet meadows.
GENERAL RANGE: Alaska to Nfld. south to Calif. and N.C.; Eurasia.
SUBSPECIFIC TAXA: Forma **variegata** (Parnell) Druce, with white variegated leaves, is occasionally cultivated and found near old habitations.

Phalaris canariensis L. Fig. p. A18.
COMMON NAME(S): Canary-Grass, Birdseed-Grass. Graines d'oiseaux.
FREQUENCY: Very rare; occasionally found where birdseed is spread or waste areas. Not persisting.
HABITAT: Waste areas, roadsides and similar open areas.
GENERAL RANGE: African and Canary Islands; introduced in Canada from B.C. to N.S.

Phalaris minor Retz. Fig. p. A18.
FREQUENCY: Rare; reported from Saint John, St. John Co. Not persisting.
HABITAT: Waste and ballast areas.
GENERAL RANGE: European; introduced in Canada from B.C. and N.B.

PHLEUM L. Timothy. Phléole or Fléole

1a Glume awns about 1.5 mm long; inflorescence slender cylindric, mostly over (usually considerably over) 3 cm long; base of culm bulb-like. **P. pratense**

1b Glume awns 2-3 mm long; inflorescence ovoid to short cylindric, mostly less than 3 cm long; base of culm not bulb-like. **P. alpinum**

Phleum alpinum L. Fig. p. A18.
COMMON NAME(S): Mountain-Timothy, Alpine Timothy. Phléole alpine.
FREQUENCY: Very rare; only known from the Restigouche River, Restigouche Co.
HABITAT: Wet meadows and rocky shores.
GENERAL RANGE: Alaska to Nfld. south in mtns. to N.H. and Maine; Mexico; S. America; Greenland; Iceland; Eurasia.

++

SYNONYMS: **P. commutatum** Gaudin var. **americana** (Fourn.) Hulten.
SPECIAL NOTE: For more information see H. Hinds, 1983. The Rare Vascular Plants of New Brunswick, Syllogeus No. 50.

Phleum pratense L. Fig. p. A18. Map p. B15.
COMMON NAME(S): Common Timothy. Mil.
FREQUENCY: Common throughout the Province.
HABITAT: Fields and similar open areas.
GENERAL RANGE: Eurasian; introduced across Canada, Greenland and elsewhere.

PHRAGMITES Trin

Phragmites australis (Cav.) Trin. Fig. p. A18. Map p. B16.
COMMON NAME(S): Reed, Cane Grass. Roseau commun.
FREQUENCY: Uncommon and scattered especially along the east coast.
HABITAT: Marshes, often brackish, shores and wet ditches.
GENERAL RANGE: Dist. Mack., and B.C. to N.S. south to Calif. and Mexico, La. and Md.; Eurasia.
SYNONYMS: **P. communis** Trin. var. **berlandieri** (Fourn.) Fern.

POA L.

Bluegrass, Meadow-Grass. Pâturin

1a Plants annual or short-lived perennial by rooting at the nodes; densely tufted with soft, bright green leaves, without long persistent basal leaves; panicle branches 1-2; callus at base of lemma without tuft of crinkled hairs (do not confuse with hairs on keel of lemma); lemma with 5 distinct nerves pubescent at base. **P. annua**

1b Plants perennial, creeping or tufted, the bases often with persistent old leaves or dry leaf sheaths; leaves dark green, bluegreen or glaucous; panicle branches usually 2 or more; callus at base of lemma with or without a tuft of crinkled hairs; lemma glabrous to pubescent on nerves. 2

2a(1b) Lemmas with only 3 distinct nerves (1 keel and 2 marginal nerves). 3

2b(1b) Lemmas with 5 distinct nerves (1 keel, 2 marginal and 2 intermediate nerves). 4

3a(2a) Marginal nerves of lemma glabrous, keel nerve pubescent at base; widely spreading lower branches of panicle 4-8 per node; ligule 0.7-2.2(3) mm long. **P. alsodes**

3b(2a) Marginal nerves of lemma pubescent. 5

4a(2b) Panicle branches 1-3 per node, floriferous mostly beyond middle; lemmas glabrous except for webbed callus below base; upper ligules 0.6-1.5(3) mm long. **P. saltuensis**

4b(2b) Panicle branches 3-5 per node, floriferous to below middle; lemma keel glabrous or appressed pubescent; upper ligules 1.5-8 mm long. 6

5a(3b) Panicle branches ascending, usually 2-3 per node; inflorescence narrowly elongate; plants often blue-green. 7

++

5b(3b) Panicle branches ascending-spreading, usually 4-6 per node; inflorescence broadly ovate in outline; plants dark green. 8

6a(4b) Upper ligules 4-8 mm long, acuminate; lemma margins and keel glabrous or keel appressed pubescent; lower sheaths and upper culm often scabrous; plants not rhizomatous. **P. trivialis**

6b(4b) Upper ligules about 1 mm long, +/- truncate; lower sheaths glabrous or rarely pubescent; plants rhizomatous. **P. pratensis**

7a(5a) Exposed culm nodes usually 2 or more; culm strongly compressed, 2-edged, decumbent at base from slender rhizomes; spikelets usually purple tinged; lemma with blunt, bronze tip; rachilla glabrous; leaves strongly ascending, to 4 mm broad; panicle dense, branches usually 2 per node, bearing spikelets nearly to base. **P. compressa**

7b(5a) Exposed culm nodes none or 1; culm not flattened, tufted; not rhizomatous; spikelets green; lemma acuminate to acute, not bronzed at tip; rachilla puberulent; leaves softer textured, not strongly ascending, to 2 mm broad; panicle lax, bearing spikelets toward the tips. **P. glaucantha**

8a(5b) Upper ligules (2.3)2.6-5 mm long, tapering to rounded summit; anthers 0.8-1.2 mm long; rachilla glabrous. **P. palustris**

8b(5b) Upper ligules 0.2-0.7(1) mm long, truncate; anthers 1.2-1.5 mm long; rachilla +/- puberulent. **P. nemoralis**

Poa alsodes Gray Fig. p. A18. Map p. B16.
COMMON NAME(S): Grove Meadow-Grass. Pâturin des bosquets.
FREQUENCY: Uncommon throughout most of the Province and rare in the eastern section.
HABITAT: Rich deciduous woods.
GENERAL RANGE: Ont. to N.S. south to Tenn. and N. Eng.

Poa annua L. Fig. p. A19. Map p. B16.
COMMON NAME(S): Annual Bluegrass, Low Speargrass. Canche naine.
FREQUENCY: Scattered throughout the Province.
HABITAT: Fields, cultivated ground and similar open areas.
GENERAL RANGE: Eurasian; introduced across Canada and Greenland.

Poa compressa L. Fig. p. A19. Map p. B16.
COMMON NAME(S): Wiregrass, Canada Bluegrass.
FREQUENCY: Common throughout the Province.
HABITAT: Dry sandy or rocky soils.
GENERAL RANGE: Eurasian; introduced across Canada.

Poa glaucantha Gaudin Fig. p. A19. Map p. B16.
COMMON NAME(S): Glaucous Poa.
FREQUENCY: Rare; five localities are known in the Province,; Martin Head, St. John Co., mouth of Pt. Wolfe River, Albert Co., Sisson Gorge, Victoria Co., and Tetagouche Falls, Gloucester Co.
HABITAT: Calcareous ledges.
GENERAL RANGE: Alaska and B.C. to Nfld. south to Ore. and Colo.

SYNONYMS: **P. glauca** Vahl. ssp. **glaucantha** (Gaudin) Lindm.
SPECIAL NOTE: For more information see H. Hinds, 1983. The Rare Vascular Plants of New Brunswick. Syllogeus No. 50.

Poa nemoralis L. Fig. p. A19.
COMMON NAME(S): Wood Meadow-Grass. Foin à vaches.
FREQUENCY: Rare; reported from near Fredericton, York Co. and Belledune Point, Gloucester Co. (UNB).
HABITAT: Partly shaded river banks and ledges.
GENERAL RANGE: Alaska and B.C. to Nfld. south to Calif. and Va.; Greenland; Iceland; Eurasia.

Poa palustris L. Fig. p. A19. Map p. B16.
COMMON NAME(S): Fowl Meadow-Grass, Swamp Meadow-Grass. Pâturin des marais.
FREQUENCY: Common throughout the Province.
HABITAT: Wet meadows and similar open areas.
GENERAL RANGE: Alaska to Nfld. south to Calif. and N.C.; Greenland; Eurasia.

Poa pratensis L. Fig. p. A19. Map p. B16.
COMMON NAME(S): Kentucky Bluegrass, Junegrass. Foin à vaches.
FREQUENCY: Common throughout the Province.
HABITAT: Meadows, fields and similar open areas.
GENERAL RANGE: Mostly Eurasian; introduced across Canada.
SPECIAL NOTE: Var. **angustifolia** (L.) Gaud. is reported from the Province (Scoggin 1978-79). These probably represent plants growing in less than optimum conditions and not the introduced European variety.

Poa saltuensis Fern. & Wieg. Fig. p. A19. Map p. B16.
COMMON NAME(S): Forest Meadow-Grass, Two-rayed Poa.
FREQUENCY: Scattered throughout except along the eastern section.
HABITAT: Mostly moist deciduous open woods.
GENERAL RANGE: Ont. to N.S. south to Iowa and R.I.

(**Poa subcaerulea** Sm.)
SPECIAL NOTE: Reported from Eastern N.B. by Fernald, 1950. This species is now considered to be mainly Eurasian, reaching N. America only in Alaska. The N.B. specimens are referred to the broad concept of **P. pratensis**.

Poa trivialis L. Fig. p. A19. Map p. B16.
COMMON NAME(S): Rough-stalked Meadow-Grass. Pâturin rude.
FREQUENCY: Uncommon; reported from the lower St. John River and Kenebecasis Rivers and from near St. Nicholas River, Kent Co.
HABITAT: Wet open ground.
GENERAL RANGE: Eurasian; introduced across Canada.

++

PUCCINELLIA Parl. Alkali-Grass, Goose-Grass

1a Anthers usually greater than 1.5 mm long, spikelets to 12 mm long, 4-9 flowered; leaves to 3.5 mm broad; culms to over 5 dm tall. **P. americana**

1b Anthers less than 1.2 mm long; spikelets to 9 mm long, 2-6 flowered. 2

2a(1b) Plants producing overground runners with short branches between nodes in autumn; lemmas to 3.2 mm long, glabrous; palea keel entire or with few spinules; anthers about 0.6 mm long. **P. ambigua**

2b(1b) Plants +/- tufted or cespitose, usually lacking overground runners. 3

3a(2b) Panicle branches smooth. 4

3b(2b) Panicle branches scabrous. 5

4a(3a) Involute leaves and culms conspicuously glaucous, stiffly erect; culms to 3 dm tall. **P. laurentiana**

4b(3a) Flat leaves and culms not conspicuously glaucous; culms usually not over 2 dm tall. **P. langeana**

5a(3b) Flat leaves to 6 mm broad; lemmas obtuse. 6

5b(3b) Involute leaves to 3 mm broad; lemmas long hairy below middle, acute; palea-keel spinulose above middle, long hairy towards base. **P. lucida**

6a(5a) Panicle broad, its branches spreading to deflexed; lemma summit ciliolate. **P. distans**

6b(5a) Panicle narrow its branches erect or spreading; lemmas serrulate, but not ciliolate. 7

7a(6b) Panicle long exserted; spikelets to 5 mm long; lemmas thick coriaceous to 2.5 mm long, their midribs +/- excurrent. **P. fasciculata**

7b(6b) Panicle with base included in upper leaf sheath; spikelet to 7 mm long; lemmas thin, hyaline above to 3 mm long, their midribs not excurrent. **P. macra**

Puccinellia ambigua Soer. Fig. p. A19.
FREQUENCY: Reported only from Grande Anse, Gloucester Co. (Sorensen, 1953).
HABITAT: Sandy openings in salt marshes.
GENERAL RANGE: Hudson Bay to Gaspé Pen. and Nfld. south to N.B. and P.E.I.
SPECIAL NOTE: According to Sorensen, 1953, all the stoloniferous specimens of **P. paupercula** (Holm) Fern. & Weath. & var. **alaskana** Scribn. & Merr. should be called **P. ambigua**. The non-stoloniferous specimens are **P. langeana** (Berl.) Soer. ssp. **alaskana** (Scribn. & Merr.) Sorensen.

Puccinellia americana Soer. Fig. p. A19.
FREQUENCY: Locally common along the Bay of Fundy and east coast, as at Waterside, Albert Co.
HABITAT: Salt marshes.
GENERAL RANGE: Gaspé Pen. to N.S. south along the coast to Del.
SYNONYMS: **P. maritima** of eastern Canada reports, not **P. maritima** (Huds.) Parl. which is a native of Greenland and Europe.

Puccinellia distans (L.) Parl. Fig. p. A19.
FREQUENCY: Scattered along the eastern shore and occasionally inland as at Fredericton, York Co.

++

HABITAT: Alkaline or saline areas along the coast or inland along salted roadsides and walkways.
GENERAL RANGE: Eurasian; introduced across Canada.

(**Puccinellia fasciculata** (Torr.) Bickn.) Fig. p. A19.
FREQUENCY: Unreported, but to be expected in our area.
HABITAT: Saltmarshes.
GENERAL RANGE: Nev., Utah, and Ariz.; P.E.I. and N.S. south along the coast to Va.; inland to N.Y.; Europe.

Puccinellia langeana (Berlin) Sorensen Map p. B16.
FREQUENCY: Scattered along coastal areas of the Province.
HABITAT: Saltmarshes and brackish tidal flats.
GENERAL RANGE: Alaska to Baffin Is. south to Man., Que. to N.S.; Greenland; Asia.
SUBSPECIFIC TAXA: Our plants are referable to ssp. **alaskana** (Scribn. & Merr.) Sorensen.
SYNONYMS: **P. paupercula** (Holm.) Fern. & Weath. var. **alaskana** (Scribn. & Merr.) Fern. & Weath.

Puccinellia laurentiana Fern. & Weath.
FREQUENCY: Uncommon; reported from Charlotte, Queens, Kent, Gloucester and Restigouche Counties.
HABITAT: Coastal sands, saltmarshes and inland salt springs.
GENERAL RANGE: Ont. to Gaspe Pen. south to P.E.I. and N.B.; Greenland.

Puccinellia nuttalliana (Schultes) Hitchc.
FREQUENCY: Very rare; reported only from Restigouche Co.
HABITAT: Coastal sands and saltmarshes.
GENERAL RANGE: Alaska and B.C. to N.B. south to Calif., Mex., Tex. and Minn.
SYNONYMS: **P. lucida** Fern. & Weath.

SCHIZACHNE Hack.

Schizachne purpurascens (Torr.) Swallen Fig. p. A20. Map p. B16.
COMMON NAME(S): False Melic-Grass, Purple Oat-Grass.
FREQUENCY: Mostly concentrated in the west central and northern parts of the Province.
HABITAT: Open, often rocky, woods.
GENERAL RANGE: Alaska to Nfld. south to Mexico and Pa.; Eurasia.
SUBSPECIFIC TAXA: Var. **pubescens** Dore, with the upper surface of the leaf blades sparsely pilose instead of glabrous is reported from near Woodstock, Carleton Co. (UNB). Its range is concentrated in southern Quebec and eastern and southern Ontario. Forma **albicans** Fern., with spikelets and basal sheaths pale rather than purplish or bronze occurs rarely with the species as near Kedgewick, Restigouche Co.

++

SCHIZACHYRIUM Nees Beardgrass. Bardon

Schizachyrium scoparium (Michx.) Nash Fig. p. A20. Map p. B16.
COMMON NAME(S): Bluestem, Broom-or Prairie-Beardgrass.
FREQUENCY: Rare; reported from the St. John River drainage system in Madawaska, Carleton, York and Queens Counties.
HABITAT: Gravel strands, rocky shores and occasionally along railways.
GENERAL RANGE: Ont. to N.S. south to Ariz. and Fla.
SYNONYMS: **Andropogon scoparius** Michx. var. **septentrionalis** Fern.

SECALE L. Rye

Secale cereale L. Fig. p. A20.
COMMON NAME(S): Rye. Seigle cultivé.
SPECIAL NOTE: Eurasian; cultivated and rarely escaped to roadsides, railways and similar open areas. Not persisting.

SETARIA Beauv. Foxtail. Sétaire

1a Bristles relatively inconspicuous, extending only 1-2 mm beyond spikelets; fertile lemma falling free from the glumes and sterile lemma at maturity; leaves to over 3 cm broad. **S. italica**

1b Bristles conspicuous, extending 3-6 mm beyond spikelets; fertile lemma falling with the glumes and sterile lemma at maturity; leaves to about 12 mm broad. 2

2a(1b) Bristles yellowish to golden-brown; fertile lemma conspicuously cross-wrinkled, upper half exposed in fruit; sheath margins glabrous. **S. glauca**

2b(1b) Bristles greenish to purplish; fertile lemma smooth, nearly concealed by second glume; sheath margins ciliate. **S. viridis**

Setaria glauca (L.) Beauv. Fig. p. A20. Map p. B17.
COMMON NAME(S): Pigeon-Grass, Yellow Foxtail. Foin sauvage.
FREQUENCY: Uncommon in the southern part of the Province.
HABITAT: Fields,roadsides and similar open areas.
GENERAL RANGE: Eurasian; introduced across Canada.

Setaria italica (L.) Beauv.
COMMON NAME(S): German or Hungarian Millet. Millet des oiseaux.
FREQUENCY: Rare; occasionally introduced as in Fredericton, York Co. Not persisting.
HABITAT: Waste areas.
GENERAL RANGE: Eurasian; occasionally introduced in B.C. and from Ont. to N.S. except Nfld.

Setaria viridis (L.) Beauv. Fig. p. A20. Map p. B17.
COMMON NAME(S): Bottle-Grass, Green Foxtail. Sétaire verdâtre.

++

FREQUENCY: Scattered throughout the Province.
HABITAT: Fields, roadsides and similar open areas.
GENERAL RANGE: Eurasian; introduced across Canada.
SUBSPECIFIC TAXA: Small plants about 10-30 cm high, with spreading branches at base and leaf blades only 2-5 mm wide, have been named var. **weinmannii** (Roemer & Schultes) Heuffel (also called var. **breviseta** (Döll) Hitchc.). It has been collected mostly in the northeastern part of the Province.

SPARTINA Schreb. Cord- or Marsh-Grass

1a Leaves tightly involute to base when fresh or dried, about 1 mm in diameter (3 mm when flattened); glumes unawned; spikes (1)2-4; culms slender, wiry, from rhizomes 1-3 mm thick. **S. patens**
1b Leaves flat when fresh, +/- involute at least towards tip when dried, about 4-5 mm broad when flattened; glumes awned or not; spikes 6-30 (or more); culms robust from rhizomes 3-8 mm thick. 2
2a(1b) Leaves harshly scabrous on margins; glumes awned; rhizome rigid, covered with hard, brownish or purplish scales; spike often purple tinged. **S. pectinata**
2b(1b) Leaves glabrous; glumes unawned; rhizome flaccid, covered with soft, light-coloured scales; spikes usually yellowish tinged. **S. alterniflora**

Spartina alterniflora Loisel. Fig. p. A20. Map p. B17.
COMMON NAME(S): Salt-water Cord-Grass. Herbe salée.
FREQUENCY: Scattered throughout along the coast.
HABITAT: Salt marshes and coastal shores.
GENERAL RANGE: Que. to Nfld. south to N.J.; introduced in Europe.
SYNONYMS: **S. alterniflora** var. **pilosa** (Merr.) Fern., and var. **glabra** (Muhl.) Fern.

Spartina patens (Ait.) Muhl. Fig. p. A20. Map p. B17.
COMMON NAME(S): Salt-meadow Grass, High-water Grass. Musotte.
FREQUENCY: Scattered throughout in coastal areas.
HABITAT: Salt marshes.
GENERAL RANGE: Que. to Nfld. south to Fla. and Tex.; Europe; Africa.
SYNONYMS: **S. patens** var. **monogyna** (Curtis) Fern.

Spartina pectinata Link Fig. p. A20. Map p. B17.
COMMON NAME(S): Fresh-water Cord-Grass, Slough-Grass. Chaume.
FREQUENCY: Common throughout most of the Province but becoming uncommon towards the north.
HABITAT: Shores and wet ditches.
GENERAL RANGE: Dist. Mack. and B.C. to Nfld. south to Oreg., N. Mex. and N.C.
SYNONYMS: **S. pectinata** var. **suttiei** (Farw.) Fern.

++

SPHENOPHOLIS Scribn.

Sphenopholis obtusata (Michx.) Scribn. Fig. p. A20. Map p. B17.
COMMON NAME(S): Slender Wedge-Grass.
FREQUENCY: Scattered throughout except rare in the eastern lowlands.
HABITAT: Calcareous ledges and shores.
GENERAL RANGE: Alaska to Nfld. south to Oreg., Tex. and Fla.
SUBSPECIFIC TAXA: Plants with leaf blades above and sheaths pilose have been named **S. intermedia** (Rydb.) Rydb. var. **pilosa** Dore. It occurs throughout the range of species in the Province.
SYNONYMS: **S. intermedia** (Rydb.) Rydb.

SPOROBOLUS R. Br. Drop-seed

1a Plants perennial, robust, usually over 50 cm high; panicle more than 5 cm long; lemma and palea glabrous; glumes distinctly unequal; palea not prolonged beyond lemma. **S. asper**

1b Plants annual, slender, usually less than 50 cm high, maturing and conspicuous only in late summer and fall, often reddish or purplish tinged; panicle usually less than 5 cm long and hidden in upper leaf sheaths; lemma and palea pubescent; glumes +/- unequal; palea prolonged beyond lemma, tapering to a slender beak. **S. vaginiflorus**

Sporobolus asper (Michx.) Kunth Fig. p. A20. Map p. B17.
COMMON NAME(S): Tall Drop-seed.
FREQUENCY: Very rare and possibly extirpated from the Province. Formally known from near the mouth of the Mactaquac River, York Co. and from Southern Head, Grand Manan Is., Charlotte Co. The Mactaquac Headpond drowned the first site and the second has not been confirmed since 1934.
SPECIAL NOTE: For more information see H. Hinds, 1983. The Rare Vascular Plants of New Brunswick. Syllogeus No. 50.

Sporobolus neglectus Nash
SPECIAL NOTE: Reported from the Province by Fernald, 1950, but probably reannotated to **S. vaginiflorus**. No specimens at GH.

Sporobolus vaginiflorus (Torr.) Wood Fig. p. A20. Map p. B17.
COMMON NAME(S): Poverty-Grass.
FREQUENCY: Locally common in the southwest of the Province and appears to be rapidly spreading.
HABITAT: Sandy roadsides, railways and similar open areas.
GENERAL RANGE: Nebr. and Ont. to N.S. south to Ariz. and Ga.
SUBSPECIFIC TAXA: Our material is all var. **inequalis** Fern. with the palea prolonged well beyond the lemma.

++

TORREYOCHLOA Church

Torreyochloa pallida (Torr.) Trin. Fig. p. A20.
COMMON NAME(S): Fernald's Manna-Grass.
FREQUENCY: Scattered throughout the Province.
HABITAT: Shallow quiet waters.
GENERAL RANGE: B.C. to Nfld. south to Mo., Tenn., and Va.; Asia.
SUBSPECIFIC TAXA: Our material is referable to var. **fernaldii** (Hitchc.) Dore.
SYNONYMS: **Glyceria fernaldii** (Hitchc.) St. John, **Puccinellia fernaldii** (Hitchc.) E. Voss.

TRISETUM L.

1a Lemma awn divergent, 4-5 mm long; panicle drooping. **T. melicoides**
1b Lemma awn straight, less than 1 mm long; panicle strict, congested. **T. triflorum**

Trisetum melicoides (Michx.) Scribn. Fig. p. A20. Map p. B17.
HABITAT: Calcareous ledges and shores.
GENERAL RANGE: Aggregate species Ont. to Nfld. south to Wisc., N.Y. and Maine.
SUBSPECIFIC TAXA: Two varieties are recognized in our area.

1a Leaf blades and sheaths glabrous. var. **melicoides**
1b Leaf blades and sheaths pilose. var. **majus**

1. Var. **melicoides**
FREQUENCY: Widely scattered in the Province except absent from the westcentral portion.
2. Var. **majus** (Gray) Hitchc.
FREQUENCY: Uncommon and mostly found in the western portion of the Province.

Trisetum triflorum (Bigel.) Love & Love Fig. p. A20. Map p. B17.
COMMON NAME(S): False Oats.
HABITAT: Rock crevices.
SUBSPECIFIC TAXA: Two varieties are recognized in the Province.
1a Glume sides and keel pilose. var. **triflorum**
1b Glume glabrous or +/- scabrous; keel sometimes hispid. var. **molle**

1. Var. **triflorum**
FREQUENCY: Very rare; known only from Gordon Falls, Point Wolfe escarpment and near the Kinnie Brook-Alma River confluence, Albert Co. (UNB).
SYNONYMS: **T. spicatum** (L.) Richter var. **pilosiglume** Fern.

2. Var. **molle** (Hulten) Love & Love
FREQUENCY: Found throughout the Province wherever rock crevices provide a suitable habitat.
SYNONYMS: **T. spicatum** (L.) Richter var. **molle** (Michx.) Beal.
GENERAL RANGE: Alaska to Nfld. south to New Eng. and mtns. of N.C.; S. America.

++

TRITICUM L. Wheat. Blé

Triticum aestivum L. Fig. p. A20. Map p. B17.
FREQUENCY: Uncommon; occasionally introduced, but not persisting.
HABITAT: Roadsides, fields, railways near grain elevators and similar open areas.
GENERAL RANGE: Eurasian; introduced but not persisting across Canada.

ZIZANIA L. Wild Rice

1a Pistillate lemmas whitish, nearly linear, very gradually tapering to scabrous awn, the ridges narrow, +/- scabrous at least above; staminate florets less than 1.5 mm wide. **Z. aquatica**
1b Pistillate lemmas often purplish, lanceolate, more abruptly narrowed to awn which has longer and denser pubescence, scabrous only between the broader, rounded, lustrous and smooth ridges; staminate florets 1.5-2 mm thick. **Z. palustris**

Zizania aquatica L.
FREQUENCY: Very rare; reported from the estuary of the Miramichi River, South Nelson, Northumberland Co.
SUBSPECIFIC TAXA: Our material is referable to var. **brevis** Fassett which is smaller in nearly every respect compared to the species. It is known from eastern Ont. to south-central Que. and the estuary of the Miramichi River, N.B.

Zizania palustris L. Fig. p. A21. Map p. B18.
COMMON NAME(S): Northern Wild Rice.
FREQUENCY: Common in the lower drainage system of the St. John River and scattered elsewhere in the southern part of the Province.
HABITAT: Shallow still or slow moving waters.
GENERAL RANGE: Ont. to N.S. south Ohio and Fla.
SUBSPECIFIC TAXA: Var. **interior** (Fassett) Dore with leaves to 3 cm broad and ligules to 1.5 cm long has been reported along the St. John River. No specimens have been seen.
SYNONYMS: **Z. aquatica** L. var. **angustifolia** Hitchc.

VENTENATA Koeler

Ventenata dubia (Leers) Cosson
SPECIAL NOTE: European; reported from a newly seeded roadside, Fredericton, York Co. Not persisting. (UNB).

CYPERACEAE (Sedge Family)

1a Flowers unisexual; achene enclosed in a sac (perigynia). **Carex**
1b Flowers bisexual; achene not enclosed. 2

++

2a(1b) Spikelets several, their scales 2-ranked and laterally compressed. 3

2b(1b) Spikelet scales spirally imbricated, not laterally compressed. 4

3a(2a) Inflorescence partly axillary; bristles present at base of achene; stems terete, hollow, jointed; style persistent. **Dulichium arundinaceum**

3b(2a) Inflorescence terminal; bristles wanting; stems +/- 3-edged, solid, not jointed. **Cyperus**

4a(2b) Inflorescence a single terminal spikelet. 5

4b(2b) Inflorescence of several spikelets. 6

5a(4a) Basal leaves absent (one species with capillary culms from base); cauline sheaths bladeless or with blades less than 2 cm long; scales yellowish-brown, greenish or reddish-purple. 7

5b(4a) Basal leaves present; scales lead-coloured; achene subtended by numerous, long, silky bristles. **Eriophorum**

6a(4b) Spikelets normally with one fertile flower subtended by (2)3 or more sterile scales. 8

6b(4b) Spikelets normally several flowered, without sterile scales. 9

7a(5a) Bristles obsolete to usually shorter than or rarely surpassing achene, not long and silky; achenes most often with a persistent tubercle at base of style; involucral bracts reduced to scales. **Eleocharis**

7b(5a) Bristles silky, greatly exceeding achene; achenes without persistent tubercle; at least the lowermost involucral bract with a blunt awn shorter than spkikelet. **Scirpus**

8a(6a) Achenes terete, broadly truncate at base; styles 3-cleft, deciduous; bristles wanting. **Cladium mariscoides**

8b(6a) Achenes lenticular, cuneate at base; styles 2-cleft, basal portion persisting as a pointed beak; bristles often present. **Rhyncospora**

9a(6b) Leaves capillary; achene trigonous, with persistent bulbous tubercle; densely tufted annuals. **Bulbostylis capillaris**

9b(6b) Leaves flat, channeled or absent, not capillary; achenes not capped by persistent tubercle; perennials. 10

10a(9b) Perianth bristles numerous, white to copper coloured, long, straight and silky; floral scales membranous; inflorescence unbranched. **Eriophorum**

10b(9b) Perianth bristles, if present, less than 8 or rudimentary, in some species several times longer than achene and then crinkled; not straight and never white; inflorescence branched. **Scirpus**

BULBOSTYLIS (Kunth) Clarke

Bulbostylis capillaris (L.) Nees Fig. p. A21. Map p. B18.

FREQUENCY: Locally common and apparently spreading from the southwest quarter of the Province.

HABITAT: Dry roadsides, railways and similar open areas.

GENERAL RANGE: Minn. and Ont. to N.B. south to Tex. and Fla.

SYNONYMS: **B. capillaris** var. **crebra** Fern.

++

CAREX L. Sedge. Laiche

KEY TO SECTIONS

1a Spike 1. 2
1b Spikes 2 or more. 3
2a(1a) Achenes lenticular or plano-convex; stigmas 2. 4
2b(1a) Achenes trigonous; stigmas 3. 5
3a(1b) Achenes lenticular or plano-convex; stigmas 2. 6
3b(1b) Achenes trigonous; stigmas 3. 7
4a(2a) Culms solitary or few from fibrous creeping rhizomes or stolons. **DIOICAE** (**C. gynocrates**)
4b(2a) Culms densely tufted. **STELLULATAE** (**C. exilis**)
5a(2b) Pistillate scales persistent; staminate scales with margins united nearly to middle; leaves soft and lax. **POLYTRICHOIDEA** (**C. leptalea**)
5b(2b) Pistillate scales soon deciduous; culms stiff and wiry. **ORTHOCERATES** (**C. pauciflora**)
6a(3a) Spikes of 2 kinds, terminal staminate (at least below). 8
6b(3a) Spikes essentially alike. 9
7a(3b) Staminate scales with margins united at base. **PHYLLOSTACHYA** (**C. backii**)
7b(3b) Staminate scales loosely ascending with margins free to base. 10
8a(6a) Bract at base of inflorescence distinctly sheathing. **BICOLORES** (**C. aurea**)
8b(6a) Bract nearly or quite sheathless. 11
9a(6b) Culms branching. **CHORDORRHIZAE** (**C. chordorrhiza**)
9b(6b) Culms simple, erect or ascending. 12
10a(7b) Style joined with achene. 13
10b(7b) Style continuous with achene. 14
11a(8b) Pistillate scales acuminate to long-awned. **CRYPTOCARPAE**
11b(8b) Pistillate scales obtuse to acute. **ACUTAE**
12a(9b) Some or all spikes gynaecandrous. 15
12b(9b) Some or all spikes androgynous. 16
13a(10a) Achenes only obscurely 3-angled, with rounded or convex sides. 44
13b(10a) Achenes more sharply 3-angled. 17
14a(10b) Perigynia +/- leathery, scarcely inflated. **PALUDOSAE**
14b(10b) Perigynia mostly thin and papery, usually strongly inflated. 18
15a(12a) Perigynia narrowly to broadly wing-margined, distinctly beaked, not thickened at base. **OVALES**
15b(12a) Perigynia at most thin edged, corky-thickened at base. 19
16a(12b) Spikes in mostly simple, compact or interrupted heads. **BRACTEOSAE**
16b(12b) Spikes in branching spike-like heads. 20
17a(13b) Leaves pubescent beneath; lower bracts leaf-like. 21
17b(13b) Leaves glabrous. 22
18a(14b) Awns of pistillate scales equalling or longer than blades. **PSEUDOCYPERAE**
18b(14b) Awns wanting or smooth, less than half as long as body of pistillate scales. 23
19a(15b) Perigynia thin-edged, not puncticulate, distinctly beaked and spongy at base; plants usually densely caespitose. **STELLULATAE**
19b(15b) Perigynia with rounded margins. 24

++

20a(16b) Leaf-sheath loose, often cross-puckered or dotted ventrally; culms soft, sharply angled; perigynia corky at base, at least 4 mm long. **VULPINAE**

20b(16b) Leaf sheaths tight around slender culms; perigynia not corky nor inflated. 25

21a(17a) Lowest bract of inflorescence with prolonged closed sheath; perigynia glabrous. 26

21b(17a) Lowest bract of inflorescence sheathless or short-sheathing. 27

22a(17b) Achene completely filling the short-beaked, mature perigynia; bracts +/- bladeless. 28

22b(17b) Achene not completely filling mature perigynia; bracts usually with well developed blades. 29

23a(18b) Staminate spikes 1; perigynia 1 or 2 cm long. 30

23b(18b) Staminate spikes usually more than 1; perigynia at most 1 cm long. **VESICARIAE**

24a(19b) Perigynia short-beaked or +/- beakless, to 4 mm long, puncticulate; plants rhizomatous or stoloniferous. **HELIONASTES**

24b(19b) Perigynia with long slender bidentate beak; plants caespitose. **DEWEYANAE**

25a(20b) Leaves to 6 mm broad; inner band of leaf-sheath conspicuously cross-puckered; scales awned. **MULTIFLORAE** (C. **vulpinoidea**)

25b(20b) Leaves usually less than 3 mm broad; inner band of leaf-sheath not cross-puckered; scales not awned. **PANICULATAE**

26a(21a) Terminal spike staminate throughout; culms pilose. **SYLVATICAE**

26b(21a) Terminal spike pistillate above the staminate base; culms glabrous. **GRACILLIMAE** (C. **gracillima**)

27a(21b) Perigynia short-beaked or beakless. **VIRESCENTES** (C. **hirtifolia**)

27b(21b) Perigynia long-beaked, pubescent. **TRIQUETRAE**

28a(22a) Staminate spike not overtopped by pistillate spikes, to 8 mm long, +/- sessile; leaves involute-filiform; culms capillary, densely tufted. **ALBAE** (C. **eburnea**)

28b(22a) Staminate spike or flowers overtopping pistillate spikes or flowers. **DIGITATAE**

29a(22b) Bract at base of inflorescence sheathless or short sheathing. 31

29b(22b) Bract at base of inflorescence with prolonged tubular sheath. 32

30a(23b) Perigynia inflated, to 8 mm thick, coarsely nerved, persistent. **LUPULINAE**

30b(23b) Perigynia hardly inflated, to 3 mm thick, finely many-nerved, soon deciduous. **FOLLICULATAE**

31a(29a) Perigynia pubescent, beaked. **HIRTAE**

31b(29a) Perigynia glabrous or scabrous margined. 33

32a(29b) Mature perigynia divergent to reflexed, nerved, the beak conspicuously bidentate. **EXTENSAE**

32b(29b) Mature perigynia ascending, the beak (when present) +/- entire or if deeply notched, the perigynia not strongly nerved. 34

33a(31b) Perigynia scabrous-papillate, abruptly narrowed to beak; leaves to over 1.5 cm broad, scabrous above; scabrous culms from long stout scaly rhizomes. **ANOMALAE** (C. **scabrata**)

33b(31b) Perigynia glabrous, beakless or short-beaked. 35

+++

34a(32b) Terminal spike pistillate above staminate base; lateral spikes loosely ascending or drooping; perigynia nerved, beakless. **GRACILLIMAE** (C. **gracillima**)

34b(32b) Terminal spike completely staminate or with a few perigynia at tip. 36

35a(33b) Terminal spike usually sessile or short-peduncled; roots without yellowish felt. **ATRATAE**

35b(33b) Terminal spike usually long-peduncled; roots covered with a yellowish felt. **LIMOSAE**

36a(34b) Perigynia beak nearly half to as long as body. 37

36b(34b) Perigynia beakless or with a short beak less than ¼ the length of the body. 38

37a(36a) Perigynium-beak about as long as globose-ovoid nerveless body; plant base strongly fibrillose. **LONGIROSTRES** (C. **sprengelii**)

37b(36a) Perigynium-beak shorter than body; plant base not fibrillose. 39

38a(36b) Pistillate scales with purplish-brown or black margins; plants long stoloniferous. **PANICEAE** (C. **livida**)

38b(36b) Pistillate scales whitish or green to reddish-brown. 40

39a(37b) Pistillate scales with purplish-brown or black margins; plants loosely stoloniferous. **PANICEAE** (C. **vaginata**)

39b(37b) Pistillate scales pale brown to green or white. 41

40a(38b) Pistillate scales reddish-brown; plants loosely stoloniferous; perigynia to about 3.5 mm long, often resin-dotted; leaves stiff, glaucous, often folded. **GRANULARES** (C. **crawei**)

40b(38b) Pistillate scales whitish to pale brown; plants mostly non-stoloniferous. 42

41a(39b) Perigynia distinctly nerved at least towards base; plant base reddish-purple to dark-brown; lower sheaths without blades. **SYLVATICAE**

41b(39b) Perigynia few nerved or nerveless; plant base drab or brown; lowest sheaths with elongate blades. **CAPILLARES** (C. **capillaris**)

42a(40b) Perigynia with spongy stipe to 1 mm long, glabrous or hispidulous. **LAXIFLORAE**

42b(40b) Perigynia without spongy stipe, nearly beakless, sessile. 43

43a(42b) Pistillate scales acuminate to cuspidate; perigynia to 4 mm long, with elevated nerves. **GRANULARES**

43b(42b) Pistillate scales often rough-awned; perigynia to 6 mm long, with impressed nerves. **OLIGOCARPAE**

44a(13a) Pistillate scales smooth. **MONTANAE**

44b(13a) Pistillate scales rough-cuspidate. **MITRATAE** (C. **caryophyllea**)

ACUTAE

1a Perigynia distinctly nerved, at least dorsally, stipitate. 2

1b Perigynia nerveless or lightly few-nerved at base dorsally (C. **haydenii** sometimes faintly 1-2 nerved dorsally). 3

2a(1a) Stolons short ascending or none, stems densely tufted; scales with central green portion about as wide as the darker, brownish margins; leaves mostly overtopping spikelets. C. **lenticularis**

2b(1a) Stolons long, horizontal; stems not densely tufted; scales with slender green stripe scarcely if at all broader than darker margins; leaves mostly shorter than culms; perigynia often marked with purple-brown to nearly black. **C. nigra**

3a(1b) Pistillate spikes spreading or nodding, the upper ones sessile; perigynia tapering to beak, twisted when mature, their scales purple-margined, shorter than perigynia. **C. torta**

3b(1b) Pistillate spikes erect or ascending; mature perigynia not twisted. 4

4a(3b) Terminal staminate spike 1, less than 2 cm long, mostly around 1 cm long; culm and leaf margins scabrous; lowest bract rarely equalling uppermost spike; rare plants of subalpine peaty turf. **C. bigelowii**

4b(3b) Terminal staminate spikes 2 or more, mostly 2-6 cm long. 5

5a(4b) Perigynia inflated, brownish at maturity, suborbicular to obovoid, lowest ones often much exceeded by spreading scales; spikes cylindric, not tapering (or only slightly) at base; lower leaf sheaths very slightly or not at all filamentous; base of stem with shiny, dark purple or black, acute leafless sheaths with lighter stripes. **C. haydenii**

5b(4b) Perigynia flattened, greenish or straw-coloured at maturity; scales shorter than to only slightly exceeding perigynia. 6

6a(5b) Lower leaf sheaths mostly strongly filamentous; culms sharply triangular throughout, with strongly concave scabrous sides; lowermost leaves reduced to pointed and bladeless sheaths; membranous ventral surface of sheaths usually minutely scabrous (glabrous in var. **strictior**); plants densely or loosely clumped, forming stools or tussocks in wet meadows. **C. stricta**

6b(5b) Lower leaf sheaths not filamentous; culms smooth throughout or scabrous near top on the angles and with flattish sides; lower part of stem crowded with old leaves, thus base of stem often thickened to 10 mm and often spongy; membraneous ventral surface of sheaths glabrous; stems often arising singly. **C. aquatilis**

ATRATAE

1a Pistillate spikes +/- sessile. 2

1b Pistillate spikes on slender peduncles 1-4 cm long; perigynia to 3.5 cm long; leaves less than 5 mm broad; scales acute to acuminate, dark purplish-red. **C. atratiformis**

2a(1a) Inflorescence crowded, 1-3 cm long; pistillate scales obtuse to subacute. **C. media**

2b(1a) Inflorescence looser, 3-10 cm long; pistillate scales pointed, often long awned. **C. buxbaumii**

BRACTEOSAE

1a Leaf sheaths loose, septate-nodulose and often mottled with green and white, inner surface nerveless, thin and friable; leaves to 1 cm wide;

perigynia pale green, plano-convex, nerveless on inner face, 3-4.5 mm long, 1.5-2.5 mm broad, tapering gradually to beak. **C. sparganioides**

1b Leaf sheaths tight, not predominantly septate-nodulose on back; leaves up to 4.5 mm broad. 2

2a(1b) Perigynia spongy thickened below middle. 3

2b(1b) Perigynia of uniform thickness, not spongy thickened; scales reddish brown, short pointed or mucronate, shorter than the nerveless, widely spreading perigynia; inflorescence +/- loose; lower spikes remote; leaves to 3 mm broad. **C. pairaei**

3a(2a) Culms +/- stiffly ascending; leaves to 3 mm broad; perigynia 3-4.5 mm long, mostly nerveless, deep green, those of the lowest spike (6)9-12, abruptly tapering to beak; stigmas strongly coiled; less densely tufted than the next. **C. flaccidula**

3b(2a) Culms mostly densely tufted but weak and reclining, diffusely spreading; leaves to 2 mm wide; perigynia pale green, 2.8-3.5 mm long, more gradually tapering to beak and usually less serrate on beak, those of the lowest spike 5-12; stigmas elongate, flexuous or recurved. **C. sylvicola**

CRYPTOCARPAE

1a Pistillate spikes widely spreading or drooping. 2

1b Pistillate spikes erect, subsessile or short peduncled; perigynia faintly nerved at base; scales often dark purple with green midnerves, 2-3 times as long as perigynia; leaves to 9 mm broad; cespitose plants of saline marshes with long stolons. **C. salina**

2a(1a) Densely cespitose plants of damp fresh water habitats. 3

2b(1a) Long stoloniferous plants of salt marshes; pistillate scales long awned; leaves to 12 mm broad. **C. paleacea**

3a(2a) Sheaths of leaves scabrous-hispidulous on ventral side; bodies of most or all pistillate scales on lower part of spikelet truncate or tapered at summit. **C. gynandra**

3b(2a) Sheaths of leaves smooth on ventral side; bodies of most pistillate scales shallowly lobed at summit on each side of base of awn. **C. crinita**

DEWEYANAE

1a Inflorescence tinged light brown, less than 5 mm broad, scarcely interrupted; spikes lance cylindric; bracts inconspicuous; perigyynia 0.8-1.2 mm wide; scales oblong; leaves to 2.5 mm wide; often densely cespitose. **C. bromoides**

1b Inflorescence light green, greater than 5 mm wide, +/- interrupted; spikes ovoid or ovoid-cylindric; bracts conspicuous; perigynia 1.3-1.6 mm wide; scales ovate; leaves to 5 mm wide. **C. deweyana**

CYPERACEAE

++

DIGITATAE

1a Terminal spike androgynous; pistillate scales awned; perigynia nerveless, glabrous; plants densely cespitose. C. **pedunculata**

1b Terminal spike completely staminate; pistillate scales blunt; perigynia pubescent, 2-ribbed and several nerved; plants loosely cespitose with slender rootstocks. C. **concinna**

EXTENSAE

1a Perigynia horizontally spreading, mostly greenish at maturity, (1.5)-2.0-2.5-(3.0 mm) long with mostly straight beak 1 mm long or less. C. **viridula**

1b Perigynia mostly reflexed, yellowish green to brownish at maturity 3-6 mm long with deflexed beak ½ or more times as long as body. 2

2a(1b) Leaves bluegreen (glaucous); staminate spike long peduncled. Lower spike remote; pistillate scales mostly blunt; culms obtusely angled; perigynia inflated, ovoid. C. **lepidocarpa**

2b(1b) Leaves not glaucous; staminate spike sessile or short pedunculate; spikes more crowded; pistillate scales acuminate; culms acutely angled; perigynia less inflated, more ovoid. 3

3a(2b) Pistillate scales shiny brown or reddish-brown, conspicuous; perigynia (4.5)-5-6.2 mm long; beak usually at least minutely serrulate towards tip and conspicuously bent; leaves to 5 mm broad. C. **flava**

3b(2b) Pistillate scales pale and inconspicuous; perigynia (3)-3.5-4.5-(5) mm long with smooth nearly straight beak; leaves to 3.5 mm broad. C. **cryptolepis**

FOLLICULATAE

1a Leaves broad, 6-17 mm wide; lowest bract 5-10 mm wide with sheaths prolonged at summit; perigynia 10-15 mm long, usually slightly exceeding awned scales. C. **folliculata**

1b Leaves slender, 2-5 mm wide; lowest bract 1-3 mm wide with sheaths concave at summit; perigynia 8-13 mm long, often twice as long as awnless scales. C. **michauxiana**

GRANULARES

1a Staminate spike overtopping uppermost pistillate spike; perigynia inconspicuously nerved or nerveless; plants stoloniferous, to 3 dm tall. C. **crawei**

1b Staminate spike +/- sessile, not fully overtopping uppermost pistillate spike; perigynia strongly nerved; plants densely cespitose to 8 dm tall and with short rhizomes. C. **granularis**

++

HELEONASTES

1a Lowest bract bristle-like, prolonged to several times the length of subtended spikelet. 2

1b Lowest bract absent or up to 3 times longer than its spikelet. 3

2a(1a) Perigynia conspicuously nerved on both ventral and dorsal surfaces (3.3-4 mm long); beak very short, smooth; perigynia distinctly plano-convex; staminate flowers basal. **C. trisperma**

2b(1a) Perigynia with few nerves ventrally, nerveless or with few nerves dorsally (2.0-2.7 mm long); beak serrulate; scales often fulvous tinged. **C. brunnescens**

3a(1b) Spikes staminate at summit with up to 3 finely nerved, biconvex, minutely beaked perigynia less than 3 mm long (or the terminal spike with up to 6 perigynia). **C. disperma**

3b(1b) Spikes pistillate at summit, staminate at base; perigynia plano-convex. 4

4a(3b) Inflorescence compact, usually even lowest spikes overlapping. 5

4b(3b) Inflorescence interrupted; only upper spikelets crowded. 6

5a(4a) Inflorescence less than 2 cm long, ovate; perigynia (2.5) 2.8-3.5 mm long, smooth-margined, attenuate to tip but essentially beakless. **C. tenuiflora**

5b(4a) Inflorescence longer than 2 cm; perigynia 1.7-2.6 mm long, with a small serrulate beak. 7

6a(4b) Culms smooth; terminal spike with prolonged staminate base; perigynia oval, distinctly stipitate, 2.5-3.3 mm, half as wide, lightly nerved on both faces, with a smooth short beak; scales reddish-brown, ovate, nearly covering perigynia. **C. mackenziei**

6b(4b) Culms scabrous above, terminal spike not prolonged at base. 8

7a(5b) Culms to 3 dm high, usually much overtopping involute or narrow leaves (0.5-1.5 mm broad); perigynia abruptly beaked, 1.3-1.9 mm long, faintly nerved to nerveless; plants of saline or brackish shores and saltmarshes. **C. glareosa**

7b(5b) Culms to 6 dm high, usually slightly overtopped by the pale green, soft, flat leaves (2.5-4 mm broad) perigynia ovate 2.2-2.6 mm long, nerved on both sides and tapering to a rough-margined beak; plants of wet woods, alluvial thickets. **C. arcta**

8a(6b) Beak very short; leaves 2-4 mm wide, glaucous when fresh; spikes 10-30 flowered; perigynia 1.8-3 mm long, only very slightly if at all serrulate at base of beak, appressed spreading. **C. canescens**

8b(6b) Beak to .5 mm long; leaves 1-2.5 mm wide, not glaucous; spikes 5-10 flowered; perigynia loosely spreading, distinctly and minutely serrulate at base of beak. **C. brunnescens**

HIRTAE

1a Nerves of perigynia hidden by dense pubescence; perigynia not over 5 mm long. 2

1b Nerves of perigynia conspicuous under sparse pubescence; perigynia to 8 mm long. 3

++

2a(1a) Leaves mostly filiform-convolute, to 2 mm broad, plants mostly of boggy habitats. C. **lasiocarpa**

2b(1a) Leaves flat, 2-5 mm broad, scabrous; plants mostly of wet marshes and meadows. C. **lanuginosa**

3a(1b) Leaf sheaths (especially at mouth) and backs of scales pilose; perigynia beak at least half as long as body; culms obtusely angled, +/- smooth; unknown in N.B., but reported from P.E.I. and N.S. C. **hirta**

3b(1b) Leaf sheaths and back of scales glabrous; perigynia beak less than half as long as body; culm sharply angled, scabrous; widespread. C. **houghtonii**

LAXIFLORAE

1a Bracts of fertile stems bladeless; bases of plants and staminate scales strongly dark reddish; terminal spike long-peduncled; leaves of the sterile shoots elongate, 10-25 mm wide; perigynia narrowly obovoid, 3.7-4.9 mm long with an erect or outcurved beak; plants of rich deciduous woodlands. C. **plantaginea**

1b Bracts of fertile stems with normal flat green blades not more than 10 mm wide; staminate spike whitish to purplish brown. 2

2a(1b) Perigynia with only 2 main side nerves, otherwise nerveless or obscurely 5-7 nerved; pistillate scales obtuse or short awned; beak of perigynia slightly bent, about as long as contracted base; angles of bract sheath serrulate. C. **leptonervia**

2b(1b) Sides of perigynia each with 7 or more conspicuous nerves; pistillate scales awned; abrupt-curved beak shorter than stipe; spikes slender and loosely flowered, with basal perigynia separated from next by approximately its own length; angles of bract-sheath smooth or slightly roughened. C. **ormostachya**

LIMOSAE

1a Pistillate scales lanceolate, about ½ as wide and twice as long as perigynia, often dark purple-brown; lowest bract usually surpassing terminal spike; perigynia 3-4 mm long; stems loosely cespitose from short rhizomes. C. **paupercula**

1b Pistillate scales ovate, about as wide and as long as perigynia; lowest bract not surpassing terminal spike. 2

2a(1b) Culms glabrous, obtusely triangular; pistillate scales usually purple-brown to blackish, enfolding base of perigynia; perigynia to 3.5 mm long. C. **rariflora**

2b(1b) Culms scabrous, sharply triangular; pistillate scales usually straw-colored to golden-brown, not enfolding base of perigynia; perigynia to 4.5 mm long. C. **limosa**

++

LUPULINAE*

1a Perigynia 6-10 mm long, 6-13 veined. **C. retrorsa** (not in the Lupulinae but easily mistaken)

1b Perigynia 10-20 mm long, 13-25 veined. 2

2a(1b) Sheath of uppermost leaf absent or less than 1.5 cm long; beak of perigynia 0.2-4.2 mm long, poorly defined; achenes ellipsoid to obovoid; perigynia lanceolate to ovoid, convex to base, 1-12 per spike, ascending and forming an ovoid to obovoid spike. **C. intumescens**

2b(1b) Sheath of uppermost leaf usually 1.7 cm or longer; beak of perigynia 4.5-10 mm long; achenes rhombic or triangular; staminate peduncle shorter than to equalling the uppermost pistillate spike. **C. lupulina**

MONTANAE

1a Culms of various lengths, some or all pistillate spikes crowded among base of leaves. 2

1b Culms all elongate. 3

2a(1a) Perigynia glabrous or nearly so, 3.2-4.7 mm long, beak 0.9-1.7 mm long, all or mostly all hidden among the harsh, firm leaf bases; leaves 2.5-5 mm wide; pistillate scales long tapering. **C. tonsa**

2b(1a) Perigynia distinctly hairy (**C. deflexa** rarely subglabrous); culms of varying lengths with some hidden among leaf bases. 4

3a(1b) Staminate spike inconspicuous; pistillate spike usually crowded; lowest pistillate bract of elongate culms leaf-like, often exceeding staminate spike; perigynia 2.5-3 mm long, short beaked; scales shorter than perigynia, acute; blades 1-3 mm wide; rhizomes freely branching. **C. deflexa**

3b(1b) Staminate spike conspicuous. 5

4a(2b) Bract of lowest pistillate spike foliaceous, exceeding staminate spike; pistillate scales shorter than perigynia, acute; perigynia 2.5-3 mm long, th short beak about 0.5 mm long; plants loosely cespitose. **C. deflexa**

4b(2b) Bract scale-like, not foliaceous; pistillate scales lance-ovate, tapering to long tip; perigynia 3.2-4.7 mm long, beak 0.9-1.7 mm long; plants densely cespitose. **C. umbellata**

5a(3b) Perigynia body (excluding beak and spongy base) +/- as long as thick, blunt angled. 6

5b(3b) Perigynia body longer than wide, distinctly 3-angled. 7

6a(5a) Plants apparently not stoloniferous, densely cespitose. **C. communis**

6b(5a) Plants with long horizontal stolons. 8

7a(5b) Lower pistillate spikes remote, not overlapping. **C. novae-angliae**

7b(5b) Lower pistillate spikes crowded or overlapping. 9

8a(6b) Perigynia subglobose, beak nearly as long as body. **C. lucorum**

8b(6b) Perigynia obtusely 3-angled, beak 1/4 - 1/5 as long as body. **C. caryophyllea** (not in the Montanae but easily mistaken)

*Based on A.A. Reznicek & P.W. Ball, Can. J. Bot. 52: 2387-2399.

++

9a(7b) Body of mature perigynia +/- exceeding their scales; beak about 0.4-0.7 mm long; stems stiff and erect, surpassing leaves. **C. peckii**

9b(7b) Body of mature perigynia +/- as long as their scales or slightly shorter; beak of perigynia about 0.5-1.4 mm long; stems weak and often loosely spreading or arching and usually shorter than leaves. **C. emmonsii**

OLIGOCARPAE

1a Much of inflorescence scabrous; perigynia to 4 mm long; leaves to 5 mm broad. **C. conoidea**

1b Much of inflorescence smooth; perigynia to 5 mm long; leaves to 1 cm broad. **C. amphibola**

OVALES

1a Inflorescence mostly over 4 cm long (1.5 to 8 cm) spikes +/- strung out from each other, at least the lowest spike not reaching the base of the next one above, usually nodding and flexuous. 2

1b Inflorescence to about 4 cm long, erect, not nodding; spikes mostly overlapping except sometimes the lowest. 3

2a(1a) Sheaths at summit with knob-like auricles on either side; inner band of sheath greatly prolonged; perigynia usually silvery, distinctly 5-7 nerved on upper face, 3.5-5 mm long, 2-3 mm broad; leaves glaucous and stiff; plants of maritime sands and rocks. **C. silicea**

2b(1a) Sheaths without knob-like auricles; leaves not glaucous and stiff. 4

3a(1b) Mature perigynia 2.6-7.5 times as long as wide. 5

3b(1b) Mature perigynia less than 2.6 times as long as wide. 6

4a(2b) Pistillate scales equalling or surpassing the tips of perigynia and essentially completely concealing their bodies. 7

4b(2b) Pistillate scales shorter than perigynia, but sometimes longer than body of perigynia. 8

5a(3a) Perigynia thin and scale-like, barely distended over achene. 9

5b(3a) Perigynia relatively thick, obviously distended over achene, 0.6-1.1 mm broad, 3.2-4.2 mm long; inflorescence stiffly erect, spikes crowded. **C. crawfordii**

6a(3b) Perigynia rhombic to suborbicular, 3-4 mm long, 2-3.4 mm broad; spikes 5-10, ovoid with rounded base; pistillate scales much shorter than perigynia, hyaline and obtuse. **C. cumulata**

6b(3b) Perigynia ovate-lanceolate (3.5) 4-7 mm long. 10

7a(4a) Perigynia green at maturity, 3-4 mm long, orbicular to short elliptic, abruptly contracted to (0.5)-0.7-1.0 mm beak; white nerves conspicuous on both faces, 5-(7) on ventral face. **C. argyrantha**

7b(4a) Perigynia brownish at maturity, at least at base, body ovate usually tapering to beak, finely nerved below, nerveless or short nerved (up to 5 nerves) above, 4-5 mm long, 1.9-2.7 mm wide; spikes markedly tapering at base, achene 1.3 mm wide, 1.7 mm long. **C. aenea**

+++

8a(4b) Perigynia less than 2 mm broad. 11

8b(4b) Perigynia 2.5-3 mm broad, 4-5 mm long; scales tipped by awl-like point; leaf blades 1-2.5 mm wide; plants of estuaries and salt marshes. **C. hormathodes**

9a(5a) Perigynia 3-5(5.5) mm long, at most 1.5 mm broad with wing abruptly narrowed above base; leaf blades 2.5-8 mm wide, abundant and spreading on the usually numerous sterile culms; inner band of leaf-sheaths firm, green-striate. **C. tribuloides**

9b(5a) Perigynia 4-7 mm long, 1.2-2.5 mm broad with wing extending continuously to base; leaf blades 1-3 mm wide, few and ascending on the infrequent sterile culms. **C. scoparia**

10a(6b) Inflorescence nodding or arching; scales long acuminate (awl-like), perigynia to 5.5 mm long; plants of brackish marshes and meadows. **C. hormathodes**

10b(6b) Inflorescence stiffly erect, usually more compact; scales acute or blunt; plants of fresh habitats. 12

11a(8a) Perigynia thin and scale-like, only slightly distended over achene, to 5 mm long, sharply nerved below, obscurely nerved above; leaves to 8 mm broad. 13

11b(8a) Perigynia thin, but firm, distinctly swollen over achene; leaves 1-3 mm wide; perigynia ascending, without spreading tips, 2.6-4 (4.5) mm long, 1.5-1.8(2) mm broad; inflorescence often nodding above the lowest spike; spikes ovoid or ellipsoid. **C. tenera**

12a(10b) Scales nearly as broad and long as perigynia, the latter nerveless or lightly nerved below. 14

12b(10b) Scales mostly shorter and narrower than perigynia. 15

13a(11a) Tips of perigynia +/- appressed-ascending; spikes longer than thick (top shaped) with more than 30 perigynia, 8-12 mm long; perigynia mostly broadest near the middle; inside band of leaf sheath usually firm at summit, white area extending only a short way below summit. **C. tribuloides**

13b(11a) Tips of perigynia loosely spreading-ascending giving a rough appearance to the spikes which are subglobose, the lowermost usually separated in the flexuous inflorescence 5-8 mm long with 15-30 perigynia; inner band of leaf sheath very friable at summit with white area extending a long way below summit. **C. projecta**

14a(12a) Basal bract of inflorescence conspicuously dilated (to 2 mm wide) usually with a prolonged awl-like tip (to 6 cm); perigynia full and leathery, green towards tip, brown at base, 4-5 mm long, 2-3 mm wide with a firm, narrow wing, nerveless or short nerved on upper face; achene 1.8-2.1 mm wide, plants of dry soils. **C. adusta**

14b(12a) Basal bract not especially dilated or prolonged; perigynia membranous, 1.8-2.1. mm wide, 3.8-4.5 mm long; achene 1 mm or less wide; scales ovate, reddish brown with broad hyaline margins; perigynia sharply nerved on both faces, the beak very slender, reddish tipped; plants of wet meadows and pastures. **C. leporina**

15a(12b) Perigynia thin and scale-like, barely distended over achene, wing thin and continuous to base, nerved on both surfaces, 1.2-2.6 mm broad, 4-7 mm long. **C. scoparia**

15b(12b) Perigynia conspicuously distended over achene. 16

16a(15b) Perigynia nerveless or obscurely nerved at base above, 1-2 mm broad, 2-4 mm long; pistillate scales brown, slightly exceeding the **body** of the perigynia; leaf blades 1.5-4.5 mm wide; all spikes strongly overlapping; culms densely cespitose. **C. bebbii**

16b(15b) Perigynia distinctly nerved on upper face; 1.7-2 mm wide, 3.5-5 mm long; scales deep brown or fulvous; leaf blades to 4 mm wide; spikes appearing bicolorous green-brown; achene about 1 mm wide. 17

17a(16b) Leaves to 4.0 mm broad; tight sheaths with hyaline inner band scarcely prolonged at summit; sterile culms rare or wanting; spikes crowded; perigynia to 5 mm long, their beaks appressed; scales reddish-brown to deep brown nearly equalling beak of perigynia. **C. tincta**

17b(16b) Leaves to 6.5 mm broad; loose +/- mottled sheaths with inner hyaline band prolonged at summit; sterile culms frequent with leaves crowded near summit; perigynia to 4 mm long, their beaks spreading or slightly recurved; scales pale yellowish brown about as long as or slightly shorter than body of perigynia; spikes usually less crowded. **C. normalis**

PANICEAE

1a Perigynia beakless or with a very short tip. 2

1b Perigynia with a long bidentate beak, faintly nerved, 3-5 mm long; pistillate spikes peduncled and loosely flowered; scales shorter than perigynia; leaves to 5 mm broad. **C. vaginata**

2a(1a) Leaves whitened, at most 3.5 mm broad, folded or involute; perigynia glaucous, granular, finely nerved, beakless, but with a short straight point and entire orifice, 2.2-4.6 mm long. **C. livida**

2b(1b) Leaves green or slightly blue-green, flat when fresh or becoming revolute; perigynia 2.5-5 mm long, stramineus or purplish, shining and very faintly nerved. **C. panicea**

PANICULATAE

1a Inflorescence to 10 cm, flexuous, +/- interrupted, especially below; summit of leaf sheaths copper tinged; perigynia stramineous to brown, nearly concealed by their reddish-brown scales. **C. prairea**

1b Inflorescence to 5(8) cm, stiffly erect, dense, dark brown; summit of leaf sheaths pale green or slightly discolored, often red dotted; perigynia lustrous olivaceous to dark brown, widely divergent and conspicuous at maturity. **C. diandra**

++

PSEUDO-CYPEREAE

1a Perigynia ascending-spreading, inflated, membranous, +/- round in cross section; leaves inconspicuously nodulose, to 1 cm broad with ligules as broad as long. 2

1b Perigynia reflexed (at least the lower ones), flattened triangular in cross section, not inflated, with close and very strong nerves; leaves more strongly septate-nodulose, to 1.5 cm broad with ligules prolonged. 3

2a(1a) Perigynia 8-10 nerved, conspicuously inflated, ovoid, shiny; leaves lax, to 7 mm broad; culm blunt angled, smooth or slightly scabrous above; achene rough-papillate. **C. lurida**

2b(1a) Perigynia 15-20 nerved, inconspicuously inflated, lanceolate; leaves firm, to 1.5 cm broad; culm sharply angled, +/- strongly scabrous above; achene smooth. **C. hystricina**

3a(1b) Perigynia 4-5.5(6.5) mm long with teeth straight or slightly divergent (0.5-1(1.3) mm long. **C. pseudo-cyperus**

3b(1b) Perigynia 5.5-7 mm long with teeth curved divergent (1) 1.2-2.2 mm long. **C. comosa**

STELLULATAE*

1a Spikes usually solitary; leaves involute; anthers 2.0-3.6 mm long. **C. exilis**

1b Spikes 2-8; leaves flat or plicate; anther 0.6-2.2(2.35) mm long. 2

2a(1b) Widest leaves 2.8-5 mm wide. 3

2b(1b) Widest leaves 0.8-2.7 mm wide; terminal spikes partly or wholly pistillate; anthers 0.6-2.2(2.35) mm long. 4

3a(2a) Lower perigynia of spikes mostly 1.1-1.6 times as long as wide, mostly 2.1-3.0 mm wide; distance between lowest 2 spikes 1.5-18.0 mm. **C. atlantica**

3b(2a) Lower perigynia of spikes (1.5)1.7-3.0 times as long as wide, mostly 1.2-2.0 mm wide; longer pistillate scales 1.4-2.2 mm long; infructescences mostly 15-30 mm long, lowest 2 spikes 1.3-9.5 mm distant. **C. wiegandii**

4a(2b) Terminal spikes without a distinct clavate base of staminate scales, staminate portion less than 1 mm in length, often unisexual; anthers (1.0)1.2-2.2(2.35) mm long; beak white at tip; calciphilous. **C. sterilis**

4b(2b) Terminal spikes with a distinct clavate base 1.0-11.5 mm long of staminate scales; anthers 0.6-1.6(2.) mm long. 5

5a(4b) Lower perigynia 2.0-3.0 mm wide. **C. atlantica**

5b(4b) Lower perigynia 0.8-1.95 mm wide. 6

6a(5b) Lower perigynia mostly 1.9-3.0(3.8) mm long, 1.0-2.0(2.2) times as long as wide; beaks mostly 0.4-0.95(1.25) mm long, mostly 0.2-0.5 times as long as body. 7

6b(5b) Lower perigynia (2.65)2.85-4.6 mm long, 0.8-2.1 mm wide, 1.8-3.6 times as long as wide; beaks mostly 0.85-2.0 mm long, mostly (0.4)0.45-0.86 times as long as body, serrulate on margins. 8

7a(6a) Perigynia mostly nerveless over achene or occasionally up to 6 nerves on upper surface, 1.95-3.0(3.3) mm long, 1.1-1.8 mm wide, beak conspicuously

*Taken in part from Reznicek & Ball, 1980

setulose-serrulate, dark reddish brown 0.5-0.95 mm long. 0.18-0.44 times as long as body; perigynia often +/- convexly tapering from widest point to beak, forming a "shoulder"; calciphilous. **C. interior**

7b(6a) Perigynia (0)1-12 nerved over achene on upper surface; beak more sparsely serrulate with definite spaces between the often single teeth, 0.45-1.25 mm long, 0.23-0.5(0.63) times as long as body; perigynia mostly cuneate or even concavely tapered from widest point to beak; oxyphilous. 9

8a(6b) Perigynia strongly nerved, fusiform, plump, about 3 mm long and .8 mm wide, dark stramineus at maturity; beak to 1 mm long, barely notched; plants of damp calcareous shores. **C. josselynii**

8b(6b) Perigynia strongly to weakly nerved or nerveless above, ovate; beak conspicuously bidentate; plants of acid habitats. **C. echinata**

9a(7b) Widest leaves 1.6-2.7 mm wide; infructescence mostly 18-45 mm long; perigynia 2.3-3.8 mm long, 1.5-3.0 mm wide, 1.1-1.7 times as long as wide; beak usually 0.6-1.25 mm long. **C. atlantica**

9b(7b) Widest leaves (0.65)0.8-1.6 mm wide, infructescence mostly 8-20 mm long; perigynia 1.9-3.0 mm long, 1.3-2.0 mm wide, (1.5)1.3-1.7(1.9) times as long as wide; beak 0.45-0.95 mm long. **C. howei**

SYLVATICAE

1a Leaf sheaths and blades +/- hairy; pistillate spikes 4-8 mm thick, up to 2.5 cm long; perigynia 3.5-5 mm long, finely nerved; beak about half as long as body; scales chestnut brown. **C. castanea**

1b Leaf sheaths and blades glabrous (lowermost bladeless sheaths sometimes finely hairy); spikes mostly 1.5-8 cm long, 2-6 mm thick. 2

2a(1b) Perigynia 3-5 mm long, definitely 3-angled, strongly stipitate and prominently nerved at base; achene sessile; basal leaves 5-10 mm wide; ligule obscure, hyaline; pistillate scales mostly cuspidate. **C. arctata**

2b(1b) Perigynia 4.5-10 mm long, obscurely 3-angled, tapering and faintly nerved at base; achene stipitate; basal leaves 3-4 mm wide; ligule well developed, longer than wide; pistillate scales ciliolate at tip, rarely cuspidate. **C. debilis**

VESICARIAE

1a Perigynia beak notched (less than 0.4 mm long), **not** definitely slender toothed. 2

1b Perigynia beak slender toothed. 3

2a(1a) Stigmas 2; achenes lenticular; perigynia essentially nerveless, 3 mm long; plants of wet, sandy or gravelly soil. **Carex saxatilis**

2b(1a) Stigmas 3; achenes trigonous; perigynia several nerved, 4-7 mm long; plants of grassy bogs and wet meadows. **C. oligosperma**

3a(1b) Perigynia (4) 4.5-7 mm thick, thin and papery; achenes deeply notched on one angle; pistillate spike thick cylindric. **C. tuckermanii**

3b(1b) Perigynia 2.5-3.5 (4) mm thick; achenes not notched. 4.

++

4a(3b) Perigynia horizontally spreading or at least the lowest reflexed at maturity, 7-12 mm long; lowest pistillate bract 3 or more times as long as the entire inflorescence; staminate spike usually one, with base slightly if at all elevated above summit of the crowded pistillate spikelets; without long horizontal stolons. **C. retrorsa**

4b(3b) Perigynia ascending or spreading (4) 4.5-7.5 mm long; lowest pistillate bract less than 3 times as long as inflorescence; staminate spikes mostly 2-4, usually well elevated above pistillate spikes. 5.

5a(4b) Culms slender, sharp and usually scabrous angled below lowest pistillate bract and between spikelets; leaves 3-5 (6) mm broad, weakly or not septate and nodulose; plants cespitose. **C. vesicaria**

5b(4b) Culms thick and spongy at base, blunt angled, more or less smooth below the lowest pistillate bract; leaves 1.5-10.5 mm broad; perigynia more or less widely spreading (ascending when immature), strongly +/- septate and nodulose, plants rhizomatous. 6

6a(5b) Leaves involute, glaucous, widest blades 1.5-4 (-7.2) mm wide; rare. **C. rostrata**

6b(5b) Leaves flat or channeled, not glaucous, widest blades (4.5-) 5-12 mm wide; widespread. **C. utriculata**

VIRESCENTES

1a Perigynia glabrous, +/- nerveless. **C. pallascens**

1b Perigynia pubescent, finely nerved. **C. swanii**

Carex adusta Boott Fig. p. A21. Map p. B18.
FREQUENCY: Uncommon.
HABITAT: Dry acid soils.
GENERAL RANGE: B.C. to Nfld. south to Minn. and Maine.

Carex aenea Fern. Fig. p. A21. Map p. B18.
FREQUENCY: Scattered throughout the Province.
HABITAT: Open, dry disturbed soil.
GENERAL RANGE: Alaska to Nfld. south to Mont. and New Eng.
SPECIAL NOTE: Named by M. Fernald from material collected in Kent Co.

Carex albolutescens Schwein.
SPECIAL NOTE: No voucher-specimens have been seen. Material reported from Bass River, Kent Co. is **C. silicea**, the relevant collection in NBM.

Carex amphibola Steud. Fig. p. A21. Map p. B18.
FREQUENCY: Very rare; collected near Hatfield Pt., Queens Co., Woodstock, Carleton Co., and Sussex, Kings Co.
HABITAT: Moist shaded woods.
GENERAL RANGE: Minn. and Ont. to N.B. south to Tex. and Ga.

++

SUBSPECIFIC TAXA: Our material is referable to var. **turgida** Fern.
SPECIAL NOTE: For more information see H. Hinds, 1983. The Rare Vascular Plants of New Brunswick. Syllogeus No. 50.

Carex aquatilis Wahl. Fig. p. A21. Map p. B18.
FREQUENCY: Locally common throughout the Province.
HABITAT: Wet open areas.
GENERAL RANGE: Alaska to Nfld. south to Calif. and N.J.; Greenland; Eurasia.
SUBSPECIFIC TAXA: Var. **altior** (Rydle.) Mack., with leaves to 8 mm broad, culms scabrous at summit, acutely triangular rather than leaves to about 5 mm broad, culms mostly smooth, obtusely triangular as in var. **aquatilis**, is recorded from Kings and Victoria Counties.

Carex arcta Boott Fig. p. A21. Map p. B18.
FREQUENCY: Uncommon and mostly confined to the southern half of the Province.
HABITAT: Shores and wet woods.
GENERAL RANGE: Alaska to Lab. and N.B. south to Calif. and New Eng.
SPECIAL NOTE: The apparent hybrid with **C. castanea** (**C. X knieskernii** Dew.) is reported from the Tobique R., Victoria Co.

Carex arctata Boott Fig. p. A21. Map p. B18.
FREQUENCY: Common throughout the Province.
HABITAT: Banks, wood edges, clearings.
GENERAL RANGE: Ont. to Nfld. south to Minn. and Conn.

Carex argyrantha Tuckerm.
SPECIAL NOTE: Reported from the Province. (Scoggan 1978-79). No voucher specimens have been seen. Most N.B. reports are based on **C. aenea**.

Carex atlantica Bailey Fig. p. A22.
FREQUENCY: Rare; collected near the Caanan River, northwest corner of Westmorland Co., near Scotch Settlement, Kings Co., and from near the mouth of Little Salmon River, St. John Co.
HABITAT: Wet and often boggy open ground.
GENERAL RANGE: Mich. to N.S. south to Tex. and Fla.

Carex atratiformis Britt. Fig. p. A22. Map p. B18.
FREQUENCY: Uncommon in the northern part of the Province.
HABITAT: Chiefly calcareous ledges, slopes and banks.
GENERAL RANGE: Ont. to Nfld. south to Mich. and n. New Eng.

Carex aurea Nutt. Fig. p. A22. Map p. 18.
FREQUENCY: Common throughout the Province.
HABITAT: Damp, often calcareous meadows and shores.
GENERAL RANGE: Alaska to Nfld. south to Calif. and Pa.

++

Carex backii Boott Fig. p. A22.
FREQUENCY: Very rare; known only from Mt. Zachie-Jonah, Albert Co. and McLeod Hill, York Co.
HABITAT: Shaded ledges and rocky soil.
GENERAL RANGE: B.C. to N.B. south to Oreg. and N.J.
SPECIAL NOTE: For more information see H. Hinds, 1983. The Rare Vascular Plants of New Brunswick. Syllogeus No. 50.

Carex bebbii Boott Fig. p. A22. Map p. B18.
FREQUENCY: Uncommon.
HABITAT: Damp open ground.
GENERAL RANGE: Alaska to Nfld. south to Oreg. and N.J.

Carex bigelowii Torr. Fig. p. A22.
FREQUENCY: Very rare; known only from Mt. Carleton, Northumberland Co.
HABITAT: Peaty rock crevices.
GENERAL RANGE: Great Bear L. to Nfld. south to Great Slave L. and N.B.; Greenland; Iceland; Eurasia.
SPECIAL NOTE: For more information see H. Hinds, 1983. The Rare Vascular Plants of New Brunswick. Syllogeus No. 50.

Carex bromoides Schkuhr Fig. p. A22. Map p. B18.
FREQUENCY: Scattered throughout the Province.
HABITAT: Wet woods, swamps and bogs.
GENERAL RANGE: Wisc. to N.S. south to La. and Fla.

Carex brunnescens (Pers.) Poir. Fig. p. A22. Map p. B19.
SUBSPECIFIC TAXA: Two intergrading varieties occur in our range.

1a Spikes +/- congested 1-10 mm apart; inflorescence 1.5-3.5 cm long; leaves and stems mostly erect. var. **brunnescens**
1b Spikes open 1-2.5 cm apart; inflorescence 2-7 cm long; leaves and stems arching. var. **sphaerostachya**

1. Var. **brunnescens** Fig. p. A22. Map p. B19.
FREQUENCY: Uncommonly scattered.
HABITAT: Mostly heights of land, mountain summits and headlands.
GENERAL RANGE: Alaska to Nfld. south to Oreg. and N.C.; Greenland; Iceland; Eurasia.

2. Var. **sphaerostachya** (Tuck.) Kuk. Map p. B19.
FREQUENCY: Common throughout the Province.
HABITAT: Low or upland woods, borders of bogs and swamps.
GENERAL RANGE: B.C. to Nfld. south to Oreg. and N.C.

Carex buxbaumii Wahl. Fig. p. A22. Map p. B19.
FREQUENCY: Scattered throughout except rare in the eastern lowlands.
HABITAT: Shores, meadows, and marly bogs.
GENERAL RANGE: Alaska to Nfld. south to Calif. and N.C.; Greenland; Eurasia.

++

Carex canescens L. Fig. p. A22. Map p. B19.
FREQUENCY: Common throughout the Province.
HABITAT: Sphagnum bogs and boggy swamps and shores.
GENERAL RANGE: Alaska to Nfld. south to Calif. and Va.; Greenland; Iceland; Eurasia.
SYNONYMS: **C. curta** Good.
SUBSPECIFIC TAXA: Var. **subloliacea** (Laestad.) Hartman, with perigynia to 2 mm long; beak mostly smooth and spikes to about 7 mm long instead of perigynia to 3 mm long, beak often serrulate and spikes to 12 mm long as in var. **canescens**, occurs scattered in the range of the species.
SPECIAL NOTE: The apparent hybrid with **C. mackenziei** (**C. X pseudohelvola** Kihlm.) is reported from Bathurst, Gloucester Co. and Grand Manan, Charlotte Co.

Carex capillaris L. Fig. p. A22. Map p. B19.
FREQUENCY: Uncommon; occurring rarely along the Bay of Fundy and more commonly along the westcentral and northern borders.
HABITAT: Rocky shores, moist ground of wood edges, roads, clearings and similar mostly calcareous areas.
GENERAL RANGE: Alaska to Nfld. south to Oreg. and Vt.; Greenland; Iceland; Eurasia.

Carex caryophyllea Latourr.
FREQUENCY: Very rare; found only near the mouth of Pt. Wolfe River, Fundy National Park, Albert Co. (UNB).
HABITAT: Grassy meadow; probably introduced in ships ballast. This large colony appears to be steadily increasing.
GENERAL RANGE: Eurasian; this is apparently the only authenticated introduction in Canada. Also known in North America from s. Me. to e. N.Y. and D.C.

Carex castanea Wahl. Fig. p. A23. Map p. B19.
FREQUENCY: Locally common and scattered in lime rich areas of the Province.
HABITAT: Wet meadows, roadside ditches and similar usually calcareous areas.
GENERAL RANGE: Man. to Nfld. south to Minn. and Conn.
SPECIAL NOTE: The apparent hybrid with **C. arctata** (**C. X knieskernii** Dew.) has been reported from the Tobique R., Victoria Co.

Carex chordorrhiza L.f. Fig. p. A23. Map p. B19.
FREQUENCY: Rare; known only from the St. John River drainage.
HABITAT: Sphagnum bogs and marshy often calcareous areas.
GENERAL RANGE: Alaska to Nfld. south to B.C. and Vt.

Carex communis Bailey Fig. p. A23. Map p. B19.
FREQUENCY: Common in suitable habitats throughout the Province.
HABITAT: Woods and rocky slopes.
GENERAL RANGE: Man. to N.S. south to Ark. and Ga.

Carex comosa Boott Fig. p.A23.
FREQUENCY: Very rare; known only from the Jolicure Lakes, Westmorland Co. and from Lakeville, Carleton Co.

HABITAT: Marshy shores.
GENERAL RANGE: B.C. to N.S. south to Calif. and Fla.

Carex concinna R. Br. Fig. p. A23. Map p. B19.
FREQUENCY: Rare; known only from a few sites along the Restigouche R., Restigouche and Madawaska Counties, and at Sisson Gorge and Plaster Rock, Victoria Co. This latter station is the southernmost for the species.
HABITAT: Calcareous ledges, shores and open slopes.
GENERAL RANGE: Alaska to Nfld. south to B.C., Colo., and N.B.
SPECIAL NOTE: For more information see H. Hinds, 1983. The Rare Vascular Plants of New Brunswick. Syllogeus No. 50.

Carex conoidea Schkuhr Fig. p. A23. Map p. B19.
FREQUENCY: Uncommon; centered in the southern part of the Province.
HABITAT: Damp grassy areas.
GENERAL RANGE: Minn. to Nfld. south to Iowa and N.C.

Carex crawei Dewey Fig. p. A23.
FREQUENCY: Very rare; known only from two sites on the Restigouche R., Restigouche Co. Reports from Ingleside, Kings Co. revised by Boivin to **C. conoidea**, the relevant collection in DAO.
HABITAT: Limestone pavement along the river.
GENERAL RANGE: B.C. to Nfld. south to Wash. and Ala.
SPECIAL NOTE: For more information see H. Hinds, 1983. The Rare Vascular Plants of New Brunswick. Syllogeus No. 50.

Carex crawfordii Fern. Fig. p. A23. Map p. B19.
FREQUENCY: Common throughout the Province.
HABITAT: Roadsides, sandy shores, clearings and similar open areas.
GENERAL RANGE: Alaska to Nfld. south to Wash. and Tenn.

Carex crinita Lam. Fig. p. A23. Map p.B19.
FREQUENCY: Common throughout the Province.
HABITAT: Wet sandy soils of roadside ditches, and open swampy ground.
GENERAL RANGE: Man. to Nfld. south to La. and Ga.

Carex cryptolepis Mack. Fig. p. A23.
FREQUENCY: Rare; several scattered locations known in central and southern parts of the Province as at Lake George near Magundy, York Co., and Kerr Lake, Charlotte Co.
HABITAT: Sandy or mucky shores, ditches, swales; occasionally in bogs.
GENERAL RANGE: Ont. to Nfld. south to Wisc. and Pa.
SYNONYMS: **C. flava** L. var. **fertilis** Peck

Carex cumulata (Bailey) Mack. fig. p. A23. Map p. B20.
FREQUENCY: Common in the Grand Lake area of Sunbury and Queens Counties; scattered elsewhere in the southern and eastcentral part of the Province.

++

HABITAT: Dry or moist disturbed soils such as the coal spoils area.
GENERAL RANGE: Man. to N.S. south to Mich. and N.J.

Carex debilis Michx. Fig. p. A23. Map p. B20.
FREQUENCY: Common throughout the southern half of the Province.
HABITAT: Wet fields, roadside ditches, open bottomland woods and shores.
GENERAL RANGE: Ont. to Nfld. south to Tex. and Fla.
SUBSPECIFIC TAXA: Our material is referable to var. **rudgei** Bailey.

Carex deflexa Hornem. Fig. p. A24. Map. p. B20.
FREQUENCY: Common throughout the Province.
HABITAT: Open woodlands.
GENERAL RANGE: Alaska to Nfld. south to B.C. and Mass.; Greenland.

Carex demissa Hornem.
SPECIAL NOTE: Reported from s. N.B. (Scoggan 1978-79). This record could not be authenticated.

Carex deweyana Schw. Fig. p. A24. Map p. B20.
FREQUENCY: Common throughout the Province.
HABITAT: Open woods, especially beech-maple stands.
GENERAL RANGE: Dist. Mack. to Nfld. south to Idaho and Pa.

Carex diandra Schrank Fig. p. A24. Map p. B20.
FREQUENCY: Uncommon; unknown from most of the southern part of the Province.
HABITAT: Boggy lowlands especially in calcareous or alkaline areas.
GENERAL RANGE: Alaska to Nfld. south to Calif. and N.J.; Iceland; Eurasia.

Carex disperma Dewey Fig. p. A24. Map p. B20.
FREQUENCY: Common throughout the Province.
HABITAT: Boggy woods and thickets.
GENERAL RANGE: Alaska to Nfld. south to Calif. and Pa.

Carex eburnea Boott Fig. p. A24. Map p. B20.
FREQUENCY: Rare; known from one site on the Pt. Wolfe R., St. John Co. and from several sites in the northwest part of the Province.
HABITAT: Moist calcareous ledges.
GENERAL RANGE: Alaska to Nfld. south to B.C. and Va.

Carex echinata Murray Fig. p. A24. Map p. B20.
FREQUENCY: Common throughout the Province.
HABITAT: Acidic wetlands such as sphagnum bogs, peaty or sandy lake or river shores, boggy meadows, roadside ditches and wet pastures.
GENERAL RANGE: B.C. to Nfld. south to Calif. and S.C.
SYNONYMS: **C. stellulata** Good.; **C. muricata** L. & vars., **C. leersii** Willd. & vars.

++

Carex emmonsii Dewey Fig. p. A24.
SPECIAL NOTE: Reported by Boivin 1966-67, and Scoggan, 1978-79 from the Province. No documenting specimens have been seen.
SYNONYMS: **C. nigromarginata** Schwein. var. **minor** (Boott) Gleason.

Carex exilis Dewey Fig. p. A24. Map p. B20.
FREQUENCY: Rare, known only from two Bay of Fundy locations and two marly bogs in Victoria Co.
HABITAT: Sphagnum bogs often in calcareous areas.
GENERAL RANGE: Ont. to Nfld. south to Minn. and Del.

C. flaccidula Steudel. Fig. p. A24. Map p. B20.
FREQUENCY: Uncommon; scattered in the St. John and Kennebecasis River valleys.
HABITAT: Rich wooded slopes.
GENERAL RANGE: Man. to N.S. south to Kans. and S.C.
SYNONYMS: **C. convoluta** Mack.

Carex flava L. Fig. p. A24. Map p. B20
FREQUENCY: Common throughout the Province
HABITAT: Shores, wet meadows.
GENERAL RANGE: Alaska to Nfld. south to Ohio and N.J.
SPECIAL NOTE: The apparent hybrid with **C. viridula** (**C. X subviridula** (Kür.) Fern.) is reported from Madawaska Co.

Carex folliculata L. Fig. p. A24. Map p. B21.
FREQUENCY: Uncommon; confined to the southern half of the Province.
HABITAT: Swampy woods.
GENERAL RANGE: Ont. to Nfld. south to La. and Fla.

Carex glareosa Wahl. Fig. p. A24.
FREQUENCY: Very rare; known only from Dalhousie, Restigouche Co. and Ste. Cecile, Shippegan Is., Gloucester Co.
HABITAT: Brackish marshes.
GENERAL RANGE: Alaska to Baffin Is. south to n. Man. and ne. N.B.
SUBSPECIFIC TAXA: Our material is referable to var. **amphigena** Fern.
SYNONYMS: **C. amphigena** (Fern.) Mack.; **C. bipartita** Bell. var. **amphigena** (Fern.) Polunin.
SPECIAL NOTE: For more information see H. Hinds, 1983. The Rare Vascular Plants of New Brunswick. Syllogeus No. 50.

Carex gracillima Schw. Fig. p. A25. Map p. B21.
FREQUENCY: Uncommonly scattered across the central portion of the Province.
HABITAT: Mostly in rich woodlands or edges of woods and streams.
GENERAL RANGE: Man. to Nfld. south to Mo. and N.C.

Carex granularis Muhl. Fig. p. A25. Map p. B21.
FREQUENCY: Uncommonly scattered around the Province as at Mactaquac Reach, York Co., and shore of Restigouche R. near Slate Mtn.
HABITAT: Shores and grassy meadows, often in calcareous areas.

GENERAL RANGE: Sask. to N.S. south to Kans. and Fla.
SUBSPECIFIC TAXA: Our material is referable to var. **haleana** (Olney) Porter.

Carex gynandra Schwein.
FREQUENCY: Common throughout the Province.
HABITAT: Wet Clearings, roadside ditches, swampy, often coniferous, woods.
GENERAL RANGE: Ont. to Nfld. south to Wisc., Tenn. and N.C.
SYNONYMS: **C. crinita** var. **gynandra** Schwein. & Torr.

Carex gynocrates Wormsk. Fig. p. A25. Map p. B21.
FREQUENCY: Rare; known only from a few localities in the northern portion of the Province.
HABITAT: Sphagnum bogs in calcareous areas.
GENERAL RANGE: Alaska to Nfld. south to B.C. and Pa.; Greenland; Asia.

Carex haydenii Dewey Fig. p. A25. Map p. B21.
FREQUENCY: Uncommon; known mainly from the area between Grand Lake, and Mactaquac Dam with an outlier in Madawaska Co.
HABITAT: Wet meadows and shores.
GENERAL RANGE: Ont. to St. Pierre & Miquelon south to Nebr. and N.J.

Carex hirtifolia Mack. Fig. p. A25. Map p. B21.
FREQUENCY: Rare; collected in a few places in the southern half of the Province.
HABITAT: Dry rich woods and rocky floodplains.
GENERAL RANGE: Ont. to N.S. south to Kans. and Md.
SPECIAL NOTE: For more information see H. Hinds, 1983. The Rare Vascular Plants of New Brunswick. Syllogeus No. 50.

Carex hormathodes Fern. Fig. p. A25. Map p. B21.
FREQUENCY: Common all along the coast.
HABITAT: Brackish marshes and sands.
GENERAL RANGE: Eastern Que. to N.S. south to Va.

Carex houghtonii Torr. Fig. p. A25. Map p. B21.
FREQUENCY: Uncommonly scattered across the Province.
HABITAT: Sandy clearings, roadsides and banks.
GENERAL RANGE: Alta. to Nfld. south to Minn. and New Eng.

Carex howei Mack. fig. p. A25. Map p. B21.
FREQUENCY: Very rare; only known from Southern Head, Grand Manan Is., Charlotte Co.(UNB)
HABITAT: Acid bogs and similar wet areas.
GENERAL RANGE: Mich. to s. Que. and N.S. south to Tex. and Fla.
SYNONYMS: **C. atlantica** Bailey ssp. **capillacea** Reznicek.

Carex hystricina Muhl. Fig. p. A25. Map p. B21.
FREQUENCY: Fairly common in the mid St. John River valley and uncommon elsewhere in the southern part of the Province.
HABITAT: Muddy shores and swamps.
GENERAL RANGE: B.C. to Nfld. south to Calif. and Va.

++

Carex interior Bailey Fig. p. A25. Map p. B21.
FREQUENCY: Uncommon; found mostly in calcareous areas throughout.
HABITAT: Bogs and wet meadows.
GENERAL RANGE: Yukon and B.C. to Nfld. south to Calif. and Del.

Carex intumescens Rudge Fig. p. A25.
FREQUENCY: Common throughout the Province.
HABITAT: Alluvial woods and swamps.
GENERAL RANGE: Man. to Nfld. south to Tex. and Fla.

Carex josselynii (Fern.) Mac.
FREQUENCY: Very rare; known only from the mouth of Michaud Brook, Madawaska Co. (UNB).
HABITAT: Springy and rocky calcareous shores.
GENERAL RANGE: Northern Me. and nw. N.B.
SYNONYMS: **C. echinata** Murr. according to some authors.

Carex lacustris Willd. Fig. p. A25. Map p. B21.
FREQUENCY: Uncommon; most collections from the southeast part of the Province but found as far north as Shippigan Is., Gloucester Co.
HABITAT: Marshy lowlands, ditches and swamps.
GENERAL RANGE: Alta. to Nfld. south to Idaho and Va.

Carex lanuginosa Michx. Fig. p. A26. Map p. B22.
FREQUENCY: Scattered throughout the Province.
HABITAT: Wet meadows, roadside ditches and similar wet areas.
GENERAL RANGE: B.C. to Nfld. south to Calif. and Va.

Carex lasiocarpa Ehrh. Fig. p. A26. Map p. B22.
FREQUENCY: Scattered throughout but apparently rarer northward.
HABITAT: Boggy open areas.
GENERAL RANGE: Alaska to Nfld. south to Wash. and N.J.: Eurasia.
SUBSPECIFIC TAXA: Our plants are refered to var. **americana** Fern.
GENERAL RANGE: Ont. to Nfld. south to Mich. and Mass.; Asia.
SYNONYMS: **C. lanuginosa** Ehrh. var. **americana** (Fern.) Boivin.

Carex lenticularis Michx. Fig. p. A26. Map p. B22.
FREQUENCY: Common throughout the Province.
HABITAT: Lake and river shores.
GENERAL RANGE: B.C. to Nfld south to Calif. and Mass.
SUBSPECIFIC TAXA: Var. **blakei** Dewey, with the terminal spike staminate only at base instead of normally entirely staminate, occurs scattered throughout the Province and elsewhere from Ont. to Nfld.

Carex lepidocarpa Tausch Fig. p. A26. Map p. B22.
FREQUENCY: Very rare; known only from Petit Rocher and Belledune, Gloucester Co.
HABITAT: Calcareous bogs.
GENERAL RANGE: Ont. to Nfld. south to N.B. and N.S.; Europe.

++

SYNONYMS: C. flava L. var. lepidocarpa (Tausch) Godr.
SPECIAL NOTE: For more information see H. Hinds, 1983. The Rare Vascular Plants of New Brunswick. Syllogeus No. 50.

Carex leporina L. Fig. p. A26.
FREQUENCY: Very rare; known only from Mactaquac Provincial Park and Fredericton, York Co.
HABITAT: Damp thickets, wet meadows and roadside ditches.
GENERAL RANGE: Eurasian; introduced in B.C., Nfld., P.E.I., N.B. and N.S.

Carex leptalea Wahl. Fig. p. A26. Map p. B22.
FREQUENCY: Probably common throughout the Province.
HABITAT: Bogs, coniferous or mixed hardwood swamps, and wet grassy fields.
GENERAL RANGE: Alaska to Nfld. south to Calif., Tex. and Fla.

Carex leptonervia Fern. Fig. p. A26. Map p. B22.
FREQUENCY: Probably common throughout the Province.
HABITAT: Damp upland woods and clearings.
GENERAL RANGE: Minn. and Ont. to Nfld. south to Wisc. and N.C.

Carex limosa L. Fig. p. A26. Map p. B22.
FREQUENCY: Uncommon; mostly collected in the southeast half of the Province.
HABITAT: Sphagnum bogs often surrounding bog ponds.
GENERAL RANGE: Alaska to Nfld. south to Calif. and Del.; Iceland; Eurasia.
SPECIAL NOTE: The apparent hybrid with C. **paupercula** (C. X **sublimosa** Lepage) is occasionally reported with the parents.

Carex livida (Wahl.) Willd. Fig. p. A26.
FREQUENCY: Very rare; known only from Miscou Island, Gloucester Co. and near mouth of Upsalquitch R., Restigouche Co.
HABITAT: Calcareous bogs and meadows.
GENERAL RANGE: Alaska and B.C. to Nfld. south to Calif. and N.J.; Iceland; Eurasia.
SPECIAL NOTE: For more information see H. Hinds, 1983. The Rare Vascular Plants of New Brunswick. Syllogeus No. 50.
SUBSPECIFIC TAXA: Our plants are all var. **radicaulis** Paine.

Carex lucorum Link Map p. B22.
FREQUENCY: Uncommon; scattered mainly in the southern half of the Province.
HABITAT: Sandy open heathy barrens and dry pine or mixed woods.
GENERAL RANGE: Ont. to N.S. south to Mich., Tenn. and N.C.
SYNONYMS: C. **pensylvanica** Lam. var. **distans** Peck.

Carex lupulina Muhl. Fig. p. A26. Map p. B22.
FREQUENCY: Uncommon in the southern part of the Province.
HABITAT: Borders of wet woods, swampy shores.
GENERAL RANGE: Minn. and Ont. to N.S. south to Tex. and Fla.

++

Carex lurida Wahl. Fig. p. A26. Map p. B22.
FREQUENCY: Common in the southern half of the Province.
HABITAT: Swampy areas.
GENERAL RANGE: Minn. and Ont. to N.S. south to Mexico and Fla.

Carex mackenziei Krecz. fig. p. A26. Map p. B22.
FREQUENCY: Scattered along the coastal areas of the Province.
HABITAT: Saline or brackish marshes.
GENERAL RANGE: Alaska to Nfld. south to Me.; Greenland; Iceland; Scandinavia; Asia.
SYNONYMS: C. **norvegica** Willd.
SPEICAL NOTE: The apparent hybrid with **C. canescens** (C. **X pseudohelvola** Kihlm.) is reported from Bathurst, Gloucester Co. and Grand Manan, Charlotte Co.

Carex media R. Br. Fig. p. A27. Map p. B22.
FREQUENCY: Very rare; known only from Pemouet Brook and Restigouche R., Restigouche Co., and Third Lake, Madawaska Co.
HABITAT: Shores and wet meadows, usually calcareous.
GENERAL RANGE: Alaska to Labr. south to B.C., Idaho, n. Mich. and N.B.
SYNONYMS: C. **vahlii** Mack.
SPECIAL NOTE: For more information see H. Hinds, 1983. The Rare Vascular Plants of New Brunswick. Syllogeus No. 50.

Carex michauxiana Boeckl. Fig. p. A27. Map p. B23.
FREQUENCY: Uncommon; scattered locations in central, eastern and southwestern parts of the Province.
HABITAT: Peat bogs and boggy meadows.
GENERAL RANGE: Ont. to Nfld. south to Mich. and Mass.; Asia.

Carex nigra (L), Reich. Fig. p. A27. Map p. B23.
FREQUENCY: Locally common; probably more common than collections seem to indicate.
HABITAT: Peaty coastal headlands and meadows and occasionally introduced inland as at Fredericton, York Co.
GENERAL RANGE: Ont. to Nfld. south along the coast to R.I.; Greenland; Iceland; Eurasia.
SYNONYMS: C. **acuta** L.; C. **goodenowii** Gay.

Carex normalis Mack. Fig. p. A27.
FREQUENCY: Very rare; known only from near the mouth of Limestone Stream on Aroostook R., Victoria Co. (UNB).
HABITAT: Open rocky woods.
GENERAL RANGE: Man. to N.B. south to Okla., and N.C.

Carex novae-angliae Schw. Fig. p. A27. Map p. B23.
FREQUENCY: Common in the southern half of the Province and rare northward.
HABITAT: Open, dryish woodlands.
GENERAL RANGE: Ont. to Nfld. south to Wisc. and Conn.

++

Carex oligosperma Michx. Fig. p. A27. Map p. B23.
FREQUENCY: Locally common; scattered throughout.
HABITAT: Acid bogs and swamps.
GENERAL RANGE: Alta. to Nfld. south to Minn. and Pa.

Carex ormostachya Wieg. Fig. p. A27. Map p. B23.
FREQUENCY: Uncommon.
HABITAT: Mostly in rich hardwoods.
GENERAL RANGE: Ont. to N.S. south to Mass. and Pa.
SYNONYMS: **C. laxiflora** Lam. var. **ormostachya** (Wieg.) Gl.

Carex pairaei Schultz Fig. p. A27.
FREQUENCY: Very rare; known only from the city of Saint John, St. John Co.
HABITAT: Fields and roadsides.
GENERAL RANGE: Eurasian; introduced in Canada in N.B.; also in Pa. and N.C.

Carex paleacea Wahl. Fig. p. A27. Map p. B23.
FREQUENCY: Common along the coast.
HABITAT: Salt marshes and brackish meadows.
GENERAL RANGE: Hudson Bay to Nfld. south to Man. and Mass.
SPECIAL NOTE: The apparent hybrid with **C. limosa** (**C. X sublimosa** Lepage) occurs rarely with the parents.

Carex pallescens L. Fig. p. A27. Map p. B23.
FREQUENCY: Probably common throughout the Province.
HABITAT: Moist meadows, clearings and edges of woods.
GENERAL RANGE: Ont. to Nfld. south to Mich. and N.J.; Iceland; Eurasia.
SUBSPECIFIC TAXA: Our material is referable to var. **neogaea** Fern.

Carex panicea L. Fig. p. A27.
FREQUENCY: Very rare; known only from Bass River, Kent Co.
HABITAT: Meadows and grasslands.
GENERAL RANGE: Eurasian; introduced in Ont., N.B. and Nfld.

Carex pauciflora Lightf. Fig. p. A28. Map p. B23.
FREQUENCY: Scattered throughout the Province.
HABITAT: Sphagnum bogs.
GENERAL RANGE: Alaska to Nfld. south to Wash. and Conn.; Eurasia.

Carex paupercula Michx. Fig. p. A28. Map p. B23.
HABITAT: Sphagnum bogs.
SUBSPECIFIC TAXA: Two varieties occur in our range.

1a Culms to about 8 dm tall, scabrous for about 1 dm below inflorescence, and usually with paler scales. var. **pallens**
1b Culms often shorter, mostly smooth; scales darker margined. var. **irrigua**

1. Var. **pallens** Fern.

FREQUENCY: Uncommonly scattered throughout.
GENERAL RANGE: Ont. to Nfld. south to Minn. and Pa.
2. Var. **irrigua** (Wahl.) Fern.
FREQUENCY: Common throughout the Province.
GENERAL RANGE: Alaska to Nfld. south to Wash. and Mass.

Carex peckii Howe Fig. p. A28. Map p. B23.
FREQUENCY: Mostly uncommon except rare in the eastern lowlands.
HABITAT: rocky slopes, clearings and dry woods, often calcareous.
GENERAL RANGE: Alaska to N.S. south to B.C. and N.J.
SYNONYMS: **C. nigromarginata** Schw. var. **elliptica** (Boott) Gl.

Carex pedunculata Muhl. Fig. p. A28. Map. p. B23.
FREQUENCY: Fairly common throughout except rare in the eastern lowlands.
HABITAT: Mostly deciduous or mixed woods in fertile soils.
GENERAL RANGE: Man. to Nfld. south to S. Dak. and Ga.

Carex pensylvanica Lam.
SPECIAL NOTE: Reported from N.B. by Scoggan, 1978-79 and Boivin, 1966-67. These may refer to **C. lucorum** (**C. pensylvanica** var. **distans**). No voucher specimens have been seen.

Carex plantaginea Lam. Fig. p. A28. Map p. B23.
FREQUENCY: Very rare; known from a few sites in the westcentral part of the Province.
HABITAT: Rich, usually calcareous, hardwoods.
GENERAL RANGE: Ont. to N.S. south to Minn. and Ga.
SPECIAL NOTE: for more information see H. Hinds, 1983. The Rare Vascular Plants of New Brunswick. Syllogeus No. 50.

Carex prairea Dewey Fig. p. 28. Map p. B24.
FREQUENCY: Rare; known only from a few sites in the western and southern parts of the Province.
HABITAT: Alkaline boggy areas.
GENERAL RANGE: B.C. to N.S. south to Nebr. and N.J.

Carex projecta Mack. Fig. p. A28. Map p. B24.
FREQUENCY: Probably common throughout the Province.
HABITAT: Meadows, clearings and swampy ground.
GENERAL RANGE: Man. to Nfld. south to Minn. and W. Va.
SPECIAL NOTE: The type specimen was collected in Kouchibouguac by Fowler.

Carex pseudocyperus L. Fig. p. A28. Map p. B24.
FREQUENCY: Scattered throughout the Province.
HABITAT: Rich bottomland soils.
GENERAL RANGE: Alta. to Nfld. south to N. Dak. and Pa.; Eurasia; Africa.

Carex rariflora (Wahl.) Sm. Fig. p. A28.
FREQUENCY: Very rare; known only from sphagnum bogs of Shippegan and Miscou Islands, Gloucester Co. (UNB).
HABITAT: Sphagnum bogs.
GENERAL RANGE: Alaska to Nfld. south to N.B., Mt. Katahdin, Maine; Greenland; Iceland; Eurasia.

Carex retrorsa Schw. Fig. p. A28. Map p. B24.
FREQUENCY: Common throughout the Province.
HABITAT: Mostly alluvial woods and wet meadows.
GENERAL RANGE: Dist. Mack. to Nfld. south to Oreg. and N.J.

Carex rostrata Stokes Fig. p. A28.
FREQUENCY: Very rare; reported from Canoe Landing Lake, Northumberland Co. and Bass River, Kent Co.
HABITAT: Wet boggy shores and meadows.
GENERAL RANGE: Alaska to Nfld. south to Minn., Mich. and Wisc., Greenland; Iceland; Eurasia.

Carex salina Wahl. Fig. p. A29. Map p. B24.
HABITAT: Brackish marshes and coastal bogs.
GENERAL RANGE: Aggregate species Man. to Nfld. south to Mass.
SUBSPECIFIC TAXA: Two varieties are found in our range.

1a Culms to about 3 dm tall, obtusely angled; leaves to about 4 mm wide; pistillate spikes to about 3 cm long. var. **salina**.
1b Culms 6-9 dm tall, +/- acutely angled; leaves to about 9 mm wide; pistillate spikes to about 8 cm long. var. **kattegatensis**.

1. Var. **salina**.
FREQUENCY: Very rare; known only Shippegan Island, Gloucester Co. (UNB).
2. Var. **kattegatensis** (Fries) Allmq.
FREQUENCY: Uncommonly scattered along the coast.

Carex saxatilis L. Fig. p. A29.
FREQUENCY: Very rare; known only from shore of Kennebecasis R. near Rothesay, Kings Co.
HABITAT: Wet rocky shores.
GENERAL RANGE: Sask. to Nfld. south to Wash. and Maine; Greenland; Iceland; Eurasia.
SUBSPECIFIC TAXA: Our material is referable to var. **miliaris** (Michx.) Bailey.

Carex scabrata Schw. Fig. p. A29. Map p. B24.
FREQUENCY: Probably common throughout.
HABITAT: Shaded springs, rills and streams.
GENERAL RANGE: Ont. to N.S. south to Mo. and S.C.

Carex scoparia Schkuhr Fig. p. A29. Map p. B24.
FREQUENCY: Common except in the northern part of the Province.
HABITAT: Wood edges, open disturbed areas.
GENERAL RANGE: B.C. to Nfld. south to Oreg. and Fla.

Carex silicea Olney Fig. p. A29. Map p. B24.
FREQUENCY: Common in its preferred habitat.
HABITAT: Coastal sands.
GENERAL RANGE: Que. to Nfld. south to Md.

Carex sparganioides Muhl. Fig. p. A29. Map p. B24.
FREQUENCY: Rare; found in Woodstock and Meductic, Carleton Co. and Robinsonville, Restigouche Co.
HABITAT: Rich hardwoods usually in calcareous areas.
GENERAL RANGE: Ont. to N.B. south to Ohio and Pa.
SUBSPECIFIC TAXA: Our material is referable to var. **cephaloidea** (Dew.) Carey.
SYNONYMS: **C. cephaloidea** (Dew.) Dewey
SPECIAL NOTE: For more information see H. Hinds, 1983. The Rare Plants of New Brunswick. Syllogeus No. 50.

Carex sprengelii Dewey Fig. p. A29. Map p. B24.
FREQUENCY: Rare; known only from a few locations in Queens, Carleton, Victoria and Restigouche Counties.
HABITAT: Banks, slopes and alluvial bottomland in calcareous areas.
GENERAL RANGE: B.C. to N.B. south to Mont. and N.J.
SPECIAL NOTE: For more information see H. Hinds, 1983. The Rare Plants of New Brunswick. Syllogeus No. 50.

Carex sterilis Willd.
FREQUENCY: Very rare; found only in Mactaquac Provincial Park, York Co. (UNB).
HABITAT: Wet calcareous meadows.
GENERAL RANGE: Alta. to Nfld. south to Ind. and N.J.
SYNONYMS: **C. muricata** L. var. **sterilis** (Carey) Gl.; **C. elachycarpa** Fern.

Carex stipata Muhl. Fig. p. A29. Map p. B24.
FREQUENCY: Common throughout the Province.
HABITAT: Wet sandy organic soils.
GENERAL RANGE: Alaska to Nfld. south to Calif. and Fla.

Carex stricta Lam. Fig. p. A29. Map p. B25.
HABITAT: Wet meadows.
GENERAL RANGE: Aggregate species from Man. to N.S. south to Colo. and Fla.
SUBSPECIFIC TAXA: Two intergrading varieties are recognized from our area.

1a Plants with elongate underground stolons, scarcely tussock forming; ventral surface of sheaths minutely scabrous; leaves pale or glaucous.
var. **strictior**

++

1b Plants very densely tufted forming tussocks or stools, horizontal stolons short or wanting; ventral surface of sheath glabrous; leaves usually darker. var. **stricta**

1. Var. **strictior** (Dew.) Carey
FREQUENCY: Uncommon and more coastal than the species.
2. Var. **stricta**
FREQUENCY: Scattered throughout except rare in coastal and northern parts of the Province.

Carex swanii (Fern.) Mack. Fig. p. A29.
FREQUENCY: Very rare; only known from Ross Island, Charlotte Co.
HABITAT: Wood edges and dry open fields.
GENERAL RANGE: Wisc. and Ont. to N.S. south to Ark. and N.C.
SPECIAL NOTE: For more information see H. Hinds, 1983. The Rare Plants of New Brunswick. Syllogeus No. 50.

Carex sylvicola Webber & Ball Fig. p. A29.
FREQUENCY: Common in its preferred habitat. Absent from the eastern lowlands.
HABITAT: Rich deciduous woods.
GENERAL RANGE: Man. to N.S. south to La. and Ga.
SYNONYMS: **C. rosea** Schkuhr

Carex tenera Dewey Fig. p. A29. Map p. B25.
FREQUENCY: Uncommonly scattered in the southern part of the Province.
HABITAT: Moist open ground and woodlands.
GENERAL RANGE: B.C. to Nfld. south to Mont. and N.C.

Carex tenuiflora Wahl. Fig. p. A29 Map p. B25.
FREQUENCY: Rare throughout the Province.
HABITAT: Sphagnum bogs and boggy meadows in calcareous areas.
GENERAL RANGE: Alaska to Nfld. south to B.C. and Maine; Eurasia.
SPECIAL NOTE: For more information see H. Hinds, 1983. The Rare Plants of New Brunswick. Syllogeus No. 50.

Carex tincta Fern. Fig. p. A30. Map p. B25.
FREQUENCY: Locally common especially in the Saint John River drainage system.
HABITAT: Moist meadows, roadside ditches, woodland borders and clearings.
GENERAL RANGE: B.C. to Alta., Mich., Que., N.B. and P.E.I. south to N.Y. and N. Eng.

Carex tonsa (Fern.) Bickn. Fig. p. A30. Map p. B25.
FREQUENCY: Common in its preferred habitat and especially common in the eastern lowlands.
HABITAT: Dry sandy areas, dunes, railways, and open woods.
GENERAL RANGE: Alta. to Lab. south to Minn. and Va.

++

Carex torta Boott Fig. p. A30. Map p. B25.
FREQUENCY: Common throughout the Province.
HABITAT: Rocky margins of streams.
GENERAL RANGE: Minn. and Ont. to N.S. south to Ark. and Ga.

Carex tribuloides Wahl. Fig. p. A30. Map p. B25.
FREQUENCY: Probably common except towards the northeast part of the Province.
HABITAT: Moist woodland, banks and shores.
GENERAL RANGE: Nebr., Mich. and Ont. to N.S. south to Okla. and Fla.

Carex trisperma Dewey Fig. p. A30. Map p. B25.
FREQUENCY: Common throughout the Province.
HABITAT: Boggy woods and clearings.
GENERAL RANGE: Dist. Mack. to Nfld. south to Minn. and Va.
SUBSPECIFIC TAXA: Var. **billingsii** Knight, with reduced perigynia to 3.3 mm long, 1-2 per spike instead of 4 mm long, to about 5 in each spike, as in var. **trisperma**, occurs rarely with the species.

Carex trisperma X **C. canescens**
SPECIAL NOTE: A specimen with sterile, malformed pollen and intermediate characteristics is reported from the edge of Sunken Island Bog near Sackville, Westmorland Co. (UNB).

Carex tuckermanii Dewey Fig. p. A30. Map p. B25.
FREQUENCY: Uncommonly scattered throughout except rarer towards the north.
HABITAT: Alluvial woods and borders.
GENERAL RANGE: Minn. and Ont. to N.S. south to Iowa and N.J.

Carex umbellata Schkuhr Fig. p. A30. Map p. B25.
FREQUENCY: Scattered throughout.
HABITAT: Dry sandy fields, roadsides, railways and similar open areas.
GENERAL RANGE: B.C. to Nfld. south to Mo. and Va.

Carex utriculata Boott Fig. p. A28 (See **C. rostrata**), Map p. B24.
FREQUENCY: Widespread throughout the Province.
HABITAT: Wet meadows, pastures, ditches and boggy pond margins.
GENERAL RANGE: Alaska to Nfld. south to Baja Calif., N. Mex., Nebr., Ind. and Va.
SYNONYMS: **C. rostrata** var. **utriculata** (Boott) Bailey.

Carex vaginata Tausch Fig. p. A30. Map p. B25.
FREQUENCY: Rare; reported from a few localities in the northern part of the Province.
HABITAT: Calcareous peat bogs and meadows.
GENERAL RANGE: Alaska to Nfld. south to B.C. and Vt.; Greenland; Iceland; Eurasia.
SPECIAL NOTE: For more information see H. Hinds, 1983. The Rare Plants of New Brunswick. Syllogeus No. 50.

++

Carex vesicaria L. Fig. p. A30. Map p. B25, B26.
HABITAT: Shores, swamps, wet meadows.
GENERAL RANGE: Aggregate species B.C. to Nfld. south to Calif. and Del.; Eurasia.
SUBSPECIFIC TAXA: An extremely variable species divided by Fernald, 1950, into 5 varieties. Only var. **jejeuna** Fern., having perigynia subglobose to globose-ovoid bodies abruptly rounded to beak instead of slenderly ovoid, gradually tapering to beak as in the species, appears worth separate recognition in the Province. It occurs occasionally with the species.
SPEICAL NOTE: The apparent hybrid with **C. saxatilis** var. **miliaris** (**C. X mainensis** Porter) is reported from near Rothesay, Kings Co.

Carex viridula Michx. Fig. p. A30. B26.
FREQUENCY: Scattered throughout except rare in the eastern lowlands.
HABITAT: Shores, often rocky and calcareous.
GENERAL RANGE: Alaska to Nfld. south to Calif. and N.J.; Greenland; Asia.

Carex vulpinoidea Michx. Fig. p. A30. Map p. B26.
FREQUENCY: Common throughout the Province, except rare in the eastern lowlands.
HABITAT: Wet shores, ditches, clearings and meadows.
GENERAL RANGE: B.C. to Nfld. south to Oreg. and Fla.

Carex wiegandii Mack. Fig. p. A31. Map p. B26.
FREQUENCY: Rare; reported from a few eastern lowland locations and one locality in Madawaska Co.
HABITAT: Sphagnum bogs, boggy thickets and shores.
GENERAL RANGE: Ont. to Nfld. south to Pa. and Mass.
SPECIAL NOTE: For more information see H. Hinds, 1983. The Rare Vascular Plants of New Brunswick. Syllogeus 50.

CLADIUM P. Br. Twig-rush

Cladium mariscoides (Muhl.) Torr. Fig. p. A31. Map p. B26.
FREQUENCY: Very rare; found only in the southwest corner of the Province at Scotch Lake, York Co., Potters, Bonaparte, Eel Bk. and David Watt Pond, Charlotte Co.
HABITAT: Boggy shores.
GENERAL RANGE: Ont. to Nfld. south to Minn. and Fla.
SPECIAL NOTE: For more information see H. Hinds, 1983. The Rare Vascular Plants of New Brunswick. Syllogeus 50.

CYPERUS L. Galingale, Umbrella-Sedge. Souchet

1a Spikelets pinnately disposed along the rachis, but **not** all from the same point; scales 2.3-3 mm long, yellowish, without green midrib; leaves often yellowish green; achenes trigonous, stigmas 3. **C. esculentus**
1b Spikelets spreading radially on a short rachis forming a sub-capitate spike 2

++

2a(1b) Stigmas 2; achenes lenticular. 3

2b(1b) Stigmas 3; achenes trigonous. 4

3a(2a) Scales mostly dark towards base, sometimes throughout; style branches united to half their length, +/- deciduous at maturity. **C. rivularis**

3b(2a) Scales mostly dark towards tip and sides; style branches free nearly to base, +/- persistant and conspicuous. **C. diandrus**

4a(2b) Scales with 7-13 conspicuous nerves. 5

4b(2b) Scales with 3-5 mostly inconspicuous nerves, concentrated near midrib, sides nerveless; achenes often replaced by bulblets. **C. dentatus**

5a(4a) Scales reddish-brown, with narrow aristate tip outwardly curved; stamens 2; aromatic annuals (fragrance of **Melilotus**) with fibrous roots. **C. aristatus**

5b(4a) Scales greenish-gray, erect at the tip, obtuse or mucronate; stamens usually 3; perennials with corm-like thickenings at base of culm. **C. filiculmis**

Cyperus aristatus Rottb. Fig. p. A31. Map p. B26.
COMMON NAME(S): Awned Cyperus. Souchet courbé.
FREQUENCY: Rare; locally common in the Grand Lakes area west to Fredericton and south to Belleisle Bay, Kings Co.
HABITAT: Sandy shores.
GENERAL RANGE: B.C. to N.B. south to Mexico and Fla.
SYNONYMS: **C. inflexus** Muhl.

Cyperus dentatus Torr. Fig. p. A31. Map p. B26.
FREQUENCY: Locally common in the southwest part of the Province.
HABITAT: Sandy and rocky shores.
GENERAL RANGE: Que. to N.S. south to Ind. and N.C.

Cyperus diandrus Torr. Fig. p. A31. Map p. B26.
FREQUENCY: Very rare; reported from the Fredericton area, York Co. and from the NW Miramichi R. near Stuart Bk., Northumberland Co. (UNB).
HABITAT: Muddy or sandy shores.
GENERAL RANGE: Ont. to N.B. south to N. Mex. and S.C.
SPECIAL NOTE: For more information see H. Hinds, 1983. The Rare Vascular Plants of New Brunswick. Syllogeus 50.

Cyperus esculentus L. Fig. p. A31. Map p. B26.
COMMON NAME(S): Yellow Nut-grass or Sedge. Amande de terre.
FREQUENCY: Locally common on the St. John River drainage system and rare on the SW Miramichi R., Northumberland Co.
HABITAT: Sandy shores and occasionally adjacent fields.
GENERAL RANGE: Ont. to N.S. south to Wash., Mexico and Fla.; tropical America; Eurasia.

Cyperus filiculmus Vahl Fig. p. A31.
FREQUENCY: Very rare; known only from Grand Lake, Queens Co. (UNB).
HABITAT: Sandy or gravelly shores.

GENERAL RANGE: Ont. to N.S. south to Tex. and Fla.
SPECIAL NOTE: For more information see H. Hinds, 1983. The Rare Vascular Plants of New Brunswick. Syllogeus 50.

Cyperus rivularis Kunth fig. p. A31.
FREQUENCY: Very rare; known only from the NW Miramichi River near Stuart Bk., Northumberland Co.
HABITAT: Sandy or muddy shores, often tidally influenced.
GENERAL RANGE: Ont. to N.B. south to Calif. and Ga.; S. America.
SPECIAL NOTE: For more information see H. Hinds, 1983. The Rare Vascular Plants of New Brunswick. Syllogeus 50.

DULICHIUM Pers.

Dulichium arundinaceum (L.) Britt. Fig. p. A31. Map p. B26.
COMMON NAME(S): Three-way Sedge
FREQUENCY: Commonly scattered throughout the Province.
HABITAT: Wet shores and similar low areas
GENERAL RANGE: B.C. to Nfld. south to Calif. and Fla.

ELEOCHARIS R. Br. Spike-Rush. Eleocharide

1a Culms 3-angled; spikelet subulate-lanceolate, about as thick as top of culm; when submerged producing numerous elongate capillary sterile floating culms; achenes compressed-trigonous; stigmas 2. **E. robbinsii**
1b Culms terete; spikelets thicker than top of culm. 2
2a(1b) Achene tubercle conical, appearing as a continuation of the body of the achene, not separated by a constriction; summit of leaf sheath without a prominant tooth; stigmas 3; tuber bearing. 3
2b(1b) Achene tubercle distinct from achene and +/- separated from body of achene by a narrow constriction; leaf sheath often with a prominent tooth at apex; stigmas 2 or 3. 4
3a(2a) Culms less than 1 dm tall, densely tufted, mat forming annual; spikelets 2-3 mm long; scales green or pale brown; achenes to 1.5 mm long. **E. parvula**
3b(2a) Culms to 4 dm tall; perennial with creeping rhizomes; spikelets 4-7 mm long; scales tinged with purple or brown; achenes to 2.5 mm long. **E. pauciflora**
4a(2b) Achenes 3-sided, styles 3 cleft. 5
4b(2b) Achenes 2-sided, styles 2-3 cleft. 6
5a(4a) Achenes white or pearly with prominent longitudinal and vertical reticulations; basal scales of spikelet fertile; capillary culms (usually sterile) often forming a turf in or near the water's edge. **E. acicularis**
5b(4a) Achenes colored, not as above; basal scales sterile. 7
6a(4b) Summit of leaf sheath with thin and membranous loose white tip; spikelets to 9 mm long; tufted perennial or with culms scattered along delicate stolons. **E. olivacea**
6b(4b) Summit of leaf sheath with a firm, close orifice. 8

++

7a(5b) Plant tufted, without rhizomes; achenes greenish yellow or olive with lanceolate tubercle several times as long as broad and not over ¼ the width of achene; culms very unequal in length. **E. intermedia**

7b(5b) Plants with rhizomes; achenes orange or golden-brown, bluntly angled and conspicuously honeycombed; tubercle low deltoid, about as high as wide; culms elongate, +/- of equal size. 9

8a(6b) Densely tufted annuals with soft culms; anthers less than 0.7 mm long; basal scales of spikelets sterile. 10

8b(6b) Rhizomatous perennials with stiff culms; anthers longer than 1 mm long. 11

9a(7b) Culms angular-cylindric, 6-8 angled, about 0.3-0.5 mm thick; achene yellow to golden-brown, lustrous, cross ridges +/- apparent. **E. tenuis** var. **borealis**

9b(7b) Culms 4-5 angled, about 0.2-0.3 mm thick; achenes yellow to olive-drab with coarse and deep honeycomb reticulations. **E. tenuis**

10a(8a) Tubercle nearly as broad as achene; scales brown or reddish brown; spikelets oval, +/- rounded at summit. **E. obtusa**

10b(8a) Tubercle less than 2/3 as wide as mature achene; scales purplish brown; spikelets ovate, more pointed at summit and usually with fewer flowers. **E. ovata**

11a(8b) Basal scales of spikelet usually 2 or 3 not encircling spikelet; culm stout, to 4 mm thick; tubercle very slender-conic, only slightly thicker at base than at tip; summit of leaf sheath often black; plants of fresh habitats. **E. smallii**

11b(8b) Basal scale solitary, nearly encircling spikelet; culms slender, rarely over 0.7 mm thick; tubercle much wider at base than at middle; summit of leaf sheath light brown or reddish; plants of both, fresh and saline habitats. **E. erythropoda**

Eleocharis acicularis (L.) R. & S. Fig. p. A31. Map p. B26.
COMMON NAME(S): Needle Spike-rush.
FREQUENCY: Probably common throughout the Province.
HABITAT: Shallow water, wet shores and similar wet areas.
GENERAL RANGE: Alaska to Nfld. south to Calif. and Fla.; Greenland; Iceland; Eurasia.

Eleocharis erythropoda Steud. Fig. p. A31. Map p. B26.
COMMON NAME(S): Bald Spike-rush.
FREQUENCY: Scattered throughout the Province.
HABITAT: Wet shores, stream margins, riverbanks, bogs and brackish marshes.
GENERAL RANGE: Alta. to N.S. south to N. Mex. and Va.; Hawaii; Asia.
SYNONYMS: **E. calva** Torr.; **E. halophila** Fern. & Brack.

Eleocharis intermedia (Muhl.) Schultes Fig. p. A31.
FREQUENCY: Very rare; known from Dawsonville, Restigouche Co., Parent Ferry, York Co., and Pearl Settlement, Victoria Co. (UNB).
HABITAT: Wet, muddy soil.
GENERAL RANGE: Ont. to N.B. south to Minn., Iowa and Md.
SPECIAL NOTE: For more information see H. Hinds, 1983. The Rare Vascular Plants of New Brunswick. Syllogeus 50.

++

Eleocharis obtusa (Willd.) Schultes Fig. p. A32. Map p. B27.
COMMON NAME(S): Blunt Spike-rush.
FREQUENCY: Commonly scattered towards the southern part of the Province and rarer northwards.
HABITAT: Wet shores and muddy pools.
GENERAL RANGE: B.C. to N.S. south to Calif. and Fla.; Hawaii.

Eleocharis olivacea Torr.
FREQUENCY: Very rare; only known from Kendricks Lake, Charlotte Co. (UNB).
HABITAT: Wet peat.
GENERAL RANGE: Ont. to N.B. south to Minn., Ohio and Fla.

Eleocharis ovata (Roth) R.&S. Fig. p. A32. Map p. B27.
FREQUENCY: Common throughout the Province.
HABITAT: Muddy shores and pools.
GENERAL RANGE: Ont. to N.B. south to Minn., Ind. and Conn.; Hawaii; Eurasia.

Eleocharis parvula (R. & S.) Link Fig. p. A32. Map p. B27.
COMMON NAME(S): Low Spike-rush.
FREQUENCY: Uncommonly scattered in coastal areas and occasionally inland.
HABITAT: Brackish and fresh shores.
GENERAL RANGE: B.C. to Nfld. south to Calif. and Fla.; S. America; W.I.; Eurasia; Africa.

Eleocharis pauciflora (Lightf.) Link Fig. p. A32. Map p. B27.
FREQUENCY: Rare; known from only a few stations mostly in the western part of the Province. Unknown from the eastern part.
HABITAT: Calcareous shores, ledges and bogs.
GENERAL RANGE: Yukon to Nfld. south to Calif. and N.J.; Greenland; Iceland; Eurasia.
SYNONYMS: **E. quinquiflora** (Hartm.) Schwarz.
SPECIAL NOTE: For more information see H. Hinds, 1983. The Rare Vascular Plants of New Brunswick. Syllogeus 50.

Eleocharis robbinsii Oakes Fig. p. A32. Map p. B27.
FREQUENCY: Rare; known from only a few stations in the southwest part of the Province.
HABITAT: Shallow water.
GENERAL RANGE: Ont. to N.S. south to Wisc., Ind., N.Y. and Fla.
SPECIAL NOTE: For more information see H. Hinds, 1983. The Rare Vascular Plants of New Brunswick. Syllogeus 50.

Eleocharis smallii Britt. Fig. p. A32. Map p. B27.
COMMON NAME(S): Small's Spike-rush.
FREQUENCY: Common throughout the Province.
HABITAT: Wet open areas and shallow water.
GENERAL RANGE: Alaska to Nfld. south to Oreg. and N. Eng.; Greenland; Eurasia.
SYNONYMS: **E. palustris** (L.) R. & S.

Eleocharis tenuis (Willd.) Schultes Fig. p. A32. Map p. B27.
HABITAT: Wet sands, gravels and peats.
GENERAL RANGE: Aggregate species B.C. to Nfld. south to Mont., Tenn. and S.C.
SUBSPECIFIC TAXA: Two varieties are recognized in our area. These are separated in the main key.

1. Var. **tenuis**
FREQUENCY: Common in the southern part of the Province.
2. Var. **borealis** (Svenson) Gl. Map p. B27.
FREQUENCY: Uncommonly scattered throughout the Province.
SYNONYMS: **E. elliptica** Kunth

ERIOPHORUM L. Cotton-grass. Linaigrette

1a Spikelet single, not surrounded by leafy bracts; culm leaves with bladeless sheaths. 2
1b Spikelets usually more than 1 (sometimes closely congested); inflorescence subtended by 1 or more leafy bracts; culm leaves usually blade bearing. 3
2a(1a) Lowermost scales of spikelet +/- 5 nerved, reddish brown at base, gray-brown above. **E. chamissonis**
2b(1a) Lowermost scales 1-veined, dark gray in the centre and hyaline on the margins. **E. vaginatum**
3a(1b) Involucral bract usually solitary, shorter than inflorescence; cauline leaves channeled, 1-1.5 mm wide. 4
3b(1b) Involucral bracts 2 or more; cauline leaves flat, at least in lower half, to 6 mm wide. 5
4a(3a) Uppermost cauline leaf blade shorter than its sheath; scales drab or blackish. **E. gracile**
4b(3a) Uppermost cauline leaf blade longer than its sheath; scales brown or straw-color. **E. tenellum**
5a(3b) Fertile scales brownish to reddish with inconspicuous midnerve; spikelets with short pedicels, +/- congested and upright at maturity (August); bristles tawny or copper colored. **E. virginicum**
5b(3b) Fertile scales olive-green to blackish with +/- conspicuous midnerve; spikelets with longer pedicels and therefore nodding at maturity (July); bristles usually white. 6
6a(5b) Midvein of fertile scales prominent, +/- scabrous and enlarged at tip; summit of leaf sheath and base of involucre not conspicuously darkened; bristles white to light brown. **E. viridi-carinatum**
6b(5b) Midvein of fertile scales obscure, not extending to the thin apex; summit of leaf sheath and base of involucre darkened; bristles white. **E. angustifolium**

Eriophorum angustifolium Honckeny Fig. p. A32. Map p. B27.
FREQUENCY: Common in the southeast half of the Province; rare elsewhere.
HABITAT: Acid bogs, boggy meadows and shores.
GENERAL RANGE: Alaska to Nfld. south to Oreg. and N. Eng.; Greenland; Iceland; Eurasia.

++

Eriophorum chamissonis C.A. Mey. Fig. p. A32. Map p. B27.
FREQUENCY: Rare; known from a few east coastal locations and one inland station near Arthurette, Victoria Co.
HABITAT: Bogs, often calcareous.
GENERAL RANGE: Alaska to Nfld. south to Oreg., Utah, Colo., and Minn.; Siberia; Kamchatka.
SUBSPECIFIC TAXA: Forma **albidum** (Nyl.) Fern., with whitish instead of reddish-brown bristles, occurs rarely with the species.

Eriophorum gracile Koch Fig. p. A32.
FREQUENCY: Very rare; known only from near Bass R., Kent Co. and from Ste. Marie-Sur-Mer, Gloucester Co. (UNB).
HABITAT: Peat bogs and boggy ditches.
GENERAL RANGE: Alaska to Labr. south to Calif., Colo., Del. and P.E.I.: Eurasia.
SPECIAL NOTE: For more information see H. Hinds, 1983. The Rare Vascular Plants of New Brunswick. Syllogeus 50.

Eriphorum tenellum Nutt. Fig. p. A32. Map p. B27.
FREQUENCY: Locally common in the southeastern part of the Province.
HABITAT: Peat bogs.
GENERAL RANGE: Ont. to Nfld. south to Ill. and N.J.

Eriophorum vaginatum L. Fig. p. A32. Map p. B27.
FREQUENCY: Locally common throughout most of the Province.
HABITAT: Peat bogs and boggy meadows.
GENERAL RANGE: Alaska to Nfld. south to Alta., Sask., Man. and N.J.; Greenland; Eurasia.
SUBSPECIFIC TAXA: Our material is referable to ssp. **spissum** (Fern.) Hulten.

Eriophorum virginicum L. Fig. p. A33. Map p. B28.
FREQUENCY: Probably scattered throughout the Province.
HABITAT: Bogs and similar peaty open areas.
GENERAL RANGE: Ont. to Nfld. south to Minn. and Ga.

Eriophorum viridi-carinatum (Engelm.) Fern. Fig. p. A33. Map p. B28.
FREQUENCY: Uncommonly scattered throughout except rare in the eastern lowlands.
HABITAT: Wet peaty meadows, often in calcareous areas.
GENERAL RANGE: Alaska to Nfld. south to B.C. and N.Y.

RHYNCOSPORA Vahl Beak-Rush

1a Plants with slender rhizomes, thus culms usually arising singly; perianth bristles barbed upwards. **R. fusca**

1b Plants cespitose; perianth bristles barbed downwards. 2

2a(1b) Bristles about 9-12, villous at base; inflorescence white to pale brown on long peduncles; achenes greenish-brown with obscure markings; leaves to 2.5 mm broad. **R. alba**

2b(1b) Bristles 6(9), not villous at base. 3

3a(2b) Spikelets several, dark brown, spreading in fan-like clusters; achenes broadly obovoid 1.1-1.7 mm long (excluding tubercle) with prominent lighter margins; leaves flat, to 3.5 mm broad. **R. capitellata**

3b(2b) Spikelets 1-4 (7), light brown, in narrow, ascending clusters, not fan-shaped; achenes oblong-ellipsoid or narrowly obovate 1.8-2.4 mm long, with less prominent margins; leaves involute-filiform. **R. capillacea**

Rhyncospora alba (L.) Vahl Fig. p. A33. Map p. B28.
COMMON NAME(S): Beak-rush.
FREQUENCY: Commonly scattered in the southeast half of the Province.
HABITAT: Wet peat bogs and similar open areas.
GENERAL RANGE: Alaska to Nfld. south to Calif. and N.C.; W.I.; Eurasia.
SUBSPECIFIC TAXA: Forma **leviseta** Gale, with perianth-bristles smooth instead of barbed, is reported from Miscou Is., Gloucester Co. and Sterling Lake Bog, York Co.

Rhyncospora capillacea Torr. Fig. p. A33.
FREQUENCY: Very rare; reported from near Crocks Pt., York Co. and gorge of the Aroostook R., Victoria Co.
HABITAT: Crevices of wet calcareous ledges, bogs and shores.
GENERAL RANGE: Alta. to Nfld. south to S. Dak. and Va.
SPECIAL NOTE: For more information see H. Hinds, 1983. The Rare Vascular Plants of New Brunswick. Syllogeus 50.

Rhyncospora capitellata (Michx.) Vahl Fig. p. A33. Map p. B28.
FREQUENCY: Uncommon; reported mostly from the eastern lowlands.
HABITAT: Damp shores, rock crevices and ledges.
GENERAL RANGE: Ont. to N.S. south to Wisc., Tex. and Fla.

Rhyncospora fusca (L.) Ait. f. Fig. p. A33.
FREQUENCY: Rare; reported from Fox Lake Creek, Westmorland Co. and a few localities in Charlotte Co.
HABITAT: Springy or boggy ground.
GENERAL RANGE: Sask. and Ont. to N.S. south to Minn., Mich. and Del.

SCIRPUS L. Bulrush. Scirpe

1a Involucral bract 1 or apparently none; culm leafy only towards base. 2

1b Involucral bracts 2-several, leaf-like; culm leafy to above middle of culm. 3

2a(1a) Involucral bract consisting of the slightly modified lowest scale of spikelet. 4

2b(1a) Involucral bract very different from scales in form, size and texture. 5

3a(1b) Spikelets large, 1-4 cm long; scales awned. 6

3b(1b) Spikelets small, 2-8 (18) mm long, usually with long peduncles and pedicels. 7

4a(2a) Culm many ridged, +/- terete, smooth; spikelets to 6 mm long; lower scales awn-tipped; bristles barbless, about twice as long as achenes; basal aphyllopodia enlarged. **S. caespitosus**

4b(2a) Culm trigonous, scabrous on angles at least above. 8

5a(2b) Spikelets lustrous purple-brown to castaneous, alternating on opposite sides of spike; achenes lance-fusiform, long beaked and stipitate; bristles wanting. **S. rufus**

5b(2b) Spikelets solitary or in clusters; achenes short beaked, not stipitate; bristles usually present. 9

6a(3a) Bracts +/- 5; achenes sharply trigonous with slightly concave sides; bristles +/- 6, strong, persistent, distinctly barbed, as long or slightly longer than achene. **S. fluviatilis**

6b(3a) Bracts 2 (3); achenes plano-convex to trigonous with convex sides; bristles 0-6, weak and deciduous, obscurely barbed, scarcely half as long as achene. **S. maritimus**

7a(3b) Spikelets mostly on individual pedicels. 10

7b(3b) Spikelets sessile in dense, globose clusters of from 3-15 or more. 11

8a(4b) Stems from creeping rhizomes; bristles white, flat, smooth, several times the length of achene (resembling a miniature **Eriophorum**); bract trigonous or cylindrical. **S. hudsonianus**

8b(4b) Stems cespitose; bristles terete, ciliate, shorter than to slightly surpassing achene; bract flat, scabrous. **S. clintonii**

9a(5b) Spikelets several, usually distinctly pedicelled (congested in **S. acutus** var. **congestus**); culm stout and naked above basal sheaths, 5-20 mm thick at base, often over 1 m tall. 12

9b(5b) Spikelets 1-few, crowded, +/- sessile, appearing lateral. 13

10a(7a) Bristles slightly longer than achene; scales orange-brown, midrib green with short point; involucre not blackened at base. **S. lineatus**

10b(7a) Bristles many times longer than achene. 14

11a(7b) Lower sheaths reddish; bristles barbed for most of their length; styles 2-cleft; achenes lenticular. **S. microcarpus**

11b(7b) Lower sheaths completely green. 15

12a(9a) Mature spikelets ovate, +/- twice as long as wide, to 10 mm long, many on extended rays; scales +/- shiny orange-brown (reddish-brown when immature), round-ovate, to 3 mm long, with red papillae confined mainly to the vicinity of the greenish midrib, otherwise glabrous; mature achenes about 1.6-2.1 (2.4) mm long (average 1.85) including apiculus, nearly as long as scales and the tip +/- visible when mature. **S. validus**

12b(9a) Mature spikelets ovoid-cylindric, about 2.5 or more times as long as wide, to 20 mm long; scales duller, pale or whitish brown, broadly ovate, to 4 mm long, densely red papillose on upper half and often also puberulent, midrib not prominent; mature achene 2.2-2.7 mm long (average 2.5) including apiculus, much shorter than scales and completely hidden. **S. acutus**

13a(9b) Tufted annuals with soft, terete or obscurely 3-angled dull green culms; plants of tide influenced shores. **S. smithii**

++

13b(9b) Rhizomatous perennials, not tufted. 16

14(10b) Plant 120-150 cm tall; culm 1.8-4 mm wide at summit; base of **involucels** and spikelets pale-brownish; leaves to 1 cm broad. **S. pedicellatus**

14b(10b) Plant 60-110 cm tall; culm 1-1.8 mm at summit; base of **involucels** and spikelets greenish-black; leaves to 5 mm broad. **S. atrocinctus**

15a(11b) Bristles crinkly, many times longer than achene and scales at maturity, smooth; plants tufted, not stoloniferous. **S. cyperinus**

15b(11b) Bristles +/- straight, less than twice as long as achene, inconspicuously barbed on upper half only; plants solitary or a few together from strong scaly stolons. **S. georgianus**

16a(13b) Culms terete; spikelets, ascending; leaves numerous, reduced to capillary strands in aquatic plants or few and rigid or wanting in emersed colonies; achene trigonous, beak +/- as long as body; involucral bract with straight callous tip. **S. subterminalis**

16b(13b) Culms trigonous, especially apparent at upper sheath; spikelets usually more than 1. 17

17a(16b) Spikelets lanceolate to slenderly ovoid; midrib of scales +/- greenish; basal leaf sheaths often strongly fibrillose on the inner side; tip of involucral bract blunt, slightly deflected; achenes trigonous; styles 3-cleft; leaves more than half as tall as culm. **S. torreyi**

17b(16b) Spikelets ovoid to subcylindric; midrib of scales brown; basal leaf sheaths not fibrillose; tip of involucral bract acute to acuminate; achenes plano-convex; styles usually 2-cleft; leaves less than half as tall as culm. **S. americanus**

Scirpus acutus Muhl. Fig. p. A33. Map p. B38.
COMMON NAME(S): Great Bulrush. Scirpe aigu.
FREQUENCY: Common, especially along the coast, but also scattered inland.
HABITAT: Wet shores and meadows.
GENERAL RANGE: Alaska to Nfld. south to Calif. and Ga.; tropical America; Eurasia.
SYNONYMS: **S. lacustris** L. ssp. **glaucus** (Sm.) Hartm.

Scirpus americanus Pers. Fig. p. A33. Map p. B28.
COMMON NAME(S): Three-square. Jonc des chaisiers.
FREQUENCY: Common on the east coast of the Province. Uncommon elsewhere.
HABITAT: Brackish or fresh shores and marshes.
GENERAL RANGE: Alaska to Nfld. south to Calif. and Fla.; S. America; Eurasia.

Scirpus atrocinctus Fern. Fig. p. A33. Map p. B28.
COMMON NAME(S): Black-girded Wool-grass.
FREQUENCY: Common throughout the Province.
HABITAT: Wet meadows, swamps and similar open areas.
GENERAL RANGE: B.C. to Labr. south to Wash. and N.J.

Scirpus caespitosus L. Fig. p. A33. Map p. B28.
COMMON NAME(S): Tufted Club-rush. Scirpe gazonnant.
SUBSPECIFIC TAXA: Two varieties may be recognized in our range.

++

1a Culms soft and flexuous; basal sheaths membranous, black to dark gray. var. **delicatulus**

1b Culms more wiry, erect and +/- straight; basal sheaths straw-colored or brownish-gray. var. **callosus**

1. Var. **delicatulus** Fern.
FREQUENCY: Very rare; known only from near the mouth of Little Salmon R., St. John Co. Reported from n. N.B. by Roland and Smith, 1969.
HABITAT: Wet calcareous ledges.
GENERAL RANGE: Gaspé to Nfld. south to Me. and N.S.
2. Var. **callosus** Bigel.
FREQUENCY: Common along the coast but rare inland.
HABITAT: Peat bogs.
GENERAL RANGE: Alaska to Labr. south to Utah, N.C. and N.Eng.

Scirpus clintonii Gray Fig. p. A33. Map p. B28.
COMMON NAME(S): Clinton's Club-rush.
FREQUENCY: Uncommonly scattered throughout the Province.
HABITAT: Calcareous ledges and shores.
GENERAL RANGE: Alta. to N.B. south to N.Y. and Me.

Scirpus cyperinus (L.) Kunth Fig. p. A34. Map p. B28.
COMMON NAME(S): Common Wool-grass. Scirpe souchet.
FREQUENCY: Common mostly in the southeast half of the Province.
HABITAT: Swampy ground and wet meadows.
GENERAL RANGE: Ont. to Nfld. south to Okla. and N.C.
SYNONYMS: **S. cyperinus** var. **pelius** Fern.

Scirpus fluviatilis (Torr.) Gray Fig. p. A34. Map B28.
COMMON NAME(S): River-Bulrush.
FREQUENCY: Rare; reported from a few locations on the lower St. John and Kennebecasis Rivers and College Bridge Lake, Westmorland Co.
HABITAT: Shallow water.
GENERAL RANGE: Alta. to N.B. south to Calif. and Va.; Asia.
SPECIAL NOTE: For more information see H. Hinds, 1983. The Rare Vascular Plants of New Brunswick. Syllogeus 50.

Scirpus georgianus Harper Fig. p. A33. Map p. B28.
FREQUENCY: Probably common throughout the Province.
HABITAT: Swampy ground and wet meadows.
GENERAL RANGE: Alta. to Nfld. south to Oreg. and Ga.
SYNONYMS: **S. atrovirens** Willd. var. **georgianus** (Harper) Fern.

Scirpus hudsonianus (Michx.) Fern. Fig. p. A34. Map p. B29.
FREQUENCY: Uncommonly scattered throughout the Province.
HABITAT: Marly or springy shores, bogs and boggy meadows.
GENERAL RANGE: Alaska to Nfld. south to B.C. and N. Eng.; Eurasia.
SYNONYMS: **Eriophorum alpinum** L.

++

Scirpus lineatus Michx. Fig. p. A34.
FREQUENCY: Very rare; known only from near Havelock, Westmorland Co. (UNB).
HABITAT: Calcareous roadside ditches and banks.
GENERAL RANGE: Ont. to N.B.; Oreg., Iowa, Me. south to Tex. and Va.
SYNONYMS: **S. pendulus** Muhl.

Scirpus maritimus L. Fig. p. A34. Map p. B29.
COMMON NAME(S): Salt-marsh Bulrush.
FREQUENCY: Common along the coast.
HABITAT: Brackish marshes and shores.
GENERAL RANGE: Alaska to Côte Nord south to Calif., Tex. and Va.; Eurasia.
SYNONYMS: **S. paludosus** Mack.; **S. maritimus** var. **paludosus** (Nels.) Kuk.
SUBSPECIFIC TAXA: Material from Bathurst Bay, Gloucester Co., Katy's Cove, Charlotte Co., and Miramichi Bay, Northumberland Co. approach var. **tuberosus** (Desf.) R. & S., with spikelets less than 6 mm wide, lanceolate, rather than more than 6 mm wide, ovate as in var. **maritimus**.

Scirpus microcarpus Presl Fig. p. A34. Map p. B29.
FREQUENCY: Probably common throughout the Province.
HABITAT: Damp sandy shores, swamp margins, river banks and marshes.
GENERAL RANGE: Alaska to Labr. south to Calif. and W. Va.; Asia.
SYNONYMS: **S. rubrotinctus** Fern.

Scirpus pedicellatus Fern. Fig. p. A34. Map p. B29.
FREQUENCY: Locally common only in the lower St. John R. drainage system.
HABITAT: Wet meadows and shores.
GENERAL RANGE: Ont. to N.S. south to Minn., Iowa and N.J.

Scirpus rufus (Huds.) Schrad. Fig. p. A34. Map p. B29.
COMMON NAME(S): Red Club-rush.
FREQUENCY: Rare; mostly reported from the Chaleur Bay area and Grand Manan Island.
HABITAT: Brackish marshes.
GENERAL RANGE: Alaska to e. Que., Nfld. and Maritime Provinces; Eurasia.
SUBSPECIFIC TAXA: Our material is referable to var. **neogaeus** Fern.

Scirpus smithii Gray Fig. p. A34.
FREQUENCY: Very rare; only known from near the mouth of Stuart Bk. on the NW Miramichi R., Northumberland Co. (UNB).
HABITAT: Muddy shore.
GENERAL RANGE: Minn. to N.B. south to Ill. and Va.; Asia.
SPECIAL NOTE: For more information see H. Hinds, 1983. The Rare Vascular Plants of New Brunswick. Syllogeus No. 50.

Scirpus subterminalis Torr. Fig. p. A34. Map p. B29.
COMMON NAME(S): Swaying-rush.
FREQUENCY: Scattered in the southern half of the Province.
HABITAT: Shallow quiet waters.
GENERAL RANGE: Alaska and B.C.; Ont. to Nfld. south to Oreg., Mich. and Ga.

Scirpus torreyi Olney Map p. B29.
COMMON NAME(S): Torrey's Bulrush.
FREQUENCY: Uncommon in the southern half of the Province.
SYNONYMS: **S. subterminalis** Torr. var. **cylindricus** (Torr.) Koyama

Scirpus validus Vahl Fig. p. A34. Map p. B29.
FREQUENCY: Common throughout the Province.
HABITAT: Shores, marshes, shallow water and similar open areas.
GENERAL RANGE: B.C. to Nfld. south throughout N. America; S. America; Hawaii; W.I.
SYNONYMS: **S. lacustris** L. ssp. **validus** (Vahl) Koyama; **S. validus** var. **crebra** Fern.

ARACEAE (Arum Family)

1a Leaves compound; summit of spadix flowerless. **Arisaema**
1b Leaves simple; spadix flowered to summit. 2
2a(1b) Leaves iris-like, narrow and long lanceolate; spadix appearing lateral from base of leaf-like spathe. **Acorus americanus**
2b(1b) Leaves expanded; spathe not leaf-like. 3
3a(2b) Spathe open at maturity, ovate, with white inner surface; leaves about 1 dm wide; plants semi-aquatic with long rhizomes rooting at the nodes. **Calla palustris**
3b(2b) Spathe nearly closed at maturity, with mottled purple-brown inner surface; leaves short petioled, about 6 dm long, thick, fleshy and with a skunk-like odor. **Symplocarpus foetidus**

ACORUS L. Sweetflag, Calamus-root

Acorus americanus Fig. p. A35.
COMMON NAME(S): Flagroot. Belle-Angélique or Redote.
FREQUENCY: Common in the southern half of the Province and rare northwards.
HABITAT: Marshes and shallow water.
GENERAL RANGE: B.C. to N.S. south to Wash. and Fla.
SYNONYMS: **A. calamus** L.

ARISAEMA Mart. Jack-in-the-pulpit

Arisaema triphyllum (L.) Schott Fig. p. A34. Map p. B29.
COMMON NAME(S): Jack-in-the-pulpit. Petit Prêcheur.
SPECIAL NOTE: Two well marked subspecies occur in our range.

1a Spathe white-ridged lengthwise (difficult to see in pressed material), the tube gradually curving into the narrowly ovate-acuminate hood; leaflets green beneath, narrowly ovate; plants of mucky acid soils, cedar swamps; blooming June-July ssp. **stewardsonii**

+++

1b Spathe shallowly ridged, not white striped lengthwise, usually dark purple, the summit of the tube with a distinct flange; leaflets glaucous beneath, broadly ovate; tip of spathe broadly ovate, abruptly acute, caudate; plants of rich, circumneutral soil of hardwoods and floodplains; blooming May-June. ssp. **triphyllum**

1. Ssp. **triphyllum**
FREQUENCY: Probably common throughout the Province.
GENERAL RANGE: Man. to N.S. south to Ky. and Ga.
SYNONYMS: **A. atrorubens** (Ait.) Blume
2. Ssp. **stewardsonii** (Britt.) Huttleston
FREQUENCY: Common in the southern part of the Province.
COMMON NAME(S): Stewardson's Jack-in-the-pulpit.
GENERAL RANGE: Minn. to Que. and N.S. south to Pa. and N.C.
SYNONYMS: **A. stewardsonii** Britt.

CALLA L. Water-Arum

Calla palustris L. fig. p. A35. Map p. B29.
COMMON NAME(S): Wild Calla. Calla des marais, Choucalle des marais.
FREQUENCY: Scattered throughout the Province except rarer towards the northern part.
GENERAL RANGE: Alaska to Nfld. south to Colo. and Fla.; Eurasia.

SYMPLOCARPUS Salisb. Skunk-cabbage. Symplocarpe

Symplocarpus foetidus (L.) Nutt. Fig. p. A35. Map p. B30.
COMMON NAME(S): Eastern Skunk-cabbage. Tabac du Diable or Chou puant.
FREQUENCY: Rare; scattered in a few localities in the southern part of the Province such as Lake Utopia, Charlotte Co.
HABITAT: Alder thickets and swampy woods.
GENERAL RANGE: Man. to N.S. south to Iowa and Ga.; Asia.
SPECIAL NOTE: For more information see H. Hinds, 1983. The Rare Vascular Plants of New Brunswick. Syllogeus No. 50.

LEMNACEAE (Duckweed Family)

1a Thalli without roots, minute (to 1.5 mm). **Wolffia arhiza**
1b Thalli with roots, usually larger than 1.5 mm long. 2
2a(1b) Thalli with 1 root, green on both surfaces; nerves 5 or less. **Lemna**
2b(1b) Thalli with 2 or more roots, dark green above, purple below; nerves often more than 5; larger. **Spirodela polyrhiza**

++

LEMNA L. Duckweed, Duck's-meat. Lenticule

1a Thalli to 5 mm long (usually less), rounded, soon becoming free floating. **L. minor**

1b Thalli to 1 cm long including narrow attachment stalks, oblong to broadly lanceolate and usually submerged. **L. trisulca**

Lemna minor L. Fig. p. A35. Map p. B30.
COMMON NAME(S): Duckweed. Lentille d'eau. Merde de grenouille.
FREQUENCY: Most specimens from the south of the Province but probably more widely distributed and undercollected.
HABITAT: Still waters.
GENERAL RANGE: Alaska to N.S. south to Calif. and Fla.; Uruguay; Eurasia.

Lemna trisulca L. Fig p. A35. Map p. B30.
COMMON NAME(S): Star-Duckweed.
FREQUENCY: Uncommon; collected only in the southern part of the Province.
HABITAT: Quiet waters.
GENERAL RANGE: Alaska to N.S. south to Calif. and Fla.; Uruguay; Eurasia.

SPIRODELA Schleid.

Spirodela polyrhiza (L.) Schleid. Fig. p. A35. Map p. B30.
COMMON NAME(S): Water-flaxseed. Lentille d'eau.
FREQUENCY: Uncommonly scattered mostly in the southern part of the Province.
HABITAT: Quiet waters and borders of streams.
GENERAL RANGE: B.C. to N.S. south to tropical America; Eurasia.
SPECIAL NOTE: For more information see H. Hinds, 1983. The Rare Vascular Plants of New Brunswick. Syllogeus 50.

WOLFFIA Horkel Water-meal

Wolffia arhiza Wimm.
SPECIAL NOTE: Material in aquaria at UNB supposedly collected from pond near Sheffield, Sunbury Co. No field specimens seen.
GENERAL RANGE: Minn. to N.B. south to Tex., Fla. and tropical America.

XYRIDACEAE (Yellow-eyed Grass Family)

XYRIS L. Yellow-eyed Grass

Xyris montana Ries. Fig. p. A35. Map p. B30.
FREQUENCY: Uncommon in the southern part of the Province and along the eastern shore.

HABITAT: Peaty open areas and margins of bog ponds.
GENERAL RANGE: Mich. to Ont. and Nfld. south to Pa. and N.J.

(**X. caroliniana** Walt.)
SPECIAL NOTE: Material from Charlotte Co. collected by J. Vroom, 1881, and revised by Weatherby (NBM) to this species, has been revised to **X. montana** by this author.

ERIOCAULACEAE (Pipewort Family)

ERIOCAULON L. Pipewort

1a Mature heads +/- depressed hemispheric with pale-hyaline involucre usually conspicuous; bracts and perianth glabrous or with few white swollen hairs; scapes +/- 4-angled to 1 dm tall; plants of estuarine waters and shores. **E. parkeri**

1b Mature heads +/- depressed-spherical with drab-blackish involucre +/- concealed; bracts and perianth thickly beset with white swollen hairs; scapes with up to 7 ribs, commonly over 2 dm tall; plants of fresh waters and shores. **E. septangulare**

Eriocaulon parkeri Robins. Fig. p. A35.
COMMON NAME(S): Parker's Pipewort.
FREQUENCY: Very rare; reported from the estuary of the Miramichi R., Northumberland Co.
HABITAT: Brackish muddy shores.
GENERAL RANGE: Ottawa and St. Lawrence R. estuaries and Miramichi R. estuary south to Me. and Va.
SYNONYMS: **E. septangulare** var. **parkeri** (Rob.) Boivin & Cayouette.
SPECIAL NOTE: For more information see H. Hinds, 1983. The Rare Vascular Plants of New Brunswick. Syllogeus 50.

Eriocaulon septangulare With. Fig. p. A35. Map p. B30.
COMMON NAME(S): White-buttons, Duckgrass.
FREQUENCY: Common throughout most of the Province.
HABITAT: Shallow water of mostly ponds and lakes.
GENERAL RANGE: Ont. to Nfld. south to Ind. and Del.; Ireland; Scotland.

PONTEDERIACEAE (Pickerelweed Family)

1a Mature leaves cordate; inflorescence spicate; flowers showy, blue with yellow centre; stem emergent. **Pontederia cordata**

1b Leaves linear (in our species); inflorescence of inconspicuous single, yellowish flowers from spathe; stem submerged. **Heteranthera dubia**

HETERANTHERA R. & P. Mud-plantain

Heteranthera dubia (Jacq.) MacM. Fig. p. A35. Map B30.
COMMON NAME(S): Water-stargrass.
FREQUENCY: Rare; reported from near the mouth of the Nashwaak R. and Pickards Padou, York Co. Easily overlooked.
HABITAT: Shallow quiet water.
GENERAL RANGE: Wash. and Ont. to N.B. south to Calif. and Fla.; tropical America.
SPECIAL NOTE: For more information see H. Hinds, 1983. The Rare Vascular Plants of New Brunswick. Syllogeus 50.

PONTEDERIA L.

Pontederia cordata L. Fig. p. A35. Map p. B30.
COMMON NAME(S): Pickerelweed.
FREQUENCY: Common in the southwest quarter of the Province.
HABITAT: Shallow water and muddy shores.
GENERAL RANGE: Ont. to N.S. south to Tex. and Fla.; tropical America.

JUNCACEAE (Rush Family)

1a Inflorescence umbellate to spicate; capsule unilocular with 3 large seeds; plants often pilose. **Luzula**
1b Inflorescence cymose to capitate; capsule +/- trilocular with many minute seeds; plants glabrous. **Juncus**

JUNCUS L. Rush, Bog-Rush. Jonc.

1a Stem leaves reduced to basal sheaths; inflorescence lateral, appearing to burst from the side of the stem. 2
1b At least some of the stem leaves bearing blades; inflorescence terminal with no involucral bract appearing to be an elongation of the stem beyond the inflorescence. 3
2a(1a) Stems densely tufted, not from creeping rhizomes; stamens 3. **J. effusus**
2b(1a) Stems arising from creeping rhizomes; stamens 6. 4
3a(1b) Leaves septate-nodulose. 5
3b(1b) Leaves not septate-nodulose. 6
4a(2b) Capsule tapering to a point; sepals green or pale brown when mature; anthers ½ as long as filaments or shorter; rhizome less than 2 mm in diameter; involucral bract usually more than ½ as long as stem below inflorescence. **J. filiformis**
4b(2b) Capsule blunt; sepals deep brown to purple-brown; anthers as long as filaments or longer; involucral bract usually less than ½ as long as stem below inflorescence. **J. balticus**

5a(3a) Flowers occurring singly or in 2's or 3's, often replaced by fascicles of reduced leaves. 7

5b(3a) Flowers in clusters of 2 or more. 8

6a(3b) Leaf blades flat. 9

6b(3b) Leaf blades terete, channeled on upper side. 10

7a(5a) Stems erect or ascending; flowers mostly sessile along branches of an open branched inflorescence; anthers much longer than filaments. **J. pelocarpus**

7b(5a) Stems repent or floating; flowers on axillary or terminal filiform peduncles; anthers +/- as long as filaments. **J. subtilis**

8a(5b) Flowers in hemispherical or narrower heads. 11

8b(5b) Flowers in dense globose heads; sepals and petals lance-subulate, subequal, reddish-brown, to 4 mm long; upper leaves surpassing inflorescence; leaf sheaths with yellowish membranous auricles; stems to +/- 6 dm tall, mostly shorter. **J. nodosus**

9a(6a) Inflorescence 1/3 to 2/3 entire height of plant; leaf sheaths gradually tapering to summit; sepals acuminate, spreading, longer than capsules, green (brown with age), hyaline margined; tufted annuals with fibrous roots. **J. bufonius**

9b(6a) Inflorescence much less than 1/3 height of plant. 12

10a(6b) Inflorescence simple, bearing 1 or 2 heads with up to 4 flowers. 13

10b(6b) Inflorescence with more than 4 flowers. 14

11a(8a) Seeds with tails at both ends. 15

11b(8a) Seeds tailless, blunt or with dark points at both ends. 16

12a(9b) Leaf sheaths extending +/- halfway up the stem; petals and sepals obtuse; rhizome horizontal; maritime plants. **J. gerardi**

12b(9b) Leaf sheaths confined to base or lower third of stem; sepals and petals acute; rhizome short and erect. 17

13a(10a) Inflorescence many times overtopped by upper involucral bracts; capsules long beaked, less than 4 mm long, dark chestnut-colored; seeds few, less than 1.5 mm long, short tailed; stems densely tufted; plants of subalpine summits. **J. trifidus**

13b(10a) Involucral leaf rudimentary or up to 3 cm long; capsules acute to acuminate, short beaked, to 7 mm long, pale brown; seeds numerous, to 2.5 mm long including tails; stems solitary or few together from slender rhizomes; plants of bogs and similar wet places. **J. stygius**

14a(10b) Involucral bract +/- flexuous, up to 2.5 dm long; capsules chestnut-colored; seeds reddish with minute white tails. **J. greenei**

14b(10b) Involucral bract +/- stiffly erect, 1-6 cm long; capsules greenish to straw-colored; seeds pale brown with tails about ½ as long as body. **J. vaseyi**

15a(11a) Most heads with many flowers each, subglobose or hemispherical; mature capsules equal to or longer than perianth; plants usually over 30 cm tall; seeds 1.3-1.8 mm long. **J. canadensis**

15b(11a) Most heads with 2-5(7) flowers each, narrower than hemispherical; mature capsules greatly exceeding perianth; seeds 0.7-1 mm long; plants usually less than 30 cm tall. 18

16a(11b) Plants stout, 3-10 dm tall; solitary cauline leaf overtopping inflorescence; plants often in shallow water and with dense capillary leaves arising from rhizome. **J. militaris**

16b(11b) Plants more slender, 2-6 dm tall (rarely over 5 dm); cauline leaves not overtopping inflorescence; plants often of shores, but not normally in water; capillary leaves lacking. 19

17a(12b) Auricles and margins of leaf sheath white-hyaline and fragile; auricles prolonged (0.5)1-3(5) mm long; leaf sheaths often some shade of red; perianth 3-4.5 mm long. **J. tenuis**

17b(12b) Auricles and margins of leaf sheaths firm, yellowish or amber coloured; auricles gradually rounded, the free portion up to 0.5 mm long; basal leaf sheaths drab-yellowish brown, never reddish, nor white-hyaline; inflorescence usually more congested; perianth 4-6 mm long. **J. tenuis** var. **uniflorus**

18a(15b) Inflorescence usually divaricate or widely spreading; perianth (1.5) 2-2.5 mm long; sepals +/- blunt with hyaline rounded summit and hyaline margins about ½ the width of the darker central portion; seeds with white tails about 1/10 as long as body. **J. brachycephalus**

18b(15b) Inflorescence usually narrowly erect or ascending; perianth 2.5-3.5 mm long, sepals sharp-pointed with narrower hyaline margins (about ½ the width of the central portion); seeds with white tails about ½ as long as body. **J. brevicaudatus**

19a(16b) Perianth equal to or longer than capsules; petals 1.5-2.5 mm long, blunt (often apiculate), slightly shorter than sepals; inflorescence at least twice as long as wide, branches ascending; heads relatively few, 1-10 flowered; anthers (0.3)0.4-(0.5) mm long. **J. alpinus**

19b(16b) Perianth shorter than capsules; petals 2-3 mm long, short pointed, equal to or slightly longer than sepals; inflorescence less than twice as long as wide, branches usually more spreading; heads ample, usually many flowered; anthers (0.5)-0.6-0.7-(0.9). **J. articulatus**

Juncus alpinus Vill. Fig. p. A36. Map p. B30.

FREQUENCY: Uncommonly scattered throughout the Province.

HABITAT: Calcareous shores and springy areas.

GENERAL RANGE: Aggregate species from Alaska to Nfld. south to Wash. and Vt.; Greenland; Iceland; Eurasia.

SUBSPECIFIC TAXA: Our material is mostly var. **rariflorus** Hartm., with usually smaller, more elongate heads and one or more flowers long pedicelled above the others, while var. **alpinus** has more rounded heads with sessile flowers.

SPECIAL NOTE: **Juncus alpinus** and **J. articulatus** appear to intergrade where they occur together. Their hybrid, **J. X alpiniformis** Fern., has been reported from St. Andrews, Charlotte Co. Other apparent hybrids of **J. alpinus** include **J. X nodosiformis** Fern. with **J. nodosus**, and **J. X gracilescens** Herm. with **J. brevicaudatus** reported from Petit Rocher, Gloucester Co., the type locality.

SYNONYMS: **J. alpinoarticulatus** Chaix.

++

Juncus articulatus L. Fig. p. A36. Map p. B30.
FREQUENCY: Locally common and scattered throughout except rarer in the northern part of the Province.
HABITAT: Sandy or muddy shores, ditches and similar wet areas.
GENERAL RANGE: Alaska to Nfld. south to Calif. and N.C.; Iceland; Eurasia.

Juncus balticus Willd. Fig. p. A36. Map p. B30.
FREQUENCY: Common especially on or near the coast.
HABITAT: Meadows, ditches, and brackish marshes.
GENERAL RANGE: Alaska to Nfld. south to Calif. and Pa.; Iceland; Eurasia.
SUBSPECIFIC TAXA: Our material is referable to var. **littoralis** Engelm.

Juncus brachycephalus (Engelm.) Fern. Fig. p. A36. Map p. B30.
FREQUENCY: Rare; known only from a few localities which have escaped drowning by hydroelectrical developments along the St. John River in Carleton, Victoria and Madawaska Counties.
HABITAT: Calcareous shores.
GENERAL RANGE: Ont. to N.S. south to Ill. and N.J.
SPECIAL NOTE: For more information see H. Hinds, 1983. The Rare Vascular Plants of New Brunswick. Syllogeus 50.

Juncus brevicaudatus (Engelm.) Fern. Fig. p. A36. Map p. B30.
FREQUENCY: Common throughout the Province.
HABITAT: Wet open areas.
GENERAL RANGE: Alta. to Nfld. south to Minn. and N.C.

Juncus bufonius L. Fig. p. A36. Map p. B31.
COMMON NAME(S): Toad Rush. Jonc des crapauds.
FREQUENCY: Scattered throughout most of the Province. Often locally common.
HABITAT: Lagoon beaches, roadside ditches, wet cultivated fields and shores.
GENERAL RANGE: Alaska to Nfld. south to Calif. and Fla.; Greenland; Iceland; Eurasia.
SUBSPECIFIC TAXA: Var. **halophilus** Buch. & Fern., with uppermost flowers mostly in pairs or in threes and petals +/- obtuse instead of flowers solitary and petals acuminate as in the species, occurs rarely in brackish marshes and sands.

Juncus canadensis Gay Fig. p. A36. Map p. B31.
FREQUENCY: Common towards the southern part of the Province, rarer northwards.
HABITAT: Shores, ditches and similar wet areas.
GENERAL RANGE: Minn. and Ont. to Nfld. south to La. and Ga.

Juncus effusus L. Fig. p. A36. Map p. B31.
COMMON NAME(S): Soft Rush, Taper Rush. Jonc épars.
HABITAT: Peaty open areas, ditches and pond margins.
SUBSPECIFIC TAXA: Our material consists of 7 +/- intergrading varieties.

1a Inflorescence crowded, less than 2 cm in diameter. 2
1b Inflorescence diffuse, usually more than 2 cm in diameter. 3

++

2a(1a) Inflorescence +/- spherical; culm fluted or coarsely ribbed below inflorescence; involucral bract conspicuously dilated at base. var. **conglomeratus**

2b(1a) Inflorescence less compact, usually irregularly lobed; culm finely, many striate below inflorescence; involucral bract scarcely dilated at base. var. **compactus**

3a(1b) Culms coarsely fluted below inflorescence with 12-30 conspicuous ridges. 4

3b(1b) Culms +/- smooth or with 30-70 weak ridges below inflorescence. 5

4a(3a) Perianth segments and capsule subequal, not longer than 3 mm. var. **costulatus**

4b(3a) Outer perianth segments (sepals) 3 mm or longer, overtopping petals and capsule. var. **pylaei**

5a(3b) Sepals usually less than 2.5 mm long, +/- spreading at base. 6

5b(3b) Sepals usually longer than 2.5 mm long; sepals and subequal petals with appressed tips, as long as or slightly shorter than capsule. var. **solutus**

6a(4a) Culm 3 mm or less thick at summit of 3-12 mm long upper sheath; capsule rarely notched at tip. var. **decipiens**

6b(4a) Culm usually more than 3 mm thick at summit of 6-17 cm long upper sheath; branches of inflorescence to over 5 cm long; capsule notched at tip. var. **effusus**

1. Var. **compactus** Lej. & Court.
FREQUENCY: Uncommon; most frequently found on the east coast.
GENERAL RANGE: Introd. in B.C.; Ont. to Nfld. south to Mich. and N.Y.
2. Var. **conglomeratus** (L.) Engelm.
FREQUENCY: Very rare; known only from Cross Creek, York Co.
GENERAL RANGE: Nfld., N.B. and N.S.; Europe.
3. Var. **costulatus** Fern.
FREQUENCY: Uncommon.
GENERAL RANGE: Que. to N.S. south to N. Eng., N.Y. and S.C.
4. Var. **decipiens** Buch.
FREQUENCY: Rare.
GENERAL RANGE: Ont. to Nfld. south to Mich. and N. Eng.
5. Var. **effusus**
FREQUENCY: Uncommon; known from a few locations in the southeast half of the Province.
GENERAL RANGE: Nfld., P.E.I. and N.B.; Eurasia.
6. Var. **pylaei** (Laharpe) Fern.
FREQUENCY: Common throughout the Province.
GENERAL RANGE: Ont. to Nfld. south to Minn. and W. Va.
7. Var. **solutus** Fern. & Wieg.
FREQUENCY: Common towards the south of the Province; rarer northwards.
GENERAL RANGE: Ont. to Nfld. south to Tex. and Fla.

Juncus filiformis L. Fig. p. A36. Map p. B31.
FREQUENCY: Common throughout the Province.
HABITAT: Shores, wet ditches and edges of marshes.
GENERAL RANGE: Alaska to Nfld. south to Oreg. and W. Va.; Greenland; Iceland; Eurasia.

++

Juncus gerardii Loisel. Fig. p. A36. Map p. B31.
COMMON NAME(S): Black Grass.
FREQUENCY: Commonly scattered along the coast.
HABITAT: Brackish or saline sandy areas and marshes.
GENERAL RANGE: B.C. to Nfld. south along the coast to Va. and Fla.; Greenland; Eurasia.

Juncus greenei Oakes & Tuckerm. Fig. p. A36.
FREQUENCY: Very rare; reported from Bass River, Kent Co. and Cape Jourimain, Westmorland Co.
HABITAT: Wet meadows.
GENERAL RANGE: Ont. to N.S. south to Minn. and N.J.
SPECIAL NOTE: For more information see H. Hinds, 1983. The Rare Vascular Plants of New Brunswick. Syllogeus 50.

Juncus militaris Bigel. Fig. p. A37. Map p. B31.
FREQUENCY: Uncommon; known only from the southern part of the Province.
HABITAT: Shallow water or margins of ponds and streams.
GENERAL RANGE: Ont. to Nfld. south to Mich. and Md.

Juncus nodosus L. Fig. p. A37. Map p. B31.
FREQUENCY: Uncommonly scattered throughout the Province; rarer towards the south.
GENERAL RANGE: Alaska to Nfld. south to Calif. and Va.

Juncus pelocarpus E. Meyer Fig. p. A37. Map p. B31.
FREQUENCY: Mostly common, but rare in the northwest half of the Province.
HABITAT: Bog margins; and muddy or peaty shores.
GENERAL RANGE: Ont. to Nfld. south to Minn. and Del.

Juncus stygius L. Fig. p. A37. Map p. B31.
FREQUENCY: Rare; known only from a few localities in the southeast of the Province.
HABITAT: Boggy open ground.
GENERAL RANGE: Alaska to Nfld. south to B.C. and Maine; Eurasia.
SUBSPECIFIC TAXA: Our material is all ssp. **americanus** (Buch.) Hulten.
SPECIAL NOTE: For more information see H. Hinds, 1983. The Rare Vascular Plants of New Brunswick. Syllogeus 50.

Juncus subtilis E. Meyer Fig. p. A37.
FREQUENCY: Very rare; reported from a locality 30 miles north of Saint John on St. John R. and at Gaugus L., Northumberland Co.
HABITAT: Shallow water and wet shores.
GENERAL RANGE: Ont. to Nfld. south to N.B. and Maine; Greenland.
SPECIAL NOTE: For more information see H. Hinds, 1983. The Rare Vascular Plants of New Brunswick. Syllogeus 50.

Juncus tenuis Willd. Fig. p. A37. Map p. B31.
COMMON NAME(S): Path Rush.
FREQUENCY: Common throughout the Province.

++

HABITAT: Open often disturbed and compacted soils.
GENERAL RANGE: Alaska to Nfld. south to Calif. and Fla.; S. America; Europe; Africa; Australia; New Zealand; Japan.
SUBSPECIFIC TAXA: Var. **uniflorus** (Farw.) (**J. dudleyi** Wieg.) is separated from var. **tenuis** in the main key. It is commonly found on less disturbed ground mostly in the northwest half of the Province.

Juncus torreyi Coville
SPECIAL NOTE: One large colony recently identified, is reported from a ditch along Trans-Canada Highway north of Shogomoc R. bridge. Not included in main key. (UNB)

Juncus trifidus L. Fig. p. A37.
FREQUENCY: Very rare; known only from the summit of Big Bald Mtn., Northumberland Co.
HABITAT: Rock crevices and peaty heathland.
GENERAL RANGE: Hudson Bay, Baffin Is. and n. Nfld. south to n. N.Y. and n. N.Eng.; Greenland; Iceland; Eurasia.
SPECIAL NOTE: For more information see H. Hinds, 1983. The Rare Vascular Plants of New Brunswick. Syllogeus 50.

Juncus vaseyi Engelm. Fig. p. A37. Map p. B31.
FREQUENCY: Uncommonly scattered along the eastern part of the Province.
HABITAT: Rock crevices and wet sandy shores and dune hollows.
GENERAL RANGE: B.C. to Labr. south to Idaho and Maine.
SYNONYMS: **J. greenei** var. **vaseyi** (Engelm.) Boivin.

LUZULA DC. Woodrush

1a Flowers mostly solitary (1-3) on drooping or arcuate pedicels; mature capsules longer than perianth; rhizomes well developed 2
1b Flowers in compact spikes, usually on erect peduncles to 6 cm; mature capsules equal to or shorter than perianth; rhizomes lacking. 3
2a(1a) Inflorescence simple, usually unbranched from base; mature perianth 2.5-4 mm long; seed conspicuously appendaged. **L. acuminata**
2b(1a) Inflorescence with long peduncles branched near tip; mature perianth 2-2.5 mm long; seeds without appendage; capsules dark purple-brown at maturity. **L. parviflora**
3a(1b) Perianth segments unequal, usually less than 2.5 mm long; style shorter than ovary; seed appendage (caruncle) less than ¼ as 0.8-1.1 mm long seed. **L. pallescens**
3b(1b) Perianth segments +/- equal; style as long as or longer than ovary; caruncle up to ¼ as long as 1.1-1.3 mm long seed. **L. multiflora**

Luzula acuminata Raf. Fig. p. A37. Map p. B32.
FREQUENCY: Probably common throughout the Province.
HABITAT: Mixed or hardwood areas, rocky slopes, banks and clearings.
GENERAL RANGE: Alta. to Nfld. south to S. Dak. and Ga.

Luzula multiflora (Retz.) Lejeune Fig. p. A37. Map p. B32.
FREQUENCY: Common throughout the Province.
HABITAT: Open woods, fields, banks and meadows.
GENERAL RANGE: Alaska to Nfld. south to Calif. and Va.; Greenland; Iceland; Eurasia.
SUBSPECIFIC TAXA: Var. **acadiensis** Fern. differs from var. **multiflora** in having perianth longer than capsule and most or all spikes sessile and pale brown instead of perianth usually shorter than capsule and most spikes reddish brown on ascending rays. It is rare in the south of the Province.
SYNONYMS: **L. campestris** L.

Luzula pallescens (Wahl.) Bess. Fig. p. A37.
FREQUENCY: Very rare; reported from St. Andrews, Charlotte Co. and Dalhousie, Restigouche Co.
HABITAT: Clearings, open woods and meadows.
GENERAL RANGE: Eurasia; introd. in Canada in Sask., Ont., Que., Nfld., N.B. and N.S.
SYNONYMS: **L. campestris** (L.) DC. var. **pallescens** Wahlenb.

Luzula parviflora (Ehrh.) Desv. Fig. p. A37. Map p. B32.
FREQUENCY: Common in the north of the Province and rare southwards. Unknown in the eastern lowlands.
HABITAT: Damp coniferous or mixed woods, cool ravines and banks.
GENERAL RANGE: Alaska to Nfld. south to Calif. and New Eng.; Greenland; Eurasia.
SUBSPECIFIC TAXA: Our material is referable to ssp. **melanocarpa** (Michx.) Hamet-Ahti.

LILIACEAE (Lily Family)

1a Leaves mainly basal, plants scapose. 2
1b Leaves mainly cauline. 3
2a(1a) Inflorescence umbellate or single flowered (if 2-3 flowered, not 2-ranked). 4
2b(1a) Inflorescence not as above. 5
3a(1b) Leaves scale-like, their function replaced by clusters of needle-like branches; flowers greenish-white, inconspicuous; fruit a bright red berry. **Asparagus officinalis**
3b(1b) Leaves various, not scale-like. 6
4a(2a) All parts of plant with strong onion or garlic odor and taste; flowers sometimes replaced by small bulblets, **Allium tricoccum** flowers appear after leaves have withered in spring). **Allium**
4b(2a) Plants not with onion or garlic taste or odor. 7
5a(2b) Leaves in 2 opposite ranks. 8
5b(2b) Leaves not 2-ranked. 9
6a(3b) Leaves mainly whorled. 10
6b(3b) Leaves mainly alternate. 11
7a(4b) Scapes 1-3 flowered. 12

++

7b(4b) Scapes several yellowish flowered; berries blue; leaves broadly lanceolate, dark green. **Clintonia borealis**

8a(5a) Inflorescence densely racemose; scape very sticky; flowers small, white. **Tofieldia glutinosa**

8b(5a) Inflorescence an irregular corymb; scape not sticky; flowers large, yellow or orange. **Hemerocallis**

9a(5b) Perianth campanulate, segments fused; inflorescence narrowly racemose, one sided; flowers very fragrant. **Convallaria majalis**

9b(5b) Perianth widely spreading, segments separate; inflorescence open racemose, not one sided; flowers not fragrant. 13

10a(6a) Leaves 3 per whorl, pinnately veined; perianth with green outer and coloured inner segments; fruit a berry. **Trillium**

10b(6a) Some leaves more than 3 per whorl, parallel veined; perianth not as above. 14

11a(6b) Stem climbing by stipular tendrils; inflorescence umbelliform; flowers with foetid odor. **Smilax rotundifolia**

11b(6b) Plants not as above. 15

12a(7a) Leaves linear, dark green; flowers bright blue. **Scilla sibirica**

12b(7a) Leaves lanceolate, mottled; flowers yellow. **Erythronium americanum**

13a(9b) Style one; perianth segments greenish striped below, lanceolate, non-glandular; stem to 2 dm tall. **Ornithogallum umbellatum**

13b(9b) Styles 3; perianth segments greenish white, obovate, glandular near base; stem to about 6 dm tall. **Zigadenus elegans**

14a(10b) Plants with less than 3 whorls of leaves; stems flocculent pubescent; flowers small with elongate, recurved styles free nearly to base; fruit a dark purple berry. **Medeola virginiana**

14b(10b) Plants with more than 3 whorls of leaves; flowers large with styles fused; fruit a loculicidal capsule. **Lilium**

15a(11b) Flowers borne in axils of ordinary stem leaves. 16

15b(11b) Flowers in terminal inflorescence. 17

16a(15a) Stem branched from near base; pedicels with conspicuous angle, usually single flowered. **Streptopus**

16b(15a) Stems unbranched; pedicels not angled, often 2 flowered. **Polygonatum pubescens**

17a(15b) Tall plants (to 2 m tall); perianth segments 6; fruit capsular. 18

17b(15b) Small acid woodland plants (less than 2 dm tall); perianth segments 4, white; inflorescence racemose; fruit a 1-2 seeded dryish berry. **Maianthemum canadense**

18a(17a) Leaves broad, plaited; greenish-yellow flowers in panicles. **Veratrum viride**

18b(17a) Leaves narrowly lanceolate, flat; perianth segments orange-red with purple-brown spots; inflorescence racemose. **Lilium lancifolium**

ALLIUM Onion, Leek, Garlic. Ait.

1a Leaves linear, tubular or +/- flat, present at flowering. 2

++

1b Leaves flat, broadly lanceolate, 1-3 dm long, 2-6 cm wide with slender petioles, usually shrivelling before the whitish flowers appear on stalks 1.5-4 dm high. **A. tricoccum**

2a(1a) Leaves tubular; flowers mostly roseate, well developed in a hemispherical or subglobose umbel. 3

2b(1a) Leaves flattened; flowers few or none, mostly replaced by small whitish or purplish bulblets. **A canadense**

3a(2a) Leaves slender, 1-2 mm thick, +/- equalling or overtopping the flowering scape; bulbs crowded. **A. schoenoprasum**

3b(2a) Leaves larger, 2-4 mm thick, rarely reaching as high as the flowering scape; bulbs few or solitary. **A. schoenoprasum** ssp. **sibiricum**

Allium canadense L. Fig. p. A37. Map p. B32.
COMMON NAME(S): Wild Garlic, Canada Garlic. Ail du Canada.
FREQUENCY: Rare; restricted to the drainage system of the St. John River. Sometimes locally common as above Crocks Point, York Co.
HABITAT: Banks, rocky shores and adjacent meadows.
GENERAL RANGE: Minn. to Ont. and N.S. south to Tex. and Fla.
SPECIAL NOTE: For more information see H. Hinds, 1983. The Rare Vascular Plants of New Brunswick. Syllogeus 50.

Allium schoenoprasum L. Fig. p. A38. Map p. B32.
COMMON NAME(S): Chives, Ciboulette or Brulotte.
FREQUENCY: Scattered throughout except in the eastern lowlands.
HABITAT: Rocky or gravelly shores and strands.
GENERAL RANGE: Alaska to Nfld. south to Oreg. and N. Eng.; Eurasia.
SUBSPECIFIC TAXA: Our material is chiefly ssp. **sibiricum** (L.) Celak. Ssp. **schoenoprasum**, the cultivated chive, occurs uncommonly as an introduction usually not far from former habitations. These are separated above.

Allium tricoccum Ait. Fig. p. A38. Fig. p. B32.
COMMON NAME(S): Ramp, Wild Leek. Ail sauvage.
FREQUENCY: Rare; populations are found in three areas; the Carleton Co. Eel and Meduxnekeag River valleys, the Petitcodiac River valley, Westmorland Co. and Trout Brook Valley, Queens Co.
HABITAT: Rich hardwoods and alluvial bottomslands.
GENERAL RANGE: Man. to N.S. south to Iowa and Ga.
SPECIAL NOTE: For more information see H. Hinds, 1983. The Rare Vascular Plants of New Brunswick. Syllogeus 50.

ASPARAGUS L. Asparagus. Asperge

Asparagus officinalis L. Fig. p. A38. Map p. B32.
COMMON NAME(S): Garden asparagus. Asperge commun.
Garden asparagus. Asperge commun.

++

FREQUENCY: Scattered mostly towards the east and west of the Province.
HABITAT: Roadsides, sandy meadows and thickets.
GENERAL RANGE: European; introduced throughout Canada except not reported from Sask.

CLINTONIA Raf.

Clintonia borealis (Ait.) Raf. Fig. p. A38. Map p. B32.
COMMON NAME(S): Clintonia, Bluebead-Lily, Corn-Lily. Lis boréale
FREQUENCY: Common throughout the Province.
HABITAT: Coniferous or mixed woods.
GENERAL RANGE: Man. to Nfld. south to Minn. and Ga.

CONVALLARIA L.

Convallaria majalis L. Fig. p. A38. Map p. B32.
COMMON NAME(S): Lily-of-the-valley. Muguet.
FREQUENCY: Scattered mostly in the southern part of the Province.
HABITAT: Moist open woods, thickets and waste areas.
GENERAL RANGE: Eurasian; introduced in Canada from Ont. to Nfld.

ERYTHRONIUM L. Dog's-tooth-Violet. Fawn-Lily

Erythronium americanum Ker. Fig. p. A38. Map p. B32.
COMMON NAME(S): Trout-Lily, Yellow Adder's-tongue. Ail doux.
FREQUENCY: Common in the western half of the Province.
HABITAT: Mostly rich woods.
GENERAL RANGE: Ont. to N.S. south to Okla. and Fla.

HEMEROCALLIS L. Day-Lily. Lis d'un jour

1a Perianth orange, without fragrance. **H. fulva**
1b Perianth yellow, fragrant. **H. lilioasphodelus**

Hemerocallis fulva L. Fig. p. A38. Map p. B32.
COMMON NAME(S): Orange Day-Lily. Lis jaune, Lis d'un jour.
FREQUENCY: Common throughout the Province.
HABITAT: Vicinity of old house sites and waste areas.
GENERAL RANGE: Eurasian; introduced but rarely spreading by seed. Established in Canada from Ont. to N.S.

Hemerocallis lilioasphodelus L.
COMMON NAME(S): Yellow Day-lily.
FREQUENCY: Rare; recorded from Carleton Co.
HABITAT: Vicinity of old house sites. Persisting, but rarely setting seeds.

++

GENERAL RANGE: Asiatic; introduced in Canada from Ont. to N.B.
SYNONYMS: **H. flava** L.

LILIUM L. Lily. Lis

1a Leaves mainly whorled; perianth unspotted. **L. canadense**
1b Leaves alternate or scattered; perianth variously spotted. 2
2a(1b) Flowers erect, perianth segments clawed, pubescent within towards base and pustulate on veins above; upper leaves often with axillary bulblets. **L. bulbiferum**
2b(1b) Flowers nodding; perianth segments recurved, not clawed, conspicuously purple-brown spotted within; stem scabrous below and arachnoid above; leaves broad based with long acuminate blades. **L. lancifolium**

Lilium bulbiferum L.
COMMON NAME(S): Orange or Saffron Lily. Lis safrane.
SPECIAL NOTE: European; rarely found as a garden escape as near Woodstock, Carleton Co. and near Moncton, Westmorland Co.
SYNONYMS: **L. croceum** Chaix.

Lilium canadense L. Fig. p. A38. Map p. B33.
COMMON NAME(S): Canada Lily.
FREQUENCY: Scattered across the Province from Madawaska Co. to Kent Co. and south.
HABITAT: Alluvial meadows and moist open woods.
GENERAL RANGE: Ont. to N.S. south to Ohio and Va.
SUBSPECIFIC TAXA: Forma **rubrum** Britt., with reddish flowers, occurs rarely with the species.

Lilium lancifolium Thunb. Fig. p. A38.
COMMON NAME(S): Tiger-Lily. Martagon.
SPECIAL NOTE: Asiatic; rarely escaped from cultivation as near Kouchibouguac, Kent Co.
SYNONYMS: **L. tigrinum** L.

MAIANTHEMUM Weber Wild Lily-of-the-valley

Maianthemum canadense Desf. Fig. p. A38. Map p. B33.
COMMON NAME(S): Wild Lily-of-the-valley. Muguet.
FREQUENCY: Common throughout the Province.
HABITAT: Moist, usually acid, woods.
GENERAL RANGE: Man. to Nfld. south to Iowa and Ga.

MEDEOLA L.

Medeola virginiana L. Fig. p. A38. Map p. B33.
COMMON NAME(S): Indian Cucumber-root. Concombre sauvage, Jarnotte.

FREQUENCY: Scattered throughout the Province.
HABITAT: Rich, mostly acid, woods.
GENERAL RANGE: Minn. and Ont. to N.S. south to La. and Fla.

ORNITHOGALUM L. Star-of-Bethlehem

Ornithogalum umbellatum L. Fig. p. A38.
COMMON NAME(S): Nap-at-noon, Sleepy Dick. Dame des onze heures.
SPECIAL NOTE: European; rarely escaped from cultivation as at Fredericton, York Co., and Woodstock, Carleton Co.

POLYGONATUM Mill. Solomon's-seal. Sceau-de-Salomon.

Polygonatum pubescens (Willd.) Pursh Fig. p. A39. Map p. B33.
FREQUENCY: Probably scattered throughout except rare in the eastern lowlands.
HABITAT: Rich deciduous woods.
GENERAL RANGE: Ont. to N.S. south to Iowa and Ga.

SCILLA L. Squill

Scilla sibirica Andr.
SPECIAL NOTE: Asiatic; occasionally escaped from cultivation as in Woodstock, York Co.

SMILACINA Desf. False Solomon's-Seal

1a Leaves caudate tipped, membranous with wavy margins; inflorescence a panicle (compound raceme). **S. racemosa**

1b Leaves acute tipped, coriaceous with straight margins; inflorescense racemose. 2

2a(1b) Cauline leaves less than four, glabrous; plants of boggy habitats. **S. trifolia**

2b(1b) Cauline leaves more than 6, minutely pubescent beneath; plants of the vicinity of shores and similar exposed habitats. **S. stellata**

Smilacina racemosa (L.) Desf. Fig. p. A39. Map p. B33.
COMMON NAME(S): False Spikenard, False Solomon's-seal. Faux Sceau-de-Salomon.
FREQUENCY: Uncommonly scattered especially in the southern and western part of the Province.
HABITAT: Open deciduous woods, slopes and clearings.
GENERAL RANGE: Man. to N.S. south to Calif. and Ga.

Smilacina stellata (L.) Desf. Fig. p. A39. Map p. B33.
COMMON NAME(S): Star-flowered False Solomon's-seal. Smilacine étoilée.

FREQUENCY: Widely distributed throughout the Province.
GENERAL RANGE: Alaska to Nfld. south to Calif. and Va.

Smilacina trifolia (L.) Desf. Fig. p. A39. Map p. B33.
COMMON NAME(S): Three-leaved False Solomon's-seal.
FREQUENCY: Scattered throughout the Province.
HABITAT: Bogs, wet meadows and woods, usually in sphagnum moss.
GENERAL RANGE: Yukon to Nfld. south to Alta., Minn. and N.J.

SMILAX L. Greenbrier, Catbrier

Smilax herbacea L. Fig. p. A39. Map p. B33.
COMMON NAME(S): Carrion-flower, Jacob's-ladder. Raisin de couleuvre.
FREQUENCY: Locally common in the west and southwest of the Province.
HABITAT: Alluvial woods and meadows.
GENERAL RANGE: Ont. to N.B. south to Mo. and Ga.

STREPTOPUS Michx. Twisted-stalk

1a Leaves amplexicaule, glaucous beneath; perianth segments greenish-white, widely spreading from near middle. **S. amplexifolius**
1b Leaves mostly sessile, those subtending branches subamplexicaule; perianth segments pinkish to purple, recurved from near tips. **S. roseus**

Streptopus amplexifolius (L.) DC. Fig. p. A39. Map p. B33.
COMMON NAME(S): Liverberry, Scootberry, White Mandarin.
FREQUENCY: Scattered throughout the Province.
HABITAT: springy coniferous or mixed woods.
GENERAL RANGE: Alaska to Nfld. south to Calif. and N.C.; Greenland; Eurasia.
SUBSPECIFIC TAXA: Our material is all referable to var. **americanus** Schultes.

Streptopus roseus Michx. Fig. p. A39. Map p. B33.
COMMON NAME(S): Rose Twisted-stalk, Rose Mandarin.
FREQUENCY: Common throughout the Province.
HABITAT: Bottomlands, coniferous and mixed woods.
GENERAL RANGE: Alaska to Nfld. south to Minn. and Ga.
SUBSPECIFIC TAXA: Our material is all referable to var. **perspectus** Fassett.

TOFIELDIA Huds. False Asphodel

Tofieldia glutinosa (Michx.) Pers. Fig. p. A39. Map p. B33.
COMMON NAME(S): Sticky False Asphodel.
FREQUENCY: Uncommon; scattered in western and northern parts of the Province.
HABITAT: Mostly calcareous ledges and shores.
GENERAL RANGE: Alaska to Nfld. south to Calif. and Ga.

++

TRILLIUM L. Trillium, Wakerobin. Trille

1a Leaves petiolate; ovary obscurely 3-lobed to rounded; petals undulate-margined, white with red markings near base. **T. undulatum**

1b Leaves sessile; ovary 6-angled. 2

2a(1b) Peduncles +/- erect, held above leaves; petals dark maroon (aberrant forms with creamy white, pink or greenish petals), 2.5-6 cm long; filaments ¼-½ as long as anthers. **T. erectum**

2b(1b) Peduncles bent horizontal to declinate below the leaves at maturity; petals white, less than 3 cm long; filaments +/- as long as anthers. **T. cernuum**

Trillium cernuum L. Fig. p. A39. Map p. B33.
COMMON NAME(S): Nodding Trillium.
FREQUENCY: Uncommonly scattered throughout the Province but locally common.
HABITAT: Alluvial bottomland and rich deciduous woods.
GENERAL RANGE: Sask. to Nfld. south to Iowa and Ga.

Trillium erectum L. Fig. p. A39. Map p. B33.
COMMON NAME(S): Purple Trillium, Stinking Benjamin.
FREQUENCY: Scattered throughout except rare in the eastern lowlands.
HABITAT: Hardwood slopes and rich woods.
GENERAL RANGE: Man. to N.S. south to Tenn. and Ga.
SUBSPECIFIC TAXA: Numerous color and monstrous forms have been named. Forma **luteum** Louis-Marie, with yellowish petals, and forma **viridiflorum** (Hook.) Peattie, with greenish petals, occur frequently throughout the range of the species.

Trillium undulatum Willd. Fig. p. A39. Map p. B34.
COMMON NAME(S): Painted Trillium.
FREQUENCY: Common throughout the Province.
HABITAT: Mainly coniferous woods.
GENERAL RANGE: Ont. to N.S. south to Wisc. and Ga.

UVULARIA L. Bellwort

Uvularia sessilifolia L. Fig. p. A39. Map p. B34.
COMMON NAME(S): Little Merrybells, Sessile-leaved Bellwort.
FREQUENCY: Locally common throughout, except rare in the northern part of the Province.
HABITAT: Deciduous woods and moist meadows.
GENERAL RANGE: Man. to N.S. south to Ark. and Ga.

VERATRUM L. False Hellebore. Vérâtre or Varaire

Veratrum viride Ait. Fig. p. A40. Map p. B34.
COMMON NAME(S): White Hellebore, Indian Poke. Tabac du diable.
FREQUENCY: Mostly common in the western half of the Province.

HABITAT: Alluvial meadows and woods.
GENERAL RANGE: Alaska to Labr. and N.B. south to Tenn. and Ga.

ZIGADENUS Michx.

Zigadenus elegans Pursh Fig. p. A40.
COMMON NAME(S): White Camass, Alkali-Grass.
FREQUENCY: Very rare; known only from Belledune Point, Gloucester Co. and Heron Is., Restigouche Co.
HABITAT: Coastal headlands and inland edges of salt marshes.
GENERAL RANGE: Alaska to N.B. south to Oreg. and mts. of N.C.
SYNONYMS: **Z. glaucus** Nutt.
SPECIAL NOTE: For more information see H. Hinds, 1983. The Rare Vascular Plants of New Brunswick. Syllogeus 50.

IRIDACEAE (Iris Family)

1a Flowers less than 1 cm wide; sepals and petals similar; stems winged; plants without rhizomes. **Sisyrinchium**
1b Flowers 6-12 cm wide; sepals and petals dissimilar; stems wingless; plants with stout rhizomes. **Iris**

IRIS L. Iris. Fleur-de-lis

1a Flowers yellow to cream. **Iris pseudacorus**
1b Flowers blue or violet with white markings. 2
2a(1b) Dwarf plants mainly of coastal headlands and other exposed habitats; petals narrow, inconspicuous, involute or tubular, bristle-tipped; capsule broadly ellipsoid, blunt; seeds pyriform; leaves up to 1.5 cm broad. **I. setosa**
2b(1b) Larger plant of moist or wet fresh habitats; petals oblanceolate to obovate, rounded at summit; capsule thick-cylindric to ellipsoid, beaked; seeds not pyriform; leaves to 3 cm broad. **I. versicolor**

Iris pseudacorus L. Fig. p. A40.
COMMON NAME(S): Common Yellow Iris.
SPECIAL NOTE: European; escaped locally from cultivation as in Sackville, Westmorland Co. Also reported elsewhere in Canada from B.C., Man., Ont. to N.S.

Iris setosa Pallas Fig. p. A40. Map p. B34.
COMMON NAME(S): Beachhead-Iris or -Flag.
FREQUENCY: Scattered along the Bay of Fundy and east coastal areas and one locality on Grand Lake, Queens Co.
HABITAT: Headlands, dune hollows and banks.
GENERAL RANGE: Eastern Que. to Nfld. south along the coast to Me.
SUBSPECIFIC TAXA: Our material is all referable to var. **canadensis** Foster.

+++

Iris versicolor L. Fig. p. A40. Map p. B34.
COMMON NAME(S): Blue Flag. Clajeux, Iris sauvage.
FREQUENCY: Locally common throughout the Province.
HABITAT: Meadows, shores, marshes and similar wet, open areas.
GENERAL RANGE: Man. to Nfld. south to Minn. and Va.

SISYRINCHIUM L. Blue-eyed Grass. Bermudienne

Sisyrinchium montanum Greene Fig. p. A40.
FREQUENCY: Common throughout the Province.
HABITAT: Meadows, shores, and similar moist open ground.
GENERAL RANGE: Yukon to Nfld. south to Calif. and mts. of Va.; Greenland.
SYNONYMS: **S. montanum** var. **crebrum** Fern.; **S. bermudiana** L. var. **crebrum** (Fern.) Boivin.

ORCHIDACEAE (Orchis Family)

1a Entire lip petal inflated, pouch-like (only basal portion of lip saccate in **Goodyera** & **Epipactis**); scapes usually single flowered. 2
1b Entire lip petal not inflated, scapes single or multiple flowered. 3
2a(1a) Leaf solitary. **Calypso bulbosa**
2b(1a) Leaves 2 or more. **Cypripedium**
3a(1b) Plants without chlorophyll, saprophytes. **Corallorhiza**
3b(1b) Plants with chlorophyll, autotrophs 4
4a(3b) Plants single flowered. 5
4b(3b) Plants several to many flowered. 6
5a(4a) Leaves reduced to bladeless sheaths. **Arethusa bulbosa**
5b(4a) Leaves with well developed blades. **Pogonia ophioglossoides**
6a(4b) Leaves all or mostly basal (small inconspicuous bracts may occur on the flowering stem). 7
6b(4b) Plants with well developed cauline leaves. 8
7a(6a) Leaves narrowly lanceolate, less than 2.5 cm broad; lip not saccate. **Spiranthes**
7b(6a) Leaves usually broader (if less than 2.5 cm broad, lip saccate). 9
8a(6b) Plant with a single major cauline leaf. 10
8b(6b) Plants with more than one major leaf. 11
9a(7b) Plants with a single basal leaf. 12
9b(7b) Plants with more than one basal leaf. 13
10a(8a) Flowers minute, 3-4 mm wide, greenish. **Malaxis**
10b(8a) Flowers larger, to 8 mm wide, yellowish-white; leaf sheath often extending 1/3 the length of scape. **Platanthera clavellata**
11a(8b) Plants with 2 opposite sessile leaves. **Listera**
11b(8b) Plants with alternate leaves. 14
12a(9a) Flowers small; white lip petal 6-8 mm long, 3 lobed with purple spots; leaf orbicular to oblong. **Amerorchis rotundifolia**
12b(9a) Flowers larger; magenta lip petal to 1.5 cm long (lip uppermost) with yellow-tipped hairs; leaf grass-like. **Calopogon tuberosus**
13a(9b) Flowers yellowish-green. **Liparis loeselii**

++

13b(9b) Flowers white or pink. 15

14a(11b) Floral bracts 2 to 4 times the length of the flowers; leaves +/- gradually reduced in size towards summit. 16

14b(11b) Floral bracts shorter than to slightly longer than flowers; leaves abruptly reduced in size towards summit. **Platanthera**

15a(13b) Flowers pink; leaves oblong-ovate, erect or ascending. **Galearis spectabilis**

15b(13b) Flowers white; leaves depressed. 17

16a(14a) Lip petal spurless; basal leaves reduced to bladeless sheaths. **Epipactis helleborine**

16b(14a) Lip petal spurred; basal leaves with well developed, large blades. **Coeloglossum viride**

17a(15b) Leaves 2, appearing opposite, orbicular to elliptic. **Platanthera**

17b(15b) Leaves several, ovate to lanceolate. **Goodyera**

AMERORCHIS Hulten

Amerorchis rotundifolia (Banks) Hulten Fig. p. A40. Map p. B34.
COMMON NAME(S): Small Round-leaved Orchis.
FREQUENCY: Very rare; known only from Shea Lake, Victoria Co., Belledune, Gloucester Co. and near First Lake, Restigouche Co.
HABITAT: Calcareous bogs.
GENERAL RANGE: Alaska to Nfld. south to B.C., Alta., n. N. Eng. and N.B.; Greenland.
SYNONYMS: **Orchis rotundifolia** Banks.
SPECIAL NOTE: For more information see H. Hinds, 1983. The Rare Vascular Plants of New Brunswick. Syllogeus 50.

ARETHUSA L.

Arethusa bulbosa L. Fig. p. A40. Map p. B34.
COMMON NAME(S): Arethusa, Swamp-pink. Aréthuse.
FREQUENCY: Uncommonly scattered throughout the Province.
HABITAT: Sphagnous bogs.
GENERAL RANGE: Sask. to Nfld. south to Ind. and N.J.
SUBSPECIFIC TAXA: Forma **albiflora** Rand & Red. with white flowers, is reported from Arthurette Bog, Victoria Co.

CALOPOGON R. Br.

Calopogon tuberosus (L.) BSP. Fig. p. A40. Map p. B34.
COMMON NAME(S): Grass-pink.
FREQUENCY: Uncommonly scattered in the southern half of the Province.
HABITAT: Sphagnous boggy areas.

GENERAL RANGE: Man. to Nfld. south to Tex. and Fla.
SYNONYMS: **C. pulchellus** (Salisb.) R. Br.

CALYPSO Salisb.

Calypso bulbosa (L.) Oakes Fig. p. A40. Map p. B34.
COMMON NAME(S): Calypso, Venus'-slipper.
FREQUENCY: Uncommonly scattered in the western part of the Province.
HABITAT: Mostly arborvitae swamps and similar moist, mossy habitats.
GENERAL RANGE: Alaska to Nfld. south to Calif. and N. Eng.; Eurasia.

COELOGLOSSUM Hartman

Coeloglossum viride (L.) Hartman Fig. p. A40. Map p. B34.
COMMON NAME(S): Frog-Orchis, Long-bracted Green Orchis.
FREQUENCY: Uncommonly scattered throughout except rare in the northeast and southwest of the Province.
HABITAT: Rich woods and meadows, often calcareous.
GENERAL RANGE: Alaska to Nfld. south to Wash. and N.C.; Iceland; Eurasia.
SUBSPECIFIC TAXA: Our material is all referable to var. **virescens** (Muhl.) Luer.
SYNONYMS: **Habenaria viridis** (L.) R. Br., **C. viride** ssp. **bracteatum** (Muhl.) Hulten.

CORALLORHIZA Chat. Coral-root

1a Lateral lobes near base of lip prominent; perianth 5-10 mm long; sepals and petals 3 nerved; capsule yellowish-brown or brownish; scape to 6 dm tall with sheaths extending beyond middle. **C. maculata**

1b Lateral lobes near base of lip wanting or inconspicuous; perianth 4-5.5 mm long; sepals and petals 1 nerved; capsule greenish; scape to 3 dm tall with sheaths towards base. **C. trifida**

Corallorhiza maculata Raf. Fig. p. A40. Map p. B34.
COMMON NAME(S): Spotted Coral-root.
FREQUENCY: Scattered throughout the Province.
HABITAT: Mostly dryish coniferous or mixed woods.
GENERAL RANGE: B.C. to Nfld. south to Calif. and N.C.
SUBSPECIFIC TAXA: Forma **flavida** (Peck) Farw., with lip unspotted, is reported from Allandale, York Co. as is forma **punicea** (Bartl.) Weath. & Adams (the whole plant suffused with reddish purple).

Corallorhiza trifida Chat. Fig. p. A41. Map p. B35.
COMMON NAME(S): Early or Pale Coral-root.
FREQUENCY: Scattered throughout the Province but rare in the eastern lowlands.

HABITAT: Seepage areas in rich coniferous or mixed woods.
GENERAL RANGE: Alaska to Nfld. south to Oreg. and Ga.; Greenland; Iceland; Eurasia.

CYPRIPEDIUM L. Lady's-slipper, Moccasin-flower. Sabot de la Vierge

1a Leaves 2, basal; scapose stem with single pale pink to crimson-pink or pure white flower; lip open along full length above. **C. acaule**
1b Leaves more than 2, cauline; lip open only at base. 2
2a(1b) Lip yellow; sepals sharply acute; stems finely pubescent, to 8 dm tall. **C. calceolus**
2b(1b) Lip white with pink or purple markings; sepals obtuse or rounded; stems conspicuously hirsute, to 10 dm tall. **C. reginae**

Cyrpripedium acaule Ait. Fig. p. A41. Map p. B35.
COMMON NAME(S): Stemless or Pink Lady's-slipper.
FREQUENCY: Uncommon throughout except rare in the northern part of the Province.
HABITAT: Boggy heathland and acid coniferous or mixed woods.
GENERAL RANGE: Alta. to Nfld. south to Minn. and Ga.
SUBSPECIFIC TAXA: Forma **albiflorum** Rand & Redf., with white lip and yellow sepals and petals, occurs frequently with and sometimes without the normal pink form.

(**Cypripedium arietinum** R. Br.)
COMMON NAME(S): Ram's-head Lady's-slipper.
SPECIAL NOTE: Not reported from the Province but occurring very close to the northwestern border of Carleton Co. and may be expected there or in Victoria Co. It differs from the other Cypripedia in having a purplish veined lip which tapers to a conical tip and the lateral sepals are separate and spreading.

Cypripedium calceolus L. Fig. p. A41.
COMMON NAME(S): Yellow Lady's-slipper.
SUBSPECIFIC TAXA: Two varieties are recognized in the Province.

1a Sepals usually greenish-yellow streaked with madder purple, 4-7 cm long; lateral petals 5-8 cm long, coloured like sepals; lip petal less than 3 cm long; flowers only faintly scented; plants of subacid, rich, well drained to moist woods and cedar swamps. var. **pubescens**
1b Sepals madder purple, 2-3.9 cm long; lateral petals madder purple 3-5 cm long; lip petal greater than 3 cm long; flowers pungently and agreeably fragrant; plants of circumneutral moist meadows, bogs and bog margins as well as exposed calcarious ledges. var. **parviflorum**

1. Var. **parviflorum** (Salisb.) Fern. Map p. B35 (see **C. pubescens**).
FREQUENCY: Rare; recorded from a few locations in Victoria and Restigouche Counties.
GENERAL RANGE: B.C. to Nfld. south to Tex. and Ga.
SYNONYMS: **C. parviflorum** Salisb.
2. Var. **pubescens** (Willd.) Correll Map p. B35 (see **C. pubescens**).

FREQUENCY: Uncommonly scattered throughout the Province but rare in the eastern lowlands.
GENERAL RANGE: Sask. to N.S. south to Mo. and Ga.
SYNONYMS: **C. pubescens** Willd.

Cypripedium reginae Walt. Fig. p. A41. Map p. B35.
COMMON NAME(S): Showy Lady's-slipper.
FREQUENCY: Rare and possibly endangered in the Province. Several previously prosperous populations have been decimated by pickers and collectors as well as habitat disruption.
HABITAT: Calcareous bogs and boggy meadows, edges of arborvitae swamps.
GENERAL RANGE: Man. to Nfld. south to N. Dak. and Ga.

EPIPACTIS Sw. Helleborine

Epipactis helleborine (L.) Crantz Fig. p. A41. Map p. B35.
COMMON NAME(S): Helleborine.
FREQUENCY: Locally common in the Meduxnekeag River valley, rare near Poley Mtn., Kings Co. and near Albert Mines, Albert Co.
HABITAT: Rich mostly calcareous deciduous or mixed woods.
GENERAL RANGE: Eurasian; introduced in Canada in B.C., Ont., Que., and N.B.

GALEARIS Raf.

Galearis spectabilis (L.) Raf. Fig. p. A41. Map p. B35.
COMMON NAME(S): Showy Orchis. Orchide admirable.
FREQUENCY: Rare; scattered along the Meduxnekeag, Eel and St. John R. valleys, Carleton Co.
HABITAT: Rich, often calcareous, deciduous woods and bottomlands.
GENERAL RANGE: Ont. to N.B. south to Kans. and Ga.
SYNONYMS: **Orchis spectabilis** L.

GOODYERA R. Br. Rattlesnake-plantain

1a Leaf blades 4-6 cm long, only midvein outlined in white or entirely pale green above; perianth 6-9 mm long; plants 20-50 cm tall. **G. oblongifolia**
1b Leaf blades to 4 cm long, usually with pale green or white lines throughout; perianth 2.5-5.5 mm long; plants 5-35 cm tall. 2
2a(1b) Cauline bracts (6) 7-10 (14); beak of lip petal about ¼ length of lip. **G. pubescens**
2b(1b) Cauline bracts 2-5(6); beak of lip petal about ½ length of lip. 3
3a(2b) Basal portion of mature lip petal deeply saccate, nearly as deep as long, tip portion strongly recurved; perianth about 4 mm long; racemes +/- one sides; leaves 1-3 cm long, widest near base, tapering to acute tip. **G. repens**

3b(2b) Basal portion of mature lip petal only shallowly saccate, longer than deep, tip portion only slightly recurved; perianth about 5 mm long; racemes loosely spiral; leaf blades 2-5(6) cm long, tapering from near middle to each end. **G. tesselata**

Goodyera oblongifolia Raf. Fig. p. A41. Map p. B35.
COMMON NAME(S): Giant Rattlesnake-plantain.
FREQUENCY: Rare; reported only in the northern third of the Province.
HABITAT: Damp arborvitae or other coniferous or mixed woods.
GENERAL RANGE: Alaska to N.S. south to Calif., Mexico, Minn. and Me.
SPECIAL NOTE: For more information see H. Hinds, 1983. The Rare Vascular Plants of New Brunswick. Syllogeus 50.

Goodyera pubescens (Willd.) R. Br. Fig. p. A41. Map p.B35.
COMMON NAME(S): Downy Rattlesnake-plantain.
FREQUENCY: Very rare; reported only from Petitcodiac, Westmorland Co. Not seen since 1881 (UNB).
HABITAT: Dry or moist coniferous or mixed woods.
GENERAL RANGE: Ont. to N.B. south to Tenn. and Fla.
SPECIAL NOTE: For more information see H. Hinds, 1983. The Rare Vascular Plants of New Brunswick. Syllogeus 50.

Goodyera repens (L.) R. Br. Fig. p. A41. Map p. B35.
COMMON NAME(S): Dwarf Rattlesnake-plantain.
SUBSPECIFIC TAXA: Two varieties are found in the Province.

1a Leaves dark green without white markings along the veins. var. **repens**
1b Leaves with prominant white lines outlining the major veins. var. **ophioides**

1. Var. **ophioides** Fern.
FREQUENCY: Scattered throughout but most common in the northern half of the Province.
GENERAL RANGE: Alaska to Nfld. south to Tenn. and S.D.
2. Var. **repens**
FREQUENCY: Very rare; reported only from Debec, Carleton Co. (UNB).
GENERAL RANGE: B.C. to N.B. south to s. Alta. and Thunder Bay, Ont.; Eurasia.

Goodyera tesselata Lodd. Fig. p. A41. Map p. B35.
COMMON NAME(S): Checkered Rattlesnake-plantain.
FREQUENCY: Uncommonly scattered throughout the Province.
HABITAT: Mostly dry coniferous woods and occasionally in arborvitae bogs.
GENERAL RANGE: Ont. to Nfld. south to Minn. and Conn.

LIPARIS Richard Twayblade

Liparis loeselii (L.) Richard Fig. p. A41. Map p. B36.
COMMON NAME(S): Bog or Yellow Twayblade.

FREQUENCY: Rare; reported from the east coast in Kent Co., near Fredericton, York Co. and from Grand Manan Is., Charlotte Co.
HABITAT: Damp alder thickets, meadows, and low sandy areas.
GENERAL RANGE: Wash. and B.C., Sask. to N.S. south to Nebr. and N.J.; Eurasia.

LISTERA R. Br. Twayblade

1a Stem distance from leaves to base of inflorescence about 2-3 times length of leaves. 2
1b Stem distance from leaves to base of inflorescence about ½ as long as leaves. 3
2a(1a) Rachis and pedicels glabrous; lip purplish-green, 4-5 mm long with 2 forward projecting horn-like teeth at base, 1.5 mm long, not auricled; leaves +/- truncate or subcordate at base. **L. cordata**
2b(1a) Rachis and pedicels glandular; lip dull red, 10 mm long, without conspicuous horns, auricled at base; leaves rounded at base. **L. australis**
3a(1b) Pedicels and ovaries glandular; lip with narrow neck at base. 4
3b(1b) Pedicels and ovaries glabrous; lip sessile with basal auricles, slightly narrowed at middle, not at base, cleft to 1/3 of its length. **L auriculata**
4a(3a) Lip auricles distant from base of long neck by about 1-1.5 mm; summit notch of lip shallow, about equal to neck or less. **L. convallarioides**
4b(3a) Lip auricles +/- reaching base of short neck; summit notch of lip deeper, 2-3 times the length of neck. **L. X veltmanii**

Listera auriculata Wieg. Fig. p. A42. Map p. B36.
COMMON NAME(S): Auricled Twayblade.
FREQUENCY: Scattered throughout but rare in the eastern lowlands.
HABITAT: Alder thickets, banks, and arborvitae swamps.
GENERAL RANGE: Ont. to Nfld. south to Minn. and N. Eng.
SPECIAL NOTE: The hybrid with **L. convallarioides** (L. X **veltmanii** Case) has been reported from near Clair, and near Summit Depot, Madawaska Co.

Listera australis Lindl. Fig. p. A42. Map p. B36.
COMMON NAME(S): Southern Twayblade.
FREQUENCY: Very rare; found only in Kouchibouguac National Park, Kent Co.
HABITAT: In sphagnum at edge of bog with stunted black spruce.
GENERAL RANGE: Ont. to N.S. south to La. and Fla.
SPECIAL NOTE: For more information see H. Hinds, 1983. The Rare Vascular Plants of New Brunswick. Syllogeus 50.

Listera convallarioides (SW.) Nutt. Fig. p. A42. Map p. B36.
COMMON NAME(S): Broad-lipped Twayblade.
FREQUENCY: Rare; known only from the northern part of the Province.
HABITAT: Peaty or mossy woods, arborvitae swamps, calcareous bogs.
GENERAL RANGE: B.C. to Nfld. south to Calif. and N.C.; Asia.

Listera cordata (L.) R. Br. Fig. p. A42. Map p. B36.
COMMON NAME(S): Heartleaf Twayblade.

++

FREQUENCY: Probably common throughout the Province.
HABITAT: Moist banks, coniferous swamps and bogs.
GENERAL RANGE: Alaska to Nfld. south to Calif. and N.C.; Greenland; Iceland; Eurasia.

MALAXIS Soland. ex. Sev. Adder's-mouth.

1a Lip entire, pointed; pedicels 1.5-3 mm long; raceme slender, elongate; leaf sheathing base of stem. **M. brachypoda**
1b Lip 2-lobed with indistinct central tooth; pedicels 4-10 mm long; raceme thick, rounded; leaf +/- sheathing bottom ½ of stem. **M. unifolia**

Malaxis brachypoda (Gray) Fern. Fig. p. A42.
COMMON NAME(S): White Adder's-tongue.
FREQUENCY: Very rare; reported only on Grand Manan Is., Charlotte Co.
HABITAT: Grassy hummocks in wet ground at edge of field.
GENERAL RANGE: Alaska to Nfld. south to Calif., Tenn. and N.J.; Eurasia.
SPECIAL NOTE: For more information see H. Hinds, 1983. The Rare Vascular Plants of New Brunswick. Syllogeus 50.
SYNONYMS: **M. monophyllos** (L.) Sw. var. **brachypoda** (Gray) Morris.

Malaxis unifolia Michx. Fig. p. A42. Map p. B37.
COMMON NAME(S): Green Adder's-tongue.
FREQUENCY: Probably scattered through the Province.
HABITAT: Alder thickets, clearings, edges of woods, and shores.
GENERAL RANGE: Man. to Nfld. south to Tex. and Fla.

PLATANTHERA L. C. Rich Rein-Orchis, Fringed Orchis

1a Leaves 1 or 2. 2
1b Leaves more than 2, mostly cauline. 3
2a(1a) Leaf 1. 4
2b(1a) Leaves 2, basal, +/- orbicular; lip entire. 5
3a(1b) Lip entire or toothed, not 3-parted. 6
3b(1b) Lip fringed. 7
4a(2a) Leaf sheathing lower ¼-1/3 of stem; lip broadest towards 3-toothed apex; spur much exceeding lip. **P. clavellata**
4b(2a) Leaf sheathing base of stem; lip broadest at base, tapering to apex; spur equalling lip or only slightly longer. **P. obtusata**
5a(2b) Scape naked (rarely with 1 bract); flowers sessile yellowish-green; spur 0.9-2.5 cm long, tapering to rounded tip. **P. hookeri**
5b(2b) Scape with 1-6 bracts below inflorescence; whitish-green; flowers with short pedicels; spur 0.8-4.5 cm long, with parallel sides and slightly enlarged tip. **P. orbiculata**

++

6a(3a) Lip broadened towards toothed apex, bearing a conspicuous tubercle dorsally and small lateral lobes near base. **P. flava**

6b(3a) Lip narrowed towards entire apex, without tubercle or side lobes. 8

7a(3b) Lip deeply three-parted; spur 1-4 cm long. 9

7b(3b) Lip fringed, but not 3-parted; flowers white; spur (15)18-20(27) mm long. **P. blephariglottis**

8a(6b) Flowers greenish to greenish-yellow; lip lanceolate, usually less than 1.5 mm wide at base. **P. hyperborea**

8b(6b) Flowers white to greenish white; lip usually abruptly expanded near base, greater than 2 mm wide. **P. dilatata**

9a(7a) Lateral lobes of lip divided more than half way to base; flowers yellowish, creamy or greenish. 10

9b(7a) Lateral lobes of lip divided less than half way to base; flowers pink-purple. 11

10a(9a) Sepals (6)6.5-9(10) mm long; lateral petals broadly obovate, erose or denticulate at apex; spur 3-4 cm long; flowers creamy-white. **P. leucophaea**

10b(9a) Sepals 3.5-5(6) mm long; lateral petals linear-oblong or lanceolate, entire at apex; spur 1-1.7 cm long; flowers yellowish-green. **P. lacera**

11a(9b) Inflorescence 2.5-4.5 cm in diameter; perianth 4-7 mm long; lip 6-16 mm broad. **P. psychodes**

11b(9b) Inflorescence 5-9 cm in diameter; perianth 9-12 mm long; lip 1.8-3 cm broad. **P. grandiflora**

Platanthera blephariglottis (Willd.) Lindl. Fig. p. A42. Map p. B35.
COMMON NAME(S): White Fringed Orchis.
FREQUENCY: Uncommon mostly in the southern part of the Province.
HABITAT: Sphagnous bogs and meadows.
GENERAL RANGE: Ont. to Nfld. south to Miss. and Fla.
SYNONYMS: **Habenaria blephariglottis** (Willd.) Hook.

Platanthera clavellata (Michx.) Luer Fig. p. A42. Map p. B35.
COMMON NAME(S): Green Woodland Orchis.
FREQUENCY: Fairly common in the southern half of the Province.
HABITAT: Roadside ditches, river banks and open arborvitae swamps.
GENERAL RANGE: Ont. to Nfld. south to Tex. and Fla.
SYNONYMS: **Habenaria clavellata** (Michx.) Spreng.

Platanthera dilatata (Pursh) Lindl. Fig. p. A42. Map p. B36.
COMMON NAME(S): Bog-candle, Scent-bottle.
FREQUENCY: Scattered throughout the Province.
HABITAT: Springy areas, arborvitae bogs, and sedge meadows.
GENERAL RANGE: Alaska to Nfld. south to Calif. and N.J.; Asia.
SYNONYMS: **Habenaria dilatata** (Pursh) Hook.
SPECIAL NOTE: **P. X media** (Rydb.) Luer, the apparent hybrid with **P. hyperborea** is reported from near the mouth of Cross Creek, York Co.

Platanthera flava (L.) Lindl. Fig. p. A42.
COMMON NAME(S): Pale Green Orchis.
FREQUENCY: Very rare; reported from near the Oromocto River, Sunbury Co.
HABITAT: Swampy woods.
GENERAL RANGE: Ont. to N.S. south to Tex. and Fla.
SUBSPECIFIC TAXA: Our material is referable to var. **herbiola** (R. Br.) Luer.
SPECIAL NOTE: For more information see H. Hinds, 1983. The Rare Vascular Plants of New Brunswick. Syllogeus 50.
SYNONYMS: **Habenaria flava** (L.) R. Br. var. **herbiola** (R. Br.) Arnes & Correll.

Platanthera grandiflora (Bigel.) Lindl. Fig. p. A43 (see **P. psycodes**). Map p. B36.
COMMON NAME(S): Large Purple-fringed Orchis.
FREQUENCY: Uncommonly scattered mostly in the southern half of the Province.
HABITAT: Rocky shores, wet meadows, and similar wet open areas.
GENERAL RANGE: Ont. to Nfld. south to Tenn. and N.C.
SYNONYMS: **Habenaria fimbriata** (Ait.) R. Br., **H. psycodes** var. **grandiflora** (Bigel.) Gray, **H. grandiflora** (Bigel.) Torr.
SUBSPECIFIC TAXA: Forma **leucopheopsis** Boivin, with white flowers, has been reported on Grand Manan Is. Charlotte Co.

Platanthera hookeri (Torr.) Lindl. Fig. p. A42. Map p. B36.
COMMON NAME(S): Hooker's Orchis.
FREQUENCY: Probably scattered throughout the Province in its preferred habitat.
HABITAT: Deciduous or mixed woods.
GENERAL RANGE: Man. to Nfld. south to Iowa and N. Eng.
SYNONYMS: **Habenaria hookeri** Torr.

Platanthera hyperborea (L.) Lindl. Fig. p. A42. Map p. B36.
COMMON NAME(S): Northern Green Orchis.
FREQUENCY: Common throughout the Province.
HABITAT: Mostly coniferous woods, occasionally in calcareous sedge meadows, bogs, and roadside ditches.
GENERAL RANGE: Alaska to Nfld. south to Calif. and N. Eng.; Greenland; Iceland; Asia.
SYNONYMS: **Habenaria hyperborea** (L.) R. Br.

Platanthera lacera (Michx.) G. Don. Fig. p. A42. Map p. B36.
COMMON NAME(S): Ragged Orchis.
FREQUENCY: Scattered in the southern third of the Province.
HABITAT: Moist meadows, roadsides, clearings and similar grassy open areas.
GENERAL RANGE: Ont. to Nfld. south to Tex. and Fla.
SPECIAL NOTE: **P. X andrewsii** (M. White) Luer, the apparent hybrid with **P. psycodes**. occurs where the parents occur together as on Grand Manan Is., Charlotte Co.
SYNONYMS: **Habenaria lacera** (Michx.) Lodd.

(**Platanthera leucophaea** (Nutt.) Lindl.)
COMMON NAME(S): Prairie White Fringed Orchis.
SPECIAL NOTE: Known from Crystal Bog near Island Falls, Aroostook Co., Maine not far from the N.B. border. To be looked for in calcareous fens. Specimens reported by

Boivin, 1966-67, from Stoneridge, York Co. appear to be pale flowered **P. grandiflora**. The report from Mactaquac Provincial Park (Hinds, 1983) is **P. lacera**.
SYNONYMS: **Habenaria leucophaea** Nutt.

Platanthera obtusata (Banks) Lindl. Fig. p. A43. Map p. B36.
COMMON NAME(S): Blunt-leaf Orchis, One-leaf Rein-Orchis.
FREQUENCY: Scattered throughout the Province.
HABITAT: Arborvitae bogs, mossy hummocks of coniferous swamps and woods.
GENERAL RANGE: Alaska to Nfld. south to Oreg. and N. Eng.; Norway; Asia.
SYNONYMS: **Habenaria obtusata** (Pursh) Richards.

Platanthera orbiculata (Pursh) Lindl. Fig. p. A43. Map p. B36.
SUBSPECIFIC TAXA: Two well marked varieties occur in our range.

1a Spur (0.8) 1.8-2.3 (2.7) cm long; leaves generally orbicular, to 2 dm long; lip narrowly lanceolate, 1-1.5 cm long; upper sepal +/- rounded. var. **orbiculata**

1b Spur (2.5) 3-4.3 (4.5) cm long; leaves usually more elliptical, to 2.5 dm long; lip linear, 1.5-2.2 cm long; upper sepal broadly ovate. var. **macrophylla**

1. Var. **orbiculata**
COMMON NAME(S): Round-leaved Orchis.
FREQUENCY: Uncommonly scattered in many parts of the Province; rarer in the northwest and southwest counties.
HABITAT: Dryish or swampy deciduous, coniferous or mixed woods.
GENERAL RANGE: Alaska to Nfld. south to Oreg. and S.C.
SYNONYMS: **Habenaria orbiculata** (Pursh) Torr.

2. Var. **macrophylla** (Goldie) Luer
COMMON NAME(S): Large Round-leaved Orchis.
FREQUENCY: Rare; reported from near College Bridge, Westmorland Co., Fredericton, York Co.; near Clair, Madawaska Co. and Lower Woodstock, Carleton Co.
HABITAT: Rich, usually deciduous or sometimes mixed with hemlock, old woods.
GENERAL RANGE: Ont. to Nfld. south to Wisc. and Conn.
SYNONYMS: **Habenaria macrophylla** Goldie, **H. orbiculata** var. **macrophylla** (Goldie) Boivin. **Platanthera macrophylla** (Goldie) Reddoch & Reddoch.
SPECIAL NOTE: For more information see Hinds, 1983. The Rare Vascular Plants of New Brunswick. Syllogeus 50.

Platanthera psycodes (L.) Lindl. Fig. p. A43. Map p. B36.
COMMON NAME(S): Small Purple Fringed Orchis, Soldier's-plume.
FREQUENCY: Scattered throughout the Province; rarer in the northern counties.
HABITAT: Rocky shores, moist meadows, roadsides, borders of woods.
GENERAL RANGE: Man. to Nfld. south to Iowa and Ga.
SPECIAL NOTE: **P. X andrewsii** (M. White) Luer, the apparent hybrid with **P. lacera**, is reported from near Sussex, Queens Co. and from Grand Manan Is., Charlotte Co.
SYNONYMS: **Habenaria psycodes** (L.) Spreng.

++

POGONIA Juss.

Pogonia ophioglossoides (L.) Juss. Fig. p. A43. Map p. B37.
COMMON NAME(S): Snakemouth, Pogonia, Beard-flower.
FREQUENCY: Scattered mostly in the southern half of the Province north to Miscou Is., Gloucester Co.
HABITAT: Sphagnous bogs and shores.
GENERAL RANGE: Ont. to Nfld. south to Tex. and Fla.
SUBSPECIFIC TAXA: Forma **albiflora** Rand & Redf., with white instead of pink flowers, occurs occasionally with the species.

SPIRANTHES Richard Ladies'-tresses

1a Flowers in +/- single loose spiral (sometimes difficult to see in mounted specimens); leaf blades less than 5 cm long and 2/5 or more as broad, often withered at flowering time. **S. lacera**

1b Flowers in 2 or more rows in a more crowded spike; leaf blades over 4.5 cm long and less than 2/5 as broad, usually persisting at flowering. 2

2a(1b) Lip with broad yellow stripe; leaves relatively short and broad, 7-23 mm wide, 3-15 cm long, all basal; cauline bracts 1-2; plants 7-20 cm tall; flowering in July to early August. **S. lucida**

2b(1b) Lip white or creamy, without yellow stripe; leaves +/- grasslike, often well developed on lower stem; cauline bracts 3-6; plants mostly taller; flowering from August to early September. 3

3a(2b) Lip constricted below expanded tip and +/- sharply deflexed; lateral sepals partly united with dorsal sepals and lateral petals to form a hood. **S. romanzoffiana**

3b(2b) Lip oblong, ruffled but not constricted below tip, not deflexed; lateral sepals free. **S. cernua**

Spiranthes cernua (L.) Richard
COMMON NAME(S): Nodding Ladies' tresses. Screw-auger.
SPECIAL NOTE: Plants reported from near Sussex, Kings Co. have been annotated to **S. romanzoffiana** by this author. Doubtfully a member of our flora.

Spiranthes lacera Raf. Fig.p. A43. Map p. B37.
COMMON NAME(S): Northern Slender Ladies'-tresses.
FREQUENCY: Scattered mostly in the southern half of the Province.
HABITAT: Sandy clearings, alder thickets, and similar dryish open areas.
GENERAL RANGE: Sask. to NS. south to Minn. and N.C.
SPECIAL NOTE: The apparent hybrid with **S. romanzoffiana**, is reported from Kouchibouguac National Park growing with both parents in a sandy roadside ditch.

Spiranthes lucida (Eat.) Ames Fig. p. A43. Map p. B37.
COMMON NAME(S): Broad-leaved Ladies'-tresses.
FREQUENCY: Uncommonly scattered in the southern half of the Province. Locally common as at the confluence of the Renous and SW Miramichi Rivers, Northumberland Co.

HABITAT: Springy shores, meadows and thickets, often calcareous.
GENERAL RANGE: Ont. to N.S. south to Minn., Mo., and N.C.

Spiranthes romanzoffiana Cham. Fig. p. A43. Map p. B37.
COMMON NAME(S): Hooded Ladies'-tresses.
FREQUENCY: Scattered throughout the Province.
HABITAT: Damp fields, ditches, and shores.
GENERAL RANGE: Alaska to Nfld. south to Calif. and N. Eng.; Ireland; Scotland.

CLASS II **MAGNOLIOPSIDA** (DICOTYLEDONAE)

SALICACEAE (Willow Family)

1a Aments long and pendulous, their scales laciniate; leaves (in our species) generally ovate; stamens 6-60; buds with numerous scales. **Populus**
1b Aments mostly short and stiff, their scales entire; leaves generally lanceolate to linear; stamens 7 or less, usually 2; buds with solitary scale. **Salix**

POPULUS L. Poplar, Aspen, Cottonwood, Peuplier

1a Petioles flattened in cross-section; buds not sticky. 2
1b Petioles round in cross-section; buds large and sticky. 3
2a(1a) Leaves finely toothed with 4-8 teeth per cm, glabrous from the first. **P. tremuloides**
2b(1a) Leaves coarsely toothed with about 1.5 teeth per cm, white tomentose when young, becoming glabrous. **P. grandidentata**
3a(1b) Leaves and twigs +/- glabrous; fragrant. **P. balsamifera**
3b(1b) Leaves and twigs tomentose, not fragrant. 4
4a(3b) Leaves permanently white-tomentose beneath, coarsely toothed and usually lobed. **P. alba**
4b(3b) Leaves less tomentose beneath except on veins, coarsely toothed but not lobed. **P. X jackii**

Populus alba L. Fig. p. A44. Map p. B38.
COMMON NAME(S): White or Silver-leaved Poplar.
SPECIAL NOTE: Eurasian; established from plantings throughout much of the Province. Also introduced across most of Canada. The similar **P. canescens** (Aiton) Am., with toothed but not lobed leaves, has been collected in the south of the Province.

Populus balsamifera L. Fig. p. A44.
COMMON NAME(S): Balsam-Poplar. Liard.
FREQUENCY: Common throughout the Province.
HABITAT: Shores, edges of moist fields, alluvial bottomlands.

++

GENERAL RANGE: Alaska to Nfld. south to Oreg. and Pa.
SPECIAL NOTE: Balm-of-Gilead, **P. X jackii** Sarg. (syn. **P. X gileadensis** Rouleau), the sterile hybrid with **P. deltoides**, has been reported in the Province.

Populus grandidentata Michx. Fig. p. A44. Map p. B38.
COMMON NAME(S): Large-toothed Aspen. Grand Tremble, Peuplier à grandes dents.
FREQUENCY: Common throughout most of the Province except scarcer in the northern counties.
HABITAT: Dry, mostly sandy woods and recent burns.
GENERAL RANGE: Man. to N.S. south to Minn. and N.C.

Populus tremuloides Michx. Fig. p. A44.
COMMON NAME(S); Trembling or Quaking Aspen. Tremble
FREQUENCY: Common throughout much of the Province.
HABITAT: Burns, clearings, and similar disturbed wooded areas.
GENERAL RANGE: Alaska to Nfld. south to Calif., Mexico, Tenn. and Va.

SALIX L. Willow. Osier, Saule

1a Mature leaf base definitely broadly rounded to cordate; leaf tip acute to short acuminate (**S. lucida** sometimes has rounded leaf bases, but with long acuminate tips). 2

1b Leaves +/- gradually narrowed to base. 3

2a(1a) Young leaves and branchlets glabrous or pubescent, not fragrant; styles 0.7-1.2 mm long; mature leaves ovate to lanceolate, stipules often large and persistant. 4

2b(1a) Young leaves and branchlets glabrous, fragrant of balsam; styles 0.3-0.5 mm long; leaves broadly lanceolate to ovate; stipules wanting or minute. **S. pyrifolia**

3a(1b) Leaf margin distinctly revolute and blades +/- permanently pubescent beneath. 5

3b(1b) Leaf margin neither distinctly revolute nor blades hairy beneath. 6

4a(2a) Leaves pale green beneath; styles 0.2-0.5 mm long; capsules 4-7 mm long. **S. eriocephala**

4b(2a) Leaves strongly whitened beneath; styles 0.6-1.2 mm long; capsules 6-11 mm long. **S. myricoides**

5a(3a) Twigs, blades and capsules white-woolly; leaves oblanceolate or narrowly oblong. **S. candida**

5b(3a) Pubescence more lustrous, silvery-velvety, not as dense. 7

6a(3b) Mature leaf tip abruptly long acuminate to caudate (young leaves often acute); capsules glabrous. 8

6b(3b) Mature leaf tip acute to short acuminate; capsules glabrous or pubescent. 9

7a(5b) Leaves +/- oblanceolate, 1/3 to 1/5 as long as wide, acute to blunt at tip; capsule beak twice as long as plump base; style +/- obsolete; low shrubs of mostly upland sites. **S. humulis**

7b(5b) Leaves linear-lanceolate to broadly lanceolate 1/6 to 1/10 as long as wide. 10

8a(6a) Young petioles with conspicuous glands on upper side near base of blade; stamens +/- 5. **S. lucida**

8b(6a) Young petioles without conspicuous glands. 11

9a(6b) Mature leaves usually less than 1.5 cm wide, linear to linear lanceolate or narrowly oblanceolate. 12

9b(6b) Mature leaves usually more than 1.5 cm wide, lanceolate to elliptic or broadly obovate-oval. 13

10a(7b) Leaves gradually tapering from below the middle to tip, 5-15(20) mm wide, 5-12(17) cm long, densely and closely +/- permanently silvery-satiny beneath, nearly obscuring the lower surface; stipules linear-lanceolate or wanting; capsules +/- sessile, 6-8 mm long; bracts black-tipped; shrub or small tree to 8 m high; branchlets yellowish to reddish-brown puberulent, becoming glabrous. **S. viminalis**

10b(7b) Leaves tapering from near or above middle to tip, 8-15(22) mm wide, 4-10(13) cm long, +/- loosely silky-pubescent and rugose-glaucescent beneath, the lower surface clearly visible through the pubescence, becoming glabrate; stipules minute or wanting; capsule pedicels slightly longer than blackish bracts; branchlets mostly reddish-brown, often with a white waxy coating in nature which tends to disappear when pressed, sometimes thinly pubescent, becoming glabrous; shrub or small tree to 5 m high. **S. pellita**

11a(8b) Flowers and capsules grouped in loose whorls along the rachis of ament; leaves with a +/- continuous marginal vein; stamens 3-7. **S. nigra**

11b(8b) Flowers and capsules spirally arranged on rachis of ament; leaves without a continuous marginal vein; stamens 2. 14

12a(9a) Leaves +/- opposite except on new sprouts, finely rugose-reticulate on both sides, purple tinged; aments often paired; filaments often fused; capsules pubescent, ovoid, 2-3 mm long, sessile. **S. purpurea**

12b(9a) Leaves regularly alternate. 15

13a(9b) Leaves irregularly, coarsely toothed or entire. 16

13b(9b) Leaves regularly, finely toothed, darkening when pressed. 17

14a(11b) Leaves whitened and usually slightly silky pubescent beneath; ovary nearly sessile; twigs not brittle at base. **S. alba**

14b(11b) Leaves green on both sides, promptly glabrous beneath; ovary short pedicelled; twigs brittle and easily snapping off at base. **S. fragilis**

15a(12b) Petioles 5-15 mm long; blade whitened beneath, silvery or fulvous hairy at first, usually becoming glabrate; bracts linear-lanceolate to narrowly oblanceolate or spatulate, brownish; style 0.1-0.4 mm long. **S. petiolaris**

15b(12b) Petioles usually less than 3 mm long; blade not whitened beneath, glabrate or permanently silvery-silky, nearly entire or sharply and remotely denticulate; bracts lanceolate, yellowish, mostly deciduous; stigma +/- sessile. **S. exigua**

16a(13a) Blades usually less than 6 cm long (2.5-5.5(6)) cm long, 0.6-2.5 cm wide, glabrous from the first, whitened beneath, abruptly acute or rounded at tip, margin entire, slightly revolute; twigs glabrous; glabrous capsules and young leaves purplish; bog plants to about 1 m high. **S. pedicellaris**

16b(13a) Blades usually greater than 6 cm long, usually pilose or tomentose becoming glabrate; twigs glabrous or pilose. 18

++

17a(13b) Leaves glabrous from the first, slightly paler beneath; rare plant of gypsum outcrops. **S. myrtillifolia**

17b(13b) Leaves silky beneath and glaucous; plant mostly of river bars. **S. sericea**

18a(16b) Leaves dull, dark green above, veins conspicuously raised and surface sparingly to heavily pilose below (fulvous hairs rare), becoming glabrate; bracts yellowish-brown, sparingly pilose; pistillate aments expanding with young leaves, bracted; branchlets +/- pilose. **S. bebbiana**

18b(16b) Leaves bright, +/- lustrous green above, glaucous and glabrous below or with scattered fulvous hairs, veins not especially evident below; aments bractless, expanding well before the leaves appear; bracts dark brown, long villous; branchlets +/- smooth and lustrous. **S. discolor**

Salix alba L. Fig. p. A44. Map p. B38.
COMMON NAME(S): White Willow. Saule blanc.
SPECIAL NOTE: Eurasian; cultivated throughout the Province and persisting near human habitations. Naturalized in Canada from Man. to N.S. **Salix X rubens** Schrank, the hybrid with **S. fragilis** has been reported from Caissie Pt., Kent Co.

Salix bebbiana Sarg. Fig. p. A44.
COMMON NAME(S): Long-beaked Willow. Chaton or Petit Minou.
FREQUENCY: Common throughout the Province.
HABITAT: Old fields, roadsides, clearings and similar moist open areas.
GENERAL RANGE: Alaska to Nfld. south to Ariz. and Md.; Siberia.

Salix candida Fluegge ex Willd. Fig. p. A44. Map p. B38.
COMMON NAME(S): Hoary Willow. Saule chenu.
FREQUENCY: Rare; most common in the extreme northeast in Gloucester Co.
HABITAT: Calcareous or alkaline bogs and springy areas.
GENERAL RANGE: Alaska to Nfld. south to B.C., Idaho, and N.J.

Salix discolor Muhl. Fig. p. A44.
COMMON NAME(S): Pussy-Willow. Petit Minou or Chaton.
FREQUENCY: Common throughout the Province.
HABITAT: Swampy edges of woods, shores, damp roadsides and similar open areas.
GENERAL RANGE: B.C. to Nfld. south to Idaho and Md.
SPECIAL NOTE: The hybrid with **S. humilis** appears occasionally where the parent species occur together.

Salix eriocephala Michx. Fig. p. A44.
COMMON NAME(S): Red-tipped Willow.
FREQUENCY: Common throughout the Province.
HABITAT: Gravelly or sandy shores.
GENERAL RANGE: Man. to Nfld. south to Kans. and N.C.
SYNONYMS: **S. rigida** Muhl., **S. cordata** Michx. var. **rigida** (Muhl.) Carey.
SPECIAL NOTE: The apparent hybrid with **S. sericea** appears ocasionally where the parents occur together as near Ludlow, Northumberland Co.

SALICACEAE

++

Salix exigua Nutt. Fig. p. A44. Map p. B38.
COMMON NAME(S): Sandbar Willow.
FREQUENCY: Common throughout most of the Province.
HABITAT: River bars and beaches, alluvial shores.
GENERAL RANGE: Alaska to N.S. south to Okla. and Md.
SYNONYMS: **S. interior** Rowlee, **S. exigua** ssp. **interior** (Rowlee) Cronq.

Salix fragilis L.
COMMON NAME(S): Crack-Willow.
SPECIAL NOTE: Eurasian; occasionally planted and persisting as near Bathurst, Gloucester Co., Campbellton, Restigouche Co. and Kouchibouguac, Kent Co. Elsewhere across Canada it has escaped in from Alta. to N.S. **S. X rubens** Schrank, the hybrid with **S. alba** is reported from Cassie Pt., Kent Co.

Salix humilis Marsh. Fig. p. A45. Map p. B38.
COMMON NAME(S): Prairie Willow.
FREQUENCY: Common throughout the Province.
HABITAT: Mostly sandy soils of old fields, open woods and clearings.
GENERAL RANGE: Sask. to Nfld. south to Wisc. and N. Eng.
SPECIAL NOTE: The hybrid with **S. discolor** occurs rarely with the parents.

Salix lucida Muhl. Fig. p. A45.
COMMON NAME(S): Shining Willow.
FREQUENCY: Common throughout the Province.
HABITAT: Wet depressions, gravelly shores, wet roadsides, and similar disturbed areas.
GENERAL RANGE: Sask. to Nfld. south to S. Dak. and Md.
SPECIAL NOTE: **S. X schneideri** Boivin, the hybrid with **S. nigra** has been collected on the shore of Grand Lake, Queens Co., and near Westfield, Kings Co.
SYNONYMS: **S. lucida** var. **angustifolia** Anderss. and var. **intonsa** Fern.

Salix myricoides Muhl. Fig. p. A44. Map p. B38.
FREQUENCY: Uncommonly scattered on the St. John, SW Miramichi and Restigouche Rivers.
HABITAT: Sandy or gravelly shores and adjacent thickets.
GENERAL RANGE: Ont. to Nfld. south to Ill. and Maine.
SYNONYMS: **S. glaucophylloides** Fern., **S. X myricoides** (Muhl.) Carey.

Salix myrtillifolia Anderss. Fig. p. A45.
FREQUENCY: Very rare; reported only from near Wilson Brook, Albert Mines, Albert Co.
HABITAT: Gypsum cliffs and adjacent woods.
GENERAL RANGE: Alaska to Nfld. south to B.C. and N.B.; mtns of Calif. east to Colo.
SPECIAL NOTE: For more information see H. Hinds, 1983. The Rare Vascular Plants of New Brunswick. Syllogeus 50.

Salix nigra Marsh. Fig. p. A45. Map p. B38.
COMMON NAME(S): Black Willow

++

FREQUENCY: Locally common in the lower St. John R. valley and rare in the upper SW Miramichi R. valley.
HABITAT: Shores, alluvial woods and meadows.
GENERAL RANGE: Ont. to N.B. south to Mexico, Tex. and Fla.
SPECIAL NOTE: **S. X schneideri** Boivin the apparent hybrid with **S. lucida** has been found on the shores of Grand Lake, Queens Co., and Westfield, Kings Co.

Salix pedicellaris Pursh Fig. p. A45. Map p. B38.
COMMON NAME(S): Bog Willow.
FREQUENCY: Uncommonly scattered mostly towards the south of the Province; rarer northwards.
HABITAT: Bogs and wet shrubby meadows.
GENERAL RANGE: Yukon and B.C. to Nfld. south to Oreg. and N.J.
SYNONYMS: **S. pedicellaris** var. **hypoglauca** Fern.

Salix pellita Anderss. Fig. p. A45. Map p. B38.
COMMON NAME(S): Silky Willow. Saule satiné.
FREQUENCY: Common throughout the Province.
HABITAT: Alluvial shores and adjacent thickets.
GENERAL RANGE: Sask. to Nfld. south to Mich. and N. Eng.
SPECIAL NOTE: This species appears to intergrade with several introduced hybrid taxa involving the **S. viminalis** group which are difficult to separate.

Salix petiolaris Sm. Fig. p. A45. Map p. B38.
COMMON NAME(S): Slender Willow.
FREQUENCY: Common in the southern half of the Province and rarer northwards.
HABITAT: Wet meadows and shores.
GENERAL RANGE: B.C. to N.S. south to Nebr. and N.J.
SPECIAL NOTE: The apparent hybrid with **S. discolor**, was collected near Sunpoke Lake, Sunbury Co. (UNB).
SYNONYMS: **S. petiolaris** var. **gracilis** Anderss.

Salix purpurea L. Fig. p. A45.
COMMON NAME(S): Purple Osier, Basket Willow.
SPECIAL NOTE: Eurasian; rarely introduced but spreading as near Jolicure Lakes, Westmorland Co. Known in Canada from Ont. to N.S.

Salix pyrifolia Anderss. Fig. p. A44.
COMMON NAME(S): Balsam Willow.
FREQUENCY: Common throughout the Province.
HABITAT: Edges of bogs and wet thickets.
GENERAL RANGE: B.C. to Nfld. south to Minn. and N. Eng.
SYNONYMS: **S. balsamifera** Barratt ex Hook.

Salix sericea Marsh. Fig. p. A45. Map p. B38.
COMMON NAME(S): Silky Willow.
FREQUENCY: Locally common especially in the south and southcentral part of the Province.

HABITAT: Gravelly shores and streambanks.
GENERAL RANGE: Ont. to N.S. south to Mo. and S.C.

Salix viminalis L. Fig. p. A45. Map p. B39.
COMMON NAME(S): Osier. Osier blanc.
SPECIAL NOTE: This species and hybrids with the European **S. caprea** L. and **S. cinerea** L. and also the native **S. pellita**, are frequently found near inhabited as well as in more remote areas. Some forms appear to be sterile pistillate clones while others appear to be fertile and spreading mostly throughout the southern half of the Province.

JUGLANDACEAE (Walnut Family)

JUGLANS L. Walnut. Noyer.

Juglans cinerea L. Fig. p. A46. Map p. B39.
COMMON NAME(S): Butternut, White Walnut. Arbre à noix longue.
FREQUENCY: Common and native in the St. John R. valley and upper SW Miramichi R. valley. Rarely planted elsewhere.
HABITAT: Rich deciduous woods.
GENERAL RANGE: N. Dak. and Ont. to N.B. south to Ark. and Ga.

MYRICACEAE (Wax-Myrtle Family)

1a Leaves entire or slightly serrate towards apex; stipules wanting. **Myrica**
1b Leaves pinnatifid, +/- fern-like; stipules subcordate. **Comptonia**

COMPTONIA L'Her.

Comptonia peregrina (L.) Coult. Fig. p. A45. Map p. B39.
COMMON NAME(S): Sweet-fern.
FREQUENCY: Common throughout the southeastern half of the Province.
HABITAT: Disturbed roadsides, clearings and similar dryish open areas.
GENERAL RANGE: Ont. to N.S. south to Minn. and Ga.
SYNONYMS: **Myrica asplenifolia** L.

MYRICA L.

1a Leaves light green, dull above; catkins borne at summit of previous year's branchlets; fruit a 2-winged nutlet. **M. gale**
1b Leaves dark-green, shiny above; catkins borne on old wood +/- below leafy tips; fruit a wax-covered globular nutlet. **M. pensylvanica**

Myrica gale L. Fig. p. A45. Map p. B39.
COMMON NAME(S): Sweet Gale. Bois-sent-bon or Piment royal.
FREQUENCY: Common throughout the Province.
HABITAT: Wet shores, borders of woods and bogs, wet shrubby thickets.
GENERAL RANGE: Alaska to Nfld. south to Oreg. and N.C.; Europe; Asia.

Myrica pensylvanica Loisel. Fig. p. A45. Map p. B39.
COMMON NAME(S): Bayberry, Candleberry.
FREQUENCY: Common along the east coast and rarer on the Bay of Fundy coast.
HABITAT: Sandy fields, clearings, shores and dunes near the coast.
GENERAL RANGE: Que. to Nfld. south along the coast to N.C.

BETULACEAE (Birch Family)

1a Fruit consisting of small, flat winged or margined nutlets; involucre wanting; pistillate ament scales woody and enlarged in fruit. 2

1b Fruit consisting of a foliaceous involucre enclosing or subtending a single wingless nut; pistillate ament scales deciduous. 3

2a(1a) Bracts of fruiting cone-like aments thick, woody, persistent; stamens 4. **Alnus**

2b(1a) Bracts of fruiting aments thin, +/- falling with the nutlets; stamens 2. **Betula**

3a(1b) Mature involucre inflated, enclosing the small nut; tree with deeply fissured dark bark. **Ostrya**

3b(1b) Mature involucre closely enfolding nut and prolonged into a slender beak (in our species); fastigiate shrub. **Corylus**

ALNUS Ehrh. Alder. Aulne or Aune

1a Cones on peduncles to 2 cm long; buds sessile; leaves finely, mostly simply serrate; pistillate aments enclosed in bud over winter; bushy shrub to about 4 m tall. **A. viridis**

1b Cones sessile or on peduncles less than 1 cm long; buds stalked; leaves coarsely, doubly serrate; pistillate aments naked over winter; shrubs or small trees to 10 m tall. 2

2a(1b) Leaves rhombic to obovate, widest at or above middle, finely, mostly simply serrate on margin, cuneate to rounded at base, green below and glutinous when young, cross-veins between main veins weakly developed. **A. serrulata**

2b(1b) Leaves mostly ovate, widest near or below middle, mostly coarsely and doubly serrate on margin, broadly rounded or cordate at base, becoming whitened beneath, cross-veins between main veins prominent. **A. incana**

Alnus incana (L.) Moench. Fig. p. A46.
COMMON NAME(S): Speckled Alder. Verne or Aulne blanchâtre.
FREQUENCY: Common throughout the Province.
HABITAT: Wet, mostly acid ground, old fields, and streambanks.

GENERAL RANGE: Sask. to Nfld. south to Iowa and W. Va.
SUBSPECIFIC TAXA: Our material is referable to ssp. **rugosa** (Du Roi) Clausen.
SYNONYMS: **A. rugosa** (Du Roi) Spreng. var. **americana** (Regel) Fern.

Alnus serrulata (Ait.) Willd.
FREQUENCY: Very rare; known only from below the dam of the Grand Falls of the St. Croix R., Charlotte Co. (UNB).
HABITAT: Upper gravelly shore and edge of dry acid woods.
GENERAL RANGE: N.B. and N.S. south to La. and Fla.
SYNONYMS: **A. rugosa** (DuRoi)preng. var. **serrulata** (Ait.) Winkler.

Alnus viridis (Chaix.) DC. Fig. p. A46.
COMMON NAME(S): Green or Mountain-Alder. Aulne vert.
FREQUENCY: Probably common throughout the Province.
HABITAT: Sandy or gravelly banks, rocky shores and moist slopes.
GENERAL RANGE: Ont. to Nfld. south to N.Y. and N. Eng.
SUBSPECIFIC TAXA: Our material is referable to ssp. **crispa** (Ait.) Turrill.
SYNONYMS: **A. crispa** (Ait.) Pursh, **A. crispa** var. **mollis** Fern.

BETULA L. Birch. Bouleau

1a Leaves with 9-11 pairs of primary veins; bark yellowish brown or silvery-grey becoming darker in age; pistillate aments oval, sessile. **B. alleghaniensis**

1b Leaves with 7 or less primary veins; pistillate aments cylindrical, mostly stalked. 2

2a(1b) Leaves +/- rounded at tip, 0.5-2.5 cm long; low shrubs with dark, close bark. 3

2b(1b) Leaves acute or acuminate at tip, mostly over 5 cm long; trees. 4

3a(2b) Twigs glandular warty, finely pubescent; leaves rarely over 3 cm long, green beneath. **B. glandulosa**

3b(2b) Twigs not glandular warty, heavily pubescent; leaves whitened beneath. **B. pumila**

4a(2b) Leaves usually coarsely doubly serrate, +/- rounded or cordate at base, acute to short acuminate at tip, pubescent beneath at least when young; staminate aments 2-several. 5

4b(2b) Leaves usually simply serrate, truncate to broadly tapering at base, +/- long acuminate at tip, glabrous beneath or pubescent in angles of main veins; staminate aments usually solitary. 6

5a(4a) Leaves ovate to broadly lanceolate, +/- rounded or tapering at base; bracts of fruiting aments with sub-quadrangular or sub-rhomboidal side-lobes, the median lobe tapering to tip; young twigs pubescent. **B. papyrifera**

5b(4a) Leaves cordate at base; bracts of fruiting aments with smooth curving outer margins, the median lobe long and parallel-sided, rounded at tip; young twigs glabrous. **B. cordifolia**

6a(4b) Leaves +/- triangular in outline with long acuminate tips; bracts of pistillate aments 1.5-4 mm long, horizontally divergent from central axis. 7

++

6b(4b) Leaves usually more ovate in outline with shorter acuminate tips; bracts of pistillate aments 5-7 mm long, ascending from central axis; leaves dull-green above, 6-8 cm long. **B. X caerulea**

7a(6a) Leaves with caudate tips; bark chalky or ash-white, not flaking in layers; twigs wiry, ascending at tips; trunks often 2-several. **B. populifolia**

7b(6a) Leaves not caudate tipped; bark lustrous, creamy or pinkish-white, often flaking in layers; twigs drooping at tip; trunk single. **B. pendula**

Betula alleghaniensis Britt. Fig. p. A46.
COMMON NAME(S): Yellow Birch. Merisier.
FREQUENCY: Common throughout the Province.
HABITAT: Cool, moist forests on acid soils.
GENERAL RANGE: Ont. to Nfld. south to Iowa and N.C.
SYNONYMS: **B. lutea** Michx. f.
SUBSPECIFIC TAXA: Var. **macrolepis** Fern., with pistillate ament bracts more than 8 mm long, occurs occasionally with the species.

Betula cordifolia Regel Fig. p. A46.
FREQUENCY: Scattered throughout the Province.
HABITAT: Mostly spruce forests.
GENERAL RANGE: Ont. to Nfld. south to Iowa. and N.C.
SYNONYMS: **B. papyrifera** var. **cordifolia** (Regel) Fern.
SPECIAL NOTE: Hybrid swarms between this species and **B. populifolia** form a confusing series of fairly common taxa variously called **B. X caerulea-grandis** Blanch., **B. X caerulea** Blanch. and **B. X caerulea** Blanch. nm. **grandis** (Blanch.) Boivin. **Betula cordifolia** and/or **B. papyrifera** also hybridize with **B. pumila** producing **B. X sandbergii** Britt. nm. **fernaldii** Lepage and with **B. glandulosa** to produce an excedingly complex series of rare taxa variously called **B. minor** (Tuckerm.) Fern., **B. borealis** Spach, **B. X neoborealis** Lepage, **B. X ungavensis** Lepage, **B. dutillyii** Lepage and **B. saxophila** Lepage.

Betula glandulosa Michx. Fig. p. A46.
COMMON NAME(S): Dwarf or Scrub Birch.
FREQUENCY: Very rare; known only from the summits of Big Bald and Carleton Mtns., Northumberland Co.
HABITAT: Rocky and peaty barrens of subalpine summits.
GENERAL RANGE: Alaska to Nfld. south to Calif. and Maine; Greenland.
SPECIAL NOTE: For more information see H. Hinds, 1983. The Rare Vascular Plants of New Brunswick. Syllogeus 50. Also see **B. cordifolia** concerning hybrid taxa.

Betula papyrifera Marsh. Fig. p. A46.
COMMON NAME(S): White, Canoe or Paper-Birch. Bouleau blanc or Bouleau à canot.
FREQUENCY: Common throughout the Province.
HABITAT: Slopes, ridges and shores.
GENERAL RANGE: Alaska to Nfld. south to Calif. and n. N. Eng.
SYNONYMS: **B. papyrifera** var. **pensilis** Fern., and var. **macrostachya** Fern.
SPECIAL NOTE: See **B. cordifolia** concerning possible hybrid taxa.

Betula pendula Roth
COMMON NAME(S): European Birch.
SPECIAL NOTE: Eurasian; occasionally escaped from cultivation as near Sackville, Westmorland Co., and St. Andrews, Charlotte Co. Elsewhere known in Canada from Man., Ont., P.E.I., and N.S.

Betula populifolia Marsh. Fig. p. A46. Map p. B39.
COMMON NAME(S): Grey, Fire, or Oldfield Birch. Bouleau rouge.
FREQUENCY: Common in the southeast half of the Province and rare northwards.
HABITAT: Dry sandy upland mixed forests, sand plain forests, old fields and dry peaty clearings.
GENERAL RANGE: Ont. to N.S. south to Ind. and Va.
SPECIAL NOTE: See **B. cordifolia** concerning possible hybrid taxa.

Betula pumila L. Fig. p. A46. Map p. B39.
COMMON NAME(S): Low or Swamp Birch.
FREQUENCY: Rare in the southwest with two known localities; one near Sunpoke Lake, Sunbury Co., and the other on Whitehead Is., Charlotte Co. Fairly common in the northeast as on Miscou Is., Gloucester Co.
HABITAT: Boggy thickets and wet shrubby meadows.
GENERAL RANGE: Ont. to Nfld. south to Oreg. and N.J.
SPECIAL NOTE: See **B. cordifolia** concerning possible hybrid taxa.

CORYLUS L. Hazel, Hazelnut, Filbert. Noisetier, Coudrier

Corylus cornuta Marsh. Fig. p. A47.
COMMON NAME(S): Beaked Hazelnut.
FREQUENCY: Common throughout the Province.
HABITAT: Open acid woodlands, borders of woods.
GENERAL RANGE: B.C. to Nfld. south to Calif. and Ga.

OSTRYA Scop. Ironwood, Hop-Hornbeam

Ostrya virginiana (Mill.) Koch Fig. p. A47. Map p. B39.
COMMON NAME(S): American Hop-Hornbeam, Leverwood. Bois dur, Bois de fer, Bois à levier.
FREQUENCY: Locally common in the southern half of the Province.
HABITAT: Rich, mostly deciduous woods.
GENERAL RANGE: Man. to N.S. south to Tex. and Fla.

FAGACEAE (Beech Family)

1a Mature fruit (acorn), 1 seeded with a cup-like basal involucre formed of many coalesced scales; leaves lobed. **Quercus**

1b Mature fruit a 1-2 seeded, 3-sided, nut enclosed in a soft spiny husk; leaves sharply toothed. **Fagus grandifolia**

FAGUS L. Beech. Hêtre

Fagus grandifolia Ehrh. Fig. p. A47.
FREQUENCY: Common throughout much of the Province.
HABITAT: Rich, mature woods.
GENERAL RANGE: Ont. to N.S. south to Tex. and Fla.

QUERCUS L. Oak. Chêne

1a Lobes of leaves with rounded apex.
1b Lobes of leaves with pointed apex; involucral cup mostly shallow, scales appressed. **Q. rubra**
2a(1a) Involucral cup enclosing about 2/3 of acorn, fringed by loose, shaggy scales; fruit +/- sessile. **Q. macrocarpa**
2b(1a) Involucral cup not fringed, enclosing less than ½ of acorn; fruiting peduncle 3-8 cm long. **Q. robur**

Quercus macrocarpa Michx. Fig. p. A47. Map p. B39.
COMMON NAME(S): Mossy-cup Oak.
FREQUENCY: Scattered in the lower St. John R. valley and very rare or planted elsewhere in the Province.
HABITAT: Swampy woods and bottomlands.
GENERAL RANGE: Sask. to N.B. south to Wyo. and N.C.

Quercus robur L. Fig. p. A47.
COMMON NAME(S): English Oak.
SPECIAL NOTE: European; only known from St. Andrews, Charlotte Co. where it is thoroughly naturalized. Known elsewhere in Canada from P.E.I. and N.S.

Quercus rubra L. Fig. p. A47. Map p. B39.
COMMON NAME(S): Northern Red Oak. Chêne rouge.
FREQUENCY: Scattered throughout the Province except rare in the northernmost counties.
HABITAT: Sandy or gravelly acid woodlands.
GENERAL RANGE: Ont. to N.S. south to Ia. and N.C.
SUBSPECIFIC TAXA: Our material is referable to var. **borealis** (Michx. f.) Fern.

ULMACEAE (Elm Family)

ULMUS L. Elm. Orme

1a Leaves usually greater than 8 cm long, unequal at base, doubly serrate. 2

++

1b Leaves less than 8 cm long; +/- equal at base, simple or slightly double serrate; flowers in short-pedicellate dense clusters; spring blooming. **U. pumila**

2a(1a) Leaf buds glabrous or slightly puberulent; flowers with pedicels to 2 cm long; samara ciliate, elliptic; leaves not aromatic when dried. **U. americana**

2b(1a) Leaf buds densely fulvous pubescent; flowers sessile or with pedicels less than 1 cm long; samara entire, nearly circular; leaves fragrant when dried. **U. rubra**

Ulmus americana L. Fig. p. A47. Map p. B40.
COMMON NAME(S): American or White Elm. Orme blanc.
FREQUENCY: Common throughout most of the Province.
HABITAT: Native on alluvial soils mostly of river bottoms but extensively planted elsewhere.
GENERAL RANGE: Sask. to N.S. south to Tex. and Fla.

Ulmus pumila L.
COMMON NAME(S): Siberian or Dwarf Elm.
SPECIAL NOTE: Asiatic; introduced and spreading to fields and along railways in Fredericton. Also reported from Kouchibouguac National Park, Kent Co. Known in Canada also from Ont.

Ulmus rubra Muhl. Fig. p. A47.
COMMON NAME(S): Slippery or Red Elm. Orme rouge. Orme gras.
SPECIAL NOTE: Reports from St. Andrews, Charlotte Co. may refer to the Eurasian **U. glabra** Hud. Documenting specimens have not been seen. Known elsewhere from N. Dak. and Ont. to Que. south to Tex. and Fla.

CANNABINACEAE (Hemp Family)
(Alternate name **MORACEAE**)

1a Stems climbing; leaves deeply 3-7 lobed; fruit the "hop" of commerce. **Humulus lupulus**

1b Stems upright; leaves compound of 3-7 narrowly lanceolate, pubescent leaflets; fruit a nutlet. **Cannabis sativa**

CANNABIS L.

Cannabis sativa L. fig. p. A47. Map p. B40.
COMMON NAME(S): Marijuana, Hemp. Chanvre.
SPECIAL NOTE: Asiatic; Previously cultivated for hemp rope. Occasionally still cultivated (illegally) and rarely escaped to beaches, shores and similar places of recreation. Known in Canada also from Alta. to N.B.

++

HUMULUS L. Hop. Houblon

Humulus lupulus L. Fig. p. A47. Map p. B40.
COMMON NAME(S): Common Hop.
SPECIAL NOTE: Eurasian; extensively introduced throughout the Province and across Canada.

URTICACEAE (Nettle Family)

1a Leaves opposite. 2
1b Leaves alternate; flowers in cymes from upper leaf axils; plant with stringing hairs. **Laportea canadensis**
2a(1a) Plant with stinging hairs. **Urtica**
2b(1a) Plant without stinging hairs. 3
3a(2b) Stems watery, +/- translucent; flowers in axillary cymes or panicles; weak-stemmed, glabrous, annuals usually less than 4 dm tall. **Pilea pumila**
3b(2b) Stems not watery; flowers in long, erect, axillary spikes; erect, pubescent perennials to 9 dm tall. **Boehmeria cylindrica**

BOEHMERIA Jacq. False Nettle

Boehmeria cylindrica (L.) Sw. Fig. p. A47.
COMMON NAME(S): Bog-Hemp.
FREQUENCY: Very rare; known from Ash Swamp near McGowans Corner, Sunbury Co. Previously also known from the mouth of the Shogomoc R., York Co. but destroyed by the Mactaquac hydroelectrical development.
HABITAT: Wet alluvial woods.
GENERAL RANGE: Minn. and Ont. to N.B. south to Tex. and Fla.
SPECIAL NOTE: For more information see H. Hinds, 1983. The Rare Vascular Plants of New Brunswick. Syllogeus 50.

LAPORTEA Gaud. Wood-Nettle

Laportea canadensis (L.) Wedd. Fig. p. A47. Map p. B40.
COMMON NAME(S): Ortie du Canada.
FREQUENCY: Common throughout most of the Province.
HABITAT: Alluvial woods.
GENERAL RANGE: Sask. to N.S. south to Okla. and Fla.

PILEA Lindl. Richweed, Clearweed, Coolwort

Pilea pumila (L.) Gray Fig. p. A48.
COMMON NAME(S): Petite Ortie.
FREQUENCY: Uncommon; known only from the St. John R. valley in York and Carleton Counties.

++

HABITAT: Alluvial woods and similar moist shaded areas.
GENERAL RANGE: Ont. to P.E.I. south to Tex. and Fla.

URTICA L. Nettle. Ortie.

1a Leaves blunt-tipped, oval or elliptic; stipules less than 5 mm long; soft-stemmed, tap-rooted annuals to 5 dm tall. **U. urens**

1b Leaves ovate to lanceolate, acute to acuminate at apex; stipules to 1.5 cm long; tough-stemmed, rhizomatous perennials to over 1 mm tall. **U. dioica**

Urtica dioica L. fig. p. A48. Map p. B40.
COMMON NAME(S): Stinging Nettle.
SUBSPECIFIC TAXA: Our material can be divided into two subspecies and three varieties.

1a Leaves mostly cordate at base, coarsely toothed (11-14 teeth per margin); upper stem and petioles with stout bristles and pilose; dioecious. ssp. **dioica**

1b Leaves mostly rounded or only slightly cordate at base, more finely toothed (11-38 teeth per margin); stem and petioles without stout bristles. ssp. **gracilis** 2

2a(1b) Leaves usually less than 3 times as long as broad, averaging 17 teeth per margin, more than 3 times as long as petiole. var. **gracilis**

2b(1b) Leaves usually more than 3 times as long as broad, averaging 25 teeth per margin, 3 or less times as long as petiole. var. **procera**

1. Ssp. **dioica**
SPECIAL NOTE: Eurasian; rarely introduced as in Fredericton, York Co. and Chatham, Northumberland Co.

2. Ssp. **gracilis** (Ait.) Seland. var. **gracilis** (Ait.) Tayl. & MacB.
FREQUENCY: Scattered throughout most of the Province.
HABITAT: Rich, moist deciduous woods.
GENERAL RANGE: Alaska to Nfld. south to Oreg. and W. Va.
SYNONYMS: **U. gracilis** Ait.

3. Ssp. **gracilis** var. **procera** (Muhl.) Weddell
FREQUENCY: Uncommonly scattered throughout much of the Province.
HABITAT: Waste places, beaches, roadsides, and clearings.
GENERAL RANGE: N. Dak. and Sask. to N.S. south to La. and N.C.
SYNONYMS: **U. procera** Muhl.

Urtica urens L. Fig. p. A48.
COMMON NAME(S): Burning Nettle, Dog-Nettle.
SPECIAL NOTE: Eurasian; rarely introduced near habitations. Also reported from across most of Canada.

++

SANTALACEAE (Sandalwood Family)

1a Flowers white, all perfect, borne in terminal corymbs or panicles; fruit a nut. **Comandra umbellata**

1b Flowers bronze, purplish or greenish, borne in cymules in axils of middle leaves, only the central one perfect; fruit juicy. **Geocaulon lividum**

COMANDRA Nutt. Bastard-Toadflax

Comandra umbellata (L.) Nutt. Fig. p. A48. Map p. B40.
FREQUENCY: Scattered along the east coast and inland along the Restigouche R. and Grand Lake, Queens Co.
HABITAT: Dry sandy soils in the open or partly shaded.
GENERAL RANGE: Yukon to Nfld. south to Ariz. and Ga.
SYNONYMS: **C. richardsiana** Fern.

GEOCAULON Fern.

Geocaulon lividum (Richards.) Fern. Fig. p. A48. Map p. B40.
COMMON NAME(S): Northern Comandra.
FREQUENCY: Uncommonly scattered mostly in the northeastern part of the Province.
HABITAT: Sphagnous bogs and sandy coniferous woods.
GENERAL RANGE: Alaska to Nfld. south to Wash. and N. Eng.

LORANTHACEAE (Mistletoe Family)

ARCEUTHOBIUM Bieb. Dwarf Mistletoe

Arceuthobium pusillum Peck. Fig. p. A48. Map p. B40.
FREQUENCY: Probably scattered throughout the Province.
HABITAT: Usually parasitic on Black Spruce in boggy areas.
GENERAL RANGE: Sask. to Nfld. south to Minn. and N.J.

ARISTOLOCHIACEAE (Birthwort Family)

ASARUM L. Wild Ginger. Asarette

Asarum canadense L. Fig. p. A48. Map p. B40.
COMMON NAME(S): Wild Ginger. Gingembre sauvage.
FREQUENCY: Locally common in the St. John R. valley above Fredericton and in the Restigouche R. valley. Also reported from the Tobique R. valley.
HABITAT: Rich, deciduous and calcareous woods and bases of ledges.
GENERAL RANGE: Man. to N.B. south to Kans. and N.C.

POLYGONACEAE (Buckwheat Family)

1a Basal leaves large, to over 4 dm wide, cordate, on stout, edible petioles; achenes broadly 3-winged, to 1 cm long; stems hollow. **Rheum rhaponticum**

1b Basal leaves smaller, usually less than 1.5 dm wide, rarely cordate; achenes trigonous or lenticular, not broadly winged. 2

2a(1b) Sepals all alike, often petaloid. 3

2b(1b) Sepals not all alike, the inner 3 enlarged and often with 1 or more conspicuous tubercles on the midnerve, the outer 3 often reflexed, not enlarged in fruit. **Rumex**

3a(2a) Leaves linear-filiform, jointed near base and early deciduous; pedicels jointed near base; flowers in loose terminal racemes which occur in the axils of ocreolae. **Polygonella articulata**

3b(2a) Leaves broader, persistent; pedicels jointed at summit; flowers clustered in axils of leaves or ocreolae, in spike-like panicles, or in corymbed or panicled racemes. 4

4a(3b) Achene +/- included in closely appressed and enlarged calyx (some species with linear to linear-lanceolate leaves often produce late season fruit that are slightly exserted from the calyx). **Polygonum**

4b(3b) Achene exserted or loosely embraced by the shriveling calyx; smooth annuals with hastate, cordate or deltoid leaves. **Fagopyrum**

FAGOPYRUM Mill. Buckwheat. Sarrasin

Fagopyrum esculentum Moench Fig. p. A48.
COMMON NAME(S): Buckwheat. Sarrasin or Blé noir.
SPECIAL NOTE: Asiatic; cultivated and very occasionally self sowing, but scarcely persisting. Introduced across much of Canada.
SYNONYMS: **F. sagittatum** Gilib.

Fagopyrum tataricum (L.) Gaertn. Fig. p. A48. Map p. B40.
COMMON NAME(S): India-wheat. Sarrasin de Tartarie.
SPECIAL NOTE: Asiatic; cultivated and often more persistent than the previous species.

POLYGONELLA Michx.

Polygonella articulata (L.) Meisn. Fig. p. A48.
COMMON NAME(S): Jointweed.
FREQUENCY: Very rare; recorded from Westfield, Kings Co. and Bethel, Charlotte Co.
HABITAT: Sandy roadsides or railways.
GENERAL RANGE: Minn. to Ont. and N.S. south to Iowa and N.C.
SPECIAL NOTE: For more information see H. Hinds, 1983. The Rare Vascular Plants of New Brunswick. Syllogeus 50.

++

POLYGONUM L. Knotweed, Smartweed. Renouée

1a Leaves elliptic to ovate, truncate, cordate, sagittate or hastate at base. 2

1b Leaves +/- linear to lanceolate, tapering to petiole. 3

2a(1a) Leaves ovate, +/- cordate at base; flowers not in heads. 4

2b(1a) Leaves sagittate or hastate at base; flowers in headlike clusters. 5

3a(1b) Flowers few in axils of leaves or bracts. 6

3b(1b) Flowers in cylindrical spikes or small separated groups without bracts or leaves. 7

4a(2a) Stems stout, shrubby, bushy-branched, 1-2.5 m tall; plants monoecious. 8

4b(2a) Stems slender, usually climbing or trailing. 9

5a(2b) Basal lobes of leaves directed directly back; achenes trigonous; peduncles glabrous. **P. sagittatum**

5b(2b) Basal lobes directed outward; achenes lenticular; peduncles glandular-bristly. **P. arifolium**

6a(3a) Achenes roughened (often in lines), not glossy. 10

6b(3a) Achenes smooth, glossy to sublustrous (**P. fowleri** with achenes sometimes minutely puncticulate). 11

7a(3b) Ocreae usually fringed with bristles at summit. 12

7b(3b) Ocreae usually lacking apical bristles (surface may be glabrous or hairy). 13

8a(4a) Leaves truncate to slightly cuneate at base. **P. cuspidatum**

8b(4a) Leaves cordate to rounded at base. 14

9a(4b) Ocreae reflexed-bristly at base; outer calyx lobes obscurely keeled. **P. cilinode**

9b(4b) Ocreae glabrous. 15

10a(6a) Stems erect; to 1 m high; plants of brackish marshes. **P. ramosissimum**

10b(6a) Stems low; depressed or trailing, to 3.5 dm high; plants of non-saline habitats. 16

11a(6b) Plants +/- erect; leaves lanceolate, oblong or linear, to 6 cm long; normal, early achenes included within calyx. 17

11b(6b) Plants prostrate to ascending; leaves linear-oblanceolate; achenes usually strongly exsert from calyx. 18

12a(7a) Sepals punctate with sessile glands. 19

12b(7a) Sepals not glandular. 20

13a(7b) Racemes 1-4, terminal, very dense; perennials with slender trailing rhizomes; habit often aquatic with floating leaves. 21

13b(7b) Racemes both terminal and axillary, usually numerous; mostly annuals without rhizomes; habit terrestrial with erect stems. 22

14a(8b) Flowers greenish to greenish-white; fruiting calyx wing-angled; ocreae not bristly at summit; achenes trigonous; glabrous perennials to 3.5 m tall. **P. sachalinense**

14b(8b) Flowers roseate; fruiting calyx not winged; ocreae with divergent bristly-ciliate flange; achenes lenticular; soft-hairy annual with fibrous roots. **P. orientale**

15a(9b) Calyx pubescent, as long as or longer than pedicel, scarcely winged; achenes dull. **P. convolvulus**

15b(9b) Calyx glabrous, shorter than pedicel, the outer lobes broadly winged; achenes shiny. **P. scandens**

16a(10b) Plant homophyllous; achene with 1 concave side about ½ as wide as the other two. 23

16b(10b) Plant heterophyllous; achene with +/- equal concave sides; perianth divided about 2/3 way to base with overlapping segments. **P. aviculare**

17a(11a) Pedicels included within sheathing ocreae; calyx about 1.3-1.8 mm long; leaves +/- rugose-veiny, especially when dry, oblanceolate; undistended normal achene sublustrous, about 2 mm long. **P. prolificum**

17b(11a) Pedicels exserted from sheathing ocreae; calyx about 3 mm long; leaves flat with inconspicuous side veins; achene glossy, to 4 mm long. **P. ramosissimum**

18a(11b) Achene broadly ovoid, abruptly beaked, sublustrous but often minutely puncticulate, 3-4 mm long; ocreae 3-4(6) mm long, +/- veinless; leaves elliptic to obovate, 4-10 mm broad, broadly rounded. **P. fowleri**

18b(11b) Achene ovate-lanceolate, gradually tapering above, glossy, undistended 4-4 7 mm long, distended 5.7-7 mm long; ocreae to about 1 cm or more, veiny; leaves linear to oblanceolate, 2-7 mm broad, acutish. **P. oxyspermum**

19a(12a) Achenes dull, 2-3.5 mm long; calyx greenish to purplish-tipped; plant parts acrid; annuals with fibrous roots. **P. hydropiper**

19b(12a) Achenes shiny, 3-4 mm long; calyx white; plant not acrid; annuals or short-lived perennials often with creeping rootstock. **P. punctatum**

20a(12b) Stem and peduncles stipitate-glandular; leaves scabrous; spikes arching or recurved. **P. careyi**

20b(12b) Stem and peduncles without obvious stalked glands. 23

21a(13a) Peduncles pubescent and often glandular; spikes narrowly cylindrical, to over 1.5 dm long, 1.5 cm thick; ocreolae ciliate and pubescent; ocreae in terrestrial forms without divergent flange. **P. coccineum**

21b(13b) Peduncles and ochreolae glabrous; spikes broadly cylindrical or ovoid, to 4 cm long, 2 cm thick; ochrea of terrestrial forms with divergent flange. **P. amphibium**

22a(13b) Racemes nodding; outer sepals in fruit strongly 3-nerved and anchor-shaped; peduncles and axis of inflorescence with inconspicuous glands; achene 1.8-2.2 mm long. **P. lapathifolium**

22b(13b) Racemes erect; outer sepals in fruit with inconspicuous, irregularly forked nerves; peduncle and axis of inflorescence with stipitate glands; achene 2.2-3.5 mm long. **P. pensylvanicum**

23a(16a) Perianth constricted above achene, 3.5-4 mm long, white margined, the outer lobes +/- cucullate; achene about 2.5 mm long olivaceous. **P. achoreum**

23b(16a) Perianth not constricted above, 2-3 mm long, roseate to purple margined, the outer lobes scarcely overlapping in fruit; achene 1.8-2.5 mm long, dark brown, to 4 mm long, greenish and shiny in late season achenes and prominently exserted. **P. arenastrum**

24a(20b) Perennials with subligneous forking and elongate rootstocks; aquatic or subaquatic with decumbent rooting bases; leaves lanceolate, to 2.5 cm wide; spikes bright pink, not crowded. **P. hydropiperoides**

24b(20b) Annuals with fibrous roots and without elongate subligneous rootstocks. 25

++

25a(24b) Spikes slender, 3-5 mm thick, each flower +/- distinctly visible; ocreae usually long ciliate; leaves linear-lanceolate to linear-oblong, not dark mottled, usually less than 8 mm wide; riparian. **P. minus**

25b(24b) Spikes dense, 4-11 mm thick, crowded and overlapping; ocreae shortly ciliate, sometimes eciliate; leaves lanceolate, 3-15 cm long, 0.5-3 cm broad, usually dark mottled; ruderal. **P. persicaria**

Polygonum achoreum Blake Fig. p. A48. Map p. B40.
COMMON NAME(S): Leathery Knotweed.
FREQUENCY: Locally common in the lower St. John R. drainage system.
HABITAT: Disturbed ground, roadsides and cultivated fields.
GENERAL RANGE: Alaska to Nfld. south to Idaho and N.Y.

Polygonum amphibium L. Fig. p. A48. Map p. B40.
COMMON NAME(S): Water-Smartweed, Amphibious Bistort. Renouée amphibie.
FREQUENCY: Common in the southern half of the Province and rarer towards the north.
HABITAT: Shallow water and mostly muddy shores, ditches and wet meadows.
GENERAL RANGE: Alaska to Nfld. south to Utah and N.J.; Iceland; Eurasia.
SPECIAL NOTE: Our material is referable to var. **stipulaceum** (Coleman) Fern.
SYNONYMS: **P. natans** (Michx.) Eat.

Polygonum arenastrum Jord. Fig. p. A48. Map p. B41.
FREQUENCY: Common in the southern half of the Province.
HABITAT: Disturbed soils of paths, parking lots and roadsides.
GENERAL RANGE: European; introduced transcontinentally.
SYNONYMS: **P. aviculare** L. of many early reports, **P. neglectum** Bess.

Polygonum arifolium L. Fig. p. A49. Map p. B41.
COMMON NAME(S): Halberd-leaved Tearthumb.
FREQUENCY: Rare; known from only a few sites in the southern half of the Province.
HABITAT: Swampy areas.
GENERAL RANGE: Minn. to Ont. and N.S. south to Ohio and Fla.
SUBSPECIFIC TAXA: Our material is referable to var. **pubescens** (Keller) Fern.
SPECIAL NOTE: For more information see H. Hinds, 1983. The Rare Vascular Plants of New Brunswick. Syllogeus 50.

Polygonum aviculare L. Map p. B41.
COMMON NAME(S): Prostrate Knotweed.
FREQUENCY: Locally common mostly towards the southern half of the Province.
HABITAT: Waste places, roadsides, gardens and cultivated fields.
GENERAL RANGE: Widespread throughout North America and probably elsewhere, the complete range not yet known.
SYNONYMS: **P. monspeliense** Pers.

Polygonum careyi Olney Fig. p. A49. Map p. B41.
COMMON NAME(S): Carey's Knotweed.
FREQUENCY: Scattered in the southwest half of the Province.
HABITAT: Rich cultivated ground, clearings and swamp margins.
GENERAL RANGE: Ont. to N.B. south to Minn. and Del.

++

Polygonum caespitosum Blume
SPECIAL NOTE: Asiatic; rarely introduced as near Mouth-of-Meswick, York Co. (UNB). Also known in Canada from B.C., Ont. and Que.
SPECIFIC TAXA: Our material is referable to var. **longisetum** (de Bruyn) Stewart.

Polygonum cilinode Michx. Fig. p. A49. Map p. B41.
COMMON NAME(S): Fringed Black Bindweed.
FREQUENCY: Common throughout the Province.
HABITAT: Rocky slopes, clearings, and dry banks.
GENERAL RANGE: Sask. to Nfld. south to Minn. and N.C.

Polygonum coccineum Muhl. Fig. p. A49.
COMMON NAME(S): Water-Smartweed.
FREQUENCY: Very rare; known only from the Canal near Lake Utopia, Charlotte Co.
HABITAT: Quiet waters and adjacent shores.
GENERAL RANGE: Dist. Mack. to N.S. south to Calif. and N.C.
SYNONYMS: **P. amphibium** var. **emersum** Michx.

Polygonum convolvulus L. Fig. p. A49.
COMMON NAME(S): Black Bindweed. Renouée liseron or Chevrier.
FREQUENCY: Common throughout the Province.
HABITAT: Disturbed ground of roadsides, fields, old habitations, and estuary beaches.
GENERAL RANGE: Eurasian; introduced from Alaska through all Provinces south throughout much of U.S.A.; Greenland.

Polygonum cuspidatum Sieb. & Zucc. Fig. p. A49.
COMMON NAME(S): Japanese Knotweed. Fleeceflower. Bambou, Canne, Jonc de Saint-Joseph.
FREQUENCY: Uncommonly scattered in the southern half of the Province.
HABITAT: Old habitations and waste areas.
GENERAL RANGE: Asiatic; previously frequently cultivated and persisting but not spreading by seed.

Polygonum fowleri Robins. Fig. p. A49. Map p. B41.
COMMON NAME(S): Fowler's Knotweed.
FREQUENCY: Locally common in the extreme southwest of the Province.
HABITAT: Sea beaches and sandy coastal marshes.
GENERAL RANGE: Alaska and B.C., n. Ont. and Que. south to Wash. on the west coast, and to N.S. on the east coast; Asia.
SYNONYMS: **P. buxifolium** Nutt. ex Bong.

Polygonum hydropiper L. Fig. p. A49. Map p. B41.
COMMON NAME(S): Common Smartweed. Curage.
FREQUENCY: Probably scattered throughout the Province and locally common.
HABITAT: Moist open soils.
GENERAL RANGE: Probably all or mostly introduced from Eurasia; Alaska and B.C. to Nfld. south to Calif. and Ala.

++

Polygonum hydropiperoides Michx. Fig. p. A49. Map p. B41.
COMMON NAME(S): False Water-pepper, Mild Smartweed.
FREQUENCY: Uncommonly scattered in the southwest of the Province and locally common.
HABITAT: Shallow water and adjacent mostly muddy shores.
GENERAL RANGE: Alaska to B.C.; Ont. to N.S. south to Mexico and Fla.; S. America.

Polygonum lapathifolium L. Fig. p. A49. Map p. B41.
COMMON NAME(S): Willow-weed.
SUBSPECIFIC TAXA: Two well marked varieties occur in our range.

1a Leaves white-tomentose beneath, usually less than 1 dm long; spikes erect, to about 4 cm long. var. **salicifolium**
1b Leaves mostly glabrous beneath except for sessile glands, to 2.5 dm long; spikes often drooping, to about 8 cm long. var. **lapathifolium**

1. Var. **salicifolium** Sibth.
FREQUENCY: Rare; known from a few scattered sites across the central part of the Province.
HABITAT: Gravelly shores.
GENERAL RANGE: B.C. to Nfld. south to Calif. and Pa.
SYNONYMS: **P. incanum** Willd.

2. Var. **lapathifolium**
FREQUENCY: Scattered throughout most of the Province but rare in the northern counties.
HABITAT: Shores, damp clearings, roadside ditches, cultivated fields.
GENERAL RANGE: B.C. to Nfld. south to Calif. and N.J.; Greenland; Iceland; Eurasia.
SPECIAL NOTE: Probably at least partly introduced in our range.
SYNONYMS: **P. lapathifolium** var. **ovatum** A. Braun, **P. lapathifolium** var. **prostratum** Wimm., **P. scabrum** Moench.

Polygonum minus Huds.
COMMON NAME(S): Lesser Persicaria.
FREQUENCY: Rare; known from a few sites along the St. John R. in York and Carleton Counties.
HABITAT: Gravel shores.
GENERAL RANGE: European; reported as introduced in Canada only in N.B. (UNB). Known also from Mass. to Ind. and Pa.

Polygonum orientale L. Fig. p. A49.
COMMON NAME(S): Prince's-plume, Kiss-me-over-the-garden-gate. Bâton de Saint-Jean, Monte-au-ciel.
SPECIAL NOTE: Asiatic; cultivated and persistent near old habitations.

++

Polygonum oxyspermum Mey. & Bunge Map p. B41.
FREQUENCY: Rare; known from only a few localities in the east and northeast of the Province.
HABITAT: Sandy and gravelly coastal shores.
GENERAL RANGE: Que. to Nfld. south along the coast to N.S.; Europe; Asia.
SUBSPECIFIC TAXA: Our plants are referable to ssp. **raii** (Bab.) Webb. & Chater.
SYNONYMS: **P. raii** Bab.
SPECIAL NOTE: For more information see H. Hinds, 1973. The Rare Vascular Plants of New Brunswick. Syllogeus 50.

Polygonum pensylvanicum L. Fig. p. A49. Map p. B41.
COMMON NAME(S): Pinkweed. Persicaire glanduleuse.
FREQUENCY: Scattered throughout the southern part of the Province.
HABITAT: Cultivated ground, clearings and shores.
GENERAL RANGE: Ont. to N.S. south to Tex. and Fla.
SYNONYMS: **P. pensylvanicum** var. **laevigatum** Fern.

Polygonum persicaria L. Fig. p. A50. Map p. B41.
COMMON NAME(S): Lady's-thumb, Heart's-ease, Black Heart. Fer à cheval, Persicaire.
FREQUENCY: Common throughout the Province.
HABITAT: Cultivated lands, roadsides, clearings.
GENERAL RANGE: Eurasian; introduced across much of N. America and Greenland.
SUBSPECIFIC TAXA: Var. **ruderale** (Salisb.) Meisn., with +/- prostrate stem, smaller more rhombic-lanceolate leaves, and shorter spikes, is recorded from gravelly beaches of Grand Lake, Queens Co.

Polygonum prolificum (Small) Robins.
SPECIAL NOTE: Reported by Munroe, 1979, from Kouchibouguac National Park, but annotated by this author to **P. ramosissimum**. Known from B.C. to N.S. (except N.B. and P.E.I.) south to Wash., Tex. and Va.
SYNONYMS: **P. ramosissimum** var. **prolificum** Small.

Polygonum punctatum (Small) Robins. Fig. p. A50.
COMMON NAME(S): Water-Knotweed.
FREQUENCY: Very rare; known only from Grand Manan Is., Charlotte Co.
HABITAT: Sandy or muddy shores.
GENERAL RANGE: B.C. to N.S. south to Calif. and Fla.; tropical America; s. Asia.
SUBSPECIFIC TAXA: Our plants are referable var. **confertiflorum** (Meisn.) Fassett.

Polygonum ramosissimum Michx. Fig. p. A50. Map p. B42.
COMMON NAME(S): Bushy Knotweed.
FREQUENCY: Uncommonly scattered along the east coast and rare on the Bay of Fundy.
HABITAT: Sandy areas of brackish marshes and ditches.
GENERAL RANGE: Alaska and B.C. to N.S. south to Calif. and Del.

Polygonum sachalinense Schmidt.
COMMON NAME(S): Sachaline, Giant Knotweed.
SPECIAL NOTE: Asiatic; rarely cultivated and persisting near old habitations as in Saint John. Also known in Canada from B.C. and from Ont. to N.S.

Polygonum sagittatum L. Fig. p. A50. Map p. B42.
COMMON NAME(S): Arrow-leaved Tearthumb. Cratte-cul.
FREQUENCY: Common throughout most of the Province.
HABITAT: River bottomlands, wet ditches and meadows.
GENERAL RANGE: Man. to Nfld. south to Tex. and Fla.

Polygonum scandens L. Fig. p. A50. Map p. B42.
COMMON NAME(S): Climbing False Buckwheat. Renouée grimpante.
FREQUENCY: Locally common on the SW Miramichi, St. John and St. Croix River valleys.
HABITAT: Alluvial meadows and thickets.
GENERAL RANGE: Sask. to Nfld. south to Tex. and Fla.

RHEUM L.

Rheum rhaponticum L. Fig. p. A50 (fruit).
COMMON NAME(S): Rhubarb. Rhubarbe.
FREQUENCY: Scattered throughout the southern half of the Province.
SPECIAL NOTE: Asiatic; persisting near old habitations throughout the southern half of the Province, and across much of Canada.

RUMEX L. Dock, Sorrel. Doche

1a Basal leaves and often others with basal lobes; leaves acid tasting; flowers unisexual. 2
1b Basal leaves cuneate, truncate or cordate at base; leaves not acid tasting; flowers perfect. 3
2a(1a) Leaves sagittate; sepals greatly enlarged in fruit. **R. acetosa**
2b(1a) Leaves hastate; sepals not greatly enlarged in fruit. **R. acetosella**
3a(1b) Valves of fruit without enlarged grains; lower leaves acute at base, flat-margined or merely undulate; valves to 7 mm long. **R. longifolius**
3b(1b) Valves of fruit with enlarged grains (at least one). 4
4a(3b) Valve margin entire, dentate or undulate. 5
4b(3b) Valve margin with bristle-like or spinose teeth. 6
5a(4a) Valves each with one well-developed grain. 7
5b(4a) Valves only one with a small grain; inflorescence open, upper whorls +/- distant; valves to 3 mm long; pedicels to 5 mm long. **R. sanguineus**
6a(4b) Perennials; well-developed grain one (3 in ssp. **transiens**); basal leaves broadly to narrowly ovate, long-petioled, often red-veined, crenulate; plants of non-saline habitats. **R. obtusifolius**
6b(4b) Annuals or biennials; well-developed grains 3; basal leaves lanceolate; stems soft, hollow; plants of brackish or saline soil. **R. maritimus**
7a(5a) Ascending stems with axillary branches or leaf tufts; leaves linear-lanceolate, tapering to both ends; fruiting pedicels filiform, curved; valves triangular, 3-6 mm long, truncate at base. **R. salicifolius**
7b(5a) Erect stems usually without axillary branches or leaf tufts; leaves broader and truncate or cordate at base. 8

8a(7b) Leaves oblong-lanceolate, with +/- flat margins; valves to 8 mm long; base of callus grains separated from base of valve; stem to 2.5 m tall; plants of swamps and shallow water. **R. orbiculatus**

8b(7b) Leaves lanceolate, with strongly wavy margins; valves to 6 mm long; callus grains with base even with base of valve or projecting below; stem to 1 m; plants of mostly dry waste areas, roadsides. **R. crispus**

Rumex acetosa L. Fig. p. A50. Map p. B42.
COMMON NAME(S): Garden- or Meadow-Sorrel, Sourdock. Grande Oseille.
FREQUENCY: Locally common in the Fredericton and Saint John area.
HABITAT: Roadsides, fields and similar open areas.
GENERAL RANGE: Eurasian; introduced and naturalized from Alaska to Nfld.
SUBSPECIFIC TAXA: Our plants are referable to ssp. **thyrsiflorus** (Fingerh.) Hayek.

Rumex acetosella L. Fig. p. A50.
COMMON NAME(S): Sheep-Sorrel, Common Sorrel. Oseille or Surette.
FREQUENCY: Common throughout the Province.
HABITAT: Dry disturbed soils.
GENERAL RANGE: Eurasian; naturalized across N. America.

Rumex crispus L. Fig. p. A50. Map p. B42.
COMMON NAME(S): Yellow or Curly-leaf Dock. Patience crépue, Reguette.
FREQUENCY: Probably common throughout the Province.
HABITAT: Roadsides, shores and waste areas.
GENERAL RANGE: Eurasian; introduced across Canada.

Rumex longifolius DC. Fig. p. A50. Map p. B42.
COMMON NAME(S): Long-leaved Dock.
FREQUENCY: Scattered throughout much of the Province.
HABITAT: Wet fields, river margins and waste areas.
GENERAL RANGE: Eurasian; introduced from Alaska and B.C. to Nfld.; Greenland.

Rumex maritimus L.
COMMON NAME(S): Golden Dock. Rumex fausse-persicaire.
SUBSPECIFIC TAXA: Two well marked varieties are found within our range.

1a Basal leaves +/- cordate at base; bristles about as long as body of valve; grain rounded on both ends. var. **persicarioides**

1b Basal leaves tapering to base; bristles longer than width of body of valve; grain acuminate at tip, nearly acute at base. var. **fueginus**

1. Var. **persicarioides** (L.) R.S. Mitchell. Fig. p. A51.
HABITAT: Salt marshes and coastal shores.
FREQUENCY: Very rare; known only from Point Sapin, Kent Co. and Miscou Is., Gloucester Co.
GENERAL RANGE: Wash. to Calif. on the west coast; E. Que. to N.S. and south to Mass.
SYNONUMS: **R. persicarioides** L.
SPECIAL NOTE: For more information see Hinds, 1983. The Rare Vascular Plants of New Brunswick. Syllogeus 50.

2. Var. **fueginus** (Phil.) Dusen Fig. p. A50. Map p. B42.
FREQUENCY: Rare; scattered along the east coast and Bay of Fundy.
HABITAT: Salt marshes and coastal shores.
GENERAL RANGE: Alaska to N.S. south to Calif., Ark. and N.Y.; S. America; Eurasia.

Rumex obtusifolius L. Fig. p. A50. Map p. B42.
COMMON NAME(S): Red-veined or Bitter Dock. Patience rouge, Rumex sanguin.
FREQUENCY: Scattered throughout much of the Province.
HABITAT: Shores and similar open wet areas.
GENERAL RANGE: Eurasian; introduced from Alaska and B.C.; Ont. to Nfld., N.B., P.E.I. and N.S.

Rumex orbiculatus Gray. Fig. p. A51. Map p. B42.
COMMON NAME(S): Great Water-Dock.
FREQUENCY: Scattered throughout much of the Province except rare in the central section.
HABITAT: Shores and similar wet ground.
GENERAL RANGE: B.C. to Nfld. south to Nebr. and N.J.

Rumex salicifolius Weinm. Fig. p. A51. Map p. B42.
COMMON NAME(S): Narrow-leaved Dock.
FREQUENCY: Scattered over much of the Province and often locally common.
HABITAT: Coastal sands, shores, and gravel roadsides.
GENERAL RANGE: Alaska to Nfld. south to Calif. and N. Engl.; Eurasia.
SUBSPECIFIC TAXA: Our plants are referable to ssp. **triangulivalvis** Danser.
SYNONYMS: **R. triangulivalvis** (Danser) Rech. f.

Rumex sanguineus L. Fig. p. A51.
COMMON NAME(S): Bloody or Red-veined Dock.
SPECIAL NOTE: Eurasian; probably introduced in ballast near St. John, but not persisting.

CHENOPODIACEAE (Goosefoot Family)

1a Stems leafless, fleshy and jointed; flowers opposite, embedded in stem. **Salicornia europaea**
1b Stems leafy, not jointed; leaves mostly alternate. 2
2a(1b) Leaves sessile, entire, linear to linear-oblong. 3
2b(1b) Leaves broader, often toothed or lobed, petioled. 4
3a(2a) Leaves spine-tipped, filiform; flowers +/- solitary, axillary. **Salsola kali**
3b(2a) Leaves not spine-tipped. 5
4a(2b) Plants stellate-pubescent; leaves entire, lanceolate; sepals scarious at maturity, +/- erect. **Axyris amaranthoides**
4b(2b) Plants smooth, farinose or glandular-pubescent. 6
5a(3b) Leaves fleshy, cylindrical to plano-convex; sepals (at least one) carinate or cucullate on the back, but scarcely winged. **Suaeda**

5b(3b) Leaves hardly fleshy, +/- flattened, linear to linear-lanceolate; fruiting sepals bearing a short dorsal wing; flowers in dense axillary or terminal spikes. **Kochia scoparia**

6a(4b) Fruit concealed by 2 enlarged, opposed bracteoles. **Atriplex**

6b(4b) Fruit enclosed by the persistent calyx. **Chenopodium**

ATRIPLEX L. Orache. Arroche*

Note: The "lower leaves" in **Atriplex** are the earlier-formed leaves on the central axis in the middle to lower part of the plant. They often differ considerably in size and form from the later-developing "upper leaves", and they frequently drop off before the bracteoles and seed are fully mature.

1a Plants whitish or greyish-green, densely scaly; bracteoles cartilaginous and hardened in the lower half; seed transversely elliptical, light brown; leaf venation appearing (magnified 12X) as a conspicuous, dark green pattern (Kranztypus) when the leaf surface is scraped with a sharp blade. **A. lacinata**

1b Plants green, glabrous to sparsely mealy; bracteoles herbaceous, succulent or spongy-inflated, not hardened in the lower half; seed ovate to orbicular or round-elliptic, dark brown or black; leaf venation appearing (magnified 12X) as the normal dicotyledonous type and not showing a dark green reticulate pattern when the leaf surface is scraped with a sharp blade. 2

2b(1b) Lower leaves linear or ovate-lanceolate to linear-lanceolate. 3

2b(1b) Lower leaves triangular 6

3a(2a) Lower leaves linear, margins without a pair of basal lobes, entire or with a few irregular teeth in the mid to apical portions; bracteoles thick-spongy and usually bi-tuberculate. **A. littoralis**

3b(2a) Lower leaves ovate-lanceolate to linear-lanceolate, margins with a pair of basal lobes (sometimes only one lobe of the pair is developed), otherwise entire or variously toothed throughout; bracteoles thick-spongy and bi-tuberculate, or thin-herbaceous and without tubercles. 5

4a(3b) Bracteoles thick-spongy and bi-tuberculate, broadly triangular to ovate-triangular, base truncate; seeds round ellipetic, wider than long. **A. subspicata**

4b(3b) Bracteoles thin-herbaceous and without tubercles, rhombic or ovate, base cuneate, obtuse or rounded; seeds orbicular, not wider than long. 5

5a(4b) Bracteoles rhombic with margins united almost to the middle, lateral angles pointed and often well-developed; radicle of brown seeds subbasal to median and up-pointing; basal lobes of lower leaves often strongly developed and forward-curving; plants symmetrical with opposing branches usually of almost equal length; non-halophytic ruderal species widespread in disturbed ground. **A. patula**

*Contributed by P.M. Taschereau

5b(4b) Bracteoles ovate with margins free to the base, lateral angles pointed or rounded but never well developed; radicle of brown seeds basal and out-pointing; basal lobes of lower leaves never strongly developed, out-pointing; plants asymmetrical with opposing branches usually of unequal length; halophytic saltmarsh species restricted to coastal habitats. **A. acadiensis**

6a(2b) Stalked bracteoles present in some of the leaf axils; margins of bracteoles irregularly denticulate to laciniate, the lateral angles often developed into 1-3 short or long teeth; upper leaves ovate-lanceolate, cuneate-based, frequently with a pair of forward-curving basal lobes. **A. franktonii**

6b(2b) Stalked bracteoles absent; margins of bracteoles entire or with a few minute teeth, lateral angles rounded or short-pointed but never clearly developed into teeth; upper leaves ovate-lanceolate or triangular, cuneate-based or broadly obtuse to truncate; with or without a pair of out-pointing basal lobes. 7

7a(6b) Inflorescence with leafy bracts to the tip, glomerules loose, irregularly spaced; bracteoles thick and spongy, margins united up to the middle; seeds more than 2.5 mm wide, usually not distinctly dimorphic, dark brown to black, irregularly biconvex; radicle median and more or less strongly up-pointing. **A. glabriuscula**

7b(6b) Inflorescence with leafy bracts only at the base, glomerules tight, contiguous or irregularly spaced; bracteoles thin to slightly thickened and spongy, margins united only at the base; seeds mostly small glossy-black but also some larger, dull brown, flattened and disc-shaped; radicle subbasal, obliquely up-pointing to out-pointing. **A. prostrata**

Atriplex acadiensis Taschereau Fig. p. A51.
FREQUENCY: Commonly scattered along the east coast and Bay of Fundy.
HABITAT: Salt marshes, particularly in the **Spartina patens** zone; also in the halophytic vegetation fringing sandy and cobble beaches in protected bays and inlets.
GENERAL RANGE: Coastal areas of P.Q., N.B., P.E.I., N.S., and Maine.

Atriplex franktonii Taschereau Fig. p. A51.
FREQUENCY: Common along the east coast.
HABITAT: Sandy coastal beaches, and occasionally in salt marshes.
GENERAL RANGE: Coastal areas of P.Q., N.B. and N.S. especially along the Northumberland Strait; coasts of Cape Breton Island, and shores of Bras d'Or Lake, Cape Breton.

Atriplex glabriuscula Edmondston Fig. p. A51.
FREQUENCY: Absent on the Northumberland Straits; frequent in the Passamaquoddy Bay area, Charlotte Co., and other parts of the Bay of Fundy.
HABITAT: On, or slightly above the strandline along exposed cobble or sand beaches.
GENERAL RANGE: Coastal areas from NFLD. to N. Eng. and coasts of W. and N.W. Europe.
SYNONYM: **A. patula** var. **oblanceolata** (Vict. & Rousseau) Boivin.

Atriplex laciniata L. Fig. p. A51.
FREQUENCY: Scattered along the Northumberland Straits.
HABITAT: Upper parts of more or less protected sand or sand and cobble beaches.

++

GENERAL RANGE: European; introduced in N. America on beaches bordering the Northumberland Straits and Gulf of St. Lawrence in P.Q., N.B., P.E.I., and N.S.
SYNONYM: **A. sabulosa** Rouy

Atriplex littoralis L.
FREQUENCY: Rare; collected in Richibucto, Kent Co., and Cape Tormentine, Westmorland Co.
HABITAT: Sheltered sandy beaches, muddy areas of salt marshes.
GENERAL RANGE: European; introduced in N.B., P.E.I. and N.S., and sporadically south along the coast to Mass.
SYNONYM: **A. patula** var. **littoralis** (in part) in Fernald, Gray's Man., ed. 8, 1950.

Atriplex patula L. Fig. p. A51.
COMMON NAMES: Common Orache, Spreading Orache, Spearscale, Spreading Atriplex.
FREQUENCY: Scattered inland and in coastal areas throughout the cultivated and settled parts of the province.
HABITAT: Weed of disturbed ground, margins of lawns and sidewalks, along paths and roadsides. Somewhat salt tolerant, but not a halophyte. When present on the coast, confined to the uppermost beach margins, on fallen soil banks, or around fishing shacks.
GENERAL RANGE: Cosmopolitan anthropophyte.

Atriplex subspicata (Nutt.) Rydberg Fig. p. A52.
FREQUENCY: Common on the east coast and Bay of Fundy shores.
HABITAT: Salt marsh margins and upper areas of quiet beaches.
GENERAL RANGE: P.Q. along the coast to Mass.; inland in saline habitats in W. Canada, C. and W. U.S.A.

Atriplex prostrata Boucher ex DC. Fig. p. A52. Map p. B43.
FREQUENCY: Scattered along the east coast and Bay of Fundy, and inland.
HABITAT: Salt marshes, sea beaches; inland about buildings, in wast places, and around salt springs.
GENERAL RANGE: Widely distributed throughout Canada, U.S., Europe from Scand. southward; Asia, n. Africa, Australia.
SYNONYM: **A. patula** var. **hastata**, in Fernald, **Gray's Man.** ed. 8, 1950; **A. hastata**, in Aellen, **Flora Eur**. 1:97, 1964; **A. triangularis** Willdenow.

AXYRIS L.

Axyris amaranthoides L. Fig. p. A52.
COMMON NAME(S): Russian Pigweed. Ansérine de Russie.
SPECIAL NOTE: Eurasian; introduced but scarcely spreading as in Fredericton, York Co.

CHENOPODIUM L. Goosefoot, Pigweed. Ansérine

1a Plants glandular aromatic; leaves +/- truncate at base, sinuate-pinnatifid with rounded lobes. **C. botrys**

++

1b Plants not glandular aromatic. 2

2a(1b) Seeds all horizontal in the enfolding perianth (**C. glaucum** may have both horizontal and vertical seeds in the same cluster). 3

2b(1b) Seeds mostly vertical (a few sometimes horizontal, especially in **C. rubrum**). 4

3a(2a) Sepals rounded on back. 5

3b(2a) Sepals elevated on the back along the midline. 6

4a(2b) Flowers confluent, in large spherical glomerules, red and strawberry-like at maturity. **C. capitatum**

4b(2b) Flowers not fleshy or confluent at maturity. 7

5a(3a) Leaves densely white-farinose beneath, 1-2.5 cm long, sinuate margined; seeds smooth. **C. glaucum**

5b(3a) Leaves green, glabrous or sparingly farinose beneath. 8

6a(3b) Seed and pericarp +/- smooth surfaced. **C. album**

6b(3b) Seed and pericarp conspicuously reticulate-roughened. **C. macrocalycium**

7a(4b) Leaves all triangular-hastate, entire or shallowly sinuate; seeds about 1.5 mm broad; stigmas to 1 mm long; perennial plants of non-saline, ruderal habitats. **C. bonus-henricus**

7b(4b) Basal leaves rhombic-ovate or obovate, +/- coarsely toothed, the upper lanceolate and subentire; seeds about 1 mm broad; stigmas minute; annual plants of saline soils. **C. rubrum**

8a(5b) Leaves entire; flowers in branched axillary panicles; sepals with conspicuous apiculate tips; seeds black, shiny, exposed at maturity. **C. polysperum**

8b(5b) Leaves variously toothed above basal lobes. 9

9a(8b) Flowers and mature seeds prominently mixed in glomerules (except very late season); pericarp usually separable. 10

9b(8b) Flowers all nearly the same stage of development in glomerules; pericarp strongly adherent. 11

10a(9a) Seeds with rounded margins, 1.5-2.2 mm broad; pericarp black, thick, attached; sepal edges meeting or overlapping, covering more than ½ of mature fruit. **C. hybridum**

10b(9a) Seeds with margins obtuse to +/- acute, 1.5-2.5 mm broad; pericarp whitish to yellow-green, thin, delicate, cellular-reticulate, readily separable or attached; sepal margins meeting or separated by as much as their width, exposing more than ½ of the mature fruit. **C. gigantospermum**

11a(9b) Seeds with margins acute, 1.2-1.5 mm broad; sepals keeled, farinose; inflorescence usually shorter than leaves. **C. murale**

11b(9b) Seeds with margins rounded, 1.0-1.2 mm broad; sepals not keeled nor farinose; inflorescence usually longer than the leaves. **C. urbicum**

Chenopodium album L. Fig. p. A52.

COMMON NAME(S): Lamb's-quarters, Pigweed, Poulette grasse, Chou gras.

FREQUENCY: Common throughout the Province.

HABITAT: Cultivated and similarly disturbed ground, high tide line of estuarine and lagoon beaches.

GENERAL RANGE: Eurasian; introduced and established across N. America.

++

Chenopodium bonus-henricus L. Fig. p. A52.
COMMON NAME(S): Good King Henry. Ansérine bon-henri.
SPECIAL NOTE: Eurasian; introduced and reported from Bathurst, Gloucester Co., Chatham and Newcastle, Northumberland Co. Known in Canada from Ont. to N.S.

Chenopodium botrys L. Fig. p. A52.
COMMON NAME(S): Jerusalem-oak, Feather-geranium. Herbe à printemps.
SPECIAL NOTE: Eurasian; Reported only from Fredericton, York Co. Known in Canada also from B.C.; Ont. to N.S.

Chenopodium capitatum (L.) Aschers. Fig. p. A52.
COMMON NAME(S): Strawberry-Blite, Indian-paint. Blette, Épinard-Fraise.
FREQUENCY: Rare; known from only a few localities in York and Carleton Counties.
HABITAT: Clearings, burns, and waste areas.
GENERAL RANGE: Alaska to N.S. south to Calif. and N.J.; Europe.

Chenopodium gigantospermum Aellen Fig. p. A52. Map p. B43.
COMMON NAME(S): Maple-leaved Goosefoot.
FREQUENCY: Rare; known only from Mt. Zachie-Jonah, Albert Co. and near Fredericton, York Co.
HABITAT: Moist rocky woods, clearings and waste places.
GENERAL RANGE: Yukon to N.S. south to Calif. and Va.; Eurasia.
SYNONYMS: **C. hybridum** L. var. **gigantospermum** (Aellen) Rouleau.

Chenopodium glaucum L. Fig. p. A52.
COMMON NAME(S): Oak-leaved Goosefoot. Patte d'oie glauque.
SPECIAL NOTE: Eurasian; reported from brackish shores and waste areas, Dalhousie, Restigouche Co., Miscou Island, Gloucester Co., and Fredericton, York Co. Not previously known from the Atlantic Provinces.

Chenopodium macrocalycium Aellen.
FREQUENCY: Locally common in the Passamaquoddy Bay and northeast coastal areas; rarely occurring inland as in Sheffield, Sunbury Co.
HABITAT: Coastal beaches, cultivated ground and river gravels inland.
GENERAL RANGE: Probably originating from South or Central America; coastal areas of N.B. and N.S. south along the coast to Fla.

Chenopodium murale L. Fig. p. A52.
COMMON NAME(S): Nettle-leaf Goosefoot.
SPECIAL NOTE: Eurasian; introduced in Saint John in ballast but not persisting. Known also in Canada from B.C.; Ont. to N.B.

Chenopodium polyspermum L. Fig. p. A52.
SPECIAL NOTE: Eurasian; introduced, but not spreading, in St. Andrews, Charlotte Co. Known also in Canada from Sask.; Ont. to N.B.

Chenopodium rubrum L. Fig. p. A52.
COMMON NAME(S): Coast-Blite, Fat Hen, French Spinach, Ansérine rouge.

++

FREQUENCY: Rare; scattered mostly along the coast.
HABITAT: Salt-marshes and saline soils.
GENERAL RANGE: Yukon and B.C. to N.S. south to Calif. and N.J.; Eurasia.

Chenopodium urbicum L. Fig. p. A53.
SPECIAL NOTE: Eurasian; introduced, but not persisting, in Kent, St. John, and Westmorland Counties. Known also in Canada from B.C.; Ont. to N.S.

KOCHIA Roth

Kochia scoparia (L.) Schrad. Fig. A53.
COMMON NAME(S): Summer-cypress, Red Belvedere. Petits soldats.
SPECIAL NOTE: Eurasian; introduced near grain elevators in Fredericton, York Co. Also known in Canada from B.C. to N.S.

SALICORNIA L. Samphire, Glasswort, Saltwort.

Salicornia europaea L. Fig. p. A53.
COMMON NAME(S): Chickenclaws. Corail.
FREQUENCY: Common along the coast and rare inland.
HABITAT: Salt marshes and inland salt springs.
GENERAL RANGE: Alaska to Nfld. south to Calif. and Ga.; Eurasia.
SYNONYMS: S. **europaea** vars. **simplex** (Pursh) Fern. and **prostrata** (Pall.) Fern.

SALSOLA L.

Salsola kali L. Fig. p. A53.
SUBSPECIFIC TAXA: Two varieties are recognized within our range.

1a Principal leaves flattened on one side to about 3 cm long; bracteal leaves to about 1.5 cm long, harshly spinous at tip, dilated at base. var. **kali**
1b Principal leaves filiform, to about 7 cm long; bracteal leaves to about 8 mm long; weakly spinous at tip. var. **tenuifolia**

1. Var. **kali**
COMMON NAME(S): Saltwort. Soude.
FREQUENCY: Uncommonly scattered along the coast.
HABITAT: Sandy seashores.
GENERAL RANGE: Que. to Nfld. south along the coast to Ga.; Eurasia.

2. Var. **tenuifolia** Tausch
COMMON NAME(S): Russian Thistle. Chardon de Russie.
FREQUENCY: Introduced in Fredericton, York Co. and St. Stephens, Charlotte Co.
HABITAT: Waste areas and cultivated fields.
GENERAL RANGE: Asian; introduced across Canada.
SYNONYMS: S. **iberica** Sennen & Pau.

++

SUAEDA Forsk. Sea-Blite

1a Stems reclining to prostrate at base, only flowering tips ascending; at least one sepal strongly keeled or cucullate on back; mature seeds 1.2-1.5 mm long; leaves not whitened, the lower about 2 cm long becoming purplish-red. **S. americana**

1b Stems usually more erect; all sepals rounded or obscurely cucullate on back; mature seeds 1.5-2 mm long; leaves often whitened, 3-5 mm long. **S. maritima**

Sueda americana (Pers.) Fern.
FREQUENCY: Uncommonly scattered on the east coast.
HABITAT: Salt marshes.
GENERAL RANGE: Yukon to N.S. south along the coast to s. Maine.

Sueda maritima (L.) Dumort Fig. p. A53.
FREQUENCY: Common along the coast.
HABITAT: Sandy shores and salt marshes.
GENERAL RANGE: Alaska to Nfld. south to Calif. and Fla.

AMARANTHACEAE (Amaranth Family)

AMARANTHUS L. Amaranth, Water-hemp. Amarante, Acnide

1a Flowers mainly in small axillary clusters; whitish stem diffusely branched; upper leaves usually less than 3 cm long. **A. albus**

1b Flowers mainly in large, dense, elongate, leafy-bracted, terminal panicles; stem seldom much branched, often reddish at base; upper leaves 10 cm or larger. **A. retroflexus**

Amaranthus albus L. Fig. p. A53.
COMMON NAME(S): Tumbleweed. Herbe-à-cochon.
FREQUENCY: Rare; collected in a few scattered localities in the south of the Province.
HABITAT: Waste areas or cultivated ground.
GENERAL RANGE: B.C. to N.S. south to Mexico and S. America; Eurasia.

Amaranthus retroflexus L. Fig. p. A53.
COMMON NAME(S): Green Amaranth, Pigweed, Redroot. Canne à pied rouge.
FREQUENCY: Scattered throughout most of the Province.
HABITAT: Cultivated and waste ground.
GENERAL RANGE: Tropical America; probably introduced throughout Canada.

++

AIZOACEAE (Carpetweed Family)

MOLLUGO L.

Mollugo verticillata L. Fig. p. A53. Map p. B43.
COMMON NAME(S): Carpetweed. Mollugine.
FREQUENCY: Locally common in the St. John, Miramichi and St. Croix River valleys.
HABITAT: Open ground, especially sandy river shores.
GENERAL RANGE: Tropical America; introduced in B.C. and from Ont. to N.S.

PORTULACACAE (Purslane Family)

1a Leaves 2, basal; flowers in racemes. **Claytonia caroliniana**
1b Leaves many, cauline; flowers axillary or terminal. 2
2a(1b) Plant succulent; leaves spathulate to obovate, alternate or sub-opposite, 1-3 cm long; petals yellow, slightly shorter than sepals; calyx fused to base of ovary; capsule circumscissile; coarse plants of weedy habit. **Portulaca oleracea**
2b(1b) Plant not succulent; leaves oblanceolate to spathulate-obovate, 5-13 mm long, opposite; petals whitish, exceeding sepals; calyx free of ovary; capsule opening lengthwise; slender annuals of springs, pools and wet ditches. **Montia fontana**

CLAYTONIA L. Spring-beauty

Claytonia caroliniana Michx. Fig. p. A53. Map p. B43.
FREQUENCY: Scattered throughout most of the Province but rarer in the eastern lowlands and towards the northeast.
HABITAT: Rich, mostly deciduous or mixed woods.
GENERAL RANGE: Ont. to Nfld. south to Tenn. and N.C.

MONTIA L. Miner's Lettuce

Montia fontana L. Fig. p. A53.
COMMON NAME(S): Water-Blinks.
FREQUENCY: Rare; recorded from Loggieville, Northumberland Co., Shediac, Westmorland Co. and Wolf Is., Charlotte Co.
HABITAT: Springy, usually coastal shores.
GENERAL RANGE: Alaska and B.C. to Nfld. south to Calif. and Maine; Greenland; Iceland; Eurasia.
SYNONYMS: **M. lamprosperma** Cham.
SPECIAL NOTE: For more information see H. Hinds, 1983. The Rare Vascular Plants of New Brunswick. Syllogeus 50.

PORTULACA L. Purslane. Pourpier

Portulaca oleracea L. Fig. p. A53. Map p. B43.
COMMON NAME(S): Common Purslane, Pusley. Pourpier potager.
FREQUENCY: Scattered in the southern part of the Province.
HABITAT: Cultivated ground.
GENERAL RANGE: Eurasian; introduced across Canada.

CARYOPHYLLACEAE (Pink Family)

1a Sepals separate. 2
1b Sepals fused, 5-lobed. 3
2a(1a) Leaves narrowly linear or awl-like. 4
2b(1a) Leaves narrowly lanceolate or broader. 5
3a(1b) Corolla wanting; calyx lobes green; leaves linear-subulate, connate at base; prostrate annual. **Scleranthus annuus**
3b(1b) Corolla present; petals long-clawed at base. 6
4a(2a) Stipules wanting; styles alternating with sepals, 4-5; leaves linear-filiform or subulate. **Sagina**
4b(2a) Stipules scaly-membranaceous; styles opposite sepals, usually 3. 7
5a(2b) Petals entire or shallowly notched. 8
5b(2b) Petals deeply cleft. 9
6a(3b) Styles 2. 10
6b(3b) Styles 3-5. 11
7a(4b) Leaves opposite; styles and valves of capsule 3. **Spergularia**
7b(4b) Leaves whorled; styles and valves of capsule 5. **Spergula arvensis**
8a(5a) Plants succulent, often densely tufted; petals and stamens inserted on a conspicuous 10-lobed stamineal disk; halophytic plants of sandy beaches. **Honkenya peploides**
8b(5a) Plants not succulent nor densely tufted; stamineal disk wanting; plants of non-saline habitats. 12
9a(5b) Capsule cylindric; plants usually hairy; styles normally 5. **Cerastium**
9b(5b) Capsule ovoid or ellipsoid; plants usually glabrous; styles normally 3. **Stellaria**
10a(6a) Flowers less than 1 cm long; white; calyx turbinate. **Gypsophila paniculata**
10b(6a) Flowers at least 2 cm long, pinkish; calyx terete or 5-angled. 13
11a(6b) Calyx-lobes leaf-like, 2-3 cm long; styles opposite the purplish-red petals; plant silky pubescent. **Agrostemma githago**
11b(6b) Calyx-lobes not prolonged; styles alternate with the petals. 14
12a(8b) Ovules and young seeds with a pale, spongy appendage (strophiole); stem retrorsely puberulent; leaves oval to narrowly oblong, puberulent, to 3.5 cm long; petals 3.5-8 mm long; rhizomatous perennials. **Moehringia lateriflora**
12b(8b) Ovules and young seeds without strophiole; stem branches cinereous-puberulent; leaves ovate, acuminate, ciliate, 2-7 mm long; petals less than 4 mm long; annuals. **Arenaria serpyllifolia**
13a(10b) Petals crowned with an appendage at summit of claw; calyx terete; stout perennial. **Saponaria officinalis**

13b(10b) Petals unappendaged; calyx 5-angled; annuals. **Vaccaria pyramidata**
14a(11b) Calyx bracteate; styles 2. **Dianthus**
14b(11b) Calyx naked at base; styles 3-5. 15
15a(14b) Capsule with as many teeth as styles, entire. **Lychnis**
15b(14b) Capsule teeth twice as many as styles, or teeth bifid. **Silene**

AGROSTEMMA L. Corn-Cockle

Agrostemma githago L. Fig. p. A53.
COMMON NAME(S): Purple Cockle. Nielle des blés.
SPECIAL NOTE: Eurasian; introduced with grain as near grain elevators, Fredericton, York Co.

ARENARIA L. Sandwort. Sabline

Arenaria serpyllifolia L. Fig. p. A54.
COMMON NAME(S): Thyme-leaved Sandwort.
FREQUENCY: Rare; known only from Bass River, Kent Co.
HABITAT: Dry, +/- sterile ground.
GENERAL RANGE: Eurasian; introduced across much of southern Canada.

CERASTIUM L. Mouse-ear Chickweed

1a Plant densely white-woolly throughout. **C. tomentosum**
1b Plant villous or pilose but not densely white-woolly. 2
2a(1b) Petals shorter than or only slightly exceeding sepals; basal offshoots herbaceous; axillary tufts uncommon. **C. fontanum**
2b(1b) Petals greatly exceeding sepals; basal offshoots becoming hard, dry and overlapping; axillary tufts common. **C. arvense**

Cerastium arvense L. Fig. p. A54. Map p. B43.
COMMON NAME(S): Field Chickweed. Mouron d'alouette.
FREQUENCY: Probably scattered throughout the Province.
HABITAT: Gravelly or rocky exposed slopes, and calcareous ledges.
GENERAL RANGE: Alaska to Nfld. south to Calif. and Ga.; Greenland; Eurasia.

Cerastium fontanum Baumg. Fig. p. A54.
COMMON NAME(S): Common Mouse-ear Chickweed. Mouron à oreille de souris.
FREQUENCY: Common throughout the Province.
HABITAT: Cultivated ground and waste places.
GENERAL RANGE: Eurasian; introduced across Canada and Greenland.
SUBSPECIFIC TAXA: Our material is referable to ssp. **triviale** (Link) Jalas.
SYNONYMS: **C. vulgatum** L., **C. holosteoides** Fries.

++

Cerastium tomentosum L.
COMMON NAME(S): Dusty Miller, Snow-in-Summer. Barbette, Argentine.
SPECIAL NOTE: European; cultivated and rarely escaped as in Kouchibouguac National Park, Kent Co. and also in B.C., Man., Ont. to N.S.

DIANTHUS L. Pink, Carnation. Oeillet

Dianthus armeria L. Fig. p. A54.
COMMON NAME(S): Deptford Pink.
SPECIAL NOTE: Eurasian; rarely introduced as in Prince William, York Co. Also known in Canada from B.C. and from Ont. to N.S.

Dianthus deltoides L.
COMMON NAME(S): Maiden-Pink.
SPECIAL NOTE: Eurasian; a garden escape as in the Acadia Forest Exp. Station, Sunbury Co. Also known in Canada from B.C. and from Ont. to N.S.

GYPSOPHILA L.

Gypsophila paniculata L.
COMMON NAME(S): Baby's-breath.
SPECIAL NOTE: Eurasian; cultivated and rarely escaped as in St. Stephen, Charlotte Co. Also known in Canada from B.C. to N.B.

HONKENYA Ehrh.

Honkenya peploides (L.) Ehrh. Fig. p. A54. Map p. B43.
COMMON NAME(S): Seabeach-Sandwort. Sabline faux-peplus.
FREQUENCY: Scattered but locally common along the coast.
HABITAT: Coastal sands.
GENERAL RANGE: Strait of Belle Isle along the coast to Md.
SUBSPECIFIC TAXA: Our plants are referable to ssp. **robusta** (Fern.) Hulten.
SYNONYMS: **Arenaria peploides** L. var. **robusta** Fern.

LYCHNIS L. Campion. Lychnide

1a Calyx glabrous, cylindrical, red nerved; petals usually pink, laciniate. **L. flos-cuculi**
1b Calyx hairy, at least on nerves, slender tubular; petals scarlet, 2-lobed. **L. chalcedonica**

Lychnis chalcedonica L. Fig. p. A54.
COMMON NAME(S): Scarlet Lychnis. Maltese-cross, London Pride. Croix de Jerusalem, Lampette de Chalcedoine.

++

SPECIAL NOTE: Asiatic; cultivated and occasionally escaped as in Fredericton, York Co. and across much of Canada and N. America.

Lychnis flos-cuculi L. Fig. p. A54.
COMMON NAME(S): Ragged-Robin, Cuckoo-flower. Fleur-de-coucou.
SPECIAL NOTE: European; rarely introduced to roadsides and meadows as in Saint John, near the Eel R., Carleton Co. and in Nigadoo, Gloucester Co.

MOEHRINGIA L.

Moehringia lateriflora (L.) Fenzl. Fig. p. A54.
COMMON NAME(S): Grove-Sandwort.
FREQUENCY: Common along the coast; rarer inland and towards the north.
HABITAT: Sandy thickets, dry heath and open woodlands.
GENERAL RANGE: Alaska to Nfld. south to Calif. and Md.; Eurasia.
SYNONYMS: **Arenaria lateriflora** L.

SAGINA L. Pearlwort

1a Annuals without basal rosettes or sterile offshoots; sepals and capsule valves normally 5 (rarely 4); seeds usually stipitate-glandular. **S. decumbens**
1b Perennials with basal rosettes of leaves and sterile shoots; mat forming. 2
2a(1b) Sepals and valves of capsule 4 (rarely 5); sepals finally spreading; axillary leaf fascicles frequent. **S. procumbens**
2b(1b) Sepals and valves of capsule 5 (rarely 4); sepals remaining appressed; leaf fascicles uncommon. **S. saginoides**

Sagina decumbens (Ell.) T. & G. Fig. p. A54.
SPECIAL NOTE: Probably introduced from eastern U.S.; known only from Saint John, St. John Co. Also known in Canada from B.C., Alta., Sask. and from Ont. to N.B.

Sagina nodosa (L.) Fensz Fig. p. A54. Map p. B43.
COMMON NAME(S): Knotted Pearlwort.
FREQUENCY: Rare; found only in a few localities in the northeast and in the southwest.
HABITAT: Moist rocky, gravelly or peaty ground.
GENERAL RANGE: Dist. Mack. to Nfld. south to Alta. Minn. and Maine; Greenland; Iceland; Eurasia.
SUBSPECIFIC TAXA: Plants with the upper part of the stem minutely glandular pubescent occur in Charlotte Co. and may be called var. **pubescens** Mert. & Koch.

Sagina procumbens L. Fig. p. A54. Map p. B43.
COMMON NAME(S): Birdseye.
FREQUENCY: Scattered throughout the Province.
HABITAT: Rocky beaches, seepage areas in fresh or brackish places.
GENERAL RANGE: Alaska to Nfld. south to Ohio and Del.; Greenland; Iceland; Eurasia.

++

Sagina saginoides (L.) Karst. Fig. p. A54.
SPECIAL NOTE: Reported from N.B. by Boivin, 1966-67.
GENERAL RANGE: Alaska and B.C., Alta. south to Calif. and Mex.; Keewatin to Gaspé Pen. south to Nfld. and N.B.; Greenland, Iceland; Eurasia

SAPONARIA L.

Saponaria officinalis L. Fig. p. A55.
COMMON NAME(S): Bouncing-Bet, Soapwort. Herbe à savon.
SPECIAL NOTE: Eurasian; naturalized on sand bars, ditches and waste areas in scattered localities mostly in the southern half of the Province, and also across much of Canada.

SCLERANTHUS L. Knawel

Scleranthus annuus L. Fig. p. A55.
COMMON NAME(S): Annual Knawel. German Knotgrass. Gravelle, Herbe aux alouettes.
SPECIAL NOTE: Eurasian; introduced in waste areas as in Kouchibouguac National Park and vicinity. Also known from B.C. to Sask.; Ont. to N.S.

SILENE L. Catchfly, Campion

1a Styles 3 (rarely 4); flowers usually perfect. 2

1b Styles 5, or flowers male only; flowers usually white (rarely pink) opening at night, fragrant; calyx teeth narrowly triangular, acuminate; capsule with 10 +/- erect teeth or 5 bifid teeth, ovoid, narrow mouthed, stems densely and softly hairy, +/- glandular above; annuals or short-lived perennials. **S. alba**

2a(1a) Calyx glabrous. 3

2b(1a) Calyx glandular-hispid. 4

3a(2a) Capsule completely included in inflated, globose calyx; stem without glutinous bands; petals white, deeply bifid; leaves to 3 cm broad, +/- glaucous; stout rhizomatous perennial. **S. vulgaris**

3b(2a) Capsule summit +/- exposed at maturity; stem with glutinous bands; petals pink (often wanting), +/- emarginate; leaves 2-12 mm wide; slender annuals. **S. antirrhina**

4a(2b) Inflorescence appearing subsecundly racemose; calyx pale green to white, to 3 cm long, greatly inflated, ovoid, with prominent network of green veins and subulate teeth; flowers fragrant, opening at night; petal limb to 10 mm long, bifid. **S. noctiflora**

4b(2b) Inflorescence cymose with +/- equal branching; calyx to 10 mm long, ovoid in fruit, with inconspicuous network of veins and triangular acute teeth; petal limb to 6 mm long, entire or emarginate; flowers not especially fragrant, opening in daytime. **S. gallica**

++

Silene alba (P. Mill.) Krause Fig. p. A55. Map p. B43.
COMMON NAME(S): White Cockle, Campion, Evening-Lychnis. Oeillet de Dieu, Floquet.
FREQUENCY: Uncommonly scattered throughout the Province.
HABITAT: Fields and waste places.
GENERAL RANGE: Eurasian; introduced across much of N. America.
SYNONYMS: **Lychnis alba** Mill.

Silene antirrhina L. Fig. p. A55.
COMMON NAME(S): Sleepy Catchfly.
FREQUENCY: Rare; probably introduced from further west; known from St. Leonard, Madawaska Co., Fredericton, York Co. and St. Stephen, Charlotte Co.
HABITAT: Dry roadways and railways, open woods and fields.
GENERAL RANGE: B.C. to N.B. south to Mexico and S. America.
SUBSPECIFIC TAXA: The Fredericton material is var. **apetala** Farw., without petals.

Silene gallica L. Fig. p. A55.
COMMON NAME(S): Small-flowered Catchfly.
SPECIAL NOTE: European; introduced in Sackville, Westmorland Co. and also in B.C., Ont. and N.S.

Silene noctiflora L. Fig. p. A55. Map p. B43.
COMMON NAME(S): Night-flowering Catchfly, Sticky Cockle, Apprapemouche, Fleur de nuit.
SPECIAL NOTE: European; introduced in waste ground in scattered localities in the Province. Also known from much of N. America.
SYNONYMS: **Melandrium noctiflora** (L.) Fries.

Silene vulgaris (Moench) Garcke Fig. p. A55.
COMMON NAME(S): Bladder-Campion. Pétards, Péteux.
SPECIAL NOTE: Eurasian; introduced in waste areas and shores throughout much of the Province and across N. America.
SYNONYMS: **S. cucubalus** Wibel.

SPERGULA L. Spurrey

Spergula arvensis L. Fig. p. A55. Map p. B43.
COMMON NAME(S): Corn-spurrey, Stickwort. Grippe.
SPECIAL NOTE: Eurasian; introduced and common throughout the Province in fields and waste areas. Known also from across much of N. America.

SPERGULARIA J. & C. Presl Sand-Spurrey

1a Leaves with fascicles of smaller leaves in their axils; capsules +/- as long as stipitate-glandular calyx; seeds dark brown, reticulate sculptured and papillate; stipules acuminate, much longer than wide. **S. rubra**

1b Leaves usually without fascicles; capsules as long as or longer than calyx; seeds dull or light reddish brown, not sculptured, smooth or +/- papillate and/or wing-margined. 2

2a(1b) Free length of stipule wider than long; leaves blunt-tipped; seeds often partially wing-margined, .8-1.4 mm long; capsules to 5.2 mm long; plants essentially glabrous. **S. canadensis**

2b(1b) Free length of stipule about as long or longer than wide; leaves mucronate; seeds papillate or smooth, wingless or the lowermost seeds in some capsules wing-margined; plants glabrous or glandular pubescent. **S. marina**

Spergularia canadensis (Pers.) Don. Fig. p. A55. Map p. B44.
FREQUENCY: Uncommonly scattered along the coast.
HABITAT: Salt lagoons and estuaries.
GENERAL RANGE: Alaska to Nfld. south along the coast to N.Y.

Spergularia marina (L.) Griseb. Fig. p. A55. Map p. B44.
FREQUENCY: Scattered along the coast.
HABITAT: Salt lagoons and estuaries and salt springs inland.
GENERAL RANGE: B.C. to Nfld. south to Calif. and Fla.; S. America; Iceland; Eurasia.

Spergularia rubra (L.) J. & C. Presl Fig. p. A56. Map p. B44.
SPECIAL NOTE: European; introduced and common throughout much of the Province in mostly dry sandy soils. Known also in Canada from B.C.; Ont. to Nfld.

STELLARIA L. Chickweed, Starwort.

1a Leaves (at least the lower and middle) distinctly petioled, +/- ovate; stems not angled, often rooting at the nodes; calyx usually pilose. **S. media**

1b Leaves +/- sessile, mostly narrower; stems angled, sometimes weakly so; calyx glabrous. 2

2a(1b) Petals significantly shorter than sepals or wanting. 3

2b(1b) Petals +/- as long or longer than sepals. 4

3a(2a) Flowers in axillary cymes; longer leaves 1.1-2.2 cm long, 2-4 times as long as wide; sepals acuminate; pedicels 4-10 mm long; seeds strongly papillate. **S. alsine**

3b(2a) Flowers solitary in forks of the stem, appearing lateral or axillary; longer leaves .7-7 cm long, (2)5-10 times as long as wide; sepals blunt or acute; pedicels 10-25 mm long; seed obscurely rugose. **S. calycantha**

4a(2b) Flowers solitary in leaf axils or in few-flowered leafy-bracted terminal cymes. 5

4b(2b) Flowers in well-developed lateral or terminal, scarious-bracted cymes. 6

5a(4a) Leaves lanceolate or elliptic lanceolate, +/- acute, scarcely fleshy, to 2 cm long; capsule well exserted from calyx; seeds papillate; fresh habitats. **S. crassifolia**

5b(4a) Leaves oval or elliptic, +/- blunt and fleshy when fresh, to 1.5 cm long; capsule inserted or barely exserted from calyx; salt marsh habitat. **S. humifusa**

6a(4b) Leaves stiff, crowded, +/- ascending, pale glaucous, lance-attenuate; capsule black and shining. **S. longipes**

6b(4b) Leaves well spaced, membranaceous, pale to dark green, linear to lanceolate; capsule stramineus to dark fuscous. 7

7a(6b) Sepals not ciliate, obscurely nerved, to 4.5 mm long; petals +/- equaling sepals; leaves linear, acute at base and summit; inflorescence sparce, of terminal and axillary cymes; fruiting pedicels ascending; seeds +/- smooth. **S. longifolia**

7b(6b) Sepals usually ciliate, strongly 3-nerved, to 5.5 mm long; petals longer than sepals; leaves lanceolate to lance-attenuate, broadest near base; inflorescence many flowered, of terminal cymes; fruiting pedicels divergent; capsules pale brown; seeds coarsely tuberculate. **S. graminea**

Stellaria alsine Grimm Fig. p. A56. Map p. B44.
FREQUENCY: Uncommonly scattered in the southern half of the Province.
HABITAT: Wet banks and similar springy areas.
GENERAL RANGE: Que. to Nfld. south to Pa. and Md.

Stellaria calycantha (Ledeb.) Bong. Fig. p. A56. Map p. B44.
FREQUENCY: Common throughout the Province.
HABITAT: Wet seepage areas.
GENERAL RANGE: Alaska to Nfld. south to Calif. and N. Eng.; Greenland; Iceland; Eurasia.

Stellaria crassifolia Ehrh. Fig. p. A56. Map p. B44.
FREQUENCY: Rare; reported only from Dalhousie, Restigouche Co., Miramichi, Northumberland Co. and Shediac, Westmorland Co.
HABITAT: Wet banks near the coast.
GENERAL RANGE: Alaska to Nfld. south to B.C., Idaho and N.S.; Iceland; Eurasia.
SPECIAL NOTE: For more information see H. Hinds, 1983. The Rare Vascular Plants of New Brunswick. Syllogeus 50.

Stellaria graminea L. Fig. p. A56.
COMMON NAME(S): Common Stitchwort. Mouron des champs.
FREQUENCY: Common throughout the Province.
HABITAT: Fields and similar grassy areas.
SPECIAL NOTE: Eurasian; introduced in fields and similar grassy areas throughout the Province and across much of N. America.

Stellaria humifusa Rottb. Fig. p. A56. Map p B44.
FREQUENCY: Uncommonly scattered along the coast.
HABITAT: Salt marshes and meadows.
GENERAL RANGE: Alaska to Nfld. south to Oreg. and Maine; Greenland; Iceland; Eurasia.

Stellaria longifolia Muhl. Fig. p. A56. Map p. B44.
FREQUENCY: Uncommon or overlooked and probably scattered throughout the Province.
HABITAT: Damp meadows and shores.
GENERAL RANGE: Alaska to Nfld. south to Calif. and S.C.; Eurasia.

Stellaria longipes Goldie Fig. p. A55. Map p. B44.
FREQUENCY: Rare; only known from a few sites on Bay de Chaleur, Gloucester and Restigouche Counties.
HABITAT: Sand plains and gravelly banks near the coast.
SPECIAL NOTE: For more information see H. Hinds, 1983. The Rare Vascular Plants of New Brunswick. Syllogeus 50.

Stellaria media (L.) Cyrillo Fig. p. A55. Map p. B44.
COMMON NAME(S): Common Chickweed. Mouron des oiseaux.
SPECIAL NOTE: Eurasian; introduced in cultivated ground mostly in the south of the Province. Also known across much of Canada.

VACCARIA Moench

Vaccaria pyramidata Medic. Fig. p. A56.
COMMON NAME(S): Cowherb, Cow-Cockle. Saponaire des vaches.
SPECIAL NOTE: Eurasian; introduced but not persisting in waste areas, Campbellton, Restigouche Co. Also known in Canada from B.C. to N.S.
SYNONYMS: **Saponaria vaccaria** L.

CERATOPHYLLACEAE (Hornwort Family)

CERATOPHYLLUM L. Hornwort, Coon-tail. Cornifle

1a Leaves 1-2 forked, their segments with teeth on +/- wide base of green tissue; fruit wingless with 1 terminal and 2 basal spines. **C. demersum**
1b Leaves 2-4 forked, their segments entire or with teeth on little or no base of green tissue; fruit narrowly winged with 8-14 spines. **C. muricatum**

Ceratophyllum demersum L. Fig. p. A56. Map p. B44.
FREQUENCY: Apparently confined to the St. John River valley.
HABITAT: Still or slowly moving waters.
GENERAL RANGE: Dist. Mack. to N.S. south to Calif., Mexico and Fla.; C. America.

Ceratophyllum muricatum Cham. Fig. p. A56.
FREQUENCY: Very rare; known only from Sunpoke Lake, Sunbury Co. and Great Pond, Grand Manan, Charlotte Co.
HABITAT: Still waters.
GENERAL RANGE: Mich. and Ont. to N.S. south to Mexico and Fla.; C. America.
SYNONYMS: **C. echinatum** Gray.
SPECIAL NOTE: For more information see H. Hinds, 1983. The Rare Vascular Plants of New Brunswick. Syllogeus 50.

++

NYMPHAEACEAE (Water-lily Family)

1a Leaves peltate, to 1 dm long; flowers dull purple, less than 3 cm broad, thickly glutinous beneath when fresh. **Brasenia schreberi**

1b Leaves cordate. 2

2a(1b) Petals white or roseate; sepals green or purplish, widely spreading at anthesis; leaves orbicular, usually purple below. **Nymphaea odorata**

2b(1b) Petals inconspicuous and stamen-like; sepals yellow, often tinged with red or green, inflexed; leaves elliptical. **Nuphar luteum**

BRASENIA Schreb.

Brasenia schreberi Gmel. Fig. p. A57.
COMMON NAME(S): Water-shield, Purple Wen-dock.
FREQUENCY: Uncommonly scattered mostly in the southwest of the Province.
HABITAT: Ponds and backwaters.
GENERAL RANGE: Alaska and B.C. to N.S. south to Calif. and Fla.; C. America; Asia; Africa; Australia.

NUPHAR Sm. Yellow Pond- or Water-lily, Cow-lily.

Nuphar luteum (L.) Sibth. & Sm. Fig. p. A57. Map p. B45.
SUBSPECIFIC TAXA: Two subspecies occur within our range.

1a Stigmatic disk greenish; flower over 1 dm broad; anthers longer than filaments; floating leaf-blades to 4 dm long. ssp. **variegatum**

1b Stigmatic disk reddish; flower to 6 cm broad; anthers shorter than filaments; floating leaf-blades to 2 dm long. ssp. **pumilum**

1. Ssp. **pumilum** (Timm) E.O. Beal
COMMON NAME(S): Small Pond-lily.
FREQUENCY: Probably scattered throughout the Province.
HABITAT: Lakes, ponds and quiet backwaters.
GENERAL RANGE: Man. to Nfld. south to Minn. and N.J.
SYNONYMS: **N. microphyllum** (Pers.) Fern., **N. X rubrodiscum** (Morong) Fern.

2. Ssp. **variegatum** (Dur.) E.O. Beal
COMMON NAME(S): Beaver-Root, Bobber, Pond-Poppy, Bullhead-lily. Pied-de-cheval.
FREQUENCY: Common throughout the Province.
HABITAT: Common in bog and other ponds, lakes and quiet backwaters.
GENERAL RANGE: Yukon and B.C. to Nfld. south to Idaho and Del.
SYNONYMS: **N. variegatum** Dur., **N. advena** (Soland.) R. Br.

++

NYMPHAEA L. Water-lily. Lis d'eau.

Nymphaea odorata Ait. Fig. p. A57. Map p. B45.
COMMON NAME(S): Fragrant Water-lily. Nénuphar blanc.
FREQUENCY: Uncommonly scattered principally in the southwest portion of the Province with one population on Miscou Is., Gloucester Co.
HABITAT: Lakes, ponds and similar +/- still waters.
GENERAL RANGE: Man. to Nfld. south to Tex. and Fla.

RANUNCULACEAE (Crowfoot Family)

1a Flowers regular with 5 spurred petals; fruit a follicle. **Aquilegia**
1b Flowers spurless. 2
2a(1b) Flowers regular. 3
2b(1b) Flowers irregular, white and purplish, the upper petaloid sepal helmut-shaped; inner 2 petals with small spurs; fruit follicular. **Aconitum bicolor**
3a(2a) Perianth small and inconspicuous. 4
3b(2a) Perianth conspicuous. 5
4a(3a) Flowers greenish in panicles, usually unisexual; fruit an achene. **Thalictrum**
4b(3a) Flowers white in racemes, bisexual; fruit a berry. **Actaea**
5a(3b) Stem-leaves opposite or whorled; sepals petaloid. 6
5b(3b) Stem-leaves alternate or leaves all basal. 7
6a(5a) Leaves trifoliate, leaflets toothed; stems climbing; sepals 4; styles plumose. **Clematis**
6b(5a) Leaves simple, deeply parted nearly to midrib; stems not climbing; sepals 5 or more; styles not plumose. **Anemone**
7a(5b) Leaves alternate; main stem sometimes creeping. 8
7b(5b) Leaves all basal; plants with elongate rhizomes. 9
8a(7a) Sepals and petals both present (sepals sometimes yellowish); fruit an achene; petals with nectar-pit at base. **Ranunculus**
8b(7a) Sepals present, petals wanting or inconspicuous. 10
9a(7b) Leaves trifoliate; fruit stipitate follicles; flowers white; filiform rhizomes golden yellow. **Coptis trifolia**
9b(7b) Leaves 3-lobed; fruit achenous; flowers blue, pink or white; pedicels and petioles long hairy. **Hepatica**
10a(8b) Leaves coarsely toothed, oblong ovate to broadly cordate or reniform; petaloid sepals 5; flowers several, axillary or terminal. **Caltha palustris**
10b(8b) Leaves palmately divided; flowers usually solitary; sepals to 15. **Trollius europeus**

ACONITUM L. Monkshood. Aconite.

Aconitum bicolor Schultes
COMMON NAME(S): Garden Monkshood or Wolfsbane. Capuce de moine, Char de Vénus.

++

SPECIAL NOTE: Believes to be a hybrid of **A. napallus** X **A. variegatum.** Commonly cultivated and persisting near old habitations. Plants rarely produce seed. Also known in Canada from Ont. to Nfld., N.B. and N.S.

ACTAEA L. Baneberry.

1a Pedicels swollen, reddish; fruit white with dark spot; leaflets +/- glabrous. **A. pachypoda**

1b Pedicels filiform; fruit normally red (white in f. **neglecta** (Gillman) Robins); leaflets pubescent beneath. **A. rubra**

Actaea pachypoda Ell. Fig. p. A57. Map p. B45.
COMMON NAME(S): White Baneberry, Doll's-eyes.
FREQUENCY: Common in the western half of the Province and rarer elsewhere.
HABITAT: Mostly rich often calcareous woods.
GENERAL RANGE: Ont. to N.S. south to Okla. and Ga.

Actaea rubra (Ait.) Willd. Fig. p. A57. Map p. B45.
COMMON NAME(S): Red Baneberry. Pain de couleuvre.
FREQUENCY: Common throughout much of the Province.
HABITAT: Mixed or deciduous woods, bottomlands and thickets.
GENERAL RANGE: Alaska and Nfld. south to Calif. and N.J.
SUBSPECIFIC TAXA: Forma **neglecta** (Gillman) Robins, with white instead of red fruit, and filiform pedicels, occurs occasionally with the species.

ANEMONE L. Anemone, Wind-flower.

1a Basal and involucral leaves finely dissected. **A. multifida**

1b Leaves otherwise. 2

2a(1b) Involucral leaves +/- sessile. 3

2b(1b) Involucral leaves petioled. 4

3a(2a) Leaf divisions blunt; flowers usually solitary; achenes long-woolly in elongate head. **A. parviflora**

3b(2a) Leaf divisions acute; flowers usually 2 or more; achenes +/- glabrous in globose head. **A. canadensis**

4a(2b) Flowers usually solitary; plants to about 3 dm tall; achenes finely pubescent; stems from creeping rhizomes. **A. quinquifolia**

4b(2b) Flowers usually 2 or more; plants generally over 3 dm tall; achenes woolly; stems tufted. **A. riparia**

Anemone canadensis L. Fig. p. A57. Map p. B45.
COMMON NAME(S): Canada Anemone.
FREQUENCY: Common in the St. John, lower Restigouche and Nepisiquit R. valleys. Uncommon or rare elsewhere.
HABITAT: Shores, meadows and moist thickets.
GENERAL RANGE: Dist. Mack. and B.C. to N.S. south to N. Mex. and N.J.

++

Anemone multifida Poir. Fig. p. A57. Map p. B45.
COMMON NAME(S): Bird's-foot Anemone, Cut-leaved Anemone.
FREQUENCY: Uncommon on the St. John and Restigouche Rivers.
HABITAT: Gravel strands and ledge crevices near shore.
GENERAL RANGE: Aggregate species from Alaska and B.C. to Nfld. south to Calif., Nebr. and Maine; S. America.
SUBSPECIFIC TAXA: Two well marked varieties occur in our range.

1a Sepals bright red, to 2 cm long; stems several flowered, to 7 dm tall; Restigouche R. only. var. **richardsiana**
1b Sepals yellowish white within, yellowish, greenish or purplish outside, to 1 cm long; stems 1-3 flowered, to about 3 dm tall. var. **multifida**

Anemone parviflora Michx. Fig. p. A57.
COMMON NAME(S): Small-flowered Anemone.
FREQUENCY: Very rare; known from a few sites on the Restigouche R., and one locality near Albert Mines, Albert Co.
HABITAT: Wet calcareous ledges and shores.
GENERAL RANGE: Alaska and B.C. to Nfld. south to Oreg. and N.B.
SPECIAL NOTE: For more information see H. Hinds, 1983. The Rare Vascular Plants of New Brunswick. Syllogeus 50.

Anemone quinquifolia L. Fig. p. A57. Map p. B45.
COMMON NAME(S): American Wood-Anemone.
FREQUENCY: Scattered in the southern half of the Province.
HABITAT: Moist woods and thickets.
GENERAL RANGE: B.C. to N.S. south to Calif. and N.C.

Anemone riparia Fern. Fig. p. A57. Map p. B45.
COMMON NAME(S): Tall Thimbleweed, Riverbank Anemone.
FREQUENCY: Scattered in the St. John and Restigouche R. valleys; rare elsewhere.
HABITAT: Calcareous ledges and shores.
GENERAL RANGE: B.C. to Nfld. south to Minn. and N. Eng.

AQUILEGIA L. Columbine. Ancolie

1a Spurs hooked; flowers white, pink, blue or purple; follicle beaks to 1 cm long. **A. vulgaris** L.
1b Spurs straight or slightly recurved; flowers reddish; follicle beaks over 1 cm long. **A. canadensis** L.

Aquilegia canadensis L. Fig. p. A57.
COMMON NAME(S): Wild Columbine. Glands or Gants de Notre-Dame.
FREQUENCY: Very rare; reported only from Dalhousie, Restigouche Co.
HABITAT: Rocky woods and slopes.
GENERAL RANGE: Sask. to Nfld. south to Tex. and Fla.
SPECIAL NOTE: Collected by Titus in 1945. Specimen destroyed in NSAC fire.

++

Aquilegia vulgaris L. Fig. p. A57. Map p. B45.
COMMON NAME(S): European Columbine. Ancolie des jardins.
SPECIAL NOTE: Eurasian; cultivated and spreading to roadsides and waste areas throughout the Province and elsewhere in Canada from B.C. to Nfld., P.E.I., N.B. and N.S.

CALTHA L. Marsh-Marigold. Populage

Caltha palustris L. Fig. p. A58. Map p. B45.
COMMON NAME(S): King-cup, Cowslip. Souci d'eau.
FREQUENCY: Scattered and uncommon in the southwest and common in the northeastern third of the Province.
HABITAT: Shallow water and swampy ground.
GENERAL RANGE: Alaska to Nfld. south to Oreg. and S.C.; Iceland; Eurasia.

CLEMATIS L. Clematis. Clematite

1a Flowers many in cymose panicles, whitish. **C. virginiana**
1b Flowers solitary, purplish or reddish blue. **C. occidentalis**

Clematis occidentalis (Hornem.) DC. Fig. p. A58. Map p. B45.
COMMON NAME(S): Purple Clematis.
FREQUENCY: Scattered throughout much of the Province.
HABITAT: Mostly calcareous ledges, rocky slopes and open woods.
GENERAL RANGE: Ont. to N.B. south to Oreg. and Md.
SYNONYMS: **C. verticillaris** DC.

Clematis virginiana L. Fig. p. A58. Map p. B46.
COMMON NAME(S): Virgin's-bower. Herbe aux gueux.
FREQUENCY: Common throughout the Province.
HABITAT: Bottomlands and thickets, borders of woods.
GENERAL RANGE: Man. to N.S. south to Kans. and Ga.

COPTIS Salisb. Goldthread. Coptide

Coptis trifolia (L.) Salisb. Fig. p. A58.
FREQUENCY: Common throughout the Province.
HABITAT: Moist, mossy coniferous or mixed woods.
GENERAL RANGE: Alta. to Nfld. south to Iowa and N.C.; Greenland.
SUBSPECIFIC TAXA: Our plants are referable to ssp. **groenlandica** (Oeder) Hult.
SYNONYMS: **C. groenlandica** (Oeder) Fern.

++

DELPHINIUM L. Delphinium, Larkspur. Pied d'alouette

Delphinium elatum L.
SPECIAL NOTE: Eurasian; commonly planted and rarely persisting near habitations as in Fredericton, York Co. Not included in family key.

HEPATICA Mill. Hepatica, Liverleaf. Trinitaire

Hepatica nobilis P. Mill. Fig. p. A58. Map p. B46.
COMMON NAME(S): Round-leaved Hepatica.
FREQUENCY: Rare; scattered in York and Carleton Counties and one location in each of Restigouche, Kings and Charlotte Counties.
HABITAT: Dryish mixed or hardwoods and slopes.
GENERAL RANGE: Man. to N.S. south to Mo. and Fla.
SUBSPECIFIC TAXA: Our plants are referable to var. **obtusa** (Pursh) Steyermark.
SYNONYMS: **H. americana** (DC.) Ker-Gawl.

RANUNCULUS L. Buttercup, Crowfoot. Renoncule

1a Stem leaves simple, neither lobed nor divided. 2
1b Stem leaves variously lobed or divided. 3
2a(1a) Stem leaves cordate-ovate, crenate; plants of brackish habitats. **R. cymbalaria**
2b(1a) Stem leaves linear to narrowly lanceolate; stem rooting at the nodes; plants of fresh shores. **R. reptans**
3a(1b) Aquatic plants; leaves finely dissected. 4
3b(1b) Terrestrial plants; leaves otherwise. 5
4a(3a) Flowers yellow, extending well beyond the leaves; nutlets smooth. 6
4b(3a) Flowers white, only slightly surpassing leaves; nutlets wrinkled. 7
5a(3b) Stems creeping; basal leaf solitary, reniform, 3-4 cleft. **R. lapponicus**
5b(3b) Stems +/- erect (runners sometimes creeping); basal leaves usually several. 8
6a(4a) Flowers small, petals 3.5-5 mm long; fruit, including beak, 1.5-2 mm long, not wing-margined; submerged leaf segments with rounded tips; stipules inconspicuous. **R. gmelinii**
6b(4a) Flowers larger, petals 6-17 mm long; fruit, including beak, 2.5-3.5 mm long with corky wing-margin; ultimate leaf segments usually long acuminate; broad scarious stipules conspicuous. **R. flabellaris**
7a(4b) Leaf broadly ovate in general outline, usually collapsing out of water; petiole usually longer than narrow stipular sheaths; achene pedicels not recurved. **R. aquatilis**
7b(4b) Leaf orbicular in general outline, +/- keeping its shape out of water, sessile above or within expanded stipular sheath of petiole; achene beak to 0.5 mm long; mature pedicles recurved. **R. circinatus**
8a(5b) Petals relatively small and inconspicuous, 1.5-5 mm long, rarely longer than the sepals. 9

8b(5b) Petals larger, 6-17 mm long, conspicuously longer than the sepals. 10

9a(8a) Basal leaves usually unlobed, crenate; receptacle bristly; stem leaves sessile. **R. abortivus**

9b(8a) Basal leaves variously lobed or divided. 11

10a(8b) Receptacle usually villous. 12

10b(8b) Receptable glabrous; basal leaves deeply divided into 3 sessile divisions; achenes with stout recurved beaks about 0.6 mm long. **R. acris**

11a(9b) Achenes minutely beaked, turgid, without a sharp border; basal leaves deeply 3-parted; sepals reflexed; petals pale yellow slightly longer than sepals. **R. sceleratus**

11b(9b) Achenes conspicuously beaked, flattened and with sharp border. 13

12a(10a) Basal leaves compound. 14

12b(10a) Basal leaves shallowly lobed to 3-parted, the terminal segment sometimes stalked; achenes orbicular with a row of tubercles near margin and recurved, horizontal beak; sepals tightly reflexed; petals pale yellow; annuals. **R. sardous**

13a(11b) Stem with soft spreading hairs; achenes in globose head, beaks hooked at tip; basal leaves palmately cleft to deeply 3-parted; plants of rich woods. **R. recurvatus**

13b(11b) Stem with stiff spreading hairs; achenes in long ovoid head, beaks straight; basal leaves soon withering, parted into 3-5 3-lobed leaflets; plants of marshes and wet meadows. **R. pensylvanicus**

14a(12a) Stigma covering one side of the short, recurved style; achene to 2.5 mm long with a short triangular curved beak; fresh leaves often white mottled. **R. repens**

14b(12a) Stigma terminating the long +/- straight style; achene to 4.5 mm long with straight or curved beak to 3 mm long; basal leaves to 2 dm broad. **R. septentrionalis**

Ranunculus abortivus L. Fig. p. A58. Map p. B46.
COMMON NAME(S): Kidney-leaf Buttercup.
FREQUENCY: Scattered throughout the Province.
HABITAT: Bottomlands and moist woodlands.
GENERAL RANGE: Alaska to Nfld. south to Wash., Colo., Tex. and Fla.; Cuba.
SYNONYMS: **R. abortivus** var. **acrolasius** Fern.

Ranunculus acris L. Fig. p. A58.
COMMON NAME(S): Common Buttercup. Bouton d'or.
FREQUENCY: Common throughout the Province.
HABITAT: Pastures, wet ditches and waste areas.
GENERAL RANGE: Eurasian; common throughout Canada.
SYNONYMS: **R. acris** var. **latisectus** Beck.

Ranunculus aquatilis L. Fig. p. A58. Map p. B46.
COMMON NAME(S): White Water-Crowfoot. Herbe aux écrevisses.
HABITAT: Calcareous or brackish slow or still waters.
SUBSPECIFIC TAXA: Three varieties occur within our range.

++

1a Receptacle +/- glabrous; achenes usually more than 1.5 mm long, glabrous and mostly smooth. var. **calvescens**

1b Receptacle pilose; achenes 1.5 or less mm long; pubescent. 2

2a(1b) Petals to 6 mm long; stamens usually less than 8; achenes about 1 mm long; stem to 1 mm thick. var. **eradicatus**

2b(1b) Petals to 8 mm long; stamens usually more than 10; achenes to 1.5 mm long; stem to 2.5 mm thick. var. **capillaceus**

1. Var. **calvescens** (W. Drew) Benson
FREQUENCY: Very rare; only one collection identified from near Eel R., Restigouche Co. (UNB).
GENERAL RANGE: Ont. to Nfld. south to Mich. and Pa.
SYNONYMS: **R. trichophyllus** Chaix. var. **calvescens** Drew.

2. Var. **capillaceus** (Thuill.) DC.
FREQUENCY: Scattered throughout the Province.
GENERAL RANGE: Alaska to Nfld. south to Calif., Tex. and Del.
SYNONYMS: **R. trichophyllus** Chaix.

3. Var. **eradicatus** Laest.
FREQUENCY: Very rare; material from St.-Joseph-de-Madawaska appears to belong here (UNB).
GENERAL RANGE: Alaska to Nfld. and e. Que.
SYNONYMS: **R. trichophyllus** var. **eradicatus** (Loest.) Drew.

Ranunculus circinatus Sibth.
COMMON NAME(S): Stiff Water-Crowfoot.
FREQUENCY: Very rare; known only from N. Rusagonis Stream at Hanwell, York Co. and from Middle Tetagouche Lake, Restigouche Co. (UNB).
HABITAT: Calcareous quiet or slow moving waters.
GENERAL RANGE: Yukon and B.C. to Nfld. south to Mexico, S. Dak. and Mass.; Europe.
SUBSPECIFIC TAXA: Our plants are referable to var. **longirostris** Godron.

Ranunculus cymbalaria Pursh. Fig. p. A58. Map p. B46.
COMMON NAME(S): Seaside Crowfoot.
FREQUENCY: Scattered along the coast and inland to salt springs in Kings Co.
HABITAT: Coastal marshes, shores and salt springs inland.
GENERAL RANGE: Alaska to Nfld. south to N. Mex. and N.J.; Greenland; Eurasia.

Ranunculus flabellaris Raf. Fig. p. A58. Map p. B46.
COMMON NAME(S): Yellow Water-Crowfoot or -Buttercup.
FREQUENCY: Rare; known from a few localities in Sunbury and Queens Counties and one each from Gloucester and Westmorland Counties.
HABITAT: Shallow quiet waters and muddy shores.
GENERAL RANGE: B.C. to N.B. south to Calif. and N.C.

++

Ranunculus gmelinii DC. Map p. B46.
COMMON NAME(S): Small Yellow Water-Crowfoot or -Buttercup.
FREQUENCY: Rare; reported from a few localities mostly in the eastern part of the Province.
HABITAT: Wet meadows, ponds and streams.
GENERAL RANGE: Alaska to Nfld. south to Oreg., N. Mex., Mich. and Maine; Eurasia.
SUBSPECIFIC TAXA: Our plants are referable to var. **hookeri** (Don) Benson.

Ranunculus lapponicus L. Fig. p. A58.
COMMON NAME(S): Lapland Buttercup.
FREQUENCY: Very rare; known only from near Shea Lake, Victoria Co.
HABITAT: Cool, mossy cedar swamp in calcareous area.
GENERAL RANGE: Alaska to Labr. south to B.C. and Maine; Greenland; Spitsbergen; Eurasia.
SPECIAL NOTE: For more information see H. Hinds, 1983. The Rare Vascular Plants of New Brunswick. Syllogeus 50.

Ranunculus pensylvanicus L. f. Fig. p. A58. Map p. B46.
COMMON NAME(S): Bristly Crowfoot.
FREQUENCY: Scattered throughout except rare in the southeast.
HABITAT: Muddy shores and moist meadows.
GENERAL RANGE: B.C. to Nfld. south to Wash., Ariz. and Del.

Ranunculus recurvatus Poir. Fig. p. A59. Map p. B46.
COMMON NAME(S): Hooked Buttercup.
FREQUENCY: Locally common mostly in the south and west of the Province.
HABITAT: Rich woods.
GENERAL RANGE: Ont. to Nfld. south to Tex. and Fla.

Ranunculus repens L. Fig. p. A59.
COMMON NAME(S): Creeping Buttercup. Bassinet.
FREQUENCY: Common throughout the Province.
HABITAT: Mostly moist meadows, ditches and similar open areas.
GENERAL RANGE: Eurasian; introduced across N. America.
SUBSPECIFIC TAXA: Var. **glabratus** DC., with stem and petioles +/- glabrous, occurs occasionally with the species as in St. Andrews, Charlotte Co.

Ranunculus reptans L. Fig. p. A59. Map p. B46.
COMMON NAME(S): Creeping Spearwort.
FREQUENCY: Scattered throughout most of the Province.
HABITAT: Damp shores and shallow water.
GENERAL RANGE: Alaska to Nfld. south to Calif. and Mass.; Greenland; Iceland; Eurasia.

Ranunculus sardous Crantz Fig. p. A59.
SPECIAL NOTE: Eurasian; introduced in wharf-ballast in Saint John but not persisting. Known in Canada also from Ont.

++

Ranunculus sceleratus L. Fig. p. A59.
COMMON NAME(S): Cursed Crowfoot, Celeryleaf Buttercup. Mort aux vaches, Herbe de feu.
FREQUENCY: Rare; known from Dalhousie and Charlo, Restigouche Co., Saints Rest, Saint John, and Milltown, Charlotte Co.
HABITAT: Wet ditches, low damp ground, upper edge of salt marsh.
GENERAL RANGE: Alaska to N.B. south to Calif. and Fla.; Eurasia.
SUBSPECIFIC TAXA: Our plants are referable to var. **multifidus** Nutt.
SPECIAL NOTE: For more information see H. Hinds, 1983. The Rare Vascular Plants of New Brunswick. Syllogeus 50.

Ranunculus septentrionalis Poir. Fig. p. A59. Map p. B46.
COMMON NAME(S): Northern Crowfoot, Swamp Buttercup.
FREQUENCY: Common especially towards the northwest of the Province, rare in the southern half.
HABITAT: Wet bottomlands and swampy ground.
GENERAL RANGE: Man. to P.E.I. south to Tex. and Va.

THALICTRUM L. Meadow-rue. Pigamon

1a Stem leaves below inflorescence sessile (immediately ternately divided above expanded stipular base); largest leaflets usually widest above the middle, entire to 3-lobed, glabrous to pubescent beneath; anthers less than 3 mm long blunt, their white or yellowish filaments expanded upwards; carpels both glandular and nonglandular pubescent; plants flowering in late July to early August. **T. pubescens**

1b Stem leaves below inflorescence petioled below ternate division; largest leaflets with (3) 5-7 lobes, usually widest at or above middle, usually minutely glandular pubescent beneath; anthers to 4 mm long, acuminate, their +/- purplish filaments filiform; carpels glandular pubescent; plant usually flowering in late May to early June. **T. venulosum**

Thalictrum pubescens Pursh Fig. p. A59.
COMMON NAME(S): King-of-the-meadow, Tall Meadow-rue. Grand Pigamon.
FREQUENCY: Common throughout the Province.
HABITAT: Meadows, thickets and swamps.
GENERAL RANGE: Ont. to Nfld. south to Tenn. and Ga.

Thalictrum venulosum Trel. Fig. p. A59. Map p. B46.
COMMON NAME(S): Frontier Meadow-rue.
FREQUENCY: Scattered and locally common in the St. John R. valley and uncommon near the mouth of the Restigouche R., Restigouche Co.
HABITAT: Calcareous shores and alluvial meadows.
GENERAL RANGE: Ont. to N.S. south to Minn. and Vt.
SUBSPECIFIC TAXA: Our plants are referable to var. **confine** (Fern.) Boivin.
SYNONYMS: **T. confine** Fern.

TROLLIUS L. Globe-flower

Trollius europaeus L.
SPECIAL NOTE: European; cultivated and rarely persisting near habitations as in Lakewood, near Saint John, St. John Co.

BERBERIDACEAE (Barberry Family)

1a Plants herbaceous; fruit globose; leaves ternately compound, glaucous; flowers greenish-purple. **Caulophyllum thalictroides**
1b Plants shrubby, spiny; fruit ellipsoid; leaves simple; flowers yellowish. **Berberis**

BERBERIS L. Barberry. Epine-vinette

1a Leaves entire; nodal spines usually simple; berries dryish. **B. thunbergii**
1b Leaves finely bristle-toothed; nodal spines ternate; berries juicy. **B. vulgaris**

Berberis thunbergii DC. Fig. p. A59. Map p. B47.
COMMON NAME(S): Japanese Barberry.
SPECIAL NOTE: Asiatic; cultivated and frequently escaped to thickets, pastures, and open woods in scattered locations in the south of the Province. Also known in Canada from Ont., P.E.I. and N.S.

Berberis vulgaris L. Fig. p. A59. Map p. B47.
COMMON NAME(S): Common Barberry.
SPECIAL NOTE: Eurasian; formally cultivated and rarely escaped to thickets and pastures as in Shediac, Westmorland Co. Known previously across much of southern Canada but irradicated because of its being an alternate host of the wheat rust.

CAULOPHYLLUM Michx. Blue Cohosh. Léontice faux-pigamon

Caulophyllum thalictroides (L.) Michx. Fig. p. A59. Map p. B47.
COMMON NAME(S): Papoose-root. Graines à chapelet.
FREQUENCY: Uncommon to rare in scattered localities towards the west and south of the Province.
HABITAT: Rich deciduous woods and alluvial bottomlands.
GENERAL RANGE: Man. to N.S. south to Mo. and S.C.; Asia.

PAPAVERACEAE (Poppy Family)
(INCLUDING FUMARIACEAE)

1a Flowers regular; sap usually brightly colored or milky. (Subfamily **Papaveroideae**) 2

+++

1b Flowers irregular; sap watery. (Subfamily **Fumarioideae**) 3

2a(1a) Leaf solitary, basal; petals at least 8; capsule solitary. **Sanguinaria canadensis**

2b(1a) Leaves more than one, at least some cauline; petals 4-6; capsules usually several. 4

3a(1b) Two outer sepals spurred or saccate at base. 5

3b(1b) One outer sepal spurred or saccate at base. 6

4a(2b) Flowers yellow; fruit ellipsoid, dehiscent by longitudinal valves. **Chelidonium majus**

4b(2b) Flowers red, white or purple; fruit broader, dehiscent by small apical valves surrounding the disk. **Papaver**

5a(3a) Leaves all basal. **Dicentra cucullaria**

5b(3a) Leaves cauline, alternate; plant climbing. **Adlumia fungosa**

6a(3b) Ovary and fruit subglobose. **Fumaria officinalis**

6b(3b) Ovary and fruit elongate. **Corydalis sempervirens**

ADLUMIA Raf.

Adlumia fungosa (Ait.) Greene Fig. p. A59.
COMMON NAME(S): Climbing Fumitory, Mountain-fringe, Canary Vine.
FREQUENCY: Rare; collected in Hampstead, Queens Co. and St. Stephen, Charlotte Co. (NBM).
HABITAT: Wet woods and rocky slopes especially where recently burned.
GENERAL RANGE: Man. to N.S. south to Tenn. and N.C. Introd. in B.C. and Alta.
SPECIAL NOTE: These reports may represent escapes from cultivations since the plant was formally cultivated. Not collected in recent years.

CHELIDONIUM L.

Chelidonium majus L. Fig. p. A60.
COMMON NAME(S): Celandine, Swallowwort. Grande Eclaire, Herbe aux verrues.
SPECIAL NOTE: Eurasian; scattered in and about towns in the Province and elsewhere in Canada in B.C., and from Ont. to N.S.

CORYDALIS Medic. Corydalis. Corydale

Corydalis sempervirens (L.) Pers. Fig. p. A60. Map p. B47.
COMMON NAME(S): Pale Corydalis, Rock-Harlequin. Corydalle toujours verte.
FREQUENCY: Scattered throughout much of the Province.
HABITAT: Rock outcrops and clearings especially recently disturbed as from burning or bulldozing.
GENERAL RANGE: Alaska to Nfld. south to B.C, Mont. and Ga.

PAPAVERACEAE

++

DICENTRA Bernh.

Dicentra cucullaria (L.) Bernh. Fig. p. A60. Map p. B47.
COMMON NAME(S): Dutchman's-breeches. Cullottes de Hollandais.
FREQUENCY: Scattered throughout much of the Province but rare in the eastern lowlands.
HABITAT: Rich, often calcareous rocky woods.
GENERAL RANGE: Ont. to N.S. south to Kans. and Ga.

FUMARIA L. Fumitory, Earth-smoke. Fumeterre

Fumaria officinalis L. Fig. p. A60.
SPECIAL NOTE: Eurasian; rarely escaped from cultivation as in Chatham, Northumberland Co., or from ballast waste as on Campobello Is., Charlotte Co. Known elsewhere in Canada from all the Provinces.

PAPAVER L. Poppy. Pavot

1a Stem leaves cordate-clasping, strongly glaucous, glabrous; flowers white, red, or purple. **P. somniferum**
1b Stem leaves otherwise; flowers usually scarlet; plant bristly-hirsute. 2
2a(1b) Flowers pale scarlet; capsule about 25 mm long expanded upward; stigmatic disc 6-10 rayed. **P. dubium**
2b(1b) Flowers dark scarlet; capsule about 7 mm long +/- globose; stigmatic disc 8-20 rayed. **P. rhoeas**

Papaver dubium L. Fig. p. A60.
COMMON NAME(S): Pale Red Poppy.
SPECIAL NOTE: European; probably introduced in wharf-ballast and not persisting as in St. Stephens, Charlotte Co. and Buctouche, Kent Co. Reported in Canada also from N.S.

Papaver rhoeas L. Fig. p. A60.
COMMON NAME(S): Corn-Poppy. Coquelicot.
SPECIAL NOTE: Eurasian; locally escaped from cultivation as in Kent and Kings Counties. Known elsewhere across much of Canada.

Papaver somniferum L.
COMMON NAME(S): Common or Opium Poppy. Oeillette.
SPECIAL NOTE: Eurasian; cultivated and rarely escaping to waste areas as in Blacks Harbour, Charlotte Co. Known elsewhere across much of s. Canada.

SANGUINARIA L.

Sanguinaria canadensis L. Fig. p. A60. Map p. B47.
COMMON NAME(S): Bloodroot. Sang-dragon, Sanguinaire.

FREQUENCY: Scattered throughout much of the Province but rare in the eastern half.
HABITAT: Rich woods and alluvial thickets.
GENERAL RANGE: Man. to N.S. south to Tex. and Fla.

BRASSICACEAE (Mustard Family)
(Alternate name **CRUCIFERAE**)

1a Capsules wide, 1-4 times as long as wide. 2
1b Capsules slender, more than 4 times as long as wide. 3
2a(1a) Capsules strongly flattened. 4
2b(1a) Capsules not flattened or if flattened, less than 6 mm wide. 5
3a(1b) Principal leaves deltoid-cordate, coarsely dentate; petals white. **Alliaria officinalis**
3b(1b) Principal leaves otherwise. 6
4a(2a) Capsules flattened parallel to septum, the septum therefore as wide as the fruit; basal leaves in rosette. 7
4b(2a) Capsules flattened at right angles to septum, the septum therefore much narrower than width of fruit; basal leaves not in rosette or wanting. 8
5a(2b) Capsule transversely divided into 2 cells; plants of marine sandy shores. **Cakile edentula**
5b(2b) Capsule longitudinally divided into 2 cells; plant habitat otherwise. 9
6a(3b) Median stem leaves lobed, deeply divided or coarsely dentate. 10
6b(3b) Median stem leaves not lobed, divided or coarsely toothed. 11
7a(4a) Plants with basal rosettes, stems to 4 dm tall; seeds wingless. 12
7b(4a) Plants without well developed basal rosettes, stems to 7 dm tall; seeds winged. **Berteroa incana**
8a(4b) Capsule triangular-obcordate. **Capsella bursa-pastoris**
8b(4b) Capsule otherwise. 13
9a(5b) Capsule beakless; leaves basal, linear-subulate, entire; plant usually submerged. **Subularia aquatica**
9b(5b) Capsule beaked. 14
10a(6a) Main stem leaves bipinnate or bipinnatifid, the ultimate segments oblong or linear. 15
10b(6a) Main stem leaves 1-pinnate. 16
11a(6b) Beak of capsule 6-16 mm long. **Brassica**
11b(6b) Beak of capsule 0.5-3(5) mm long. 17
12a(7a) Petals rounded or emarginate; perennials or biennials. **Draba**
12b(7a) Petals bifid; annuals. **Eriophila verna**
13a(8b) Capsule +/- indehiscent, usually wider than long, emarginate, reticulate; leaves 1-2 pinnately parted. **Coronopus didymus**
13b(8b) Capsule dehiscent, as long as or longer than wide, smooth. 18
14a(9b) Cauline leaves toothed or lobed. 19
14b(9b) Cauline leaves entire. 20
15a(10a) Plant +/- pubescent with forked or stellate hairs; leaves 1-3-pinnate with very numerous small segments. **Descurainia sophia**
15b(10a) Plant glabrous or hairs simple. **Rorippa sylvestris**
16a(10b) Leaves ternate or palmately divided or lobed. **Dentaria**

16b(10b) Leaves pinnately divided or lobed. 21

17a(11b) Median cauline leaves remotely and sharply denticulate; flowers scented, purple, white or pink. **Hesperis matronalis**

17b(11b) Median cauline leaves entire or low sinuate-dentate. 22

18a(13b) Seed 1-2 per capsule; fruit deeply notched. **Thlaspi arvense**

18b(13b) Seeds several in each capsule; fruit shallowly notched. **Lepidium**

19a(14a) Petals yellow. 23

19b(14a) Petals white; basal leaves long petioled, the oblong blade to 3 dm long, coarsely dentate; capsules usually falling early; seeds not developing. **Armoracia rusticana**

20a(14b) Cauline leaves cuneate at base, not clasping; petals white, bifid. **Berteroa incana**

20b(14b) Cauline leaves sagittate, clasping; petals yellow. 24

21a(16b) Seeds in 2 rows in each locule. 25

21b(16b) Seeds in 1 row in each locule. 26

22a(17b) Capsule flattened. **Arabis**

22b(17b) Capsule terete or quadriangular. 27

23a(19a) Capsule indehiscent, coarsely and irregularly verrucose. **Bunias orientalis**

23b(19a) Capsule dehiscent, smooth. **Rorippa islandica**

24a(20b) Capsule smooth, dehiscent, several seeded. **Camelina sativa**

24b(20b) Capsule reticulate and pitted, indehiscent, 1-2 seeded. **Neslia paniculata**

25a(21a) Leaves divided into 3-11 +/- entire leaflets; petals white; capsule often falcate; plants aquatic. **Nasturtium microphyllum**

25b(21a) Leaves coarsely few-toothed or pinnatifid; plants terrestrial. **Diplotaxis**

26a(21b) Beak of capsule 6-16 mm long. 28

26b(21b) Beak of capsule 0.5-3(4) mm long. 29

27a(21b) Flowering stem with rosette of basal leaves; petals creamy yellow; pubescence simple. **Arabis glabra**

27b(22b) Flowering stem without basal rosette; basal leaves early deciduous; petals clear yellow; pubescence 2-4 pronged. **Erysimum**

28a(26a) Capsule indehiscent, constricted at intervals; petals conspicuously veined, yellow, pink or purple. **Raphanus raphanistrum**

28b(26a) Capsule dehiscent, not constricted at intervals; petals not conspicuously veined, yellow fading to white. 30

29a(26b) Lower flowers in axils of leaf-like bracts; stem retrorsely pubescent with simple hairs. **Erucastrum gallicum**

29b(26b) Lower flowers without bracts. 31

30a(28b) Valves of fruit each with 3 parallel veins of about equal strength; beak flattened, 2-edged; cauline leaves not clasping. **Sinapis**

30b(28b) Valves of fruit with a prominent midnerve, other nerves obscure, not parallel; beak slender, terete or angular, not 2-edged; cauline leaves clasping in some species. **Brassica**

31a(29b) Petals white. **Cardamine**

31b(29b) Petals yellow. 32

32a(31b) Cauline leaves clasping; stem ribbed or angled; each capsule valve 1-nerved. **Barbarea**

32b(31b) Cauline leaves not clasping; stem not ribbed or angled; each capsule valve 1-3 nerved. **Sisymbrium**

BRASSICACEAE

++

ALLIARIA Ehrh.

Alliaria petiolata (Bieb.) Cavara & Grande Fig. p. A60.
COMMON NAME(S): Garlic-Mustard. Alliaire.
SPECIAL NOTE: Eurasian; introduced mostly in the St. John R. valley in alluvial woods and thickets. Known elsewhere in Canada from B.C., and from Ont. to N.B.
SYNONYMS: **A. officinalis** Andrz.

ARABIS L. Rock-Cress. Arabette

1a	Flowering and fruiting pedicels inclined at 45° or more from main axis of inflorescence. 2
1b	Flowering and fruiting pedicels inclined at less than 45° from main axis of inflorescence. 3
2a(1a)	Cauline leaves glabrous, +/- entire; stem erect from base, to 9 dm high; flowers usually pink or purple; native. **A. divaricata**
2b(1a)	Cauline leaves stellate pubescent, usually irregularly toothed; stem decumbent at base; flowers white, showy; alien garden escape. **A. caucasica**
3a(1b)	Stem and leaves mostly glabrous above base. 4
3b(1b)	Stem and leaves usually hirsute to summit. **A. hirsuta**
4a(3a)	Mature capsules terete, less than 1.3 mm wide; petals creamy-yellowish or greenish-white, 3-6 mm long. **A. glabra**
4b(3a)	Mature capsules flattened, more than 1.5 mm wide; petals white. **A. drummundii**

Arabis caucassica Schlecht.
COMMON NAME(S): Wall Rock-Cress.
SPECIAL NOTE: Eurasian; cultivated and rarely escaped to disturbed dry soil as in Saint John, St. John Co. Known elsewhere in Canada from Ont., Que. and Nfld.

Arabis divaricarpa Nels. Fig. p. A60. Map p. B47.
FREQUENCY: Very rare; known from only three localities, along the Restigouche R., and Eel R., Restigouche Co.
HABITAT: Rocky banks and calcareous ledges.
GENERAL RANGE: Alaska to N.B. south to Calif. and Vt.
SPECIAL NOTE: For more information see H. Hinds, 1983. The Rare Vascular Plants of New Brunswick. Syllogeus 50.

Arabis drummondii Gray Fig. p. A60. Map p. B47.
COMMON NAME(S): Drummond's Rock-Cress.
FREQUENCY: Uncommon; widely scattered throughout much of the Province.
HABITAT: Ledges, gravels and rocky thickets.
GENERAL RANGE: Alaska to Nfld. south to Calif. and Del.

++

Arabis glabra (L.) Bernh. Fig. p. A60. Map p. B47.
COMMON NAME(S): Tower-Mustard. Tourette.
SPECIAL NOTE: Eurasian; introduced in clearings, on roadsides, and ledges mostly in the northwest of the Province. Known elsewhere in Canada from Yukon and B.C. to the Gaspé.

Arabis hirsuta (L.) Scop. Fig. p. A60. Map p. B47.
FREQUENCY: Uncommon; scattered along the Restigouche R., Restigouche Co., the St. John R. in Kings, York and Carleton Counties, and in the Caldonia Mtns., Albert and St. John Counties.
HABITAT: Calcareous ledges and gravels.
GENERAL RANGE: Alaska to N.S. south to Calif., N. Mex. and Ga.
SPECIAL NOTE: Some of our material approaches var. **adpressipilis** (Hopkins) Rollins, with stem appressed-pubescent or strigose with mostly 2-forked hairs and leaves +/- glabrous instead of leaves and stem spreading-hirsute with mostly simple hairs as in the usual var. **pycnocarpa** (Hopkins) Rollins.

ARMORACIA Gaertn., Mey., & Scherb.

Armoracia rusticana (Lam.) Gaertn., Mey., & Scherb. Fig. p. A61.
COMMON NAME(S): Horseradish. Raifort.
SPECIAL NOTE: Eurasian; cultivated and escaping to fields and roadsides as in the St. John R. valley and probably elsewhere. Known elsewhere in Canada from all the provinces, Yukon and Dist. Mack.

BARBAREA R. Br. Winter-Cress

Barbarea orthoceras Ledeb. Fig. p. A61. Map p. B47.
FREQUENCY: Uncommon; found mostly in the northern half of the Province with one population on Grand Manan Island, Charlotte Co.
HABITAT: Streambanks, sandy beaches, gravel river strands and rocky shores.
GENERAL RANGE: Alaska to Nfld. south to Calif. and N.H.; Greenland; Eurasia.
SPECIAL NOTE: Some of our populations have a weedy character and may be introduced.

Barbarea vulgaris R. Br. Fig. p. A61.
COMMON NAME(S): Common Winter-Cress, Yellow Rocket. Cresson d'hiver, Herbe de Sainte-Barbe.
SPECIAL NOTE: Eurasian; introduced and common in fields, disturbed ground and similar areas throughout the Province. Known elsewhere in Canada from all the provinces except Sask.
SPECIAL NOTE: some of our material appears to be intermediate between this species and the last and perhaps represents hybridization.

++

BERTEROA DC.

Berteroa incana (L.) DC. Fig. p. A61.
COMMON NAME(S): Hoary Alyssum.
SPECIAL NOTE: Eurasian; introduced on railways, dry fields and waste areas as near Buctouche, Kent Co., and St. Stephen, Charlotte Co. Known elsewhere in Canada from all the Provinces except P.E.I.

BRASSICA L. Mustard, Turnip. Moutard or Navette

1a	Cauline leaves petiolate or sessile. 2
1b	Cauline leaves clasping stem. 3
2a(1a)	Plant +/- glabrous, glaucous; upper leaves sessile or short petioled; capsules ascending 1.5-4 cm long; pedicels slender 9-18 mm long. **B. juncea**
2b(1a)	Plant hirsute, green; capsules erect, +/- appressed, 1-2 cm long; pedicels thick .3-5 mm long; leaves all petioled. **B. nigra**
3a(1b)	Open flowers overtopping the buds of the inflorescence; petals 6-10 mm long; leaves (at least the lowest) setose-ciliate, bright green to glaucous; roots rarely thickened. **B. rapa**
3b(1b)	Open flowers not overtopping the buds of the inflorescence; petals 10-18 mm long; leaves glabrous, glaucous; roots often thickened. **B. napus**

Brassica juncea (L.) Czern. Fig. p. A61.
COMMON NAME(S): Chinese or Leaf-Mustard. Moutarde de l'Inde.
SPECIAL NOTE: Asiatic; introduced occasionally in grains. Known elsewhere in Canada from all the Provinces.

Brassica napus L.
COMMON NAME(S): Turnip, Winter Rape. Navette.
SPECIAL NOTE: Eurasian; ocasionally escaped to fields and waste areas and persisting after cultivation. Known elsewhere in Canada from the remaining provinces.
SYNONYMS: **B. campestris** L.

Brassica nigra (L.) Koch Fig. p. A61.
COMMON NAME(S): Black Mustard. Moutarde noire.
SPECIAL NOTE: European; introduced to open fields and waste areas throughout much of the Province. Known elsewhere in Canada from all Provinces except Man. and Alta.

Brassica rapa L. Fig. p. A61. Map p. B47.
COMMON NAME(S): Bird's Rape. Navette.
SPECIAL NOTE: Eurasian; occasionally cultivated and introduced throughout much of the Province and across Canada.

CAKILE Mill. Sea-Rocket. Caquillier

Cakile edentula (Bigel.) Hook. Fig. p. A61.
COMMON NAME(S): Sea Kale.

FREQUENCY: Common along the coast.
HABITAT: Coastal beaches.
GENERAL RANGE: B.C. and Ont. to N.S. south to Calif. on the Pacific coast and to S.C. on the Atlantic.

CAMELINA Crantz

1a Stem base with both long simple and shorter stellate pubescence; capsule valves with obscure midrib; seeds brownish red. **C. microcarpa**
1b Stem base with sparse mostly stellate pubescence; capsule valves with prominent midrib; seeds light orange. **C. sativa**

Camelina microcarpa Andrz. Fig. p. A61.
COMMON NAME(S): Small-seeded False Flax.
SPECIAL NOTE: Eurasian; rarely introduced along roadsides and waste areas. Known elsewhere in Canada from all the Provinces.

Camelina sativa (L.) Crantz. Fig. p. A61.
COMMON NAME(S): Dutch Flax. Gold-of-Pleasure. Sésame d'Allemagne.
SPECIAL NOTE: Eurasian; rarely introduced along roadsides and cultivated fields as near Bass River, Kent Co. Known elsewhere in Canada from all the Provinces.

CAPSELLA Medic. Shepherd's-purse. Tabouret

Capsella bursa-pastoris (L.) Medic. Fig. p. A61.
SPECIAL NOTE: Eurasian; commonly introduced to cultivated fields and waste areas throughout the Province and across Canada.

CARDAMINE L. Bitter Cress

1a Petals conspicuous, 6-13 mm long; flowers white or pink; leaflets of lower leaves rounded to oblong, petiolulate. **C. pratensis**
1b Petals inconspicuous, 1.5-3 mm long, white; leaflets of lower leaves usually linear to oval. 2
2a(1b) Terminal leaflet of cauline leaves +/- dentate, usually much wider than laterals, all leaflets decurrent on the rachis; plants of swamps and wet woods and similar wet habitats. **C. pensylvanica**
2b(1b) Terminal leaflet of cauline leaves usually entire, about as wide as laterals, 1-3 mm wide, not decurrent; plant of dry rocky or sandy habitats. **C. parviflora**

Cardamine parviflora L. Fig. p. A61.
FREQUENCY: Very rare; known only from Schampers Cluff, Kings Co. and Grand Manan Island, Charlotte Co.
HABITAT: Ledges, sandy soil and dry woods.

GENERAL RANGE: B.C. to Nfld. south to Oreg., Tex. and Fla.; Eurasia.
SUBSPECIFIC TAXA: Our plants are referable to var. **arenicola** (Britt.) Schulz.

Cardamine pensylvanica Muhl. Fig. p. A62. Map p. B47.
FREQUENCY: Common throughout the Province.
HABITAT: Mostly wet soil along streams.
GENERAL RANGE: Alaska to Nfld. south to Calif., Tex. and Fla.

Cardamine pratensis L. Fig. p. A62. Map p. B48.
COMMON NAME(S): Cuckoo-flower, Lady's-smock.
SPECIAL NOTE: Eurasian; introduced and locally common in the southern part of the Province. Known elsewhere in Canada from Ont. to Nfld. and N.S.

CONRINGIA Adans.

Conringia orientalis (L.) Dumort. Fig. p. A62.
COMMON NAME(S): Hare's-ear-Mustard.
SPECIAL NOTE: Eurasian; rarely introduced in waste places as in Fredericton, York Co. Known elsewhere in Canada from all the Provinces.

CORONOPUS Trew Swine-Cress, Wart-Cress.

1a Fruit longer than pedicel, notched at base only, narrowed at summit; petals longer than sepals; plant glabrous. **C. squamatus**

1b Fruit shorter than pedicel, notched at both ends; petals none or shorter than sepals; plant pubescent. **C. didymus**

Coronopus didymus (L.) Sm. Fig. p. A62.
COMMON NAME(S): Lesser Swine-Cress.
SPECIAL NOTE: European or S. American; rarely introduced as in Bathurst, Gloucester Co. and Saint John, St. John Co. Known elsewhere in Canada from B.C., Alta., Que., Nfld. and N.S.

Coronopus squamatus (Forsk.) Aschers.
SPECIAL NOTE: European; rarely introduced and not persisting as in Saint John, St. John Co. Known elsewhere in Canada from Ont. and N.S.
SYNONYMS: **C. procumbens** Gilib.

DENTARIA L. Toothwort, Pepper-root. Dentaire

1a Rhizome with annual constrictions. 2

1b Rhizome +/- continuous, without annual constrictions, coarsely toothed; central leaflet of cauline leaves broadly ovate to elliptic, coarsely blunt toothed. **D. diphylla**

++

2a(1a) Rhizome +/- smooth, easily separated into bright yellow, fusiform tubers; rachis of inflorescence hirsute; cauline leaves usually 3, whorled, each 3-pinnate or pinnatifid into narrow oblanceolate segments. **D. laciniata**

2b(1a) Rhizome toothed, not easily separated at constrictions; rachis of inflorescence glabrous; cauline leaves usually 2, each 3-pinnate into broadly oblanceolate, deeply sharp-toothed segments. **D. maxima**

Dentaria diphylla Michx. Fig. p. A62. Map p. B48.
COMMON NAME(S): Snicroûte, Carcajou.
FREQUENCY: Common in the south and rare in the eastern part of the Province.
HABITAT: Rich mostly deciduous woods.
GENERAL RANGE: Ont. to N.S. south to Mich. and S.C.

Dentaria laciniata Muhl. Fig. p. A62. Map p. B48.
COMMON NAME(S): Cutleaf Toothwort.
FREQUENCY: Very rare; known only from Lower Queensbury, York Co., Perth-Andover, Victoria Co. and Upper Woodstock, Carleton Co. (UNB).
HABITAT: Rich hardwoods.
GENERAL RANGE: Ont. to N.B. south to Kans., La. and Fla.
SPECIAL NOTE: For more information see H. Hinds, 1983. The Rare Vascular Plants of New Brunswick. Syllogeus 50.

Dentaria maxima Nutt. Fig. p. A62. Map p. B48.
FREQUENCY: Locally common mostly in the southwestern part of the Province.
HABITAT: Rich, often calcareous rocky, deciduous woods.
GENERAL RANGE: Wisc. to Ont. and N.B. south to Tenn. and Pa.

DESCURAINIA Webb & Berthelot Tansy-Mustard

1a Leaves once-pinnate into entire, lobed or deeply pinnatifid obtuse segments; capsule to 10 mm long, appressed ascending. **D. richardsonii**

1b Leaves 2-3-pinnate into linear or narrowly oblanceolate acute segments; capsule to 3 cm long, loosely ascending. **D. sophia**

Descurainia richardsonii (Sweet) Schulz
COMMON NAME(S): Moutarde tanaisie grise.
FREQUENCY: Very rare; only known from Atholville, Restigouche Co. (UNB).
HABITAT: River banks.
GENERAL RANGE: Alaska to Gaspé Pen. south to Mont., N. Mex., Minn. and Maine.

Descurainia sophia (L.) Webb Fig. p. A62. Map p. B48.
COMMON NAME(S): Sagesse des chirurgiens.
SPECIAL NOTE: Eurasian; uncommonly introduced in waste areas. Known elsewhere in Canada from all provinces.

++

DIPLOTAXIS DC. Wall-Rocket

1a Leaves deeply pinnately parted; capsule with a short stalk at base; pedicel nearly as long as capsule; stem glabrous at base. **D. tenuifolius**

1b Leaves coarsely toothed to pinnatifid; capsule without a short stalk at base; pedicel shorter than capsule; stem +/- hirsute at base. **D. muralis**

Diplotaxis muralis (L.) DC. Fig. p. A62.
COMMON NAME(S): Wall-Rocket or Mustard, Stinkweed.
SPECIAL NOTE: European; rarely introduced as near Bass River, Kent Co. and not persisting. Known elsewhere in Canada from all Provinces.

Diplotaxis tenuifolia (L.) DC.
SPECIAL NOTE: European; rarely introduced in waste places in the southern part of the Province and not persisting. Known elsewhere in Canada from Ont., Que., and N.S.

DRABA L. Draba. Drave

1a Stem leaves usually more than 20, crowded especially towards base; basal leaves usually withering at flowering; biennial. **D. incana**

1b Stem leaves less than 10; basal rosette well developed; perennials. 2

2a(1b) Raceme elongate with single flowers or fruit in axils of middle stem leaves; capsule densely stellate hirsute. **D. lanceolata**

2b(1b) Raceme looser, usually confined to summit of stem; capsule glabrous. 3

3a(2b) Capsule strongly twisted, about ¼ as wide as long; stem leaves acute at base. **D. arabisans**

3b(2b) Capsule plane, not strongly twisted, about 1/3 to 2/5 as wide as long; stem leaves obtuse to rounded at base. **D. glabella**

Draba arabisans Michx. Fig. p. A62. Map p. B48.
COMMON NAME(S): Rock-Cress Draba. Drave arabette.
FREQUENCY: Rare; known from Campobello Is., Charlotte Co., Martin Head, St. John Co., and Two Brooks, Restigouche Co.
HABITAT: Crevices of dry calcareous ledges.
GENERAL RANGE: Ont. to Nfld. and N.B.
SPECIAL NOTE: For more information see H. Hinds, 1983. The Rare Vascular Plants of New Brunswick. Syllogeus 50.

Draba glabella Pursh Map p. B48.
FREQUENCY: Rare; known from one station in each of Restigouche, Gloucester, Kings and St. John Counties.
HABITAT: Calcareous ledges and shores.
SUBSPECIFIC TAXA: Two varieties are recognized in our range.

1a Fruit lanceolate, +/- acute at summit, to 3 mm broad; seeds to 1 mm long; stem leaves at most 5. var. **glabella**

++

1b Fruit elliptic to oblong-ovate, obtuse at summit, to 5 mm broad; seeds to 1.3 mm long; stem leaves at least 5. var. **megasperma**

1. Var. **glabella**
GENERAL RANGE: Alaska to Nfld. south to N.Y.; Greenland; Eurasia.
SPECIAL NOTE: For more information see H. Hinds, 1983. The Rare Vascular Plants of New Brunswick. Syllogeus 50.

2. Var. **megasperma** (Fern. & Knowlt.) Fern.
FREQUENCY: Very rare; known only from Belledune Point, Gloucester Co.
HABITAT: Calcareous sands.
GENERAL RANGE: S. Lab. to Nfld. south to Gaspé Pen. and N.B.

Draba incana L. Fig. p. A62.
FREQUENCY: Very rare; known only from the Grande Plaine, Miscou Is., Gloucester Co.
HABITAT: Dry sands.
GENERAL RANGE: Man., N.B. and P.E.I.
SUBSPECIFIC TAXA: Our plants are referable to var. **confusa** (Ehrh.) Lilj.
SPECIAL NOTE: For more informatin see H. Hinds, 1983. The Rare Vascular Plants of New Brunswick. Syllogeus 50.

Draba lanceolata Royle Fig. p. A63.
FREQUENCY: Very rare; only known from McLeod Hill near Fredericton, York Co.
HABITAT: Calcareous ledges.
GENERAL RANGE: Alaska to N.B. south to Nev., Mich. and N. Eng.; Greenland; Asia.
SYNONYMS: **D. cana** Rydb.
SPECIAL NOTE: For more information see H. Hinds, 1983. The Rare Vascular Plants of New Brunswick. Syllogeus 50.

EROPHILA DC.

Erophila verna (L.) Chev. Fig. p. A63.
COMMON NAME(S): Whitlow-grass.
SPECIAL NOTE: Eurasian; known only near Point Wolfe River, Albert Co. in gravels of camping area. Known elsewhere in Canada from B.C., Ont. and Que.
SYNONYMS: **Draba verna** L.

ERUCASTRUM Presl

Erucastrum gallicum (Willd.) Schulz Fig. p. A63. Map p. B48.
COMMON NAME(S): Dog-Mustard. Moutarde des chiens.
SPECIAL NOTE: Eurasian; occasionally introduced in fields and waste areas and locally common. Known elsewhere in Canada from all the provinces.

++

ERYSIMUM L. Wallflower, Treacle-Mustard. Vélar

1a Pedicels thickening towards tip, less than 0.6 mm long; upper surface of leaf with 2-pronged appressed hairs. **E. inconspicuum**
1b Pedicels uniformily filiform, usually over 4 mm long; upper surface of leaves with 3-4-pronged hairs. 2
2a(1b) Capsules appressed on pedicels to 5 mm long; upper surface of leaves with 4-pronged hairs; petals sulphur yellow, to 8 mm long; perennials. **E. hieracifolium**
2b(1b) Capsules divergent on pedicels to 8 mm long; upper surface of leaves with 3-pronged hairs; petals pale yellow, to 5 mm long, annuals. **E. cheiranthoides**

Erysimum cheiranthoides L. Fig. p. A63. Map p. B48.
COMMON NAME(S): Wormseed-Mustard. Herbe au chantre.
SPECIAL NOTE: Eurasian; commonly introduced to roadsides and waste areas throughout the Province and across Canada.

Erysimum hieracifolium L. Fig. p. A63.
SPECIAL NOTE: European; rarely introduced as near Rothesay, Kings Co., Jaquet R., Restigouche Co., and Brockway, York Co. Known elsewhere in Canada from Sask., and from Ont. to N.S.

Erysimum inconspicuum (Wats.) MacK. Fig. p. A63.
FREQUENCY: Very rare; known only from St. Leonard, Madawaska Co. (CAN).
HABITAT: Dry open soil.
GENERAL RANGE: E. Que. and Nfld. south to N.B. and N.S.
SUBSPECIFIC TAXA: Our plants are referable to var. **coarctatum** (Fern.) Rossbach.

HESPERIS L. Rocket

Hesperis matronalis L. Fig. p. A63. Map p. B48.
COMMON NAME(S): Dame's-Violet. Julienne des Dames.
SPECIAL NOTE: Eurasian; cultivated and commonly escaped to thickets and waste areas especially in the south and west of the Province. Known elsewhere in Canada from all the provinces.

LEPIDIUM L. Pepperwort, Peppergrass. Passerage.

1a Stem leaves auriculate or clasping. **L. campestre**
1b Stem leaves not clasping. 2
2a(1b) Capsule 5-7 mm long. **L. sativum**
2b(1b) Capsule 2-4 mm long. 3
3a(2b) Petals equalling or longer than sepals; capsule usually wider than long. **L. virginicum**
3b(2b) Petals wanting or rudimentary; capsule as long or longer than wide. 4

4a(3b) Upper leaves acute; raceme dense with 9-15 fruit per cm; capsule about as long as wide. **L. densiflorum**

4b(3b) Upper leaves obtuse to rounded at tip; raceme loose with 6-10 fruit per cm; capsule usually longer than wide. **L. ruderale**

Lepidium campestre (L.) R. Br. Fig. p. A63. Map p. B48.
COMMON NAME(S): Cow-Cress. Cresson des champs.
SPECIAL NOTE: Eurasian; frequently introduced especially in the south of the province. Known elsewhere in Canada from B.C., Alta., Ont. to N.S. and Nfld.

Lepidium densiflorum Schrad. Fig. p. A63. Map p. B48.
SPECIAL NOTE: Introduced from western provinces and commonly spreading in disturbed areas throughout the Province.

Lepidium ruderale L. Fig. p. A63.
COMMON NAME(S): Cresson puant.
SPECIAL NOTE: Eurasian; rarely introduced in fields and waste areas as in St. Andrews, Charlotte Co. Known elsewhere in Canada from Sask. to N.S.

Lepidium sativum L. Fig. p. A63.
COMMON NAME(S): Garden-Cress. Cresson alénois.
SPECIAL NOTE: Asiatic; cultivated and rarely introduced in waste areas as in Fredericton, York Co. Known elsewhere in Canada from all provinces except Nfld.

Lepidium virginicum L. Fig. p. A63.
COMMON NAME(S): Poor-man's-pepper.
FREQUENCY: Very rare; known only from Fredericton, York Co. (UNB).
HABITAT: Dry open soil.
SPECIAL NOTE: Native to e. USA; introduced from Ont. to Nfld. and N.S.

NASTURTIUM R. Br. Watercress. Cresson

Nasturtium microphyllum (Boenn.) Reichenb. Fig. p. A64. Map p. B49 (as **N. officinale**).
COMMON NAME(S): Watercress. Cresson de fontaine.
SPECIAL NOTE: Eurasian; rarely introduced to wet ditches and calcareous or alkaline streams as in Saint John, St. John Co., Dalhousie, Restigouche Co., and Penobsquis, Kings Co. Known elsewhere in Canada from all the provinces.
SYNONYMS: **N. officinale** R. Br. var. **microphyllum** (Boeuer.) Thell.

NESLIA Desv.

Neslia paniculata (L.) Desv. Fig. p. A64.
COMMON NAME(S): Ball-Mustard.
SPECIAL NOTE: Eurasian; rarely introduced on railways and in waste areas as in Fredericton, York Co., and St. Stephen, Charlotte Co. Known elsewhere in Canada from all the provinces.

++

RAPHANUS L.

1a Mature capsule scarcely constricted between the 2-3 seeds; petals pink-purple to white. **R. sativus**

1b Mature capsule strongly constricted between the 4-10 seeds; petals usually yellow, turning whitish. **R. raphanistrum**

Raphanus raphanistrum L. Fig. p. A64. Map p. B49.
COMMON NAME(S): Wild Radish, Jointed Charlock. Rave Sauvage.
SPECIAL NOTE: Eurasian; commonly introduced in fields and waste areas throughout much of the Province and across Canada.

Raphanus sativus L. Fig. p. A64.
COMMON NAME(S): Radish. Radis.
SPECIAL NOTE: Eurasian; cultivated and rarely introduced as in Fredericton, York Co. Not persisting.

RORIPPA Scop.

1a Petals 3-4 mm long; capsules linear-cylindric 1-2.5 cm long; rhizomatous perennials. **R. sylvestris**

1b Petals 1.7-2 mm long; capsules ellipsoid to +/- globose, 4-10 mm long; taprooted annual or biennial. **R. palustris**

Rorippa palustris (L.) Bess. Fig. p. A64. Map p. B49.
COMMON NAME(S): Yellow Cress. Cresson palustre.
FREQUENCY: Scattered through the Province.
HABITAT: Gravelly or muddy shores and occasionally a weed in fields.
GENERAL RANGE: Alaska to Nfld. south to Calif., Tex. and Fla.
SUBSPECIFIC TAXA: Several subspecies and varieties have been reported within our range. Their characteristics appear to intergrade and they are not recognized in this work.
SYNONYMS: **R. islandica** (Oeder) Borbas.

Rorippa sylvestris (L.) Bess. Map p. B49.
COMMON NAME(S): Creeping Yellow Cress.
SPECIAL NOTE: European; frequently introduced mostly on shores and wet meadows in the western part of the Province. Known elsewhere in Canada from all the provinces.

SINAPIS L.

1a Capsule bristly, beak as long as or longer than body; leaves all pinnatifid. **S. alba**

1b Capsule glabrous (rarely slightly bristly), beak 1/3 to nearly as long as body; leaves rhombic to oblong, merely dentate. **S. arvensis**

++

Sinapis alba L. Fig. p. A64.
COMMON NAME(S): White Mustard. Moutarde blanche.
SPECIAL NOTE: Eurasian; rarely introduced in waste areas and cultivated fields. Known elsewhere in Canada from all the Provinces.
SYNONYMS: **Brassica hirta** Moench.

Sinapis arvensis L. Fig. p. A64.
COMMON NAME(S): Charlock. Moutarde d'été.
SPECIAL NOTE: Eurasian; occasionally introduced in cultivated fields, shores and waste areas.
SYNONYMS: **Brassica kaber** (DC.) Wheeler.

SISYMBRIUM L.

1a	Capsules appressed to main axis of inflorescence. **S. officinale**
1b	Capsules spreading. 2
2a(1b)	Upper leaves divided into filiform segments; capsules 5-10 cm long, only slightly thicker than pedicels. **S. altissimum**
2b(1b)	Upper leaves with broader lobes or segments; capsules 1.2-3.5 cm long, considerably thicker than pedicels. **S. loeselii**

Sisymbrium altissimum L. Fig. p. A64. Map p. B49.
COMMON NAME(S): Tumble-Mustard. Moutarde roulante.
SPECIAL NOTE: Eurasian; introduced in fields and waste areas especially on the east coast. Known elsewhere in Canada from all the provinces.

Sisymbrium loeselii L. Fig. p. A64.
SPECIAL NOTE: Eurasian; rarely introduced in waste areas as in Fredericton, York Co. Known elsewhere in Canada from all the provinces except N.S., P.E.I. and Nfld.

Sisymbrium officinale (L.) Scop. Fig. p. A64. Map p. B49.
COMMON NAME(S): Hedge-Mustard. Herbe au chantre.
SPECIAL NOTE: Eurasian; commonly introduced in waste areas especially in the south of the province.

SUBULARIA L.

Subularia aquatica L.
COMMON NAME(S): Awlwort.
FREQUENCY: Rare; probably overlooked because of its inconspicuousness; reported from States Lake, Restigouche Co., Oromocto Lake, York Co., and from Lake Utopia, Charlotte Co.
HABITAT: Muddy shores.
GENERAL RANGE: Alaska to Nfld. south to Calif., Wyo., Ont., and n. New Eng.; Greenland; Iceland; Eurasia.

++

THLASPI L.

Thlaspi arvense L. Fig. p. A64.
COMMON NAME(S): Field Penny-Cress, Frenchweed. Cennes.
SPECIAL NOTE: Eurasian; introduced in fields and waste areas over much of the Province and across Canada.

SARRACENIACEAE (Pitcher-plant Family)

SARRACENIA L. Pitcher-plant

Sarracenia purpurea L. Map p. B49.
COMMON NAME(S): Pitcher-plant. Petits cochons, Herbe-crapaud.
FREQUENCY: Common throughout much of the Province.
HABITAT: Acid bogs and boggy margins of ponds.
GENERAL RANGE: Great Bear L. and B.C. to Nfld. south to s. cent. Sask. and Del.
SUBSPECIFIC TAXA: Forma **heterophylla** (Eat.) Boivin, with greenish yellow flowers instead of deep maroon, occurs occasionally with the species.

DROSERACEAE (Sundew Family)

DROSERA L. Sundew. Rossolis

1a Leaf blade rounded, slightly broader than long; seeds spindle-shaped, striate. **D. rotundifolia**

1b Leaf blade spathulate, much longer than wide; seeds oblong, papillate; stipules mostly free, setaceous. **D. intermedia**

Drosera intermedia Hayne Fig. p. A64. Map p. B49.
COMMON NAME(S): Narrow-leaved Sundew.
FREQUENCY: Scattered throughout much of the Province.
HABITAT: Acid bogs and boggy ponds.
GENERAL RANGE: B.C., Minn. and Ont. to Nfld. south to Tex. and Fla.; Eurasia.

Drosera rotundifolia L. A65. Map p. B49.
COMMON NAME(S): Round-leaved Sundew. Rossolis à feuilles rondes.
FREQUENCY: Common throughout the Province.
HABITAT: Acid bogs and moist acid roadside ditches.
GENERAL RANGE: Alaska to Nfld. south to Calif., Minn. and Fla.; Greenland; Iceland; Eurasia.

PODOSTEMACEAE (Riverweed Family)

PODOSTEMUM Michx.

Podostemum ceratophyllum Michx. Fig. p. A65. Map p. B49.
COMMON NAME(S): Threadfoot.
FREQUENCY: Rare and probably overlooked; reported from Quarryville, Northumberland Co., and Eel R., Carleton Co.
HABITAT: Submerged rocks in fast water.
GENERAL RANGE: Ont. to N.S. south to Ark. and Ga.
SPECIAL NOTE: For more information see H. Hinds, 1983. The Rare Vascular Plants of New Brunswick. Syllogeus 50.

CRASSULACEAE (Orpine Family)

1a Leaves opposite; flowers solitary in leaf axils; small annual herbs of muddy shores. **Crassula**
1b Leaves mostly alternate; flowers in terminal cymes; perennials. **Sedum**

CRASSULA L.

Crassula aquatica (L.) Schoenl. Fig. p. A65. Map p. B49.
COMMON NAME(S): Pigmyweed.
FREQUENCY: Uncommon but easily overlooked; populations centered in the estuary of the Miramichi R., in the Grand Lake area, Queens Co., Oromocto and St. John Rivers below Fredericton, and on Grand Manan Is., Charlotte Co.
HABITAT: Fresh or tidal muddy shores.
GENERAL RANGE: Alaska to Nfld. south to Mex., La. and Md.; Iceland; Eurasia.
SYNONYMS: **Tillaea aquatica** L.

SEDUM L. Stonecrop, Orpine. Orpin

1a Leaves +/- cylindrical, 3-6 mm long, 1.5-2.5 mm wide, entire; flowers yellow. **S. acre**
1b Leaves flat, longer and wider, the upper ones dentate. 2
2a(1b) Leaves acute at summit, entire or obscurely dentate near tip; flowers unisexual, greenish-yellow changing to purple, plants of ledges along Bay of Fundy. **S. rosea**
2b(1b) Leaves rounded at summit, coarsely dentate; flowers creamy, yellowish, pink or purple; plants of roadsides, waste areas and open woodlands. **S. telephium**

Sedum acre L. Fig. p. A65.
COMMON NAME(S): Mossy Stonecrop, Wallpepper, Love-Entangle. Vermiculaire, Orpin brûlant.

+++

SPECIAL NOTE: Eurasian; rarely introduced and spreading from cultivation as on roadside, Sheffield, Sunbury Co. Known elsewhere in Canada from B.C., Alta., Ont., Que., P.E.I. and N.S.

Sedum rosea (L.) Scop. Fig. p. A65. Map p. B50.
COMMON NAME(S): Roseroot. Orpin rosat.
FREQUENCY: Scattered along the Bay of Fundy coast.
HABITAT: Moist ledge crevices.
GENERAL RANGE: Ungava-Labr. to Baffin Is. south along the coast to Que., Nfld., N.B. and N.S.; Greenland; Iceland; Spitsb.; Eurasia.

Sedum telephium L. Fig. p. A65. Map p. B49.
COMMON NAME(S): Live-forever, Garden-Orpine. Vit-toujours.
SPECIAL NOTE: Eurasian; formally cultivated and now widespread across the Province. Known elsewhere in Canada from B.C., Man., and from Ont. east.
SYNONYMS: **S. purpureum** Schultes

SAXIFRAGACEAE (Saxifrage Family)

1a Woody shrubs; fruit a berry. **Ribes**
1b Plants herbaceous; fruit a capsule or follicle. 2
2a(1b) Basal rosettes well developed; petals present. 3
2b(1b) Basal rosettes wanting; petals not developed. 4
3a(2a) Flower one per scape; ovary one celled; leaves entire. **Parnassia glauca**
3b(2a) Flowers several per scape. 5
4a(2b) Stems prostrate; leaves opposite, ovate to rotund, short petioled; flowers 4-merous. **Chrysosplenium americanum**
4b(2b) Stems erect; leaves alternate, acuminate at both ends; flowers 5-merous. **Penthorum sedoides**
5a(3b) Leaves long petioled, the blade +/- rotund to ovate, cordate at base; inflorescence mostly unbranched. 6
5b(3b) Leaves sessile or cuneately narrowed to margined petiole; inflorescence usually branched. **Saxifraga**
6a(5a) Petals fringed; capsule beak obtuse; leaves less than 5 cm long. **Mitella nuda**
6b(5a) Petals entire; capsule beak acute; leaves greater than 5 cm long. **Tiarella cordifolia**

CHRYSOSPLENIUM L. Golden Carpet, Golden Saxifrage. Dorine

Chrysosplenium americanum Schwein. Fig. p. A65. Map p. B50.
COMMON NAME(S): Water-mat, Water-carpet. Dorine d'Amérique.
FREQUENCY: Widespread throughout the Province.
HABITAT: Swamps, seepage areas, shallow shaded streams and wet ditches.
GENERAL RANGE: Minn. and Ont. to N.S. south to Iowa and Ga.

SAXIFRAGACEAE

++

MITELLA L. Mitrewort, Bishop's-cap. Mitrelle

Mitella nuda L. Fig. p. A65. Map p. B50.
FREQUENCY: Scattered throughout the Province.
HABITAT: Cool, mossy, usually coniferous, woods.
GENERAL RANGE: Alaska to Nfld. south to Mont. and Pa.; Asia.

PARNASSIA L. Grass-of-Parnassus, Bog-stars.

Parnassia glauca Raf. Fig. p. A65. Map p. B50.
FREQUENCY: Uncommon to rare; reported from in a few localities in the western and northern part of the Province.
HABITAT: Calcareous rocky shores and marly bogs.
GENERAL RANGE: Sask. to Nfld. south to S. Dak., Ohio, and N. Eng.

PENTHORUM L.

(Sometimes included in the **Crassulaceae**)

Penthorum sedoides L. Fig. p. A65. Map p. B50.
COMMON NAME(S): Ditch-Stonecrop.
FREQUENCY: Uncommon; scattered populations concentrated in the southwest of the Province and especially in the Jemseg area of Sunbury and Queens Counties.
HABITAT: Muddy or gravelly shores.
GENERAL RANGE: Nebr. and Man. to N.B. south to Tex. and Fla.

RIBES L. Current, Gooseberry. Gadellier, Groseillier

(Sometimes put in a separate family, **GROSSULARIACEAE**)

1a Flowers solitary or in small clusters of 2-4; stems weakly spiny. **R. hirtellum**
1b Flowers in racemes. 2
2a(1b) Ovary and fruit glandular-hispid. 3
2b(1b) Ovary and fruit not as above. 4
3a(2a) Stems spiny; racemes +/- drooping; berry dark purple to black. **R. lacustris**
3b(2a) Stems not spiny; racemes ascending; berry dark red. **R. glandulosum**
4a(2b) Leaves resinous-dotted beneath; berry black. 5
4b(2b) Leaves without resinous glands. 6
5a(4a) Calyx densely pubescent; bracts at base of pedicel minute. **R. nigrum**
5b(4a) Calyx +/- glabrous; bracts at base of pedicel conspicuous, persistent. **R. americanum**
6a(4b) Middle leaf lobe ovate; stems +/- erect; calyx greenish-yellow; anthers widely separated. **R. rubrum**
6b(4b) Middle leaf lobe deltoid; stems weakly ascending; calyx purple-tinged; anthers not widely separated. **R. triste**

++

Ribes americanum P. Mill. Fig. p. A65. Map p. B50.
COMMON NAME(S): Wild Black Currant. Cassis sauvage.
FREQUENCY: Scattered mostly in the southern half of the Province. Rare towards the northeast.
HABITAT: Rich, mostly calcareous soils of moist woodlands and thickets.
GENERAL RANGE: Alta. to N.S. south to Mont., N. Mex., Mo. and Del.

Ribes glandulosum Grauer. Fig. p. A66. Map p. B50.
COMMON NAME(S): Skunk-Currant.
FREQUENCY: Common throughout the Province.
HABITAT: Rocky slopes, moist woods and thickets.
GENERAL RANGE: Alaska to Nfld. south to B.C., Alta. and N.C.

Ribes hirtellum Michx. Fig. p. A66. Map p. B50.
COMMON NAME(S): Canada or Bristly Gooseberry. Fausse-épine.
FREQUENCY: Common throughout much of the Province.
HABITAT: Rocky woods and clearings.
GENERAL RANGE: B.C. to Nfld. south to Nebr. and W. Va.
SUBSPECIFIC TAXA: Sinnott (1985) does not recognize var. **calcicola** (Fern.) Fern. nor var. **saxosum** (Hook.) Fern. in his revision of the subg. Grossularia and contends that the leaf characters are "not consistent enough with regard to habitat preference and morphology to warrant recognition at any taxonomic level." However, in N.B. plants fitting the description of var. **calcicola**, with leaves densely soft-pubescent beneath, are found only in the extreme northeast and southwest coastal areas.

Ribes lacustre (Pers.) Poir. Fig. p. A66.
COMMON NAME(S): Bristly or Swamp Currant. Groseillier sauvage.
FREQUENCY: Common throughout much of the Province.
HABITAT: Moist woods, thickets and swamps.
GENERAL RANGE: Alaska to Nfld. south to Calif., Tenn. and N.Y.

Ribes nigrum L. Fig. p. A66. Map p. B50.
COMMON NAME(S): Black Currant. Cassis.
SPECIAL NOTE: Eurasian; Formerly and occasionally presently cultivated and rarely escaped to thickets in scattered localities throughout the Province. Known elsewhere in Canada from B.C., and from Ont. to N.S.

Ribes rubrum L. Fig. p. A66. Map p. B50.
COMMON NAME(S): Red or Garden Currant. Gadellier cultivé.
SPECIAL NOTE: Formerly cultivated and spreading to open woods and thickets throughout much of the Province. Known elsewhere in Canada from B.C., Man. and from Ont. to N.S. excl. Nfld.

Ribes triste Pallas Fig. p. A66. Map p. B50.
COMMON NAME(S): Wild Red Currant.
FREQUENCY: Scattered throughout much of the Province.
HABITAT: Mostly rocky moist woods and thickets.
GENERAL RANGE: Alaska to Nfld. south to Oreg. and Va.; Asia.

SAXIFRAGA L. Saxifrage

1a Stem leafy or bracteate above the basal rosette; basal leaves stiff, sessile, finely serrate, the teeth whitened with lime secretions. **S. paniculata**

1b Stem scapose; basal leaves soft, entire or crenate; petiole wing-margined to base. **S. virginiensis**

Saxifraga paniculata P. Mill. Fig. p. A66. Map p. B50.
COMMON NAME(S): Livelong Saxifrage.
FREQUENCY: Rare; collected in Kings, St. John and Albert Counties.
HABITAT: Moist calcareous ledges.
GENERAL RANGE: Great Slave L. to Nfld. south to mts. of N.Y. and N. Eng.; Greenland; Iceland; Europe.
SYNONYMS: **S. aizoon** Jacq. var. **neogaea** Butters.
SPECIAL NOTE: For more information see H. Hinds, 1983. The Rare Vascular Plants of New Brunswick. Syllogeus 50.

Saxifraga virginiensis Michx. Fig. p. A66.
COMMON NAME(S): Early Saxifrage.
FREQUENCY: Very rare; reported only from the south end of Keswick Ridge, York Co.
HABITAT: Rocky woods and ledge crevices.
GENERAL RANGE: Man. to N.B. south to Mo. and Ga.
SPECIAL NOTE: For more information see H. Hinds, 1983. The Rare Vascular Plants of New Brunswick. Syllogeus 50.

TIARELLA L. False Mitrewort, Foamflower.

Tiarella cordifolia L. Fig. p. A66. Map p. B50.
FREQUENCY: Scattered mostly in the western part of the Province.
HABITAT: Rich woods.
GENERAL RANGE: Ont. to N.S. south to Mich., Tenn. and N.C.

HAMAMELIDACEAE (Witch-hazel Family)

HAMAMELIS L.

Hamamelis virginiana L. Fig. p. A66. Map p. B50.
COMMON NAME(S): Witch-hazel. Café du diable.
FREQUENCY: Locally common in the southwest of the Province and rare elsewhere.
HABITAT: Moist, mostly acid, woods and shore thickets.
GENERAL RANGE: Minn. to N.S. south to Mo. and Ga.

++

ROSACEAE (Rose Family)

1a Stem +/- woody at least at the base. 2

1b Stem herbaceous, dying to the ground in winter. 3

2a(1a) Leaves compound. 4

2b(1a) Leaves simple. 5

3a(1b) Leaves compound. 6

3b(1b) Leaves simple. 7

4a(2a) Fruit dry. 8

4b(2a) Fruit fleshy. 9

5a(2b) Ovary apparently superior. 10

5b(2b) Ovary inferior. 11

6a(3a) Basal leaves palmately compound. 12

6b(3a) Basal leaves pinnately compound. 13

7a(3b) Leaves palmately lobed; flowers greenish, numerous in panicled corymbs. **Alchemilla**

7b(3b) Leaves unlobed, cordate-orbicular; flowers white, solitary. **Dalibarda repens**

8a(4a) Flowers white in dense terminal panicles; sepals without alternating bracts. **Sorbaria sorbifolia**

8b(4a) Flowers yellow in open leafy cymes; sepals alternating with bracts. **Potentilla fruticosa**

9a(4b) Ovary +/- inferior. 14

9b(4b) Ovary superior; fruit of several to many 1-seeded fleshy drupelets; flowers in racemes. **Rubus**

10a(5a) Fruit fleshy, either a single large drupe or several small drupelets in a head. 15

10b(5a) Fruit dry. 16

11a(5b) Stems usually with long thorns; fruit apple-like, to about 2.5 cm thick, with 1-5 bony nutlets. **Crataegus**

11b(5b) Stems usually without thorns; fruit berry-like with 5-10 small seeds. 17

12a(6a) Petals white; fruit of small achenes scattered on the surface of a fleshy receptacle (strawberry). **Fragaria**

12b(6a) Petals yellow or dark purple; fruit a cluster of dry achenes. 18

13a(6b) Flowers numerous, congested in dense panicles, spikes or heads. 19

13b(6b) Flowers solitary or several in open racemes, corymbs or cymes. 20

14a(9a) Stems usually prickly; petals large, often pink (rarely white, reddish-purple or yellow), emarginate; solitary or several clustered; stipules adnate to petiole for more than ½ their length. **Rosa**

14b(9a) Stems not prickly; petals small, white, entire; very numerous in flat-topped or domed clusters; stipules attached at base only, early deciduous. **Sorbus**

15a(10a) Fruit a single drupe per flower; petals white. **Prunus**

15b(10a) Fruit a cluster of several, dryish drupes per flower; petals pink. **Rubus odoratus**

16a(10b) Flowers densely clustered; plants shrubby. 21

16b(10b) Flowers solitary; plants suffruticose; style plumose in fruit. **Dryas integrifolia**

+++

17a(11b) Flowers, solitary, in racemes or umbel-like clusters. 22

17b(11b) Flowers in flat or dome-shaped compound corymbs; locules as many as styles; fruit pulpy, astringent-tasting. **Aronia**

18a(12b) Leaves all basal, trifoliate; flowers yellow; achenes 2-6. **Waldsteinia fragarioides**

18b(12b) Leaves cauline as well as sometimes basal with (3-)5-(7) leaflets; flowers various; achenes more than 10. **Potentilla**

19a(13a) Flowers in dense spikes, white; pistil solitary. **Sanguisorba canadensis**

19b(13a) Flowers in panicles, white or pink; pistils 5-10. **Filipendula**

20a(13b) Inflorescence a spike-like raceme; calyx-tube armed with hooked bristles. **Agrimonia**

20b(13b) Inflorescence cymose or corymbose; calyx-tube not bristly. 23

21a(16a) Leaves palmately 3-5 lobed; inflorescence racemose, dome-shaped, as long as or wider than long; follicles bladdery. **Physocarpus opulifolius**

21b(16a) Leaves not lobed, coarsely serrate; inflorescence paniculate, longer than wide; follicles not inflated. **Spiraea**

22a(17a) Flowers white, more than 3 cm wide, pink-tinged; fruit an apple. **Malus sylvestris**

22b(17a) Flowers white, usually less than 3 cm wide; fruit berry-like with 10 small seeds. **Amelanchier**

23a(20b) Style elongate, jointed, becoming hooked in fruit; leaves differing markedly in shape from base to summit of stem. **Geum**

23b(20b) Style short, inconspicuous, not jointed nor hooked in fruit; leaves similar, except in size, from base to summit of stem. **Potentilla**

AGRIMONIA L. Agrimony. Agrimoine

1a Floral axis with scattered long +/- straight hairs about as long as or longer than width of axis; terminal leaflet of mid stem leaves usually without teeth or small lobes at base; furrows of hypanthium without strigose hairs. **A. gryposepala**

1b Floral axis covered by short and long mostly ascending hairs as long as or shorter than width of axis; terminal leaflet of mid stem leaves often with teeth or small lobes at base; furrows of hypanthium minutely strigose. (N.B. specimens are both glandular on the floral axis and on the hypanthium) **A. striata**

Agrimonia gryposepala Wallr. Fig. p. A66. Map p. B51.
FREQUENCY: Scattered mostly in the western half of the Province.
HABITAT: Rich woods and thickets.
GENERAL RANGE: B.C. to N.S. south to Kans. and N.C.

Agrimonia striata Michx. Fig. p. A66. Map p. B51.
FREQUENCY: Common throughout much of the Province.
HABITAT: Borders of woods and thickets often in disturbed areas.
GENERAL RANGE: B.C. to Nfld. south to Ariz., Nebr. and N.J.

++

ALCHEMILLA L. Lady's-mantle

Alchemilla filicaulis Buser Fig. p. A67.
COMMON NAME(S): Lady's-mantle, Dew-cup.
SPECIAL NOTE: Probably of European origin. Reported from St. Andrews, Charlotte Co., and on islands in the Restigouche R., Restigouche Co.
SYNONYMS: **A. vulgaris** L. var. **filicaulis** (Buser) Fern. and Wieg.

AMELANCHIER Medic. Juneberry, Serviceberry, Shadbush. Petite Poires, Poirier

SPECIAL NOTE: The species of this genus are all interfertile. With the clearing of extensive areas of the Province many of the species were brought into close proximity and intraspecific hybridization and introgression has produced a puzzling complex of characters. Some of the hybrids are exceedingly common and may be self perpetuating. Only the more common and more evident hybrids are included in the key.

1a	Plant in flower; young foliage present. 2
1b	Plant not in flower; mature fruit and leaves present. 3
2a(1a)	Flowers 1-5 on each branchlet; ovary summit persistently tomentose. 4
2b(1a)	Flowers in racemes usually with more than 5 flowers. 5
3a(1b)	Most racemes with 1-5 fruit, occasionally with more. 6
3b(1b)	Most racemes with 5-20 fruit. 7
4a(2a)	Very young leaves folded, deep green or bronze; petioles 5-15 mm long; style divided half way to base; flowers 2-5. 8
4b(2a)	Very young leaves revolute and glabrous (petioles slightly silky), light green to slightly reddish; petioles 2-7 mm long; style divided nearly to base; flowers mostly 1-3; ovary elongate; sepals erect. **A. bartramiana**
5a(2b)	Young foliage bronze to purple with little green showing. 9
5b(2b)	Young foliage grey-green, occasionally slightly reddish-green. 10
6a(3a)	Leaves persistently pubescent beneath especially along base of midrib and petiole; leaf tip tending towards acute, leaves ovate-elliptic to obovate, thick and lustrous, rounded to broadly cuneate at base, finely dentate 2/3 way to base; petioles permanently silky, 5-15 mm long; fruit +/- globose often with some pubescence on surface especially when young; young leaves brownish-green, silky pubescent. **A. X quinti-martii**
6b(3a)	Leaves glabrous from the first, narrowly elliptic, well grown at anthesis, apex usually short acuminate, base rounded or subcordate, margins finely and sharply serrate with 7-9 teeth per cm.; petals 8-10 mm long; sepals reflexed; fruit glabrous except summit +/- tomentose, slightly obovate. **A. X neglecta**
7a(3b)	Tip of floral leaves acuminate; some leaves occasionally acute to barely rounded and apiculate; general shape ovate to oblong, longer than wide; veins forking near margin. 11
7b(3b)	Tip of leaf rounded to apiculate or acute with an abrupt point. 12

8a(4a) Young leaves usually bronze or purple, smooth or pubescent when very young; flowers mostly more than 3 in smooth open racemes; ovary summit densely tomentose. A. X **neglecta**

8b(4a) Young leaves green, sparsely pilose on upper surface, densely silky-pubescent beneath; raceme and petioles usually permanently pubescent; flowers 2-3; veins often entering marginal teeth; leaves coarsely serrate; tall non-stoloniferous shrubs. A. X **quinti-martii**

9a(5a) Racemes open and pendulous, 12-20 flowered; leaf finely serrate; tall shrubs or small trees; petioles and racemes glabrous; leaves mostly well grown at flowering. A. **arborea** var. **laevis**

9b(5a) Racemes tightly fewer flowered, ascending. 13

10a(5b) Ovary summit glabrous or nearly so; expanding leaves tomentose. 14

10b(5b) Ovary summit densely pubescent at anthesis. 15

11a(7a) Leaves with 4-5 (3-7) very sharp teeth per cm, short acuminate to apiculate at apex, eventually glabrous, bronze-purple when young; sepals sharply reflexed; ovary summit densely tomentose especially when young. A. X **wiegandii**

11b(7a) Leaves with 5-11 teeth per cm, apex mostly long acuminate. 16

12a(7b) Leaves of flowering shoots finely serrate with 5-11 teeth per cm 5 mm from tip. 17

12b(7b) Leaves more coarsely dentate with 3-5 teeth per cm. 18

13a(9b) Ovary summit tomentose. 19

13b(9b) Ovary summit +/- glabrous; leaves usually glabrous; petals 9-12 mm long. A. X **intermedia**

14a(10a) Petals 3-10 mm long, oval to elliptic; racemes ascending, dense; leaves oblong-elliptic to broadly elliptic-subrotund. 20

14b(10a) Petals 10-20 mm long, narrowly cuneate to oblong; racemes less strict, more open; larger shrubs or trees to 20 m; leaves ovate to slightly obovate, acuminate, cordate to rounded at base. A. **arborea**

15a(10b) Leaves glabrous or nearly so from the first; sepals erect or ascending; low shrubs to about 1 m high. 21

15b(10b) Leaves tomentose beneath from the first, becoming glabrate. 22

16a(11b) Fruit with sepals reflexed; tall shrubs or small trees. 23

16b(11b) Fruit with sepals erect or ascending. 24

17a(12a) Ovary summit tomentose. 25

17b(12a) Ovary summit +/- glabrous; leaves oblong-elliptic; sepals erect to ascending. 26

18a(12b) Principal leaf veins or their primary branches extending +/- directly into teeth. 27

18b(12b) Principal veins branching before margin, many not entering teeth; leaves and petioles glabrous; leaves rounded or subtruncate at apex; racemes loose and open, with 1-3 of lowermost pedicels subtended by long petioled, large leaves; low shrub to about 1 m. A. **sanguinea** var. **gaspensis**

19a(13a) Young leaves tomentose; sepals reflexed. A. X **wiegandii**

19b(13a) Young leaves essentially glabrous. A. X **neglecta**

20a(14a) Petals 7-10 mm long, racemes 2.5-6 cm long; fastigiate tall shrubs; petioles and pedicels usually tomentose. A. **canadensis**

++

20b(14a) Petals 3-7 mm long; racemes 1-4 cm long; petioles and pedicels soon mostly glabrous; stoloniferous or suckering low shrub. **A. lucida**

21a(15a) Leaves with lateral veins +/- parallel, but forming a network short of the margin; lower 1-3 pedicels subtended by long petioled leaves; sepals 1.5-3.5 mm long; petals 6-9 mm long. **A. sanguinea** var. **gaspensis**

21b(15a) Leaves with lateral veins prominent to margin; only lowest pedicel with leaf; sepals 3-5 mm long. **A. fernaldii**

22a(15b) Raceme dense, erect, silky pubescent at anthesis, 1.5-5 cm long; lowest pedicels 7-17 mm long; petals obovate-oblong 7-10 mm long; leaves unfolding with the flowers, entire or more commonly toothed upper 2/3 only. 28

22b(15b) Raceme more open, glabrous to slightly silky, soon arching, 4-7 cm long; lowest pedicels .7-2.5 cm long; petals spatulate to linear 1-1.5 cm long; leaves mostly unfolded at anthesis, usually toothed to the base; arched ascending, straggling shrub with 1-several trunks 1-3 m high. **A. sanguinea**

23a(16a) Mature leaves and petioles glabrous; fruit glabrous, juicy and sweet; young leaves purplish. 29

23b(16a) Mature leaves glabrescent, pubescent when young and retaining some pubescence on midrib below and on petioles; fruit variously tomentose, dry and insipid; raceme of (6)8-15 fruit; pedicels not much over 1.5 cm long; hypanthium 2.5-3 mm wide. **A. arborea**

24a(16b) Leaves mostly glabrous at maturity, purplish and slightly pubescent when young; sepals glabrous; leaf tip +/- acuminate. **A. X intermedia**

24b(16b) Leaves green and densely pubescent when young, more rounded to apiculate at tip and slightly hairy when mature; sepals pubescent above. **A. canadensis**

25a(17a) Veins prominent beneath to teeth; racemes loose and spreading, glabrous; sepals erect or spreading; leaves rounded to acute at apex, sometimes acuminate, toothed to base, usually glabrous from the first. **A. fernaldii**

25b(17a) Veins not prominent near margins; leaves rounded to truncate apiculate at apex, toothed along upper 2/3 of margin, tomentose beneath when young, nearly glabrous when mature; racemes relatively compact, erect; sepals reflexed from near middle. **A. stolonifera**

26a(17b) Fruiting raceme with glabrous rachis and pedicels; mostly low .2-1.5 m high stoloniferous or suckering shrub forming circular colonies; pedicles 1-2.5 cm long; mature leaves glabrous, dark green and shining. **A. lucida**

26b(17b) Fruiting raceme tomentose; pedicels not much over 2 cm long; leaves scarcely half grown at anthesis and then copiously white felted beneath; mature leaves eventually glabrate; alder-like tall shrubs. **A. canadensis**

27a(18a) Racemes densely silky-tomentose, becoming glabrate, erect; leaves toothed to slightly below middle or rarely entire; veins mostly branching before margin; lowest pedicel 1.5-2.0 cm in fruit; sepals revolute from middle after anthesis. **A. humulis**

27b(18a) Racemes essentially glabrous or slightly silky, +/- loose and open, soon arching; lowest pedicel to 2.5 cm long in fruit; mature leaf blade elliptic to suborbicular, 3-6 cm long, toothed to base; uppermost veins mostly straight to margin without branching; sepals recurving from near middle. **A. sanguinea**

28a(22a) Some unfolded leaves usually available to show rounded-apiculate to obtusely pointed apex, tomentum white, veins obscure near margin; racemes 1.5-4 cm long; shrubs of dry, sterile, rocky, sandy or peaty acid substrates to 1.5 m high. **A. stolonifera**

28b(22a) Unfolding leaves short acuminate-apiculate to acute, dull reddish-green to bronze; tomentum usually yellowish or greyish; veins prominent to margin, but mostly branching once before entering teeth; racemes 4-5 cm long; sepals usually more pubescent; plants of rocky shores or banks in circumneutral soils. **A. humilis**

29a(23a) Leaves rounded to cuneate at base, narrowly elliptic, slightly pubescent when very young, but nearly glabrous and +/- fully expanded at flowering; fruit slightly obovate, +/- tomentose at summit but sometimes glabrous. **A. X neglecta**

29b(23a) Leaves rounded to cordate at base, glabrous from the first; fruit globose; lowest pedicels 2.5-5 cm long. **A. arborea** var. **laevis**

Amelanchier arborea (Michx. f.) Fern. Fig. p. A67. Map p. B51.
SUBSPECIFIC TAXA: Two varieties occur within our range. These are separated in the above key.

1. Var. **arborea**
FREQUENCY: Scattered in the southern part of the Province and rare northwards.
HABITAT: Deciduous woods and thickets especially in circumneutral soils.
GENERAL RANGE: Ont. to N.B. south to Minn., Okla., La. and Fla.

2. Var. **laevis** (Wieg.) Ahles
FREQUENCY: Very common in the southern part of the Province, becoming rarer northwards.
HABITAT: Borders of deciduous or mixed woods, hedgerow thickets and swamps.
GENERAL RANGE: Ont. to Nfld. south to Iowa, Del. and Ga.
SYNONYMS: **A. laevis** Wieg.

Amelanchier bartramiana (Tausch) Roemer Fig. p. A67.
COMMON NAME(S): Bartram's Shadbush. Mountain Juneberry. Amélanchier boréale, Petites Poires.
FREQUENCY: Common throughout much of the Province.
HABITAT: Mostly cool coniferous, peaty or boggy areas.
GENERAL RANGE: Ont. to Nfld. south to Minn., Pa. and Mass.

Amelanchier canadensis (L.) Medic. Fig. p. A67. Map p. B51.
FREQUENCY: Uncommonly scattered in the southern part of the Province.
HABITAT: Borders of swamps and streams.
GENERAL RANGE: Ont. to N.S. south to Ga.

Amelanchier fernaldii Wieg. Fig. p. A67. Map p. B51.
COMMON NAME(S): Fernald's Serviceberry.
FREQUENCY: Rare; collected in a few scattered localities.
HABITAT: Mostly calcareous thickets and shores.
GENERAL RANGE: Que. to Nfld. south to N.B. and N.S.

++

Amelanchier humilis Wieg. Fig. p. A67. Map p. B51.
FREQUENCY: Common throughout much of the Province except unknown along the east coast.
HABITAT: Rocky or sandy shores, ledges and hedgerows.
GENERAL RANGE: Minn. and S. Dak. to Ont. and N.B. south to Ohio and Vt.

Amelanchier X intermedia Spach Fig. p. A67. Map p. B51.
FREQUENCY: Scattered commonly in the southern and eastern portions of the Province.
HABITAT: Thickets, shores and boggy lowlands.
GENERAL RANGE: Minn. and Ont. to Nfld. south to N.C.
SPECIAL NOTE: Considered to be the fertile hybrid between **A. arborea** and **A. canadensis**.

Amelanchier lucida Fern.
HABITAT: Dry open areas such as pastures, old fields and heaths.
GENERAL RANGE: N.B. and N.S.
FREQUENCY: Rare; plants fitting the description of this species have been collected near Snake Creek, Geary, Sunbury Co., and near Dorchester, Westmorland Co. (UNB).

Amelanchier X neglecta Egglest. Fig. p. A67. Map p. B51.
FREQUENCY: Scattered throughout the Province.
HABITAT: Moist thickets, swamps and borders of woods.
GENERAL RANGE: Ont. to Nfld. south to New York and N. Eng.
SPECIAL NOTE: Considered to be the hybrid between **A. bartramiana** and **A. arborea** var. **laevis**.

Amelanchier X quinti-martii Louis-Marie Map p. B51.
FREQUENCY: Rare; mostly recorded in the eastern lowlands.
HABITAT: Wet woods and heathy barrens.
GENERAL RANGE: Ont. to N.B. south to northern N. Eng.
SPECIAL NOTE: Considered to be the hybrid of **A. bartramiana** and **A. arborea** or **A. humilis**.

Amelanchier sanguinea (Pursh) DC. Fig. p. A67. Map p. B51.
FREQUENCY: Uncommonly scattered throughout much of the Province.
HABITAT: Dry rocky or gravelly soil mostly in calcareous areas.
GENERAL RANGE: Ont. to N.B. south to Iowa, Ohio and N.C.
SUBSPECIFIC TAXA: Var. **gaspensis** Wieg. is separated from var. **sanguinea** in the key above and occurs very sparingly scattered throughout the Province on cliffs, ledges, gravelly beaches, or in alluvial woods. It has been recorded from Ont. to Que. south to Minn., Mich., Me. and N.B. SYNONYMS: **A. gaspensis** (Wieg.) Fern. & Weath.

Amelanchier stolonifera Wieg. Fig. p. A67. Map p. B51.
FREQUENCY: Scattered mostly throughout the southern half of the Province.
HABITAT: Acid rock outcrops and dry sandy or peaty areas.
GENERAL RANGE: Ont. to Nfld. south to Minn., Mich. and Va.

Amelanchier X wiegandii Neilsen Fig. p. A67. Map p. B52.
COMMON NAME(S): Wiegand's Serviceberry or Shadbush.

+++

FREQUENCY: Scattered throughout much of the Province.
HABITAT: Moist woods and streambanks.
GENERAL RANGE: Ont. to Nfld. south to Minn., N.Y. and northern N. Eng.
SPECIAL NOTE: Considered to be the fertile hybrid between **A. arborea** and **A. sanguinea.**

ARONIA Medic. Chokeberry.

1a Plant +/- glabrous at flowering, completely so at maturity; calyx-lobes deltoid. **A. melanocarpa**
1b Plant, in parts, white-tomentose at flowering, usually persisting until maturity; calyx-lobes longer than wide. **A. prunifolia**

Aronia melanocarpa (Michx.) Ell. Fig. p. A71. Map p. B54.
COMMON NAME(S): Black Chokeberry. Aronia noir.
FREQUENCY: Scattered throughout much of the Province and especially common in the southwest.
HABITAT: Peaty barrens, low areas of old fields, and wet thickets.
GENERAL RANGE: Ont. to N.S. south to Tex. and Fla.
SYNONYMS: **Aronia arbutifolia** (L.) Pers. var. **nigra** (Willd.) Seymour; **Pyrus arbutifolia** (L.) L. f. var. **nigra** Willd.; **Aronia nigra** (Willd.) Koehne.
SUBSPECIFIC TAXA: Forma **rubriflora**, forma nova, with flowers red from the beginning is reported from near Turtle Creek, Albert Co. (UNB).

Aronia prunifolia (Marsh.) Rehd. Fig. p. A71. Map p. B54.
COMMON NAME(S): Purple Chokeberry.
FREQUENCY: Scattered throughout the southern half and eastern coastal area of the Province.
HABITAT: Heath barrens, bog edges and sand plains.
GENERAL RANGE: Ont. to Nfld. south to Mich., Tex. and Fla.
SYNONYMS: **A. arbutifolia** (L.) Pers. var. **atropurpurea** (Britt.) Seymour; **Pyrus arbutifolia** (L.) L. f. var. **atropurpurea** (Britt.) Robins.

CRATAEGUS L. Hawthorn, Red Haw. Aubépine, Pommettes, Ceneles

1a Veins of large leaf blades entering sinuses as well as points of lobes; blades 3-6 lobed, +/- glabrous; fruit 6-8 mm thick. 2
1b Veins of blades entering only the leaf lobes, not the sinuses. 3
2a(1a) Styles and nutlets 2; leaves deeply lobed. **C. monogyna**
2b(1a) Style and nutlet 1; leaves more shallowly lobed. **C. curvisepala**
3a(1b) Plants in flower. 4
3b(1b) Plants not in flower. 5
4a(3a) Anthers pink, red or purple. 6
4b(3a) Anthers white or creamy, often drying brownish. 7
5a(3b) Floreal leaves mostly obtusely attenuate, rounded or cordate at base. 8

5b(3b) Floreal leaves mostly acutely attenuate at base. 9

6a(4a) Anthers 20. 10

6b(4a) Anthers 10 or less. 11

7a(4b) Anthers 20. 9

7b(4b) Anthers 10. 12

8a(5a) Leaves thick, +/- villous at least on veins beneath, lobes obtuse to acute, not acuminate. 12

8b(5a) Leaves mostly thinner, glabrous or becoming glabrate; lobes acuminate, often recurved at the tip. 13

9a(5b, 7a,10b) Floreal leaves widest above the middle, narrowly attenuate at base, unlobed or shallowly lobed above the middle, pubescent on veins beneath; inflorescence and calyx densely pubescent; anthers +/- 20, red or white; fruit yellowish. **C. punctata**

9b(5b, 7a,10b) Floreal leaves widest near or below the middle. 14

10a(6a) Floreal leaves widest below the middle. **C. flabellata** var. **grayana**

10b(6a) Floreal leaves widest above the middle. 9

11a(6b) Floreal leaves widest near or above the middle. 15

11b(6b) Floreal leaves widest below the middle. **C. flabellata**

12a(7b,8a) Mature floreal leaves usually longer than 7 cm, fruit oblong or obovate. 16

12b(7b,8a) Mature floreal leaves usually less than 6 cm long; calyx lobes glandular-serrate; fruit dark crimson, subglobose or short oblong; flesh thin, dry; nutlets 3-4, rounded at ends. **C. chrysocarpa**

13a(8b) Mature floreal leaves broadest near middle, narrowed at attenuate base, not deeply lobed, elliptic to obovate, glabrous beneath; inflorescence and calyx-tube glabrous; fruit subglobose, scarlet; flesh thick, dry, mealy; nutlets 3, with wide, shallow, irregular ventral depression, often obscure. **C. brainerdii**

13b(8b) Mature floreal leaves broadest near base, ovate to rhombic, often villous along veins beneath at least when young, short appressed villous above; inflorescence and calyx-tube glabrous or villous; fruit oblong or subglobose, crimson; flesh thick, succulent; nutlets 3-5, without ventral depression. **C. flabellata**

14a(9b) Leaves mostly pale yellowish-green, usually shallowly lobed above the middle; calyx lobes pectinate. **C. succulenta**

14b(9b) Leaves dull or dark green, usually conspicuously lobed above the middle; calyx lobes entire or shallowly lobed; fruit to 1.2 cm thick, subglobose, short-oblong to obovate. 17

15a(11a) Calyx lobes entire or shallowly glandular-serrate. 17

15b(11a) Calyx lobes pectinate; leaves +/- pale yellowish-green, shallowly lobed above the middle. **C. succulenta**

16a(12a) Leaves mostly widest near the middle; gradually narrowed to broadly attenuate at base; calyx-lobes +/- entire; fruit bright carmine red; nutlets 3 or rarely 2, acute and narrowed at base, rounded and broad at apex; flesh thick, sweet and mealy. **C. jonesae**

++

16b(12a) Leaves widest near base, broadly cuneate or rounded at base; calyx-lobes with large red stipitate glands, persistent; fruit orange-red with pale dots; flesh thin, subacid, dry and mealy; nutlets usually 5. **C. submollis**

17a(14b, 15a) Calyx lobes mostly entire; flowers 2-2.3 cm wide; inflorescence pubescent; stamens 10; anthers pink; leaves 7.5-10 cm long +/- permanently pubescent on veins beneath. **C. jonesae**

17b(14b, 15a) Calyx lobes glandular-serrate; flowers 1.6-1.8 cm wide; inflorescence glabrous; stamens 5-15; anthers pink or rarely whitish; leaves 3-5 cm long, glabrous beneath +/- scabrate above. **C. brainerdii**

Crataegus brainerdii Sarg. Fig. p. A67.
FREQUENCY: Scattered uncommonly towards the southeast and northeast of the Province.
HABITAT: Clearings, rocky shores and borders of woods.
GENERAL RANGE: Ont. to N.S. south to Mich., N.Y. and N. Eng.
SUBSPECIFIC TAXA: Our plants are referable to var. **scabrida** (Sarg.) Egglest.
SYNONYMS: **C. scabrida** Sarg.

Crataegus chrysocarpa Ashe Fig. p. A67. Map p. B52.
FREQUENCY: Scattered throughout much of the Province.
HABITAT: Streamside thickets and old fields.
GENERAL RANGE: B.C. to N.S. south to Idaho, Wisc. and N. Eng.
SYNONYMS: **C. rotundifolia** Moench.; **C. coccinea** L. (in part).

Crataegus curvisepala Lindm. Fig. p. A68.
SPECIAL NOTE: European; rarely introduced from cultivation as in Fredericton, York Co. and St. Andrews, Charlotte Co. Known elsewhere in Canada from Ont. and Que.
SYNONYMS: **C. oxyacantha** L.

Crataegus flabellata (Bosc ex Spach) K. Koch Fig. p. A68. Map p. B52.
FREQUENCY: Scattered commonly in the southern half of the Province.
HABITAT: Hedgerow thickets, shores and clearings.
GENERAL RANGE: Ont. to N.S. south to N.Y. and N. Eng.
SYNONYMS: **C. macrosperma** Ashe.
SUBSPECIFIC TAXA: Var. **grayana** (Egglest.) Palmer, separated in the above key, occurs rarely throughout the range of the species. SYNONYMS: **C. grayana** Egglest.

Crataegus jonesae Sarg.
FREQUENCY: Rare; 5 localities are known in the southcentral and southwest of the Province.
HABITAT: Open woods, shore thickets and hedgerows.
GENERAL RANGE: Que. and N.B. south to Maine.

++

Crataegus monogyna Jacq. Fig. p. A68. Map p. B52.
COMMON NAME(S): English Hawthorn.
SPECIAL NOTE: European; occasionally escaped cultivation mostly in the southern half of the Province. Known elsewhere in Canada from B.C., and from Ont. to N.S.

Crataegus punctata Jacq.
FREQUENCY: Very rare; known only from near the Hermitage, Fredericton, York Co.
HABITAT: Open shaded hillside, thickets, pastures and open rocky areas.
GENERAL RANGE: Ont. to N.B. south to Iowa and Ky.

Crataegus succulenta Schrad. ex Link Fig. p. A68. Map p. B52.
FREQUENCY: Most commonly found in the lower St. John River valley; rare elsewhere.
HABITAT: Borders of woods and pastures mostly in moist alluvial soil.
GENERAL RANGE: Man. to N.S. south to Iowa, N.Y. and N. Eng.
SUBSPECIFIC TAXA: Our plants are referable to var. **macracantha** (Lodd.) Egglest.
SYNONYMS: **C. macracantha** Lodd.

Crataegus submollis Sarg. Fig. p. A68 (as **C. mollis**)
FREQUENCY: Rare; Scattered along the St. John R. drainage from Victoria to Kings County and on the St. Croix R. near Milltown, Charlotte Co.
HABITAT: Rich alluvial thickets and pastures.
GENERAL RANGE: Ont. to N.B. south to Mass.

DALIBARDA Kalm False Violet, Dalibarda

Dalibarda repens L. Fig. p. A68. Fig. p. A52.
COMMON NAME(S): Robin-run-away.
FREQUENCY: Scattered across the central portion of the Province.
HABITAT: Shaded, moist coniferous woods.
GENERAL RANGE: Ont. to N.S. south to Mich., Ohio and N.C.

DRYAS L. Dryas, Mountain Avens

Dryas integrifolia Vahl Fig. p. A68.
FREQUENCY: Very rare; known only from Wilson Brook, Albert Co.
HABITAT: Gypsum cliffs and talus.
GENERAL RANGE: Alaska and B.C. to Nfld. and N.B.; Greenland; Siberia.
SPECIAL NOTE: For more information see H. Hinds, 1983. The Rare Vascular Plants of New Brunswick. Syllogeus 50.

FILIPENDULA Mill.

1a Petals pink; terminal leaflet deeply 7-9 lobed, pilose on veins beneath. **F. rubra**

1b Petals white or creamy; terminal leaflet 3-5 lobed, white tomentose beneath. **F. ulmaria**

++

Filipendula rubra (Hill) Robins. Fig. p. A68.
COMMON NAME(S): Queen-of-the-prairie.
SPECIAL NOTE: Native to eastern U.S.A.; occasionally escaped from cultivation in the south of the Province. Known elsewhere in Canada from Ont. to Nfld. south to N.S.

Filipendula ulmaria (L.) Maxim. Fig. p. A68. Map p. B52.
COMMON NAME(S): Queen-of-the-meadow.
SPECIAL NOTE: Eurasian; formerly cultivated and often escaped to roadsides, waste areas and former house sites. Known elsewhere in Canada from Ont. to Nfld. south to N.S.

FRAGARIA L. Strawberry. Fraisier

1a Apical tooth of terminal leaflet usually exceeding the two adjacent teeth; achenes superficial on receptacle; calyx-lobes spreading or reflexed on fruit; petals less than 1 cm long; flowering stem usually as long as or exceeding leaves; inflorescence raceme-like with unequal branches. **F. vesca**

1b Apical tooth of terminal leaflet usually shorter than the two adjacent teeth; achenes sunken in shallow pits of receptacle; calyx-lobes appressed around young fruit; petals usually longer than 1 cm; flowering stem mostly shorter than leaves; inflorescence umbel-like with +/- equal branches. **F. virginiana**

Fragaria vesca L. Fig. p. A68. Map p. B52.
COMMON NAME(S): Woodland Strawberry. Fraisier à vaches.
FREQUENCY: Locally common in the lower St. John R. valley and rare northwestward.
HABITAT: Calcareous rocky woods and clearings.
GENERAL RANGE: B.C. to Nfld. south to Calif. and Va.
SUBSPECIFIC TAXA: Our plants are referable to ssp. **americana** (Porter) Staudt. Forma **landonii** Boivin, with whitish or yellowish fruit, occurs rarely with the species as in Kingsclear, York Co. (UNB).

Fragaria virginiana Dcne. Fig. p. A68.
COMMON NAME(S): Wild strawberry.
FREQUENCY: Common throughout much of the Province.
HABITAT: Old fields, borders of woods, roadsides and ditches.
GENERAL RANGE: Alaska to Nfld. south to Calif., Tenn. and Ga.
SUBSPECIFIC TAXA: Our plants are referable to ssp. **glauca** (S. Wats.) Staudt, with stem and petiole glabrous or with appressed-ascending hairs rather than spreading hairs as in the species.

GEUM L. Avens. Benoîte

1a Sepals and petals purplish or petals varying to yellowish; flowers nodding; upper and lowermost part of style setose. **G. rivale**

1b Sepals green; petals white, yellow or orange. 2

++

2a(1b) Plant in flower. 3

2b(1b) Plant in fruit. 4

3a(2a) Petals white. 5

3b(2a) Petals orange or deep yellow. 6

4a(2b) Receptacle glabrous or minutely pubescent. 7

4b(2b) Receptacle long-hirsute. 8

5a(3a) Petals much shorter than calyx-lobes; peduncles hirsute with long spreading hairs +/- hiding shorter pubescence; basal leaves pinnately compound with pinnately lobed and incised leaflets. **G. lacinatum**

5b(3a) Petals +/- equalling to longer than calyx-lobes; peduncles puberulent with longer hairs scattered or wanting; basal leaves mostly trifoliate. **G. canadense**

6a(3b) Petals and calyx-lobes 4-5 mm long; terminal portion of style minutely pubescent; peduncles slender, not conspicuously enlarged towards summit; terminal segment of all basal leaves cordate-reniform or suborbicular, often deeply lobed. 9

6b(3b) Petals and calyx-lobes to 1 cm long; terminal portion of style conspicuously hirsute; peduncles stout, enlarged towards summit; terminal segment of some or all basal leaves cuneate-obovate to oblanceolate, incised. **G. aleppicum**

7a(4a) Fruiting heads globose, 1.7-2.5 cm in diameter; styles drab or brownish, not all reflexed, not glandular; achenes +/- glabrous; peduncles stout with crowded divergent or reflexed hairs 1-2 mm long; some or all basal leaves pinnately compound, the segments pinnately lobed and incised. **G. laciniatum**

7b(4a) Fruiting heads ovoid, 1.2-1.8 cm in diameter; achenes hirsute; styles usually purplish, minutely glandular at base, mostly all reflexed at maturity; peduncles slender, minutely puberulent, often with scattered longer hairs; all basal leaves with terminal segments cordate-reniform or suborbicular, often deeply lobed. **G. macrophyllum**

8a(4b) Fruiting heads obovoid at maturity, the styles all +/- tightly reflexed; basal leaves mostly pinnately divided with 5-9 incised leaflets. 10

8b(4b) Fruiting heads globose at maturity, the styles loosely ascending to spreading or tardily reflexed; peduncles slender; basal leaves mostly ternately compound (rarely pinnately compound with 1-2 additional pairs of much smaller leaflets, or simple trilobed). **G. canadense**

9a(6a) Style base minutely glandular; cauline leaves mostly simple, incised serrate with acute teeth, deeply 3-lobed (rarely 3-parted), the terminal lobe rounded at summit. **G. macrophyllum**

9b(6a) Style base glabrous; cauline leaves usually ternately compound (upper leaves sometimes 3-lobed), incised-crenate with blunt teeth, the terminal leaflet acute to broadly acuminate at summit. **G. urbanum**

10a(8a) Peduncles stout, enlarged towards summit; cauline leaves pinnately compound with mostly more than 3 lanceolate to rhombic leaflets, serrate with acute teeth; styles drab to brownish; calyx-lobes lanceolate or lance-ovate, 5-9 mm long. **G. aleppicum**

++

10b(8a) Peduncles slender, not enlarged towards summit; cauline leaves ternately compound (rarely 3-lobed), the oblanceolate to narrowly rhombic leaflets incised-crenate with blunt teeth; styles purplish; calyx-lobes broadly deltoid, 2.5-5 mm long. **G. urbanum**

Geum aleppicum Jacq. Fig. p. A68. Map p. B52.
FREQUENCY: Common mostly in the western part of the Province.
HABITAT: Damp woods and thickets.
GENERAL RANGE: Alaska and B.C. to Nfld. south to Calif., N. Mex. and N.J.; Eurasia.
SYNONYMS: **G. aleppicum** var. **strictum** (Ait.) Clausen

Geum canadense Jacq. Fig. p. A69. Map p. B52.
FREQUENCY: Mostly confined to the St. John R. drainage system with one report from the Miramichi R. near Newcastle, Northumberland Co.
HABITAT: Rich woods and thickets.
GENERAL RANGE: N. Dak. to N.S. south to Tex., Ala. and S.C.
SUBSPECIFIC TAXA: Var. **camporum** (Rydb.) Fern. & Weath., with leaves mostly strigose-pilose above and peduncles finely pubescent or with scattered long hairs occurs rarely in the same range as the species.

Geum laciniatum Murr. Fig. p. A69. Map p. B52.
FREQUENCY: Scattered throughout most of the Province.
HABITAT: Stream banks, damp meadows and thickets.
GENERAL RANGE: Ont. to N.S. south to Kans. and N.C.

Geum macrophyllum Willd. Fig. p. A69. Map p. B52.
FREQUENCY: Scattered over much of the Province.
HABITAT: Rich woods, thickets and meadows.
GENERAL RANGE: Alaska and B.C. to Nfld. south to Calif., Mex., Minn. and N. Eng.

Geum rivale L. Fig. p. A69. Map p. B53.
COMMON NAME(S): Water- or Purple-Avens, Chocolate-Root. Benoîte des ruisseaux.
FREQUENCY: Scattered over much of the Province and locally common.
HABITAT: Wet meadows and shores.
GENERAL RANGE: B.C. to Nfld. south to Wash., N. Mex., Ind. and N.J.; Iceland; Europe; Asia.

Geum urbanum L.
COMMON NAME(S): Wood-Avens, Herb-Bennet.
SPECIAL NOTE: Eurasian; introduced and spreading especially in the southwest of the Province as in Fredericton, York Co., St. Stephen, Charlotte Co., and Woodstock, Carleton Co. (UNB). Known elsewhere in Canada from Ont.

++

MALUS P. MILL. Apple, Crab Apple. Pommier.

Malus sylvestris (L.) P. Mill. Fig. p. A71. Map p. B54.
COMMON NAME(S): Apple. Pommier.
SPECIAL NOTE: Eurasian; commonly spread from cultivation throughout the southern half of the Province and rare northwards. Known elsewhere in Canada from all the Provinces except Sask.
SYNONYMS: **Pyrus malus** L., **Malus malus** (L.) Britt., **Malus pumila** Mill.

PHYSOCARPUS Maxim.

Physocarpus opulifolius (L.) Maxim. Fig. p. A69. Map p. B53.
COMMON NAME(S): Ninebark. Bois à sept écorces.
SPECIAL NOTE: Native in Canada in Ont. and Que.; cultivated and occasionally escaped in the southern part of the Province. Known elsewhere in Canada from Man. and N.S.

POTENTILLA L. Cinquefoil, Five-finger. Potentille or Quinte-feuille

1a Basal leaves with 3 leaflets. 2
1b Basal leaves with more than 3 leaflets. 3
2a(1a) Leaflets +/- toothed along entire margin, silky-tomentose to hirsute beneath; achenes glabrous; flowers red, purple or yellow. 4
2b(1a) Leaflets 3(5) toothed only at summit, nearly glabrous; flowers white, in erect terminal cymes; achenes densely hirsute. **P. tridentata**
3a(1b) Lowest leaves digitately compound. 5
3b(1b) Lowest leaves pinnately compound. 6
4a(2a) Leaflets silky-tomentose, +/- short petiolulate; petals red or purple, longer than calyx-lobes. **P. atrosanguinea**
4b(2a) Leaflets +/- hirsute, sessile; petals yellow, equal to or shorter than calyx-lobes. **P. norvegica**
5a(3a) Flowers solitary on long filiform pedicels from leaf axils; stem finally repent; leaves +/- appressed hirsute. **P. simplex**
5b(3a) Flowers numerous in +/- leafy cymes; stem erect. 7
6a(3b) Stem woody, to about 1 m high, shredding bark; leaflets 5-7, entire; achenes long hairy. **P. fruticosa**
6b(3b) Stem herbaceous, generally much less than 1 m high; leaflets 7-many, variously serrate; achenes glabrous. 8
7a(5b) Leaflets densely white tomentose beneath, nearly pinnatifid; stem depressed to ascending. **P. argentea**
7b(5b) Leaflets variously hairy beneath, but not densely white tomentose, nor deeply toothed; stem erect. 9
8a(6b) Stem erect, glandular villous; leaflets 7-11, oval to ovate, doubly serrate, glandular hirtellus beneath; petals pale yellow, slightly longer than calyx-lobes. **P. arguta**
8b(6b) Stem ascending from decumbent base or reduced to runners; leaflets simple serrate, not glandular beneath; petals bright yellow or purple. 10

9a(7b) Flowers 1.5-2.5 cm broad; petals pale yellow, showy; cyme stiffly erect above principal leaves; leaflets with narrowly deltoid teeth. **P. recta**

9b(7b) Flowers 8-10 mm broad; petals clear yellow; cyme diffuse, leafy; leaflets with +/- ovate teeth. **P. intermedia**

10a(8b) Stems reduced to runners; leaflets usually more than 9; flowers solitary on long pedicels; petals yellow. **P. anserina**

10b(8b) Stems ascending; leaflets less than 7; flowers in few-flowered, leafy cymes; petals red-purple. **P. palustris**

Potentilla anserina L. Fig. p. A69. Map p. B53.
COMMON NAME(S): Silverweed. Argentine, Richette.
FREQUENCY: Common along the Bay of Fundy and eastern coastal areas, and uncommon along streams inland.
HABITAT: Sandy or gravelly shores.
GENERAL RANGE: Alaska to Nfld. south to Calif., N. Mex., Iowa and N. Eng.; Greenland; Iceland; Eurasia.

Potentilla argentea L. Fig. p. A69. Map p. B53.
COMMON NAME(S): Silvery Cinquefoil.
SPECIAL NOTE: Eurasian; commonly introduced throughout much of the Province. Known elsewhere in Canada from all the Provinces.

Potentilla arguta Pursh Fig. p. A69. Map p. B53.
COMMON NAME(S): Tall Cinquefoil.
FREQUENCY: Locally common in the St. John River valley and in the northcentral area.
HABITAT: Rocky or alluvial banks and meadows.
GENERAL RANGE: Alaska to N.B. south to Oreg., Ariz., Mo. and W. Va.

Potentilla atrosanguinea Lodd.
COMMON NAME(S): Himalayan Cinquefoil.
SPECIAL NOTE: Asiatic; introduced in a strawberry bed, Kouchibouguac, Kent Co. Not reported elsewhere in Canada.

(**Potentilla egedii** Wormsk.)
SPECIAL NOTE: Listed from the Province by Scoggan, 1978-79 (as **P. egedii** var. **groenlandica** (Tratt.) Polunin) and Boivin, 1966-67 (as **P. anserina** var. **rolandii** (Boivin) Boivin). All N.B. specimens seen by this author are referable to **P. anserina** with achenes grooved on the back.

Potentilla fruticosa L. Fig. p. A69. Map p. B53.
COMMON NAME(S): Shrubby Cinquefoil, Buckbrush, Gold-Withy.
FREQUENCY: Most common in the southern and western part of the Province and rare northeastward.
HABITAT: Rocky or gravelly, rarely boggy areas, often calcareous.
GENERAL RANGE: Alaska to Nfld. south to Calif., N. Mex., S. Dak. and N.J.; Eurasia.

Potentilla intermedia L. Fig. p. A69. Map p. B53.
SPECIAL NOTE: Possibly of hybrid origin between **P. norvegica** and **P. argentea**.

Reported from Fredericton, York Co., Dalhousie, Restigouche Co. and from near Lac Unique, Madawaska Co. Known elsewhere in Canada from Ont. to N.S.

Potentilla norvegica L. Fig. p. A70. Map p. B53.
COMMON NAME(S): Rough Cinquefoil. Potentille de Montpellier.
SPECIAL NOTE: Possibly both native and introduced. Common throughout the Province in fields, waste areas and similar open areas. Known elsewhere in Canada from Yukon and Dist. Mack. to Nfld. south to Calif., Mex., Tex. and N.C.; Greenland; Iceland; Eurasia.

Potentilla palustris (L.) Scop. Fig. p. A70. Map p. B53.
COMMON NAME(S): Marsh-Five-Finger. Comaret, Argentine rouge.
FREQUENCY: Scattered throughout much of the Province.
HABITAT: Edges of fresh and salt marshes and bogs.
GENERAL RANGE: Alaska to Nfld. south to Calif., Iowa, Pa. and N.J.; Greenland; Eurasia.

Potentilla recta L. Fig. p. A70. Map p. B53.
COMMON NAME(S): Rough-fruited Cinquefoil, Sulphur Cinquefoil.
SPECIAL NOTE: Eurasian; occasionally introduced in old fields, waste areas and railways in the southern half of the Province. Known elsewhere in Canada from all the Provinces.

Potentilla simplex Michx. Fig. p. A69. Map p. B53 (as **P. canadensis**).
COMMON NAME(S): Old-field-Cinquefoil.
FREQUENCY: Common throughout the southern half of the Province.
HABITAT: Mostly old fields, roadsides and young mixed woods.
GENERAL RANGE: Ont. to Nfld. south to Minn., Okla., Tenn. and Ga.
SYNONYMS: **P. canadensis** L.

Potentilla tridentata Ait. Fig. p. A70. Map p. B54.
COMMON NAME(S): Three-toothed Cinquefoil.
FREQUENCY: Scattered throughout much of the Province.
HABITAT: Exposed rocky or sandy headlands, mountain tops and shores.
GENERAL RANGE: Dist. Mack. to Nfld. south to N. Dak., N.Y. and Ga.; Greenland.

PRUNUS L. Plum, Cherry. Prunier, Cerisier.

1a Flowers and fruit many, in terminal racemes. 2
1b Flowers and fruit solitary or in umbel-like clusters. 3
2a(1a) Leaves lance-oblong to oblong-ovate, thick, shiny above, taper pointed, dull below and often with midrib fringed with rusty hairs, finely serrate with blunt incurved teeth; calyx-lobes persistent on fruit; trees. **P. serotina**
2b(1a) Leaves ovate to obovate, abruptly short acuminate, glabrous or few white haired in axils beneath, finely sharp-serrate with erect or ascending teeth; calyx-lobes early deciduous; shrubs. **P. virginiana**
3a(1b) Leaves obscurely low dentate, narrowly oblanceolate; low depressed colonial shrub. **P. pumila** L.

++

3b(1b) Leaves conspicuously dentate, ovate to oblong-lanceolate; erect shrubs or small trees. 4

4a(3b) Calyx-lobes glandular-serrate; petals often pink tinged; fruit light red to yellowish; twigs often spiny. **P. nigra**

4b(3b) Calyx-lobes entire, glandular. 5

5a(4b) Leaves pubescent on both sides, ovate-oblong to obovate, short acuminate, coarsely and usually doubly serrate with blunt teeth; calyx-lobes subacute, reflexed; flowers 2-3 cm wide. **P. avium**

5b(4b) Leaves +/- glabrous, lanceolate to oblong-lanceolate, long acuminate, finely and sharply serrate with incurved teeth; calyx-lobes rounded, erect; flowers 1-1.5 cm wide. **P. pensylvanica**

Prunus avium L. Fig. p. A70.
COMMON NAME(S): Sweet Cherry. Massard, Cerisier de France.
SPECIAL NOTE: Eurasian; occasionally cultivated and spreading to borders of woods and roads as in Westcock, Westmorland Co. Known elsewhere in Canada from Ont., Que., and N.S.

Prunus cerasus L. Fig. p. A70.
COMMON NAME(S): Sour Cherry. Grottier.
SPECIAL NOTE: Asiatic; occasionally cultivated and rarely escaped as in Fredericton, York Co. Known elsewhere in Canada from B.C., Ont., P.E.I., and N.S.

Prunus nigra Ait. Fig. p. A70. Map p. B54.
COMMON NAME(S): Canada Plum. Guignier or Prunier sauvage.
SPECIAL NOTE: Probably mostly introduced from further west. Formerly cultivated and scattered throughout most of the Province in fencerows and former house sites. Known elsewhere in Canada from Man. to N.S.

Prunus padus L.
COMMON NAME(S): European Bird-Cherry.
SPECIAL NOTE: Material collected on Sugar Island above Fredericton, York Co. has been identified as this species. Known elsewhere in Canada from Ont.

Prunus pensylvanica L. f. Fig. p. A70.
COMMON NAME(S): Bird-, Fire-, or Pin-Cherry. Cerises d'été or Petit Merisier.
FREQUENCY: Common throughout much of the Province.
HABITAT: Clearings, burned areas and young mixed forest.
GENERAL RANGE: Dist. Mack. and B.C. to Nfld. south to Mont., Tenn. and N.C.

Prunus pumila L. Fig. p. A70. Map p. B54.
COMMON NAME(S): Sand-Cherry. Cerisier de sable, Minel du Canada, or Ragouminier.
FREQUENCY: Scattered throughout much of the Province except the eastern and southeastern part.
HABITAT: Sandy beaches and similar open areas.
GENERAL RANGE: Ont. to southern Lab. and N.B. south to Wisc., N.Y. and Mass.
SUBSPECIFIC TAXA: Our plants are referable to var. **depressa** (Pursh) Gleason.

++

Prunus serotina Ehrh. Fig. p. A70. Map p. B54.
COMMON NAME(S): Black or Rum-Cherry. Cerisier d'automne.
FREQUENCY: Locally common throughout much of the southern half of the Province and rare northward.
HABITAT: Dry to moist woods and border thickets.
GENERAL RANGE: N. Dak. to Minn., Ont., Que., N.B. and N.S. south to Mex., Tex. and Fla.

Prunus virginiana L. Fig. p. A70.
COMMON NAME(S): Choke-Cherry. Cerisier à grappes.
FREQUENCY: Common throughout much of the Province.
HABITAT: Thickets, shores and borders of dry woods.
GENERAL RANGE: Dist. Mack. and B.C. to Nfld. south to Calif., N. Mex.

ROSA L. Rose. Rosier.

1a Leaflets stipitate-glandular (not minutely glandular) over entire undersurface; pedicels glandular-bristly. 2
1b Leaflets not stipitate-glandular beneath. 3
2a(1a) Prickles pubescent; leaflets with single teeth and conspicuously raised veins beneath; glands yellowish; calyx-tube glabrous. **R. rugosa**
2b(1a) Prickles glabrous, usually curved; leaflets with double teeth and veins not conspicuously raised beneath; glands reddish; calyx-tube glabrous or glandular-hispid. 4
3a(1b) Branchlets usually with prominent prickles in pairs near the nodes; small prickles and bristles scattered or wanting. 5
3b(1b) Branchlets usually without conspicuous prickles in pairs near the nodes; prickles scattered with bristles or without. 6
4a(2b) Styles glabrous; prickles +/- similar, not mixed with glandular bristles; sepals falling as fruit ripens. **R. micrantha**
4b(2b) Styles villous; prickles usually mixed with glandular bristles especially on flowering stems; sepals usually persistent on fruit. **R. rubiginosa**
5a(3a) Pedicel and often calyx-tube glandular hispid. 7
5b(3a) Pedicel and usually calyx-tube glabrous; flowers usually double; leaflets +/- pubescent on both surfaces; calyx lobes entire, erect on fruit. **R. cinnamomea**
6a(3b) Pedicels and often calyx-tube glandular-hispid. 8
6b(3b) Pedicels and usually calyx-tube glabrous. 9
7a(5a) Leaflets finely toothed (averaging about 0.5 mm high near centre of leaflet) narrowly elliptic, acute at both ends, softly pubescent on midrib beneath and leaf rachis; stipules very narrow, hardly widened distally, often involute. **R. palustris**
7b(5a) Leaflets more coarsely toothed (averaging about 1 mm high near centre of leaflet); internodal prickles various, sometimes wanting. 10

8a(6a) Branchlets very densely covered with thin prickles and red bristles; leaflets glabrous; calyx-lobes entire. **R. nitida**

8b(6a) Branchlets not as above; leaflets +/- pubescent beneath; calyx-lobes usually pinnatifid or lobed. 11

9a(6b) Prickles stout, usually curved; calyx-lobes pinnatifid. **R. canina**

9b(6b) Prickles slender, +/- straight, sometimes wanting. 12

10a(7b) Nodal prickles +/- straight, slender, terete, rarely wanting, with base less than ½ the length of prickle; internodal prickles frequent especially at base; flowers mostly single on one-year stems; stipules firm, trough-like, scarcely widened upward, adnate portion 0.5-2 mm wide; leaflets 3-5 or 7, 1-1.5 cm wide, glabrous or pubescent, dentate mostly above middle. **R. carolina**

10b(7b) Nodal prickles usually down-curved or down-slanting. 13

11a(8b) Flowers nodding, usually double with erect overlapping petals; leaf-rachis without broad-based prickles; leaflets thin, not especially rugose. **R. centifolia**

11b(8b) Flowers on upright pedicels, usually single or semi-double; leaf-rachis with broad-based prickles; leaflets thick and rugose. **R. gallica**

12a(9b) Leaflets 7-11, less than 2 cm long, ovate to orbicular; base of pedicel without a bract. **R. spinosissima**

12b(9b) Leaflets 3-7, more than 2 cm long, ovate to elliptic; base of pedicel bearing a dilated bract. 14

13a(10b) Calyx lobes pinnatifid or lobed. 15

13b(10b) Calyx lobes entire, stipitate-glandular; flowers corymbose on branches from old wood; flattened base of prickles usually longer than ½ the length of the prickle; internodal prickles rare; stipules often glandular-dentate, widened upward, attached portion 3-10 mm wide; leaflets 7-9 glabrous, shiny, dentate upper 3/4 of margin, 1-3 cm wide. **R. virginiana**

14a(12b) Leaflets usually biserrate, glandular-puberulent beneath; stem densely beset with prickles and bristles from base to summit; petioles, leaf-rachises and margin of sepals usually glandular; calyx-lobes erect and forming a beak on the mature fruit. **R. acicularis**

14b(12b) Leaflets variously pubescent, simple serrate, not glandular beneath; stem often glabrous; prickles, if present, usually confined to base; calyx-lobes erect or reflexed on mature fruit. **R. blanda**

15a(13a) Flowers nodding, usually double; leaf-rachis without broad-based prickles. **R. centifolia**

15b(13a) Flowers erect, usually single; leaf-rachis with or without broad-based prickles. 16

16a(15b) Leaflets thin, nearly glabrous or pubescent on midrib beneath; petals 8-25(-30) mm long; leaf rachis without prickles. **R. canina**

16b(15b) Leaflets very thick, abundantly pubescent beneath; petals (25-)30-45 mm long; leaf rachis with broad-based prickles. **R. gallica**

Rosa acicularis Lindl. Fig. p. A71. Map p. B54.

COMMON NAME(S): Prickly Rose. Eglantier.

FREQUENCY: Very rare; known from St. Leonard, Madawaska Co., Pabineau Falls, Gloucester Co. and Northumberland Co., probably on the Nepisiquit R.

HABITAT: Rocky banks.

++

GENERAL RANGE: Alaska to northern N.B. south to Idaho, N. Mex., S. Dak., and Maine; Eurasia.
SUBSPECIFIC TAXA: Our plants are referable to ssp. **sayi** (Schwein.) W.H. Lewis.
SYNONYMS: **R. acicularis** var. **bourgeauiana** Crepin.

Rosa blanda Ait. Fig. p. A71. Map p. B54.
COMMON NAME(S): Meadow Rose. Rosier agréable.
FREQUENCY: Scattered throughout much of the Province and locally common along the St. John River.
HABITAT: Meadows, gravelly shores, rocky slopes.
GENERAL RANGE: Sask. to N.S. south to Nebr., Ind. and Pa.
SUBSPECIFIC TAXA: Var. **glabra** Crepin, with sepals +/- reflexed rather than sepals erect as in var. **blanda**, occurs frequently throughout the range of the species.

Rosa canina L.
COMMON NAME(S): Dog-Rose.
SPECIAL NOTE: Eurasian; cultivated and rarely escaping to fields, roadsides and waste areas. Known elsewhere in Canada from B.C., Ont., and N.S.

Rosa carolina L. Fig. p. A71. Map p. B54.
FREQUENCY: Scattered throughout the Province except rare in the northwest.
HABITAT: Sandy and rocky open areas.
GENERAL RANGE: Ont. to N.S. south to Tex. and Fla.

Rosa x centifolia L.
COMMON NAME(S): Cabbage-Rose.
SPECIAL NOTE: Asiatic; formerly cultivated and frequently escaped to roadsides and fields near present or former habitations. Known elsewhere in Canada from Ont. and P.E.I.

Rosa cinnamomea L. Fig. p. A71. Map p. B54.
COMMON NAME(S): Cinnamon-Rose. Rosier cannelle.
SPECIAL NOTE: Eurasian; cultivated and escaped to roadsides and fields throughout much of the Province. Known elsewhere in Canada from Ont. to N.S.

Rosa eglanteria L. Fig. p. A71. Map p. B55.
COMMON NAME(S): Sweet-Brier or Eglantine. Eglantier or Cébreur.
SPECIAL NOTE: Eurasian; cultivated and escaped to fields and waste areas in a few scattered areas. Known elsewhere in Canada from B.C., Ont., Que., Nfld., P.E.I. and N.S.

Rosa gallica L.
COMMON NAME(S): French Rose.
SPECIAL NOTE: Eurasian; occasionally cultivated and escaped to roadsides and thickets in the southwest of the Province.

Rosa multiflora Thunb.
SPECIAL NOTE: Asiatic; cultivated in the southern portions of the Province and rarely escaping to clearings and shores as in Milltown, Charlotte Co. (UNB). Known elsewhere in Canada from southern Ont. Not included in species key.

Rosa nitida Willd. Fig. p. A71. Map p. B55.
COMMON NAME(S): Bristly Rose. Rosier aciculaire.
FREQUENCY: Scattered throughout much of the Province.
HABITAT: Acid bogs and marshes, wet thickets, and stream margins.
GENERAL RANGE: Ont. to Nfld. south to southern N. Eng.

Rosa palustris Marsh. Fig. p. A71. Map p. B55.
COMMON NAME(S): Swamp Rose.
FREQUENCY: Uncommon to rare in the southern part of the Province.
HABITAT: Wet shores, marshes and swamps.
GENERAL RANGE: Ont. to N.S. south to Minn., Ark., and Fla.
SPECIAL NOTE: Many investigated citations of this species in the Province are misidentified and frequently turn out to be **R. virginiana**. Authenticated specimens are known from Scotchtown, Queens Co., and Pt. Wolfe R., Kings Co.

Rosa rugosa Thunb. Fig. p. A71. Map p. B55.
COMMON NAME(S): Salt-spray Rose.
SPECIAL NOTE: Asiatic; introduced near old habitations and in seaside thickets mostly in the southern part of the Province. Known elsewhere in Canada from Ont. to Nfld. south to N.B., P.E.I. and N.S.

Rosa spinosissima L. Fig. p. A71.
COMMON NAME(S): Scotch or Burnet-Rose.
SPECIAL NOTE: Eurasian; cultivated and rarely escaped to roadside thickets as in Woodstock, Carleton Co. Known elsewhere in Canada from Ont., Que., P.E.I. and N.S.

Rosa virginiana Mill. Fig. p. A71, Map p. B55.
FREQUENCY: Common throughout the southeast half of the Province and rare northwestward.
HABITAT: Shores, meadows, roadside ditches and clearings.
GENERAL RANGE: Ont. to Nfld. south to Mo. and Ala.

RUBUS L. Raspberry, Blackberry, Bramble. Framboisier, Ronce

1a Leaves simple. 2
1b Leaves compound. 3
2a(1a) Stems woody, to over 2 m tall; flowers bisexual, in cymose clusters; petals rose-purple, to 2.5 cm long; many parts stipitate-glandular; fruit dryish; leaf lobes acuminate. **R. odoratus**
2b(1a) Stem +/- herbaceous from creeping scaly rhizomes; flowers unisexual, usually solitary; petals white, 1-1.8 cm long; fruit juicy, yellowish when ripe; leaf lobes rounded. **R. chamaemorus**

3a(1b) Stems rarely over 4 dm long, without prickles or bristles, subherbaceous, without primocane and floricane development; stipules broad, +/- free from petiole; elongate runners frequently tip-rooting; fruit red. **R. pubescens**

3b(1b) Stems usually longer than 4 dm, mostly well provided with prickles and/or bristles, +/- woody with development of primocanes and floricanes; stipules linear or setaceous, adnate to petiole. 4

4a(3b) Primocane leaves densely white-tomentose beneath; ripe fruit easily separating from receptacle; raspberries. 5

4b(3b) Primocane leaves glabrous or variously hairy, but not white-tomentose beneath; fruit black, not easily separating from receptacle; blackberries. 6

5a(4a) Canes from rhizomes and stolons not tip-rooting, bristly or +/- glabrous; primocane leaves uncommonly with 5 pinnate leaflets; inflorescence +/- racemiform, open, not especially crowded at summit of shoot; calyx often bristly; fruit usually red; drupelets dull, densely tomentose. 7

5b(4a) Canes from crowns, arching, tip-rooting, with hooked prickles but no bristles nor glands, often heavily glaucous; primocane leaves often with 5 digitate leaflets; inflorescence more cymiform or umbelliform in a +/- compact cluster; fruit purple-black, shiny or glaucous; tomentum mainly separating glabrous drupelets. **R. occidentalis**

6a(4b) Broad-based curved prickles rare or absent; primocane bristles numerous, straight; **OR** prickles fewer, straight, acicular. 8

6b(4b) Broad-based curved or sometimes +/- straight stout prickles common; **OR** primocanes with prickles mixed with bristles; OR primocanes tall and arching with more slender and often few prickles. 9

7a(5a) Plants glandless; pedicels relatively short; drupelets not easily separating. **R. idaeus**

7b(5a) Plants stipitate-glandular; pedicels slender, more elongate; drupelets easily separating. **R. strigosus**

8a(6a) Primocanes and floricanes trailing or very low-arching and trailing, usually tip-rooting; flowering shoots erect from +/- prostrate floricane; leaflets 3(5); inflorescence usually glandless, few to several flowered. 10

8b(6a) Canes erect to arched-ascending, not usually tip-rooting nor with trailing tips; leaflets 5(3); inflorescence usually stipitate-glandular, racemose or corymbiform, many flowered. 11

9a(6b) Leaves essentially glabrous beneath; inflorescence without stipitate-glands or sometimes with only a few on pedicels. 12

9b(6b) Leaves +/- soft pubescent beneath at least along midrib and in axils of main veins. 13

10a(8a) Canes closely hispid or bristly without strong prickles (bristles often glandular when young); leaves flat, coriaceous, lustrous, usually evergreen. **R. hispidus**

10b(8a) Canes with strong mostly scattered prickles, glandless; leaflets with veins impressed above and very prominent and rib-like beneath, their margins plicate when fresh, not evergreen; flowers 3-15 with ascending or spreading non-glandular pedicels. **R. plicatifolius**

11a(8b) Cane bristles soft, pale, very crowded, +/- 1000 per dm, often glandular when young. **R. setosus**

11b(8b) Cane prickles harsher, less crowded, 50-500 per dm, stiff, acicular, often darker, usually not glandular. **R. vermontanus**

12a(9a) Prickles crowded, 10-100 per dm; primocanes mostly slender, lax; inflorescence corymbiform. **R. elegantulus**

12b(9a) Prickles more scattered or absent, 0-10 thin prickles per dm; primocanes stout, erect; inflorescence racemose. **R. canadensis**

13a(9b) Axis of inflorescence conspicuously stipitate-glandular; terminal leaflet of primocane leaves narrowly ovate to obovate-oblong, not more than 3/5 as broad as long; densely pubescent beneath, inflorescence usually long-cylindric, 1-2.5 dm long. **R. allegheniensis**

13b(9b) Axis of inflorescence glandless or with inconspicuous scattered sessile glands +/- concealed in hairs; terminal leaflet of primocane leaves broadly ovate to subrotund, 3/4 to about as broad as long, less densely pubescent beneath; inflorescence usually shorter, fewer-flowered, to 8 cm long. **R. pensylvanicus**

Rubus allegheniensis Porter Fig. p. A72. Map p. B55.
FREQUENCY: Locally common mostly in the southern half of the Province.
HABITAT: Thickets and clearings.
GENERAL RANGE: Ont. to N.S. south to Minn., Mo., and N.C.

Rubus canadensis L. Fig. p. A72. Map p. B55.
FREQUENCY: Common in the southern half of the Province and rare northwards.
HABITAT: Clearings and rocky slopes.
GENERAL RANGE: Ont. to Nfld. south to Tenn. and Ga.

Rubus chamaemorus L. Fig. p. A72. Map p. B55.
COMMON NAME(S): Bake-apple, Baked-apple-berry, Cloudberry, Mûres blanches, Ronce petit-mûrier.
FREQUENCY: Occasional in the southwest and northeast of the Province.
HABITAT: Acid bogs.
GENERAL RANGE: Alaska and B.C. to Nfld. south to Long Is.; Greenland; Spitzbergen; Eurasia.

Rubus elegantulus Blanch. Map p. B55.
FREQUENCY: Common throughout the southern half of the Province and rare northwards.
HABITAT: Clearings, roadsides and edges of woods.
GENERAL RANGE: Que. to Nfld. south to N.J.
SYNONYMS: **R. canadensis** var. **elegantulus** (Blanch.) Farw.

(**Rubus flagellaris** Willd. Fig. p. A72.)
SPECIAL NOTE: Material collected by Blanchard from Edmundston, Madawaska Co. (GH) was originally named **R. villosus** with a question mark in pencil and with = **procumbens** beneath. **R. flagellaris** was written in pencil above the label. This material agrees with the description of **R. plicatifolius**. I have seen no undoubted **R. flagellaris** which is more southern in distribution.

++

Rubus hispidus L. Fig. p. A72. Map p. B55.
FREQUENCY: Scattered and locally common in the southern part of the Province and uncommon northwards.
HABITAT: Clearings and open woods.
GENERAL RANGE: Ont. to N.S. south to Wisc., Ill. and N.C.
SYNONYMS: **R. mananensis** Bailey.

Rubus idaeus L. Fig. p. A72. Map p. B55.
COMMON NAME(S): Garden Raspberry. Ronce du mont Ida, Framboisier.
SPECIAL NOTE: Eurasian; frequently cultivated and occasionally escaped as near Milltown, Charlotte Co. (UNB). Known elsewhere in Canada from Ont. to Nfld. and N.S.

Rubus occidentalis L. Fig. p. A72. Map p. B55.
COMMON NAME(S): Black Raspberry. Thimbleberry. Mûrier.
FREQUENCY: Rare; found in a very few locations in Queens, York, and Carleton Counties.
HABITAT: Open woods and thickets often in calcareous areas.
GENERAL RANGE: Ont. to N.B. south to Minn., Okla. and Ga.
SPECIAL NOTE: For more information see H. Hinds, 1983. The Rare Vascular Plants of New Brunswick. Syllogeus 50.

Rubus odoratus L. Map p. B56.
COMMON NAME(S): Purple-flowering Raspberry. Calottes, Chapeaux rouges.
SPECIAL NOTE: Probably introduced from further west; cultivated and persisting near former habitations in the south of the Province.
GENERAL RANGE: Ont. and Que. south to Tenn. and Ga.

Rubus pensilvanicus Poir. Fig. p. A72. Map p. B56.
FREQUENCY: Uncommon in the southcentral and southwest of the Province.
HABITAT: Clearings and roadsides.
GENERAL RANGE: Ont. to Nfld. south to Minn., Okla., Ark. and Ala.

Rubus plicatifolius Blanch.
FREQUENCY: Very rare; found only in Edmundston and near Green R., Madawaska Co. (UNB, GH).
HABITAT: Grassy banks and roadside ditches.
GENERAL RANGE: Que. to N.S. south to N.Y. and Conn.
SPECIAL NOTE: See **R. flagellaris**.

Rubus pubescens Raf. Fig. p. A72. Map p. B56.
FREQUENCY: Common throughout much of the Province.
HABITAT: Moist thickets, woods and shores.
GENERAL RANGE: Mackenzie R. delta to Nfld. south to Wash., S. Dak. and N.J.
SUBSPECIFIC TAXA: Var. **pilosifolius** Hill, with leaves densely pilose beneath, rather than +/- glabrous as in the species, has been recorded from the Wolf Is., Charlotte Co.

++

(**Rubus recurvicaulis** Blanch.)
SPECIAL NOTE: Reports from N.B. have been referred to other species.

Rubus setosus Bigel. Map p. B56.
FREQUENCY: Scattered in the southern half of the Province.
HABITAT: Mostly damp thickets and open woods.
GENERAL RANGE: Ont. to N.S. south to Wisc., W. Va. and Md.

Rubus strigosus Michx. Map p. B56.
COMMON NAME(S): Red Raspberry, Framboisier.
FREQUENCY: Common throughout much of the Province.
HABITAT: Clearings, burns and similar open areas.
GENERAL RANGE: Alaska to Nfld. south to Calif., Mex., Tenn. and N.C.

Rubus vermontanus Blanch. Fig. p. A72. Map p. B56.
FREQUENCY: Scattered in the southeast half of the Province.
HABITAT: Thickets and clearings.
GENERAL RANGE: Ont. to N.S. south to Minn., Pa., and N. Eng.
SYNONYMS: **R. weatherbyi** Bailey.

SANGUISORBA L. Burnet

Sanguisorba canadensis L. Fig. p. A72. Map p. B56.
COMMON NAME(S): Canada Burnet. Herbe à pisser. Pimprenelle du Canada.
FREQUENCY: Locally common in the northeast extremity of the Province in Gloucester Co. and rare in Fundy National Park, Albert Co.
HABITAT: Bogs and wet open ground.
GENERAL RANGE: Que. to Nfld. south to Ill., Ohio and Ga.

SORBARIA (Sev.) A. Braun.

Sorbaria sorbifolia (L.) A. Br. Fig. p. A72. Map p. B56.
COMMON NAME(S): False Spiraea.
SPECIAL NOTE: Asiatic; formerly cultivated and escaped to roadsides and waste areas. Known elsewhere in Canada from Alta., Sask., and Ont. to N.S.

X **SORBARONIA** Schneid.

Sorbaronia X **jackii** Rehd.
SPECIAL NOTE: This is the apparent hybrid between **Sorbus americana** and **Aronia prunifolia**. It has been collected with the parents in w. Saint John, St. John Co. and in Memramcook, Westmorland Co. It is a fastigiate shrub to about 2 m tall with leaf characteristics combining aspects of both parents.

++

SORBUS L. Mountain-Ash. Sorbier

1a Leaflets abruptly +/- acute tipped, terminal tooth not prolonged, margins +/- parallel; inflorescence, leaf-axes and undersides of leaflets white-tomentose; winter buds usually densely pubescent. **S. aucuparia**

1b Leaflets more acuminate tipped, terminal tooth +/- prolonged, margins not parallel; inflorescence and leaf-axes white-tomentose at first becoming glabrate in fruit; winter buds glabrous or sparsely pilose. 2

2a(1b) Lateral leaflets broadest near middle, abruptly acute, dull above, serrate to slightly below middle, with 3-45 teeth, about 2-3 times as long as wide; inner bud scales rusty villous; flowers about 10 mm wide; petals orbicular. **S. decora**

2b(1b) Lateral leaflets broadest below middle, long acuminate, +/- shiny above, serrate nearly to base, with 50-75 teeth; about 3.5-5 times as long as wide; bud scales glabrous or merely ciliate; flowers 5-6 mm wide; petals obovate, cuneate at base. **S. americana**

Sorbus americana Marsh. Fig. p. A73. Map p. B56.
COMMON NAME(S): American Mountain-Ash. Cormier.
FREQUENCY: Common throughout much of the Province.
HABITAT: Damp acid woods.
GENERAL RANGE: Ont. to Nfld. south to Ill., Tenn. and Ga.
SYNONYMS: **Pyrus americana** (Marsh.) DC.

Sorbus aucuparia L. Fig. p. A73. Map p. B56.
COMMON NAME(S): European Mountain-Ash, Rowan-Tree.
SPECIAL NOTE: Eurasian; escaped from cultivation in scattered locations mostly in the south of the Province. Known elsewhere in Canada from B.C., Sask., and from Ont. to N.S.
SYNONYMS: **Pyrus aucuparia** (L.) Gaertn.

Sorbus decora (Sarg.) Schneid. Fig. p. A73. Map p. B56.
COMMON NAME(S): Showy Mountain-Ash.
FREQUENCY: Scattered throughout much of the Province.
HABITAT: Rocky woods and shores.
HABITAT: Borders of woods, rocky slopes and shores.
GENERAL RANGE: Sask. to Nfld. south to Minn., Ohio and Mass.
SYNONYMS: **Pyrus decora** (Sarg.) Hzl.

SPIRAEA L. Spiraea. Spirée

1a Leaves densely tomentose beneath; panicle narrow, long tapering to summit; petals roseate. **S. tomentosa**

1b Leaves +/- glabrous beneath; panicle open-pyramidal; petals white or rarely pink-tinged. **S. latifolia**

Spiraea latifolia (Ait.) Borkh. Fig. p. A73. Map p. B56.
COMMON NAME(S): Meadow-sweet. Thé du Canada.
FREQUENCY: Common throughout much of the Province.
HABITAT: Wet or dryish meadows and shores.
GENERAL RANGE: Man. to Nfld. south to N. Dak., Ohio, and Del.
SYNONYMS: **S. alba** Du Roi var. **latifolia** (Ait.) Dippel

Spiraea tomentosa L. Fig. p. A73. Map p. B56.
COMMON NAME(S): Steeple-bush, Hardhack. Reine des prés.
FREQUENCY: Common in the southern half of the Province and unknown in the north.
HABITAT: Wet acid meadows, fields, roadside ditches and shores.
GENERAL RANGE: Que. to N.S. south to Miss. and N.C.

WALDSTEINIA Willd.

Waldsteinia fragarioides (Michx.) Tratt. Fig. p. A73. Map p. B57.
COMMON NAME(S): Barren Strawberry.
FREQUENCY: Very rare; known only from the valley of Eel R., Carleton Co. and from Second Westcock, Westmorland Co.
HABITAT: Thickets, borders of woods and shores.
GENERAL RANGE: Ont. to N.B. south to Minn., Tenn. and Ga.

FABACEAE (Pulse or Pea Family)
(Alternate name **LEGUMINOSAE**)

1a Plant woody (tree or shrub). 2
1b Plant herbaceous. 3
2a(1a) Flowers terminal, blue-violet, small, in dense spike-like racemes; pod plump, to 8 mm long, 1-2 seeded; leaves glandular-dotted; stems to 4 m high. **Amorpha fruticosa**
2b(1a) Flowers several from scaly buds or in axillary racemes, white, yellow or rose-purple; pods larger, several seeded; leaves not glandular dotted; trees or shrubs. 4
3a(1b) Plants vine-like. 5
3b(1b) Plants erect, not vine-like. 6
4a(2b) Flowers yellow on long, downy pedicels; fastigiate shrubs or small trees. **Caragana arborescens**
4b(2b) Flowers white or rose-purple in pendulus axillary racemes; trees. **Robinia**
5a(3a) Leaves evenly 1-pinnate, terminated by a tendril. 7
5b(3a) Leaves odd-pinnate, not terminated by a tendril. 8
6a(3b) Leaflets 3. 9
6b(3b) Leaflets 5-many. 10
7a(5a) Styles filiform, bearded with a tuft or ring near tip; stem +/- round. **Vicia**

++

7b(5a) Style widened towards tip, bearded along inner side; stem usually angled or winged. **Lathyrus**

8a(5b) Leaflets 3; flowers purplish to white; calyx lobes 4; stems slender. **Amphicarpaea bracteata**

8b(5b) Leaflets 5-7; flowers purple-brown; calyx lobes 5; stems stout. **Apios americana**

9a(6a) Leaflets denticulate. 11

9b(6a) Leaflets entire. 12

10a(6b) Leaves palmately compound. **Lupinus polyphyllus**

10b(6b) Leaves pinnately compound. 13

11a(9a) Inflorescence long racemose, 5-20 cm long; corolla white or yellow; stem 1-3 m high. **Melilotus**

11b(9a) Inflorescence capitate or short racemose; stem usually less than 1 m high. 14

12a(9b) Pod indehiscent, flattened, divided into several 1-seeded joints covered by hooked hairs which cling to clothing, etc.; flowers purple in dense racemes. **Desmodium**

12b(9b) Pod dehiscent, not jointed, cylindric or angled; flowers yellow. **Coronilla scorpioides**

13a(10b) Inflorescence umbellate or capitate. 15

13b(10b) Inflorescence racemose or spicate. 16

14a(11b) Pod well exserted from calyx-tube, curved or coiled; filaments not dilated at summit. 17

14b(11b) Pod +/- enclosed by persistent calyx or only shortly exserted; filaments dilated at summit. **Trifolium**

15a(13a) Leaflets 11-21; petals roseate; pods linear, 3-7 jointed. **Coronilla varia**

15b(13a) Leaflets 5-11; petals yellow to crimson; pods not jointed. 18

16a(13b) Pods flattened, divided into lobed joints; flowers over 2 cm long; leaflets about 3 cm long. **Hedysarum alpinum**

16b(13b) Pods not flattened nor jointed; flowers less than 2 cm long; leaflets 5-25 mm long. 19

17a(14a) Pods only slightly curved, not at all spirally coiled, 10-16 mm long, thinly veined, not spiny; corolla yellow. **Trigonella corniculata**

17b(14a) Pods spirally coiled at least at tip, thickly veined or spiny; corolla yellow or purple. **Medicago**

18a(15b) Inflorescence conspicuously bracteate, short-pedunculate; plant appressed-pubescent; corolla pale yellow or crimson. **Anthyllis vulneraria**

18b(15b) Inflorescence without bracts, long pedunculate; corolla yellow, often reddish tinged. **Lotus**

19a(16b) Plants +/- acaulescent; peduncles leafless; pods inflated, erect, lance-ovoid, 2-2.5 cm long. **Oxytropis campestris**

19b(16b) Plants with +/- long leafy stems; peduncles leafy; pods not especially inflated, reflexed, ovoid to lance-oblong or elliptic to about 1.7 mm long. **Astragalus**

++

AMORPHA L.

Amorpha fruticosa L. Fig. p. A73.
COMMON NAME(S): False Indigo, Indigo-bush.
SPECIAL NOTE: Eastern U.S.A.; introduced from old nursery stock in Upper Woodstock, Carleton Co. (UNB). Known elsewhere in Canada from Que. and Ont.

AMPHICARPAEA Ell.

Amphicarpaea bracteata (L.) Fern. Fig. p. A73. Map p. B57.
COMMON NAME(S): Hog-Peanut.
FREQUENCY: Scattered mostly in the southwestern part of the Province on the St. John, St. Croix and upper Miramichi R. systems.
HABITAT: Shores and moist alluvial woods.
GENERAL RANGE: Sask. to N.S. south to Mont., Tex. and Fla.
SYNONYMS: **Amphicarpa bracteata** (L.) Fern.

ANTHYLLIS L.

Anthyllis vulneraria L. Fig. p. A73.
COMMON NAME(S): Lady's-fingers.
SPECIAL NOTE: European; introduced and spreading to roadsides and waste areas near Newcastle, Northumberland Co. Known elsewhere in Canada from Ont. and Que.

APIOS Medic.

Apios americana Medic. Fig. p. A73. Map p. B57.
COMMON NAME(S): Groundnut. Patates en chapelet.
FREQUENCY: Scattered mostly in the southern half of the Province and extending north to Kouchibouguac National Park, Kent Co.
HABITAT: Alluvial woods, thickets and shores.
GENERAL RANGE: Ont. to N.S. south to Colo., Tex. and Fla.
SYNONYMS: **A. tuberosa** Moench.

ASTRAGALUS L. Milk-Vetch, Locoweed. Astragale

1a Pods sessile, rupturing calyx-tube at maturity, elliptic, about 1 cm long; flowers 6-8 mm long, the wings and keel much shorter than standard. **A. eucosmus**

1b Pods stipitate within calyx-tube, oblong to lance-oblong. 2

2a(1b) Pod falcate and deeply grooved on lower side, 10-13 mm long; stipe usually longer than calyx-tube; calyx-tube about 2 mm long; flowers about 1 cm long, the broad and prominent keel about equalling the standard; long white hairs lacking in inflorescence and on calyx-tube. **A. alpinus**

++

2b(1b) Pod straight, not deeply grooved below, 10-19 mm long; stipe 1-3 mm long, about as long as or slightly shorter than calyx-tube; calyx-tube 3-3.5 mm long; flowers about 9-12 mm long, the keel petals much shorter than the standard; long white hairs about 1.5 mm long often present in inflorescence and on calyx-lobes. **A. robbinsii**

Astragalus alpinus L. Fig. p. A73. Map p. B57.
COMMON NAME(S): Brunet's Milk-Vetch.
FREQUENCY: Uncommon along the St. John R. above Fredericton, York Co. and on the Restigouche R., Restigouche Co.
HABITAT: Gravel strands, calcareous shores and ledges.
GENERAL RANGE: Que. to Nfld. south to Vt. and Me.
SUBSPECIFIC TAXA: Our plants are referable to var. **brunetianus** Fern.

Astragalus eucosmus Robins. Fig. p. A73. Map p. B57.
FREQUENCY: Rare; reported from near Crocks Pt., York Co., near mouth of Little R., Victoria Co., confluence of Upsalquitch and Restigouche Rivers and near Two Brooks on the Restigouche R., Restigouche Co.
HABITAT: Calcareous ledges and shores.
GENERAL RANGE: Alaska and B.C. to Nfld. south to N.B. and northern Me. Also known from Colo.
SPECIAL NOTE: For more information see H. Hinds, 1983. The Rare Vascular Plants of New Brunswick. Syllogeus 50.

Astragalus robbinsii (Oakes) Gray.
COMMON NAME(S): Robbins' Milk-Vetch.
FREQUENCY: Very rare; known only from Plaster Rock, Victoria Co. (UNB).
HABITAT: Gypsum ledges.
GENERAL RANGE: Straits of Belle I., Lab., e. Que. n. N.B. and w. Nfld.
SUBSPECIFIC TAXA: Our plants are referable to var. **fernaldii** (Rydb.) Barneby.
SYNONYMS: **A. fernaldii** (Rydb.) H. F. Lewis.

CARAGANA Lam. Pea-shrub

Caragana arborescens Lam.
COMMON NAME(S): Siberian Pea-shrub.
SPECIAL NOTE: Asiatic; commonly cultivated and occasionally escaping to hedgerows and borders of woods as in Fredericton, York Co. and St. Andrews, Charlotte Co. Known elsewhere in Canada from Alta. to Que.

CORONILLA L. Crown-Vetch

1a Leaflets more than 10; flowers 10-15, roseate. **C. varia**
1b Leaflets 3 or sometimes reduced to one; flowers 3-5, yellow. **C. scorpioides**

++

Coronilla scorpioides (L.) Koch
SPECIAL NOTE: European; reported from St. Stephen, Charlotte Co. and probably introduced in ballast waste. Known elsewhere in Canada from N.S.

Coronilla varia L. Fig. p. A73.
SPECIAL NOTE: European; recently cultivated and escaping to waste areas and rocky or sandy banks as near Minto, Sunbury Co. and Mactaquac Dam area, York Co. Known elsewhere in Canada from B.C., Man., and Ont. to N.S.

DESMODIUM Desv. Tick-trefoil, Tick-clover.

1a Flowering stem terminal, originating from a cluster of leaves; terminal leaflet about as long as wide, abruptly acuminate. **D. glutinosum**
1b Flowering stems axillary and terminal; terminal leaflet usually longer than wide, acute to obtuse at apex. **D. canadense**

Desmodium canadense (L.) DC. Fig. p. A74. Map p. B57.
COMMON NAME(S): Beggar-ticks.
FREQUENCY: Scattered throughout much of the western half of the Province.
HABITAT: Shores and alluvial thickets.
GENERAL RANGE: Man. to NS. south to Okla., Ohio and Va.

Desmodium glutinosum (Muhl.) Wood Fig. p. A73.
FREQUENCY: Very rare; formerly known from Bull Island near Woodstock, Carleton Co., and at Sprague's Falls on the St. Croix R. Both sites now destroyed by hydroelectric developments.
HABITAT: Rich rocky woods.
GENERAL RANGE: Ont. to N.S. south to Minn., Mex., Tenn. and Fla.
SPECIAL NOTE: For more information see H. Hinds, 1983. The Rare Vascular Plants of New Brunswick. Syllogeus 50.

HEDYSARUM L. Sainfoin

Hedysarum alpinum L. Fig. p. A75. Map p. B57.
FREQUENCY: Uncommonly scattered along the St. John River in York and Carleton Counties; rare on the Tobique near Plaster Rock, Victoria Co.; and locally common on the Restigouche R., Restigouche Co.
HABITAT: Calcareous ledges and rocky shores.
GENERAL RANGE: Alaska to Nfld. south to B.C., Alta., Mont. and Me.; Eurasia.

LATHYRUS L. Vetchling, Wild Pea. Gesse

1a Leaflets 2. 2
1b Leaflets 4-12. 3
2a(1a) Corolla yellow; stem wingless. **L. pratensis**

++

2b(1a) Corolla purplish to roseate or white; stem winged. **L. sylvestris**

3a(1b) Leaflets acute tipped; stipules oblique, linear to lanceolate with one basal lobe; stem winged. **L. palustris**

3b(1b) Leaflets +/- obtuse tipped; stipules broadly ovate, hastate with 2 basal lobes; stem wingless. **L. japonicus**

Lathyrus japonicus Willd. Fig. p. A74. Map p. B57.
COMMON NAME(S): Beach-Pea. Pois de mer.
FREQUENCY: Scattered in coastal areas of the Province.
HABITAT: Sandy or gravelly coastal beaches.
GENERAL RANGE: Alaska to Nfld. south to Calif., and from Man. to N.J.
SUBSPECIFIC TAXA: Smooth plants have been called var. **glabra** (Ser.) Fern. and hairy ones var. **pellitus** Fern. They have similar distributions in the Province.

Lathyrus palustris L. Fig. p. A74. Map p. B57.
COMMON NAME(S): Marsh Pea, Vetchling. Pois de marais.
FREQUENCY: Scattered throughout much of the Province.
HABITAT: Damp meadows and thickets.
GENERAL RANGE: Alaska and B.C. to Nfld. south to Calif., Tenn. and N.C.; Eurasia.
SUBSPECIFIC TAXA: Glabrous plants with +/- wingless stems have been separated as var. **myrtifolius** (Muhl.) Gray, reported from Ont. to N.S. Pubescent plants with winged stems have been called var. **pilosus** (Cham.) Ledeb. and occur mostly in coastal areas of the Province. Glabrous plants with stout winged stems refer to var. **palustris** which occurs both inland and on the coast from B.C.; Man. to Nfld. south to Oreg., Minn. and N. Eng.; Asia.

Lathryus pratensis L. Fig. p. A74. Map p. B57.
COMMON NAME(S): Yellow Vetchling. Gesse des prés.
SPECIAL NOTE: Eurasian; escaped from gardens to roadsides and damp meadows. Known elsewhere in Canada from B.C., and from Ont. to Nfld. south to N.B., P.E.I. and N.S.

Lathryus sylvestris L.
COMMON NAME(S): Everlasting or Perennial Pea. Gesse des bois.
SPECIAL NOTE: Eurasian; rarely escaped from gardens. Known elsewhere in Canada from B.C. and from Ont. to N.S.

LOTUS L. Deer-Vetch

1a Stem hollow; heads 5-12(15)-flowered. **L. pedunculatus**

1b Stem solid (sometimes hollow at base); heads (1)2-7-flowered. **L. corniculatus**

Lotus corniculatus L. Fig. p. A74. Map p. B57.
COMMON NAME(S): Birdsfoot-Trefoil.
SPECIAL NOTE: Eurasian; cultivated and escaped to roadsides and fields. Known elsewhere in Canada from B.C. to Nfld. south to N.B. and N.S.

++

Lotus pedunculatus Cav.
SPECIAL NOTE: Eurasian; rarely introduced as in St. Stephen, Charlotte Co. Known elsewhere in Canada from Sask., and Man. to N.S.

MEDICAGO L. Medick. Luzerne

1a Plants low, branching near base; corolla yellow; pod slightly coiled at summit or spiny; annuals. 2
1b Plants erect or ascending, to 1 m high; corolla blue to violet; pod coiled 2-3 turns, not spiny; perennials. **M. sativa**
2a(1a) Pod spiny, spirally coiled. 3
2b(1a) Pod not spiny, slightly incoiled at summit, heavily veined, often stipitate-glandular. **M. lupulina**
3a(2a) Leaflets broadly obcordate, about as broad as long, dark blotched above; marginal vein of pod with 3 conspicuous grooves. **M. arabica**
3b(2a) Leaflets cuneate-obovate or -oblong, longer than broad, not dark blotched above; marginal vein of pod very wide, not 3-grooved. **M. polymorpha**

Medicago arabica (L.) Huds. Fig. p. A74.
COMMON NAME(S): Spotted Medick.
SPECIAL NOTE: Eurasian; rarely introduced in ballast as in Saint John (NBM). Known elsewhere in Canada from B.C.

Medicago lupulina L. Fig. p. A74. Map p. B57.
COMMON NAME(S): Black Medick. Lupuline.
SPECIAL NOTE: Eurasian; commonly introduced to roadsides and waste areas throughout the Province. Known elsewhere in Canada from B.C. to Nfld. south to N.B., P.E.I. and N.S.

Medicago polymorpha L. Fig. p. A74.
COMMON NAME(S): Bur-Clover.
SPECIAL NOTE: Eurasian; rarely introduced as in Buctouche, Kent Co., and Saint John, St. John Co. (NBM). Known elsewhere in Canada from B.C., Sask., Ont. and N.S.

Medicago sativa L. Fig. p. A75. Map p. B58.
COMMON NAME(S): Alfalfa, Lucerne. Luzerne.
SPECIAL NOTE: Eurasian; commonly cultivated and escaped to fields and waste areas throughout much of the Province. Known elsewhere in Canada from Dist. Mack. and B.C. to N.S.

MELILOTUS Mill. Sweet Clover, Melilot. Mélilot

1a Pods pubescent, black at maturity; corolla yellow fading to whitish when dried; standard, keel and wings +/- equal. **M. altissima**
1b Pods glabrous, brown at maturity; corolla white or yellow; standard slightly longer than wings and keel. 2

++

2a(1b) Pods 2.5-3.5 mm long, strongly nerved; corolla 5-6 mm long, yellow fading to white when dried. **M. officinalis**

2b(1b) Pods 3-4 mm long, obscurely nerved; corolla white 4-5 mm long. **M. alba**

Melilotus alba Desr. Fig. p. A75. Map p. B58.
COMMON NAME(S): White Melilot. Trèfle d'odeur.
SPECIAL NOTE: Eurasian; very common weed along roadsides and in waste areas throughout the Province. Known elsewhere in Canada from Yukon and B.C. to Nfld., N.B., P.E.I. and N.S.

Melilotus altissima Thuill.
SPECIAL NOTE: Eurasian; rarely escaped to roadsides and waste areas as in Saint John, St. John Co. Known elsewhere in Canada from Ont. and N.S.

Melilotus officinalis (L.) Lam. Fig. p. A75. Map p. B58.
COMMON NAME(S): Yellow Sweet Clover. Trèfle d'odeur jaune.
SPECIAL NOTE: Eurasian; scattered in fields and waste areas and along roadsides throughout the Province. Known elsewhere in Canada from Yukon and B.C. to Nfld., N.B., P.E.I., and N.S.

OXYTROPIS DC. Stemless Locoweed

Oxytropis campestris (L.) DC. Fig. p. A75. Map p. B58.
COMMON NAME(S): St. John River Oxytropis.
FREQUENCY: Scattered along the St. John and Restigouche Rivers.
HABITAT: Gravelly river strands.
GENERAL RANGE: Man. to Baffin Is., Que., Nfld., N.B. and N.S.
SUBSPECIFIC TAXA: Our plants are referable to var. **johannensis** Fern.
SYNONYMS: **O. johannensis** Fern.

ROBINIA L. Locust. Robinier

1a Branchlets, leaflets and pods +/- glabrous; stipules spiny; corolla white; pods wing-margined. **R. pseudo-acacia**

1b Branchlets, leaflets and pods glandular hispid; stipules usually setaceous; corolla pinkish; pods wingless. **R. viscosa**

Robinia pseudo-acacia L. Fig. p. A75. Map p. B58.
COMMON NAME(S): Black or Yellow Locust, False Acacia. Acacia.
SPECIAL NOTE: Native to eastern U.S.; cultivated and occasionally escaped in the southern half of the Province. Known elsewhere in Canada from B.C., and from Ont. to N.S.

++

Robinia viscosa Vent. Fig. p. A75. Map p. B58.
COMMON NAME(S): Clammy Locust.
SPECIAL NOTE: Native to eastern U.S.; cultivated and occasionally escaped in the lower St. John River valley. Known elsewhere in Canada from Ont. to N.S.

TRIFOLIUM L. Clover, Trefoil. Trèfle

1a Calyx inflated in fruit; standard below the keel; corolla pink. **T. resupinatum**
1b Calyx not inflated in fruit; standard above keel. 2
2a(1b) Heads 1-5-flowered; corolla creamy white 5-6 mm long; stem diffusely branched, prostrate. **T. ornithopodioides**
2b(1b) Heads 10-many-flowered. 3
3a(2b) Corolla hidden by persistent long hairs of calyx-lobes. **T. arvense**
3b(2b) Corolla not hidden. 4
4a(3b) Corolla yellow, sometimes brownish after flowering. 5
4b(3b) Corolla not yellow. 6
5a(4a) Terminal leaflet +/- sessile; or all petiolules equal; stipules as long as petiole. **T. aureum**
5b(4a) Terminal leaflet conspicuously stalked; stipules about ½ as long as petiole. 7
6a(4b) Flowers distinctly pedicelled in head; corolla white or pinkish, brownish in age. 8
6b(4b) Flowers +/- sessile in head; corolla purplish or roseate. 9
7a(5b) Leaflets deeply notched at summit, veins prominent; flowers 2.5-3.5 mm long, 5-15 per head standard faintly veined. **T. dubium**
7b(5b) Leaflets only slightly notched at summit, veins inconspicuous; flowers 3.5-4.5 mm long, 20-30 per head, standard strongly veined. **T. campestre**
8a(6a) Stems prostrate, creeping; peduncles erect, long; calyx-teeth shorter than calyx-tube. **T. repens**
8b(6a) Stems erect; peduncles relatively short; calyx-teeth as long as or longer than calyx-tube. **T. hybridum**
9a(6b) Heads usually +/- pedunculate; leaflets narrowly elliptic or oblong; stipules lanceolate, awl-tipped; calyx glabrous outside, pubescent within; corolla purplish. **T. medium**
9b(6b) Heads sessile in axil of reduced leaves; leaflets oval or obovate; stipules ovate, linear-tipped; calyx tawny pubescent outside, glabrous within; corolla roseate. **T. pratense**

Trifolium aureum Pollich Fig. p. A75. Map p. B58.
COMMON NAME(S): Yellow or Hop Clover. Trèfle jaune.
SPECIAL NOTE: Eurasian; commonly introduced along roadsides throughout the Province. Known elsewhere in Canada from all the Provinces.
SYNONYMS: **T. agrarium** L.

++

Trifolium arvense L. Fig. p. A75. Map p. B58.
COMMON NAME(S): Rabbit-foot Clover, Stone Clover. Trèfle des champs.
SPECIAL NOTE: Eurasian; introduced on roadsides, fields and waste areas mostly on the east coast and in the southwest. Known elsewhere in Canada from B.C. and from Ont. to N.S.

Trifolium campestre Schreb. Fig. p. A75. Map p. B58.
COMMON NAME(S): Low Hop Clover.
SPECIAL NOTE: Eurasian; commonly introduced on roadsides and waste areas throughout much of the Province. Known elsewhere in Canada from all provinces.
SYNONYMS: **T. procumbens** L.

Trifolium dubium Sibth. Fig. p. A75.
COMMON NAME(S): Suckling Clover.
SPECIAL NOTE: European; rarely introduced in the south of the Province as in Saint John, St. John Co. Known elsewhere in Canada from B.C. and from Ont. to N.S.

Trifolium hybridum L. Fig. p. A75.
COMMON NAME(S): Alsike Clover. Trèfle Alsike.
SPECIAL NOTE: Eurasian; commonly introduced throughout much of the Province. Known elsewhere in Canada from all provinces.

Trifolium medium L. Fig. p. A75. Map p. B58.
COMMON NAME(S): Zigzag Clover.
SPECIAL NOTE: Eurasian; rarely introduced along roadsides, railways and shores in Victoria and Restigouche Counties. Known elsewhere in Canada from B.C. and from Ont. to N.S.

Trifolium ornithopodioides L.
COMMON NAME(S): Bird's-foot Fenugreek.
SPECIAL NOTE: European; locally introduced from ship ballast in Saint John, St. John Co. but not persisting.
SYNONYMS: **Trigonella ornithopodioides** (L.) DC.

Trifolium pratense L. Fig. p. A76.
COMMON NAME(S): Red Clover. Trèfle rouge.
SPECIAL NOTE: Eurasian; cultivated and escaped to meadows, roadsides and similar open areas throughout the Province. Known elsewhere in Canada from all the provinces.

Trifolium repens L. Fig. p. A76. Map p. B58.
COMMON NAME(S): White Clover. Trèfle blanc.
SPECIAL NOTE: Eurasian; cultivated and escaped to roadsides and waste areas throughout the Province. Known elsewhere in Canada from all the provinces.

Trifolium resupinatum L.
COMMON NAME(S): Reversed Clover.
SPECIAL NOTE: Asiatic; reported on ballast waste, Saint John, St. John Co. by Fowler, 1885. Not persisting. Known elsewhere in Canada from Que.

TRIGONELLA L.

Trigonella corniculata L.
SPECIAL NOTE: European; probably introduced in ship ballast in St. Stephen, Charlotte Co. and not persisting. Known elsewhere in Canada from Ont. and N.S.

VICIA L. Vetch, Tare. Vesce

1a Flowers +/- sessile in axils of leaves. 2

1b Flowers on long pedicels or peduncles. 3

2a(1a) Calyx-tube several times longer than calyx-lobes; flowers in clusters of about 4-6; perennials. **V. sepium**

2b(1a) Calyx-tube nearly as long as calyx-lobes; flowers usually 1 or 2; annuals. **V. sativa**

3a(1b) Peduncle 1-6(9) flowered; leaflets 4-12; pod .6-1.3 cm long; flowers .3-.8 cm long. 4

3b(1b) Peduncle many-flowered; leaflets 16-24; pod 2-3 cm long; flowers 1-1.5 cm long. **V. cracca**

4a(3a) Pod hirsute, 6-10 mm long; flowers 3-4 mm long; calyx-lobes +/- equal; stipules with 2 simple lobes; tendrils usually branched. **V. hirsuta**

4b(3a) Pod glabrous, 10-13 mm long; flowers 7-8 mm long; calyx-lobes unequal; stipules 2-lobed, outer simple, inner trifid; tendrils usually unbranched. **V. tetrasperma**

Vicia cracca L. Fig. p. A76.
COMMON NAME(S): Cow or Tufted Vetch, Canada-pea. Jargeau.
SPECIAL NOTE: Eurasian; introduced to roadsides, fields and similar open areas throughout much of the Province. Known elsewhere in Canada from all the provinces.

Vicia hirsuta (L.) S.F. Gray Fig. p. A76.
COMMON NAME(S): Hairy Vetch.
SPECIAL NOTE: Eurasian; rarely introduced to roadsides and waste areas. Known elsewhere in Canada from B.C., Alta., and from Ont. to N.S.

Vicia sativa L. Fig. p. A76. Map p. B58.
COMMON NAME(S): Spring-Vetch. Vesce cultivée.
SPECIAL NOTE: Eurasian; occasionally cultivated and escaped to roadsides and waste areas mostly in the southern half of the Province.
SUBSPECIFIC TAXA: Var. **nigra** (L.) Ehrh., with purple corolla less than 18 mm long and seeds about 3 mm broad, occurs rarely with the species that has bicolored reddish and purple corolla and seeds about 5 mm broad.

++

Vicia sepium L. Map p. B59.
COMMON NAME(S): Hedge Vetch.
SPECIAL NOTE: Eurasian; introduced on roadsides and in old fields and waste areas in a few scattered locations. Known elsewhere in Canada from Ont. to Nfld., N.B., P.E.I., and N.S.

Vicia tetrasperma (L.) Moench Fig. p. A76.
COMMON NAME(S): Four-seeded Vetch. Lentillon, Cicérole.
SPECIAL NOTE: Eurasian; introduced on roadsides and waste areas. Known elsewhere in Canada from B.C., and from Ont. to N.S.

LINACEAE (Flax Family)

Linum L. Flax. Lin

1a Stem leaves opposite; corolla white, yellow at base, 5 mm long; stem to 3 dm tall. **L. catharticum**
1b Stem leaves alternate; corolla normally pale blue, about 1 cm long; stem usually greater than 5 dm tall. **L. usitatissimum**

Linum catharticum L. Fig. p. A76.
COMMON NAME(S): Fairy-Flax.
SPECIAL NOTE: Eurasian; probably introduced in ship ballast in St. Andrews, Charlotte Co. and not persisting. Known elsewhere in Canada from Ont. to N.S.

Linum usitatissimum L. Fig. p. A76.
COMMON NAME(S): Common Flax. Lin.
SPECIAL NOTE: Eurasian; cultivated and occasionally escaped to fields, railways and waste areas in the south of the Province. Known elsewhere in Canada from all the provinces.

OXALIDACEAE (Wood-Sorrel Family)

OXALIS L. Wood-Sorrel, Lady's-Sorrel. Oxalide

1a Leaves all basal; flowers solitary, white, reddish veined. **O. acetosella**
1b Leaves cauline; flowers several per inflorescence, yellow. 2
2a(1b) Plants with many hairs blunt, septate, flattened; stipules lacking; plants rhizomatous. **O. stricta**
2b(1b) Plants with hairs nonseptate nor flattened, sharp pointed; stipules present, inconspicuous; plants not rhizomatous. **O. dillenii**

Oxalis acetosella L. Fig. p. A76. Map p. B59 (as **O. montana**).
COMMON NAME(S): Wood-Sorrel. Surette, Paine de lièvre.
FREQUENCY: Common throughout much of the Province.

HABITAT: Coniferous or mixed woods.
GENERAL RANGE: Ont. to Nfld. south to Minn., Tenn. and N.C.
SYNONYMS: **O. montana** Raf.

Oxalis dillenii Jacq.
SPECIAL NOTE: Natural range mostly western and European; reported from the Province by G. Eiten, 1963, as ssp. **filipes** (Small) Eiten.

Oxalis stricta L. Fig. p. A76.
COMMON NAME(S): Yellow Sorrel. Pain d'oiseau.
FREQUENCY: Common throughout much of the Province.
HABITAT: Roadsides, fields and similar open areas.
GENERAL RANGE: Sask. to N.S. south to Wash., Ariz., Tenn. and Ga.

GERANIACEAE (Geranium Family)

1a Leaves palmately cleft or compound; fertile stamens +/- 10. **Geranium**
1b Leaves pinnately compound; fertile stamens 5. **Erodium**

ERODIUM L'Hér. Storksbill.

Erodium cicutarium (L.) L'Hér. Fig. p. A77.
COMMON NAME(S): Alfilaria, Filaria, Pin-Clover. Aeguillettes, Herbe à la fourchette.
SPECIAL NOTE: Eurasian; introduced in gardens, fields and waste areas in scattered localities. Known elsewhere in Canada from all provinces except Nfld.

GERANIUM L. Cranesbill, Wild Geranium

1a Leaves palmately decompound, at least the terminal leaflet stalked, triangular-ovate in general outline. **G. robertianum**
1b Leaves merely lobed, pentagonal or +/- orbicular in general outline. 2
2a(1b) Petals about twice as long as sepals, purple; fruiting pedicels deflexed. **G. pratense**
2b(1b) Petals about as long as sepals, pink-purple; fruiting pedicels erect. **G. bicknellii**

Geranium bicknellii Britt. Fig. p. A77. Map p. B59.
COMMON NAME(S): Bicknell's Wild Geranium.
FREQUENCY: Uncommon; mostly southern and locally common.
HABITAT: Disturbed soils, burns, and clearings.
GENERAL RANGE: Yukon to Nfld. south to Calif. and Mass.
SUBSPECIFIC TAXA: Our plants are referable to var. **longipes** (Wats.) Fern.

++

Geranium pratense L. Fig. p. A77. Map p. B59.
COMMON NAME(S): Meadow Crane's-bill.
SPECIAL NOTE: Eurasian; cultivated and escaped to roadsides and similar open areas. Known elsewhere in Canada from Man. to Nfld., N.B., P.E.I. and N.S.

Geranium robertianum L. Fig. p. A77. Map p. B59.
COMMON NAME(S): Herb-Robert. Herbe à Robert, Herbe à l'esquinancie.
FREQUENCY: Uncommon; scattered in the extreme south of the Province.
HABITAT: Rocky woods and wet ledges.
GENERAL RANGE: B.C.; Ont. to Nfld. south to Ill., Ohio, and Md.; Eurasia.

POLYGALACEAE (Milkwort Family)

POLYGALA L. Polygala, Milkwort.

1a Flowers few (1-4), large, subtended by leaves in terminal cluster; corolla showy, rose-purple (rarely white); upper leaves ovate. **P. paucifolia**
1b Flowers small, numerous, in compact or elongate racemes. 2
2a(1b) Stems several from subligneus perennial rootstock; leaves lanceolate-elliptic; inflorescence slender, tapering to tip; corolla white. **P. senega**
2b(1b) Stems simple from slender, annual, tap-rooted base; leaves linear to narrowly elliptic; inflorescence dense, short cylindric, rounded at summit; corolla pinkish purple. **P. sanguinea**

Polygala paucifolia Willd. Fig. p. A77. Map p. B59.
COMMON NAME(S): Fringed Polygala, Gaywings, Bird-on-the-wing.
FREQUENCY: Rare; recorded in a few scattered localities in Sunbury, Queens, and Charlotte Counties.
HABITAT: Mostly moist acid mixed woods.
GENERAL RANGE: Sask. to Gaspé Pen. south to Minn., Tenn. and Ga.
SPECIAL NOTE: For more information see H. Hinds, 1983. The Rare Vascular Plants of New Brunswick. Syllogeus 50.

Polygala sanguinea L. Fig. p. A77. Map p. B59.
FREQUENCY: Rare; recorded from one locality each in Charlotte, Sunbury and Queens Counties.
HABITAT: Open woods, clearings (often old burn sites), and sterile roadsides.
GENERAL RANGE: Ont. to N.S. south to Minn., Okla., Tenn. and S.C.

Polygala senega L. Fig. p. A77. Map p. B59.
COMMON NAME(S): Seneca-Snakeroot. Sénéca.
FREQUENCY: Uncommon; scattered and locally common in York, Carleton, Victoria and Restigouche Counties along the western portion of the Province.
HABITAT: Calcareous woods and shores.
GENERAL RANGE: Alta. to N.B. south to S. Dak., Tenn. and Ga.

++

CALLITRICACEAE (Water-Starwort Family)

CALLITRICHE Water-Starwort

1a Leaves all submerged, linear, 1-nerved, truncate or slightly notched at summit, opposite pairs not connected at base; fruit nearly orbicular, conspicuously winged around entire margin, deeply furrowed between carpels. **C. hermaphroditica**

1b Leaves usually of 2 forms (broad 3-5 nerved floating and narrow 1-nerved submerged), opposite leaf pairs connected at base by narrow wing; fruit obovate, wingless or narrowly winged, shallowly furrowed between carpels. 2

2a(1b) Fruit +/- as broad as long, its margins mostly rounded and wingless, usually widest above the middle, notch wide angled, marginal furrow narrow, shallow, surface pits irregularly arranged; floating leaves subtruncate or tapering to obtuse summit. **C. heterophylla**

2b(1b) Fruit longer than broad, its margin unequally narrow winged, widest at middle, marginal furrow wide, V-shaped, surface pits in +/- vertical rows, notch narrow; floating leaves oblanceolate to obovate tapering to rounded apex. **C. verna**

Callitriche hermaphroditica L. Fig. p. A78. Map p. B60.
FREQUENCY: Uncommon; scattered mostly in the south of the Province.
HABITAT: Quiet mostly calcareous or brackish waters.
GENERAL RANGE: Alaska to Nfld. south to Calif., Colo., Mich. and Vt.; Greenland; Iceland; Eurasia.

Callitriche heterophylla Pursh Fig. p. A78. Map p. B60.
FREQUENCY: Common throughout much of the Province.
HABITAT: Quiet waters and shores.
GENERAL RANGE: Alaska and B.C. to Nfld. south to Calif., Tex., Ohio and Maine; S. America.

Callitriche verna L. Fig. p. A78. Map p. B60.
FREQUENCY: Common throughout much of the Province.
HABITAT: Quiet waters and shores.
GENERAL RANGE: Alaska to Nfld. south to Calif., Nebr. and Va.; Greenland; Iceland; Eurasia.
SYNONYMS: **C. palustris** L.

EMPETRACEAE (Crowberry Family)

1a Flowers solitary in leaf axils; fruit fleshy, black. **Empetrum nigrum**

1b Flowers several in terminal heads, purplish in mass; fruit dry. **Corema conradii**

++

COREMA Don Broom-Crowberry

Corema conradii Torr. Fig. p. A78.
COMMON NAME(S): Poverty-grass.
FREQUENCY: Very rare; collected by G.F. Matthew (no date, probably 1800's) in Portland (now Saint John), St. John Co. (CAN). Probably extirpated from Province.
HABITAT: Barrens.
GENERAL RANGE: Magdalen Is., Que., Nfld., P.E.I., N.B. and N.S. south to N.Y., Cape Cod and N.J.
SPECIAL NOTE: Collections by G. Hay at CAN have been annotated to **Empetrum nigrum** by this author.

EMPETRUM L. Crowberry. Camarine

Empetrum nigrum L. Fig. p. A78. Map p. B60.
COMMON NAME(S): Black Crowberry, Curlewberry. Corbigeau.
FREQUENCY: Locally common along the coast and occasionally inland.
HABITAT: Exposed headlands, windswept peaty and rocky barrens.
GENERAL RANGE: Alaska and B.C. to Nfld. south to Calif., Minn., N.Y. and Long Is.; Greenland; Iceland; Spitzbergen; Eurasia.

EUPHORBIACEAE (Spurge Family)

1a Plants with milky sap; leaves +/- sessile; flowers in cyathia. 2
1b Plants without milky sap; leaves petioled; flowers in axillary heads or short spikes. 3
2a(1a) Plants erect; leaves more than 2 cm long, not oblique at base. **Euphorbia**
2b(1a) Plants prostrate or +/- ascending; leaves usually less than 1.5 cm long, oblique at base. **Chamaesyce**
3a(1b) Leaves opposite, short-petioled; pistillate bracts minute; capsule 2-locular. **Mercurialis annua**
3b(1b) Leaves alternate, long-petioled; pistillate bracts conspicuous, foliaceous, palmately cleft; capsule 3-locular. **Acalypha rhomboidea**

ACALYPHA L. Three-seeded Mercury, Copperleaf. Ricinelle

Acalypha rhomboidea Raf. Fig. p. A77. Map p. B59.
FREQUENCY: Locally common along the St. John R. in York, Queens and Kings Counties.
HABITAT: Gravelly shores, roadsides and waste areas.
GENERAL RANGE: Ont. to N.S. south to Minn., Okla., Ala. and Fla.

CHAMAESYCE L. Spurge

1a Capsules strigose; stems villous; seeds with low transverse ridges. **C. maculata**

++

1b Capsules glabrous. 2

2a(1b) Leaves entire; involucral glands without petal-like appendages; seeds smooth. **C. polygonifolia**

2b(1b) Leaves denticulate; involucral glands with white petal-like appendages; seeds transversely ridged. 3

3a(2b) Stem glabrous; leaves serrulate especially near apex. 4

3b(2b) Stem +/- hirsute; leaves serrulate along entire margin. **C. vermiculata**

4a(3a) Seeds smooth or slightly wrinkled. **C. serpyllifolia**

4b(3a) Seeds with 5-6 sharp cross-ribs. **C. glyptosperma**

Chamaesyce glyptosperma (Engelm.) Small Fig. p. A77.
FREQUENCY: Rare; reported in Victoria and Carleton Counties.
HABITAT: Rail and roadways and similar dry open areas.
GENERAL RANGE: B.C. to N.B. south to Calif., Mexico, Ohio and N.Y.
SPECIAL NOTE: For more information see H. Hinds, 1983. The Rare Vascular Plants of New Brunswick. Syllogeus 50.
SYNONYMS: **Euphorbia glyptosperma** Engelm.

Chamaesyce maculata (L.) Small Fig. p. A77.
FREQUENCY: Very rare; known only from St. Stephen, Charlotte Co.
HABITAT: Railway.
GENERAL RANGE: Ont. to N.S. south to N. Dak., Tex. and Fla.
SYNONYMS: **Euphorbia supina** Raf.

Chamaesyce polygonifolia (L.) Small Fig. p. A77. Map p. B59.
COMMON NAME(S): Seaside Spurge.
FREQUENCY: Rare; reported from a few east coastal areas in Gloucester and Kent Counties and on Grand Manan Is. in the Bay of Fundy, Charlotte Co.
HABITAT: Coastal sandy and gravelly shores and dune hollows.
GENERAL RANGE: Wisc., Mich. and Ont. to N.S. south along the coast to Ga.
SPECIAL NOTE: For more information see H. Hinds, 1983. The Rare Vascular Plants of New Brunswick. Syllogeus 50.
SYNONYMS: **Euphorbia polygonifolia** L.

Chamaesyce serpyllifolia (Pers.) Small Fig. p. A77.
FREQUENCY: Rare; reported from York, Victoria and Madawaska Counties along the St. John R.
HABITAT: Sandy and alluvial open areas.
GENERAL RANGE: B.C. to N.B. south to Mexico, Tex. and Mo.
SPECIAL NOTE: Probably introduced in eastern Canada from further west.
SYNONYMS: **Euphorbia serpyllifolia** Pers.

Chamaesyce vermiculata (Raf.) House Fig. p. A78. Map p. B59.
FREQUENCY: Locally common on the St. John R. drainage system.
HABITAT: Dry sandy and gravelly shores, roadways and waste areas.
GENERAL RANGE: Ont. to N.S. south to Wisc., Pa. and N.J.; introd. westward to B.C., Ariz. and N. Mex.
SYNONYMS: **Euphorbia vermiculata** Raf.

++

EUPHORBIA L. Spurge

1a Primary rays of umbel 3; seeds pitted on outside face, furrowed on inside; leaves ovate to obovate. **E. peplus**

1b Primary rays of umbel 5 or more; seeds smooth or reticulate. 2

2a(1b) Primary rays of umbel 5; leaves serrulate; capsules smooth; seeds conspicuously reticulate. **E. helioscopia**

2b(1b) Primary rays of umbel usually more than 5; leaves entire; seeds smooth. 3

3a(2b) Leaves very numerous, +/- linear, 1-3 mm wide, 1-2 cm long; stems densely tufted from horizontal rhizomes. **E. cyparissias**

3b(2b) Leaves less crowded, 4-12 mm wide, 3-7 cm long; stem scattered from deep roots. **E. esula**

Euphorbia cyparissias L. Fig. p. A78. Map p. B59.
COMMON NAME(S): Cypress Spurge. Rhubarbe des pauvres, Euphorbe cyprès.
SPECIAL NOTE: Eurasian; formerly cultivated and mostly introduced near old habitations, roadsides and waste areas in the south of the Province. Known elsewhere in Canada from all provinces except Alta.

Euphorbia escula L. Fig. p. A79.
COMMON NAME(S): Wolf's-milk. Leafy Spurge. Embranchée.
SPECIAL NOTE: Eurasian; formerly cultivated and locally escaped to fields and roadsides as near Woodstock, Carleton Co., and St. Andrews, Charlotte Co. Known elsewhere in Canada from B.C. to N.S.

Euphorbia helioscopia L. Fig. p. A79.
COMMON NAME(S): Wartweed. Revéille-matin.
SPECIAL NOTE: Eurasian; formerly cultivated and locally escaped as in Fredericton, York Co. Known elsewhere in Canada from B.C. to Alta. and from Ont. to N.S.

Euphorbia peplus L.
COMMON NAME(S): Petty Spurge.
SPECIAL NOTE: Eurasian; rarely cultivated and occasionally escaped or persisting as near Richibucto, Kent Co. Known elsewhere in Canada from B.C.; Sask. to all eastern provinces.

MERCURIALIS L. Mercury

Mercurialis annua L. Fig. p. A79.
COMMON NAME(S): Boys-and-Girls.
SPECIAL NOTE: European; probably introduced in ship ballast in Newcastle, Northumberland Co. and in Buctouche, Kent Co. Probably not persisting. Known elsewhere in Canada from Ont. to N.S.

ANACARDIACEAE (Cashew Family)

1a Leaflets 3, ovate to rhombic or elliptic; fruit smooth, grayish white; straggling or low climbing shrubs. **Toxicodendron**

1b Leaflets many, lanceolate to narrowly oblong; fruit red hispid; tall shrub or small tree. **Rhus typhina**

RHUS L. Sumac

Rhus typhina L. Fig. p. A78. Map p. B60.
COMMON NAME(S): Staghorn-Sumac, Velvet Sumac. Vinaigrier.
FREQUENCY: Common in the southwest half of the Province and rarer north and eastwards.
HABITAT: Dry edges of woods and rocky slopes.
GENERAL RANGE: Ont. to N.S. south to Minn., Ky., and N.C.

TOXICODENDRON P. Mill. Poison Ivy, Herbe à la puce

1a Stems mostly trailing, without aerial roots; leaflets conspicuously toothed, broadly ovate to subrotund, acute, midrib pubescence not tufted. **T. rydbergii**

1b Stems often climbing by aerial rootlets; leaflets +/- entire, the terminal one ovate-acuminate, midrib pubescence tufted. **T. radicans**

Toxicodendron radicans (L.) Kuntze Map p. B60.
FREQUENCY: Uncommon; populations concentrated in the lower St. John R. drainage system.
HABITAT: Swampy woods or thickets.
GENERAL RANGE: Ont. to NS. south to Ky., and Fla.
SYNONYMS: **Rhus radicans** L.

Toxicodendron rydbergii (Small) Greene Fig. p. A79. Map p. B60.
FREQUENCY: Common in the St. John R. drainage system and scattered elsewhere throughout the Province.
HABITAT: Rocky or sandy woods and shores.
GENERAL RANGE: B.C. to N.S. south to N. Mex., Ariz., Kans. and Va.
SYNONYMS: **Rhus radicans** L. var. **rydbergii** (Small) Rehd.

AQUIFOLIACEAE (Holly Family)

1a Leaves membranous, finely mucronate, +/- entire, usually darkening when dried; petals and stamens free; flowers usually solitary; fruiting pedicels longer than 1 cm. **Nemopanthus mucronata**

1b Leaves coriaceous, acute but not finely mucronate, regularly fine toothed, not darkening when dried; petals slightly united at base; stamens adnate to corolla tube; flowers in small clusters; fruiting pedicels shorter than 1 cm. **Ilex verticillata**

ILEX L. Holly. Houx

Ilex verticillata (L.) Gray Fig. p. A79. Map p. B60.
COMMON NAME(S): Black alder, Winterberry. Apalanche.
FREQUENCY: Common in the south and eastern parts of the Province, rare elsewhere.
HABITAT: Swampy or boggy open woods and thickets.
GENERAL RANGE: Ont. to Nfld. south to Minn., Tenn. and Ga.

NEMOPANTHUS Raf.

Nemopanthus mucronata (L.) Trel. Fig. p. A79.
COMMON NAME(S): Mountain-Holly. Catberry. Faux Houx.
FREQUENCY: Common throughout much of the Province.
HABITAT: Damp acid woods, bogs, thickets and shores.
GENERAL RANGE: Ont. to Nfld. south to Minn., Ohio and Va.

CELASTRACEAE (Staff-tree Family)

1a Leaves alternate; vines. **Celastrus scandens**
1b Leaves opposite; shrubs. **Euonymus**

CELASTRUS L. Staff-tree, Shrubby Bittersweet.

Celastrus scandens L. Fig. p. A79. Map p. B60.
COMMON NAME(S): Climbing Bittersweet. Beurreau des arbres.
FREQUENCY: Formerly inhabiting several islands in the St. John R. in Carleton and York Counties and extirpated due to flooding by the Mactaquac Hydroelectric Development.
HABITAT: Thickets, streambanks and woods.
GENERAL RANGE: Sask. to N.B. (formerly) south to Okla., Ala. and Ga.
SPECIAL NOTE: For more information see H. Hinds, 1983. The Rare Vascular Plants of New Brunswick. Syllogeus 50.

EUONYMUS L. Spindle-tree

1a Leaves linear to linear-oblong, +/- entire and revolute margined; +/- whorled; habit procumbent to spreading, to about 1 m high. **E. nanus**
1b Leaves ovate to oblong-lanceolate, crenately serrate margined, opposite; shrub or small tree to about 5 m high. **E. europaeus**

Euonymus europaeus L.
COMMON NAME(S): Spindle-tree.
SPECIAL NOTE: Eurasian; cultivated and locally escaped to hedgerows and open fields in St. Andrews, Charlotte Co. Known elsewhere in Canada from Ont. and Que.

Euonymus nanus Bieb.
SPECIAL NOTE: Asiatic; cultivated and locally escaped to woods and thickets in Lincoln, York Co. Not reported elsewhere in Canada.

ACERACEAE (Maple Family)

ACER L. Maple. Erable

1a Leaves pinnately compound; wings of fruit +/- parallel; flowers apetalous, the staminate in sessile clusters, the pistillate in drooping racemes. **A. negundo**

1b Leaves simple. 2

2a(1b) Leaf sinuses rounded. 3

2b(1b) Leaf sinuses pointed, +/- sharply angled. 4

3a(2a) Leaves with lobes ending in very fine acute or hair-like points; petals conspicuous; fruit wings +/- horizontally divergent; inflorescence peduncled; pedicels glabrous, erect-ascending; sap of petiole milky. **A. platanoides**

3b(2a) Leaves with lobes acute, rounded or ending in blunt narrowed points; petals none; wings of fruit divergent at 120° or less; inflorescence sessile; pedicels pubescent, drooping; petiole sap not milky. **A. saccharum**

4a(2b) Leaves green above and silvery or bluish-green beneath. 5

4b(2b) Leaves green above and beneath. 6

5a(4a) Terminal lobe of leaves at least 2/3 the length of the blade; branchlets first ascending then drooping or arching, unpleasant smelling when broken; flowers appearing much before the leaves; fruit maturing in spring. **A. saccharinum**

5b(4a) Terminal lobe of leaves less than 2/3 the length of the blade; branchlets upright or spreading, not unpleasant when broken. 7

6a(4b) Branchlets green with white stripes; leaves mostly more than 10 cm long, finely doubly-serrate with 7-12 teeth per cm; inflorescence racemose, drooping; petals conspicuous. **A. pensylvanicum**

6b(4b) Branchlets not striped; leaves mostly less than 10 cm long (longer on vigorous shoots), coarsely and +/- simply serrate with 2-3 teeth per cm; inflorescence a slender erect panicle; petals inconspicuous. **A. spicatum**

7a(5b) Leaves heavy, prominently ribbed, glabrous, lobes ovate, crenately or bluntly serrate; flowers and fruit in drooping racemes 6-15 cm long. **A. pseudoplatanus**

7b(5b) Leaves thinner, not prominently ribbed, +/- pubescent beneath at least when young, lobes triangular in general outline, +/- sharply serrate; flowers and fruit in dense clusters mostly less than 6 cm long. **A. rubrum**

ACERACEAE

++

Acer ginnala Maxim.
COMMON NAME(S): Amur Maple.
SPECIAL NOTE: Asiatic; cultivated and very locally escaped to hedgerows and open woods in Lincoln, Sunbury Co. Known elsewhere in Canada from Ottawa, Ont. Not included in key

Acer negundo L. Fig. p. A79. Map p. B60.
COMMON NAME(S): Box-Elder, Manitoba Maple. Erable à Giguere.
SPECIAL NOTE: Native to U.S.A. and questionably north to Man. and Alta. to Ont.; introduced across much of southern Canada.
SUBSPECIFIC TAXA: In var. **violaceum** (Kirsch.) Jaeg. the twigs are purplish beneath a glaucous bloom while in var. **negundo** the twigs are green. Both varieties are scattered throughout eastern and southwestern parts of the Province.

Acer pensylvanicum L. Fig. p. A79.
COMMON NAME(S): Striped Maple, Moosewood. Bois d'original, Bois barré.
FREQUENCY: Common throughout much of the Province.
HABITAT: Cool moist woods.
GENERAL RANGE: Ont. to N.S. south to Mich., Tenn. and Ga.

Acer platanoides L. Map p. B60.
COMMON NAME(S): Norway Maple. Erable de Norvège.
SPECIAL NOTE: European; cultivated and commonly escaped in or near towns and villages mostly in the southern part of the Province. Known elsewhere in Canada from Ont. to Nfld., N.B. and N.S.

Acer pseudoplatanus L.
COMMON NAME(S): Sycamore Maple.
SPECIAL NOTE: European; cultivated and escaped in Westcock, Westmorland Co. Known elsewhere in Canada from N.S.

Acer rubrum L. Fig. p. A79.
COMMON NAME(S): Red, Scarlet, Soft, or Swamp Maple. Plaine, Plaine rouge.
FREQUENCY: Common throughout much of the Province.
HABITAT: Swamps, alluvial lowlands and moist uplands.
GENERAL RANGE: Ont. to Nfld. south to Tex. and Fla.
SUBSPECIFIC TAXA: Var. **trilobum** Koch. with fertile branch leaves 3-lobed and cuneate or rounded at base rather than 5-lobed and subcordate at the base as in var. **rubrum**, occurs scattered throughout the Province.

Acer saccharinum L. Fig. p. A79. Map p. B60.
COMMON NAME(S): Silver, White, Soft, or River Maple. Plaine blanche.
FREQUENCY: Common along the St. John R. and rare elsewhere.
HABITAT: Alluvial bottomlands.
GENERAL RANGE: Ont. to N.B. south to S. Dak., Tenn. and Fla. Extensively planted and occasionally escaped elsewhere.

Acer saccharum Marsh. Fig. p. A79.
COMMON NAME(S): Sugar or Rock Maple. Erable à sucre.
FREQUENCY: Rich, often rocky or upland woods.
GENERAL RANGE: Ont. to N.S. south to Tex., Ark. and Ga.

Acer spicatum Lam. Fig. p. A79.
COMMON NAME(S): Mountain Maple. Plaine bâtard, Fouéreux.
FREQUENCY: Common throughout much of the Province.
HABITAT: Cool rocky woods and bottomlands.
GENERAL RANGE: Sask. to Nfld. south to Iowa, Tenn. and Ga.

HIPPOCASTANACEAE (Buckeye Family)

AESCULUS L. Horse-Chestnut, Buckeye. Marronnier

Aesculus hippocastanum L. Fig. p. A79. Map p. B61.
COMMON NAME(S): Horse-Chestnut. Marronnier d'Inde.
SPECIAL NOTE: European; cultivated and occasionally escaped as in St. Andrews, Charlotte Co. Known elsewhere in Canada from Ont. to N.S.

BALSAMINACEAE (Touch-me-not Family)

IMPATIENS L. Touch-me-not. Balsam, Jewelweed. Impatiente

1a Leaves +/- opposite or whorled; flowers pinkish-purple to 4 cm long in cymes or umbel-like clusters; stem to 2 m high. **I. glandulifera**
1b Leaves alternate; flowers yellow to orange; stems to about 1.5 m high. 2
2a(1b) Leaves finely sharp toothed; flowers lemon yellow in numerous erect racemes; spur not curved. **I. parviflora**
2b(1b) Leaves coarsely blunt-toothed; flowers pale yellow or orange; spur curved. 3
3a(2b) Spur bent at right angles to pale yellow saccate sepal, about 4-5(8) mm long. **I. pallida**
3b(2b) Spur strongly recurved parallel to orange saccate sepal, about 7-10 mm long. **I. capensis**

Impatiens capensis Meerb. Fig. p. A79.
COMMON NAME(S): Spotted Touch-me-not, Snapweed. Chou sauvage.
FREQUENCY: Scattered throughout much of the Province.
HABITAT: Seepage areas and swampy woods.
GENERAL RANGE: Dist. Mack. and B.C. to Nfld. south to Minn., Okla., Ala. and Fla.
SUBSPECIFIC TAXA: Various colour forms have been named. Those reported from the Province include f. **citrina** (Weath.) Fern. & Schub., with flowers lemon-yellow with crimson spots and f. **immaculata** (Weath.) Fern. & Schub., with flowers +/- unspotted.

++

Impatiens glandulifera Royle Fig. p. A80.
COMMON NAME(S): Himalayan Balsam.
SPECIAL NOTE: Asiatic; probably introduced in ballast waste and locally common along the Bay of Fundy in Charlotte and St. John Counties. Known elsewhere in Canada from B.C., Ont., Que. and N.S.

Impatiens pallida Nutt. Fig. p. A80.
COMMON NAME(S): Pale Touch-me-not.
FREQUENCY: Very rare; known only from Upper Woodstock, Carleton Co., Salmon R., Victoria Co., and the lower Restigouche R., Restigouche Co.
HABITAT: Rich calcareous woods and meadows.
GENERAL RANGE: Ont. to Nfld. south to Kans., Tenn. and Ga.
SPECIAL NOTE: For more information see H. Hinds, 1983. The Rare Vascular Plants of New Brunswick. Syllogeus 50.

Impatiens parviflora DC.
SPECIAL NOTE: Asiatic; introduced and spreading in vacant lots and similar shaded areas in Fredericton, York Co. (UNB). Known elsewhere in Canada from Que., P.E.I. and N.S.

RHAMNACEAE (Buckthorn Family)

RHAMNUS L. Buckthorn. Nerprun

1a Leaves +/- entire, twigs pubescent; shrub or small tree to 7 m tall. **R. frangula**

1b Leaves serrate; twigs glabrous at maturity. 2

2a(1b) Branches spine-tipped; leaves with 2-3 pairs of lateral veins, subopposite, clustered near ends of shoots; twigs rugose from leaf scars; shrub or small tree to 6 m tall. **R. cathartica**

2b(1b) Branches not spine-tipped; leaves with up to 7 pairs of lateral veins, alternate; twigs not rugose; shrub to 1 m tall. **R. alnifolia**

Rhamnus alnifolia L'Hér. Fig. p. A80. Map p. B61.
COMMON NAME(S): Alder-leaved Buckthorn.
FREQUENCY: Scattered throughout the Province except rare in the eastern lowlands.
HABITAT: Swamps and bog margins especially in calcareous areas.
GENERAL RANGE: B.C. to Nfld. south to Calif., Nebr. and N.J.

Rhamnus cathartica L. Fig. p. A80.
COMMON NAME(S): Common Buckthorn. Nerprun.
SPECIAL NOTE: Eurasian; occasionally introduced in hedgerows and open woods in the south of the Province. Known elsewhere in Canada from Sask. to N.S.

Rhamnus frangula L. Fig. p. A80. Map p. B61.
COMMON NAME(S): Alder-Buckthorn. Nerprun bourdaine.
SPECIAL NOTE: Eurasian; locally introduced and aggressively spreading to open woods as in the Fredericton area, York Co. Known elsewhere in Canada from Man. to N.S.

++

VITACEAE (Grape Family)

1a Leaves simple; petals partly fused. **Vitis**

1b Leaves palmately 5-foliate; petals separate. **Parthenocissus**

PARTHENOCISSUS Planch. Virginia Creeper, Woodbine. Vigne-Vierge

1a Plants often high climbing by means of tendrils with adhesive disks; cymes with definite central axis; leaves dull above. **P. quinquefolia**

1b Tendrils rarely with adhesive disks; plants usually not high climbing; cymes dichotomously branched, the central axis poorly defined; leaves glossy above. **P. inserta**

Parthenocissus inserta (Kerner) Fritsch Map p. B61.
SPECIAL NOTE: Native from Man. to Que. south to Calif., N. Mex., Kans. and N. Eng.; sparingly introduced in N.B., P.E.I. and N.S.

Parthenocissus quinquefolia (L.) Planch. Fig. p. A80. Map p. B61.
SPECIAL NOTE: Native in woods and thickets from Ont. and Que. south to Mex., Tex. and Fla.; cultivated and escaped in scattered localities throughout N.B., Nfld., P.E.I. and N.S.

VITIS L. Grape. Raisin or Vigne

1a Leaves densely rusty-tomentose beneath; berries to 2 cm thick. **V. labrusca**

1b Leaves +/- glabrous beneath; berries usually less than 1 cm thick. **V. riparia**

Vitis labrusca L.
COMMON NAME(S): Fox-Grape.
SPECIAL NOTE: Native to eastern U.S.A.; rarely escaped as in Milltown, Charlotte Co. Known elsewhere in Canada from Ont. Possibly a hardy cultivar derived from **V. labrusca**.

Vitis riparia Michx. Fig. p. A80. Map p. B61.
COMMON NAME(S): Riverbank or Frost Grape. Vigne or Raisin sauvage.
FREQUENCY: Scattered along the St. John R. system and rare on the Restigouche and Miramichi Rivers.
HABITAT: Alluvial thickets and banks.
GENERAL RANGE: Man. to N.S. south to N.Mex., Tex., Tenn. and Va.

TILIACEAE (Linden Family)

TILIA L. Linden, Basswood. Tilleul

1a Lower leaf surface mostly glabrous; sterile stamens (staminodia) in front of each petal. **T. americana**

1b Lower leaf surface pubescent; staminodia absent. 2

2a(1b) Lower leaf surface sparingly pubescent; cymes about 3-flowered. **T. platyphyllos**

2b(1b) Lower leaf surface glabrous except for hairy tufts in axils of main veins; cymes up to 15-flowered. 3

3a(2b) Flowering cymes pendant; tertiary leaf veins prominent beneath; hairs on underside of leaf whitish. **T. europaea**

3b(2b) Flowering cymes ascending; tertiary veins obscure; hairs on underside of leaf reddish-brown. **T. cordata**

Tilia americana L. Fig. p. A80. Map p. B61.
COMMON NAME(S): Basswood, Whitewood. Bois blanc.
FREQUENCY: Scattered in the southwest part of the Province sparingly north along the Southwest Miramichi R., Northumberland Co.
HABITAT: Rich deciduous woods.
GENERAL RANGE: Man. to N.B. south to Tex., Ark., Tenn. and Del.

Tilia cordata P. Mill.
COMMON NAME(S): Littleleaf Linden. Tilleul à petites feuilles.
SPECIAL NOTE: European; introduced and locally spreading as in Lincoln, York Co., and near Anagance, Kings Co. Not reported as established elsewhere in Canada.

Tilia europaea L. Map p. B61.
COMMON NAME(S): European Linden.
SPECIAL NOTE: European; planted and locally spreading as in St. Andrews, Charlotte Co. Also reported from Kent, Kings and York Counties Known elsewhere in Canada from Ont. to N.S.
SYNONYMS: **T. X vulgaris** Hayne (**T. cordata** X **T. platyphyllos**)

Tilia platyphyllos Scop. Map p. B61.
COMMON NAME(S): Bigleaf Linden.
SPECIAL NOTE: Eurasian; planted and locally spreading as in Belleisle Bay, Kings Co. Also reported from Gloucester, Northumberland, and Charlotte Co. Known elsewhere in Canada from Ont. and Que.

MALVACEAE (Mallow Family)

1a Carpels loosely united in a ring, separating in fruit; anthers terminal on stamen-column. 2

++

1b Carpels united to form a compound ovary, not separating in fruit; anthers borne along length of stamen-column; calyx subtended by involucre of several bracts. **Hibiscus trionum**

2a(1a) Involucre wanting; leaves unlobed, broadly ovate, cordate at base, entire or shallowly crenate; corolla yellow; fruiting carpels long beaked, 2-9 seeded. **Abutilon theophrasti**

2b(1a) Involucre of 1-3 bractlets; stigma with filiform branches. 3

3a(2b) Involucral bractlets free, usually 3, linear or lanceolate. 4

3b(2b) Involucral bractlets united, 6-9 cleft. 5

4a(3a) Involucral bractlets united near base; fruit central axis longer than the 20 or more carpels forming a narrow cone. **Lavatera thuringiaca**

4b(3a) Involucral bractlets free to base; fruit central axis not surpassing the 10-20 carpels. **Malva**

5a(3b) Flowers usually less than 30 mm in diamter, some on conspicuous pedicels or peduncles; plants velvety; leaves 3-5 lobed; stem less than 1.5 mm tall. **Althaea officinalis**

5b(3b) Flowers at least 30 mm in diameter, subsessile in spikes; plants rough hairy; leaves 5-7 lobed; stems to 3 m tall. **Alcea rosea**

ABUTILON Mill. Indian Mallow

Abutilon theophrasti Medic. Fig. p. A80.
COMMON NAME(S): Velvet-leaf, Butter-Print. Mauve jeune, Mauve des Indes.
SPECIAL NOTE: Asiatic; probably introduced in grain to gardens and waste areas as in Cross Creek, Penniac and Keswick Ridge, York Co. Known elsewhere in Canada from B.C.; Sask. to N.S.

ALCEA L. Hollyhock. Rose trémière, Passerose

Alcea rosea L. Fig. p. A80.
SPECIAL NOTE: Eurasian; cultivated and rarely persisting in old gardens as near Woodstock, Carleton Co. Known elsewhere in Canada from B.C.; Man. to N.S.
SYNONYMS: **Althaea rosea** Cav.

ALTHAEA L. Guimauve

Althaea officinalis L. Fig. p. A80.
COMMON NAME(S): Marshmallow.
SPECIAL NOTE: Eurasian; rarely cultivated and scarcely persisting as near Woodstock, Carleton Co. Known elsewhere in Canada from Ont. and Que.

++

HIBISCUS L. Rose-Mallow. Ketmie

Hibiscus trionum L. Fig. p. A80.
COMMON NAME(S): Flower-of-an-hour. Oeil de faisan.
SPECIAL NOTE: Eurasian; rarely introduced in waste areas as in St. Stephen, Charlotte Co. Known elsewhere in Canada from Sask. to N.S.

LAVATERA L. Tree-Mallow

Lavatera thuringiaca L.
SPECIAL NOTE: European; reported from the Province by Boivin, 1966-67. Known elsewhere in Canada as a garden escape from Sask. to Ont.

MALVA L. Mallow. Mauve

1a Upper leaves deeply cleft or lobed half way or more to centre. 2
1b Upper leaves shallowly lobed. 3
2a(1a) Upper leaves 5-7 parted, the segments deeply pinnatifid; bractlets linear to narrowly lanceolate; stem rough pubescent with simple hairs. **M. moschata**
2b(1a) Upper leaves deeply lobed, the segments shallowly lobed or merely toothed; bractlets oblong to ovate; plant stellate pubescent. **M. alcea**
3a(1b) Stem +/- prostrate or ascending. 4
3b(1b) Stem erect. 5
4a(3a) Petals about twice as long as sepals; fruiting carpels 12-15, rounded or only faintly ridged on back; stem +/- stellate pubescent, branching from rootstock. **M. neglecta**
4b(3a) Petals less than twice as long as sepals; fruiting carpels 8-11, conspicuously reticulate on back; stem slightly pubescent to glabrous, branching above base. **M. rotundifolia**
5a(3b) Fruiting pedicels less than twice as long as calyx; petals rose-purple to 2.5 cm long; calyx lobes obtuse or rounded. **M. sylvestris**
5b(3b) Fruit +/- sessile or pedicel less than twice as long as calyx; petals about 6 mm long, whitish or blue-tinged; calyx lobes acuminate. **M. verticillata**

Malva alcea L. Fig. p. A81.
SPECIAL NOTE: European; cultivated and locally escaped to borders of fields as in Woodstock, Carleton Co. Known elsewhere in Canada from Ont. to N.S.

Malva moschata L. Fig. p. A81. Map p. B61.
COMMON NAME(S): Musk-Mallow. Mauve musquée.
SPECIAL NOTE: European; cultivated and escaped to fields and waste areas throughout much of the Province. Known elsewhere in Canada from B.C.; Man. to N.S.

Malva neglecta Wallr.
COMMON NAME(S): Cheeses. Fromagère.

SPECIAL NOTE: Eurasian; introduced in waste areas and gardens as in Fredericton, York Co. and near Hatfield Pt., Kings Co. Known elsewhere in Canada from B.C. to Nfld., P.E.I. and N.S.

Malva rotundifolia L. Fig. p. A81.
COMMON NAME(S): Cheeses. Fromagère.
SPECIAL NOTE: European; introduced and known only from waste grain, Fredericton, York Co. Known elsewhere in Canada from B.C.; Sask. to P.E.I.

Malva sylvestris L. Fig. p. A81.
COMMON NAME(S): High Mallow.
SPECIAL NOTE: European; occasionally introduced but not persisting as in Fredericton, York Co. Known elsewhere in Canada from B.C., Alta.; Ont. and Que.

Malva verticillata L. Fig. p. A81.
SPECIAL NOTE: Asiatic; rarely introduced in waste areas. Known elsewhere in Canada from Alta. to N.S.

HYPERICACEAE (St. John's-wort Family)
(Alternate names **CLUSIACEAE, GUTTIFERAE**)

1a Petals pale purplish or pinkish; stamens in fascicles of 3 alternating with large glands; leaves often purplish tinged, scarcely reduced in inflorescence. **Triadenum fraseri**

1b Petals yellow or orange; stamens fascicled or not, inter-staminal glands wanting; leaves not purplish tinged, usually much reduced in inflorescence. **Hypericum**

HYPERICUM L. St. John's-wort. Millepertuis

1a Styles +/- divergent, separated to base; stigmas capitate; lateral leaf veins conspicuous or leaves linear. 2

1b Styles united below forming beak on capsule (separating when capsule opens); stigmas minute, not capitate; leaves elliptic, lateral veins obscure; stems 3-5 dm tall from creeping rhizomes. **H. ellipticum**

2a(1a) Stems to 10 dm, often 2-3 arising from subligneus perennial base; stamens very numerous, fascicled; petals black-dotted near margins; inflorescence very diffuse. **H. perforatum**

2b(1a) Stems to about 7 dm (most species much lower), arising from herbaceous annual roots; stamens fewer, not fascicled; petals not black-dotted. 3

3a(2b) Inflorescence strict, +/- congested, often flat-topped, branches relatively short and erect; capsule summit tapered. 4

3b(2b) Inflorescence diffuse, branches relatively long, spreading to ascending; capsule summit rounded. 5

++

4a(3a) Upper leaves 1-3 nerved throughout, tapering from near or above middle to base, linear to narrowly oblanceolate, 1-6 mm broad, rounded at tip; sepals linear-lanceolate, 2-4.5 mm long. 6

4b(3a) Upper leaves (3)5-7 nerved at base, rounded from below middle to base, clasping around stem, lanceolate, 5-15 mm broad, acute at tip; sepals lance-attenuate, 4-7 mm long. **H. majus**

5a(3b) Uppermost flowers with reduced leaves at base; leaves green below. **H. boreale**

5b(3b) Uppermost flowers with scale-like bracts at base; leaves pale below. **H. mutilum**

6a(4a) Leaves 2-6 mm wide, 3-5 veined; mature capsules 3-5 mm long; seeds often undeveloped. **H. dissimulatum**

6b(4a) Leaves 1-4 mm wide, 1-3 veined; mature capsules 5-6.5 mm long. **H. canadense**

Hypericum boreale (Britt.) Bickn. Fig. p. A81. Map p. B61.
FREQUENCY: Scattered throughout much of the Province except in Madawaska and Restigouche Counties.
HABITAT: Shores and shallow water.
GENERAL RANGE: Ont. to Nfld. south to Iowa, Ohio and Va.
SUBSPECIFIC TAXA: Forma **callitrichoides** Fassett, with long, unbranched, sterile stems and rounded leaves occurs submerged or on muddy shores occasionally in the range of the species.
SYNONYMS: **H. mutilum** L. ssp. **boreale** (Britt.) Gillett.

Hypericum canadense L. Fig. p. A81. 62.
FREQUENCY: Common in the southern and central portions of the Province but rare or unknown from western and northern portions.
HABITAT: Shores and meadows.
GENERAL RANGE: Man. to Nfld. south to Iowa, Ala. and Ga.

Hypericum dissimulatum Bickn. Map p. B62.
FREQUENCY: Rare; known from Grand Manan Is., near Waweig, Charlotte Co. and near Sunpoke L., Sunbury Co.
HABITAT: Shores and damp open areas.
GENERAL RANGE: Ont. to N.S. south to eastern N.Y.
SPECIAL NOTE: Probably a hybrid between **H. canadense** and **H. mutilum** or **H. boreale**.

Hypericum ellipticum Hook. Fig. p. A81.
FREQUENCY: Common throughout much of the Province.
HABITAT: Wet open areas.
GENERAL RANGE: Ont. to Nfld. south to N.Dak., Ohio and Md.

Hypericum majus (Gray) Britt. Fig. p. A81. Map p. B62.
FREQUENCY: Uncommonly scattered in the south of the Province and rare in the north.
HABITAT: Damp open areas.
GENERAL RANGE: B.C. to N.S. south to Wash., Nebr. and Del.

+++

Hypericum mutilum L. Fig. p. A81. Map p. B62.
FREQUENCY: Uncommonly scattered in the southern half of the Province.
HABITAT: Damp open areas.
GENERAL RANGE: Man. to N.S. south to Tex. and Fla.

Hypericum perforatum L. Fig. p. A81.
COMMON NAME(S): Common St. John's Wort.
SPECIAL NOTE: Eurasian; commonly introduced on roadsides and waste areas throughout much of the Province. Known elsewhere in Canada from B.C.; Ont. to Nfld., P.E.I. and N.S.

TRIADENUM Raf.

Triadenum fraseri (Spach) Gleason Fig. p. A81. Map p. B62 (as **Hypericum virginicum**).
COMMON NAME(S): Marsh St. John's wort.
FRQUENCY: Common throughout much of the Province.
HABITAT: Marshy or swampy open or partly shaded areas.
GENERAL RANGE: Sask. to Nfld. south to Neb., Mich., and N. Eng.
SYNONYMS: **Hypericum virginicum** L. var. **fraseri** (Spach) Fern.

ELATINACEAE (Waterwort Family)

Elatine L. Waterwort.

1a Carpels usually 2; seeds +/- straight, their pits with rounded ends, not diminishing in size towards each end, longitudinal ribs straight; sepals, petals and stamens 2. **E. minima**
1b Carpels usually 3; seeds mostly curved, their pits with angled ends, diminishing in size towards each end, longitudinal ribs zigzag or obscure; sepals, petals and stamens 3. **E. americana**

Elatine americana (Pursh) Arn. Fig. p. A82.
FREQUENCY: Rare; found on the lower Miramichi R., Northumberland Co., the lower St. John drainage in Sunbury and Kings Co. and in the Jolicure Lakes area, Westmorland Co.
HABITAT: Muddy often tidal influenced shores and shallow water.
GENERAL RANGE: Man. to N.B. south to Okla., Mo., and Va.
SYNONYMS: **E. triandra** Schkuhr var. **americana** (Pursh) Fassett.

Elatine minimum (Nutt.) Fisch. & Mey. Fig. p. A82. Map p. B62.
FREQUENCY: Rare; found on the lower Miramichi R., Northumberland Co., and scattered in the southwestern corner of the Province.
HABITAT: Muddy or sandy shores and shallow water.
GENERAL RANGE: Ont. to Nfld. south to Minn., Mich. and Md.
SPECIAL NOTE: For more information see H. Hinds, 1983. The Rare Vascular Plants of New Brunswick. Syllogeus 50.

CISTACEAE (Rockrose Family)

1a Leaves needle or scale-like, less than .5 mm broad; flowers bright yellow, conspicuous in mass; plants bushy, heather-like; leafy basal offshoots wanting. **Hudsonia tomentosa**

1b Leaves broader and longer, neither scale nor needle-like; flowers inconspicuous in diffuse panicles; leafy basal offshoots present. **Lechea**

HUDSONIA L. Hudsonia. Hudsonie

Hudsonia tomentosa Nutt. Fig. p. A82. Map p. B62.
COMMON NAME(S): Beach-heather.
FREQUENCY: Locally common in coastal areas of Gloucester, Northemberland and Kent Counties and inland on the summit of Big Bald Mtn., Northumberland Co.
HABITAT: Coastal dunes, sand spits, open sandy woods and gravelly sub-alpine summits.
GENERAL RANGE: Alta. to Labr. south to Minn., Ind. and N.C.

LECHEA L. Pinweed

1a Leaves appressed pilose over entire undersurface; inner and outer sepals subequal. **L. maritima**

1b Leaves sparsely pubescent on midrib and margins beneath; outer sepals about one half as long as inner. **L. intermedia**

Lechea intermedia Leggett Fig. p. A82. Map p. B62.
FREQUENCY: Scattered in southcentral and eastern parts of the Province.
HABITAT: Dry sandy open areas.
GENERAL RANGE: Man. to N.S. south to S. Dak., Nebr., N.Y. and Vt.

Lechea maritima Leggett Fig. p. A82.
COMMON NAME(S): Seaside Pinweed.
FREQUENCY: Rare; known from Portage and Fox Islands, Northumberland Co., Kouchibouguac National Park, Buctouche Bar and Southbeach, Richibucto, Kent Co.
HABITAT: Coastal grassy plains and dunes.
GENERAL RANGE: N.B. and Me. south along the coast to Va.
SPECIAL NOTE: For more information see H. Hinds, 1983. The Rare Vascular Plants of New Brunswick. Syllogeus 50.

VIOLACEAE (Violet Family)

VIOLA L. Violet

1a Leaves and flower stalks rising directly from rootstock or from runners. 2

1b Leaves and flower stalks rising from axils of upright stems. 3

2a(1a) Plants purple-blue or violet flowered except albinos, not stoloniferous. 4
2b(1a) Plants normally white flowered. 5
3a(1b) Stipules small, entire or finely fringed; native perennials. 7
3b(1b) Stipules large and leaflike, divided near the base; introduced annuals or short-lived perennials without rhizomes. 8
4a(2a) Lateral petals bearded. 9
4b(2a) Lateral petals not bearded; sinus of leaf narrow, with basal lobes often overlapping; leaves pubescent above, glabrous below. **V. selkirkii**
5a(2b) Leaves lanceolate to elliptic, long tapering to base, crenulate, +/- obtuse at apex. **V. lanceolata**
5b(2b) Leaves kidney-shaped to broadly ovate, cordate. 6
6a(5b) Leaves kidney-shaped, rounded at tip; plants not stoloniferous; lateral petals glabrous. **V. renifolia**
6b(5b) Leaves broadly ovate, more pointed at tip; plants stoloniferous. 10
7a(3a) Flowers yellow or white within, pale violet on back; stipules mostly entire. 11
7b(3a) Flowers violet (white only in albinos); stipules toothed or fringed. 12
8a(3b) Petals 2-3 times as long as sepals, variously whitish, yellowish and violet; flowers large. **V. tricolor**
8b(3b) Petals as large as or slightly larger than sepals, pale yellow and relatively small. **V. arvensis**
9a(4a) Leaves cordate, coarsely serrate along entire margin. 13
9b(4a) Leaves oblong-ovate, usually sharply incised or toothed towards the subcordate or truncate base. **V. fimbriatula**
10a(6b) Blades completely glabrous; petioles mostly hirsutulus; flowers very fragrant; lateral petals slightly bearded or beardless; stolons often bearing flowers and leaves; cleistogamous peduncles erect. **V. macloskeyi**
10b(6b) Blades with some pubescence especially when young; flowers only slightly fragrant or nonfragrant; runners flowerless; cleistogamous flowers on prostrate peduncles; lateral petals heavily bearded. **V. incognita**
11a(7a) Flowers yellow; stipules narrowly ovate, persistent; capsules glabrous. **V. pubescens**
11b(7a) Flowers white within, pale violet on back; stipules slender tipped, white and deciduous after flowering; capsule puberulent. **V. canadensis**
12a(7b) Leaves orbicular-cordate, mostly glabrous and membranous. 14
12b(7b) Leaves ovate-subcordate, less than 3 cm wide, puberulent and subcoreaceous, gradually narrowed to blunt apex. **V. adunca**
13a(9a) Spurred petal glabrous and shorter than laterals; lateral petals with hairs enlarged towards tip; flowers violet with darker veins towards centre, on erect peduncles usually longer than leaves; leaves and sepals glabrous. **V. obliqua**
13b(9a) Spurred petal bearded and as long as laterals; lateral petals with hairs narrowed towards tip. 15
14a(12a) Leaves usually not over 3 cm wide, acuminate and blunt at apex, smooth or with scattered pubescence on upper surface; often sprawling with long internodes; stipules narrowly lanceolate, sub-entire or short toothed only at base; petals mostly dark blue. **V. labradorica**

++

14b(12a) Leaves up to 4 cm wide, more rounded or abrupt blunt tipped; stipules more ovate-lanceolate and bristly serrate most of their length with some teeth up to 2 cm long; flowers pale blue or violet; more upright growing.
V. conspersa

15a(13b) Plants mostly glabrous; leaves broadly ovate; sepals blunt tipped, nonciliate; often densely clumped plants of shores and calcareous bogs.
V. nephrophylla

15b(13b) Plants pubescent, especially the youngest leaves, petioles, and pedicels. 16

16a(15b) Sepals glabrous; leaves narrowly ovate, angular, 1-1.5 times as long as broad; plants of gravelly and sandy shores and rock crevices.
V. novae-angliae

16b(15b) Sepals closely ciliate to tip; leaves ovate to reniform, round-tipped, about as broad as long or broader; plants of moist open woodlands and meadows.
V. septentrionalis

Viola adunca Sm. Fig. p. A82. Map p. B62.
COMMON NAME(S): Early Blue Violet. Violette à éperon crochu.
FREQUENCY: Uncommon; scattered in northern and southern parts of the Province.
HABITAT: Rocky slopes, shores, sand plains and pine woods.
GENERAL RANGE: B.C. to N.B. south to Calif. and N. Eng.
SPECIAL NOTE: This species is part of a confusing complex which includes **V. conspersa** and **V. labradorica**. They are not always clear cut and appear to intergrade in some localities. Populations in the northeast near the coast and on Heron Island, Restigouche and Gloucester Counties are essentially stemless.

Viola arvensis Murr. Fig. p. A82.
COMMON NAME(S): Wild Pansy. Pansée des champs.
SPECIAL NOTE: Eurasian; rarely introduced in fields and similar waste areas as in Woodstock, Carleton Co. Known elsewhere in Canada from all the provinces.

Viola canadensis L. Fig. p. A82.
COMMON NAME(S): Canada-Violet, Tall White Violet.
FREQUENCY: Very rare; known only from Carleton and Victoria Counties.
HABITAT: Rich calcareous woods.
GENERAL RANGE: Ont. to N.S. south to Utah and Pa.
SPECIAL NOTE: For more information see H. Hinds, 1983. The Rare Vascular Plants of New Brunswick. Syllogeus 50.

Viola conspersa Reichenb. Fig. p. A82. Map p. B62.
COMMON NAME(S): American Dog Violet. Violette décombante.
FREQUENCY: Common through much of the Province except rare in the northeast.
HABITAT: Damp mostly deciduous woods.
GENERAL RANGE: Man. to N.S. south to Tenn. and Ga.

Viola fimbriatula Sm. Fig. p. A82.
COMMON NAME(S): Fringed Violet.
FREQUENCY: Very rare; known only from Grand Manan Island, Charlotte Co.

HABITAT: Dry, open ground.
GENERAL RANGE: Ont. to P.E.I. south to Okla. and Fla.
SYNONYMS: **V. sagittata** Ait. var. **ovata** (Nutt.) T. & G.

Viola incognita Brainerd Fig. p. A82.
COMMON NAME(S): Large-leaved White Violet. Violette méconnue.
FREQUENCY: Common throughout much of the Province.
HABITAT: Damp woods, thickets and meadows.
GENERAL RANGE: Ont. to Nfld. south to N.Dak. and Del.

Viola labradorica Schrank Fig. p. A82. Map p. B62.
FREQUENCY: Uncommon; populations are found in Albert, Restigouche and Gloucester Counties.
HABITAT: Cool moist rock or ledge crevices.
GENERAL RANGE: Yukon and B.C. to Nfld. south to Calif. and N. Eng.
SYNONYMS: **V. adunca** Sm. var. **minor** (Hook.) Fern.

Viola lanceolata L. Fig. p. A83. Map p. B62.
COMMON NAME(S): Lance-leaved Violet.
FREQUENCY: Common in the southcentral and southwest of the Province.
HABITAT: Moist shores and thickets.
GENERAL RANGE: Ont. to St-Pierre and Miquelon south to Tex. and Fla.

Viola macloskyi Lloyd Fig. p. A83.
COMMON NAME(S): Northern White Violet.
FREQUENCY: Common throughout much of the Province.
HABITAT: Wet open areas, thickets and woods.
GENERAL RANGE: Dist. Mack. and B.C. to Nfld. south to Calif. and S.C.
SUBSPECIFIC TAXA: Our plants are referable to ssp. **pallens** (Banks ex DC.) Baker. **V. X sublanceolata** House, the hybrid with **V. lanceolata** has been collected at Douglas Harbour, Queens Co.
SYNONYMS: **V. pallens** (Banks ex DC.) Brainerd.

Viola nephrophylla Greene Fig. p. A83. Map p. B63.
FREQUENCY: Uncommon; scattered in southcentral and northern areas of the Province.
HABITAT: Gravelly shores.
GENERAL RANGE: Dist. Mack. and B.C. to Nfld. south to Calif. and N.Eng.

Viola novae-angliae House Fig. p. A83. Map p. B63.
COMMON NAME(S): New England Violet.
FREQUENCY: Uncommon in the western half of the Province.
HABITAT: Ledge crevices, gravelly or rocky shores and wet rocks.
GENERAL RANGE: Ont. to N.B. south to Minn. and Maine.

Viola obliqua Hill Fig. p. A83.
COMMON NAME(S): Blue Marsh Violet.
FREQUENCY: Common throughout much of the Province.
HABITAT: Wet meadows and swamps.

++

GENERAL RANGE: Ont. to Nfld. south to Nebr., Tenn. and Ga.
SPECIAL NOTE: **V. X melissaefolia** Greene, the hybrid with **V. septentrionalis** occurs occasionally in the southern part of the Province (Map p. B63).
SYNONYMS: **V. cucullata** Ait.

Viola pubescens Ait. Fig. p. A83. Map p. B63.
COMMON NAME(S): Smooth Yellow Violet.
FREQUENCY: Common throughout much of the Province except rare in the east and central portions.
HABITAT: Moist deciduous woods.
GENERAL RANGE: Man. to N.S. south to Okla. and N.C.
SUBSPECIFIC TAXA: Our plants are referable to var. **eriocarpa** (Schwein.) Russell.
SYNONYMS: **V. eriocarpa** Schwein., **V. pensylvanica** Michx.

Viola renifolia Gray Fig. p. A83. Map p. b63.
COMMON NAME(S): Kidney-leaved Violet.
FREQUENCY: Scattered throughout much of the Province.
HABITAT: Moist cool woods and rocky slopes.
GENERAL RANGE: Alaska-Yukon and B.C. to Nfld. south to Wash. and Conn.
SUBSPECIFIC TAXA: Var. **brainerdii** (Greene) Fern., with blades and petioles glabrous rather than villous, occurs occasionally with the species.

Viola selkirkii Pursh Fig. p. A83. Map p. B63.
COMMON NAME(S): Great-spurred Violet.
FREQUENCY: Uncommon; mostly scattered in the southern and western parts of the Province.
HABITAT: Rich deciduous woods and rocky slopes.
GENERAL RANGE: Alaska-Yukon and B.C. to Nfld. south to Minn. and N.Eng.

Viola septentrionalis Greene Fig. p. A83. Map p. B63.
COMMON NAME(S): Northern Blue Violet.
FREQUENCY: Common in the southern half of the Province.
HABITAT: Moist clearings and open woods.
GENERAL RANGE: B.C. to Nfld. south to Wash. and Va.

Viola tricolor L.
COMMON NAME(S): Pansy. Pensée.
SPECIAL NOTE: Eurasian; rarely introduced as in Fredericton, York Co., Woodstock, Carleton Co. and Kouchibouguac National Park, Kent Co. Known elsewhere in Canada from B.C. to N.S.

THYMELAEACEAE (Mezereum Family)

1a Calyx yellowish, its lobes indistinct; styles exserted; drupes green, yellowish or red in clusters of 3-4; stem 1-3.3 m high. **Dirca palustris**

1b Calyx pinkish-purple with 4 spreading lobes, styles included; drupes bright red, usually densely clustered; stems 3-9 dm high. **Daphne mezereum**

++

DAPHNE L. Mezereum

Daphne mezereum L. Fig. p. A83.
COMMON NAME(S): Daphne. Bois gentil, Bois joli.
SPECIAL NOTE: Eurasian; rarely introduced in rich woods as in Shediac, Kent Co., Saint John, St. John Co., and Kingshurst, Kings Co. Known elsewhere in Canada from Ont. and all the Atlantic Provinces.

DIRCA L.

Dirca palustris L. Fig. p. A83. Map p. B63.
COMMON NAME(S): Leatherwood, Wicopy. Bois de plomb.
FREQUENCY: Rare; reported from eastern Carleton Co., Keswick Ridge, and Curry Mtn., York Co. and Weldon, Albert Co.
HABITAT: Rich deciduous or mixed woods.
GENERAL RANGE: Minn. and Ont. to N.S. south to La. and Fla.

ELAEAGNACEAE (Oleaster Family)

1a Leaves alternate; flowers pedicelled, the male and bisexual with 4 stamens; plants monoecious. **Elaeagnus angustifolius**
1b Leaves opposite; flowers +/- sessile, the male with 8 stamens; plants dioecious. **Shepherdia canadensis**

ELAEAGNUS L. Oleaster. Chalef

Elaeagnus angustifolia L.
COMMON NAME(S): Russian Olive, Oleaster. Olivier de Bohême.
SPECIAL NOTE: Eurasian; introduced and locally spreading from cultivation in Chaleur Provincial Park, Restigouche Co. Known elsewhere in Canada from B.C., Alta., Man., and Ont.

SHEPHERDIA Nutt.

Shepherdia canadensis (L.) Nutt. Fig. p. A83. Map p. B63.
COMMON NAME(S): Soapberry, Bitter Berries, Soopolalie.
FREQUENCY: Uncommon but locally abundant on the northeast coast, the Restigouche R., Restigouche Co., Victoria Co. on the Tobique and St. John Rivers, and near Albert Mines, Albert Co.
HABITAT: Calcareous ledges, slopes and open woods.
GENERAL RANGE: Alaska and Alta. to Nfld. south to Oreg., N. Mex., Ohio and N. Eng.

++

LYTHRACEAE (Loosestrife Family)

1a Calyx cylindric; petals usually 6; leaves mostly opposite; stems erect, herbaceous; plants of wet meadows, roadside ditches and similar open areas. **Lythrum salicaria**

1b Calyx +/- globose; petals usually 5; leaves usually in whorls of 3 or 4; stems spongy-based, semi-woody, arching and sometimes rooting at tip; plants of shallow water of ponds and streams. **Decodon verticillatus**

DECODON Gmel. Swamp Loosestrife

Decodon verticillatus (L.) Ell.
COMMON NAME(S): Water-willow, Water-oleander.
FREQUENCY: Very rare; known only from Pocowogamis Lake, York Co. (UNB)
HABITAT: Shallow water and boggy pond margins.
GENERAL RANGE: Minn. and Ont. to Que. and N.S. south to La. and Fla.
SUBSPECIFIC TAXA: Our plants appear to be intermediate between var. **laevigatus** T. & G., with smooth pedicels and var. **verticillatus** with leaves soft pubescent beneath.

LYTHRUM L. Loosestrife

Lythrum salicaria L. Fig. p. A84. Map p. B63.
COMMON NAME(S): Spiked or Purple Loosestrife. Salicaire, Bouquet violet.
SPECIAL NOTE: Eurasian; introduced and rapidly spreading to wet roadsides and similar open areas. Known elsewhere in Canada from all provinces except Sask.

ONAGRACEAE (Evening-Primrose Family)

1a Fruit bristly; calyx-lobes and petals 2; leaves opposite, long petioled; stem unbranched below inflorescence. **Circaea**

1b Fruit not bristly; calyx-lobes and petals 4 or if minute or wanting, plants creeping. 2

2a(1b) Plant prostrate, rooting at nodes; calyx-lobes persistent in fruit; petals minute or wanting; leaves opposite. **Ludwigia palustris**

2b(1b) Plant usually erect; calyx-lobes not persistent in fruit; petals conspicuous; leaves opposite and/or alternate. 3

3a(2b) Flowers yellow, calyx-tube prolonged beyond ovary; seeds without long hairs. **Oenothera**

3b(2b) Flowers pinkish or white; calyx-tube slightly if at all prolonged beyond ovary; seeds with tuft of long hairs on one end. **Epilobium**

CIRCAEA L. Enchanter's Nightshade

1a Leaves oblong-ovate, +/- twice as long as wide; fruit with 5 longitudinal ridges; stigma shallowly lobed. **C. lutetiana**

1b Leaves ovate, less than twice as long as wide; fruit not ridged; stigma deeply lobed. 2

2a(1b) Upper internodes much shorter than lower; fruit regularly maturing, 1-celled, 1-seeded; rhizome tuberous-thickened; anthers 0.2-0.3 mm long. **C. alpina**

2b(1b) Upper internodes not conspicuously shorter than lower; fruit rarely maturing, 2-3-celled, 1-seeded; rhizome slender, not tuberous-thickened; anthers 0.5-0.8 mm long. **C. X intermedia**

Circaea alpina L. Fig. p. A84. Map p. B63.
FREQUENCY: Common throughout much of the Province.
HABITAT: Cool, moist woods.
GENERAL RANGE: Alaska and Dist. Mack.; B.C. to Nfld. south to Calif. and Ga.; Eurasia.

Circaea X intermedia Ehrh. Map p. B63.
FREQUENCY: Scattered throughout much of the Province.
HABITAT: Rich often alluvial woods.
GENERAL RANGE: Que. to N.S. south to W. Va. and Va.
SPECIAL NOTE: This is the fertile hybrid between **C. alpina** and **C. lutetiana** ssp. **canadensis**.
SYNONYMS: **C. canadensis** Hill.

Circaea lutetiana (L.) Aschers. & Magnus Fig. p. A84. Map p. B64.
FREQUENCY: Scattered mostly in the southern half of the Province.
HABITAT: Rich, usually deciduous woods and thickets.
GENERAL RANGE: Man. to N.S. south to Okla. and Ga.; Asia.
SUBSPECIFIC TAXA: Our plants are referable to ssp. **canadensis** (L.) Aschers. & Magnus.
SYNONYMS: **C. quadrisulcata** (Maxim) Franch. & Sav.

EPILOBIUM L. Willow-herb

1a Petals showy, 1-2 cm long; stem usually over 1 m tall. 2

1b Petals 3-9 mm long; stem less than 1 m tall. 3

2a(1a) Leaves opposite, serrate; petals purple, notched; stem villous throughout. **E. hirsutum**

2b(1a) Leaves alternate, entire; petals pinkish to magenta or rarely white, entire; stem glabrous. **E. angustifolium**

3a(1b) Basal margins of leaves long decurrent as a ridge on the stem; leaves denticulate. 4

3b(1b) Basal margins of leaves not decurrent; leaves usually entire or wavy-margined and often revolute. 5

4a(3a) Leaves elongate-lanceolate, tapering to base and tip, distinctly petioled, closely and irregularly denticulate; inflorescence +/- intricately branched; seeds beakless, tuft of hairs cinnamon-coloured. **E. coloratum**

4b(3a) Leaves generally broader, oblong, elliptic or ovate, sub-sessile, remotely low dentate; inflorescence usually less intricately branched; seeds +/- beaked below tuft of white hairs or tuft wanting. **E. ciliatum**

5a(3b) Plant villous throughout with short horizontally spreading hair. **E. strictum**

5b(3b) Plant glabrous or puberulent with incurved hairs. 6

6a(5b) Leaves about twice as long as broad, wavy-margined or minutely denticulate; stems often several from creeping basal stolons. **E. hornemannii**

6b(5b) Leaves more than twice as long as broad, often revolute-margined. 7

7a(6b) Most leaves (10-)15-30 times as long as wide, evenly fine hairy above; median leaves usually alternate. **E. leptophyllum**

7b(6b) Most leaves (5-)6-12(-16) times as long as wide, +/- glabrous above; median leaves usually opposite. **E. palustre**

Epilobium angustifolium L. Fig. p. A84.
COMMON NAME(S): Fireweed, Great Willow-herb. Bouguets rouges.
FREQUENCY: Common throughout much of the Province.
HABITAT: Clearings, thickets and open woods, especially in burned over areas.
GENERAL RANGE: Alaska to Baffin Is. and Labr. south to Calif. and N.C.; Greenland; Iceland; Eurasia.
SUBSPECIFIC TAXA: Forma **albiflorum** (Dum.) Haussk. with +/- white petals occurs occasionally with the normal roseate or purplish form.

Epilobium ciliatum Raf.
SUBSPECIFIC TAXA: Two well marked subspecies are found within our range.

1a Seeds with a terminal tuft of white hairs; plants of fresh shores or springy ground. ssp. **glandulosum**

1b Seeds without tuft of hairs; plants of brackish tidal mud flats. ssp. **ciliatum** var. **ecomosum**

1. ssp. **glandulosum** (Lehm.) Hock & Raven
COMMON NAME(S): Glandular Willow-herb.
FREQUENCY: Common throughout much of the Province.
GENERAL RANGE: Alaska and Yukon to Nfld. south to Calif., N. Mex., Ohio and Md.
SYNONYMS: **E. adenocaulon** Hausskn. var. **occidentale** Trel., **E. watsonii** Barbey var. **occidentale** (Trel.) C.L. Hitchc., **E. glandulosum** Lehm.

2. ssp. **ciliatum** var. **ecomosum** (Fassett) Boivin
COMMON NAME(S): Bald Willow-herb.
FREQUENCY: Very rare; known only from the estuary of the Miramichi R., Northumberland Co. (UNB).
GENERAL RANGE: Estuary of the St. Lawrence and Miramichi Rivers.
SYNONYMS: **E. adenocaulon** Hausskn. var. **ecomosum** (Fasset) Munz., **E. ecomosum** (Fassett) Fern.

++

Epilobium coloratum Biehler Fig. p. A84. Map p. B64.
COMMON NAME(S): Purple-veined Willow-herb.
FREQUENCY: Uncommon; reported from Kent, York, Carleton and Victoria Counties.
HABITAT: Open seepy areas.
GENERAL RANGE: S. Dak. and Ont. to N.S. south to Kans. and Ga.

Epilobium hirsutum L. Fig. p. A84.
COMMON NAME(S): Great Hairy Willow-herb. Epilobe hérissé.
SPECIAL NOTE: Eurasian; introduced in Saint John, St. John Co. Known elsewhere in Canada from Ont. to N.S.

Epilobium hornemannii Reichenb. Fig. p. A84. Map p. B64.
COMMON NAME(S): Hornmann's Willow-herb.
FREQUENCY: Uncommon; scattered in the northwest, central highlands and Bay of Fundy areas.
HABITAT: Cool, moist slopes, ledges and shores.
GENERAL RANGE: Northern N. America south to Colo. in the west and N.H. and Me. in the east; Eurasia.
SYNONYMS: **E. alpinum** L. var. **nutans** (Hornem.) Hook.

(**Epilobium lactiflorum** Hausskn.)
SPECIAL NOTE: Reports from Falls Brook, St. John Co. (Hinds, 1983) have been reannotated to **E. leptophyllum**.

Epilobium leptophyllum Raf. Fig. p. A84. Map p. B64.
COMMON NAME(S): Narrow-leaved Willow-herb.
FREQUENCY: Scattered throughout much of the Province.
HABITAT: Wet ditches, seepage areas and marshes.
GENERAL RANGE: Alta. to N.S. south to Colo., Ohio, and Va.
SPECIAL NOTE: Sometimes combined with the following species.

Epilobium palustre L. Fig. p. A84. Map p. B64.
COMMON NAME(S): Swamp Willow-herb.
FREQUENCY: Common throughout much of the Province.
HABITAT: Wet ditches and boggy areas.
GENERAL RANGE: Alaska to Labr. south to Wyo., Ont., and Ct.; Eurasia.
SPECIAL NOTE: Of the several species and varieties named in the **E. palustre** complex, e.g., **E. nesophilum** Fern., **E. palustre** vars. **grammadophyllum** Haussk., **labradoricum** Haussk., **oliganthum** (Michx.) Fern., and **lapponicum** Wahl. only **E. leptophyllum** appears to be significantly distinct.

Epilobium strictum Muhl. Fig. p. A84. Map p. b64.
FREQUENCY: Uncommon; scattered mostly in the southeast half of the Province.
HABITAT: Wet meadows, boggy swales and marshes.
GENERAL RANGE: Minn. and Ont. to N.S. south to Ill. and Va.

ONAGRACEAE

+++

LUDWIGIA L. False Loosestrife

Ludwigia palustris (L.) Ell. Fig. p. A85. Map p. B64.
COMMON NAME(S): Water-purslane.
FREQUENCY: Common in the southwest third of the Province and uncommon in the estuary of the Miramichi River, Northumberland Co.
HABITAT: Shallow water and wet shores.
GENERAL RANGE: B.C. to N.S. south to Calif., Mex., and Fla.; S. America; Eurasia; Africa.

OENOTHERA L. Evening-Primrose. Onagre

1a Ovary four-angled; mature capsule winged, tapering at base. 2
1b Ovary and capsule +/- cylindric, not winged nor especially tapering at base. 3
2a(1a) Inflorescence nodding; calyx-tube 4-8 mm long; capsule stipe 2-4 mm long; leaves less than 6 cm long, obtuse; stem to 6 dm tall. **O. perennis**
2b(1a) Inflorescence erect; calyx-tube 5-15 mm long; capsule subsessile; leaves to 1 dm long, acute; stem to 1 m high. **O. fruticosa**
3a(1b) Elongate appendages at tip of bud touching at least at base, distinctly terminal, not subtended by an auricle. **O. biennis**
3b(1b) Elongate appendages at tip of bud separated at base, subterminal, the actual end of calyx-lobes represented by an evident auricle. **O. parviflora**

Oenothera biennis L. Fig. p. A85.
COMMON NAME(S): Evening-Primrose.
FREQUENCY: Uncommon; scattered in the lower St. John R. valley, Grand Lake areas of Queens Co., and on the east coast.
HABITAT: Roadsides, waste areas, sandy and gravelly shores and old fields.
GENERAL RANGE: B.C. to Nfld. south to Wash., Tex. and Fla.
SUBSPECIFIC TAXA: Many varieties and forms have been proposed but because of the extreme variability of this species and the possibility of hybridization, these are not considered in this work.

Oenothera fruticosa L.
FREQUENCY: Very rare; collected near St. Andrews, Charlotte Co.
HABITAT: Old fields, edges of thickets and roadsides.
GENERAL RANGE: Eastern U.S.A.; probably introduced and not persisting. Known elsewhere in Canada from Man., Ont. and N.S.
SUBSPECIFIC TAXA: Our plants are referable to ssp. **glauca** (Michx.) Straleg.
SYNONYMS: **O. tetragona** Roth.

Oenothera parviflora L. Fig. p. A85. Map p. B64.
FREQUENCY: Common throughout much of the Province.
HABITAT: Sandy or gravelly shores, headlands, talus and waste areas.
GENERAL RANGE: B.C. to Nfld. south to Mont., Ill. and N.J.

SUBSPECIFIC TAXA: Var. **oakesiana** (Robbins) Fern. differs from the species in having leaves narrowly lanceolate and the pubescence mostly of minute closely appressed hairs rather than leaves more broadly lanceolate and pubescence at least partly of long spreading hairs. It has been reported from Moncton, Westmorland Co. and Oak Point, Kings Co. SYNONYMS: **O. cruciata** (S. Wats.) Munz var. **stenopetala** (Bickn.) Fern.

Oenothera perennis L. Fig. p. A85. Map p. B64.
COMMON NAME(S): Sundrops.
FREQUENCY: Common throughout much of the Province.
HABITAT: Moist to dry open areas.
GENERAL RANGE: Man. to Nfld. south to Mo. and Ga.

HALORAGACEAE (Water-Milfoil Family)

1a Leaves all alternate, the emersed foliaceous; stamens 3; carpels united into angled indehiscent fruit. **Proserpinaca palustris**

1b Leaves alternate or opposite, the emersed (when present) bract-like; stamens 4 or 8; carpels separated into 4 distinct nutlets. **Myriophyllum**

MYRIOPHYLLUM L. Water-Milfoil. Myriophylle

1a Stems mostly unbranched, from creeping base, short and scape-like, leafless or scale-leaved. **M. tenellum**

1b Stems elongate; submerged leaves pinnately dissected. 2

2a(1b) Leaves usually less than 1 cm long, filiform throughout, with 3-7 pairs of divisions; flowering scape emergent; uppermost staminate flowers and bracts alternate; stems often intricately branched from near the base and reddish-tinged. **M. alterniflorum**

2b(1b) Leaves usually longer than 1 cm long, with 5-13 pairs of divisions; bracts and flowers whorled; plants often long trailing. 3

3a(2b) Flowers and fruit in axils of submerged or emergent leaves. 4

3b(2b) Flowers and fruit on emergent branches. 5

4a(3a) Fruit tuberculate roughened; plants not amphibous; winter buds well developed in fall. **M. farwellii**

4b(3a) Fruit +/- smooth; plants amphibous; winter buds absent. **M. humile**

5a(3b) Bracts of uppermost flowers pectinate to pinnatifid, as long as or longer than flowers or fruit; leaves with 9-12 pairs of divisions; winter buds common along the stem. **M. verticillatum**

5b(3b) Bracts of uppermost flowers entire to serrulate or spinose-dentate. 6

6a(5b) Bracts shorter than or equalling the flowers or fruit, entire to serrulate; stamens usually 8; fruit about 2.5 mm long, not conspicuously beaked nor ridged. **M. exalbescens**

6b(5b) Bracts greatly exceeding flowers or fruit, spinulose-dentate; stamens 4; fruit about 1.5 mm long, conspicuously beaked and dorsally ridged. **M. heterophyllum**

++

Myriophyllum alterniflorum DC. Fig. p. A85.
FREQUENCY: Scattered throughout much of the Province.
HABITAT: Shallow water of lakes and rivers.
GENERAL RANGE: Alaska-Dist. Mack. and B.C. to Nfld. south to Minn. and Conn.; Greenland; Iceland; Europe.

Myriophyllum exalbescens Fern. Fig. p. A85. Map p. B64.
FREQUENCY: Rare; known only from Madawaska and Restigouche Counties.
HABITAT: Mostly alkaline waters.
GENERAL RANGE: Alaska to Nfld. south to Mich., Kans., and W. Va.

Myriophyllum farwellii Morong Fig. p. A85. Map p. B64.
FREQUENCY: Rare; five populations are known in York, Westmorland and Northumberland Counties.
HABITAT: Acid ponds and slow streams.
GENERAL RANGE: B.C.; Ont. to N.S. south to Minn. and N. Eng.

Myriophyllum heterophyllum Michx. Fig. p. A85.
FREQUENCY: Very rare; known from Mill Bk., Queens Co. and mouth of Grant Bk., Hatfield Pt., Kings Co.
HABITAT: Streams and ponds.
GENERAL RANGE: S. Dak. and Ont. to N.B. south to N. Mex. and Fla.
SPECIAL NOTE: For more information see H. Hinds, 1983. The Rare Vascular Plants of New Brunswick. Syllogeus 50.

Myriophyllum humile (Raf.) Morong Fig. p. A85.
FREQUENCY: Very rare; known only from the mouth of Little Digdeguash River on Spednic Lake.
HABITAT: Shallow acid waters and shores.
GENERAL RANGE: Minn. and Ont. to N.S. south to Pa. and Md.

Myriophyllum tenellum Bigel. Fig. p. A85. Map p. B65.
FREQUENCY: Uncommon; widely scattered throughout much of the Province and probably often overlooked.
HABITAT: Shallow water of sandy or muddy bottomed lakes and ponds.
GENERAL RANGE: Minn. and Ont. to Nfld. south to Pa. and Long Is.
SPECIAL NOTE: For more information see H. Hinds, 1983. The Rare Vascular Plants of New Brunswick. Syllogeus 50.

Myriophyllum verticillatum L. Map p. B65.
FREQUENCY: Uncommon; widely scattered throughout central and southern parts of the Province.
HABITAT: Quiet waters of lakes and streams.
GENERAL RANGE: Alaska to Nfld. south to Calif., Tex. and Md.

++

PROSERPINACA L. Mermaid-weed. Proserpinie

Proserpinaca palustris L. Fig. p. A85. Map p. B65.
FREQUENCY: Rare; known only from the extreme southwest corner of the Province in York and Charlotte Counties.
HABITAT: Shallow acid waters and shores.
GENERAL RANGE: Minn. and Ont. to N.S. south to Tex. and Fla.; Mex., C. America; W.I.
SPECIAL NOTE: For more information see H. Hinds, 1983. The Rare Vascular Plants of New Brunswick. Syllogeus 50.

HIPPURIDACEAE (Mare's-tail Family)

HIPPURIS L. Mare's-tail. Hippuride

Hippuris vulgaris L. Fig. p. A85. Map p. b65.
COMMON NAME(S): Queue de cheval.
FREQUENCY: Locally common; scattered throughout much of the Province.
HABITAT: Damp shores and shallow waters.
GENERAL RANGE: Alaska to Nfld. south to Calif., N. Mex. and N.Y.; S. America; Greenland; Iceland; Eurasia.

ARALIACEAE (Ginseng Family)

1a Plants completely herbaceous from a deeply buried tuber; stem less than 1.5 dm tall; umbel solitary, terminal; leaves palmately compound with 3-5 leaflets. **Panax trifolius**

1b Plants +/- woody at base, not tuber bearing; stem usually over 3 dm tall; umbels 3 or more; leaves alternate or basal, twice or thrice compound, the segments pinnate. **ARALIA**

ARALIA L. Sarsaparilla

1a Inflorescence a raceme-like panicle of umbels; leaflets cordate-ovate; stem to 3 m high, much branched. **A. racemosa**

1b Inflorescence a corymb of umbels; leaflets oblong-ovate; stem less than 1 m high. 2

2a(1b) Stem prickly at base, leafy throughout; terminal leaflet long stalked. **A. hispida**

2b(1b) Stem smooth, bearing a single long petioled ternate-pinnate leaf and a shorter naked scape with 3-7 umbels. **A. nudicaulis**

Aralia hispida Vent. Fig. p. A86.
COMMON NAME(S): Bristly Sarsaparilla. Salsepareille.
FREQUENCY: Common throughout much of the Province.
HABITAT: Mostly open woods and clearings.
GENERAL RANGE: Alta. to Nfld. south to Minn. and N.C.

++

Aralia nudicaulis L. Fig. p. A86.
COMMON NAME(S): Wild Sarsaparilla.
FREQUENCY: Common throughout much of the Province.
HABITAT: Mostly deciduous woods and clearings.
GENERAL RANGE: B.C. to Nfld. south to Wash. and Ga.

Aralia racemosa L. Fig. p. A86. Map p. B65.
COMMON NAME(S): Spikenard. Grande Salsepareille, Anis sauvage.
FREQUENCY: Uncommon; scattered throughout much of the Province.
HABITAT: Rich deciduous woods.
GENERAL RANGE: Man. to N.S. south to Kans. and Ga.

PANAX L. Ginseng

Panax trifolius L. Fig. p. A86. Map p. b65.
COMMON NAME(S): Dwarf Ginseng, Groundnut. Petit Ginseng.
FREQUENCY: Uncommon; widely scattered in the southern half of the Province.
HABITAT: Rich deciduous woods.
GENERAL RANGE: Ont. to N.S. south to Nebr. and Ga.

APIACEAE (Parsley Family)
(Alternate name UMBELLIFERAE)
(Note: Mature fruit needed for positive identification)

1a Fruit covered with spreading prickles or bristles. 2
1b Fruit without spreading bristles or prickles (bristles closely appressed in **Osmorhiza**). 3
2a(1a) Leaflets 3-5; flowers greenish. **Sanicula**
2b(1a) Leaflets many, linear; flowers white. 4
3a(1b) Leaves simple. 5
3b(1b) Leaves (at least some) compound. 6
4a(2b) Involucre of large, pinnately dissected bracts. **Daucus carota**
4b(2b) Involucre of small, linear bracts. **Torilis nodosa**
5a(3a) Blades orbicular with crenate margins, long petioled; stem creeping. **Hydrocotyle americana**
5b(3a) Blades ovate, sessile or perfoliate, entire margined; stem erect. **Bupleurum rotundifolium**
6a(3b) Leaves much dissected or decompound, the leaflets not immediately obvious, being deeply lobed or divided into ovate, oblong, linear or filiform divisions less than 2 cm wide. 7
6b(3b) Leaves organized into distinct and separate leaflets of essentially uniform shape, these often more than 2 cm wide. 8
7a(6a) Bulblets developed in leaf axils. **Cicuta bulbifera**

7b(6a) Bulblets not developed. 9

8a(6b) Leaves 3-foliolate or once pinnate. 10

8b(6b) Leaves 2-3 times compound. 11

9a(7b) Bracts at base of umbellets filiform or absent. 12

9b(7b) Bracts linear or wider. 13

10a(8a) Upper leaf sheaths conspicuously dilated to over 1 cm wide. **Heracleum**

10b(8a) Upper leaf sheaths not conspicuously dilated, less than 1 cm wide. 14

11a(8b) Fruit strongly flattened parallel to broad commissure, lateral ribs conspicuously raised; sheath of upper leaves greatly dilated to over 1 cm broad. **Angelica**

11b(8b) Fruit flattened laterally, perpendicular to narrow commissure, or +/- terete, wingless; upper sheaths not conspicuously expanded. 15

12a(9a) Fruit flattened dorsally, parallel to broad commissure, lateral ribs prominently thin winged. **Conioselinum chinense**

12b(9a) Fruit flattened laterally, perpendicular to narrow commissure or nearly terete, lateral ribs not winged. 16

13a(9b) Leaf-sheaths entire; fruit broadly ovoid, barely constricted at commissure, ribs prominent, +/- undulate. **Conium maculatum**

13b(9b) Leaf-sheaths villous-ciliate; fruit linear, constricted at commissure, ribs obsolete. **Anthriscus cerefolium**

14a(10b) Leaves with 3 leaflets. **Cryptotaenia canadensis**

14b(10b) Leaves with 5 or more leaflets. 17

15a(11b) Principal leaf veins directly ending in a sinus and also with a branch to a tooth. **Cicuta**

15b(11b) Principal leaf veins directly ending on a tooth. 18

16a(12b) Fruit flattened laterally, perpendicular to narrow commissure, obtuse at summit. **Carum carvi**

16b(12b) Fruit +/- terete in cross section, acute at summit. **Aethusa cynapium**

17a(14b) Leaflets lanceolate to linear. **Sium suave**

17b(14b) Leaflets oblong to ovate, obovate or subrotund. 19

18a(15b) Leaflets lustrous, +/- sessile, coarsely serrate; ribs of fruit strong and conspicuously elevated. **Ligusticum scothicum**

18b(15b) Leaflets dull green (often variegated in **Aegopodium**), some or all petiolulate; ribs of fruit inconspicuous. 20

19a(17b) Flowers yellow; lateral ribs of fruit conspicuously winged. **Pastinaca sativa**

19b(17b) Flowers white; lateral ribs of fruit conspicuous but not winged. **Pimpinella saxifraga**

20a(18b) Flowers yellow; fruit without stylopodium. **Zizia aurea**

20b(18b) Flowers white; fruit with conspicuous stylopodium. 21

21a(20b) Leaves thrice compound, the ultimate divisions coarsely dentate to lobed; fruit linear-cylindric to narrowly clavate. **Osmorhiza**

21b(20b) Leaves 1-2-ternate, the ultimate divisions sharply serrate; fruit oblong-ovate, flattened laterally. **Aegopodium podagraria**

APIACEAE

++

AEGOPODIUM L.

Aegopodium podagraria L. Fig. p. A86.
COMMON NAME(S): Goutweed. Herbe aux goutteux.
SPECIAL NOTE: Eurasian; cultivated and persisting near former or present habitations. Known elsewhere in Canada from B.C. and Man. to Que. and all Atlantic Provinces. Var. **variegatum** Bailey, with white streaked leaves, is our most common form.

AETHUSA L.

Aethusa cynapium L. Fig. p. A86.
COMMON NAME(S): Fool's-Parsley.
SPECIAL NOTE: Eurasian; rarely introduced in waste areas and cultivated ground as in Fredericton, York Co. Known elsewhere in Canada from B.C.; Ont. to N.S.

ANGELICA L. Angelica. Angelique

1a Involucels to 1 mm wide, persistent, often reddish margined; fruit only slightly flattened with thick corky dorsal ribs scarcely differing from lateral ribs; plant strictly coastal. **A. lucida**

1b Involucels inconspicuous, less than .5 mm wide, deciduous; fruit strongly flattened with broadly winged lateral ribs and low dorsal ribs; plants not strictly coastal. 2

2a(1b) Uppermost leaf sheaths +/- tubular, veins inconspicuous; pedicels finely scabrous; fruit cross-section showing 1-3 oil tubes in each interval between ribs, seed closely attached to outer coat. **A. sylvestris**

2b(1b) Uppermost leaf sheaths inflated, veins conspicuous; pedicels heavily scabrous; fruit cross-section showing 25-30 oil tubes surrounding the loose seed. **A. atropurpurea**

Angelica atropurpurea L. Fig. p. A86. Map p. B65.
COMMON NAME(S): Alexanders.
FREQUENCY: Locally common in the northwest third of the Province.
HABITAT: Wet meadows and thickets.
GENERAL RANGE: Northern Ont. to Labr. south to Ill. and Md.

Angelica lucida L. Fig. p. A86. Map p. B65.
COMMON NAME(S): Sea-coast Angelica. Céloplèvre brillante.
FREQUENCY: Scattered along the west Bay of Fundy coast and the east coast.
HABITAT: Coastal headlands, moist meadows, and thickets.
GENERAL RANGE: Alaska to Labr. south to Calif. and the west coast and N.Y. on the east coast.
SYNONYMS: **Coelopleurum lucidum** (L.) Fern.

++

Angelica sylvestris L.
COMMON NAME(S): Woodland Angelica.
SPECIAL NOTE: Eurasian; introduced and rapidly spreading along roadsides and waterways as in the Fredericton area, York Co. and Saint John, St. John Co. (UNB). Known elsewhere in Canada from N.S.

ANTHRISCUS Bernh. Beak-Chervil

Anthriscus sylvestris (L.) Hoffm. Fig. p. A86.
COMMON NAME(S): Cow-parsley. Persil sauvage, Petrosine.
SPECIAL NOTE: Eurasian; rarely escaped in waste areas and roadsides as in St. Andrews, Charlotte Co. and near Long Point, Queens Co. Known elsewhere in Canada from Ont. to N.S.

BUPLEURUM L.

Bupleurum rotundifolium L. Fig. p. A86.
COMMON NAME(S): Hare's-ear.
SPECIAL NOTE: Eurasian; rarely introduced probably as ballast waste as in Saint John, St. John Co. Known elsewhere in Canada from B.C., Ont., and Que.

CARUM L. Caraway. Anis, Carvi

Carum carvi L. Fig. p. A86.
COMMON NAME(S): Caraway. Anis Canadien.
SPECIAL NOTE: Eurasian; introduced to roadsides and fields throughout the Province. Known elsewhere in Canada from all the provinces.

CICUTA L. Water-Hemlock. Cicutaire

1a Leaflets mostly ovate-lanceolate; leaf axils not bulblet-bearing. **C. maculata**

1b Leaflets linear to narrowly lanceolate; leaf axils bulblet-bearing. **C. bulbifera**

Cicuta bulbifera L. Fig. p. A86. Map p. B65.
COMMON NAME(S): Bulblet-bearing Water-Hemlock.
FREQUENCY: Common throughout much of the Province.
HABITAT: Marshes and swampy thickets.
GENERAL RANGE: Dist. Mack.- B.C. to Labr. south to Oreg. and Va.

++

Cicuta maculata L. Fig. p. A87. Map p. B65.
COMMON NAME(S): Spotted Cowbane, Musquash-root. Carotte à Moreau.
FREQUENCY: Scattered mostly in southern and eastern portions of the Province.
HABITAT: Wet meadows and thickets.
GENERAL RANGE: Alaska-Dist. Mack.-B.C. to N.S. south to Tex. and N.C.

CONIOSELINUM Hoffm.

Conioselinum chinense (L.) BSP. Fig. p. A87. Map p. B65.
COMMON NAME(S): Hemlock-parsley.
FREQUENCY: Widely separate populations occur along the Bay of Fundy and in the northern third of the Province; rare elsewhere.
HABITAT: Cool wet woods and ravines.
GENERAL RANGE: Alaska-B.C. to Nfld. south to Calif. and N.C.; Asia.

CONIUM L.

Conium maculatum L. Fig. p. A87.
COMMON NAME(S): Poison-Hemlock. Cigue.
SPECIAL NOTE: Eurasian; rarely introduced in waste places as in Tracey Mills, Carleton Co. Known elsewhere in Canada from B.C.; Sask. to N.S.

CRYPTOTAENIA DC.

Cryptotaenia canadensis (L.) DC. Fig. p. A87.
COMMON NAME(S): Wild Chervil, Honewort. Cerfeuil sauvage.
FREQUENCY: Very rare; known from near Sussex, Kings Co., Woodstock, Carleton Co., mouth of the Tobique and Salmon Rivers, Victoria Co.
HABITAT: Rich deciduous woods.
GENERAL RANGE: Man. to N.B. south to Tex. and Ga.
SPECIAL NOTE: For more information see H. Hinds, 1983. The Rare Vascular Plants of New Brunswick. Syllogeus 50.

DAUCUS L. Carrot. Carotte

Daucus carota L. Fig. p. A87.
COMMON NAME(S): Wild Carrot, Queen Anne's-lace. Carotte sauvage.
SPECIAL NOTE: Eurasian; uncommonly introduced along roadsides and fields usually in calcareous areas as near Woodstock, Carleton Co. Known elsewhere in Canada from B.C. to Labr., P.E.I. and N.S.

++

HERACLEUM L.

1a Leaves 3-foliolate; style base not extending above deep notch of fruit; plant +/- woolly. **H. lanatum**

1b Leaves once pinnately compound; style base extending above shallow notch of fruit; plant pubescent and +/- scabrous. **H. spondyleum**

Heracleum lanatum Michx. Fig. p. A87. Map p. B65.
COMMON NAME(S): Cow-parsnip. Berce.
FREQUENCY: Common throughout much of the Province.
HABITAT: Wet meadows, thickets and shores.
GENERAL RANGE: Alaska-Yukon to Labr. south to Calif. and Ga.; Asia.

Heracleum spondylium L.
COMMON NAME(S): Hogweed.
SPECIAL NOTE: Eurasian; rarely introduced to roadsides and waste areas as in St. Andrews, Charlotte Co. (UNB). Known elsewhere in Canada from Ont. to Nfld., and N.S.

HYDROCOTYLE L. Water-Pennywort

Hydrocotyle americana L. Fig. p. A87. Map p. B66.
FREQUENCY: Scattered throughout much of the Province except rare in the north.
HABITAT: Moist woods, meadows and thickets.
GENERAL RANGE: Minn. to Ont. and N.S. south to Tenn. and N.C.

LIGUSTICUM L. Lovage

Ligusticum scothicum L. Fig. p. A87. Map p. B66.
COMMON NAME(S): Scotch Lovage. Livêche écossaise, Persil de mer.
FREQUENCY: Common along the Bay of Fundy and east coast.
HABITAT: Coastal headlands, margins of salt marshes and ledges.
GENERAL RANGE: Ont. to Labr. south to N.Y. and Conn.; Greenland; Iceland; Europe.

OSMORHIZA Raf. Sweet Cicely. Osmorhize

Note: Style and its swollen base (stylopodium) here called style complex.

1a Base of pedicels with involucel of narrow bractlets (often falling late in season); style complex +/- straight and parallel. 2

1b Base of pedicels without involucel; style complex less than 1 mm long (usually much smaller) and +/- divergent. 3

2a(1a) Style complex at most 1.5 mm long. **O. claytoni**

2b(1a) Style complex at least 2 mm long. **O. longistylis**

++

3a(1b) Fruit as long as or usually shorter than pedicel, +/- 1.5 cm long (from tip of style complex to bristly base), narrowed portion of apex less than 1 mm long; stem leaflets +/- acute tipped, thin, light green. **O. depauperata**

3b(1b) Fruit usually longer than pedicel, about 2 cm long, narrowed portion of apex more than 1 mm long; stem leaflets +/- acuminate tipped, thicker, dark green. **O. chilensis**

Osmorhiza chilensis H. & A. Fig. p. A87. Map p. B66.
FREQUENCY: Scattered in the eastern Bay of Fundy region, central highlands and northern counties.
HABITAT: Moist to dryish woodlands.
GENERAL RANGE: Alaska-B.C. to Nfld. south to Calif. and Maine; S. America.

Osmorhiza claytoni (Michx.) Clarke Fig. p. A87. Map p. B66.
FREQUENCY: Common throughout much of the Province except uncommon to rare in the eastern lowlands.
HABITAT: Moist woods and clearings.
GENERAL RANGE: Ont. to Nfld. south to S. Dak., Ala. and N.C.

Osmorhiza depauperata Phil. Fig. p. A87. Map p. B66.
FREQUENCY: Uncommon; collected only in the northernmost counties.
HABITAT: Moist woods and clearings.
GENERAL RANGE: Alaska-Dist. Mack. and B.C. to Labr. south to Calif. and Vt.; S. America.
SYNONYMS: **O. obtusa** (Coult. & Rose) Fern.

Osmorhiza longistylis (Torr.) DC. Fig. p. A87. Map p. B66.
FREQUENCY: Uncommonly scattered in the western part of the Province with one outlier in Restigouche County.
HABITAT: Moist woods and clearings.
GENERAL RANGE: Alta. to N.S. south to N.Mex. and Ga.

PASTINACA L.

Pastinaca sativa L. Fig. p. A88.
COMMON NAME(S): Parsnip. Panais.
SPECIAL NOTE: European; introduced and locally common along roadsides and in fields in the southern half of the Province. Known elsewhere in Canada from all provinces.

PIMPINELLA L.

Pimpinella saxifraga L. Fig. p. A88.
COMMON NAME(S): Burnet-Saxifrage.
SPECIAL NOTE: Eurasian; introduced and rapidly spreading along roadsides and waste areas as near Fredericton, York Co. Known elsewhere in Canada from Ont., Nfld., and N.S.

++

SANICULA L. Sanicle, Black Snakeroot. Sanicle

1a Styles at least 1 mm long, conspicuous, divergent and exceeding bristles of the mature fruit. 2

1b Styles less than 1 mm long, inconspicuous and shorter than bristles of the mature fruit; fruit beak +/- longer than bristles; sepals lance-subulate, exceeding the petals; leaves 3-parted, lateral leaflets often deeply lobed. **S. trifoliata**

2a(1a) Sepals lance-subulate, 1-2 mm long; fruit +/- sessile, 4-6 mm long; anthers greenish white. **S. marilandica**

2b(1a) Sepals ovate-lanceolate to ovate, obtuse or subacute, 0.5-1 mm long; fruit stipitate, 2.5-4 mm long; anthers bright yellow. **S. gregaria**

Sanicula gregaria Bickn. Fig. p. A88. Map p. B66.
FREQUENCY: Very rare; known from only a few localities in Carleton and Kings Counties.
HABITAT: Rich woods.
GENERAL RANGE: Ont. to N.S. south to Minn., and Fla.
SPECIAL NOTE: For more information see H. Hinds, 1983. The Rare Vascular Plants of New Brunswick. Syllogeus 50.

Sanicula marilandica L. Fig. p. A88. Map p. B66.
FREQUENCY: Common throughout much of the Province.
HABITAT: Woods, thickets, meadows and shores.
GENERAL RANGE: B.C. to Nfld. south to N. Mex. and Fla.

Sanicula trifoliata Bickn. Fig. p. A88. Map p. B66.
FREQUENCY: Very rare; known only from Carleton and Kings Counties.
HABITAT: Rich hardwoods.
GENERAL RANGE: Ont. to N.B. south to Tenn. and N. Eng.
SPECIAL NOTE: For more information see H. Hinds, 1983. The Rare Vascular Plants of New Brunswick. Syllogeus 50.

SIUM L.

Sium suave Walt. Fig. p. A88. Map p. B66.
COMMON NAME(S): Water-parsnip. Berle.
FREQUENCY: Common throughout much of the Province except rare in the northwest third.
HABITAT: Wet meadows, shores, swamps and shallow water.
GENERAL RANGE: Alaska to Nfld. south to Calif. and Fla.; Asia.

TORILIS Adans.

Torilis nodosa (L.) Gaertn. Fig. p. A88.
COMMON NAME(S): Hedge-parsley.
SPECIAL NOTE: Eurasian; rarely introduced in cultivated ground as in Fredericton, York Co. and St. Stephen, Charlotte Co. Not persisting. Not known elsewhere in Canada.

++

ZIZIA Koch

Zizia aurea (L.) Koch Fig. p. A88. Map p. B66.
COMMON NAME(S): Golden Alexanders. Zizie dorée.
FREQUENCY: Scattered in the St. John R. drainage system and rare elsewhere.
HABITAT: Damp meadows, thickets and shores.
GENERAL RANGE: Sask. to N.B. and N.S. south to Tex. and Ga.

CORNACEAE (Dogwood Family)

CORNUS L. Dogwood, Cornel. Cornouiller

1a	Plant subherbaceous, to about 2 dm high; flowers surrounded by involucre of 4 petal-like white expanded bracts; fruit red. **C. canadensis**
1b	Plants woody shrubs usually more than 2 m high; flowers without involucre of expanded bracts; fruit white or blue. 2
2a(1b)	Leaves and branches alternate; stems yellowish; fruit deep blue. **C. alternifolia**
2b(1b)	Leaves and branches opposite. 3
3a(2b)	Pith of 1 or 2 year branches white; leaves tapering at base. 4
3b(2b)	Pith of 1 or 2 year branches brownish; leaves tapering or rounded at base, often with appressed reddish-brown pubescence beneath. **C. amomum**
4a(3a)	Branches speckled with purple; leaves oval or rotund with 6-8 pair of lateral veins, softly white-downy beneath; fruit blue. **C. rugosa**
4b(3a)	Branches reddish; leaves lanceolate to ovate with 3-4 pairs of lateral veins, minutely appressed pubescent beneath. **C. sericea**

Cornus alternifolia L.f. Fig. p. A88. Map p. B66.
COMMON NAME(S): Pagoda-Dogwood, Green Osier.
FREQUENCY: Scattered throughout much of the Province except rare in the northeast third.
HABITAT: Partly shaded dry woods.
GENERAL RANGE: Man. to N.S. south to Mo. and Ga.

Cornus amomum Mill. Fig. p. A88. Map p. B66.
COMMON NAME(S): Silky Dogwood, Red Willow. Cornouiller soyeax.
FREQUENCY: Rare; only known from the St. Croix and Eel Rivers in York and Carleton Counties.
HABITAT: Rocky shores and wet thickets along streams.
GENERAL RANGE: Ont. to N.B. south to N.Dak., Okla., and Ga.

Cornus canadensis L. Fig. p. A88.
COMMON NAME(S): Bunchberry, Dwarf Cornel. Quatre-temps, Rougets.
FREQUENCY: Common throughout much of the Province.
HABITAT: Heaths, bog edges and mixed woods.
GENERAL RANGE: Alaska and Yukon to Labr. south to Calif. and N.J.; Greenland; Asia.

Cornus rugosa Lam. Fig. p. A88. Map p. B67.
COMMON NAME(S): Round-leaved Dogwood. Bois de calumet.
FREQUENCY: Scattered throughout much of the Province.
HABITAT: Rocky slopes and dry woods.
GENERAL RANGE: Man. to N.S. south to Iowa and Va.

Cornus sericea L. Fig. p. A89.
COMMON NAME(S): Red Osier. Hart rouge.
FREQUENCY: Common throughout much of the Province.
HABITAT: Low woods, meadows and shores.
GENERAL RANGE: Alaska to Labr. south to Calif., N. Mex. and N. Eng.
SYNONYMS: **C. stolonifera** Michx.

PYROLACEAE (Wintergreen Family)

1a Plant without chlorophyll; leaves scale-like. 2
1b Plant with normal green leaves. 3
2a(1a) Petals separate; fruit erect; plant blackening when dried. **Monotropa**
2b(1a) Petals united, the flower urn-shaped; fruit nodding; plant not blackening when dried. **Pterospora andromedea**
3a(1b) Leaves congested near summit of stem, lanceolate; inflorescence corymbose to subumbellate. **Chimophila umbellata**
3b(1b) Leaves basal, oval to elliptic or subrotund; inflorescence racemose or flowers solitary. 4
4a(3b) Flowers solitary; petals widely spreading. **Moneses uniflora**
4b(3b) Flowers racemose; petals slightly incurved. 5
5a(4b) Raceme one-sided; petals greenish yellow; style long-exsert, straight. **Orthilia secunda**
5b(4b) Raceme spiral; petals whitish to pink; style short-straight to long-recurved. **Pyrola**

CHIMAPHILA Pursh Prince's-Pine, Pipsissewa

Chimaphila umbellata (L.) Spreng. Fig. p. A89. Map p. B67.
COMMON NAME(S): Prince's-Pine, Pipsissewa. Herbe à clé.
FREQUENCY: Uncommon; scattered in the central and northeastern portions of the Province.
HABITAT: Mostly coniferous or mixed forests.
GENERAL RANGE: Man. to Nfld. south to Minn. and Ga.
SUBSPECIFIC TAXA: Our plants are referable to ssp. **cisatlantica** (Blake) Hulten.

MONESES Salisb.

Moneses uniflora (L.) Gray. Fig. p. A89. Map p. B67.
COMMON NAME(S): One-flowered Pyrola.

++

FREQUENCY: Scattered throughout much of the Province.
HABITAT: Cool, mossy, mostly coniferous woods.
GENERAL RANGE: Alaska to Labr. south to Calif. and N. Eng.; Eurasia.

MONOTROPA L.

1a Flower solitary; plant white or pinkish, +/- glabrous, blackening. **M. uniflora**

1b Flowers several in bracted raceme; plant yellowish or brownish, usually pubescent. **M. hypopithys**

Monotropa hypopithys L. Fig. p. A89. Map p. B67.
COMMON NAME(S): Pinesap, False Beech-drops. Monotrope de Pin, Sucepin.
FREQUENCY: Uncommon; scattered in the southern half of the Province and rare northwards.
HABITAT: Coniferous forests.
GENERAL RANGE: Alaska and B.C. to Nfld. south to Calif., Mex., and Fla.; Eurasia.
SYNONYMS: **Hypopithys monotropa** Crantz.

Monotropa uniflora L. Fig. p. A89. Map p. B67.
COMMON NAME(S): Indian-pipe, Convulsion-root. Ghost-flower.
FREQUENCY: Common throughout much of the Province.
HABITAT: Shaded mostly deciduous woods.
GENERAL RANGE: Alaska and B.C. to Labr. south to Calif., Mex. and Fla.; C. America; Asia.

ORTHILIA Raf.

Orthilia secunda (L.) House Fig. p. A89. Map p. B67.
COMMON NAME(S): One-sided Pyrola or Wintergreen.
FREQUENCY: Common throughout much of the Province but rare along the Bay of Fundy.
HABITAT: Mostly mixed woods.
GENERAL RANGE: Alaska to Labr. south to Calif. and Va.; Greenland; Iceland; Eurasia.
SYNONYMS: **Pyrola secunda** L.
SUBSPECIFIC TAXA: Ssp. **obtusata** (Turcz.) Bocher, with about 10 or less creamy white flowers and leaves +/- rounded at summit instead of as many as 20 greenish yellow flowers and leaves mostly acute at summit, occurs occasionally throughout with the species.

PTEROSPORA Nutt.

Pterospora andromedea Nutt. Fig. p. A89. Map p. B67.
COMMON NAME(S): Pine-drops, Giant Bird's-nest.
FREQUENCY: Very rare; found on Curry Mtn. and Keswick Ridge, York Co. and along the Restigouche R., Restigouche Co.

HABITAT: Mature White Pine forests.
GENERAL RANGE: B.C., Alta., Sask., and Ont. to N.B. south to Calif., Mex. and Vt.
SPECIAL NOTE: For more information see H. Hinds, 1983. The Rare Vascular Plants of New Brunswick. Syllogeus 50.

PYROLA L. Pyrola, Wintergreen

1a Style straight, 0.5-1.3 mm long; stigma broadly peltate, 5-lobed, without a subterminal collar. **P. minor**
1b Style recurved, 4-7 mm long; stigma with subterminal collar. 2
2a(1b) Bracts below the flowers 1-5, ovate-lanceolate, +/- clasping stem; sepals longer than broad. 3
2b(1b) Bracts below the flowers 0-3, narrowly lanceolate, not clasping stem; sepals about as long as wide. 4
3a(2a) Bracts of inflorescence single nerved near acuminate tip; anthers purplish-red; petals pink or crimson-red; leaves rounded or cordate at base. **P. asarifolia**
3b(2a) Bracts of inflorescence 3-nerved near acute tip; anthers yellowish; flowers whitish; leaves usually cuneate at base. **P. americana**
4a(2b) Leaf blade nearly orbicular, usually shorter than petiole; corolla greenish-white. **P. chlorantha**
4b(2b) Leaf blade oblong to elliptic, usually longer than petiole; corolla white or creamy. **P. elliptica**

Pyrola americana Sweet Fig. p. A89. Map p. B67.
COMMON NAME(S): Round-leaved Pyrola. Muguet des Bois.
FREQUENCY: Uncommon; scattered in the southwest and northeast portions of the Province.
HABITAT: Moist woods, bogs and barrens.
GENERAL RANGE: Man. to Nfld. south to Minn. and N.C.; Greenland; Eurasia.
SYNONYMS: **P. rotundifolia** L.

Pyrola asarifolia Michx. Fig. p. A89. Map p. B67.
COMMON NAME(S): Pink Pyrola or Wintergreen.
FREQUENCY: Scattered throughout much of the Province but rare in the south and east portions.
HABITAT: Mostly coniferous woods.
GENERAL RANGE: Alaska to Nfld. south to Oreg., N. Mex. and N. Eng.; Asia.

Pyrola chlorantha Sw. Fig. p. A89. Map p. B67.
FREQUENCY: Uncommon; scattered throughout much of the Province but rare towards the south.
HABITAT: Mostly coniferous woods.
GENERAL RANGE: Alaska to Labr. south to Calif., Ariz. and Pa.; Eurasia.
SYNONYMS: **P. virens** Schweigger.

++

Pyrola elliptica Nutt. Fig. p. A89. Map p. B67.
COMMON NAME(S): Shinleaf.
FREQUENCY: Common throughout much of the Province.
HABITAT: Dry to moist mostly mixed woods.
GENERAL RANGE: B.C. to Nfld. south to Idaho and N. Eng.; Japan.

Pyrola minor L. Fig. p. A89. Map p. B67.
FREQUENCY: Rare; recorded along the Bay of Fundy, central highlands and in the northwest of the Province.
HABITAT: Cool moist woods.
GENERAL RANGE: Alaska to Labr. south to Calif., N. Mex. and N. Eng.; Greenland; Iceland; Eurasia.

ERICACEAE (Heath Family)

1a	Plants with erect stems usually over 1 dm high. 2
1b	Plants with low, weak, +/- horizontal stems usually less than 1 dm high. 3
2a(1a)	Ovary superior; fruit capsular. 4
2b(1a)	Ovary inferior; fruit fleshy. 5
3a(1b)	Leaves tapering at base; fruit fleshy. 6
3b(1b)	Leaves rounded to cordate at base, evergreen; flowers pink or white, very fragrant. **Epigaea repens**
4a(2a)	Leaves rusty-tomentose beneath, margins inrolled; petals white, separate. **Ledum groenlandicum**
4b(2a)	Leaves not as above; petals +/- united. 7
5a(2b)	Leaves resinous-dotted beneath; berry fruit with 10 hard, seeds. **Gaylussacia**
5b(2b)	Leaves not resinous-dotted beneath; berry fruit with numerous small seeds. **Vaccinium**
6a(3b)	Leaves wintergreen flavored; ovary wholly or partly superior; calyx fleshy. **Gaultheria**
6b(3b)	Leaves not wintergreen flavored; ovary inferior; calyx not fleshy. 8
7a(4b)	Leaves brown resinous beneath, minutely denticulate; flowers white in elongate 1-sided racemes. **Chamaedaphne calyculata**
7b(4b)	Leaves not as above; flowers white or pinkish. 9
8a(6b)	Stem +/- prostrate, long trailing; leaves oblanceolate; corolla constricted above. **Arctostaphylos uva-ursi**
8b(6b)	Stem +/- erect; leaves lanceolate; corolla not constricted above. **Vaccinium**
9a(7b)	Flowers appearing before the leaves, corolla lobed nearly to base; leaves minutely white pubescent and also with scattered rufous hairs beneath, margins ciliate. **Rhododendron canadense**
9b(7b)	Flowers appearing after the leaves develop; corolla shallowly lobed; leaves not as above. 10
10a(9b)	Corolla open saucer-shaped; leaves blunt-tipped; anthers without terminal pores. **Kalmia**

++

10b(9b) Corolla nearly globose; leaves apiculate, whitened beneath, margins inrolled; anthers with terminal pores. **Andromeda polifolia**

ANDROMEDA L. Andromeda

Andromeda polifolia L. Fig. p. A90.
COMMON NAME(S): Bog-rosemary. Androméde glauque.
FREQUENCY: Common throughout much of the Province.
HABITAT: Peat bogs and boggy shores.
GENERAL RANGE: Man. to Labr. south to Minn. and N.J.; Greenland.
SUBSPECIFIC TAXA: Our plants are referable to var. **glaucophylla** (Link.) DC.

ARCTOSTAPHYLOS Adans. Bearberry, Manzanita

Arctostaphylos uva-ursi (L.) Spreng. Fig. p. A90. Map p. B68.
COMMON NAME(S): Bearberry, Sandberry, Kinnikinnick. Bousserole, Raisin d'ours.
FREQUENCY: Uncommon; scattered in southcentral and eastern portions of the Province.
HABITAT: Dry ledges and sands.
GENERAL RANGE: Alaska to Labr. south to Calif., N. Mex. and Va.
SUBSPECIFIC TAXA: Ssp. **coactilis** (Fern. & Macbr.) Love & Kapoor, with branchlets white-tomentose, occurs occasionally with the species.

CALLUNA Salisb.

Calluna vulgaris (L.) Hull
COMMON NAME(S): Scotch Heather, Ling.
SPECIAL NOTE: European; reported from Sunbury, Albert and Westmorland Counties but scarcely established. Known elsewhere in Canada from B.C., Que., Nfld., St-Pierre and Miquelon and N.S.

CHAMAEDAPHNE Moench

Chamaedaphne calyculata (L.) Moench Fig. p. A90. Map p. B68.
COMMON NAME(S): Leather-leaf, Cassandra. Faux Bleuets, Petit-daphné.
FREQUENCY: Common throughout much of the Province.
HABITAT: Peat bogs and margins of acid bog ponds.
GENERAL RANGE: Alaska to Nfld. south to B.C., Man., Iowa and Ga.; Eurasia.
SUBSPECIFIC TAXA: Our plants are referable to var. **angustifolia** (Ait.) Rehd.

EPIGAEA L. Trailing Arbutus, Ground Laurel.

Epigaea repens L. Fig. p. A90. Map p. B68.
COMMON NAME(S): Mayflower. Fleur de Mai.

FREQUENCY: Common throughout much of the Province except rare in the western portion.
HABITAT: Cool, moist, peaty woods.
GENERAL RANGE: Man. to Nfld. south to Iowa and Fla.

GAULTHERIA L. Wintergreen

1a Plants erect; leaves 25-35 mm long, often reddish tinged; flowers 5-merous; fruit red. **G. procumbens**

1b Plants trailing; leaves less than 10 cm long; flowers 4-merous; fruit white. **G. hispidula**

Gaultheria hispidula (L.) Muhl. Fig. p. A90. Map p. B68.
COMMON NAME(S): Creeping Snowberry, Moxieplum. Oeufs de perdrix.
FREQUENCY: Scattered throughout much of the Province.
HABITAT: Cool, moist coniferous woods.
GENERAL RANGE: B.C. to Nfld. south to Idaho and N.C.
SYNONYMS: **Chiogenes hispidula** (L.) Torr. & Gray

Gaultheria procumbens L. Fig. p. A90. Map p. B68.
COMMON NAME(S): Checkerberry, Teaberry. Thé des bois.
FREQUENCY: Common in the eastern lowlands and uncommon to rare elsewhere.
HABITAT: Mostly dry mixed or coniferous woods.
GENERAL RANGE: Man. to Nfld. south to Minn. and Ga.

GAYLUSSACIA HBK. Huckleberry.

1a Plants variously pubescent and stipitate-glandular; leaves oval, rounded and conspicuously apiculate at apex, glossy above. **G. dumosa**

1b Plants +/- glabrous except for sessile resin glands; leaves elliptic, abruptly sharp pointed, pale and dull above. **G. baccata**

Gaylussacia baccata (Wang.) Koch Fig. p. A90. Map p. B68.
COMMON NAME(S): Black Huckleberry. Gueules noires.
FREQUENCY: Common in the southern and eastern portions of the Province and rare elsewhere.
HABITAT: Edges of dry woods, sandy shores and clearings.
GENERAL RANGE: Ont. to Nfld. south to La. and Ga.

Gaylussacia dumosa (Andr.) T. & G. Fig. p. A90. Map p. B68.
COMMON NAME(S): Dwarf Huckleberry.
FREQUENCY: Uncommon or overlooked; scattered in the south and east of the Province mostly near the coast.
HABITAT: Sphagnous bogs and wet barrens.
GENERAL RANGE: Nfld. south along the coast to Miss. and Fla.
SUBSPECIFIC TAXA: Our plants are referable to var. **bigeloviana** Fern.

++

KALMIA L. Laurel

1a Leaves flat, glabrous beneath; flowers at base of new growth. **K. angustifolia**

1b Leaves inrolled and white-pubescent beneath; flowers at summit of new growth. **K. polifolia**

Kalmia angustifolia L. Fig. p. A90.
COMMON NAME(S): Lambkill, Sheep-Laurel. Crevard de moutons.
FREQUENCY: Common throughout much of the Province.
HABITAT: Rocky barrens, clearings, and old pastures.
GENERAL RANGE: Ont. to Nfld. south to Mich. and Ga.

Kalmia polifolia Wang. Fig. p. A90.
COMMON NAME(S): Pale or Bog Laurel.
FREQUENCY: Scattered throughout much of the Province.
HABITAT: Peat Bogs and similar open areas.
GENERAL RANGE: Alaska to Labr. south to Oreg. and N.J.

LEDUM L. Labrador-tea. Ledon

Ledum groenlandicum Oeder Fig. p. A90.
COMMON NAME(S): Labrador-tea.
FREQUENCY: Scattered throughout much of the Province.
HABITAT: Peat bogs and peaty lowlands.
GENERAL RANGE: Alaska to Labr. south to Oreg. and N.J.; Greenland.

RHODODENDRON L. Rhododendron, Rosebay. Rhodora

Rhododendron canadense (L.) Torr. Fig. p. A90. Map p. B68.
COMMON NAME(S): Rhodora.
FREQUENCY: Common throughout much of the Province but uncommon to rare in the western and northern portions of the Province.
HABITAT: Sphagnous bogs, peaty lowlands and barrens.
GENERAL RANGE: Que. and Nfld. south to Pa. and N.J.
SUBSPECIFIC TAXA: Forma **albiflora** (Rand & Redf.) Rehd., with white flowers, occurs rarely with the species. (UNB).

VACCINIUM L. Blueberry, Bilberry, Cranberry. Airelle or Atocas

1a Corolla shallowly 5-lobed or toothed; pedicels usually less than 1 cm long. 2

1b Corolla deeply 4-lobed or cleft below the middle, the lobes recurved; pedicels 1-4 cm long (cranberries). 3

++

2a(1a) Corolla +/- constricted at or near summit; leaves deciduous; fruit blue or black. 4

2b(1a) Corolla campanulate, not constricted above; leaves leathery, evergreen; fruit red. **V. vitis-idaea**

3a(1b) Pedicel bracts foliaceous, 2-4 mm long; fruit 10-20 mm thick, red; leaves blunt. **V. macrocarpon**

3b(1b) Pedicel bracts scale-like, reddish, to about 1.5 mm long; fruit 6-10 mm thick, brownish-red, dotted; leaves acute. **V. oxycoccus**

4a(2a) Leaves entire. 5

4b(2a) Leaves variously toothed; flowers 5-merous. 6

5a(4a) Leaves downy; flowers 5-merous. **V. myrtilloides**

5b(4a) Leaves glabrous, leathery with conspicuous veins; flowers 4-merous. **V. uliginosum**

6a(4b) Low shrubs usually less than 5 dm high forming colonies. 7

6b(4b) Tall shrubs to 4 m high with several stems; leaves to 8 cm long, 4 cm broad. **V. corymbosum**

7a(6a) Leaves oblanceolate; corolla pink; anthers long awned at base of tubules. **V. caespitosum**

7b(6a) Leaves lanceolate; corolla usually white; anthers not awned at base of tubules. 8

8a(7b) Leaves averaging 10.6 mm long, 3.6 mm wide; corolla 3-3.5 mm long; rare subalpine. **V. boreale**

8b(7b) Leaves averaging 16.5 mm long, 6 mm wide; corolla 5-7 mm long; common throughout. **V. angustifolium**

Vaccinium angustifolium Ait. Fig. p. A91.
COMMON NAME(S): Low or Late Sweet Blueberry. Bleuets.
FREQUENCY: Common throughout much of the Province.
HABITAT: Acid barrens, bogs and clearings.
GENERAL RANGE: Man. to Labr. south to Iowa and Va.

Vaccinium boreale Hall & Aalders.
FREQUENCY: Very rare; collected only on Big Bald Mtn., Northumberland Co.
HABITAT: Peaty barrens.
GENERAL RANGE: Que. and Labr. south to N.Y. and Me.
SPECIAL NOTE: For more information see H. Hinds, 1983. The Rare Vascular Plants of New Brunswick. Syllogeus 50.

Vaccinium caespitosum Michx. Fig. p. A91. Map p. B68.
COMMON NAME(S): Dwarf Bilberry.
FREQUENCY: Uncommon; widely scattered throughout much of the Province.
HABITAT: Rocky shores, ledge crevices and clearings.
GENERAL RANGE: Alaska to Labr. south to Calif. and N. Eng.

Vaccinium corymbosum L. Fig. p. A91.
COMMON NAME(S): Highbush-Blueberry.
FREQUENCY: Very rare; recorded only on Grand Lake and Spednic Lake, York Co., and Tower Hill, Charlotte Co.

HABITAT: Swamps and thickets near shore.
GENERAL RANGE: Ont. to N.B. south to Wisc. and N.J.

Vaccinium macrocarpon Ait. Fig. p. A91. Map p. B68.
COMMON NAME(S): Large or American Cranberry. Gros Atocas.
FREQUENCY: Common in the south and eastern portion of the Province and rare towards the northwest.
HABITAT: Peat bogs and wet lowlands mostly near the coast.
GENERAL RANGE: Minn. and Ont. to Nfld. south to Ark. and N.C.
SYNONYMS: **Oxycoccus macrocarpus** (Ait.) Pers.

Vaccinium myrtilloides Michx. Fig. p. A91.
COMMON NAME(S): Sour-top or Velvet-leaf Blueberry. Bleuets.
FREQUENCY: Common throughout much of the Province.
HABITAT: Clearings, thickets and peat bogs.
GENERAL RANGE: Dist. Mack. to Labr. south to Mont. and Va.

Vaccinium oxycoccus L. Fig. p. A91. Map p. B68.
COMMON NAME(S): Bog Cranberry.
FREQUENCY: Scattered in the southeastern half of the Province and uncommon elsewhere.
HABITAT: Peat bogs and boggy lowlands.
GENERAL RANGE: Ont. to Labr. south to N.B., P.E.I. and N.S.; Greenland.
SYNONYMS: **Oxycoccus quadripetalus** Gilib. var. **microphyllus** (Lange) Porsild.
SUBSPECIFIC TAXA: Our plants are referable to var. **intermedium** Gray.

Vaccinium uliginosum L. Fig. p. A91.
COMMON NAME(S): Alpine Bilberry, Bog-Blueberry.
FREQUENCY: Very rare; reported only from Bald Mtn., Northumberland Co.
HABITAT: Rocky barrens, alpine summits and peat bogs.
GENERAL RANGE: Alaska to Labr. south to Calif., Mich. and Vt.; Greenland; Iceland; Eurasia.
SUBSPECIFIC TAXA: Our plants are referable to ssp. **pubescens** (Wormsk.) Young.
SPECIAL NOTE: For more information see H. Hinds, 1983. The Rare Vascular Plants of New Brunswick. Syllogeus 50.

Vaccinium vitis-idaea L. Fig. p. A91. Map p. B68.
COMMON NAME(S): Rock-Cranberry, Mountain-Cranberry. Berris, Graines rouges, Pommes de terre.
FREQUENCY: Common along the Bay of Fundy and scattered elsewhere towards the northeast.
HABITAT: Rocky barrens and peaty headlands and clearings.
GENERAL RANGE: Alaska to Nfld. south to N. Eng.; Greenland; Iceland; Eurasia.
SUBSPECIFIC TAXA: Our plants are referable to ssp. **minus** (Lodd.) Hulten.

PRIMULACEAE (Primrose Family)

1a Leaves all basal; plants scapose; inflorescence umbellate. **Primula**

++

1b Leaves cauline as well as sometimes basal, opposite, alternate or whorled; plants not scapose; flowers in racemes or axillary. 2

2a(1b) Leaves opposite or +/- whorled. 3

2b(1b) Leaves alternate; flowers white, 5-merous, in terminal racemes. **Samolus valerandi**

3a(2a) Leaves mostly opposite or in several whorls; flowers 5-6 merous. 4

3b(2a) Leaves lanceolate, in a single terminal whorl (alternate scale leaves on lower stem); flowers +/- 7-merous. **Trientalis borealis**

4a(3a) Flowers yellow. **Lysimachia**

4b(3a) Flowers white, pink or red. 5

5a(4b) Plants fleshy, erect or ascending; flowers about 3 mm wide, calyx petaloid pink or white; perennials. **Glaux maritima**

5b(4b) Plants not fleshy, trailing; flowers 10-12 mm wide, scarlet or brick red; annuals. **Anagallis arvensis**

ANAGALLIS L. Pimpernel

Anagallis arvensis L. Fig. p. A91.
COMMON NAME(S): Scarlet Pimpernel.
SPECIAL NOTE: Eurasian; occasionally cultivated and rarely escaped to sandy open areas. Known elsewhere in Canada from B.C. and Alta.; Ont. to N.S.

GLAUX L.

Glaux maritima L. Fig. p. A91. Map p. B68.
COMMON NAME(S): Sea-Milkwort, Saltwort.
FREQUENCY: Common along the coast and Bay of Fundy.
HABITAT: Coastal salt marshes.
GENERAL RANGE: Alaska to Nfld. south to Calif. on the west coast and to Va. on the east coast.

LYSIMACHIA L. Loosestrife

1a Stem prostrate; leaves +/- orbicular; flowers axillary. **L. nummularia**

1b Stem erect; leaves lanceolate to elliptic or ovate. 2

2a(1b) Flowers large, showy, uniformly yellow. 3

2b(1b) Flowers smaller, corolla dark streaked or dotted. 4

3a(2a) Leaves +/- sessile, pubescent at least on veins beneath. 5

3b(2a) Leaves with long, ciliate-margined petioles, glabrous. **L. ciliata**

4a(2b) Flowers in racemes; leaves opposite. 6

4b(2b) Flowers solitary in leaf axils; leaves whorled. **L. quadrifolia**

5a(3a) Flowers whorled in upper leaf axils; corolla-lobe margins glandular-ciliate; calyx-lobes green throughout, to about 1 cm long. **L. punctata** L.

5b(3a) Flowers in terminal leafy panicles; corolla-lobe margins entire; calyx-lobes dark-margined, to about 5 mm long. **L. vulgaris**

++

6a(4a) Open racemes terminal; pedicles 9-17 mm long; leaf axils often bearing elongate bulblets. **L. terrestris**

6b(4a) Dense racemes from axils of narrowly lanceolate midstem leaves; pedicels to about 3 mm long; plants not bulblet bearing. **L. thyrsiflora** L.

Lysimachia ciliata L. Fig. p. A91. Map p. B69.
COMMON NAME(S): Fringed Loosestrife.
FREQUENCY: Scattered throughout much of the Province but uncommon to rare on the east coast.
HABITAT: Wet thickets, shores and meadows.
GENERAL RANGE: B.C. to N.S. south to Oreg. and Fla.
SYNONYMS: **Steironema ciliatum** (L.) Raf.

Lysimachia nummularia L. Fig. p. A91. Map p. B69.
COMMON NAME(S): Moneywort, Creeping Jenny. Monnayère.
SPECIAL NOTE: European; introduced in moist lawns, thickets and roadsides in scattered localities in the southern half of the Province. Known elsewhere in Canada in B.C. and from Ont. to Nfld., P.E.I. and N.S.

Lysimachia punctata L. Fig. p. A92.
COMMON NAME(S): Garden-Loosestrife.
SPECIAL NOTE: Eurasian; cultivated and occasionally escaped to roadsides and waste areas as in Woodstock, Carleton Co. Known elsewhere in Canada in B.C. and from Ont. to N.S.

Lysimachia quadrifolia L. Fig. p. A92. Map p. B69.
COMMON NAME(S): Whorled Loosestrife.
FREQUENCY: Rare; known only from Grand Manan Island, Charlotte Co. and the mouth of the Nerepis R. and several localities on Belleisle Bay, Kings Co.
HABITAT: Sandy or gravelly shores, and damp woods.
GENERAL RANGE: Ont. to N.B. south to Ill. and Ga.
SPECIAL NOTE: For more information see H. Hinds, 1983. The Rare Vascular Plants of New Brunswick. Syllogeus 50.

Lysimachia terrestris (L.) BSP. Fig. p. A92. Map p. B69.
COMMON NAME(S): Swamp- or Yellow-Loosestrife, Swamp-candles.
FREQUENCY: Common throughout much of the Province except rare towards the northwest.
HABITAT: Wet thickets, bogs and shores.
GENERAL RANGE: Man. to Labr. south to Iowa and Ga.
SPECIAL NOTE: **L. X conmixta** Fern., the hybrid with **L. thyrsiflora** was collected on the shore of Grand Lake, Queens Co. (UNB).

Lysimachia thyrsiflora L. Fig. p. A92. Map p. B69.
COMMON NAME(S): Tufted Loosestrife.
FREQUENCY: Uncommonly scattered over much of the Province.
HABITAT: Swamps, wet thickets and bogs.
GENERAL RANGE: Alaska to Gaspé Pen. south to Calif. and W. Va.; Eurasia.

++

Lysimachia vulgaris L. Fig. p. A92.
COMMON NAME(S): Garden-Loosestrife. Chasse-bosses.
SPECIAL NOTE: Eurasian; cultivated and occasionally escaped to roadsides and waste areas as in Carleton, Kings and Restigouche Counties. Known elsewhere in Canada from Ont. to N.S.

PRIMULA L. Primrose, Cowslip. Primevère

1a Corolla yellow, orange or scarlet; leaves usually over 15 cm long; plant soft hairy. **P. veris**
1b Corolla pale pink or lilac to white, all with yellow centre; leaves less than 10 cm long. 2
2a(1b) Leaves beneath and often calyx heavily white-mealy, 2.5-10 cm long; capsule 9-10 mm long; seeds +/- roughened. **P. farinosa**
2b(1b) Leaves beneath and calyx glabrous or only slightly white-mealy, 1-4 cm long; capsule 5-8 mm long; seeds smooth. **P. mistassinica**

Primula farinosa L. Fig. p. A92.
COMMON NAME(S): Bird's-eye Primrose.
FREQUENCY: Very rare; known only from the vicinity of the mouth of Pt. Wolfe River, Albert Co.
HABITAT: Coastal, usually calcareous ledges.
GENERAL RANGE: Ont. to Labr. south to Maine; Europe.
SYNONYMS: **P. laurentiana** Fern.
SPECIAL NOTE: For more information see H. Hinds, 1983. The Rare Vascular Plants of New Brunswick. Syllogeus 50.

Primula mistassinica Michx. Fig. p. A92. Map p. B69.
COMMON NAME(S): Bird's-eye Primrose, Dwarf Canadian Primrose.
FREQUENCY: Uncommonly scattered in the western portion of the Province and rare in the southwest and eastern lowlands.
HABITAT: Mostly wet calcareous ledges, shores and slopes.
GENERAL RANGE: Yukon to Labr. south to B.C., Iowa, Ont. and Me.

Primula veris L.
COMMON NAME(S): English Cowslip.
SPECIAL NOTE: Eurasian; cultivated and locally escaped in damp meadows and woods as near Shampers Bluff and Greenwich, Kings Co. Known elsewhere in Canada in B.C. and from Ont. to Nfld., and N.S.

SAMOLUS L.

Samolus valerandi L. Fig. p. A92.
COMMON NAME(S): Water-Pimpernel, Brookweed.
FREQUENCY: Very rare; known from Northumberland and Kent counties.

HABITAT: Tidal estuaries above high tide.
GENERAL RANGE: Alaska, B.C., Ont. to N.S.; Ill., Mich. to Calif., Mex., and Fla.; tropical America.
SUBSPECIFIC TAXA: Our plants are referable to ssp. **parviflorus** (Raf.) Hulten.
SYNONYMS: **S. parviflorus** Raf.
SPECIAL NOTE: For more information see H. Hinds, 1983. The Rare Vascular Plants of New Brunswick. Syllogeus 50.

TRIENTALIS L. Star-flower

Trientalis borealis Raf. Fig. p. A92.
COMMON NAME(S): American Star-flower.
FREQUENCY: Common throughout much of the Province.
HABITAT: Coniferous and mixed woods.
GENERAL RANGE: Yukon and B.C. to Labr. south to Calif. and Va.

PLUMBAGINACEAE (Leadwort or Plumbago Family)

LIMONIUM Mill.

Limonium carolinianum (Walt.) Britt. Fig. p. A92. Map p. B69 (as **L. Nashii**).
COMMON NAME(S): Sea-Lavender, Marsh-Rosemary. Lavande de mer.
FREQUENCY: Scattered along the coast and Bay of Fundy.
HABITAT: Coastal salt marshes.
GENERAL RANGE: Que. to Nfld. south to Fla. and Tex.
SYNONYMS: **L. nashii** Small

OLEACEAE (Olive Family)

1a Leaves simple; corolla lilac-purple to white; fruit capsular; shrubs or small trees. **Syringa**
1b Leaves pinnately compound; corolla absent; fruit winged; trees. **Fraxinus**

FRAXINUS L. Ash. Frêne

1a Fruit winged nearly to base; calyx greatly reduced or wanting; leaves subsessile (petiolules less than 2 mm long). 2
1b Fruit winged on upper half only; calyx persisting at base of fruit; petiolules often longer than 2 mm. 3
2a(1a) Terminal buds brown to nearly black, conical; sessile bases of leaflets joined above by rusty tomentum; leaflets 11-14 cm long. **F. nigra**

++

2b(1a) Terminal buds black or blue-black, low rounded; +/- sessile or short winged petiolules not joined above by rusty tomentum; leaflets 3.5-10 cm long. **F. excelsior**

3a(1b) Terminal buds low, +/- rounded; leaf scars with upper margin deeply U-shaped; wing only slightly decurrent on swollen, seed-bearing body of fruit, abruptly acute at apex; leaflets mostly rounded at base, entire or usually only serrulate along upper ½, upper and lower surface contrasting dark and light; petiolules scarcely winged. **F. americana**

3b(1b) Terminal buds conical; leaf scars with upper margin straight or slightly down curved; wing decurrent about ½ way on body of fruit, acuminate to summit; leaflets mostly tapering at base, usually regularly serrulate nearly to base, upper and lower surfaces not strongly contrasting; petiolules winged. **F. pennsylvanica**

Fraxinus americana L. Fig. p. A92. Map p. B69.
COMMON NAME(S): White Ash. Frêne blanc, Franc-frêne.
FREQUENCY: Common in the southern third of the Province and rare northwards.
HABITAT: Rich hardwoods.
GENERAL RANGE: Ont. to N.S. south to Tex. and Fla.

Fraxinus excelsior L.
COMMON NAME(S): European Ash.
SPECIAL NOTE: Eurasian; cultivated and escaped in open woods and waste areas as in St. Andrews, Charlotte Co. Known elsewhere in Canada from Nfld., and N.S.

Fraxinus nigra Marsh. Fig. p. A92. Map p. B69.
COMMON NAME(S): Black Ash. Frêne noir, Frêne gras.
FREQUENCY: Common throughout much of the Province.
HABITAT: Swamps, and river bottomlands.
GENERAL RANGE: Man. to Nfld. south to N. Dak. and Del.

Fraxinus pennsylvanica Marsh. Fig. p. A93. Map p. B69.
COMMON NAME(S): Red Ash. Frêne rouge.
FREQUENCY: Common in the Saint John R. drainage system and scattered in the St. Croix R. valley; rare elsewhere.
HABITAT: River banks.
GENERAL RANGE: Alta. to N.S. south to Mont., Tex. and Fla.

SYRINGA L.

Syringa vulgaris L. Fig. p. A93. Map p. B69.
COMMON NAME(S): Lilac. Lilas
SPECIAL NOTE: European; extensively cultivated and persisting near former habitations mostly in the southern half of the Province. Known elsewhere in Canada from Sask., Ont. to Nfld., P.E.I. and N.S.

GENTIANACEAE (Gentian Family)

1a Leaves reduced to scales; corolla greenish. **Bartonia**

1b Leaves otherwise. 2

2a(1b) Corolla-lobes united only at base, bearing fringed scale-like appendages at base; stigma decurrent on ovary. **Lomatogonium rotatum**

2b(1b) Corolla-lobes united for most of their length, not bearing scale-like appendages at base; stigma terminal on ovary. 3

3a(2b) Corolla-lobes usually spurred, greenish purple. **Halenia deflexa**

3b(2b) Corolla not spurred, pink, white, or blue. 4

4a(3b) Corolla lobes pink, flaring from cylindric +/- colorless tube; small annuals. **Centaurium pulchellum**

4b(3b) Corolla lilac to blue (rarely white) throughout; relatively stout annuals, biennials or perennials. 5

5a(4b) Corolla plaited between the lobes, not fringed within; flowers +/- sessile, clustered at stem summit, subtended by 1-2 bracts; perennials. **Gentiana**

5b(4b) Corolla not plaited, lobes densely ciliate-fringed within; flowers pedicelled, cymose from most leaf axils, not bracted; annuals or biennials. **Gentianella amarella**

BARTONIA Muhl.

1a Stem purplish; scale leaves and branches alternate; flowers whitish. **B. paniculata**

1b Stem green; scale leaves and branches mainly opposite; flowers yellowish. **B. virginica**

Bartonia paniculata (Michx.) Muhl. Fig. p. A93.
COMMON NAME(S): Screw-stem.
FREQUENCY: Very rare; known only from Ross Island, Charlotte Co. and near the gorge of the Little Salmon R., St. John Co. (UNB).
HABITAT: Wet peaty or sandy lowlands.
GENERAL RANGE: N.B., Nfld. and N.S. south to Tex. and Fla.
SUBSPECIFIC TAXA: Our plants are referable to ssp. **iodandra** (Robins.) Gillett.
SPECIAL NOTE: For more information see H. Hinds, 1983. The Rare Vascular Plants of New Brunswick. Syllogeus 50.

Bartonia virginica (L.) BSP.
COMMON NAME(S): Bartonia.
FREQUENCY: Very rare; known only from Pt. Escuminac, Northumberland Co. (UNB).
HABITAT: Peat bogs and peaty thickets.
GENERAL RANGE: Southern Ont. to N.S.; Minn. to Ohio south to La. and Fla.
SPECIAL NOTE: For more information see H. Hinds, 1983. The Rare Vascular Plants of New Brunswick. Syllogeus 50.

GENTIANACEAE

++

CENTAURIUM Hill Centaury

Centaurium pulchellum (Sw.) Druce Fig. p. A93.
SPECIAL NOTE: Eurasian; introduced in fields and roadsides as near Cape Bruin, Westmorland Co. Not reported from elsewhere in Canada.

GENTIANA L. Gentian. Gentiane

1a Middle and upper leaves ovate, rounded or cordate at base, 1.5-3 cm broad; corolla pale blue to nearly white, lobes acute, 3-5 mm longer than plaits. **G. rubricaulis**

1b Middle and upper leaves linear, tapering to base, 0.3-1.5 cm broad; corolla +/- bright blue, lobes rounded or broadly obtuse. 1-2 mm longer than plaits. **G. linearis**

Gentiana linearis Froel. Fig. p. A93. Map p. B69.
COMMON NAME(S): Closed Gentian, Bastard Gentian.
FREQUENCY: Rare; scattered in the Miramichi drainage system with old reports from near Fredericton, York Co. and from Kent Co.
HABITAT: Damp thickets and peaty lowlands.
GENERAL RANGE: Man. to Labr. south to Nebr. and W. Va.

Gentiana rubricaulis Schwein. Fig. p. A93. Map p. B70.
FREQUENCY: Very rare; known only from a few localities in Charlotte Co.
HABITAT: Moist meadows and thickets.
GENERAL RANGE: Southern Ont. and N.B. south to Mich., Minn., Nebr. and Me.
SYNONYMS: **G. linearis** var. **rubricaulis** (Schw.) Gillett.

GENTIANELLA Moench

Gentianella amarella (L.) Borner Fig. p. A93. Map p. B70.
COMMON NAME(S): Felwort.
FREQUENCY: Very rare; known from near Campbellton, Restigouche Co., Grande Plaine, Gloucester Co. and from Saint John, St. John Co.
HABITAT: Moist dune hollows, and calcareous ledges.
GENERAL RANGE: Alaska, Yukon to Labr. south to Calif., Mex. and Vt.; Greenland; Iceland; Eurasia.
SUBSPECIFIC TAXA: Our plants are referable to ssp. **acuta** (Michx.) Gillett.

HALENIA Borkh.

Halenia deflexa (Sm.) Griseb. Fig. p. A93. Map p. B70.
COMMON NAME(S): Spurred Gentian.
FREQUENCY: Scattered throughout most of the Province except rare in the eastern lowlands.

++

HABITAT: Damp banks, woods and thickets.
GENERAL RANGE: B.C. to Labr. south to Mont., and N. Eng.; Mexico.

LOMATOGONIUM A. Br.

Lomatogonium rotatum (L.) Fries Fig. p. A93.
COMMON NAME(S): Marsh Felwort.
FREQUENCY: Very rare; known only from S. Wolf Island, Charlotte Co.
HABITAT: Wet meadows and shores.
GENERAL RANGE: Alaska to Labr. south to B.C., Idaho, N.Mex. and Maine. Not known from P.E.I. or N.S.
SPECIAL NOTE: For more information see H. Hinds, 1983. The Rare Vascular Plants of New Brunswick. Syllogeus 50.

APOCYNACEAE (Dogbane Family)

APOCYNUM L. Dogbane, Indian Hemp. Apocyn

1a Corolla pink or white with pink stripes within, campanulate, +/- 5 mm long; lobes spreading or reflexed; calyx-lobes less than ½ as long as corolla tube. 2

1b Corolla greenish white or white, tubular to urceolate, 3-6 mm long, lobes erect or slightly divergent; calyx-lobes reaching to or beyond the summit of corolla-tube. **A. sibericum**

2a(1a) Calyx-lobes about twice as long as wide, 1/3 to 1/2 as long as corolla-tube; corolla-lobes recurving; all inflorescences flowering simultaneously; flowers finally nodding; leaves spreading or declining. **A. androsaemifolium**

2b(1a) Calyx-lobes at least ½ as long as corolla-tube; corolla-lobes erect to spreading, not recurving; central inflorescence flowering much earlier than inflorescences of leafy side branches; flowers mostly erect; leaves +/- ascending. **A. X medium**

Apocynum androsaemifolium L. Fig. p. A93. Map p. B70.
COMMON NAME(S): Spreading Dogbane. Herbe à la puce.
FREQUENCY: Scattered throughout much of the Province.
HABITAT: Roadsides, dry fields and ditches.
GENERAL RANGE: Alaska, Yukon and B.C. to Nfld. south to Calif., Mex. and Ga.
SUBSPECIFIC TAXA: Our plants are referable to var. **incanum** A. DC.

Apocynum X medium Greene Map p. B70.
FREQUENCY: Uncommon in scattered localities mostly in the southern half of the Province.
HABITAT: Thickets and shores.
GENERAL RANGE: B.C. to Nfld. south to N. Mex. and Va.
SPECIAL NOTE: This is the hybrid between **A. androsaemifolium** var. **incanum** and **A. sibiricum**.

++

Apocynum sibiricum Jacq. Fig. p. A93. Map p. B70.
COMMON NAME(S): Indian Hemp.
FREQUENCY: Scattered throughout much of the Province but rare in the eastern lowlands.
HABITAT: Gravelly shores.
GENERAL RANGE: Dist. Mack. and B.C. to Nfld. south to Calif., Tex. and Fla.
SYNONYMS: **A. cannabinum** L. var. **hypericifolium** Gray.

VINCA L. Periwinkle. Pervenche

Vinca minor L. Fig. p. A93.
COMMON NAME(S): Common Periwinkle. Pervenche.
SPECIAL NOTE: Eurasian; cultivated and escaped to waste areas as on Keswick Ridge, York Co. Known elsewhere in Canada from B.C.; Ont. to N.S.

ASCLEPIADACEAE (Milkweed Family)

ASCLEPIAS L. Milkweed. Asclépiade

1a Fruiting pedicels erect; pods lance-fusiform, attenuate; corolla pink to rose-purple. **A. incarnata**
1b Fruiting pedicels deflexed; pods thick lanceolate; corolla purple to +/- green. **A. syriaca**

Asclepias incarnata L. Fig. p. A94. Map p. B70.
COMMON NAME(S): Swamp-Milkweed.
FREQUENCY: Locally common in the southwest third of the Province.
HABITAT: Shores and wet meadows.
GENERAL RANGE: Man. to N.S. south to N. Mex., La., and S.C.
SUBSPECIFIC TAXA: Ssp. **pulchra** (Ehrh. ex Wild.) Woods., with leaves broadly lanceolate to elliptic and short-pilose beneath rather than oblong-lanceolate and +/- glabrous beneath, occurs frequently with the species.

Asclepias syriaca L. Fig. p. A94. Map p. B70.
COMMON NAME(S): Common Milkweed. Herbe à coton, Petits cochons.
FREQUENCY: Scattered in the southern part of the Province and rare northwards.
HABITAT: Fields, roadsides, waste areas and thickets.
GENERAL RANGE: Man. to N.S. south to Kans. and Ga.

CONVOLVULACEAE (Morning-glory Family)

1a Stems slender, yellow or orange, attached as parasites to surrounding vegetation; leaves scale-like. **Cuscuta**

++

1b Stems stout, often high climbing or trailing; leaves green, normally expanded. 2

2a(1b) Calyx bractless at base; stigma filiform. **Convolvulus arvensis**

2b(1b) Calyx closely subtended by 2 +/- cordate bracts; stigma oval or oblong. **Calystegia sepium**

CALYSTEGIA R. Br.

Calystegia sepium (L.) R. Br. Fig. p. A94. Map p. B70 (as **Convolvulus sepium**).
COMMON NAME(S): Hedge-Bindweed, Wild Morning-glory. Gloire du matin.
FREQUENCY: Scattered throughout much of the Province and common along the east coast.
HABITAT: Edge of salt marshes, thickets, woods and waste areas.
GENERAL RANGE: B.C. to Nfld. south to Calif. and Fla.
SUBSPECIFIC TAXA: Our plants are referable to ssp. **americana** (Sims.) Brummitt.
SYNONYMS: **Convolvulus sepium** L. incl. var. **repens** (L.) Gray.

CONVOLVULUS L. Bindweed. Liseron

Convolvulus arvensis L. Fig. p. A94.
COMMON NAME(S): Field-Bindweed, Corbine. Liseron des champs, Petit liseron.
SPECIAL NOTE: Eurasian; introduced in waste areas as in Saint John, St. John Co. Known elsewhere in Canada from B.C. to N.S.

CUSCUTA L. Dodder, Love-tangle, Coral-vine. Cuscute

1a Stigma filiform; flowers 5-merous in dense head-like clusters; calyx campanulate, lobes acute, reddish; corolla short-cylindric, white or pink, lobes ovate, acute; stem reddish. **C. epithymum**

1b Stigma globose. 2

2a(1b) Corolla-lobes obtuse or rounded; calyx-lobes shorter than corolla tube, oblong to oval-ovate overlapping only at base. 3

2b(1b) Corolla-lobes acute, often inflexed; calyx-lobes +/- as long as corolla tube, broadly ovate, widely overlapping; flowers mostly 5-merous. **C. pentagona**

3a(2a) Flowers chiefly 4-merous; corolla-lobes much shorter than tube, erect to spreading, when withered mostly capping capsule. **C. cephalanthi**

3b(2a) Flowers chiefly 5-merous; corolla-lobes much shorter than tube, widely spreading, when withered surrounding capsule. **C. gronovii**

Cuscuta cephalanthi Engelm.
FREQUENCY: Rare; collected near Aroostook, Victoria Co., Neguac Provincial Park, Northumberland Co. and near the mouth of Tetagouche R., Bathurst, Gloucester Co.
HABITAT: Edge of salt marshes, and river banks.
GENERAL RANGE: B.C., Alta., Man., Que., N.B. and N.S. south to Oreg., Tex. and Va.

++

Cuscuta epithymum L. Fig. p. A94.
COMMON NAME(S): Clover-Dodder. Cuscute du thym.
SPECIAL NOTE: European; reported from Ingleside, Kings Co. and from Restigouche Co. Known elsewhere in Canada from B.C. and Ont.

Cuscuta gronovii Willd. Fig. p. A94. Map p. B70.
FREQUENCY: Common in the southern half of the Province.
HABITAT: Damp meadows and shores.
GENERAL RANGE: Man. to N.S. south to Ariz. and Fla.

Cuscuta pentagona Engelm. Fig. p. A94.
FREQUENCY: Very rare; reported from Quarryville, Northumberland Co. and Grand Lake, Queens Co.
HABITAT: River or lake shores.
GENERAL RANGE: B.C., Sask. to Nfld., south to Calif., Mex. and Fla.; tropical America.

MENYANTHACEAE (Buckbean Family)
(Often included in the **GENTIANACEAE**)

1a Leaves floating, cordate-ovate (like miniature water-lily leaves); flowers borne in umbel-like clusters near summit of petiole; corolla lobes ciliate or entire. **Nymphoides cordata**
1b Leaves trifoliate, basal; flowers racemose; corolla conspicuously fringed. **Menyanthes trifoliata**

MENYANTHES L.

Menyanthes trifoliata L. Fig. p. A94. Map p. B70.
COMMON NAME(S): Buckbean, Bogbean. Herbe à canards.
FREQUENCY: Scattered throughout much of the Province.
HABITAT: Bogs and shallow ponds.
GENERAL RANGE: Alaska, Yukon to Labr. south to Calif. and Del.; Greenland; Iceland; Eurasia.
SUBSPECIFIC TAXA: Our plants are referable to var. **minor** Raf.

NYMPHOIDES Hill

Nymphoides cordata (Ell.) Fern. Fig. p. A94. Map p. B70.
COMMON NAME(S): Floating-heart. Faux-Nymphéa.
FREQUENCY: Scattered in the southern part of the Province.
HABITAT: Shallow water of lakes, ponds and slow streams.
GENERAL RANGE: Ont. to Nfld. south to La. and Fla.

POLEMONIACEAE (Polemonium Family)

1a Leaves pinnately compound; corolla tube very short; stamens conspicuously exsert. **Polemonium van-bruntiae**

1b Leaves entire; corolla tube well developed; stamens not exserted. 2

2a(1b) Leaves alternate; plants erect; flowers inconspicuous in leafy-bracted capitate clusters; annuals. **Collomia linearis**

2b(1b) Leaves chiefly opposite; plants low, matted; flowers showy in open cymes; perennials. **Phlox subulata**

COLLOMIA Nutt.

Collomia linearis Nutt. Fig. p. A94. Map p. B71.
COMMON NAME(S): Collomia.
FREQUENCY: Rare; reported in scattered localities in Restigouche, Gloucester, Kings, St. John and Charlotte Counties. Possibly introduced from further west.
HABITAT: Railways and waste areas.
GENERAL RANGE: Alaska, Yukon and B.C. to N.S. south to Calif., N. Mex. and Wisc.
SPECIAL NOTE: For more information see H. Hinds, 1983. The Rare Vascular Plants of New Brunswick. Syllogeus 50.

PHLOX L. Phlox

Phlox subulata L.
COMMON NAME(S): Moss-Phlox, Mountain Phlox, Moss-Pink.
SPECIAL NOTE: Native to eastern U.S.A. from Mich. to southern Ont. south to Tenn. and N.C. Cultivated and occasionally spreading in cemeteries and similar dry open areas in the southern part of the Province.

POLEMONIUM L. Greek Valerian, Jacob's-ladder

Polemonium van-bruntiae Britt. Fig. p. A94.
COMMON NAME(S): Jacob's-ladder.
FREQUENCY: Very rare; reported from Hoyt, Sunbury Co. and from Trout Lake, Charlotte Co.
HABITAT: Wooded swamps, alluvial bottomlands, sphagnous bogs and mossy glades. Occasionally escaping from cultivation to damp waste areas.
GENERAL RANGE: Apparently native from N.Y. and Vt. to W. Va.
SPECIAL NOTE: Material collected by J. Britton in 1885 in woods on uninhabited side of Trout Brook, Trout Lake, Charlotte Co. may be native but this locality cannot be confirmed. For more information see H. Hinds, 1983. The Rare Vascular Plants of New Brunswick. Syllogeus 50.

++

BORAGINACEAE (Borage Family)

1a Nutlets with hooked prickles. 2

1b Nutlets otherwise. 3

2a(1a) Inflorescence bracteate to summit; calyx-lobes conspicuous, to about 4 mm long; fruiting pedicels erect. **Lappula squarrosa**

2b(1a) Inflorescence +/- bractless, at least towards summit; calyx-lobes inconspicuous, about 2 mm long; fruiting pedicels spreading to +/- recurved. 4

3a(1b) Corolla with elongate, acute lobes, tube inconspicuous, shorter than calyx; anthers with prominent appendage, erect and forming column around style. **Borago officinalis**

3b(1b) Corolla with shorter mostly obtuse or rounded lobes, tube well developed, slightly shorter to longer than calyx; anthers without appendage. 5

4a(2b) Leaves usually less than 10, elliptic-oblong, to about 8 cm wide, uppermost +/- clasping; nutlets covered by prickles. **Cynoglossum**

4b(2b) Leaves more than 10, lanceolate, to about 3 cm wide, not clasping; nutlets mostly prickly only on margins. **Hackelia deflexa**

5a(3b) Corolla irregular (tube +/- curved, lobes unequal). 6

5b(3b) Corolla regular. 7

6a(5a) Corolla about 10 mm wide; stamens conspicuously exsert; corolla-tube throat open; receptacle not pitted. **Echium vulgare**

6b(5a) Corolla about 5 mm broad; stamens included; corolla-tube throat closed by bristly-hirsute scales; receptacle pitted. **Anchusa arvensis**

7a(5b) Corolla tubular or campanulate with +/- erect lobes. 8

7b(5b) Corolla +/- broadly funnelform to salviform. 9

8a(7a) Plants prostrate, succulent, glabrous; leaves blue-green; corolla about 8 mm long; seabeaches plants. **Mertensia maritima**

8b(7a) Plants erect, not succulent, hispid-hirsute; leaves dark green; corolla 1.5 cm long; ruderal plants. **Symphytum**

9a(7b) Flowers greenish white to pale yellow; nutlets smooth, glossy white with broad basal attachment scar; stem to about 1 m high. **Lithospermum officinale**

9b(7b) Flowers blue or white; nutlets with a small basal attachment scar; stem to about 7 dm high. 10

10a(9b) Nutlets smooth, glossy brown to black; leaves alternate; stem usually branching above the base (except in **M. stricta**); corolla blue or white. **Myosotis**

10b(9b) Nutlets dull, roughened with low ridges, white to brown; leaves mostly opposite; stem branching at base; corolla white with yellow centre. **Plagiobothrys scouleri**

ANCHUSA L.

Anchusa arvensis (L.) Bieb. Fig. p. A94.

COMMON NAME(S): Bugloss. Chaudronnette.

SPECIAL NOTE: Eurasian; occasionally introduced on sandy beaches and waste areas as along the east coast. Known elsewhere in Canada from Alta. to Nfld., P.E.I. and N.S.

SYNONYMS: **Lycopsis arvensis** L.

++

BORAGO L.

Borago officinalis L. Fig. p. A95.
COMMON NAME(S): Borage. Bourrache.
SPECIAL NOTE: European; cultivated and occasionally persisting as in Fredericton, York Co. and Restigouche Co. Known elsewhere in Canada from B.C. to N.S.

CYNOGLOSSUM L. Hound's-tongue, Beggar's-lice. Langue de chien

1a Stem abundantly leafy to summit; racemes axillary and terminal; flowers reddish purple. **C. officinale**
1b Stem leafy mostly towards base; racemes terminal; flowers blue. **C. boreale**

Cynoglossum boreale Fern. Fig. p. A95. Map p. B71.
COMMON NAME(S): Northern Wild Comfrey.
FREQUENCY: Rare; reported from 1 or 2 localities in Restigouche, Gloucester, Victoria and Carleton Counties.
HABITAT: Rich, often calcareous woods and thickets.
GENERAL RANGE: B.C. to Nfld. south to Iowa and Conn.
SPECIAL NOTE: For more information see H. Hinds, 1983. The Rare Vascular Plants of New Brunswick. Syllogeus 50.

Cynoglossum officinale L. Fig. p. A95.
COMMON NAME(S): Common Hound's-tongue. Langue de chien.
SPECIAL NOTE: Eurasian; rarely introduced in waste areas as in Saint John, St. John Co. Known elsewhere in Canada from B.C. to N.S.

ECHIUM L. Viper's Bugloss. Vipérine

Echium vulgare L. Fig. p. A95. Map p. B71.
COMMON NAME(S): Blueweed, Blue Devil. Vipérine.
SPECIAL NOTE: Eurasian; introduced along roadsides and waste areas scattered throughout much of the Province. Known elsewhere in Canada from B.C. to Nfld., P.E.I. and N.S.
SUBSPECIFIC TAXA: Forma **albiflorum** Hoffm., with white or whitish flowers, and forma **roseum** Zimm. with pink flowers, occur occasionally with the species. (UNB).

HACKELIA Opiz Stickseed, Beggar's-lice

Hackelia deflexa (Wahl.) Opiz Fig. p. A95.
FREQUENCY: Very rare; known only from the Eel R. and Restigouche R., Restigouche Co.
HABITAT: Thickets, open woods and rocky shores.
GENERAL RANGE: Dist. Mack. to N.B. south to Wash., Colo., and Vt.; Eurasia.
SUBSPECIFIC TAXA: Our plants are referable to var. **americana** (Gray) Fern. & Johnson.
SYNONYMS: **H. americana** (Gray) Fern.

++

LAPPULA Moench Stickseed. Bardanette

Lappula squarrosa (Retz.) Dumort. Fig. p. A95. Map p. B71.
SPECIAL NOTE: Eurasian; occasionally introduced in waste areas and fields as in St. Andrews, Charlotte Co., and in York, Victoria and Restigouche Counties. Known elsewhere in Canada from Alaska and B.C. to Nfld., P.E.I. and N.S.
SYNONYMS: **L. myosotis** Moench

LITHOSPERMUM L. Stoneseed, Gromwell, Puccoon. Grémil

Lithospermum officinale L. Fig. p. A95.
COMMON NAME(S): Common Gromwell. Herbe aux perles, Graines de lutin.
SPECIAL NOTE: European; introduced to thickets, pastures and waste areas as near Campbellton, and Heron Island, Restigouche Co. Known elsewhere in Canada from Man., Ont. and Que.

MERTENSIA Roth Lungwort, Bluebells

Mertensia maritima (L.) S.F. Gray Fig. p. A95. Map p. B71.
COMMON NAME(S): Sea-Lungwort, Oysterleaf. Sanguine de mer.
FREQUENCY: Scattered along the Bay of Fundy and east coast.
HABITAT: Sandy or gravelly sea beaches.
GENERAL RANGE: Alaska, Yukon to Baffin Is. south on west coast to B.C. and on east coast to Mass.; Greenland; Iceland; Spitsb.; Eurasia.

MYOSOTIS L. Forget-me-not. Grémillet

1a Calyx in fruit longer than pedicels; corolla about 1.5 mm broad; inflorescence often over 3/4 total height of plant. **M. stricta**

1b Calyx in fruit shorter than pedicels; corolla usually over 2 mm broad; inflorescence rarely over ½ height of plant. 2

2a(1b) Stem and calyx strigose, hairs neither glandular nor hooked; perennials of mostly wet habitats. 3

2b(1b) Stem and calyx with divergent hooked hairs; annuals or biennials of mostly dry habitats. 4

3a(2a) Calyx-lobes about as long as tube; corolla 4-6 mm broad; style shorter than calyx-tube. **M. laxa**

3b(2a) Calyx-lobes much shorter than tube; corolla 5-10 mm broad; style equal to or longer than calyx-tube. **M. scorpioides**

4a(2b) Calyx-lobes much longer than tube; corolla spreading horizontally, 4-8 mm wide. **M. sylvatica**

4b(2b) Calyx-lobes slightly longer than tube; corolla spreading-ascending, 2-3 mm broad. **M. arvensis**

++

Myosotis arvensis (L.) Hill Fig. p. A95. Map p. B71.
SPECIAL NOTE: Eurasian; cultivated and occasionally escaped to roadsides and waste areas in the southern part of the Province. Known elsewhere in Canada from B.C. to Nfld., P.E.I. and N.S.

Myosotis laxa Lehm. Fig. p. A95. Map p. B71.
FREQUENCY: Scattered throughout much of the Province.
HABITAT: Seepage areas and shores.
GENERAL RANGE: B.C.; Ont. to Nfld. south to Calif.; Tenn. to Ga.; tropical America; Eurasia.

Myosotis scorpioides L. Fig. p. A95. Map p. B71.
SPECIAL NOTE: Eurasian; cultivated and frequently escaped throughout much of the Province. Known elsewhere in Canada from Alaska, Yukon and B.C.; Man. to Nfld., P.E.I. and N.S.

Myosotis stricta Link
SPECIAL NOTE: European; reported only from Pt. Wolfe R. camping area, Fundy National Park, Albert Co. (UNB). Known elsewhere in Canada from B.C., Alta., Ont. and Que.
SYNONYMS: **M. micrantha** Pallas

Myosotis sylvatica Hoffm. Fig. p. A95. Map p. B71.
SPECIAL NOTE: Eurasian; cultivated and escaped to wet meadows and waste areas in many parts of the Province. Known elsewhere in Canada from all provinces except Man.

PLAGIOBOTHRYS F. & M.

Plagiobothrys scouleri (Hook. & Arn.) I. M. Johnston Fig. p. A96. Map p. B71.
SPECIAL NOTE: Native to the west coast of N.A. east to southwest Man. Collected in sandy clearings near Val Comeau, Gloucester Co. and in dry compacted soil, UNB campus, Fredericton, York Co. (UNB). Not reported elsewhere east of Man.
SYNONYMS: **P. scopulorum** (Greene) I.M. Johnson, **Allocarya cognata** Greene.
SUBSPECIFIC TAXA: Our plants are referable to var. **penicillatus** (Greene) Cronq.

SYMPHYTUM L. Comfrey. Consoude

1a Leaves long decurrent; upper stem +/- winged, prickles not especially decurved. **S. officinale**
1b Leaves sessile or short decurrent; stem not winged, prickles decurved. **S. asperum**

Symphytum asperum Lepechin Fig. p. A96. Map p. B72.
COMMON NAME(S): Prickly Comfrey.
SPECIAL NOTE: Asiatic; introduced to roadsides and waste areas in Carleton, Victoria and Albert Counties. Known elsewhere in Canada from B.C.; Man. to Nfld., P.E.I. and N.S.

++

Symphytum officinale L. Fig. p. A96.
COMMON NAME(S): Common Comfrey. Langue de vache, Herbe à la coupure
SPECIAL NOTE: Eurasian; introduced to roadsides and waste areas in St. John County. Known elsewhere in Canada from B.C., Alta., Ont. to Nfld. and N.S.

VERBENACEAE (Vervain Family)

VERBENA L. Vervain. Verveine

1a Inflorescence diffuse, of elongate spikes, lax; white flowers and fruit usually well separated on peduncle. **V. urticifolia**

1b Inflorescence compact, of shorter, erect spikes; blue or pink (rarely white) corolla and fruit +/- contiguous on peduncle. **V. hastata**

Verbena hastata L. Fig. p. A96. Map p. B72.
COMMON NAME(S): Blue Vervain. Simpler's-joy. Verveine bleue.
FREQUENCY: Uncommonly scattered throughout much of the Province.
HABITAT: Shores, wet meadows and river bottoms.
GENERAL RANGE: B.C.; Sask. to N.S. south to Calif. and Fla.
SUBSPECIFIC TAXA: Forma **rosea** Cheney, with pink rather than blue corolla, is reported from Crocks Pt., York Co.

Verbena urticifolia L. Fig. p. A96. Map p. B72.
COMMON NAME(S): White Vervain, Bur-Vine. Verveine blanche.
FREQUENCY: Uncommon; scattered along the St. John R. in York, Carleton and Victoria Counties.
HABITAT: Alluvial bottomlands and thickets.
GENERAL RANGE: Sask.; Ont. to N.B. south to Tex. and Fla.

LAMIACEAE (Mint Family)
(Alternate name **LABIATAE**)

1a Flowers in terminal spikes; bracts +/- inconspicuous. 2

1b Flowers in axils of scarcely reduced leaves, in whorls, heads, lateral panicles or corymbs. 3

2a(1a) Calyx nearly regular or slightly oblique. 4

2b(1a) Calyx 2-lipped. 5

3a(1b) Stem mostly prostrate or trailing (flowering stems sub-erect); leaves rotund-cordate to nearly reniform, crenate. **Glecoma hederacea**

3b(1b) Stem erect or sometimes decumbent at base; plants often stoloniferous. 6

4a(2a) Corolla nearly regular; stamens exserted. 7

4b(2a) Corolla irregular. 8

5a(2b) Corolla upper lip wanting, pink-purple lower lip 5-lobed, 12-18 mm long. **Teucrium canadense**

5b(2b) Corolla upper lip present. 9

6a(3b) Corolla nearly regular. 10

++

6b(3b) Corolla irregular. 11

7a(4a) Corolla tube +/- equal to calyx-lobes; leaves strongly glandular-aromatic; perennials. **Mentha**

7b(4a) Corolla tube much longer than calyx-lobes; leaves glandular beneath; annuals. **Elsholtzia ciliata**

8a(4b) Calyx violet-tinged, lobes acute, very regular. 12

8b(4b) Calyx green, lobes aristate, slightly irregular. 13

9a(5b) Plant prostrate to suberect 1-5 dm tall; calyx with upper lip nearly truncate, lower lip with 3 lance-acuminate lobes; leaves mostly entire. **Prunella vulgaris**

9b(5b) Plant erect 2-8 dm tall; one calyx lobe much broader than others, conspicuously cross-veined; leaves coarsely serrate. **Dracocephalum parviflorum**

10a(6a) Calyx strongly irregular, swollen on lower side near base; flowers bluish; very aromatic annuals to about 4 dm high; leaves few toothed. **Hedeoma pulegioides**

10b(6a) Calyx +/- regular, not swollen; flowers white to pale purple; leaves regularly serrate. 14

11a(6b) Calyx upper side with conspicuous protuberance; flowers blue. **Scutellaria**

11b(6b) Calyx without protuberance. 15

12a(8a) Leaves narrowly lanceolate, sessile, glabrous; corolla pink-purple 2-3 cm long. **Physostegia virginiana**

12b(8a) Leaves dentoid-ovate, minutely pubescent, petiolate; corolla blue, about 1 cm long. **Agastache foeniculum**

13a(8b) Leaves cordate-ovate, long petiolate; corolla creamy-white with pink-purple spots. **Nepeta cataria**

13b(8b) Leaves lanceolate, rounded or truncate at base, sessile or short petiolate; corolla pinkish purple. **Stachys**

14a(10b) Flowers sessile, corolla white; plants glabrous, not aromatic. **Lycopus**

14b(10b) Flowers with pedicels to 2 mm long at flowering, corolla pinkish-purple; plants glandular-aromatic and +/- pubescent. **Mentha**

15a(11b) Flowers many in dense terminal or axillary clusters or heads. 16

15b(11b) Flowers in few-flowered whorls in leaf axils. 17

16a(15a) Heads all terminal in open corymbs; corolla whitish; leaves lance-linear. **Pycnanthemum virginianum**

16b(15a) Heads terminal and axillary; corolla crimson, pinkish-purple or white; leaves ovate to nearly lanceolate. 18

17a(15b) Upper leaves greatly reduced, deeply lobed; corolla pale pink, densely villous. **Leonurus cardiaca** L.

17b(15b) Upper leaves scarcely reduced; corolla finely pubescent, white, pink or purplish. 19

18a(16b) Calyx, corolla and often bracts of the inflorescence crimson; leaves to 1.5 dm long, acuminate, glandular-aromatic on petioles to about 2 cm long. **Monarda didyma**

18b(16b) Calyx green; corolla rose-purple to white; leaves to 4 cm long, blunt-tipped, subsessile. **Clinopodium vulgare**

++

19a(17b) Sepals spiny-tipped; lower lip of corolla with 2 yellow knobs; stem +/- stiffly erect, hispid hairy. **Galeopsis**

19b(17b) Sepals narrowly acuminate, not spiny; lower lip of corolla without yellow knobs. 20

20a(19b) Leaves dentate; corolla longer than 1 cm, upper lip erect, constricted at base, deeply concave; stem decumbent at base, subglabrous. **Lamium**

20b(19b) Leaves entire; corolla less than 1 cm long, upper lip straight to slightly spreading, not constricted at base, flat or slightly concave; stem erect, finely pubescent. **Satureja hortensis**

AGASTACHE Clayt. Giant Hyssop

Agastache foeniculum (Pursh) Ktze. Fig. p. A96.
COMMON NAME(S): Blue or Fennel Giant Hyssop. Agastache fenouil.
SPECIAL NOTE: Native from Dist. Mack. and B.C. to Ont. south to Wash. and Ill. Introduced in waste areas and roadsides as in Moore's Mills, Charlotte Co. and near St. Quentin, Restigouche Co. Known elsewhere in eastern Canada from eastern Ont., and Que.

CLINOPODIUM L.

Clinopodium vulgare L. Fig. p. A96. Map p. B72.
COMMON NAME(S): Wild Basil, Dogmint. Sarriette vulgaire.
FREQUENCY: Scattered mostly in the western half of the Province and rare in the eastern lowlands.
HABITAT: Open woods, slopes, waste areas.
GENERAL RANGE: Man. to Nfld. south N. Mex., Tenn. and N.C.; Eurasia.
SYNONYMS: **Satureja vulgaris** (L.) Fritsch var. **neogaea** Fern.

DRACOCEPHALUM L. False Dragonhead

Dracocephalum parviflorum Nutt. Fig. p. A96.
COMMON NAME(S): American Dragonhead.
SPECIAL NOTE: Native to Alaska and B.C. to Que. Probably introduced to fields and waste areas as in Scotchtown, Queens Co. (UNB). Known elsewhere in eastern Canada from N.S.
SYNONYMS: **Moldavica parviflora** (Nutt.) Britt.

ELSHOLTZIA Willd.

Elsholtzia ciliata (Thimb.) Hylander Fig. p. A96.
SPECIAL NOTE: Asiatic; rarely introduced to fields and roadsides as near Grand Falls, Victoria Co. Known elsewhere in Canada from Man. to Que.

++

GALEOPSIS L. Hemp-Nettle. Gratte, Galéopside

1a Stem hispid hairy, swollen beneath nodes; leaves ovate-lanceolate, coarsely serrate. **G. tetrahit**

1b Stem soft appressed hairy, not swollen; leaves narrowly linear to elliptic, obscurely serrate to entire. **G. ladanum**

Galeopsis ladanum L. Fig. p. A96.
COMMON NAME(S): Red Hemp-Nettle.
SPECIAL NOTE: Eurasian; introduced to roadsides and waste areas as in wharf ballast, Saint John, St. John Co. Known elsewhere in Canada from Que., Nfld., and N.S.

Galeopsis tetrahit L. Fig. p. A96.
COMMON NAME(S): Common Hemp-Nettle. Ortie royale, Gratte.
SPECIAL NOTE: Eurasian; introduced throughout the Province in waste areas. Known elsewhere in Canada from Dist. Mack. and all provinces.

GLECHOMA L. Ground Ivy

Glechoma hederacea L. Fig. p. A96.
COMMON NAME(S): Gill-over-the-ground, Run-away-Robin. Lierre terrestre.
SPECIAL NOTE: Eurasian; introduced in waste areas throughout much of the Province. Known elsewhere in Canada from all provinces.

HEDEOMA Pers. Mock Pennyroyal

Hedeoma pulegioides (L.) Pers. Fig. p. A97. Map p. B72.
COMMON NAME(S): American Pennyroyal, Pudding-grass.
FREQUENCY: Rare; scattered in the southern quarter of the Province.
HABITAT: Ledges and dry gravelly soil.
GENERAL RANGE: S. Dak. to Ont. and N.S. south to Kans. and Fla.

LAMIUM L. Dead-Nettle

1a Upper lip of corolla 1/4-1/2 as long as tube, 3-5 mm long; corolla pink or purplish; annuals or biennials. 2

1b Upper lip of corolla 2/3 as long as tube or longer; corolla white; perennials. **L. album**

2a(1a) Upper bracteal leaves +/- clasping, often wider than long; calyx-lobes about as long as tube. **L. amplexicaule**

2b(1a) Upper bracteal leaves shortly petiolate, longer than wide; calyx-lobes longer than tube. 3

3a(2b) Leaves irregularly incised-dentate, light green; corolla tube naked or slightly hairy within. **L. hybridum**

3b(2b) Leaves shallowly and regularly crenate-serrate, dark green or purplish; corolla-tube pubescent within. **L. purpureum**

++

Lamium album L. Fig. p. A97.
COMMON NAME(S): Snowflake, White Dead-Nettle. Lamier blanc, Ortie blanche, Marachemin.
SPECIAL NOTE: Eurasian; cultivated and locally introduced in the Newcastle and Chatham area, Northumberland Co. Known elsewhere in Canada from Sask. to Que.

Lamium amplexicaule L. Fig. p. A97.
COMMON NAME(S): Henbit. Pain de poule.
SPECIAL NOTE: Eurasian; rarely introduced in fields and waste areas as in Lamèque, Gloucester Co. and Fredericton, York Co. Known elsewhere in Canada from Dist. Mack. and B.C. to Labr., Nfld. and N.S.

Lamium hybridum Vill.
COMMON NAME(S): Henbit.
SPECIAL NOTE: Eurasian; rarely introduced in cultivated soil as in Fredericton, York Co. Not persisting. Known elsewhere in Canada from Que. and P.E.I.

LEONURUS L. Motherwort. Agripaume

Leonurus cardiaca L. Fig. p.A97. Map p. B72.
COMMON NAME(S): Motherwort.
SPECIAL NOTE: Eurasian; occasionally introduced in the southern part of the Province in gardens and waste areas. Known elsewhere in Canada from B.C.; Sask. to N.S.

LYCOPUS L. Water-Horehound, Bugleweed. Lycope

1a Leaves shallowly to deeply lobed, especially at base; calyx-lobes spine-tipped, greatly exceeding the mature nutlets. **L. americanus**
1b Leaves serrate to slightly below the middle, entire margined and tapering to base; calyx-lobes acute, shorter than mature nutlets. **L. uniflorus**

Lycopus americanus Muhl. Fig. p. A97. Map p. B72.
FREQUENCY: Scattered throughout much of the Province.
HABITAT: Seepage areas and river bottomlands.
GENERAL RANGE: B.C. to Nfld. south to Calif., Tex. and Fla.

Lycopus uniflorus Michx. Fig. p. A97.
FREQUENCY: Common throughout much of the Province.
HABITAT: Wet woods, bottomlands, shores and seepage areas.
GENERAL RANGE: B.C. to Nfld. south to Calif. and N.C.; Asia.

MENTHA L. Mint. Menthe

1a Bracts of inflorescence similar to ordinary foliage leaves. 2
1b Bracts of inflorescence small and inconspicuous; flowers in a spike. 3
2a(1a) Calyx broadly campanulate, 1.5-2.5 mm long with teeth broadly triangular; plants usually fertile. **M. arvense**

++

2b(1a) Calyx narrowly campanulate or tubular, 2-4 mm long with teeth narrowly triangular or subulate; plant sterile, often red-tinged. **M. X gentilis**

3a(1b) Spike 5-15 mm in diameter usually tapering to tip; leaves +/- sessile. **M. spicata**

3b(1b) Spike oblong, 12-20 mm in diameter, rounded at tip; leaves petiolate. **M. X piperita**

Mentha arvensis L. Fig. p. A97.
COMMON NAME(S): Common Mint. Menthe du Canada.
FREQUENCY: Common throughout much of the Province.
HABITAT: Shores, and similar wet open areas.
GENERAL RANGE: Alaska, Yukon to Labr. south to Calif., N. Mex. and Del.; Eurasia.
SPECIAL NOTE: Several varieties and numerous forms have been named. Ssp. **haplocalyx** Brig. appears to be the name now accepted for the majority of our specimens.
SYNONYMS: **M. arvensis** var. **villosa** (Benth.) Fern.

Mentha X gentilis L. Fig. p. A97. (**M. arvensis X M. spicata**)
COMMON NAME(S): Red Mint.
SPECIAL NOTE: European; occasionally cultivated and rarely introduced as near Yoho Lake, York Co. Known elsewhere in Canada from B.C.; Ont. to Nfld., P.E.I. and N.S.
SYNONYMS: **M. cardiaca** Baker, **M. X gentilis** nm. **cardiaca** (Baker) Boivin.

Mentha X piperata L. (**M. aquatica X M. spicata**)
COMMON NAME(S): Peppermint. Menthe poivrée.
SPECIAL NOTE: European; cultivated and rarely escaped to waste areas as in Fredericton, York Co. Known elsewhere in Canada from B.C.; Ont. to St. Pierre and Miquelon, P.E.I. and N.S.

Mentha spicata L. Fig. p. A97.
COMMON NAME(S): Spearmint. Baume.
SPECIAL NOTE: Eurasian; cultivated and occasionally introduced in wet meadows and waste areas as in Fundy National Park, Albert Co. and near Grande Digue Pt., Kent Co. Known elsewhere in Canada from B.C., Sask. to N.S.

MONARDA L. Monarda, Horsemint

Monarda didyma L. Fig. p. A97.
COMMON NAME(S): Oswego-tea, Bee-balm. Monarde écarlate, Thé d'Oswego.
SPECIAL NOTE: Native to eastern U.S.A.; cultivated and locally escaped in Carleton and Gloucester Counties. Known elsewhere in Canada from Ont. and Que.

NEPETA L. Catmint

Nepeta cataria L. Fig. p. A97.
COMMON NAME(S): Catnip. Herbe à chats, Chataire.

SPECIAL NOTE: Eurasian; occasionally cultivated and locally established as near Woodstock, Carleton Co. and in Kouchibouguac National Park, Kent Co. Known elsewhere in Canada from B.C. to Nfld., P.E.I. and N.S.

PHYSOSTEGIA Benth.

Physostegia virginiana (L.) Benth. Fig. p. A98.
COMMON NAME(S): False Dragonhead. Cataleptique.
FREQUENCY: Very rare; known only from Grand Manan Island, Charlotte Co.
HABITAT: Damp thickets, meadows and shores.
GENERAL RANGE: Man. to N.S. south to N. Dak. and N.C.
SPECIAL NOTE: Widely cultivated and possibly introduced east of Quebec City.
SYNONYMS: **Dracocephalum virginianum** L.

PRUNELLA L. Selfheal.

Prunella vulgaris L. Fig. p. A98.
COMMON NAME(S): Heal-all, Carpenter-weed. Herbe au charpentier.
FREQUENCY: Common throughout much of the Province.
HABITAT: Fields, roadsides, shores and waste areas.
GENERAL RANGE: Alaska and B.C. to Nfld. south to Calif., N. Mex. and N.C.; Eurasia.
SUBSPECIFIC TAXA: Var. **elongata** Benth., with leaves lanceolate and narrowly cuneate at base, rather than leaves ovate to ovate-oblong and broadly cuneate or rounded at base, as in the species, occurs commonly throughout the Province. Forma **candida** Fern. with white flowers has been recorded from Carleton and York Counties. SYNONYMS: **P. vulgaris** var. **lanceolata** (Bart.) Fern.

PYCNANTHEMUM Michx. Mountain-mint

Pycnanthemum virginianum (L.) Durand & Jackson
FREQUENCY: Very rare; known only from Jenkins Cove, Kings Co. (UNB).
HABITAT: Moist rock crevices near shore.
GENERAL RANGE: N. Dak. to Ont., Que. and N.B. south to Kans. and N.C.
SPECIAL NOTE: For more information see H. Hinds, 1983. The Rare Vascular Plants of New Brunswick. Syllogeus 50.

SATUREJA L. Savory, Calamint. Sarriette

Satureja hortensis L.
COMMON NAME(S): Summer-Savory. Sarriette des jardins.
SPECIAL NOTE: European; occasionally cultivated and rarely escaped as in Kouchibouguac, Kent Co. and St. Andrews, Charlotte Co. Known elsewhere in Canada from Ont. to Nfld., St. Pierre and Miquelon.

++

SCUTELLARIA L. Skullcap

1a Flowers in one-sided bracteate racemes; petioles 1-2 cm long; corolla 5-8 mm long. **S. lateriflora**
1b Flowers in axils of scarcely reduced leaves; petioles usually less than 1 cm long; corolla 15-25 mm long. 2
2a(1b) Corolla less than 15 mm long; petioles 3-10 mm long. **S. X churchilliana**
2b(1b) Corolla to 25 mm long; petioles less than 4 mm long. **S. galericulata**

Scutellaria X churchilliana Fern. Map p. B72.
COMMON NAME(S): Churchill's Skullcap.
FREQUENCY: Rare; known from York, Carleton and Restigouche Counties.
HABITAT: River gravels, and thickets.
GENERAL RANGE: Ont. to N.S. south to Maine.
SPECIAL NOTE: This is the hybrid between **S. galericulata** and **S. lateriflora**.

Scutellaria galericulata L. Fig. p. A98. Map p. B72.
COMMON NAME(S): Common Skullcap. Toque. Tertianaire.
FREQUENCY: Common throughout much of the Province.
HABITAT: Moist open areas.
GENERAL RANGE: Alaska and Yukon to Labr. south to Calif., N. Mex. and Del.; Eurasia.
SYNONYMS: **S. epilobiifolia** Hamilton

Scutellaria lateriflora L. Fig. p. A98. Map p. b72.
COMMON NAME(S): Mad-dog Skullcap.
FREQUENCY: Common throughout much of the Province.
HABITAT: Bottomlands, seepage areas and wet depressions.
GENERAL RANGE: B.C. to Nfld. south to Calif. and Ga.

STACHYS L. Hedge-Nettle. Epiaire

1a Fibrous rooted annuals, non-stoloniferous. **S. annua**
1b Stoloniferous or rhizomatous perennials. 2
2a(1b) Stem to over 1 m high, mostly unbranched, long bristly on angles, glabrous on sides; petioles 8-25 mm long; calyx bristly, but not glandular. **S. tenuifolia**
2b(1b) Stem usually less than 1 m tall, often branched near base, pubescent on sides, strigose to hirsute on angles, sessile or short petiolate; calyx glandular-pubescent, often purple tinged. **S. palustris**

Stachys annua (L.) L. Fig. p. A98.
COMMON NAME(S): Field Woundwort.
SPECIAL NOTE: European; introduced in Clifton, Kent Co. (NBM). Probably not persisting. Not previously reported from Canada.

Stachys palustris L. Fig. p. A98. Map p. B72.
COMMON NAME(S): Woundwort, Hedge-Nettle. Crapaudine.
SUBSPECIFIC TAXA: Two varieties are recognized within our range.

1a Stem angles with conspicuous pustulate-based deflexed hairs and shorter deflexed hairs on sides; calyx with glandular and nonglandular hairs +/- in equal numbers. var. **palustris**
1b Stem angles with nonpustulate or inconspicuously pustulate-based spreading hairs and dense, shorter hairs on sides; calyx with glandular hairs much shorter than nonglandular hairs. var. **pilosa**

1. Var. **palustris**
FREQUENCY: Common throughout much of the Province.
HABITAT: Mostly streambanks, ditches and waste areas.
GENERAL RANGE: Man. to Nfld., N.B. and N.S.

2. Var. **pilosa** (Nutt.) Fern.
FREQUENCY: Uncommon; scattered mostly in the northern part of the Province south to Grand Lake, Queens Co.
HABITAT: Alluvial shores.
GENERAL RANGE: Alaska, Yukon and B.C. to Nfld. south to Ariz. and N. Eng.

Stachys tenuifolia Willd. Fig. p. A98. Map p. B72.
FREQUENCY: Rare; found along the St. John R. in York and Carleton Counties.
HABITAT: Alluvial meadows, shores and thickets.
GENERAL RANGE: Man. to N.B. south to Tex. and S.C.
SUBSPECIFIC TAXA: Our plants are referable to var. **platyphylla** Fern.

TEUCRIUM L. Germander, Wood-Sage. Germandrée

Teucrium canadense L. Fig. p. A98. Map p. B73.
COMMON NAME(S): American Germander.
FREQUENCY: Rare; recorded from Grand Manan Is., Charlotte Co. and from Kent and Northumberland Counties along the east coast.
HABITAT: Coastal sands, thickets and woods.
GENERAL RANGE: Ont. to N.S. south to Tex. and Fla.

SOLANACEAE (Nightshade Family)

1a Corolla deeply lobed; anthers erect, surrounding style, opening by terminal pores; fruit a berry. **Solanum**
1b Corolla funnelform, shallowly lobed to +/- entire; anthers opening by longitudinal slits; fruit a berry or capsule. 2
2a(1b) Berry fruit surrounded by inflated papery husk. **Physalis**
2b(1b) Fruit a capsule. 3

++

3a(2b) Corolla greenish-yellow with purple veins and throat; spike-like inflorescence leafy, one-sided. **Hyoscyamus niger**

3b(2b) Corolla white or pale violet; flowers single, axillary. **Datura stramonium**

DATURA L. Jimsonweed, Thorn-apple. Pomme épineuse, Herbe aux sorciers

Datura stramonium L. Fig. p. A98.
COMMON NAME(S): Stramonium.
SPECIAL NOTE: Asiatic; rarely introduced in waste areas as in St. Andrews and St. Stephen, Charlotte Co. Not persisting. Known elsewhere in Canada from B.C. to Sask., Ont. to N.S.
SUBSPECIFIC TAXA: Var. **tatula** (L.) Torr., with pale violet rather than white corolla and is reported from Richibucto, Kent Co., and Saint John, St. John Co.

LYCOPERSICON Mill. Tomato

Lycopersicon esculentum Mill.
SPECIAL NOTE: Tropical American; frequently cultivated and occasionally escaped in the southern part of the Province. Not established. Known elsewhere in Canada from B.C., Ont. to N.S.
SYNONYMS: **Lycopersicum esculentum** Mill.

HYOSCYAMUS L. Henbane

Hyoscyamus niger L. Fig. p. A98.
COMMON NAME(S): Black Henbane. Jusquiame
SPECIAL NOTE: European; rarely introduced as ballast waste as in Miramichi, Northumberland Co., Richibucto and Bass River, Kent Co., and Saint John, St. John Co. Known elsewhere in Canada from B.C. to N.S.

PHYSALIS L. Ground-Cherry. Coqueret

Physalis alkekengi L.
COMMON NAME(S): Chinese Lantern, Winter-Cherry. Alkekenge.
SPECIAL NOTE: Asiatic; frequently cultivated and rarely introduced as in Fredericton, York Co. and Milltown, Charlotte Co. Known elsewhere in Canada from Ont.

Physalis pubescens L. Fig. p. A98.
SPECIAL NOTE: Tropical American and southern U.S.A.; rarely introduced as in Bass River, Kent Co. Known elsewhere in Canada from B.C., Man. to Que.

SOLANUM L. Nightshade. Morelle

1a	Plant viney, 1-3 m long; corolla bluish-purple; fruit red.	**S. dulcamara**
1b	Plant erect, 1-6 dm high; corolla white; fruit black.	**S. nigrum**

Solanum dulcamara L. Fig. p. A99. Map p. B73.
COMMON NAME(S): Bittersweet-Nightshade. Morelle douce-amère.
SPECIAL NOTE: Eurasian; introduced throughout much of the southern part of the Province. Known elsewhere in Canada from all provinces.
SUBSPECIFIC TAXA: Var. **villosissimum** Desv., with leaves and stems heavily pubescent rather than glabrous, occurs rarely with the species.

Solanum nigrum L.
COMMON NAME(S): Black Nightshade. Tue-chien, Morelle noire, Crève-chien.
SPECIAL NOTE: Eurasian; introduced in waste areas as in Kouchibouguac, Kent Co. and Saint John, St. John Co. Known elsewhere in Canada from all provinces.

SCROPHULARIACEAE (Figwort Family)

1a	Leaves all basal, linear, 2-5 cm long, tufted; flowers on filiform scape-like pedicels; corolla regular. **Limosella australis**
1b	Leaves not all basal. 2
2a(1b)	Corolla lobes spreading, much longer than tube. 3
2b(1b)	Corolla lobes equal to or shorter than tube. 4
3a(2a)	Corolla yellow or reddish; stamens 5; calyx 5-lobed. **Verbascum**
3b(2a)	Corolla white or bluish; stamens 2 or 4; calyx 4-lobed. **Veronica**
4a(2b)	Mid stem leaves chiefly alternate. 5
4b(2b)	Mid stem leaves opposite or whorled (floral leaves and bracts sometimes alternate). 6
5a(4a)	Corolla spurred. 7
5b(4a)	Corolla not spurred. 8
6a(4b)	Flowers +/- congested, terminal. 9
6b(4b)	Flowers single in axils of leaves or bracts. 10
7a(5a)	Flowers axillary on pedicels at least twice as long as calyx. 11
7b(5a)	Flowers in terminal bracted racemes on pedicels about as long as calyx. **Linaria**
8a(5b)	Leaves deeply lobed to dissected. **Pedicularis**
8b(5b)	Leaves entire to serrate. 12
9a(6a)	Flowers sessile in terminal spike; corolla whitish. **Chelone glabra**
9b(6a)	Flowers slender pedicelled; inflorescence racemose or paniculate. 13
10a(6b)	Fertile stamens 2; calyx lobed nearly to base. 14
10b(6b)	Fertile stamens 4; calyx-lobes equal to or shorter than tube. 15
11a(7a)	Leaves palmately 5-9-lobed, long petioled; corolla pale blue with yellow palate; stem trailing. **Cymbalaria muralis**
11b(7a)	Leaves lanceolate, short petioled; corolla blue-purple with yellow palate; stem erect. **Chaenorrhinum minus**

++

12a(8b) Corolla nearly regular, purple, spotted within; leaves ovate, crenate-dentate, mostly long petioled. **Digitalis purpurea**

12b(8b) Corolla 2-lipped, pale-greenish, +/- hidden by creamy white 3-lobed bracts; leaves linear, sessile, entire. **Castilleja septentrionalis**

13a(9b) Corolla less than 1 cm long, reddish-brown, about thrice as long as calyx; sterile stamen glabrous. **Scrophularia**

13b(9b) Corolla to 3 cm long, about 5 times as long as calyx, whitish; sterile stamen bearded. **Penstemon digitalis**

14a(10a) Plant glandular-viscid, erect or ascending, mostly simple; calyx with pair of bractlets at base; corolla yellow. **Gratiola neglecta**

14b(10a) Plant glabrous, semi-erect to reclining, branching from base; calyx without bractlets; corolla pale lavender. **Lindernia dubia**

15a(10b) Calyx 4-lobed. 16

15b(10b) Calyx 5-lobed. 17

16a(15a) Leaves to about 2 cm long, ovate, coarsely serrate, palmately veined; corolla pale white, blue or purple marked with darker purple lines. **Euphrasia**

16b(15a) Leaves pinnately veined, chiefly more than 2 cm long, linear to lanceolate. 18

17a(15b) Leaves linear, entire, sessile, 1-nerved; corolla pinkish-purple. **Agalinus**

17b(15b) Leaves lanceolate to ovate or reniform-cordate, pinnately-veined; corolla yellow to blue or pinkish. **Mimulus**

18a(16b) Calyx inflated in fruit; corolla yellow. **Rhinanthus minor**

18b(16b) Calyx not inflated in fruit; corolla white with yellow palate or rose-red. 19

19a(18b) Capsule oblique at summit, opening along longer side; corolla white with yellow palate; leaves linear to lanceolate entire or the upper with teeth or lobes near base, often darkening on drying. **Melampyrum lineare**

19b(18b) Capsule +/- symmetrical, opening along 2 sides; corolla rose-red; leaves oblong-lanceolate, serrate, not darkening on drying. **Odontites verna**

AGALINUS Raf. Gerardia

1a Calyx-lobes acuminate; leaves sharp pointed; pedicels 1-4 mm long. **A. purpurea**

1b Calyx-lobes oblong or semicircular, blunt; leaves obtuse, +/- fleshy; pedicels 2-10 mm long. **A. maritima**

Agalinus maritima (Raf.) Raf. Fig. p. A99.
COMMON NAME(S): Salt-marsh Gerardia.
FREQUENCY: Very rare; reported from Kent Co. Last collected in 1865.
HABITAT: Coastal salt marshes.
GENERAL RANGE: N.B. and N.S. south to Fla. and Tex.
SYNONYMS: **Gerardia maritima** Raf.
SPECIAL NOTE: For more information see H. Hinds, 1983. The Rare Vascular Plants of New Brunswick. Syllogeus 50.

++

Agalinus purpurea L. Fig. p. A99.
COMMON NAME(S): Small-flowered Gerardia. Gérardie appauvrie.
FREQUENCY: Very rare; known only from Ross Is., Charlotte Co. and from Quarryville on the SW Miramichi R., Northumberland Co.
HABITAT: Boggy swales and seepy stream banks.
GENERAL RANGE: Man. to N.B. south to Tex. and Fla.
SUBSPECIFIC TAXA: Our plants are referable to var. **parviflora** (Benth.) Boivin.
SYNONYMS: **Gerardia purpurea** L. var. **parviflora** Benth.
SPECIAL NOTE: For more information see H. Hinds, 1983. The Rare Vascular Plants of New Brunswick. Syllogeus 50.

CASTILLEJA Mutis Indian Paint-brush, Painted-cup. Castilléjie

Castilleja septentrionalis Lindl. Fig. p. A99. Map p. B73.
COMMON NAME(S): Northern Painted-cup. Castilléjie pâle.
FREQUENCY: Rare; recorded along the St. John R. in Carleton, Victoria and Madawaska Counties and the Restigouche R., Restigouche Co.
HABITAT: Stream banks and peaty alpine summits (not N.B.).
GENERAL RANGE: Dist. Mack. to Labr. south to Utah and mtns. of N. Eng.
SYNONYMS: **C. pallida** (L.) Spreng. ssp. **septentrionalis** (Lindl.) Scoggan.

CHAENORRHINUM (DC.) Reichenb.

Chaenorrhinum minus (L.) Lange Fig. p. A99. Map p. B73.
COMMON NAME(S): Dwarf Snapdragon.
SPECIAL NOTE: Eurasian; frequently introduced mostly in the western half of the Province to roadsides, railways and waste areas. Known elsewhere in Canada from all provinces.

CHELONE L.

Chelone glabra L. Fig. p. A99. Map p. B73.
COMMON NAME(S): Turtlehead, Galmony. Tête de tortue, Galane.
FREQUENCY: Scattered throughout much of the Province.
HABITAT: Wet ditches, meadows and stream banks.
GENERAL RANGE: Man. to Nfld. south to Mo., and Ga.
SUBSPECIFIC TAXA: Var. **dilatata** Fern. & Wieg., with the upper leaves scarcely reduced and +/- rounded at the base, is reported from the northern part of the Province.

CYMBALARIA Hill.

Cymbalaria muralis Baumg. Fig. p. A99. Map p. B73.
COMMON NAME(S): Coliseum-Ivy, Kenilworth-Ivy, Ivy-leaved Toadflax. Cymbalaire.

++

SPECIAL NOTE: European; rarely introduced to roadsides and waste areas as in Upper Woodstock, Carleton Co., St. Andrews, Charlotte Co., and Saint John, St. John Co. Known elsewhere in Canada from B.C.; Ont. to N.S.

EUPHRASIA L. Eyebright

1a Corolla 2.5-4.5 mm long, lilac or purple, rarely white; inflorescence condensed; cauline internodes 1-3 times as long as hirsute or glabrous leaves; flowering from 3rd to 8th node; calyx teeth acute and straight. **E. randii**

1b Corolla over 5 mm long; inflorescence mostly elongate; flowering from (3rd) 5th to 12th node. 2

2a(1b) Bracts of inflorescence +/- erect, cuneate at base, teeth glabrous, mostly ascending; capsule 4-6 mm long. **E. stricta**

2b(1b) Bracts of inflorescence +/- spreading, more rounded or truncate at base, teeth spreading, glabrous or with glandular hairs or bristles. 3

3a(2b) Plants flowering from 7th to 9th node; leaves +/- bristly and also often with short glandular hairs; stem to 31 cm long, sometimes purplish; corolla white with lilac or all lilac 5-8 (-10) mm long; capsule 4.5-7 mm long. **E. arctica**

3b(2b) Plants flowering from 10th to 12th node; leaves glabrous or finely ciliate; stem to 25 cm long, generally dark purplish; corolla white suffused with purple, 5-7 mm long; capsule 4-5.5 mm long. **E. nemorosa**

Euphrasia arctica Lange ex Rostr. Fig. p. A99. Map p. B73.
SPECIAL NOTE: Probably introduced from the British Isles. Uncommon in the northern part of the Province and along the Bay of Fundy. Known elsewhere in Canada from Que. to Nfld. south to P.E.I. and N.S. Also known in B.C.
SUBSPECIFIC TAXA: Our plants are referable to ssp. **borealis** (Towns.) Yeo.
SYNONYMS: **E. borealis** Towns.

Euphrasia nemorosa (Pers.) Walk. Fig. p. A99. Map p. B73.
SPECIAL NOTE: Probably introduced from Europe. Common throughout much of the Province in open ground. Known elsewhere in Canada from Que. to N.S.
SYNONYMS: **E. americana** Wetts., **E. canadensis** Sensu auctt. non Townsend.

Euphrasia randii Rob. Fig. p. A99. Map p. B73.
COMMON NAME(S): Rand's Eyebright.
FREQUENCY: Very rare; known from Campobello Is., Grand Manan Is., and Machias Seal Is., Charlotte Co.
HABITAT: Grassy openings or cliff tops on coastal headlands.
GENERAL RANGE: Que. to Nfld. south to N.B., P.E.I., N.S. and Maine.
SUBSPECIFIC TAXA: Var. **farlowii** Robins, reported from Charlotte Co., differs from the species in several features.

++

1a Plants compact, diffusely branched, cauline internodes equal to twice as long as leaves; leaves 1.5-6 mm long, +/- densely hirsute; corolla white or cream, rarely with lilac upper lip or purplish; capsule 2.5-4 mm long. var. **farlowii**

1b Plants lax, cauline internodes 1-3(-5) times as long as leaves; leaves obtuse, crenate, glabrous or sparingly hirsute; corolla purple or lilac, rarely white; capsule 3-5 mm long. var. **randii**

Euphrasia stricta J.P. Wolff ex J.F. Lehm. Fig. p. A99. Map p. B73.
SPECIAL NOTE: Probably introduced from Europe. Scattered in pastures, thickets and seepy banks, throughout much of the Province. Known elsewhere in Canada from Que., P.E.I. and N.S.
SYNONYMS: **E. officinalis** L., **E. rigidula** Jord.

GRATIOLA L. Hedge-hyssop

Gratiola neglecta Torr. Fig. p. A99. Map p. B73.
FREQUENCY: Rare; scattered along the St. John R. drainage system in Sunbury, York and Carleton Counties.
HABITAT: Muddy shores and shallow water.
GENERAL RANGE: B.C. to N.B., N.S. south to Calif. Tex. and Ga.
SPECIAL NOTE: For more information see H. Hinds, 1983. The Rare Vascular Plants of New Brunswick. Syllogeus 50.

LIMOSELLA L. Mudwort

Limosella australis R. Br. Fig. p. A100. Map p. B74.
FREQUENCY: Rare; reported from Grand Manan Is. and near St. Stephen, Charlotte Co., Kouchibouguac National Park, Kent Co., and the estuary of the Miramichi R., Northumberland Co.
HABITAT: Brackish sands or mud.
GENERAL RANGE: Que. to Nfld. south to Va.
SYNONYMS: **L. subulata** Ives.

LINARIA Mill. Toadflax. Linaire

1a Leaves +/- clasping, lance-ovate to broadly ovate; corolla yellow, about 4 cm long. **L. dalmatica**

1b Leaves linear, not clasping; flowers 3 cm long or less. 2

2a(1b) Corolla yellow with orange palate, 2-3 cm long; capsule 8-12 mm long. **L. vulgaris**

2b(1b) Corolla blue or whitish. 3

3a(2b) Corolla blue or violet (rarely white), throat open with rounded ridges; slender annual or biennial with fibrous roots. **L. canadensis**

3b(2b) Corolla pale blue or creamy white striped with violet, throat closed by yellow spotted palate; leafy perennial from creeping rootstock. **L. repens**

Linaria canadensis (L.) Dumont Fig. p. A100. Map p. B74.
COMMON NAME(S): Old-field Toadflax.
FREQUENCY: Uncommon; scattered in the southwest third of the Province.
HABITAT: Dry sandy openings.
GENERAL RANGE: B.C., Sask., Ont. to N.B., N.S. south to Calif., Mex. and Fla.

Linaria dalmatica (L.) Mill.
COMMON NAME(S): Dalmatian Toadflax.
SPECIAL NOTE: European; cultivated and rarely escaped to roadsides and waste areas as in Woodstock, Carleton Co. Known elsewhere in Canada from B.C. to Ont. and N.S.

Linaria repens (L.) Mill. Fig. p. A100. Map p. B74
COMMON NAME(S): Striped Toadflax.
SPECIAL NOTE: European; introduced and locally common on roadsides and in waste areas as in Northumberland Co. and Shediac, Westmorland Co. Known elsewhere in Canada from Nfld. and N.S.

Linaria vulgaris Hill Fig. p. A100.
COMMON NAME(S): Butter-and-eggs, Common Toadflax. Gueule de lion.
SPECIAL NOTE: Eurasian; introduced throughout much of the Province to roadsides and waste areas. Known elsewhere in Canada from all provinces.

LINDERNIA All. False Pimpernel

Lindernia dubia (L.) Pennell Fig. p. A99. Map p. B74.
FREQUENCY: Scattered throughout the southern half of the Province.
HABITAT: Muddy shores and shallow water.
GENERAL RANGE: Ont. to N.B. and N.S. south to Tex. and Ga.
SUBSPECIFIC TAXA: Var. **inundata** Pennell, with bracteal leaves rounded at summit and flowers undeveloped rather than leaves +/- acutish at summit and flowers expanding, is reported from the Miramichi R. estuary, Northumberland Co. (UNB).

MELAMPYRUM L.

Melampyrum lineare Desr. Fig. p. A100. Map p. B74.
COMMON NAME(S): Cow-wheat.
FREQUENCY: Common in the south and east of the Province.
HABITAT: Coniferous woods, heaths and rocky barrens.
GENERAL RANGE: B.C. to Nfld. south to Wash., Minn., Ga. and S.C.

++

MIMULUS L. Monkey-flower. Mimule

1a Leaves sessile, rounded or clasping at base, lanceolate; corolla purple or pinkish. **M. ringens**

1b Leaves petiolate, ovate; corolla yellow. 2

2a(1b) Plant erect, glabrous; leaves palmately veined from near base; corolla to 4 cm long, spotted with red-brown within. **M. guttatus**

2b(1b) Plant creeping at base, viscid-villous; leaves pinnately veined; corolla to 3 cm long, unspotted within. **M. moschatus**

Mimulus guttatus Fisch. ex DC. Fig. p. A100.
COMMON NAME(S): Yellow Monkey-flower. Mimule ponctué.
SPECIAL NOTE: Native to western U.S.A. and Canada east to Sask. south to Mexico; introduced in wet fields and roadside ditches near Springfield, Kings Co. (UNB).

Mimulus moschatus Dougl. Fig. p. A100. Map p. B74.
COMMON NAME(S): Muskflower. Mimule musquée.
SPECIAL NOTE: Native from B.C. to Calif. and Colo.; probably introduced in moist meadows and streamsides as along the eastern shore of the Bay of Fundy and along the east coast. Known elsewhere in eastern Canada from Ont. to Nfld., P.E.I. and N.S.

Mimulus ringens L. Fig. p. A100. Map p. B74.
COMMON NAME(S): Monkey-flower.
FREQUENCY: Common throughout much of the Province.
HABITAT: Shores and wet depressions.
GENERAL RANGE: Sask. to N.S. south to Tex. and Ga.
SUBSPECIFIC TAXA: Forma **peckii** House, with white rather than pinkish or purplish flowers, is recorded from Woodstock, Carleton Co.

ODONTITES Ludwig

Odontites verna (Bell.) Dum. Fig. p. A100.
COMMON NAME(S): Red Bartsia.
SPECIAL NOTE: Eurasian; introduced on Cape Tormentine and near Stony Creek, Westmorland Co., on Grand Manan Is., Charlotte Co., and Saint John, St. John Co. Known elsewhere in Canada from Alta., Man. to N.S.
SYNONYMS: **O. serotina** (Wettst.) Dum.

PEDICULARIS L. Lousewort, Wood-betony. Pédiculaire

1a Basal half of lower leaves divided to midrib, acuminate tipped; calyx-lobes 5, subequal; stem 4-6 dm high. **P. furbishiae**

1b Lower leaves lobed but not divided to midrib, +/- rounded to acute tipped; calyx very oblique divided into two entire lateral halves; stem 1.5-4 dm high. **P. canadensis**

++

Pedicularis canadensis L. Fig. p. A100.
COMMON NAME(S): Wood-betony, Common Lousewort, Chicken's-heads.
FREQUENCY: Very rare; reported from Curry Mtn., York Co. (now believed extirpated) and the lower St. Croix R. near Spragues Falls, Charlotte Co.
HABITAT: Open dry woods and clearings.
GENERAL RANGE: Man. to N.B. south to Mex. and Fla.
SPECIAL NOTE: For more information see H. Hinds, 1983. The Rare Vascular Plants of New Brunswick. Syllogeus 50.

Pedicularis furbishiae Wats. Fig. p. A100. Map p. B74.
COMMON NAME(S): Furbish's Lousewort, St. John River Wood-betony.
FREQUENCY: Very rare; known from three sites on or near the Saint John R. in Victoria Co.
HABITAT: River and railway banks.
GENERAL RANGE: N.B. and northern Maine. Unknown elsewhere in the world!
SPECIAL NOTE: For more information see H. Hinds, 1983. The Rare Vascular Plants of New Brunswick. Syllogeus 50.

PENSTEMON Mitchell Beard-tongue

Penstemon digitalis Nutt. Fig. p. A101.
COMMON NAME(S): Foxglove Beard-tongue.
SPECIAL NOTE: Native to eastern U.S.A.; cultivated and introduced to fields and roadsides as in Charlotte and Carleton Counties. Known elsewhere in Canada from Ont., Que. and N.S.

RHINANTHUS L. Yellow Rattle

Rhinanthus minor L. Fig. p. A101. Map p. B74.
COMMON NAME(S): Common Yellow Rattle. Claquette, Sonnette.
SPECIAL NOTE: Probably Eurasian; introduced throughout much of the Province to roadsides, old fields and clearings. Known elsewhere in Canada from all provinces.
SYNONYMS: **R. crista-galli** L.

SCROPHULARIA L. Figwort. Scrophulaire.

1a Leaves sharply serrate or incised near base, mostly acuminate at tip, cuneate to truncate at base; sterile stamen greenish-yellow. **S. lanceolata**
1b Leaves more coarsely blunt toothed, abruptly acute at tip, cordate at base; sterile stamen brown to dull purple. **S. nodosa**

Scrophularia lanceolata Pursh Fig. p. A101. Map p. B74.
FREQUENCY: Uncommon; scattered in the southwest third of the Province.
HABITAT: Open woods and old fields.
GENERAL RANGE: B.C., Sask., Ont. to N.B., N.S. south to Calif. and S.C.

++

Scrophularia nodosa L. Fig. p. A101.
SPECIAL NOTE: Eurasian; rarely introduced as in Saint John and Fairview, St. John Co. and near Sussex, Kings Co. (UNB). Known elsewhere in Canada from Que. and Nfld.

VERBASCUM L. Mullein. Molène

1a Leaves densely tomentose on both surfaces, decurrent, entire; inflorescence dense, mostly an unbranched spike; stem to 2 m high. **V. thapsus**
1b Leaves glabrous, coarsely toothed or lobed, not decurrent; inflorescence open racemose; stem to 1 m tall. **V. blattaria**

Verbascum blattaria L. Fig. p. A101.
COMMON NAME(S): Moth-Mullein.
SPECIAL NOTE: Eurasian; rarely introduced to roadsides and waste areas as in St. Andrews, Charlotte Co. and Woodstock, Carleton Co. Known elsewhere in Canada from B.C.; Ont. to N.B. and N.S.

Verbascum thapsus L. Fig. p. A101. Map p. B74.
COMMON NAME(S): Common Mullein, Flannel-plant. Tabac du diable, Bouillon blanc.
SPECIAL NOTE: Eurasian; introduced to roadsides and waste areas throughout the Province. Known elsewhere in Canada from all provinces except Sask.

VERONICA L. Speedwell. Véronique

1a Main axis terminated by opposite leaves; flowers in axillary racemes. 2
1b Main axis terminating in an inflorescence; flowers borne solitary in axils of leaves or in terminal spikes; upper bract-like leaves alternate. 3
2a(1a) Leaves linear-lanceolate, sessile, obscurely dentate; pedicels filiform, reflexed in fruit. **V. scutellata**
2b(1a) Leaves ovate to oblong or lanceolate. 4
3a(1b) Flowers crowded in spike-like inflorescence; leaves sharply serrate, 4-10 mm long; erect perennial to 1 m high. **V. longifolia**
3b(1b) Flowers in looser, raceme-like inflorescence; leaves usually less than 3 cm long. 5
4a(2b) Stem pubescent; leaves sessile or on winged petioles. 6
4b(2b) Stem +/- glabrous; leaves petiolate, lance-ovate, broadest near base. **V. americana**
5a(3b) Pedicels 7-35 mm long, often deflexed; mature capsules veiny. 7
5b(3b) Pedicels 1-5 mm long; mature capsules not veiny. 8
6a(4a) Leaves sessile or on petioles less than 2 mm long, ovate or cordate, coarsely crenate-serrate; pedicels 5-9 mm long equal to or longer than sepals. **V. chamaedrys**
6b(4a) Leaves abruptly narrowed to winged petioles or sessile, elliptic to obovate, finely serrate; pedicels 1-2 mm long, shorter than sepals. **V. officinalis**
7a(5a) Fruiting pedicels over 3 cm long, about twice as long as subtending leaves. **V. persica**

7b(5a) Fruiting pedicels less than 1 cm long, less than twice as long as subtending leaves. 9

8a(5b) Bracts of inflorescence conspicuously extended beyond fruit; plant glabrous or stipitate glandular, +/- fleshy; leaves narrowly oblong to oblanceolate, entire or shallowly dentate; corolla white. **V. peregrina**

8b(5b) Bracts of inflorescence inconspicuous, scarcely extended beyond fruit; plants pilose to closely puberulent; leaves mostly broader; corolla blue. 10

9a(7b) Style much surpassing lobes of capsule; corolla mostly bright blue. **V. polita**

9b(7b) Style shorter than lobes of capsule; corolla light blue and white. **V. agrestis**

10a(8b) Plants mostly erect at base, pilose; mid stem leaves serrate or lobed; annuals. 11

10b(8b) Plants usually decumbent at base, finely puberulent; mid stem leaves entire or low dentate; rhizomatous perennials. **V. serpyllifolia**

11a(10a) Style shorter than capsule lobes; mid stem leaves lobed. **V. verna**

11b(10a) Style as long as or longer than capsule lobes; mid stem leaves serrate. **V. arvensis**

Veronica agrestis L. Fig. p. A101.
COMMON NAME(S): Field Speedwell.
SPECIAL NOTE: Eurasian; introduced in waste areas in Fredericton, York Co., Richibucto, Kent Co., Saint John, St. John Co. and St. Andrews, Charlotte Co. Known elsewhere in Canada from B.C., Alta., Ont. to Nfld., P.E.I. and N.S.

Veronica americana (Raf.) Schwein. ex Benth. Fig. p. A101. Map p. B75.
COMMON NAME(S): American Brooklime.
FREQUENCY: Scattered throughout much of the Province.
HABITAT: Shallow water and seepage areas.
GENERAL RANGE: Alaska, Yukon and B.C. to Nfld. south to Calif., Mex. and N.C.; Asia.

Veronica arvensis L. Fig. p. A101.
COMMON NAME(S): Corn-Speedwell.
SPECIAL NOTE: Eurasian; reported from Fundy National Park, and Hillsborough, Albert Co., and St. Andrews, Charlotte Co. Known elsewhere in Canada from B.C., Ont. to Labr., Nfld., and N.S.

Veronica chamaedrys L. Fig. p. A101. Map p. B75.
COMMON NAME(S): Bird's-eye, Angel's-eye. Petit-chêne, Herbe à Thérèse.
SPECIAL NOTE: Eurasian; introduced to moist roadsides and waste areas in the western half of the Province. Known elsewhere in Canada from B.C., Ont. to Nfld., P.E.I. and N.S.

Veronica longifolia L. Map p. B75.
SPECIAL NOTE: Eurasian; occasionally introduced to roadsides and waste areas in the southern half of the Province. Known elsewhere in Canada from Alta., Sask., Ont. to Nfld., P.E.I. and N.S.

++

Veronica officinalis L. Fig. p. A101. Map p. B75.
COMMON NAME(S): Common Speedwell, Gypsyweed. Thé d'Europe.
SPECIAL NOTE: Eurasian; commonly introduced to roadsides, fields and waste areas throughout the Province. Known elsewhere in Canada from B.C., Ont. to Nfld., P.E.I. and N.S.
SYNONYMS: **V. officinalis** var. **tournefortii** (Vill.) Reichenb.

Veronica peregrina L. Fig. p. A101. Map p. B75.
COMMON NAME(S): Neckweed, Purslane-Speedwell. Véronique voyageuse.
SUBSPECIFIC TAXA: Two subspecies are recognized in the Province.

1a Plants essentially glabrous. ssp. **peregrina**
1b Plants +/- short stipitate glandular at least above. ssp. **xalapensis**

1. Ssp. **peregrina**
FREQUENCY: Very rare; known only from nursery, Acadia Exp. Station, Sunbury Co. (UNB).
HABITAT: Dry sandy open area.
GENERAL RANGE: B.C., Ont. to P.E.I. south to Tex. & Fla.

2. Ssp. **xalapensis** (HBK.) Pennell
FREQUENCY: Widely scattered in the southern half of the Province.
HABITAT: Dry sandy or gravelly and muddy areas.
GENERAL RANGE: Alaska, Yukon and B.C. to Nfld. south to C. America and Ga.

Veronica persica Poir. Fig. p. A102. Map p. B75.
COMMON NAME(S): Bird's-eye.
SPECIAL NOTE: Eurasian; occasionally introduced in the St. John R. valley. Known elsewhere in Canada from B.C. to Nfld., P.E.I. and N.S.
SUBSPECIFIC TAXA: Our plants are referable to var. **aschersoniana** (Lehm.) Boivin.

Veronica polita Fries Fig. p. A102.
COMMON NAME(S): Wayside Speedwell.
SPECIAL NOTE. Eurasian; introduced in lawns and roadsides as in Fredericton, York Co. and St. Andrews, Charlotte Co. (UNB). Known elsewhere in Canada from Man. and Ont.

Veronica scutellata L. Fig. p. A102. Map p. B75.
COMMON NAME(S): Marsh-Speedwell. Véronique en écusson.
FREQUENCY: Common throughout much of the Province.
HABITAT: Shores, and similar wet places.
GENERAL RANGE: Alaska, Yukon and B.C. to Nfld. south to Calif. and Va.; Iceland; Eurasia.

Veronica serpyllifolia L. Fig. p. A102. Map p. B75.
COMMON NAME(S): Thyme-leaved Speedwell.
SUBSPECIFIC TAXA: Two subspecies are recognized in our range.

1a Pubescence of stem and pedicels incurved, nonglandular; corolla 3-5 mm wide, pale blue and blue streaked; filaments less than 3 mm long. ssp. **serpyllifolia**

1b Pubescence of stem and pedicels both incurved and nonglandular and also spreading glandular; corolla 5-8 mm wide, deep blue-violet with darker lines; filaments to 4 mm long. ssp. **humifusa**

1. Ssp. **serpyllifolia**
SPECIAL NOTE: Eurasian; introduced in moist lawns, roadsides and waste areas throughout much of the Province. Known elsewhere in Canada from B.C., Man. to Labr., Nfld., P.E.I. and N.S.

2. Ssp. **humifusa** (Dickson) Syme
FREQUENCY: Scattered throughout much of the Province.
HABITAT: Shores and seepage areas.
GENERAL RANGE: Alaska and B.C. to Labr. south to Calif., Mex., and N.Y.; S. America; Eurasia.
SYNONYMS: **V. tenella** All.

Veronica verna L. Fig. p. A102.
COMMON NAME(S): Spring Speedwell.
SPECIAL NOTE: Eurasian; rarely introduced to gravelly or rocky roadsides as in Fundy National Park, Albert Co. (UNB). Known elsewhere in Canada from B.C. and Ont.

OROBANCHACEAE (Broom-rape Family)

1a Flowers white to violet, scapose, solitary; plants usually cespitose, conspicuously glandular-villous; parasitic on many different species. **Orobanche uniflora**

1b Flowers whitish with purple-brown stripes, numerous in large panicle; plants minutely glandular-pubescent, parasitic only on beech. **Epifagus virginiana**

EPIFAGUS Nutt.

Epifagus virginiana (L.) Bart. Fig. p. A102. Map p. B75.
COMMON NAME(S): Beech-drops.
FREQUENCY: Scattered throughout the southern part of the Province but uncommon in the eastern lowlands.
HABITAT: Parasitic on roots of beech.
GENERAL RANGE: Ont. to N.S. south to La. and Fla.

OROBANCHE L. Broom-rape

Orobanche uniflora L. Fig. p. A102. Map p. B75.
COMMON NAME(S): One-flowered Cancer-root.

+++

FREQUENCY: Uncommon in the southern half of the Province.
HABITAT: Parasitic on many different plants in damp thickets, woods and meadows.
GENERAL RANGE: Ont. to Nfld. south to Calif. and Fla.

LENTIBULARIACEAE (Bladderwort Family)

1a Calyx 2-lobed; leaves linear, to 3 cm long or finely dissected or divided; bladders usually present; corolla yellow or purple. **Utricularia**

1b Calyx 5-lobed; leaves elliptic to ovate in basal rosette, glutinous above; corolla violet. **Pinguicula vulgaris**

PINGUICULA L. Butterwort. Grassette

Pinguicula vulgaris L. Fig. p. A102.
FREQUENCY: Very rare; reported only from the Restigouche R., Restigouche Co.
HABITAT: Damp calcareous shores and banks.
GENERAL RANGE: Alaska to Baffin Is., Labr. and Nfld. south to B.C., N.Y. and Vt.; Greenland; Iceland; Eurasia.
SPECIAL NOTE: For more information see H. Hinds, 1983. The Rare Vascular Plants of New Brunswick. Syllogeus 50.

UTRICULARIA L. Bladderwort.

1a Flowers purple or violet. 2

1b Flowers yellow. 3

2a(1a) Leaves entire, linear, to 3 cm long, bladders inconspicuous; flowers solitary; corolla facing upwards; plants of sandy wet soil or shallow water. **U. resupinata**

2b(1a) Leaves finely dissected, whorled; bladders conspicuous, terminal; flowers 1-4-flowered, facing outwards; coarser plants of deeper water. **U. purpurea**

3a(1b) Leaves entire, linear, inconspicuous; bladders minute; bracts broad-based with a pair of narrow inner bractlets. **U. cornuta**

3b(1b) Leaves divided or dissected; bladders usually conspicuous; bracts without inner bractlets. 4

4a(3b) Ultimate leaf divisions flattened, +/- as wide as primary divisions; lower corolla lip +/- twice as long as upper. 5

4b(3b) Leaf divisions progressively narrower towards ultimate segments; lower corolla lip +/- as long as upper. 6

5a(4a) Bladders on specialized branches distinct from leaves; ultimate leaf segments blunt tipped, margins spinulose. **U. intermedia**

5b(4a) Bladders attached to leaves, not on separate branches; ultimate leaf segments acuminate, margins entire. **U. minor**

6a(4b) Stems mostly free floating; leaves repeatedly forked with numerous bladders; pedicels down curved in fruit. 7

6b(4b) Stems creeping on mud or other debris in shallow water; leaves 1 or rarely 2-forked with few bladders. **U. gibba**

7a(6a) Leaves with well defined, thicker central rachis, segments spinulose; bract at base of pedicel with basal lobes; cleistogamous flowers wanting; scape scaly, to 20 cm long. **U. vulgaris**

7b(6a) Leaves usually without a central rachis, margins entire; bract at base of pedicel without basal lobes; cleistogamous flowers scattered at base of scape; scape without scales, to 10 cm long. **U. geminiscapa**

Utricularia cornuta Michx. Fig. p. A102. Map p. B75.
FREQUENCY: Scattered throughout much of the Province.
HABITAT: Acid peat bogs and boggy shores.
GENERAL RANGE: Man. to Ont. and Labr. south to Tex. and Fla.

Utricularia geminiscapa Benj. Fig. p. A102. Map p. B75.
FREQUENCY: Rare; reported from Sterling Lake, York Co., near Richibucto, Kent Co., Cape Enrage, Albert Co. and Grand Manan Is., Charlotte Co.
HABITAT: Bog ponds and slow streams.
GENERAL RANGE: Que. to Nfld. south to Wisc., N.Y. and Va.
SYNONYMS: **U. clandestina** Nutt.
SPECIAL NOTE: For more information see H. Hinds, 1983. The Rare Vascular Plants of New Brunswick. Syllogeus 50.

Utricularia gibba L. Fig. p. A102.
FREQUENCY: Very rare; reported from Grand Manan Is., Charlotte Co. and Yoho Lake, York Co.
HABITAT: Shallow water and bogs.
GENERAL RANGE: Minn. and Ont. to N.B., N.S. south to Calif., Mex. and Fla.; W.I.; C. America.
SPECIAL NOTE: For more information see H. Hinds, 1983. The Rare Vascular Plants of New Brunswick. Syllogeus 50.

Utricularia intermedia Hayne Fig. p. A103. Map p. B76.
FREQUENCY: Common throughout much of the Province.
HABITAT: Ponds and slow streams.
GENERAL RANGE: Alaska to Labr. and Nfld. south to Calif. and Del.; Greenland; Iceland; Eurasia.

Utricularia minor L. Fig. p. A103. Map p. B76.
FREQUENCY: Rare; reported from the Nepisiguit and NW Miramichi Rivers, Northumberland Co.
HABITAT: Shallow pools, and boggy meadows.
GENERAL RANGE: Alaska to Labr. and Nfld. south to Calif. and N.J.; Greenland; Iceland; Eurasia.

SPECIAL NOTE: For more information see H. Hinds, 1983. The Rare Vascular Plants of New Brunswick. Syllogeus 50.

Utricularia purpurea Walt. Fig. p. A103. Map p. B76.
COMMON NAME(S): Greater Purple Bladderwort
FREQUENCY: Scattered in the southwest corner of the Province.
HABITAT: Lakes, ponds and slow streams.
GENERAL RANGE: Wisc. to Ont., Que., N.B., N.S. south to La. and Fla.; W.I.; C. America.

Utricularia resupinata Greene Fig. p. A103.
COMMON NAME(S): Lesser Purple Bladderwort
FREQUENCY: Very rare; collected by Colin Livingstone in 1986 on sandy shore of Phipp's Lake, Long Reach, King Co. This locality has not been relocated.
HABITAT: Damp shores and shallow water.
GENERAL RANGE: Wisc. to Ont., Que., N.B., N.S. south to Ill. and Fla.
SPECIAL NOTE: For more information see H. Hinds, 1983. The Rare Vascular Plants of New Brunswick. Syllogeus 50.

Utricularia vulgaris L. Fig. p. A103. Map p. B76.
FREQUENCY: Common in the south of the Province and rare northwards.
HABITAT: Quiet waters.
GENERAL RANGE: Alaska to Labr. and Nfld. south to Calif., Mex. and Va.; Eurasia.

PHRYMACEAE (Lopseed Family)

PHRYMA L. Lopseed

Phryma leptostachya L. Fig. p. A103. Map p. B76.
COMMON NAME(S): Lopseed.
FREQUENCY: Uncommon; reported from Kings, York, Carleton and Victoria Counties.
HABITAT: Rich woods.
GENERAL RANGE: Man. to N.B. south to Tex. and Fla.

PLANTAGINACEAE (Plantain Family)

1a Plants to about 4 cm high; scapes 1-flowered; leaves narrowly linear, to 5 cm long. **Littorella americana**
1b Plants over 1 dm high; scapes many flowered; leaves linear to broadly ovate, to over 1 dm long. **Plantago**

LITTORELLA Bergius

Littorella americana Fern. Fig. p. A103.
COMMON NAME(S): Shore-weed. American Littorella.

++

FREQUENCY: Uncommon and overlooked; scattered in the southwest third of the Province.
HABITAT: Shallow water and muddy shores.
GENERAL RANGE: Minn. to Ont., Que. and Nfld. south to N.Y., Me. and Vt.
SYNONYMS: **L. uniflora** (L.) Aschers. var. **americana** (Fern.) Gl.
SPECIAL NOTE: For more information see H. Hinds, 1983. The Rare Vascular Plants of New Brunswick. Syllogeus 50.

PLANTAGO L. Plantain, Ribgrass, Ribwort. Plantain, Queue de rat

1a Inflorescence usually much less than 1/3 as long as peduncle; bracts and sepals scarcely keeled. 2
1b Inflorescence 1/3 or more the length of peduncle; bracts and sepals prominently keeled. 3
2a(1a) Leaves lanceolate, 1-4 dm long; scape deeply furrowed; inflorescence short cylindric, to 8 cm long at maturity, rounded at base. **P. lanceolata**
2b(1a) Leaves elliptic to obovate or oblanceolate, 1-2 dm long; scape not furrowed; inflorescence cylindric, to 10 cm at maturity, usually tapering at base. **P. media**
3a(1b) Leaves lanceolate, thick and fleshy; corolla tube pubescent without. **P. maritima**
3b(1b) Leaves broadly ovate, not fleshy; corolla tube glabrous without. 4
4a(3b) Capsule dehiscing well below middle, 3-6 mm long; seeds 4-9 per capsule; leaf margin often dentate. **P. rugelii**
4b(3b) Capsule dehiscing from above to about the middle, 2-4 mm long; seeds 6-12(22) per capsule; leaf margins +/- entire. **P. major**

Plantago lanceolata L. Fig. p. A103. Map p. B76.
COMMON NAME(S): Ribgrass, Ripplegrass, Buckhorn, English Plantain. Bonne femme, Herbe à cinq côtes.
SPECIAL NOTE: Eurasian; uncommonly scattered mostly in the south of the Province. Known elsewhere in Canada from all provinces.
SUBSPECIFIC TAXA: Forma **composita** Farw., with compound spike is reported from Fredericton, York Co. (UNB).

Platago major L. Fig. p. A103.
COMMON NAME(S): Common Plantain, Whiteman's-foot. Grand Plantain.
SPECIAL NOTE: Eurasian; common throughout much of the Province. Known elsewhere in Canada from all provinces.

Plantago maritima L. Fig. p. A103.
COMMON NAME(S): Seaside-Plantain, Goose-tongue. Perce-pierre, Passe-pierre.
FREQUENCY: Common in coastal areas.
HABITAT: Coastal salt marshes, sands and ledges.
GENERAL RANGE: Alaska to Calif. on Pacific coast east across coastal areas of northern Canada to Ungava Bay south on the Atlantic coast to N.J.; Greenland; S. America; Iceland; Eurasia.

++

SUBSPECIFIC TAXA: Our plants are referable to var. **juncoides** (Lam.) Gray.
SYNONYMS: **P. juncoides** Lam., **P. oliganthos** Roemer & Schultes.

Plantago media L. Fig. p. A103.
COMMON NAME(S): Hoary Plantain. Plantain bâtard.
SPECIAL NOTE: Eurasian; rarely introduced to lawns and waste areas as in Bathurst, Gloucester Co. Known elsewhere in Canada from B.C., Man., Ont. to N.S.

Plantago rugelii Dcne. Fig. p. A103.
FREQUENCY: Very rare; reported from Woodstock, Carleton Co. and Saint John, St. John Co.
HABITAT: Shores and waste areas.
GENERAL RANGE: Ont. to N.S. south to Tex. and Fla.
SPECIAL NOTE: Probably introduced in the Province.

RUBIACEAE (Madder Family)

1a Principal leaves whorled; stem 4-angled; corolla tube very short; fruit dry, of 2 +/- globose 1-seeded carpels. **Galium**
1b Principal leaves opposite. 2
2a(1b) Plant shrubby, 1-3 m high; leaves deciduous; flowers white, in dense globose peduncled heads. **Cephalanthus occidentalis**
2b(1b) Plant herbaceous or evergreen; flowers axillary or terminal. 3
3a(2b) Stem trailing; leaves evergreen, round-ovate; fruit fleshy, red; corolla lobes villous above, white. **Mitchella repens**
3b(2b) Stem erect, often tufted; leaves not evergreen, obovate or spatulate; corolla white or blue with yellow eye; fruit capsular. **Hedyotis caerulea**

CEPHALANTHUS L.

Cephalanthus occidentalis L. Fig. p. A104.
COMMON NAME(S): Buttonbush. Bois noir.
FREQUENCY: Rare; only known from the St. Croix R. valley in York and Charlotte Counties.
HABITAT: Swamps and stream margins.
GENERAL RANGE: Ont. to N.B., N.S. south to Calif., Mex. and Fla.
SPECIAL NOTE: For more information see H. Hinds, 1983. The Rare Vascular Plants of New Brunswick. Syllogeus 50. The Grand Lakes reference probably refers to the Grand Lake of the St. Croix R., York Co. and the latest collections were made in 1985 (UNB).

GALIUM L. Bedstraw, Cleavers. Gaillet

1a Leaves blunt or abruptly acute-tipped. 2
1b Leaves sharply acuminate, cuspidate or mucronate-tipped. 3

++

2a(1a) Leaves linear to linear-lanceolate, broadest near base; flowers in dense terminal panicles; corolla clear white; stem short-bearded below nodes. **G. boreale**

2b(1a) Leaves elliptic to oval or obovate, broadest near or above middle. 4

3a(1b) Fruit bristly. 5

3b(1b) Fruit smooth. 6

4a(2b) Fruit bristly; stem short, erect; leaves in 2-4 whorls, broadly ovate. **G. kamtschaticum**

4b(2b) Fruit smooth. 7

5a(3a) Leaves mostly 8 per whorl, down-pointed hispid on upper surface, margins and midvein beneath, bristle-tipped; stem harshly down-pointed bristly; annuals. **G. aparine**

5b(3a) Leaves 4-6 per node, mostly smooth above, down-pointed hispid on midvein beneath but with upwardly pointed ciliate margins, bristle-tipped; stem +/- barbed hispid on angles; perennials. **G. triflorum**

6a(3b) Corolla yellow; leaves linear, +/- 6 per node. **G. verum**

6b(3b) Corolla white; leaves broader. 8

7a(4b) Inflorescence with repeated divergent branching, each branch bearing 5-many flowers; pedicels usually less than 5 mm long; corolla white, 4-parted, 2.5-3.3 mm broad; leaves less than 1.5 cm long and with stems sparingly down-pointed scabrous; fruit to 2 mm thick. **G. palustre**

7b(4b) Inflorescence only once or twice branched, each branch bearing 2-4 flowers; pedicels 5-10 mm long. 9

8a(6b) Stem smooth or slightly hairy below, erect from decumbent base. **G. mollugo**

8b(6b) Stem down-pointed scabrous, weak, with matted basal offshoots. **G. asprellum**

9a(7b) Corolla rarely to 1.5 mm broad, mostly with 3 blunt greenish white lobes; stems reclining, +/- downwardly scabrous, developing matted basal offshoots. 10

9b(7b) Corolla to 3 mm broad, usually with 4 acute white lobes; stems erect or ascending, smooth or slightly hispidulous, lacking matted basal offshoots. 11

10a(9a) Pedicels often scabrous, arching in fruit, .3-3 cm long, 1-3 per peduncle; leaves 4 per node, linear to linear-oblanceolate; fruit to 2 mm thick. **G. trifidum**

10b(9a) Pedicels smooth, straight, widely spreading in fruit, to 8 mm long; leaves 4-6 per node, oblanceolate to oblong-spatulate. 12

11a(9b) Leaves spreading or ascending, scabrous on midvein beneath, 1.5-3 cm long; inflorescence mostly terminal, not surpassed by sterile branches; fruit 2.5-3.5 mm thick. **G. obtusum**

11b(9b) Leaves mostly reflexed on lower main stem, midvein usually glabrous beneath, margins closely spreading ciliate, often +/- revolute, 0.5-1.5 cm long; peduncles to 1.7 cm long; inflorescence once branched, soon surpassed by sterile branches; fruit less than 2 mm thick. **G. labradoricum**

12a(10b) Mature fruit to 1.5 mm thick; leaves to 2 cm long; stem ascending to reclining. **G. tinctorium**

12b(10b) Mature fruit to 1 mm thick; 1-2 flowered peduncles 0.5-4 mm long; leaves to 1 cm long; stem low, intricately branched, forming dense mats. **G. brevipes**

++

Galium aparine L. Fig. p. A104.
COMMON NAME(S): Cleavers, Goosegrass. Gaillet gratteron, Herbe collante.
SPECIAL NOTE: Our plants are probably Eurasian; sparingly introduced in waste areas and fields as in Saint John, St. John Co., Petitcodiac, Westmorland Co., and Fredericton, York Co. Known elsewhere in Canada from B.C. to Nfld. & N.S.

Galium asprellum Michx. Fig. p. A104. Map p. B76.
COMMON NAME(S): Rough Bedstraw. Gaillet piquant.
FREQUENCY: Common throughout much of the Province.
HABITAT: Marshes, wet meadows and thickets.
GENERAL RANGE: Ont. to Nfld., P.E.I. and N.S. south to Nebr. and N.C.

Galium boreale L. Fig. p. A104. Map p. B76.
COMMON NAME(S): Northern Bedstraw, Crosswort.
FREQUENCY: Uncommon; scattered in southcentral and westcentral parts of the Province.
HABITAT: Open woods, thickets and meadows.
GENERAL RANGE: Alaska to Gaspé Pen. and N.S. south to Calif., Tex. and Del.; Greenland; Iceland; Eurasia.

Galium brevipes Fern. & Wieg.
FREQUENCY: Rare; reported from Spednic Lake and Big Kedron Lakes, York Co. (UNB).
HABITAT: Low depressions and shores (often calcareous).
GENERAL RANGE: Minn. and Ont. to N.B. south to Mich., N.Y. and Me.

Galium kamtschaticum Steller Fig. p. A104.
COMMON NAME(S): Northern Wild Licorice.
FREQUENCY: Very rare; known from Moose Lake, York Co. and from near Summit Depot, Restigouche Co.
HABITAT: Cool, mossy coniferous woods.
GENERAL RANGE: Alaska and B.C. to Nfld. south to mts. of N.Y. and N. Eng.; Asia.
SPECIAL NOTE: For more information see H. Hinds, 1983. The Rare Vascular Plants of New Brunswick. Syllogeus 50.

Galium labradoricum Wieg. Fig. p. A104. Map p. B76.
COMMON NAME(S): Bog Bedstraw.
FREQUENCY: Rare; reported from scattered localities in Gloucester, Restigouche, Madawaska, Carleton and Charlotte Counties
HABITAT: Cool, moist woods and bogs.
GENERAL RANGE: Dist. Mack. and Alta. to Labr. and Nfld. south to Minn. and N.J.

Galium mollugo L. Fig. p. A107. Map p. B76
COMMON NAME(S): White Bedstraw, Baby's Breath, Smooth Bedstraw. Caillelait, Rable.
SPECIAL NOTE: Eurasian; commonly introduced throughout much of the Province in fields, waste areas and roadsides. Known elsewhere in Canada from B.C., Ont. to Nfld., P.E.I. and N.S.
SUBSPECIFIC TAXA: Most of our plants are referable to var. **erectum** (Huds.) Syme, with inflorescence branches +/- erect, and leaves linear-lanceolate rather than branches of inflorescence spreading and leaves mostly oblanceolate or obovate.
SYNONYMS: **G. erectum** Huds., **G. album** Miller.

++

Galium obtusum Bigel. Fig. p. A104. Map p. B76.
FREQUENCY: Uncommon; mostly reported from along the east coast.
HABITAT: Shores and similar wet places.
GENERAL RANGE: Minn. and Ont. to N.B. and N.S. south to Ariz. and Fla.

Galium palustre L. Fig. p. A104. Map p. B76.
COMMON NAME(S): Marsh Bedstraw.
FREQUENCY: Common throughout much of the Province.
HABITAT: Wet open shores, meadows, and similar open areas.
GENERAL RANGE: Ont. to Nfld. south to Wisc. and N. Eng.; Eurasia.

Galium tinctorium L. Fig. p. A104. Map p. B77.
COMMON NAME(S): Clayton's Bedstraw.
FREQUENCY: Scattered over much of the Province.
HABITAT: Wet depressions and bottomlands.
GENERAL RANGE: Ont. to Nfld. south to Nebr., Tex. and S.C.
SYNONYMS: **G. trifidum** L. var. **tinctorium** (L.) T. & G., **G. claytonii** Michx.

Galium trifidum L. Fig. p. A104. Map p. B77.
COMMON NAME(S): Dyer's Cleavers. Gaillet trifide.
SUBSPECIFIC TAXA: Three subspecies are recognized within our range.

1a Plant glabrous throughout, +/- fleshy. ssp. **halophilum**
1b Plant +/- scabrous on stem, leaves, pedicels and peduncles. 2
2a(1b) Many flowers and fruit single on axillary pedicels, a few with 2-3 flowers per peduncle. ssp. **subbiflorum**
2b(1b) Most flowers and fruit in groups of three from summit of peduncle. ssp. **trifidum**

1. Ssp. **halophilum** (Fern. & Wieg.) Puff
FREQUENCY: Scattered along the east coast.
HABITAT: Salt marshes and brackish shores.
GENERAL RANGE: Que. to Labr. south to Mass.
SYNONYMS: **G. trifidum** var. **halophilum** Fern. & Wieg.

2. Ssp. **subbiflorum** (Fern. & Wieg.) Puff
FREQUENCY: Rare; a few scattered localities reported on the western and northwestern border of the Province.
HABITAT: Shores and bogs, usually calcareous.
GENERAL RANGE: Alaska to Nfld. south to Calif., N. Mex., Minn. and N.Y.
SYNONYMS: **G. tinctorium** var. **subbiflorum** (Wieg.) Fern.

3. Ssp. **trifidum**
FREQUENCY: Common throughout much of the Province.
HABITAT: Wet shores, marshes and meadows.
GENERAL RANGE: Alaska to Nfld. south to Calif., Tex. and Ga.; Greenland; Iceland; Eurasia.

++

Galium triflorum Michx. Fig. p. A104. Map p. B77.
COMMON NAME(S): Sweet-scented Bedstraw.
FREQUENCY: Common throughout much of the Province.
HABITAT: Bottomlands and mixed or hardwoods.
GENERAL RANGE: Alaska to Nfld. south to Calif. and Fla.; Greenland; Eurasia.

Galium verum L. Fig. p. A105.
COMMON NAME(S): Yellow or Our Lady's Bedstraw. Gaillet vrai, Grappelle, Herbe à la Vierge.
SPECIAL NOTE: Eurasian; rarely introduced to roadsides and waste areas as near Debec, Carleton Co., near St. George, Charlotte Co. and near Springhill, York Co. (UNB). Known elsewhere in Canada from B.C. to Nfld. and N.S. **G. X pomeranicum** Retz. (hybrid of **G. verum X G. mollugo** var. **erectum**), with yellowish-white flowers, is reported from Springhill, York Co. (UNB). SYNONYMS: **G. X ochroleucum** Wolff.

HEDYOTIS L.

Hedyotis caerulea (L.) Hook. Fig. p. A105. Map p. B77.
COMMON NAME(S): Bluets, Innocence, Quaker-Ladies. Houstonie-Bleue.
FREQUENCY: Locally common in the southwest quarter of the Province.
HABITAT: Lawns, meadows and similar open areas.
GENERAL RANGE: Ont. to N.B., N.S. south to Mo. and Ga.
SYNONYMS: **Houstonia caerulea** L.

MITCHELLA L.

Mitchella repens L. Fig. p. A105. Map p. B77.
COMMON NAME(S): Partridge-berry. Pain de perdrix, Mitchella rampant.
FREQUENCY: Scattered mostly in the southern half of the Province and rare northwards.
HABITAT: Coniferous or mixed woods.
GENERAL RANGE: Minn. to Ont. and Nfld. south to Tex. and Fla.

CAPRIFOLIACEAE (Honeysuckle Family)

1a Plants erect, shrubby. 2
1b Plants trailing, semiwoody or herbaceous. 3
2a(1a) Leaves pinnately compound; inflorescence cymose; fruit berry-like. **Sambucus**
2b(1a) Leaves simple. 4
3a(1b) Stems trailing, semiwoody; leaves 1-2 cm long, oval to obovate; flowers funnelform, paired on upright branches; corolla white with rose-purple markings. **Linnaea borealis**
3b(1b) Stems erect, herbaceous; leaves obovate to subrhombic, sessile, to over 2 dm long; flowers in sessile, axillary clusters, purplish-red. **Triosteum aurantiacum**

++

4a(2b) Leaves variously dentate. 5

4b(2b) Leaves +/- entire, wavy margined or shallowly lobed. 6

5a(4a) Corolla funnelform, yellowish to reddish; fruit dry capsular; low colonial shrub. **Diervilla lonicera**

5b(4a) Corolla open saucer-shaped, united only at base; fruit drupaceous; usually taller, cespitose shrubs. **Viburnum**

6a(4b) Flowers small, campanulate, hairy within, white or pinkish, clustered in leaf axils and in short terminal spikes; fruit white, 2-seeded. **Symphoricarpos albus**

6b(4b) Flowers larger, corolla +/- deeply lobed, often bilabiate, axillary, peduncles bearing paired terminal flowers, their ovaries sometimes united. **Lonicera**

DIERVILLA Duham.

Diervilla lonicera Mill. Fig. p. A105.
COMMON NAME(S): Bush-Honeysuckle. Herbe bleue, Dièreville chèvrefeuille, Chèvrefeuille d'Acadie.
FREQUENCY: Common throughout much of the Province.
HABITAT: Roadsides, field margins, mixed woods and rocky slopes.
GENERAL RANGE: Sask. to Nfld. south to Iowa and N.C.

LINNAEA Gronov.

Linnaea borealis L. Fig. p. A105.
COMMON NAME(S): Twinflower, Pink Bells. Linnée boréale
FREQUENCY: Common throughout much of the Province.
HABITAT: Mixed or coniferous woods and clearings.
GENERAL RANGE: Alaska to Nfld. south to Calif. and Md.; Greenland; Eurasia.
SUBSPECIFIC TAXA: Our plants are referable to ssp. **longiflora** (Torr.) Hulten.

LONICERA L. Honeysuckle. Chèvrefeuille

1a Branchlets solid; pith white. 2

1b Branchlets hollow; pith brown. 3

2a(1a) Leaves dark green, widest at or above middle. 4

2b(1a) Leaves light green, widest below the middle, marginally ciliate, otherwise +/- glabrous; berries red. **L. canadensis**

3a(1b) Leaves and branchlets glabrous; corolla pink (rarely white) not fading to yellow; peduncles 15-20 mm long. **L. tatarica**

3b(1b) Leaves and branchlets +/- pubescent; corolla white, fading to yellow. 5

4a(2a) Leaves spreading pilose at least beneath, appearing with flowers; bracts at least as long as ovary; corolla nearly regular; ovaries completely fused; berries blue. **L. villosa**

4b(2a) Leaves finely puberulent, well expanded before flowers; bracts minute or none; corolla strongly bilabiate; ovaries separate above middle; berries orange-yellow to deep red. **L. oblongifolia**

5a(3b) Leaves broadly oval or obovate; buds long-pointed, narrowly conical; corolla strongly bilabiate (upper lip relatively shallowly lobed); filaments completely pubescent. **L. xylosteum**

5b(3b) Leaves narrowly ovate to oblong or triangular-lanceolate; buds obtuse, ovate; corolla nearly regular; filaments pubescent only at base. 6

6a(5b) Leaves mostly pubescent only on midrib and veins beneath; corolla glabrous; peduncles glabrous or slightly pubescent. **L. X bella**

6b(5b) Leaves densely pubescent beneath; corolla pubescent, saccate at base; sepals ciliate; peduncles densely pubescent. **L. morrowii**

Lonicera X bella Zabel.
COMMON NAME(S): Belle Honeysuckle.
SPECIAL NOTE: Asiatic; introduced in woods and thickets in scattered localities throughout the Province. Known elsewhere in Canada from Ont. and Que. This is the hybrid between **L. morrowii** and **L. tatarica**.

Lonicera canadensis Bartr. Fig. p. A105.
COMMON NAME(S): Fly-Honeysuckle.
FREQUENCY: Common throughout much of the Province.
HABITAT: Damp coniferous or mixed woodlands.
GENERAL RANGE: Ont. to N.S. south to Iowa and N.C.

Lonicera morrowii Gray Fig. p. A105.
COMMON NAME(S): Morrow's Honeysuckle.
SPECIAL NOTE: Asiatic; cultivated and rarely introduced as in St. Stephen, Charlotte Co.

Lonicera oblongifolia (Goldie) Hook. Fig. p. A105. Map p. B77.
COMMON NAME(S): Swamp Fly-Honeysuckle.
FREQUENCY: Rare; reported from a few localities in Charlotte, Carleton, Victoria and Madawaska Counties.
HABITAT: Mostly calcareous bogs and swamps.
GENERAL RANGE: Sask. to N.B., N.S. south to Minn. and Pa.
SPECIAL NOTE: For more information see H. Hinds, 1983. The Rare Vascular Plants of New Brunswick. Syllogeus 50.

Lonicera tatarica L. Fig. p. A105. Map p. B77.
COMMON NAME(S): Tatarian Honeysuckle.
SPECIAL NOTE: Eurasian; introduced in open woods and thickets in scattered localities throughout the Province. Known elsewhere in Canada from Alta. to Que. and N.S.
SUBSPECIFIC TAXA: Forma **albiflora** (DC.) House, with white rather than pink flowers, occurs occasionally throughout.

Lonicera villosa (Michx.) R. & S. Fig. p. A105.
COMMON NAME(S): Mountain Fly-Honeysuckle. Chèvrefeuille velu.
FREQUENCY: Common throughout much of the Province.

++

HABITAT: Peaty or rocky barrens, swamps and bogs.
GENERAL RANGE: B.C. to Nfld. south to Wash., Nev., and Pa.
SUBSPECIFIC TAXA: Several intergrading varieties were named by Fernald. These are not recognized in this work.
SYNONYMS: **L. caerulea** L., **L. villosa** vars. **tonsa** Fern., **solonis** (Eat.) Fern., and **calvescens** (Fern. & Wieg.) Fern.

Lonicera xylosteum L. Fig. p. A105.
COMMON NAME(S): Fly-Honeysuckle.
SPECIAL NOTE: Eurasian; rarely introduced to open woods, clearings and thickets as near Fredericton, York Co. and St. Andrews, Charlotte Co. Known elsewhere in Canada from Ont. and Que.

SAMBUCUS L. Elder, Elderberry. Sureau

1a Flowers and fruit in flat-topped or domed cymes mostly with more than 3 main branches near base; pith white; berries purple-black. **S. canadensis**
1b Flowers and fruit in elongate, panicle-like cymes usually with 3 main branches near base; pith brown; berries red. **S. racemosa**

Sambucus canadensis L. Fig. p. A105. Map p. B77.
COMMON NAME(S): Elderberry. Sureau blanc.
FREQUENCY: Common in the southern half of the Province and uncommon northwards.
HABITAT: Shores, swamps and damp thickets.
GENERAL RANGE: Man. to N.S. south to Okla. and Ga.

Sambucus racemosa L. Fig. p. A106.
COMMON NAME(S): Red-berried Elder, Stinking Elder. Sureau rouge.
FREQUENCY: Common throughout much of the Province.
HABITAT: Mixed woodland, clearings and shores and thickets.
GENERAL RANGE: Alaska and B.C. to Nfld. south to Calif., N. Mex. and Ga.; Eurasia.
SUBSPECIFIC TAXA: Our plants are referable to ssp. **pubens** (Michx.) House.

SYMPHORICARPOS Duham. Snowberry

Symphoricarpos albus (L.) Blake Fig. p. A106.
COMMON NAME(S): Snowberry. Belluaine.
SPECIAL NOTE: Introduced from further west; scattered throughout much of the Province. Known elsewhere in Canada from Alta. to Nfld., P.E.I. and N.S.
SUBSPECIFIC TAXA: Our plants are referable to var. **laevigatus** (Fern.) Blake.

TRIOSTEUM L. Feverwort, Horse-Gentian.

Triosteum aurantiacum Bickn. Fig. p. A106. Map p. B77.
COMMON NAME(S): Wild Coffee, Tinker's-weed. Trioste orangé.

FREQUENCY: Rare; scattered in the St. John R. drainage system in York, Carleton and Victoria Counties.
HABITAT: Rich woods and thickets.
GENERAL RANGE: Ont. to N.B., N.S. south to Iowa and Ga.
SYNONYMS: **T. perfoliatum** L. var. **aurantiacum** (Bickn.) Wieg.

VIBURNUM L. Viburnum, Arrow-wood. Viorne

1a Leaves coarsely dentate and/or lobed. 2
1b Leaves finely dentate. 3
2a(1a) Leaves 3-lobed. 4
2a(1a) Leaves unlobed, glabrous or pubescent in axils of veins beneath; fruit blue-black. **V. recognitum**
3a(1b) Leaves broadly rounded to cordate at base; lower surface stellate pubescent. 5
3b(1b) Leaves +/- tapering at base, glabrous or brown-scurfy below. 6
4a(2a) Leaves with glands near summit of petiole; outer flowers of inflorescence enlarged. 7
4b(2a) Leaves without glands; outer flowers not enlarged. 8
5a(3a) Inflorescence with peduncles 0.5-5 cm long; leaves ovate to elliptic-lanceolate; marginal flowers not enlarged. **V. lantana**
5b(3a) Inflorescence sessile (without peduncle above subtending leaves); marginal flowers enlarged; leaves broadly ovate-subrotund. **V. lantanoides**
6a(3b) Leaves sharply serrate; cymes sessile; tall shrubs or small trees to 10 m high. **V. lentago**
6b(3b) Leaves shallowly crenate, denticulate or +/- entire; cymes with peduncles 5-25 mm long; shrub to 4 m high. **V. cassinoides**
7a(4a) Petiole grooves broad, flat-bottomed, glands near summit +/- stipitate, mostly convex-topped. **V. trilobum**
7b(4a) Petiole grooves narrow, V-shaped; glands sessile, broader than long, concave-topped. **V. opulus**
8a(4b) Leaves with at least 5 veins from base, sparingly simple pilose beneath; petioles glabrous; fruit yellow or red. **V. edule**
8b(4b) Leaves with 3 veins from base, stellate pubescent beneath; petioles pubescent; fruit dark purple-black. **V. acerifolium**

Viburnum acerifolium L. Fig. p. A106.
COMMON NAME(S): Maple-leaved Viburnum, Dockmackie.
FREQUENCY: Very rare; known only from near the Woodland Dam on the St. Croix R. (formerly Spragues Falls), Charlotte Co.
HABITAT: Dry deciduous woodland.
GENERAL RANGE: Minn. to N.B. south to Tenn. and Ga.
SPECIAL NOTE: For more information see H. Hinds, 1983. The Rare Vascular Plants of New Brunswick. Syllogeus 50.

++

Viburnum cassinoides L. Fig. p. A106.
COMMON NAME(S): Witherod, Wild Raisin. Alisier, Bleuets sains.
FREQUENCY: Common throughout much of the Province.
HABITAT: Acid swamps, clearings and thickets.
GENERAL RANGE: Ont. to Nfld. south to Wisc. and Ala.

Viburnum edule (Michx.) Raf. Fig. p. A106. Map p. B77.
COMMON NAME(S): Squashberry, Mooseberry. Pimbina
FREQUENCY: Scattered in the northwestern quarter of the Province and rare in Fundy National Park, Albert Co.
HABITAT: Cool, moist woods, ravines and mountain slopes.
GENERAL RANGE: Alaska to Nfld. south to Oreg., Colo., and N. Eng.
SYNONYMS: **V. pauciflorum** La Pylaie.

Viburnum lantana L. Fig. p. A106.
COMMON NAME(S): Wayfaring-tree, Twistwood. Mancienne.
SPECIAL NOTE: Eurasian; sparingly introduced from cultivation to thickets and roadsides as in Lincoln, York Co., and in Rockwood Park, Saint John, St. John Co. Known elsewhere in Canada from Ont. and Que.

Viburnum lantanoides Michx. Fig. p. A106.
COMMON NAME(S): Hobblebush, Mooseberry, Witch-hobble. Bois d'orignal.
FREQUENCY: Common throughout much of the Province.
HABITAT: Cool, moist mixed woods.
GENERAL RANGE: Ont. to N.S. south to Tenn. and Ga.
SYNONYMS: **V. alnifolium** Marsh.

Viburnum lentago L. Fig. p. A106. Map p. B77.
COMMON NAME(S): Sweet Viburnum, Nannyberry. Alisier, Bourdaine.
FREQUENCY: Rare; reported from the Eel R. at Benton Falls and from the St. Croix R., Charlotte Co.
HABITAT: Stream margins, swamps and thickets.
GENERAL RANGE: Sask. to N.B. south to Colo., S. Dak., and Ga.
SPECIAL NOTE: For more information see H. Hinds, 1983. The Rare Vascular Plants of New Brunswick. Syllogeus 50.

Viburnum opulus L. Fig. p. A106.
COMMON NAME(S): Guelder-Rose. Obier, Rose de Gueldre.
SPECIAL NOTE: Eurasian; cultivated and occasionally escaped to woods and thickets near present or past habitations. Known elsewhere in Canada from Ont. and N.S.

Viburnum trilobum Marsh.
COMMON NAME(S): Highbush Cranberry, Pembina. Viorne d'Amerique, Pimbina.
FREQUENCY: Common throughout much of the Province.
HABITAT: Borders of woods, thickets and shores.
GENERAL RANGE: B.C. to Nfld. south to Wash., Wyo., Pa. and N. Eng.
SYNONYMS: **V. opulus** var. **americanum** (Mill.) Ait.

Viburnum recognitum Fern. Fig. p. A106. Map p. B77.
COMMON NAME(S): Arrow-wood.
FREQUENCY: Rare; reported from the St. Croix R. valley in York and Charlotte Co. and near Good Corner, Carleton Co.
HABITAT: Stream banks and damp thickets.
GENERAL RANGE: Ont. to N.B. south to Tex. and Fla.
SYNONYMS: **V. dentatum** L. var. **lucidum** Ait.
SPECIAL NOTE: For more information see H. Hinds, 1983. The Rare Vascular Plants of New Brunswick. Syllogeus 50.

VALERIANACEAE (Valerian Family)

VALERIANA L. Valérian

1a Leaflet margins forward-pointing scabrous; basal leaves pinnately divided into several +/- equal pinnae; corolla about 4 mm long scarcely gibbous near base; stem to over 1.5 m high. **V. officinalis**
1b Leaflet margins glabrous or ciliate. 2
2a(1b) Leaflet margins glabrous; basal leaves +/- persistent, simple, 2.5-7.5 cm long; corolla 3-4 mm long, scarcely gibbous, lobes about equalling tube; bracts of inflorescence ciliate only at base; stem to 7 dm high. **V. dioica**
2b(1b) Leaflet margins spreading ciliate; basal leaves rarely persisting, simple or divided, 5-15 cm long; corolla 5-7 mm long distinctly gibbous near base, lobes shorter than tube; bracts of inflorescence ciliate; stem to about 1 m tall. **V. sitchensis**

Valeriana dioica L. Fig. p. A106.
COMMON NAME(S): Marsh Valerian, Northern Valerian.
FREQUENCY: Very rare; known only from near Summit Depot, Restigouche Co. (UNB).
HABITAT: Moist woods in calcareous area.
GENERAL RANGE: Yukon to Labr., Gaspé Pen. south to Wash. and northern N.B.
SUBSPECIFIC TAXA: Our plants are referable to ssp. **sylvatica** (Soland. ex Richards.) F.G. Meyer.

Valeriana officinalis L. Fig. p. A107. Map p. B77.
COMMON NAME(S): Common Valerian. Herbe aux chats.
SPECIAL NOTE: Eurasian; occasionally introduced from cultivation to roadsides and thickets throughout the Province and locally aggressive. Known elsewhere in Canada from B.C., Man. to N.S.

Valeriana sitchensis Bong. Fig. p. A107. Map p. B78.
FREQUENCY: Rare; reported from Charlotte, Carleton, Victoria, Madawaska, Restigouche and Gloucester Counties.
HABITAT: Calcareous swamps and bogs.
GENERAL RANGE: Alaska and B.C. to Gaspé Pen. and N.B. south to Calif., Ohio and Vt.
SUBSPECIFIC TAXA: Our plants are referable to ssp. **uliginosa** (Torr. & Gray) F. G. Meyer.
SYNONYMS: **V. uliginosa** (T.&G.) Rydb.

DIPSACACEAE (Teasel Family)

KNAUTIA L.

Knautia arvensis (L.) Coult. Fig. p. A107.
COMMON NAME(S): Blue-buttons, Gypsy's Rose, Pincushion. Scabieuse des champs, Oreille d'âne, Mirliton.
SPECIAL NOTE: Eurasian; rarely introduced to roadsides and waste areas as in St. Andrews, Charlotte Co. (UNB). Known elsewhere in Canada from all provinces except P.E.I. and N.S.

CUCURBITACEAE (Cucumber or Gourd Family)

ECHINOCYSTIS T. & G. Wild Balsam-apple

Echinocystis lobata (Michx.) T. & G. Fig. p. A107. Map p. B78.
COMMON NAME(S): Wild or Prickly Cucumber. Concombre grimpant.
FREQUENCY: Scattered throughout much of the Province but rarer towards the north.
HABITAT: Bottomlands, wet meadows, thickets and shores.
GENERAL RANGE: B.C. to N.S. south to Idaho, Ariz. and Fla.

CAMPANULACEAE (Bluebell Family)

1a Corolla regular; anthers separate. 2
1b Corolla irregular, +/- bilabiate; anthers united. **Lobelia**
2a(1a) Flowers sessile in axils of cordate-clasping leaves; corolla lobes longer than tube; capsule slender-cylindric. **Triodanis perfoliata**
2b(1a) Flowers pedicelled; corolla lobes equal to or shorter than tube; capsule obconic or turbinate to subglobose. **Campanula**

CAMPANULA L. Harebell, Bellflower. Campanule

1a Flowers small, .5-1.2 cm long, white or pale blue; stem weak, scabrous, often tangled with other vegetation; leaves linear to narrowly lanceolate. **C. aparinoides**
1b Flowers larger 1.5-5.0 cm long, usually blue. 2
2a(1b) Cauline leaves linear to linear lanceolate or spatulate, +/- glabrous. 3
2b(1b) Cauline leaves broader, ovate to lanceolate or lance-oblong, hirsute. 4
3a(2a) Stem simple, erect, stout; basal leaves not significantly different from cauline; flowers erect; corolla to 3 cm long. **C. persicifolia**
3b(2a) Stem slender, branched or simple; basal leaves (rarely present at flowering) cordate-ovate, greatly differing from cauline; flowers usually nodding; corolla less than 2 cm long. **C. rotundifolia**

++

4a(2b) Corolla inflated-campanulate (urceolate); calyx with large reflexed cordate appendages at base; basal and lower leaves ovate-oblong tapering to petiole. **C. medium**

4b(2b) Corolla campanulate, not inflated in the middle; calyx unappendaged at base; basal and lower leaves cordate, ovate to ovate-oblong. **C. rapunculoides**

Campanula aparinoides Pursh Fig. p. A107. Map p. B78.
COMMON NAME(S): Marsh-Bellflower, Bedstraw Bellflower. Campanule faux-gaillet.
FREQUENCY: Scattered in the southern half of the Province and rare northwards.
HABITAT: Wet meadows and shores.
GENERAL RANGE: Sask. to N.B., N.S. south to Colo. and Ga.
SYNONYMS: **C. uliginosa** Rydb.

Campanula medium L.
COMMON NAME(S): Canterbury-bells.
SPECIAL NOTE: European; occasionally cultivated and rarely escaped to roadsides and waste areas as near St. Andrews, Charlotte Co. Known elsewhere in Canada from B.C. and Ont.

Campanula persicifolia L.
SPECIAL NOTE: Eurasian; occasionally cultivated and rarely escaped to fields and roadsides as in Fairvale Station, and Lakeside, Kings Co., Perth-Andover, Victoria Co., and Chamcook Mtn., Charlotte Co. Known elsewhere in Canada from B.C., Ont. and Que.

Campanula rapunculoides L. Fig. p. A107.
COMMON NAME(S): Creeping Bellflower. Lygurian Bellflower. Campanule fausse-raiponce, Clochettes.
SPECIAL NOTE: Eurasian; formerly cultivated and commonly escaped to dooryards, waste areas and roadsides throughout much of the Province. Known elsewhere in Canada from all provinces except Sask.

Campanula rotundifolia L. Fig. p. A107. Map p. B78.
COMMON NAME(S): Harebell, Bluebell.
FREQUENCY: Scattered throughout much of the Province.
HABITAT: Calcareous ledges, rocky fields, roadsides and gravelly river strands.
GENERAL RANGE: Alaska to Nfld. south to Calif., Mex., and N.J.; Greenland; Iceland; Spitsbergen; Eurasia.

LOBELIA L. Lobelia. Lobélie

(Separated as the family **LOBELIACEAE** by some authors)

1a Leaves fleshy, linear, restricted to submerged basal rosette; flowers pale blue or white. **L. dortmanna**

1b Leaves distributed along the stem, not submerged nor fleshy. 2

2a(1b) Corolla to 4.5 cm long, cardinal red. **L. cardinalis**

2b(1b) Corolla to 1.3 cm long. 3

3a(2b) Capsule inflated, 4-7 mm thick when mature; corolla whitish, pinkish or pale violet; leaves ovate to obovate. **L. inflata**

3b(2b) Capsule not especially inflated, 3-4 mm thick when mature. 4

4a(3b) Corolla bright blue with white eye; leaves linear or narrowly lanceolate. **L. kalmii**

4b(3b) Corolla white to pale blue; leaves lanceolate to obovate. **L. spicata**

Lobelia cardinalis L. Fig. p. A107. Map p. B78.
COMMON NAME(S): Cardinal-flower.
FREQUENCY: Locally common in the southwest corner of the Province mainly on the Eel, Oromocto, St. Croix and Magaguadavic Rivers.
HABITAT: Shores and adjacent meadows and swamps.
GENERAL RANGE: Minn. to N.B. south to Tex. and Fla.

Lobelia dortmanna L. Fig. p. A107. Map p. B78.
COMMON NAME(S): Water-Lobelia, Water-Gladiole.
FREQUENCY: Scattered throughout much of the Province.
HABITAT: Shallow water of lakes and ponds.
GENERAL RANGE: B.C. and Dist. Mack. to Nfld. south to Oreg. on the west coast and Minn. to Pa. and N.J. on the east coast; Europe.

Lobelia inflata L. Fig. p. A107. Map p. B78.
COMMON NAME(S): Indian-tobacco. Lobélie gonflée. Herbe à l'asthme.
FREQUENCY: Common in the southern half of the Province and rare northwards.
HABITAT: Recently disturbed areas, fields, waste places.
GENERAL RANGE: Minn. to Nfld. south to Kans. and Ga.

Lobelia kalmii L. Fig. p. A107. Map p. B78.
COMMON NAME(S): Kalm's Lobelia.
FREQUENCY: Uncommonly scattered in western and northern part of the Province.
HABITAT: Calcareous shores and wet ledges.
GENERAL RANGE: B.C. to Nfld. south to Wash., Colo., and N.J.
SUBSPECIFIC TAXA: Forma **leucantha** Rouleau, with white instead of blue flowers, occurs occasionally with the species as at Two Brooks on the Restigouche R., Restigouche Co.

Lobelia spicata Lam. Fig. p. A108.
COMMON NAME(S): Pale-spike Lobelia.
FREQUENCY: Very rare; reported from Moncton, Westmorland Co. and from St. Andrews, Charlotte Co.
HABITAT: Woods, clearings, moist meadows and thickets.
GENERAL RANGE: Alta. to N.S. south to Kans. and Ga.
SPECIAL NOTE: For more information see H. Hinds, 1983. The Rare Vascular Plants of New Brunswick. Syllogeus 50.

++

TRIODANIS Raf.

Triodanis perfoliata (L.) Nieuwl. Fig. p. A107.
COMMON NAME(S): Venus's Looking-glass.
FREQUENCY: Very rare; reported only from St. Andrews, Charlotte Co.
HABITAT: Roadsides, waste areas, shores.
GENERAL RANGE: B.C., Mont. to Ont. and N.B. south to Calif., Mex. and Fla.
SYNONYMS: **Specularia perfoliata** (L.) A. DC.

ASTERACEAE (Aster Family)*
(Alternate name **COMPOSITAE**)

1a Heads radiate or discoid. 2
1b Heads all ligulate; plants usually with milky juice. 3
2a(1a) Heads radiate (marginal ray-florets ligulate, more than 5 mm long, central disk-florets tubular). 4
2b(1a) Heads discoid (florets all tubular or rays less than 4 mm long). 5
3a(1b) Leaves mostly alternate. 6
3b(1b) Leaves mostly basal, stem leaves reduced when present. 7
4a(2a) Leaves mostly alternate. 8
4b(2a) Leaves mostly opposite or basal. 9
5a(2b) Leaves mostly alternate. 10
5b(2b) Leaves whorled, opposite or basal. 11
6a(3a) Leaves +/- entire. 12
6b(3a) Leaves toothed to lobed. 13
7a(3b) Leaves +/- entire; plants rhizomatous; phyllaries +/- overlapping with basal series of bractlets; pappus bristles usually darkened. **Hieracium**
7b(3b) Leaves toothed or lobed. 14
8a(4a) Leaves +/- entire. 15
8b(4a) Leaves toothed to lobed. 16
9a(4b) Leaves mostly basal (stem leaves bract-like if present). 17
9b(4b) Leaves mostly opposite, sometimes alternate above, toothed to deeply lobed or dissected. 18
10a(5a) Leaves +/- entire (lowest leaves sometimes sparsely toothed). 19
10b(5a) Leaves toothed or lobed. 20
11a(5b) Leaves basal, toothed or lobed; receptacle +/- naked; pappus of capillary bristles; leaves deeply once 5-7 lobed, the reduced stem leaves subentire; heads creamy white, racemose or corymbose; plants +/- white-woolly. **Petasites palmatus**
11b(5b) Leaves opposite. 21
12a(6a) Pappus bristles densely plumose; achenes long-beaked. 22
12b(6a) Pappus of capillary bristles. 23
13a(6b) Heads blue, rarely pure white. 24
13b(6b) Heads yellow, orange, pink, creamy or greenish white. 25
14a(7b) Pappus-bristles plumose. 26
14b(7b) Pappus-bristles various, not plumose. 27
15a(8a) Receptacle chaffy. 28

*Based in part on Scoggan, 1978-79

15b(8a) Receptacle +/- naked. 29

16a(8b) Receptacle chaffy (at least towards the centre). 30

16b(8b) Receptacle +/- naked (occasionally short hairy). 31

17a(9a) Heads numerous, whitish, in corymbose or racemose panicles; scape with scaly bracts. **Petasites palmatus**

17b(9a) Heads solitary at tip of scape. 32

18a(9b) Leaves +/- entire (occasionally the lowest ones sparingly toothed). 33

18b(9b) Leaves toothed to deeply lobed or dissected. 34

19a(10a) Phyllaries with hooked tips in several rows forming a bur; coarse weedy plants with large ovate, cordate leaves. **Arctium**

19b(10a) Phyllaries without hooked tips. 35

20a(10b) Some phyllaries spine-tipped. 36

20b(10b) No phyllaries spine-tipped. 37

21a(11b) Leaves entire or obscurely serrate; achenes with minutely barbed awns; rays orange or yellow. **Bidens**

21b(11b) Leaves toothed to deeply lobed. 38

22a(12a) Plants glabrous; leaves +/- grasslike, clasping; heads large, solitary, yellow or purple. **Tragopogon**

22b(12a) Plants +/- spiny-hispid; leaves lanceolate to oblanceolate; heads 2 or more, yellow. **Picris echioides**

23a(12b) Achenes +/- flattened, beaked or tapered to summit. **Lactuca canadensis**

23b(12b) Achenes +/- cylindric. 39

24a(13a) Heads less than 3 cm broad, panicled; achenes +/- flattened; pappus of capillary bristles. **Lactuca biennis**

24b(13a) Heads more than 3 cm broad, +/- sessile; achenes obscurely 5-angled; pappus of minute scales. **Cicorium intybus**

25a(13b) Pappus-bristles plumose or wanting. 40

25b(13b) Pappus-bristles various, neither plumose nor wanting. 41

26a(14a) Receptacle chaffy; inner achenes slender beaked; leaves hirsute. **Hypochaeris radicata**

26b(14a) Receptable naked; achenes +/- beakless. **Leontodon autumnalis**

27a(14b) Heads several. 42

27b(14b) Heads solitary on the scape. 43

28a(15a) Receptacle flat or slightly convex; pappus of 2 readily deciduous awned scales; perennials (rarely annuals) with rhizomes or tuberous roots. **Helianthus**

28b(15a) Receptacle conical; pappus none or short +/- toothed crown. **Rudbeckia**

29a(15b) Rays white, pink, blue or purple. 44

29b(15b) Rays cream-colour to yellow or deep orange. 45

30a(16a) Rays white (rarely pink or reddish purple in **Achillea**). 46

30b(16a) Rays yellow to orange. 47

31a(16b) Rays white, pink, bluish or purplish. 48

31b(16b) Rays yellow to orange. 49

32a(17b) Scape naked; rays white to pink or purple; pappus wanting. **Bellis perennis**

32b(17b) Scape scaly or leafy-bracted; rays yellow; pappus of capillary bristles; basal leaves cordate-rotund, with deep, narrow sinus and shallowly lobed, appearing long after the flowering scape. **Tussilago farfara**

33a(18a) Pappus of capillary subplumose bristles; receptacle minutely fringed around pits; heads solitary or few. **Arnica**

33b(18a) Pappus of teeth, scales or awns. 50

34a(18b) Receptacle with pits minutely fringed; pappus of barbed capillary bristles; rays yellow. **Arnica**

34b(18b) Receptacle chaffy. 51

35a(19b) Receptacle bristly; florets showy, white, pink, blue or purplish, marginal ones mostly enlarged and ray-like; phyllaries with a terminal appendage which is toothed, pectinate or spiny; pappus of scales, much reduced or wanting. **Centaurea**

35b(19b) Receptacle naked; pappus of capillary bristles. 52

36a(20a) Heads unisexual; phyllaries of pistillate heads united to form a prickly or tubercled bur. 53

36b(20a) Heads of many perfect florets with distinct phyllaries. 54

37a(20b) Pappus of elongate bristles. 55

37b(20b) Pappus wanting or a short toothed crown. 56

38a(21b) Receptacle bristly or chaffy. 57

38b(21b) Receptacle +/- naked; leaves perfoliate or petiolate, sometimes whorled; rays wanting; pappus of capillary bristles. **Eupatorium**

39a(23b) Plants from short to elongate rhizomes and fibrous roots; phyllaries of unequal lengths; pappus sordid or tawny. **Hieracium**

39b(23b) Plants from a taproot or several strong roots, lacking rhizomes; phyllaries 1-rowed; pappus +/- white. **Crepis**

40a(25a) Pappus none; heads corymbed or panicled; phyllaries +/- equal with minute bractlets; leaves ovate to +/- rotund. **Lapsana communis**

40b(25a) Pappus-bristles plumose; heads terminating leafy stems; outer phyllaries loose and spreading; leaves +/- lanceolate. **Picris echioides**

41a(25b) Achenes +/- strongly flattened. 58

41b(25b) Achenes +/- terete. 59

42a(27a) Plants with short to elongate rhizomes and fibrous roots; phyllaries +/- overlapping and with small bractlets at base; pappus +/- sordid or tawny. **Hieracium**

42b(27a) Plants from a taproot or several strong roots; lacking rhizomes; phyllaries 1-rowed; pappus whitish. **Crepis**

43a(27b) Achenes strongly tuberculate above, long-beaked; scapes hollow, weak; pappus white; plants with taproot. **Taraxacum**

43b(27b) Achenes not tuberculate, beakless; scapes +/- solid and firm; pappus +/- sordid or tawny; plants rhizomatous. **Hieracium**

44a(29a) Peduncles +/- leafy or bracteate; rays broad and relatively few; achenes mostly several nerved; pappus copius and firm; phyllaries relatively broad with expanded tip often remaining green in 2 or more series of unequal length; mostly late summer and fall blooming. **Aster**

44b(29a) Peduncles +/- naked or with few reduced leaves; rays narrow and numerous; achenes 2-nerved; pappus scanty and fragile; phyllaries relatively long and narrow, +/- equal and without green tips; mostly spring and early summer blooming. **Erigeron**

45a(29b) Inflorescence +/- flat-topped; heads in tight clusters; leaves entire, numerous, linear to linear-lanceolate with 3 or more +/- parallel veins. **Euthamia graminifolia**

45b(29b) Inflorescence not flat-topped; heads in racemes, panicles or axillary clusters; leaves lanceolate to ovate. **Solidago**

46a(30a) Heads large, solitary at the ends of branches, the disk to 12 mm broad, rays to 13 mm long; receptacle hemispherical to conical; achenes quadrangular or subterete. **Anthemis**

46b(30a) Heads small, numerous in a terminal corymb; disk +/- 5 mm broad; rays to +/- 5 mm long; receptacle flattish; achenes strongly flattened, callous-margined. **Achillea**

47a(30b) Phyllaries papery, dry and scarious-margined; heads solitary at the ends of branches, the disk about 1.5 cm broad, the rays 1.5 cm long; receptacle hemispherical; pappus none or a minute crown; leaves deeply pinnately divided into narrow toothed or pinnatifid segments. **Anthemis tinctoria**

47b(30b) Phyllaries not dry, papery; leaves less divided. 60

48a(31a) Pappus a short crown or none; phyllaries dry, papery and scarious-margined; rays white. 61

48b(31a) Pappus of capillary bristles, phyllaries not papery. 62

49a(31b) Pappus of scales, short crown or wanting. 63

49b(31b) Pappus of capillary bristles. 64

50a(33b) Achenes laterally compressed at right angles to the phyllaries, and enfolded by the receptacular chaff; phyllaries +/- green and herbaceous, subequal or in 2 or more series of unequal length. **Helianthus**

50b(33b) Achenes compressed parallel to the phyllaries, free from the receptacular chaff; phyllaries dimorphic, 2-rowed, the outer series +/- herbaceous. 65

51a(34b) Rays white; leaves at most rather coarsely toothed, lance-ovate to ovate, petiolate; rays 4 or 5, small, roundish; disk 3 or 4 mm broad; receptacle conical; pappus consisting of fringed scales; annual. **Galinsoga**

51b(34b) Rays yellow to orange-yellow. 66

52a(35b) Plants glabrous or +/- pubescent, but not white woolly. 67

52b(35b) Plants +/- white-woolly, at least on the lower leaf surfaces. 68

53a(36a) Burs with hooked prickles, 2 flowered, 2 seeded; heads 1-few in leaf axils; leaves +/- broadly lobed. **Xanthium**

53b(36a) Burs with a single row of tubercles near summit; 1 flowered, 1 seeded; leaves +/- deeply lobed or dissected. **Ambrosia**

54a(36b) Leaves not prickly or spiny. 69

54b(36b) Leaves prickly or spiny on the margins. 70

55a(37a) Heads whitish; achenes with 10-12 pale ribs; leaves coarsely and irregularly serrate. **Erechtites hieracifolia**

55b(37a) Heads yellow or orange-yellow; achenes 5-10 grooved, glabrous or strigose; leaves often deeply lobed. **Senecio**

56a(37b) Inflorescence spike-like, racemose or paniculate; heads small and numerous. **Artemisia**

56b(37b) Inflorescence corymbiform or heads solitary on ends of branches. 71

57a(38a) Pappus of elongate barbed awns; disk yellow to orange, 1-2 cm broad; florets perfect; phyllaries smooth. **Bidens**

57b(38a) Pappus wanting or undeveloped; heads unisexual, numerous and small; mature phyllaries of pistillate heads with a single series of tubercles or erect spines near pointed apex. **Ambrosia**

58a(41a) Achenes beaked; heads relatively few flowered, 13-34(54) florets per head. **Lactuca**

58b(41a) Achenes beakless; heads many flowered, 85-250 florets per head. **Sonchus**

59a(41b) Phyllaries in a single series; heads yellow or orange; pappus whitish; plants with taproot or several strong roots, lacking rhizomes. **Crepis**

59b(41b) Phyllaries in 2 or more series. 72

60a(47b) Receptacle flat to merely convex; pappus consisting of 2 or 4 deciduous scales. **Helianthus**

60b(47b) Receptacle hemispheric or ovoid-conical, its chaff short and subtending only disk-florets; achenes 4-angled. **Rudbeckia**

61a(48a) Leaves deeply pinnately dissected into narrow segments; mature receptacle conical. **Matricaria**

61b(48a) Leaves coarsely toothed to deeply broad-lobed; receptacle flat or merely convex. 82

62a(48b) Peduncles +/- leafy or bracteate; rays broad and relatively few; achenes mostly several nerved; pappus copius and firm; phyllaries relatively broad with expanded tip often remaining green in 2 or more series of unequal length; mostly late summer or fall blooming. **Aster**

62b(48b) Peduncles +/- naked or with few reduced leaves; rays narrow and numerous; achenes 2-nerved; pappus scanty and fragile; phyllaries relatively long and narrow, +/- equal and without green tips; mostly spring and early summer blooming. **Erigeron**

63a(49a) Principal leaves merely coarsely toothed to moderately deeply lobed. **Chrysanthemum segetum**

63b(49a) Principal leaves 1-2 pinnate-pinnatifid. 73

64a(49b) Phyllaries subequal and essentially 1-rowed; heads solitary or in corymbs. **Senecio**

64b(49b) Phyllaries in 2 or more series of unequal length. 74

65a(50b) Plants essentially terrestrial (occasionally rooted in shallow water); leaves simple, incised or trifoliate; achenes flattened or angular. **Bidens**

65b(50b) Plants aquatic; submerged leaves dissected into filiform divisions; emersed leaves simple, lanceolate to ovate, serrate; achenes +/- cylindric. **Megalodonta beckii**

66a(51b) Rays remaining attached to the 4-sided marginal achenes and becoming papery; receptacle conical; phyllaries in 2 or 3 subequal series; pappus none or a small crown; leaves ovate, coarsely serrate, petiolate. **Heliopsis**

66b(51b) Rays soft, deciduous from the achenes at maturity. 75

67a(52a) Involucre 3-4 mm high; phyllaries mostly scarious, ciliate to fimbriate at summit; achenes strigose. **Conyza canadensis**

67b(52a) Involucre 6-8 mm high; phyllaries green, entire at summit; achenes pilose. **Aster laurentianus**

68a(52b) Plants bisexual, the outer (pistillate) florets of each head threadlike to summit, the central (perfect) florets coarser and distinctly flaring at summit; phyllaries whitish to brown or purplish, often with a dark spot near tip. **Gnaphalium**

68b(52b) Plant unisexual. 76

69a(54a) Phyllaries hooked at spiny tip; achenes attached basally. **Arctium**

69b(54a) Phyllaries not hooked, either terminated by a straight spine or more commonly by a broad, erose to lacerate or pectinate appendage; achenes obliquely or laterally attached. **Centaurea**

70a(54b) Heads 1-flowered, whitish to blue, very numerous in a terminal globose cluster to about 6 cm thick; pappus of short narrow scales or awns; leaves white-tomentose beneath, pinnatifid into lanceolate to oblong-triangular lobes, to about 4 dm long and 2 dm broad, at least the middle and upper ones clasping. **Echinops**

70b(54b) Heads many flowered. 77

71a(56b) Leaves crenate to coarsely incised; heads solitary at the ends of the branches; achenes of the outer flowers broadly winged, long-stipitate, the others merely 2-nerved, short stipitate. **Cotula coronopifolia**

71b(56b) Leaves deeply cut-toothed to finely pinnate or dissected. 78

72a(59b) Inflorescence a branching raceme or panicle of slender, drooping, whitish, creamy, or pink heads; pappus white to reddish brown; lower leaves often deeply cleft; root tuberous. **Prenanthes**

72b(59b) Inflorescence a corymb or panicle of erect or ascending, broad yellow heads; pappus +/- sordid or tawny; leaves merely toothed; root not tuberous. **Hieracium**

73a(63b) Ray-achenes narrowly 3-winged; annual. **Chrysanthemum coronarium**

73b(63b) Ray-achenes 5-ribbed or 5-angled; rhizomatous perennials. **Tanacetum**

74a(64b) At least the upper ovate stem-leaves cordate-clasping; lower leaves long-petioled; leaves densely white-woolly beneath, the lower ones elliptic. **Inula helenium**

74b(64b) None of the leaves cordate-clasping; heads small and numerous; in racemes, corymbs, panicles or axillary clusters. **Solidago**

75a(66b) Pappus of up to 6 +/- barbed awns; rays usually less than 1.5 cm long. **Bidens**

75b(66b) Pappus deciduous, of 2 scales; rays usually more than 5 cm long. **Helianthus**

76a(68b) Basal leaves conspicuous, tufted, persistent, the stem-leaves usually few and much reduced; inflorescence relatively small; phyllaries white to yellow-tinged or roseate. **Antennaria**

76b(68b) Basal leaves soon deciduous, the stem-leaves scarcely reduced; inflorescence many-headed and relatively large; phyllaries pearly white. **Anaphalis margaritacea**

77a(70b) Heads yellow, 3 or 4 cm high; phyllaries broad, the spines of the inner ones pinnatisect; pappus double, consisting of 10 outer smooth long awns and 10 much shorter sparsely pectinate inner ones, the obliquely attached achene also with a crown of 10 short horny teeth; plant spreading short-villous. **Cnicus**

77b(70b) Heads white to violet or purple; achenes basally attached; pappus single, consisting of naked or plumose capillary bristles. 79

78a(71b) Mature receptacle strongly conical; leaves deeply dissected into narrow segments; bruised plant with a characteristic pineapple odour. **Matricaria matricarioides**

78b(71b) Mature receptacle flat or merely somewhat convex; leaves 1-3 pinnatifid; plant with a characteristic strong scent. **Tanacetum**

79a(77b) Pappus plumose; phyllaries narrow, +/- entire, shorter than the disk; leaves not white mottled. **Circium**

79b(77b) Pappus not plumose. 80

80a(79b) Receptacle with short bristle tips on partitions of honeycomb-like pits; pappus reddish, wings of stem with fairly broad uncut portion adjoining the white-woolly stem, this to +/- 1.5 m tall. **Onopordum**

80b(79b) Receptacle setose or densely bristly; pappus white. 81

81a(80b) Stem +/- wingless; phyllaries broad-based, sharply serrate, at least the outer ones considerably surpassing the disk; leaves glabrous or slightly tomentose, +/- white-mottled along the main veins; filaments united. **Silybum**

81b(80b) Stem conspicuously spiny-winged by decurrent leaf-bases; heads usually red-purple; filaments separate; receptacle densely bristly; wings of stem toothed to base. **Carduus**

82a(61b) Leaves deeply serrate to remotely toothed or notched, nonaromatic; heads to 4 cm broad. 83

82b(61b) Leaves bipinnatifid into ovate, irregularly cut divisions, strongly aromatic; heads, including rays, less than 3 cm broad. **Tanacetum parthenium**

83a(82a) Cauline leaves acute to acuminate, sharply serrate with forward pointing teeth; corolla of tubular florets compressed and winged below; stem to 2 m tall. **Leucanthemella serotina**

83b(82a) Cauline leaves obtuse to rounded, remotely toothed or notched; corolla of tubular florets unwinged; stem usually less than 1 m tall. **Leucanthemum vulgare**

ACHILLEA L. Yarrow. Achillée

1a Leaves pinnately dissected; heads numerous, the disk 2-4 mm wide. **A. millefolium**

1b Leaves shallowly serrate to subentire; heads several, the disk 4-8 mm wide. **A. ptarmica**

Achillea millefolium L. Fig. p. A108.
COMMON NAME(S): Yarrow, Milfoil. Herbe à dindes.
FREQUENCY: Common throughout much of the Province.
HABITAT: Roadsides, waste areas, lawns, meadows and shores.
GENERAL RANGE: Alaska to Nfld. south to Calif., Mex. and Fla.; Eurasia.
SYNONYMS: Most of our plants are referable to var. **lanulosa** (Nutt.) Piper, apparently a native tetraploid (2n = 36). The morphologically similar var. **millefolium** is a hexaploid (2n = 54) believed to be introduced from Eurasia and very rare in the Atlantic Provinces. Forma **roseoides** Breitung, with ray ligules pink or reddish occurs occasionally throughout.
SYNONYMS: **A. lanulosa** Nutt.

++

Achillea ptarmica L. Fig. p. A108. Map p. B78.
COMMON NAME(S): Sneezeweed. Herbe à éternuer.
SPECIAL NOTE: Eurasian; formerly cultivated and now escaped to fields, roadsides and rocky shores throughout the Province. Often appearing as if native. Known elsewhere in Canada from all provinces except Sask. Forma **multiplex** (Reynier) Heimerl, with heads all ligulate, occurs frequently with the species which has both ligulate and disk florets.

AMBROSIA L. Ragweed

1a Mid stem leaves usually palmately 3-lobed or sometimes unlobed and lanceolate, harshly strigose; fruit +/- 4-5 mm long. **A. trifida**
1b Mid stem leaves pinnately deeply lobed, finely strigose; fruit +/- 12 mm long. **A. artemisiifolia**

Ambrosia artemisiifolia L. Fig. p. A108.
COMMON NAME(S): Common Ragweed. Petite herbe à poux.
FREQUENCY: Scattered throughout the southern part of the Province.
HABITAT: Roadsides, waste areas, beaches, and fields.
GENERAL RANGE: Native range uncertain; Dist. Mack. and B.C. to Nfld. south to Mex. and Fla.
SUBSPECIFIC TAXA: Our plants are referable to var. **elatior** (L.) Desc.

Ambrosia trifida L. Fig. p. A108. Map p. B78.
COMMON NAME(S): Giant or Great Ragweed. Grande herbe à poux.
FREQUENCY: Locally common in the Miramichi R. drainage system, Northumberland Co., in Fredericton, York Co. and St. Andrews, Charlotte Co.
HABITAT: Thickets and waste areas.
GENERAL RANGE: Native range uncertain: B.C. to N.S. south to Mex. and Fla.

ANAPHALIS DC.

Anaphalis margaritacea (L.) Clarke Fig. p. A108.
COMMON NAME(S): Pearly Everlasting. Immortelle.
FREQUENCY: Common throughout much of the Province.
HABITAT: Roadsides, dry fields, open woods and clearings.
GENERAL RANGE: Alaska and B.C. to Nfld. south to Calif., Mex., S. Dak. and N. Eng.

ANTENNARIA Gaertn. Pussy-toes, Everlasting, Ladies'-tobacco. Immortelle

1a Basal leaves less than 1.5 cm wide, 1-nerved or obscurely 3-nerved. **A. neglecta**
1b Basal leaves over 4 cm wide, 3-5 nerved. **A. plantaginifolia**

++

Antennaria neglecta Greene Fig. p. A108.
SUBSPECIFIC TAXA: Three varieties occur within our range.

1a Cauline leaves of pistillate plants flattened or inrolled and scarious at summit. 2
1b Cauline leaves of pistillate plants blunt or pointed but not flattened or inrolled nor scarious at summit. var. **attenuata**
2a(1a) New basal leaves +/- glabrous on upper surface. var. **randii**
2b(1a) New basal leaves tomentulose on upper surface. var. **neglecta**

1. Var. **attenuata** (Fern.) Cronq. Map p. B78.
FREQUENCY: Scattered throughout much of the Province.
HABITAT: Old fields, ditches, clearings, thickets, dry rocks and ledges.
GENERAL RANGE: Ont. to Nfld. south to Minn. and Va.
SYNONYMS: **A. neodioica** Greene, **A. rupicola** Fern.

2. Var. **neglecta**
FREQUENCY: Common throughout much of the Province.
HABITAT: Fields, pastures, open woods and similar open areas.
GENERAL RANGE: Yukon-B.C. to Nfld. south to Calif., Ariz. and Va.
SYNONYMS: **A. petaloidea** Fern.

3. Var. **randii** (Fern.) Cronq. Map p. B79.
FREQUENCY: Uncommon in the southern half of the Province.
HABITAT: Rocky barrens, open woods and fields.
GENERAL RANGE: Man. to N.B. and N.S. south to Minn. and Va.
SYNONYMS: **A. canadensis** Greene, **A. neodioica** var. **randii** (Fern.) Boivin.

Antennaria plantaginifolia (L.) Richards. Fig. p. A108.
COMMON NAME(S): Plantain-leaved Pussy-toes.
FREQUENCY: Very rare; reported only from near confluence of Branch Stream and Digdeguash R., Charlotte Co. (UNB).
HABITAT: Dry open woods, clearings and fields.
GENERAL RANGE: Man. to N.B. and N.S. south to Tex. and Va.
SUBSPECIFIC TAXA: Our plants are referable to var. **arnoglossa** (Greene) Cronq.

ANTHEMIS L. Dog-Fennel, Chamomile. Camomille

1a Rays yellow; leaves 1-pinnatifid, the lanceolate ultimate segments coarsely dentate; stoloniferous perennial. **Anthemis tinctoria**
1b Rays white; leaves 2-3 pinnatifid, the ultimate segments linear; annuals. 2
2a(1b) Ray florets without stigmas; leaves +/- glabrous, unpleasantly scented. **Anthemis cotula**
2b(1b) Ray florets with stigmas; leaves +/- tomentose, without unpleasant scent. **Anthemis arvensis**

++

Anthemis arvensis L. Fig. p. A108.
COMMON NAME(S): Corn-Chamomile.
SPECIAL NOTE: Eurasian; rarely introduced along roadsides and in waste areas as near Mary's Point, Albert Co. and Buctouche, Kent Co. Known elsewhere in Canada from B.C., Ont., Nfld., P.E.I. and N.S.
SUBSPECIFIC TAXA: Our plants are referable to var. **agrestis** (Wallr.) DC.

Anthemis cotula L. Fig. p. A108. Map p. B79.
COMMON NAME(S): Mayweed, Dog-Fennel. Maroute, Camomille des chiens.
SPECIAL NOTE: Eurasian; introduced to waste areas mostly in the south of the Province. Known elsewhere in Canada from all provinces and the Yukon.

Anthemis tinctoria L. Fig. p. A108. Map p. B79.
COMMON NAME(S): Yellow Chamomile. Camomille des teinturiers, Oeil de boeuf.
SPECIAL NOTE: Eurasian; cultivated and escaped to roadsides and fields scattered throughout the Province. Known elsewhere in Canada from B.C. to Nfld., and N.S.

ARCTIUM L. Burdock. Bardane

1a Heads usually long peduncled in flat-topped or convex clusters; petioles strongly angled; leaves rounded at apex. 2

1b Heads +/- sessile to long-peduncled in racemiform clusters; petioles usually hollow, only slightly angled; leaves acute at apex. 3

2a(1a) Phyllaries glabrous, equalling or surpassing the corollas; involucre 2.5-4 cm thick; petioles +/- solid. **A. lappa**

2b(1a) Phyllaries cobwebby, mostly shorter than corollas; involucre 2-3 cm thick; petioles +/- hollow. **A. tomentosum**

3a(1b) Florets longer than phyllaries; fruiting involucre 1.5-2.5 cm thick; achenes about 6 mm long. **A. minus**

3b(1b) Florets about as long as phyllaries; fruiting involucres 3.0-3.5 cm thick; achenes +/- 1 cm long. **A. nemorosum**

Arctium lappa L. Fig. p. A109. Map p. B79.
COMMON NAME(S): Great Burdock, Cukle-Buttons. Grande Bardane, Artichaut, Rhubarbe sauvage, Toques, Rapace, Catherinettes.
SPECIAL NOTE: Eurasian; occasionally introduced in waste areas. Known elsewhere in Canada from B.C., Man. to N.B. and N.S.

Arctium minus (Hill.) Bernh. Fig. p. A109. Map p. B79.
COMMON NAME(S): Common Burdock. Bardane, Tabac du diable.
SPECIAL NOTE: Eurasian; introduced to waste areas throughout much of the developed areas of the Province. Known elsewhere in Canada from all provinces.

Arctium nemorosum Lef. Fig. p. A109.
SPECIAL NOTE: European; rarely introduced to waste areas as at Ft. Folly Point, Westmorland Co. Known elsewhere in Canada from all provinces.

++

Arctium tomentosum Mill. Fig. p. A109.
SPECIAL NOTE: Eurasian; introduced to waste areas as near Chatham, Northumberland Co. Known elsewhere in Canada from Alta. to N.S.

ARNICA L. Arnica

1a Pappus white, barbellate; leaves usually narrower and more obviously denticulate; plants of mostly lime rich soils. **A. angustifolia**
1b Pappus +/- tawny, subplumose; leaves broader and less obviously denticulate; plants of mostly acid soils. **A. lanceolata**

Arnica angustifolia Vahl.
FREQUENCY: Very rare and possibly extirpated; reported only from the gorge of the Sisson R., Victoria Co. The site badly disrupted by hydroelectric developments.
HABITAT: Calcareous ledges and gravels.
GENERAL RANGE: Alaska and northern Canada south to Alta., Ont., Gaspé Pen., N.B. and Nfld. Isolated stations in S. Dak. and Minn.
SUBSPECIFIC TAXA: Our plants are referable to ssp. **lonchophylla** (Greene) Douglas & Ruyle-Douglas.
SYNONYMS: **A. lonchophylla** ssp. **chionopappa** (Fern.) Maguire.
SPECIAL NOTE: For more information see H. Hinds, 1983. The Rare Vascular Plants of New Brunswick. Syllogeus 50.

Arnica lanceolata Nutt. Fig. p. A109. Map p. B79.
COMMON NAME(S): Arnica. Arnica soyeux.
FREQUENCY: Uncommon; known from northern Northumberland, Victoria, Madawaska, Restigouche, and Gloucester Counties.
HABITAT: Wet shores, ditches, and moist meadows.
GENERAL RANGE: Alaska-Yukon and B.C., Alta., Que. and N.B. south to Calif. and Colo., N.H. and Me.
SYNONYMS: **A. mollis** Hook. var. **petiolaris** Fern.

ARTEMISIA L. Sagebrush, Wormwood. Armoise

1a Leaves +/- glabrous (sometimes sparsely silky when young). 2
1b Leaves conspicuously hairy. 3
2a(1a) Ultimate segments of leaves merely toothed. 4
2b(1a) Ultimate segments of leaves narrowly linear; basal leaves 2-3 pinnatifid or ternate. **A. campestris**
3a(1b) Leaves +/- lanceolate or lance-elliptic, entire or with a few coarse teeth or entire forward-directed teeth, white-tomentose beneath. **A. frigida**
3b(1b) Leaves deeply lobed to dissected and fern-like. 5
4a(2a) Inflorescence +/- cylindrical, dense; leaves extending out from inflorescence. **A. biennis**
4b(2a) Inflorescence bushy-branched, loose; leaves surpassed by branches of inflorescence. **A. annua**

5a(3b) Receptacle naked. 6

5b(3b) Receptacle long-hairy. 7

6a(5a) Leaf segments less than 1 mm wide, +/- 2-pinnatifid and tomentose above; involucre densely pubescent with slender pedicelled nodding heads scattered along racemose branches of panicle. **A. pontica**

6b(5a) Leaf segments greater than 1 mm wide. 8

7a(5b) Leaf segments less than 1 mm wide, densely silvery-silky on both surfaces, 2-3 ternately divided; base of stem woody. **A. frigida**

7b(5b) Leaf segments usually 2-4 mm wide, 2-3 pinnatifid; base of stem herbaceous. **A. absinthium**

8a(6b) Leaves white-tomentose above, their lobes blunt; involucre 6-7 mm high; inflorescence narrow and dense; stems matted from creeping rhizomes. **A. stelleriana**

8b(6b) Leaves +/- glabrous and green above, white-tomentose beneath, their lobes often acute; involucre 3.5-4 mm high; infloresence a leafy panicle with ascending spike-like branches. **A. vulgaris**

Artemisia absinthium L. Fig. p. A109. Map p. B79.
COMMON NAME(S): Absinthe, Absinthium, Common Wormwood. Armoise.
SPECIAL NOTE: Eurasian; cultivated and introduced to waste areas, railways and roadsides in the south and east of the Province. Known elsewhere in Canada from all provinces.

Artemisia annua L. Fig. p. A109.
COMMON NAME(S): Sweet Wormwood. Armoise annuelle.
SPECIAL NOTE: Eurasian; rarely introduced in waste places as in Chatham, Northumberland Co. Known elsewhere in Canada from Ont. to P.E.I.

Artemisia biennis Willd. Fig. p. A109. Map p. B79.
COMMON NAME(S): Biennial Wormwood.
SPECIAL NOTE: Native to western U.S.A.; introduced to waste areas scattered throughout the Province. Known elsewhere in Canada from Dist. Mack. and all provinces except Nfld.

Artemisia campestris L. Fig. p. A109. Map p. B79.
COMMON NAME(S): Sagewort Wormwood.
FREQUENCY: Locally common along the Bay of Fundy, St. John R. drainage system, and the northeast shore of Gloucester and Northumberland Counties.
HABITAT: Sandy or gravelly shores, ledges and dry slopes.
GENERAL RANGE: B.C. to Nfld. south to Calif. Mich., and Me.
SUBSPECIFIC TAXA: Our plants are referable to ssp. **caudata** (Michx.) Hall & Clements.

Artemisia frigida Willd. Fig. p. A109.
COMMON NAME(S): Prairie Sagewort.
SPECIAL NOTE: Native to western states and provinces; occasionally introduced to waste areas as in Bathurst, Gloucester Co., Sackville, Westmorland Co. and

Fredericton, York Co. One anomalous collection is from an undeveloped area, Kedgewick Forks, Restigouche Co., along a stream edge in softwoods. (UNB).
SYNONYMS: **A. ludoviciana** var. **gnaphalodes** (Nutt.) T. & G.

Artemisia pontica L. Fig. p. A109.
COMMON NAME(S): Roman Wormwood: Petite Absinthe, Plante de beauté.
SPECIAL NOTE: European; rarely introduced to waste areas and railways as in Mouth-of-Keswick, York Co. Known elsewhere in Canada from Man., Ont., Que., and N.S.

Artemisia stelleriana Bess. Fig. p. A110.
COMMON NAME(S): Dusty Miller, Beach Wormwood.
SPECIAL NOTE: Asiatic; commonly introduced in coastal areas of the Province. Known elsewhere in Canada from Ont. to N.S.; B.C.

Artemisia vulgaris L. Fig. p. A110.
COMMON NAME(S): Common Mudwort. Herbe Saint-Jean.
SPECIAL NOTE: Eurasian; commonly introduced in waste areas throughout the developed areas of the Province. Known elsewhere in Canada from all provinces.

ASTER L. Aster

1a Basal and usually the lowest leaves both cordate and petiolate (sometimes merely abruptly contracted to winged petiole in **A. ciliolatus**). 2

1b Leaves not cordate and petiolate. 3

2a(1a) Involucre glandular; inflorescence round-topped with few leaf-like bracts; phyllaries greenish, firm, imbricated; rays violet to blue; rhizomes long creeping. **A. macrophyllus**

2b(1a) Involucre not glandular. 4

3a(1b) Heads +/- discoid (rays less than 1 mm long or wanting). **A. laurentianus**

3b(1b) Heads distinctly radiate. 5

4a(2b) Middle stem leaves not wing-petioled (less than 2 mm wide closest to blade); involucre 3.5-6.5 mm high; phyllaries +/- appressed, the outer ones blunt or short-tipped with oblanceolate contrasting central bands dilated near their summits; heads short peduncled in panicles of racemose branches. **A. cordifolius**

4b(2b) Middle stem leaves wing-petioled, 4-6 mm wide; involucres 5-7 mm high; phyllaries +/- loosely ascending, linear attenuate, the lanceolate median band not strongly broadened to summit. Heads in open panicles of short loose racemes or corymbs +/- on elongate naked or 1-2 bracted slender peduncles. **A. ciliolatus**

5a(3b) Plants annual from a short taproot; leaves +/- fleshy spatulate-oblanceolate, with obtuse or rounded tips; heads few, terminating erect branchlets; phyllaries subequal, the outer herbaceous; rays 1-rowed, 4-5 mm long, much exceeding pappus; saltmarsh plants. **A. subulatus**

5b(3b) Plants perennial with fibrous roots. 6

6a(5b) Involucres glandular; achenes pubescent; inflorescence roundish-topped; rays normally violet to blue-purple; leaves entire, strongly cordate-auriculate, lanceolate, acuminate, scabrous or stiffly appressed-hairy above. **A. novae-angliae**

6b(5b) Involucres not obviously glandular. 7

7a(6b) Phyllaries hard or firm, usually pale-coriaceous at base, mostly green-tipped and often with a broad green median band, usually in several markedly unequal series; inflorescence commonly corymbiform; leaves usually firm. 8

7b(6b) Phyllaries thin and scarious or foliaceous, pliable and not markedly coriaceous toward base; leaves mostly relatively thin and soft. 9

8a(7a) Leaves linear to narrowly oblong, entire. 10

8b(7a) Leaves lanceolate to oblanceolate, ovate or oblong, entire or serrate; achenes glabrous. 13

9a(7b) Inflorescence rounded or flat-topped (heads sometimes solitary); phyllaries lacking herbaceous tips. 11

9b(7b) Inflorescence not distinctly flat or round topped. 12

10a(8a) Achenes glabrous; pappus white clavate; involucre +/- 7 mm high; with lance-subulate to lance-oblong coriaceous acutish phyllaries with a relatively long green area on back of thickened midrib; rays white, +/- 8 mm long; peduncles +/- naked. **Solidago ptarmicoides**

10b(8a) Achenes appressed-silky; phyllaries firm, with elongate green areas on back of thickened midrib, ciliate-fringed on +/- purplish margins; rays normally blue-violet, to 1.5 cm long; pappus tawny; heads 1-few on leafy-bracted peduncles; leaves scabrous, lacking lateral nerves. **A. linariifolius**

11a(9a) Pappus +/- sordid, double (a shorter outer series); achenes pubescent; heads many; rays white; leaves pinnate-veined, narrowly to broadly elliptic, up to 1.5 dm long; stem to 2 m tall. **A. umbellatus**

11b(9a) Pappus white, single; achenes glandular; stem usually less than 1 m tall. 14

12a(9b) Stem leaves auriculate-clasping at base (sometimes half-clasping). 15

12b(9b) Stem leaves not clasping at base. 16

13a(8b) Leaves +/- glaucous, thickish and firm, entire or remotely serrate, lanceolate to ovate, middle and upper ones auriculate-clasping, the lower on long winged petioles; phyllaries acute or acuminate, closely appressed; plant +/- glabrous from short rhizomes. **A. laevis**

13b(8b) Leaves green, +/- thin, sharply serrate, scabrous, rough-veiny, oblong-lanceolate, acute, sessile, the lower usually early deciduous; tips of phyllaries +/- spreading, blunt; plant +/- hispid from long subterranean stolons. **A. radula**

14a(11b) Leaves entire, linear to narrowly lanceolate, 40-75 below the inflorescence, 5-12 mm wide with revolute margins; heads solitary or few. **A. nemoralis**

14b(11b) Leaves toothed, less than 40 below the inflorescence. 17

15a(12a) Involucral bracts (at least the inner ones) long-acuminate to attenuate; leaves tapering or gradually rounded to base. **A. puniceus**

15b(12a) Involucral bracts obtuse to merely acute, sometimes acuminate. 18

16a(12b) Lobes of disk corollas comprising 45-75% of the goblet-shaped limb; leaves hairy along the midrib beneath, linear-lanceolate to sub-rhombic. **A. lateriflorus**

16b(12b) Lobes of disk corollas comprising 17-45% of the +/- funnel-shaped limb; leaves glabrous beneath. 19

17a(14b) Peduncles bearing 1-2 ovate to lanceolate bracts (often with a few smaller ones also); rays normally white; leaves oblong-lanceolate to oblanceolate or broadly oval, coarsely serrate, to about 6 cm broad. **A. acuminatus**

17b(14b) Peduncles bearing several linear to thread-like bracts; leaves lanceolate to oblong, acute or short-acuminate, low serrate to about 2.5 cm broad. **A. X blakei**

18a(15b) Involucral bracts herbaceous, usually more than 1 mm broad and sometimes expanded and foliaceous, loose or spreading to squarrose, imbricate to +/- equal; heads usually many; stem often over 2.5 mm thick; rhizome over 2 mm thick; leaves narrowly lanceolate to broadly oblanceolate; slender to stout plants of gravelly river shores, roadsides, edges of saltmarshes, etc. **A. novi-belgii**

18b(15b) Involucral bracts appressed, imbricate to +/- equal, less than 1 mm broad, slender, acute to acuminate; heads usually few; stem usually less than 2.5 mm thick; rhizome less than 2 mm thick; leaves entire, 1.5-5(9) mm wide; slender plants of cold calcareous bogs. **A. borealis**

19a(16b) Slender (stems 1-2.5(3) mm thick), rather low (to 6 dm high) plants with racemose-paniculate heads or these scattered on minutely foliose branchlets; branches +/- divergent; leaves +/- 1.5 cm broad, to 10 cm long; ligules white. 20

19b(16b) Stout (stem 3-6 mm thick), tall (to 2 m high); paniculate inflorescence elongate, leafy with numerous heads; leaves 1-2 dm long, 8-15 cm broad; ligules white (usually fading to pale blue) to blue or blue-violet. **A. lanceolatus**

20a(19a) Involucre 2.5-3.5(4) mm high; leaves sessile, linear or narrowly lanceolate, entire or slightly toothed; heads numerous in an open inflorescence with divergent, bracteate, often recurved branches +/- secund, the bracteate peduncles up to 1.5 cm long; plants mostly of shaded bottomlands. **A. vimineus**

20b(19a) Involucres 3.5-5 mm high; leaves linear or narrowly lance-linear to lance elliptic, +/- sessile; heads relatively few in an elongate, open inflorescence; plants mostly of open rocky shores. **A. tradescantii**

Aster acuminatus Michx. Fig. p. A110.
COMMON NAME(S): Whorled Wood Aster.
FREQUENCY: Common throughout much of the Province.
HABITAT: Moist woodlands and clearings.
GENERAL RANGE: Ont. to Nfld. south to Tenn. and Ga.
SUBSPECIFIC TAXA: Forma **discoideus** Ktze., with florets modified into tufts of subulate, reddish-orange chaff, is reported from near Fundy National Park, Albert Co. and from Yoho Lake, York Co. (UNB). This is presumed to be a pathological state.

Aster X blakei (Porter) House Map p. B79.
COMMON NAME(S): Blake's Aster.
FREQUENCY: Uncommon; reported from Charlotte, York, Westmorland, Kent and Gloucester Counties.

HABITAT: Edges of bogs, open woods, and damp thickets.
GENERAL RANGE: Que. to Nfld. south to N. Eng.
SPECIAL NOTE: This is the hybrid between **Aster acuminatus** and **A. nemoralis.**

Aster borealis (T. & G.) Prov. Fig. p. A110.
FREQUENCY: Very rare; reported from Grand Manan Is., Charlotte Co., Madawaska and Gloucester Counties.
HABITAT: Calcareous bogs, cedar swamps and shores.
GENERAL RANGE: Alaska-Yukon to Gaspé Pen. and N.S., Idaho and N.J.
SPECIAL NOTE: Status uncertain because of confusion with related taxa.
SYNONYMS: **A. junciformis** Rydb., **A. junceus** Sensu auctt. non Ait.

Aster ciliolatus Lindl. Fig. p. A110. Map p. B79.
FREQUENCY: Scattered throughout much of the Province.
HABITAT: Old fields, roadsides, margins of woods and thickets.
GENERAL RANGE: B.C. to N.B. and N.S. south to Mont. and N. Eng.
SPECIAL NOTE: This species is suspected of forming hybrids with the following species in the Province; **A. cordifolius**, **A. macrophyllus**, **A. puniceus**, **A. lanceolatus**, and **A. novi-belgii**.
SYNONYMS: **A. lindleyanus** T. & G.

Aster cordifolius L. Fig. p. A110. Map p. B79.
COMMON NAME(S): Heart-leaved Aster.
FREQUENCY: Common throughout much of the Province.
HABITAT: thickets, clearings, open woods and roadsides.
GENERAL RANGE: Ont. to N.S. south to Mo. and Ga.
SUBSPECIFIC TAXA: Forma **furbishiae** (Fern.) Forma nova, with varying degrees of pubescence occurs principally in the western part of the Province.
SPECIAL NOTE: The following suspected hybrids are reported for the Province:
A. cordifolius X **A. lateriflorus**, **A. macrophyllus**, and **A. puniceus**.

Aster dumosus L.
COMMON NAME(S): Bushy Aster.
FREQUENCY: Very rare and probably introduced from further west. Plants collected in Derby Jct., Northumberland Co. closely approach this species except for the rays being white (UNB). It occurred with **A. novi-belgii**, **A. lateriflorus** and their suspected hybrid.
HABITAT: Here found along a dry roadside. Normally found in dry prairies, thickets and open woods.
GENERAL RANGE: Mich. to Ont., N.Y. and Me. south to Tex. and Fla.

Aster laevis L. Fig. p. A110.
COMMON NAME(S): Smooth Aster.
FREQUENCY: Very rare and probably introduced from further west. Reported from Northumberland, Carleton, and Victoria Counties.
HABITAT: Roadsides, old fields and thickets.
GENERAL RANGE: Yukon and B.C. to Que. south to Oreg., N. Mex. and Ga.

++

Aster lanceolatus Willd. Fig. p. A110. Map p. B80.
FREQUENCY: Common in the southern half and western border of the Province.
HABITAT: Damp clearings, borders of woods, thickets and roadsides.
GENERL RANGE: Sask. to Nfld. south to Kans. and N.C.
SPECIAL NOTE: Much of our material has bluish or violet ray florets.
SYNONYMS: **A. simplex** Willd. and var. **ramosissimus** (T. & G.) Cronq., **A. paniculatus** Lam.

Aster lateriflorus (L.) Britt. Fig. p. A110. Map p. B80.
COMMON NAME(S): Calico Aster.
FREQUENCY: Common in the southern half of the Province; rarer northwards.
HABITAT: Fields, thickets and clearings.
GENERAL RANGE: Man. to N.S. south to Tex. and Fla.
SPECIAL NOTE: A suspected hybrid with **A. novi-belgii** is reported from Derby Jct., Northumberland Co. (UNB).

Aster laurentianus Fern. Fig. p. A110.
FREQUENCY: Very rare; known from Miscou Is., Tracadie and Bathurst, Gloucester Co. and Kouchibouguac National Park, Kent Co.
HABITAT: Saline shores and marshes.
GENERAL RANGE: Yukon and B.C. to N.S. south to Wash., Minn. and N.Y.
SYNONYMS: **A. brachyactis** Blake.
SPECIAL NOTE: For more information see H. Hinds, 1983. The Rare Vascular Plants of New Brunswick. Syllogeus 50.

Aster linariifolius L. Fig. p. A110. Map p. B80.
COMMON NAME(S): Spruce Aster, Savory-leaved Aster.
FREQUENCY: Locally common along the SW Miramichi R., Northumberland Co., Pabineau Falls, Gloucester Co., and rare in Kouchibouguac National Park, Kent Co.
HABITAT: Sandy roadsides, rocky banks and dry fields.
GENERAL RANGE: Minn. to N.B. south to Tex. and Fla.

Aster macrophyllus L. Fig. p. A110. Map p. B80.
COMMON NAME(S): Large-leaved Aster.
FREQUENCY: Scattered throughout much of the Province.
HABITAT: Dry woods and thickets.
GENERAL RANGE: Man. to N.S. south to Ill. and N.C.

Aster nemoralis Ait. Fig. p. A110. Map p. B80.
COMMON NAME(S): Bog Aster.
FREQUENCY: Scattered throughout the southern and eastern portions of the Province.
HABITAT: Acid bogs.
GENERAL RANGE: Ont. to Nfld. south to Mich. and N.J.

Aster novae-angliae L. Fig. p. A111.
COMMON NAME(S): New England Aster.
SPECIAL NOTE: Introduced from further south; recorded from Fredericton, York Co., Campbellton, Restigouche Co. and rapidly spreading near Milltown, Charlotte Co.

++

SUBSPECIFIC TAXA: Forma **roseus** (Desf.) Britt., with pinkish rays, is reported from Milltown, Charlotte Co.

Aster novi-belgii L. Fig. p. A111.
COMMON NAME(S): New York Aster.
FREQUENCY: Common throughout much of the Province.
HABITAT: Thickets, roadsides, gravel strands and shores.
SPECIAL NOTE: The limits of variability within the **A. novi-belgii** complex are presently being considered by several specialists of the group. There is a tremendous range of variation between the meter high, broad-leaved and many-headed forms found on roadsides and along the coast and the long, narrow-leaved and few-headed forms mostly found on the shores of our inland lakes and rivers. There is also a puzzling array of intermediate types scattered throughout the Province. No completely satisfactory disposition of the entire complex has yet been put forward. This complex includes the following taxa: **A. anticostensis** Fern., **A. crenifolius** (Fern.) Cronq., **A. foliaceus** sensu auctt. non Lindl., **A. johannensis** Fern., **A. johannensis** var. **villicaulis** (Gray) Fern., **A. longifolius** Lam., **A. tardiflorus** L. and **A. novi-belgii** var. **elodes** (Torr. & Gray) Gray, var. **litoreus** Gray, and var. **rosaceus** Rouss.

Aster pilosus Willd.
SPECIAL NOTE: Native to southern Ont. and southwest Que.; introduced along railway in Hampton, Kings Co. but not spreading (UNB). Not keyed above.

Aster praealtus Poir.
SPECIAL NOTE: Native from Man. to southern Que. south to Tex. and Tenn.; plants closely fitting the characteristics of this species are reported from near St. Andrews, Charlotte Co. and were probably introduced from hay brought from further west with C. VanHorne's cattle which were kept on nearby Ministers Island. Not keyed above.

Aster puniceus L. Fig. p. A111. Map p. B80.
COMMON NAME(S): Purple-stemmed Aster.
FREQUENCY: Common throughout much of the Province.
HABITAT: Roadside ditches, damp meadows and shores.
GENERAL RANGE: Alta. to Nfld. south to N. Dak., Ill. and Ga.
SPECIAL NOTE: Hybridizes with **A. novi-belgii**.

Aster radula Ait. Fig. p. A111. Map p. B80.
COMMON NAME(S): Rough Aster.
FREQUENCY: Common in the eastern half of the Province, and rare elsewhere.
HABITAT: Low woods, treed bogs and peaty clearings.
GENERAL RANGE: Que. to Nfld. south to W. Va. and Va.
SUBSPECIFIC TAXA: Our plants are referable to var. **strictus** (Pursh) Gray.

(**Aster subulatus** Michx.)
SPECIAL NOTE: Native from Mich. to southern Ont. and N.Y. south along the coast to Fla. and La.; Fernald reported this species from Bathurst, Gloucester Co. The material was atypical enough for him to publish it as var. **obtusifolius**. These plants were confirmed by L. Broullet to be another species Fernald named in the general area, **A. laurentianus** and not **A. subulatus** at all.

++

Aster tradescantii L. Fig. p. A111. Map p. B80.
FREQUENCY: Uncommon; scattered in the southwest corner of the Province from the vicinity of Grand Lake, Queens Co. to the St. Croix R., Charlotte Co.
HABITAT: Swamps, and rocky or gravelly shores.
GENERAL RANGE: Ont. to Nfld., N.B. and N.S. south to N.Y. and N. Eng.

Aster umbellatus Mill. Fig. p. A111.
COMMON NAME(S): Flat-topped White Aster.
FREQUENCY: Common throughout much of the Province.
HABITAT: Open woods, meadows, and thickets.
GENERAL RANGE: Ont. to Nfld. south to Nebr., Ky. and Ga.

Aster vimineus Lam. Fig. p. A111. Map p. B80.
FREQUENCY: Rare; reported from a few localities in York, Sunbury and Queens Counties (UNB).
HABITAT: Wet woods, fields and shores.
GENERAL RANGE: Mich. to Ont., Que. and N.B. south to Tex. and Ga.
SUBSPECIFIC TAXA: Our plants are referable to var. **subdumosus** Wieg.

BIDENS L. Bur-Marigold, Beggar-ticks. Bident

1a	Leaves deeply lobed or divided. 2
1b	Leaves neither lobed nor divided. 3
2a(1a)	Leaves at most only broadly lobed at base; achenes sharply 4-angled, 3-4 awned, downwardly barbed; heads discoid; outer phyllaries 2-6. **B. connata**
2b(1a)	Leaves definitely compound; achenes usually 2-awned. 4
3a(1b)	Achenes mostly convex between awns, flat; rays usually present. 5
3b(1b)	Achenes mostly concave or flat between awns; rays absent. **B. connata**
4a(2b)	Margin of outer phyllaries +/- ciliate. 6
4b(2b)	Margin of outer phyllaries smooth; achenes upwardly barbed. **B. discoidea**
5a(3a)	Heads erect, campanulate; disk-corollas 4-lobed; central nerve of achenes similar to marginal nerves, barbless; anthers included in corolla; plants of brackish river estuaries. **B. hyperborea**
5b(3a)	Heads usually nodding in fruit, hemispheric; disk-corollas 5-lobed; central nerve of achenes more prominent than marginal nerves, barbed; anthers surpassing corolla; plants of usually fresh habitats. **B. cernua**
6a(4a)	Outer phyllaries 5-8, sparsely ciliate; outer achenes 2.2-4 mm wide, 5.3-7 mm long. **B. frondosa**
6b(4a)	Outer phyllaries 10-20, copiously ciliate; outer achenes 3.5-6.5 mm wide, 6.5-11.3 mm long. **B. vulgata**

Bidens cernua L. Fig. p. A112. Map p. B80.
COMMON NAME(S): Nodding Beggar-ticks, Stick-tight, Pitchfork. Bident penché, Fourchettes.
FREQUENCY: Scattered throughout much of the Province.
HABITAT: Shores and similar wet places.

GENERAL RANGE: B.C. to N.S. south to Calif. and N.C.; Eurasia.
SUBSPECIFIC TAXA: Forma **discoidea** (Wimm. & Grab.) Briq. & Cavill., lacking ligulate florets, occurs occasionally with the rayed form.

Bidens connata Muhl. Fig. p. A112. Map p. B80.
FREQUENCY: Uncommon; reported from Fredericton, York Co. and from several east coast localities.
HABITAT: Shores and similar wet areas.
GENERAL RANGE: Ont. to N.S. south to Kans. and Va.

Bidens discoidea (T. & G.) Britt. Fig. p. A112.
FREQUENCY: Very rare; reported only from Washademoak Lake, Queens Co. (UNB).
HABITAT: Rocky, peaty or sandy shores.
GENERAL RANGE: Minn. to N.B. and N.S. south to Tex. and Ala.

Bidens frondosa L. Fig. p. A112.
FREQUENCY: Scattered in the southern half of the Province and rare in the north.
HABITAT: Wet open areas.
GENERAL RANGE: B.C., Man. to Nfld. south to Calif., La. and Va.
SUBSPECIFIC TAXA: Forma **anomala** (Porter) Fern., with awns upwardly barbed, occurs rarely with the species.

Bidens hyperborea Greene Fig. p. A112. Map p. B80.
FREQUENCY: Scattered along the east coast.
HABITAT: River estuaries.
GENERAL RANGE: Ont. to N.B. and N.S. south along the coast to N.J.
SUBSPECIFIC TAXA: Var. **arcuans** Fern., with outer achenes to 8.5 mm long instead of to 5 mm long, occurs occasionally with the species as near Newcastle, Northumberland Co. (type locality).

Bidens vulgata Greene Fig. p. A112.
COMMON NAME(S): Beggar-ticks, Stick-tight.
FREQUENCY: Scattered in the southern half of the Province.
HABITAT: Damp waste places, ditches and similar areas.
GENERAL RANGE: B.C. to N.B. and N.S. south to Calif. and Ga.
SUBSPECIFIC TAXA: Var. **schizantha** Lunell, with most leaves twice pinnate instead of with leaves once pinnate as in the species, is reported from Milltown, Charlotte Co. (UNB).

CARDUUS L. Plumeless Thistle

1a Heads large, +/- nodding, solitary at branch tips; leaf lobes and decurrent bases strongly spiny. **C. nutans**

1b Heads relatively small, clustered at branch tips; leaf lobes decurrent bases weakly spiny. **C. crispus**

++

Carduus crispus L. Fig. p. A112.
COMMON NAME(S): Welted Thistle.
SPECIAL NOTE: Eurasian; rarely introduced in waste areas as in Saint John, St. John Co. Probably not persisting. Known elsewhere in Canada from Ont. and N.S.

Carduus nutans L. Fig. p. A112.
COMMON NAME(S): Musk-Thistle.
SPECIAL NOTE: Eurasian; introduced in waste areas in Kent, Northumberland, Gloucester and Northumberland Counties. Known elsewhere in Canada from B.C. to Nfld. and N.S.

CENTAUREA L. Star Thistle, Knapweed. Centaurée

1a Middle stem leaves deeply pinnately divided. 2
1b Middle stem leaves simple, coarsely toothed, sinuate or lyrate-based (occasionally narrow lobed in **C. cyanus**). 3
2a(1a) Ultimate segments of leaves more than 4 mm wide; pappus more than 3 mm long; involucre to 2.5 cm high, the phyllaries conspicuously bicolored. **C. scabiosa**
2b(1a) Ultimate segments of leaves less than 2 mm wide; pappus less than 2 mm long or absent; involucre to 1.5 cm high, the phyllaries not conspicuously bicolored. **C. maculosa**
3a(1b) Lower leaves linear to narrowly lanceolate, persistently white-arachnoid beneath; lower leaves occasionally narrow lobed; annuals. **C. cyanus**
3b(1b) Lower leaves lanceolate to obovate, not persistently white-arachnoid beneath; perennials. 4
4a(3b) Phyllaries black fringed, without enlarged apical appendage; leaves soft pubescent. **C. montana**
4b(3b) Phyllaries with a conspicuous enlarged dark brown pectinately-fringed apical appendage; leaves rough puberulent. **C. nigra**

Centaurea cyanus L. Fig. p. A112.
COMMON NAME(S): Bachelor's-button, Bluebottle, Cornflower. Bleuet.
SPECIAL NOTE: European; cultivated and occasionally escaped to fields and waste places as in Jemseg, Queens Co. Known elsewhere in Canada from B.C., Alta. to Nfld., P.E.I. and N.S.

Centaurea maculosa Lam. Fig. p. A112.
SPECIAL NOTE: European; rarely introduced to roadsides and waste areas as in Fundy National Park, Albert Co. Known elsewhere in Canada from B.C., Ont., Que. and N.S.

Centaurea montana L.
COMMON NAME(S): Mountain-Bluet.
SPECIAL NOTE: European; cultivated and occasionally escaped to dooryards and waste areas as in Fredericton, York Co. Known elsewhere in Canada from B.C., Ont. to Nfld.

Centaurea nigra L. Fig. p. A113. Map p. B81.
COMMON NAME(S): Knapweed, Spanish-buttons. Centaurée noire.

SPECIAL NOTE: European; introduced in waste areas throughout much of the Province. Known elsewhere in Canada from B.C., Ont. to Nfld., P.E.I. and N.S.
SUBSPECIFIC TAXA: Var. **radiata** DC., with greatly enlarged marginal florets, is occasionally reported with the species.

Centaurea scabiosa L. Fig. p. A113
COMMON NAME(S): Knapweed. Centaurée Scabieuse.
SPECIAL NOTE: Eurasian; reported from the Province by Boivin, 1966-67.

CHRYSANTHEMUM L. Chrysanthemum. Chrysanthème

1a Leaves clasping stem, shallowly pinnatifid; ray achenes 2-winged; rays deep yellow. **C. segetum**
1b Leaves not clasping, pinnate-pinnatifid; ray achenes 3-winged; rays creamy yellow. **C. coronarium**

Chrysanthemum coronarium L.
COMMON NAME(S): Garland-Chrysanthemum.
SPECIAL NOTE: European; introduced from ballast waste, Saint John, St. John Co. Not persisting. Known elsewhere in Canada from Ont.

Chrysanthemum segetum L. Fig. p. A113.
COMMON NAME(S): Corn-Marigold. Marguerite dorée.
SPECIAL NOTE: Eurasian; cultivated and locally escaped as in Saint John, St. John Co. Known elsewhere in Canada from Que., P.E.I. and N.S.

CICHORIUM L. Chicory, Succory. Chicorée

Cichorium intybus L. Fig. p. A113. Map p. B81.
COMMON NAME(S): Blue Sailors.
SPECIAL NOTE: Eurasian; occasionally introduced to roadsides and waste areas scattered throughout the Province. Known elsewhere in Canada from all provinces.

CIRSIUM Mill. Plumed Thistle. Chardon

1a Heads small, numerous, unisexual; involucres 1-2 cm high; outer phyllaries with weak prickles; corollas pink-purple to white. **C. arvense**
1b Heads larger, 1-few; all flowers perfect, purple (rarely white); involucres 2-4 cm high. 2
2a(1b) Leaf bases long decurrent producing +/- spiny-winged stems; all phyllaries spine-tipped; stem rarely to 2 m tall. **C. vulgare**
2b(1b) Leaf bases sessile, not decurrent on the stem; phyllaries acute to acuminate, at most weakly short spine-tipped; stem often over 2 m high. **C. muticum**

++

Cirsium arvense (L.) Spreng. Fig. p. A113.
COMMON NAME(S): Canada Thistle. Chadron, Chaudron.
SPECIAL NOTE: Eurasian; commonly introduced to cultivated fields, roadsides and waste areas throughout much of the Province. Known elsewhere in Canada from all provinces and Dist. Mack.
SUBSPECIFIC TAXA: Forma **albiflorum** (Rand & Redf.) Hoffm., with white flowers, occurs occasionally with the species.

Cirsium muticum Michx. Fig. p. A113. Map p. B81.
COMMON NAME(S): Swamp thistle.
FREQUENCY: Scattered throughout much of the Province.
HABITAT: Swamps, bottomlands and wet meadows.
GENERAL RANGE: Sask. to Nfld. south to Tenn. and N.C.

Cirsium vulgare (Savi) Tenore Fig. p. A113.
COMMON NAME(S): Bull-Thistle. Piqueux, Gros Chardon.
SPECIAL NOTE: Eurasian; commonly introduced in cultivated fields and waste areas throughout much of the Province. Known elsewhere in Canada from B.C. to Nfld., P.E.I. and N.S.

CNICUS L.

Cnicus benedictus L. Fig. p. A113.
COMMON NAME(S): Blessed Thistle.
SPECIAL NOTE: Eurasian; formerly cultivated and rarely escaped as in Bass River, Kent Co. Known elsewhere in Canada from B.C. and N.S.

CONYZA Less.

Conyza canadensis (L.) Cronq. Fig. p. A113. Map p. B81.
COMMON NAME(S): Horseweed. Vergerette du Canada.
FREQUENCY: Probably more common than collections indicate, and found throughout much of the Province.
HABITAT: Disturbed ground and waste areas throughout the Province.
GENERAL RANGE: Dist. Mack. and B.C. to Nfld., south to Calif., Mex. and Fla.; tropical America.
SYNONYMS: **Erigeron canadensis** L.

COTULA L. Brass-buttons

Cotula coronopifolia L. Fig. p. A113.
SPECIAL NOTE: Native to S. Africa; rarely introduced in tidal mud flats as on Grand Manan Is., Charlotte Co., in Saint John, St. John Co., and on Miscou Is., Gloucester Co. Known elsewhere in Canada from B.C., Que. to N.S.

++

CREPIS L. Hawk's-beard

1a Mature achenes tawny to pale brown, 1.5-2.5 mm long; phyllaries glabrous within; receptacle glabrous. **C. capillaris**

1b Mature achenes dark purplish-brown, 2.5-4.5 mm long; phyllaries appressed strigose within; receptacle finely ciliate-fringed between achene pits. **C. tectorum**

Crepis capillaris (L.) Wallr. Fig. p. A113.
SPECIAL NOTE: European; reported as introduced in the Province by Boivin, 1966-67.

Crepis tectorum L. Fig. p. A114.
SPECIAL NOTE: Eurasian; rarely introduced along roadsides and in waste areas as in Fredericton and Lincoln, York Co. Known elsewhere in Canada from Yukon and Dist. Mack., B.C. to N.S.

ECHINOPS L. Globe-Thistle

Echinops exaltatus Schrad.
COMMON NAME(S): Russian Globe-Thistle.
SPECIAL NOTE: Siberian; cultivated and very locally escaped as near the mouth of Barnaby R., Restigouche Co. Known elsewhere in Canada from B.C., Ont. and Que.

ERECHTITES Raf.

Erechtites hieracifolia (L.) Raf. Fig. p. A114. Map p. B81.
COMMON NAME(S): Fireweed. Crève-z-yeux.
FREQUENCY: Common in the southern half of the Province and rare northwards.
HABITAT: Disturbed soils, burns, fields and waste areas.
GENERAL RANGE: Ont. to N.S. south to Tex. and Fla.
SYNONYMS: **E. hieracifolia** var. **intermedia** Fern.

ERIGERON L. Fleabane. Vergerette

1a Stem leaves cordate clasping; rays 5-10 mm long, deep pink or white. **E. philadelphicus**

1b Stem leaves not clasping. 2

2a(1b) Stems with a single head, numerous from caudex; leaves numerous linear, only slightly reduced upward; rays 4-8 mm long, white or rarely pink or rose-purple. **E. hyssopifolius**

2b(1b) Stems usually with many heads; leaves narrowly lanceolate to ovate. 3

3a(2b) All leaves except sometimes the uppermost coarsely toothed, the median broadly lanceolate, tapering to wide petiole; pubescence of stem long and spreading. **E. annuus**

3b(2b) Leaves +/- entire, mostly narrowly lanceolate to elliptic; basal and lower stem leaves sometimes shallowly toothed, tapering to narrow petioles; pubescence of stem various. 4

4a(3b) Involucre to 5 mm high with sparse long flattened +/- transparent hairs or short dense appressed white hairs; pappus of disk florets double with inner series of capillary bristles and outer series of shorter scales or bristles; rays conspicuous, usually longer than pappus. **E. strigosus**

4b(3b) Involucre to over 1 cm high, glabrous or slightly bristly and glandular viscid; pappus of disk florets a single series of capillary bristles; rays inconspicuous, +/- as long as pappus. **E. acris**

Erigeron acris L. Fig. p. A114.
FREQUENCY: Very rare; reported from Madawaska and Restigouche Counties.
HABITAT: Rocky shores and thickets.
GENERAL RANGE: Alaska and Yukon to Nfld. and N.B. south to Calif., Colo., Mich. and Maine; Eurasia.
SUBSPECIFIC TAXA: Our plants are referable to var. **kamtschaticus** (DC.) Herder.
SYNONYMS: **E. acris** var. **asteroides** (Andrz.) DC.

Erigeron annuus (L.) Pers. Fig. p. A114.
COMMON NAME(S): Daisy-Fleabane. Vergerette annuelle.
FREQUENCY: Scattered throughout much of the Province.
HABITAT: Open grassy areas, disturbed soils and waste areas.
GENERAL RANGE: B.C. to Nfld. south throughout U.S.A.

Erigeron hyssopifolius Michx. Fig. p. A114. Map p. B81.
COMMON NAME(S): Hyssop-leaved Fleabane.
FREQUENCY: Locally common in the southeast and northern parts of the Province.
HABITAT: Calcareous ledges and shores.
GENERAL RANGE: Yukon and Dist. Mack. to Nfld., N.B. and N.S. south to Mich., N.Y. and Vt.
SUBSPECIFIC TAXA: Forma **patens** Cronq., with spreading rather than appressed pubescence along the stem, is reported from Plaster Rock, Victoria Co. and from along the Restigouche R., Restigouche Co.

Erigeron philadelphicus L. Fig. p. A114. Map p. B81.
FREQUENCY: Scattered throughout much of the Province except uncommon in the eastern lowlands.
HABITAT: Moist meadows and similar open places.
GENERAL RANGE: Yukon and Dist. Mack., B.C. to Nfld. south to Calif., Tex. and Fla.

Erigeron strigosus Muhl. Fig. p. A114. Map p. B81.
COMMON NAME(S): Daisy-Fleabane, White-top. Vergerette rude.
FREQUENCY: Common throughout much of the Province.
HABITAT: Cultivated areas, roadsides and waste places.
GENERAL RANGE: B.C. to Nfld. south to Calif., Okla., Tex. and Fla.
SUBSPECIFIC TAXA: Var. **septentrionalis** (Fern. & Wieg.) Fern., with stems and lower leaves +/- glabrous instead of cinereous-strigose, is reported scattered in the western half of the Province.

++

EUTHAMIA Ell.

Euthamia graminifolia (L.) Nutt. ex Cass. Fig. p. A119. Map p. B83 (as **Solidago graminifolia**).
COMMON NAME(S): Grass-leaved Goldenrod.
FREQUENCY: Dry fields, thickets, shores and waste areas.
GENERAL RANGE: Man. to Nfld. south to N.Mex., Ky. and N.C.
SUBSPECIFIC TAXA: Most of our plants are var. **nuttallii** (Greene) Stone, with upper leaves +/- hirtellus rather than essentially glabrous.
SYNONYMS: **Solidago graminifolia** L.

EUPATORIUM L. Thoroughwort

1a Opposite bases of leaves connate around stem; flowers white; leaves lance-acuminate. **E. perfoliatum**
1b Leaves not connate at base. 2
2a(1b) Leaves whorled, lanceolate to ovate, serrate with forward directed teeth; stem purple or purple-spotted; flowers pink or purplish, rarely white. **E. maculatum**
2b(1b) Leaves opposite, broadly ovate, coarsely dentate with spreading teeth: flowers white. **E. rugosum**

Eupatorium maculatum L. Fig. p. A114.
COMMON NAME(S): Joe-Pye-Weed.
FREQUENCY: Common throughout much of the Province.
HABITAT: Damp meadows, thickets and shores.
GENERAL RANGE: B.C., Sask. to Nfld. south to Utah, Iowa and N.C.
SYNONYMS: **E. purpureum** L. var. **maculatum** (L.) Darl., **Eupatoriadelphus maculatus** (L.) King & Rob.

Eupatorium perfoliatum L. Fig. p. A114. Map p. B81.
COMMON NAME(S): Thoroughwort, Boneset.
FREQUENCY: Common in the southern half of the Province and rare northwards.
HABITAT: Damp meadows, ditches and shores.
GENERAL RANGE: Man. to N.S. south to Tex. and Fla.

Eupatorium rugosum Houtt. Fig. p. A114. Map p. B81.
COMMON NAME(S): White Snakeroot.
FREQUENCY: Uncommon the northwestern half of the Province and rare along the eastern Bay of Fundy coast.
HABITAT: Moist, cool woods and clearings.
GENERAL RANGE: Man. to N.B. and N.S. south to Tex. and Ga.

++

GALINSOGA R. & P. Quickweed.

1a Primary stem leaves +/- rhombic, coarsely serrate; disk achenes with awned pappus, densely hispid on inner face; peduncles with spreading hairs. **G. quadriradiata**

1b Primary stem leaves oblanceolate, finely serrate; disk achenes awnless, hispid at summit only; peduncles with appressed ascending hairs. **G. parviflora**

Galinsoga quadriradiata Ruiz & Pavon Fig. p. A115.
SPECIAL NOTE: Native to Central and South America; introduced and locally common in dooryards, gardens and waste areas in the southern part of the Province. Known elsewhere in Canada from all provinces except Nfld.
SYNONYMS: **G. ciliata** (Raf.) Blake.

Galinsoga parviflora Cav. Fig. p. A114.
COMMON NAME(S): Field Scabious, Joey Hooker. Galinsoge.
SPECIAL NOTE: Native of Mexico and South America; introduced and known only from Speerville, Carleton Co. (UNB). Known elsewhere in Canada from B.C., Ont. and Que.

GNAPHALIUM L. Cudweed, Everlasting

1a Inflorescence corymbiform with numerous heads terminating short branches, not conspicuously leafy-bracted; outer phyllaries dingy-white. 2

1b Inflorescence of crowded heads or spikes, conspicuously bracted; phyllaries brownish or brown-spotted. 3

2a(1a) Leaves sessile, upper surface of at least lower leaves smooth, shiny; phyllaries dull, opaque, +/- corrugated, blunt; heads to 1.5 cm broad; plant fragrant. **G. obtusifolium**

2b(1a) Leaves decurrent, upper surface dull, glandular; phyllaries shiny, with thin scarious margins, acute; heads to .8 cm broad; plant not especially fragrant. **G. viscosum**

3a(1b) Heads in small capitate clusters; pappus bristles not united at base, deciduous separately; stem +/- lax, profusely branched from base, white appressed-tomentose. **G. uliginosum**

3b(1b) Heads in distinct spikes; pappus bristles united at base, deciduous in a ring; stem mostly unbranched at base, greenish, loosely tomentose. **G. sylvaticum**

Gnaphalium obtusifolium L. Fig. p. A115. Map p. B81.
COMMON NAME(S): Catfoot.
FREQUENCY: Very rare; reported from Albert, Kent and Queens Counties.
HABITAT: Disturbed soil, fields and clearings.
GENERAL RANGE: Ont. to N.S. south to Tex. and Fla.
SPECIAL NOTE: For more information see H. Hinds, 1983. The Rare Vascular Plants of New Brunswick. Syllogeus 50.

Gnaphalium sylvaticum L. Fig. p. A115. Map p. B82.
FREQUENCY: Scattered throughout much of the Province.
HABITAT: Dirt roadways, fields and clearings.
GENERAL RANGE: Ont. to Nfld. south to N. Eng.; Iceland; Eurasia.

Gnaphalium uliginosum L. Fig. p. A115.
COMMON NAME(S): Low Cudweed.
SPECIAL NOTE: Probably Eurasian; introduced throughout the Province in disturbed soils. Known elsewhere in Canada from all provinces, the Yukon, and Dist. Mack.

Gnaphalium viscosum HBK. Fig. p. A115. Map p. B82.
COMMON NAME(S): Povertyweed, Sticky Everlasting. Gnaphale visqueuse.
FREQUENCY: Uncommon; scattered in Carleton, Northumberland, York and Sunbury Counties.
HABITAT: Clearings, and similar disturbed areas.
GENERAL RANGE: B.C., Ont. to N.S. south to Oreg., Mex., Pa. and N. Eng.

HELIANTHUS L. Sunflower. Soleil

1a Receptacle +/- flat; involucral bracts ovate, abruptly contracted above the middle to an acuminate tip; lower leaves often cordate. **H. annuus**
1b Receptacle +/- convex or low conic; involucral bracts not abruptly contracted; leaves usually not cordate. 2
2a(1b) Involucral bracts conspicuously imbricate in several series, +/- ovate to lance-ovate, appressed, sharply acute to obtuse, ciliate and mostly smooth on back; disk reddish to purplish or most often with us, yellow. **H. X laetiflorus**
2b(1b) Involucral bracts narrower, not conspicuously imbricate; disk yellow. 3
3a(2b) Main stem leaves mostly opposite. 4
3b(2b) Main stem leaves mostly alternate. 5
4a(3a) Petioles less than 1.5 cm long; leaves entire or shallow toothed, pale, soft pubescent beneath; stem +/- glabrous below inflorescence. **H. strumosus**
4b(3a) Petioles usually more than 1.5 cm long; larger leaves coarsely toothed. 6
5a(3b) Phyllaries long ciliate; petioles ciliate; leaves scabrous or +/- hirsute beneath. **H. giganteus**
5b(3b) Phyllaries rarely short ciliate, nearly glabrous; petioles usually not ciliate; leaves hispid beneath. **H. nuttallii**
6a(4b) Stem +/- glabrous below inflorescence; involucral bracts green, thin, loose and spreading; rhizome rarely tuber-bearing; stem to 1.5 m tall. **H. decapetalus**
6b(4b) Stem +/- pubescent with spreading hairs; involucral bracts dark especially near base, often recurved; rhizome usually tuber-bearing; stem to 3 m high. **H. tuberosus**

++

Helianthus annuus L. Fig. p. A115.
COMMON NAME(S): Common Sunflower. Tourne-soleil.
SPECIAL NOTE: Native to western N. America; occasionally introduced to field and waste areas in the southern half of the Province. Known elsewhere in Canada from all provinces except Nfld.

Helianthus decapetalus L. Fig. p. A115.
FREQUENCY: Very rare and possibly extirpated; reported from Crocks Pt., York Co. and Eel R., Carleton Co.
HABITAT: River banks and rich meadows.
GENERAL RANGE: Nebr., Minn. to N.B. south to Mo. and Ga.
SPECIAL NOTE: For more information see H. Hinds, 1983. The Rare Vascular Plants of New Brunswick. Syllogeus 50.

Helianthus giganteus L. Fig. p. A115.
FREQUENCY: Rare; occasionally cultivated and escaped as in Fredericton, York Co. and perhaps also native as in Albert Co.
HABITAT: Moist rich ground and clearings.
GENERAL RANGE: Ont. to N.B. and N.S. south to eastern U.S.A.
SPECIAL NOTE: For more information see H. Hinds, 1983. The Rare Vascular Plants of New Brunswick. Syllogeus 50.
SYNONYMS: **H. nuttallii** Torr. & Gray. var. **subtuberosus** (Britt.) Boivin.

Helianthus X laetiflorus Pers. (**H. rigidus** X **H. tuberosus**) Fig. p. A115.
SPECIAL NOTE: Native from B.C. to Ont. south to N. Mex. and Ind.; occasionally introduced to roadsides and waste areas as in York, Charlotte, Queens, Victoria, Kent and Northumberland Counties.

Helianthus nuttallii T. & G. Fig. p. A115.
FREQUENCY: Very rare; reported from Shediac Cape, Westmorland Co. and near Loggieville, Northumberland Co.
HABITAT: Roadsides and dooryards.
GENERAL RANGE: B.C. to Nfld., N.B. and N.S. south to Calif., Okla. and Mo.
SPECIAL NOTE: Possibly introduced from further west. For more information see H. Hinds, 1983. The Rare Vascular Plants of New Brunswick. Syllogeus 50.
SYNONYMS: **H. nuttallii** ssp. **canadensis** Long.

Helianthus strumosus L. Fig. p. A115.
FREQUENCY: Very rare; reported from Carleton and Northumberland Counties.
HABITAT: Clearings and thickets.
GENERAL RANGE: N. Dak. and Minn. to N.B. south to Okla. and Ga.

Helianthus tuberosus L. Fig. p. A115.
COMMON NAME(S): Jerusalem Artichoke. Topinambour.
SPECIAL NOTE: Native to further south and west; early introduced and cultivated in some areas; escaped to alluvial meadows and shores especially in the St. John R. valley near Fredericton, York and Sunbury Counties.

HELIOPSIS Pers.

Heliopsis helianthoides (L.) Sweet Fig. p. A116.

COMMON NAME(S): Ox-eye.

FREQUENCY: Very rare; reported from First Eel Lake and Fredericton, York Co. Probably introduced from further west or south.

HABITAT: Roadside and waste areas.

GENERAL RANGE: B.C., Sask. to N.B. south to N. Mex. and Ga.

SUBSPECIFIC TAXA: Our plants are referable to ssp. **scabra** (Dunal) Fisher.

SPECIAL NOTE: For more information see H. Hinds, 1983. The Rare Vascular Plants of New Brunswick. Syllogeus 50.

HIERACIUM L. Hawkweed. Epervière

1a Cauline leaves or bracts 0-3, +/- entire; leaves lanceolate to oblanceolate. 2

1b Cauline leaves numerous, often dentate, or if less than 3 broadly rounded to subcordate at base. 3

2a(1a) Non-stoloniferous; basal offshoots if present, not rooting. 4

2b(1a) Stoloniferous. 5

3a(1b) Phyllaries +/- the same length and with a series of small bracts at base; heads at most 2.5 cm broad; leaves mostly entire to shallowly undulate-dentate, the lower much larger than the middle and upper ones. 6

3b(1b) Phyllaries +/- imbricated, of different lengths; heads usually over 2.5 cm broad; leaves often coarsely dentate or lobed. 7

4a(2a) Slender ascending branches arising from among basal leaves; leaves finely stellate-pubescent beneath. **H. praealtum**

4b(2a) Branches rarely produced; leaves +/- glabrous to sparingly setose beneath. **H. piloselloides**

5a(2b) Heads orange-red, crowded; leaves long-setose or glabrate above, setose below. **H. aurantiacum**

5b(2b) Heads yellow. 8

6a(3a) Leaves lanceolate, thin, glabrous or thinly setose and glaucous below; peduncles filiform, glabrous or, in our range, more often finely glandular; heads with 30 or less florets. **H. paniculatum**

6b(3a) Leaves broadly oblanceolate to elliptic, thick, usually pubescent and not glaucous beneath; peduncles stout, tomentose and +/- dark glandular; heads with 40 or more florets. **H. scabrum**

7a(3b) Stem +/- leafy to inflorescence, basal leaves often absent at flowering; leaves very gradually reduced in size upwards. 9

7b(3b) Stem with leaves rapidly reduced to small bracts upwards; basal leaves usually present at flowering. 10

8a(5b) Heads usually several to many, +/- crowded. 11

8b(5b) Heads usually solitary or few on elongate peduncles. 12

9a(7a) Phyllaries +/- hispid with blackish gland-tipped or glandless hairs. 13

9b(7a) Phyllaries without blackish hairs; leaf margins revolute, entire or denticulate; blade at least marginally with numerous stiff, blunt, subconic hairs. **H. umbellatum**

10a(7b) Lower leaves elliptic or oval, rounded to cordate at base with divergent basal teeth or lobes, usually not purple mottled; scape naked or with 1-2 leaves or bracts. **H. murorum**

10b(7b) Lower leaves tapering to petiole, often purple-mottled; stem with several scattered small leaves above base. 14

11a(8a) Leaves +/- glabrous above, glaucous. **H. X floribundum**

11b(8a) Leaves setose on both sides, green. **H. caespitosum**

12a(8b) Young leaves tawny-tomentose beneath, setose above and below; scape usually less than 2.5 dm high. **H. pilosella**

12b(8b) Young leaves green and merely setose; scape 1.2-4.5 dm high, sometimes forking near or above the middle. **H. X flagellare**

13a(9a) Styles pale yellowish; phyllaries +/- obtuse; peduncles slender, to 8 cm long; leaves glabrous or slightly scabrous. **H. canadense**

13b(9a) Styles darkened by brownish scabrosities; stem stouter; branches of inflorescence stouter and shorter (peduncles to 3 cm long); leaves +/- scabrous. 15

14a(10b) Spreading hairs of peduncles glandless; basal leaves usually less than 2 cm wide; heads 1-few; phyllaries tapering to a very sharply pointed tip; plants usually less than 2 dm tall. **H. robinsonii**

14b(10b) Spreading hairs of peduncles gland-tipped; basal leaves usually over 2 cm wide; heads usually numerous; phyllaries +/- acute but not sharply long pointed; plants usually more than 2 dm tall. **H. lachenalii**

15a(13b) Midvein of leaves and stem with scattered long hairs; lower leaves to 1.8 dm long and separated by about ½ their length. **H. X dutillyanum**

15b(13b) Midvein of leaves and stem without long hairs; lower leaves to about 13 cm long and separated by about 1/3 their length. **H. kalmii**

Hieracium aurantiacum L. Fig. p. A116.
COMMON NAME(S): Devil's Paintbrush, Orange Hawkweed. Bouquets rouges.
SPECIAL NOTE: European; commonly introduced to fields and waste areas throughout much of the Province. Known elsewhere in Canada from all provinces.

Hieracium caespitosum Dumort. Fig. p. A116.
COMMON NAME(S): King Devil. Epervière des prés.
SPECIAL NOTE: European; introduced and common throughout much of the Province. Known elsewhere in Canada from B.C., Ont. to N.S.
SYNONYMS: **H. pratense** Tausch.

Hieracium canadense L. Fig. p. A116. Map p. B82.
COMMON NAME(S): Canada Hawkweed.
FREQUENCY: Scattered throughout much of the Province.
HABITAT: Disturbed ground, thickets, clearings.
GENERAL RANGE: Ont. to Nfld. south to Wash., Mont., Mich. and Maine.
SUBSPECIFIC TAXA: Var. **divaricatum** Lepage, with leaf margins ciliate, is reported from Two Brooks, Restigouche Co. (UNB).

Hieracium X **dutillyanum** Lepage Map p. B82.
FREQUENCY: Locally common in the southern half of the Province.
HABITAT: Thickets, clearings and fields.
GENERAL RANGE: Ont. and N.B.
SPECIAL NOTE: Reported as a hybrid between **H. canadense** and **H. umbellatum** var. **scabriusculum**.

Hieracium X **flagellare** Willd.
SPECIAL NOTE: European; introduced through much of the developed areas of the Province in fields and pastures. Known elsewhere in Canada from Que. to N.S.
SUBSPECIFIC TAXA: Reported to be a hybrid between **H. caespitosum** and **H. pilosella**. Plants vary in their resemblance to one parent or another. Nm. **cernuiforme** (Naegeli & Peter) Lepage, resembling **H. pilosella**, with stems about 2 dm tall, and with 1 or 2 heads, is reported from Charlotte, York, Albert and Carleton Counties. Nm. **amauracron** (Missback & Zahn) Lepage, resembling **H. caespitosum**, with stems about 4 cm tall and with up to 6 heads, is reported from St. Stephen, Charlotte Co.

Hieracium X **floribundum** Wimm. & Grab. Fig. p. A116.
COMMON NAME(S): Yellow Devil.
SPECIAL NOTE: European; introduced to lawns, fields and clearings throughout the Province. Known elsewhere in Canada from B.C., Ont. to N.S. Reported to have originated as a hybrid between **H. caespitosum** and **H. lactucella**.

Hieracium **kalmii** L. Map p. B82.
FREQUENCY: Rare; reported from York, St. John and Charlotte Counties.
HABITAT: Roadsides, railways, clearings and borders of woods.
GENERAL RANGE: Minn. to Ont., Que., N.B. and N.S. south to Ohio and N.Y.
SYNONYMS: **H. canadense** var. **kalmii** (L.) Scoggan
SUBSPECIFIC TAXA: Var. **fasciculatum** (Pursh) Lepage, with stouter stems, leaves with irregular sized teeth, and more branched inflorescence rather than slender stem, leaves regularly toothed and inflorescence more open branched, occurs uncommonly in the extreme southwest of the Province in Charlotte Co.

Hieracium **lachenalii** Gmel. Fig. p. A116.
SPECIAL NOTE: European; introduced in fields, thickets and waste areas throughout much of the Province. Known elsewhere in Canada from B.C., Ont. to Nfld., P.E.I. and N.S.
SYNONYMS: **H. vulgatum** Fries. Map p. B82.

Hieracium **paniculatum** L. Fig. p. A116.
FREQUENCY: Very rare; reported from Fredericton, York Co. and Hatfield Point, Kings Co. (UNB).
HABITAT: Open rocky woods and thickets.
GENERAL RANGE: Ont. to N.B. and N.S. south to Ala. and Ga.

+++

Hieracium pilosella L. Fig. p. A116.
COMMON NAME(S): Mouse-ear. Oreille de souris.
SPECIAL NOTE: Eurasian; introduced and spreading in low fertility fields, lawns, and pastures throughout much of the Province. Known elsewhere in Canada from B.C., Ont. to N.S.

Hieracium piloselloides Vill. Fig. p. A116.
COMMON NAME(S): King Devil.
SPECIAL NOTE: European; introduced and aggressively spreading in fields, lawns and pastures throughout the Province. Known elsewhere in Canada from Ont. to Nfld., and N.S.
SYNONYMS: **H. florentinum** All.

Hieracium praealtum Vill. ex Gochnat Map p. B82.
COMMON NAME(S): King Devil.
SPECIAL NOTE: European; introduced and spreading in York, Sunbury, Victoria, Kent and Gloucester Counties. Known elsewhere in Canada from B.C., Ont., Nfld. and P.E.I.
SUBSPECIFIC TAXA: Our plants are referable to var. **decipiens** Koch.

Hieracium robinsonii (Zahn) Fern. Fig. p. A116.
COMMON NAME(S): Robinson's Hawkweed.
FREQUENCY: Very rare; reported from Fundy National Park, Albert and St. John Counties.
HABITAT: Ledge crevices along streams.
GENERAL RANGE: Que. to Nfld., N.B. and N.S. south to N.H. and Maine.

Hieracium scabrum Michx. Fig. p. A116. Map p. B82.
COMMON NAME(S): Rough Hawkweed.
FREQUENCY: Scattered throughout much of the Province.
HABITAT: Thickets and borders of woods.
GENERAL RANGE: Minn. to Ont., N.B., P.E.I. and N.S. south to Mo. and Ga.
SUBSPECIFIC TAXA: Var. **tonsum** Fern. & St. John, with lower stem glabrescent or short hispid and leaves glabrous or the upper surface sparingly setulose, rather than stem long villous and leaves long villous on midribs beneath, is reported from the Upsalquitch R., Restigouche Co. (UNB).

Hieracium umbellatum L. Fig. p. A117. Map p. B82.
FREQUENCY: Uncommon; reported from Restigouche, York, Queens, Gloucester, Kings and St. John Counties.
HABITAT: Well drained sandy or gravelly soils of roadsides and thickets.
GENERAL RANGE: Alaska and Yukon to N.B. and P.E.I. south to Oreg., Colo. and Mich.
SUBSPECIFIC TAXA: Our plants are referable to var. **scabriusculum** (Schweinitz) Farw.

++

HYPOCHAERIS L. Cat's-ear

Hypochaeris radicata L. Fig. p. A117.
SPECIAL NOTE: Eurasian; rarely introduced to lawns and similar open areas as in Fredericton, York Co. Known elsewhere in Canada from B.C., Sask., Ont. to Nfld, N.B. and N.S.

INULA L.

Inula helenium L. Fig. p. A117.
COMMON NAME(S): Elecampane, Elf Dock. Inule aulnée.
SPECIAL NOTE: Eurasian; cultivated and occasionally introduced in the moist meadows, ditches and clearings in the southwest quarter of the Province. Known elsewhere in Canada from B.C., Ont. to Nfld. and N.S.

LACTUCA L. Lettuce. Laitue

1a Pappus white; flowers yellow. 2
1b Pappus light brown; flowers bluish to white; achene tapering to beakless or shortly stout beaked apex. **L. biennis**
2a(1a) Achene with 3-several prominent ribs per face, beak as long as body or longer; lower stem +/- prickly. **L. serriola**
2b(1a) Achene 1-ribbed with +/- 1 rib per face, beak equal to ½ as long as body. **L. canadensis**

Lactuca biennis (Moench) Fern. Fig. p. A117.
FREQUENCY: Common throughout much of the Province.
HABITAT: Clearings, old fields and waste areas.
GENERAL RANGE: Alaska and Yukon, B.C. to Nfld. south to Calif., Colo. and N.C.

Lactuca canadensis L. Fig. p. A117. Map p. B82.
COMMON NAME(S): Devil's Weed. Chicorée blanche.
FREQUENCY: Scattered in the southern half of the Province and rare northwards.
HABITAT: Moist clearings and thickets.
GENERAL RANGE: Man. to N.S. south to Tex. and Fla.
SUBSPECIFIC TAXA: Var. **longifolia** (Michx.) Farw., with leaves mostly deeply lobed instead of +/- unlobed, occurs occasionally with the species.

Lactuca serriola L. Fig. p. A117.
COMMON NAME(S): Prickly Lettuce. Laitu épineuse.
SPECIAL NOTE: Eurasian; rarely introduced along roadsides and in waste areas as near Hillsborough, Albert Co. and Moncton, Westmorland Co. Known elsewhere in Canada from B.C. to N.S.
SUBSPECIFIC TAXA: Our plants are referable to forma **integrifolia** (Bogenh.) Beck with the leaves mostly unlobed.

++

LAPSANA L.

Lapsana communis L. Fig. p. A117. Map p. B82.
COMMON NAME(S): Nipplewort. Herbe aux mamelles.
SPECIAL NOTE: Eurasian; occasionally introduced in alluvial bottomlands in the St. John and St. Croix R. drainages. Known elsewhere in Canada from B.C., Man. to Nfld., P.E.I. and N.S.

LEONTODON L. Hawkbit.

Leontodon autumnalis L. Fig. p. A117.
COMMON NAME(S): Fall Dandelion. Leontodon d'automne.
SPECIAL NOTE: Eurasian; introduced throughout much of the Province in fields, pastures and lawns. Known elsewhere in Canada from B.C., Ont. to Nfld., P.E.I. and N.S.
SUBSPECIFIC TAXA: Var. **pratensis** (Link.) Koch, with involucre and adjacent peduncle with long spreading hairs rather than +/- glabrous, occurs commonly throughout.

LEUCANTHEMELLA Tzvelev

Leucanthemella serotina (L.) Tzvelev
COMMON NAME(S): Giant or High Daisy.
SPECIAL NOTE: European; cultivated and rarely introduced to fields and roadsides as near St. George and St. Stephen, Charlotte Co. Known elsewhere in Canada from B.C., Ont. and Que.
SYNONYMS: **Chrysanthemum uliginosum** Pers., **Tanacetum serotinum** (L.) Schultz-Bip.

LEUCANTHEMUM P. Mill

Leucanthemum vulgare Lam. Fig. p. A117.
COMMON NAME(S): Ox-eye Daisy. Marguerite.
SPECIAL NOTE: Eurasian; introduced to fields and waste areas throughout much of the Province. Known elsewhere in Canada from Yukon and all provinces.
SYNONYMS: **Chrysanthemum leucanthemum** L.

MATRICARIA L. Wild Chamomile. Matricaire

1a Heads rayless; disk-corollas 4-lobed; receptacle sharply conic; bruised plant with odor of pineapple. **M. matricarioides**
1b Heads white-rayed; disk-corollas usually 5-lobed. 2
2a(1b) Rays to 1 cm long; disk to 1 cm broad; receptacle sharply conic; bruised plant with odor of pineapple. **M. chamomilla**
2b(1b) Rays often longer than 1 cm; disk to 15 mm broad; receptacle hemispheric, obtuse; plant without fragrance of pineapple. 3

3a(2b) Leaves with filiform segments to 2 cm long; heads 3-4 cm broad with rays to 2 cm long; stem +/- erect. **M. perforata**

3b(2b) Leaves with fleshy segments to .5 cm long; heads to about 3 cm broad with rays to +/- 1.2 cm long; stem and branches lax. **M. maritima**

Matricaria chamomilla L. Fig. p. A117.
COMMON NAME(S): Camomille.
SPECIAL NOTE: Eurasian; reported as introduced in waste places (Scoggan, 1978-79). Plants examined from Kouchibouguac National Park, Kent Co. and Fredericton, York Co. have been annotated to **Anthemis cotula**.

Matricaria maritima L. Fig. p. A117.
COMMON NAME(S): Scentless Chamomile. Matricaire inodore.
SPECIAL NOTE: Eurasian; introduced in fields and waste areas mostly along the coast but also inland. Known elsewhere in Canada from Dist. Mack. and B.C. to Nfld., P.E.I. and N.S.

Matricaria matricarioides (Less.) Porter Fig. p. A117.
COMMON NAME(S): Pineapple-weed. Matricaire suave, Herbe à crapaud.
SPECIAL NOTE: Eurasian; introduced in compacted soil of roadsides, paths and waste places throughout much of the Province. Known elsewhere in Canada from Yukon, Dist. Mack. and B.C. to Nfld., P.E.I. and N.S.

Matricaria perforata Merat
SPECIAL NOTE: Eurasian; introduced to shores and waste areas as in Aroostook Jct., Victoria Co. Known elsewhere in Canada from all provinces.
SYNONYMS: **Matricaria maritima** var. **agrestis** (Knaf) Wilmott.

MEGALODONTA Greene

Megalodonta beckii (Torr. ex. Spreng.) Greene Fig. p. A118. Map p. B83.
COMMON NAME(S): Water-Marigold, Beck's Beggar-ticks.
FREQUENCY: Uncommon; reported from Westmorland, Queens, Sunbury, St. John and Charlotte Counties.
HABITAT: Shallow quiet waters.
GENERAL RANGE: B.C., Sask. to N.B. and N.S. south to Oreg., Mo. and N.J.
SYNONYMS: **Bidens beckii** Torr.

ONOPORDUM L.

Onopordum acanthium L. Fig. p. A118.
COMMON NAME(S): Scotch Thistle.
SPECIAL NOTE: Eurasian; introduced in fields and waste areas as near Bass River, Kent Co. Known elsewhere in Canada from B.C., Ont. and N.S.

++

PETASITES Mill. Sweet Coltsfoot

Petasites frigidus (L.) Rydb. Fig. p. A118. Map p. B83.
FREQUENCY: Scattered throughout much of the Province.
HABITAT: Cool, moist woods.
GENERAL RANGE: Yukon to Nfld. south to Calif., Mich. and Mass.; Asia.
SUBSPECIFIC TAXA: Our plants are referable to var. **palmatus** (Ait.) Cronq.
SYNONYMS: **P. palmatus** Ait.

PICRIS L. Ox-tonque

Picris echioides L. Fig. p. A118.
SPECIAL NOTE: Eurasian; rarely introduced in waste areas as in St. Stephen, Charlotte Co. Known elsewhere in Canada from Alta., Sask., Ont. and N.S.

PRENANTHES L. Rattlesnake-root

1a Flowers pink; leaves oval to oblanceolate, usually unlobed, the lower tapering into winged petioles. **P. racemosa**
1b Flowers whitish; leaves with truncate to cordate bases, usually variously lobed or cleft. 2
2a(1b) Longer phyllaries 4-6, usually 5; heads 5-6 flowered; cauline leaves often unlobed. **P. altissima**
2b(1b) Longer phyllaries 7-10; heads 7-8 (15) flowered; cauline leaves usually lobed or pinnatifid. **P. trifoliolata**

Prenanthes altissima L. Fig. p. A118. Map p. B83.
COMMON NAME(S): White Lettuce.
FREQUENCY: Scattered throughout much of the Province.
HABITAT: Moist mixed or hardwoods.
GENERAL RANGE: Ont. to N.B., P.E.I. and N.S. south to Tenn. and Ga.
SUBSPECIFIC TAXA: Forma **integra** Rousseau, with basal leaves +/- entire rather than lobed, occurs rarely with the species as in Fredericton, York Co. (UNB).

Prenanthes racemosa Michx. Fig. p. A118. Map p. B83.
COMMON NAME(S): Glaucous Lettuce.
FREQUENCY: Uncommon; scattered in the St. John and Restigouche R. drainage systems.
HABITAT: Shores, meadows, and thickets.
GENERAL RANGE: B.C. to N.B. and N.S. south to Mont., S. Dak. and N. Eng.
SPECIAL NOTE: **P. X mainensis** Gray, the hybrid with **P. trifoliolata**, with glabrous involucre and drooping heads rather than involucre hirsute and heads mostly ascending, is recorded from the St. John R. drainage system.

Prenanthes trifoliolata (Cass.) Fern. Fig. p. A118. Map p. B83.
COMMON NAME(S): Gall-of-the-earth. Pattes d'oie.
FREQUENCY: Scattered throughout much of the Province.
HABITAT: Open woods, clearings, and thickets.
GENERAL RANGE: Que. to Nfld. south to Tenn. and N.C.

++

RUDBECKIA L. Coneflower

1a Disk dark purple-brown; pappus none; leaves simple. **R. hirta**

1b Disk yellow or grayish; pappus a short +/- toothed crown; leaves trilobed to pinnatifid. **R. laciniata**

Rudbeckia hirta L. Fig. p. A118.
COMMON NAME(S): Black-eyed Susan. Marguerite jaune.
SPECIAL NOTE: Native to central U.S.A.; introduced to roadsides, fields and waste areas throughout much of the Province. Known elsewhere in Canada from all provinces.
SUBSPECIFIC TAXA: Our plants are referable to var. **pulcherrima** Farw.
SYNONYMS: **R. serotina** Nutt.

Rudbeckia laciniata L. Fig. p. A118.
COMMON NAME(S): Cut-leaved Coneflower.
SPECIAL NOTE: Apparently native across much of southern Canada south to Ariz., Tex. and Fla. Our plants all appear to be introduced as either the type form which is rare, or as the more common horticultural form, "Golden Glow", with all or most florets ligulate.

SENECIO L. Groundsel, Ragwort. Séneçon

1a Stems white-tomentose, +/- equally leafy throughout. 2

1b Stem leafy mostly towards base, basal leaves well developed. 3

2a(1a) Leaves subentire to coarsely toothed; basal leaves absent at flowering; heads over 4 cm broad, usually less than 6 per stem; leaves lustrous green above, white tomentose beneath. **S. pseudoarnica**

2b(1a Leaves mostly deeply 1-3 pinnatifid. 4

3a(1b) Basal leaves usually truncate to cordate at base. 5

3b(1b) Basal leaves cuneate at base, oblanceolate to elliptic; heads usually few; plant +/- glabrous at flowering. **S. pauperculus**

4a(2b) Rays absent or inconspicuous; stems soft; annuals. 6

4b(2b) Rays present and conspicuous; leaves deeply 2-3 times pinnatifid +/- tomentose beneath at least when young; stem coarse, firm, to over 1 m tall. **S. jacobaea**

5a(3a) Basal leaf blades lanceolate to lance-ovate, 1.75-3.5 times as long as wide, truncate to subcondate at base; rays usually pale yellow. **S. schweinitzianus**

5b(3a) Basal leaf blades orbicular to ovate 0.8-1.5 times as long as wide; deeply cordate at base, rays deep yellow. **S. aureus**

6a(4a) Rays none; minute bracteoles black-tipped. **S. vulgaris**

6b(4a) Rays minute; bracteoles green. 7

7a(6b) Plant conspicuously glandular-hispid; bracteoles ½ length of phyllaries. **S. viscosus**

7b(6b) Plant subglabrous, scarcely glandular; bracteoles minute. **S. sylvaticus**

Senecio aureus L. Fig. p. A118. Map p. B83.
COMMON NAME(S): Golden Ragwort, Squaw-weed. Séneçon doré.
FREQUENCY: Scattered throughout much of the Province except uncommon in the eastern lowlands.
HABITAT: Damp meadows, thickets, ditches and open woods.
GENERAL RANGE: Man. to Nfld. south to S. Dak., Ark. and Fla.
SYNONYMS: **S. aureus** var. **aquilonius** Fern., var. **semicordatus** (Mackenz. & Bush) Greenm., and var. **intercursus** Fern.

Senecio jacobaea L. Fig. p. A118.
COMMON NAME(S): Tansy-Ragwort, Stinking Willie. Séneçon Jacobée, Herbe-de-Jacob.
SPECIAL NOTE: Eurasian; introduced to roadsides, old fields and waste areas especially in the eastern half of the Province. Known elsewhere in Canada from B.C., Ont. to Nfld., P.E.I. and N.S.

Senecio pauperculus Michx. Fig. p. A118. Map p. B83.
FREQUENCY: Uncommon along the St. John and Restigouche Rivers.
HABITAT: Ledges and rocky shores, often calcareous.
GENERAL RANGE: Alaska and Yukon to Nfld. south to Oreg., N. Mex., Ala. and Ga.

Senecio pseudoarnica Less. Fig. p. A119.
COMMON NAME(S): Seabeach Groundsel. Roi des champs.
FREQUENCY: Very rare; reported only from Bill's Island near Grand Harbour, Green Islands, and Long Pond beach, Grand Manan Archipelago, Charlotte Co.
HABITAT: Sandy or gravelly seabeaches.
GENERAL RANGE: Coastal areas of Alaska and B.C.; coastal areas from Labr., Nfld., N.B. and N.S.; Asia.
SPECIAL NOTE: For more information see H. Hinds, 1983. The Rare Vascular Plants of New Brunswick. Syllogeus 50.

Senecio schweinitzianus Nutt. Fig. p. A119. Map p. B83.
COMMON NAME(S): Swamp Ragwort, Robbin's Senecio. Séneçon de Robbins.
FREQUENCY: Scattered in the southwest half of the Province and rare northeastward.
HABITAT: Wet thickets, meadows and peaty swamp edges.
GENERAL RANGE: Que. to N.B., P.E.I. and N.S. south to Tenn. and N.C.
SYNONYMS: **S. robbinsii** Oakes.

Senecio sylvaticus L. Fig. p. A119.
SPECIAL NOTE: Eurasian; introduced to waste areas, upper seabeaches, clearings and open woods, especially in the southwest corner of the Province. Known elsewhere in Canada from B.C., Que. to Nfld., P.E.I. and N.S.

Senecio viscosus L. Fig. p. A119.
COMMON NAME(S): Sticky or Stinking Groundsel. Séneçon visqueux.
SPECIAL NOTE: Eurasian; introduced to waste areas, shores and railways mostly in the southcentral part of the Province. Known elsewhere in Canada from B.C., Man. to Nfld., P.E.I. and N.S.

++

Senecio vulgaris L. Fig. p. A119.
COMMON NAME(S): Common Groundsel.
SPECIAL NOTE: Eurasian; introduced in waste areas throughout much of the Province. Known elsewhere in Canada from Yukon, Dist. Mack. and all provinces.

SILYBUM Adans.

Silybum marianum (L.) Gaertn. Fig. p. A119.
COMMON NAME(S): Milk-Thistle, Lady's-Thistle. Chardon-Marie.
SPECIAL NOTE: European; rarely cultivated and escaped as in Gloucester, Victoria, York and Kent Counties. Known elsewhere in Canada from B.C., Sask., Ont., Que. and N.S.

SOLIDAGO L. Goldenrod. Verge d'Or

1a Inflorescence various but not flat-topped (corymbiform). 2

1b Inflorescence flat-topped (corymbiform), leaves glabrous, linear to linear lanceolate; rays white. **S. ptarmicoides**

2a(1a) Inflorescence either of axillary clusters or a terminal simple to branched club- or wand-shape (thyrsiform). 3

2b(1a) Inflorescence terminal and either nodding pyramidal **OR** with at least the lower branches +/- recurved and the heads oriented in one direction **OR** both. 4

3a(2a) Heads very large, involucres 7-12 mm high, phyllaries attenuate; rays 8-10; achenes glabrous. **S. macrophylla**

3b(2a) Heads smaller. 5

4a(2b) Basal and lower stem leaves largest, often present at flowering; leaves reduced upward. 6

4b(2b) Median stem leaves usually largest, crowded; lower stem leaves usually dropping off before flowering begins. 7

5a(3b) Inflorescence a compact terminal cluster of heads (sub-capitate), often sub-corymbiform; upper stem leaves reduced, the lower densely villous with long, soft curling hairs at least at base; phyllaries +/- equal in length; plants generally less than 3 dm tall; arctic-alpine species. **S. multiradiata**

5b(3b) Inflorescence otherwise; phyllaries unequal; eastern temperate or boreal species. 8

6a(4a) Leaves fleshy, smooth; plants of inland salt springs and coastal salt marshes. **S. sempervirens**

6b(4a) Leaves not fleshy nor plants of salt marshes. 9

7a(4b) Leaves triple-nerved, stems hairless to densely hairy. 10

7b(4b) Leaves not triple-nerved, usually hairy at least along the midrib of the lower surface; stem +/- hairy at least above the middle. **S. rugosa**

8a(5b) Basal and lower stem leaves largest, often present at flowering time. 11

8b(5b) Median stem leaves largest, lower stem leaves usually dropping off before flowering time. 12

9a(6b) Stems and leaves densely short hairy (grayish color); tip of inflorescence usually strongly bent to one side to nodding; open sandy soils. **S. nemoralis**

9b(6b) Stems and sometimes the leaves hairless to only sparsely hairy, but not grayish below the inflorescence. 13

10a(7a) Stems hairless below the inflorescence and sometimes within, smooth-waxy (glaucous). **S. gigantea**

10b(7a) Stem hairy at least half way to base. 14

11a(8a) Tips of phyllaries curved outward (squarrose); heads relatively large, involucre 5-9 mm high; achenes glabrous; lower leaves sharply serrate. **S. squarrosa**

11b(8a) Tips of phyllaries not usually spreading (appressed); heads relatively small. 15

12a(8b) Stems smooth, glaucous; stem leaves linear-oblanceolate tapering to subsessile base; involucre less than 5 mm high. **S. caesia**

12b(8b) Stem sparsely hairy, zigzag at the leaf nodes; stem leaves oblanceolate to obovate; involucre to 6 mm high. **S. flexicaulis**

13a(9b) Petioles of lowest stem leaves sheathing the stem; inflorescence often much longer than broad; plants of marshes and bogs and usually moist soils. **S. uliginosa**

13b(9b) Petioles not sheathing; branches of the inflorescence oriented upward and outward; heads usually secund on branches; plants of usually drier soils. **S. juncea**

14a(10b) Stem hairy to base; leaves densely short-hairy, with a few marginal teeth near the tip or entire, linear elliptic-oblanceolate; branches of the inflorescence ascending to spreading-recurved; involucre usually more than 3 mm high. **S. altissima**

14b(10b) Stem not hairy to base; involucre 2-3 mm high; leaves nearly glabrous to +/- pubescent beneath. **S. canadensis**

15a(11b) Plants hairy along entire stem; basal leaf petioles not sheathing stem. 16

15b(11b) Plants essentially glabrous except in the inflorescence. 17

16a(15a) Tips of phyllaries subulate-attenuate. **S. puberula**

16b(15a) Tips of phyllaries rounded to bluntly pointed. 18

17a(15b) Leaves thick and fleshy, margins entire; plants of salt marshes along the coast or salt springs inland. **S. sempervirens**

17b(15b) Leaves usually sparsely to coarsely serrate, not fleshy; plants of non-saline habitats, calcareous ledges or gravelly river strands. 19

18a(16b) Rays +/- white. **S. bicolor** L.

18b(16b) Rays yellow in fresh and recently opened florets. **S. hispida**

19a(17b) Petioles of lowest stem leaves sheathing the stem; plants of bogs and other wet lowlands. **S. uliginosa**

19b(17b) Petioles of lowest stem leaves not sheathing stem; plants of calcareous ledges or gravelly river strands. **S. spathulata**

Solidago altissima L. Map p. B83.

FREQUENCY: Uncommon; reported from the St. John River drainage system in Kings, Sunbury, York, Carleton and Victoria Counties and from Heron Is., Restigouche Co.

HABITAT: Alluvial meadows and sandy fields.

GENERAL RANGE: Minn. to Ont. and N.B. south to Tex. and Fla.

SYNONYMS: **S. canadensis** var. **scabra** (Muhl.) T. & G.

++

Solidago bicolor L. Fig. p. A119. Map p. B83.
COMMON NAME(S): Silverrod.
FREQUENCY: Common in the southeast half of the Province and scattered elsewhere.
HABITAT: Dry old fields, crevices of seacliffs, and open woods.
GENERAL RANGE: Mich. to Ont. and N.S. south to Ark. and Ga.

Solidago caesia L. Fig. p. A119.
COMMON NAME(S): Blue-stem Goldenrod.
FREQUENCY: Very rare; reported from Saint John, St. John County. Now apparently extirpated.
HABITAT: Rich woods and clearings.
GENERAL RANGE: Wisc. to Ont., Que., N.B., and N.S. south to Tex. and Fla.
SPECIAL NOTE: For more information see H. Hinds, 1983. The Rare Vascular Plants of New Brunswick. Syllogeus 50.

Solidago canadensis L. Fig. p. A119.
COMMON NAME(S): Canada Goldenrod. Bouquets jaunes.
FREQUENCY: Common throughout much of the Province.
HABITAT: Old fields, clearings, meadows and waste areas.
GENERAL RANGE: Alaska, Yukon, Dist. Mack. and B.C. to Nfld. south to Calif., N. Mex. and Fla.
SPECIAL NOTE: An apparent hybrid with **S. rugosa** is reported from Indian Point, St. Andrews, Charlotte Co.

Solidago flexicaulis L. Fig. p. A119. Map p. B83.
COMMON NAME(S): Zigzag Goldenrod.
FREQUENCY: Common throughout much of the Province except rare in the eastern lowlands.
HABITAT: Rich mixed or hardwoods.
GENERAL RANGE: N. Dak., Ont. to N.S. south to Kans. and N.C.

Solidago gigantea Ait. Fig. p. A119. Map p. B83.
FREQUENCY: Scattered in the southern and western portions and unknown in the north-east half of the Province.
HABITAT: Alluvial meadows.
GENERAL RANGE: Dist. Mack. and B.C. to N.S. south to Oreg., N. Mex. and Ga.

Solidago hispida Muhl. Fig. p. A120. Map p. B84.
FREQUENCY: Scattered in the western half of the Province.
HABITAT: Mostly dry rocky fields, roadsides and shores, often calcareous.
GENERAL RANGE: Sask. to Nfld. south to Ark. and Ga.
SUBSPECIFIC TAXA: Var. **tonsa** Fern., with leaves +/- glabrous is reported from limestone cliffs at the jct. of the Restigouche and Upsalquitch Rivers, Restigouche Co.

++

Solidago juncea Ait. Fig. p. A120. Map p. B84.
COMMON NAME(S): Early Goldenrod. Solidage élancée.
FREQUENCY: Scattered mostly in the western half of the Province.
HABITAT: Fields, shores, and similar open places.
GENERAL RANGE: Man. to N.S. south to Mo., Tenn. and Ga.
SPECIAL NOTE: An apparent hybrid with **S. spathulata** var. **gilmanii** is reported from Shore Is., near Crocks Pt., York Co. (UNB).

Solidago macrophylla Pursh Fig. p. A120. Map p. B84.
COMMON NAME(S): Large-leaved Goldenrod.
FREQUENCY: Scattered throughout much of the Province except rare along the western border.
HABITAT: Moist, mostly coniferous woods.
GENERAL RANGE: Ont. to Nfld. south to N.Y. and Mass.

Solidago multiradiata Ait. Fig. p. A120.
FREQUENCY: Very rare; reported from near Wilson Bk., Albert Mines. Albert Co.
HABITAT: Gypsum cliffs.
GENERAL RANGE: Alaska, Yukon, Dist. Mack. and Victoria Is. to Ungava south to Nfld., N.B. and N.S.; Siberia.
SPECIAL NOTE: For more information see H. Hinds, 1983. The Rare Vascular Plants of New Brunswick. Syllogeus 50.

Solidago nemoralis Ait. Fig. p. A120. Map p. B84.
COMMON NAME(S): Gray Goldenrod.
FREQUENCY: Scattered in the southwest quarter of the Province.
HABITAT: Poor, dry, soils of fields and roadsides.
GENERL RANGE: B.C., Man. to N.S. south to Mont., Ariz. and Ga.
SPECIAL NOTE: An apparent hybrid with **S. bicolor**, is reported from Flume Ridge near the Magaguadavic R., Charlotte Co. (UNB).

Solidago ptarmicoides (Nees) Boivin Fig. p. A111 (See **Aster ptarmicoides**)
FREQUENCY: Very rare; reported near Woodstock, Carleton Co. The site drowned by the Mactaquac hydroelectric development.
HABITAT: Rocky shores.
GENERAL RANGE: Wyo., Sask. to N.B. south to Colo., Mo. and N.Y.
SYNONYMS: **Aster ptarmicoides** (Nees) T. & G.
SPECIAL NOTE: For more information see H. Hinds, 1983. The Rare Vascular Plants of New Brunswick. Syllogeus 50.

Solidago puberula Nutt. Fig. p. A120. Map p. B84.
COMMON NAME(S): Downy Goldenrod.
FREQUENCY: Common throughout much of the Province.
HABITAT: Poor, dry fields, meadows and rocky barrens.
GENERAL RANGE: Que. to N.S. south to Miss. and Fla.
SUBSPECIFIC TAXA: An apparent hybrid with **S. bicolor**, with white rays, is reported from near the confluence of Branch Str. and Digdeguash R., Charlotte Co. (UNB). Another apparent hybrid with **S. uliginosa**, is reported from near Acton, and near Yoho Lake, York Co. (UNB).

Solidago rugosa Ait. Fig. p. A120. Map p. B84.
COMMON NAME(S): Rough-stemmed Goldenrod.
FREQUENCY: Common throughout much of the Province.
HABITAT: Damp woods, fields, and thickets.
GENERAL RANGE: Ont. to Nfld. south to Tex. and Fla.
SUBSPECIFIC TAXA: Var. **villosa** (Pursh) Fern., with lower panicle-branches shorter than their subtending leaves, rather than lower panicle-branches much longer than their subtending leaves, occurs occasionally with the species.
SPECIAL NOTE: An apparent hybrid with **S. uliginosa** is reported from the base of Big Bald Mtn., Northumberland Co. (UNB).

Solidago sempervirens L. Fig. p. A120. Map p. B84.
COMMON NAME(S): Seaside Goldenrod.
FREQUENCY: Scattered along the Bay of Fundy and common along the east coast.
HABITAT: Coastal sands and brackish marshes
GENERAL RANGE: Que. to Nfld. south along the coast to Va.
SPECIAL NOTE: **S. X erskinei** Boivin, the hybrid with **S. canadensis**, is reported from Kouchibouguac National Park, Kent Co., and from near St. Andrews, Charlotte Co.

Solidago spathulata DC. Fig. p. A120. Map p. B84.
SUBSPECIFIC TAXA: Two well-marked taxa are found within our range.

1a Heads to 9 mm high in loosely branched long inflorescence; leaves to 4 cm broad, coarsely toothed; plants usually with multiple stems to 1 m tall. ssp. **spathulata** var. **gillmanii**

1b Heads to 6 mm high in raceme-like panicle; leaves to about 1.5 cm wide, entire or serrate-dentate above the middle; plants often single stemmed or tufted, usually less than 5 dm tall. ssp. **randii** var. **racemosa**

1. Ssp. **spathulata**, var. **gillmanii** (Gray) Cronq.
FREQUENCY: Rare; reported from 4 localities along the St. John R. in York and Carleton Counties.
HABITAT: Gravelly river strands.
GENERAL RANGE: Wisc. to Ont. and N.B. south to Ind.
SYNONYMS: **S. gillmanii** (Gray) Steele, **S. racemosa** Greene var. **gillmanii** (Gray) Fern.
SPECIAL NOTE: For more information see H. Hinds, 1983. The Rare Vascular Plants of New Brunswick. Syllogeus 50.

2. Ssp. **randii** (Porter) Cronq. var. **racemosa** (Greene) Gleason
FREQUENCY: Rare; reported from 4 localities on the St. John and Restigouche R. systems in Victoria and Restigouche Counties.
HABITAT: Dry, calcareous ledges and rocky banks.
GENERAL RANGE: Wisc., Mich., Ont., Que. and N.B. south to Ind., Ky. and Va.
SYNONYMS: **S. racemosa** Greene.
SPECIAL NOTE: For more information see H. Hinds, 1983. The Rare Vascular Plants of New Brunswick. Syllogeus 50.

Solidago squarrosa Muhl. Fig. p. A120. Map p. B84.
FREQUENCY: Scattered throughout much of the Province.
HABITAT: Clearings, open woods and rocky slopes.
GENERAL RANGE: Ont. to N.B. south to Ky. and N.C.

Solidago uliginosa Nutt. Fig. p. A120. Map p. B84.
COMMON NAME(S): Bog Goldenrod.
FREQUENCY: Scattered throughout much of the Province but rare northwards.
HABITAT: Acid peat bogs, roadside ditches, wet meadows and thickets.
GENERAL RANGE: Man. to Nfld. south to Wisc., Ohio and N.H.
SUBSPECIFIC TAXA: Plants of nutrient poor habitats such as acid bogs are greatly reduced in size. Those with broad panicles have been called var. **terrae-novae** (T. & G.) Fern., while those with narrow panicles have been called var. **linoides** (T.& G.) Fern. Both varieties are reported from the Province.

SONCHUS L. Sow-Thistle. Laiteron

1a Involucre and peduncles heavily stipitate-glandular; involucre 15-22 mm broad in fruit, to over 2 cm long; flowers bright orange-yellow; upper leaf auricles rounded; achenes prominently ribbed and strongly rugose; perennials with long vertical and creeping roots. **S. arvensis**

1b Involucre and peduncles mostly glabrous (occasionally with a few gland-tipped hairs). 2

2a(1b) Mature stem leaf auricles pointed; terminal lobe of leaves (if any) deltoid; sinuses +/- acute; corollas pale yellow; annual species with tap roots. **S. oleraceus**

2b(1b) Mature stem leaf auricles rounded; terminal lobe of leaves (if any) narrowly triangular, sinuses more rounded. 3

3a(2b) Leaf margins firm spiny; achenes ribbed but not rugose; involucre 10-12 mm high; annual species with tap roots. **S. asper**

3b(2b) Leaf margins +/- soft spiny; achenes ribbed and rugose; involucre 14-22 mm broad in fruit; perennials with long vertical and creeping roots. 4

4a(3b) Phyllaries uniformly dark green to lead color; involucre broadly campanulate to hemispheric, the longest bracts 14-17 mm long. **S. arvensis**

4b(3b) Phyllaries pale-brownish, white margined; longest involucral bracts 10-15 mm. **S. arvensis** ssp. **uliginosus**

Sonchus arvensis L. Fig. p. A120.
SPECIAL NOTE: Eurasian; introduced in fields and waste areas throughout much of the Province. Known elsewhere in Canada from Dist. Mack. and all provinces.
SUBSPECIFIC TAXA: Ssp. **uliginosus** (Bieb.) Nyman, separated in the above key, is reported mostly from the northern coastal area of the Province.
SYNONYMS: **S. uliginosus** Bieb.

Sonchus asper (L.) Hill Fig. p. A121.
SPECIAL NOTE: Eurasian; introduced to waste areas, fields and roadsides throughout much of the Province. Known elsewhere in Canada from Man. to P.E.I.
SUBSPECIFIC TAXA: Forma **inermis** (Bisch.) Beck, with undivided rather than +/- pinnatifid stem-leaves, occurs frequently with the species.

Sonchus oleraceus L. Fig. p. A121.
SPECIAL NOTE: Eurasian; introduced in waste areas and cultivated ground in the southern half of the Province. Known elsewhere in Canada from Dist. Mack. and all provinces.

TANACETUM L. Tansy. Tanaisie

1a Disk florets white; leaves bipinnately divided, the divisions ovate, entire, crenate or subpinnately divided; ultimate segments rounded. **T. parthenium**

1b Disk florets yellow to orange; leaves 1-3 pinnately dissected, the leaflets lanceolate, pinnatisect to pinnately lobed, ultimate segments acute. 2

2a(1b) Heads numerous in a +/- flat-topped inflorescence; disk 5-10 mm broad; leaves glabrous; stems to 1.5 m high. **T. vulgare**

2b(1b) Heads 1-few; disk 10-18 mm broad; leaves villous; stem depressed-ascending to .5 m high. **T. bipinnatum**

Tanacetum bipinnatum (L.) Schultz-Bip. Fig. p. A121. Map p. B85.
COMMON NAME(S): Lake Huron Tansy.
FREQUENCY: Uncommon; reported along the St. John and Restigouche Rivers. Many former sites destroyed by hydroelectric and other developments.
HABITAT: Gravelly river strands or rocky banks.
GENERAL RANGE: Mich., Ont. to N.B. south to Me.
SUBSPECIFIC TAXA: Our plants are referable to ssp. **huronense** (Nutt.) Breitung.
SYNONYMS: **T. huronense** Nutt. var. **johannense** Fern.

Tanacetum parthenium (L.) Schultz-Bip. Fig. p. A121.
COMMON NAME(S): Feverfew. Grande Camomille.
SPECIAL NOTE: Eurasian; rarely escaped from cultivation as near Yoho Lake, York Co. Known elsewhere in Canada from B.C., Ont. to N.S.
SYNONYMS: **Chrysanthemum parthenium** L.

Tanacetum vulgare L. Fig. p. A121.
COMMON NAME(S): Common Tansy. Golden-buttons. Tanaisie commune, Herbe de Saint-Marc.
SPECIAL NOTE: Eurasian; introduced to roadsides, fields and waste areas throughout the Province. Known elsewhere in Canada from Dist. Mack. and all provinces.

TARAXACUM Zinn. Dandelion. Pissenlit

1a Mature achenes +/- tuberculate nearly to base, the tubercles mostly confined to upper half and without intervening flat areas. **T. latilobum**

++

1b Mature achenes mostly tuberculate above middle, occasionally with a few remote tubercles and always with broad, flat intervening areas. 2

2a(1b) Inner phyllaries +/- corniculate near tip; achenes reddish, the body to 3.5 mm long; pappus sordid or creamy; rays deep sulphur-yellow; leaves deeply dissected. **T. laevigatum**

2b(1b) Inner phyllaries not corniculate at tip; achenes olivaceous to brown or stramineous, the body to 4 mm long; pappus white; rays canary-yellow; leaves usually not as deeply dissected. **T. officinale**

Taraxacum laevigatum (Willd.) DC. Fig. p. A121. Map p. B85.
COMMON NAME(S): Red-seeded Dandelion.
SPECIAL NOTE: European; apparently introduced to fields and open woods but often appearing native when growing in calcareous ledge crevices as near Foley Mtn., Waterford, Kings Co. Also reported from East Wolf Is., Charlotte Co. and woods near Earle Wharf, Kings Co.
SYNONYMS: **Taraxacum erythrospermum** Andrz. ex Bess.

Taraxacum latilobum DC. Map p. B85.
FREQUENCY: Rare; reported from Grand Manan and Deer Islands, Charlotte Co., Saint John, St. John Co. and from Miscou Is., Gloucester Co.
HABITAT: Crevices of calcareous sea-cliffs and moist dune hollows.
GENERAL RANGE: Que., N.B. and Nfld. south to Me.
SPECIAL NOTE: For more information see H. Hinds, 1983. The Rare Vascular Plants of New Brunswick. Syllogeus 50.

Taraxacum officinale Weber Fig. p. A121.
COMMON NAME(S): Common Dandelion. Dent-de-lion.
SPECIAL NOTE: European; introduced in cultivated ground, fields and waste areas throughout the Province. Known elsewhere in Canada from Yukon, Dist. Mack. and all provinces.

TRAGOPOGON L. Goat's-beard. Salsifis

1a Flowers purple; pappus brownish. **T. porrifolius**

1b Flowers yellow; pappus whitish. 2

2a(1b) Phyllaries surpassing the rays; peduncles enlarged above; corollas sulphur-yellow; achenes to +/- 3.5 cm long; leaves not undulate-recurving. **T. dubius**

2b(1b) Phyllaries shorter than rays; peduncles scarcely enlarged; corollas canary-yellow; achenes to 2.5 cm long; leaves undulate-recurving. **T. pratensis**

Tragopogon dubius Scop.
COMMON NAME(S): Goat's-beard.
SPECIAL NOTE: European; introduced along roadsides in the north of the Province. Known elsewhere in Canada from B.C. to Que. and N.B.

++

Tragopogon porrifolius L. Fig. p. A121.
COMMON NAME(S): Salsify, Oyster-plant. Salsifis cultivé.
SPECIAL NOTE: European; escaped from cultivation to roadsides and waste areas as in Bass R., Kent Co. Known elsewhere in Canada from B.C., Alta. and Man. to Que. and N.S.

Tragopogon pratensis L. Fig. p. A121. Map p. B85.
COMMON NAME(S): Goat's-beard. Salsifis des prés.
SPECIAL NOTE: Eurasian; introduced in the southern and western portions of the Province. Known elsewhere in Canada from all provinces except Nfld.

TUSSILAGO L.

Tussilago farfara L. Fig. p. A121. Map p. B85.
COMMON NAME(S): Coltsfoot. Pas-d'âne.
SPECIAL NOTE: Eurasian; introduced on clay banks, brooksides and ditches throughout much of the Province. Known elsewhere in Canada from B.C., Ont. to Nfld., P.E.I. and N.S.

XANTHIUM L. Cocklebur, Clotbur. Lampourde

1a Leaves lanceolate, tapering at both ends, silvery-sericeous beneath; leaf axils bearing a three-pointed yellow spine. **X. spinosum**

1b Leaves ovate to reniform, +/- cordate or broadly cuneate at base, strigose beneath; leaf axils spineless. **X. strumarium**

Xanthium spinosum L. Fig. p. A121.
COMMON NAME(S): Spiny Cocklebur.
SPECIAL NOTE: European; rarely introduced as ballast waste as in Saint John, St. John Co. Known elsewhere in Canada from B.C., Sask., Ont. and Que.

Xanthium strumarium L. Fig. p. A121. Map p. B85.
FREQUENCY: Locally common in the Miramichi R. estuary, Northumberland Co. and along the St. John R. in Queens, Sunbury, and York Counties.
HABITAT: Shores and waste areas.
GENERAL RANGE: B.C. to N.B. south to Calif., N. Dak., Pa. and Mass.
SUBSPECIFIC TAXA: Our plants are referable to var. **canadense** (Mill) T. & G.).
SYNONYMS: **X. italicum** Moretti.

REFERENCES CITED

Boivin, B. 1966-67. Enumération des Plantes due Canada.

________. 1967-69. Flora of the Prairie Provinces. Phytologia 15-18.

Canadian Flora. Terminology Bulletin No. 156. Translation Bureau, Secretary of State, Hull, Qué.

Clayton, J.S., W.A. Ehrlich, D.B. Cann, J.H. Day and I.B. Marshall. 1977. Soils of Canada Vol. I. CDA, Ottawa.

Cronquest, Arthur. 1968. The Evolution and Classification of Flowering Plants. Houghton Mifflin Co., Boston.

Eiten, G. 1963. Am. Midl. Nat. 69:301.

Fernald, M.L. 1950. Gray's Manual of Botany. Eighth Edition. American Book Co., N.Y.

Fowler, J. 1885. Preliminary List of the Plants of New Brunswick. Bull. Nat. Hist. Soc. N.B. 4:8-84.

Groh, H. and C. Frankton. 1949. Canadian Weed Survey, 7th Report. Botany and Plant Pathology Div., CDA, Ottawa.

Hinds, H. 1983. The Rare Vascular Plants of New Brunswick. Syllogeus 50. National Museums of Canada, Ottawa.

Hulten, E. 1958. The amphi-atlantic plants and their phytogeographical connections. K. Sven. Vetenskapsakad. Handl., Ser. 4,7(1):1-340.

Kott, L.S. 1981. Isoetes acadiensis, a new species from eastern North America. Can. J. Bot. 59:2592-2594.

Lawson, G. 1880. Notice of new and rare plants. Trans. N.S. Inst. Sci. 6:68-75.

Manley, S. 1971. Identification of red, black and hybrid spruces. CFS Publ. 1301, DOE, Ottawa.

Monroe, D. 1979. A floristic study of Kouchibouguac National Park. Unpublished report.

Putnam, D.F. 1940. The climate of the Maritime Provinces. Can. Geog. J. 21:134-147.

Reznicek, A. and P. Ball. 1974. The taxonomy of Carex series Lupulinae in Canada. Can. J. Bot. 52:2387-2399.

________________________. 1980. The taxonomy of Carex section Stellulatae in North America north of Mexico. Contrib. U. Mich. Herb. 14:153-203.

Richards, C., F. Hyland, and L. Eastman. 1983. Check-list of the Vascular Plants of Maine. Bull. Josselyn Bot. Soc. No. 11.

Richardson, J. 1980. Ecology of Ruppia maritima L. in New Hampshire (U.S.A.) tidal marshes. Rhodora 82:403-439.

Roland, A.E. and E.C. Smith. 1969. The Flora of Nova Scotia. N.S. Museum, Halifax, N.S.

Sarvela, J., D. Britton and K. Pryer. 1981. Studies on the Gymnocarpium robertianum complex in North America. Rhodora 83:421-431.

Scoggan, H.J. 1978-79. The Flora of Canada. National Museum of National Sciences Publ. No. 7(3).

Sinnott, Q. 1985. A revision of Ribes L. subg. Grossularia (Mill.) Pers. sect. Grossularia (Mill.) Nutt. (Grossulariaceae) in North America. Rhodora 87:189-286.

Sorensen, T. 1953. A revision of the Greenland species of Puccinellia Parl. Medd. Gronl. 136(3):1-179.

Tomlinson, P.B. 1982. Anatomy of the Monocotyledons. VII Helobiae (Alismatidae). Oxford University Press.

GLOSSARY

(Illustrated terms are followed by an asterisk)

Figure 1

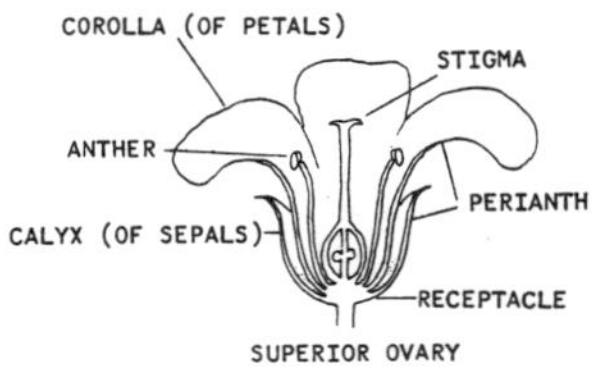

BILABIATE

Achene
A small, dry, one seeded, indehiscent fruit, e.g. individual fruit of <u>Ranunculus</u>
Acuminate
Tapering to a slender point
Acute
Forming an acute angle; sharp pointed
Ament*
A catkin, usually unisexual; dry scaly spike or spike-like raceme, e.g. inflorescence of birch or willow

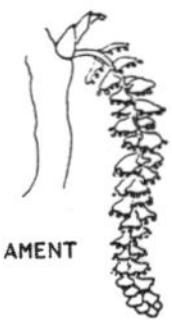
AMENT

Amplexicaule*
Clasping the stem

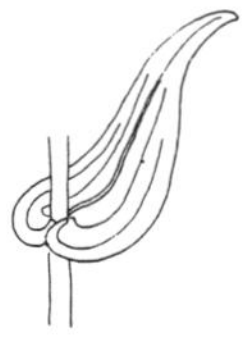
AMPLEXICAULE

Androgynous
In a bisexual spike, the staminate flowers at the apex, e.g. some Cyperaceae
Annual
Plant completing its life cycle within one year
Anthesis
Period in which flower is fully expanded and functional
Anthropophyte
Growing around human habitations
Aphyllopodia
Lower leaf sheaths lacking well developed blades, e.g. <u>Carex</u>
Apiculate*
Ending in an abrupt short point

APICULATE

Appressed
Lying close and flat against an organ, e.g. leaves against a stem
Arachnoid
Hairs or hairlike appendages interlaced as in a cobweb
Arcuate
Arched or curving into an arc of a circle
Areolate
Marked into small spaces or reticulated
Aristate
Having a bristle shaped tip or awn
Attenuate
Gradually tapering to narrow slender tips, longer than acuminate
Auricle*
A lobe-like appendage at base of organ

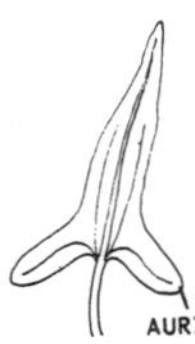
AURICLE

Autotroph
Plant able to produce organic food from inorganic sources
Awn*
A slender terminal bristle-shaped appendage

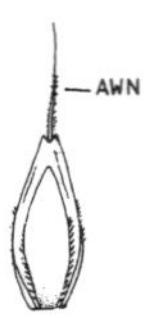

Axillary
Located in an axil or angle betweed two organs

Barbellate
Finely barbed
Bilabiate*
Having two lips, e.g. <u>Linaria vulgaris</u>
Bipinnate
Doubly or twice pinnate

Bract*
A modified or scale-like leaf subtending a flower or inflorescence

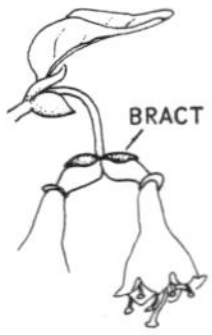

Bracteate
With bracts
Bulblet
A small bulb, especially one borne in leaf axils, e.g. <u>Lilium</u> <u>tigrinum</u>

Caducous
Falling off very easily
Calcareous
Containing carbonate of lime; of limestone; alkaline
Calciphilous
Lime loving
Callus
Hard protruberance, e.g. on fruits of <u>Rumex</u>
Campanulate*
Bell-shaped

CAMPANULATE

Capillary
Of hair-like fineness
Capitate
Arranged in a head or cluster
Capsule*
A dry usually dehiscent fruit of two or more carpels opening by valves or teeth

CAPSULE

Carpel
A simple pistil, or one member of a compound pistil
Cartilaginous
Like cartilage; firm and tough but flexible
Caruncle
Appendage near the attachment of a seed
Castaneous
Chestnut-coloured; dark brown
Catkin
See **ament**
Caudate*
With slender taillike appendage

CAUDATE

Caudex
The persistent base of stems of perennial herbs
Cauline
Pertaining to the stem
Cespitose
Forming compact tufts
Chaff
Thin dry scales, e.g. scales separating florets of many Asteraceae
Chartaceous
Papery texture
Ciliate
With fine marginal hairs
Ciliolate
Minutely ciliate

Cinereous
Ash coloured
Circumscissile*
Opening by transverse circular line, the top opening like a lid, e.g. Plantago
Circumneutral
About neutral, i.e. neither acid nor alkaline
Clavate
Clubshaped
Claw*
The narrow base or stalk of some petals
Cleistogamous
Flowers which do not open, pollinated with pollen from their own anthers, e.g. Viola
Commissure*
The surface by which one carpel joins another, as in the Apiaceae
Connate
Opposite leaves united by the base, e.g. Eupatorium perfoliatum (see perfoliate)
Connivent
Coming into contact; converging
Convolute
Rolled longitudinally; margins overlapping in the bud
Cordate*
Heartshaped; usually referring to base
Coriaceous
Leathery
Corm
Swollen base of stem that stores food material below ground
Corniculate
With hornshaped appendages
Corymb*
Broad, flat topped inflorescence equivalent to contracted raceme, the flowers or heads at perimeter open first, e.g. Achillea millifolium
Corymbose
Corymb-like
Cotyledon
Primary or seed leaf
Crenate
Margins with rounded teeth
Cucullate*
Hooded or resembling a hood
Culm
The stem of grasses and sedges
Cuneate
Wedge-shaped
Cuspidate
Tipped with a sharp firm point
Cyme*
A flattish inflorescence with central terminal flowers opening first
Cymule
A small secondary cyme of a compound cyme

Declinate
Curved downward or forward
Decompound
More than once divided
Decumbent*
Reclining with the ends ascending

CIRCUMSCISSILE

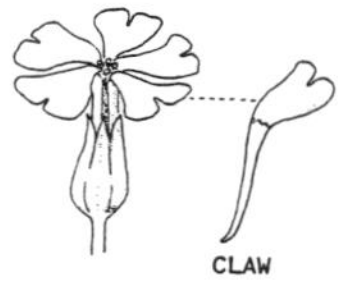
CLAW

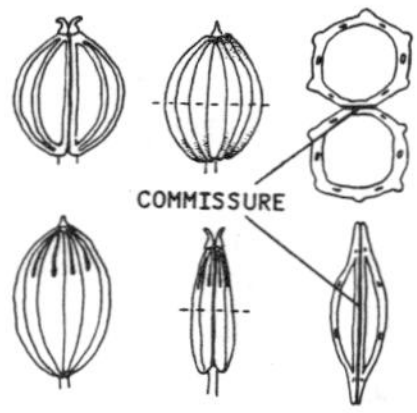
COMMISSURE

CORDATE

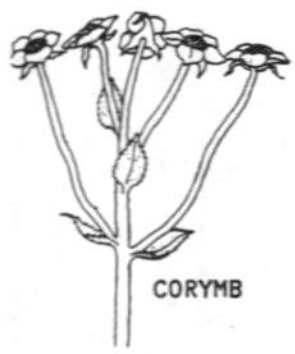
CORYMB

EMARGINATE
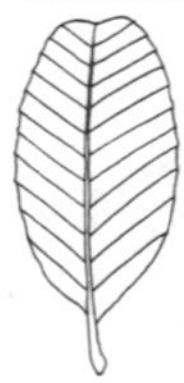

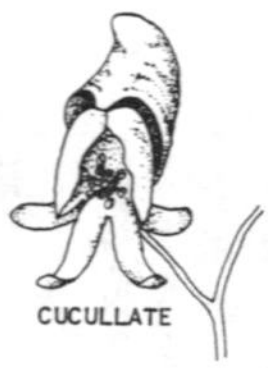
CUCULLATE

ELLIPSOID

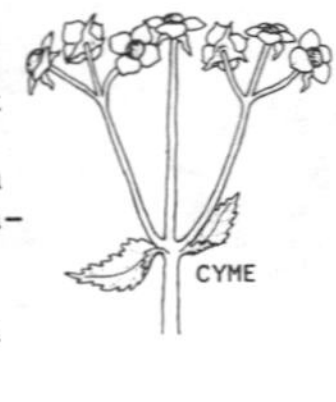
CYME

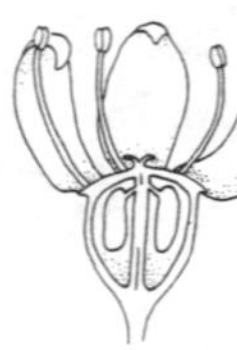
EPIGYNOUS

DECUMBENT

Decurrent
Extending downward from poin of insertion, e.g. leaf winge and extending downwards alon stem as in Cirsium vulgare
Deflexed
Turned abruptly downward
Dehiscent
Opening by valves or slits a in a capsule or anther
Deltoid
Triangular

Dentate
Toothed, usually with outwar ly directed teeth
Denticulate
Minutely toothed
Digitate
Parts diverging from commor base as the fingers of ar outstretched hand

Dimorphic
Having two forms, e.g. fertile and sterile fronds of Osmunda cinnamomea
Dioecious
Unisexual with male and female flowers on separate plants
Disarticulating
Separating at special joints when mature
Discoid
Resembling a disc; in Astereaceae composed only of tubular florets, i.e. lacking ray florets
Dissected
Divided into many segments
Divaricate
Spreading at a wide angle
Drupe
Fleshy fruit with a hard endocarp or stone , e.g. cherry
Drupaceous
Like a drupe

Emarginate*
Notched at the apex
Emergent
Appearing above the surface of the water
Ellipsoid*
Oval-shaped like an ellipse but solid
Epigynous*
Sepals, petals and stamens borne at summit of the ovary or apparently so, the ovary therefore inferior
Erose
Irregularly margined as if gnawed
Exserted
Projecting beyond surrounding parts

Falcate
Scythe-shaped
Farinose
Covered with mealy powder
Fascicle
Bundle or cluster
Fastigiate
With branches or stems erect and near together

Fibrillose
With fibers
Filiform
Thread-like
Filamentous
Composed of thread-like structures
Floreal
Bearing flowers
Floricane
Flowering stem usually second year's growth of non-flowering first-year stem, e.g. Rubus

Flexuous
Curved alternately in opposite directions
Flocculent
With fine flocks of woolly hair
Foliaceous
Leaf-like

Follicle*
Fruit of a single carpel, dehiscing by a ventral slit, e.g. milkweed "pod"
Frond
Leaf of ferns and some other Cryptogams; in Lemnaceae, the thallus-like stem which functions as foliage
Fulvous
Tawny, brownish-yellow
Fusiform*
Spindle-shaped

Gamopetalous
Petals more or less united at least above the base
Geniculate
Bent abruptly like a knee
Gibbous
Swollen on one side usually near the base
Glabrate
Nearly smooth
Glabrescent
Becoming smooth
Glabrous
Smooth, not pubescent
Gland*
A secreting structure; any protruberance having the appearance of such an organ
Glandular
Bearing glands
Glaucous
Covered with a fine bluish or whitish bloom
Glaucescent
Somewhat glaucous
Glomerules
Compact head-like cyme
Glume
Scaly bracts, specifically one of the two bracts at the base of grass spikelets
Glutinous
Having a sticky exude
Gynaecandrous
In a bisexual spike, the pistillate flowers at the apex, e.g. some Cyperaceae

HASTATE

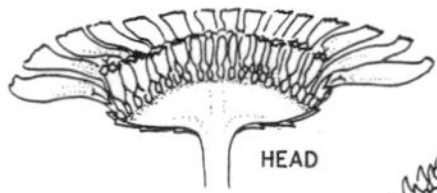
HEAD

HEAD

FOLLICLE

FUSIFORM

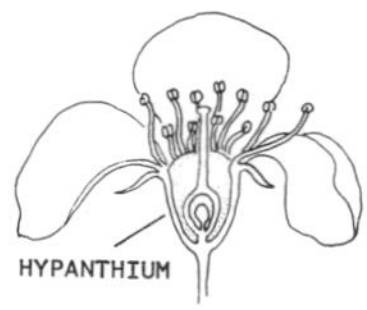
HYPANTHIUM

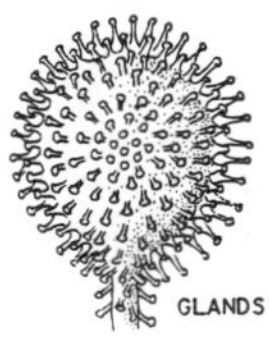
GLANDS

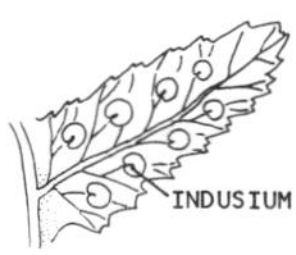
INDUSIUM

Halophytic
Of saline habitats
Hastate*
Leaf shape with two basal lobes directed outwards
Head*
Dense cluster of sessile or subsessile flowers on short axis or receptacle
Herbaceous
Plants which die back to ground at end of growing season

Heterophyllous
Having more than one kind of leaf, e.g. Polygonum
Hirsute
Pubescent with coarse stiff hairs
Hirsutulous
Slightly hirsute
Hirtellus
Minutely hirsute
Hispid
With bristly hairs or bristles
Hispidulus
Minutely hispid
Homophyllous
Having leaves of one kind

Hyaline
Transparent or translucent
Hybrid
The progeny of a cross between two species, or genetically different parents
Hypanthium*
An enlargement of the receptacle (torus) under the calyx, often simulating the calyx tube

Imbricate
Overlapping, either in width only or in width and length, e.g. involucral bracts of many Asteraceae
Imperfect
Of flowers, unisexual; lacking either stamens or pistils
Incised
Sharply and irregularly cut
Indehiscent
Not opening at maturity
Indurated
Hardened
Indusium (pl. **indusia**)*
The thin epidermal extension covering the sorus or spore bearing structure of ferns
Inferior
Lower or below; inferior ovary: one that is adnate to and below the calyx
Inflexed
Bent inward
Inflorescence
Complete flowering structure
Infructescence
Complete fruiting structure
Involucel*
A secondary involucre subtending a part of an inflorescence, e.g. umbellet of Apiaceae

Involucre*
A set of bracts commonly in a circle subtending a single flower or an infloresence

Involute
Rolled inward
Irregular Flower
Flower with dissimilar parts, i.e. with mirrorimageappearance only in one longitudinal plane of symmetry (see scape)

Joint*
Point where two parts are connected

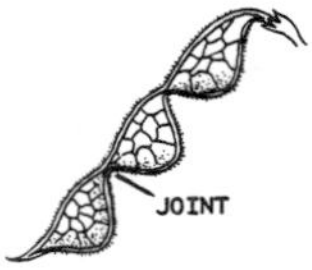

Keel*
A central dorsal ridge like the keel of a boat, also two lower united petals in Fabaceae

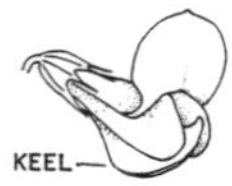

Krantztypus*
Distinctive dark green reticulate venation, most clearly seen when leaf surface scraped with a blade, e.g. Atriplex laciniata

Lacerate
Irregularly cut as if torn
Laciniate
Cut into narrow pointed lobes

Lanceolate
Lance shaped; broad toward base, narrowed at apex and several times longer than wide
Lacunae
Rows of air filled cavities on each side of the leaf midrib in certain Potamogeton
Lax
Loose and open
Lemma*
The lower (outer) of two bracts enclosing the floret in grasses

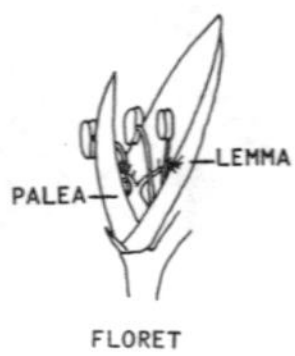

Lenticular
Having the shape of a biconvex lens, narrowly elliptic in transverse section
Lepidote
Covered with small scurfy scales
Ligneous
Woody
Ligulate
Having a ligule
Ligule*
Small, usually flat or tongue-like projections from an organ; in grasses at the summit of the leaf sheath between base of leaf blade and stem; in some Caryophyllaceae at junction of claw and blade of petals; in Asteraceae the flattened white or coloured limb of ray florets

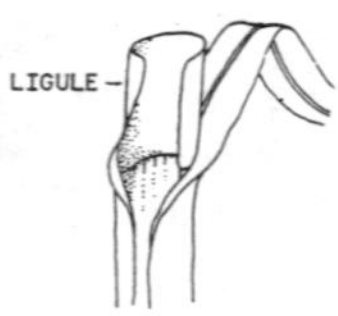

Locule
The cavity of an ovary, fruit or anther
Loculicidal capsule
Dehiscing into locule cavity along its outer median line

Lyrate*
Pinnately lobed with terminal lobe the largest

Megaspore
The larger of two kinds of spore of Isoetes and Selaginella usually giving rise to female prothallus or gametophyte
Membranous
Thin, flexible, more or less translucent
Mericarp*
Each segment of mature fruit each having the appearance of a separate fruit as in Apiaceae (see commissure)

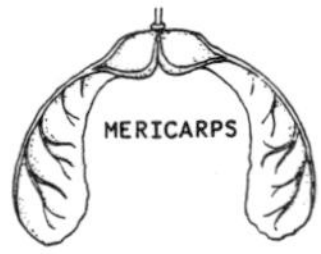

-merous
A suffix commonly referring to the number of parts in each circle of the floral organs
Monoecious
Unisexual with staminate and pistillate flowers on the same plant, often in different parts of the inflorescence, e.g. Typha
Mucronate*
Tipped with abrupt short point

Nodal glands
Paired glandular swellings at the nodes of some species of Potamogeton

Nodulose
With small knobs
Nothomorph
A hybrid produced by the combination of certain characters of the parent species, other nothomorphs being the product of a different combination of characters of the same parents

Ob-
A prefix signifying in reverse direction
Oblong
Longer than broad, the ends rounded, the sides nearly parallel
Obsolete
Rudimentary; not evident
Obtuse
Broad angled; blunt or rounded at the apex
Ocrea*
A tubular sheathing leaf-stipule as in Polygonum

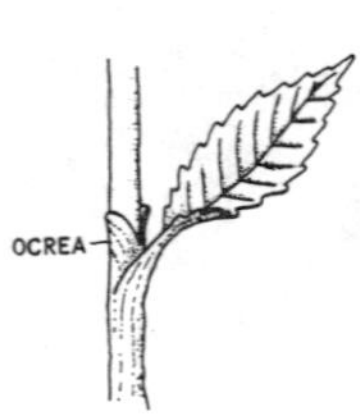

Ocreolae
Small secondary sheaths as in the inflorescence of Polygonum
Olivaceous
Olive coloured
Orbicular
Circular
Ovate*
Egg shaped

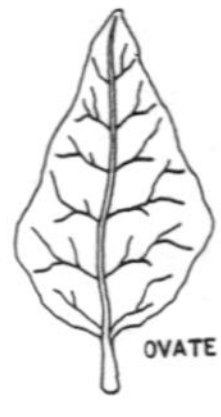

Ovoid
Solid with ovate outline
Oxyphilous
Acid loving

Palea*
The upper of the two bracts surrounding a grass floret, often wanting or rudimentary (see lemma)
Palmate
Having leaflets arising from one point near the tip of the petiole
Panicle*
A loose irregular compound inflorescence of stalked flowers or spikelets
Papillate
Bearing minute blunt projections

Pappus*
In Asteraceae, a modified calyx-limb giving a crown of hairs, bristles, awns, teeth or scales at the summit of the achene (see tubular floret)
Patent
Spreading, opening wide, diverging widely from the axis
Pectinate
Comb-like; pinnatifid with the narrow segments arranged side by side like a comb
Pedicel*
Stalk of a single flower in an inflorescence
Peduncle*
Primary flower-stalk bearing a single flower or a flower cluster
Peltate*
Round or shield shaped, the point of attachment being near the centre of the lower surface
Perennial
Living through a number of years, especially herbs which die back to ground level in Fall and send new shoots from the root in Spring

Perfect Flower
Having both stamens and pistil
Perfoliate*
Leaves so clasping the stem that the stem appears to pierce them
Perianth*
The floral envelope consisting of calyx and corolla or at least of one of these (see Figure 1)
Pericarp
The wall of the fruit
Perigynium (pl. **perigynia**)*
The inflated sac enclosing the ovary in <u>Carex</u>
Petaloid
Resembling a petal
Petiole
Leaf-stalk
Petiolulate
With a leaflet-stalk

PANICLE

PINNATIFID

PINNATISECT

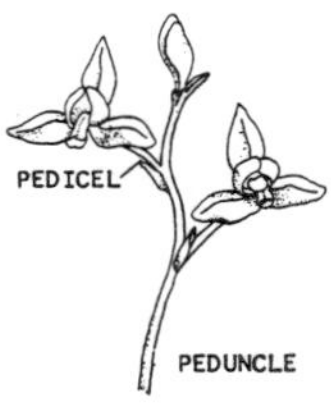

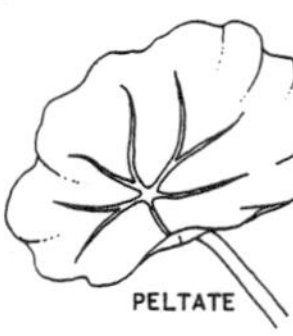
PELTATE

PERFOLIATE

RACEME

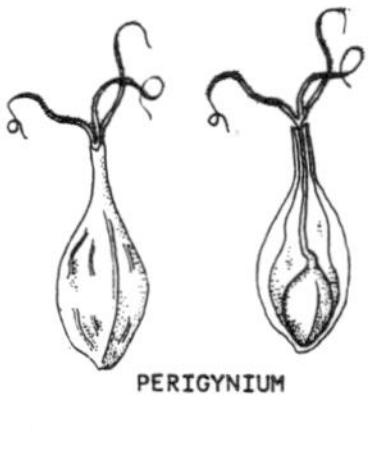
PERIGYNIUM

Phyllaries*
Involucral bracts surrounding the flowering heads in the Asteraceae (see involucre)
Phyllodium (pl. **phyllodia**)
Dilated petiole which serves the function of a leaf blade
Pilose
Hairy, especially soft, relatively long hairs
Pinna (pl. **pinnae**)
One of the primary divisions of a pinnate leaf
Pinnate
A compound leaf with leaflets arranged on both sides of a common axis
Pinnatifid*
Deeply pinnately cleft but not to mid rib or rachis
Pinnatisect*
Pinnately cleft to the mid rib or rachis

Pinnule
A secondary pinna; one of the divisions of a pinnate pinna
Pistillate
Unisexual, having pistils but no stamens
Pith
The spongy cellular tissue or parenchyma in the centre of a stem
Plano-convex
With one surface flat and the opposite surface convex
Plicate
Folded in lengthwise folds; pleated
Plumose
With long hairs; feathery
Primocane
Of <u>Rubus</u>, the first year's cane, usually without flowers, from which the second year's flowering can e or floricane develops
Puberulent
With very short hairs
Pubescent
Covered with hairs, especially short, soft hairs
Pulvinus (pl. **pulvini**)
The swollen base of a branch of an inflorescence; swollen base of a petiole
Punctate
Dotted with transluclent internal glands, coloured dots or small pits
Puncticulate
Minutely punctate
Pyriform
Pear-shaped
Pyxis
A capsule with circumscissile dehiscence, e.g. <u>Portulaca</u> (see **circumscissile**)

Raceme*
An inflorescence where each flower is borne on an individual stalk along a central stem

Rachilla
Of grasses and sedges, the axis of an individual floret as opposed to that of the spike
Rachis*
The axis of a compound leaf or of an inflorescence (see rhizome)
Radiate
Bearing ray florets; with ray florets around a central disk of tubular florets, e.g. Daisy
Radicle
Small root of the embryo
Ray floret*
The ligule or strap-shaped floret of many Asteraceae
Receptacle*
The expanded summit of a stem bearing the floral organs or bearing the crowded flowers of a head; torus (see Figure 1)
Reflexed
Bent backward abruptly
Regular flower
Radially symmetrical
Reniform*
Kidney-shaped
Repent*
Of stems, creeping and rooting at nodes (see stolon)
Resinous
Sticky; adhesive, as in the exude or resin of certain gymnosperms
Reticulations
Network pattern
Retrorse
Turned backward or downward
Revolute
Rolled backwards from the margin and covering the lower surface
Rhizomatous
Having the character of a rhizome
Rhizome*
Subterranian stem producing roots from its lower surface and sending up stems, branches and leaves from its upcurved apex
Rhomboidal
A solid with a rhombic outline (an equilateral oblique-angled figure)
Riparian
Riverine; situated on the banks of a river
Rootstock
Rhizome
Roseate
Rose coloured; pink
Rosette
Cluster of basal leaves in a circular arrangement
Ruderal
Growing in waste places or among rubbish
Rufous
Reddish-brown; rust coloured
Rugose
Wrinkled as in a leaf with sunken veins

SAMARA

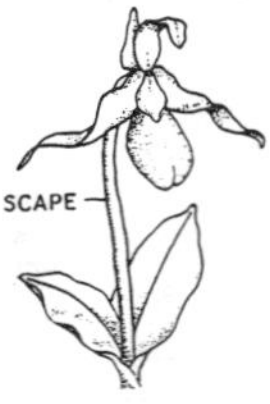
SCAPE

RAY FLORET

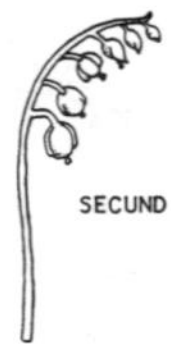
SECUND

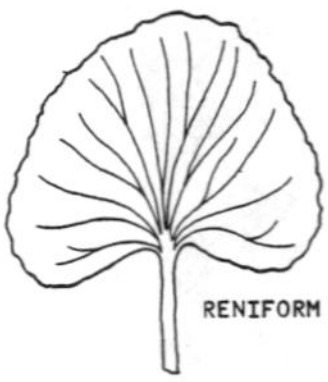
RENIFORM

SESSILE

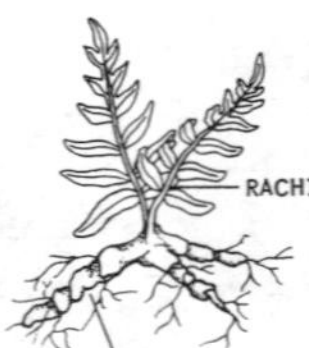

RHIZOME

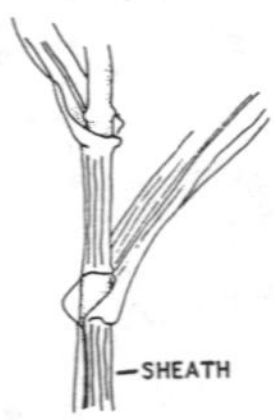
SHEATH

SILICLE

SILIQUE

SPADIX
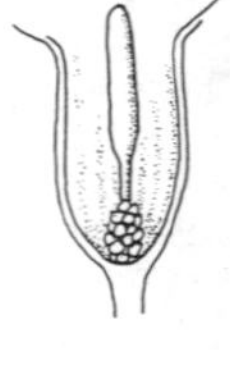

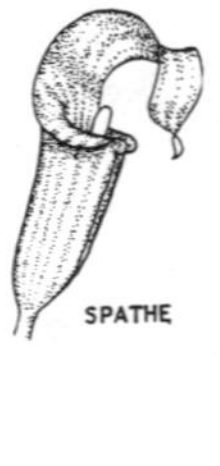
SPATHE

Saccate
With a pouch or sac, e.g. spurred sepal of Impatiens
Samara*
Indehiscent, winged usually one seeded fruit as in ash, elm, maple
Saprophyte
Plant which grows on dead organic matter
Scabrous
Roughness due to the structure of the epidermis or to the presence of short, stiff hairs
Scape*
Leafless flowering stem, the leaves usually in a basal rosette, e.g. Cypripedium acaule
Scapose
Having a scape
Scarious
Thin, dry, membraneous; not green
Scurfy
With scale-like particles
Secund*
An inflorescence in which the flowers are borne along one side of the stem
Septate-nodulose
Divided by small swollen partitions
Sericeous
Silky pubescence
Serrate
With sharp teeth directed forwards
Serrulate
Minutely serrate
Sessile*
Without a stem or petiole
Setaceous
Bristle-shaped
Setose
Bristly
Setulose
Minutely bristly
Sheath*
Tubular envelope as in lower part of leaf sheathing the stem in Poaceae, Cyperaceae and Juncaceae
Silicle*
A short silique
Silique*
A long two valved capsular fruit with a thin longitudinal internal septum bearing seeds, e.g. many Brassicaceae
Sinuate
With strongly wavy but entire margins
Sordid
Impure white colour
Sorus (pl. sori)
Cluster of sporangia or 'fruit dots' in the ferns
Spadix*
A fleshy spike of flowers as in the Araceae (see spathe)

Spathe*
Large green or coloured bract enclosing flower or inflorescence, e.g. Araceae

Spatulate*
Spoon-shaped; broadly rounded at summit tapering towards base

SPATULATE

Spicate
Arranged in a spike
Spike*
A simple inflorescence with sessile or nearly sessile flowers arranged on an elongated axis

SPIKE

Spikelet*
A small spike with flowers inconspicuous and mostly hidden by bracts, e.g. Poaceae and Cyperaceae

SPIKELET

Spinulose
With small sharp spines
Sporangia
A sac containing spores as in the ferns
Spur*
A hollow sac-like projection of a sepal or petal often carrying nectaries, e.g. Aguilegia

Squarrose
Spreading or recurved at the tip
Staminate
Unisexual, having stamens but not pistils
Stamineal disk
Disk at point of attachment of stamens
Stellate
Star-shaped
Stigma*
The part of the pistil which receives pollen, usually having a viscid or papillose surface (see Figure 1)
Stipe
The stalk of a fern frond below the leafy portion
Stipitate
With a stipe
Stipule
Bract-like or leafy appendage at the base of a petiole or sessile leaf
Stolon*
A runner or any prostrate basal branch rooting at the nodes or tip

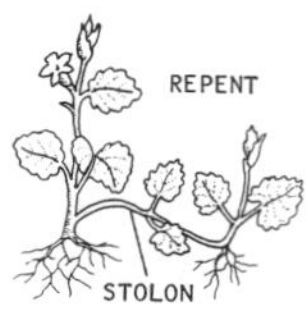

Stoloniferous
Producing stolons
Stramineous
Straw coloured; pale brownish yellow
Striate
Marked with fine longitudinal lines or stripes or veins
Strigose
With appressed stiff hairs
Strobilus (pl. **strobili**)*
A cone-shaped inflorescence with overlapping scales or bracts

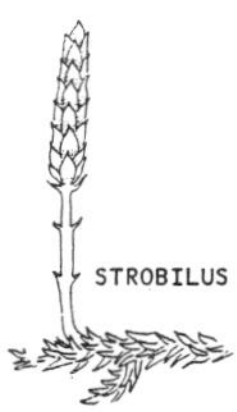
STROBILUS

Stylopodium*
Disc which runs around the base of the style, e.g. Osmorhiza

Sub-
Latin prefix signifying "nearly" or "slightly"

Subtended
To extend under so as to embrace or enfold
Subulate
Gradually tapering to a slender point
Succulent
Fleshy or juicy
Suffruticose
Low and woody
Superior Ovary*
An ovary with the perianth inserted below it (see Figure 1)
Sympetalous
A corolla with the petals more or less united above the base
Synsepalous
A calyx with the sepals united

Tepal
Sepals and petals of a similar form, not readily differentiated
Terete
Circular in transverse section
Ternate
Divided into three segments or arranged in threes
Thalloid
Resembling a thallus; a flat vegetative organ not differentiated into leaves and stems
Tomentose
Covered with matted woolly hairs
Tomentulose
Finely tomentose
Tomentum
A covering of matted woolly hairs
Trigonous
Triangular in cross section, e.g. seed of Buckwheat
Trilocular
Having three cavities or locules
Truncate
Ending abruptly with a nearly straight edge
Tuber
A thickening underground stem carrying buds or "eyes", e.g. potato
Tubercle
A small tuber or tuber-like body; also the persistent base of the style in some Cyperaceae, bacteria-containing nodules of roots of Fabaceae, and small processes on seeds of some Rumex species
Tuberculate
Bearing tubercles
Tubular floret*
Floret of many Asteraceae, regular in form having petals joined in a tube and limb of petals poorly developed

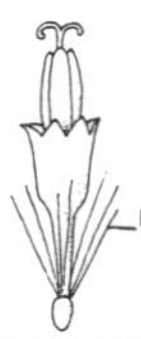

TUBULAR FLORET

Turbinate
Top-shaped

Umbel*
Flattish topped inflorescence with stalked flowers arising from one point of the common peduncle, the outer flowers

UMBEL

open first
Umbellate
Resembling an umbel
Umbelliform
In the shape of an umbel
Unilocular
Having one locule or cavity
Unisexual
Having one sex, either staminate or pistillate

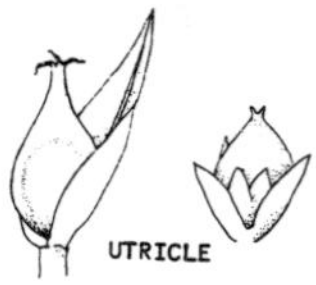
UTRICLE

URCEOLATE

Utricle*
A small bladdery one-seeded fruit
Urceolate*
Urn-shaped
Valve
One of the parts into which a dehiscent capsule splits
Verrucose
Covered with wart-like projections
Villous
Having long soft hairs

INDEX

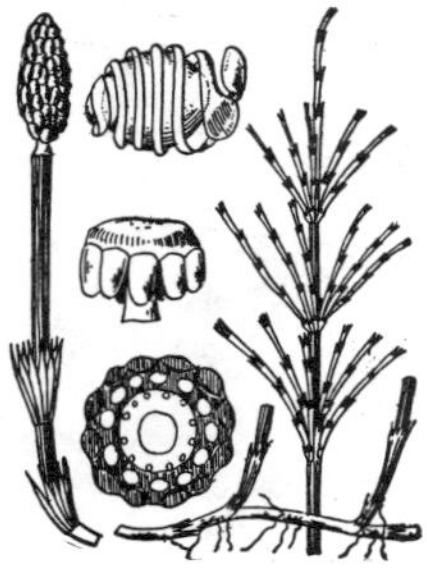

EQUISETUM ARVENSE

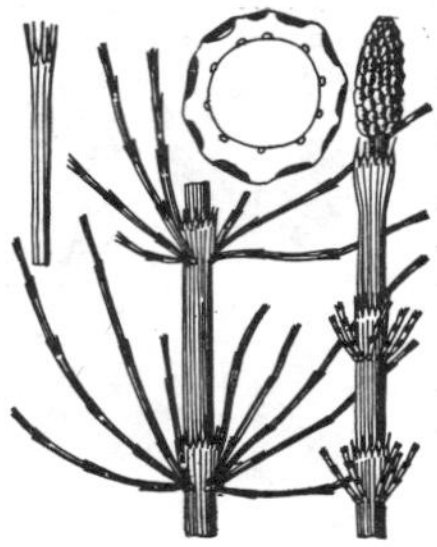

EQUISETUM FLUVIATILE

EQUISETUM HYEMALE

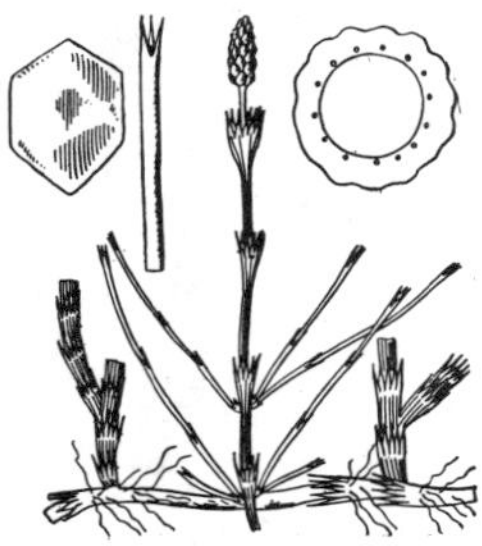

EQUISETUM X LITORALE

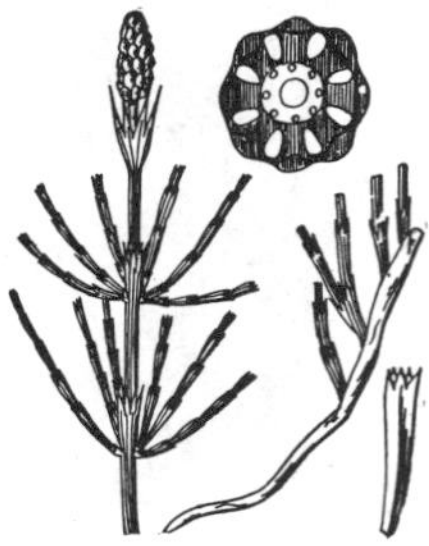

EQUISETUM PALUSTRE

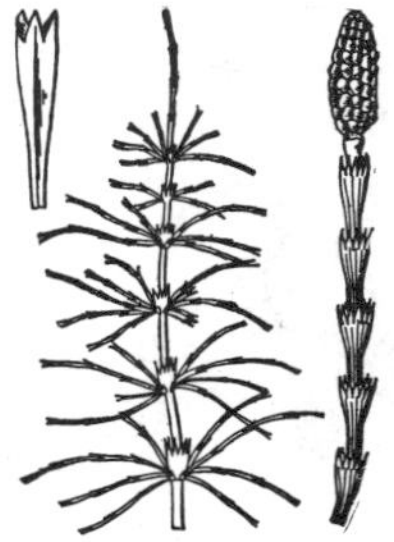

EQUISETUM PRATENSE

EQUISETUM SCIRPOIDES

EQUISETUM SYLVATICUM

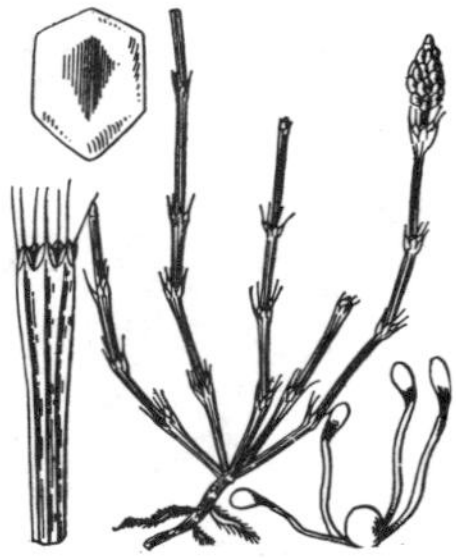

EQUISETUM VARIEGATUM

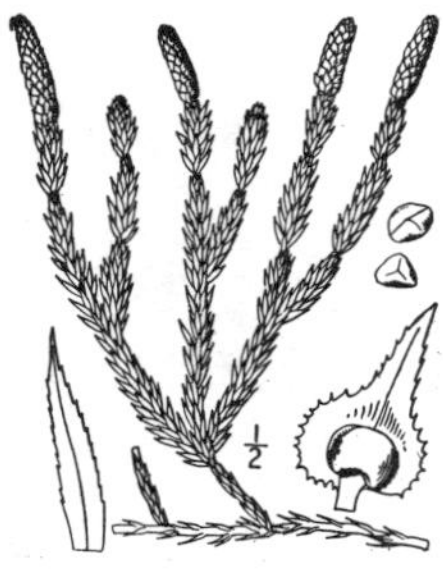

LYCOPODIUM ANNOTINIUM

LYCOPODIUM CLAVATUM

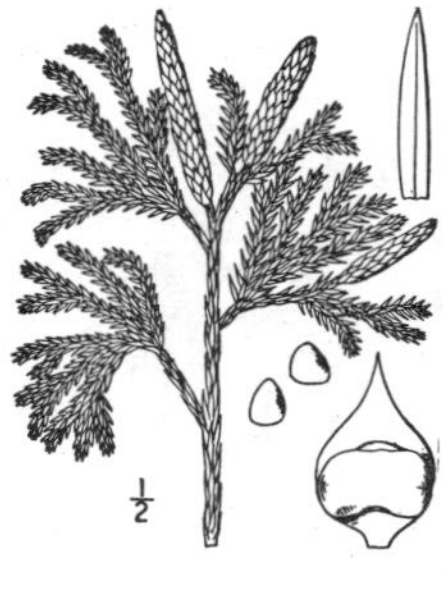

LYCOPODIUM OBSCURUM

LYCOPODIACEAE	SELAGINELLACEAE	ISOETACEAE

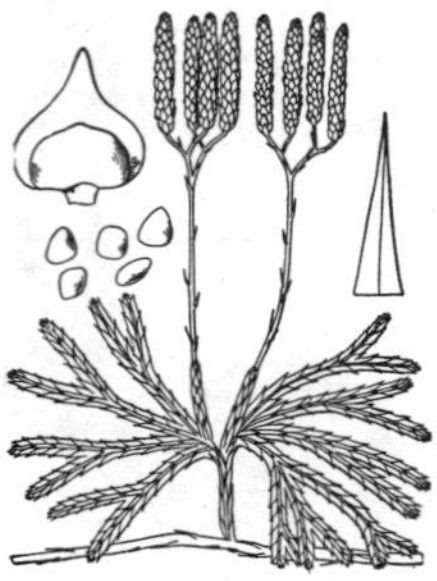

LYCOPODIUM DIGITATUM

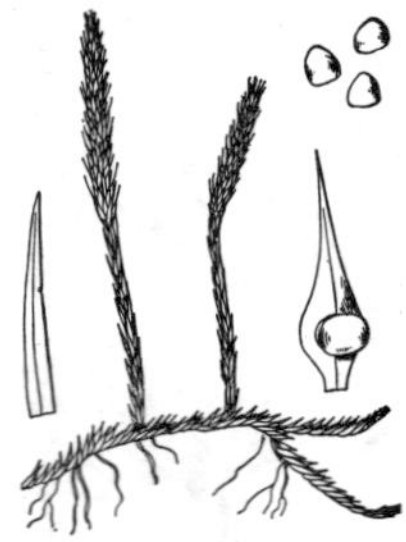

LYCOPODIUM INUNDATUM

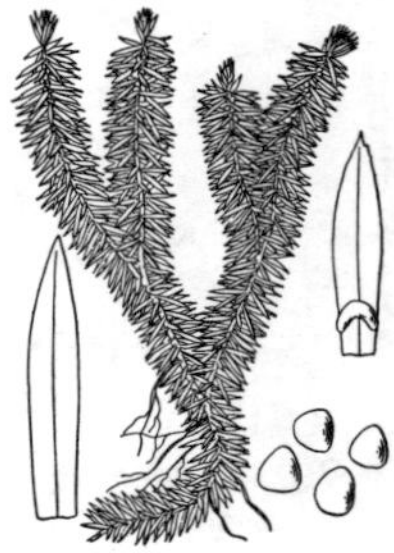

LYCOPODIUM LUCIDULUM

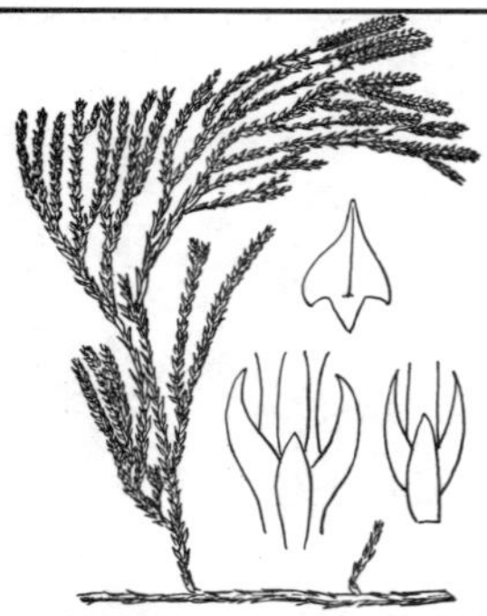

LYCOPODIUM SABINAEFOLIUM

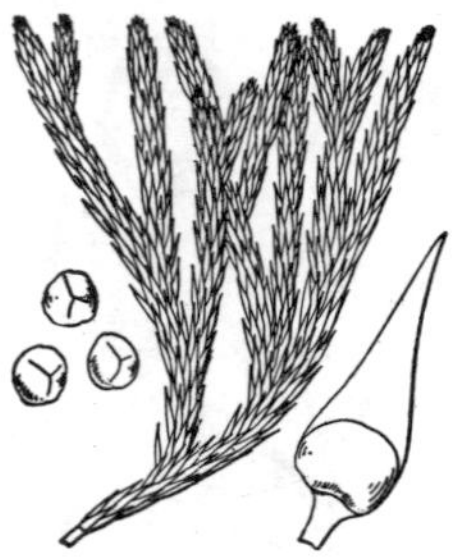

LYCOPODIUM SELAGO

LYCOPODIUM TRISTACHYM

SELAGINELLA RUPESTRIS

SELAGINELLA SELAGINOIDES

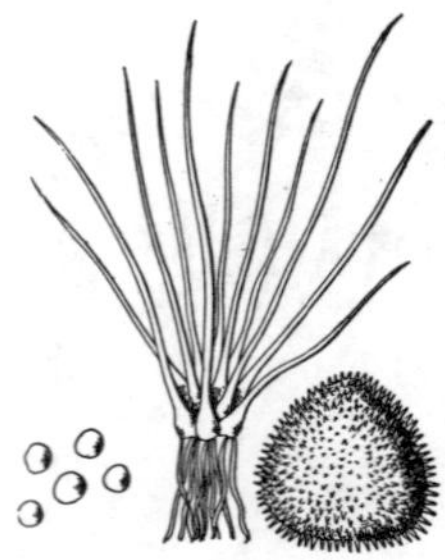

ISOETES ECHINOSPORA

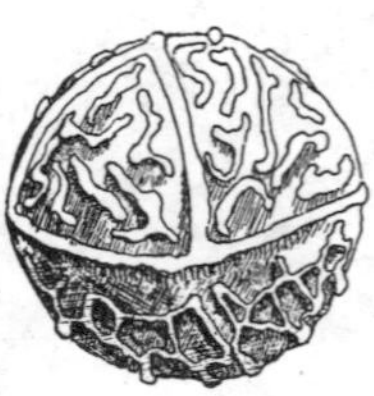

ISOETES HIEROGLYPHICA

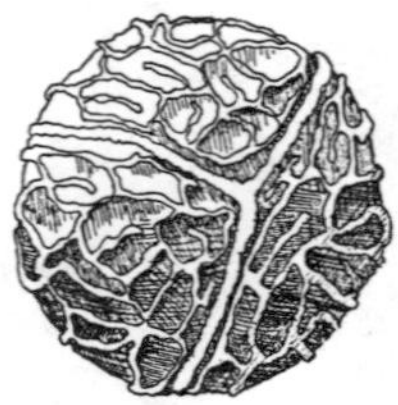

ISOETES MACROSPORA

ISOETES RIPARIA

ISOETACEAE

OPHIOGLOSSACEAE

OSMUNDACEAE

ISOETES TUCKERMANII

BOTRYCHIUM DISSECTUM

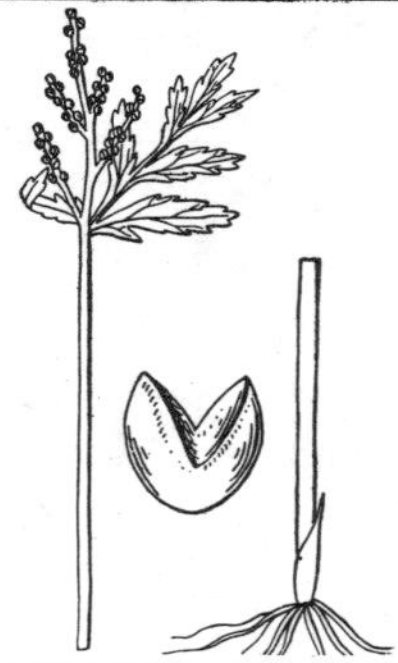

BOTRYCHIUM LANCEOLATUM

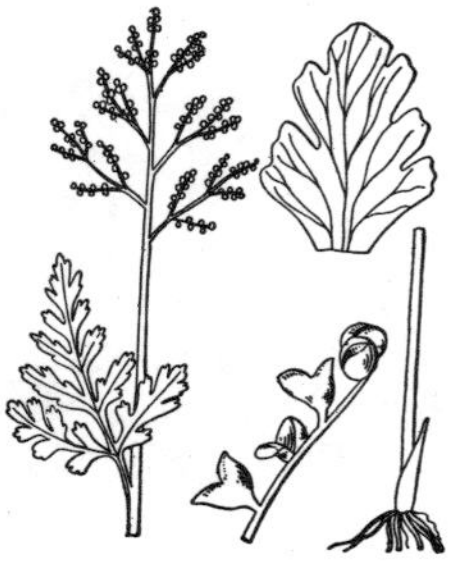

BOTRYCHIUM MATRICARIAEFOLIUM

BOTRYCHIUM MINGANENSE

BOTRYCHIUM MULTIFIDUM

BOTRYCHIUM SIMPLEX

BOTRYCHIUM VIRGINIANUM

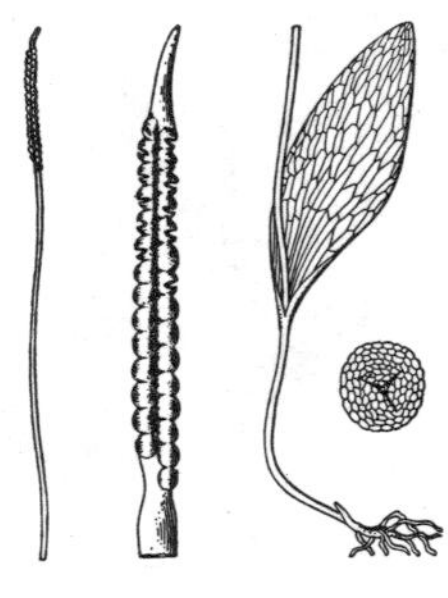

OPHIOGLOSSUM VULGATUM

OSMUNDA CINNAMOMEA

OSMUNDA CLAYTONIANA

OSMUNDA REGALIS

ADIANTUM PEDATUM

ASPLENIUM VIRIDE

ATHYRIUM FILIX-FEMINA

ATHYRIUM THELYPTEROIDES

CRYPTOGRAMMA STELLERI

CYSTOPTERIS BULBIFERA

CYSTOPTERIS FRAGILIS

DENNSTAEDTIA PUNCTILOBULA

DRYOPTERIS BOOTTII

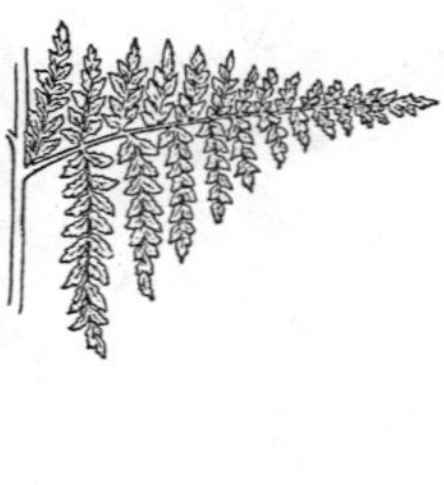
DRYOPTERIS CAMPYLOPTERA

DRYOPTERIS CRISTATA

DRYOPTERIS FRAGRANS

DRYOPTERIS GOLDIANA

DRYOPTERIS INTERMEDIA

DRYOPTERIS MARGINALIS

DRYOPTERIS SPINULOSE

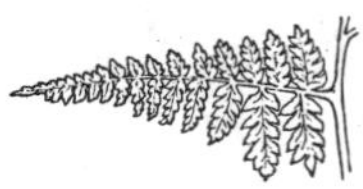

DRYOPTERIS X TRIPLOIDEA

GYMNOCARPIUM DRYOPTERIS

GYMNOCARPIUM ROBERTIANUM

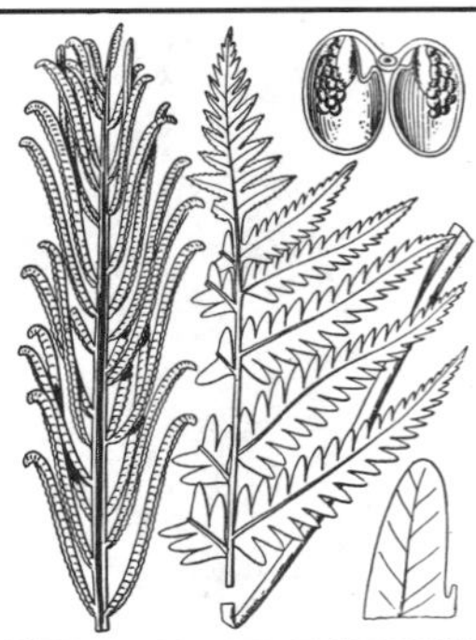

MATTEUCCIA STRUTHIOPTERIS

ONOCLEA SENSIBILIS

POLYPODIUM VIRGINIANUM

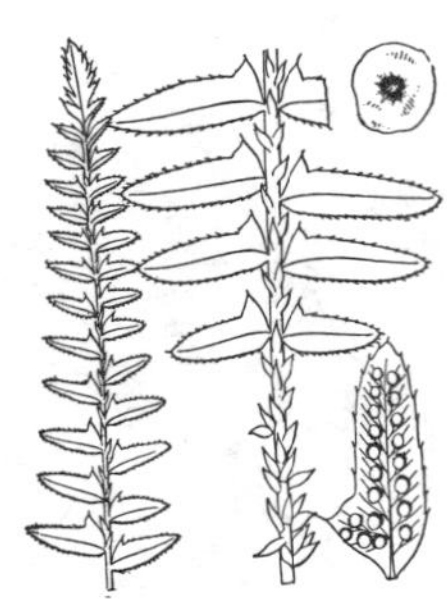

POLYSTICHUM ACROSTICHOIDES

POLYSTICHUM BRAUNII

POLYPODIACEAE | TAXACEAE | PINACEAE

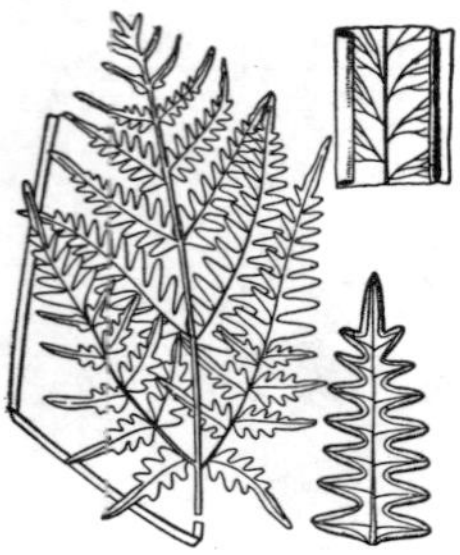

PTERIDIUM AQUILINUM

THELYPTERIS NOVEBORACENSIS

THELYPTERIS PALUSTRIS

THELYPTERIS PHEGOPTERIS

THELYPTERIS SIMULATA

WOODSIA ALPINA

WOODSIA GLABELLA

WOODSIA ILVENSIS

WOODWARDIA VIRGINICA

TAXUS CANADENSIS

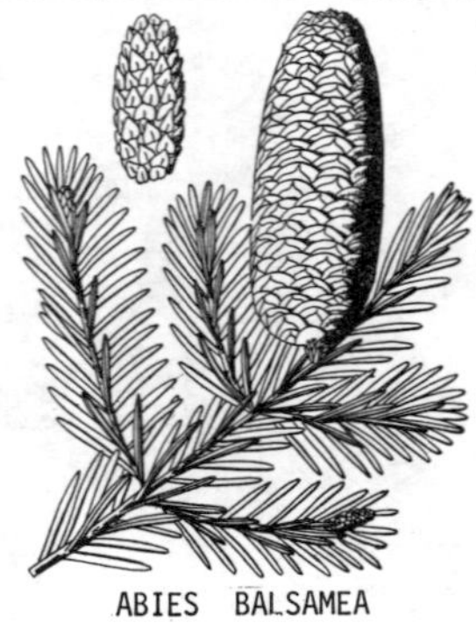

ABIES BALSAMEA

JUNIPERUS COMMUNIS

JUNIPERUS HORIZONTALIS

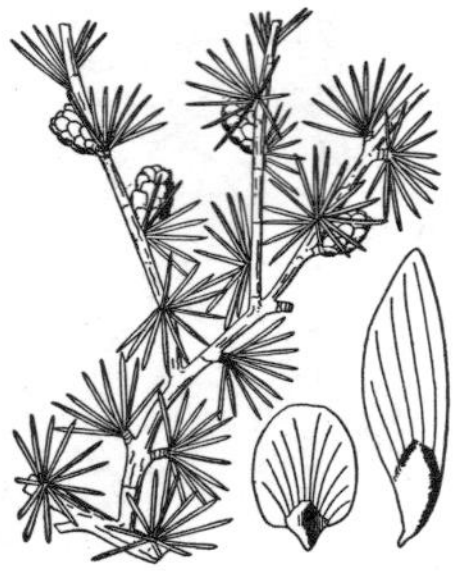

LARIX LARICINA

PICEA GLAUCA

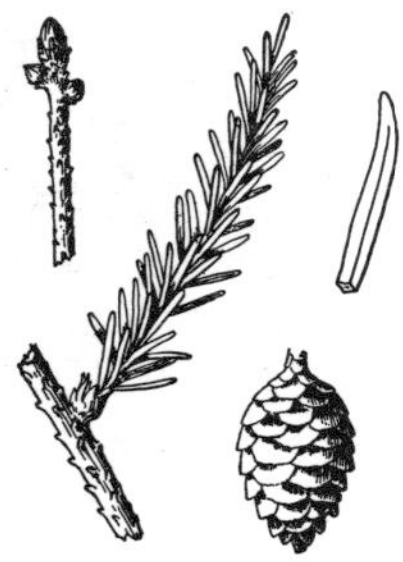

PICEA MARIANA

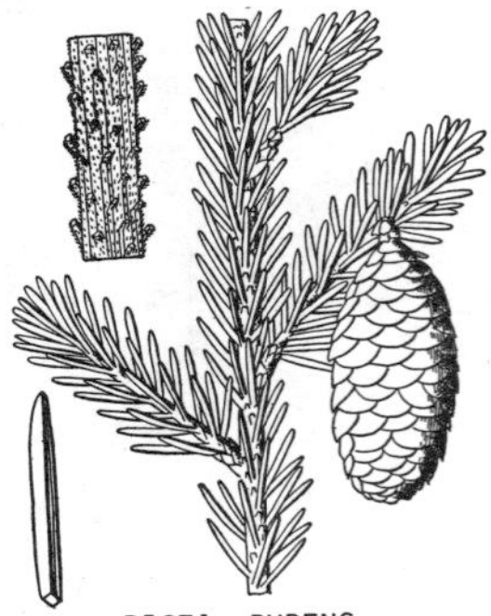

PICEA RUBENS

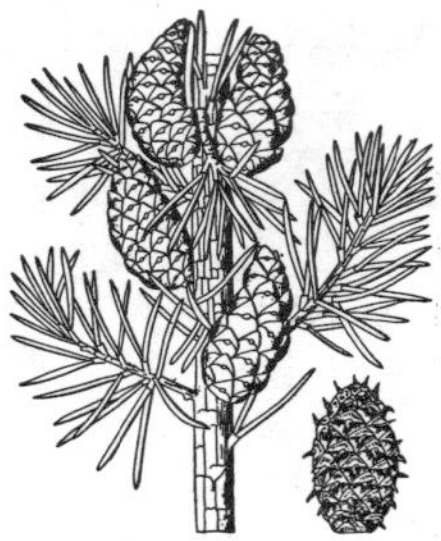

PINUS BANKSIANA

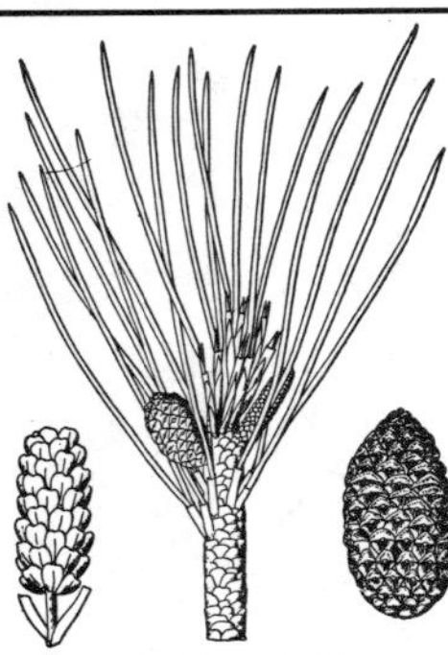

PINUS RESINOSA

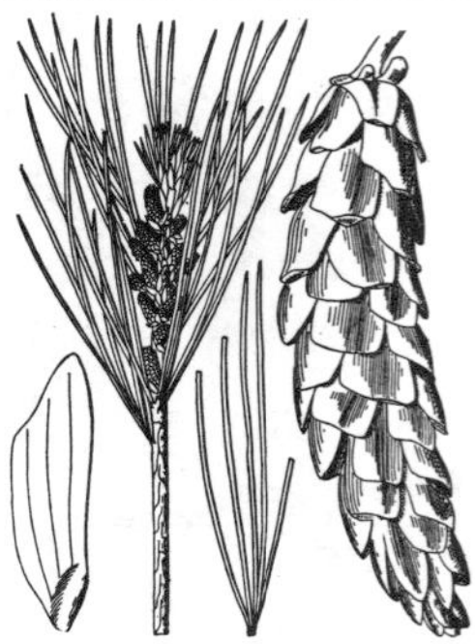

PINUS STROBUS

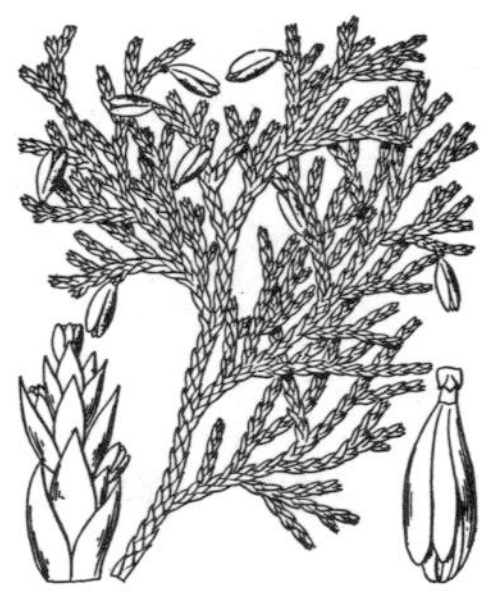

THUJA OCCIDENTALIS

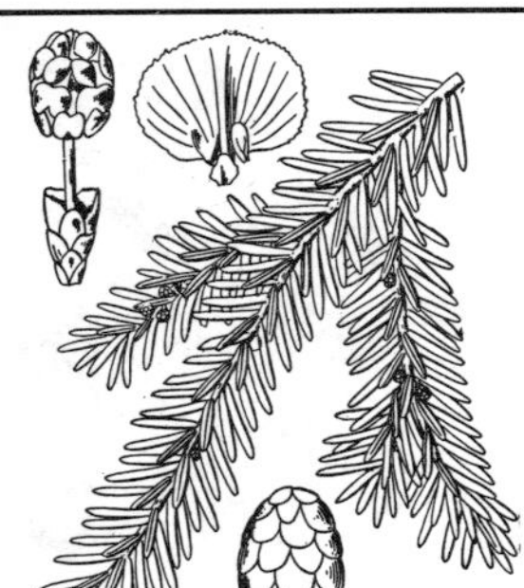

TSUGA CANADENSIS

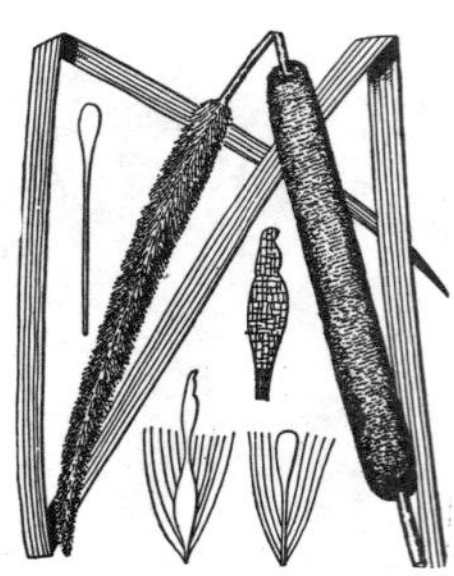

TYPHA ANGUSTIFOLIA

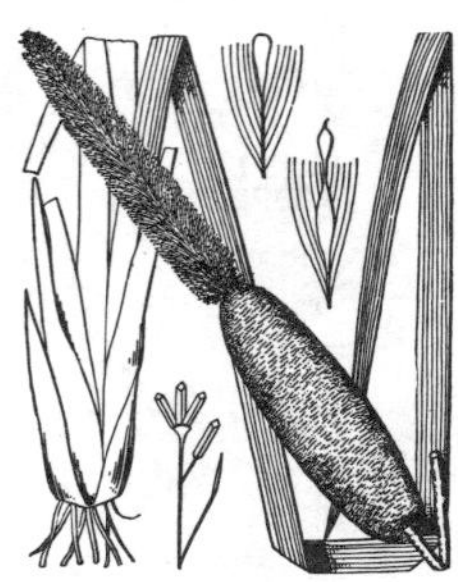

TYPHA LATIFOLIA

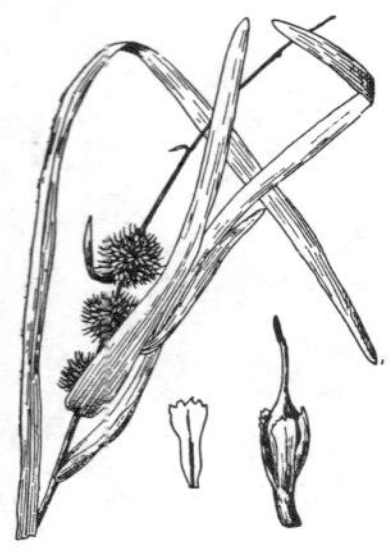

SPARGANIUM AMERICANUM

SPARGANIUM ANDROCLADUM

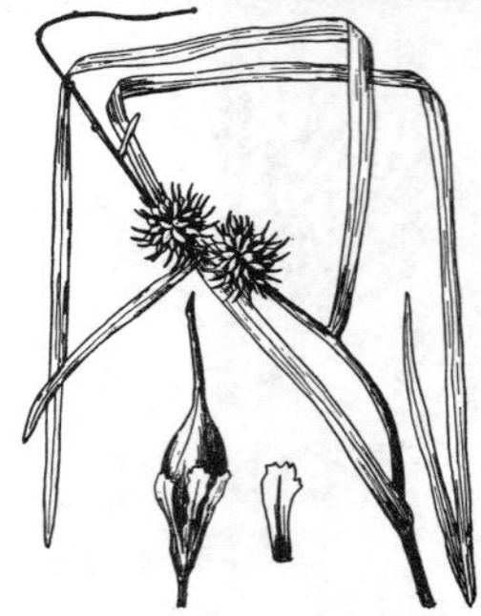

SPARGANIUM CHLOROCARPUM

SPARGANIUM EMERSUM

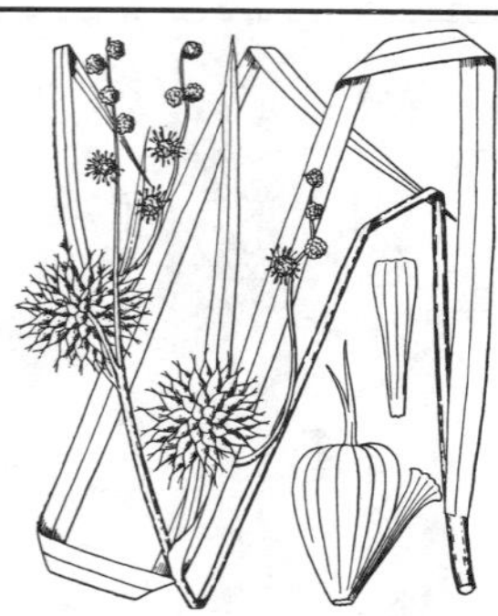

SPARGANIUM EURYCARPUM

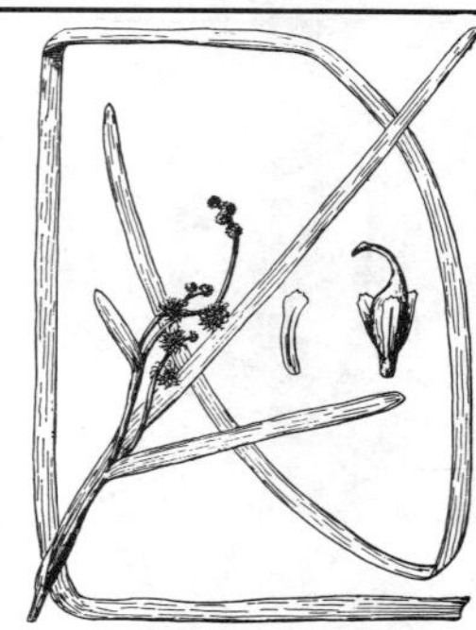

SPARGANIUM FLUCTUANS

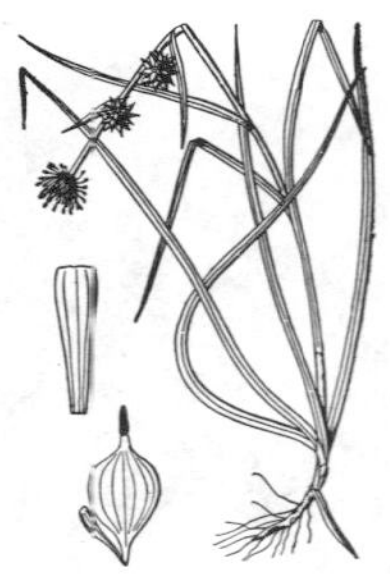

SPARGANIUM MINIMUM

POTAMOGETON ALPINUS

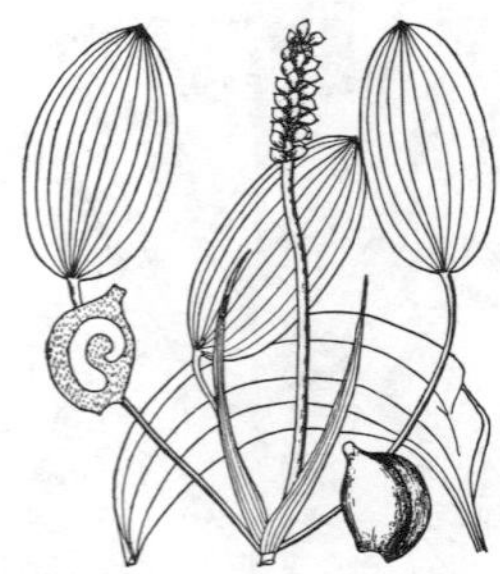

POTAMOGETON AMPLIFOLIUS

POTAMOGETON CONFERVOIDES

POTAMOGETON EPIHYDRUS

POTAMOGETON FILIFORMIS

POTAMOGETON FOLIOSUS

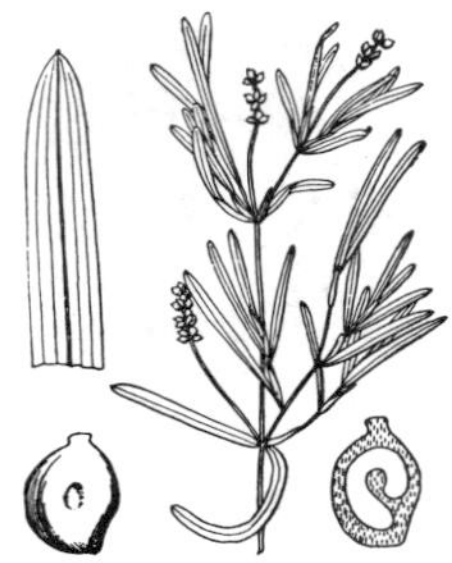

POTAMOGETON FRIESII

POTAMOGETON GRAMINEUS

POTAMOGETON NATANS

POTAMOGETON NODOSUS

POTAMOGETON OAKESIANUS

POTAMOGETON OBTUSIFOLIUS

POTAMOGETON PECTINATUS

POTAMOGETON PERFOLIATUS

POTAMOGETON PRAELONGUS

POTAMOGETON PUSILLUS

POTAMOGETON RICHARDSONII

POTAMOGETON ROBBINSII

POTAMOGETON SPIRILLUS

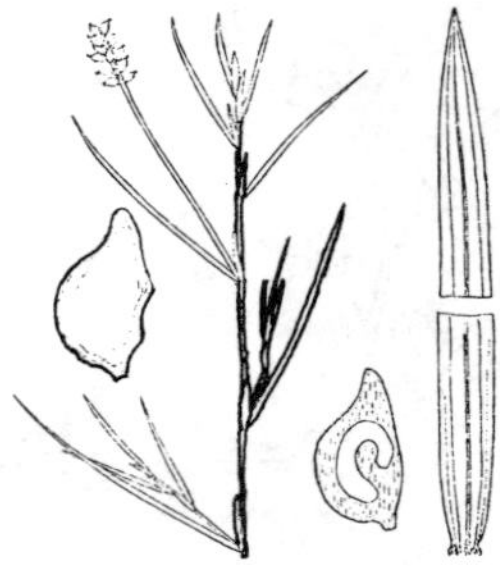

POTAMOGETON STRICTIFOLIUS

POTAMOGETON VASEYI

POTAMOGETON ZOSTERIFORMIS

RUPPIA MARITIMA

ZANICHELLIA PALUSTRIS

ZOSTERA MARINA

NAIAS FLEXILIS

SCHEUCHZERIA PALUSTRIS

TRIGLOCHIN GASPENSE

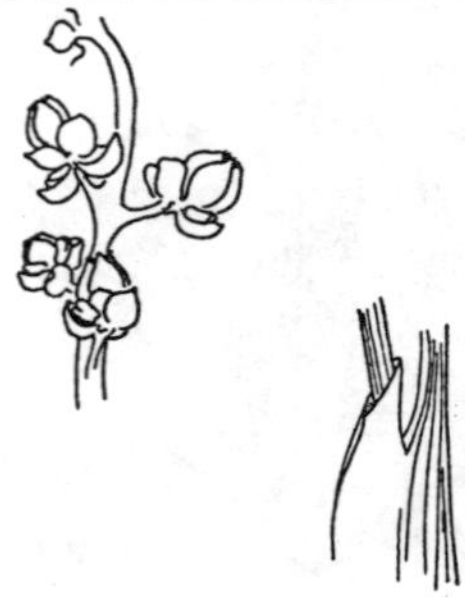

TRIGLOCHIN MARITIMA

TRIGLOCHIN PALUSTRIS

ALISMA PLANTAGO-AQUATICA

SAGITTARIA CALYCINA

SAGITTARIA CUNEATA

SAGITTARIA GRAMINEA

SAGITTARIA LATIFOLIA

ELODEA CANADENSIS

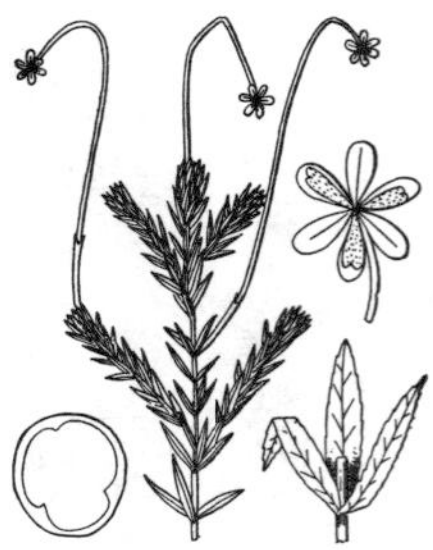

ELODEA NUTTALLII

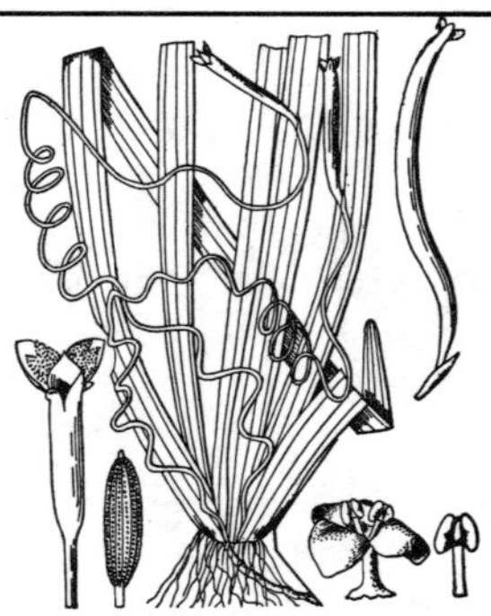

VALLISNERIA AMERICANA

AGROPYRON REPENS

AGROPYRON TRACHYCAULUM

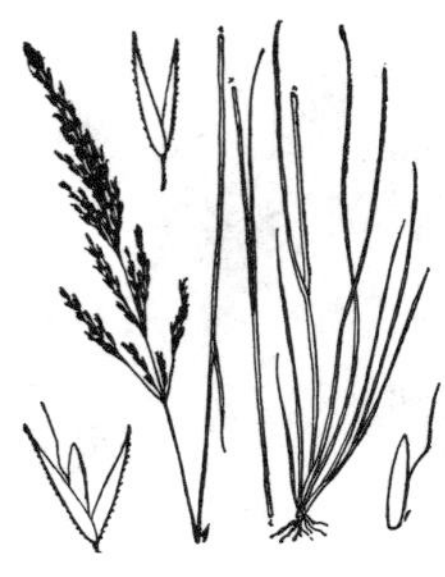

AGROSTIS CANINA

AGROSTIS CAPILLARIS

AGROSTIS MERTENSII

AGROSTIS PERENNANS

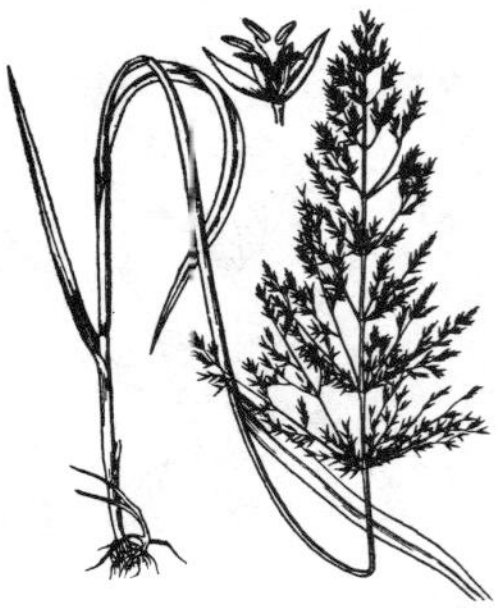

AGROSTIS STOLONIFERA

ALOPECURUS AEQUALIS

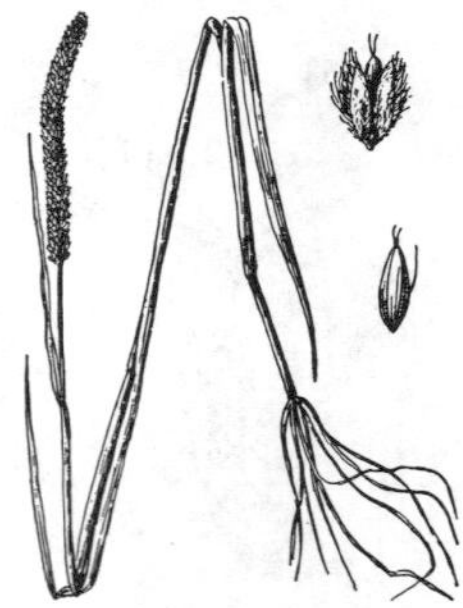

ALOPECURUS GENICULATUS

ALOPECURUS PRATENSIS

AMMOPHILA BREVILIGULATA

ANTHOXANTHUM ODORATUM

ARRHENATHERUM ELATIUS

AVENA FATUA

AVENA SATIVA

BRACHYELYTRUM ERECTUM

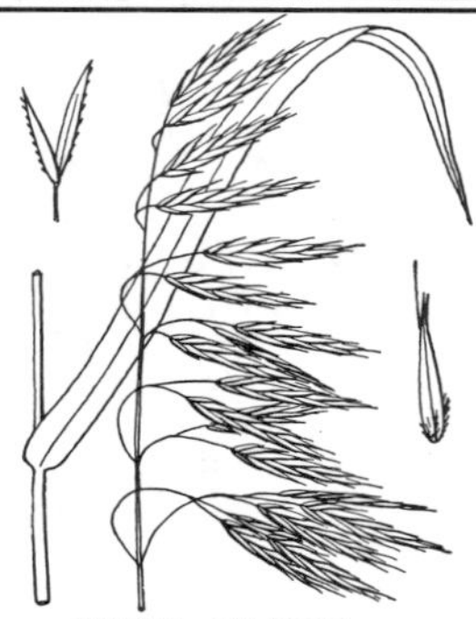

BROMUS CILIATUS

BROMUS HORDEACEUS

BROMUS INERMIS

BROMUS RACEMOSUS

BROMUS SECALINUS

BROMUS TECTORUM

CALAMAGROSTIS CANADENSIS

CALAMAGROSTIS PICKERINGII

CALAMAGROSTIS STRICTA

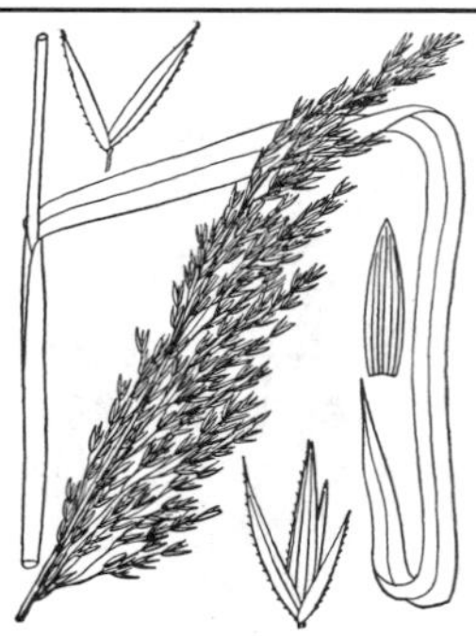

CINNA ARUNDINACEA

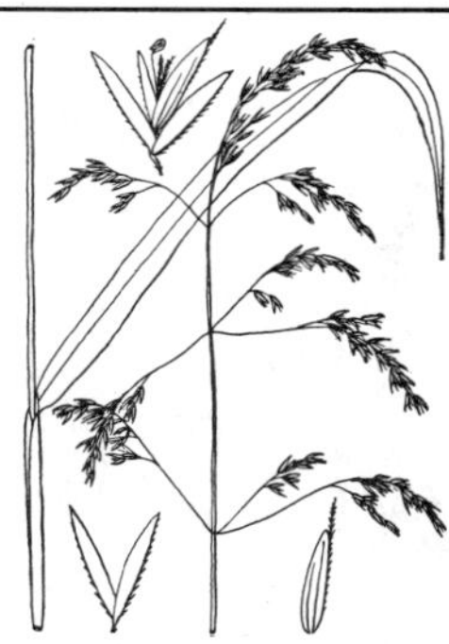

CINNA LATIFOLIA

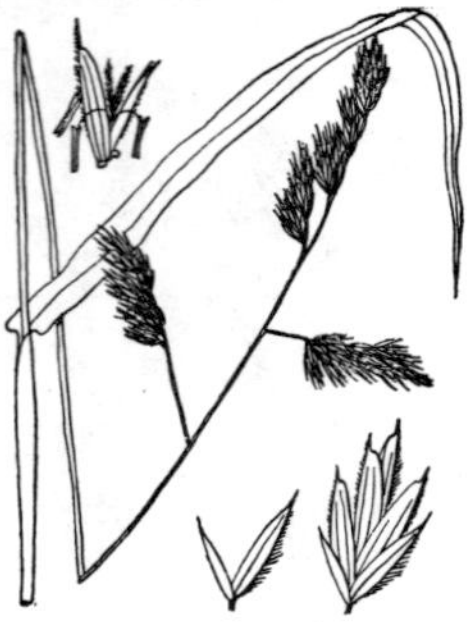

DACTYLIS GLOMERATA

DANTHONIA SPICATA

DESCHAMPSIA CAESPITOSA

DESCHAMPSIA FLEXUOSA

DICHANTHELIUM ACUMINATUM VAR. IMPLICATUM

DICHANTHELIUM ACUMINATUM VAR. LINDHEIMERI

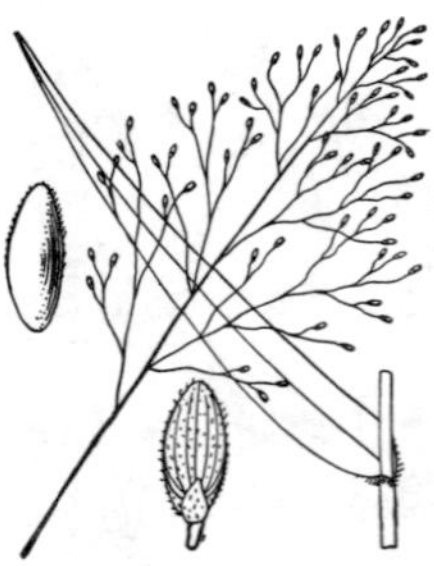

DICHANTHELIUM BOREALE

DICHANTHELIUM DEPAUPERATA

DICHANTHELIUM LINEARIFOLIUM

DICHANTHELIUM XANTHOPHYSUM

DIGITARIA ISCHAEMUM

DISTICHLIS SPICATA

ECHINOCHLOA CRUSGALLI

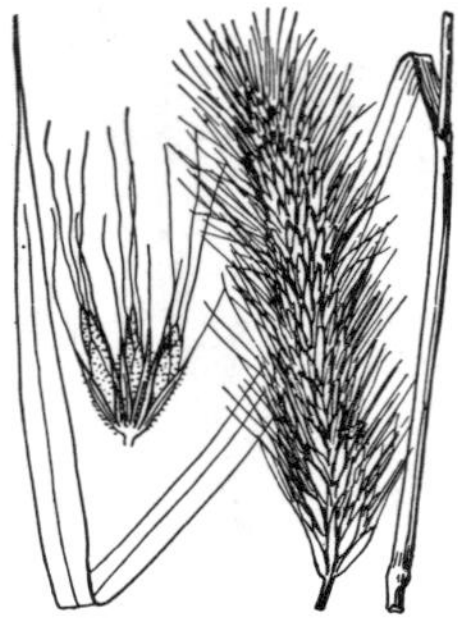

ELYMUS CANADENSIS

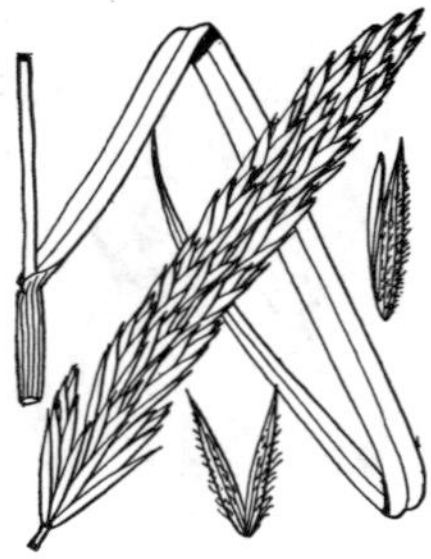

ELYMUS MOLLIS

ELYMUS VIRGINICUS

ELYMUS WIEGANDII

ERAGROSTIS FRANKII

ERAGROSTIS MINOR

ERAGROSTIS PECTINACEA

FESTUCA ARUNDINACEA

FESTUCA TENUIFOLIA

FESTUCA OBTUSA

FESTUCA PRATENSIS

FESTUCA RUBRA

GLYCERIA BOREALIS

GLYCERIA CANADENSIS

GLYCERIA GRANDIS

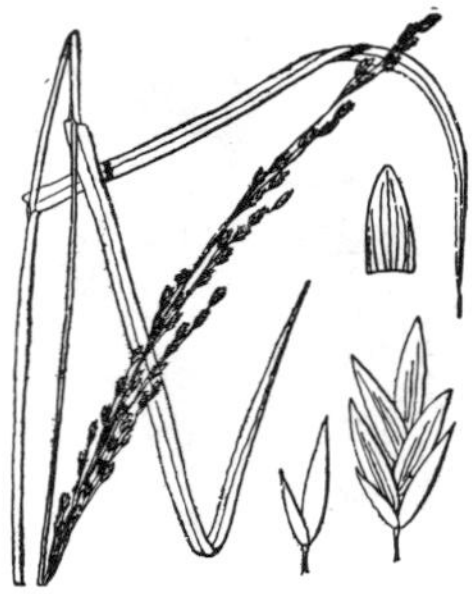

GLYCERIA MELICARIA

GLYCERIA OBTUSA

GLYCERIA STRIATA

HIEROCHLOE ODORATA

HOLCUS LANATA

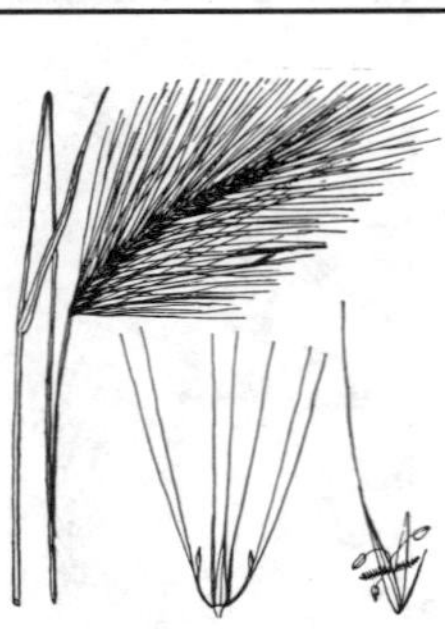

HORDEUM JUBATUM

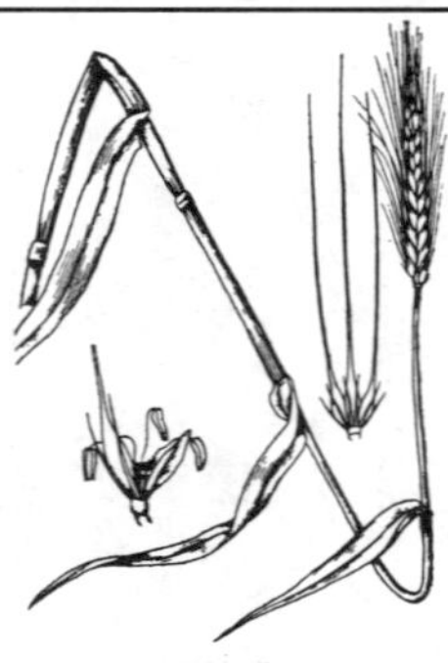

HORDEUM VULGARE

HYSTRIX PATULA

LEERSIA ORYZOIDES

LEERSIA VIRGINICA

LOLIUM PERENNE

LOLIUM TEMULENTUM

MILIUM EFFUSUM

MUHLENBERGIA FRONDOSA

MUHLENBERGIA GLOMERATA

MUHLENBERGIA MEXICANA

MUHLENBERGIA RICHARDSONIS

MUHLENBERGIA UNIFLORA

ORYZOPSIS ASPERFOLIA

ORYZOPSIS CANADENSIS

ORYZOPSIS PUNGENS

PANICUM CAPILLARE

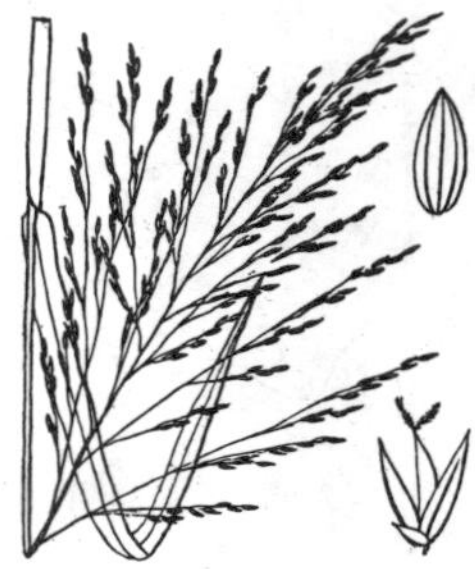

PANICUM DICHOTOMIFLORUM

PANICUM FLEXILE

PANICUM TUCKERMANII

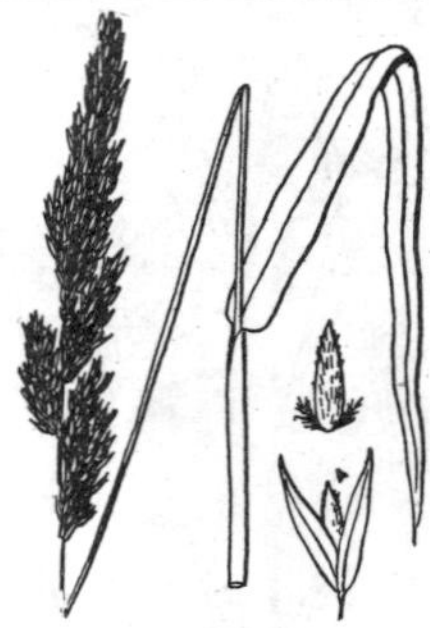

PHALARIS ARUNDINACEA

PHALARIS CANARIENSIS

PHALARIS MINOR

PHLEUM ALPINUM

PHLEUM PRATENSE

PHRAGMITES AUSTRALIS

POA ALSODES

POA ANNUA

POA COMPRESSA

POA GLAUCANTHA

POA NEMORALIS

POA PALUSTRIS

POA PRATENSIS

POA SALTUENSIS

POA TRIVIALIS

PUCCINELLIA AMBIGUA

PUCCINELLIA AMERICANA

PUCCINELLIA DISTANS

PUCCINELLIA FASCICULATA

PUCCINELLIA LAURENTIANA

SCHIZACHNE PURPURASCENS

SCHIZACHYRIUM SCOPARIUM

SECALE CEREALE

SETARIA GLAUCA

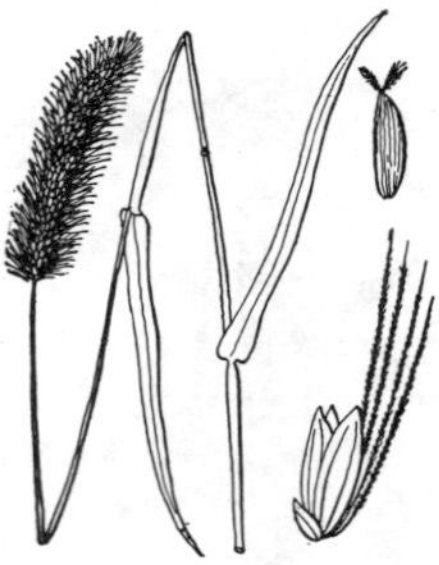

SETARIA VIRIDIS

SPARTINA ALTERNIFLORA

SPARTINA PATENS

SPARTINA PECTINATA

SPHENOPHOLIS OBTUSATA

SPOROBOLUS ASPER

SPOROBOLUS VAGINIFLORUS

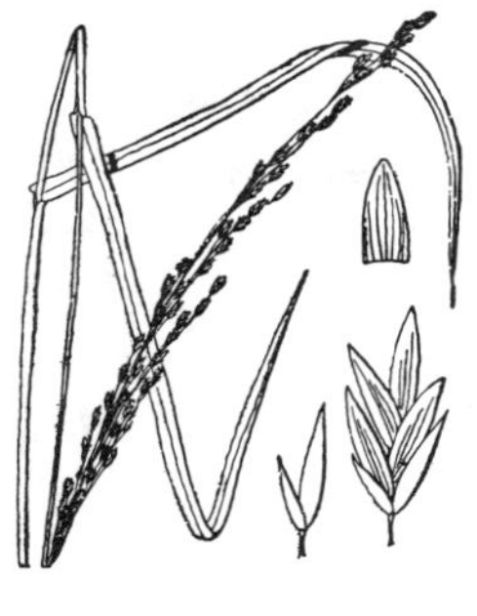

TORREYOCHLOA PALLIDA

TRISETUM MELICOIDES

TRISETUM SPICATUM

TRITICUM AESTIVUM

ZIZANIA PALUSTRIS

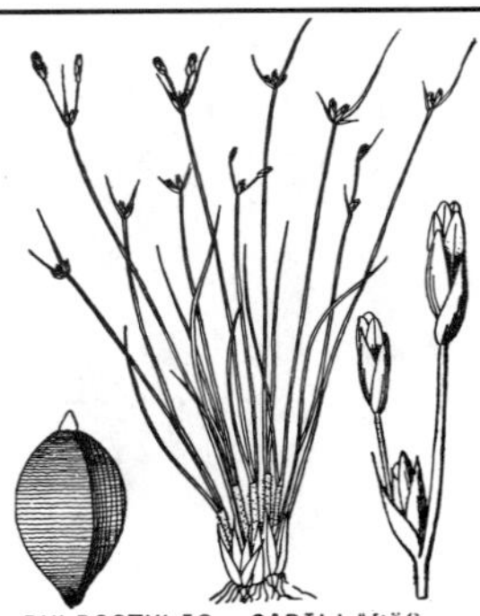

BULBOSTYLIS CAPILLARIS

CAREX ADUSTA

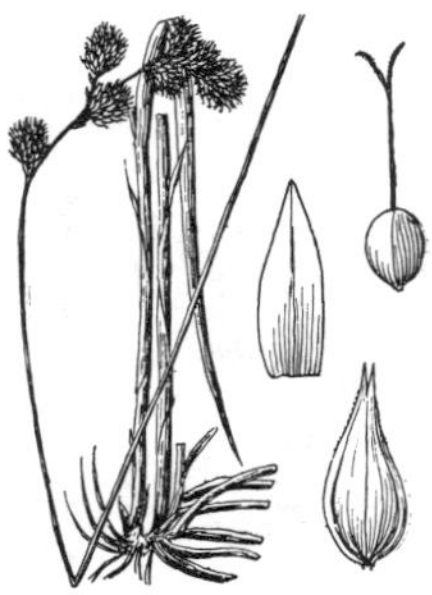

CAREX AENEA

CAREX AMPHIBOLA

CAREX AQUATILIS

CAREX ARCTA

CAREX ARCTATA

CAREX ATLANTICA

CAREX ATRATIFORMIS

CAREX AUREA

CAREX BACKII

CAREX BEBBII

CAREX BIGELOWII

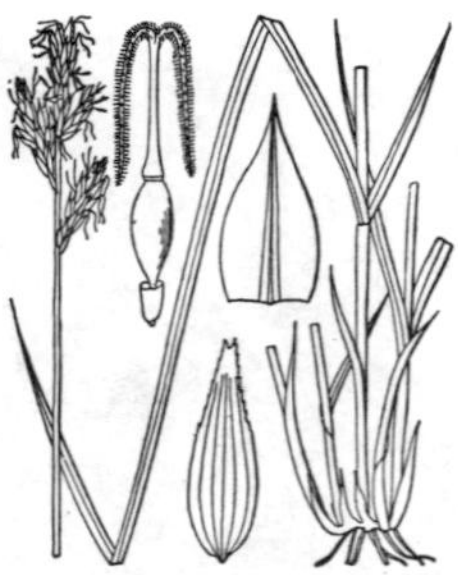

CAREX BROMOIDES

CAREX BRUNNESCENS

CAREX BUXBAUMII

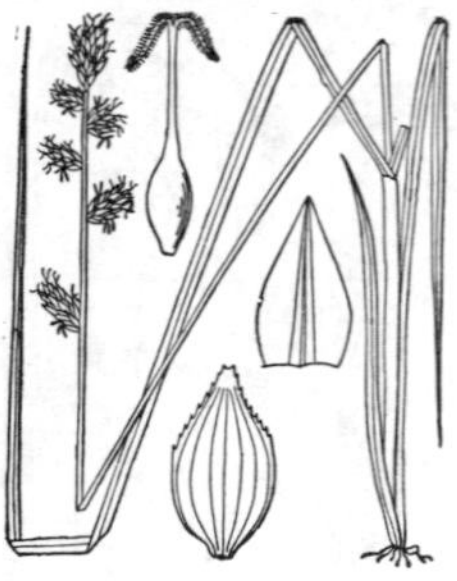

CAREX CANESCENS

CAREX CAPILLARIS

CAREX CARYOPHYLLEA

CAREX CASTANEA

CAREX CHORDORRHIZA

CAREX COMMUNIS

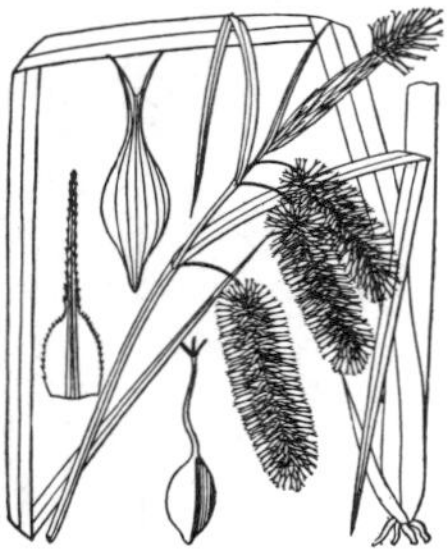

CAREX COMOSA

CAREX CONCINNA

CAREX CONOIDEA

CAREX CRAWEI

CAREX CRAWFORDII

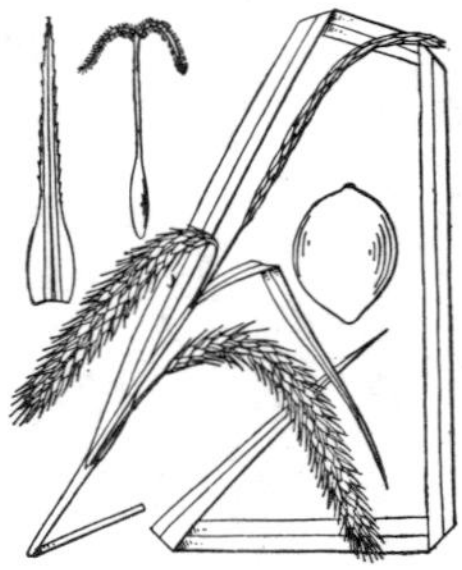

CAREX CRINITA

CAREX CRYPTOLEPIS

CAREX CUMULATA

CAREX DEBILIS

CAREX DEFLEXA

CAREX DEWEYANA

CAREX DIANDRA

CAREX DISPERMA

CAREX EBURNEA

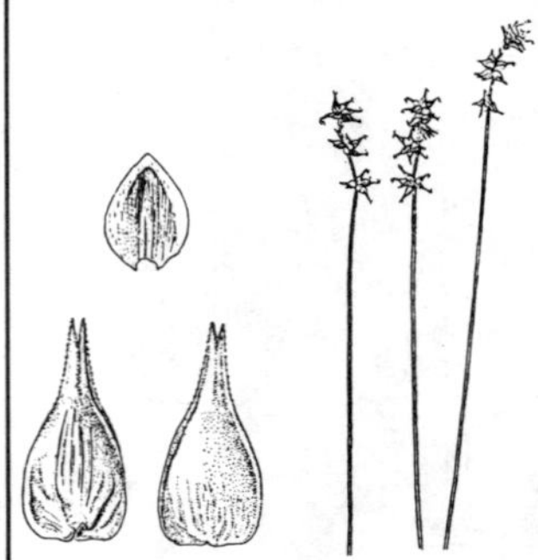

CAREX ECHINATA

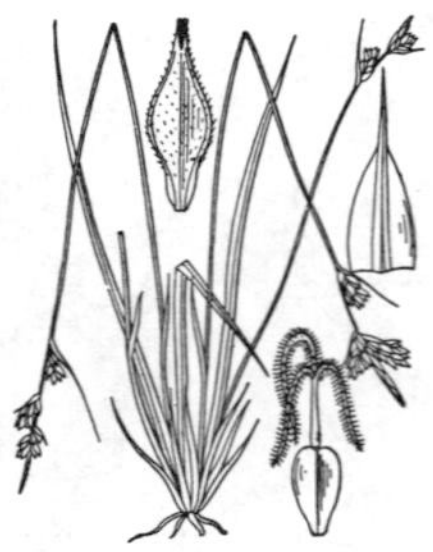

CAREX EMMONSII

CAREX EXILIS

CAREX FLACCIDULA

CAREX FLAVA

CAREX FOLLICULATA

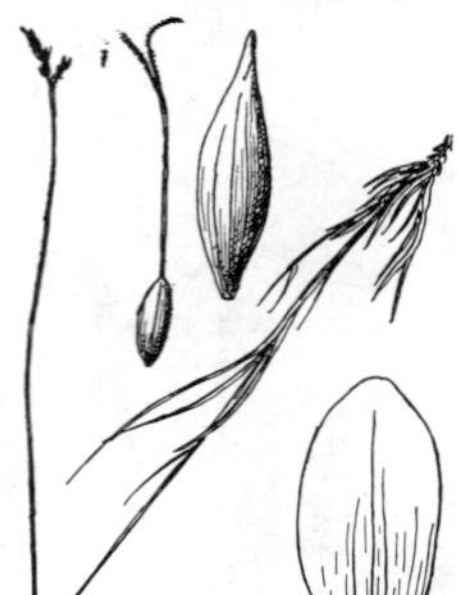

CAREX GLAREOSA

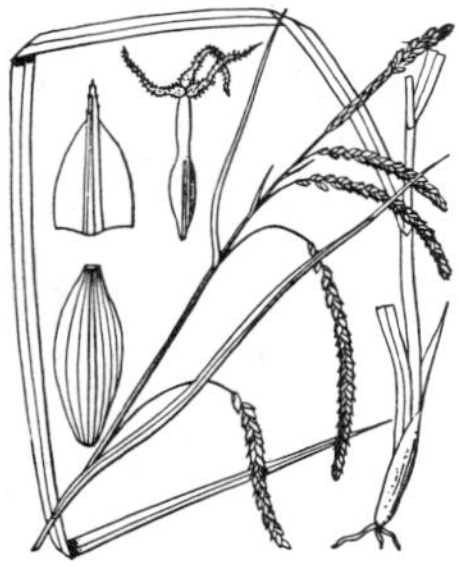

CAREX GRACILLIMA

CAREX GRANULARIS

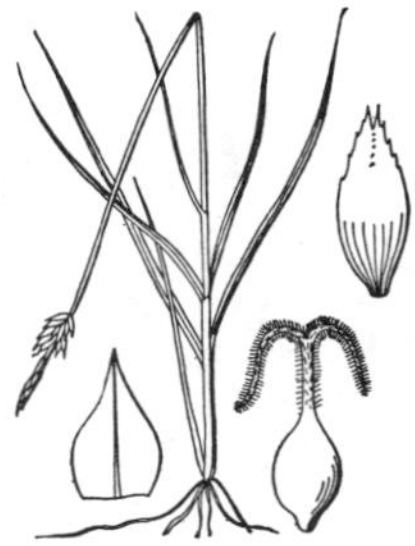

CAREX GYNOCRATES

CAREX HAYDENII

CAREX HIRTIFOLIA

CAREX HORMATHODES

CAREX HOUGHTONII

CAREX HOWEI

CAREX HYSTRICINA

CAREX INTERIOR

CAREX INTUMESCENS

CAREX LACUSTRIS

CAREX LANUGINOSA

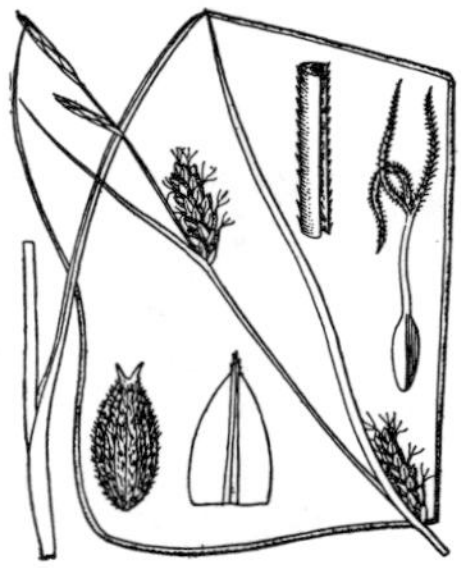

CAREX LASIOCARPA

CAREX LENTICULARIS

CAREX LEPIDOCARPA

CAREX LEPORINA

CAREX LEPTALEA

CAREX LEPTONERVIA

CAREX LIMOSA

CAREX LIVIDA

CAREX LUPULINA

CAREX LURIDA

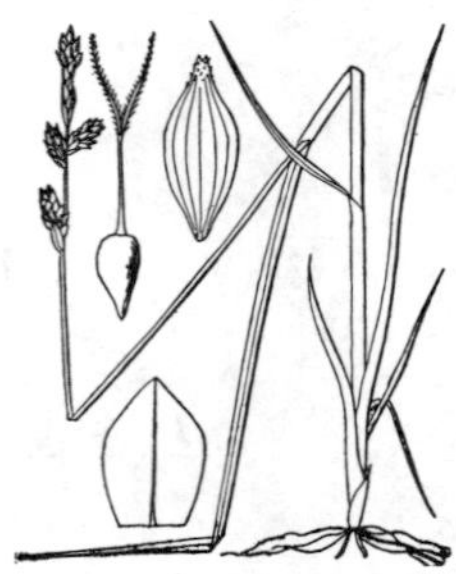

CAREX MACKENZIEI

CAREX MEDIA

CAREX MICHAUXIANA

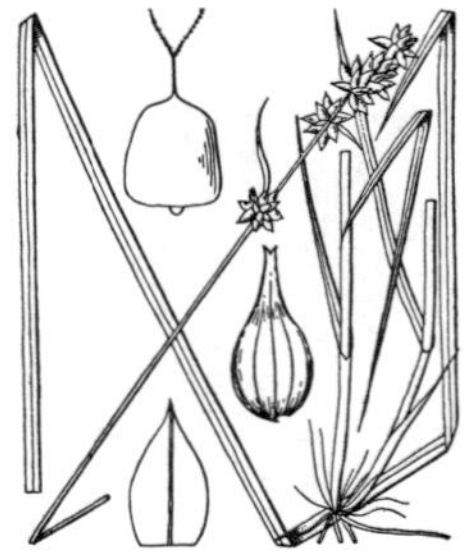

CAREX MURICATA

CAREX NIGRA

CAREX NORMALIS

CAREX NOVAE-ANGLIAE

CAREX OLIGOSPERMA

CAREX ORMOSTACHYA

CAREX PAIRAEI

CAREX PALEACEA

CAREX PALLESCENS

CAREX PANICEA

CAREX PAUCIFLORA

CAREX PAUPERCULA

CAREX PECKII

CAREX PEDUNCULATA

CAREX PENSYLVANICA

CAREX PLANTAGINEA

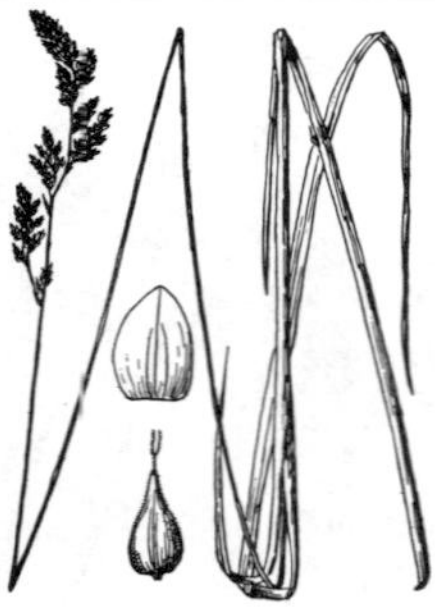

CAREX PRAIREA

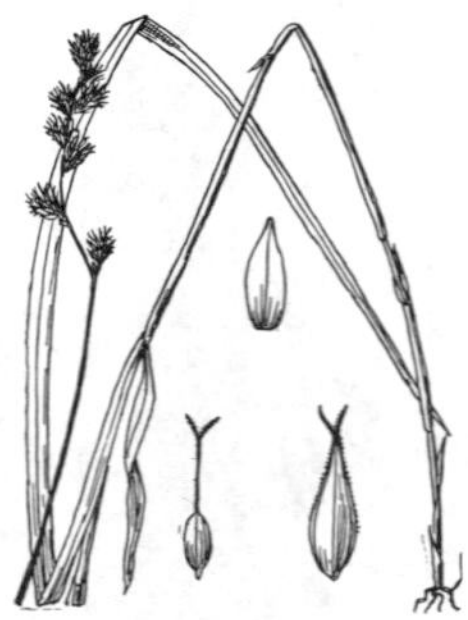

CAREX PROJECTA

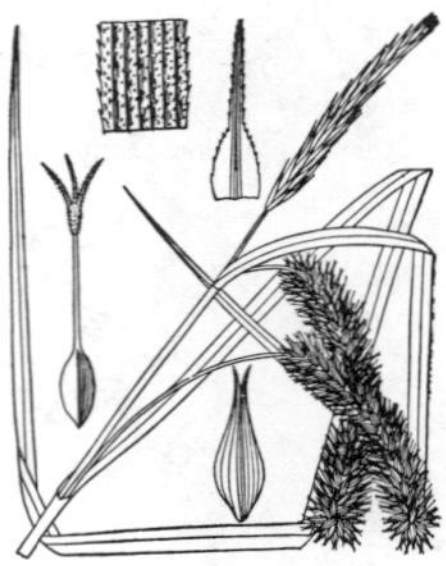

CAREX PSEUDO-CYPERUS

CAREX RARIFLORA

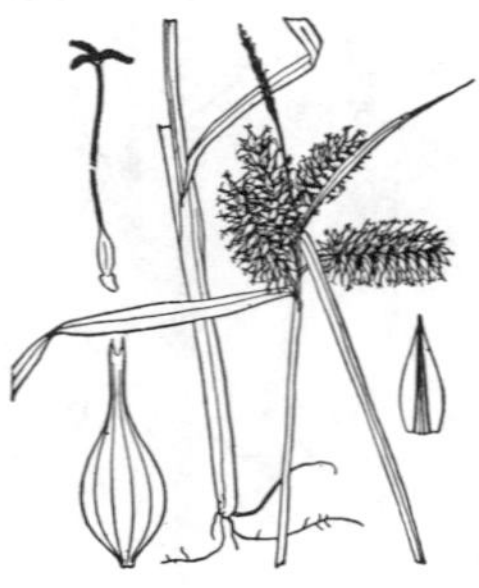

CAREX RETRORSA

CAREX ROSTRATA

CAREX SALINA

CAREX SAXATILIS

CAREX SCABRATA

CAREX SCOPARIA

CAREX SILICEA

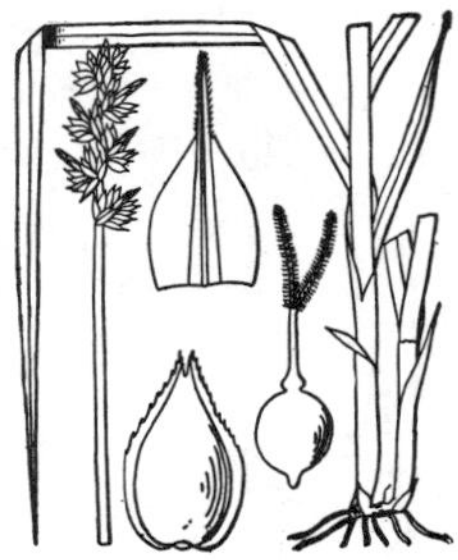

CAREX SPARGANIOIDES

CAREX SPRENGELII

CAREX STIPATA

CAREX STRICTA

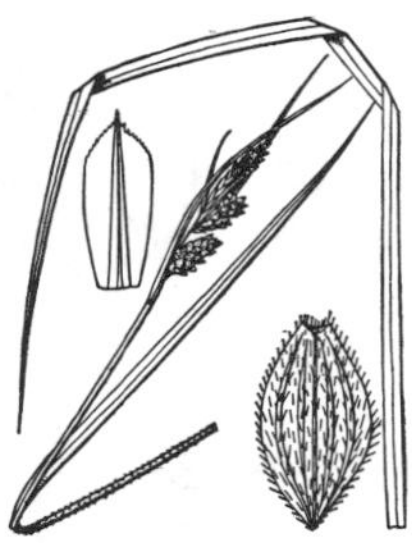

CAREX SWANII

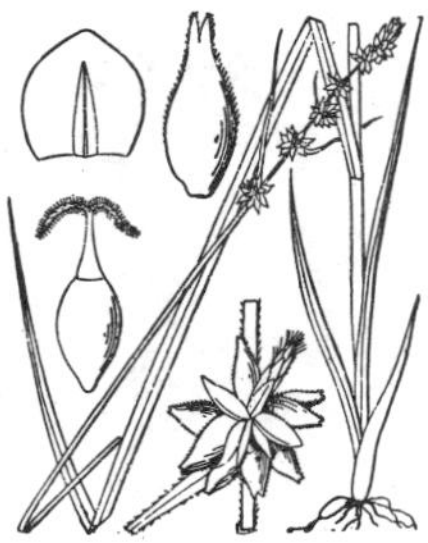

CAREX SYLVICOLA

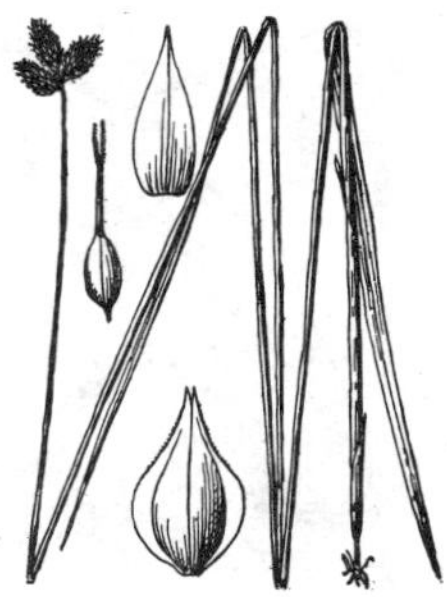

CAREX TENERA

CAREX TENUIFLORA

CAREX TINCTA

CAREX TONSA

CAREX TORTA

CAREX TRIBULOIDES

CAREX TRISPERMA

CAREX TUCKERMANII

CAREX UMBELLATA

CAREX VAGINATA

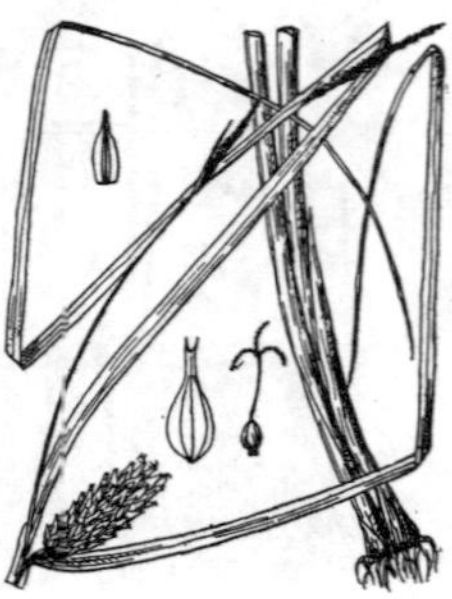

CAREX VESICARIA

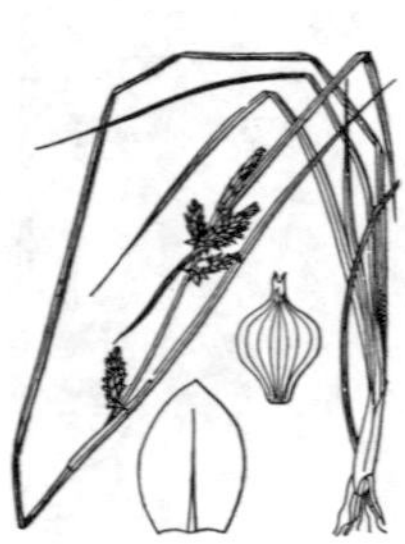

CAREX VIRIDULA

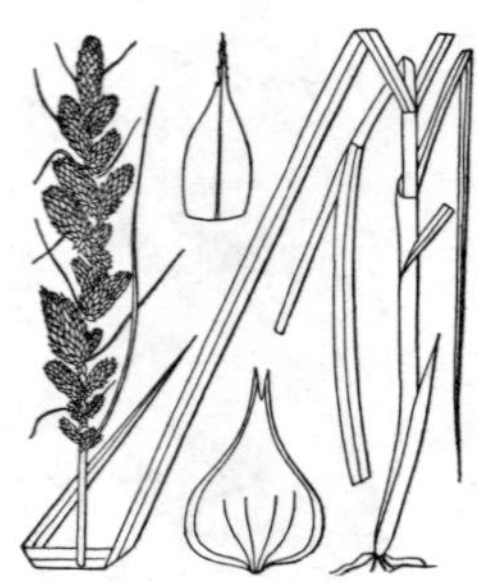

CAREX VULPINOIDEA

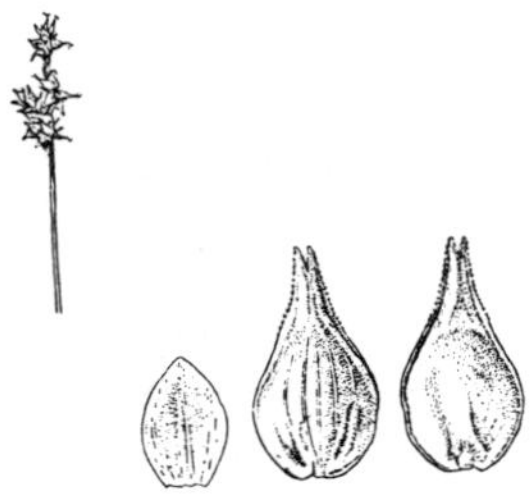

CAREX WIEGANDII

CLADIUM MARISCOIDES

CYPERUS ARISTATUS

CYPERUS DENTATUS

CYPERUS DIANDRUS

CYPERUS ESCULENTUS

CYPERUS FILICULMIS

CYPERUS RIVULARIS

DULICHIUM ARUNDINACEUM

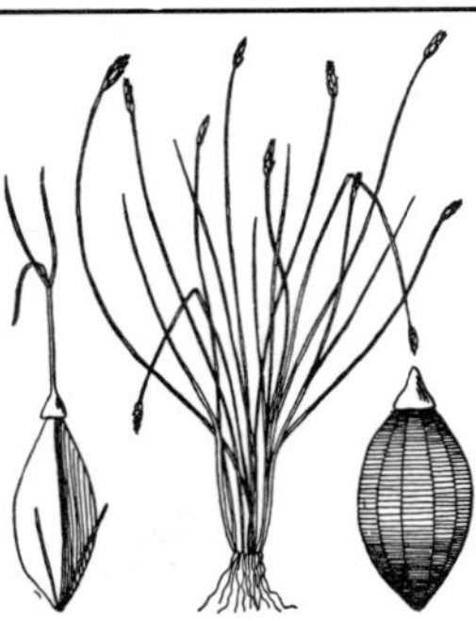

ELEOCHARIS ACICULARIS

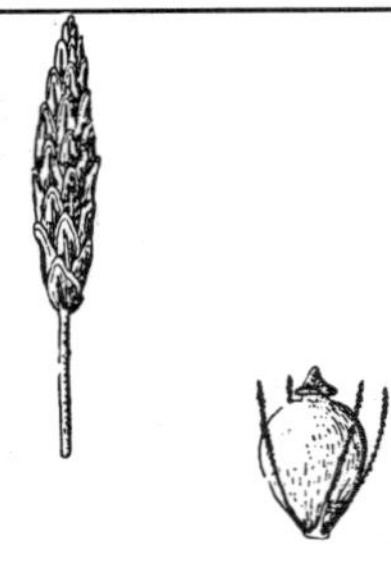

ELEOCHARIS ERYTHROPODA

ELEOCHARIS INTERMEDIA

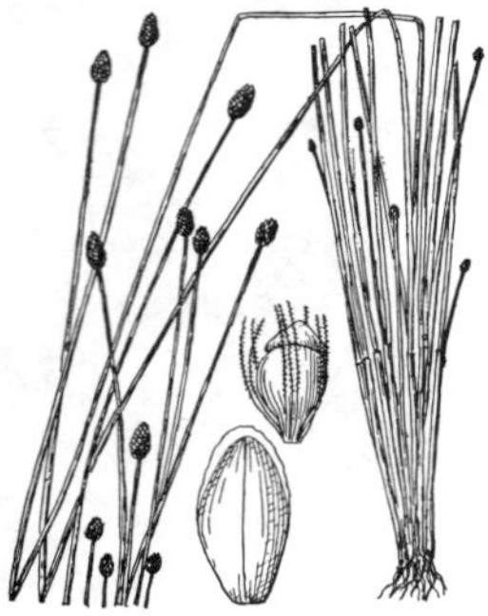

ELEOCHARIS OBTUSA

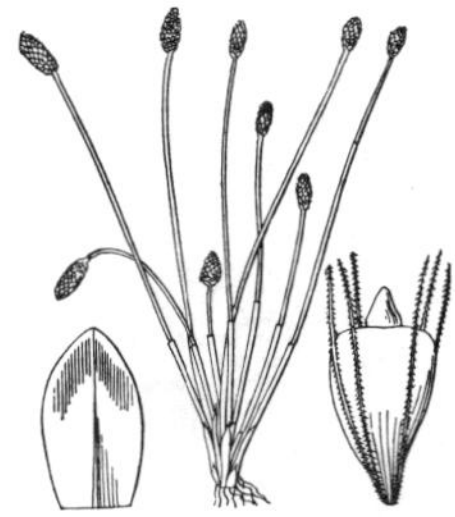

ELEOCHARIS OVATA

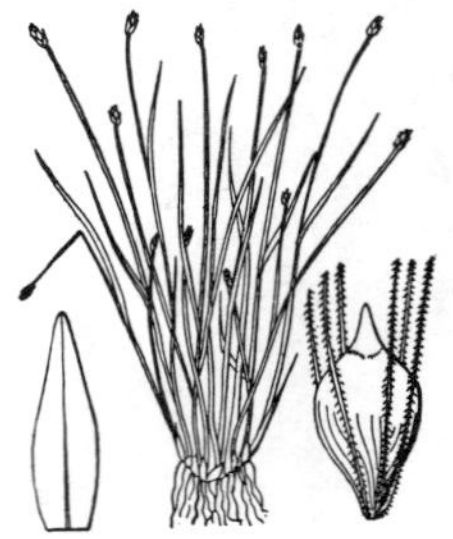

ELEOCHARIS PARVULA

ELEOCHARIS PAUCIFLORA

ELEOCHARIS ROBBINSII

ELEOCHARIS SMALLII

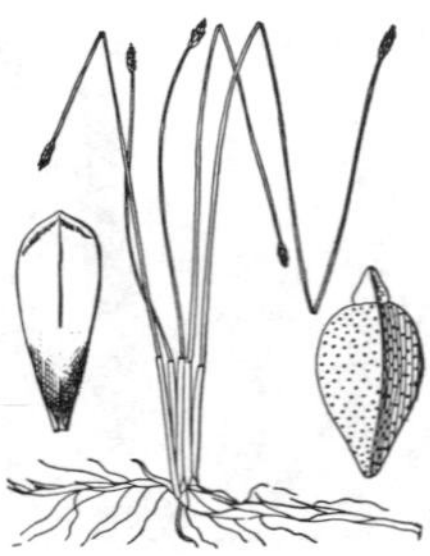

ELEOCHARIS TENUIS

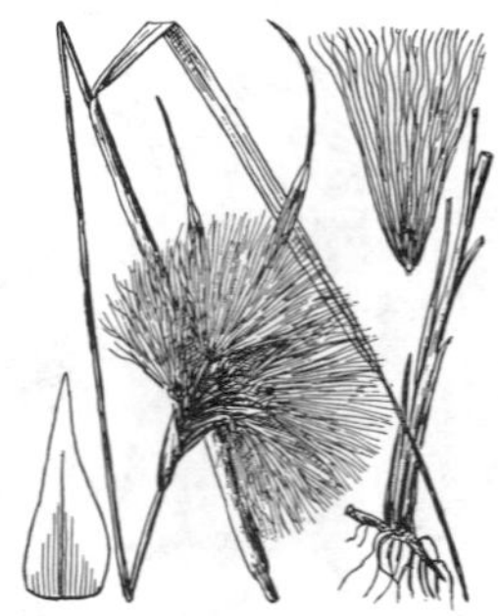

ERIOPHORUM ANGUSTIFOLIUM

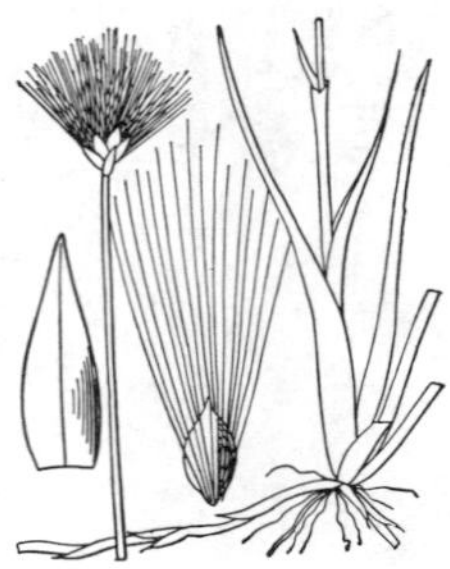

ERIOPHORUM CHAMISSONIS

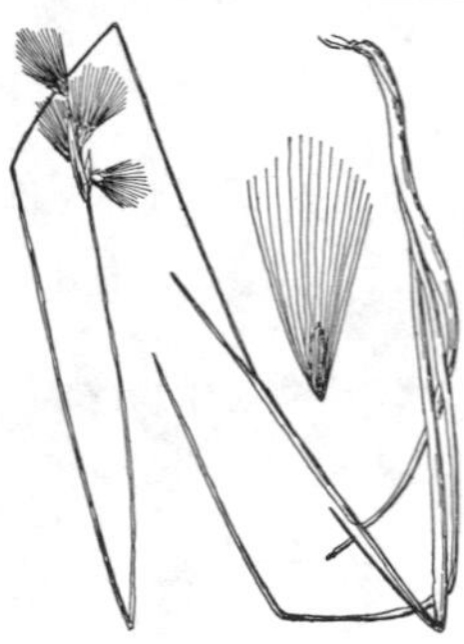

ERIOPHORUM GRACILE

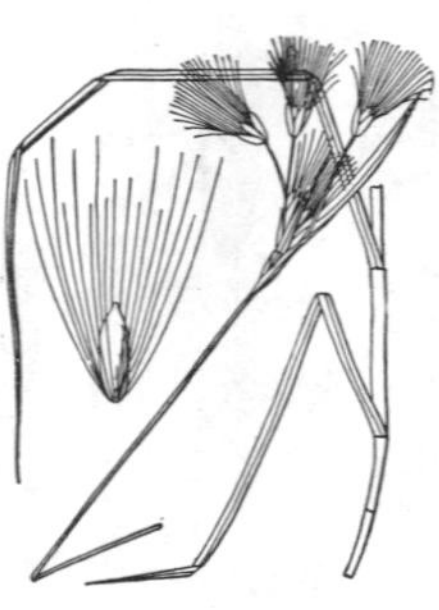

ERIOPHORUM TENELLUM

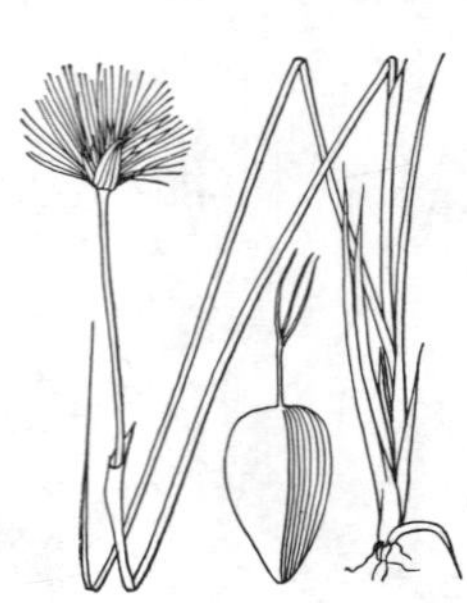

ERIOPHORUM VAGINATUM

ERIOPHORUM VIRGINICUM

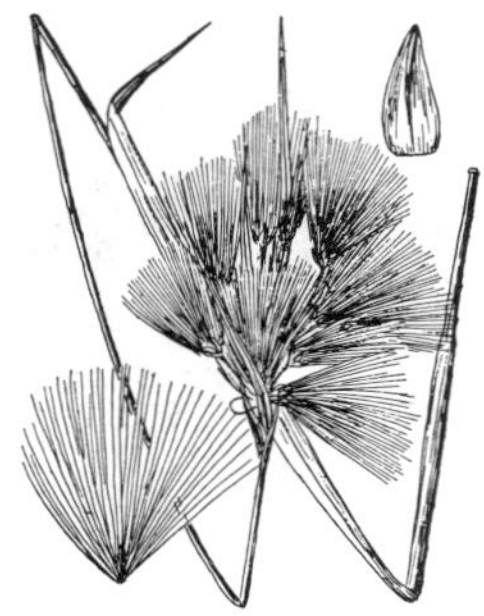

ERIOPHORUM VIRIDICARINATUM

RHYNCHOSPORA ALBA

RHYNCHOSPORA CAPILLACEA

RHYNCHOSPORA CAPITELLATA

RHYNCHOSPORA FUSCA

SCIRPUS ACUTUS

SCIRPUS AMERICANUS

SCIRPUS ATROCINCTUS

SCIRPUS ATROVIRENS

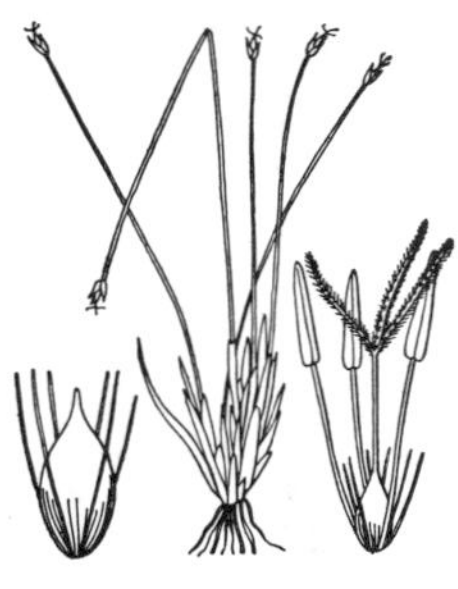

SCIRPUS CAESPITOSUS

SCIRPUS CLINTONII

SCIRPUS CYPERINUS

SCIRPUS FLUVIATILIS

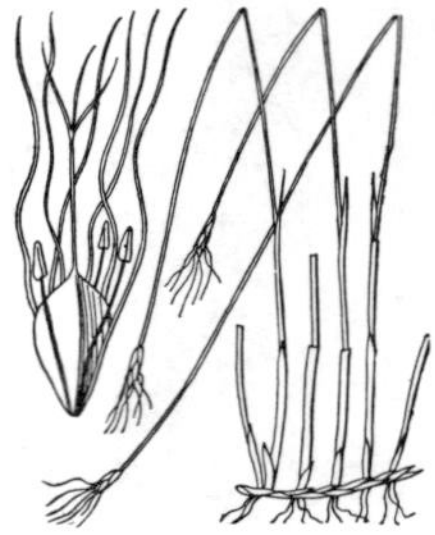

SCIRPUS HUDSONIANUS

SCIRPUS LINEATUS

SCIRPUS MARITIMUS

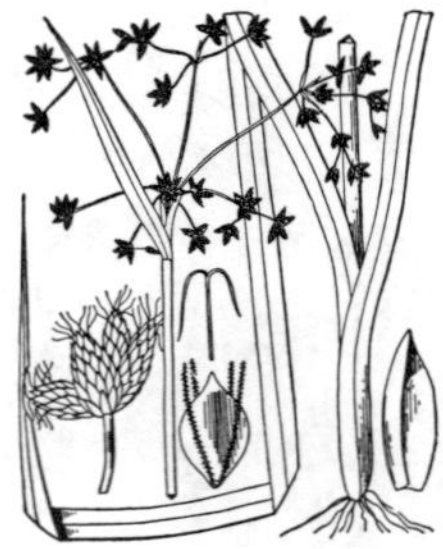

SCIRPUS MICROCARPUS

SCIRPUS PEDICELLATUS

SCIRPUS RUFUS

SCIRPUS SMITHII

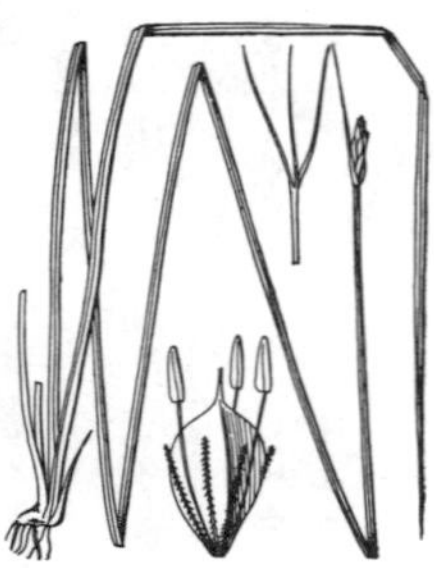

SCIRPUS SUBTERMINALIS

SCIRPUS VALIDUS

BUTOMUS UMBELLATUS

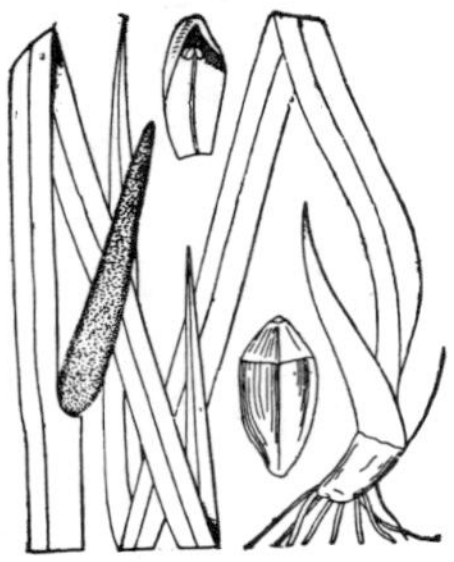

ACORUS AMERICANUS

ARISAEMA TRIPHYLLUM

CALLA PALUSTRIS

SYMPLOCARPUS FOETIDUS

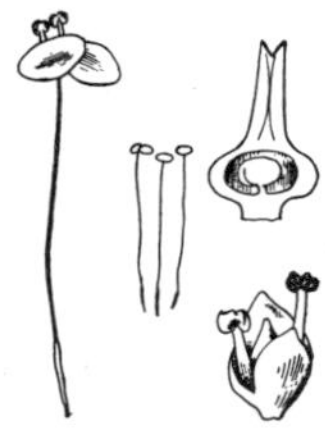

LEMNA MINOR

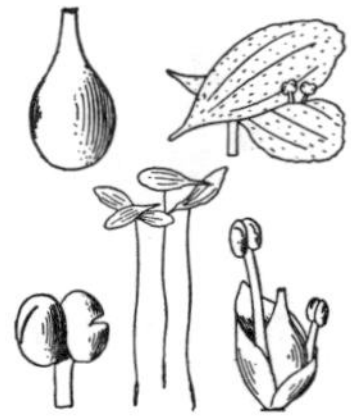

LEMNA TRISULCA

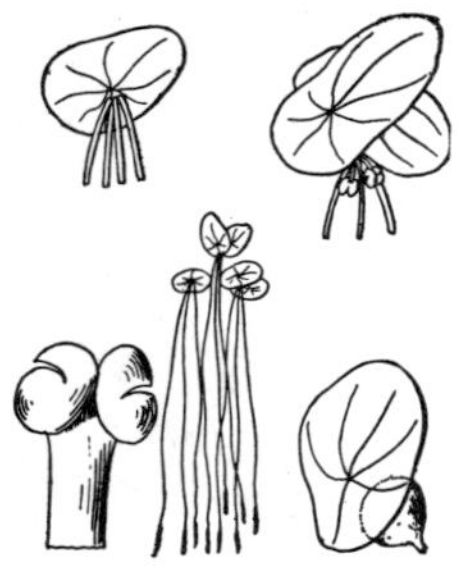

SPIRODELA POLYRHIZA

XYRIS MONTANA

ERIOCAULON PARKERI

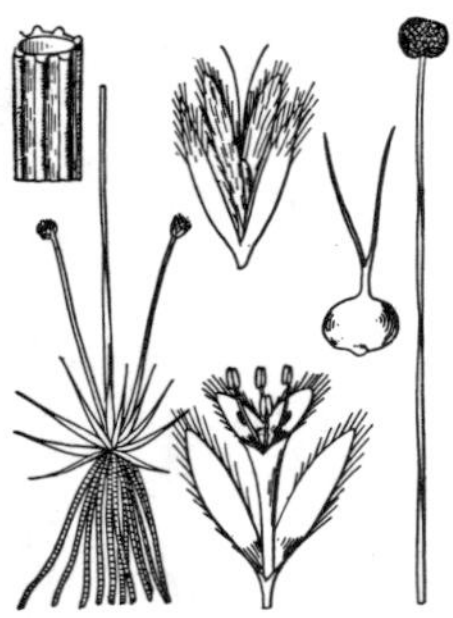

ERIOCAULON SEPTANGULARE

HETERANTHERA DUBIA

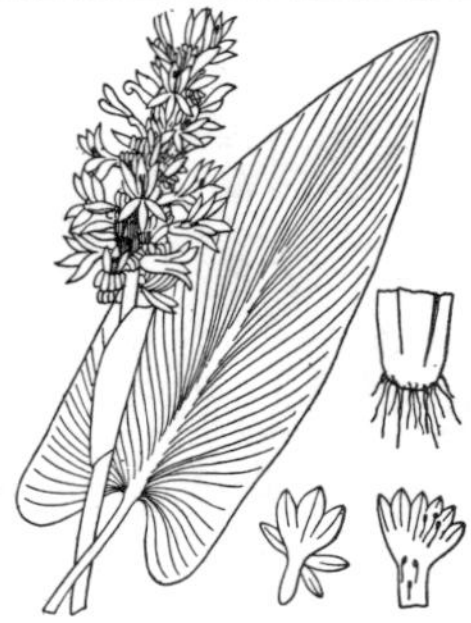

PONTEDERIA CORDATA

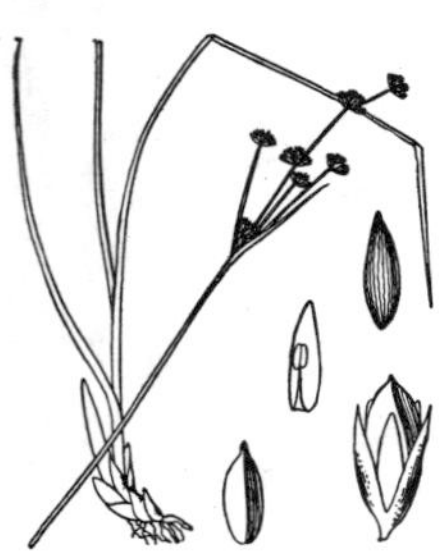

JUNCUS ALPINUS

JUNCUS ARTICULATUS

JUNCUS BALTICUS

JUNCUS BRACHYCEPHALUS

JUNCUS BREVICAUDATUS

JUNCUS BUFONIUS

JUNCUS CANADENSIS

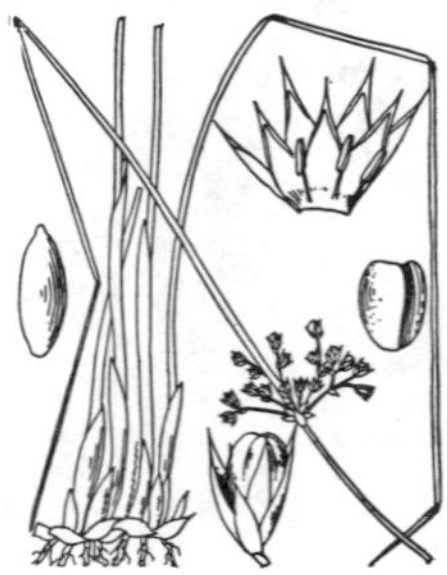

JUNCUS EFFUSUS

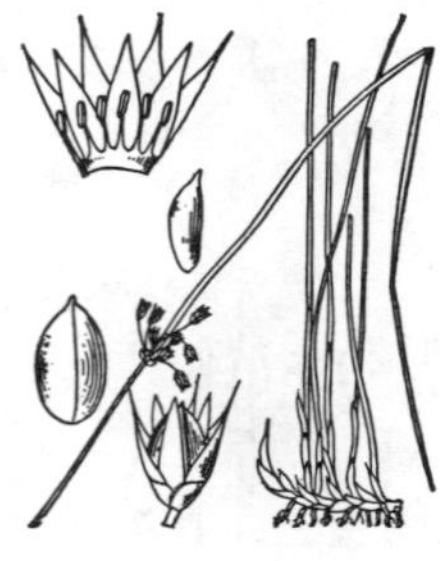

JUNCUS FILIFORMIS

JUNCUS GERARDII

JUNCUS GREENEI

JUNCUS INTERIOR

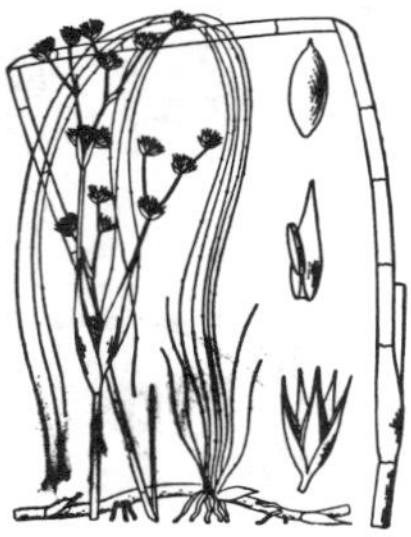

JUNCUS MILITARIS

JUNCUS PELOCARPUS

JUNCUS STYGIUS

JUNCUS SUBTILIS

JUNCUS TENUIS

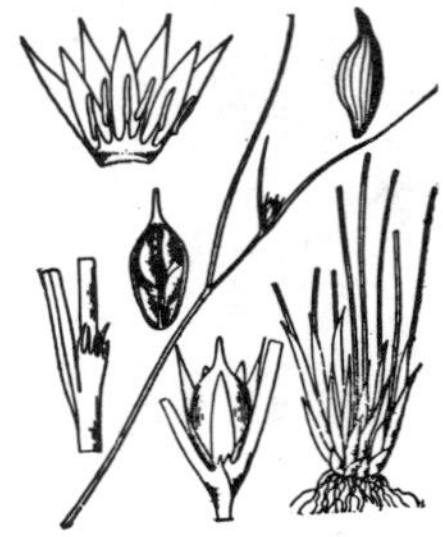

JUNCUS TRIFIDUS

JUNCUS VASEYI

LUZULA ACUMINATA

LUZULA PALLESCENS

LUZULA MULTIFLORA

LUZULA PARVIFLORA

ALLIUM CANADENSE

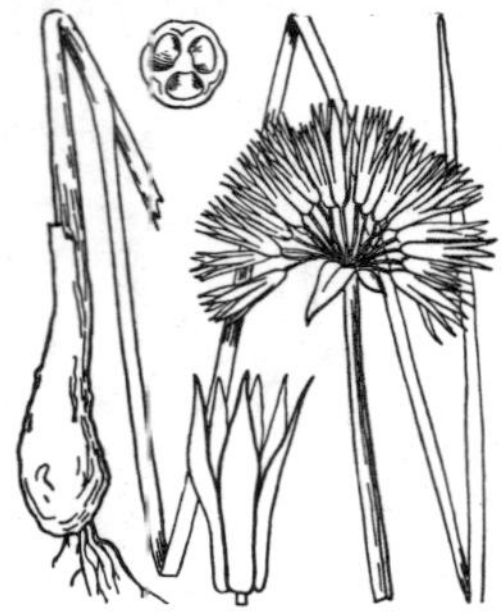

ALLIUM SCHOENOPRASUM

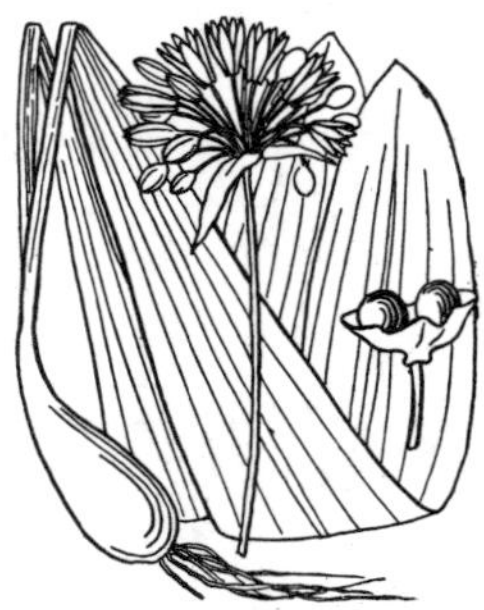

ALLIUM TRICOCCUM

ASPARAGUS OFFICINALIS

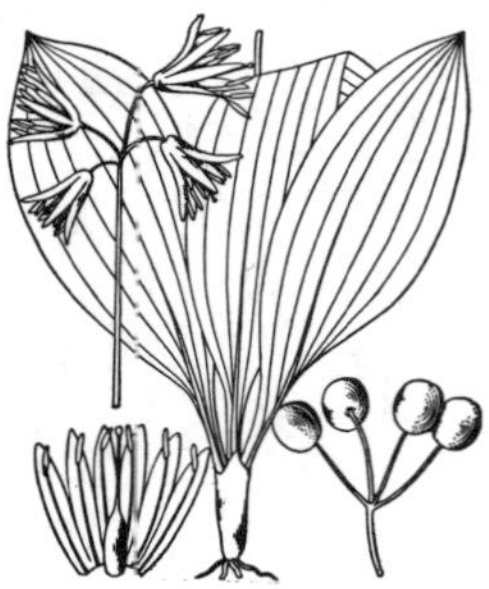

CLINTONIA BOREALIS

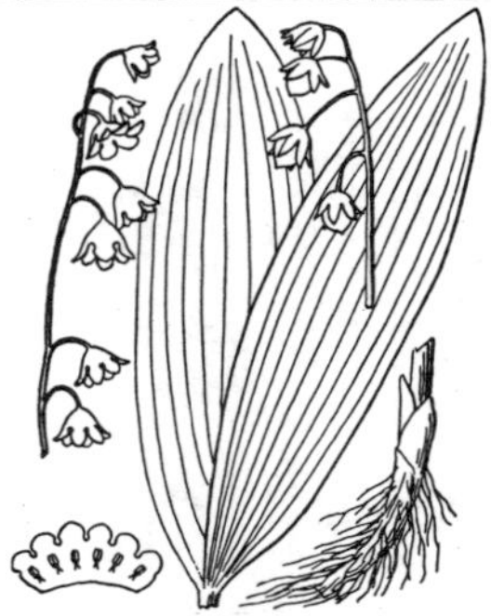

CONVALLARIA MAJALIS

ERYTHRONIUM AMERICANUM

HEMEROCALLIS FULVA

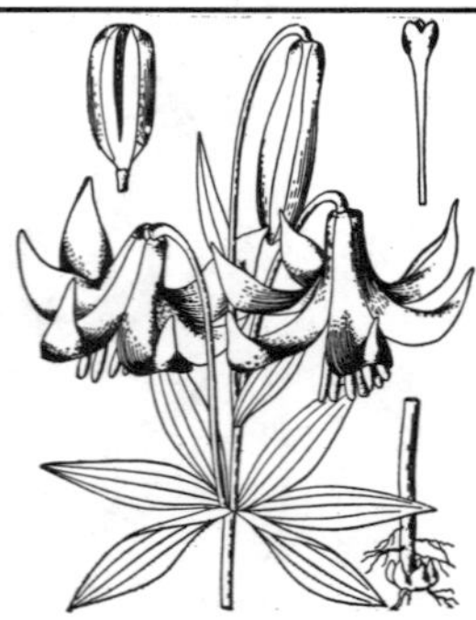

LILIUM CANADENSE

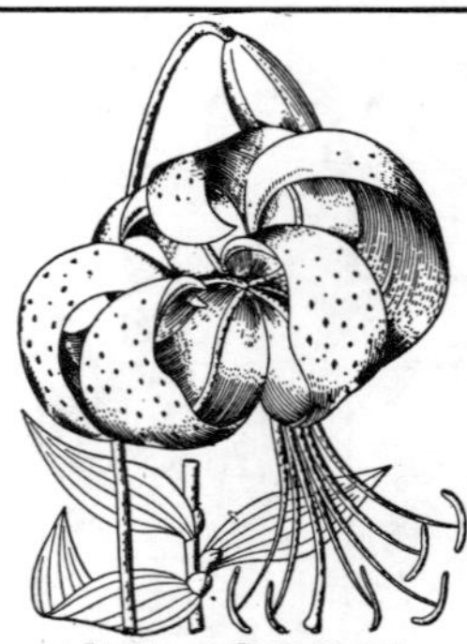

LILIUM LANCIFOLIUM

MAIANTHEMUM CANADENSE

MEDEOLA VIRGINIANA

ORNITHOGALUM UMBELLATUM

POLYGONATUM PUBESCENS

SMILACINA RACEMOSA

SMILACINA STELLATA

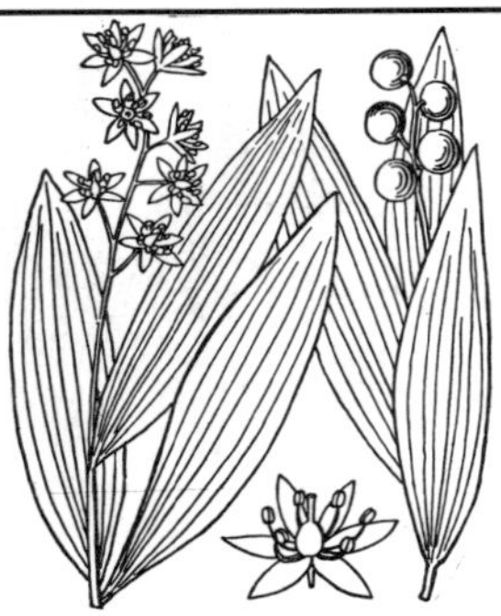

SMILACINA TRIFOLIA

SMILAX HERBACEA

STREPTOPUS AMPLEXIFOLIUS

STREPTOPUS ROSEUS

TOFIELDIA GLUTINOSA

TRILLIUM CERNUUM

TRILLIUM ERECTUM

TRILLIUM UNDULATUM

UVULARIA SESSILIFOLIA

LILACEAE	IRIDACEAE	ORCHIDACEAE
 VERATRUM VIRIDE	 ZIGADENUS ELEGANS	 IRIS GERMANICA
 IRIS PSEUDACORUS	 IRIS SETOSA	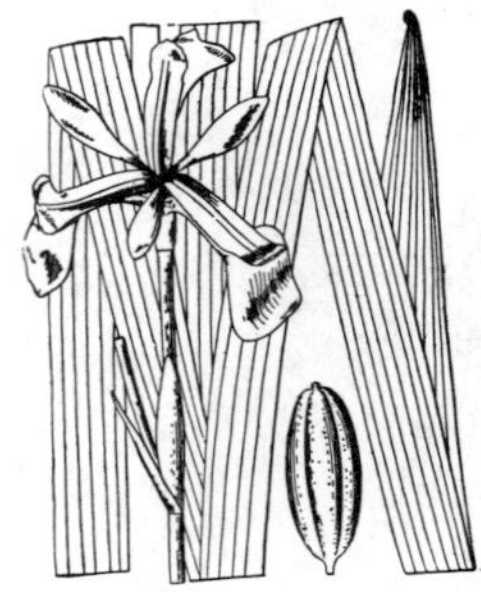 IRIS VERSICOLOR
 SISYRICHIUM MONTANUM	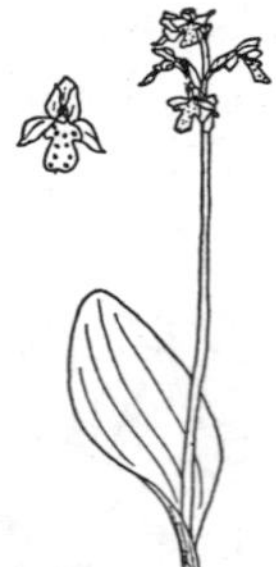 AMERORCHIS ROTUNDIFOLIA	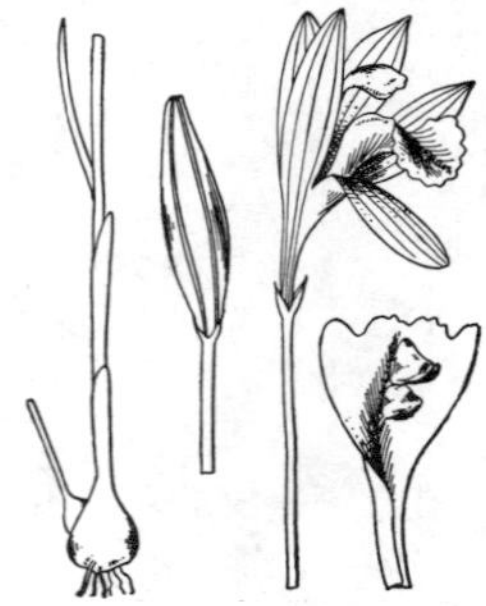 ARETHUSA BULBOSA
 CALOPOGON TUBEROSUS	 CALYPSO BULBOSA	 COELOGLOSSUM VIRIDIS

CORALLORHIZA TRIFIDA

CORALLORHIZA MACULATA

CYPRIPEDIUM ACAULE

CYPRIPEDIUM CALCEOLUS

CYPRIPEDIUM REGINAE

EPIPACTIS HELLEBORINE

GALEARIS SPECTABILIS

GOODYERA OBLONGIFOLIA

GOODYERA PUBESCENS

GOODYERA REPENS

GOODYERA TESSELATA

LIPARIS LOESELII

LISTERA AURICULATA

LISTERA AUSTRALIS

LISTERA CONVALLARIOIDES

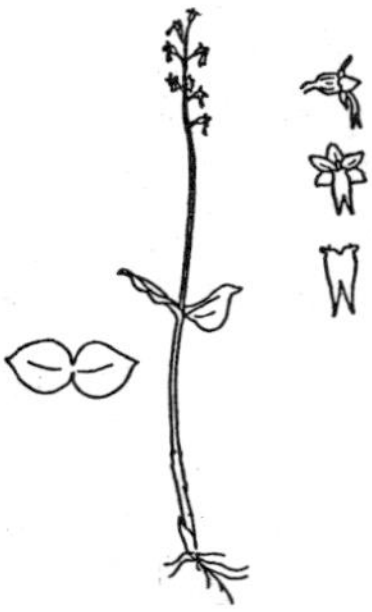

LISTERA CORDATA

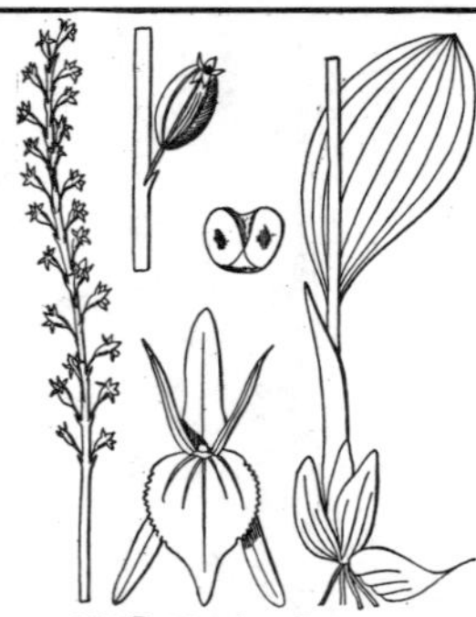

MALAXIS MONOPHYLLOS

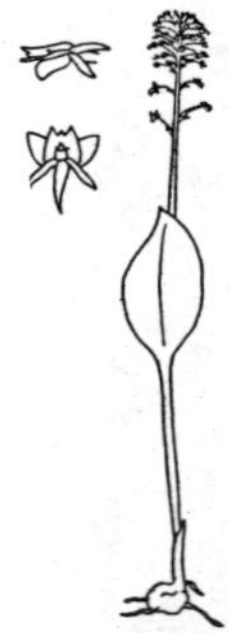

MALAXIS UNIFOLIA

PLATANTHERA BLEPHARIGLOTTIS

PLATANTHERA CLAVELLATA

PLATANTHERA DILATATA

PLATANTHERA FLAVA

PLATANTHERA HOOKERI

PLATANTHERA HYPERBOREA

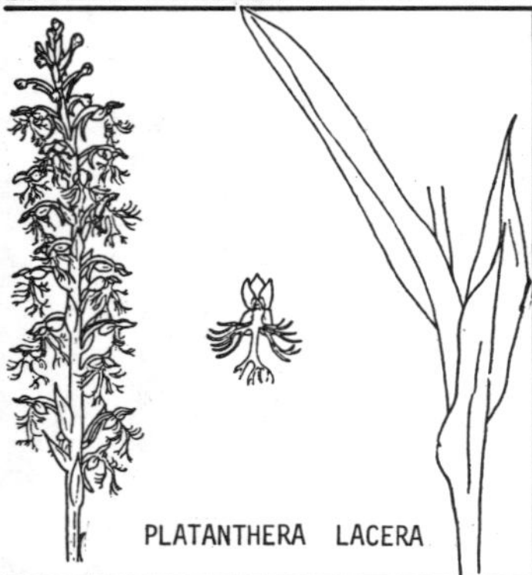

PLATANTHERA LACERA

PLATANTHERA MACROPHYLLA

PLATANTHERA OBTUSATA

PLATANTHERA ORBICULATA

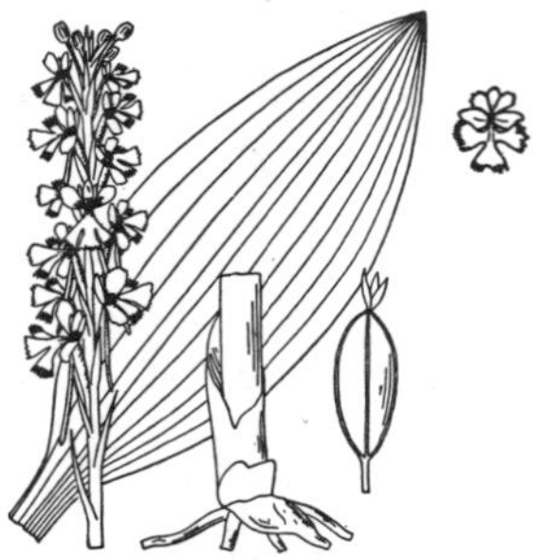

PLATANTHERA PSYCODES

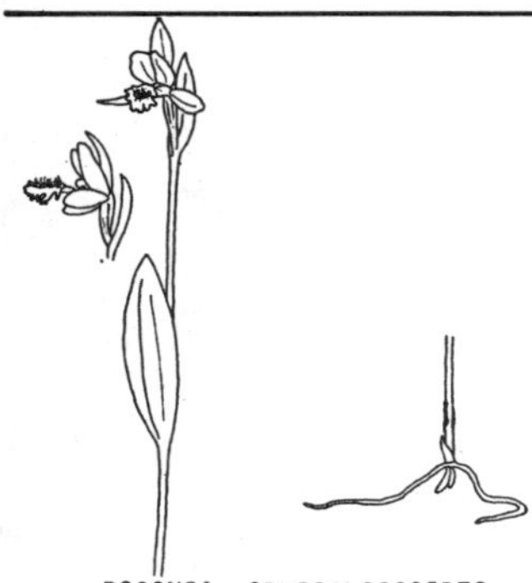

POGONIA OPHIOGLOSSOIDES

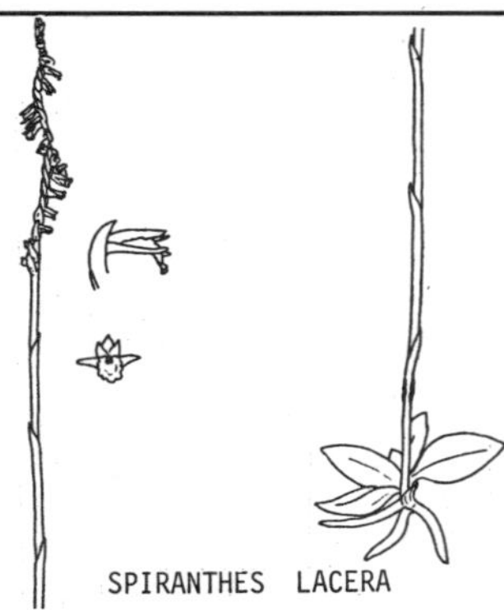

SPIRANTHES LACERA

SPIRANTHES LUCIDA

SPIRANTHES ROMANZOFFIANA

FOPULUS ALBA

POPULUS BALSAMIFERA

POPULUS GRANDIDENTATA

POPULUS TREMULOIDES

SALIX ALBA

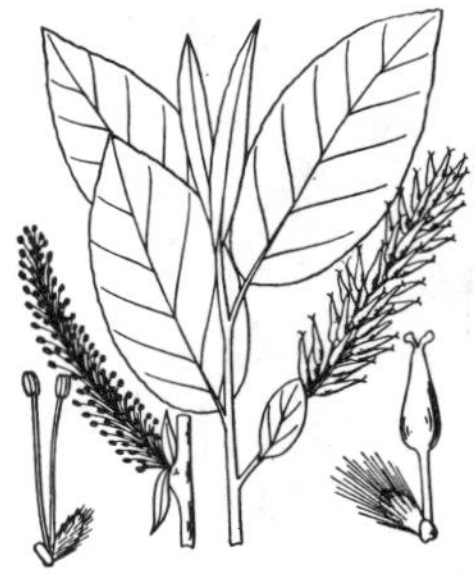

SALIX PYRIFOLIA

SALIX BEBBIANA

SALIX CANDIDA

SALIX DISCOLOR

SALIX ERIOCEPHALA

SALIX EXIGUA

SALIX GLAUCOPHYLLOIDES

SALIX HUMILIS

SALIX LUCIDA

SALIX MYRTILLIFOLIA

SALIX NIGRA

SALIX PEDICELLARIS

SALIX PELLITA

SALIX PETIOLARIS

SALIX PURPUREA

SALIX SERICEA

SALIX VIMINALIS

COMPTONIA PEREGRINA

MYRICA GALE

MYRICACEAE | JUGLANDACEAE | BETULACEAE

MYRICA PENSLYVANICA

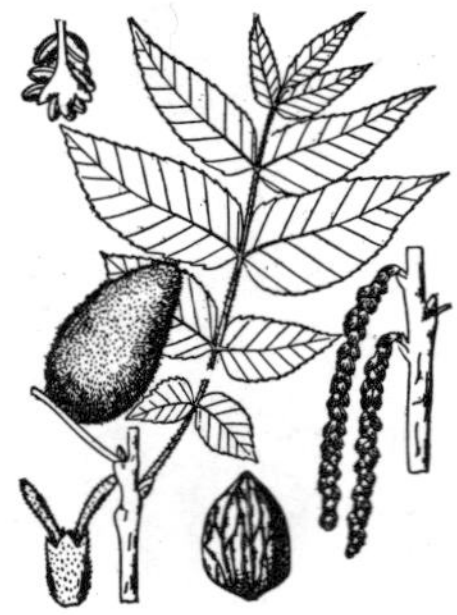

JUGLANS CINEREA

ALNUS INCANA SSP. RUGOSA

ALNUS VIRIDIS SSP. CRISPA

BETULA ALLEGHANIENSIS

BETULA X CAERULEA

BETULA CORDIFOLIA

BETULA GLANDULOSA

BETULA MINOR

BETULA PAPYRIFERA

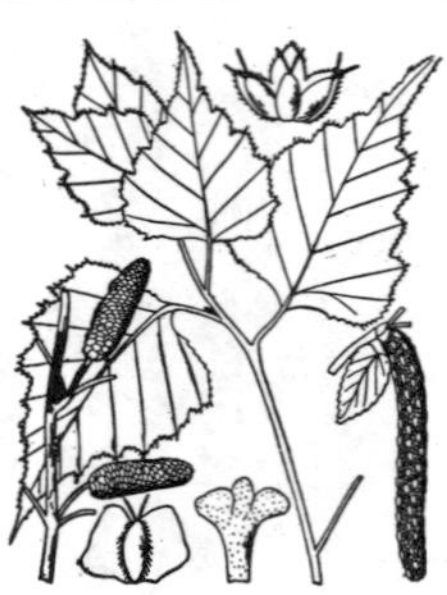

BETULA POPULIFOLIA

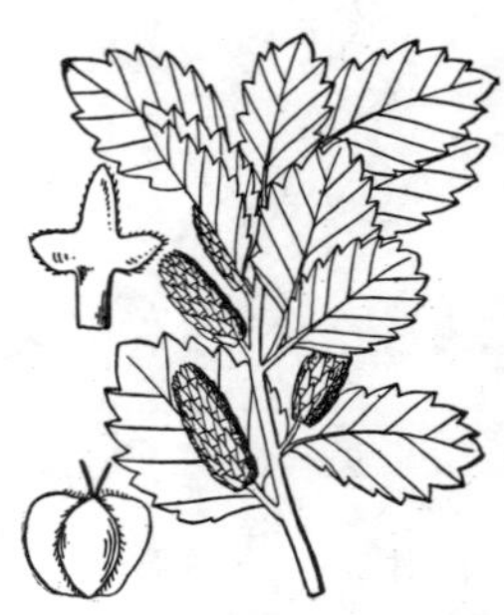

BETULA PUMILA

CORYLUS CORNUTA

OSTRYA VIRGINIANA

FAGUS GRANDIFOLIA

QUERCUS MACROCARPA

QUERCUS ROBUR

QUERCUS RUBRA

ULMUS AMERICANA

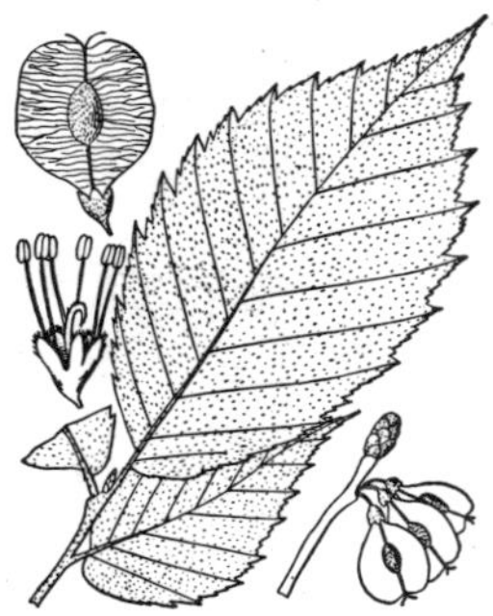

ULMUS RUBRA

CANNABIS SATIVA

HUMULUS LUPULUS

BOEHMERIA CYLINDRICA

LAPORTEA CANADENSIS

PILEA PUMILA

URTICA DIOICA

URTICA URENS

COMANDRA UMBELLATA

GEOCAULON LIVIDUM

ARCEUTHOBIUM PUSILLUM

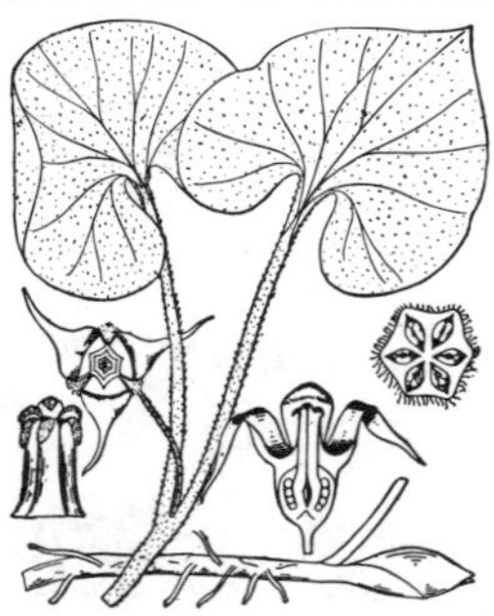

ASARUM CANADENSE

FAGOPYRUM ESCULENTUM

POLYGONUM ACHOREUM

POLYGONELLA ARTICULATA

POLYGONUM AMPHIBIUM

POLYGONUM ARENASTRUM

POLYGONUM ARIFOLIUM

POLYGONUM CAREYI

POLYGONUM CILINODE

POLYGONUM CONVOLVULUS

POLYGONUM CUSPIDATUM

POLYGONUM COCCINEUM

POLYGONUM FOWLERI

POLYGONUM HYDROPIPER

POLYGONUM HYDROPIPEROIDES

POLYGONUM LAPATHIFOLIUM

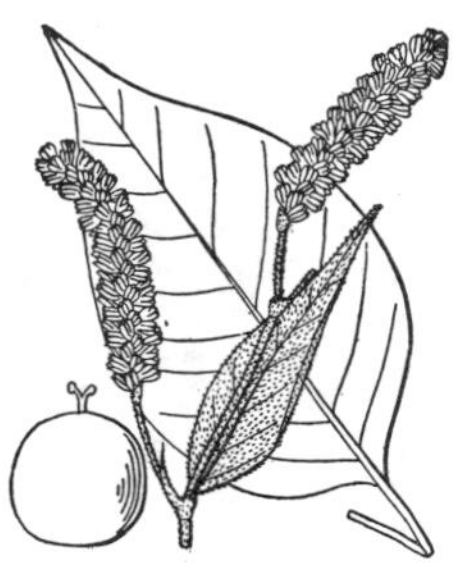

POLYGONUM ORIENTALE

POLYGONUM PENSYLVANICUM

POLYGONUM PERSICARIA

POLYGONUM PUNCTATUM

POLYGONUM RAMOSISSIMUM

POLYGONUM SAGITTATUM

POLYGONUM SCANDENS

RHEUM RHAPONTICUM

RUMEX ACETOSA

RUMEX ACETOSELLA

RUMEX CRISPUS

RUMEX LONGIFOLIUS

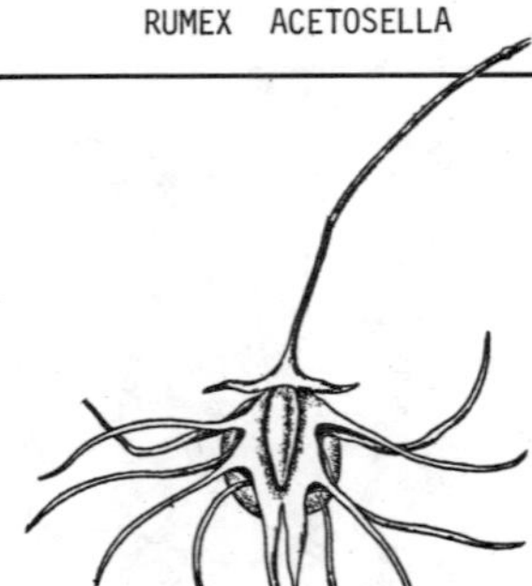

RUMEX MARITIMUS

RUMEX OBTUSIFOLIUS

RUMEX OCCIDENTALIS

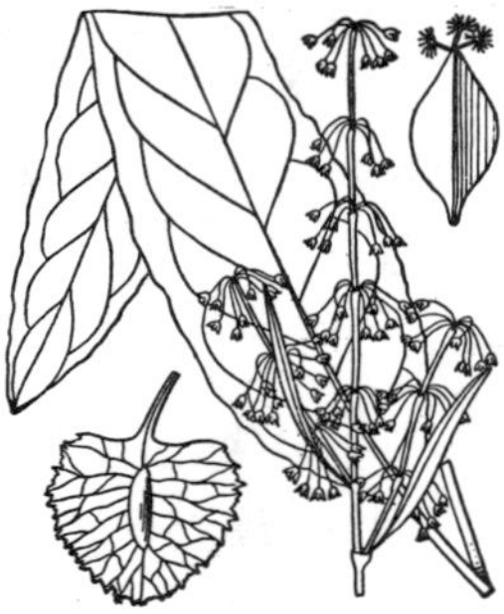

RUMEX ORBICULATUS

RUMEX PATIENTIA

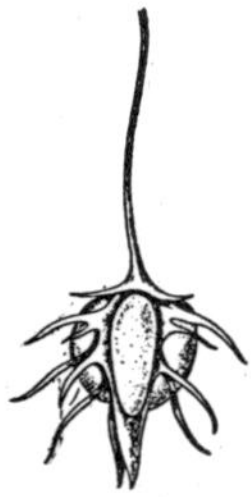

RUMEX PERSICARIOIDES

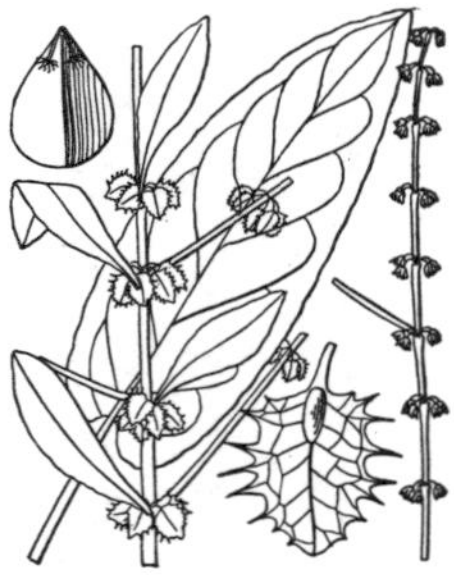

RUMEX PULCHER

RUMEX SALICIFOLIUS

RUMEX SANGUINEUS

ATRIPLEX ACADIENSIS

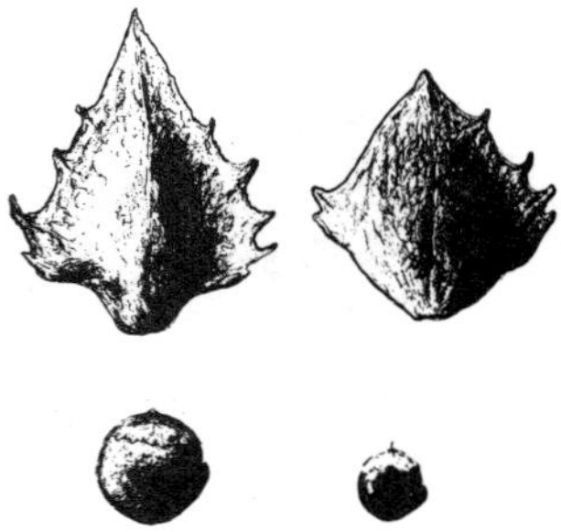

ATRIPLEX FRANKTONII

ATRIPLEX GLABRIUSCULA

ATRIPLEX LACINIATA

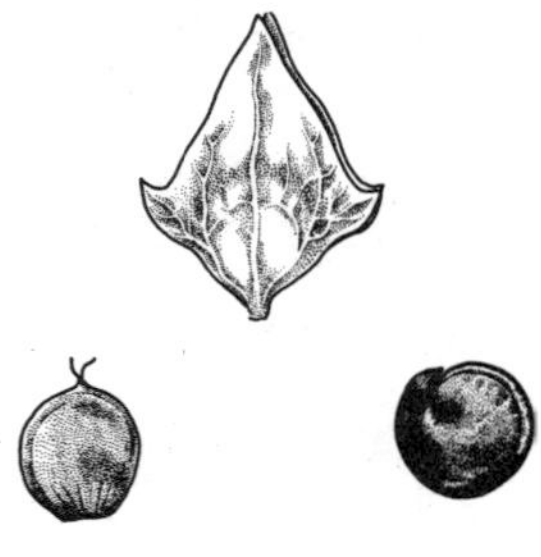

ATRIPLEX PATULA

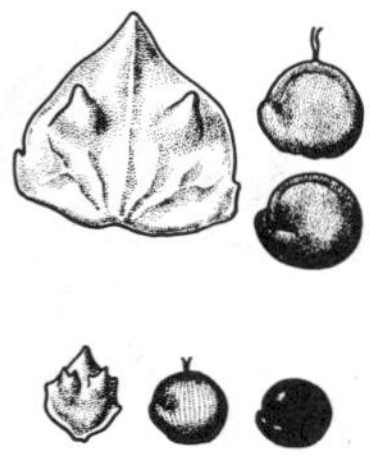

ATRIPLEX SUBSPICATA

ATRIPLEX PROSTRATA

AXYRIS AMARANTHOIDES

CHENOPODIUM ALBUM

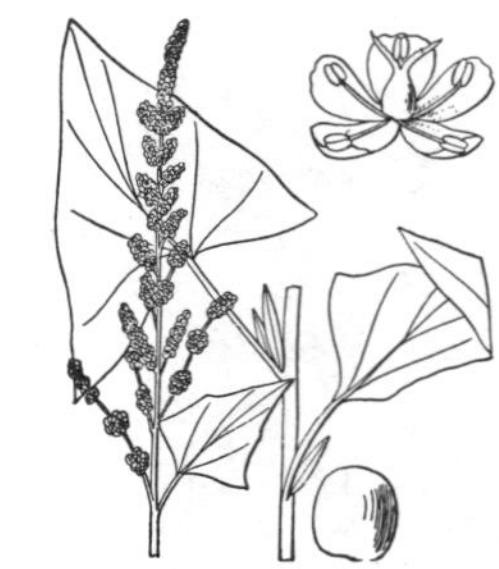

CHENOPODIUM BONUS-HENRICUS

CHENOPODIUM BOTRYS

CHENOPODIUM CAPITATUM

CHENOPODIUM GLAUCUM

CHENOPODIUM GIGANTOSPERMUM

CHENOPODIUM MURALE

CHENOPODIUM POLYSPERMUM

CHENOPODIUM RUBRUM

CHENOPODIUM URBICUM

KOCHIA SCOPARIA

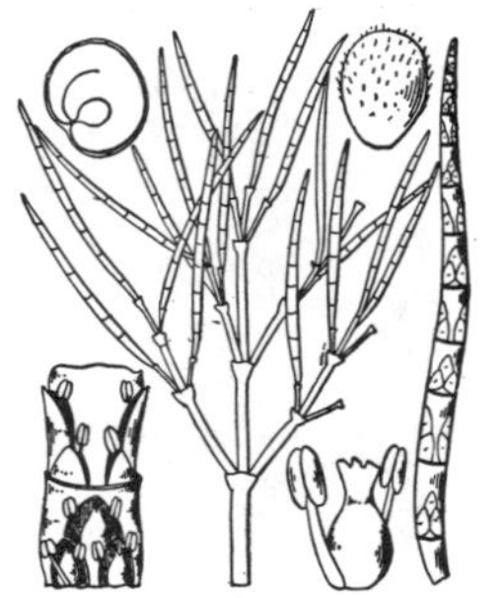

SALICORNIA EUROPAEA

SALSOLA KALI

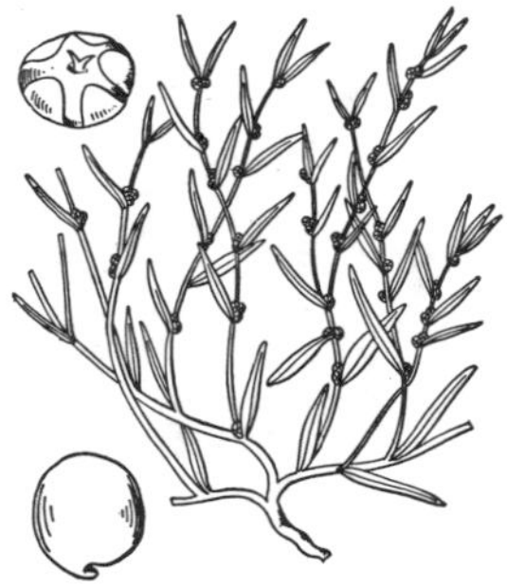

SUAEDA MARITIMA

AMARANTHUS ALBUS

AMARANTHUS RETROFLEXUS

MOLLUGO VERTICILLATA

CLAYTONIA CAROLINIANA

MONTIA FONTANA

PORTULACA OLERACEA

AGROSTEMMA GITHAGO

ARENARIA SERPYLLIFOLIA

CERASTIUM ARVENSE

CERASTIUM FONTANUM

CERASTIUM GLOMERATUM

DIANTHUS ARMERIA

HONCKENYA PEPLOIDES

LYCHNIS CHALCEDONICA

LYCHNIS FLOS-CUCULI

MOEHRINGIA LATERIFLORA

SAGINA DECUMBENS

SAGINA NODOSA

SAGINA PROCUMBENS

SAGINA SAGINOIDES

SAPONARIA OFFICINALIS

SCLERANTHUS ANNUUS

SILENE ALBA

SILENE ANTIRRHINA

SILENE ARMERIA

SILENE GALLICA

SILENE NOCTIFLORA

SILENE VULGARIS

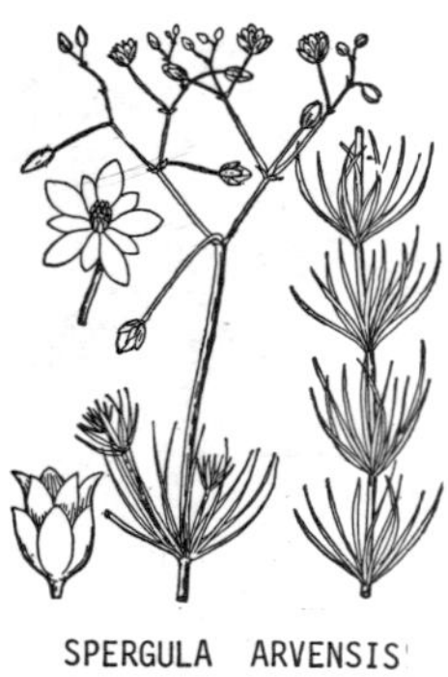

SPERGULA ARVENSIS

SPERGULARIA CANADENSIS

SPERGULARIA MARINA

SPERGULARIA RUBRA

STELLARIA ALSINE

STELLARIA CALYCANTHA

STELLARIA CRASSIFOLIA

STELLARIA GRAMINEA

STELLARIA HUMIFUSA

STELLARIA LONGIFOLIA

STELLARIA LONGIPES

STELLARIA MEDIA

VACCARIA PYRAMIDATA

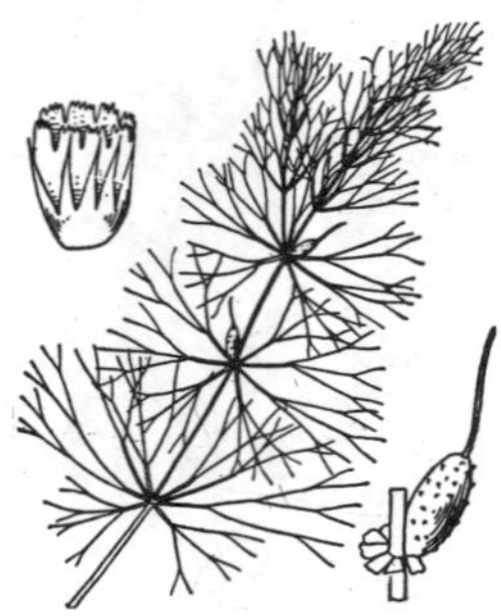

CERATOPHYLLUM DEMERSUM

CERATOPHYLLUM MURICATUM

BRASENIA SCHREBERI

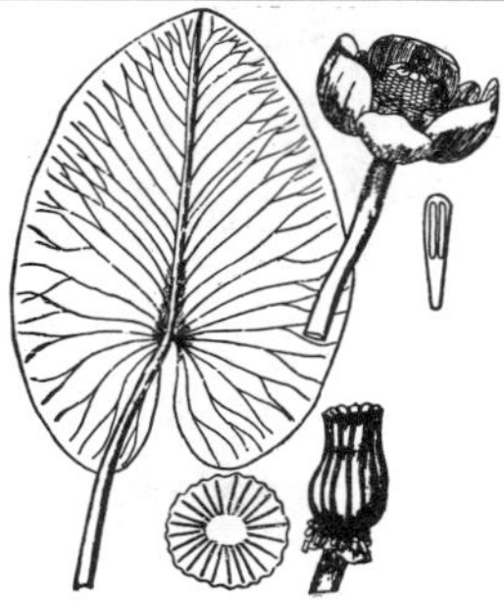

NUPHAR LUTEUM

NYMPHAEA ODORATA

ACTAEA PACHYPODA

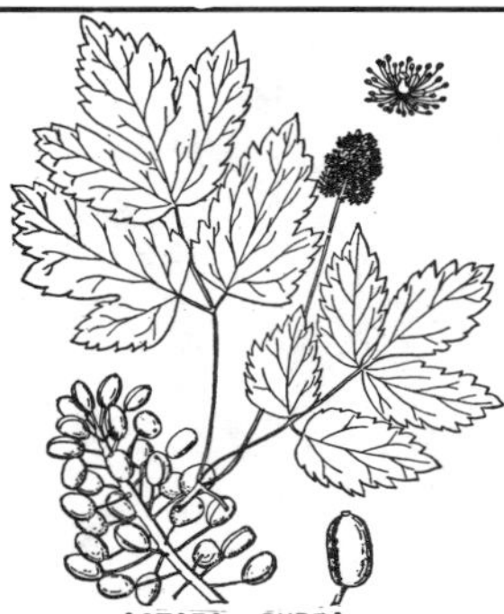

ACTAEA RUBRA

ANEMONE CANADENSIS

ANEMONE MULTIFIDA

ANEMONE PARVIFLORA

ANEMONE QUINQUEFOLIA

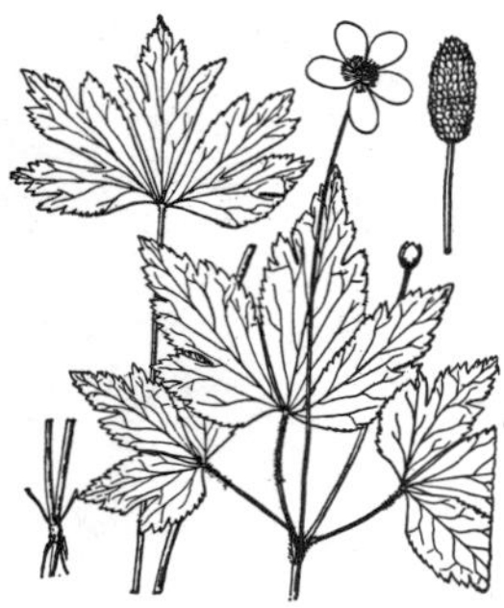

ANEMONE RIPARIA

AQUILEGIA CANADENSIS

AQUILEGIA VULGARIS

CALTHA PALUSTRIS

CLEMATIS OCCIDENTALIS

CLEMATIS VIRGINIANA

COPTIS TRIFOLIA

HEPATICA NOBILIS VAR. OBTUSA

RANUNCULUS ABORTIVUS

RANUNCULUS ACRIS

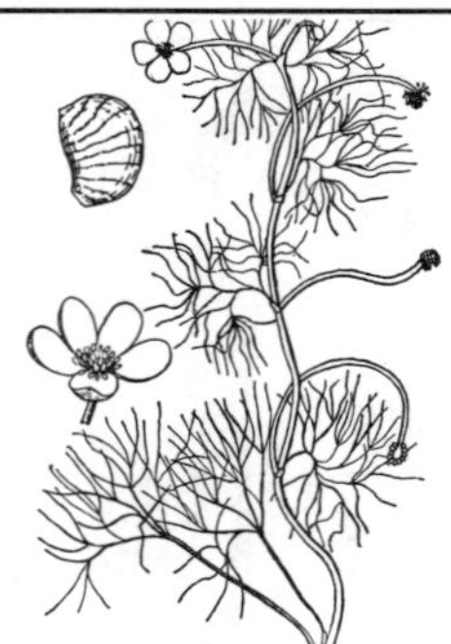

RANUNCULUS AQUATILIS

RANUNCULUS CYMBALARIA

RANUNCULUS FLABELLARIS

RANUNCULUS LAPPONICUS

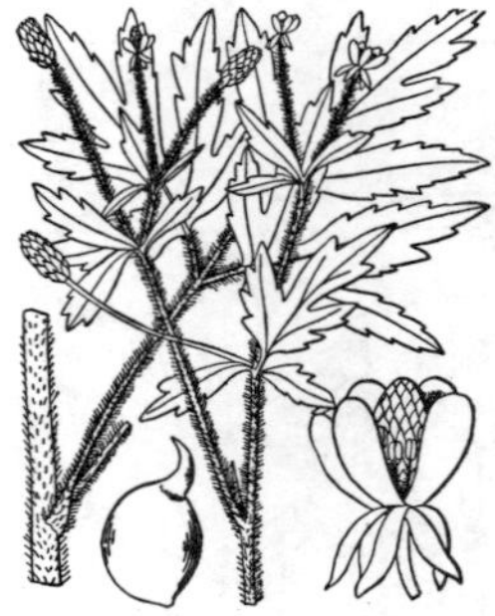

RANUNCULUS PENSYLVANICUS

RANUNCULACEAE BERBERIDACEAE PAPAVERACEAE

RANUNCULUS RECURVATUS

RANUNCULUS REPENS

RANUNCULUS REPTANS

RANUNCULUS SARDOUS

RANUNCULUS SCELERATUS

RANUNCULUS SEPTENTRIONALIS

THALICTRUM PUBESCENS

THALICTRUM VENULOSUM VAR. CONFINE

BERBERIS THUNBERGII

BERBERIS VULGARIS

CAULOPHYLLUM THALICTROIDES

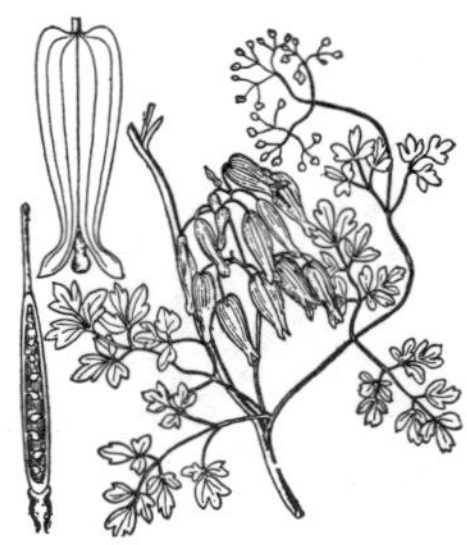

ADLUMIA FUNGOSA

CHELIDONIUM MAJUS

CORYDALIS SEMPERVIRENS

DICENTRA CUCULLARIA

FUMARIA OFFICINALIS

PAPAVER DUBIUM

PAPAVER RHOEAS

SANGUINARIA CANADENSIS

ALLIARIA PETIOLATA

ARABIS DIVARICARPA

ARABIS DRUMMONDII

ARABIS GLABRA

ARABIS HIRSUTA

ARMORACIA RUSTICANA

BARBAREA ORTHOCERAS

BARBAREA VULGARIS

BERTEROA INCANA

BRASSICA JUNCEA

BRASSICA NIGRA

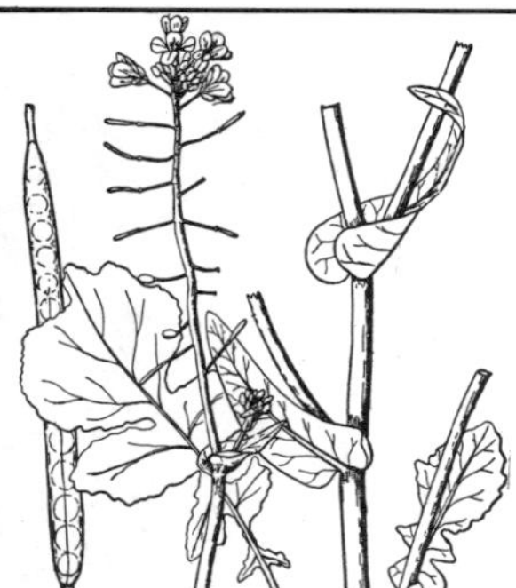
BRASSICA RAPA

CAKILE EDENTULA

CAMELINA MICROCARPA

CAMELINA SATIVA

CAPSELLA BURSA-PASTORIS

CARDAMINE PARVIFLORA

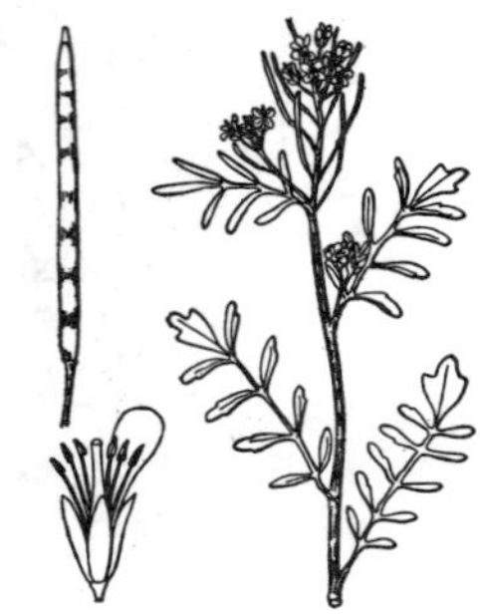

CARDAMINE PENNSYLVANICA

CARDAMINE PRATENSIS

CONRINGIA ORIENTALIS

CORONOPUS DIDYMUS

DENTARIA DIPHYLLA

DENTARIA LACINATA

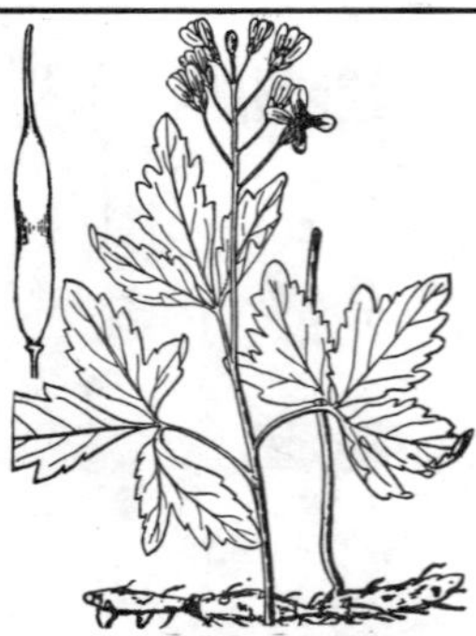

DENTARIA MAXIMA

DESCURAINIA SOPHIA

DIPLOTAXIS MURALIS

DIPLOTAXIS TENUIFOLIA

DRABA ARABISANS

DRABA INCANA

DRABA LANCEOLATA

EROPHILA VERNA

ERUCASTRUM GALLICUM

ERYSIMUM CHEIRANTHOIDES

ERYSIMUM HIERACIFOLIUM

ERYSIMUM INCONSPICUUM

HESPERIS MATRONALIS

LEPIDIUM CAMPESTRE

LEPIDIUM DENSIFLORUM

LEPIDIUM RUDERALE

LEPIDIUM SATIVUM

LEPIDIUM VIRGINICUM

NASTURTIUM MICROPHYLLUM

NESLIA PANICULATA

RAPHANUS RAPHANISTRUM

RAPHANUS SATIVUS

RORIPPA PALUSTRIS

SINAPIS ALBA

SINAPIS ARVENSIS

SISYMBRIUM ALTISSIMUS

SISYMBRIUM LOESELII

SISYMBRIUM OFFICINALE

THLASPI ARVENSE

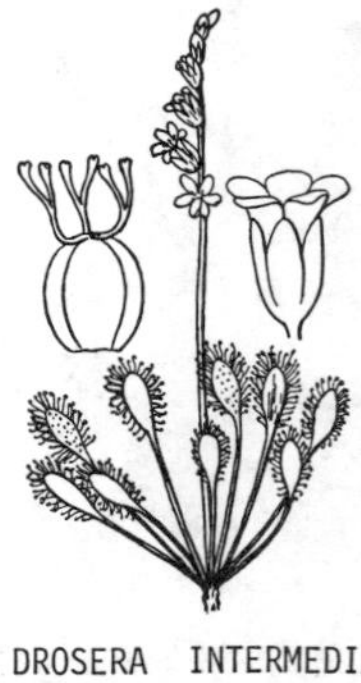

DROSERA INTERMEDIA

DROSERA ROTUNDIFOLIA

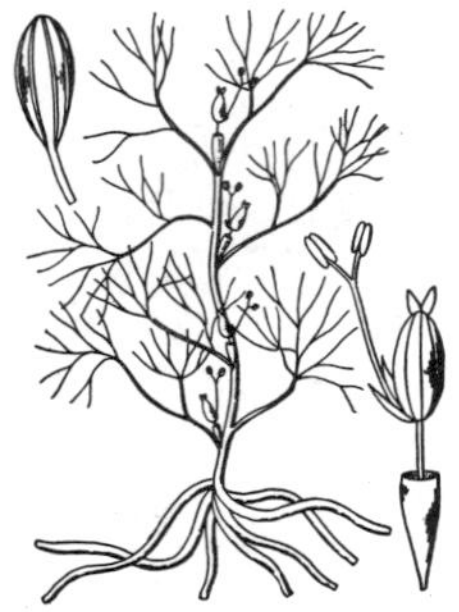

PODOSTEMON CERATOPHYLLUM

CRASSULA AQUATICA

SEDUM ACRE

SEDUM PURPUREUM

SEDUM ROSEA

SEDUM TELEPHIUM

CHRYSOSPLENIUM AMERICANUM

MITELLA NUDA

PARNASSIA GLAUCA

PENTHORUM SEDOIDES

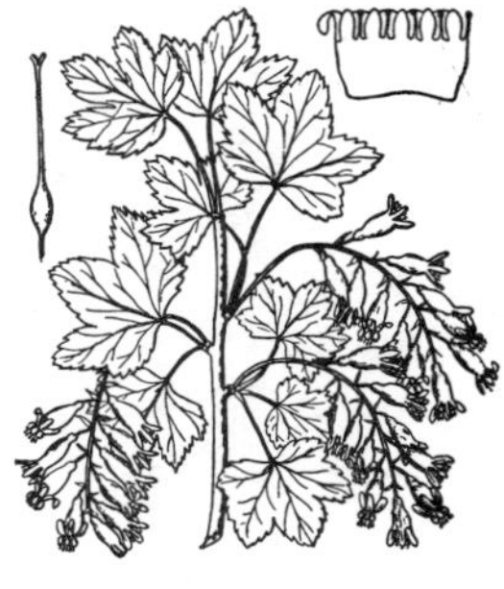

RIBES AMERICANA

SAXIFRAGACEAE | HAMAMELIDACEAE | ROSACEAE

RIBES GLANDULOSUM

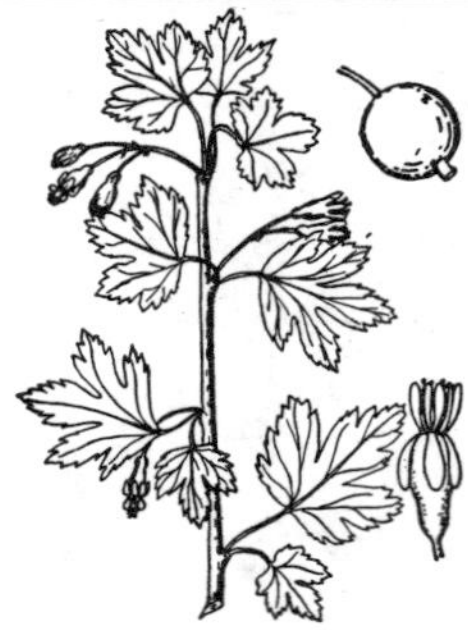

RIBES HIRTELLUM

RIBES LACUSTRE

RIBES NIGRUM

RIBES RUBRUM

RIBES TRISTE

SAXIFRAGA PANICULATA

SAXIFRAGA VIRGINIENSIS

TIARELLA CORDIFOLIA

HAMAMELIS VIRGINIANA

AGRIMONIA GRYPOSEPALA

AGRIMONIA STRIATA

ALCHEMILLA FILICAULIS

AMELANCHIER ARBOREA

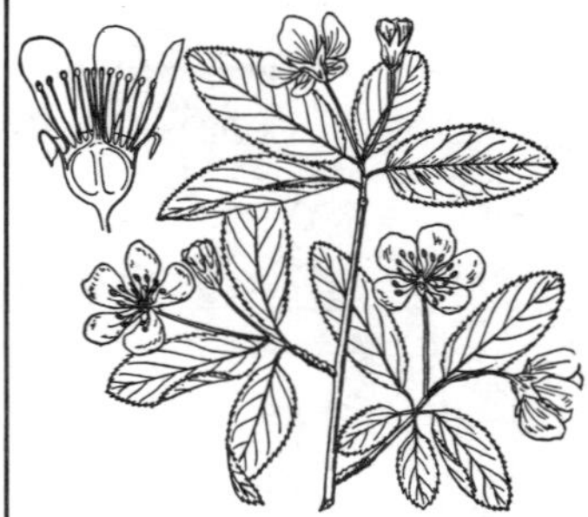

AMELANCHIER BARTRAMIANA

AMELANCHIER CANADENSIS

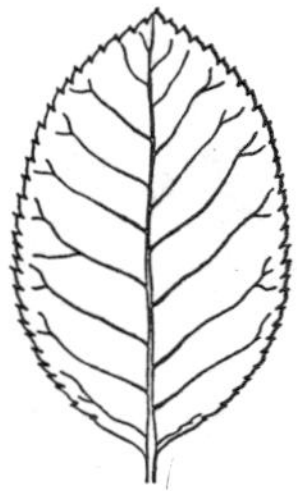

AMELANCHIER FERNALDII

AMELANCHIER HUMILIS

AMELANCHIER INTERMEDIA

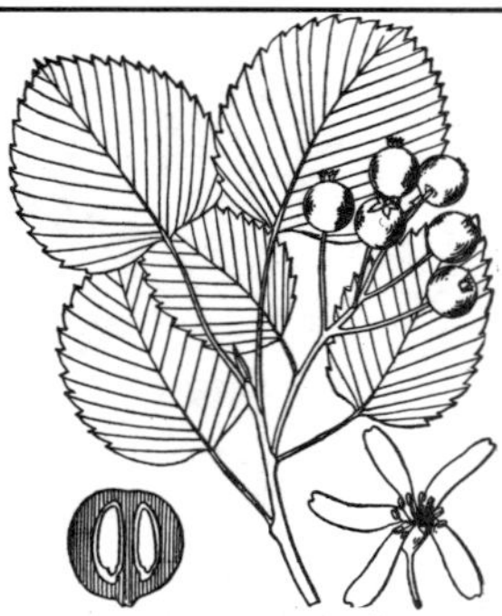

AMELANCHIER SANGUINEA

AMELANCHIER STOLONIFERA

AMELANCHIER WIEGANDII

CRATAEGUS BRAINERDII

CRATAEGUS CHRYSOCARPA

CRATAEGUS CURVISEPALA

CRATAEGUS FLABELLATA

CRATAEGUS MOLLIS

CRATAEGUS MONOGYNA

CRATAEGUS SUCCULENTA

DALIBARDA REPENS

DRYAS INTEGRIFOLIA

FILIPENDULA RUBRA

FILIPENDULA ULMARIA

FRAGARIA VESCA

FRAGARIA VIRGINIANA

GEUM ALEPPICUM

GEUM CANADENSE

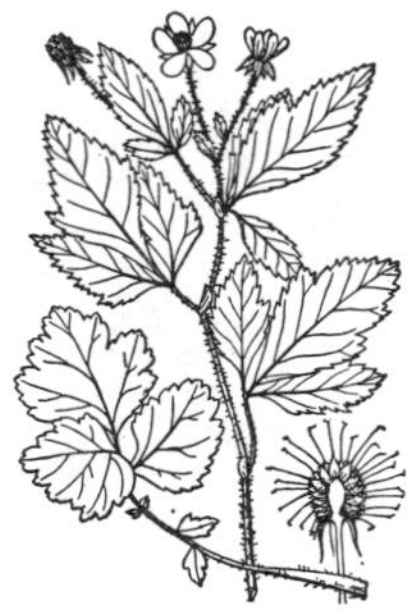

GEUM LACINIATUM

GEUM MACROPHYLLUM

GEUM RIVALE

GEUM URBANUM

PHYSOCARPUS OPULIFOLIUS

POTENTILLA ANSERINA

POTENTILLA ARGENTEA

POTENTILLA ARGUTA

POTENTILLA CANADENSIS

POTENTILLA FRUTICOSA

POTENTILLA INTERMEDIA

POTENTILLA NORVEGICA

POTENTILLA PALUSTRIS

POTENTILLA RECTA

POTENTILLA TABERNAEMONTANA

POTENTILLA TRIDENTATA

PRUNUS AVIUM

PRUNUS CERASUS

PRUNUS NIGRA

PRUNUS PENNSYLVANICA

PRUNUS PUMILA

PRUNUS SEROTINA

PRUNUS VIRGINIANA

ARONIA PRUNIFOLIA

MALUS SYLVESTRIS

ROSA ACICULARIS

ROSA BLANDA

ROSA CAROLINA

ROSA CINNAMOMEA

ROSA EGLANTERIA

ROSA NITIDA

ROSA PALUSTRIS

ROSA RUGOSA

ROSA SPINOSISSIMA

ROSA VIRGINIANA

RUBUS ALLEGHENIENSIS

RUBUS CANADENSIS

RUBUS CHAMAEMORUS

RUBUS FLAGELLARIS

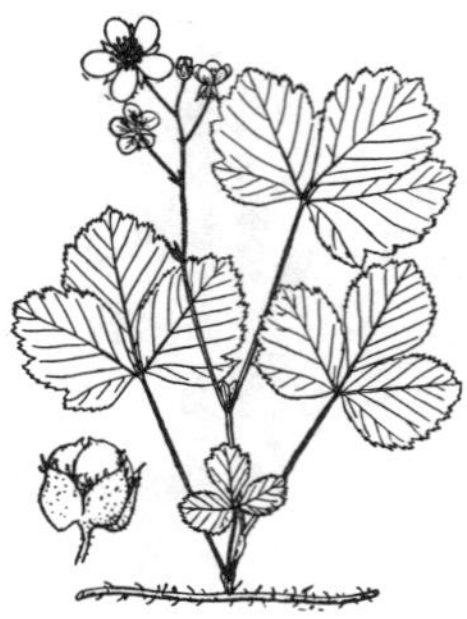

RUBUS HISPIDUS

RUBUS IDAEUS

RUBUS OCCIDENTALIS

RUBUS PENSILVANICUS

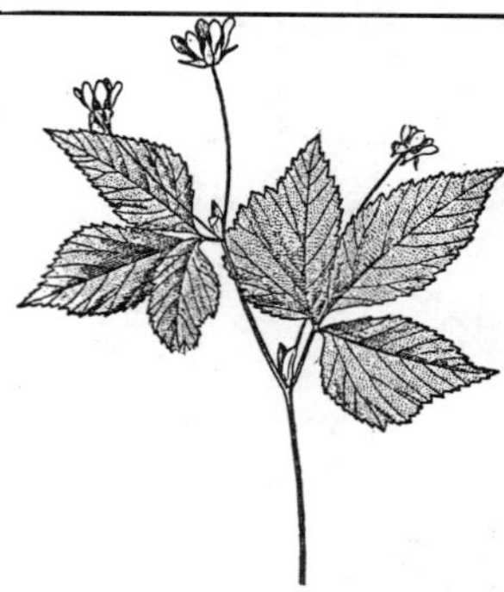

RUBUS PUBESCENS

RUBUS VERMONTANUS

SANGUISORBA CANADENSIS

SORBARIA SORBIFOLIA

SORBUS AMERICANA

SORBUS AUCUPARIA

SORBUS DECORA

SPIRAEA ALBA

SPIRAEA TOMENTOSA

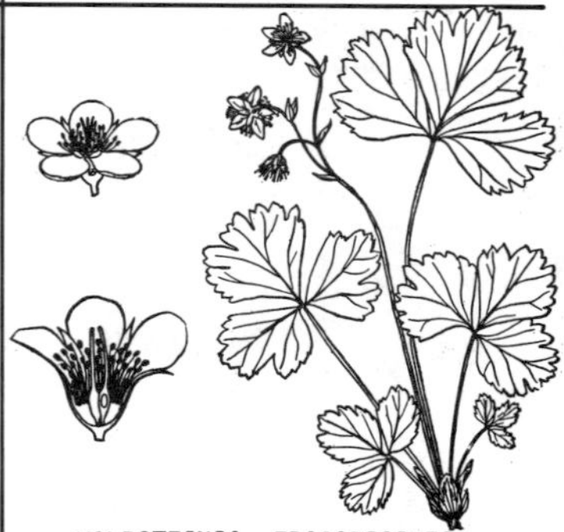

WALDSTEINIA FRAGARIOIDES

AMORPHA FRUTICOSA

AMPHICARPAEA BRACTEATA

ANTHYLLIS VULNERARIA

APIOS AMERICANA

ASTRAGALUS ALPINUS

ASTRAGALUS EUCOSMUS

CORONILLA VARIA

DESMODIUM CANADENSE

DESMODIUM GLUTINOSUM

HEDYSARUM ALPINUM

LATHYRUS JAPONICUS

LATHYRUS PALUSTRIS

LATHYRUS PRATENSIS

LOTUS CORNICULATUS

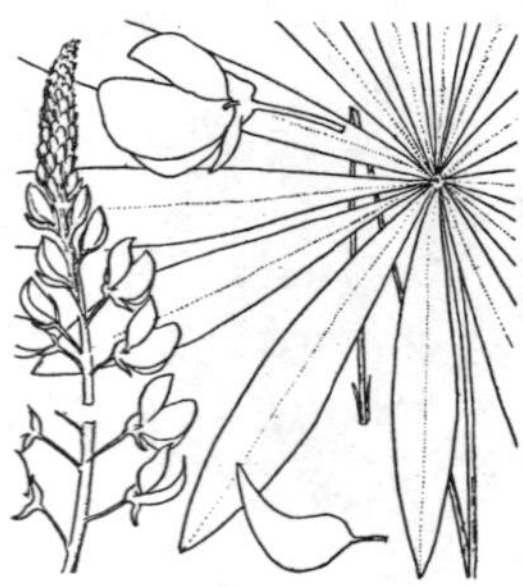

LUPINUS POLYPHYLLUS

MEDICAGO ARABICA

MEDICAGO LUPULINA

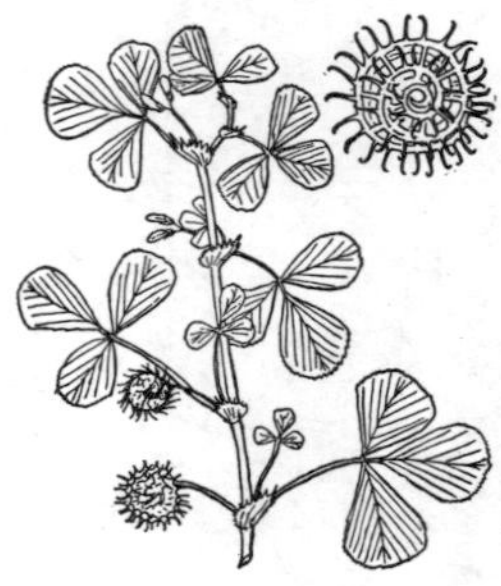

MEDICAGO POLYMORPHA

MEDICAGO SATIVA

MELILOTUS ALBA

MELILOTUS OFFICINALIS

OXYTROPIS CAMPESTRIS

ROBINIA PSEUDO-ACACIA

ROBINIA VISCOSA

TRIFOLIUM ARVENSE

TRIFOLIUM AUREUM

TRIFOLIUM CAMPESTRE

TRIFOLIUM DUBIUM

TRIFOLIUM HYBRIDUM

TRIFOLIUM MEDIUM

FABACEAE | LINACEAE | OXALIDACEAE

TRIFOLIUM PRATENSE

TRIFOLIUM REPENS

VICIA AMERICANA

VICIA CRACCA

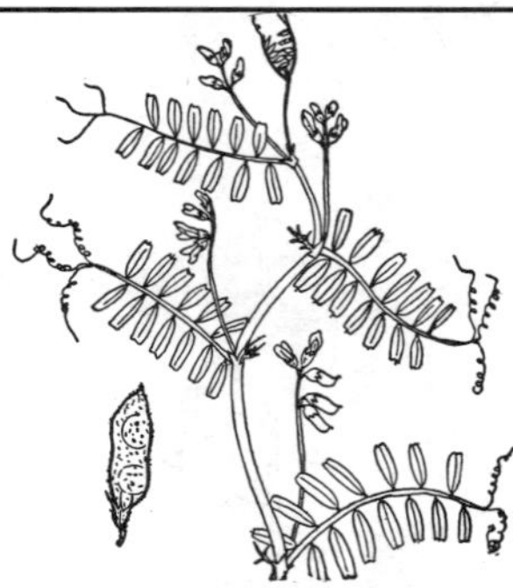

VICIA HIRSUTA

VICIA SATIVA

VICIA TETRASPERMA

LINUM CATHARTICUM

LINUM USITATISSIMUM

OXALIS ACETOSELLA

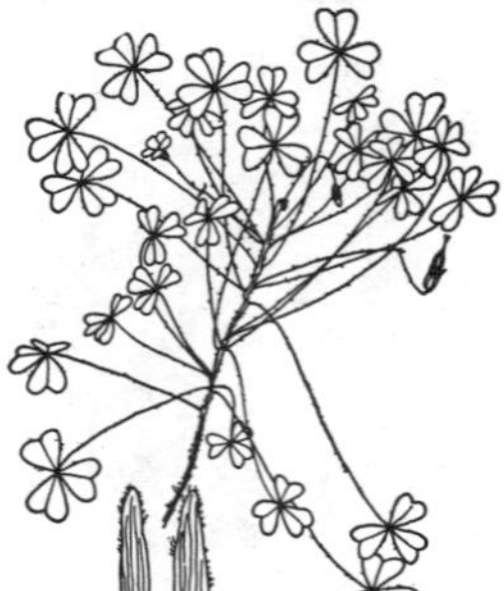

OXALIS DILLENII

OXALIS STRICTA

GERANIACEAE | POLYGALACEAE | EUPHORBIACEAE

ERODIUM CICUTARIUM

GERANIUM BICKNELLII

GERANIUM PRATENSE

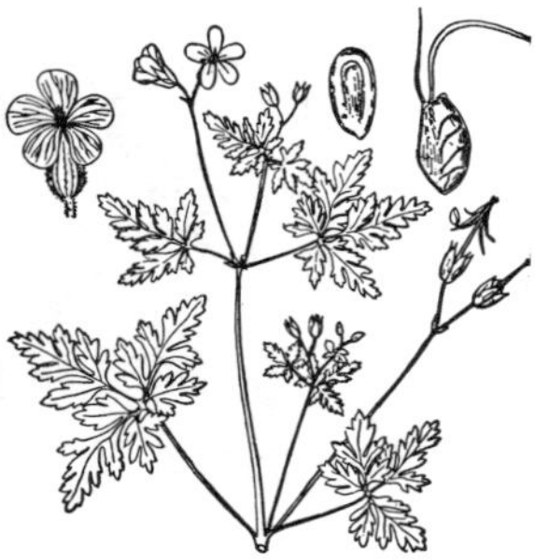
GERANIUM ROBERTIANUM

POLYGALA PAUCIFOLIA

POLYGALA SANGUINEA

POLYGALA SENEGA

ACALYPHA RHOMBOIDEA

CHAMAESYCE GLYPTOSPERMA

CHAMAESYCE MACULATA

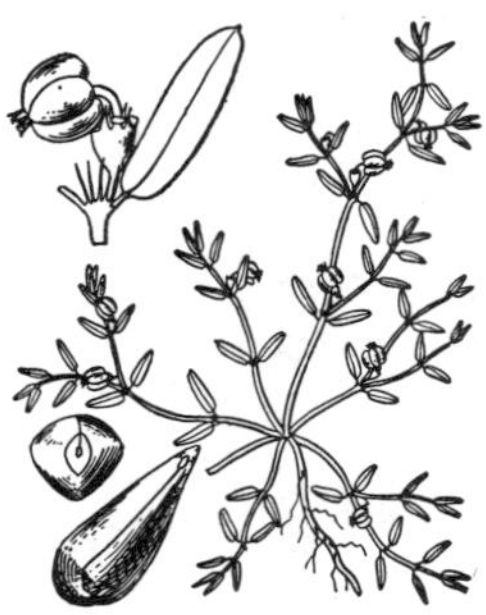
CHAMAESYCE POLYGONIFOLIA

CHAMAESYCE SERPYLLIFOLIA

CHAMAESYCE VERMICULATA

EUPHORBIA CYPARISSIAS

EUPHORBIA ESULA

EUPHORBIA HELIOSCOPIA

EUPHORBIA PEPLUS

MERCURIALIS ANNUA

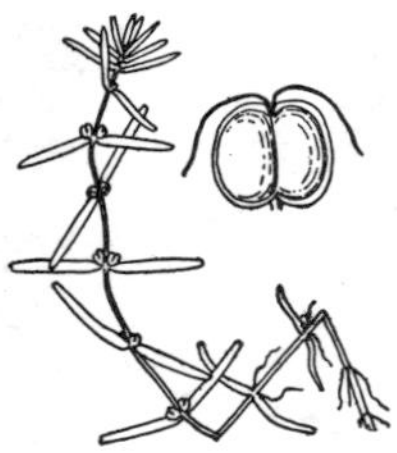

CALLITRICHE HERMAPHRODITICA

CALLITRICHE HETEROPHYLLA

CALLITRICHE VERNA

COREMA CONRADII

EMPETRUM NIGRUM

RHUS TYPHINA

TOXICODENDRON RYDBERGII

ILEX VERTICILLATA

NEMOPANTHUS MUCRONATA

CELASTRUS SCANDENS

ACER NEGUNDO

ACER PENSYLVANICUM

ACER RUBRUM

ACER SACCHARINUM

ACER SACCHARUM

ACER SPICATUM

AESCULUS HIPPOCASTANUM

IMPATIENS CAPENSIS

IMPATIENS GLANDULIFERA

IMPATIENS PALLIDA

RHAMNUS ALNIFOLIA

RHAMNUS CATHARTICA

RHAMNUS FRANGULA

PARTHENOCISSUS QUINQUEFOLIA

VITIS RIPARIA

TILIA AMERICANA

ABUTILON THEOPHRASTI

ALCEA ROSEA

ALTHAEA OFFICINALIS

HIBISCUS TRIONUM

MALVA ALCEA

MALVA MOSCHATA

MALVA ROTUNDIFOLIA

MALVA SYLVESTRIS

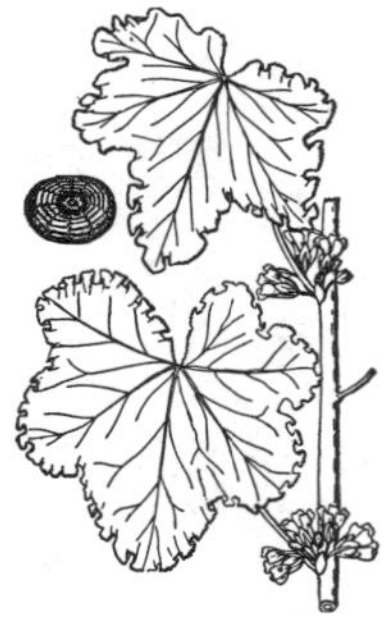

MALVA VERTICILLATA

HYPERICUM BOREALE

HYPERICUM CANADENSE

HYPERICUM ELLIPTICUM

HYPERICUM MAJUS

HYPERICUM MUTILUM

HYPERICUM PERFORATUM

TRIANDEUM VIRGINICUM

ELATINACEAE | CISTACEAE | VIOLACEAE

ELATINE MINIMA

ELATINE TRIANDRA

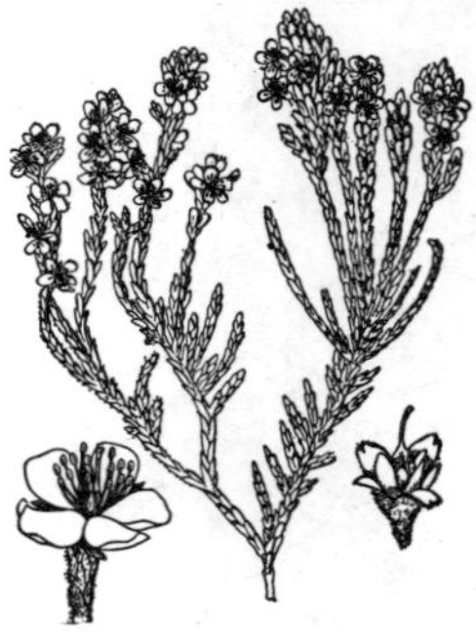

HUDSONIA TOMENTOSA

LECHEA INTERMEDIA

LECHEA MARITIMA

VIOLA ADUNCA

VIOLA ARVENSIS

VIOLA CANADENSIS

VIOLA CONSPERSA

VIOLA FIMBRIATULA

VIOLA INCOGNITA

VIOLA LABRADORICA

VIOLACEAE THYELIACEAE ELAEGNACEAE

VIOLA LANCEOLATA

VIOLA MACLOSKEYI

VIOLA NEPHROPHYLLA

VIOLA NOVAE-ANGLIAE

VIOLA OBLIQUA

VIOLA PUBESCENS VAR. ERIOCARPA

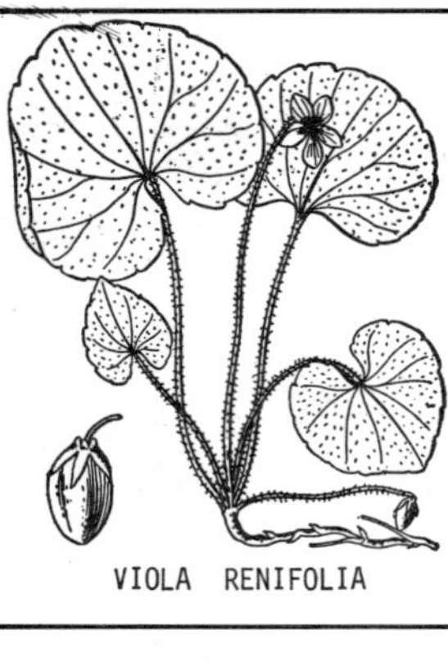

VIOLA RENIFOLIA

VIOLA SELKIRKII

VIOLA SEPTENTRIONALIS

DAPHNE MEZEREUM

DIRCA PALUSTRIS

SHEPHERDIA CANADENSIS

LYTHRUM SALICARIA

CAMISSONIA CONTORTA

CIRCAEA ALPINA

CIRCAEA LUTETIANA SSP. CANADENSIS

EPILOBIUM ANGUSTIFOLIUM

EPILOBIUM COLORATUM

EPILOBIUM CILIATUM

EPILOBIUM HIRSUTUM

EPILOBIUM HORNEMANNII

EPILOBIUM LEPTOPHYLLUM

EPILOBIUM PALUSTRE

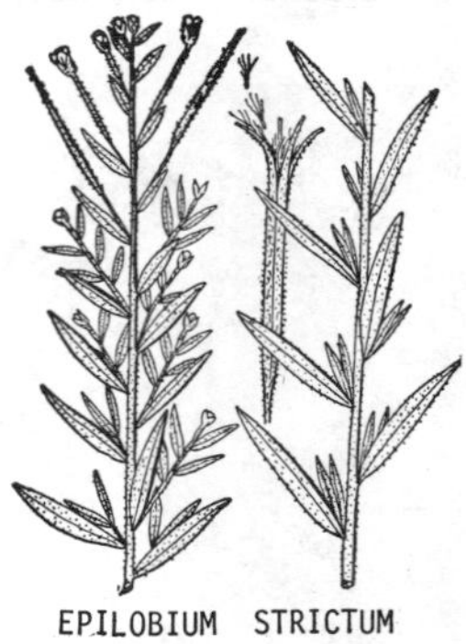

EPILOBIUM STRICTUM

ONAGRACEAE | HALORAGIDACEAE | HIPPURIDACEAE

LUDWIGIA PALUSTRIS

OENOTHERA BIENNIS

OENOTHERA PARVIFLORA

OENOTHERA PERENNIS

MYRIOPHYLLUM ALTERNIFLORUM

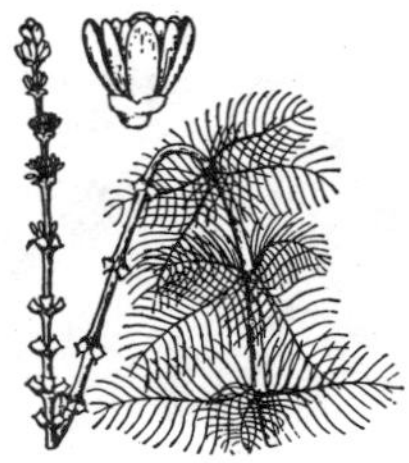

MYRIOPHYLLUM EXALBESCENS

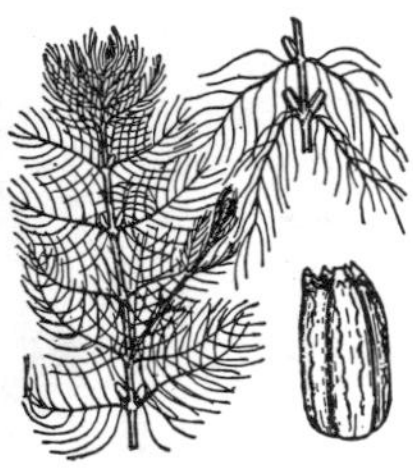

MYRIOPHYLLUM FARWELLII

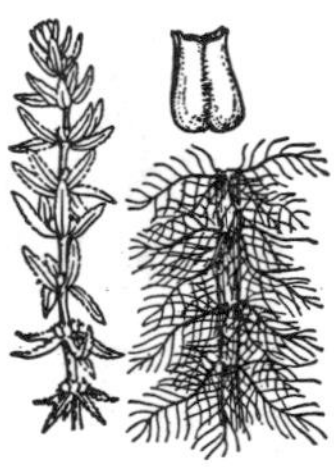

MYRIOPHYLLUM HETEROPHYLLUM

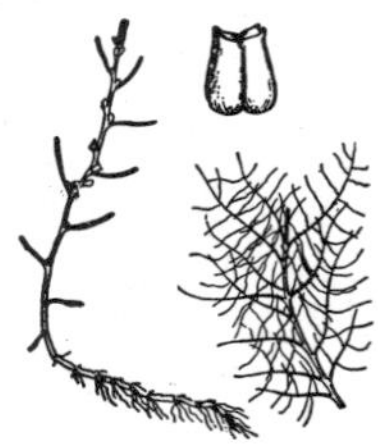

MYRIOPHYLLUM HUMILE

MYRIOPHYLLUM TENELLUM

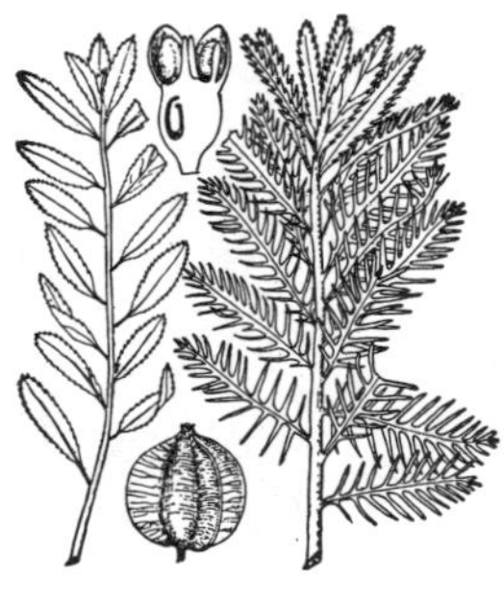

PROSERPINACA PALUSTRIS

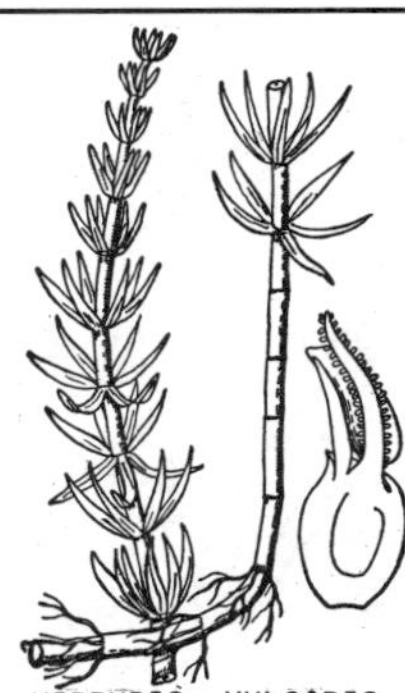

HIPPURIS VULGARIS

ARALIA HISPIDA

ARALIA NUDICAULIS

ARALIA RACEMOSA

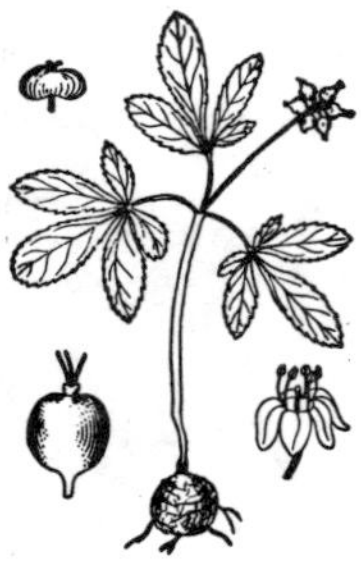

PANAX TRIFOLIUS

AEGOPODIUM PODAGRARIA

AETHUSA CYNAPIUM

ANGELICA ATROPURPUREA

ANGELICA LUCIDA

ANTHRISCUS SYLVESTRIS

BUPLEURUM ROTUNDIFOLIUM

CARUM CARVI

CICUTA BULBIFERA

CICUTA MACULATA

CONIOSELINUM CHINENSE

CONIUM MACULATUM

CRYPTOTAENIA CANADENSIS

DAUCUS CAROTA

HERACLEUM LANATUM

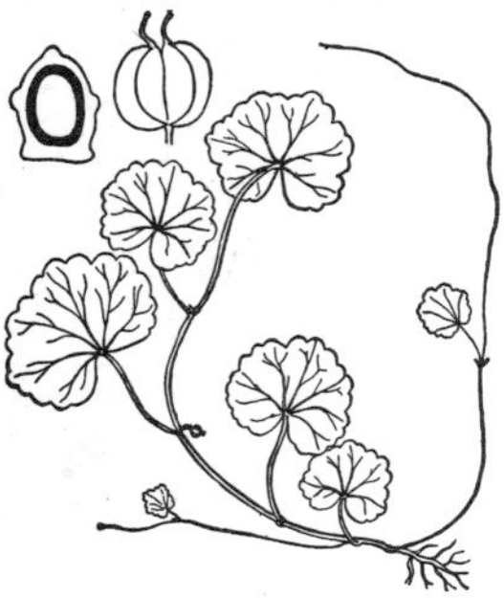

HYDROCOTYLE AMERICANA

LIGUSTICUM SCOTICUM

OSMORHIZA CHILENSIS

OSMORHIZA CLAYTONII

OSMORHIZA DEPAUPERATA

OSMORHIZA LONGISTYLIS

PASTINACA SATIVA

PIMPINELLA SAXIFRAGA

SANICULA GREGARIA

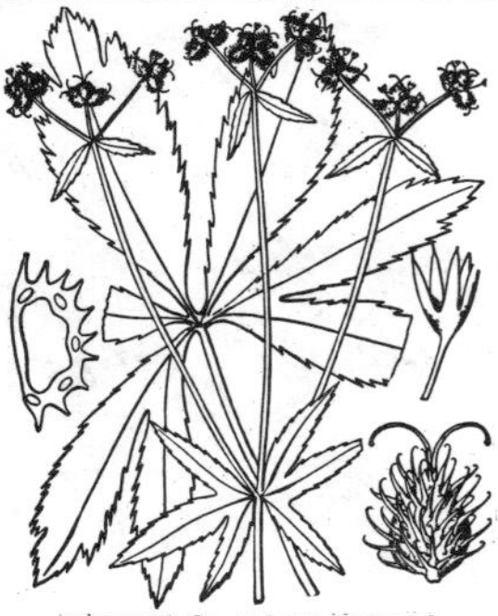

SANICULA MARILANDICA

SANICULA TRIFOLIATA

SIUM SUAVE

TORILIS NODOSA

ZIZIA AUREA

CORNUS ALTERNIFOLIA

CORNUS AMOMUM

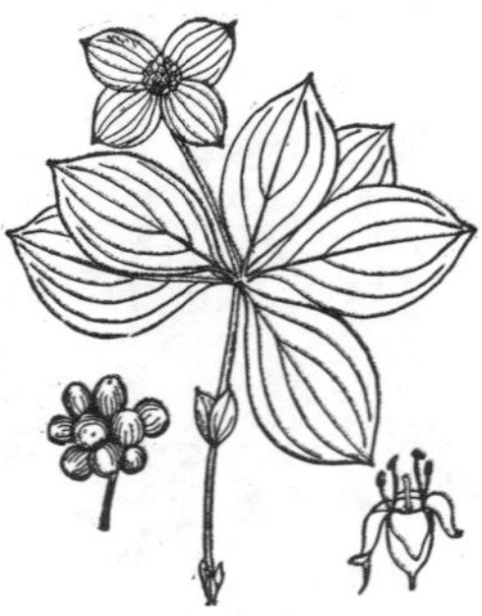

CORNUS CANADENSIS

CORNUS RUGOSA

CORNUS SERICEA

CHIMAPHILA UMBELLATA

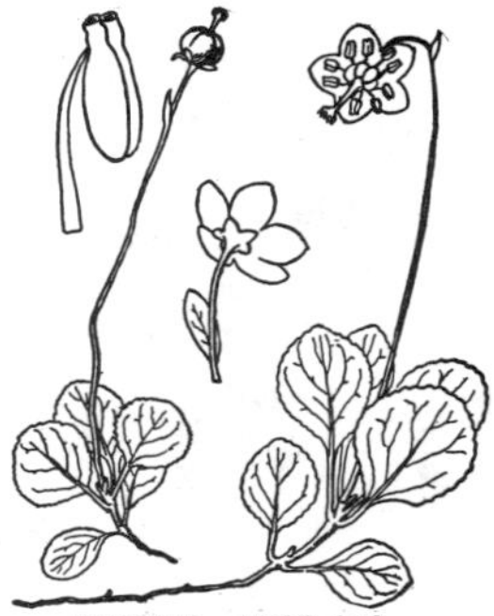

MONESES UNIFLORA

MONOTROPA HYPOPITYS

MONOTROPA UNIFLORA

ORTHILIA SECUNDA

PTEROSPORA ANDROMEDEA

PYROLA AMERICANA

PYROLA ASARIFOLIA

PYROLA CHLORANTHA

PYROLA ELLIPTICA

PYROLA MINOR

ANDROMEDA POLIFOLIA VAR. GLAUCOPHYLLA

ARCTOSTAPHYLOS UVA-URSI

CHAMAEDAPHNE CALYCULATA

EPIGAEA REPENS

GAULTHERIA HISPIDULA

GAULTHERIA PROCUMBENS

GAYLUSSACIA BACCATA

GAYLUSSACIA DUMOSA

KALMIA ANGUSTIFOLIA

KALMIA POLIFOLIA

LEDUM GROENLANDICUM

RHODODENDRON CANADENSE

VACCINIUM ANGUSTIFOLIUM

VACCINIUM CAESPITOSUM

VACCINIUM CORYMBOSUM

VACCINIUM MACROCARPON

VACCINIUM MYRTILLOIDES

VACCINIUM OXYCOCCUS

VACCINIUM ULIGINOSUM

VACCINIUM VITIS-IDAEA SSP. MINOR

ANAGALLIS ARVENSIS

GLAUX MARITIMA

LYSIMACHIA CILIATUM

LYSIMACHIA NUMMULARIA

PRIMULACEAE | PLUMBAGINACEAE | OLEACEAE

LYSIMACHIA PUNCTATA

LYSIMACHIA QUADRIFOLIA

LYSIMACHIA TERRESTRIS

LYSIMACHIA THYRSIFLORA

LYSIMACHIA VULGARIS

PRIMULA FARINOSA

PRIMULA MISTASSINICA

SAMOLUS VALERANDI SSP. PARVIFLORUS

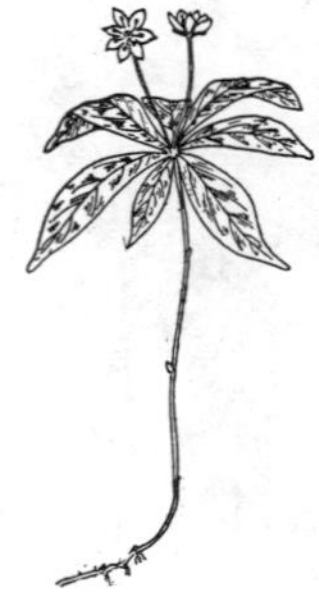

TRIENTALIS BOREALIS

LIMONIUM CAROLINIANUM

FRAXINUS AMERICANA

FRAXINUS NIGRA

OLEACEAE | GENTIANACEAE | APOCYNACEAE

FRAXINUS PENNSYLVANICA

SYRINGA VULGARIS

BARTONIA PANICULATA

CENTAURIUM PULCHELLUM

GENTIANA LINEARIS

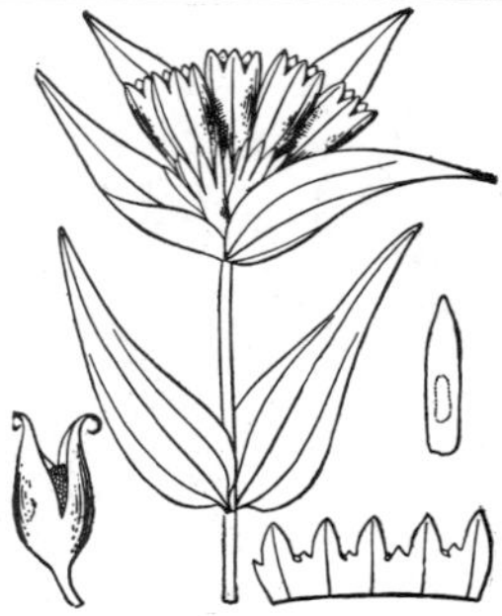

GENTIANA RUBRICAULIS

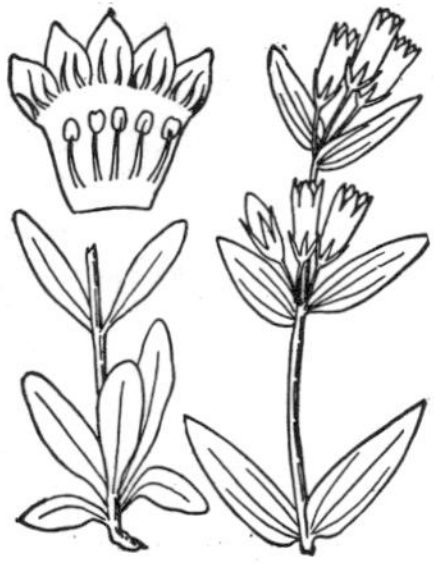

GENTIANELLA AMARELLA

HALENIA DEFLEXA

LOMATOGONIUM ROTATUM

APOCYNUM ANDROSAEMIFOLIUM

APOCYNUM SIBIRICUM

VINCA MINOR

ASCLEPIAS INCARNATA

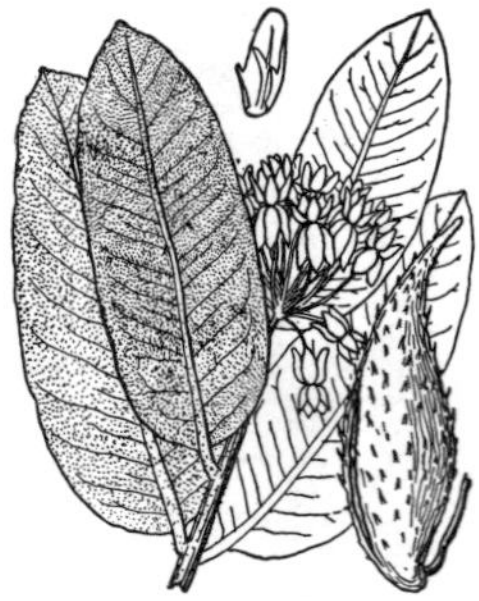

ASCLEPIAS SYRIACA

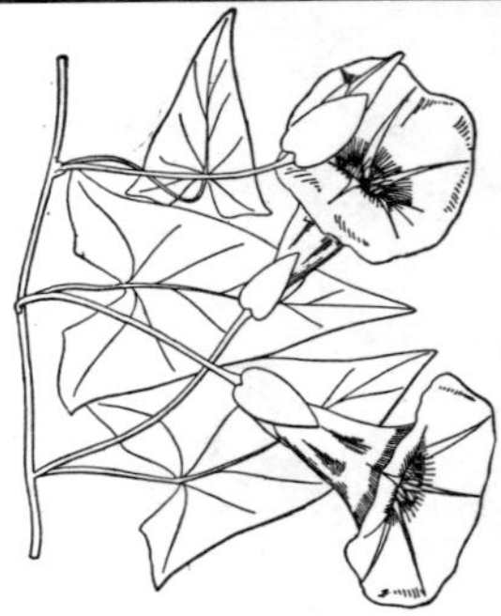

CALYSTEGIA SEPIUM

CONVOLVULUS ARVENSIS

CUSCUTA EPITHYMUM

CUSCUTA GRONOVII

CUSCUTA PENTAGONA

MENYANTHES TRIFOLIATA

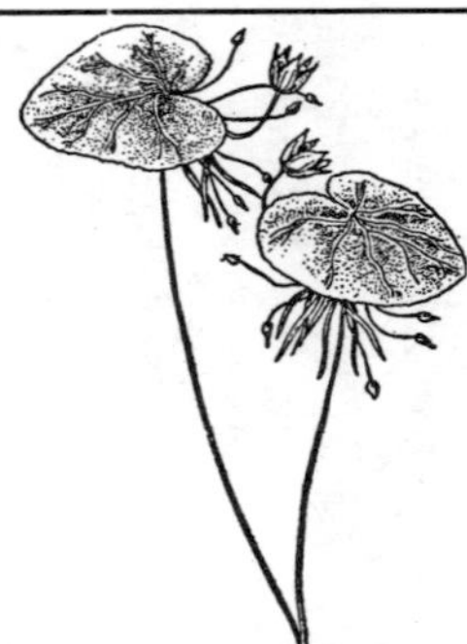

NYMPHOIDES CORDATA

COLLOMIA LINEARIS

POLEMONIUM VAN-BRUNTIAE

ANCHUSA ARVENSIS

BUGLOSSOIDES ARVENSE

BORAGO OFFICINALIS

CYNOGLOSSUM BOREALE

CYNOGLOSSUM OFFICINALE

ECHIUM VULGARE

HACKELIA DEFLEXA

LAPPULA SQUARROSA

MERTENSIA MARITIMA

MYOSOTIS ARVENSIS

MYOSOTIS LAXA

MYOSOTIS SCORPIOIDES

MYOSOTIS SYLVATICA

BORAGINACEAE VERBENACEAE LAMIACEAE

PLAGIOBOTHYS SCOULERI

SYMPHYTUM ASPERUM

SYMPHYTUM OFFICINALE

VERBENA HASTATA

VERBENA URTICIFOLIA

AGASTACHE FOENICULUM

CLINOPODIUM VULGARE

DRACOCEPHALUM PARVIFLORUM

ELSHOLTZIA CILIATA

GALEOPSIS LADANUM

GALEOPSIS TETRAHIT

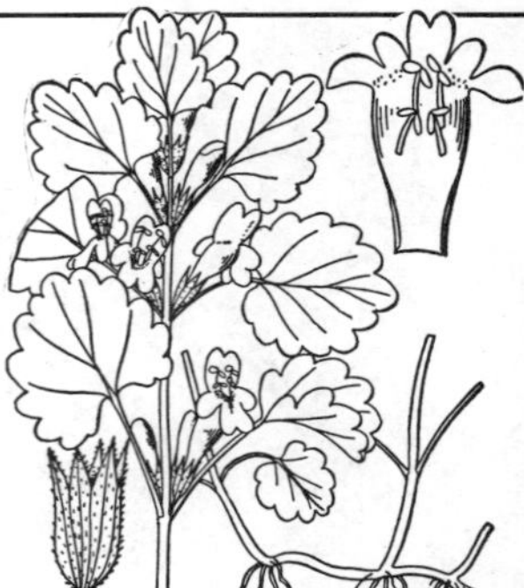

GLECHOMA HEDERACEA

HEDEOMA PULEGIOIDES

LAMIUM ALBUM

LAMIUM AMPLEXICAULE

LEONURUS CARDIACA

LYCOPUS AMERICANUS

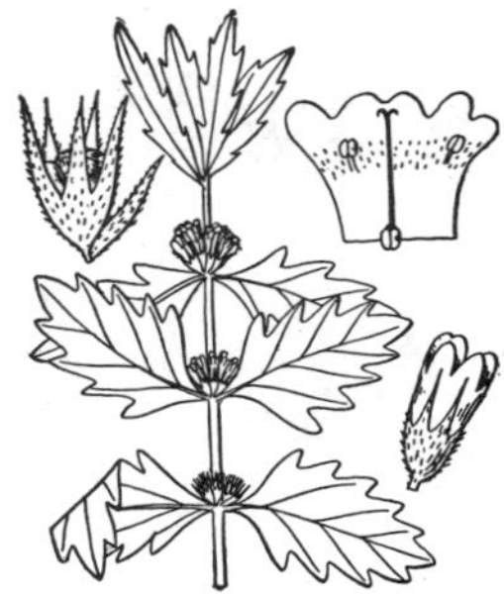

LYCOPUS EUROPAEUS

LYCOPUS VIRGINICUS

MENTHA ARVENSIS

MENTHA X GENTILIS

MENTHA SPICATA

MONARDA DIDYMA

NEPETA CATARIA

PHYSOSTEGIA VIRGINIANA

PRUNELLA VULGARIS

SCUTELLARIA GALERICULATA

SCUTELLARIA LATERIFLORA

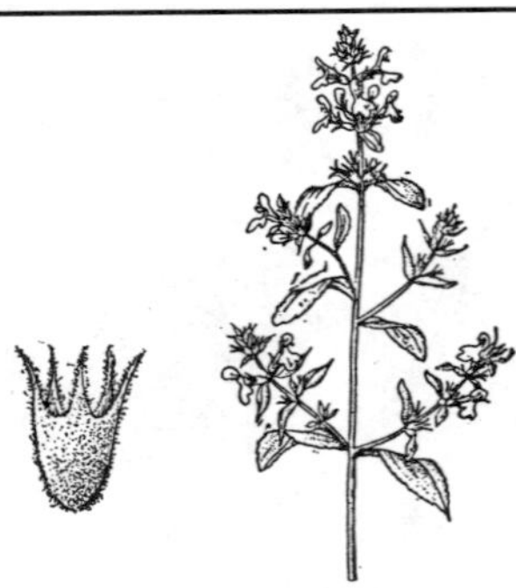

STACHYS ANNUA

STACHYS PALUSTRIS

STACHYS TENUIFOLIA

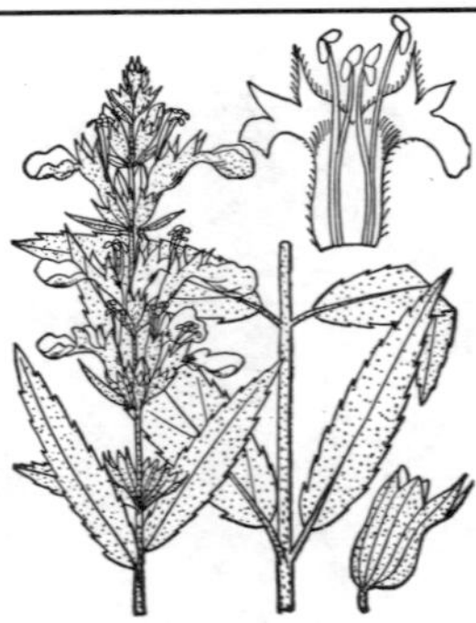

TEUCRIUM CANADENSE

DATURA STRAMONIUM

HYOSCYAMUS NIGER

PHYSALIS HETEROPHYLLA

PHYSALIS PUBESCENS

SOLANUM DULCAMARA

SOLANUM NIGRUM

AGALINIS MARITIMA

AGALINIS PURPUREA

CASTILLEJA SEPTENTRIONALIS

CHAENORRHINUM MINUS

CHELONE GLABRA

CYMBALARIA MURALIS

EUPHRASIA ARCTICA

EUPHRASIA NEMOROSA

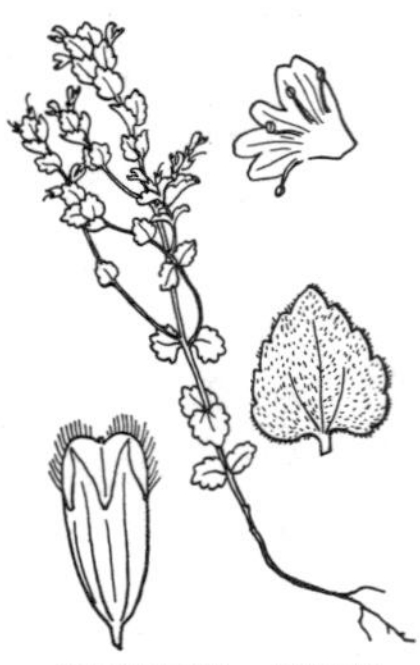

EUPHRASIA RANDII

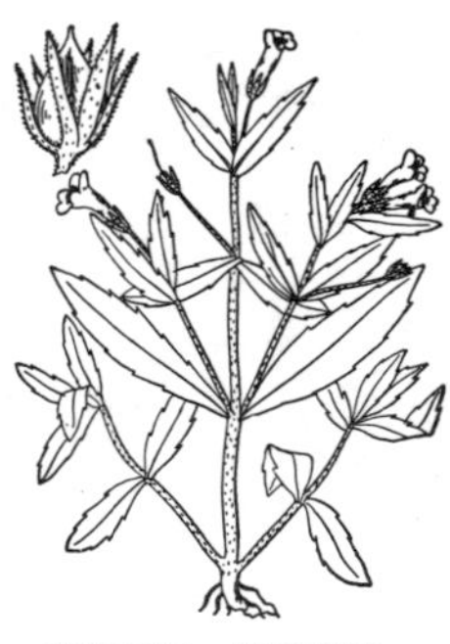

GRATIOLA NEGLECTA

LIMOSELLA AUSTRALIS

LINARIA CANADENSIS

LINARIA REPENS

LINARIA VULGARIS

LINDERNIA DUBIA

MELAMPYRUM LINEARE

MIMULUS GUTTATUS

MIMULUS MOSCHATUS

MIMULUS RINGENS

ODONTITES VERNA

PEDICULARIS CANADENSIS

PEDICULARIS FURBISHIAE

PENSTEMON DIGITALIS

RHINANTHUS MINOR

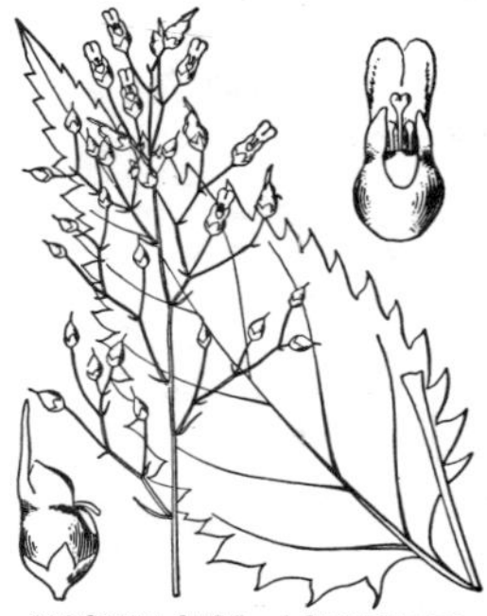

SCROPHULARIA LANCEOLATA

SCROPHULARIA NODOSA

VERBASCUM BLATTARIA

VERBASCUM THAPSUS

VERONICA AGRESTIS

VERONICA AMERICANA

VERONICA ARVENSIS

VERONICA CHAMAEDRYS

VERONICA OFFICINALIS

VERONICA PEREGRINA

SCROPHULARIACEAE | OROBANCHACEAE | LENTIBULARIACEAE

VERONICA PERSICA

VERONICA POLITA

VERONICA SCUTELLATA

VERONICA SERPYLLIFOLIA

VERONICA SPICATA

VERONICA VERNA

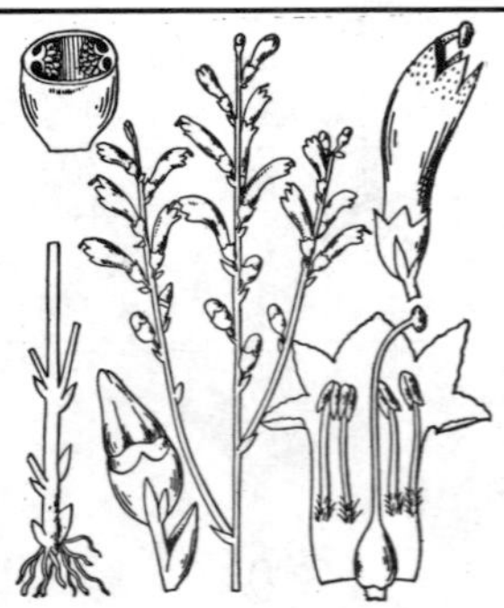

EPIFAGUS VIRGINIANA

OROBANCHE UNIFLORA

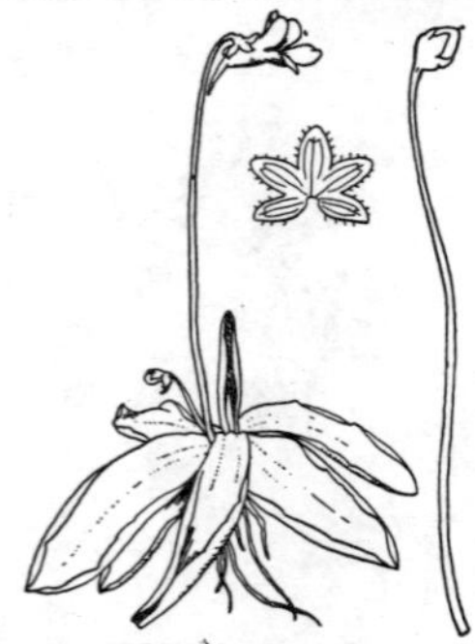

PINGUICULA VULGARIS

UTRICULARIA CORNUTA

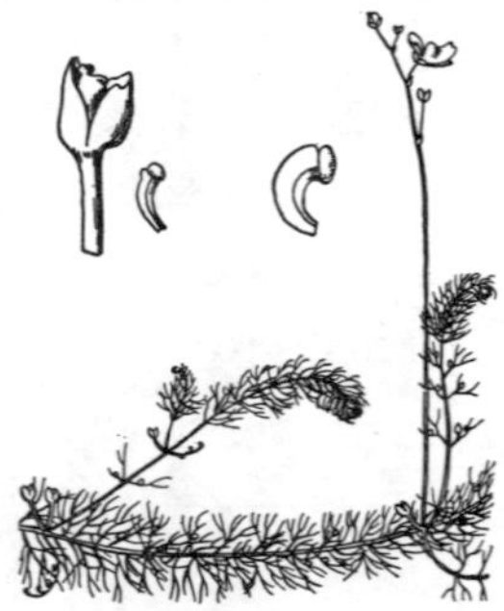

UTRICULARIA GEMINISCAPA

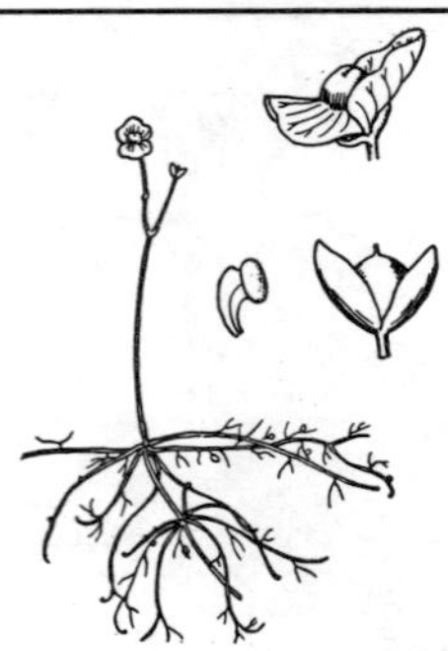

UTRICULARIA GIBBA

LENTIBULARIACEAE | VERBENACEAE | PLANTAGINACEAE

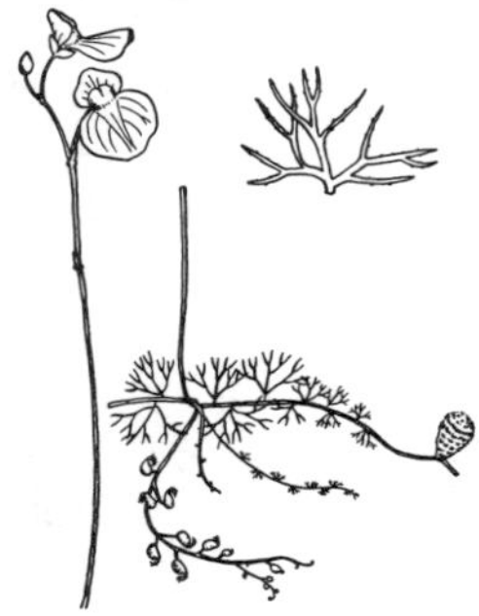
UTRICULARIA INTERMEDIA

UTRICULARIA MINOR

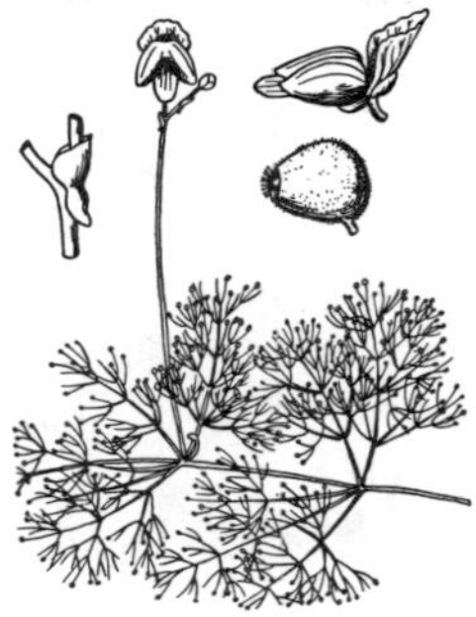
UTRICULARIA PURPUREA

UTRICULARIA RESUPINATA

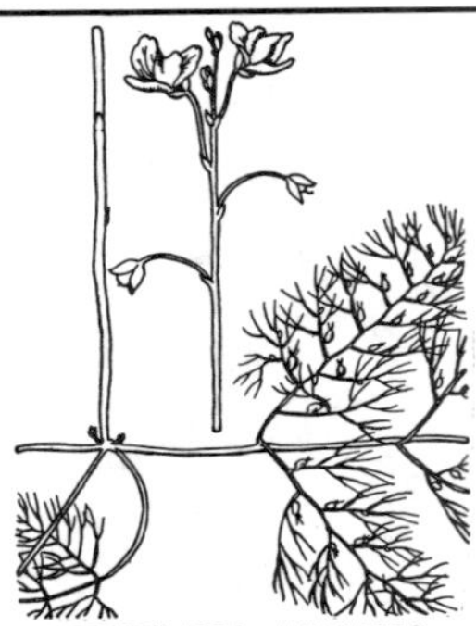
UTRICULARIA VULGARIS

PHRYMA LEPTOSTACHYA

LITTORELLA AMERICANA

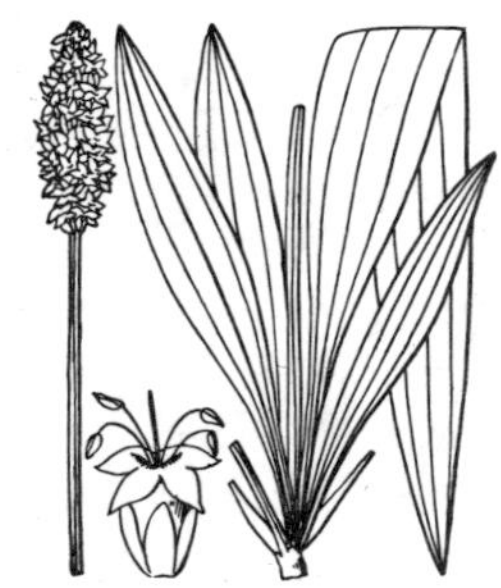
PLANTAGO LANCEOLATA

PLANTAGO MAJOR

PLANTAGO MARITIMA

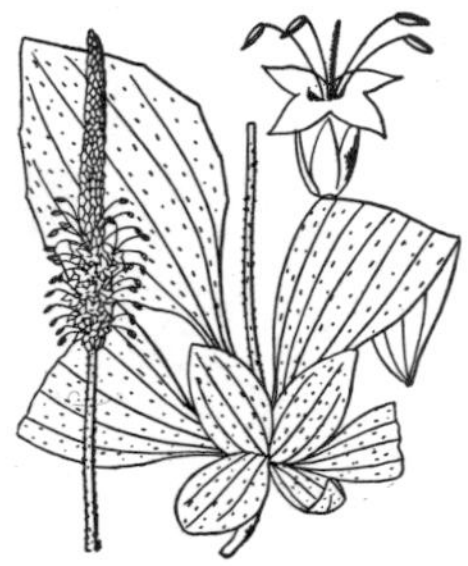
PLANTAGO MEDIA

PLANTAGO RUGELII

CEPHALANTHUS OCCIDENTALIS

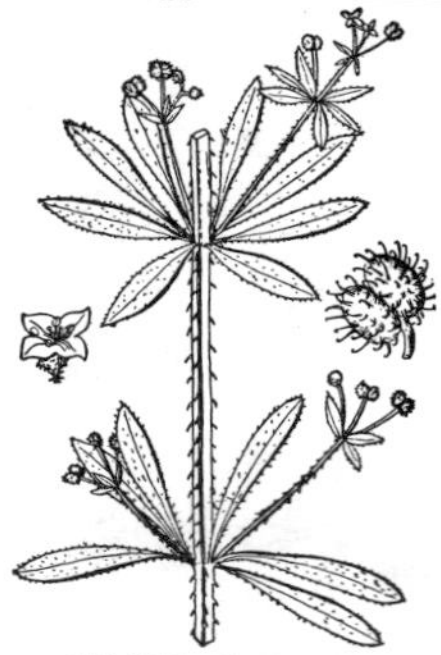

GALIUM APARINE

GALIUM ASPRELLUM

GALIUM BOREALE

GALIUM KAMTSCHATICUM

GALIUM LABRADORICUM

GALIUM MOLLUGO

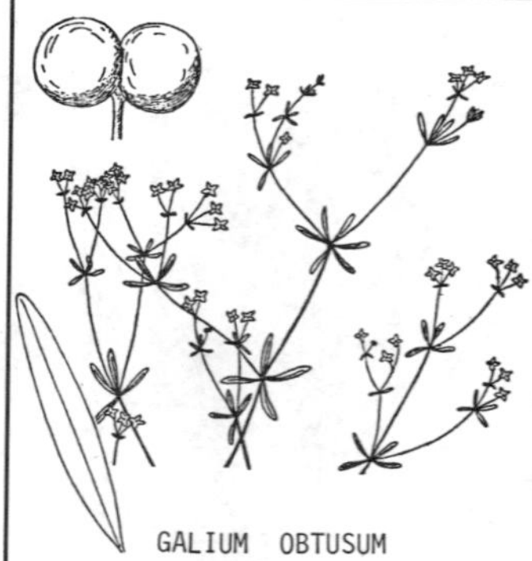

GALIUM OBTUSUM

GALIUM PALUSTRE

GALIUM TINCTORIUM

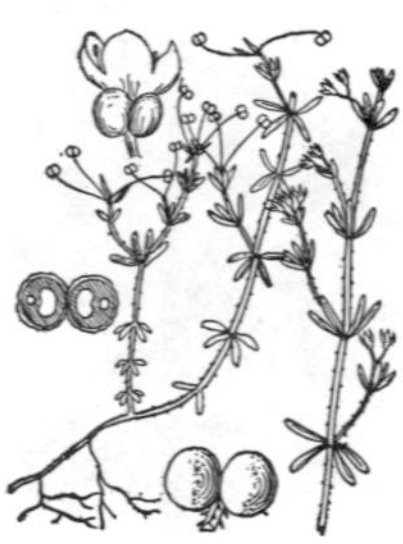

GALIUM TRIFIDUM

GALIUM TRIFLORUM

GALIUM VERUM

HEDYOTIS CAERULEA

MITCHELLA REPENS

DIERVILLA LONICERA

LINNAEA BOREALIS

LONICERA CANADENSIS

LONICERA MORROWII

LONICERA OBLONGIFOLIA

LONICERA TATARICA

LONICERA VILLOSA

LONICERA XYLOSTEUM

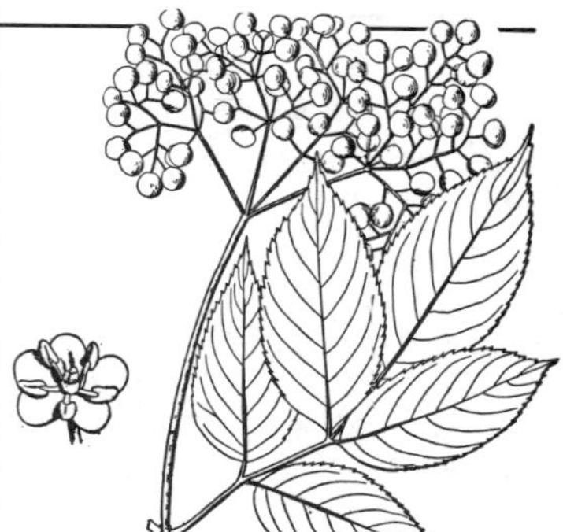

SAMBUCUS CANADENSIS

SAMBUCUS RACEMOSA

SYMPHORICARPOS ALBUS

TRIOSTEUM AURANTIACUM

VIBURNUM ACERIFOLIUM

VIBURNUM CASSINOIDES

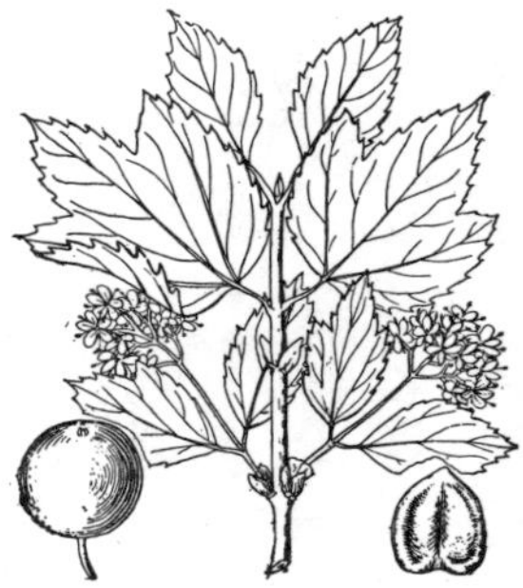

VIBURNUM EDULE

VIBURNUM LANTANA

VIBURNUM LANTANOIDES

VIBURNUM LENTAGO

VIBURNUM OPULUS

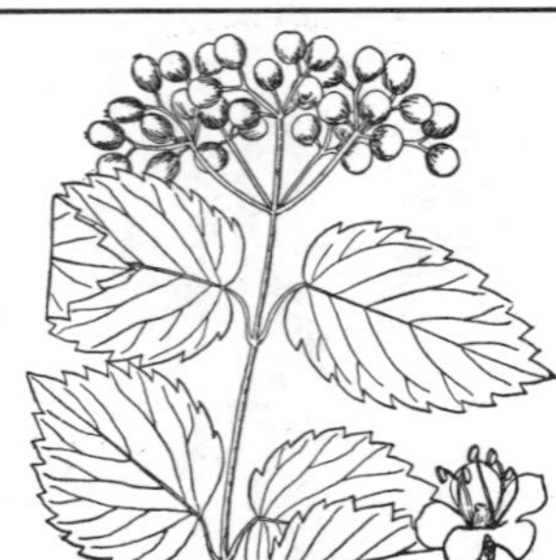

VIBURNUM RECOGNITUM

VALERIANA DIOICA SSP. SYLVATICA

VALERIANA OFFICINALIS

VALERIANA SITCHENSIS

KNAUTIA ARVENSIS

ECHINOCYSTIS LOBATA

CAMPANULA APARINOIDES

CAMPANULA RAPUNCULOIDES

CAMPANULA ROTUNDIFOLIA

TRIODANIS PERFOLIATA

LOBELIA CARDINALIS

LOBELIA DORTMANNA

LOBELIA INFLATA

LOBELIA KALMII

LOBELIA SPICATA

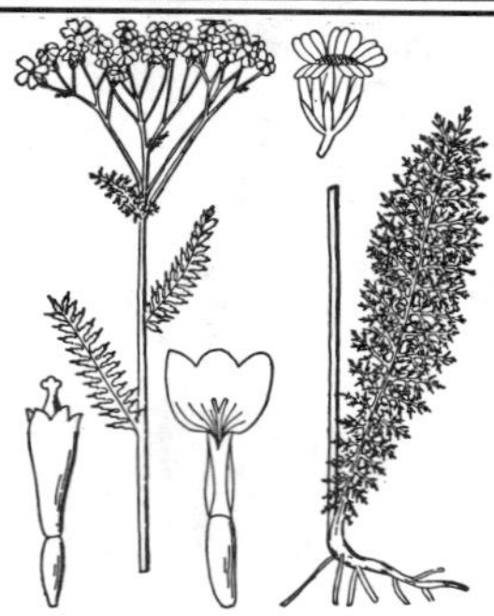

ACHILLEA MILLEFOLIUM

ACHILLEA PTARMICA

AMBROSIA ARTEMISIIFOLIA

AMBROSIA TRIFIDA

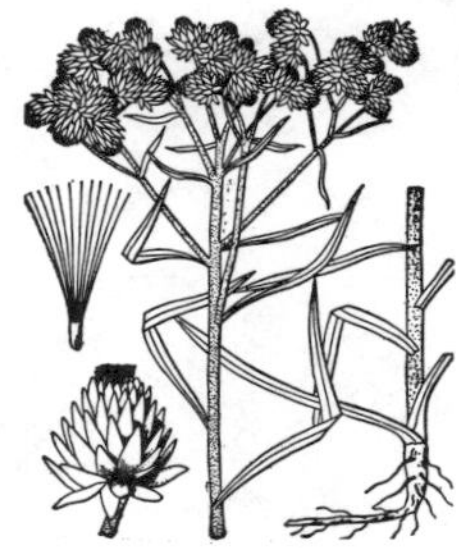

ANAPHALIS MARGARITACEA

ANTENNARIA CANADENSIS

ANTENNARIA NEGLECTA

ANTENNARIA PLANTAGINIFOLIA

ANTHEMIS ARVENSIS

ANTHEMIS COTULA

ANTHEMIS TINCTORIA

ARCTIUM LAPPA

ARCTIUM MINUS

ARCTIUM NEMOROSUM

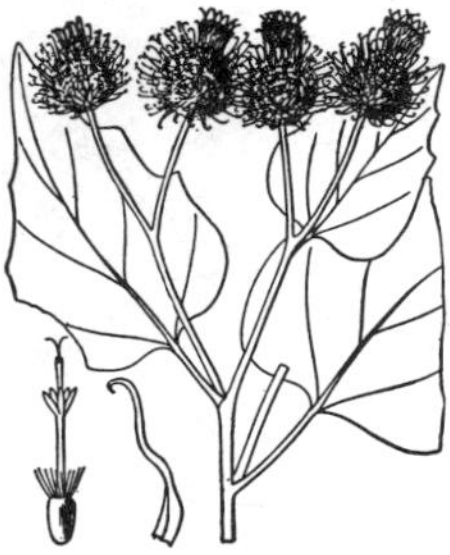

ARCTIUM TOMENTOSUM

ARNICA MOLLIS

ARTEMISIA ABSINTHIUM

ARTEMISIA ANNUA

ARTEMISIA BIENNIS

ARTEMISIA CAMPESTRIS

ARTEMISIA FRIGIDA

ARTEMISIA LUDOVICIANA

ARTEMISIA PONTICA

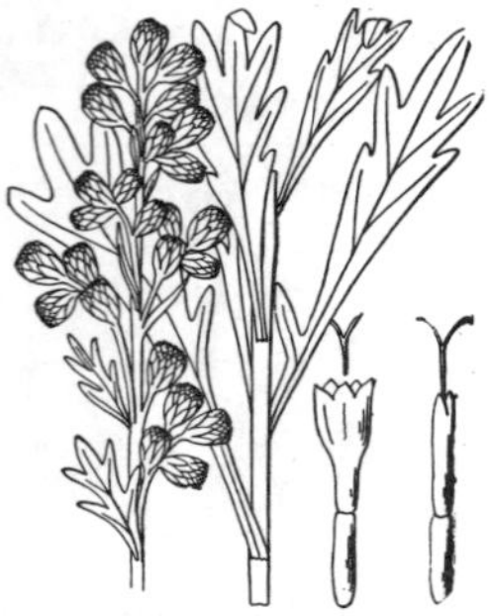

ARTEMISIA STELLERIANA

ARTEMISIA VULGARIS

ASTER ACUMINATUS

ASTER BOREALIS

ASTER CILIOLATUS

ASTER CORDIFOLIUS

ASTER LAEVIS

ASTER LATERIFLORUS

ASTER LAURENTIANUS

ASTER LINARIIFOLIUS

ASTER MACROPHYLLUS

ASTER NEMORALIS

ASTER NOVAE-ANGLIAE

ASTER NOVI-BELGII

ASTER PATENS

ASTER PRAEALTUS

ASTER PTARMICOIDES

ASTER PUNICEUS

ASTER RADULA

ASTER SIMPLEX

ASTER TARDIFLORUS

ASTER TRADESCANTII

ASTER UMBELLATUS

ASTER VIMINEUS

BELLIS PERENNIS

BIDENS CERNUA

BIDENS CONNATA

BIDENS DISCOIDEA

BIDENS FRONDOSA

BIDENS HYPERBOREA

BIDENS TRIPARTITA

BIDENS VULGATA

CARDUUS CRISPUS

CARDUUS NUTANS

CENTAUREA CYANUS

CENTAUREA MACULOSA

CENTAUREA NIGRA

CENTAUREA SCABIOSA

CHRYSANTHEMUM SEGETUM

CICHORIUM INTYBUS

CIRSIUM ARVENSE

CIRSIUM MUTICUM

CIRSIUM VULGARE

CNICUS BENEDICTUS

CONYZA CANADENSIS

COTULA CORONOPIFOLIA

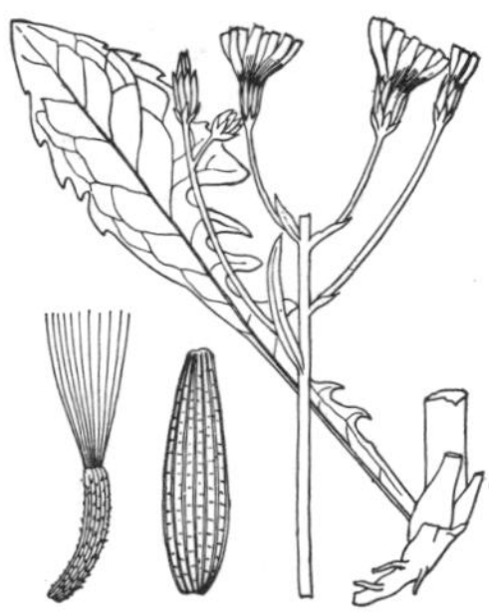
CREPIS BIENNIS

CREPIS CAPILLARIS

CREPIS TECTORUM

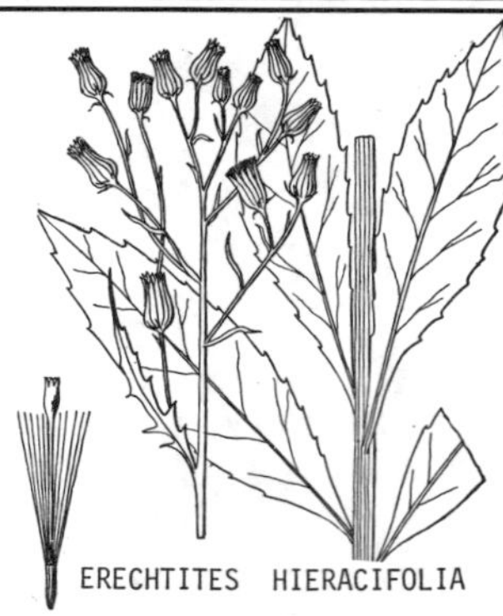

ERECHTITES HIERACIFOLIA

ERIGERON ACRIS

ERIGERON ANNUUS

ERIGERON HYSSOPIFOLIUS

ERIGERON PHILADELPHICUS

ERIGERON STRIGOSUS

EUPATORIUM MACULATUM

EUPATORIUM PERFOLIATUM

EUPATORIUM PURPUREUM

EUPATORIUM RUGOSUM

GALINSOGA PARVIFLORA

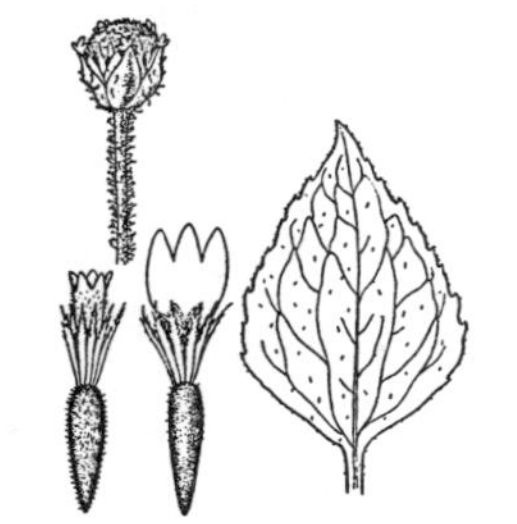

GALINSOGA QUADRIRADIATA

GNAPHALIUM OBTUSIFOLIUM

GNAPHALIUM SYLVATICUM

GNAPHALIUM ULIGINOSUM

GNAPHALIUM VISCOSUM

HELIANTHUS ANNUUS

HELIANTHUS DECAPETALUS

HELIANTHUS GIGANTEUS

HELIANTHUS LAETIFLORUS

HELIANTHUS NUTTALLII

HELIANTHUS STRUMOSUS

HELIANTHUS TUBEROSUS

HELIOPSIS HELIANTHOIDES

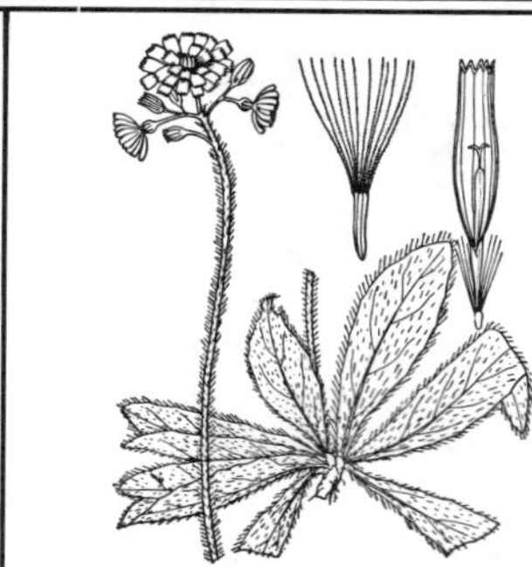

HIERACIUM AURANTIACUM

HIERACIUM CAESPITOSUM

HIERACIUM CANADENSE

HIERACIUM FLORIBUNDUM

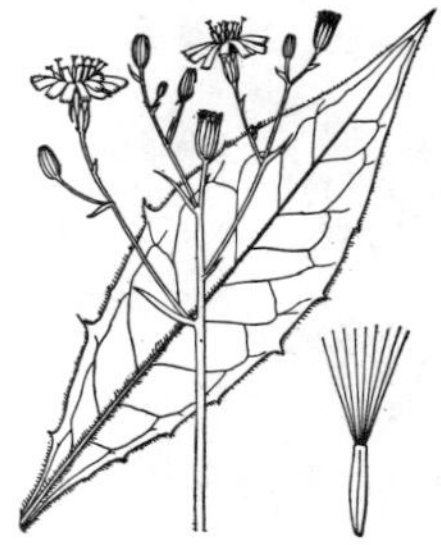

HIERACIUM LACHENALII

HIERACIUM MURORUM

HIERACIUM PANICULATUM

HIERACIUM PILOSELLA

HIERACIUM PILOSELLOIDES

HIERACIUM ROBINSONII

HIERACIUM SCABRUM

HIERACIUM UMBELLATUM

HYPOCHAERIS RADICATA

INULA HELENIUM

LACTUCA BIENNIS

LACTUCA CANADENSIS

LACTUCA SERRIOLA

LAPSANA COMMUNIS

LEONTODON AUTUMNALIS

LEUCANTHEMUM VULGARE

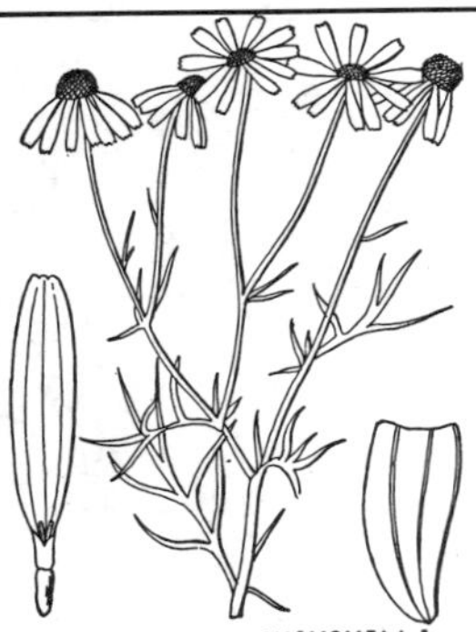

MATRICARIA CHAMOMILLA

MATRICARIA MARITIMA

MATRICARIA MATRICARIOIDES

MEGALODONTA BECKII

ONOPORDON ACANTHIUM

PETASITES FRIGIDUS

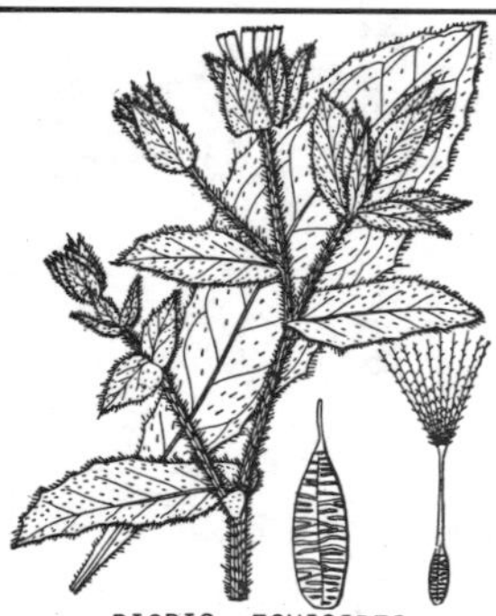

PICRIS ECHIOIDES

PRENANTHES ALTISSIMA

PRENANTHES RACEMOSA

PRENANTHES TRIFOLIOLATA

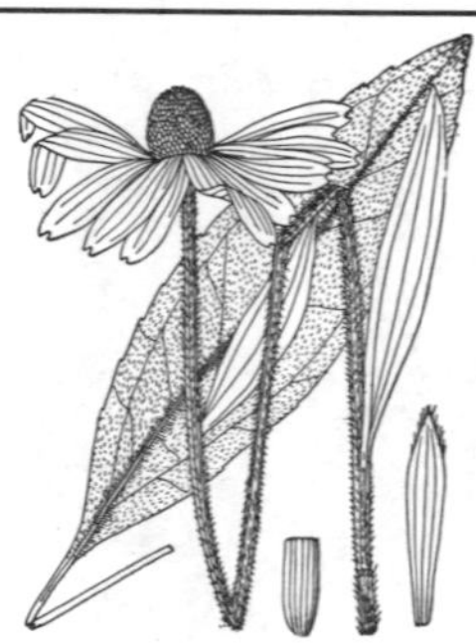

RUDBECKIA HIRTA

RUDBECKIA LACINIATA

SENECIO AUREUS

SENECIO JACOBAEA

SENECIO PAUPERCULUS

SENECIO PSEUDOARNICA

SENECIO SCHWEINITZIANUS

SENECIO SYLVATICUS

SENECIO VISCOSUS

SENECIO VULGARIS

SILYBUM MARIANUM

SOLIDAGO BICOLOR

SOLIDAGO CAESIA

SOLIDAGO CANADENSIS

SOLIDAGO FLEXICAULIS

SOLIDAGO GIGANTEA

SOLIDAGO GRAMINIFOLIA

SOLIDAGO HISPIDA

SOLIDAGO JUNCEA

SOLIDAGO MACROPHYLLA

SOLIDAGO MULTIRADIATA

SOLIDAGO NEMORALIS

SOLIDAGO PUBERULA

SOLIDAGO RUGOSA

SOLIDAGO SEMPERVIRENS

SOLIDAGO SPATHULATA

SOLIDAGO SQUARROSA

SOLIDAGO ULIGINOSA

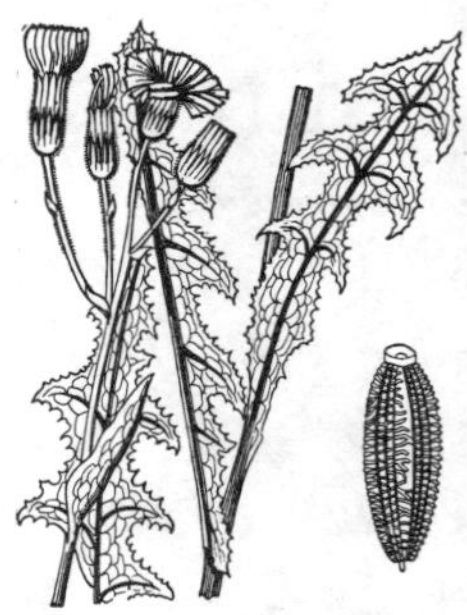

SONCHUS ARVENSIS

SONCHUS ASPER

SONCHUS OLERACEUS

TANACETUM BIPINNATUM

TANACETUM PARTHENIUM

TARAXACUM LAEVIGATUM

TANACETUM VULGARE

TARAXACUM OFFICINALE

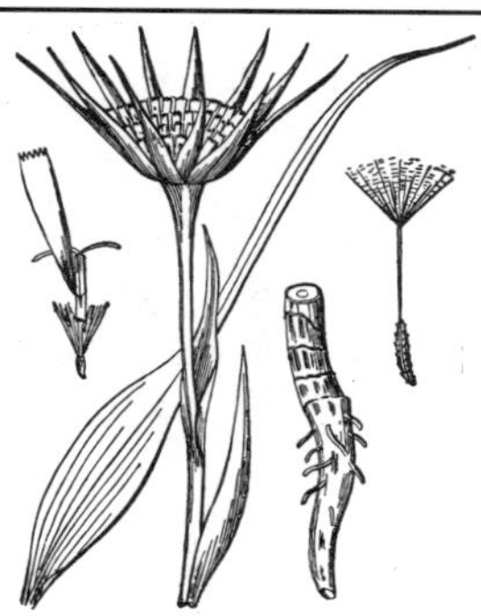

TRAGOPOGON PORRIFOLIUS

TRAGOPOGON PRATENSIS

TUSSILAGO FARFARA

XANTHIUM SPINOSUM

XANTHIUM STRUMARIUM

EQUISETUM ARVENSE

EQUISETUM FLUVIATILE

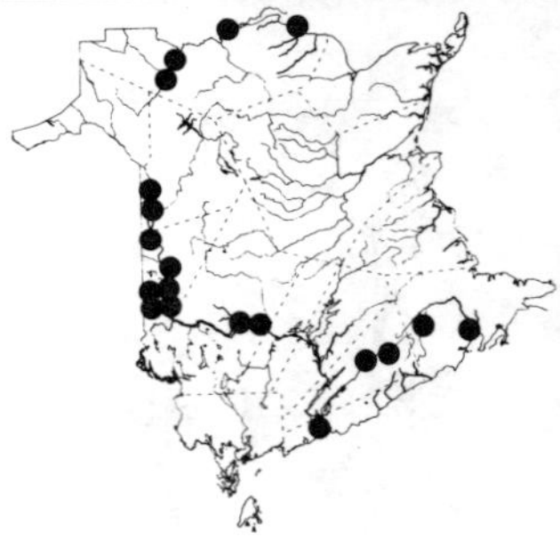

EQUISETUM HYEMALE

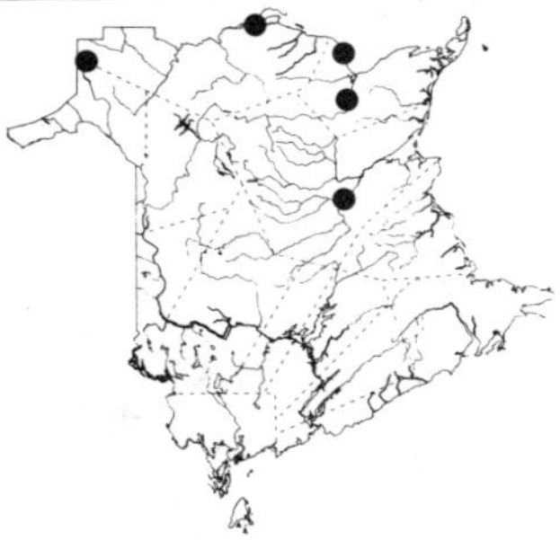

X EQUISETUM LITORALE

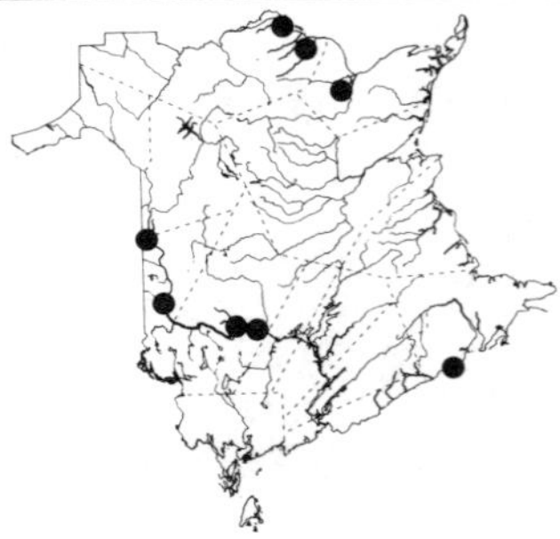

EQUISETUM PALUSTRE

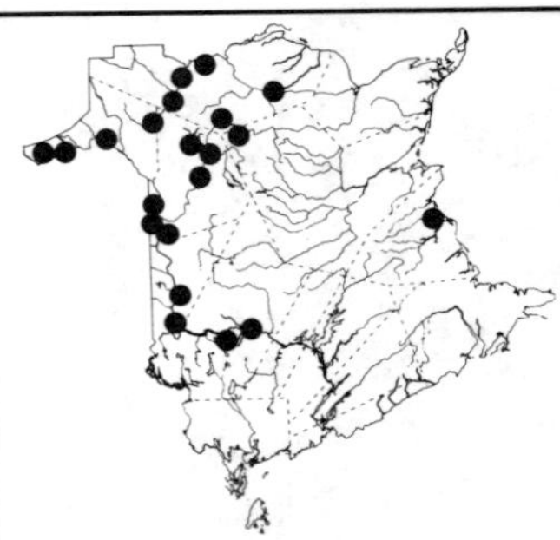

EQUISETUM PRATENSE

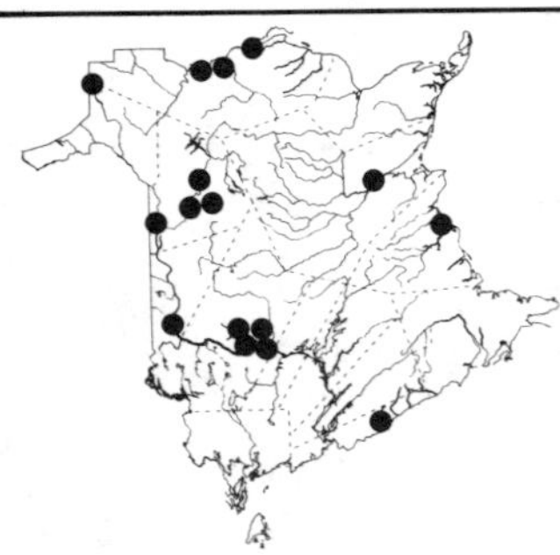

EQUISETUM SCIRPOIDES

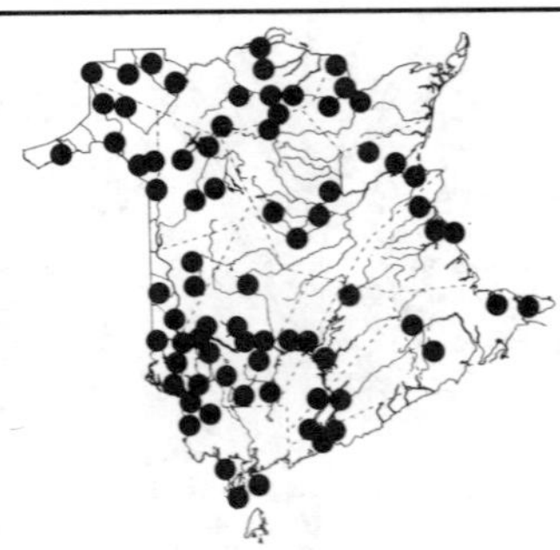

EQUISETUM SYLVATICUM

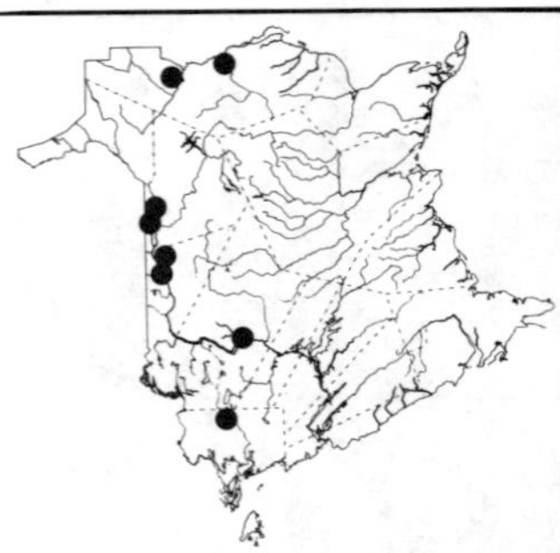

EQUISETUM VARIEGATUM

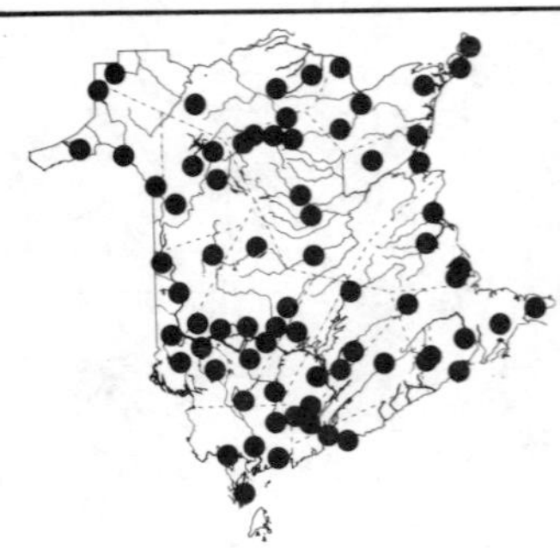

LYCOPODIUM ANNOTINUM

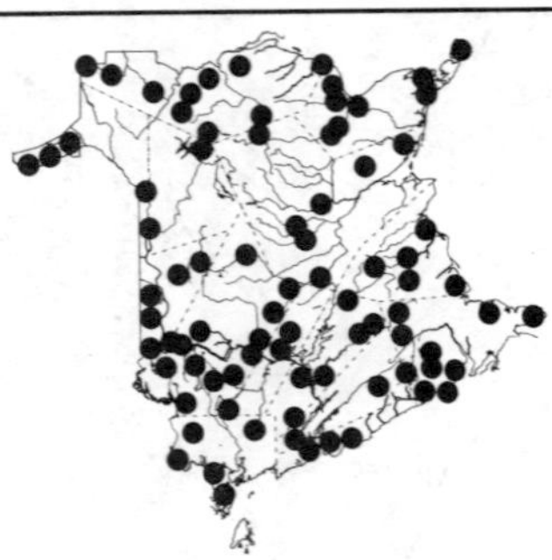

LYCOPODIUM CLAVATUM

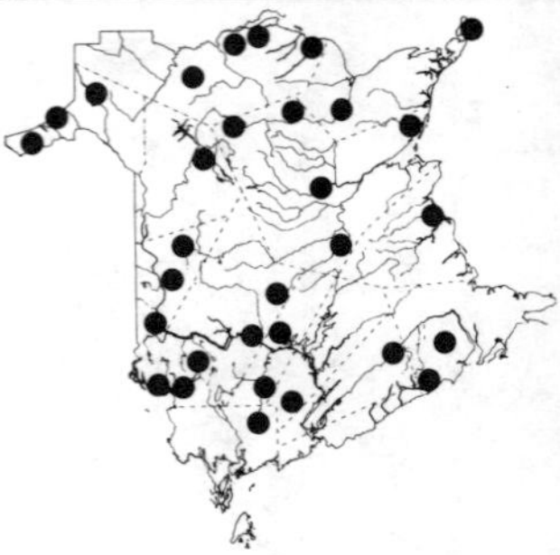

LYCOPODIUM COMPLANATUM

LYCOPODIACEAE | SELAGINELLACEAE | ISOETACEAE

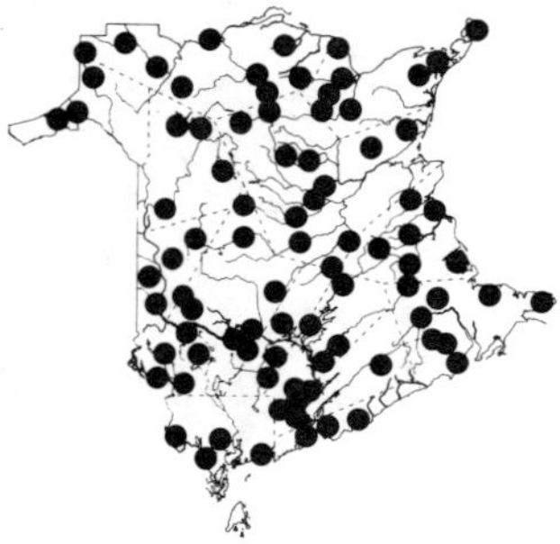

LYCOPODIUM DENDROIDEUM

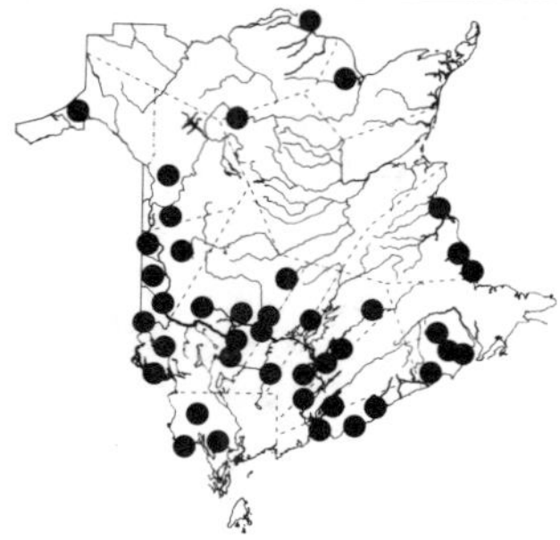

LYCOPODIUM DIGITATUM

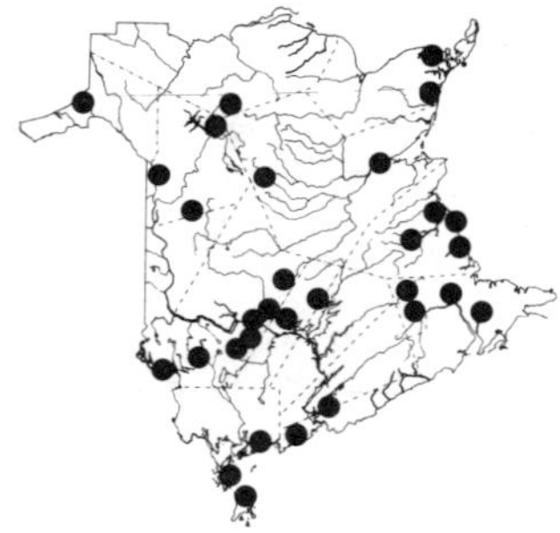

LYCOPODIUM INUNDATUM

LYCOPODIUM LUCIDULUM

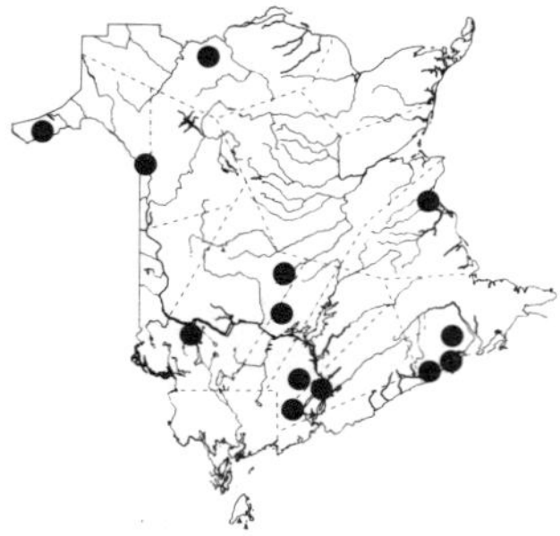

LYCOPODIUM SABINAEFOLIUM

LYCOPODIUM SELAGO

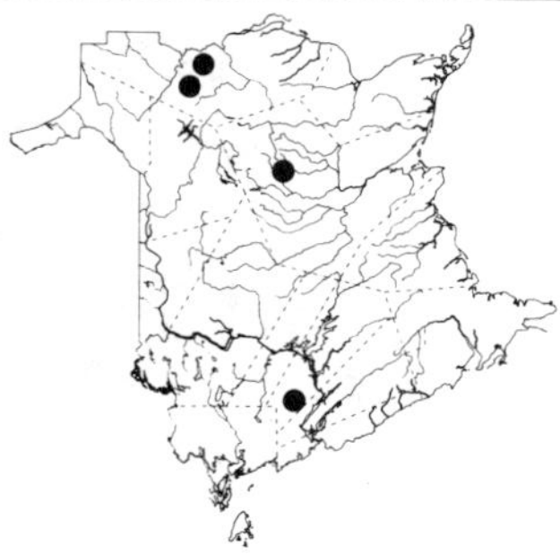

LYCOPODIUM SITCHENSE

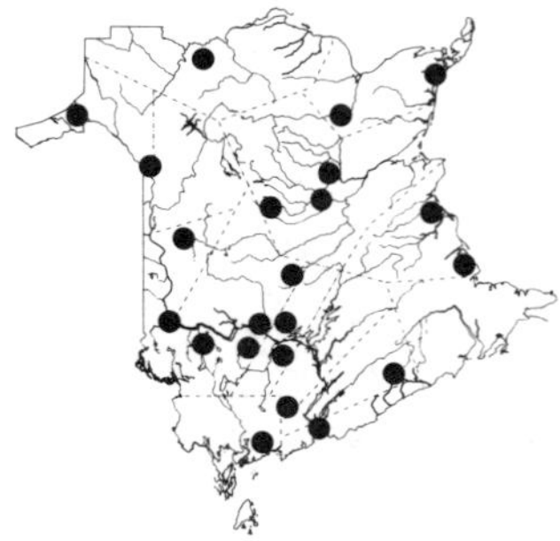

LYCOPODIUM TRISTACHYM

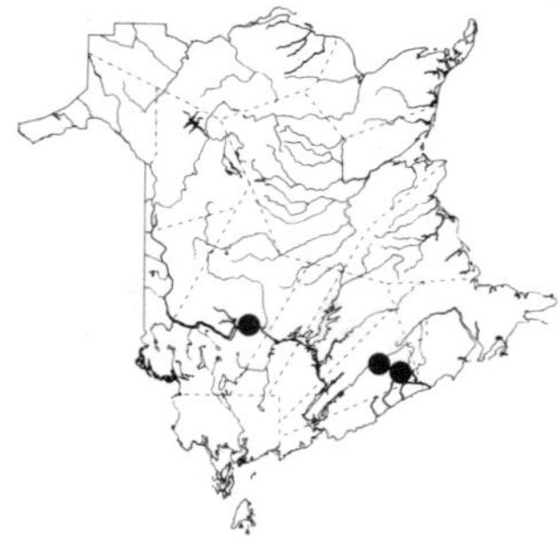

SELAGINELLA RUPESTRIS

SELAGINELLA SELAGINOIDES

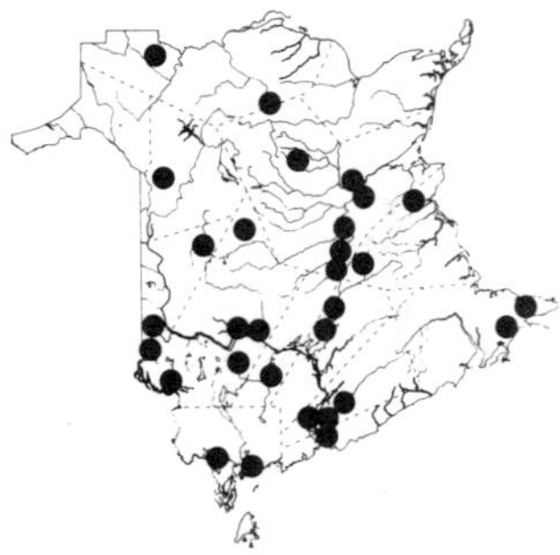

ISOETES ECHINOSPORA

X ISOETES HARVEYI

ISOETES TUCKERMANII

BOTRYCHIUM DISSECTUM

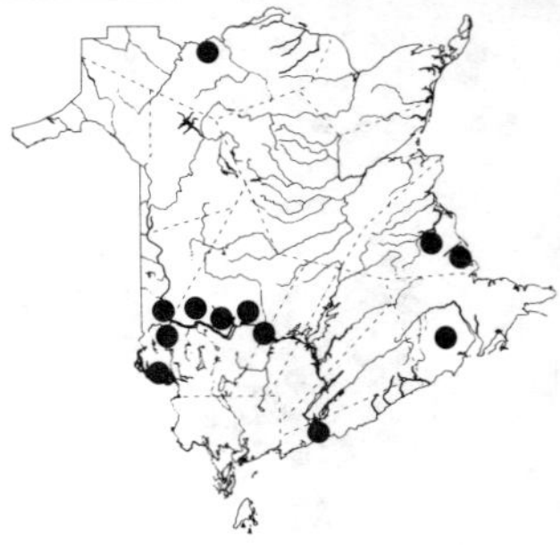

BOTRYCHIUM LANCEOLATUM

BOTRYCHIUM MATRICARIAEFOLIUM

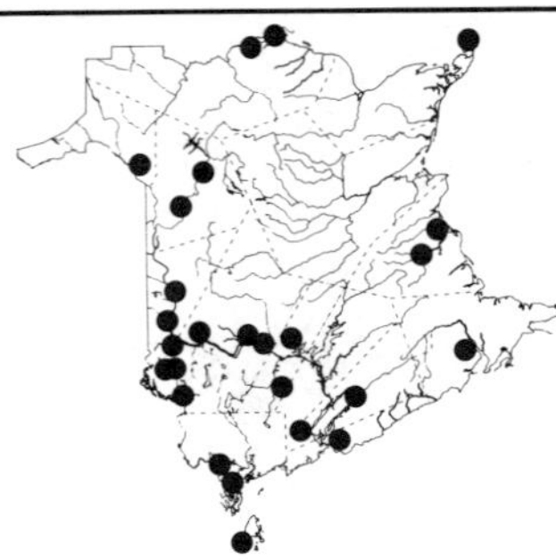

BOTRYCHIUM MULTIFIDUM

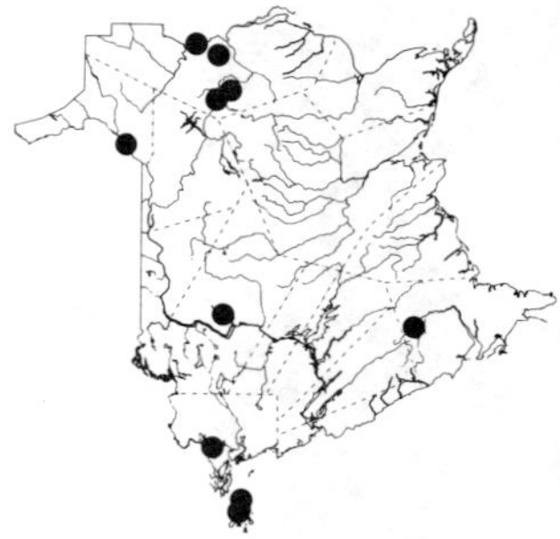

BOTRYCHIUM SIMPLEX

BOTRYCHIUM VIRGINIANUM

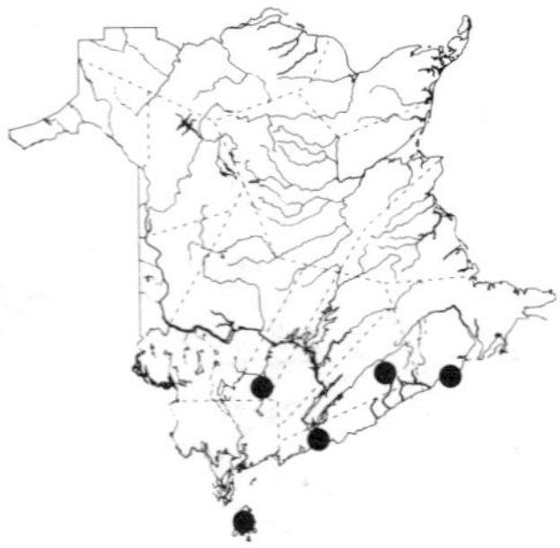

OPHIOGLOSSUM VULGATUM

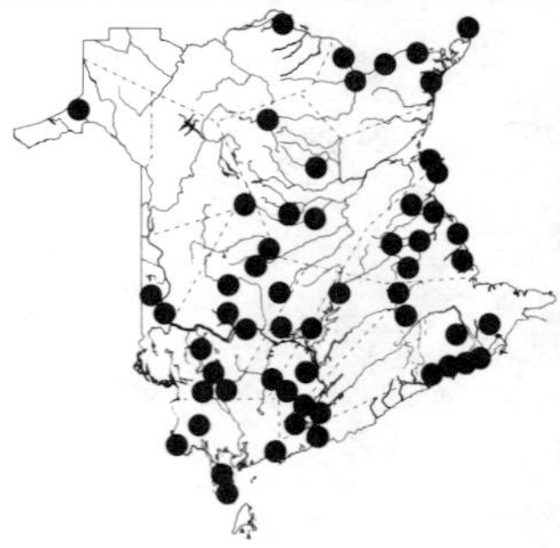

OSMUNDA CINNOMOMEA

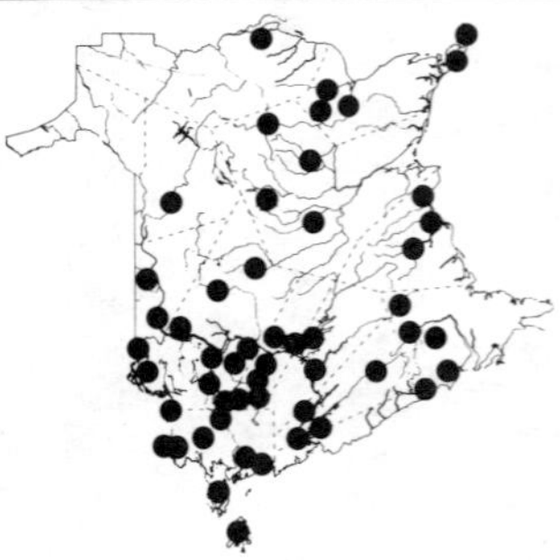

OSMUNDA REGALIS

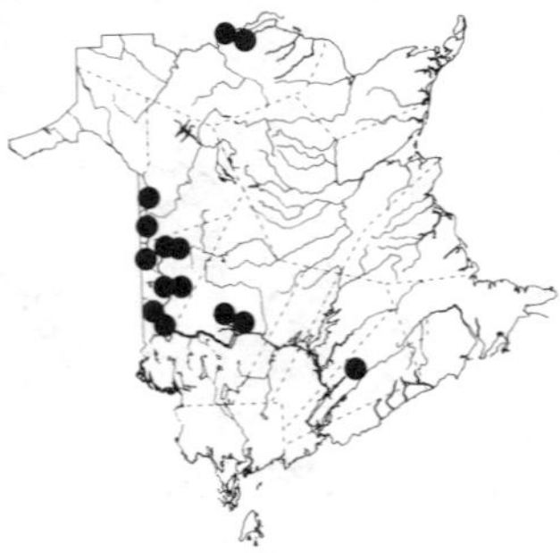

ADIANTUM PEDATUM

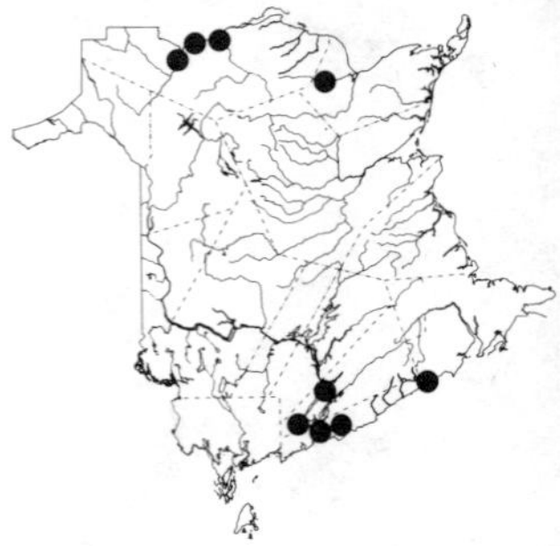

ASPLENIUM VIRIDE

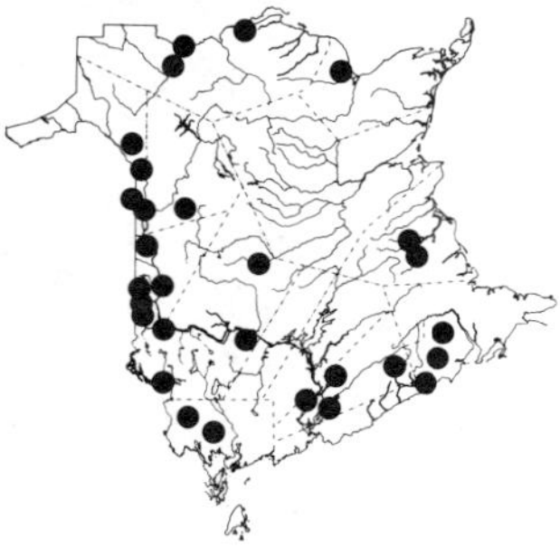
ATHYRIUM THELYPTEROIDES

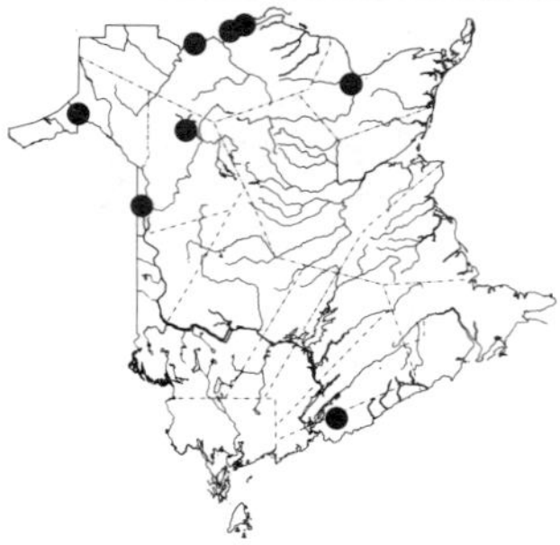
CRYPTOGRAMMA STELLERI

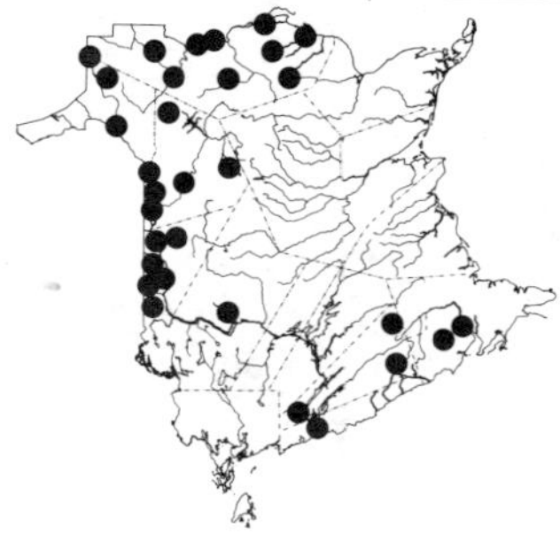
CYSTOPTERIS BULBIFERA

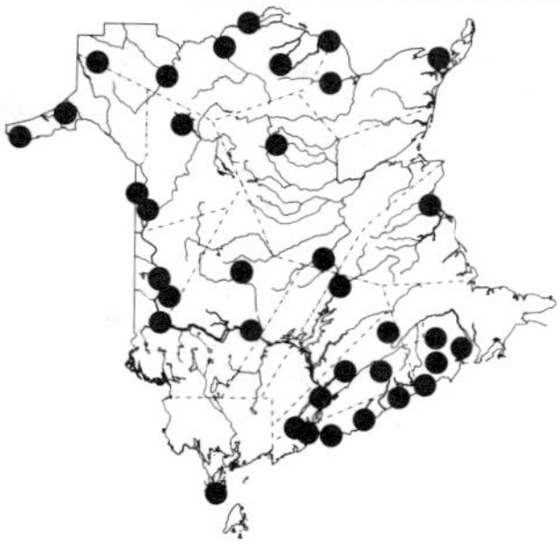
CYSTOPTERIS FRAGILIS

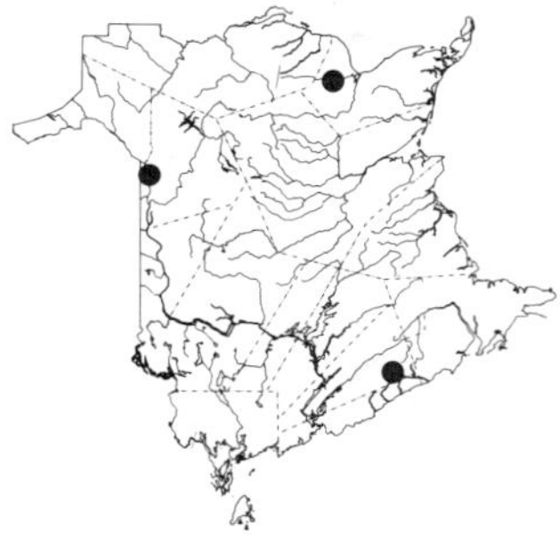
CYSTOPTERIS LAURENTIANA

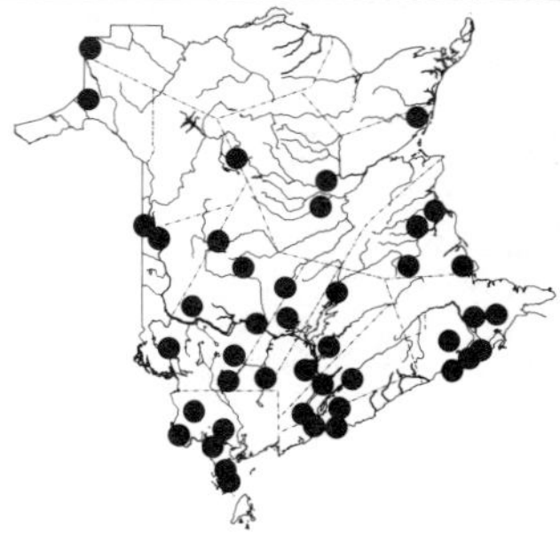
DENNSTAEDTIA PUNCTILOBULA

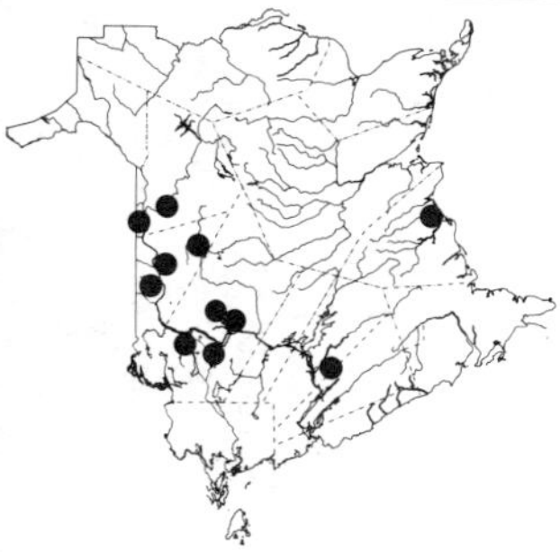
DRYOPTERIS BOOTTII

DRYOPTERIS CAMPYLOPTERA

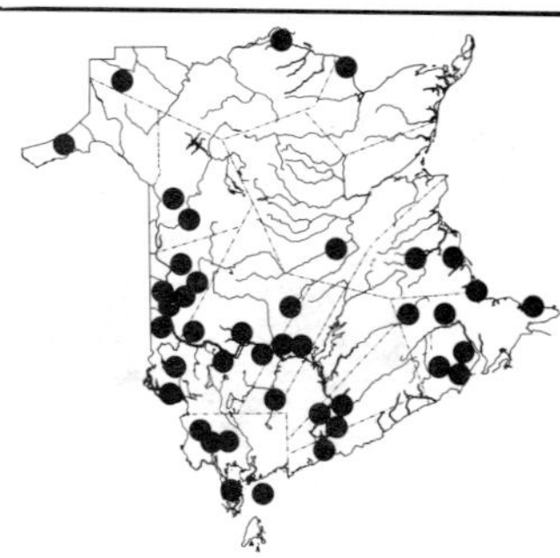
DRYOPTERIS CRISTATA

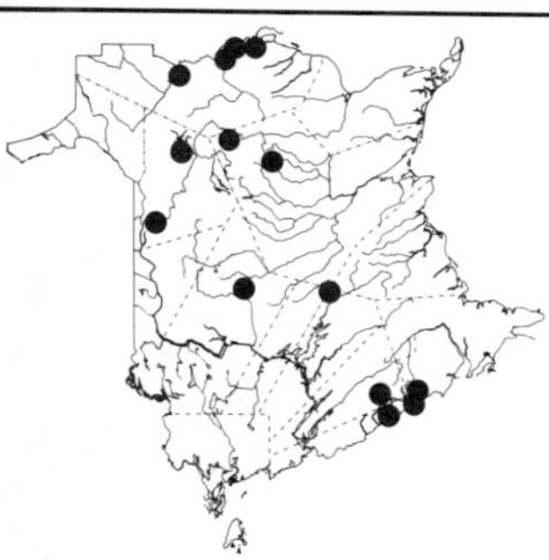
DRYOPTERIS FRAGRANS

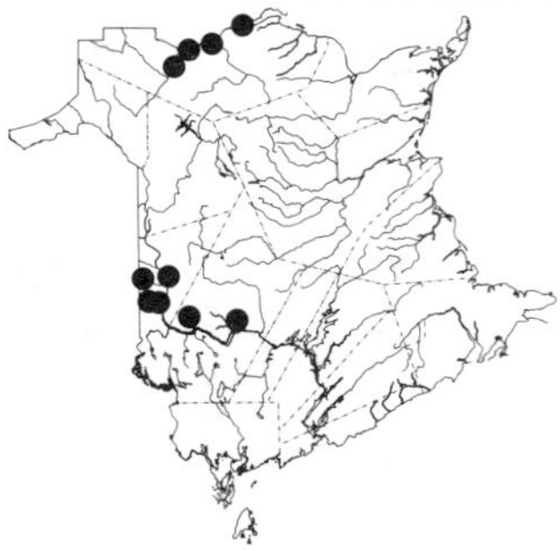
DRYOPTERIS GOLDIANA

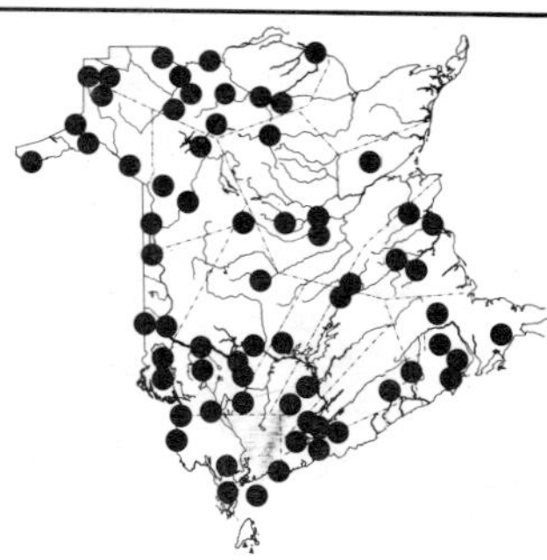
DRYOPTERIS INTERMEDIA

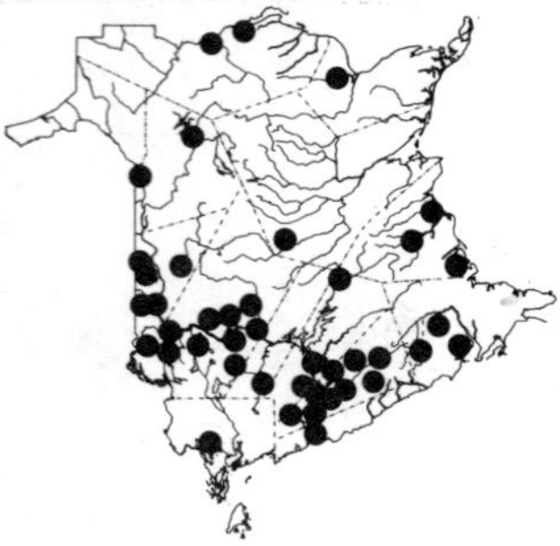

DRYOPTERIS MARGINALIS

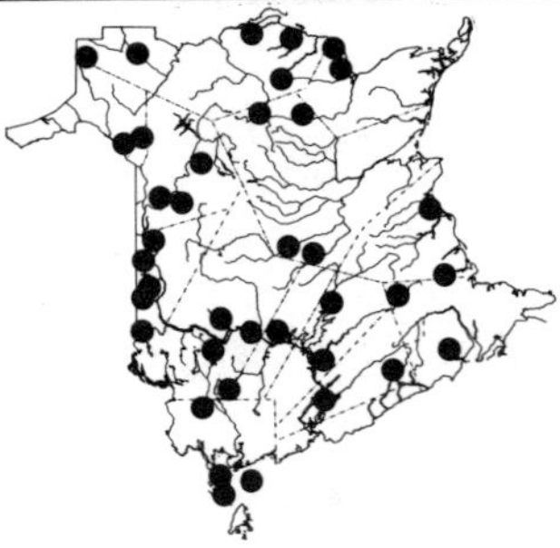

DRYOPTERIS SPINULOSE

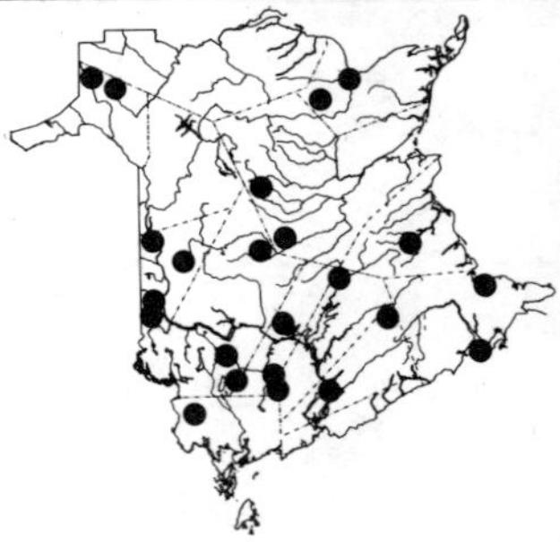

DRYOPTERIS X TRIPLOIDEA

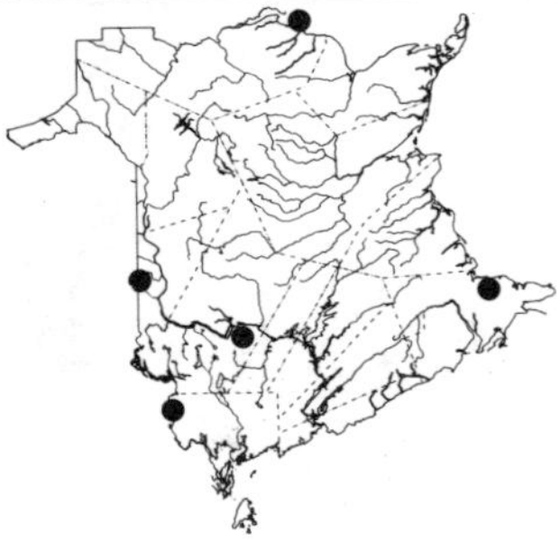

DRYOPTERIS X ULIGINOSA

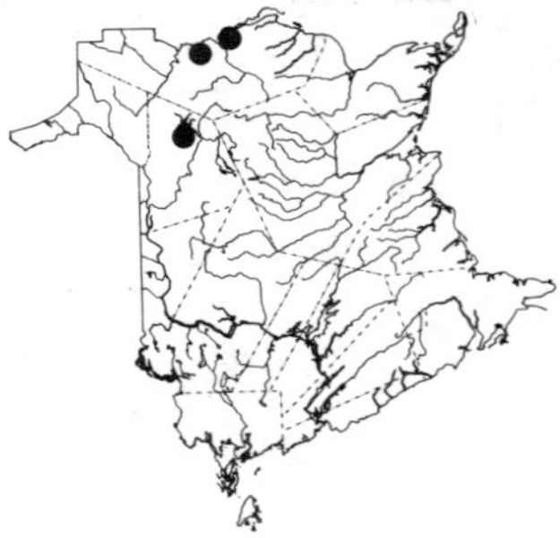

GYMNOCARPIUM ROBERTIANUM

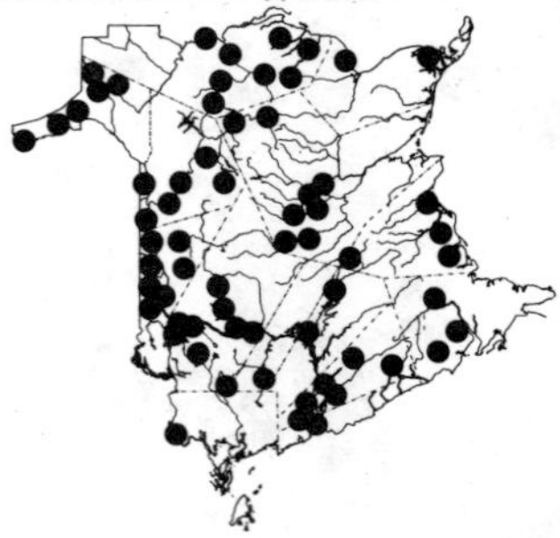

MATTEUCCIA STRUTHIOPTERIS

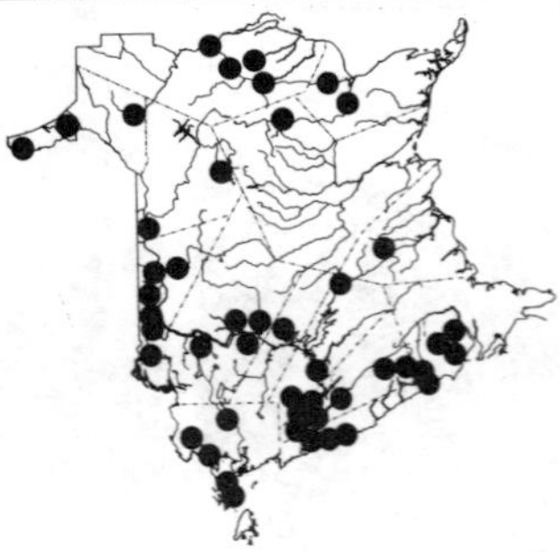

POLYPODIUM VIRGINIANUM

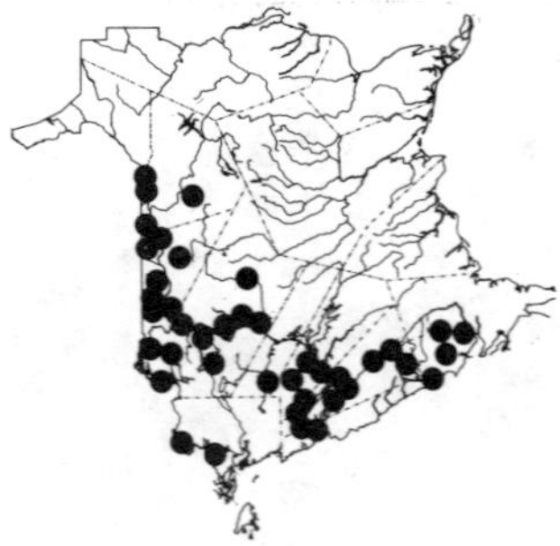

POLYSTICHUM ACROSTICHOIDES

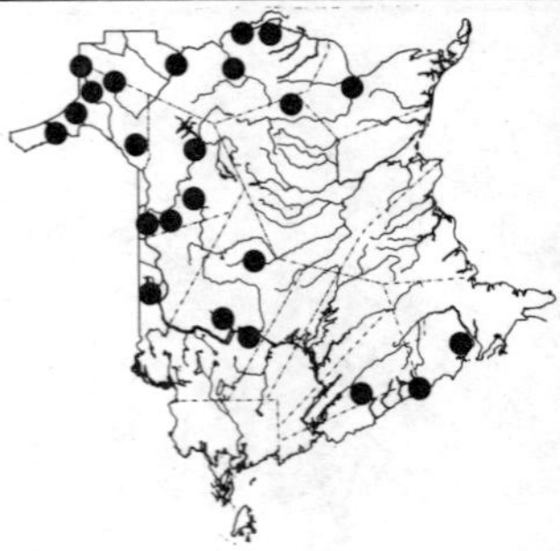

POLYSTICHUM BRAUNII

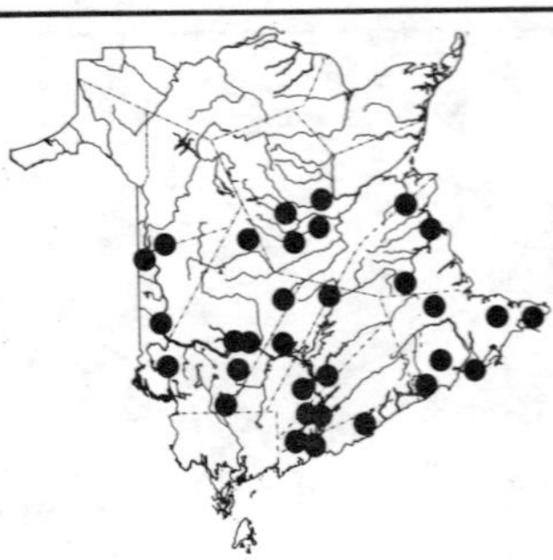

THELYPTERIS NOVEBORACENSIS

THELYPTERIS PALUSTRIS

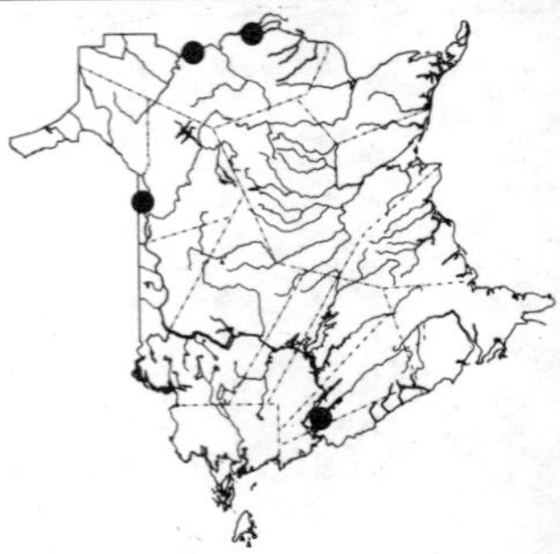

WOODSIA ALPINA

POLYPODIACEAE | TAXACEAE | PINACEAE

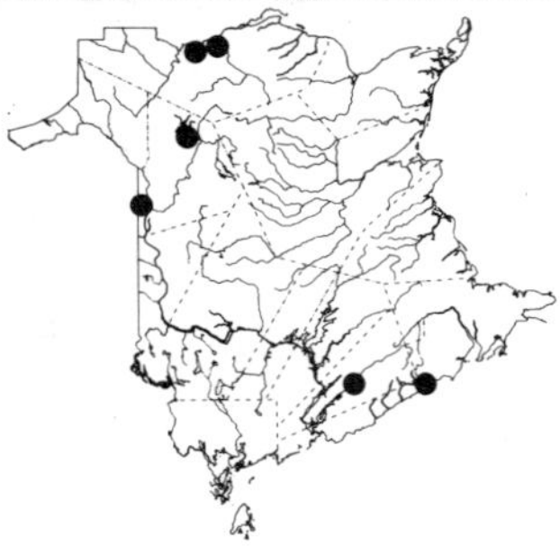

WOODSIA GLABELLA

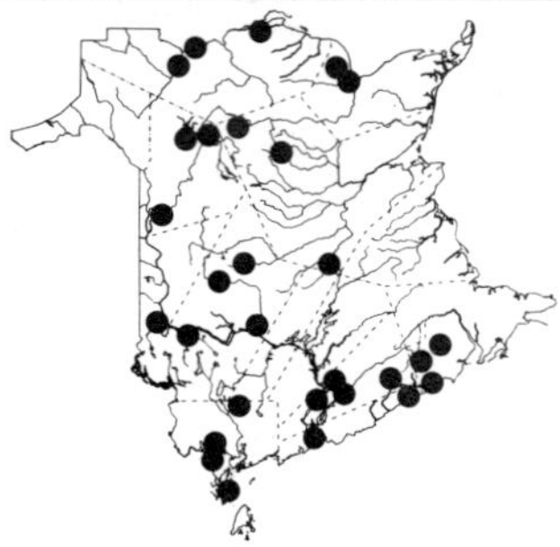

WOODSIA ILVENSIS

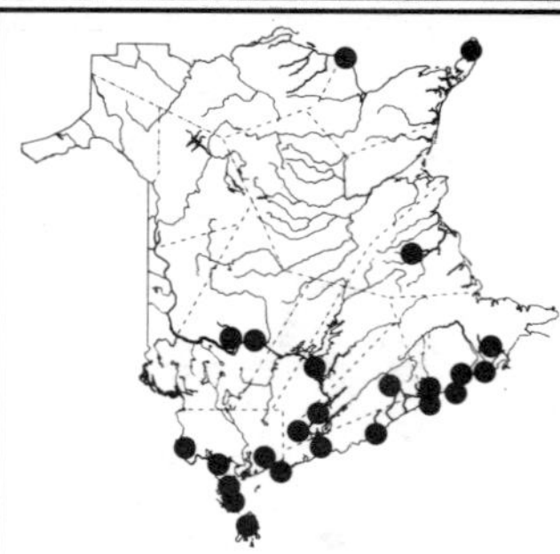

JUNIPERUS COMMUNIS

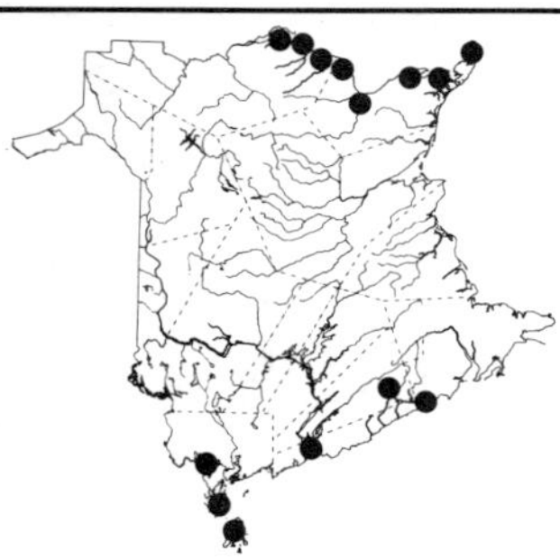

JUNIPERUS HORIZONTALIS

LARIX LARICINA

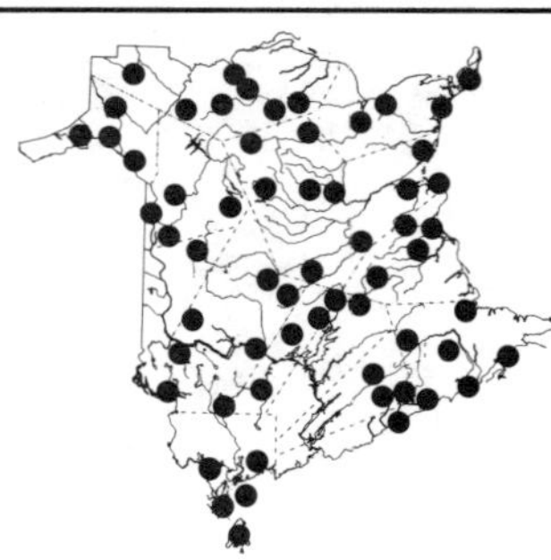

PICEA MARIANA

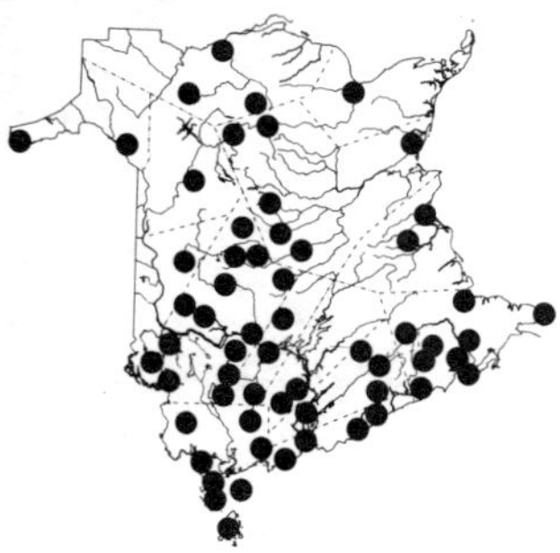

PICEA RUBENS

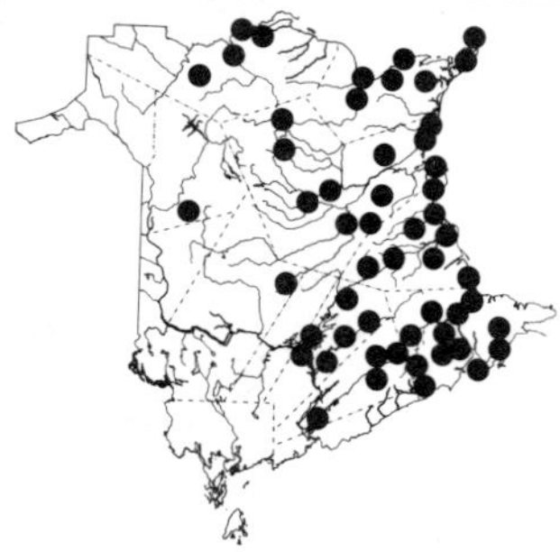

PINUS BANKSIANA

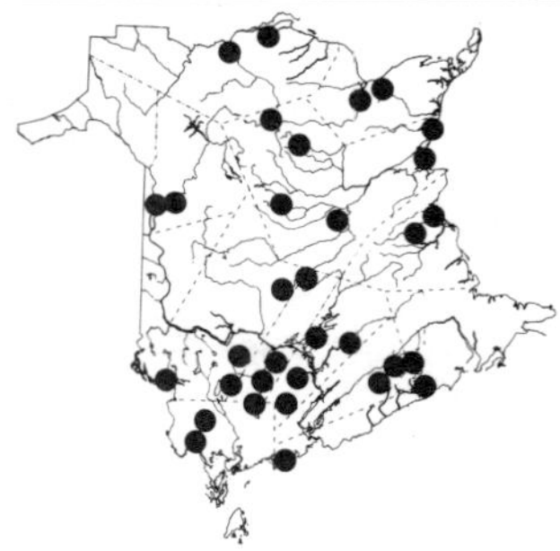

PINUS RESINOSA

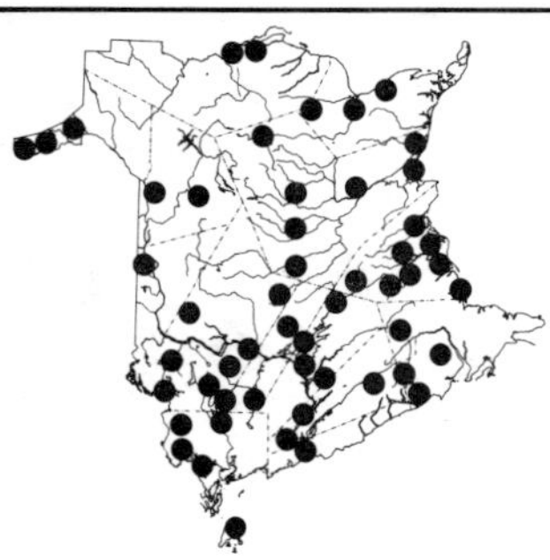

PINUS STROBUS

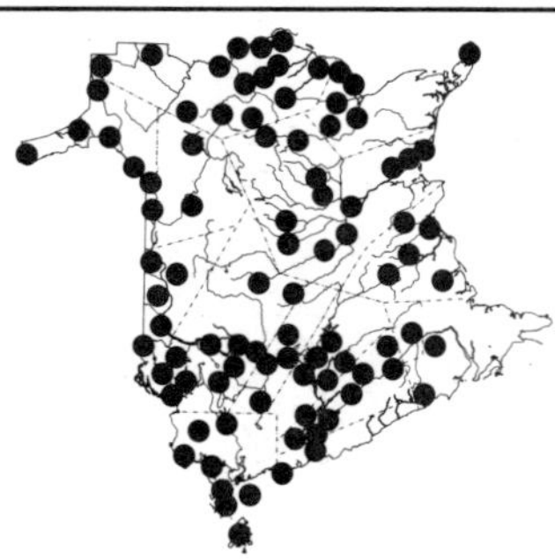

THUJA OCCIDENTALIS

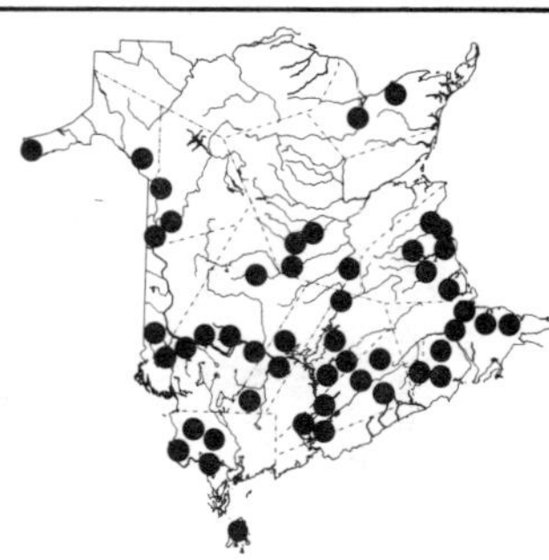

TSUGA CANADENSIS

TYPHACEAE SPARGANIACEAE POTAMOGETONACEAE

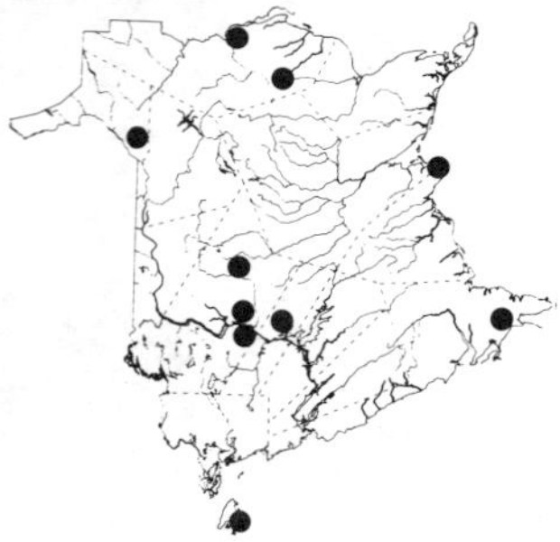

TYPHA ANGUSTIFOLIA

TYPHA LATIFOLIA

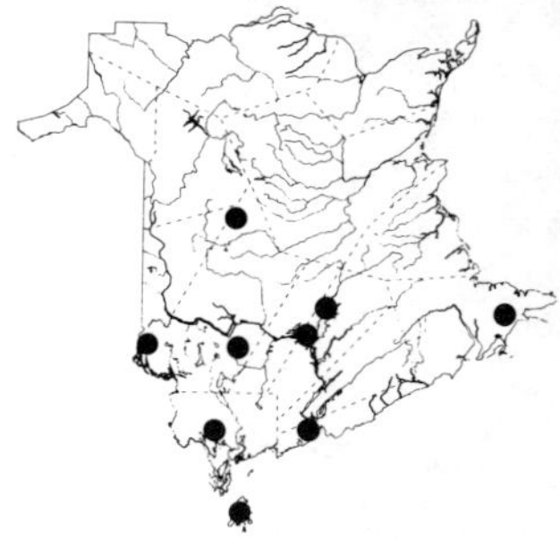

SPARGANIUM AMERICANUM

SPARGANIUM CHLOROCARPUM

SPARGANIUM EMERSUM

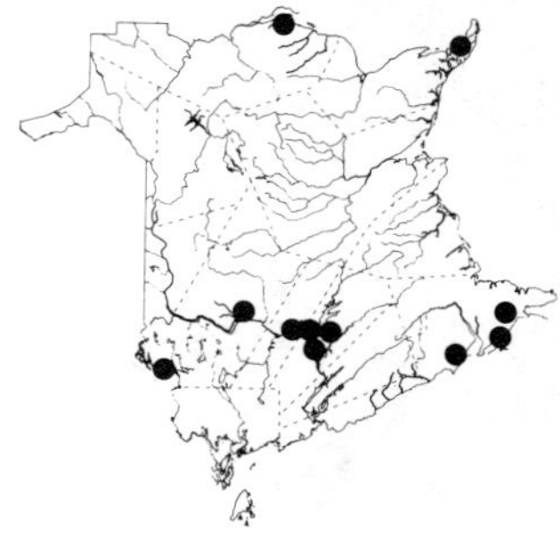

SPARGANIUM EURYCARPUM

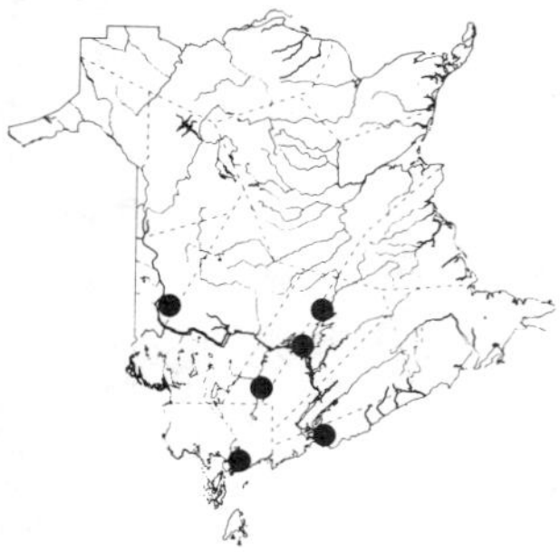

SPARGANIUM FLUCTUANS

SPARGANIUM MINIMUM

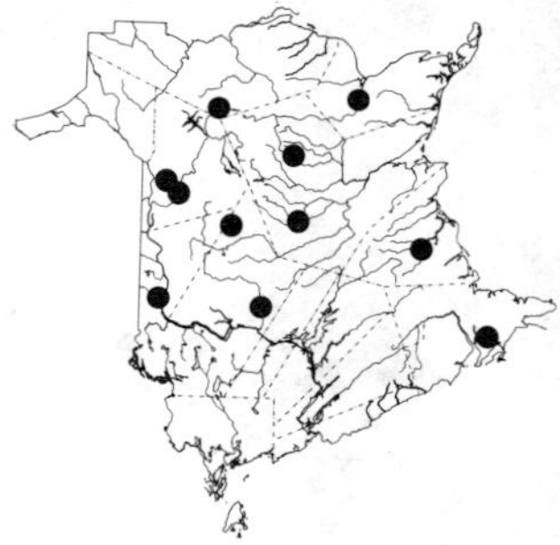

POTAMOGETON ALPINUS

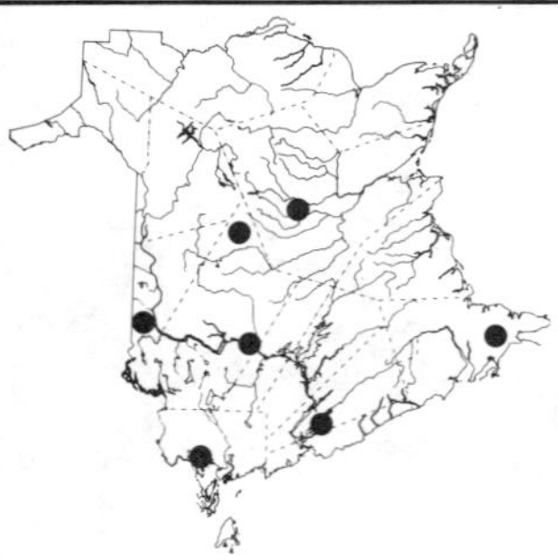

POTAMOGETON AMPLIFOLIUS

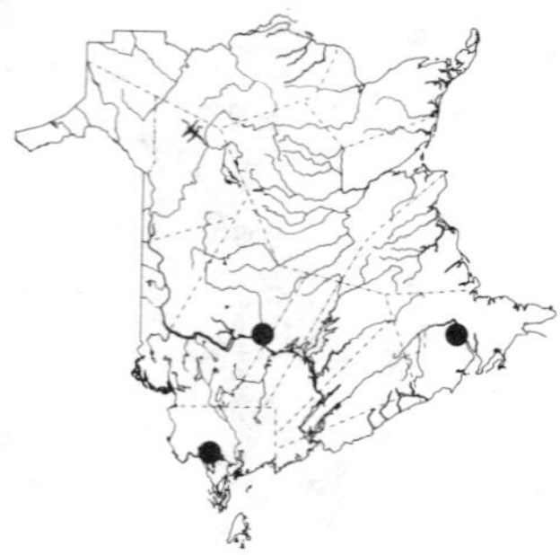

POTAMOGETON CONFERVOIDES

POTAMOGETON EPIHYDRUS

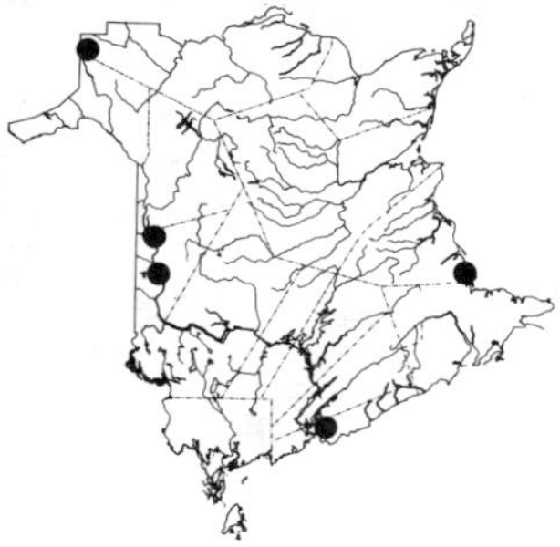

POTAMOGETON FILIFORMIS

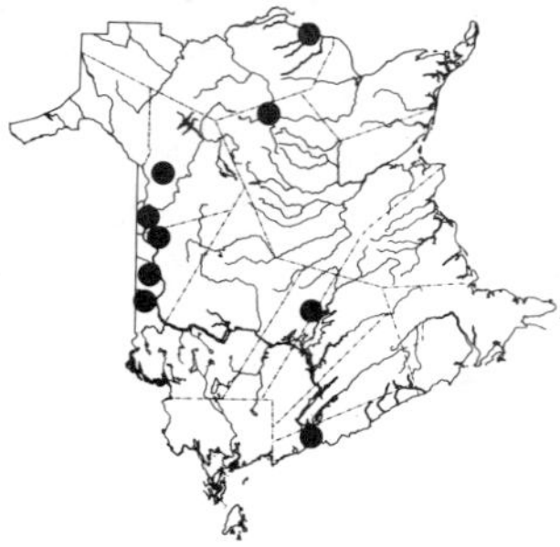

POTAMOGETON FOLIOSUS

POTAMOGETON GRAMINEUS

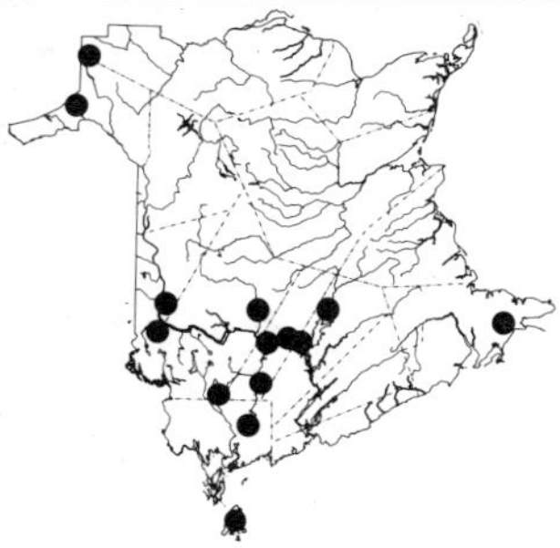

POTAMOGETON NATANS

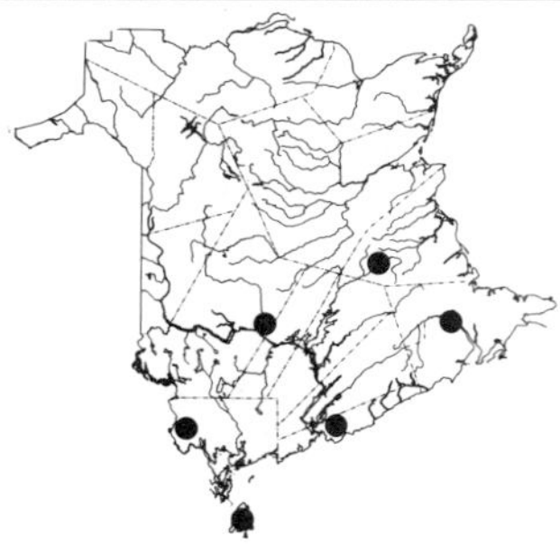

POTAMOGETON OAKESIANUS

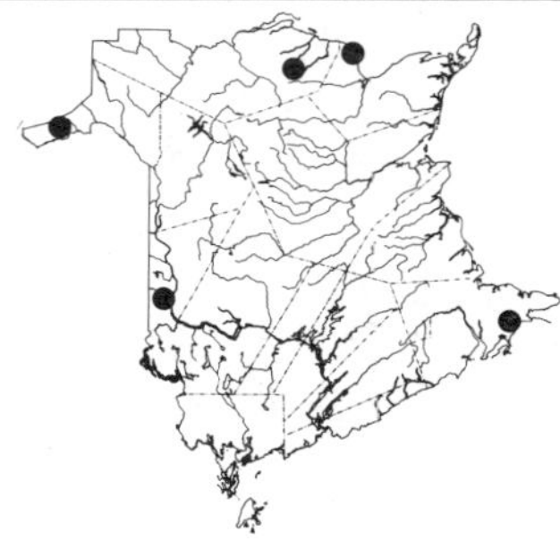

POTAMOGETON OBTUSIFOLIUS

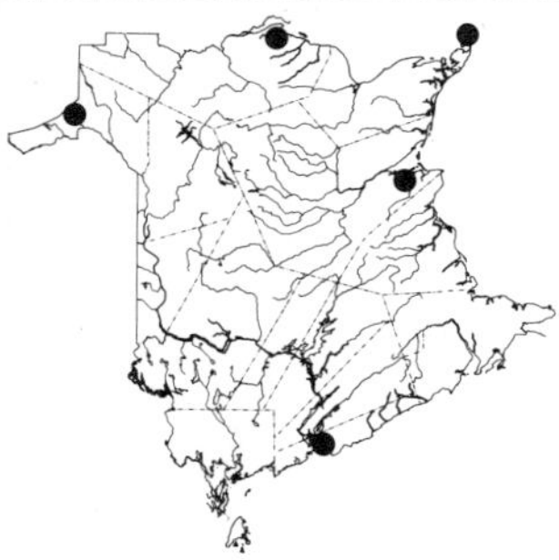

POTAMOGETON PECTINATUS

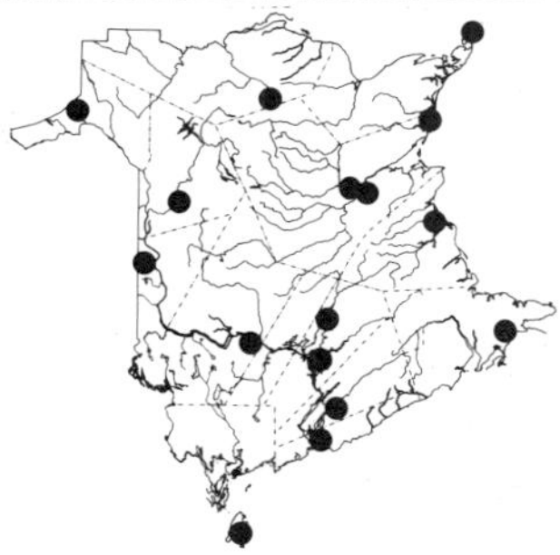

POTAMOGETON PERFOLIATUS

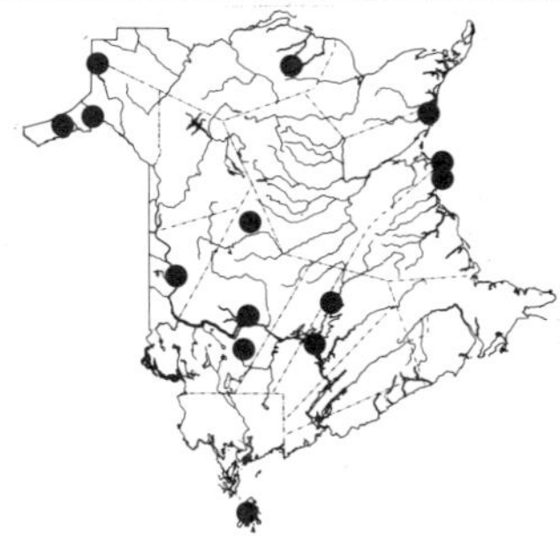

POTAMOGETON PUSILLUS

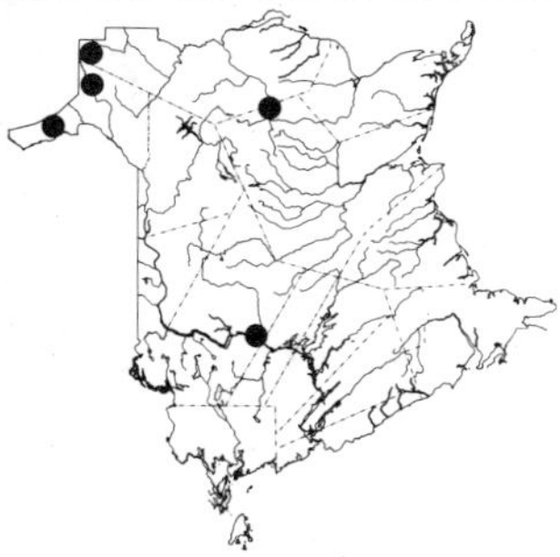

POTAMOGETON RICHARDSONII

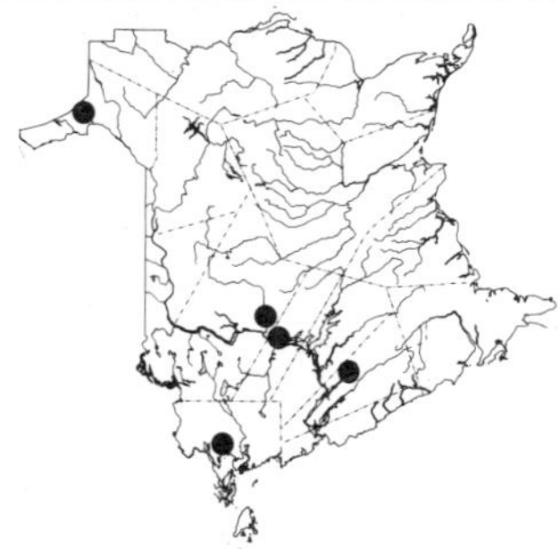

POTAMOGETON ROBBINSII

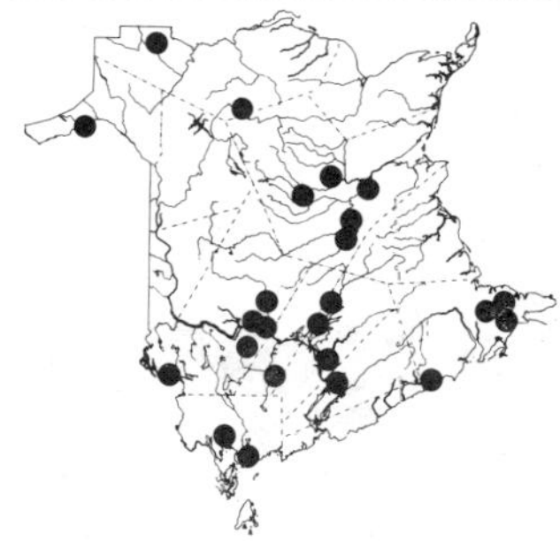

POTAMOGETON SPIRELLUS

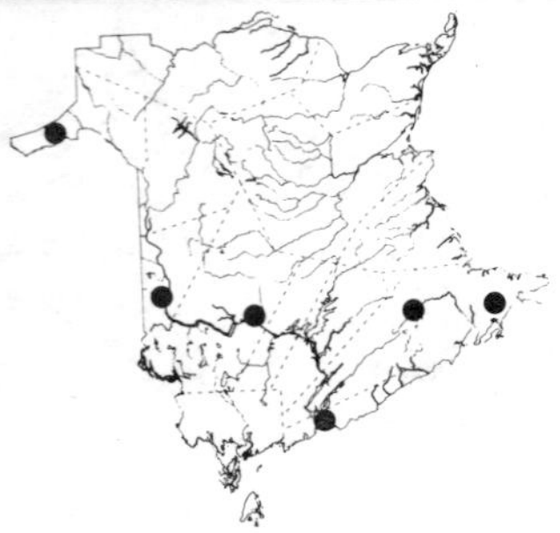

POTAMOGETON ZOSTERIFORMIS

RUPPIA MARITIMA

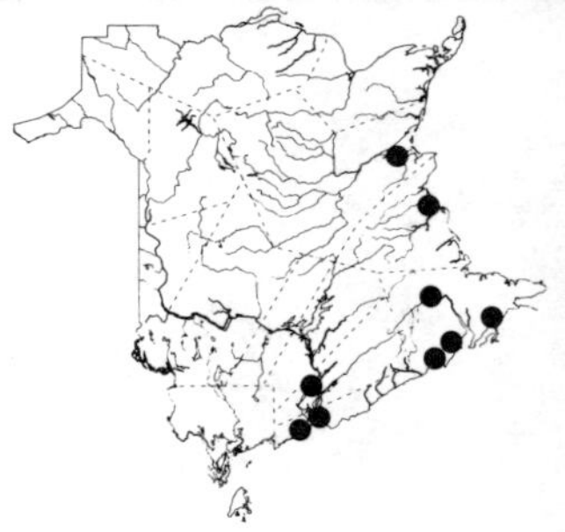

ZANICHELLIA PALUSTRIS

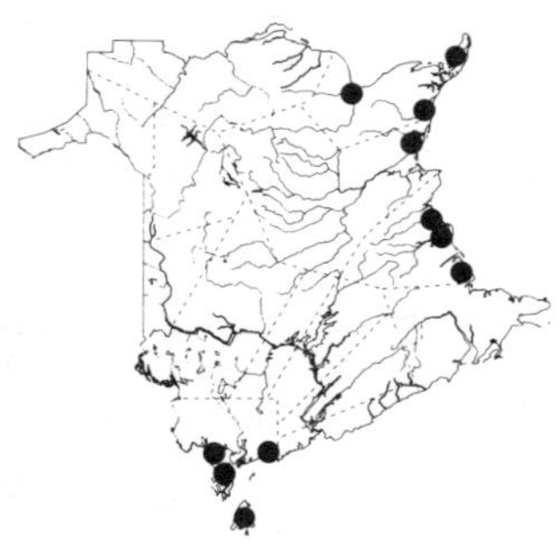

ZOSTERA MARINA
VAR. STENOPHYLLA

NAJAS FLEXILIS

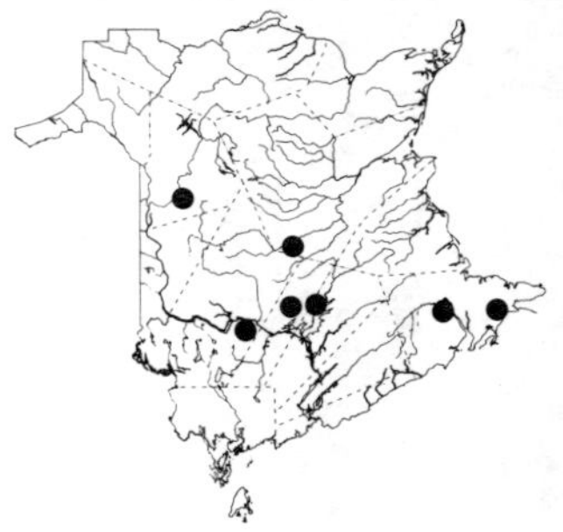

SCHEUCHZERIA PALUSTRIS

TRIGLOCHIN GASPENSE

TRIGLOCHIN MARITIMA

TRIGLOCHIN PALUSTRIS

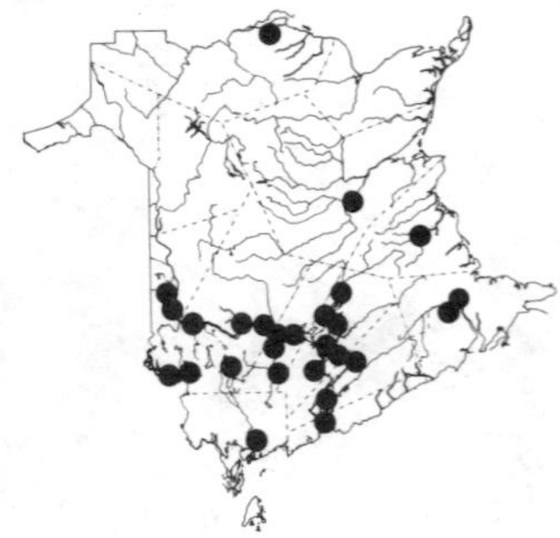

ALISMA PLANTAGO-AQUATICA

SAGITTARIA CALYCINA

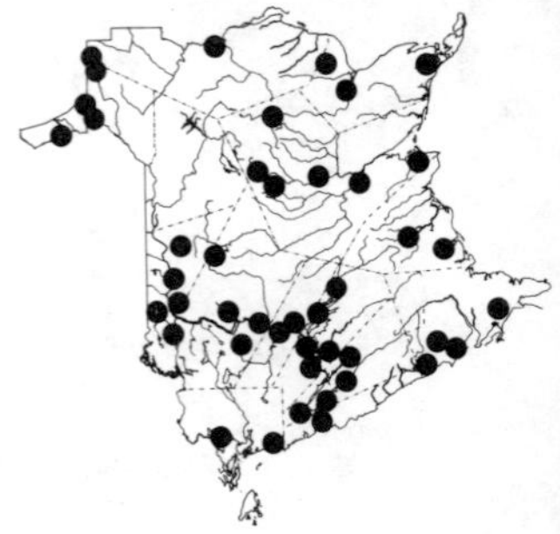

SAGITTARIA CUNEATA

ALISMACEAE | HYDROCHARITACEAE | POACEAE

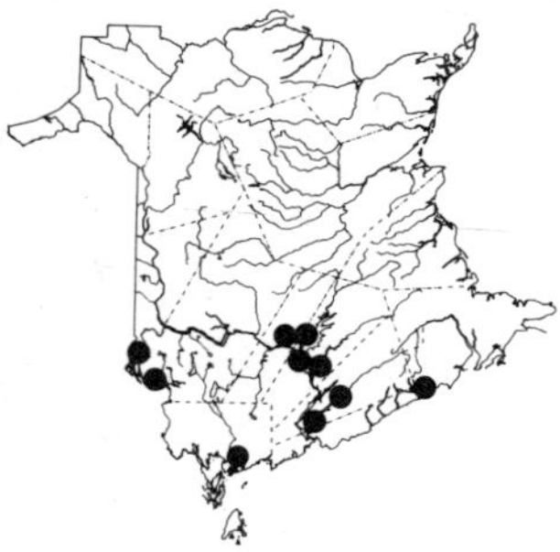

SAGITTARIA GRAMINEA

SAGITTARIA LATIFOLIA

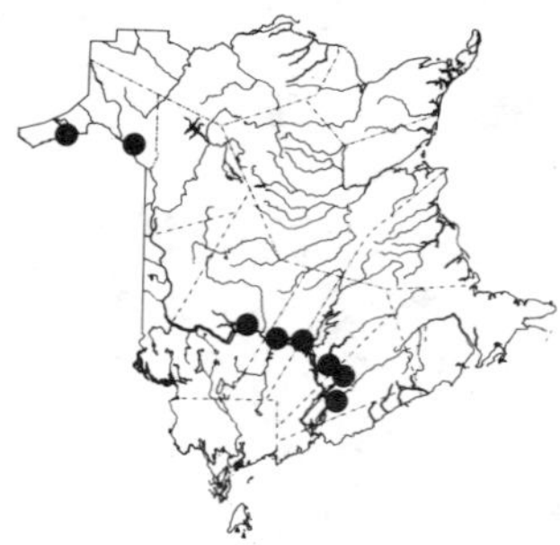

ELODEA CANADENSIS

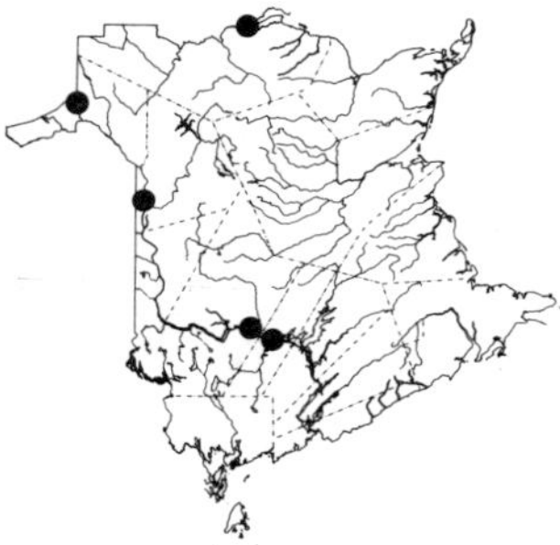

ELODEA NUTTALLII

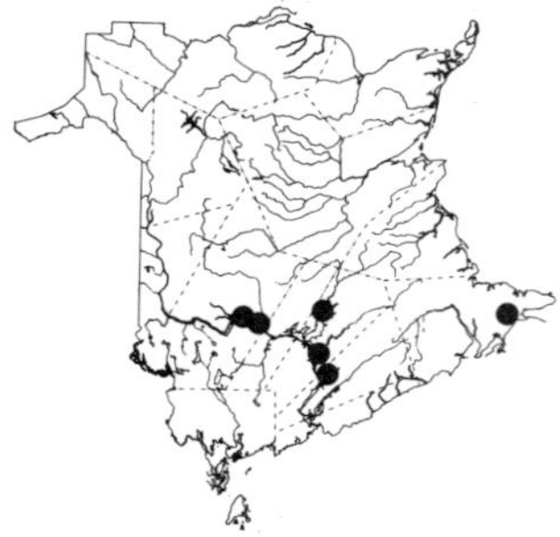

VALLISNERIA AMERICANA

AGROPYRON REPENS

AGROPYRON TRACHYCAULUM

AGROSTIS CAPILLARIS

AGROSTIS GIGANTEA

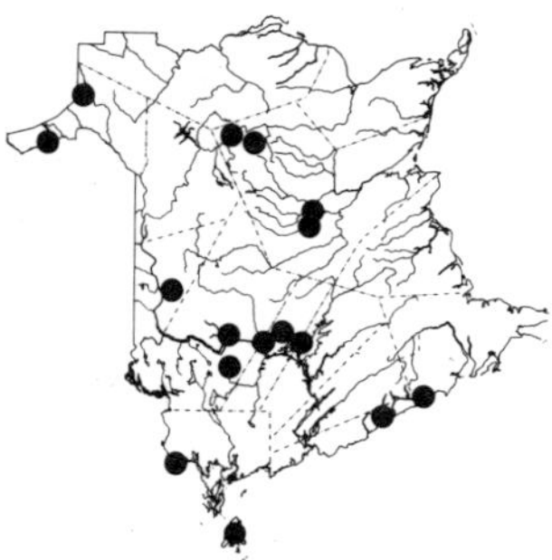

AGROSTIS PERENNANS

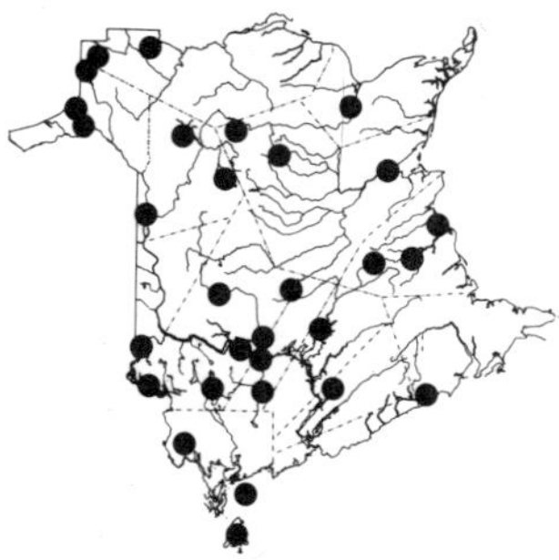

AGROSTIS SCABRA

AGROSTIS STOLONIFERA

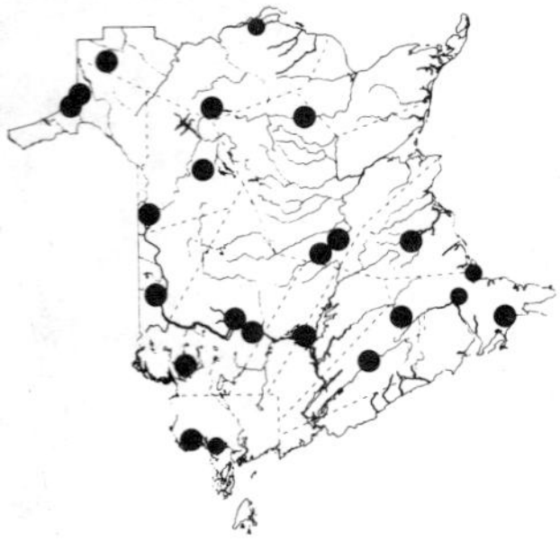

ALOPECURUS AEQUALIS

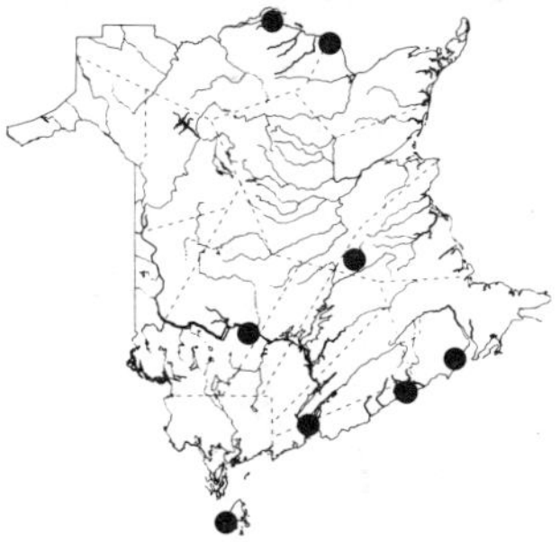

ALOPECURUS PRATENSIS

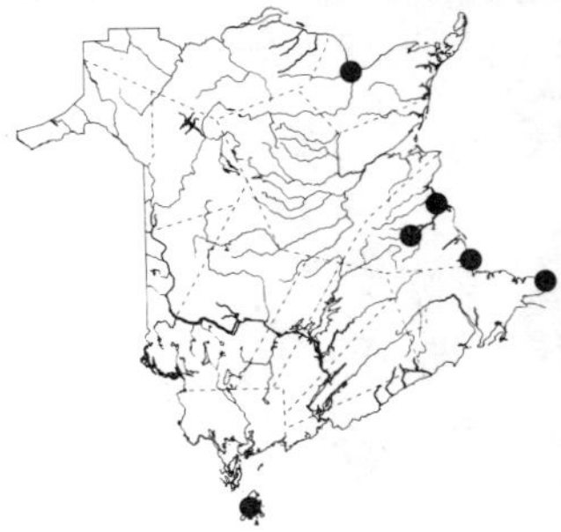

AMMOPHILA BREVILIGULATA

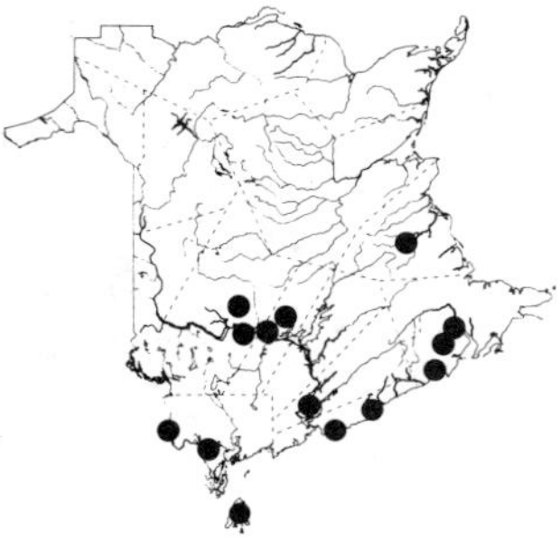

ANTHOXANTHUM ODORATUM

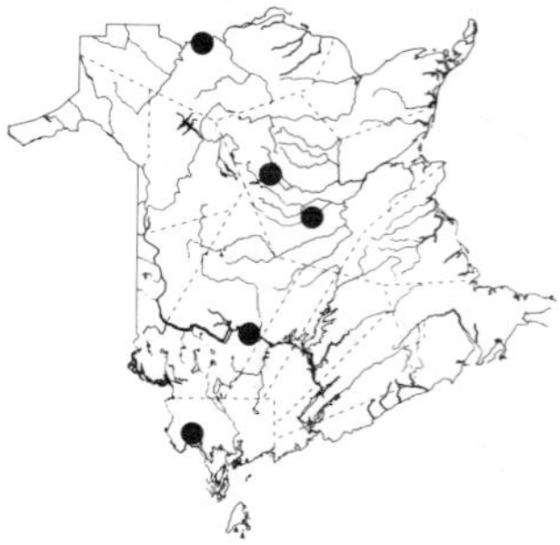

AVENA SATIVA

BRACHYELYTRUM ERECTUM

BROMUS CILIATUS

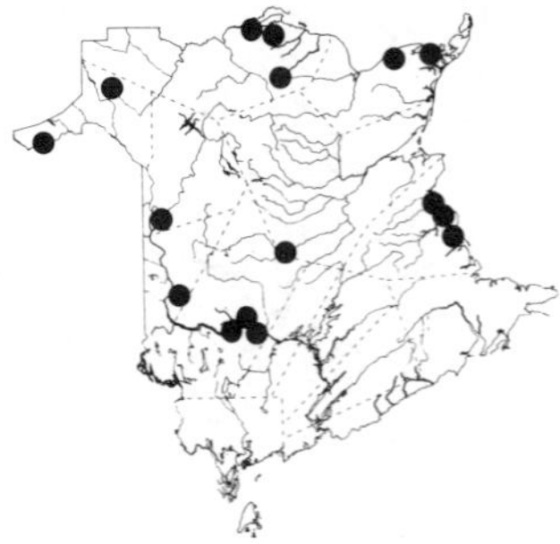

BROMUS INERMIS

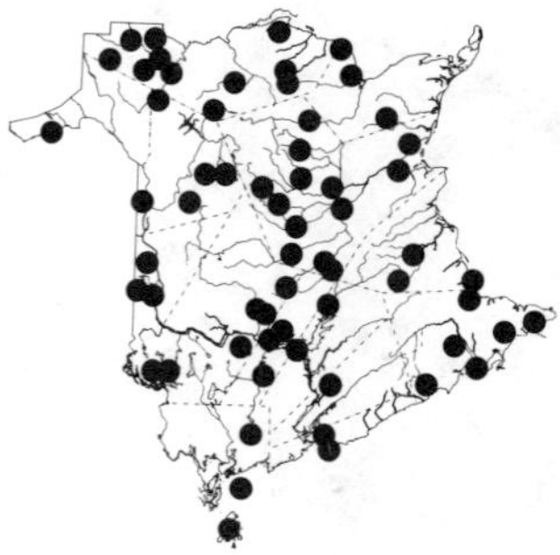

CALAMAGROSTIS CANADENSIS

CALAMAGROSTIS STRICTA

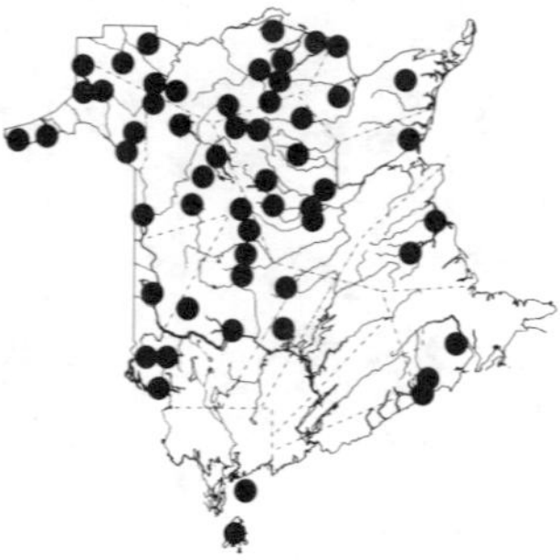

CINNA LATIFOLIA

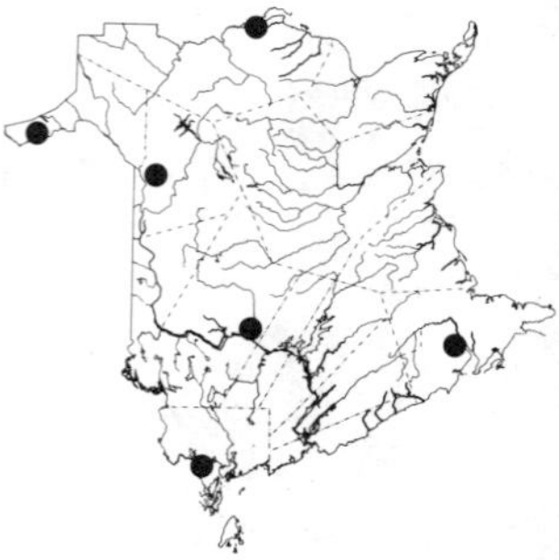

DACTYLIS GLOMERATA

DANTHONIA SPICATA

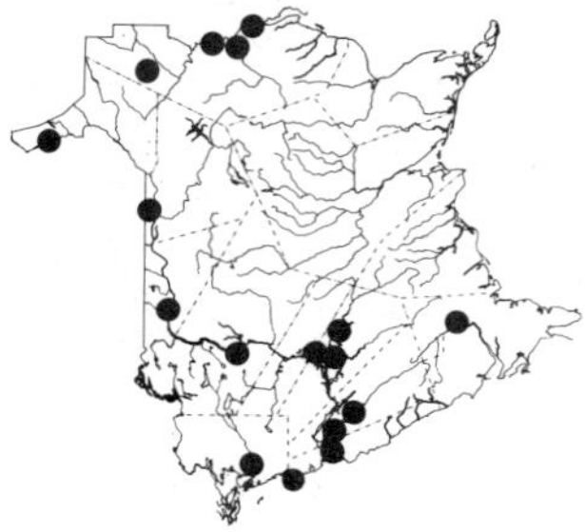

DESCHAMPSIA CAESPITOSA

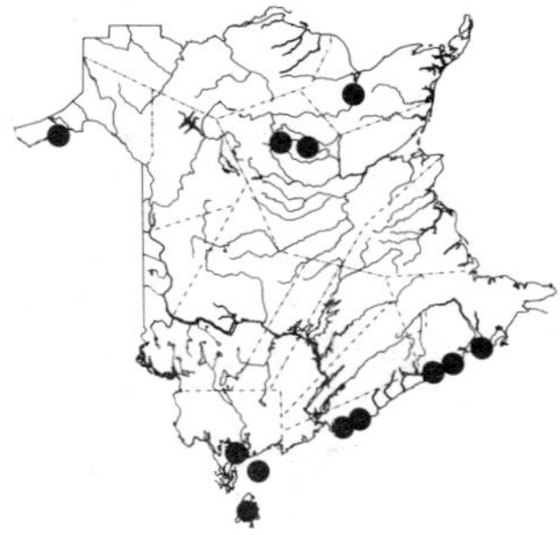

DESCHAMPSIA FLEXUOSA

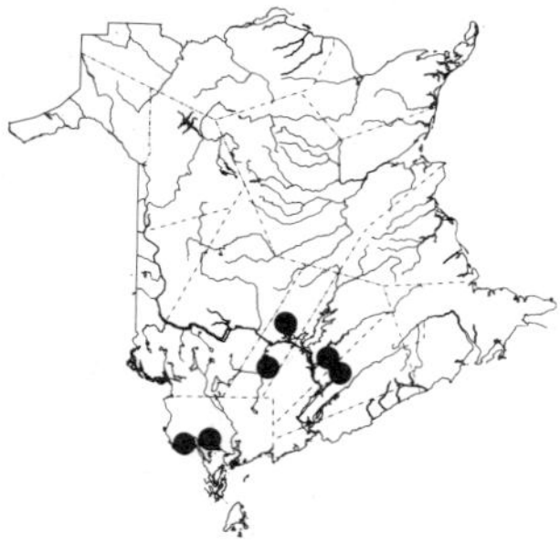

DICHANTHELIUM ACUMINATUM

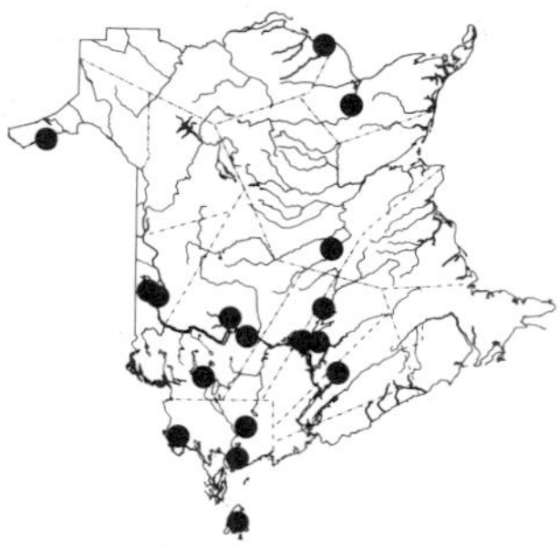

DICHANTHELIUM ACUMINATUM VAR. IMPLICATUM

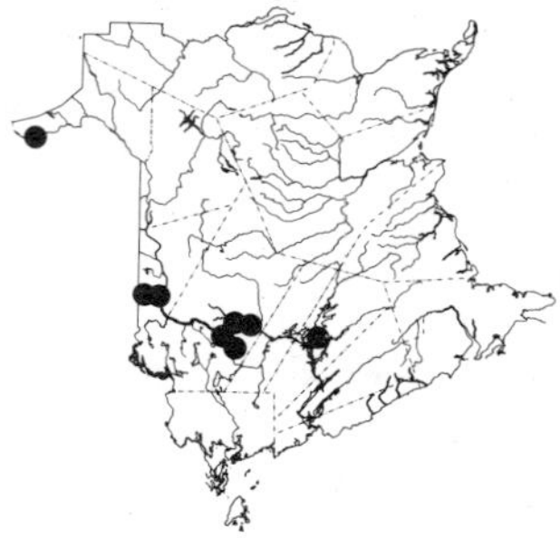

DICHANTHELIUM ACUMINATUM VAR. LINDHEIMERI

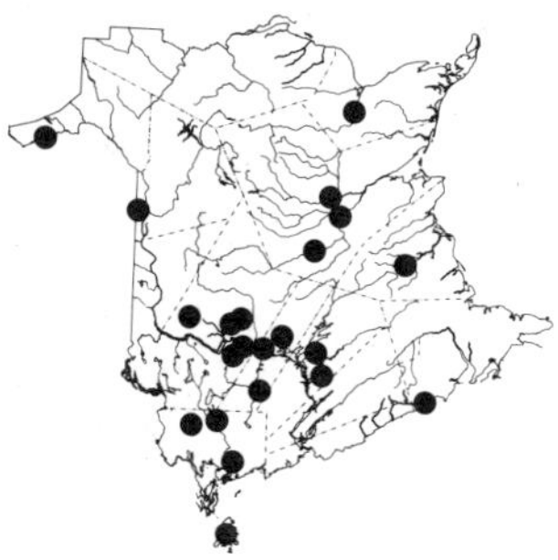

DICHANTHELIUM BOREALE

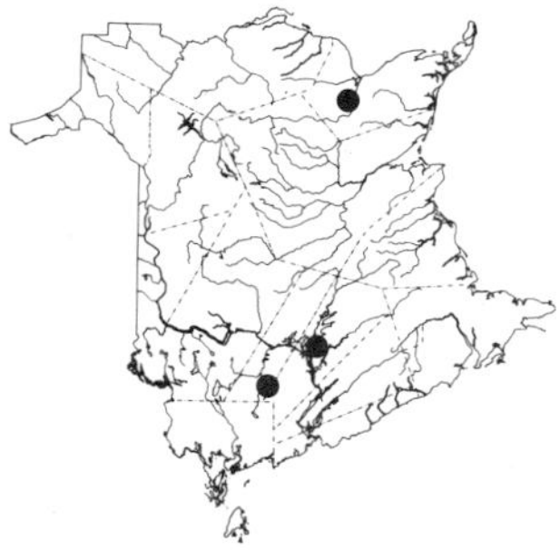

DICHANTHELIUM DEPAUPERATUM

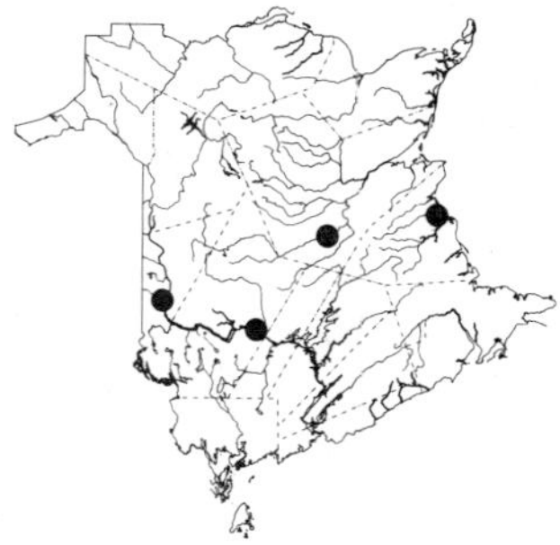

DIGITARIA ISCHAEMUM

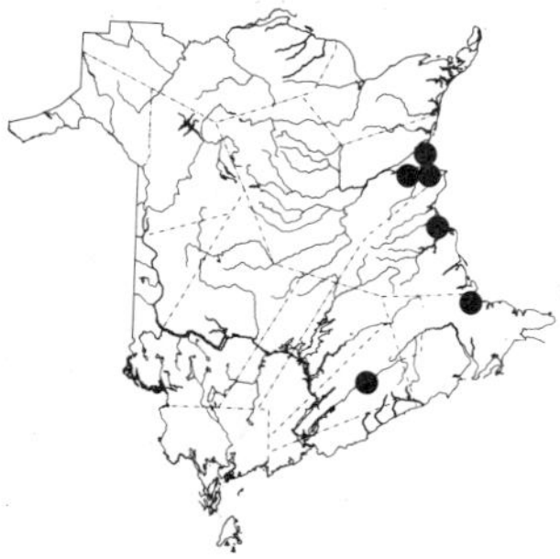

DISTICHLIS SPICATA

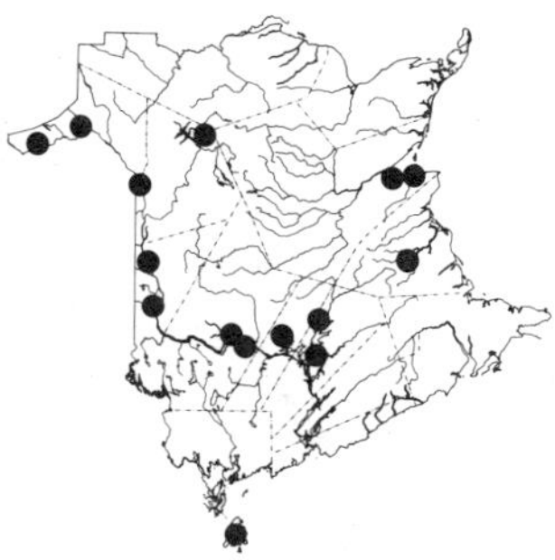

ECHINOCHLOA CRUSGALLI

ELYMUS CANADENSIS

ELYMUS MOLLIS

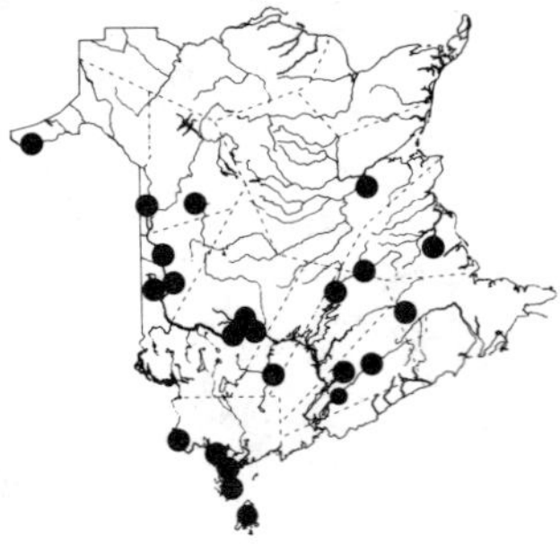

ELYMUS VIRGINICUS

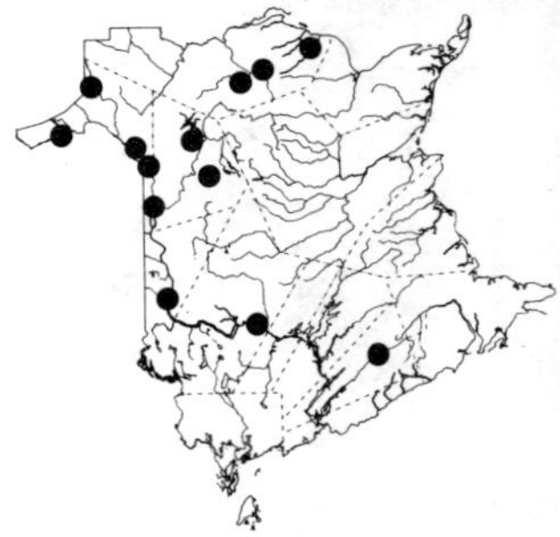

ELYMUS WIEGANDII

ERAGROSTIS PECTINACEA

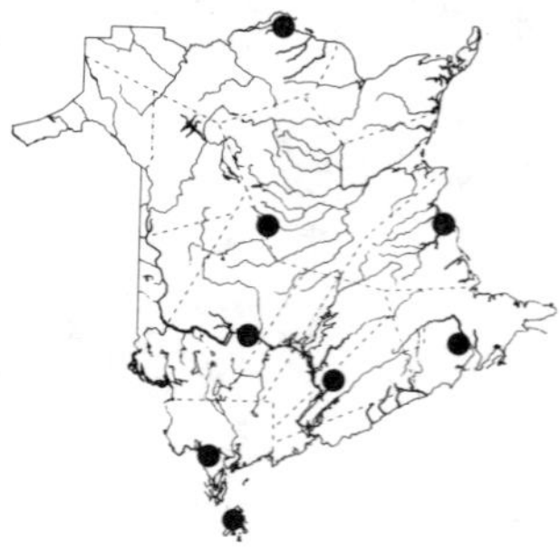

FESTUCA PRATENSIS

FESTUCA RUBRA VAR. COMMUTATA

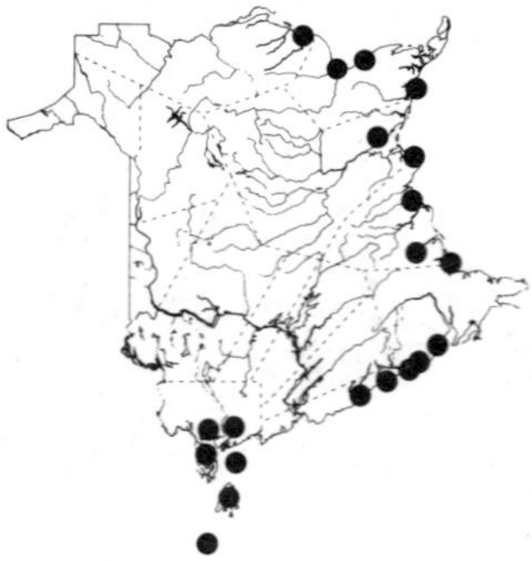

FESTUCA RUBRA VAR. JUNCEA

FESTUCA RUBRA VAR. MEGASTACHYS

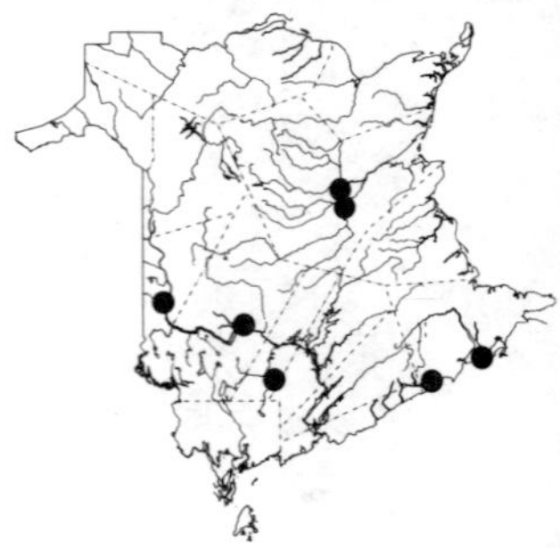

FESTUCA RUBRA VAR. RUBRA

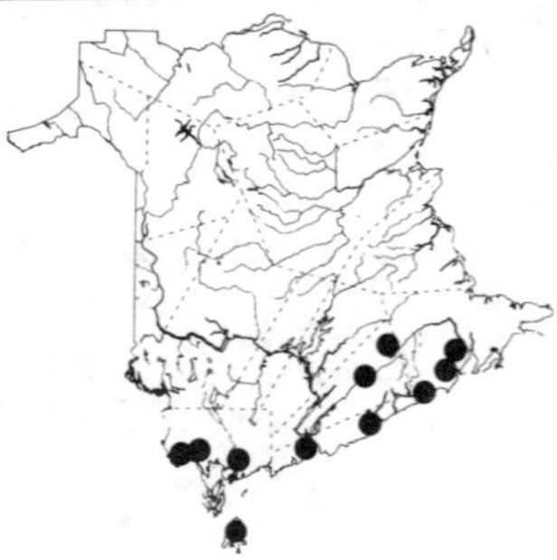

FESTUCA TENUIFOLIA

GLYCERIA BOREALIS

GLYCERIA CANADENSIS

GLYCERIA GRANDIS

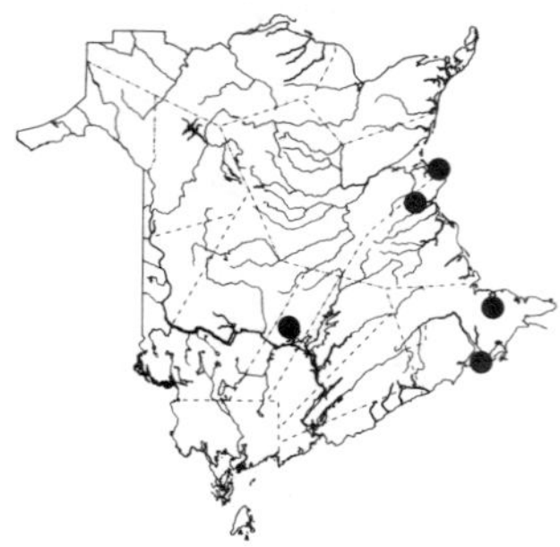

GLYCERIA X LAXA

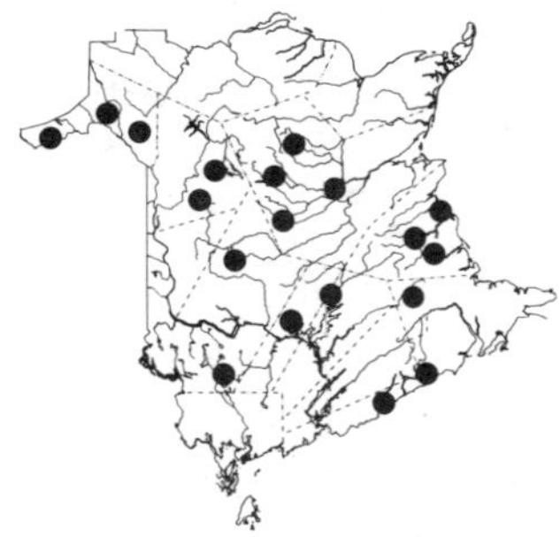

GLYCERIA MELICARIA

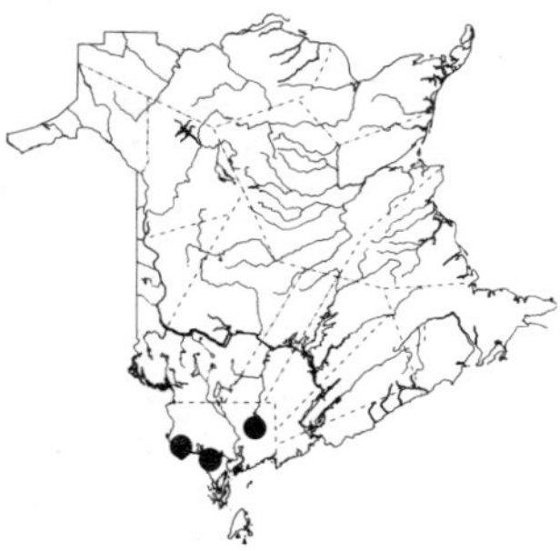

GLYCERIA OBTUSA

GLYCERIA STRIATA

HIEROCHLOE ODORATA

HORDEUM JUBATUM

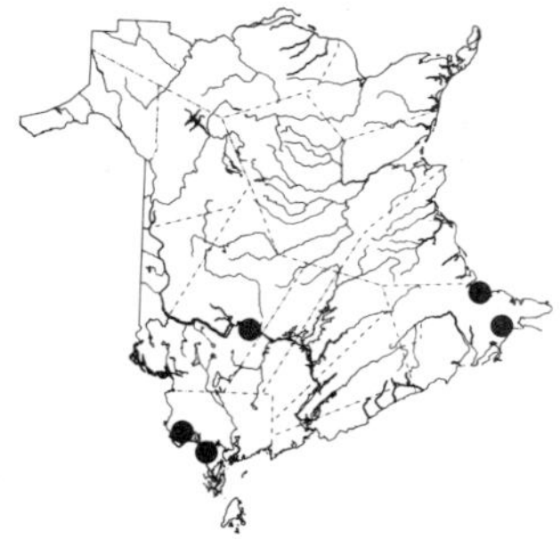

HORDEUM VULGARE

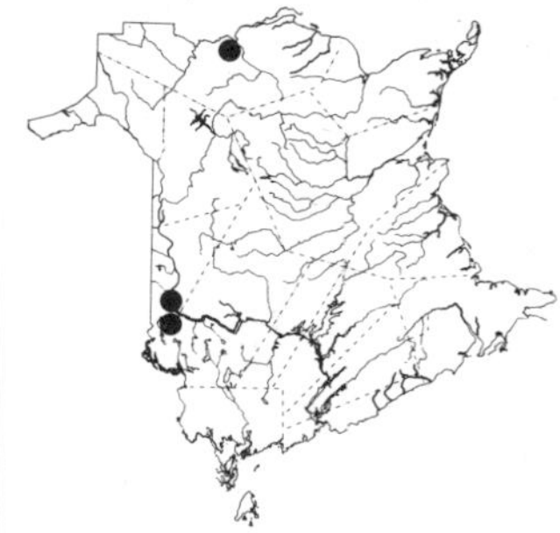

HYSTRIX PATULA VAR. BIGELOVIANA

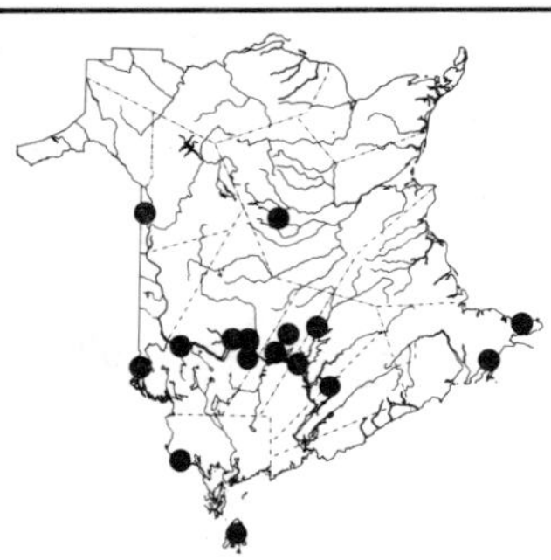

LEERSIA ORYZOIDES

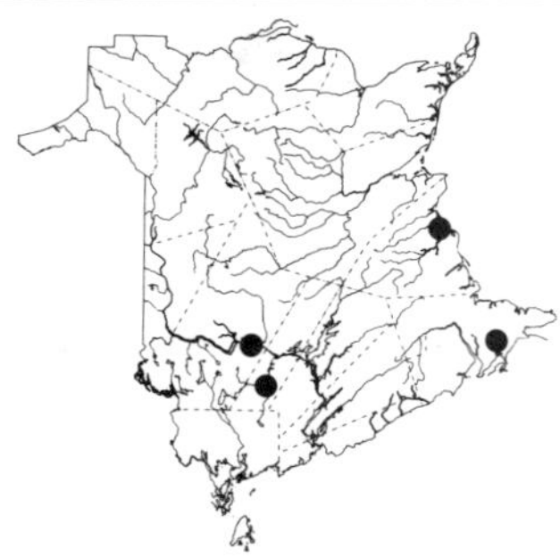

LOLIUM PERENNE

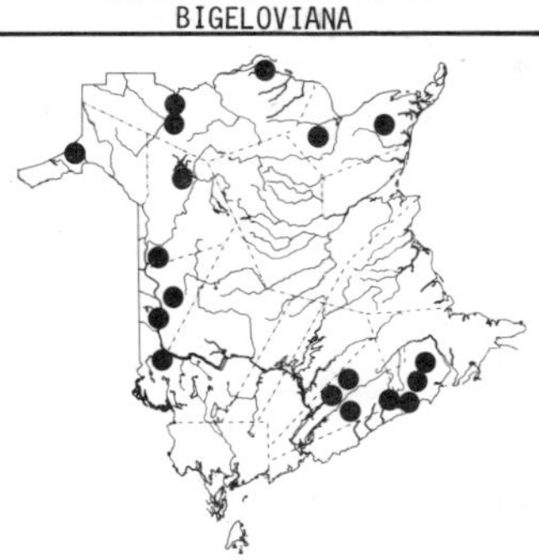

MILIUM EFFUSUM

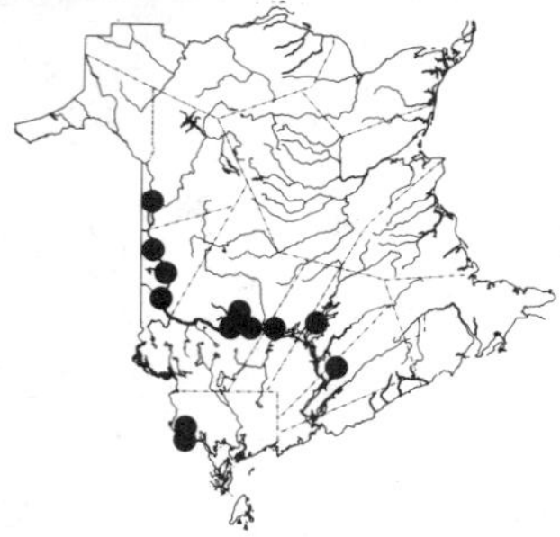

MUHLENBERGIA FRONDOSA

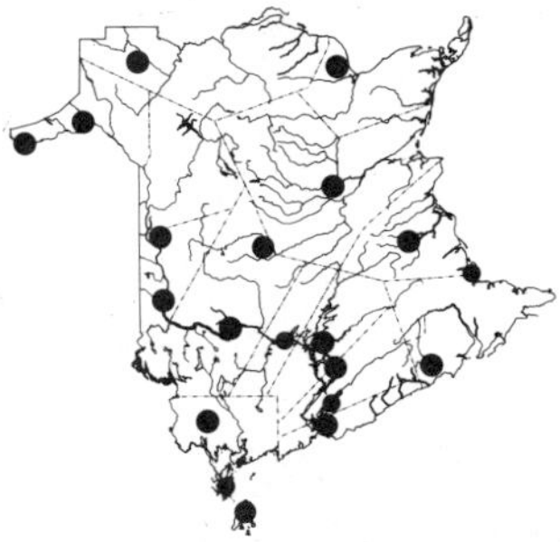

MUHLENBERGIA GLOMERATA

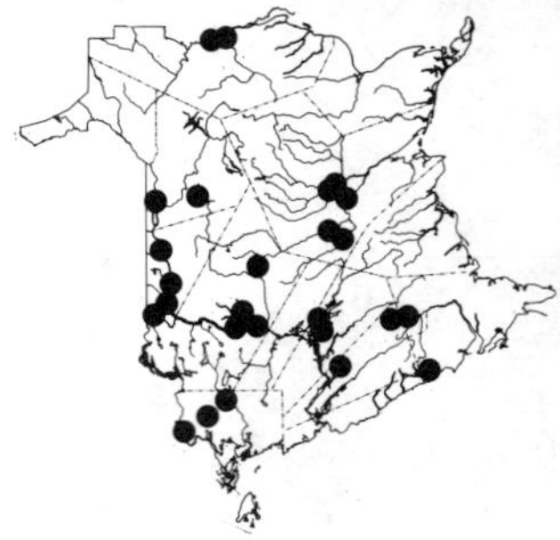

MUHLENBERGIA MEXICANA

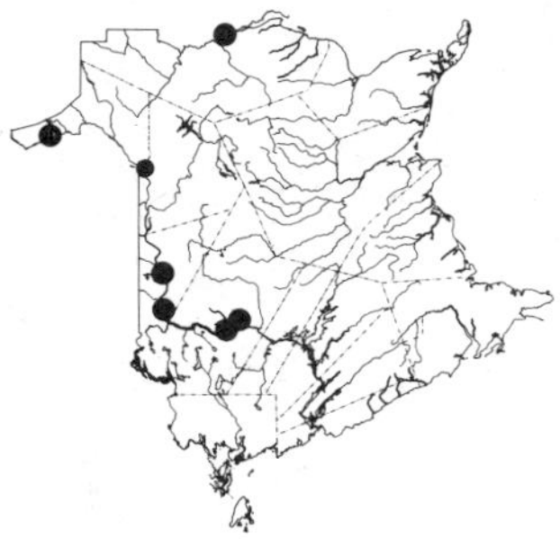

MUHLENBERGIA RICHARDSONIS

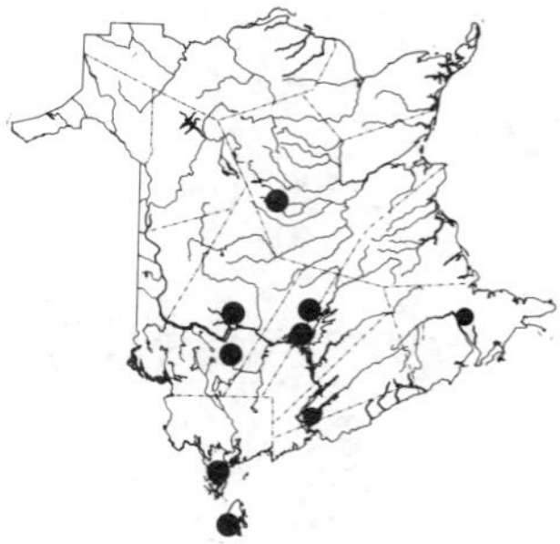

MUHLENBERGIA UNIFLORA

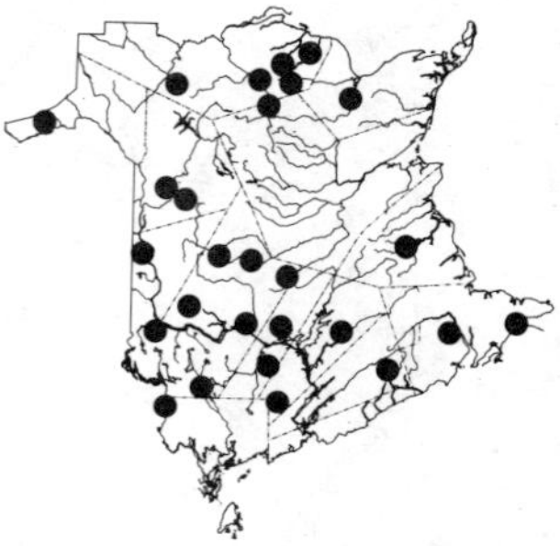

ORYZOPSIS ASPERFOLIA

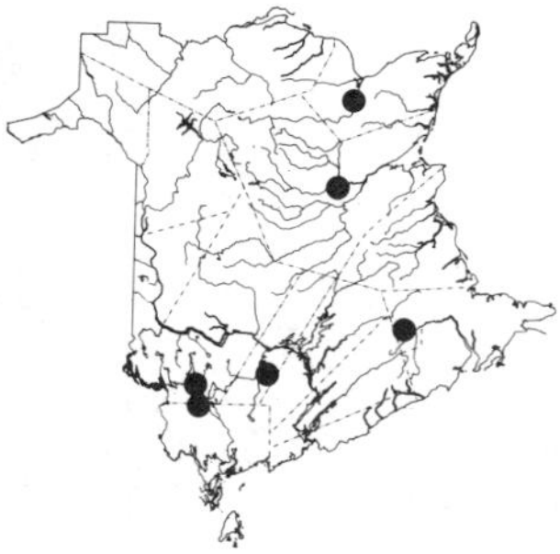

ORYZOPSIS CANADENSIS

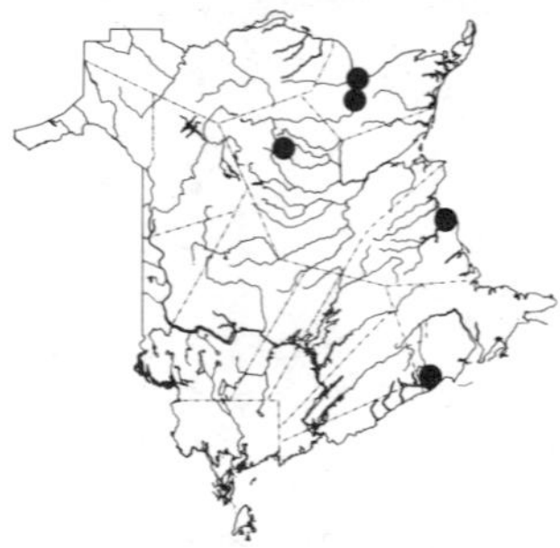

ORYZOPSIS PUNGENS

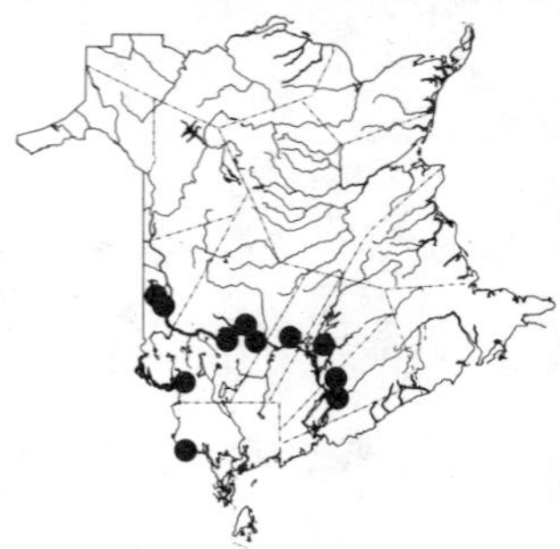

PANICUM CAPILLARE

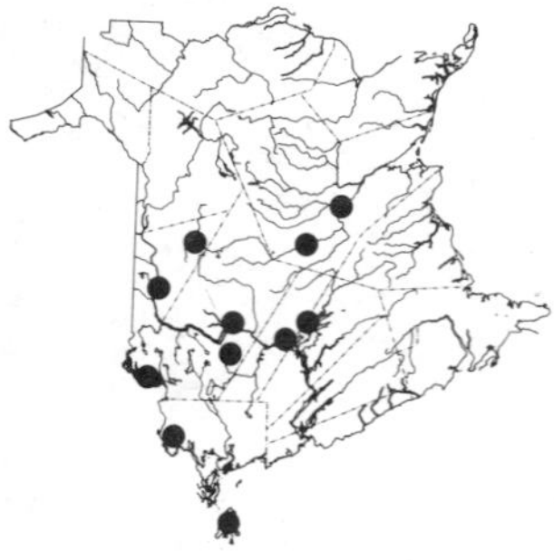

PANICUM TUCKERMANII

PHALARIS ARUNDINACEA

PHLEUM PRATENSE

PHRAGMITES AUSTRALIS

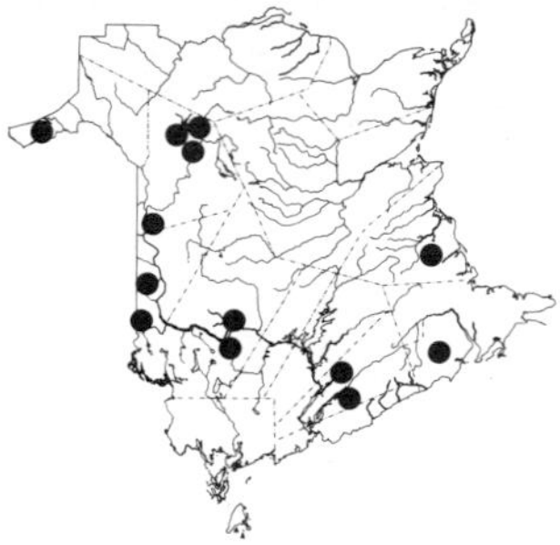

POA ALSODES

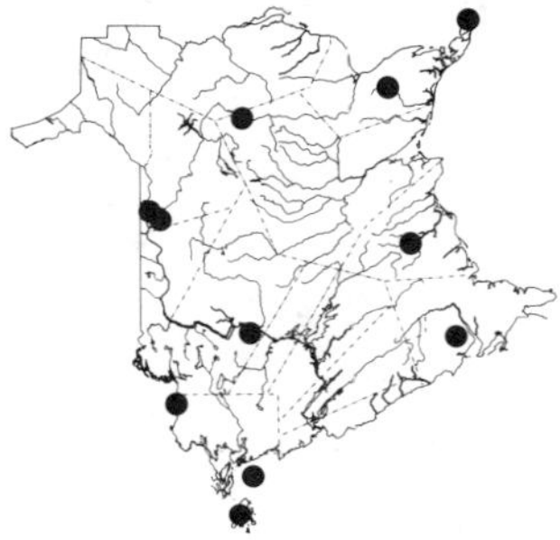

POA ANNUA

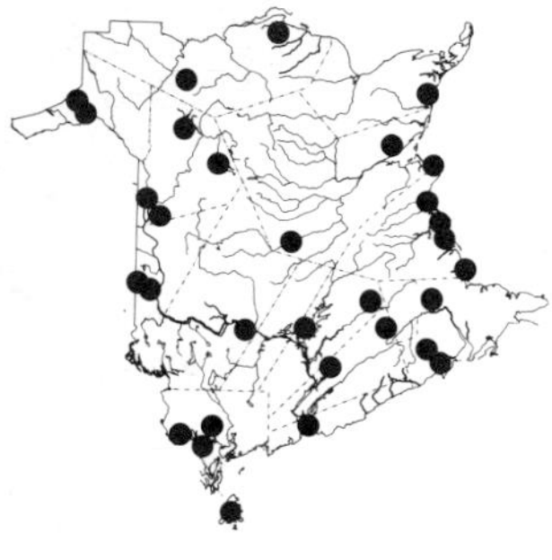

POA COMPRESSA

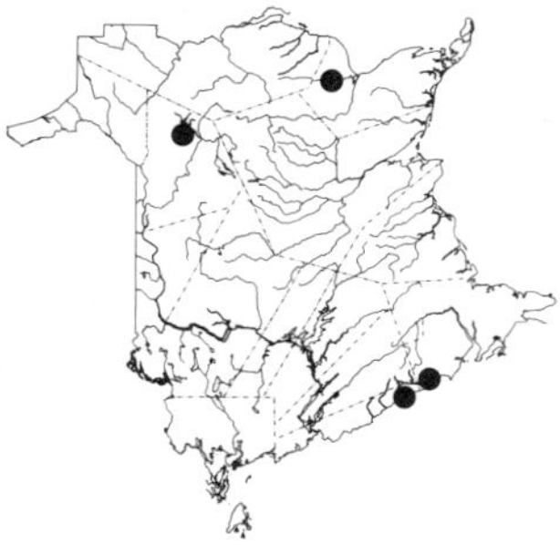

POA GLAUCANTHA

POA PALUSTRIS

POA PRATENSIS

POA SALTUENSIS

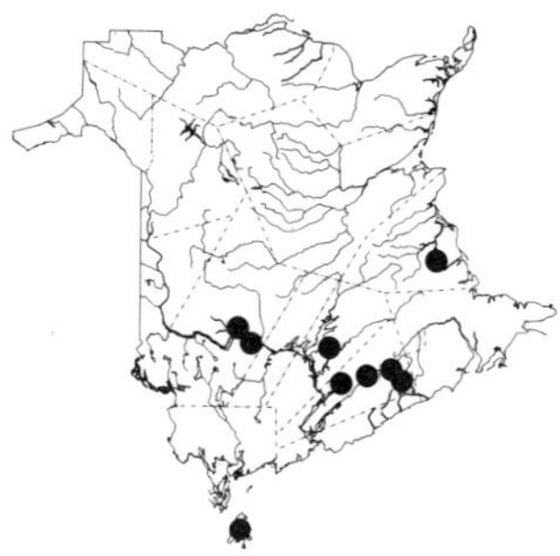

POA TRIVIALIS

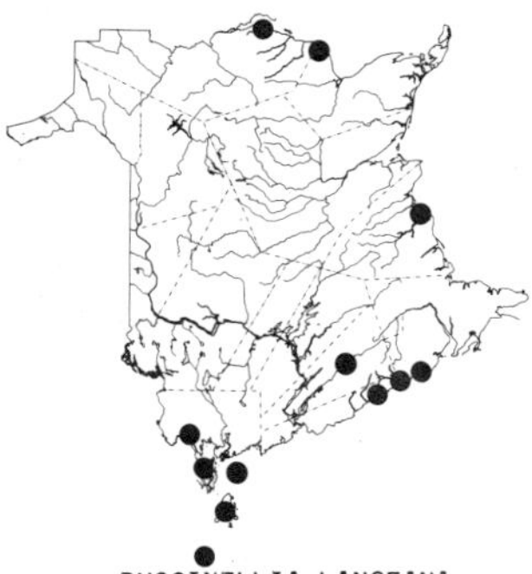

PUCCINELLIA LANGEANA

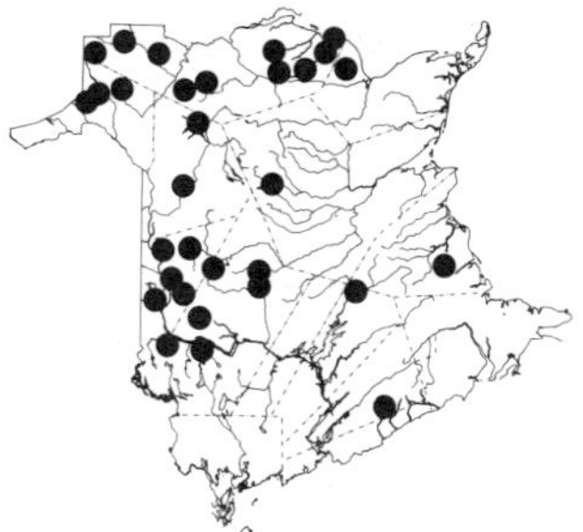

SCHIZACHNE PURPURASCENS

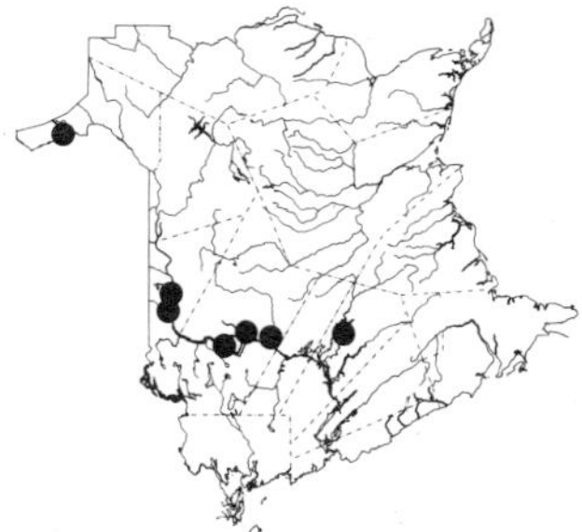

SCHIZACHYRIUM SCOPARIUM

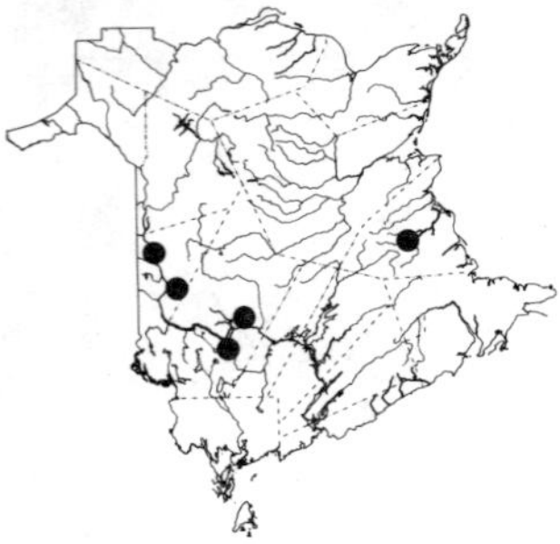

SETARIA GLAUCA

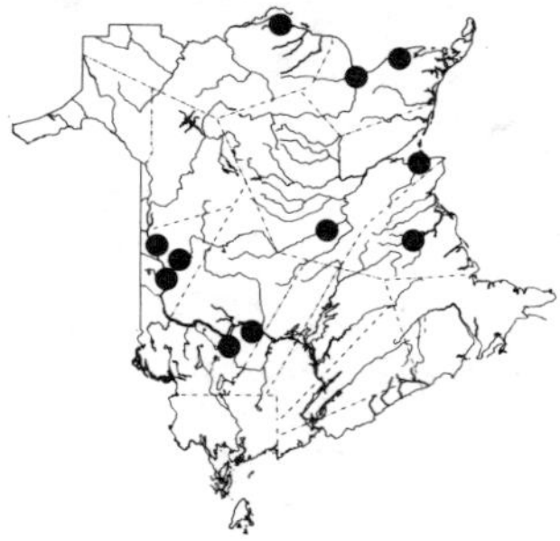

SETARIA VIRIDIS

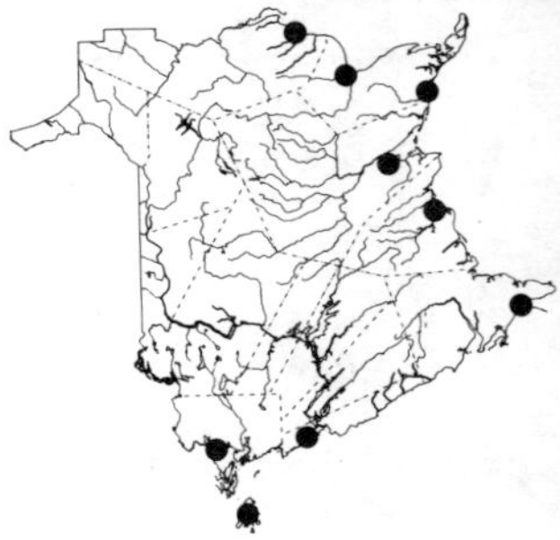

SPARTINA ALTERNIFLORA

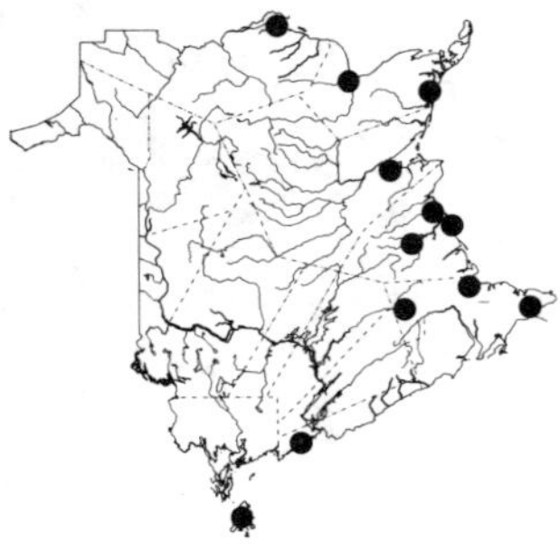

SPARTINA PATENS

SPARTINA PECTINATA

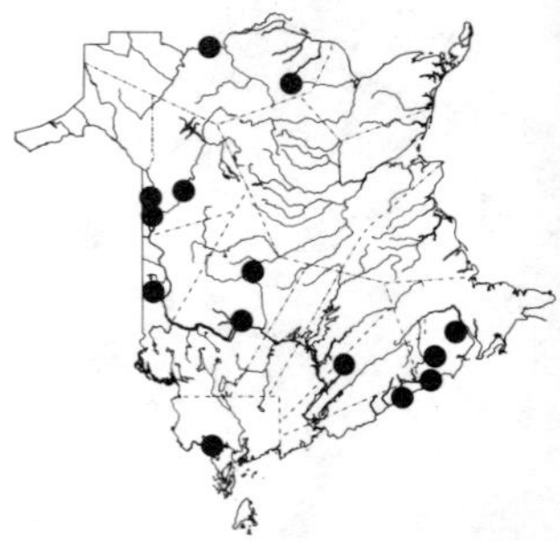

SPHENOPHOLIS OBTUSATA VAR. MAJOR

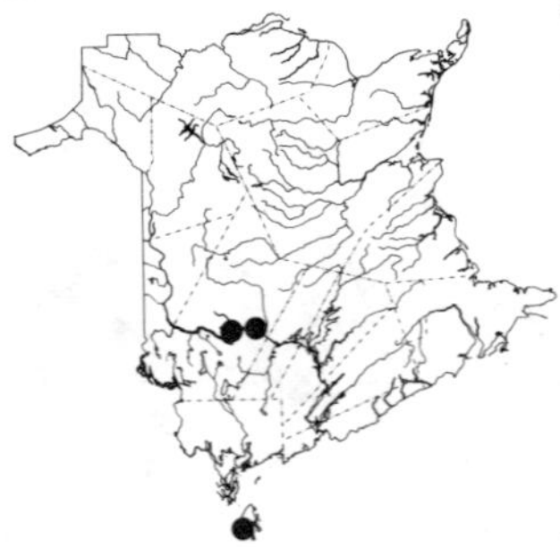

SPOROBOLUS ASPER

SPOROBOLUS VAGINIFLORUS

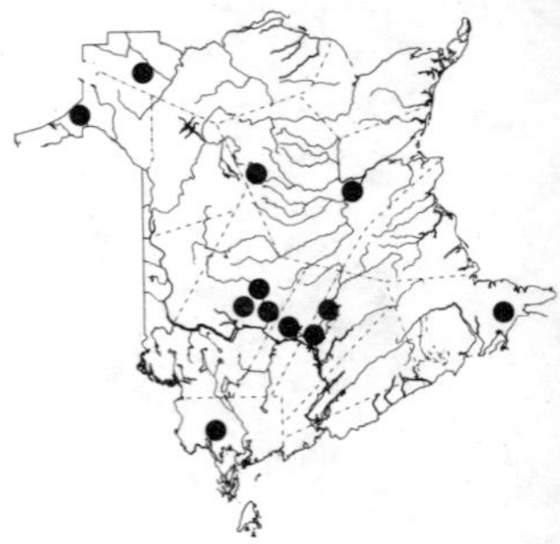

TORREYOCHLOA PALLIDA

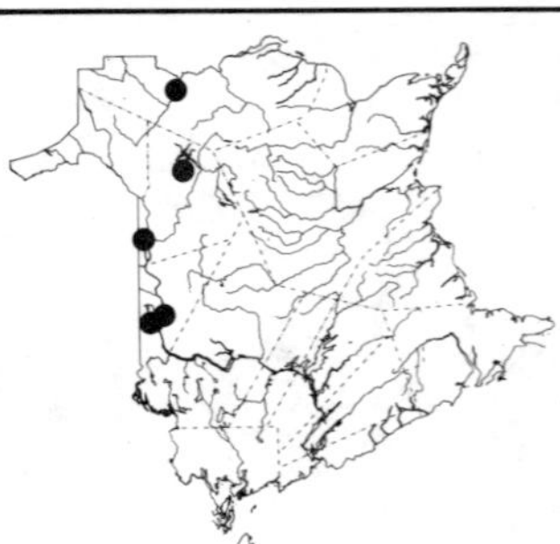

TRISETUM MELICOIDES

TRISETUM TRIFLORUM

TRITICUM AESTIVUM

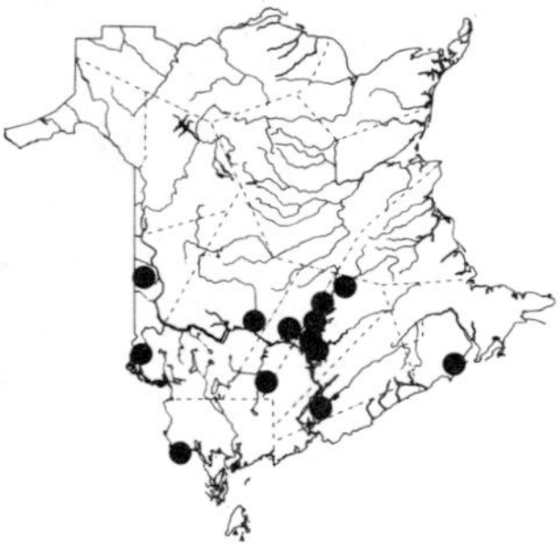

ZIZANIA PALUSTRIS

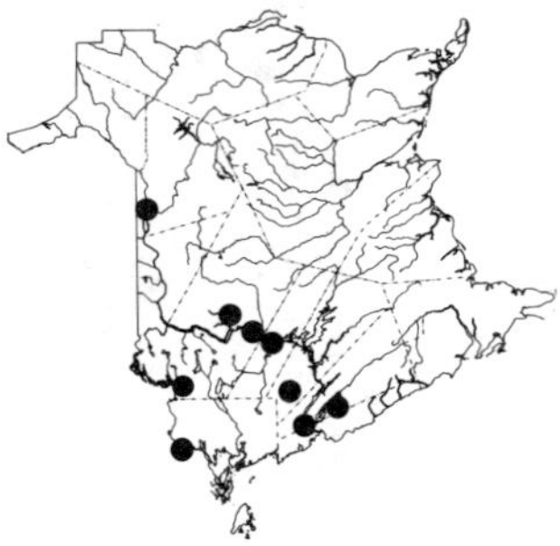

BULBOSTYLIS CAPILLARIS

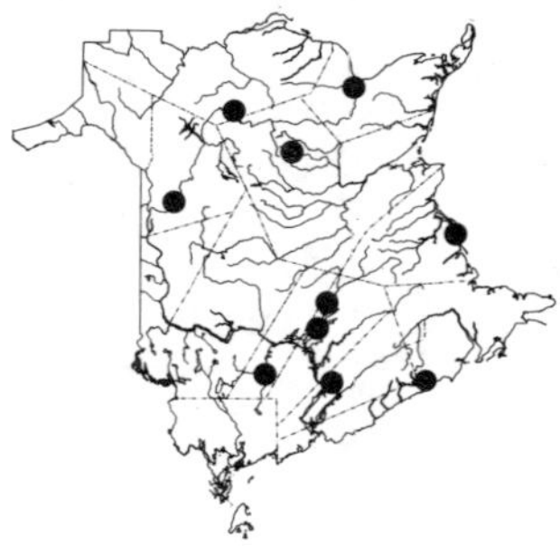

CAREX ADUSTA

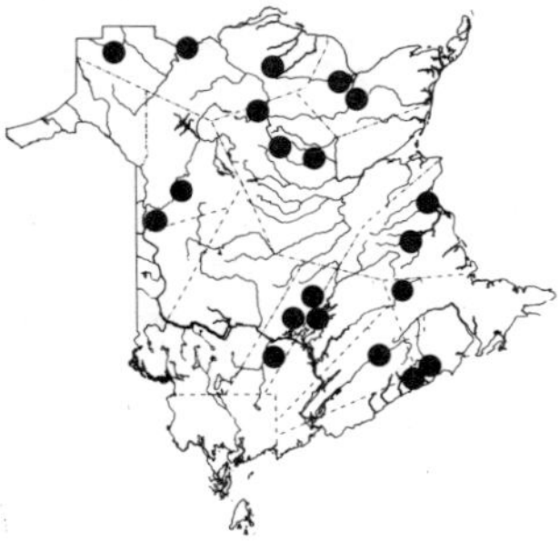

CAREX AENEA

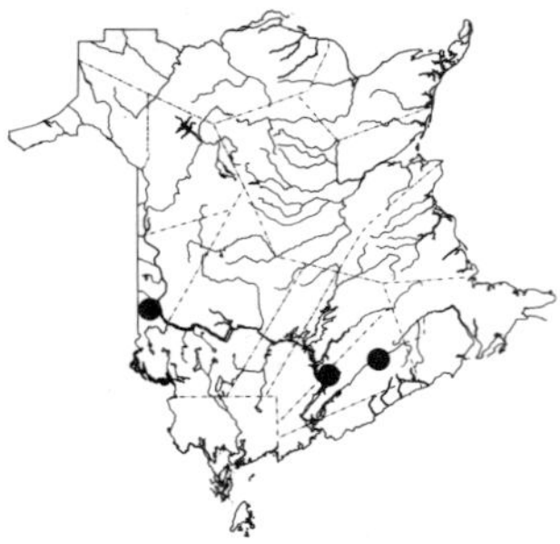

CAREX AMPHIBOLA

CAREX AQUATILIS

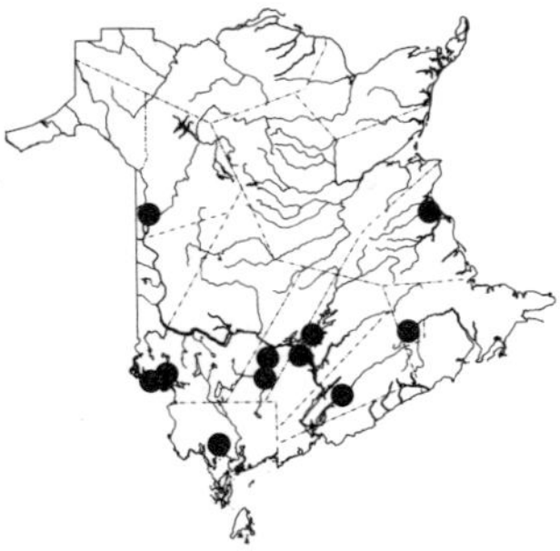

CAREX ARCTA

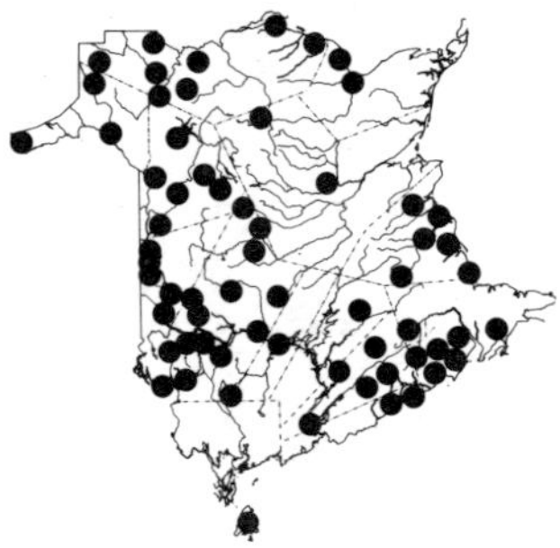

CAREX ARCTATA

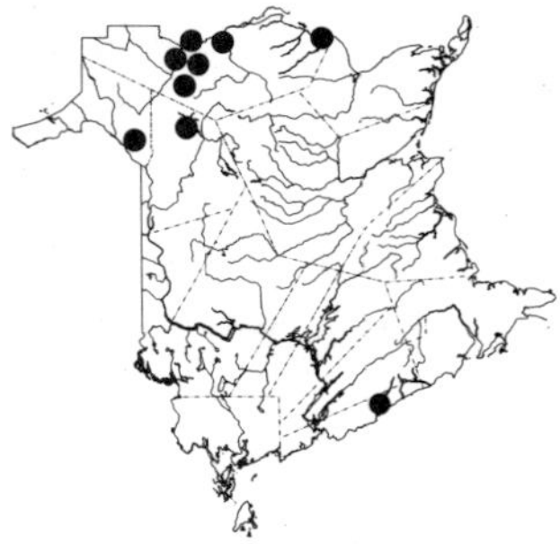

CAREX ATRATIFORMIS

CAREX AUREA

CAREX BEBBII

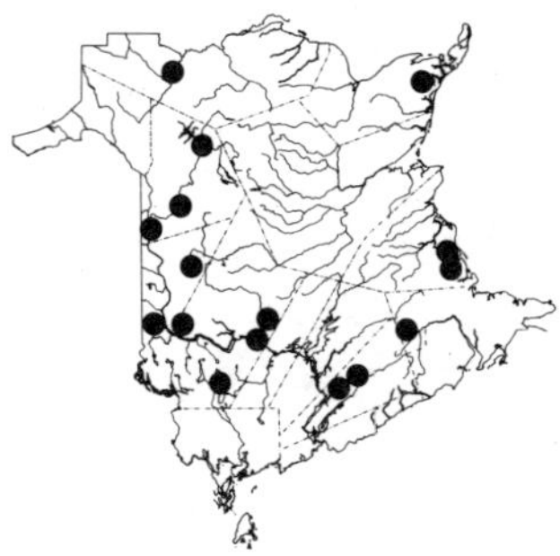

CAREX BROMOIDES

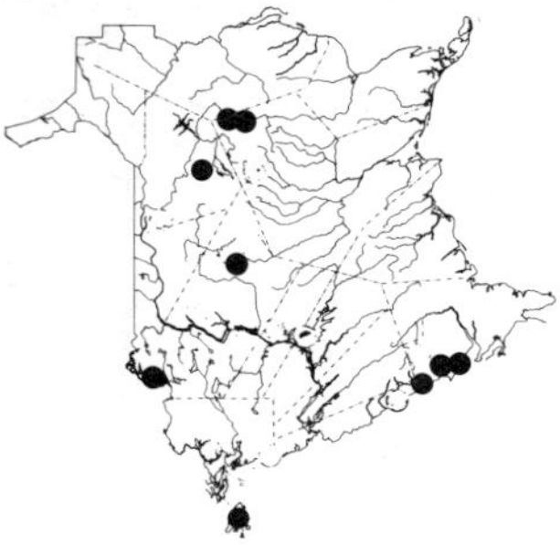

CAREX BRUNNESCENS

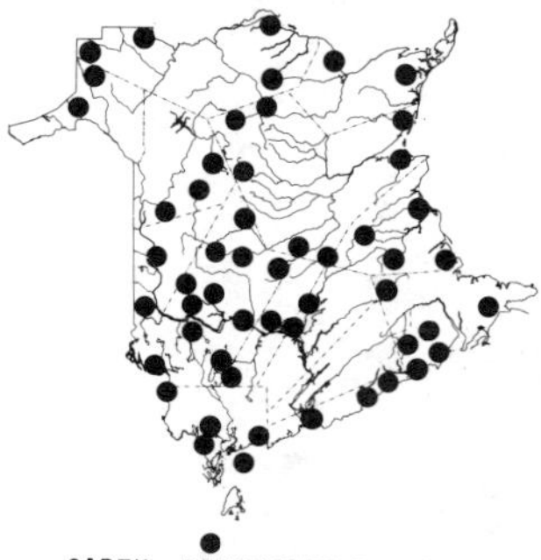

CAREX BRUNNESCENS VAR. SPHAEROSTACHYA

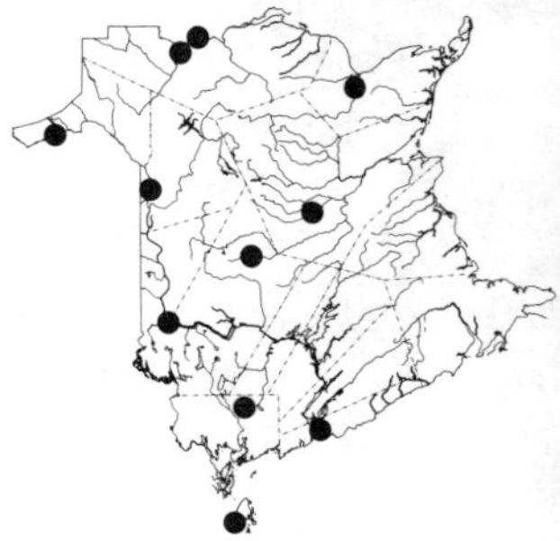

CAREX BUXBAUMII

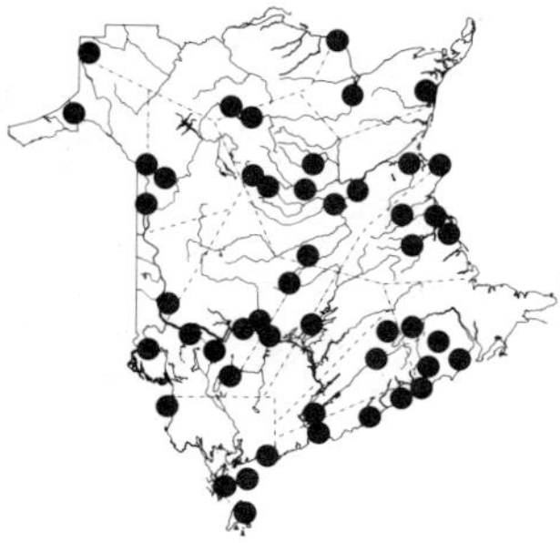

CAREX CANESCENS

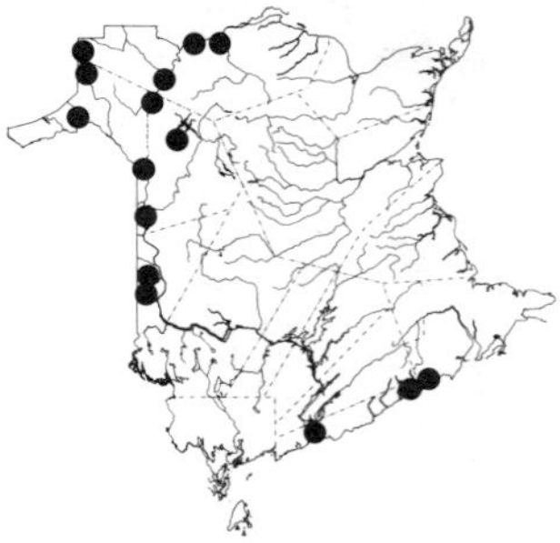

CAREX CAPILLARIS

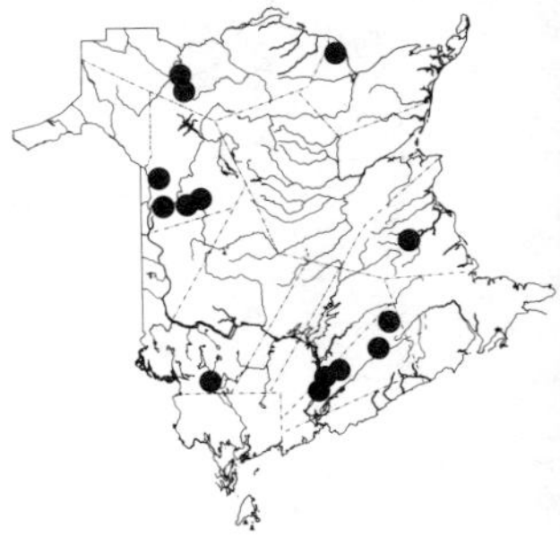

CAREX CASTANEA

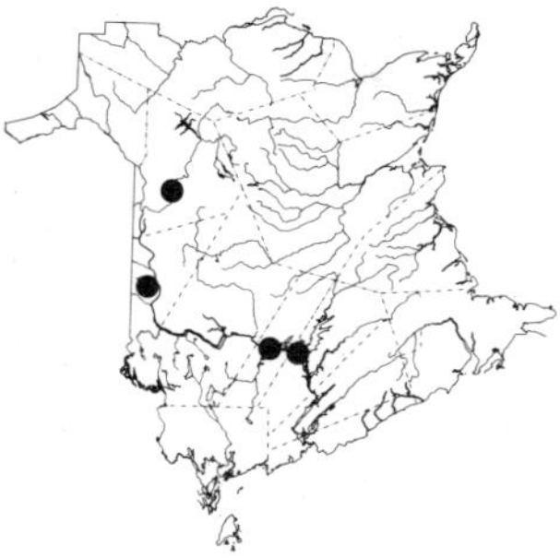

CAREX CHORDORRHIZA

CAREX COMMUNIS

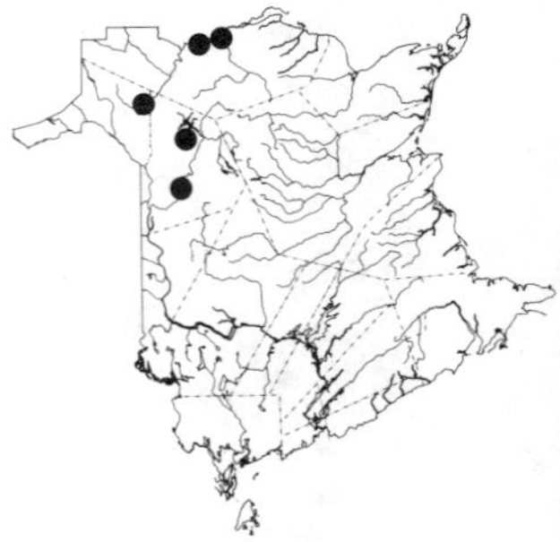

CAREX CONCINNA

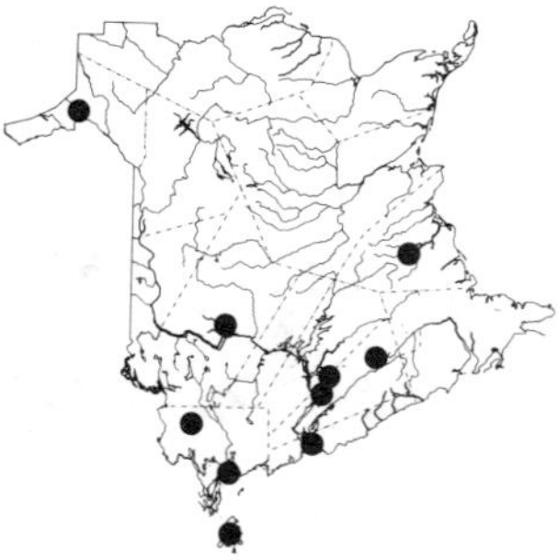

CAREX CONOIDEA

CAREX CRAWFORDII

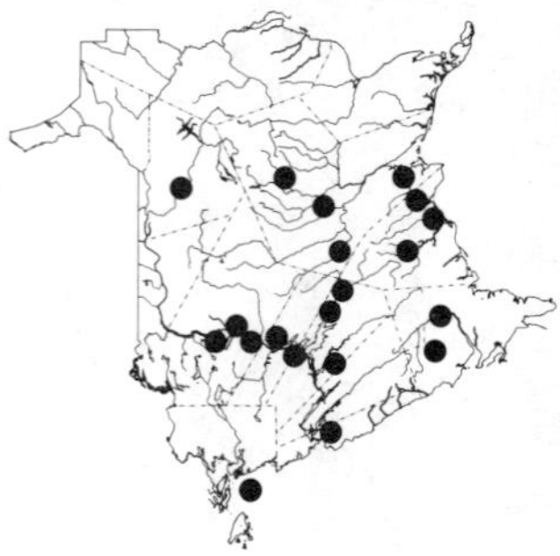

CAREX CRINITA

CAREX GYNANDRA

CAREX CUMULATA

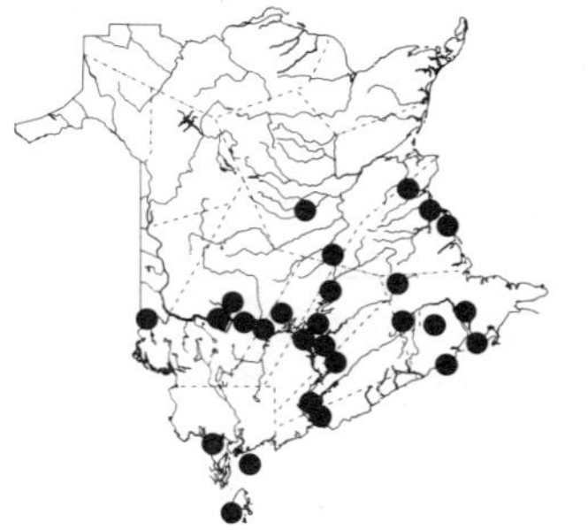

CAREX DEBILIS VAR. RUDGEI

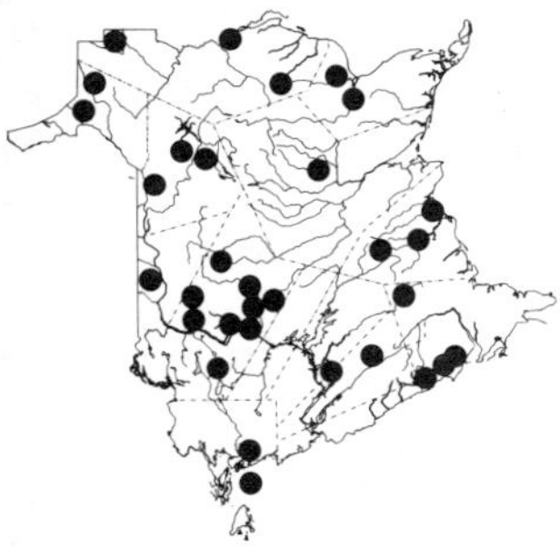

CAREX DEFLEXA

CAREX DEWEYANA

CAREX DIANDRA

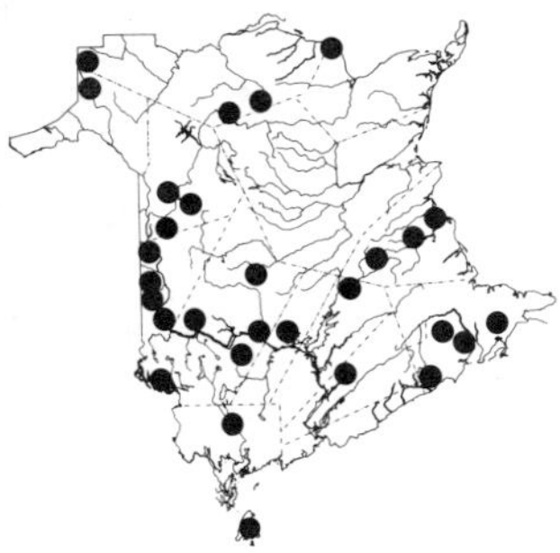

CAREX DISPERMA

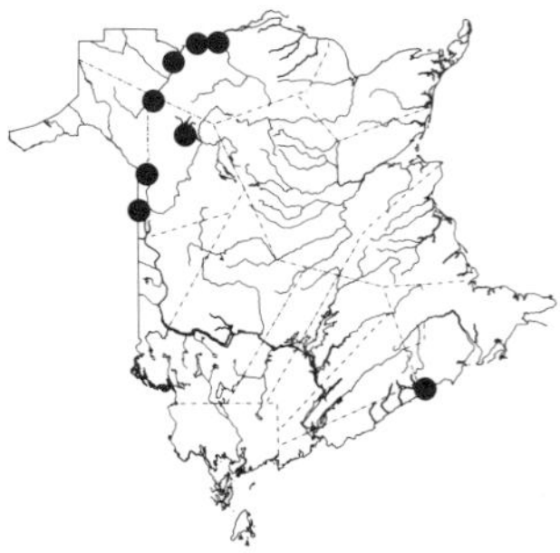

CAREX EBURNEA

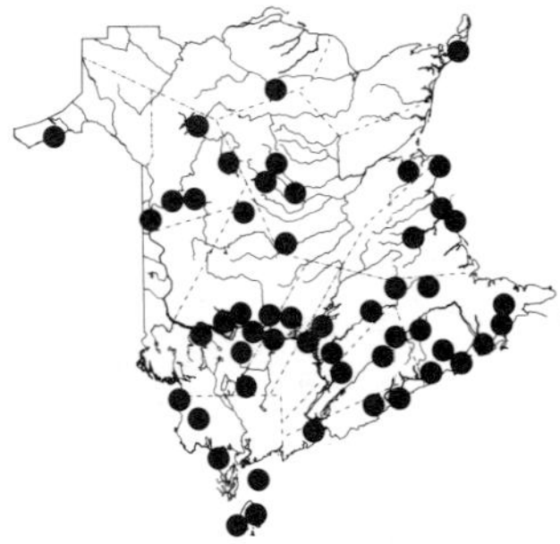

CAREX ECHINATA

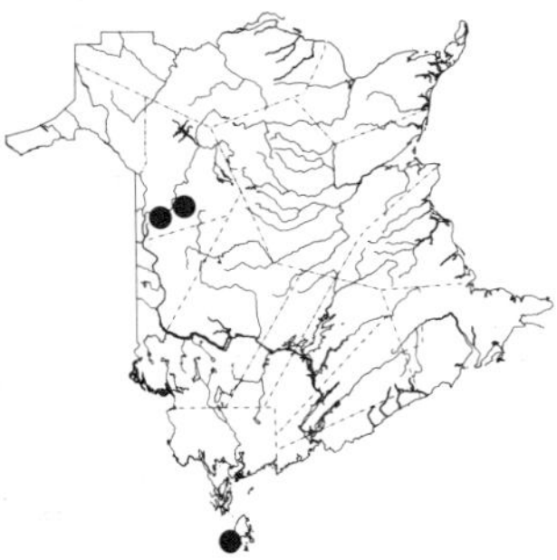

CAREX EXILIS

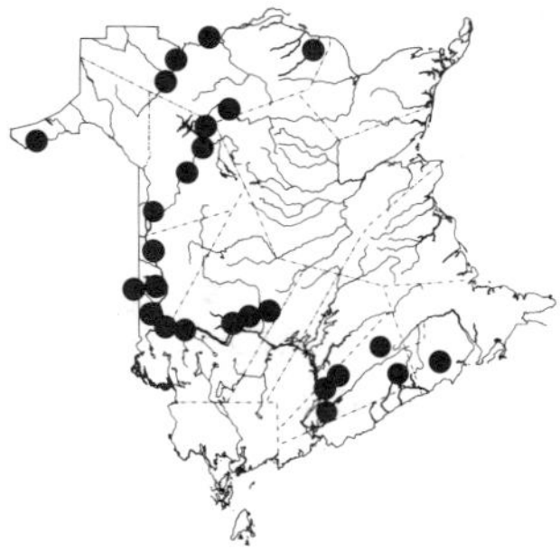

CAREX FLACCIDULA

CAREX FLAVA

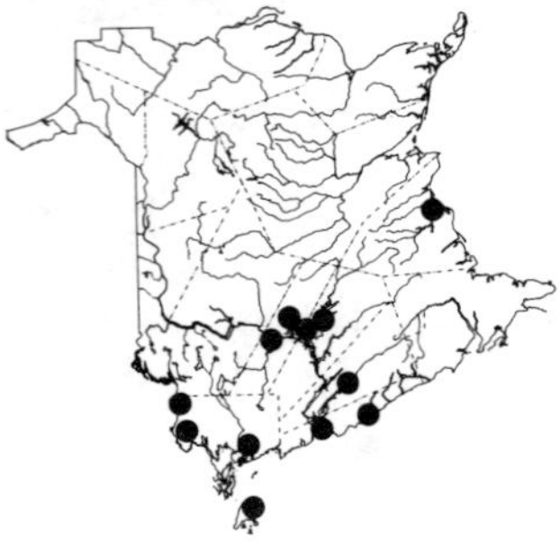

CAREX FOLLICULATA

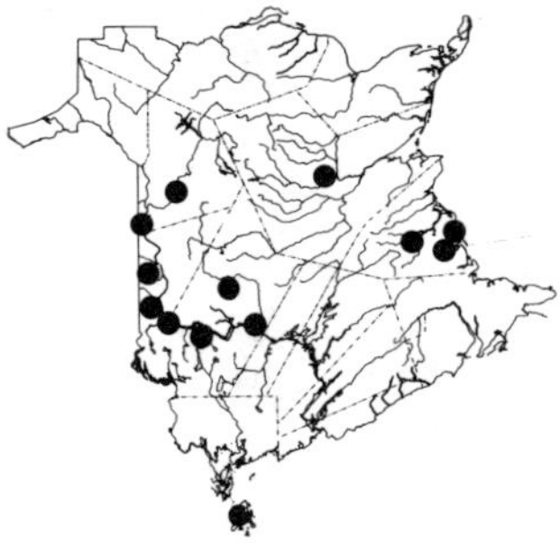

CAREX GRACILLIMA

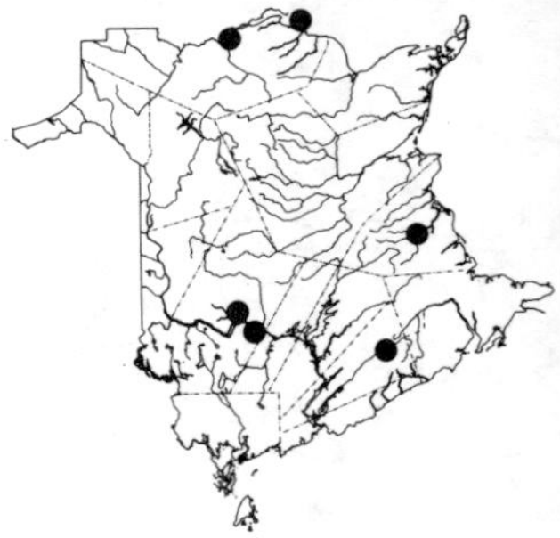

CAREX GRANULARIS VAR. HALEANA

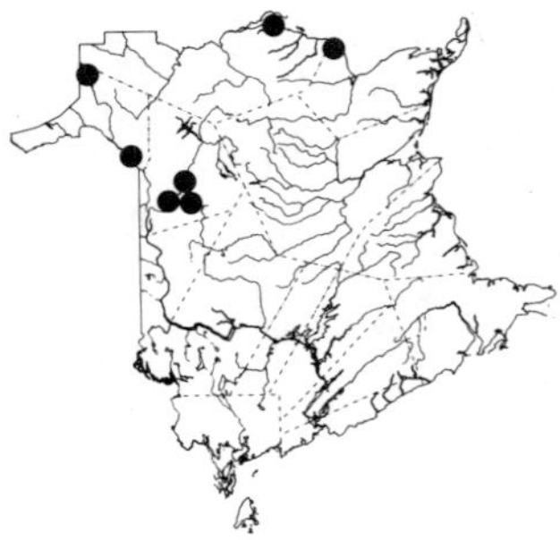

CAREX GYNOCRATES

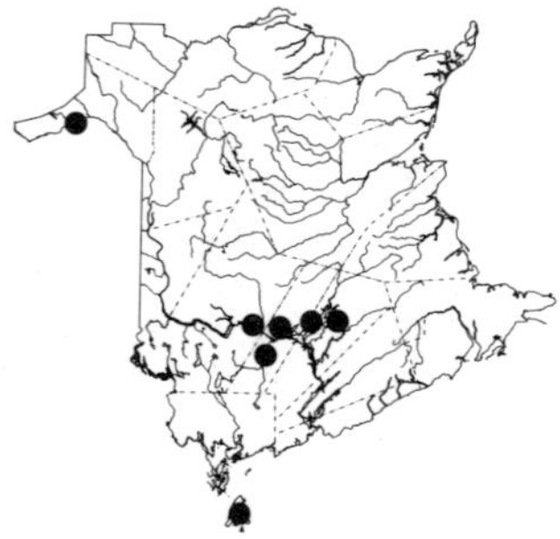

CAREX HAYDENII

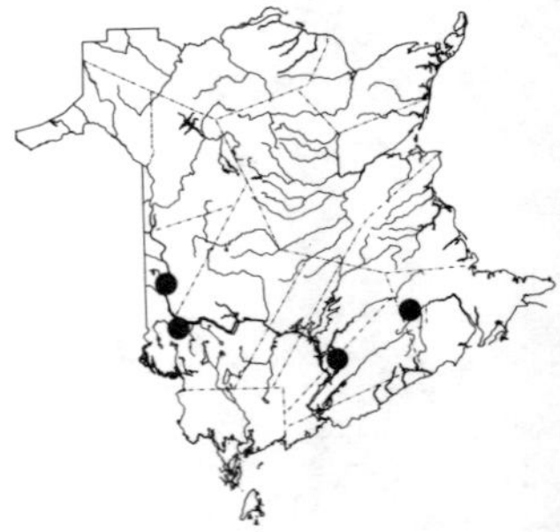

CAREX HIRTIFOLIA

CAREX HORMATHODES

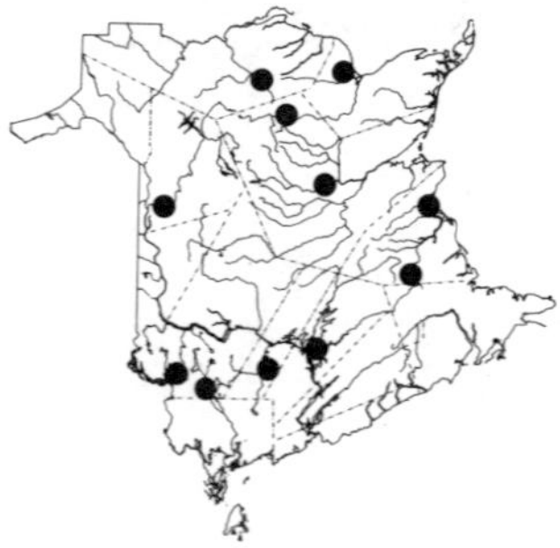

CAREX HOUGHTONII

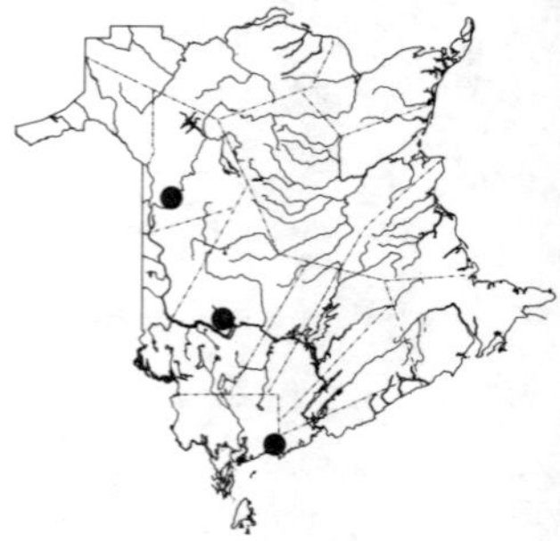

CAREX HOWEI

CAREX HYSTRICINA

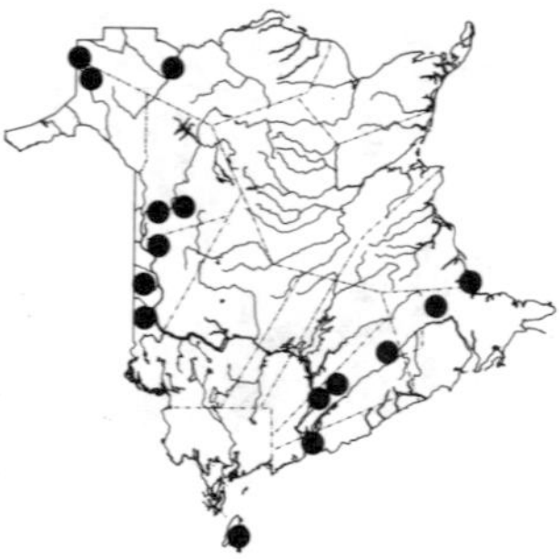

CAREX INTERIOR

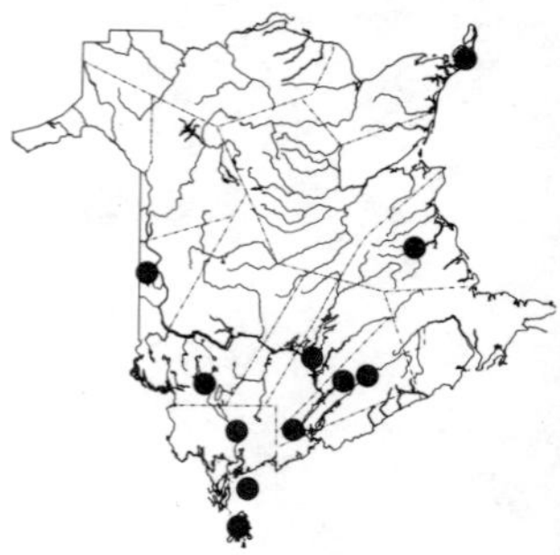

CAREX LACUSTRIS

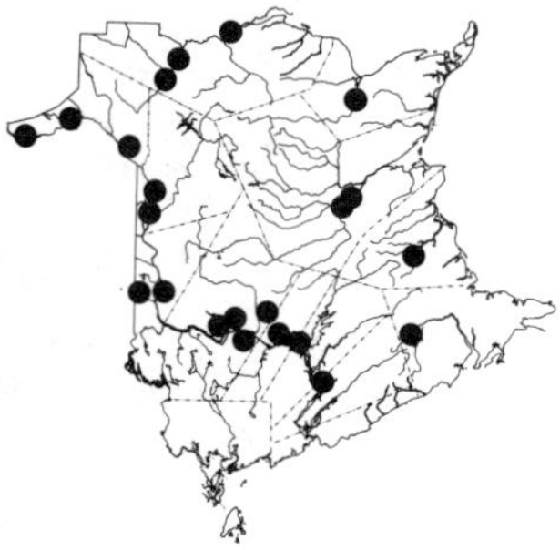

CAREX LANUGINOSA

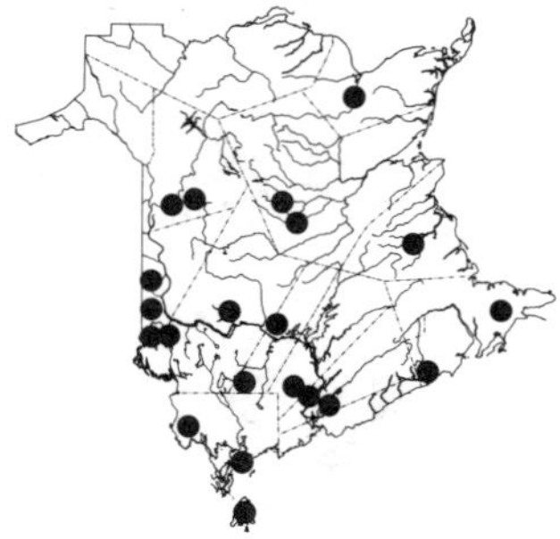

CAREX LASIOCARPA

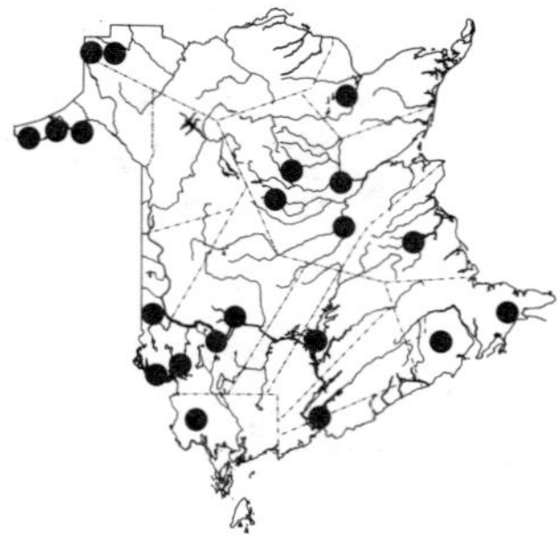

CAREX LENTICULARIS

CAREX LEPIDOCARPA

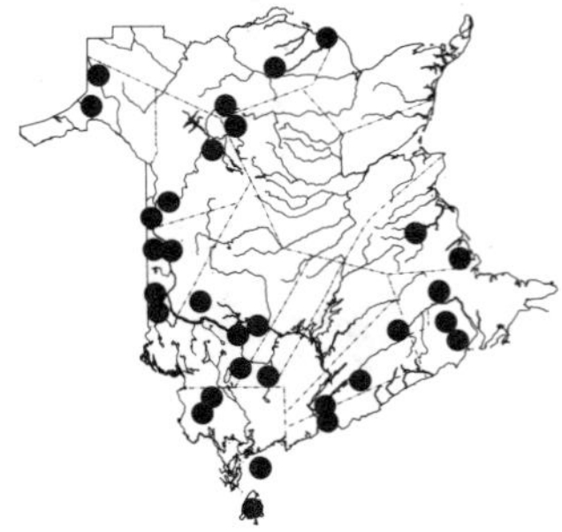

CAREX LEPTALEA

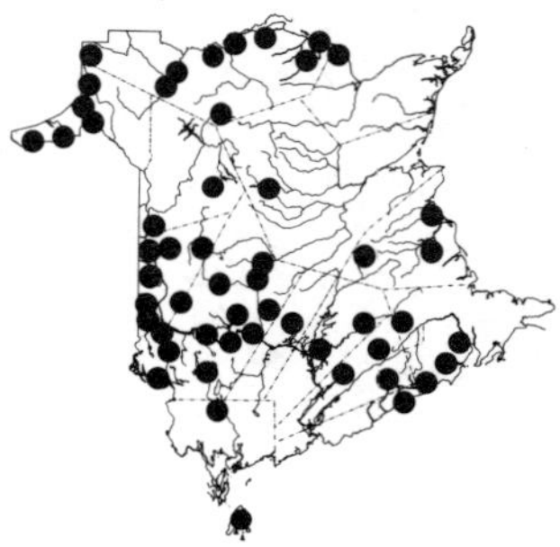

CAREX LEPTONERVIA

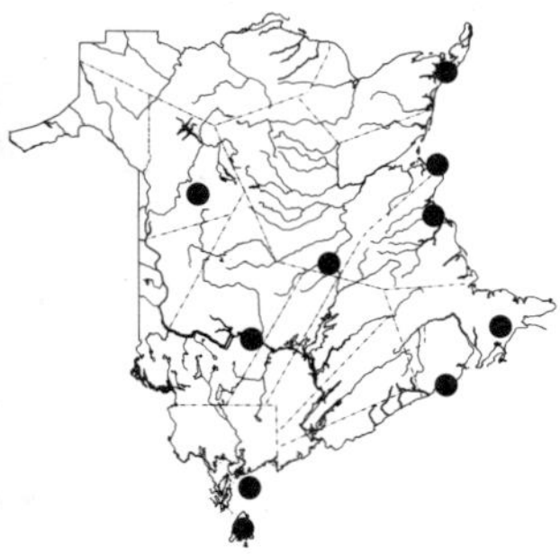

CAREX LIMOSA

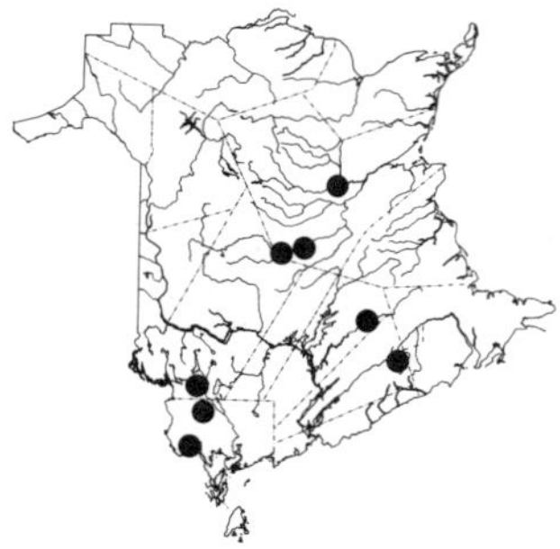

CAREX LUCORUM

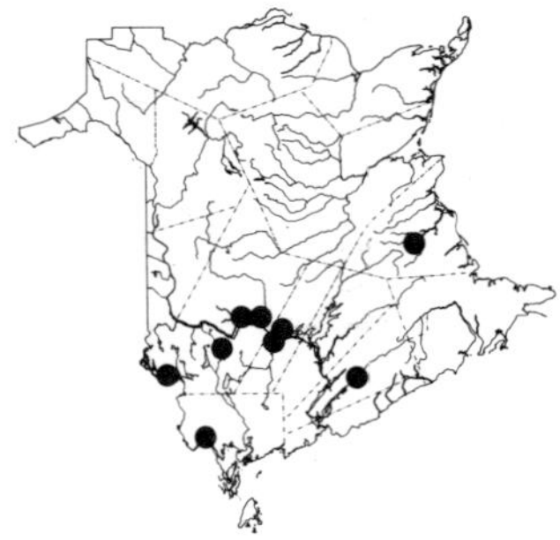

CAREX LUPULINA

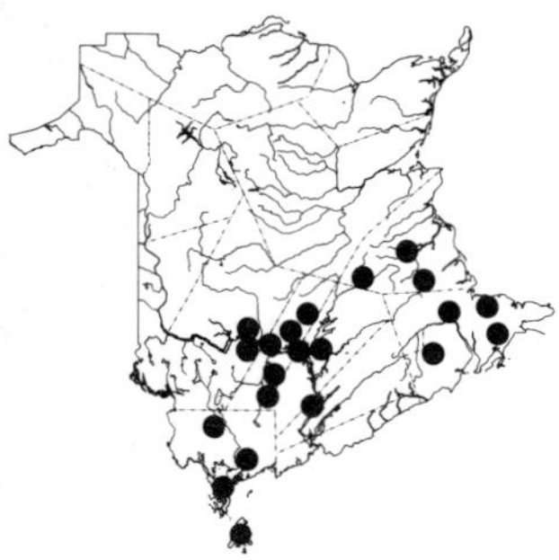

CAREX LURIDA

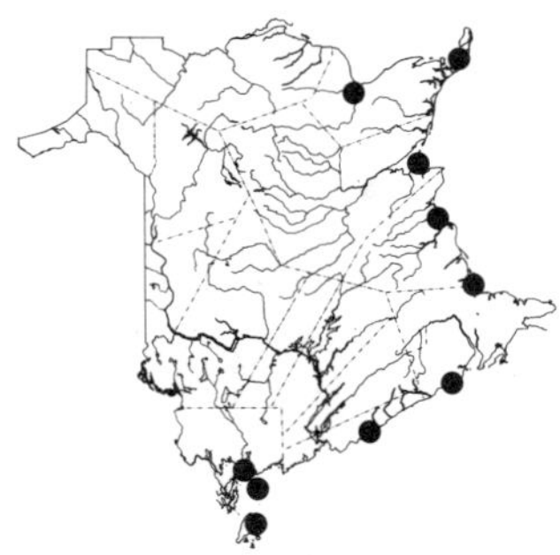

CAREX MACKENZIEI

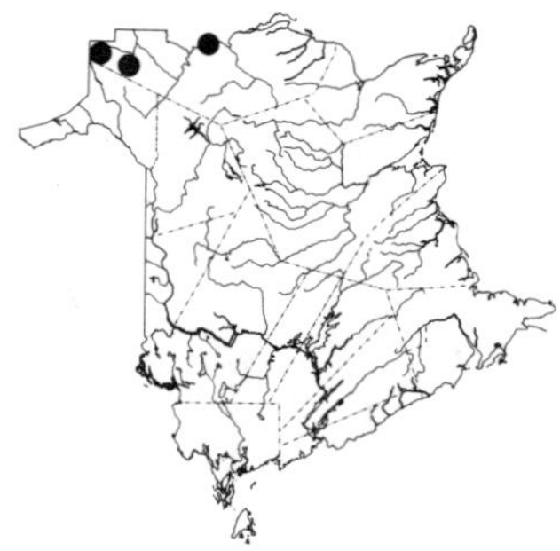

CAREX MEDIA

CAREX MICHAUXIANA

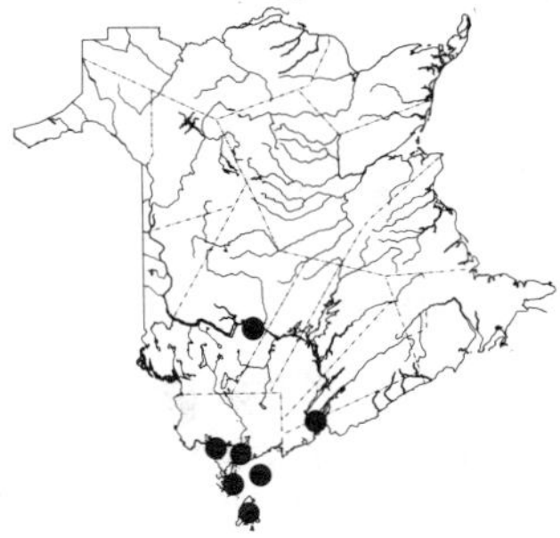

CAREX NIGRA

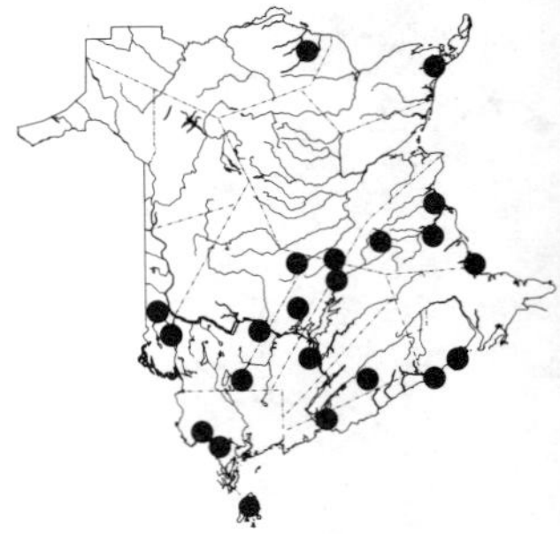

CAREX NOVAE-ANGLIAE

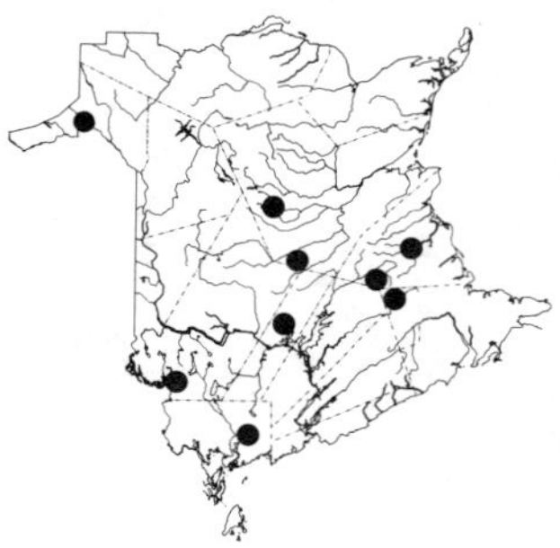

CAREX OLIGOSPERMA

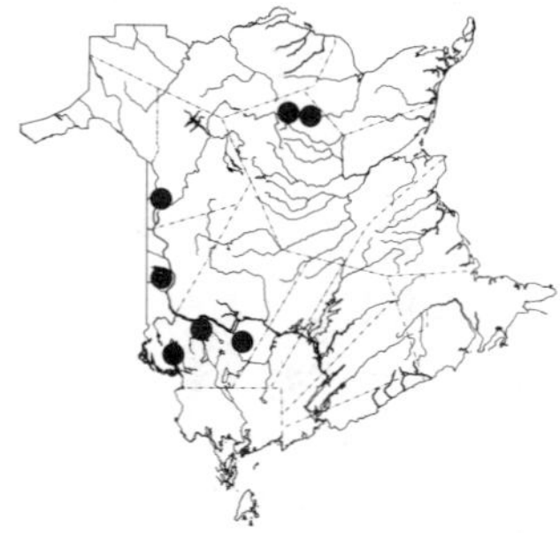

CAREX ORMOSTACHYA

CAREX PALEACEA

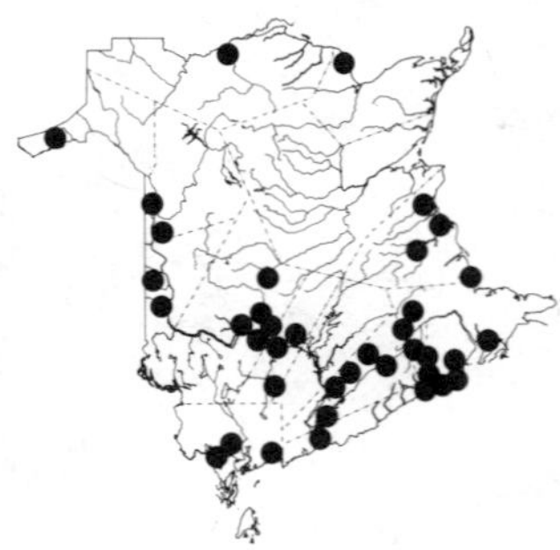

CAREX PALLESCENS VAR. NEOGAEA

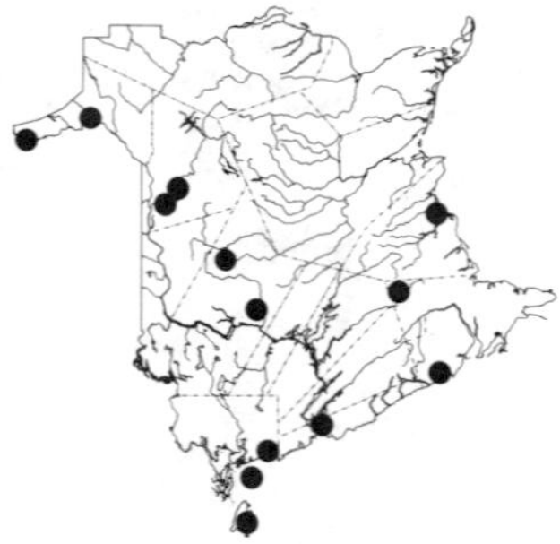

CAREX PAUCIFLORA

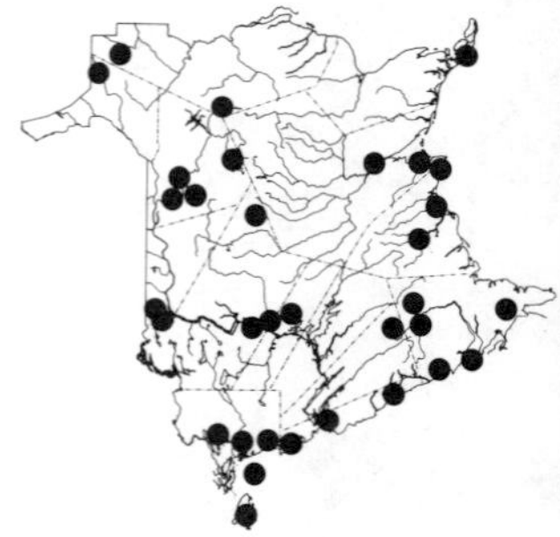

CAREX PAUPERCULA

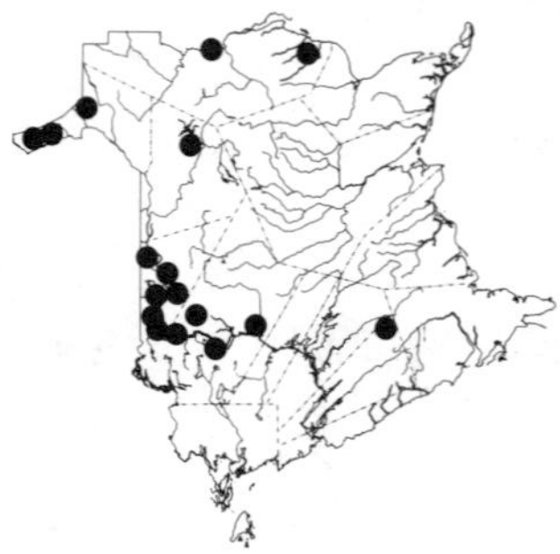

CAREX PECKII

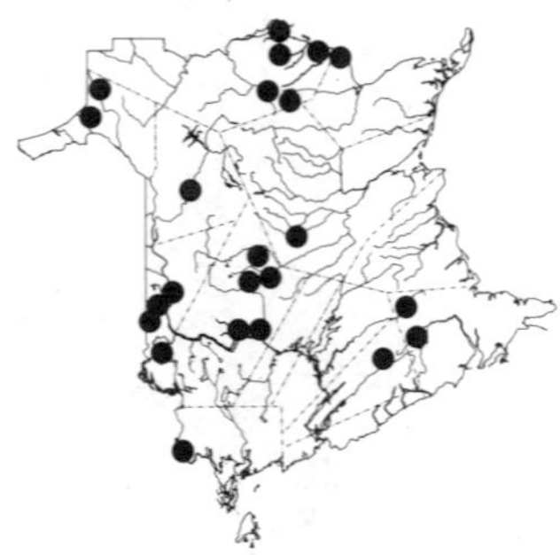

CAREX PEDUNCULATA

CAREX PLANTAGINEA

CAREX PRAIREA

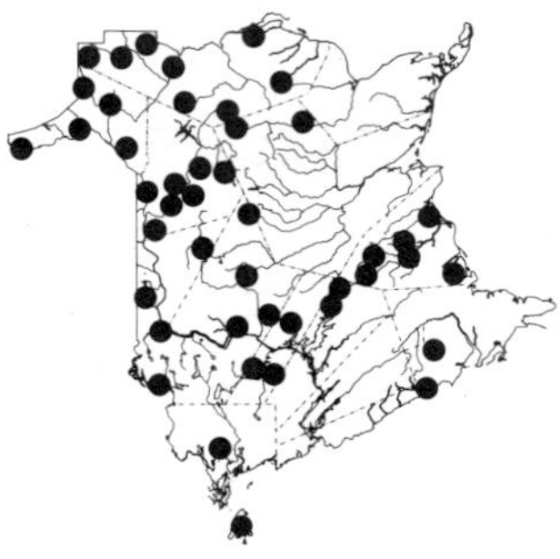

CAREX PROJECTA

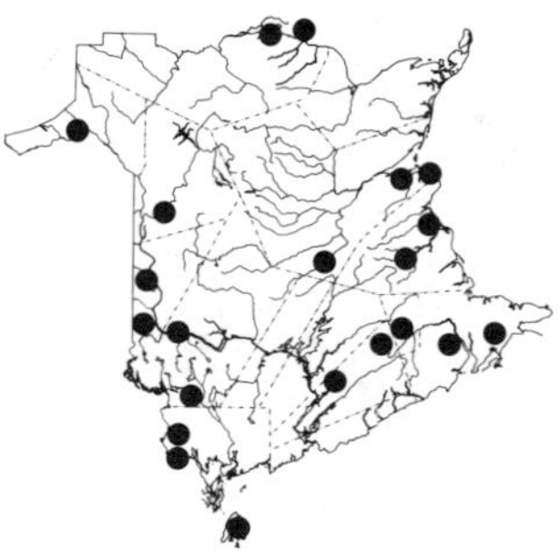

CAREX PSEUDO-CYPERUS

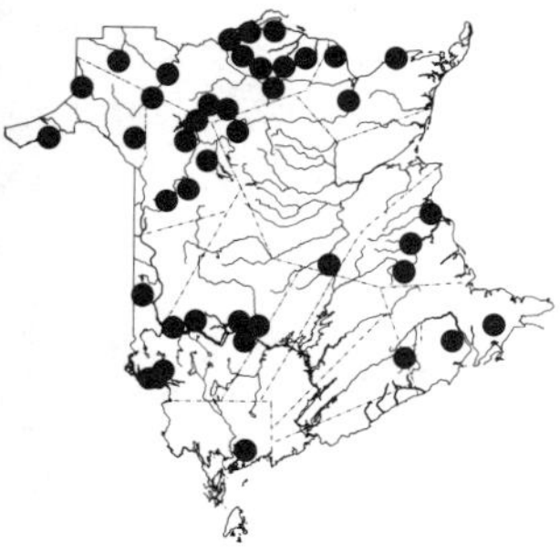

CAREX RETRORSA

CAREX UTRICULATA

CAREX SALINA VAR. KATTEGATENIS

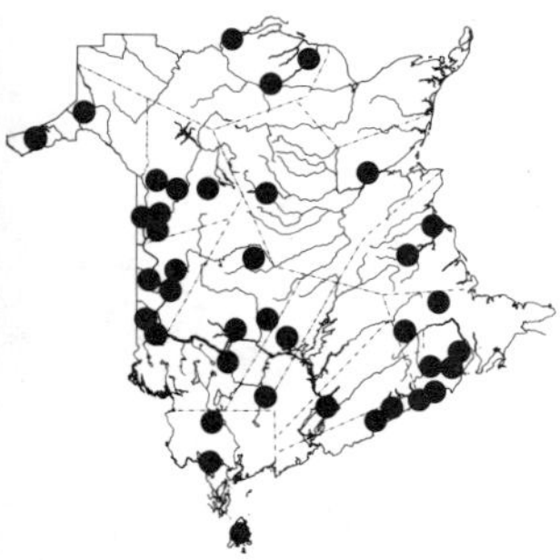

CAREX SCABRATA

CAREX SCOPARIA

CAREX SILICEA

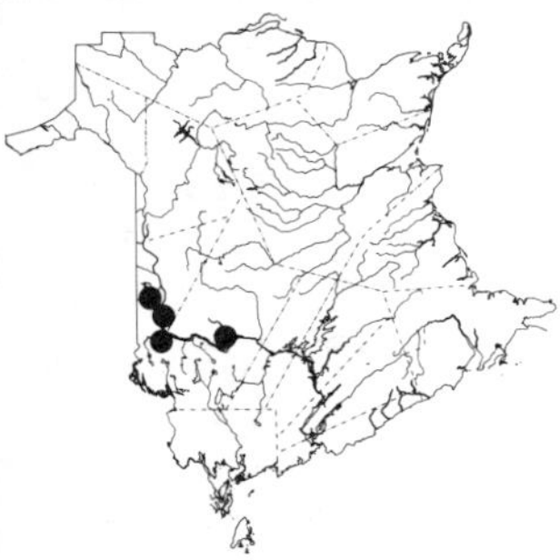

CAREX SPARGARIOIDES VAR CEPHALOIDEA

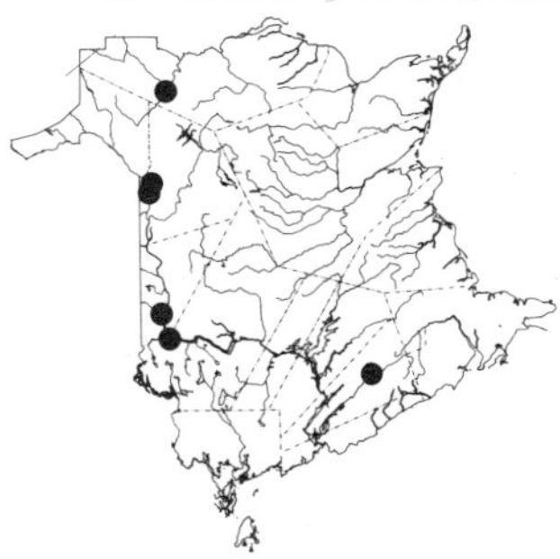

CAREX SPRENGELII

CAREX STIPATA

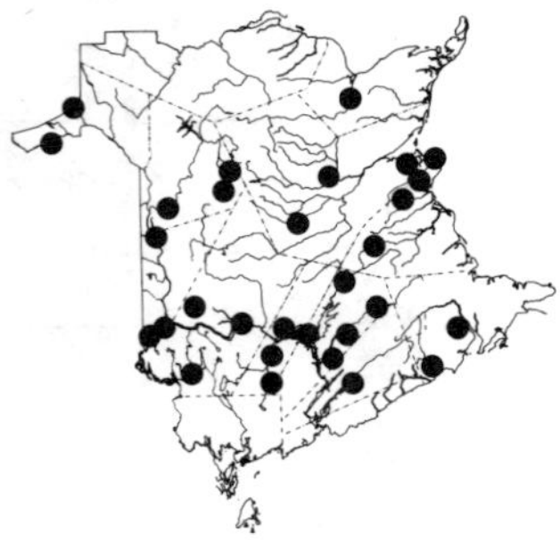

CAREX STRICTA

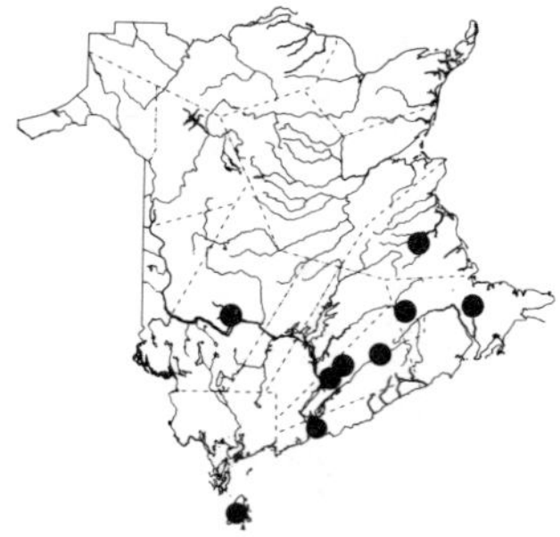

CAREX TENERA

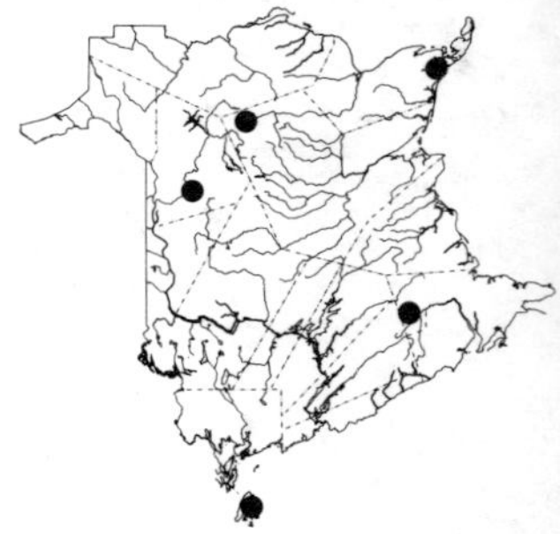

CAREX TENUIFLORA

CAREX TINCTA

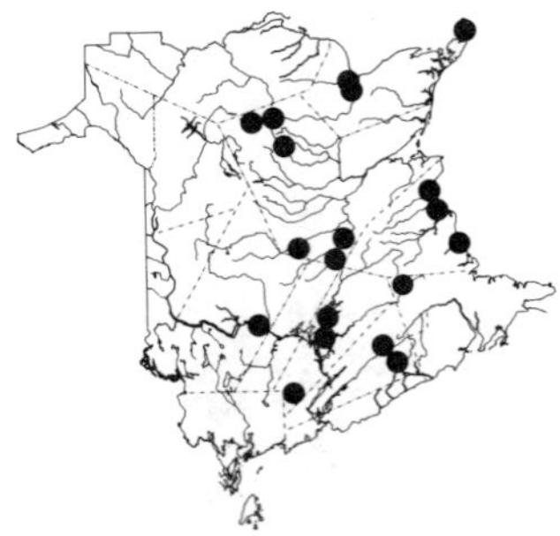

CAREX TONSA

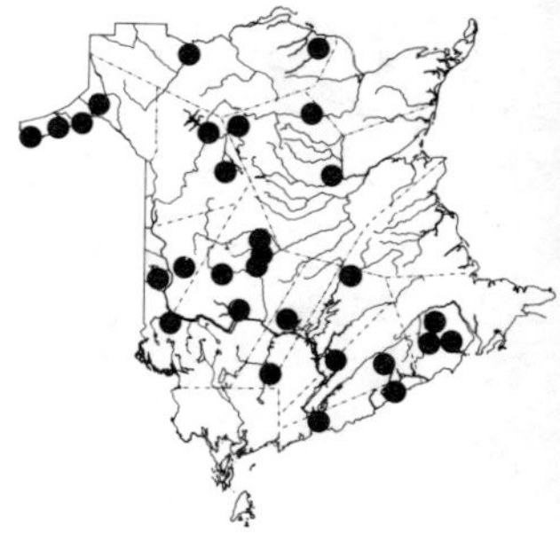

CAREX TORTA

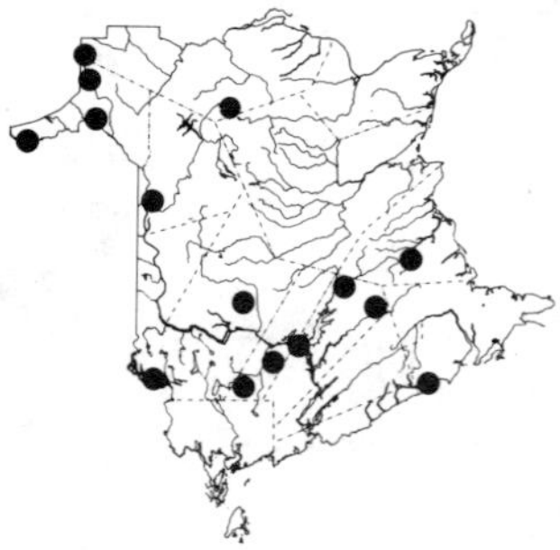

CAREX TRIBULOIDES

CAREX TRISPERMA

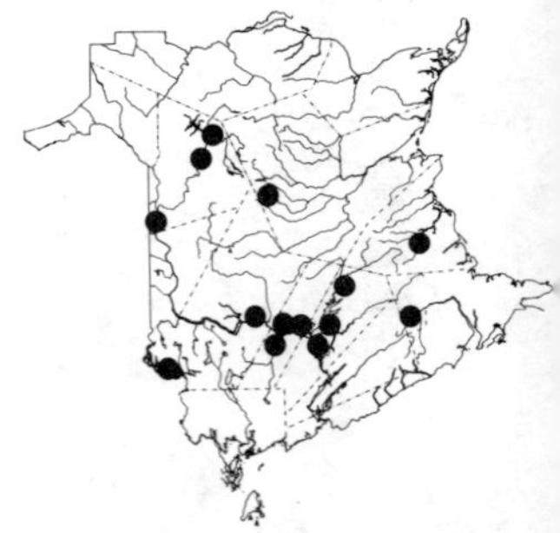

CAREX TUCKERMANII

CAREX UMBELLATA

CAREX VAGINATA

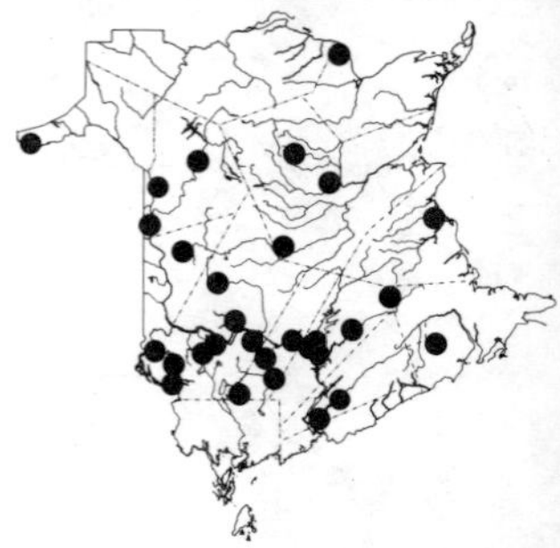

CAREX VESICARIA

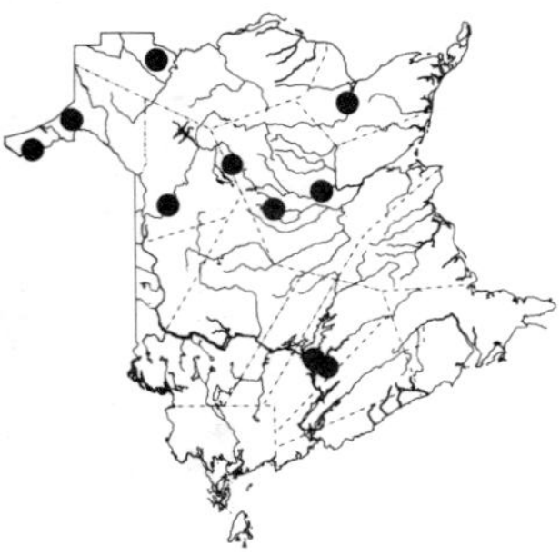
CAREX VESICARIA VAR. JEJEUNA

CAREX VIRIDULA

CAREX VULPINOIDEA

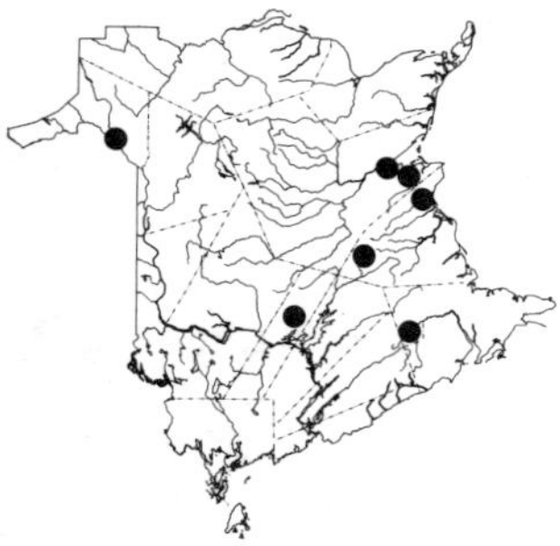
CAREX WIEGANDII

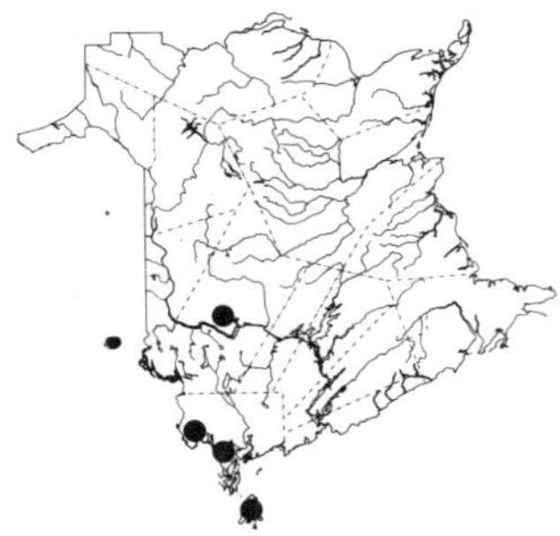
CLADIUM MARISCOIDES

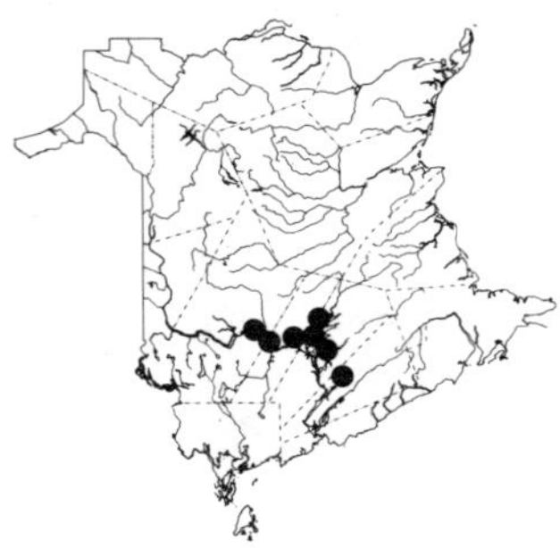
CYPERUS ARISTATUS

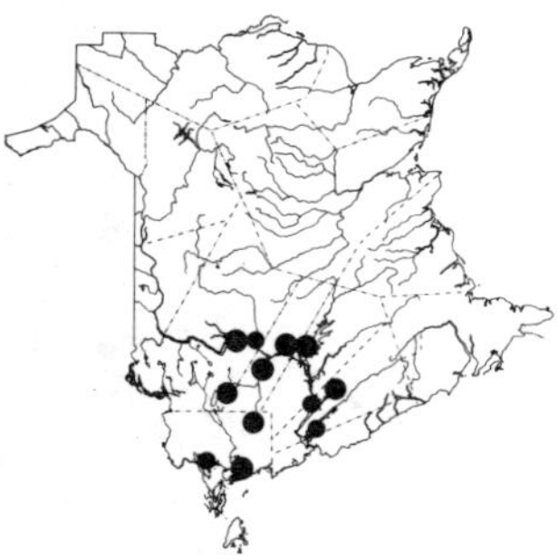
CYPERUS DENTATUS

CYPERUS DIANDRUS

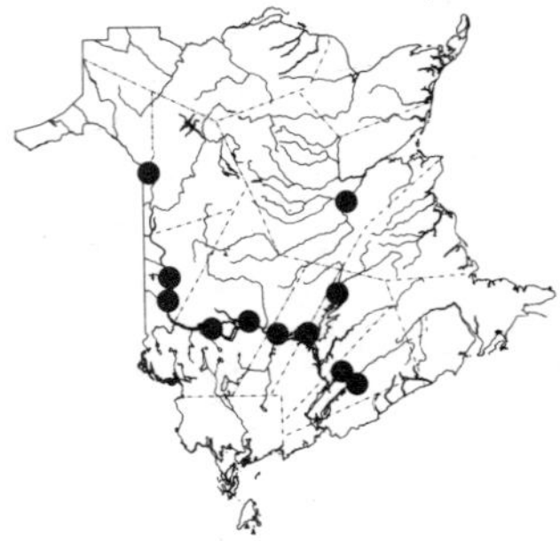
CYPERUS ESCULENTUS

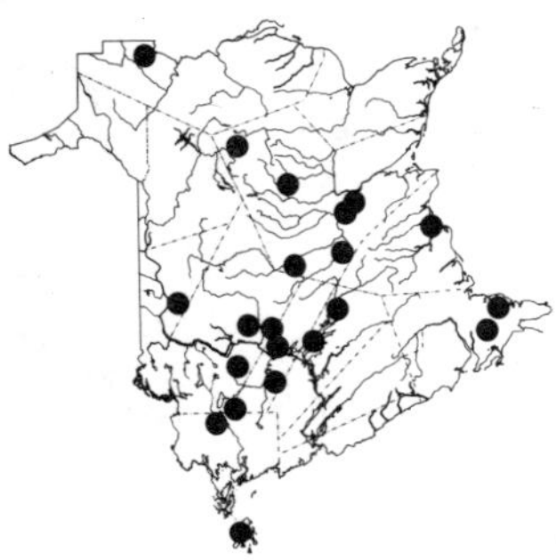
DULICHIUM ARUNDINACEUM

ELEOCHARIS ACICULARIS

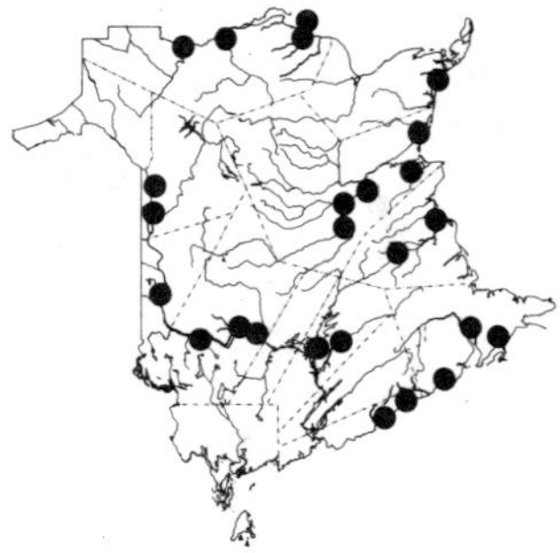
ELEOCHARIS ERYTHROPODA

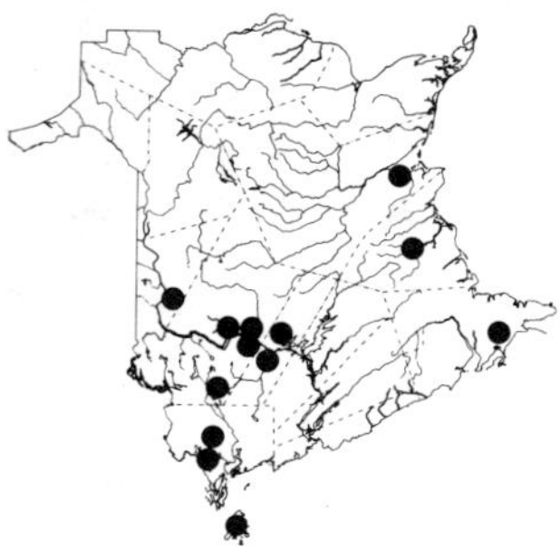

ELEOCHARIS OBTUSA

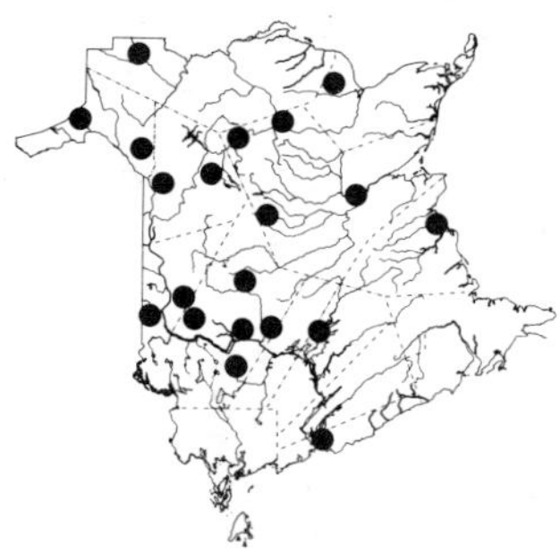

ELEOCHARIS OVATA

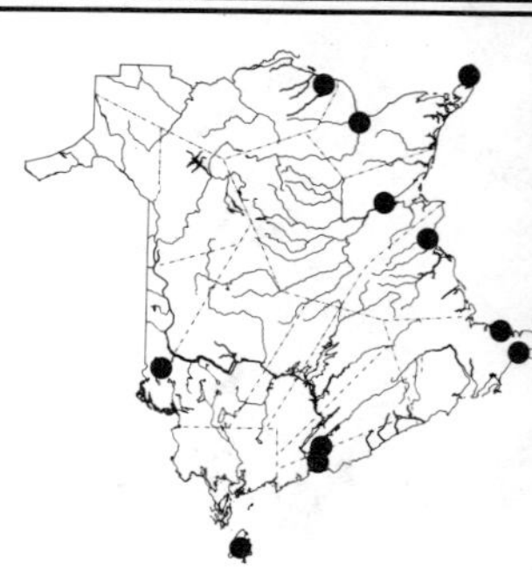

ELEOCHARIS PARVULA

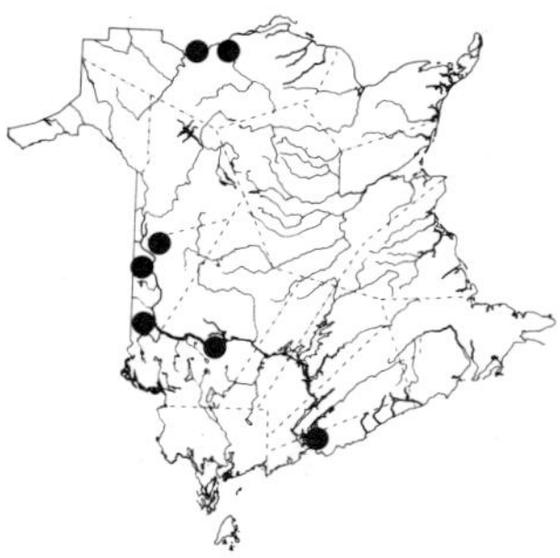

ELEOCHARIS PAUCIFLORA

ELEOCHARIS ROBBINSII

ELEOCHARIS SMALLII

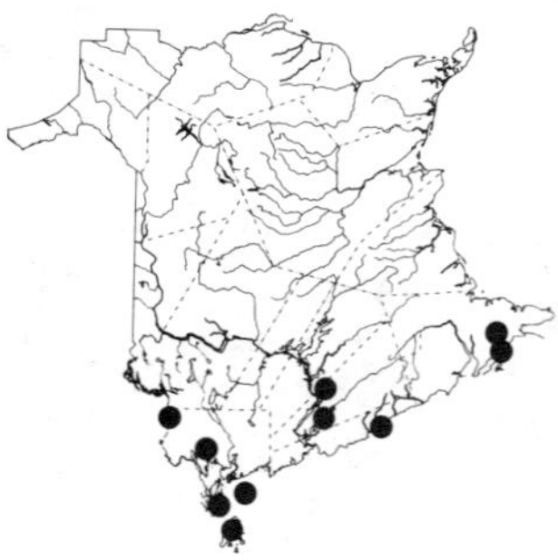

ELEOCHARIS TENUIS

ELEOCHARIS TENUIS VAR. BOREALIS

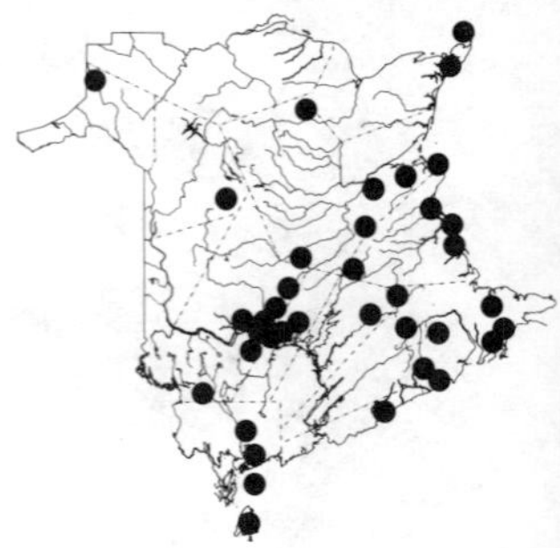

ERIOPHORUM ANGUSTIFOLIUM

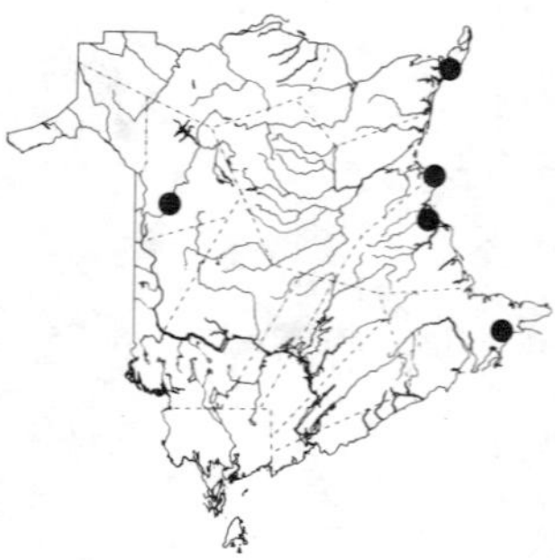

ERIOPHORUM CHAMISSONIS

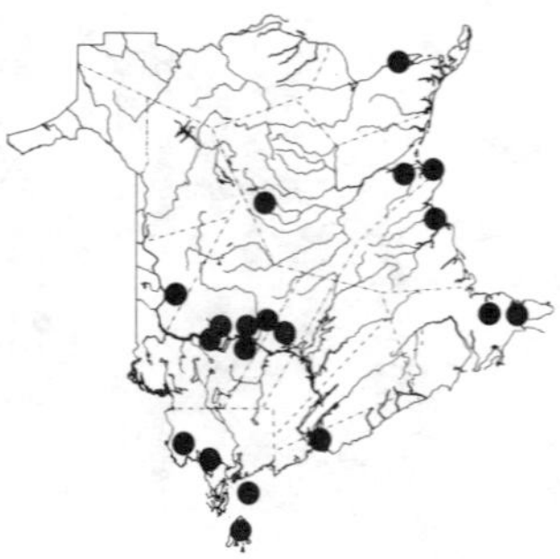

ERIOPHORUM TENELLUM

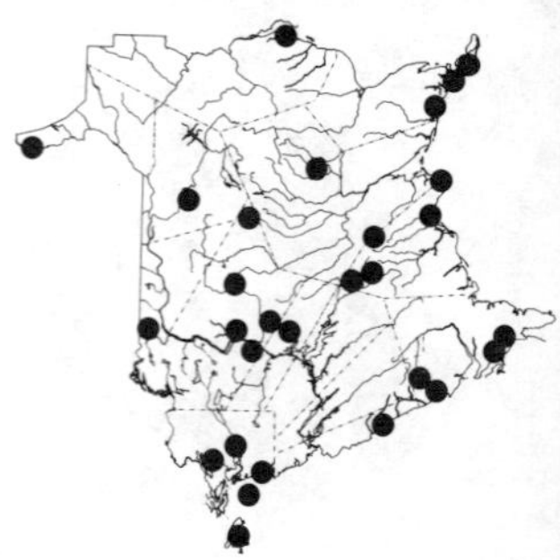

ERIOPHORUM VAGINATUM
VAR. SPISSUM

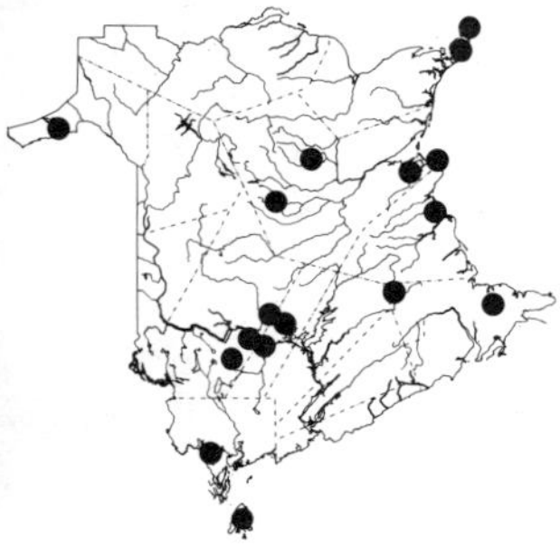

ERIOPHORUM VIRGINICUM

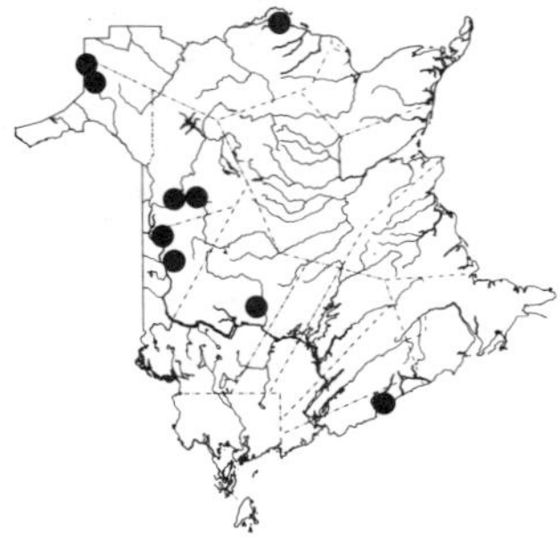

ERIOPHORUM VIRIDICARINATUM

RHYNCHOSPORA ALBA

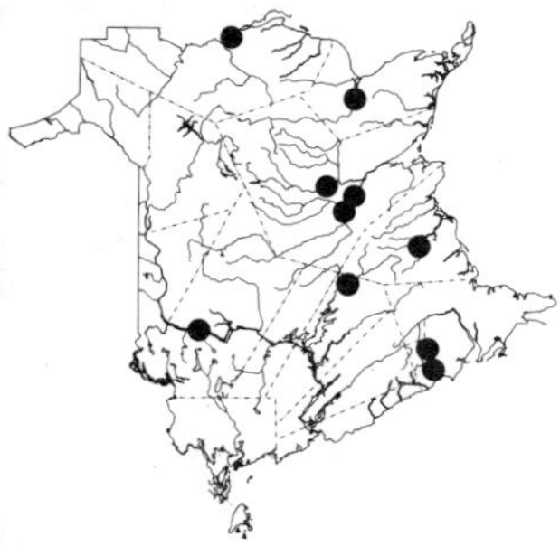

RHYNCHOSPORA CAPITELLATA

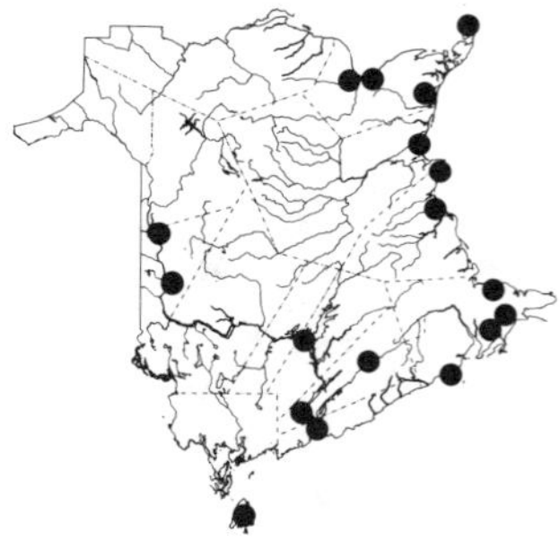

SCIRPUS ACUTUS

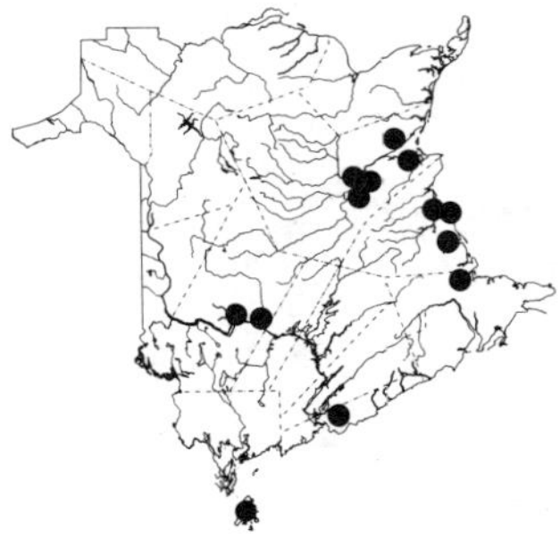

SCIRPUS AMERICANUS

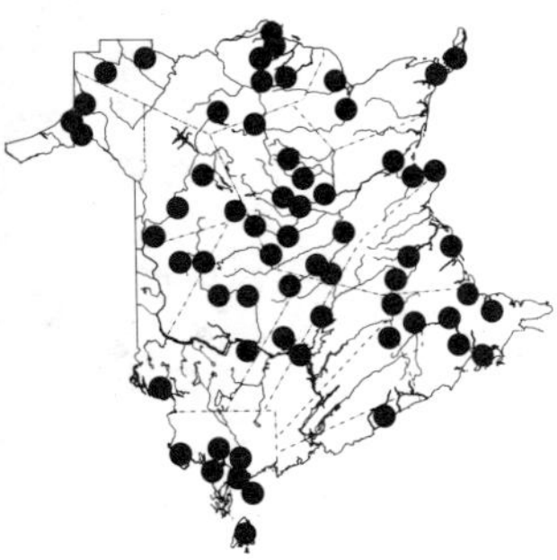

SCIRPUS ATROCINCTUS

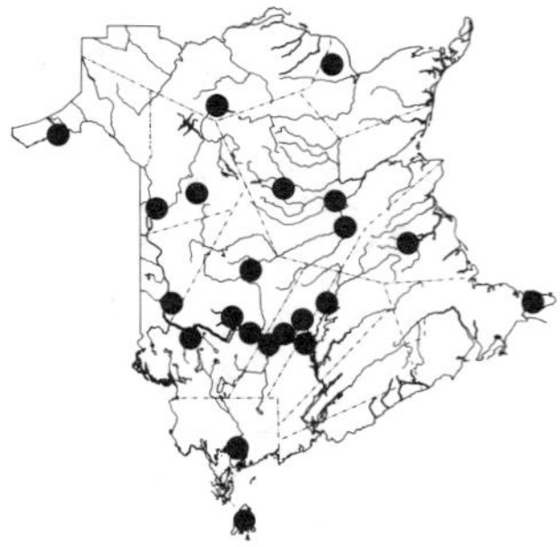

SCIRPUS ATROVIRENS VAR. GEORGIANA

SCIRPUS CAESPITOSUS

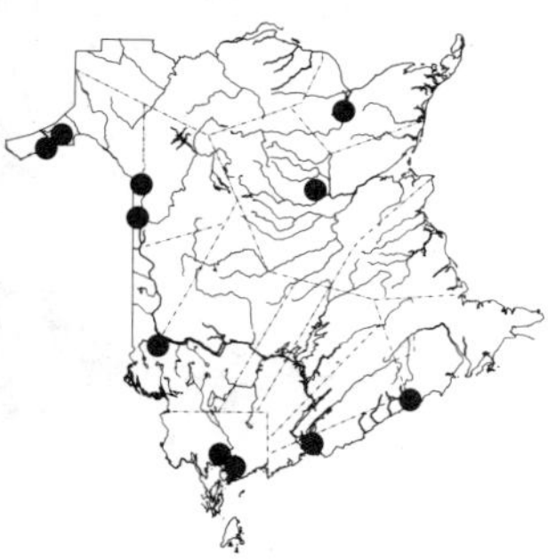

SCIRPUS CLINTONII

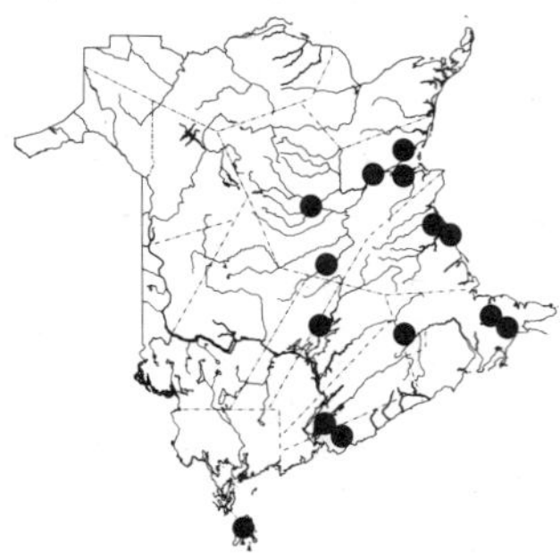

SCIRPUS CYPERINUS

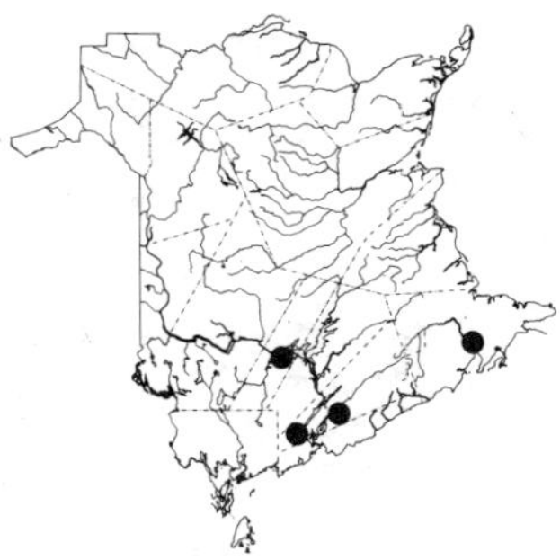

SCIRPUS FLUVIATILIS

SCIRPUS HUDSONIANUS

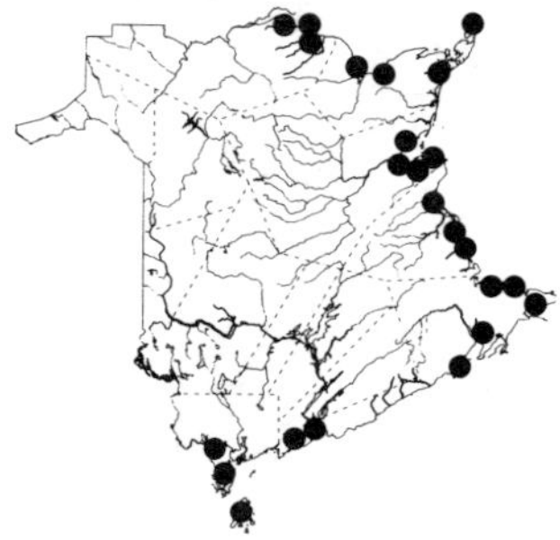

SCIRPUS MARITIMUS

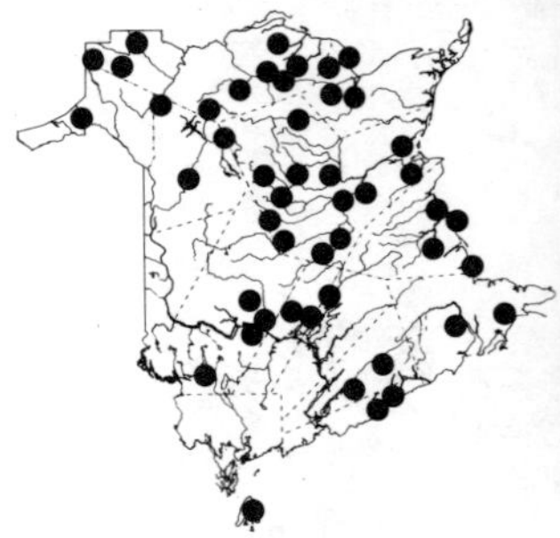

SCIRPUS MICROCARPUS

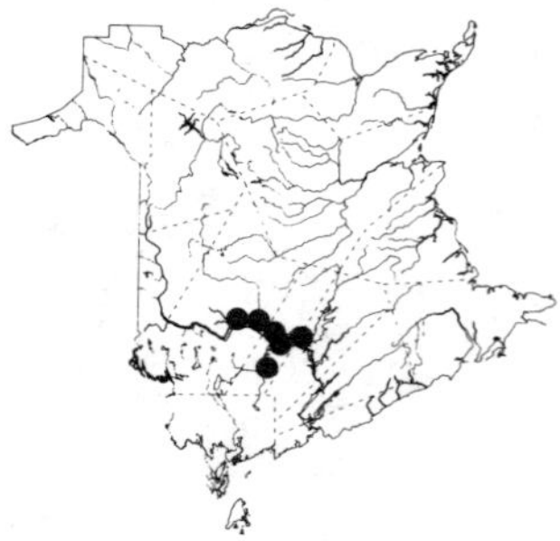

SCIRPUS PEDICELLATUS

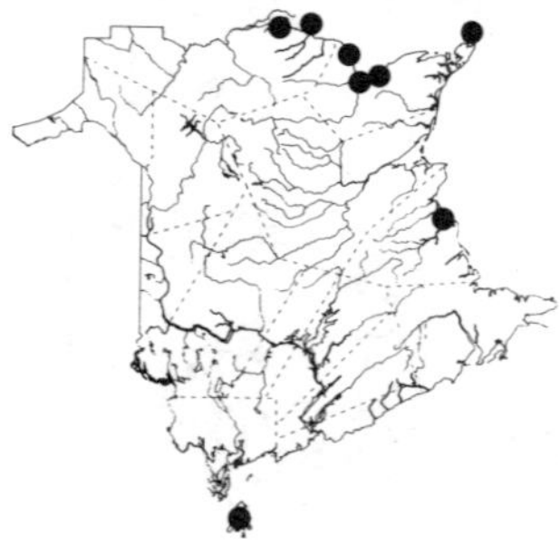

SCIRPUS RUFUS

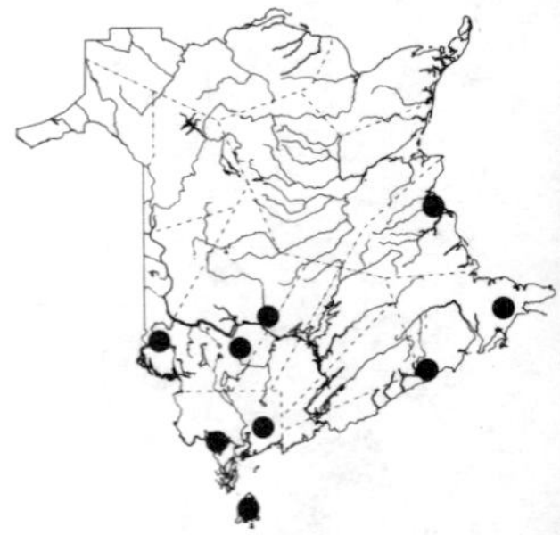

SCIRPUS SUBTERMINALIS

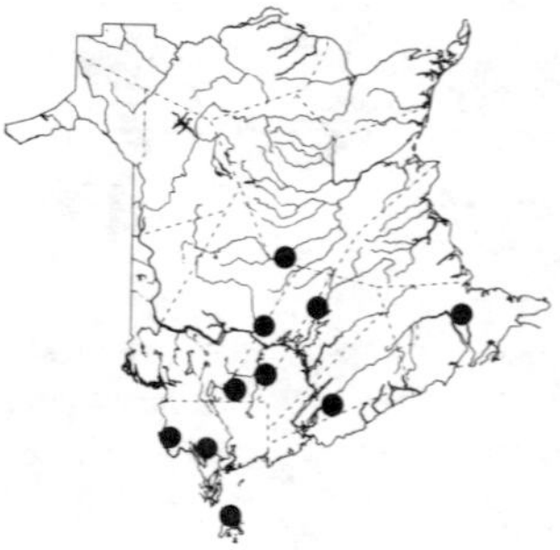

SCIRPUS TORREYI

SCIRPUS VALIDUS

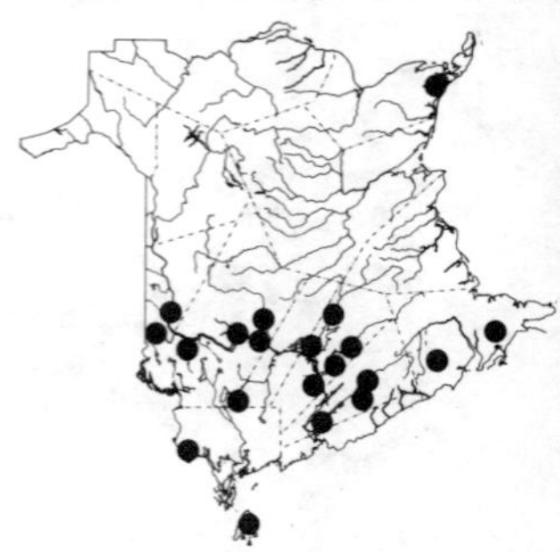

ACORUS AMERICANUS

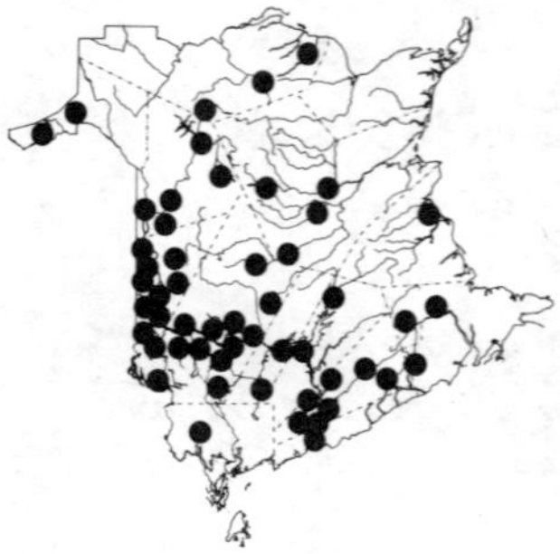

ARISAEMA TRIPHYLLUM

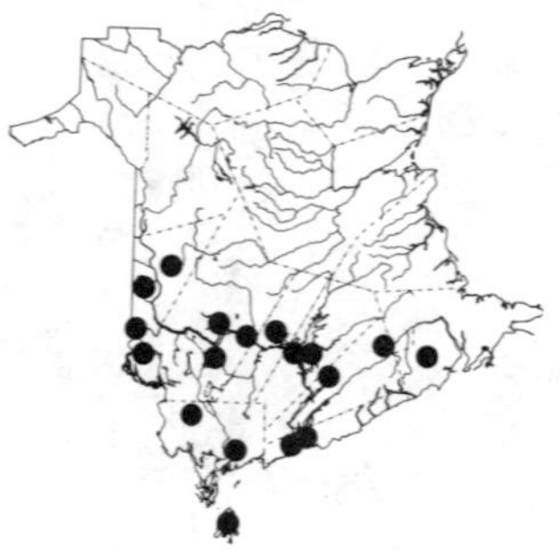

ARISAEMA TRIPHYLLUM SSP. STEWARDSONII

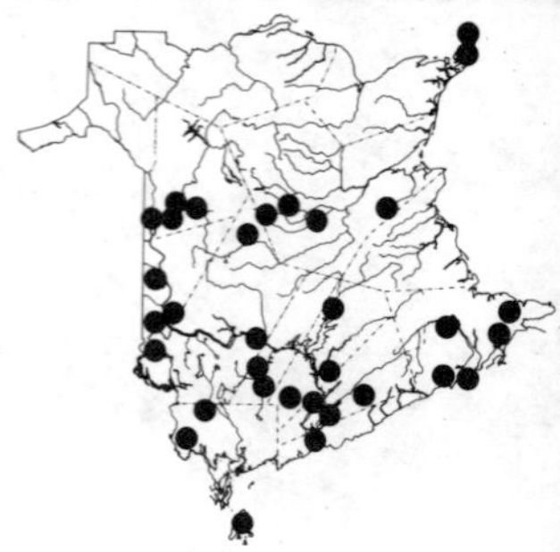

CALLA PALUSTRIS

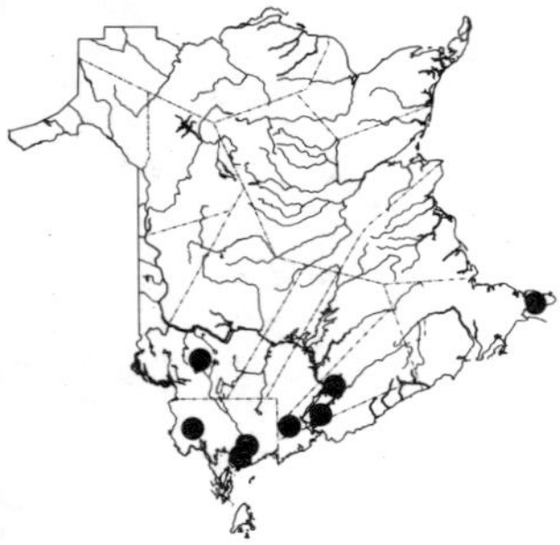

SYMPLOCARPUS FOETIDUS

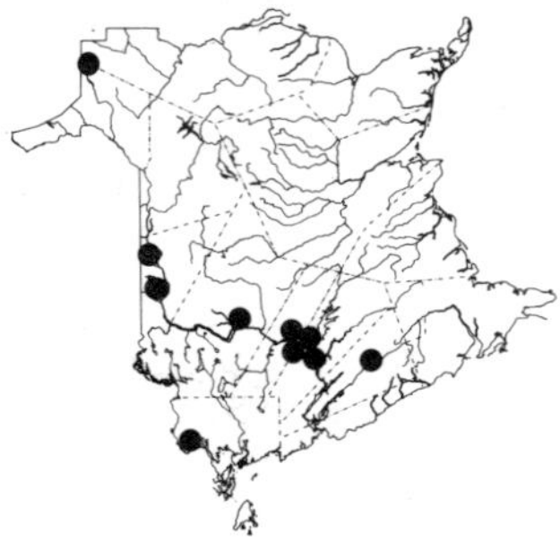

LEMNA MINOR

LEMNA TRISULCA

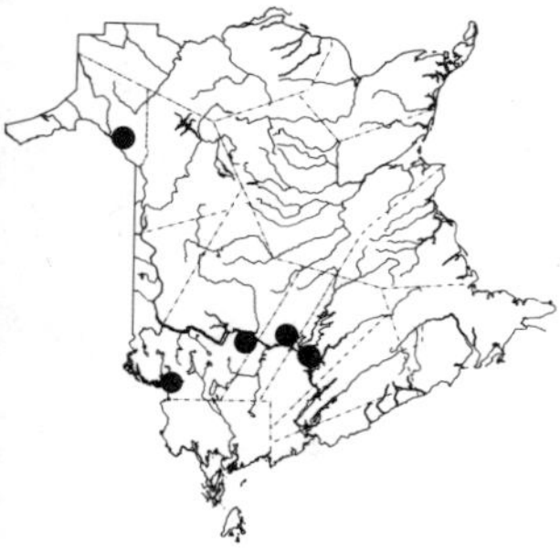

SPIRODELA POLYRHIZA

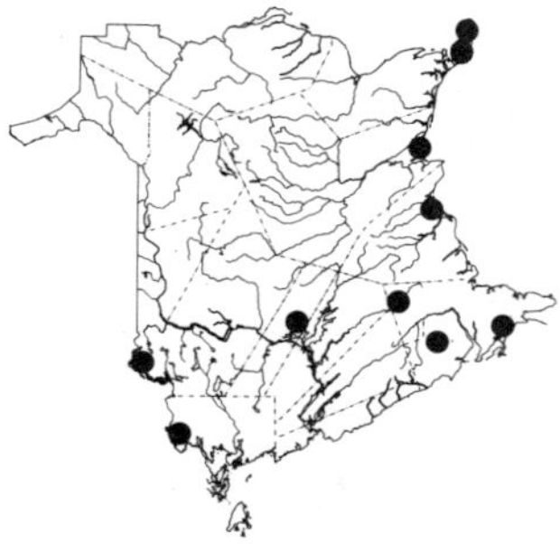

XYRIS MONTANA

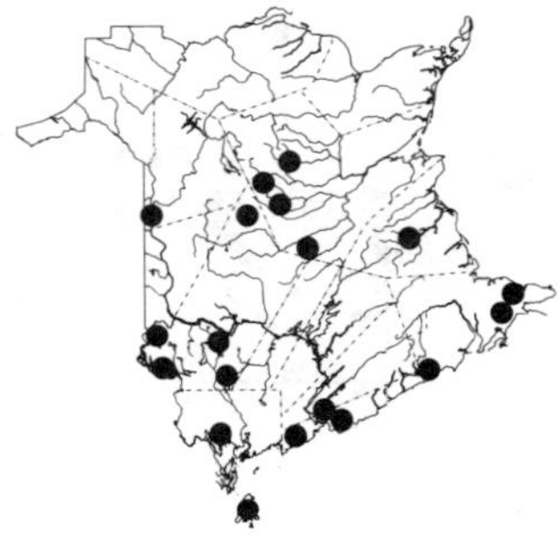

ERIOCAULON SEPTANGULARE

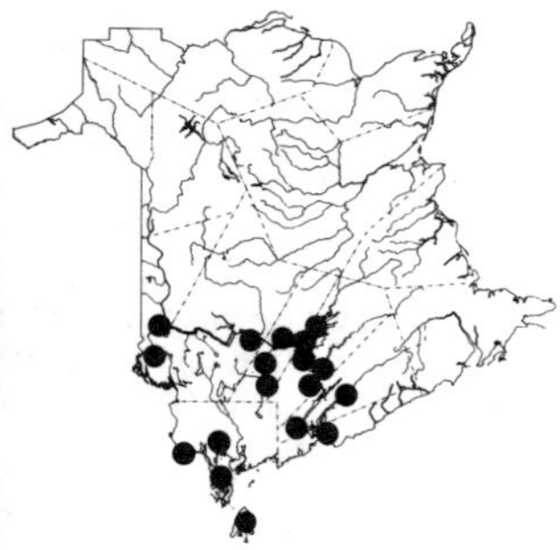

PONTEDERIA CORDATA

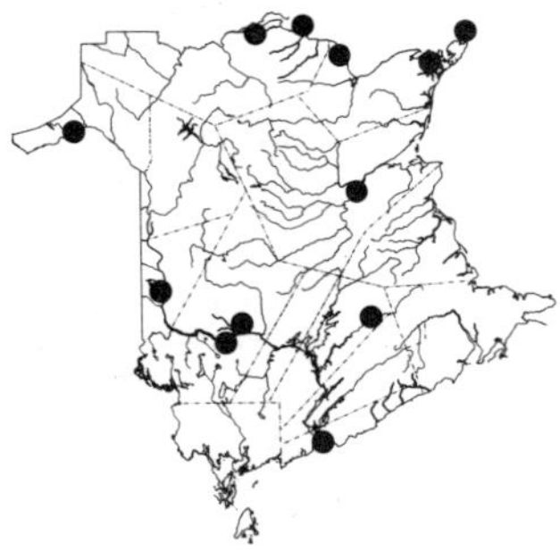

JUNCUS ALPINUS

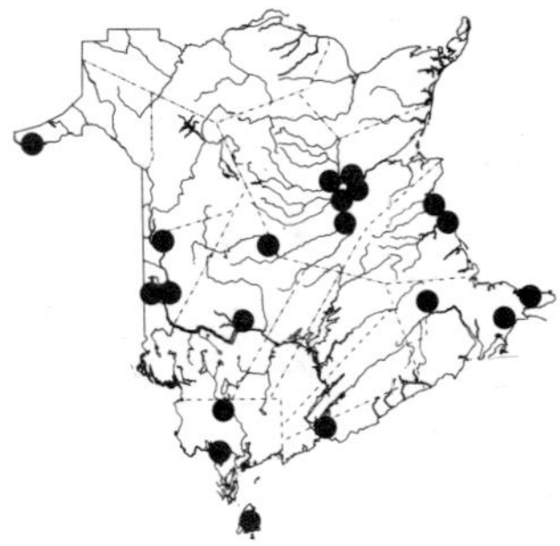

JUNCUS ARTICULATUS

JUNCUS BALTICUS VAR. LITTORALIS

JUNCUS BRACHYCEPHALUS

JUNCUS BREVICAUDATUS

JUNCUS BUFONIUS

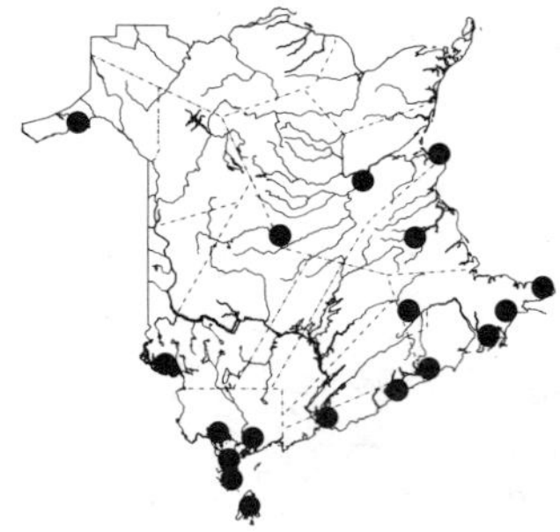

JUNCUS CANADENSIS

JUNCUS EFFUSUS

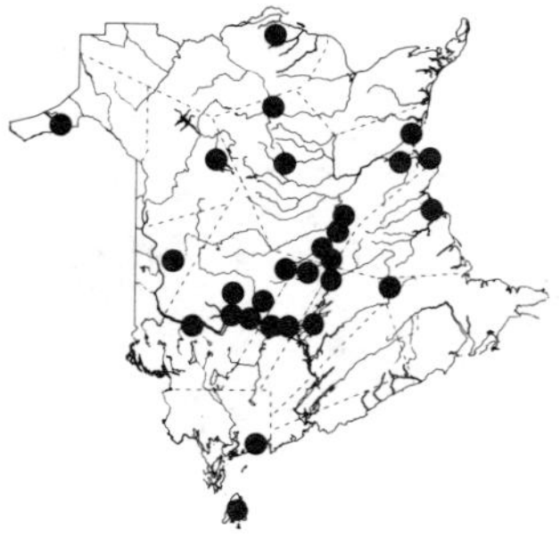

JUNCUS FILIFORMIS

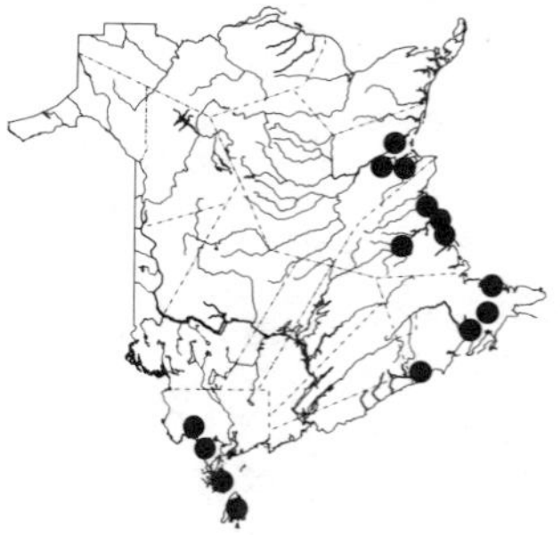

JUNCUS GERARDII

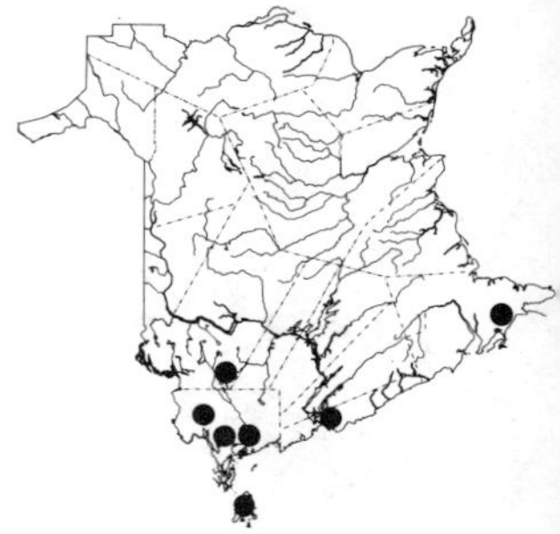

JUNCUS MILITARIS

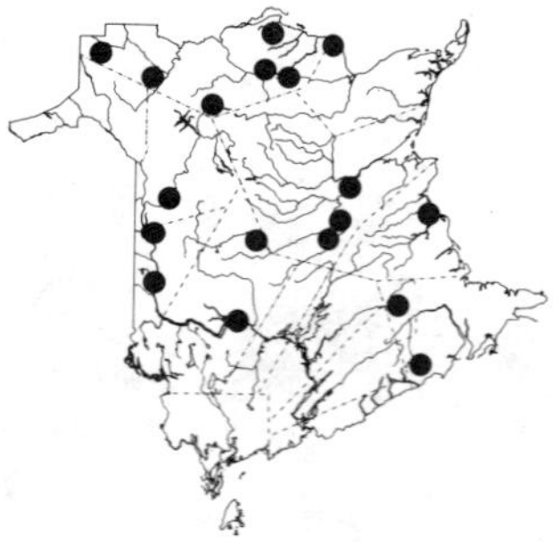

JUNCUS NODOSUS

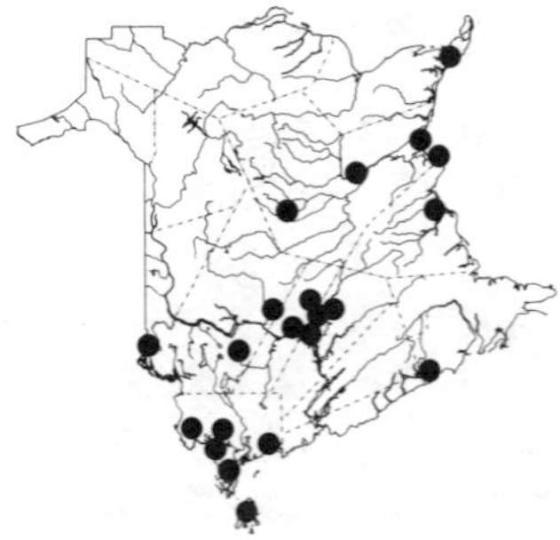

JUNCUS PELOCARPUS

JUNCUS STYGIUS

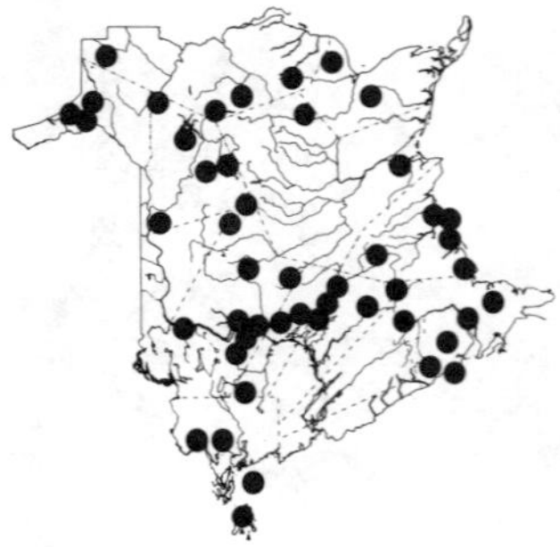

JUNCUS TENUIS

JUNCUS TENUIS VAR. UNIFLORUS

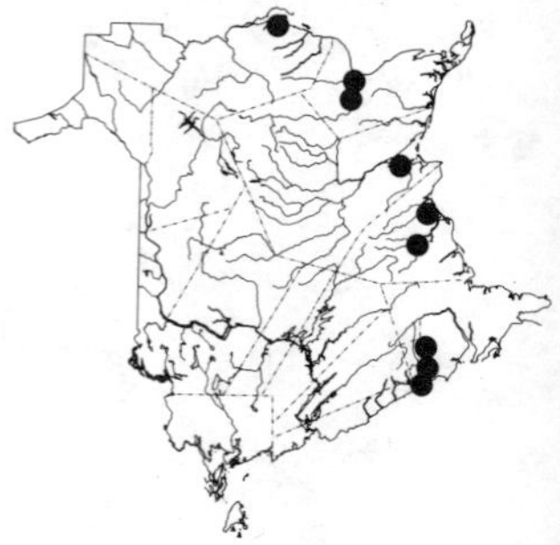

JUNCUS VASEYI

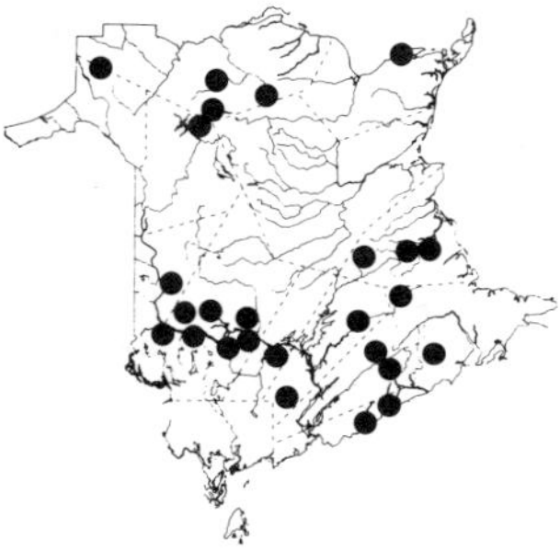

LUZULA ACUMINATA

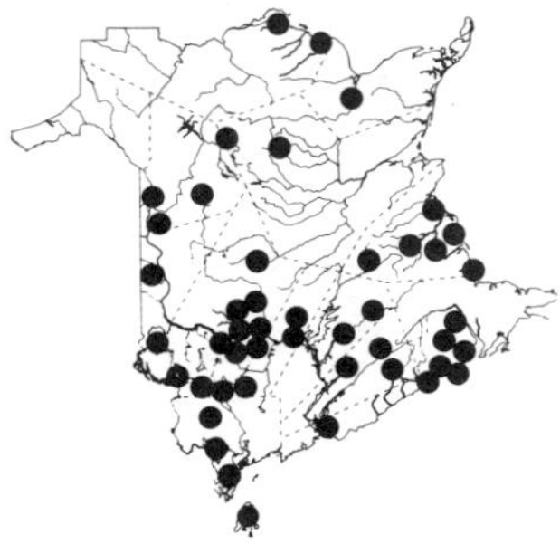

LUZULA MULTIFLORA

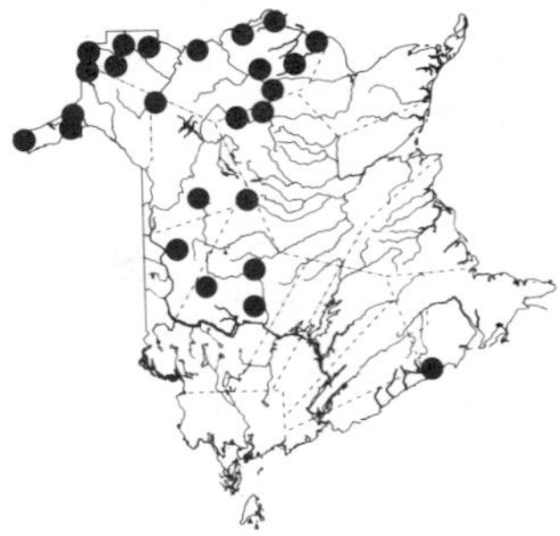

LUZULA PARVIFLORA SSP. MELANOCARPA

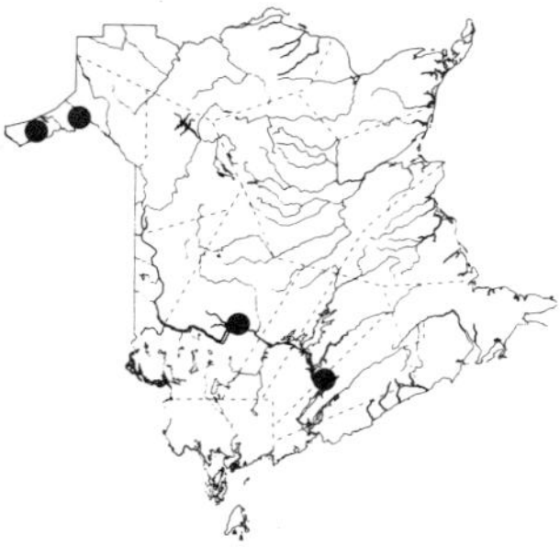

ALLIUM CANADENSE

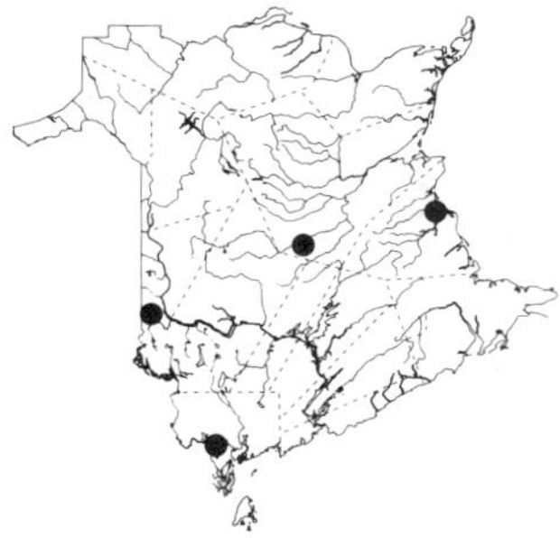

ALLIUM SCHOENOPRASUM

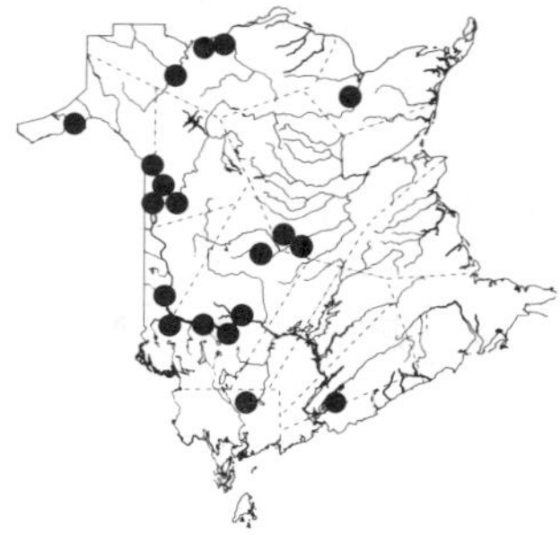

ALLIUM SCHOENOPRASUM SSP. SIBIRICUM

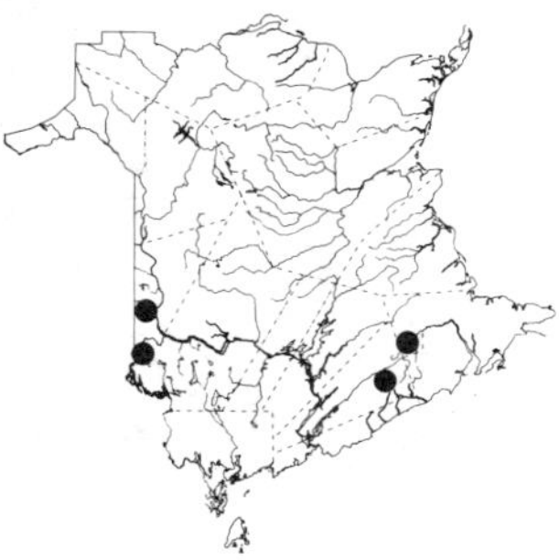

ALLIUM TRICCOCUM

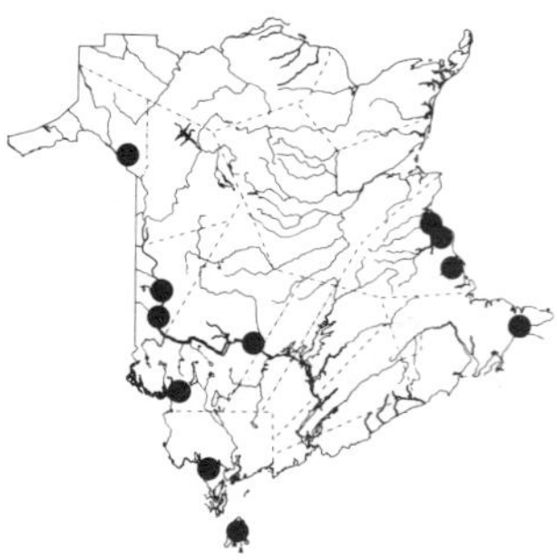

ASPARAGUS OFFICINALIS

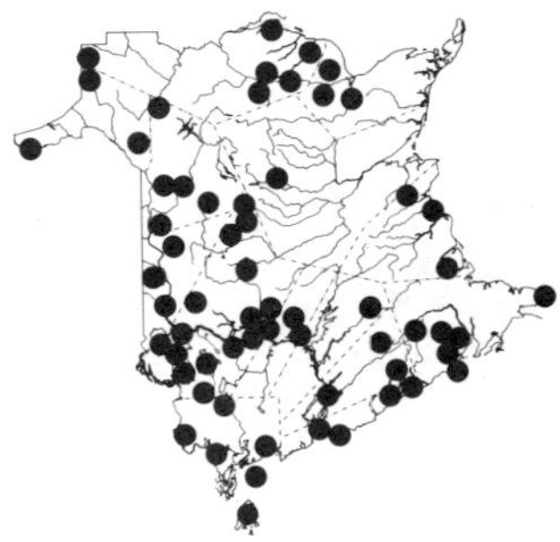

CLINTONIA BOREALIS

CONVALLARIA MAJALIS

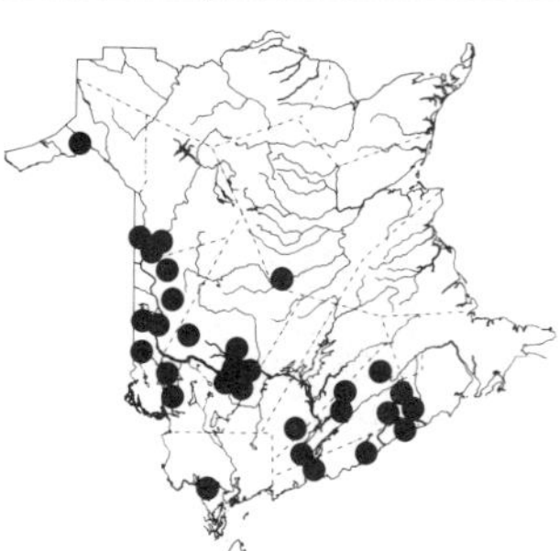

ERYTHRONIUM AMERICANUM

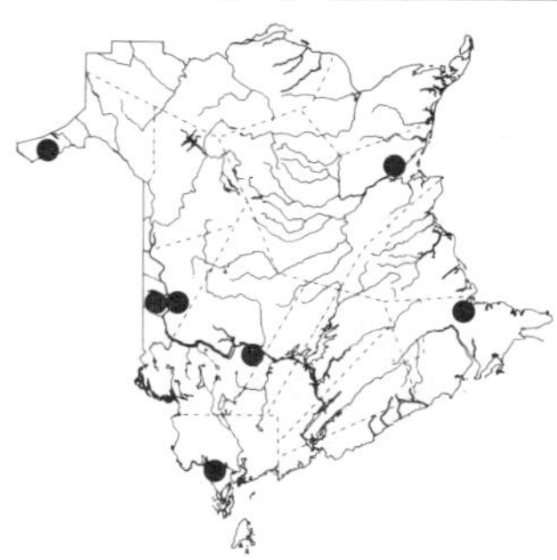

HEMEROCALLIS FULVA

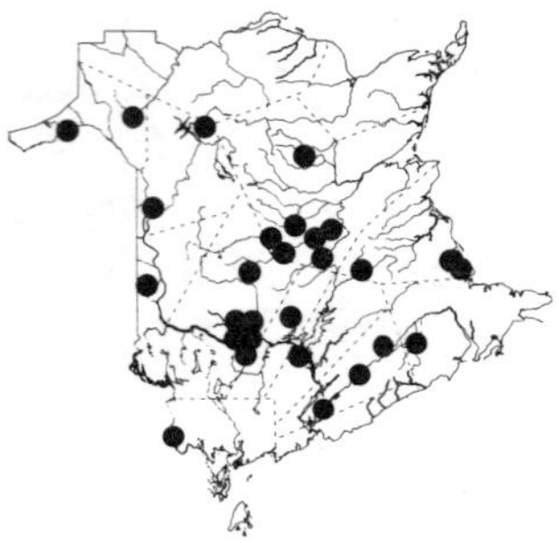

LILIUM CANADENSE

MAIANTHEMUM CANADENSE

MEDEOLA VIRGINIANA

POLYGONATUM PUBESCENS

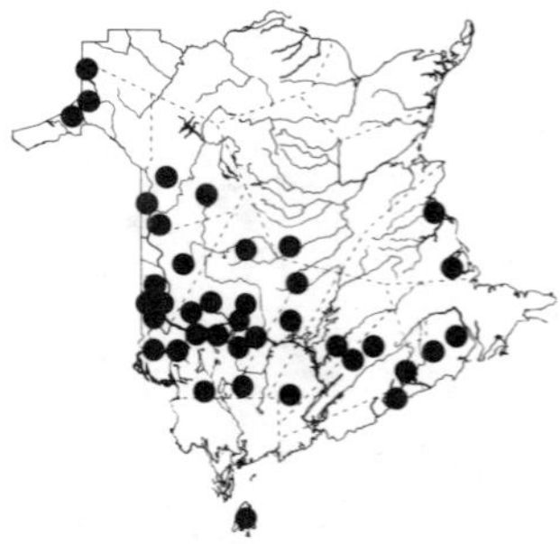

SMILACINA RACEMOSA

SMILACINA STELLATA

SMILACINA TRIFOLIA

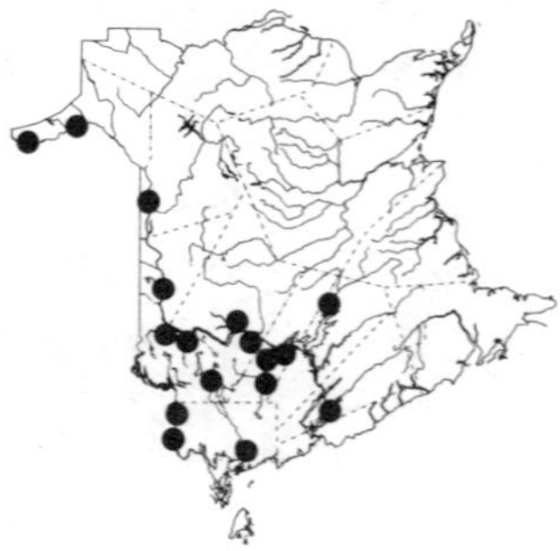

SMILAX HERBACEA

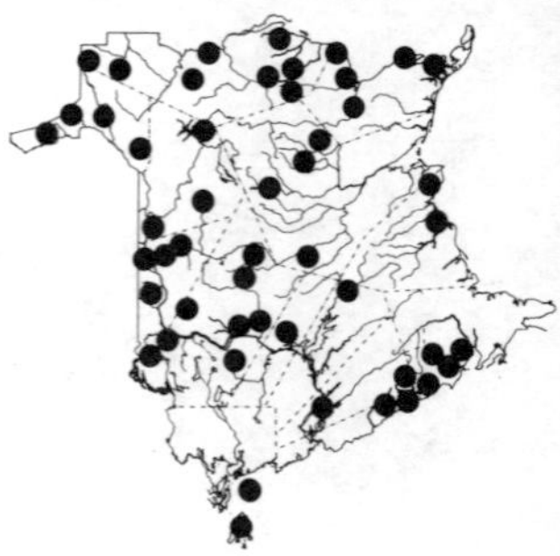

STREPTOPUS AMPLEXIFOLIUS

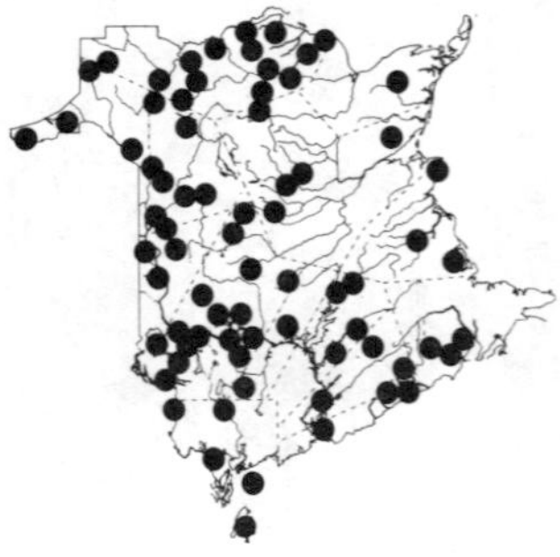

STREPTOPUS ROSEUS

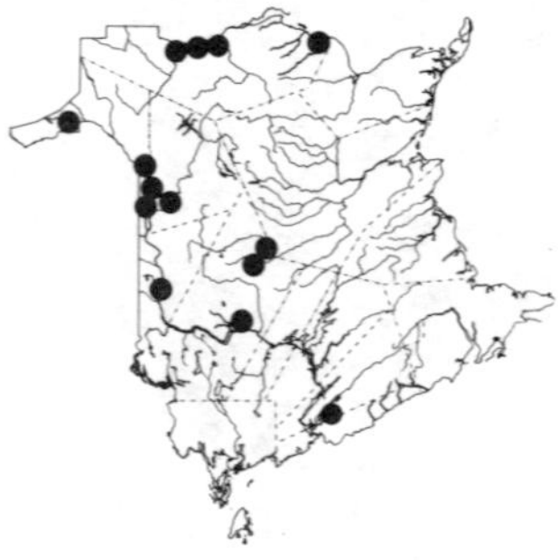

TOFIELDIA GLUTINOSA

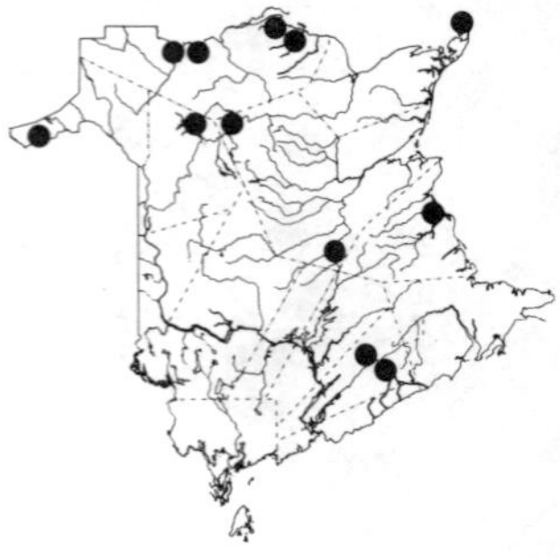

TRILLIUM CERNUUM

LILIACEAE | IRIDACEAE | ORCHIDACEAE

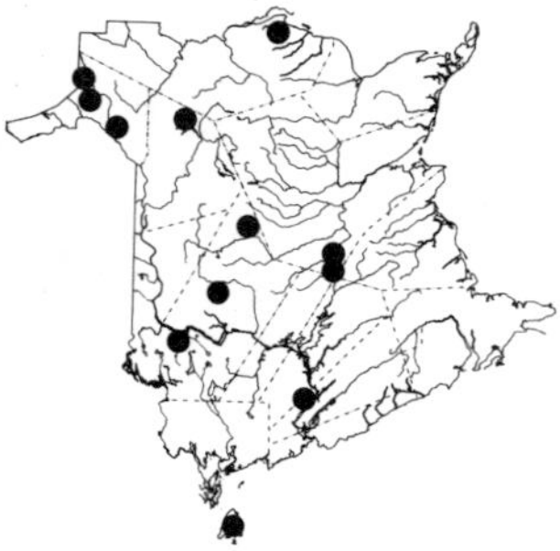

TRILLIUM ERECTUM

TRILLIUM UNDULATUM

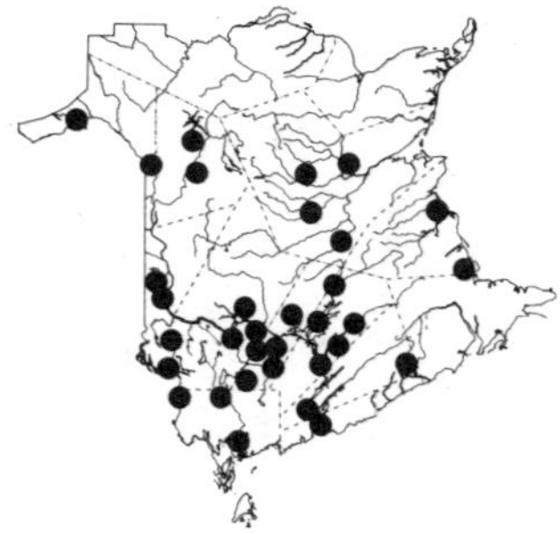

UVULARIA SESSILIFOLIA

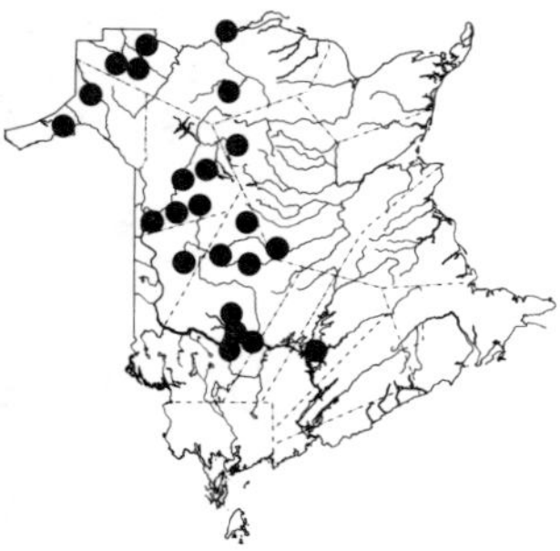

VERATRUM VIRIDE

IRIS SETOSA VAR. CANADENSIS

IRIS VERSICOLOR

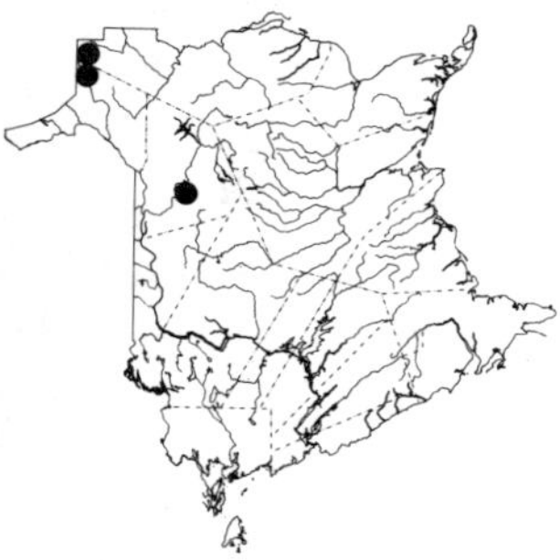

AMERORCHIS ROTUNDIFOLIA

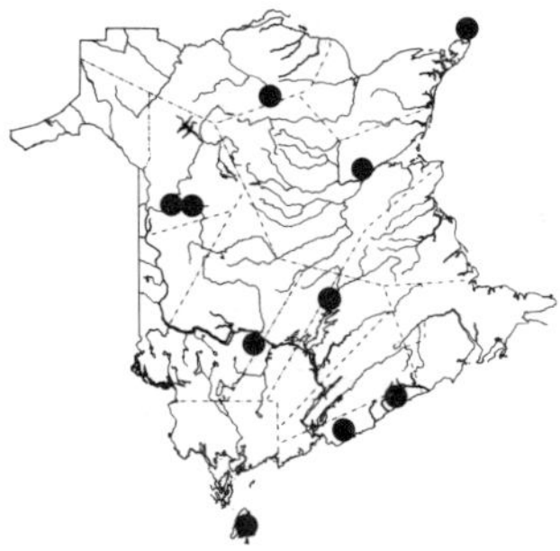

ARETHUSA BULBOSA

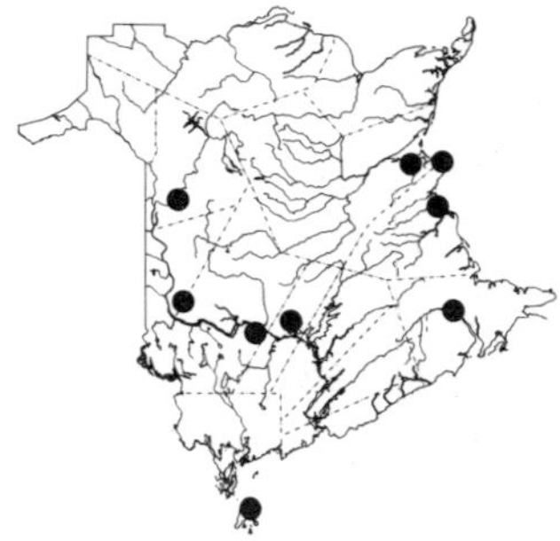

CALOPOGON TUBEROSUS

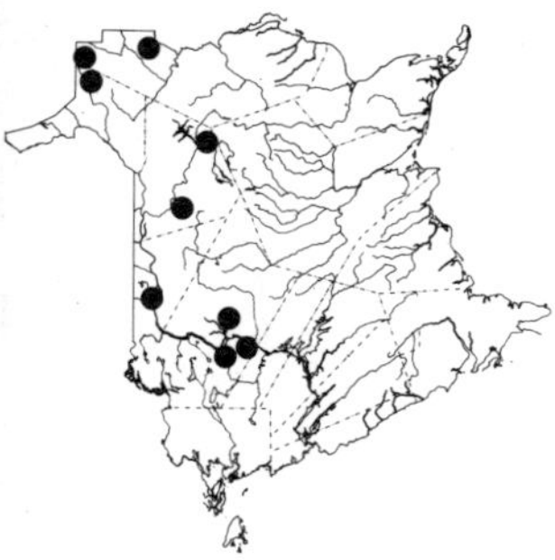

CALYPSO BULBOSA

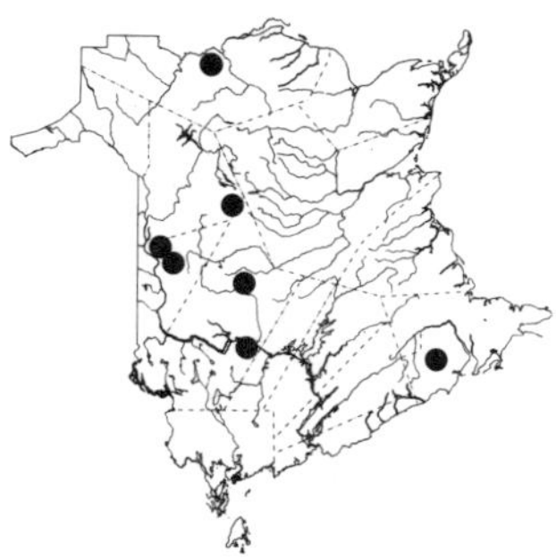

COELOGLOSSUM VIRIDE VAR. VIRESCENS

CORALLORHIZA MACULATA

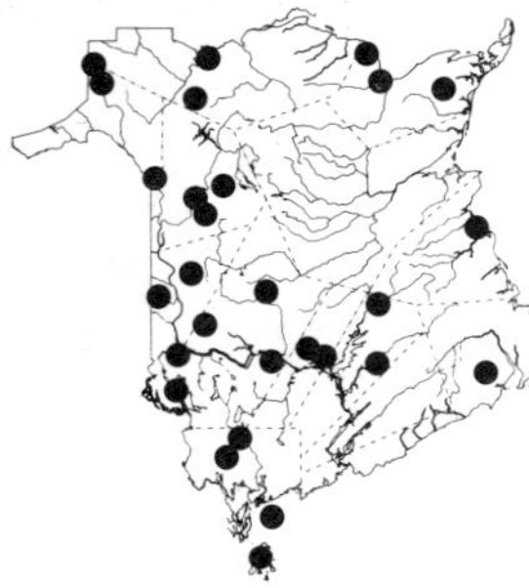

CORALLORHIZA TRIFIDA

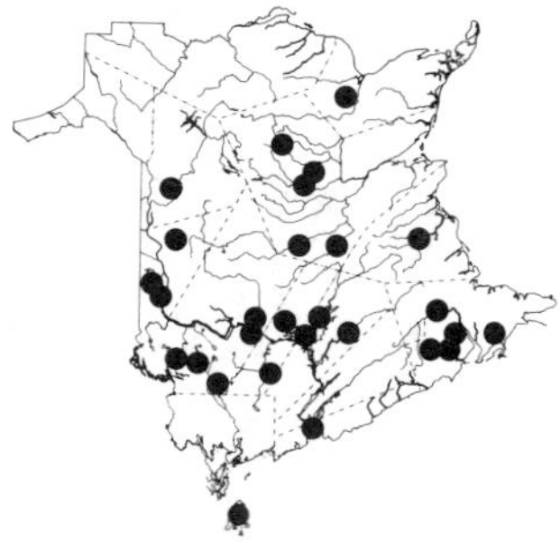

CYPRIPEDIUM ACAULE

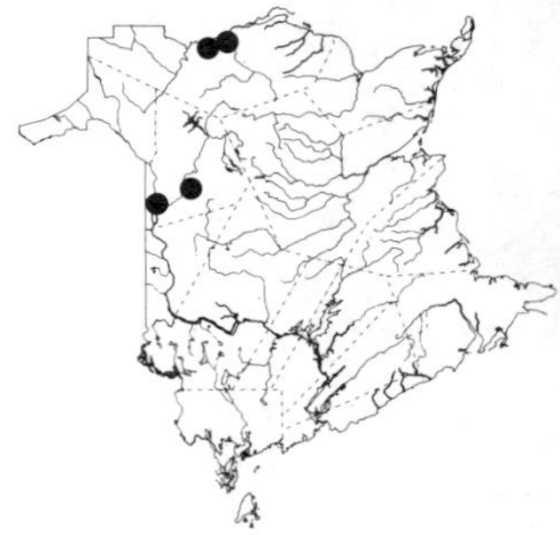

CYPRIPEDIUM PARVIFLORUM

CYPRIPEDIUM PUBESCENS

CYPRIPEDIUM REGINAE

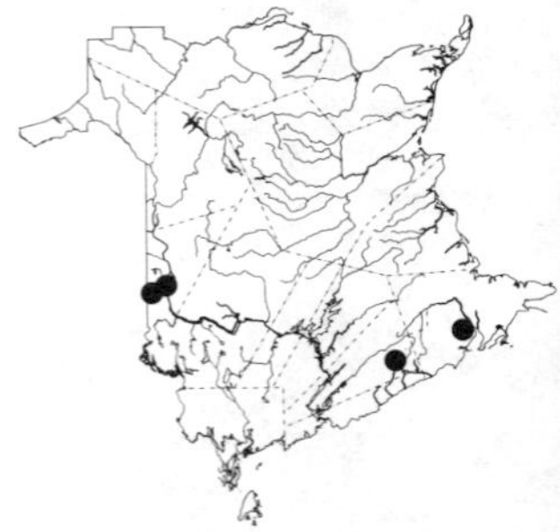

EPIPACTIS HELLEBORINE

GALEARIS SPECTABILIS

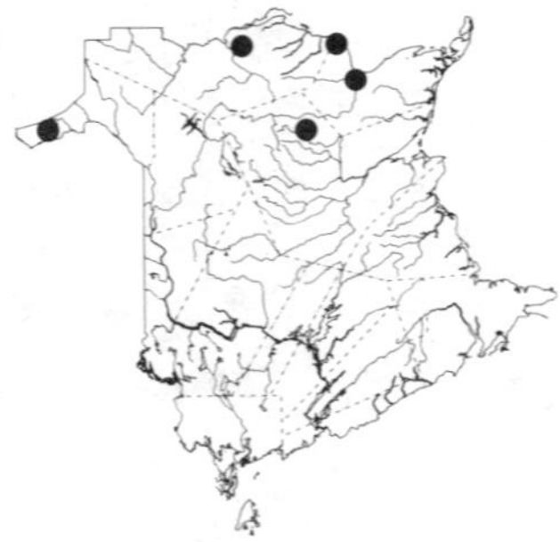

GOODYERA OBLONGIFOLIA

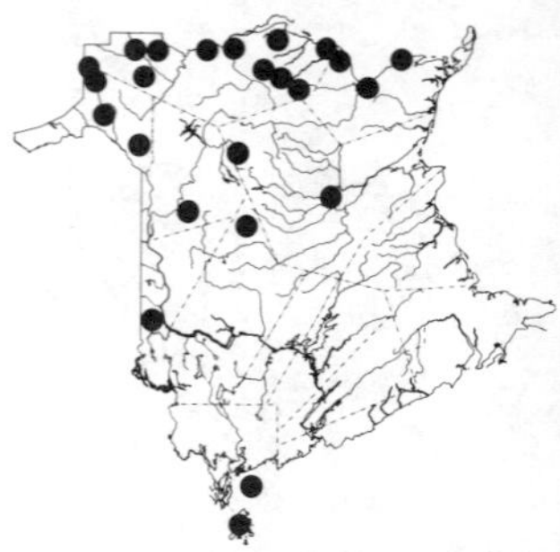

GOODYERA REPENS

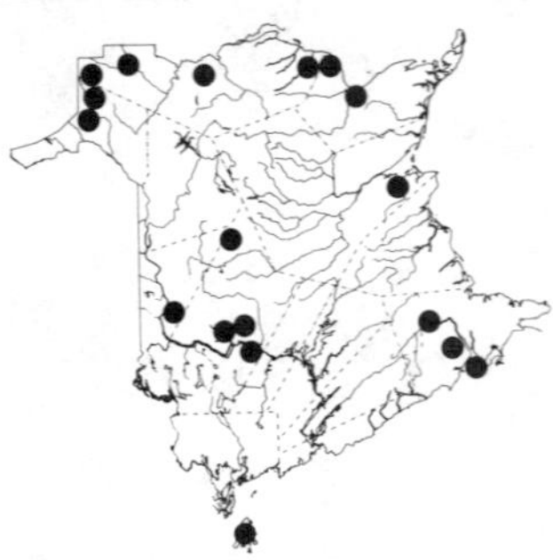

GOODYERA TESSELATA

PLATANTHERA BLEPHARIGLOTTIS

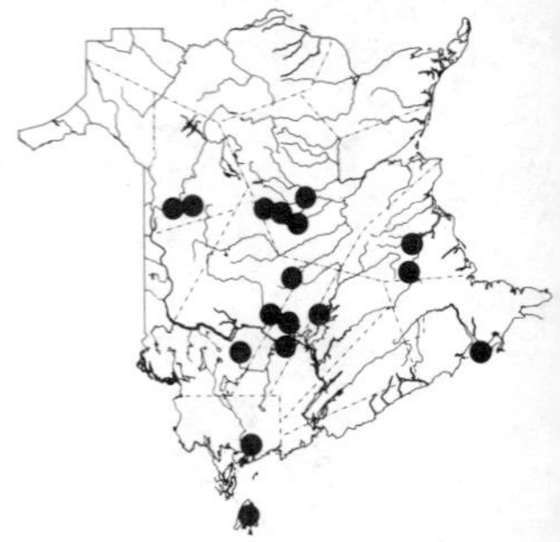

PLATANTHERA CLAVELLATA

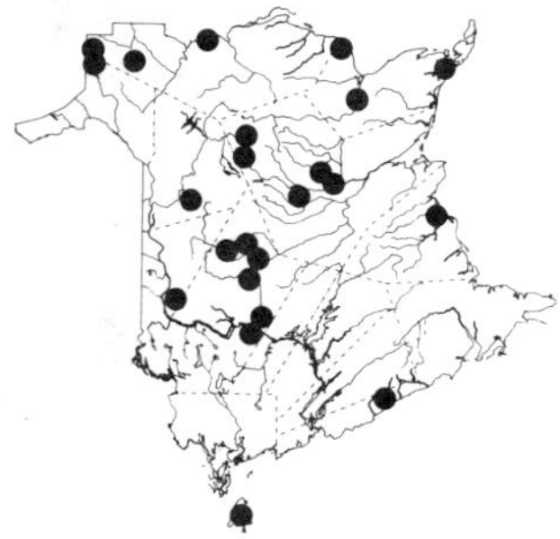

PLATANTHERA DILATATA

PLATANTHERA GRANDIFLORA

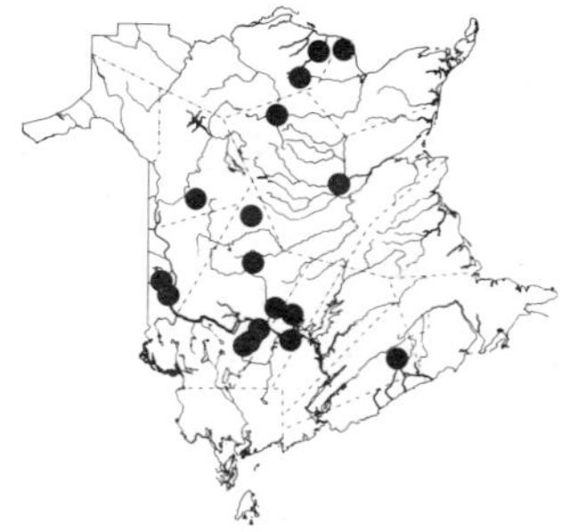

PLATANTHERA HOOKERI

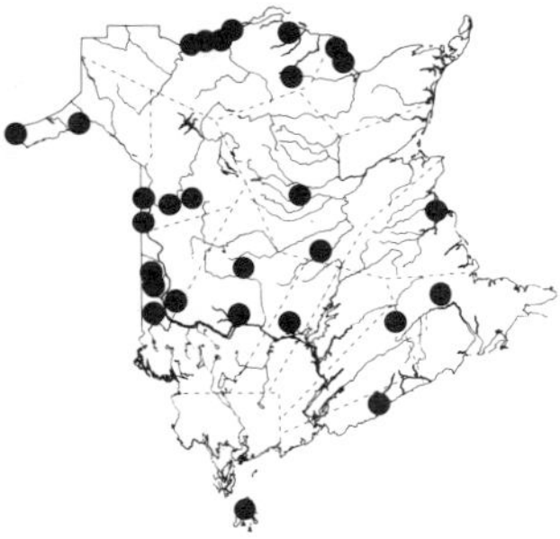

PLATANTHERA HYPERBOREA

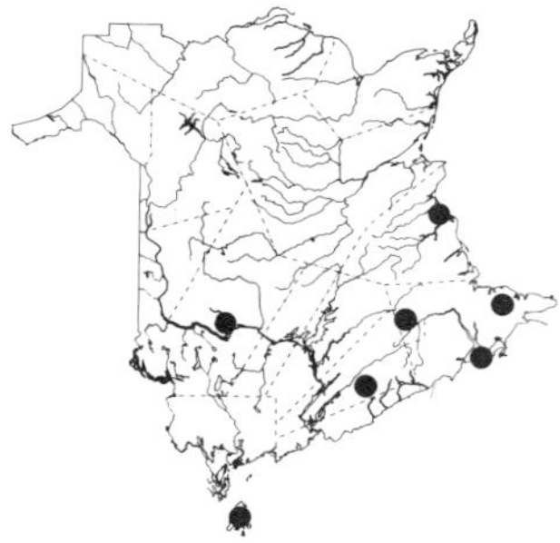

PLATANTHERA LACERA

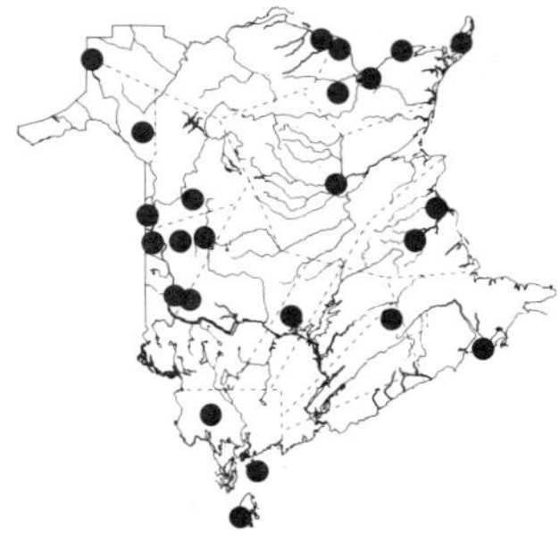

PLATANTHERA OBTUSATA

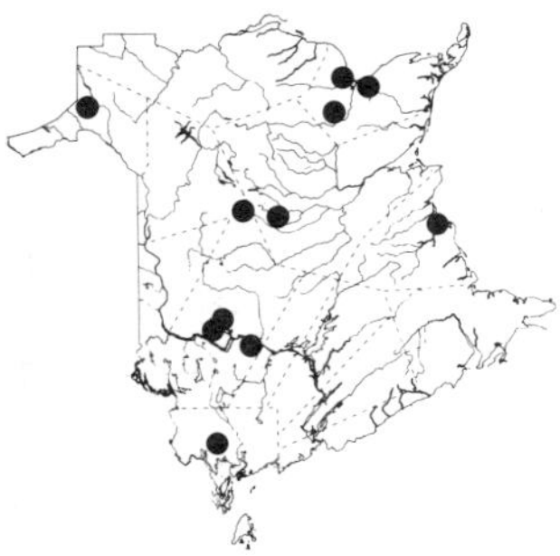

PLATANTHERA ORBICULATA

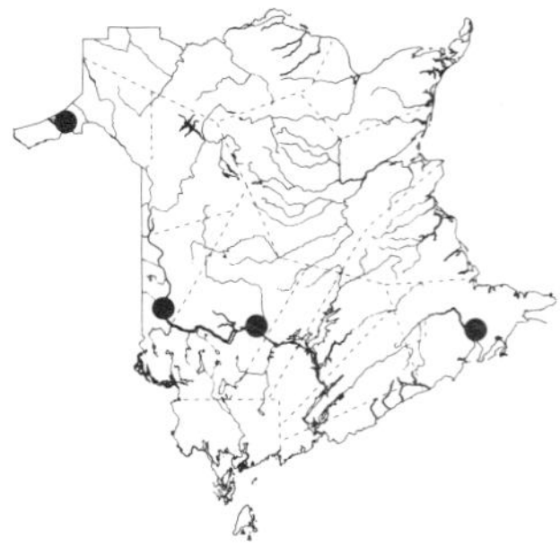

PLATANTHERA ORBICULATA VAR. MACROPHYLLA

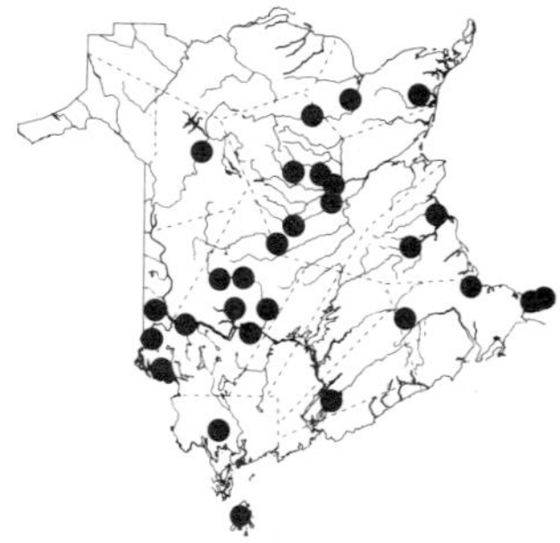

PLATANTHERA PSYCODES

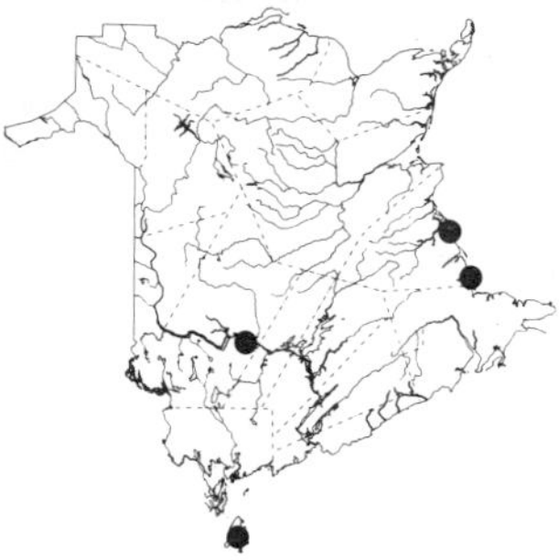

LIPARIS LOESELII

LISTERA AURICULATA

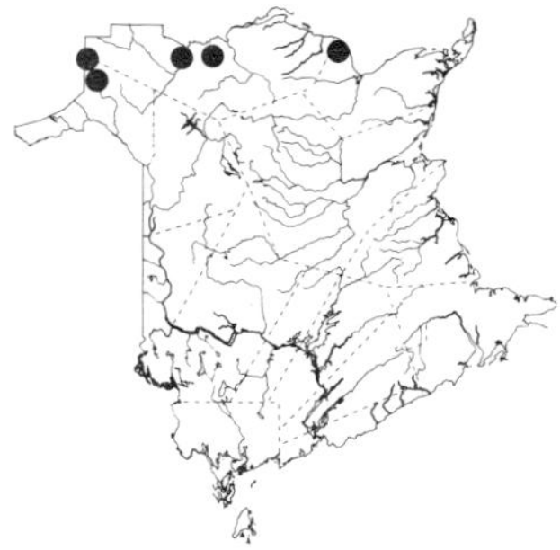

LISTERA CONVALLARIOIDES

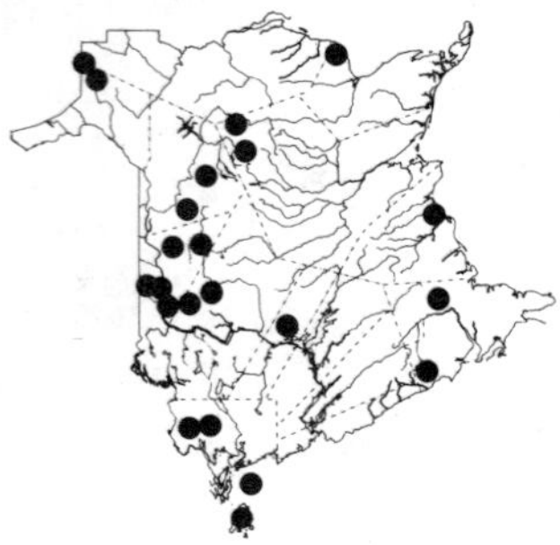

LISTERA CORDATA

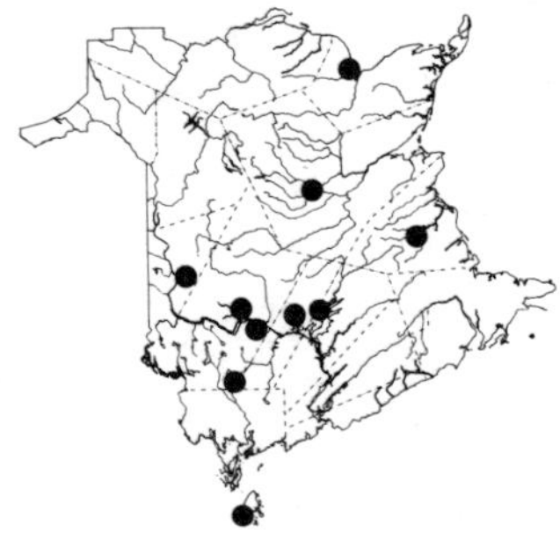

MALAXIS UNIFOLIA

POGONIA OPHIOGLOSSOIDES

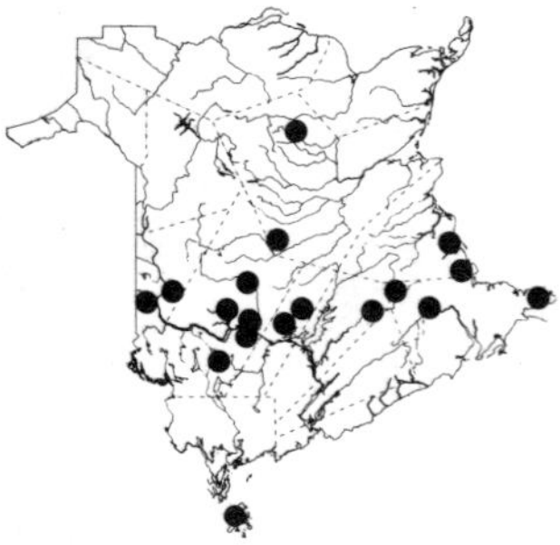

SPIRANTHES LACERA

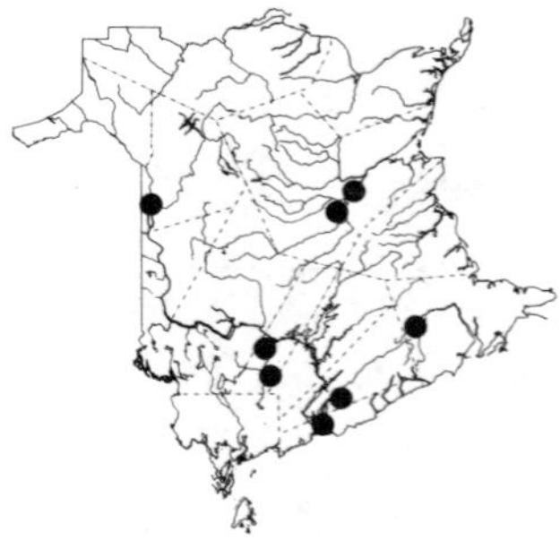

SPIRANTHES LUCIDA

SPIRANTHES ROMANZOFFIANA

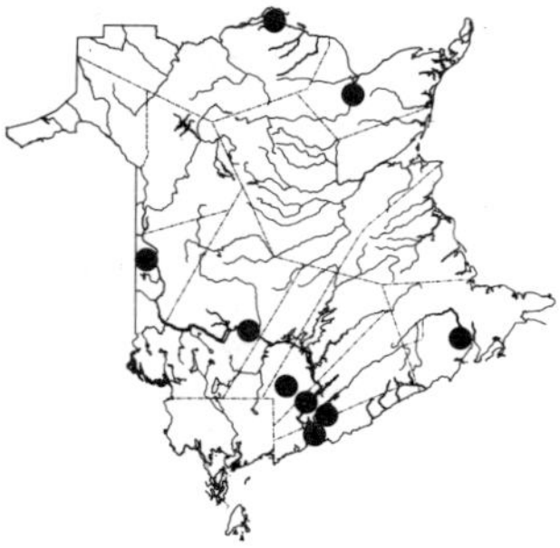

POPULUS ALBA

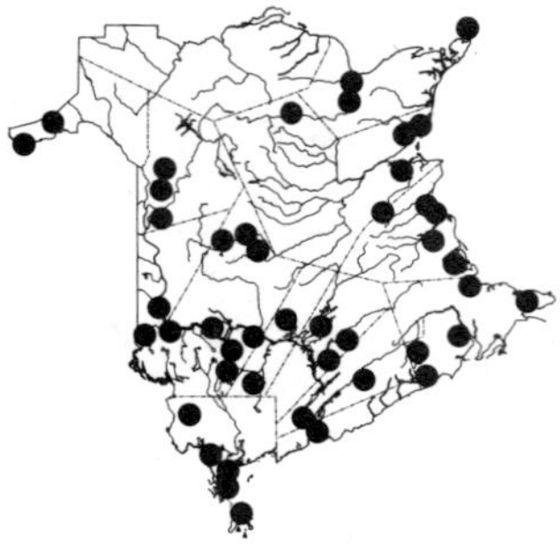

POPULUS GRANDIDENTATA

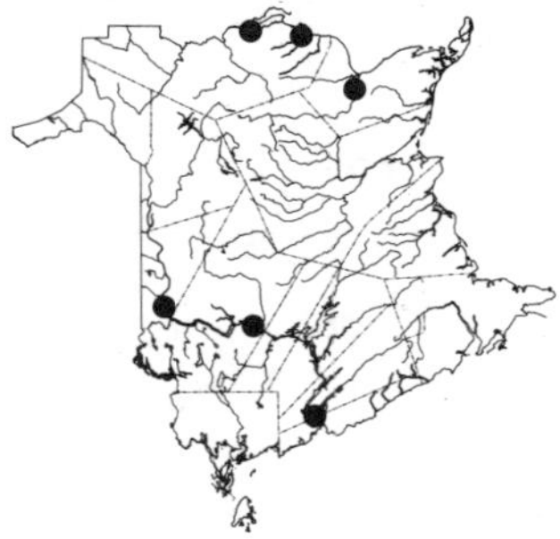

SALIX ALBA

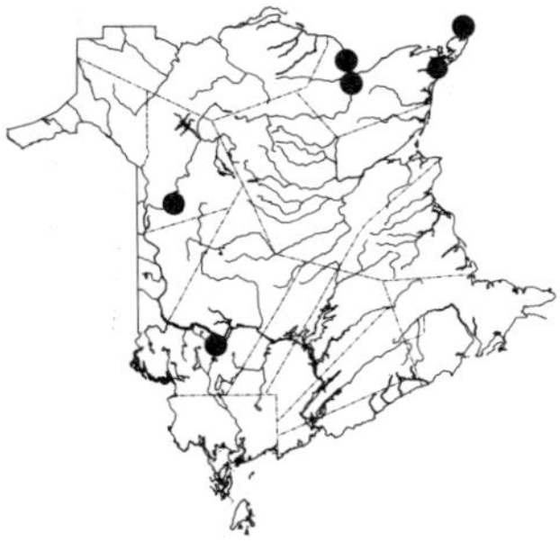

SALIX CANDIDA

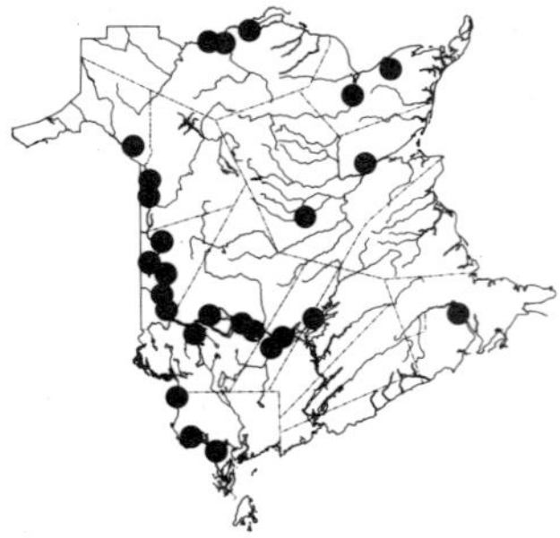

SALIX EXIGUA

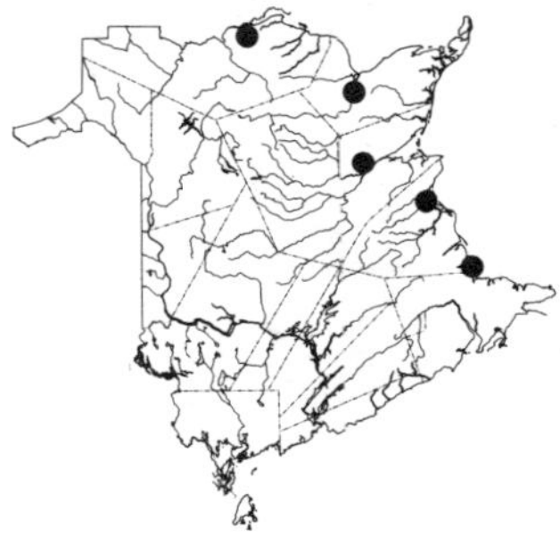

SALIX FRAGILIS

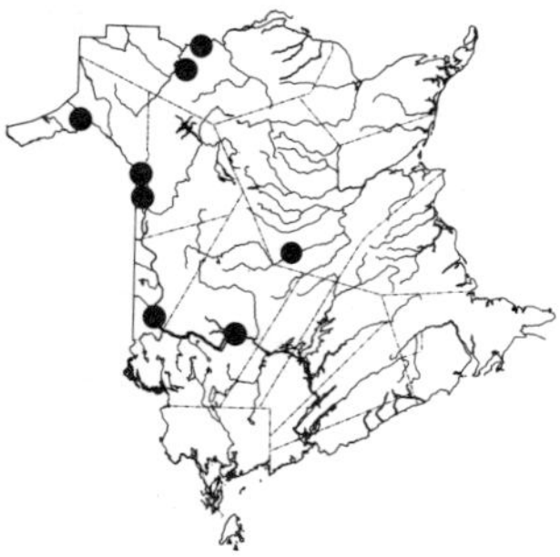

SALIX MYRICOIDES

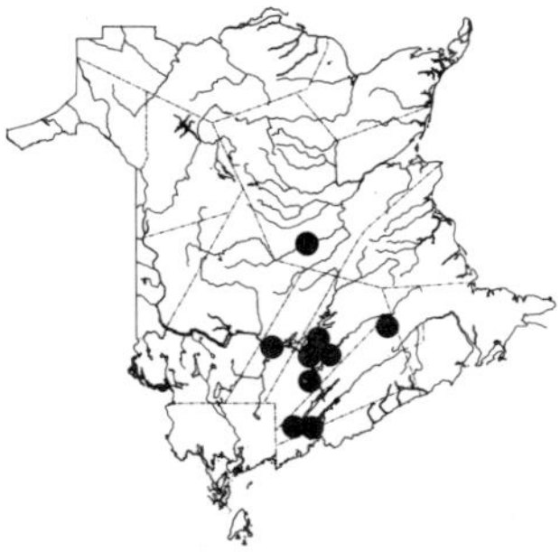

SALIX NIGRA

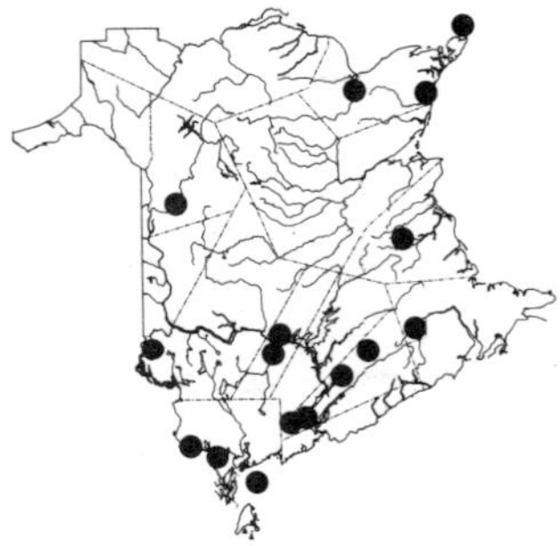

SALIX PEDICELLARIS

SALIX PELLITA

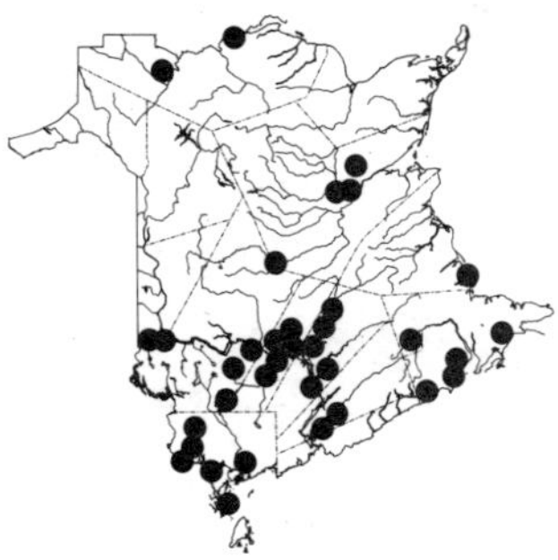

SALIX PETIOLARIS

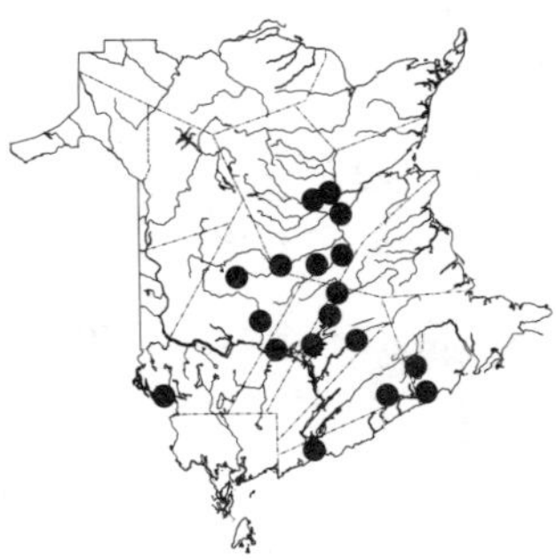

SALIX SERICEA

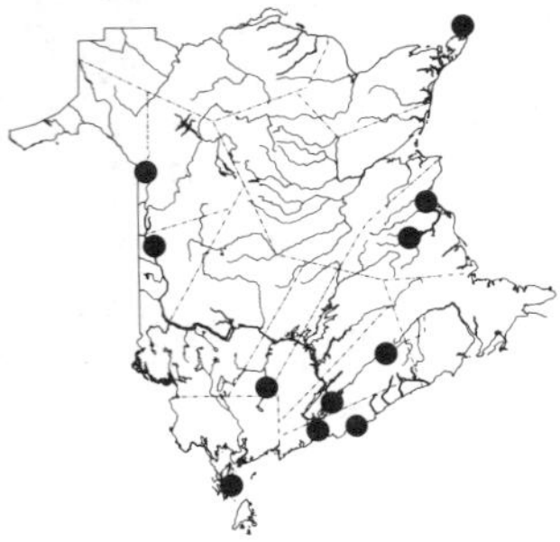

SALIX VIMINALIS

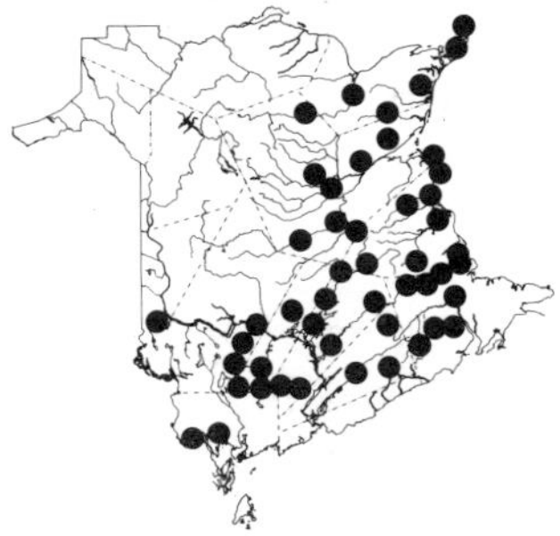

COMPTONIA PEREGRINA

MYRICA PENSYLVANICA

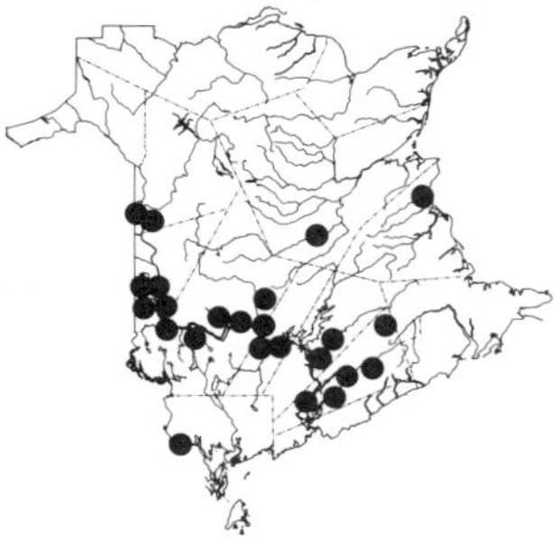

JUGLANS CINEREA

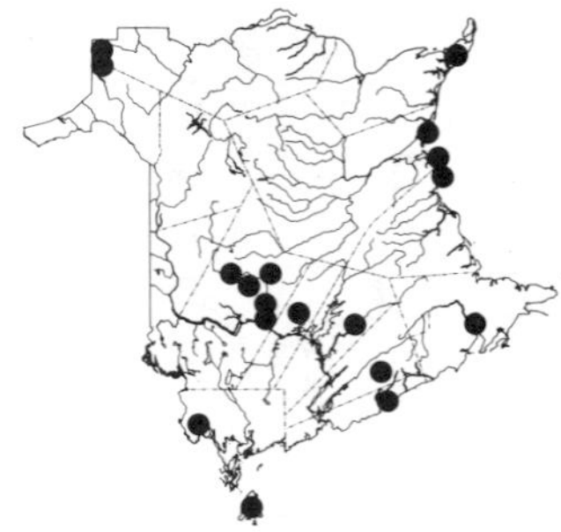

BETULA X CAERULEA

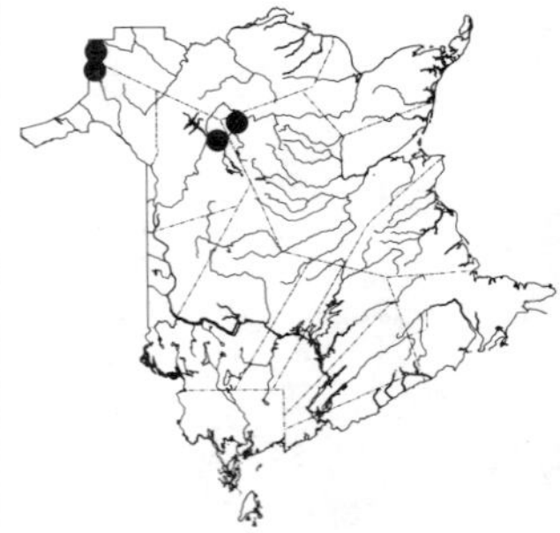

BETULA MINOR

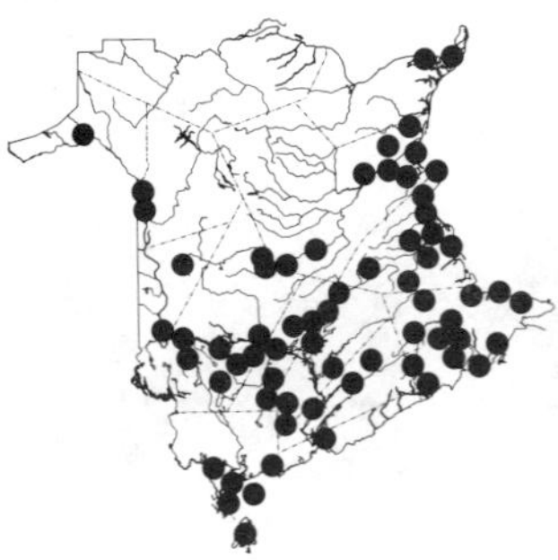

BETULA POPULIFOLIA

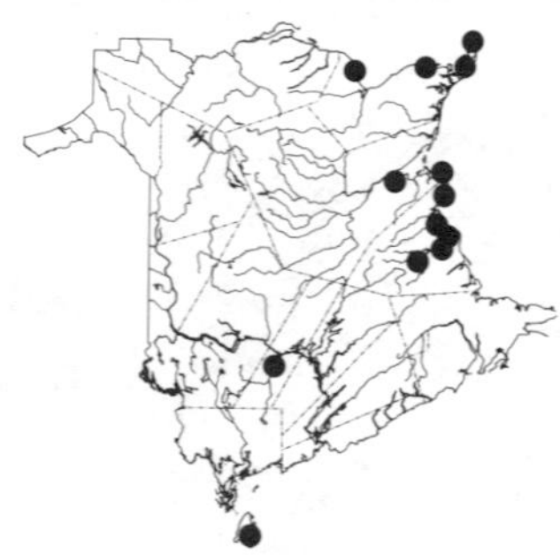

BETULA PUMILA

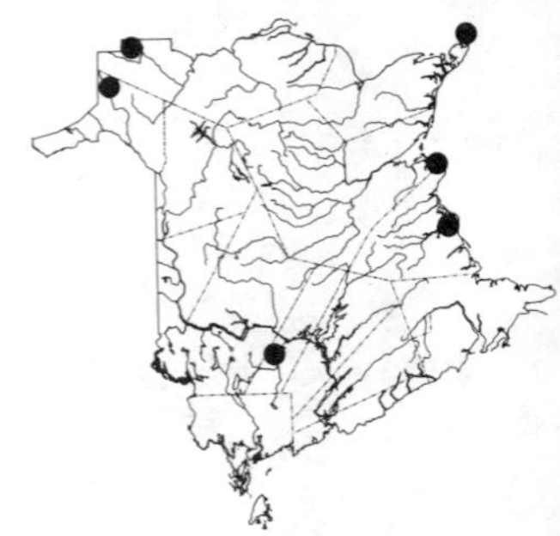

BETULA X SANDBERGII

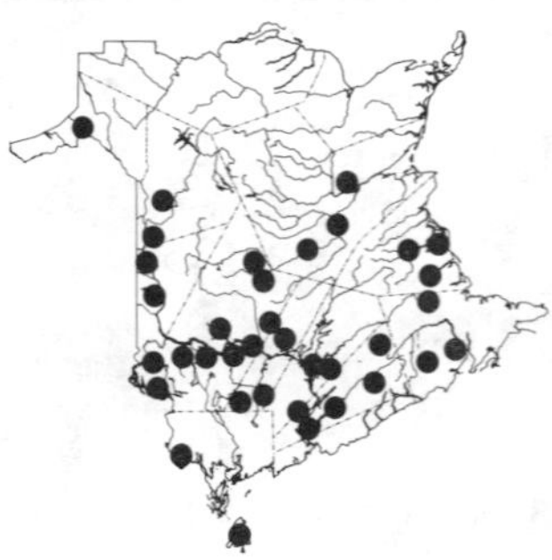

OSTRYA VIRGINIANA

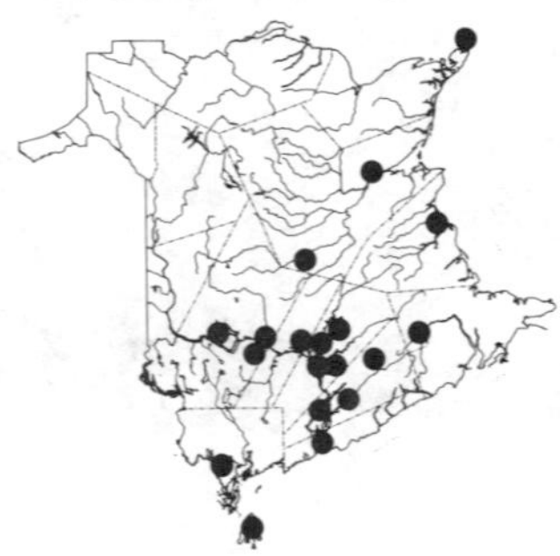

QUERCUS MACROCARPA

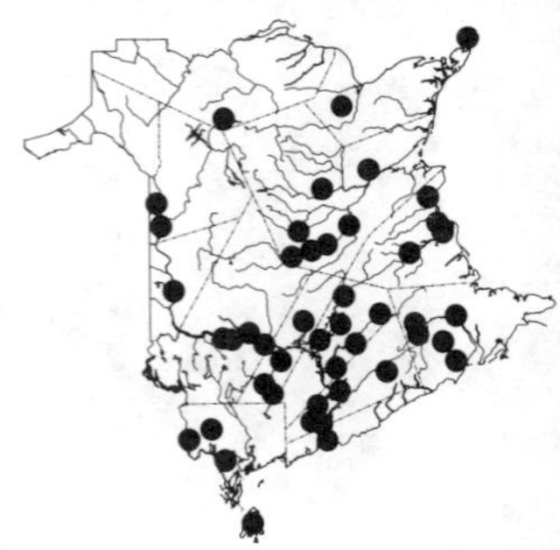

QUERCUS RUBRA

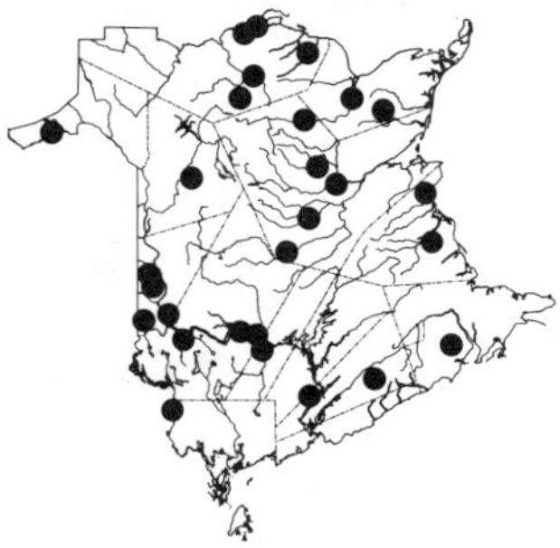

ULMUS AMERICANA

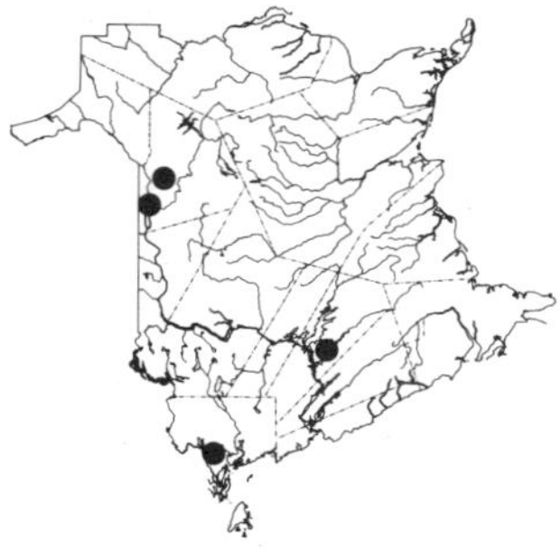

CANNABIS SATIVA

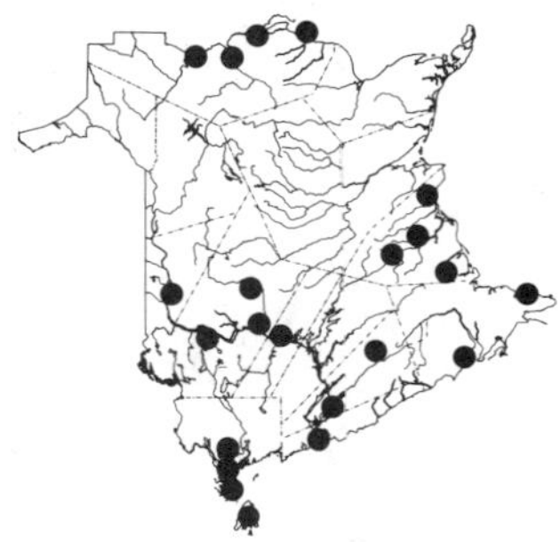

HUMULUS LUPULUS

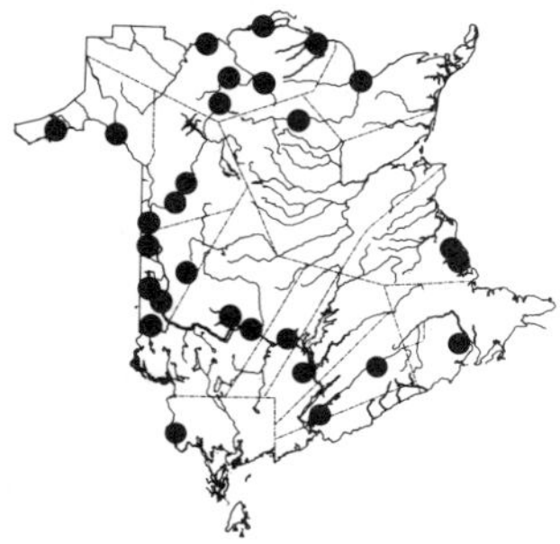

LAPORTEA CANADENSIS

URTICA DIOICA

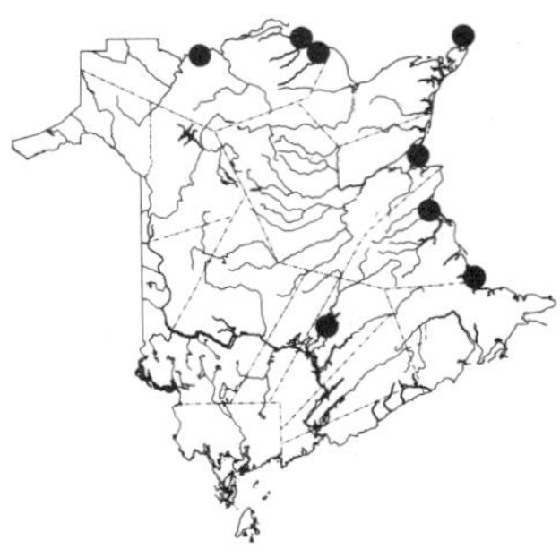

COMANDRA UMBELLATA

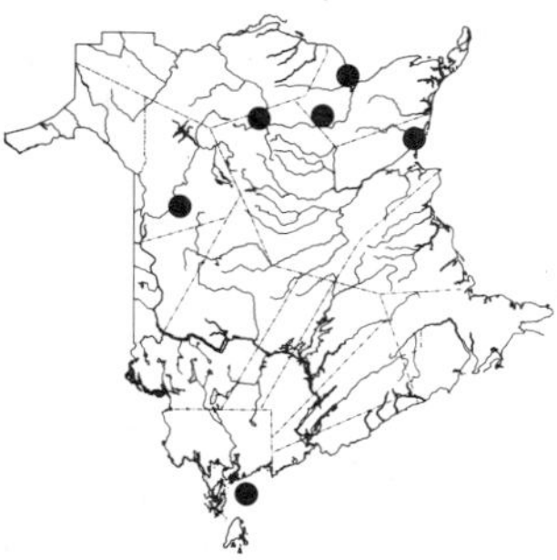

GEOCAULON LIVIDUM

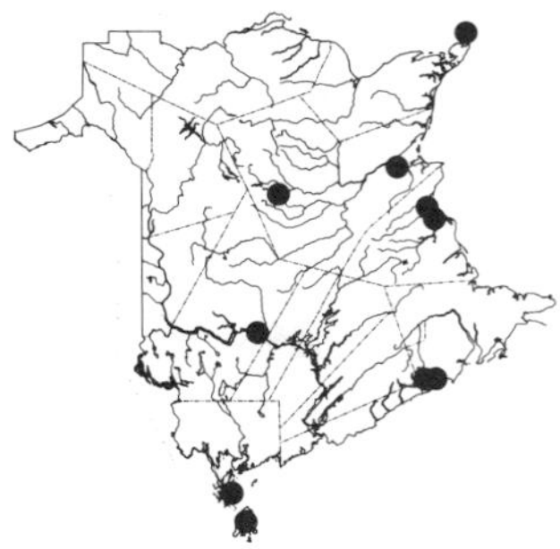

ARCEUTHOBIUM PUSILLUM

ASARUM CANADENSE

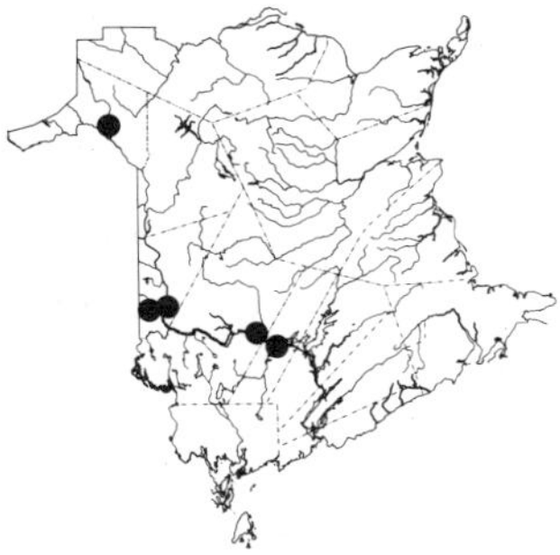

FAGOPYRUM TATARICUM

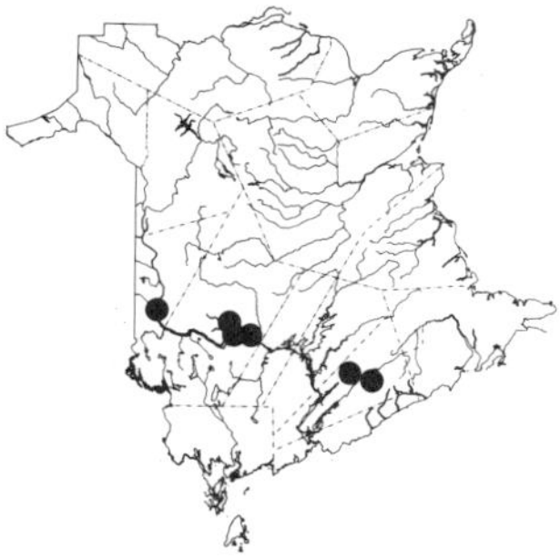

POLYGONUM ACHOREUM

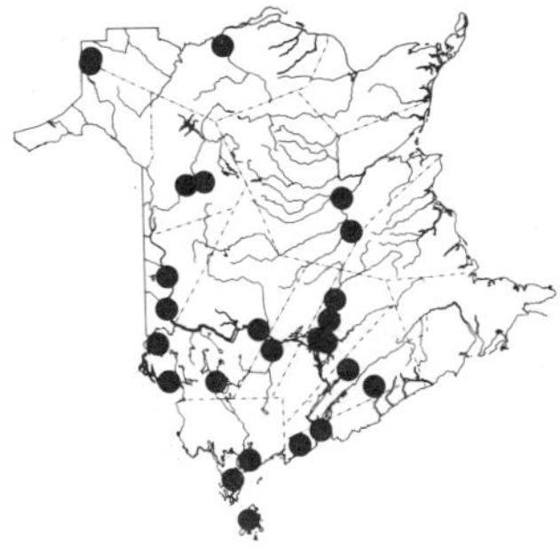

POLYGONUM AMPHIBIUM

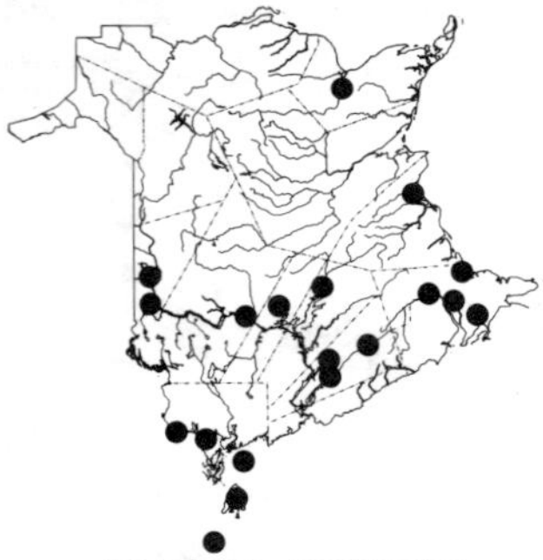
POLYGONUM ARENASTRUM

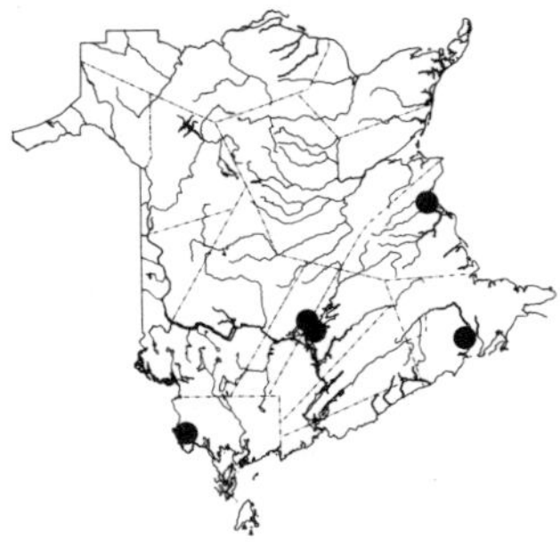
POLYGONUM ARIFOLIUM VAR. PUBESCENS

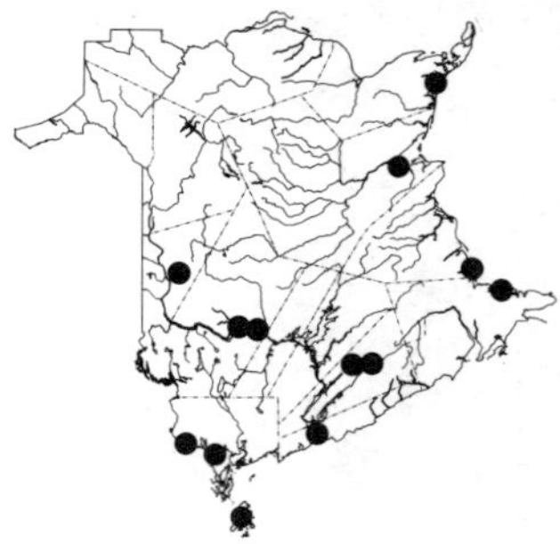
POLYGONUM AVICULARE

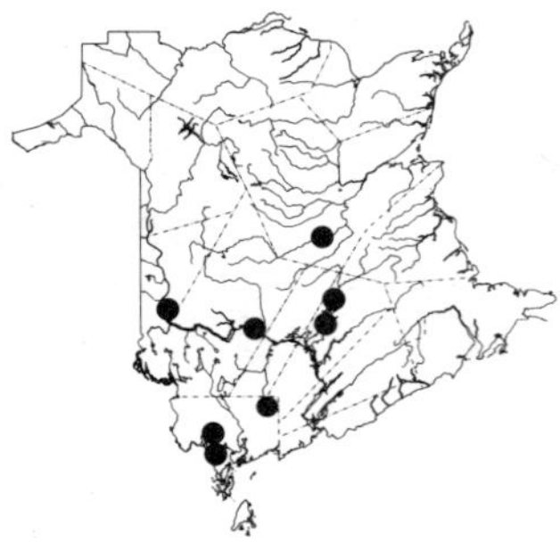
POLYGONUM CAREYI

POLYGONUM CILINODE

POLYGONUM FOWLERI

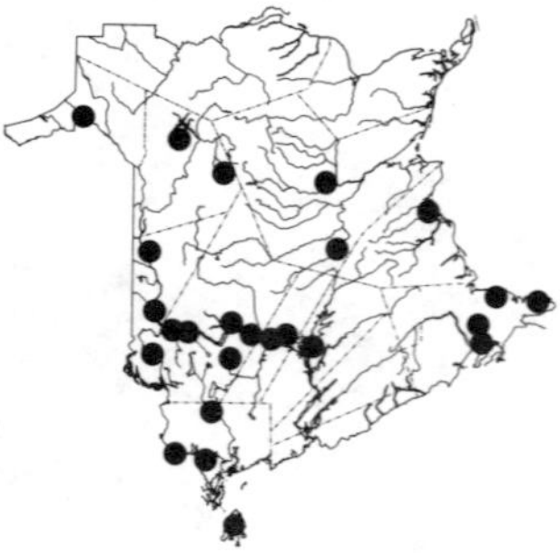
POLYGONUM HYDROPIPER

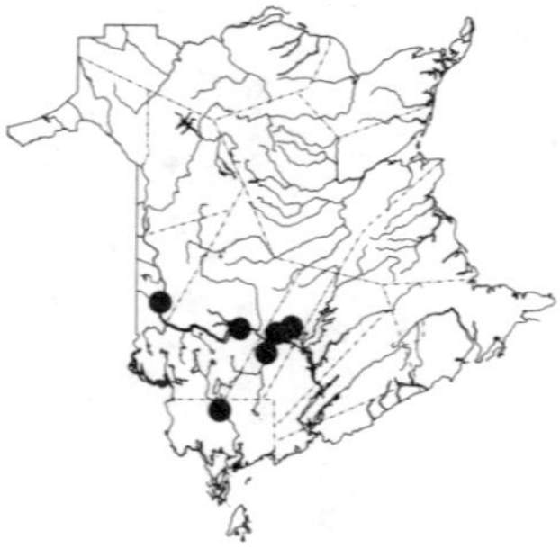
POLYGONUM HYDROPIPEROIDES

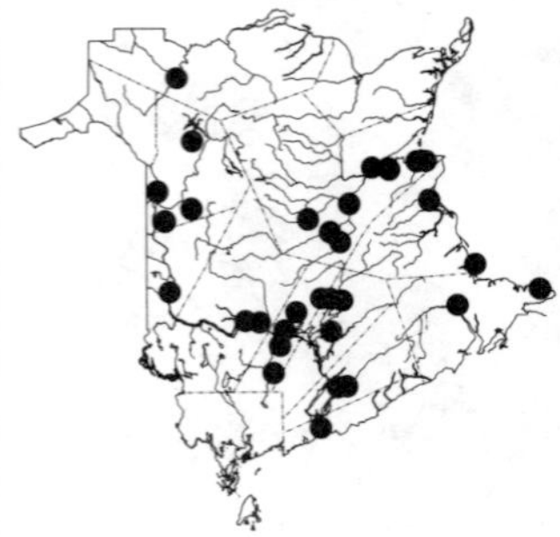
POLYGONUM LAPATHIFOLIUM

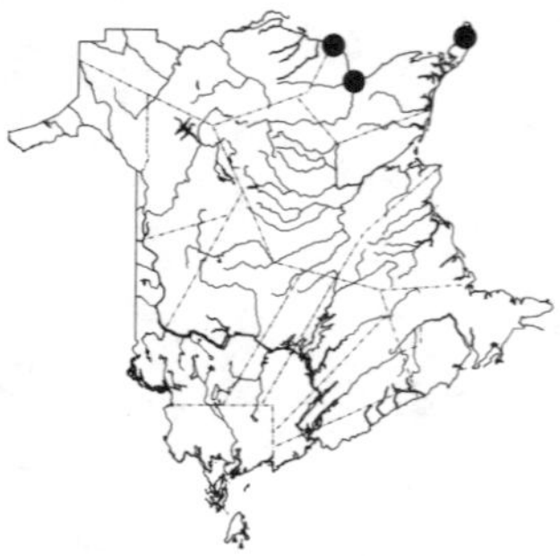
POLYGONUM OXYSPERMUM SSP. RAII

POLYGONUM PENSYLVANICUM

POLYGONUM PERSICARIA

POLYGONUM RAMOSISSIMUM

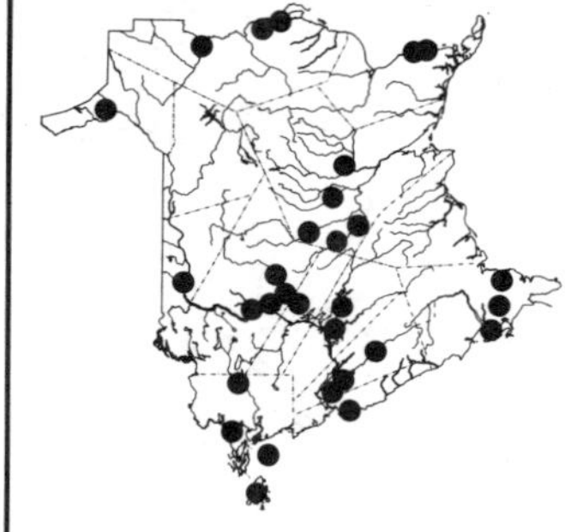

POLYGONUM SAGITTATUM

POLYGONUM SCABRUM

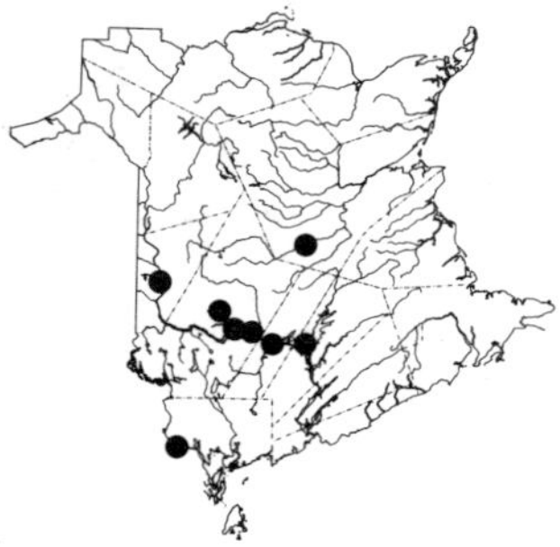

POLYGONUM SCANDENS

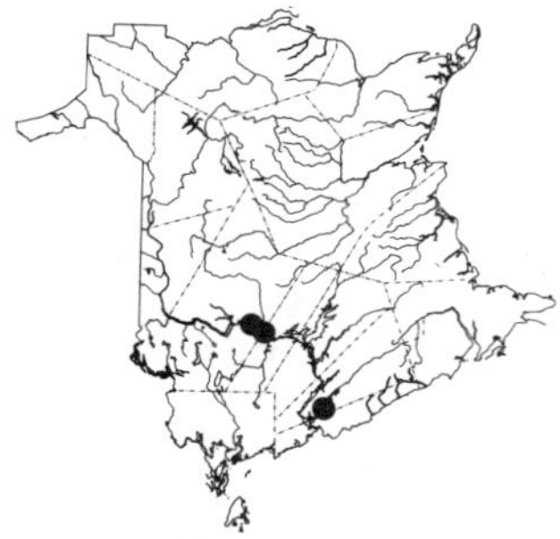

RUMEX ACETOSA

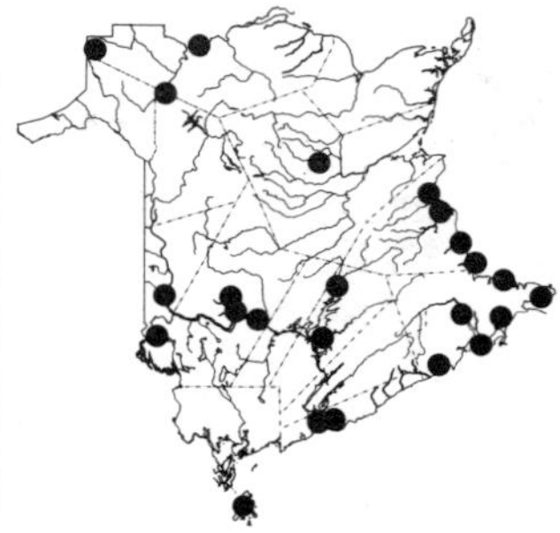

RUMEX CRISPUS

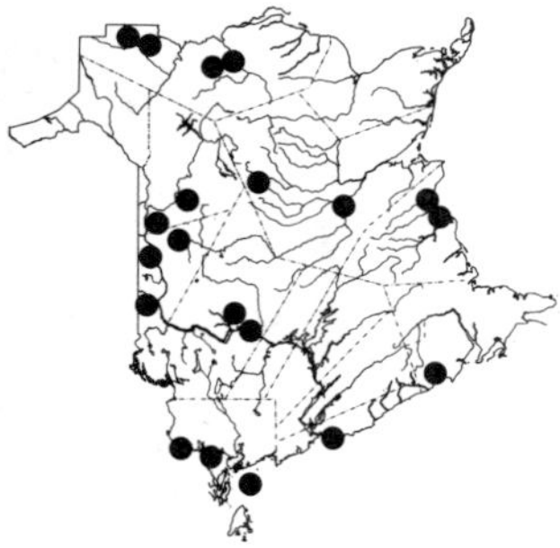

RUMEX LONGIFOLIUS

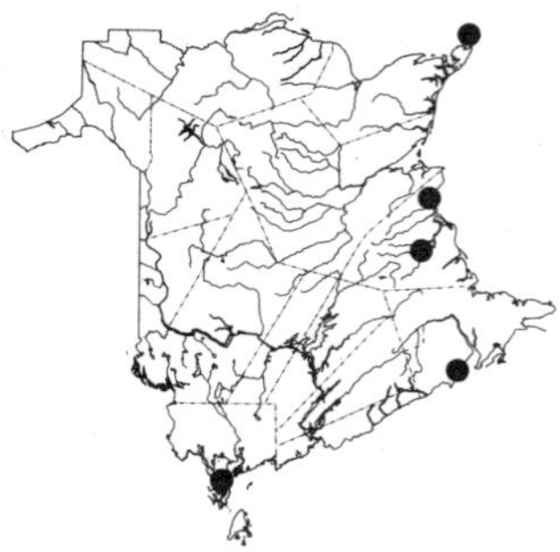

RUMEX MARITIMUS VAR. FUEGINUS

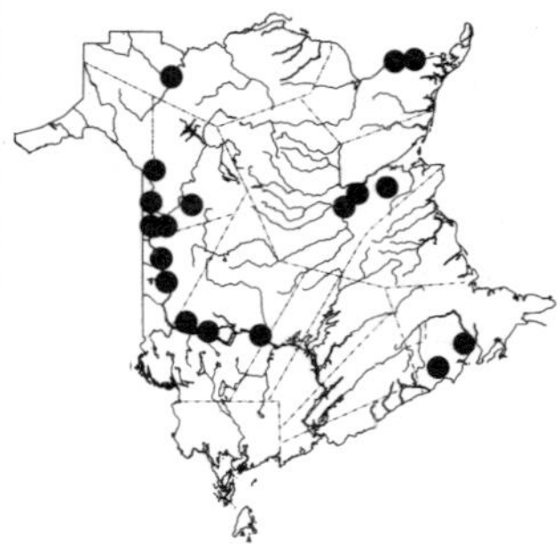

RUMEX OBTUSIFOLIUS

RUMEX ORBICULATUS

RUMEX SALICIFOLIUS

ATRIPLEX PROSTRATA

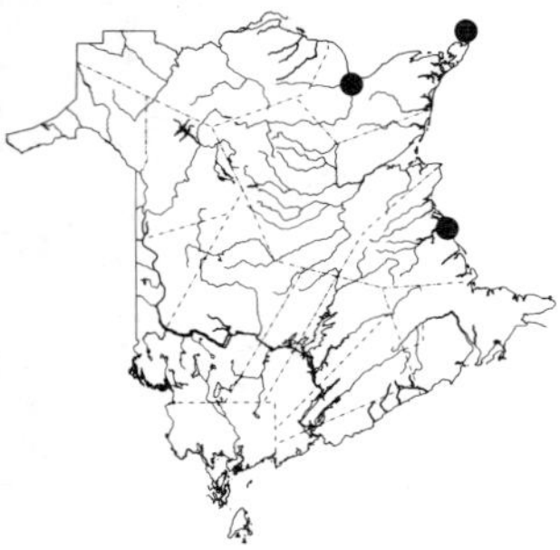

ATRIPLEX SUBSPICATA

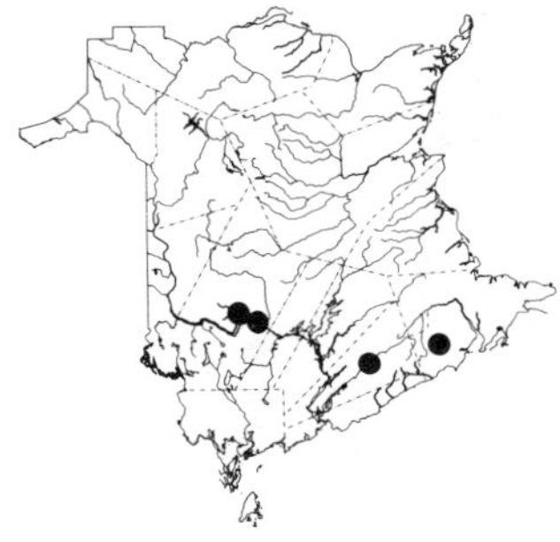

CHENOPODIUM GIGANTOSPERMUM

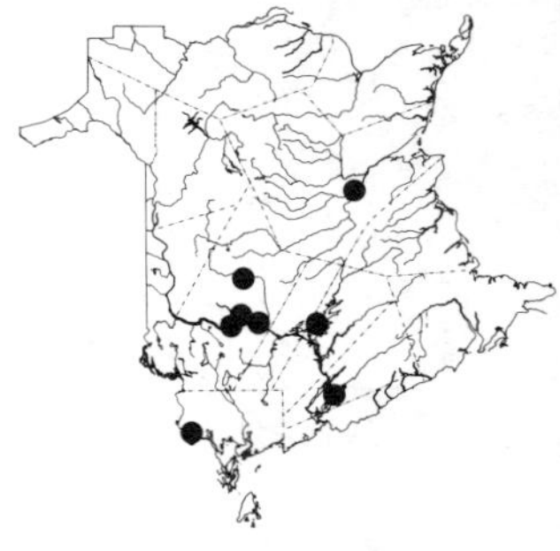

MOLLUGO VERTICILLATA

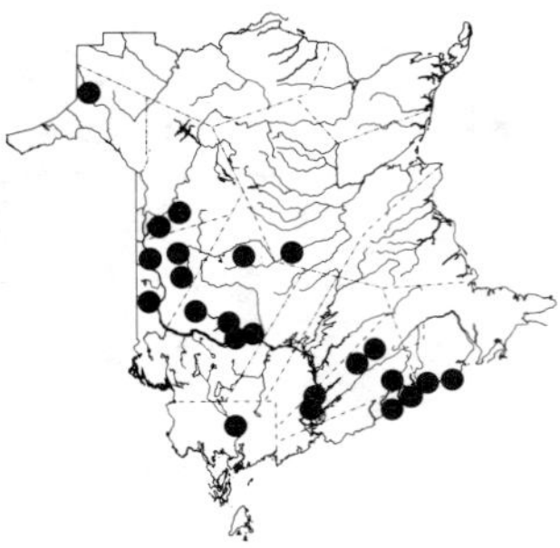

CLAYTONIA CAROLINIANA

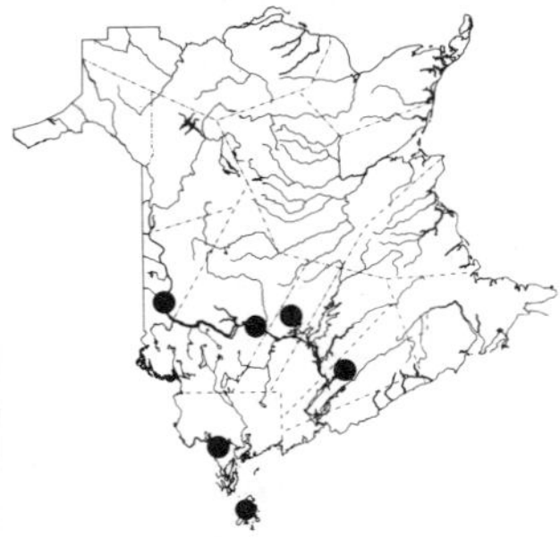

PORTULACA OLERACEA

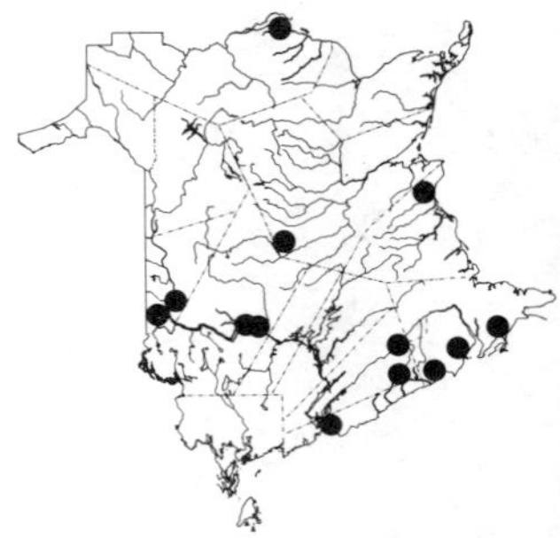

CERASTIUM ARVENSE

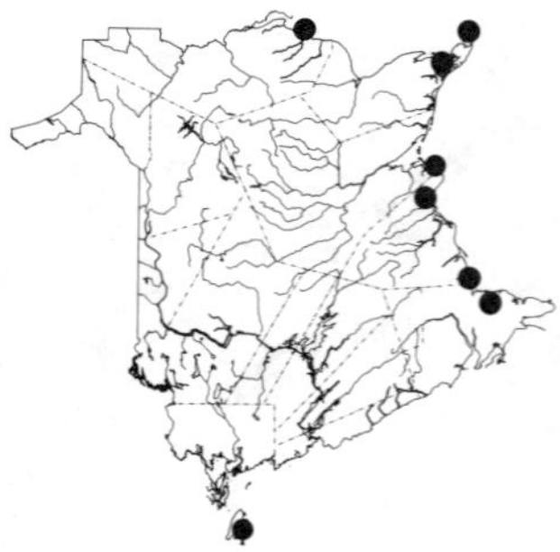

HONCKENYA PEPLOIDES

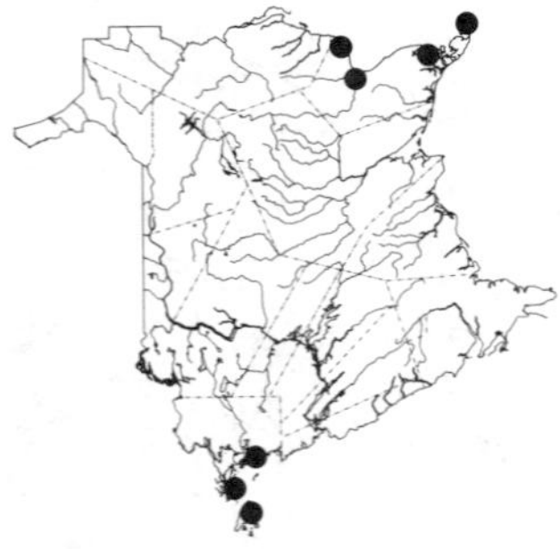

SAGINA NODOSA

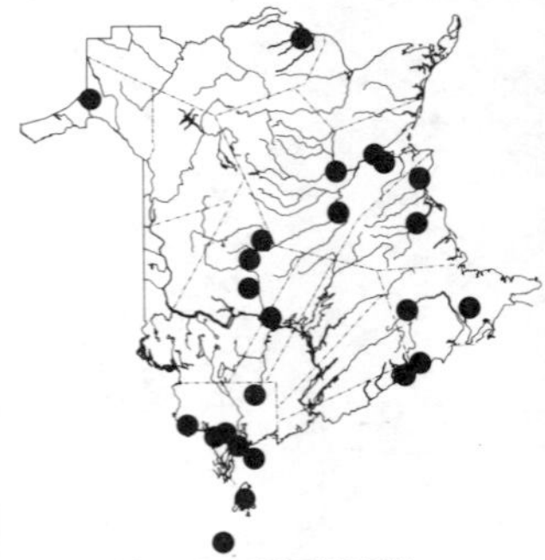

SAGINA PROCUMBENS

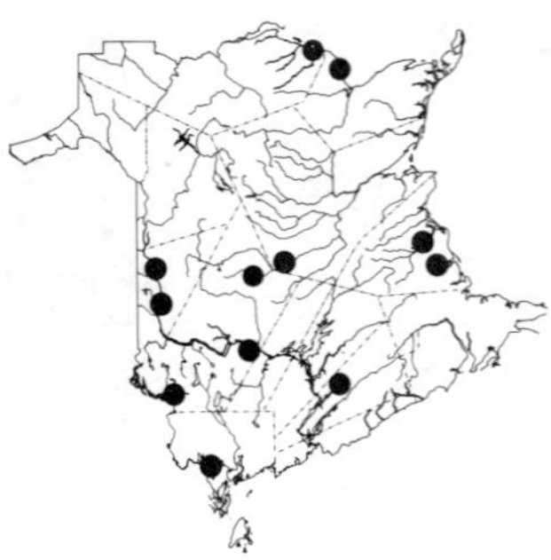

SILENE ALBA

SILENE NOCTIFLORA

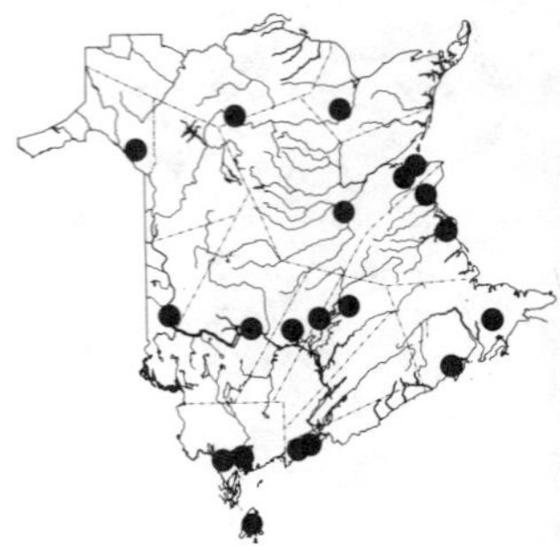

SPERGULA ARVENSIS

CARYOPHYLLACEAE | CERATOPHYLLACEAE | NYMPHAEACEAE

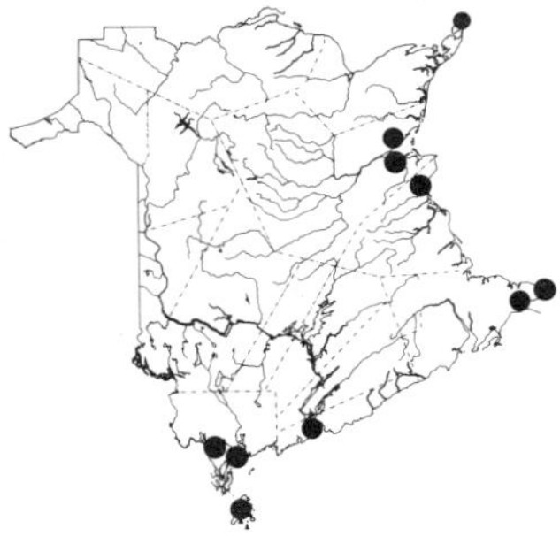

SPERGULARIA CANADENSIS

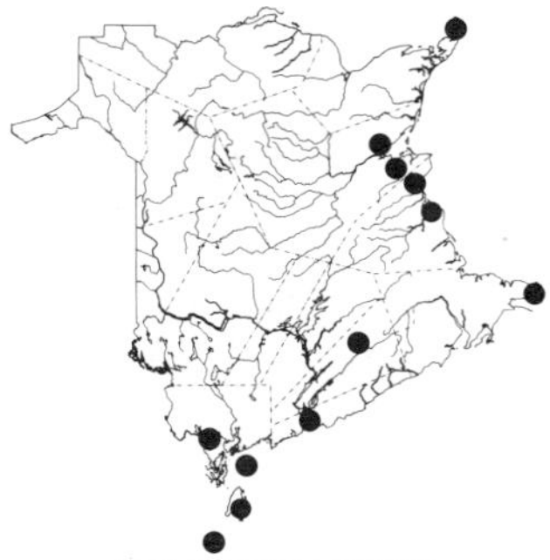

SPERGULARIA MARINA

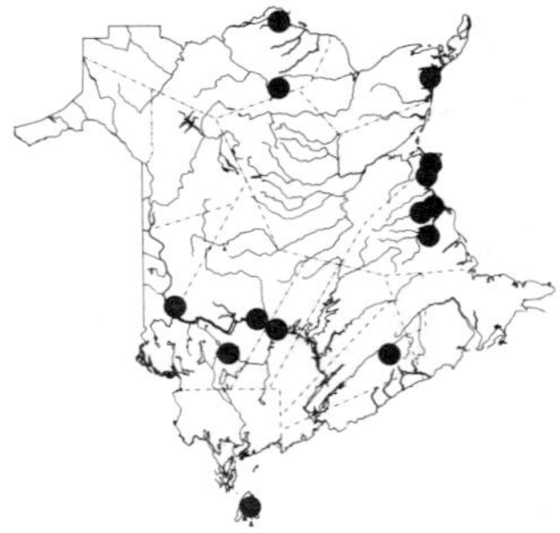

SPERGULARIA RUBRA

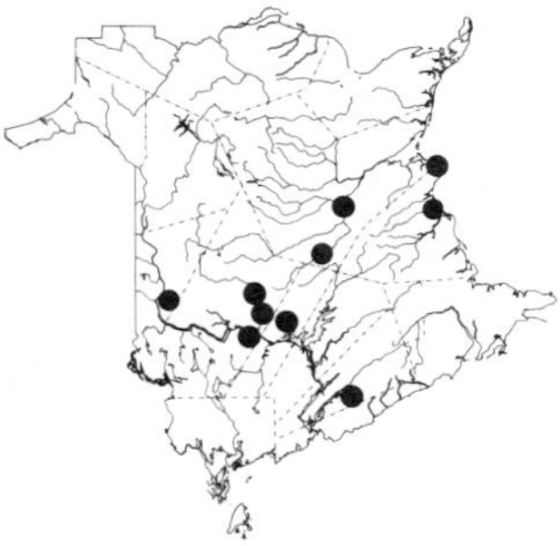

STELLARIA ALSINE

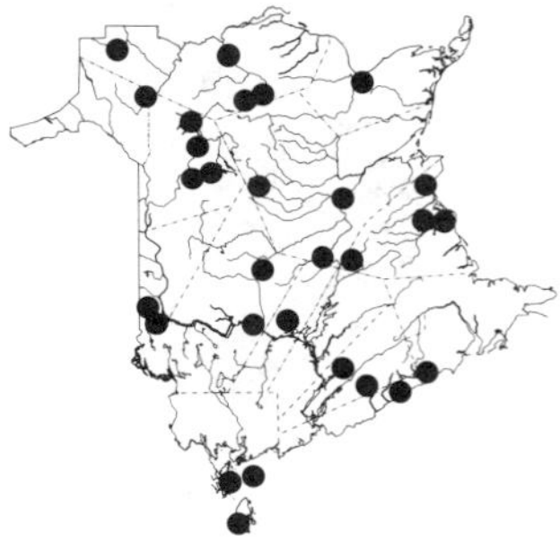

STELLARIA CALYCANTHA

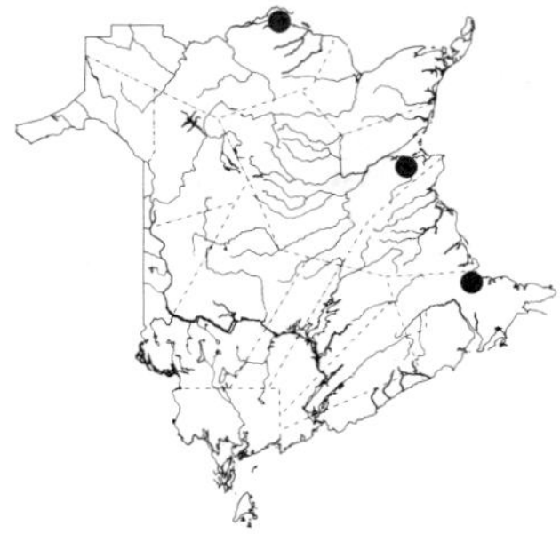

STELLARIA CRASSIFOLIA

STELLARIA HUMIFUSA

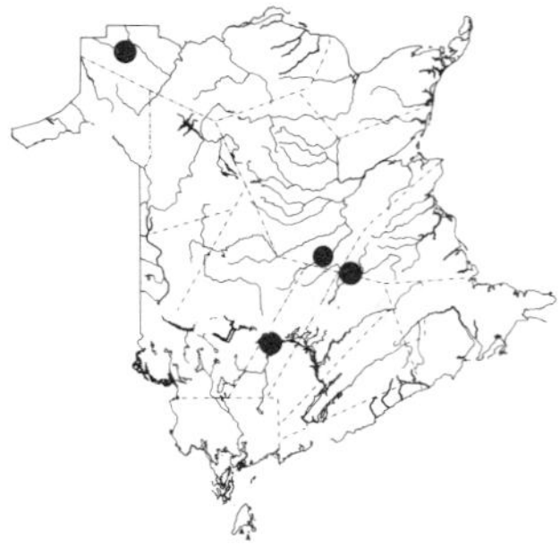

STELLARIA LONGIFOLIA

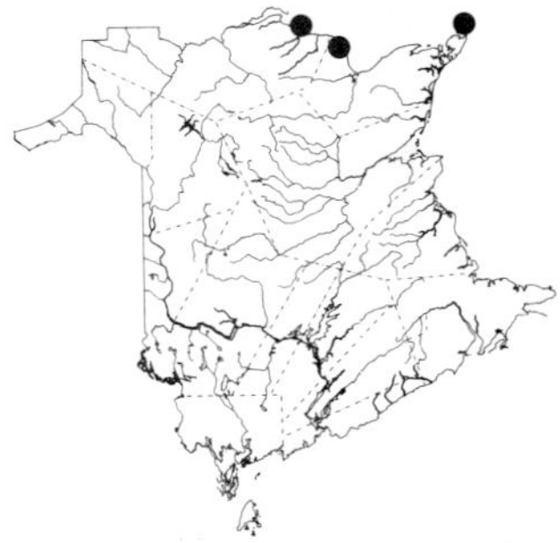

STELLARIA LONGIPES

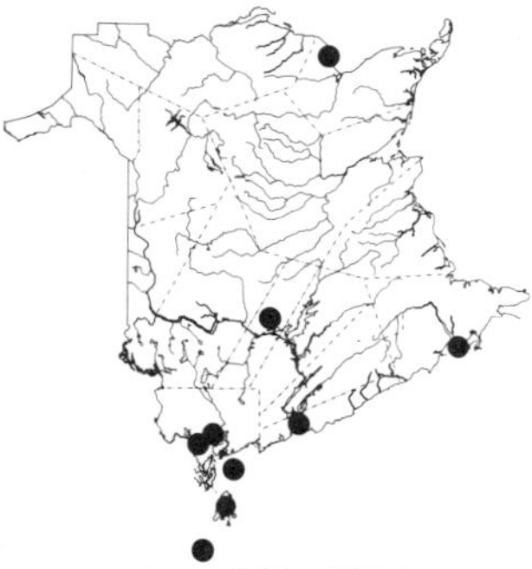

STELLARIA MEDIA

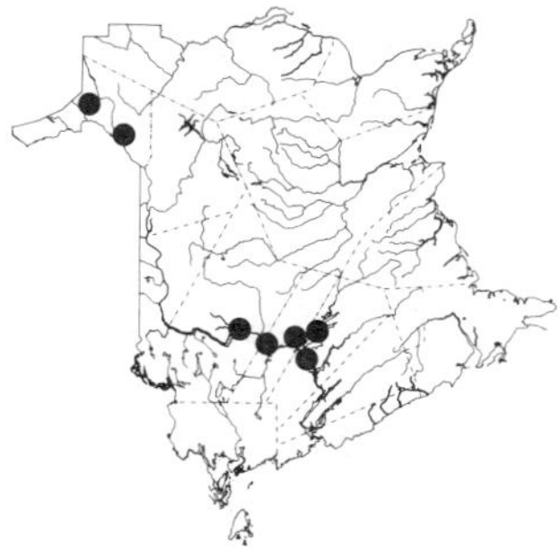

CERATOPHYLLUM DEMERSUM

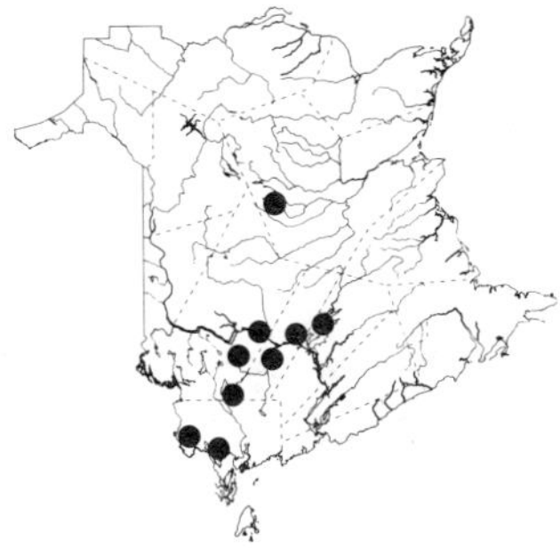

BRASENIA SCHREBERI

NUPHAR LUTEUM

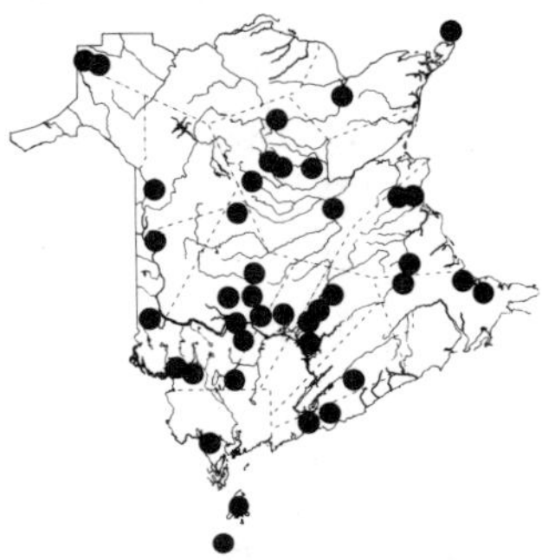

NUPHAR LUTEUM SSP. VARIEGATUM

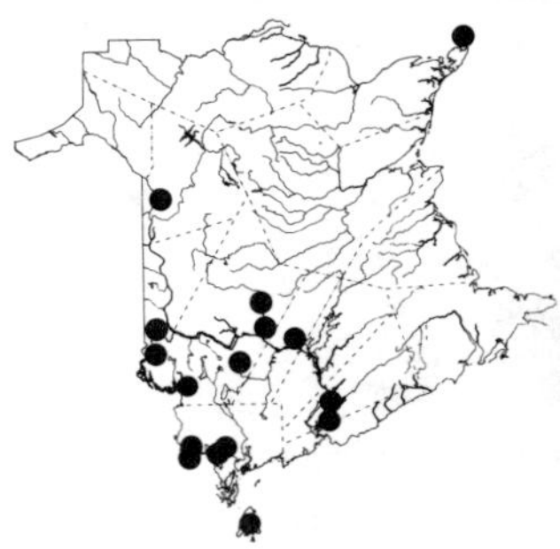

NYMPHAEA ODORATA

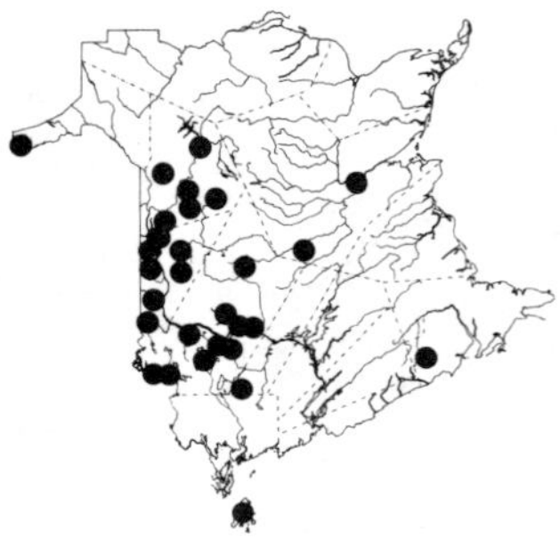

ACTAEA PACHYPODA

ACTAEA RUBRA

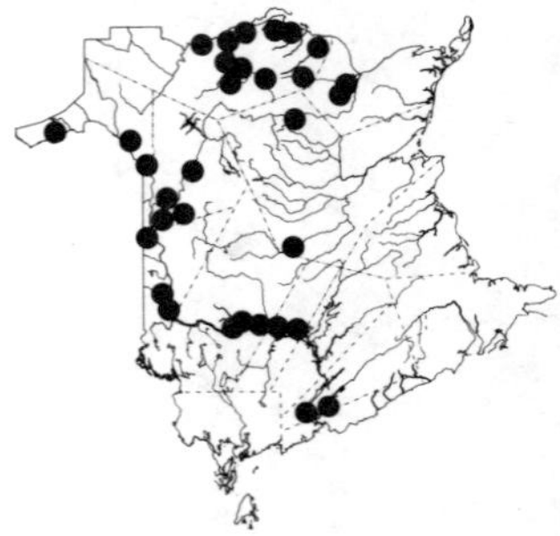

ANEMONE CANADENSIS

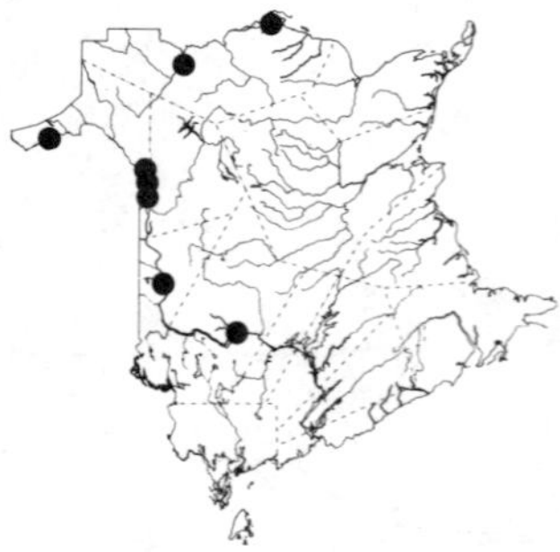

ANEMONE MULTIFIDA

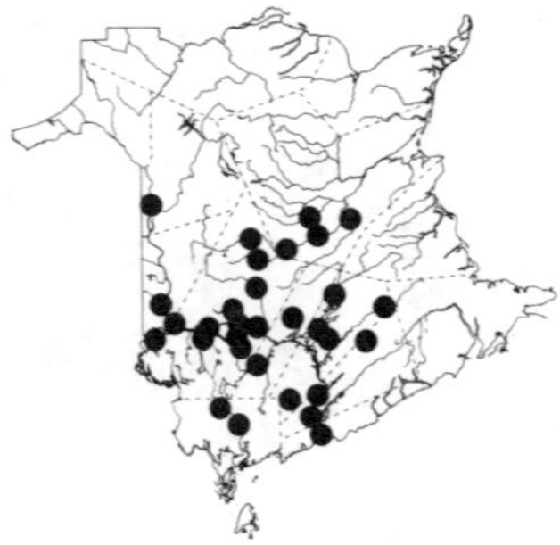

ANEMONE QUINQUEFOLIA

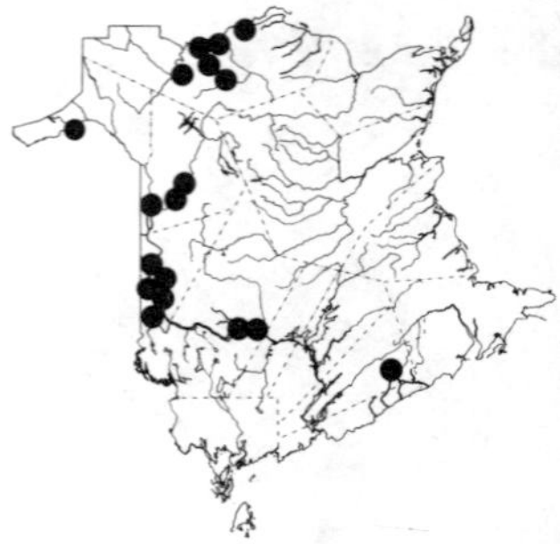

ANEMONE RIPARIA

AQUILEGIA VULGARIS

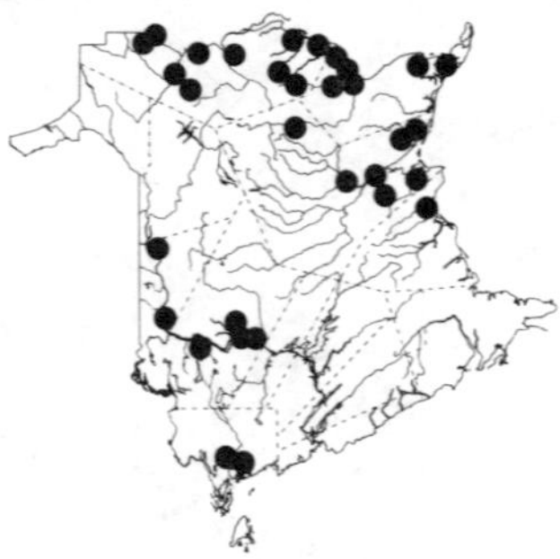

CALTHA PALUSTRIS

CLEMATIS OCCIDENTALIS

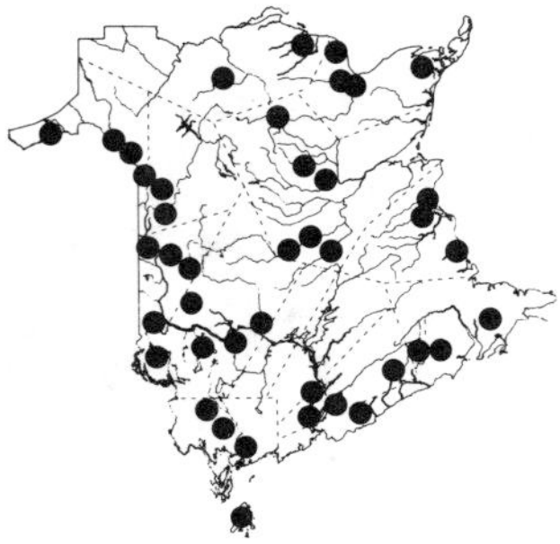

CLEMATIS VIRGINIANA

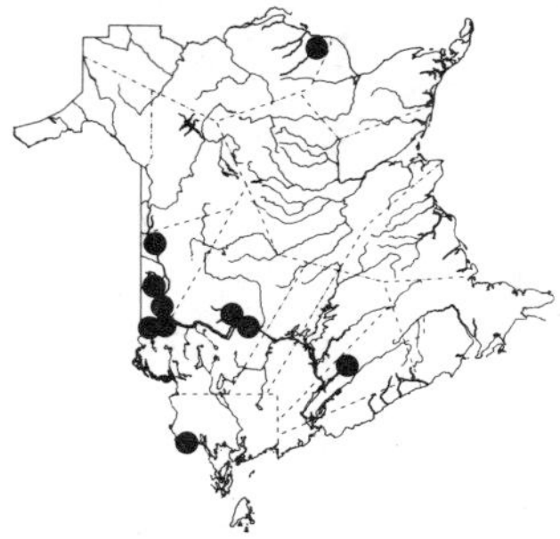

HEPATICA NOBILIS

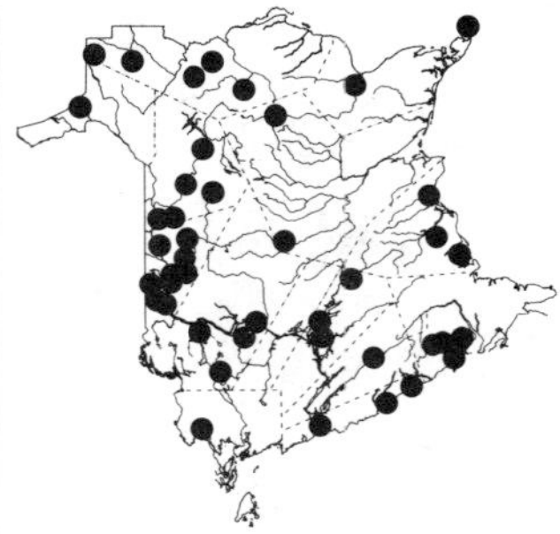

RANUNCULUS ABORTIVUS

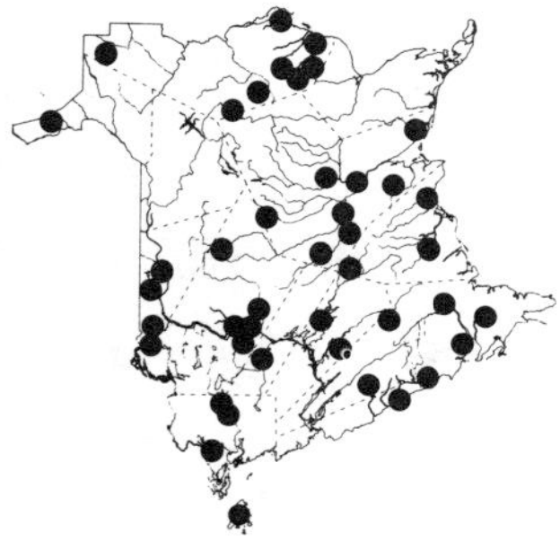

RANUNCULUS AQUATILIS

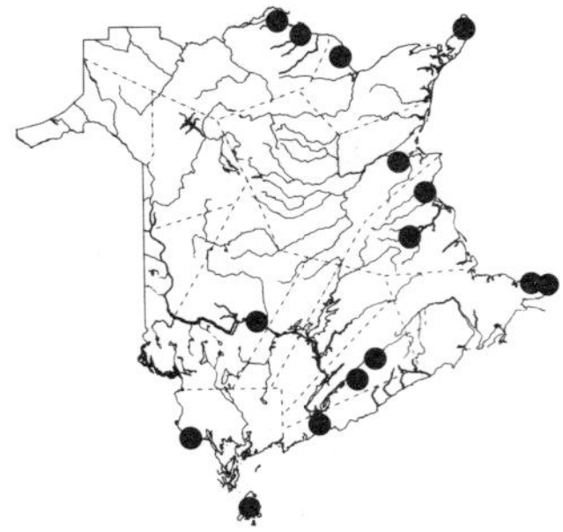

RANUNCULUS CYMBALARIA

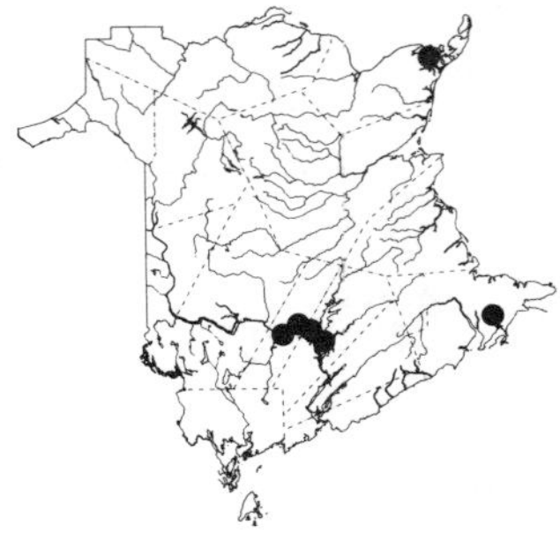

RANUNCULUS FLABELLARIS

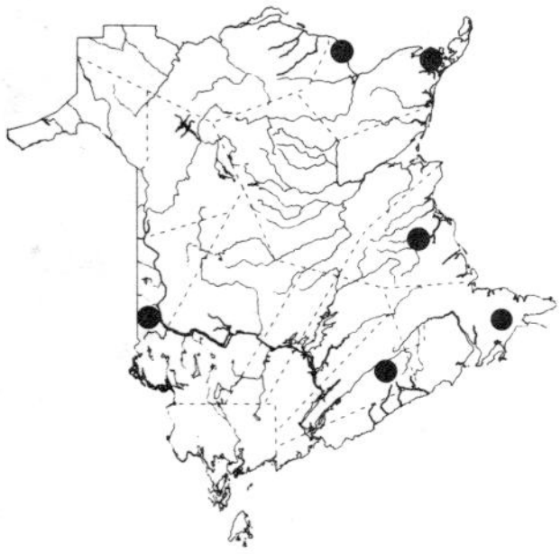

RANUNCULUS GMELINII VAR. HOOKERI

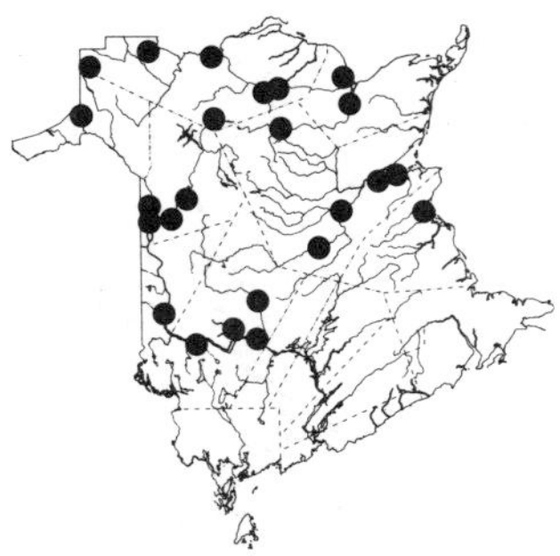

RANUNCULUS PENSYLVANICUS

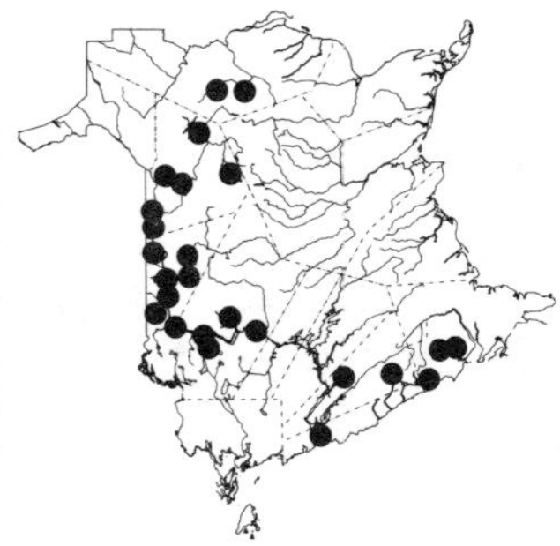

RANUNCULUS RECURVATUS

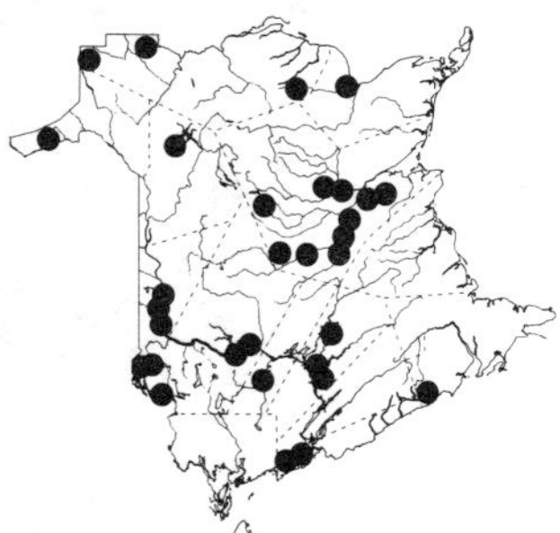

RANUNCULUS REPTANS

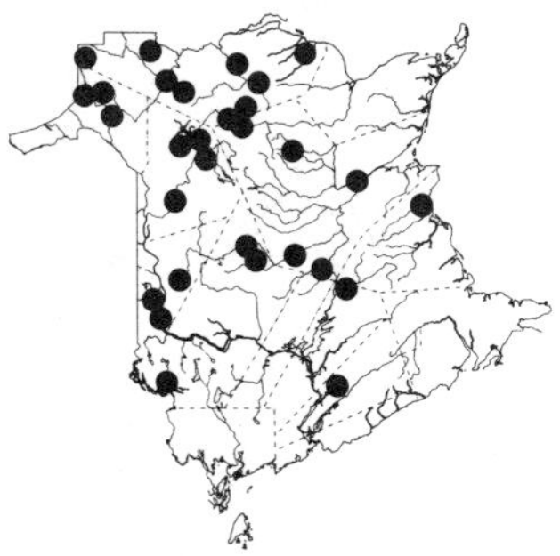

RANUNCULUS SEPTENTRIONALIS

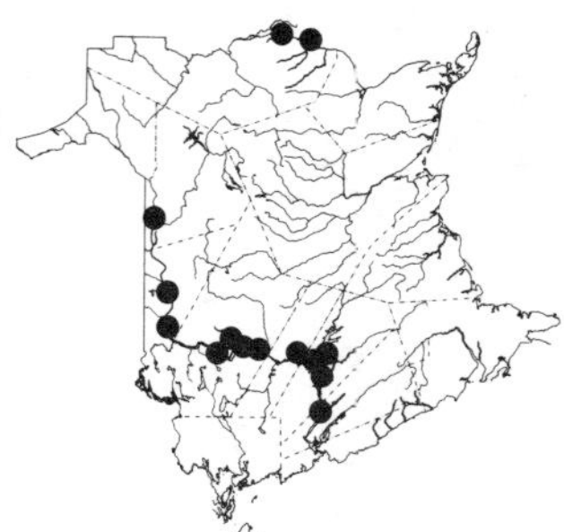

THALICTRUM VENULOSUM

BERBERIDÁCEAE	PAPAVERACEAE	BRASSICACEAE

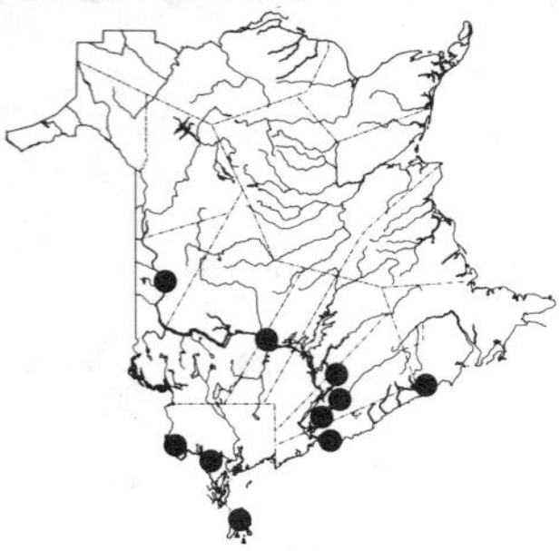

BERBERIS THUNBERGII

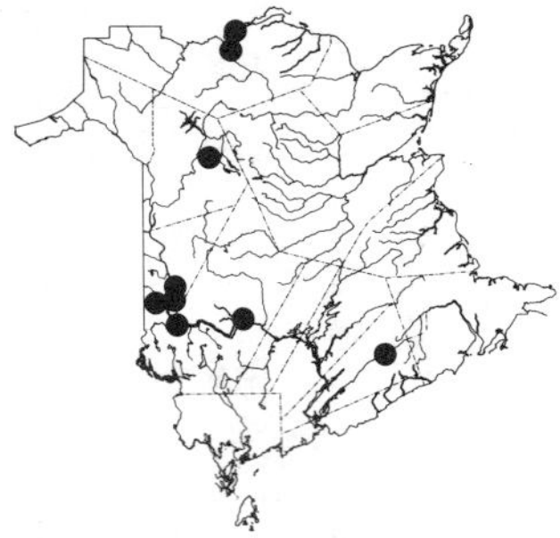

CAULOPHYLLUM THALICTROIDES

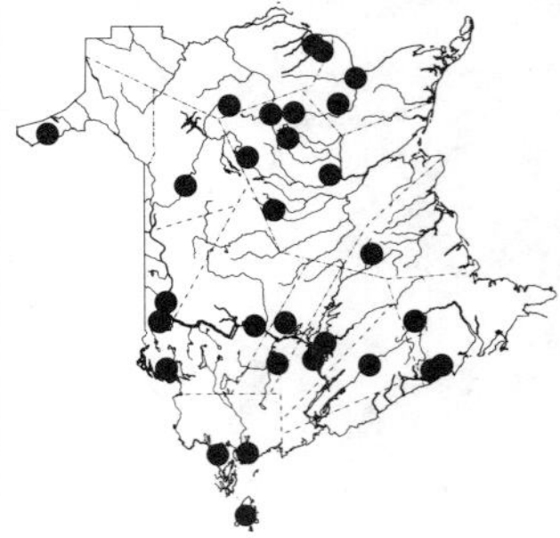

CORYDALIS SEMPERVIRENS

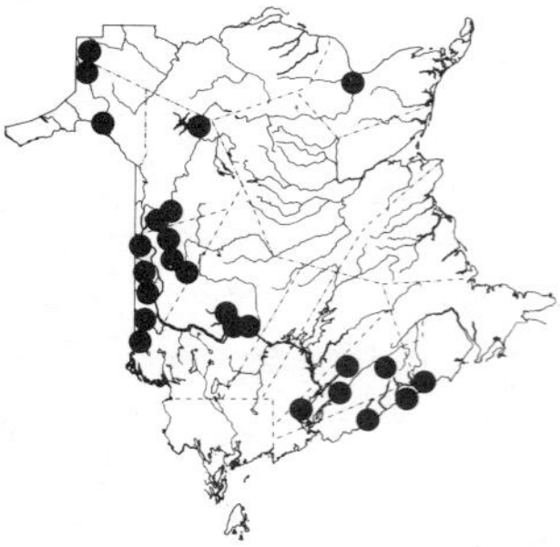

DICENTRA CUCULLARIA

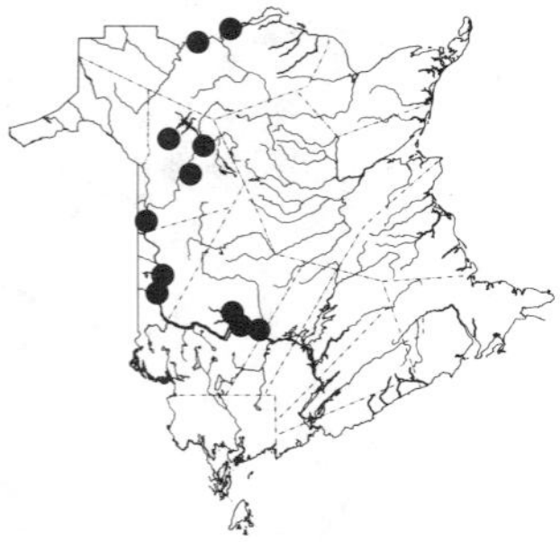

SANGUINARIA CANADENSIS

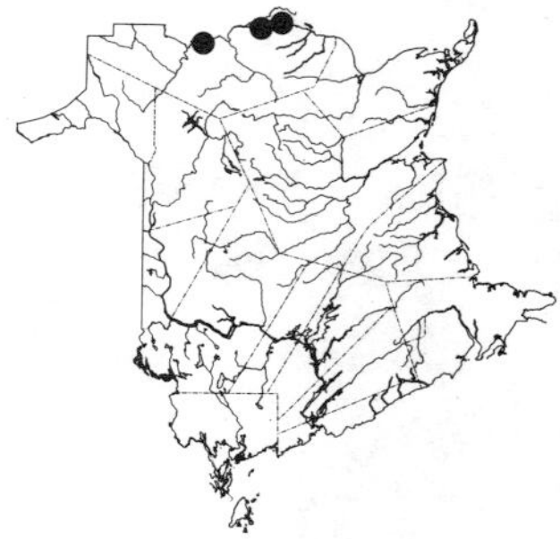

ARABIS DIVARICARPA

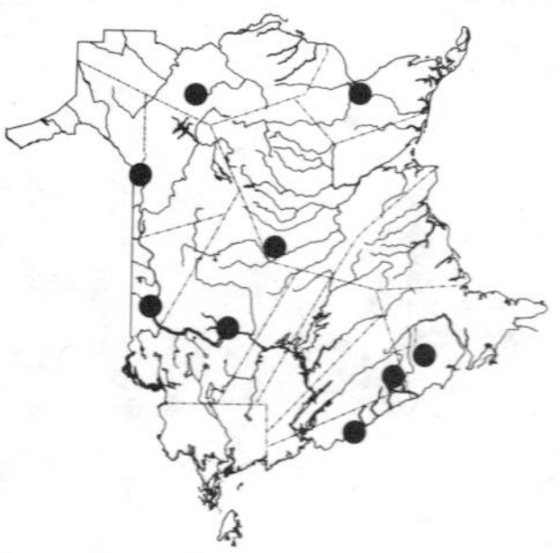

ARABIS DRUMMONDII

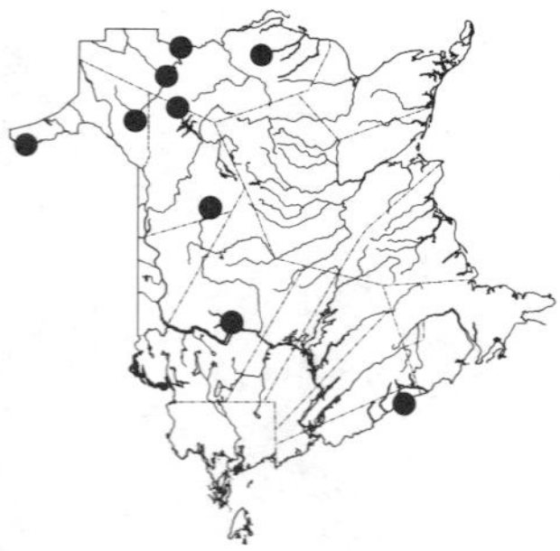

ARABIS GLABRA

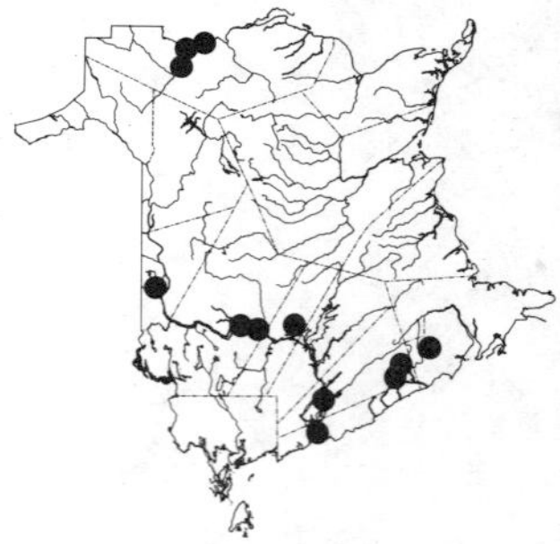

ARABIS HIRSUTA

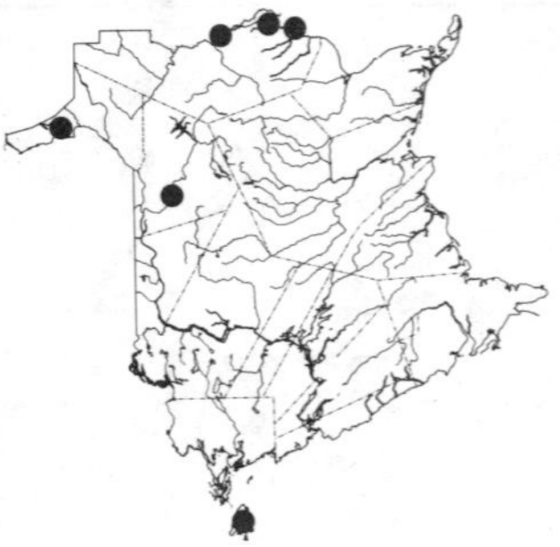

BARBAREA ORTHOCERAS

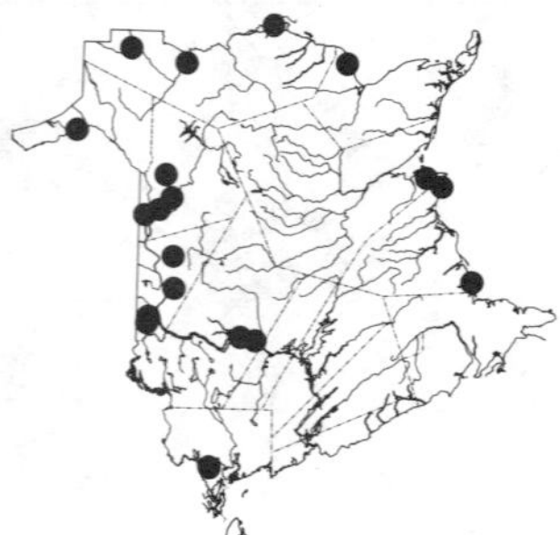

BRASSICA RAPA

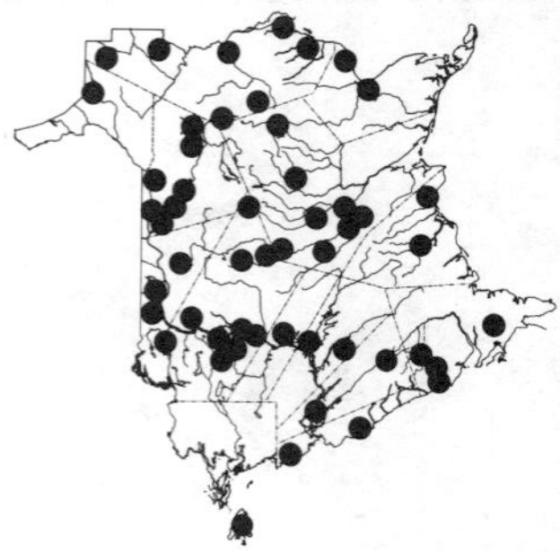

CARDAMINE PENSYLVANICA

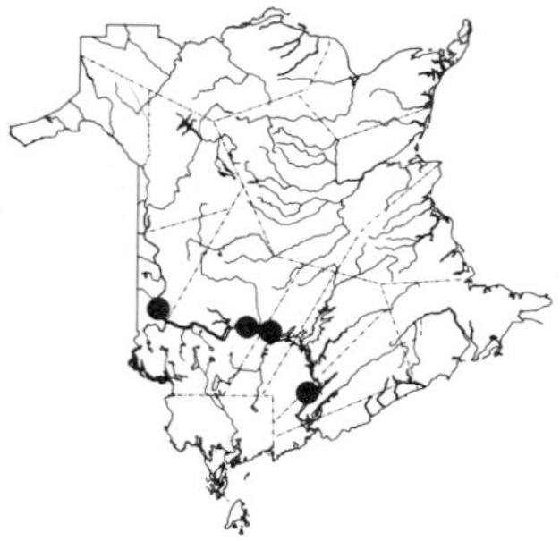

CARDAMINE PRATENSIS

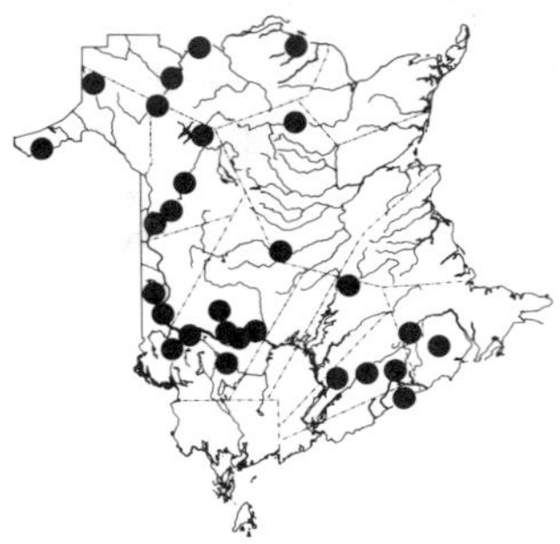

DENTARIA DIPHYLLA

DENTARIA LACINIATA

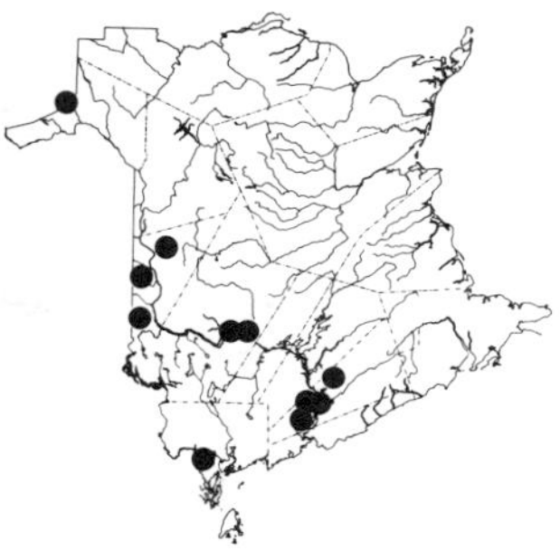

DENTARIA MAXIMA

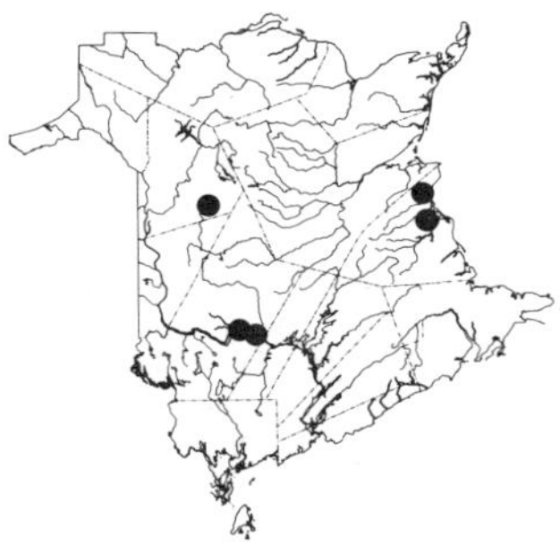

DESCURAINIA SOPHIA

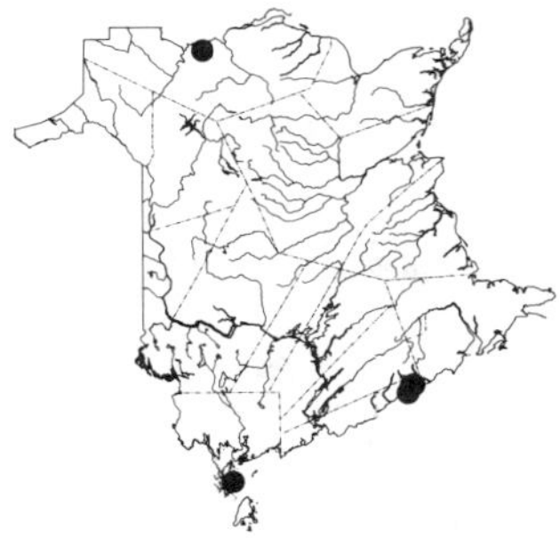

DRABA ARABISANS

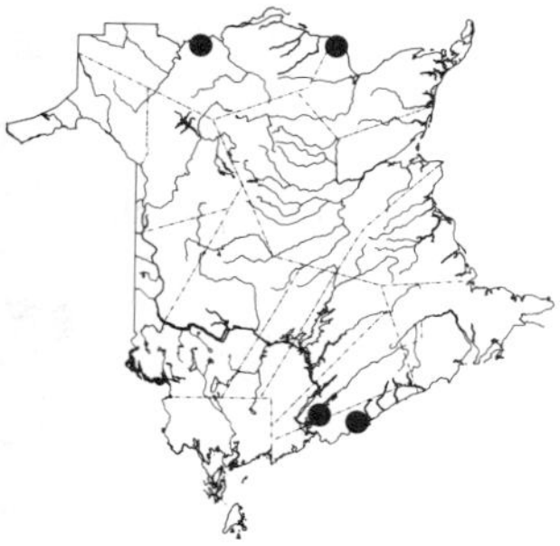

DRABA GLABELLA

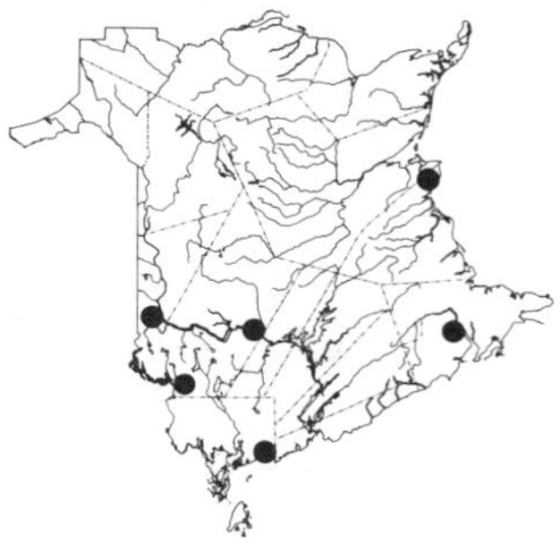

ERUCASTRUM GALLICUM

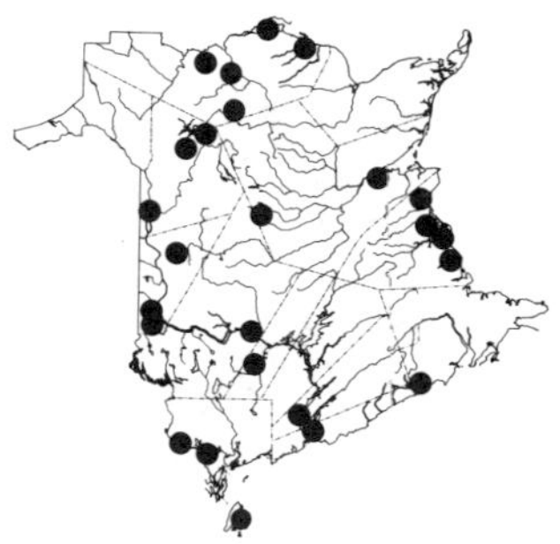

ERYSIMUM CHEIRANTHOIDES

HESPERIS MATRONALIS

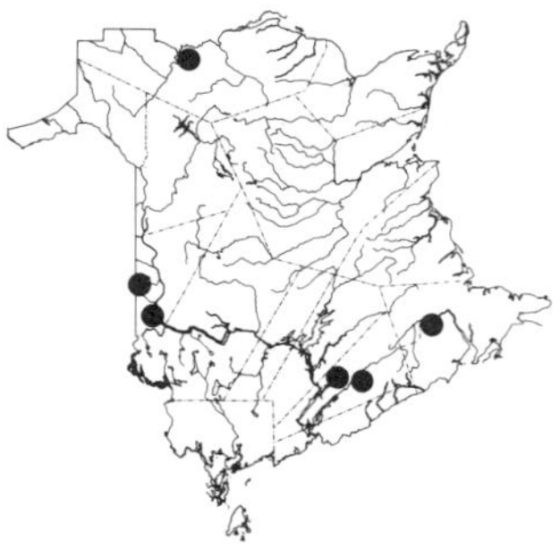

LEPIDIUM CAMPESTRE

LEPIDIUM DENSIFLORUM

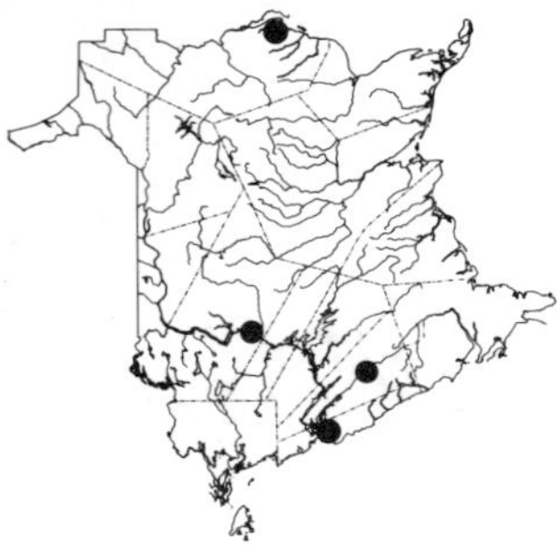

NASTURTIUM OFFICINALE

RAPHANUS RAPHANISTRUM

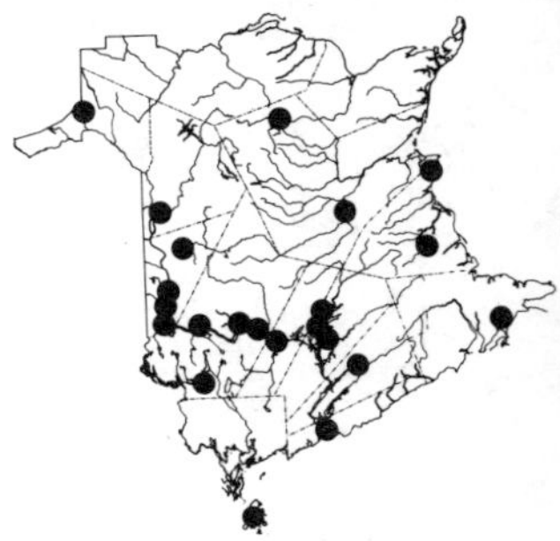

RORIPPA PALUSTRIS

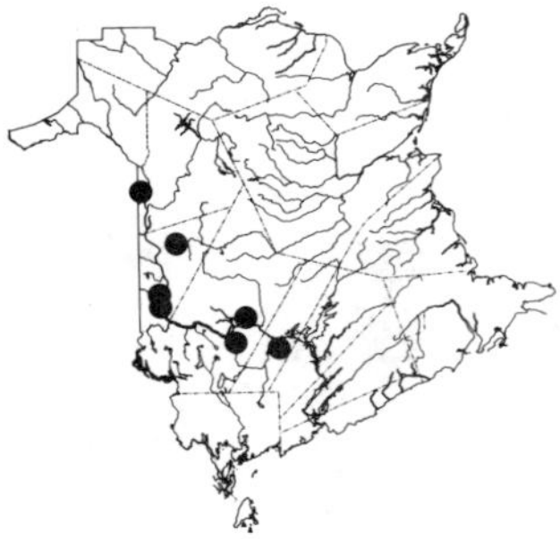

RORIPPA SYLVESTRIS

SISYMBRIUM ALTISSIMUM

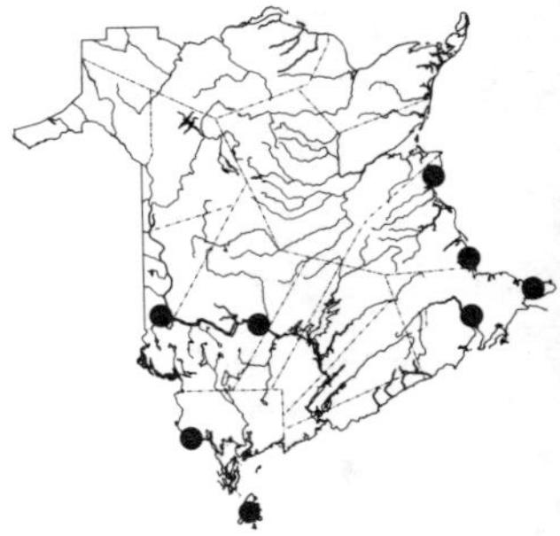

SISYMBRIUM OFFICINALE

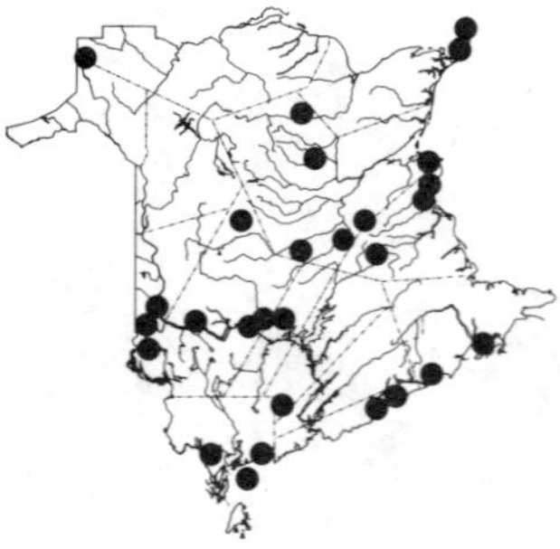

SARRACENIA PURPUREA

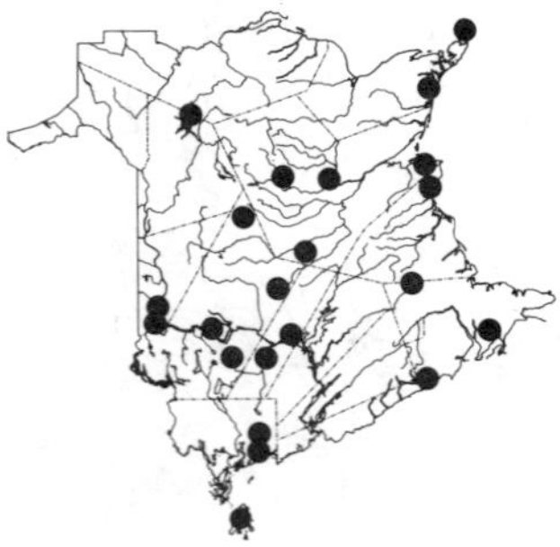

DROSERA INTERMEDIA

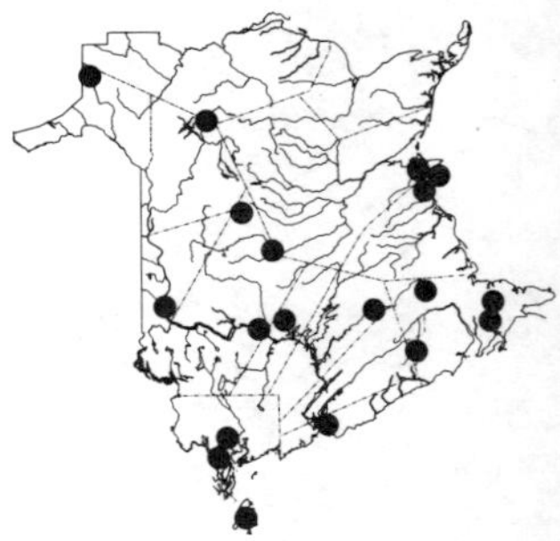

DROSERA ROTUNDIFOLIA

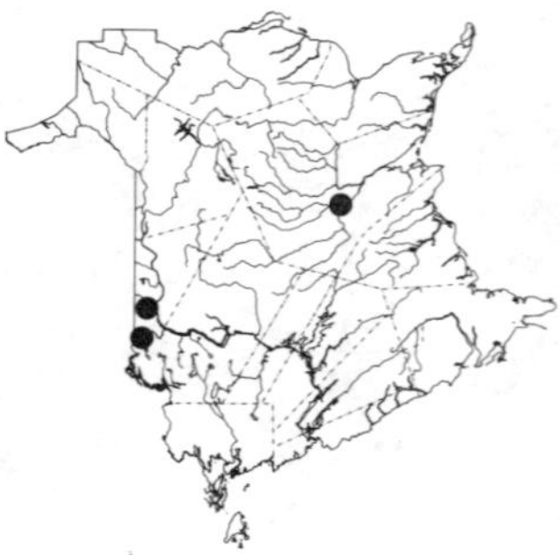

PODOSTEMON CERATOPHYLLUM

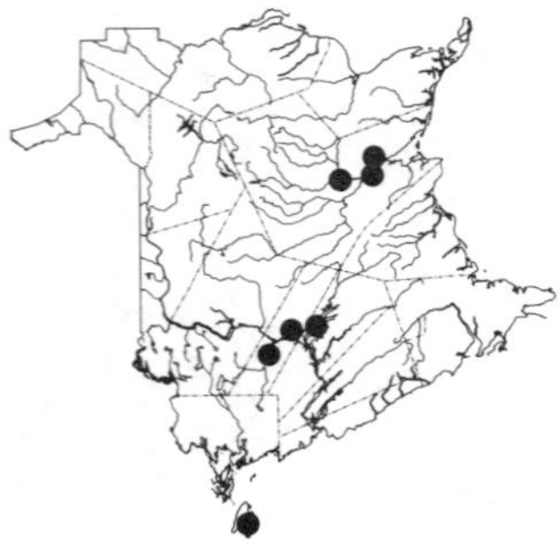

CRASSULA AQUATICA

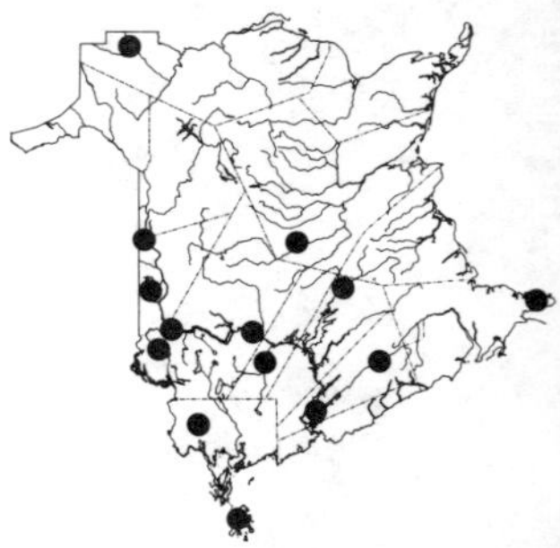

SEDUM TELEPHIUM

CRASSULACEAE SAXIFRAGACEAE HAMAMELIDACEAE

SEDUM ROSEA

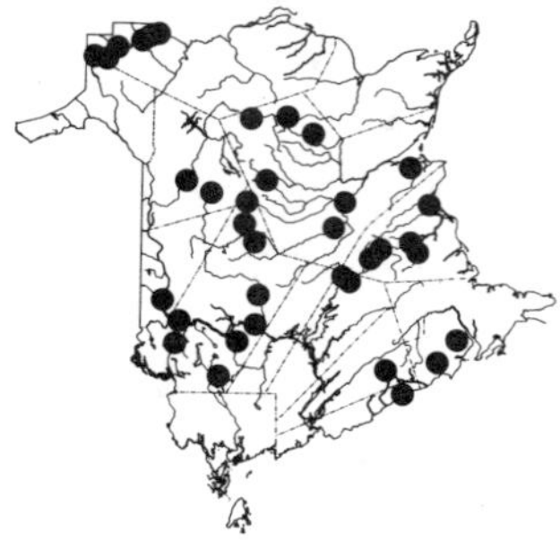

CHRYSOSPLENIUM AMERICANUM

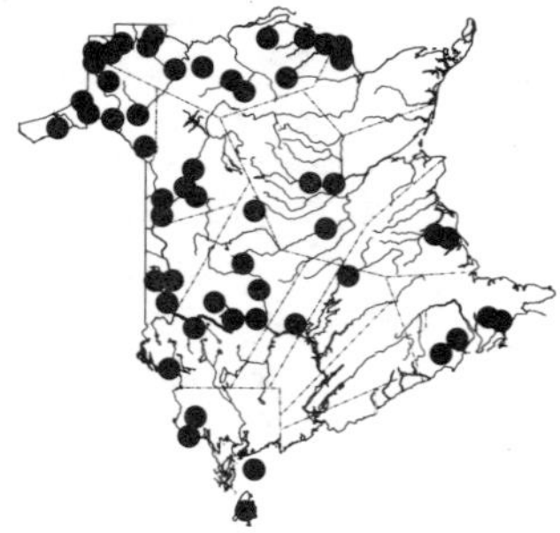

MITELLA NUDA

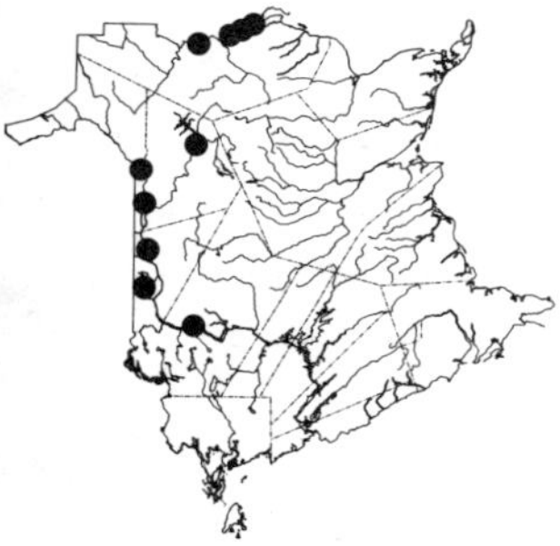

PARNASSIA GLAUCA

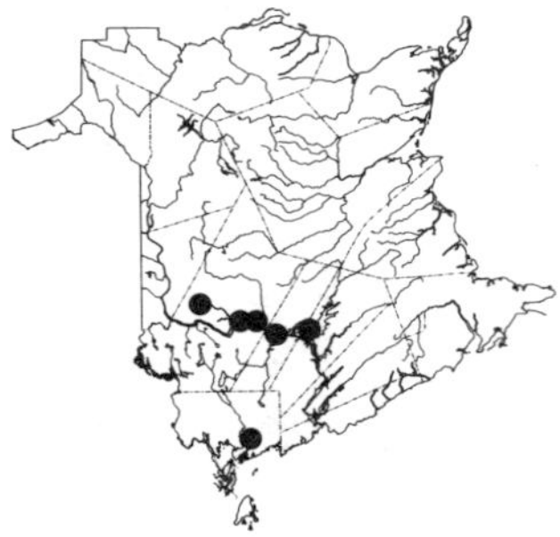

PENTHORUM SEDOIDES

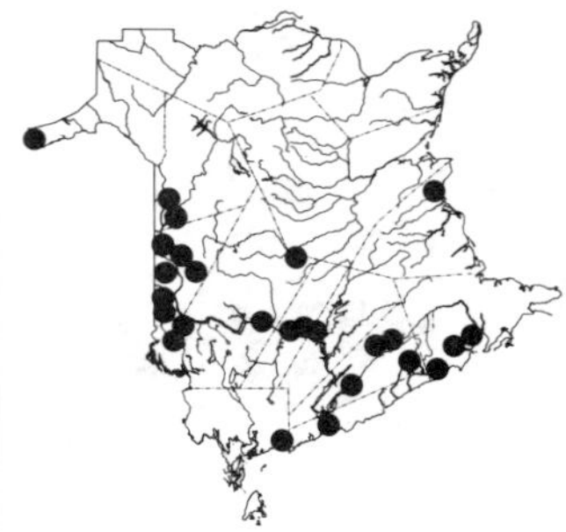

RIBES AMERICANUM

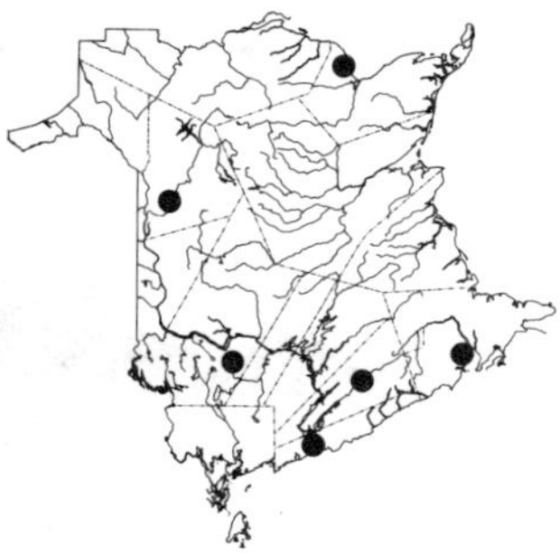

RIBES NIGRUM

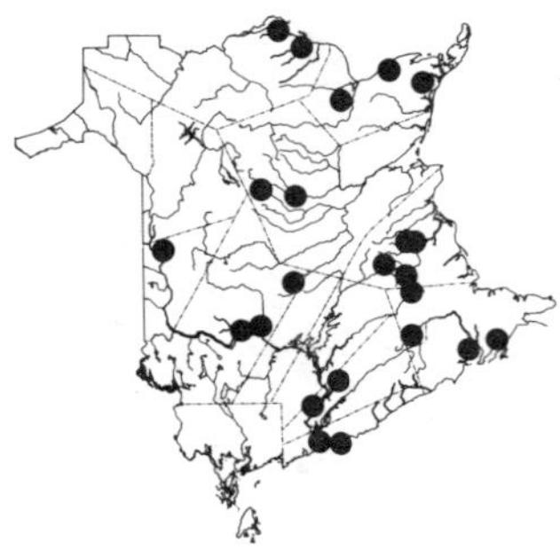

RIBES RUBRUM VAR. ALASKANUM

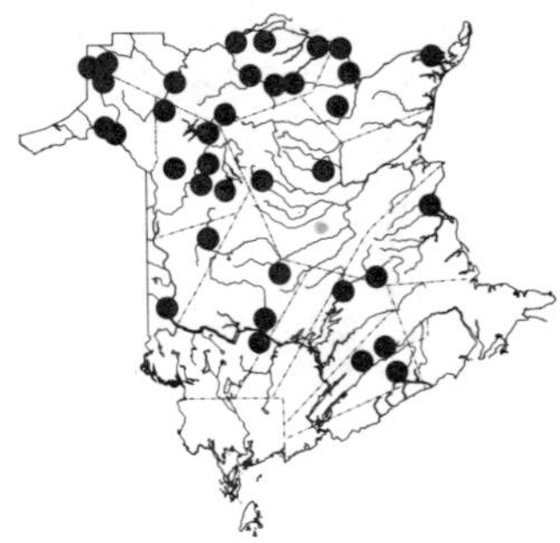

RIBES TRISTE

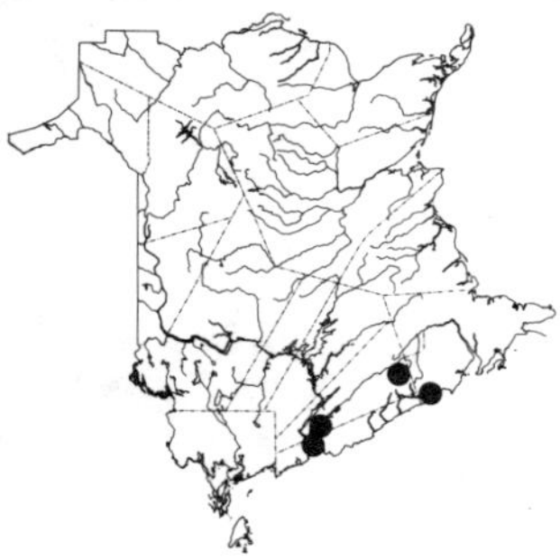

SAXIFRAGA PANICULATA

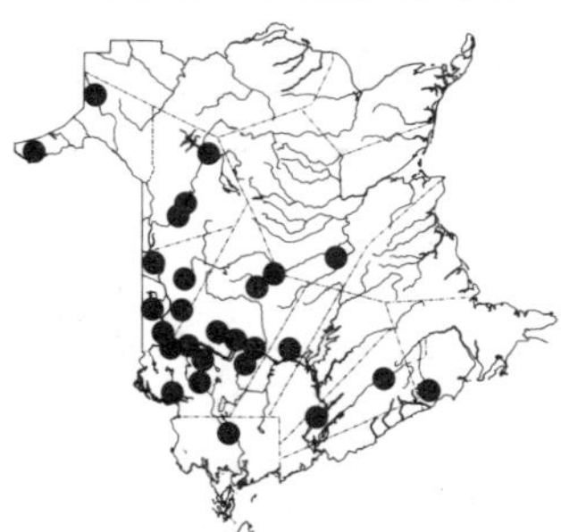

TIARELLA CORDIFOLIA

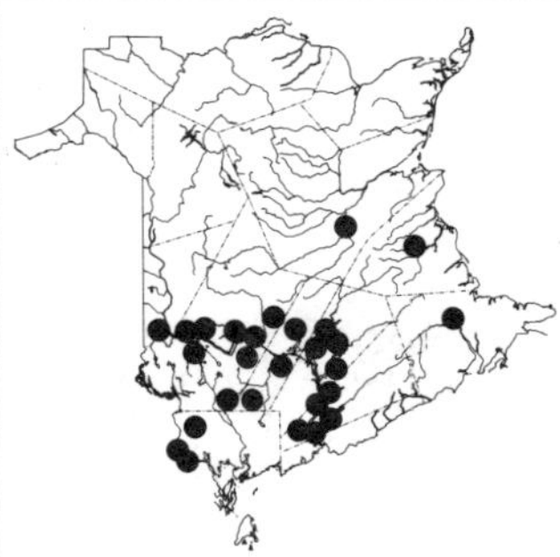

HAMAMELIS VIRGINIANA

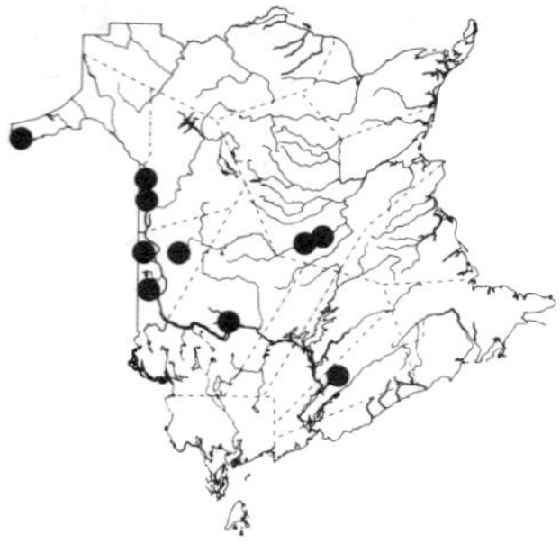

AGRIMONIA GRYPOSEPALA

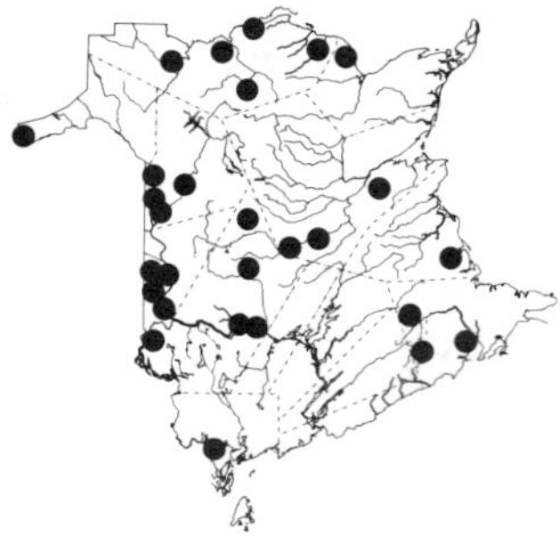

AGRIMONIA STRIATA

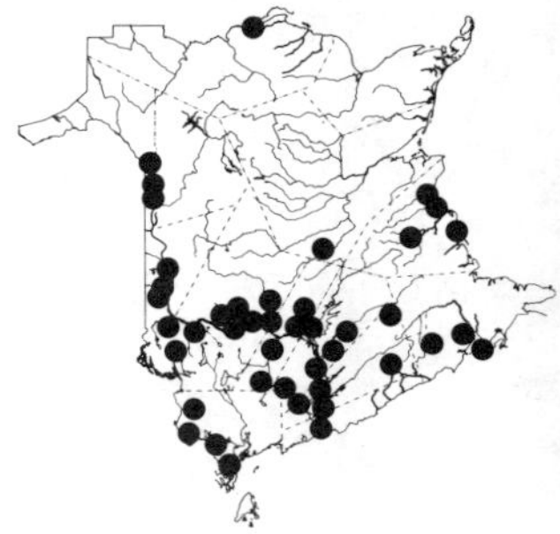

AMELANCHIER ARBOREA

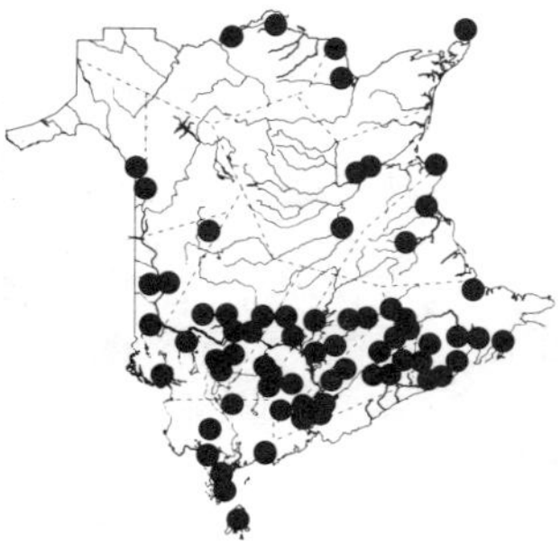

AMELANCHIER ARBOREA VAR. LAEVIS

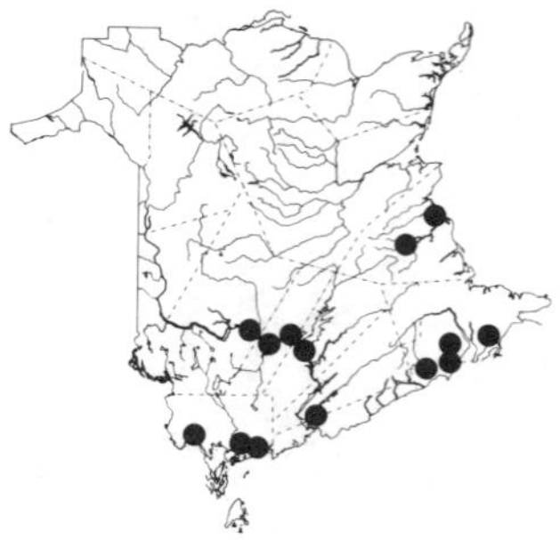

AMELANCHIER CANADENSIS

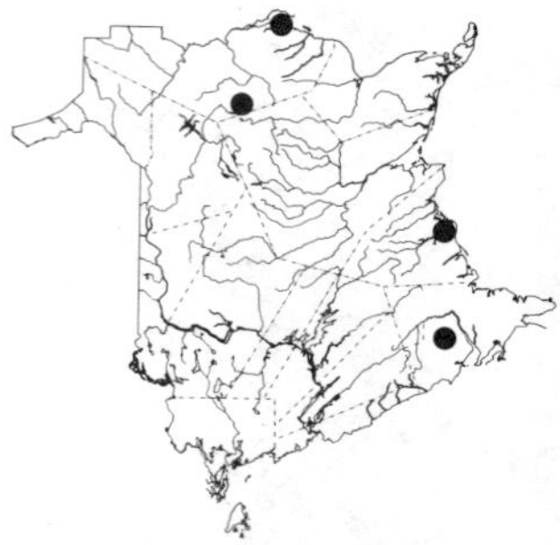

AMELANCHIER FERNALDII

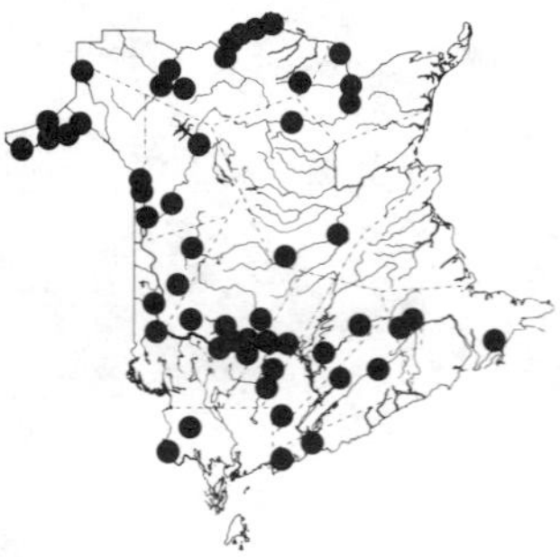

AMELANCHIER HUMILIS

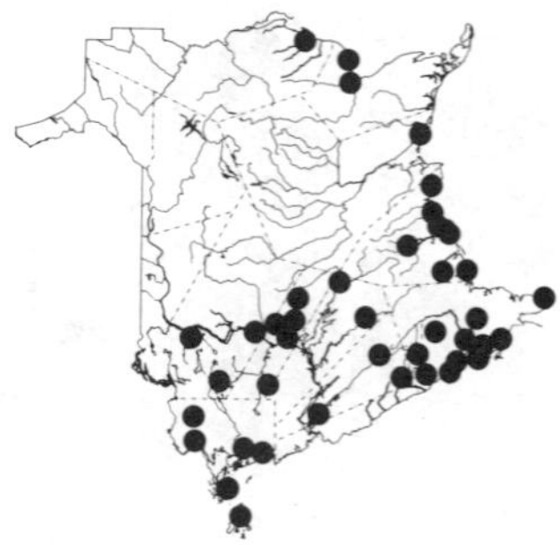

AMELANCHIER INTERMEDIA

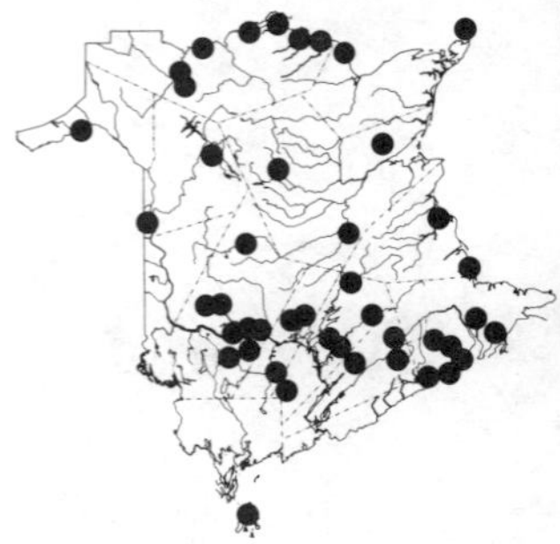

AMELANCHIER X NEGLECTA

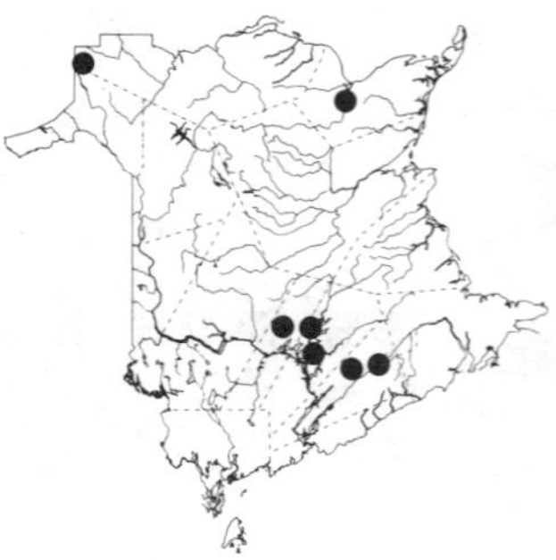

AMELANCHIER X QUINTI-MARTII

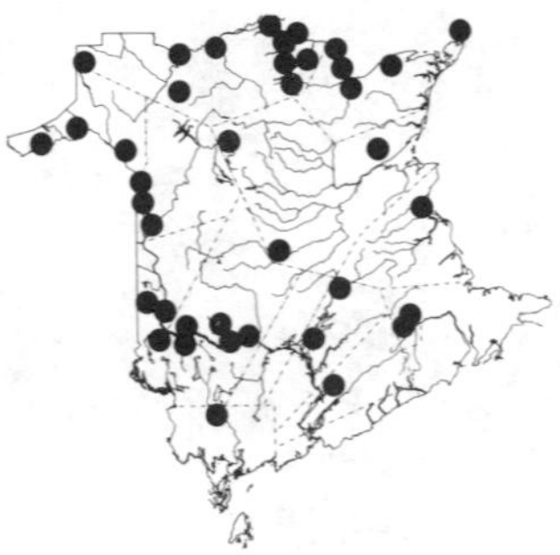

AMELANCHIER SANGUINEA

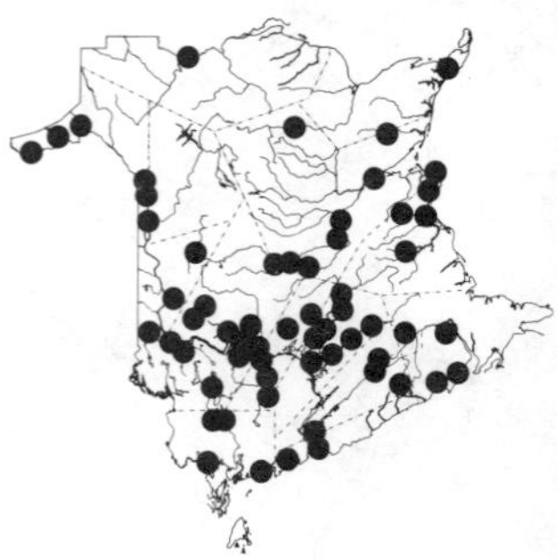

AMELANCHIER STOLONIFERA

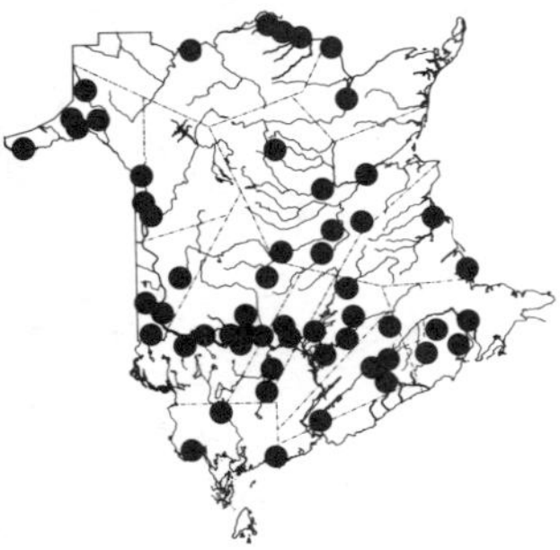

AMELANCHIER WIEGANDII

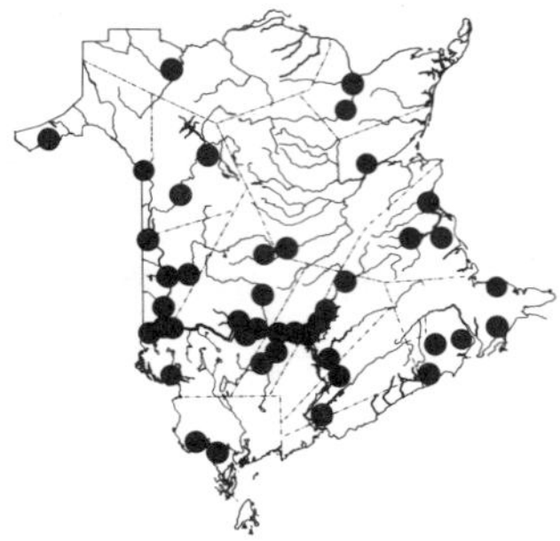

CRATAEGUS CHRYSOCARPA

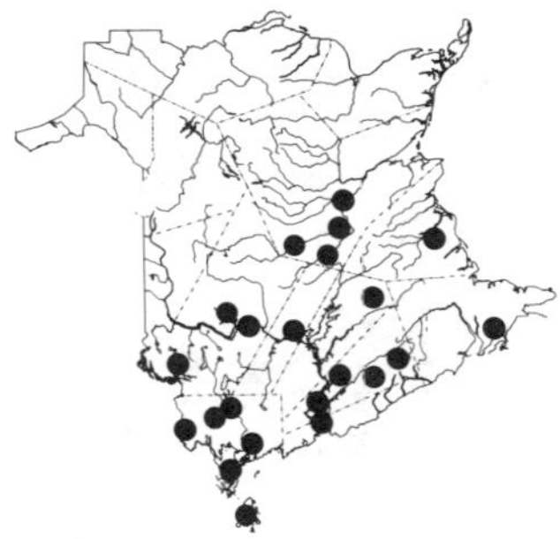

CRATAEGUS FLABELLATA

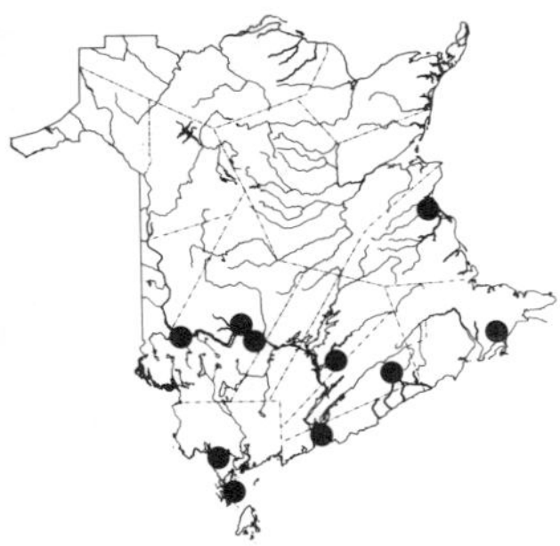

CRATAEGUS MONOGYNA

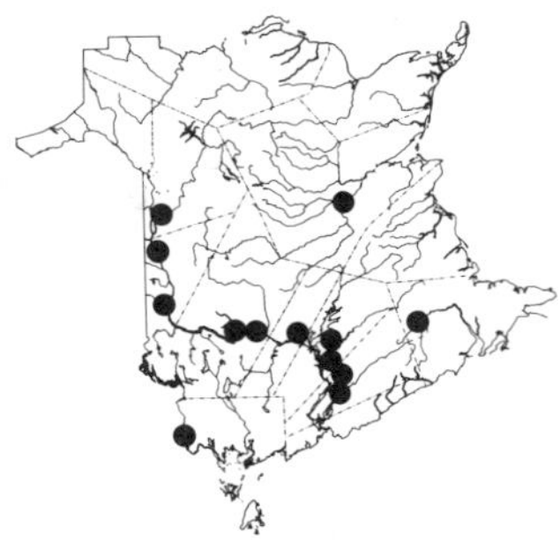

CRATAEGUS SUCCULENTA

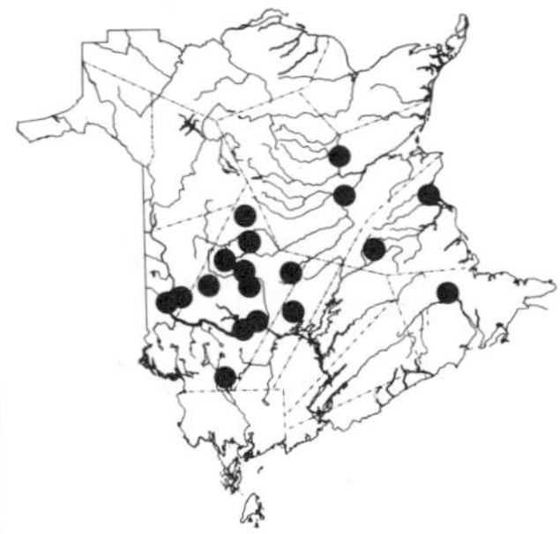

DALIBARDA REPENS

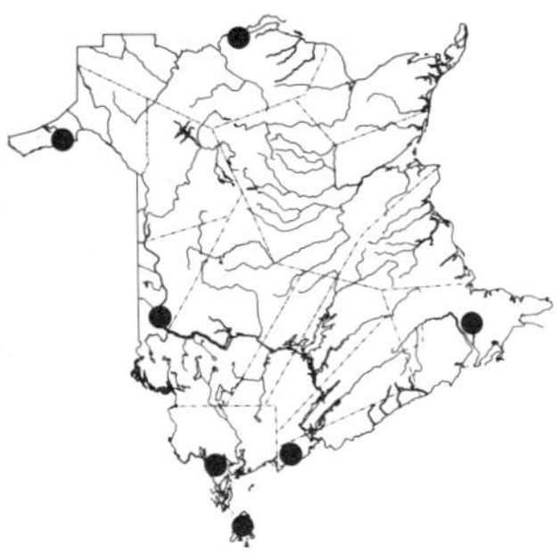

FILIPENDULA ULMARIA

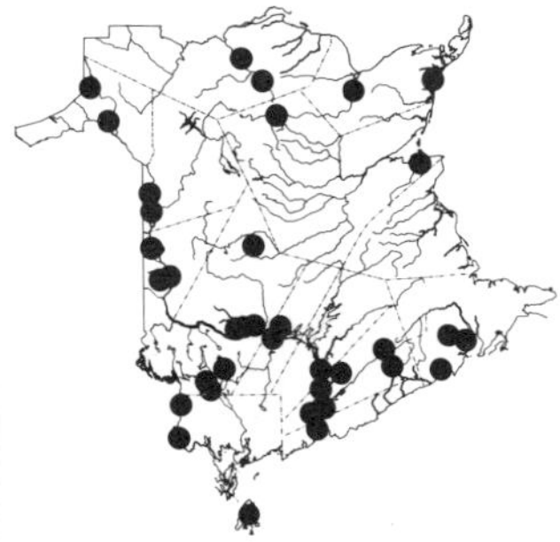

FRAGARIA VESCA

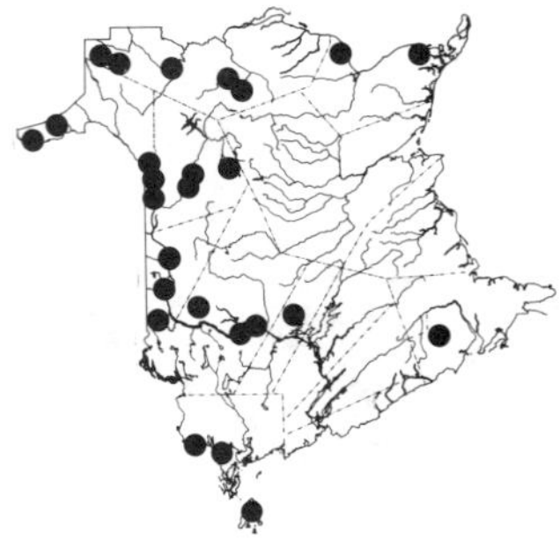

GEUM ALEPPICUM

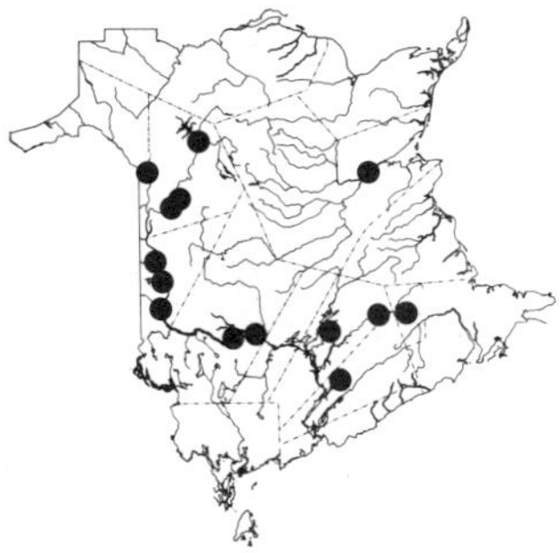

GEUM CANADENSE

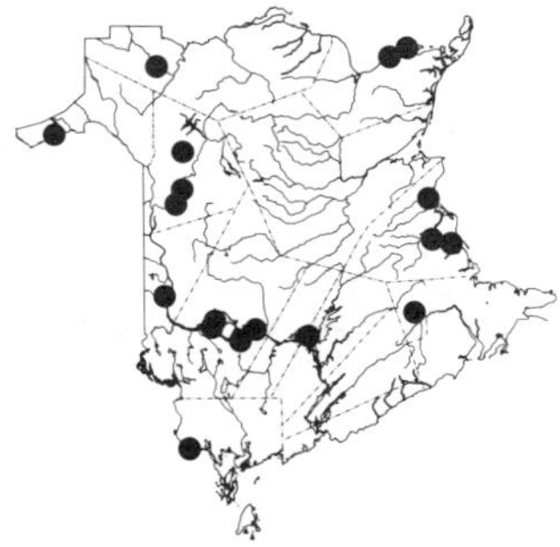

GEUM LACINIATUM

GEUM MACROPHYLLUM

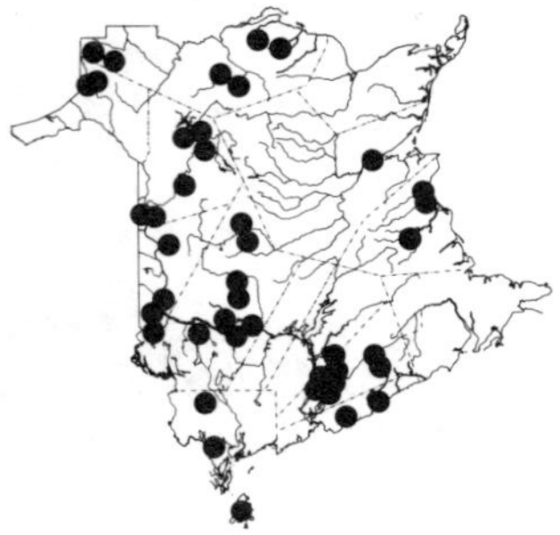

GEUM RIVALE

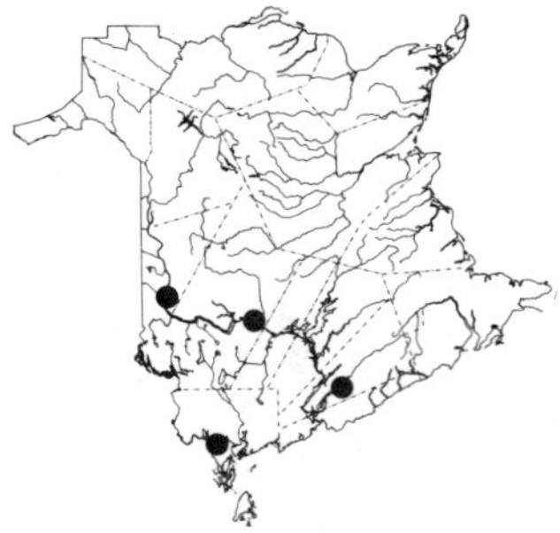

GEUM URBANUM

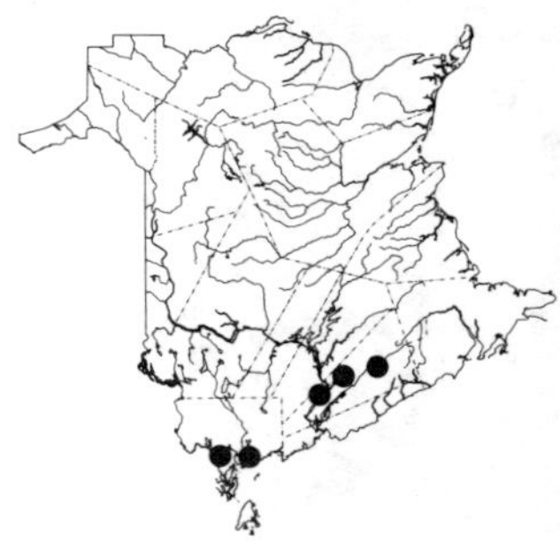

PHYSOCARPUS OPULIFOLIUS

POTENTILLA ANSERINA

POTENTILLA ARGENTEA

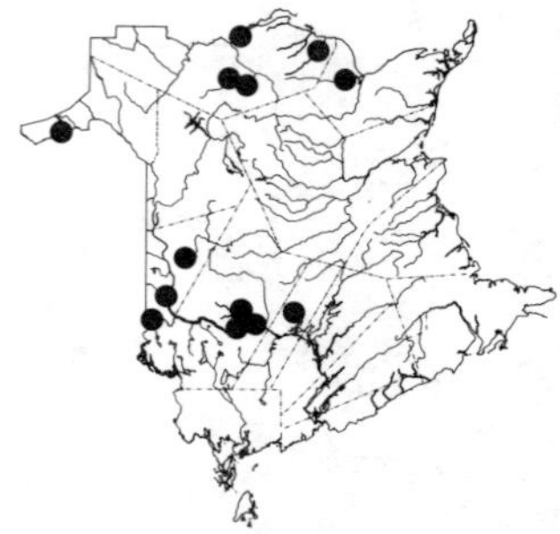

POTENTILLA ARGUTA

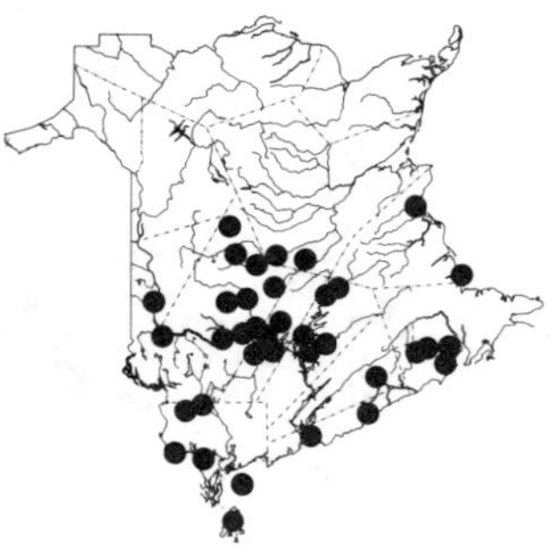

POTENTILLA SIMPLEX

POTENTILLA FRUTICOSA

POTENTILLA INTERMEDIA

POTENTILLA NORVEGICA

POTENTILLA PALUSTRIS

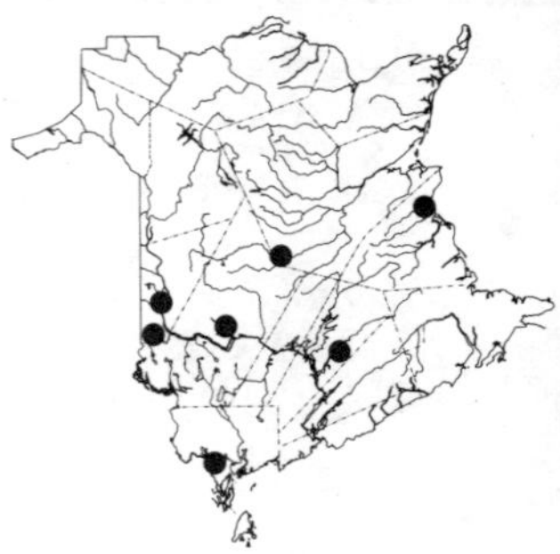

POTENTILLA RECTA

POTENTILLA TRIDENTATA

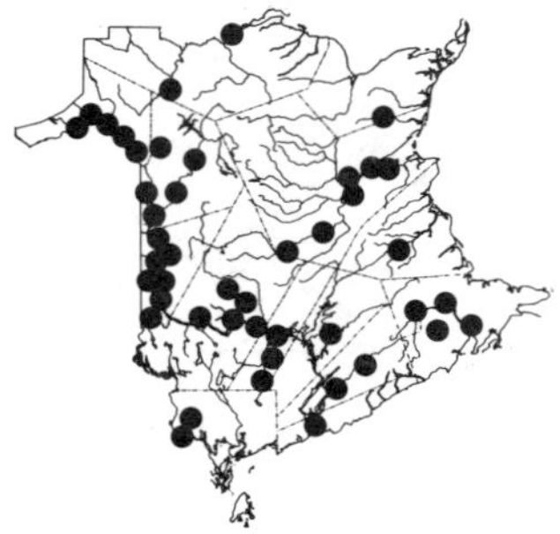

PRUNUS NIGRA

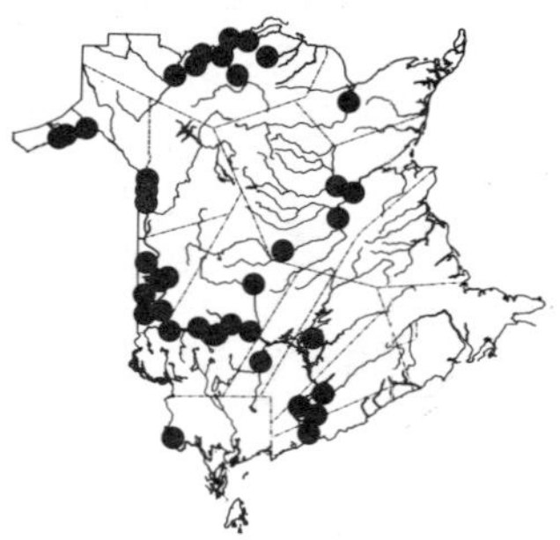

PRUNUS PUMILA

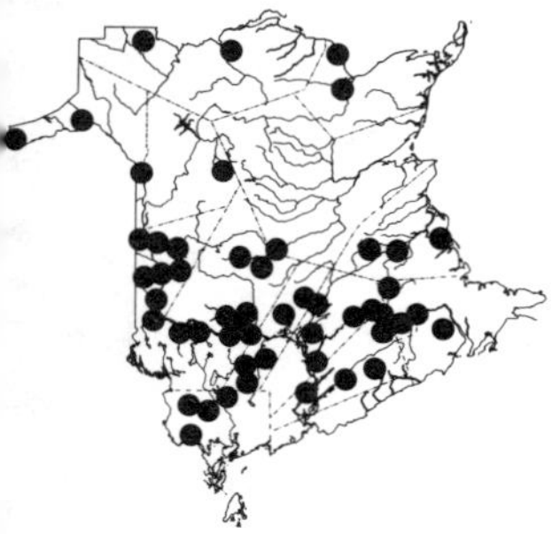

PRUNUS SEROTINA

PRUNUS VIRGINIANA

MALUS SYLVESTRIS

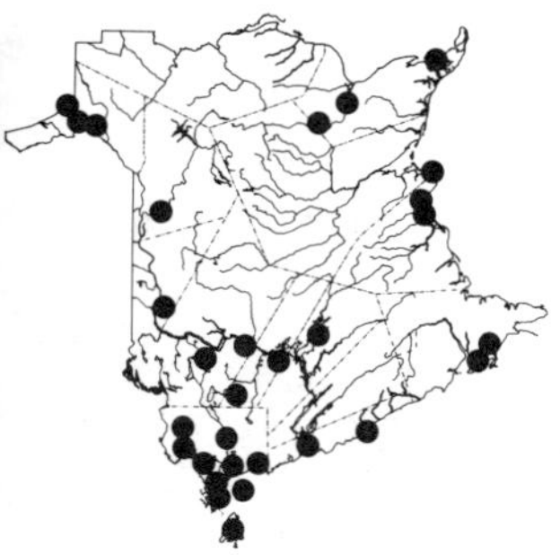

ARONIA MELANOCARPA

ARONIA PRUNIFOLIA

ROSA ACICULARIS

ROSA BLANDA

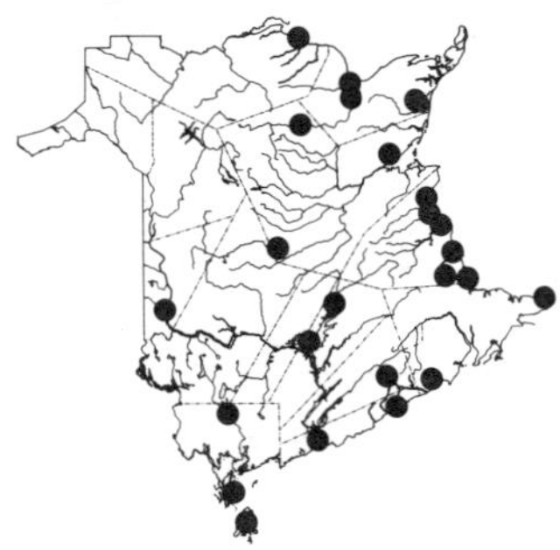

ROSA CAROLINA

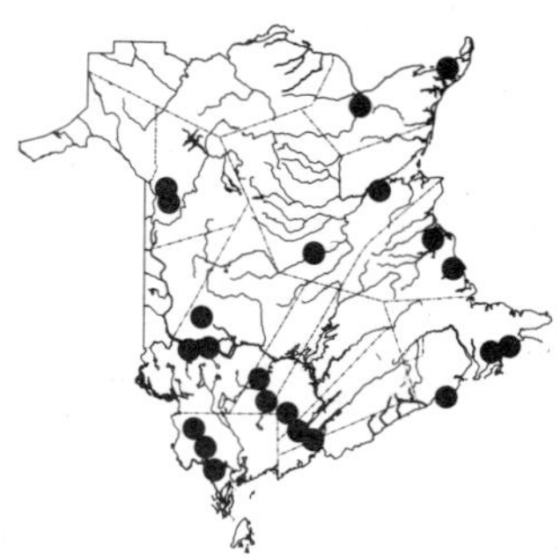

ROSA CINNAMOMEA

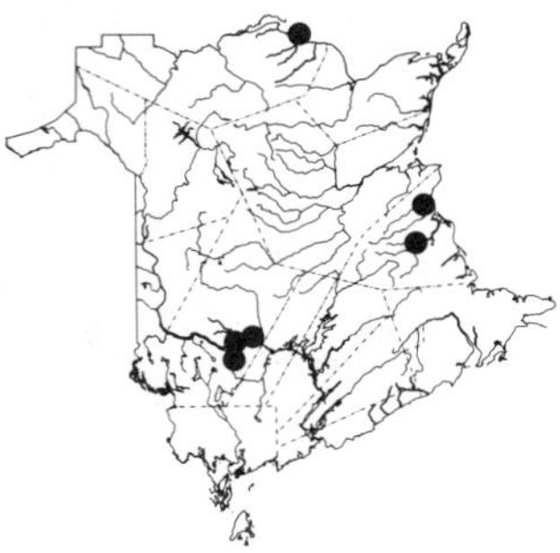
ROSA EGLANTERIA

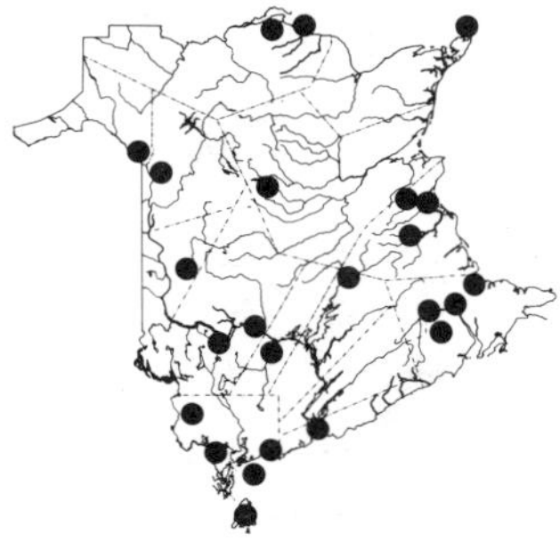
ROSA NITIDA

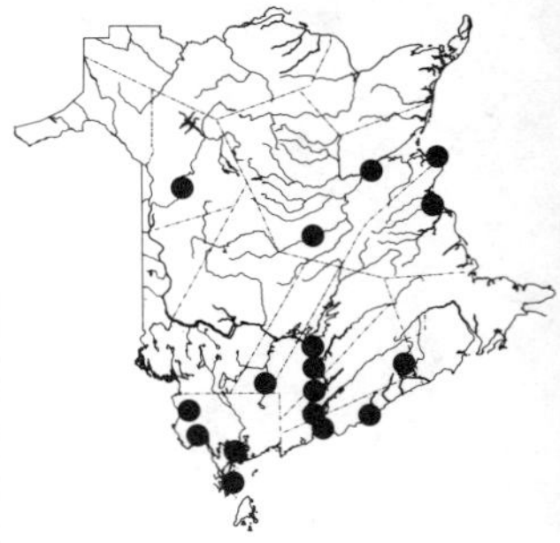
ROSA PALUSTRIS

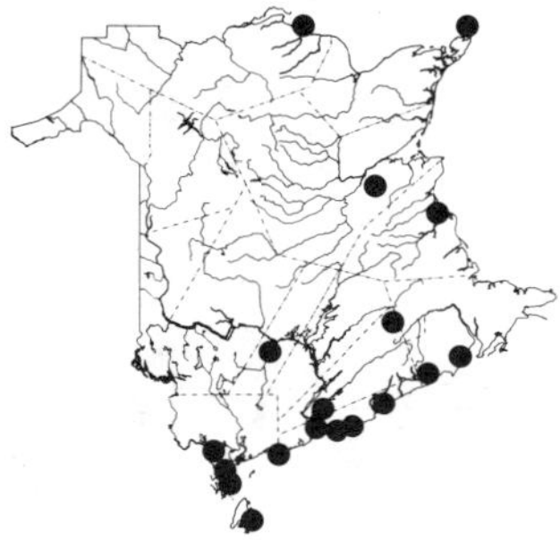
ROSA RUGOSA

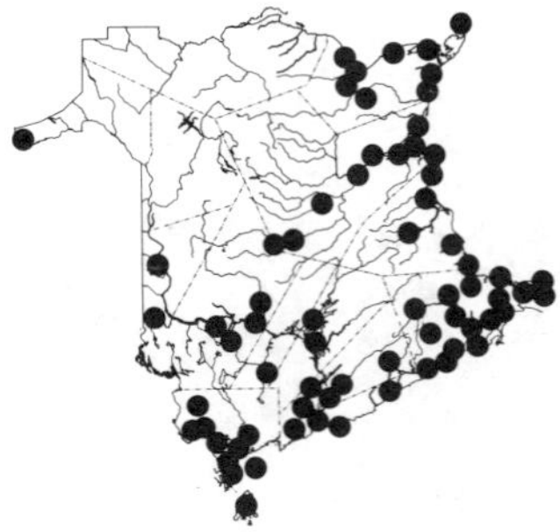
ROSA VIRGINIANA

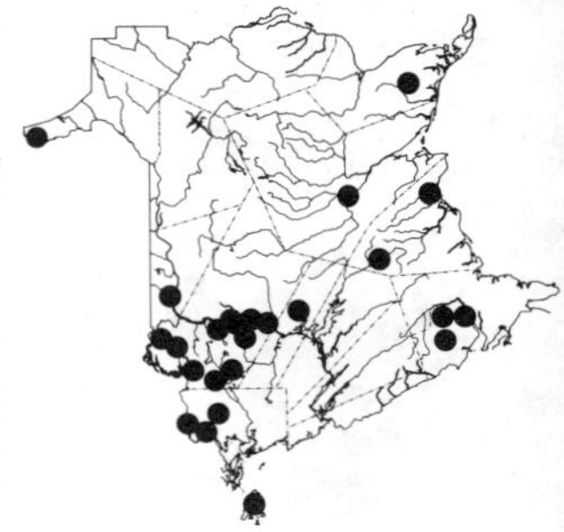
RUBUS ALLEGHENIENSIS

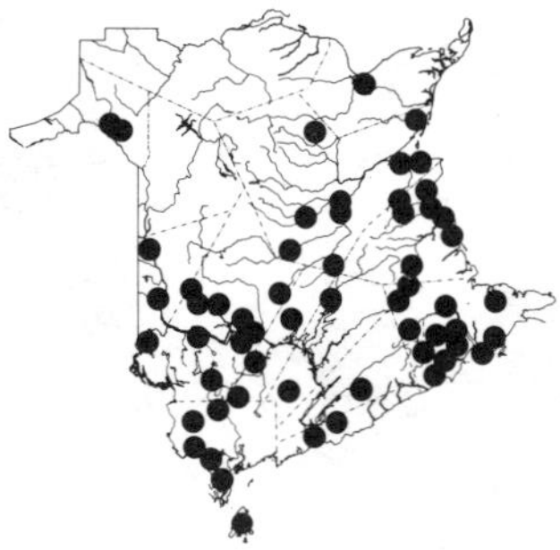
RUBUS CANADENSIS

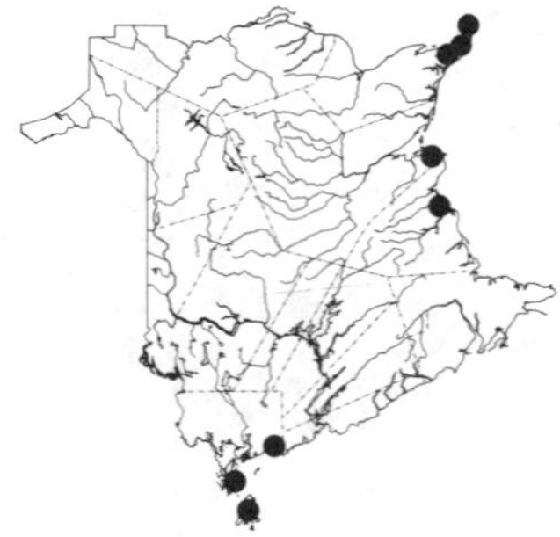
RUBUS CHAMAEMORUS

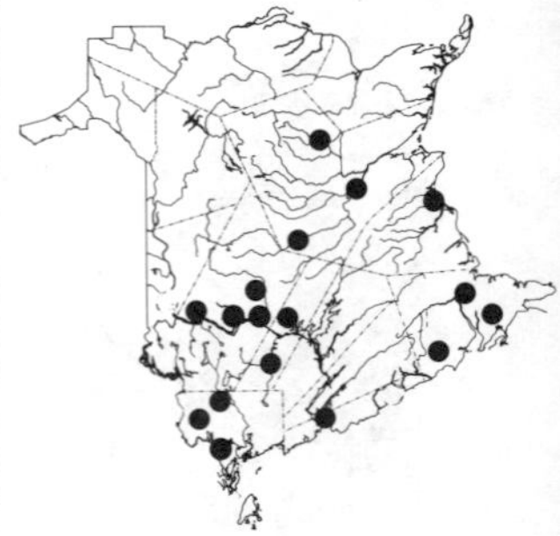
RUBUS ELEGANTULUS

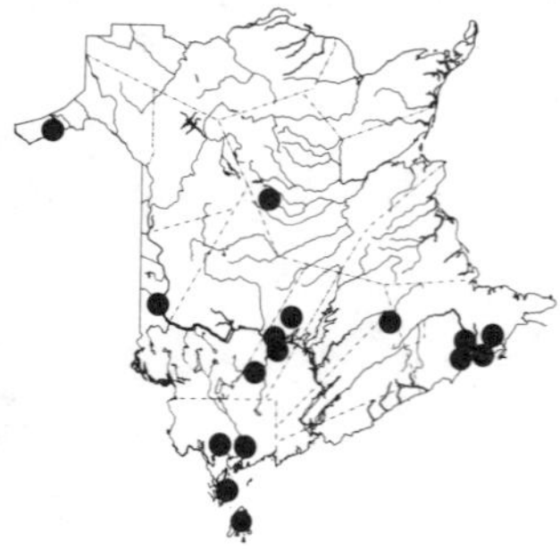
RUBUS HISPIDUS

RUBUS IDAEUS

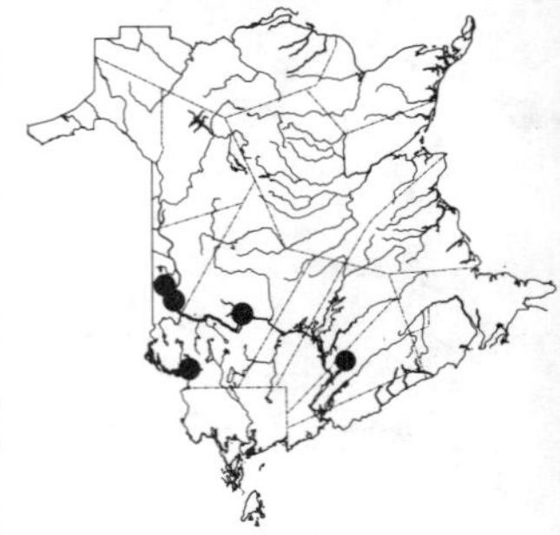
RUBUS OCCIDENTALIS

RUBUS ODORATUS

RUBUS PENSILVANICUS

RUBUS PUBESCENS

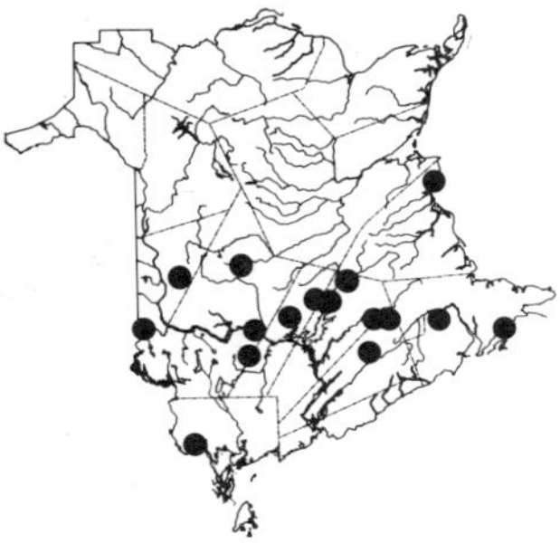

RUBUS SETOSUS

RUBUS STRIGOSUS

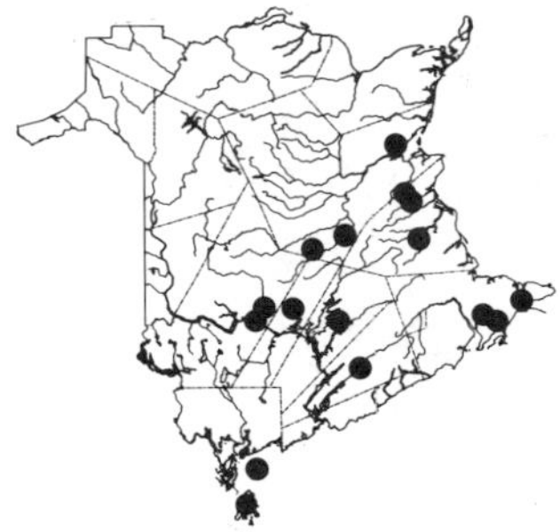

RUBUS VERMONTANUS

SANGUISORBA CANADENSIS

SORBARIA SORBIFOLIA

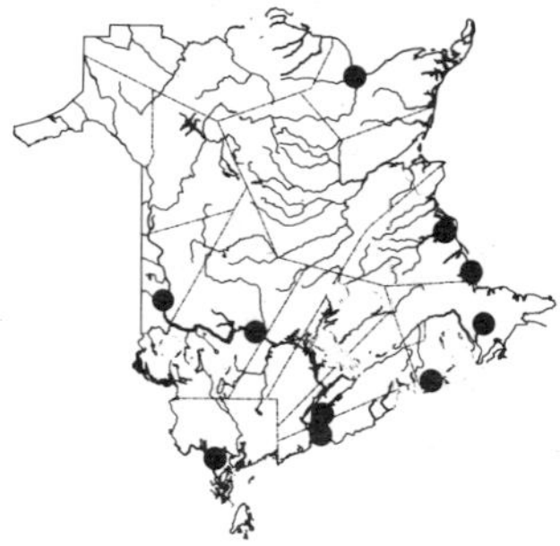

SORBUS AUCUPARIA

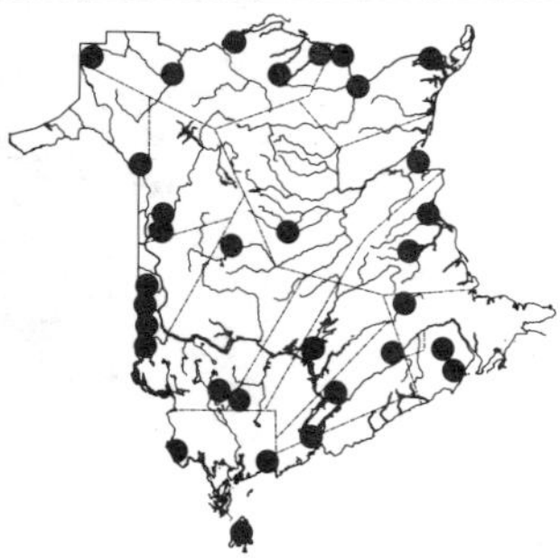

SORBUS DECORA

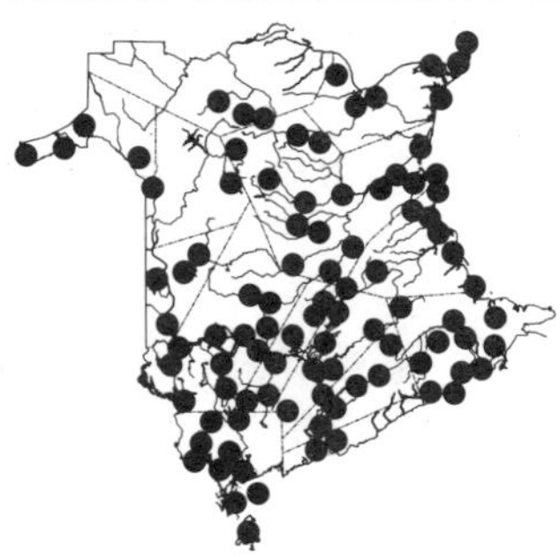

SPIRAEA LATIFOLIA

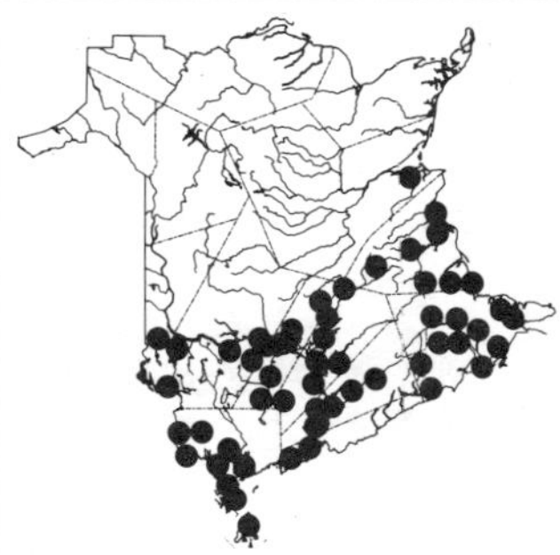

SPIRAEA TOMENTOSA

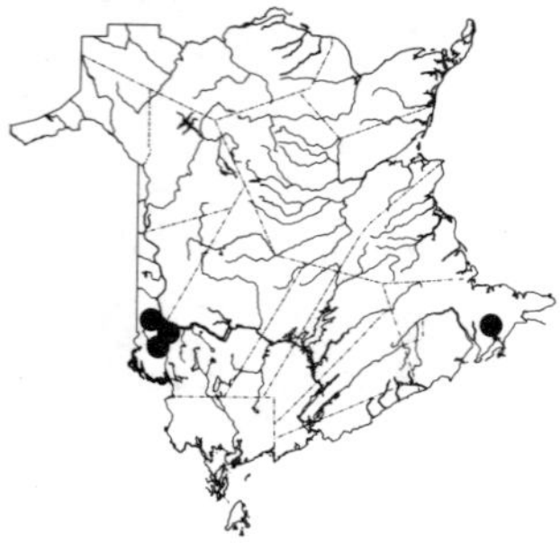

WALDSTEINIA FRAGARIOIDES

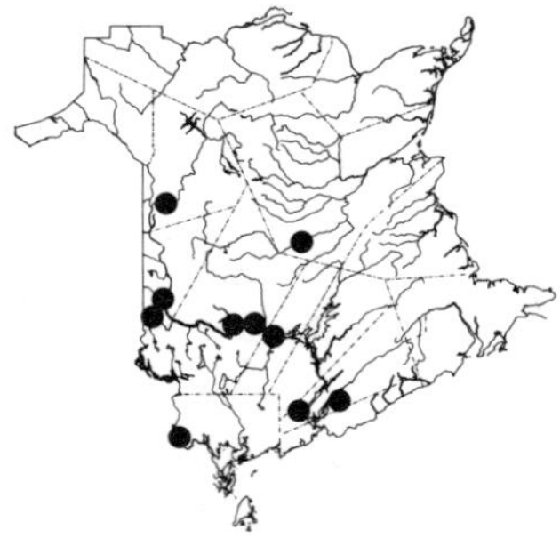

AMPHICARPAEA BRACTEATA

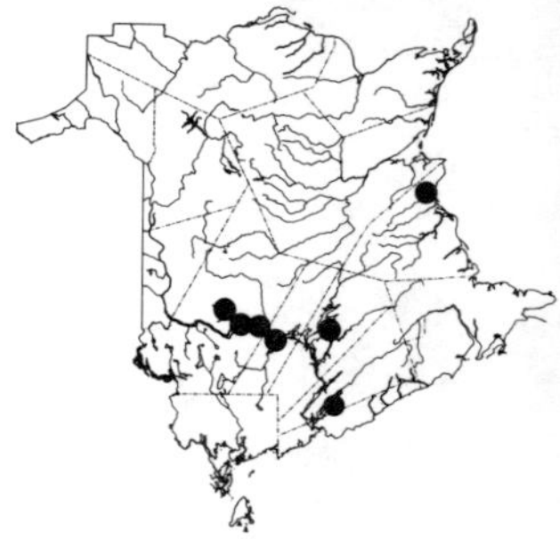

APIOS AMERICANA

ASTRAGALUS ALPINUS
VAR. BRUNETIANUS

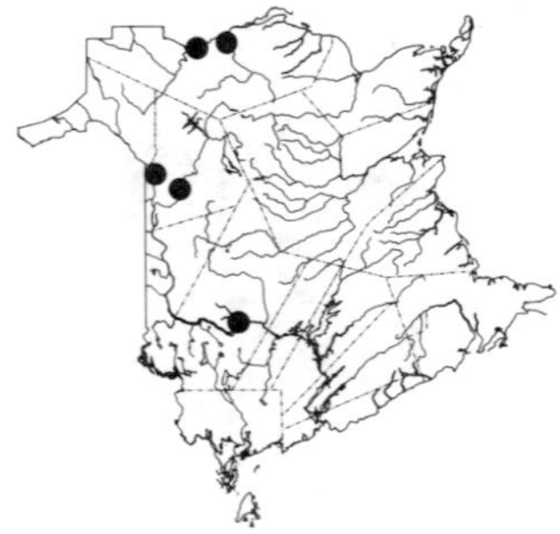

ASTRAGALUS EUCOSMUS

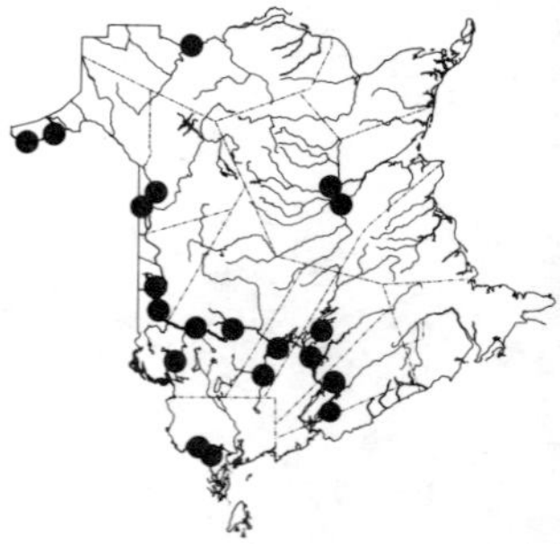

DESMODIUM CANADENSE

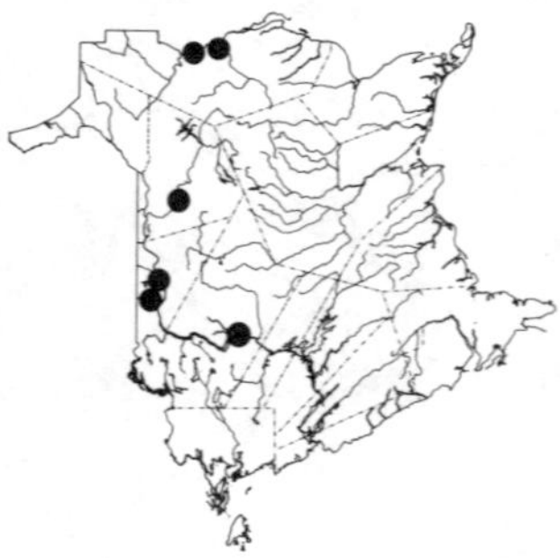

HEDYSARUM ALPINUM

LATHYRUS JAPONICUS

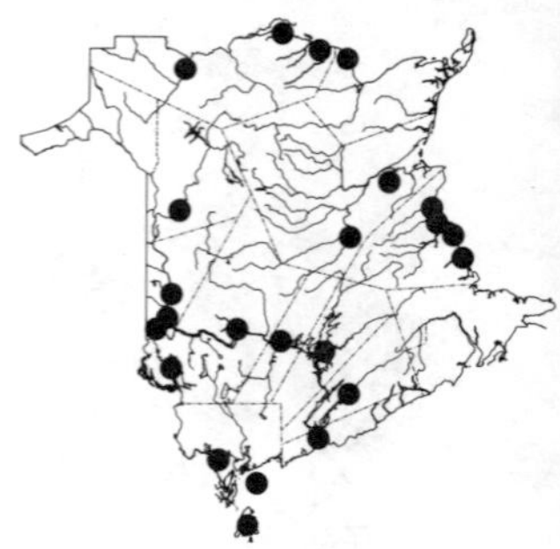

LATHYRUS PALUSTRIS

LATHYRUS PRATENSIS

LUPINUS POLYPHYLLUS

MEDICAGO LUPULINA

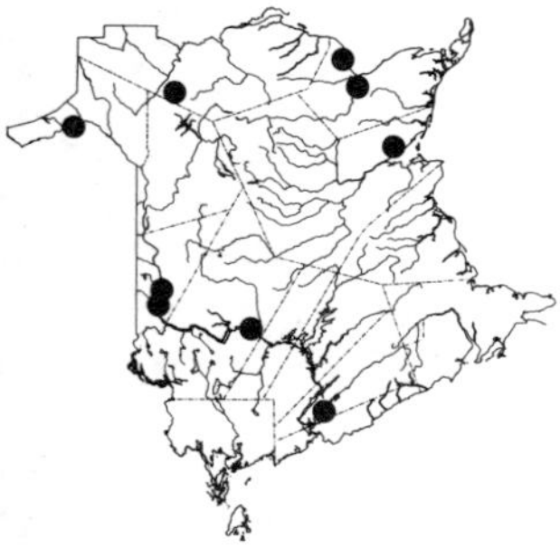

MEDICAGO SATIVA

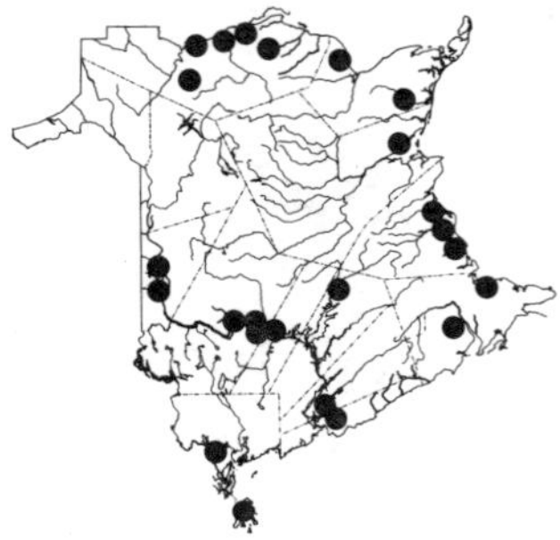

MELILOTUS ALBA

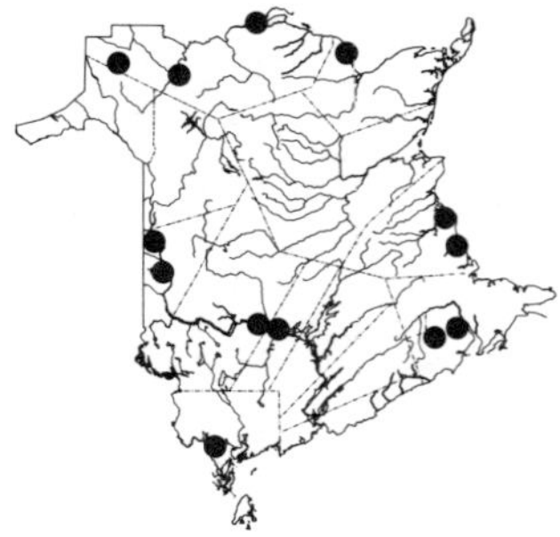

MELILOTUS OFFICINALIS

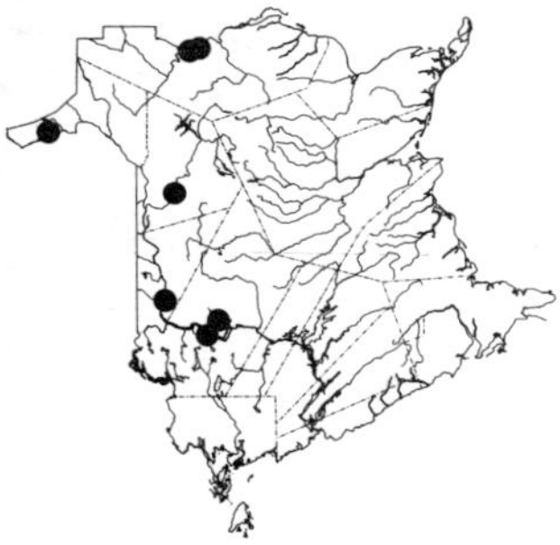

OXYTROPIS CAMPESTRIS VAR. JOHANNENSIS

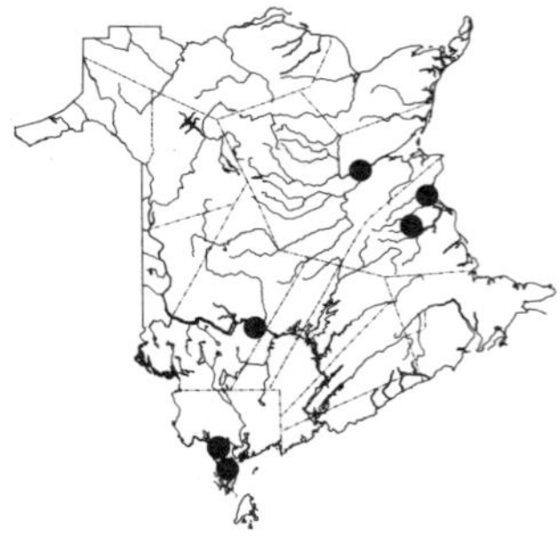

ROBINIA PSEUDO-ACACIA

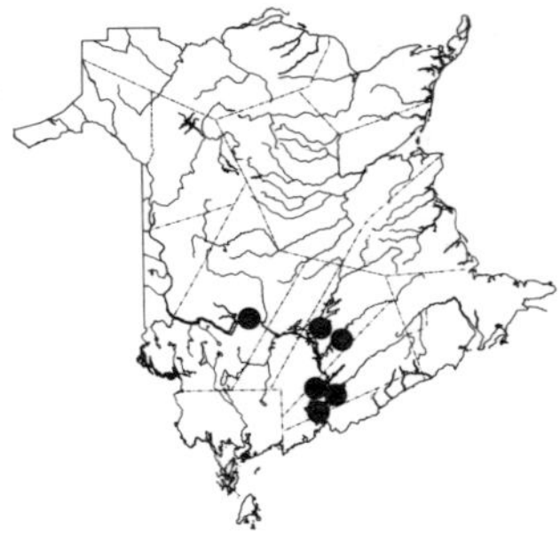

ROBINIA VISCOSA

TRIFOLIUM AUREUM

TRIFOLIUM ARVENSE

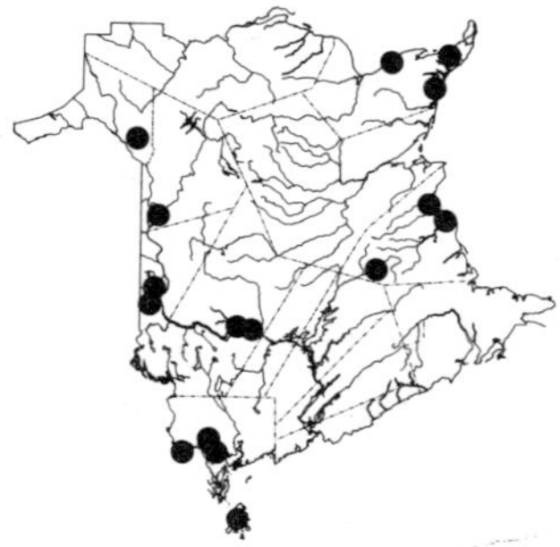

TRIFOLIUM CAMPESTRE

TRIFOLIUM MEDIUM

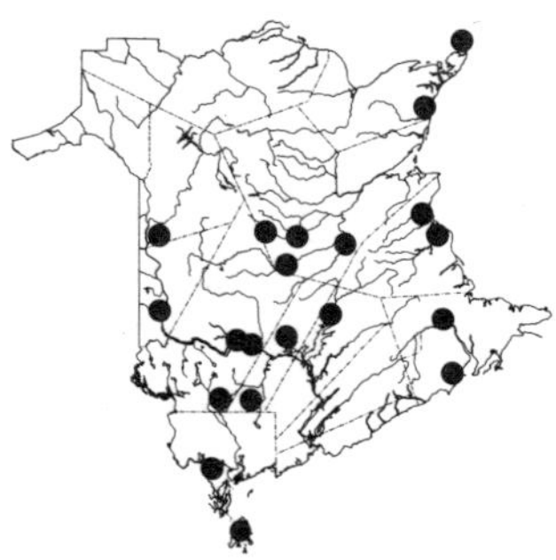

TRIFOLIUM REPENS

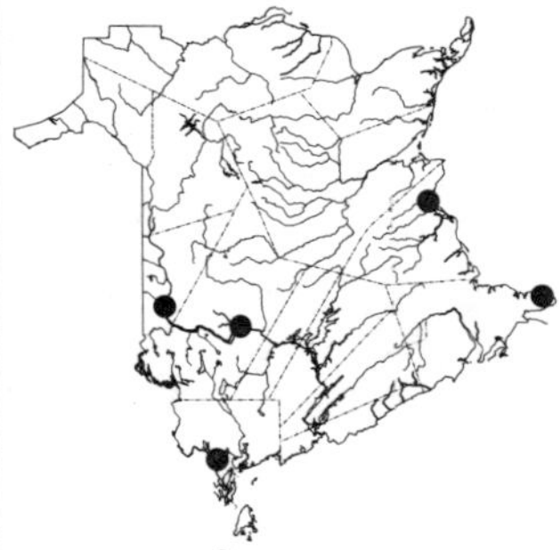

VICIA SATIVA

VICIA SEPIUM

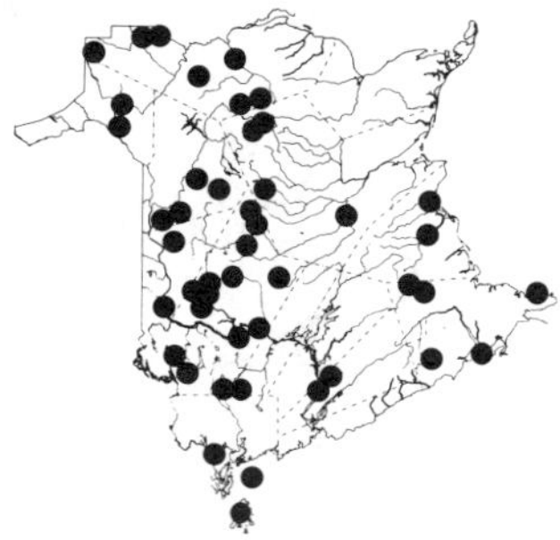

OXALIS MONTANA

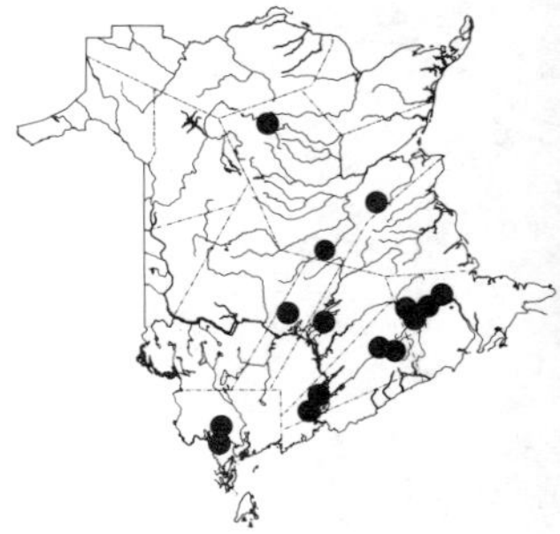

GERANIUM BICKNELLII

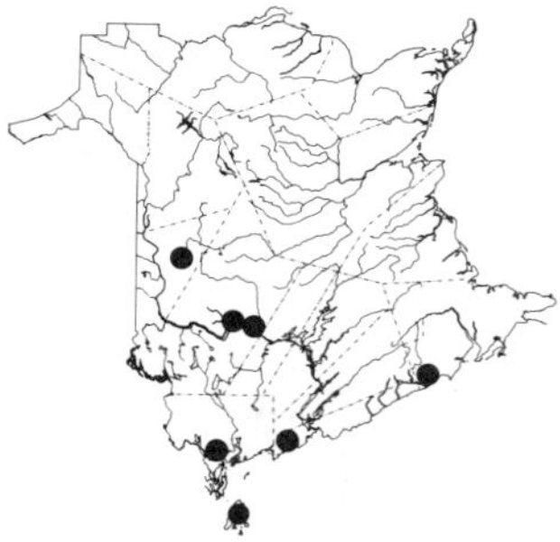

GERANIUM PRATENSE

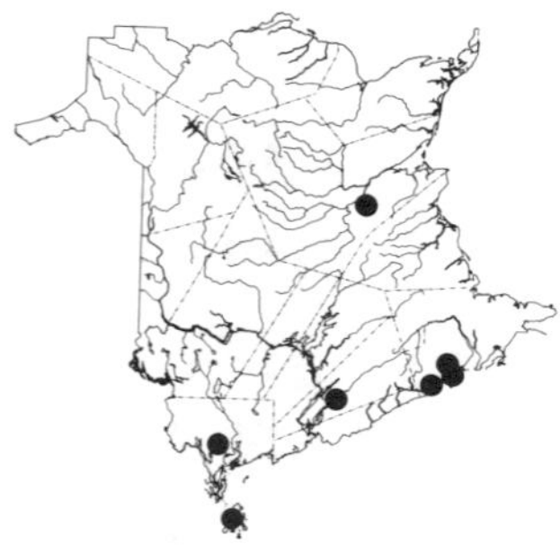

GERANIUM ROBERTIANUM

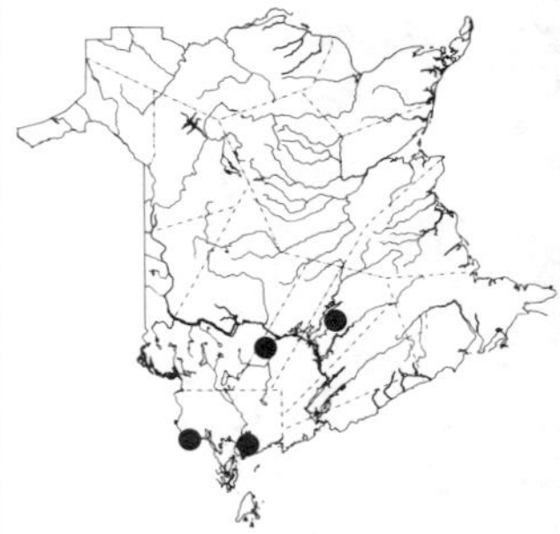

POLYGALA PAUCIFOLIA

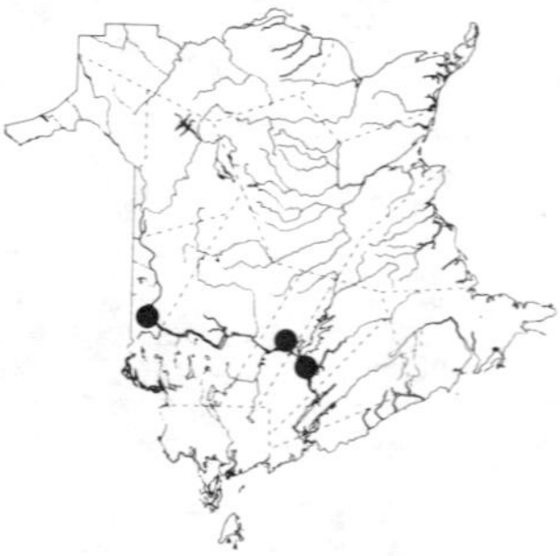

POLYGALA SANGUINEA

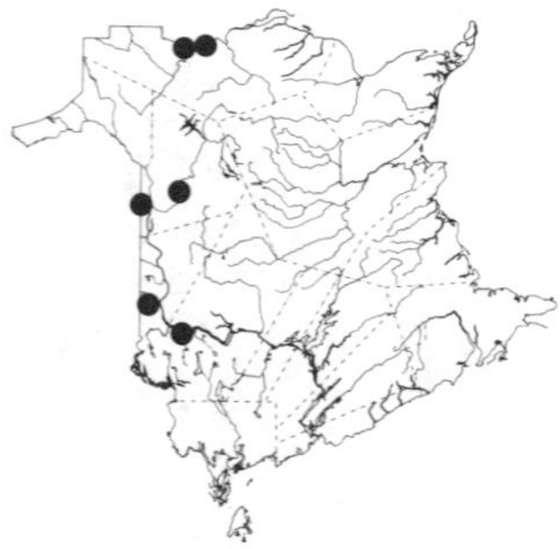

POLYGALA SENEGA

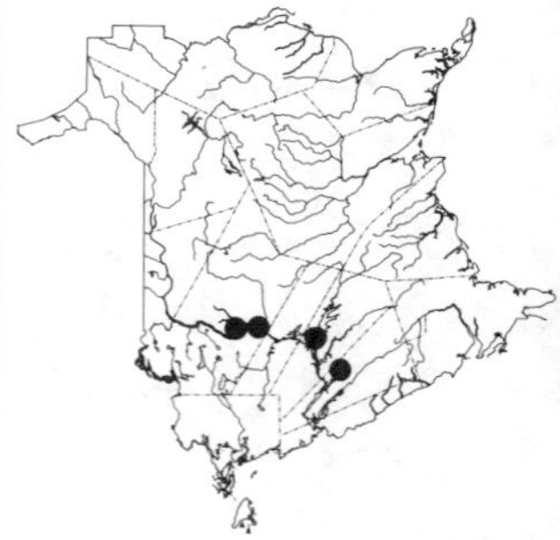

ACALYPHA RHOMBOIDEA

CHAMAESYCE POLYGONIFOLIA

CHAMAESYCE VERMICULATA

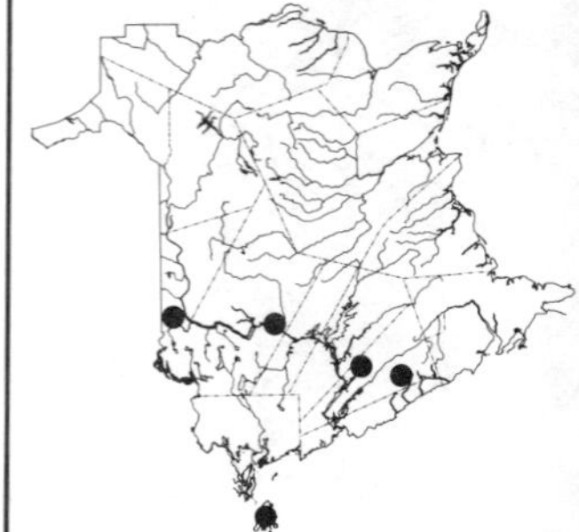

EUPHORBIA CYPARISSIAS

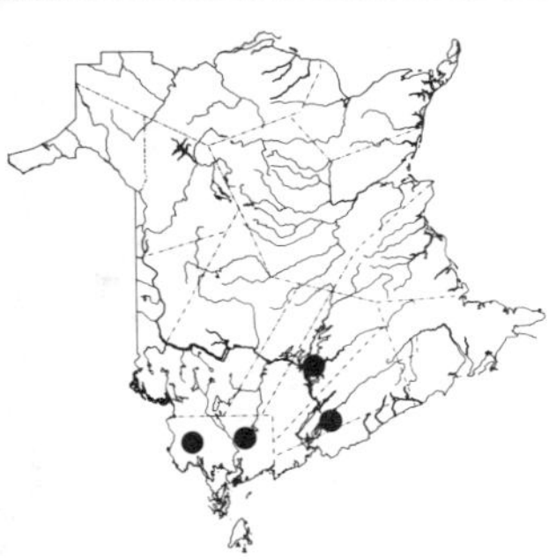

CALLITRICHE HERMAPHRODITICA

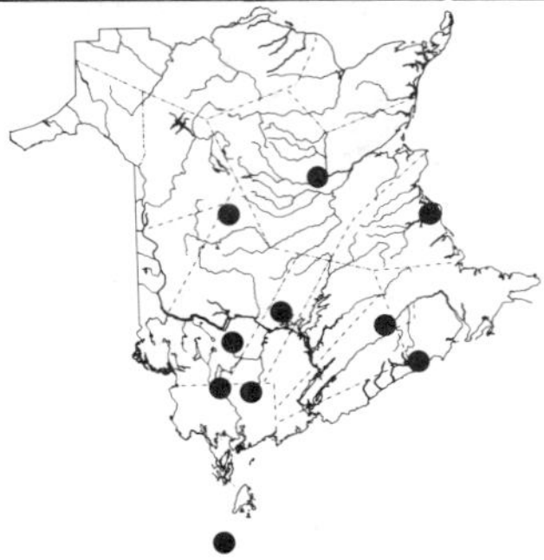

CALLITRICHE HETEROPHYLLA

CALLITRICHE VERNA

EMPETRUM NIGRUM

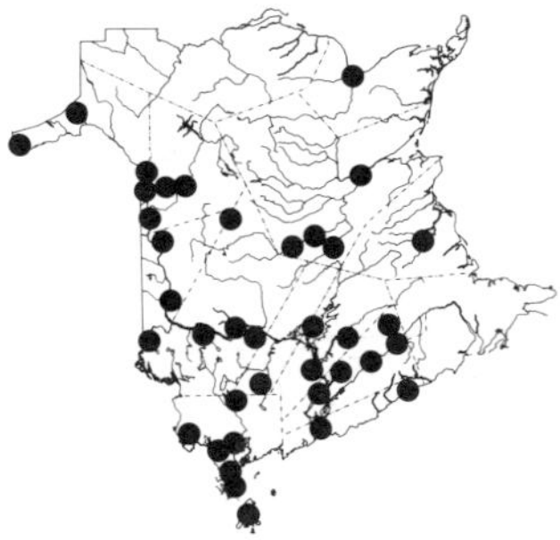

RHUS TYPHINA

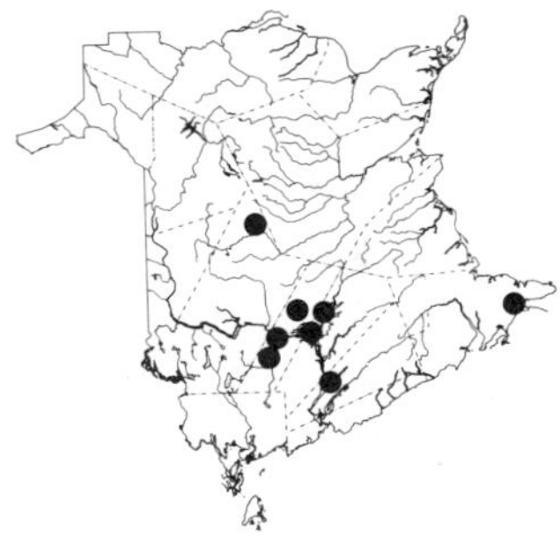

TOXICODENDRON RADICANS

TOXICODENDRON RYDBERGII

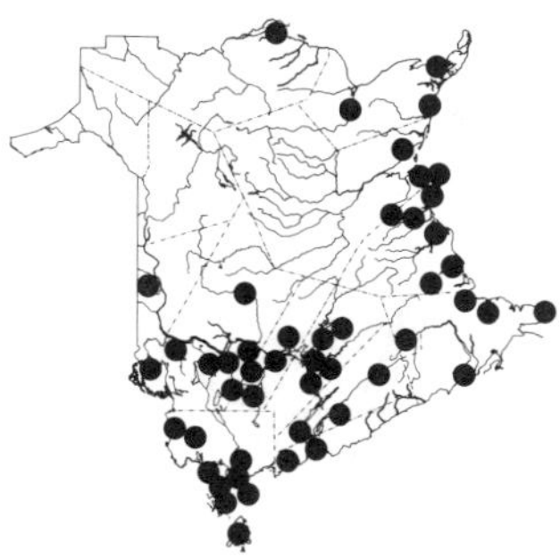

ILEX VERTICILLATA

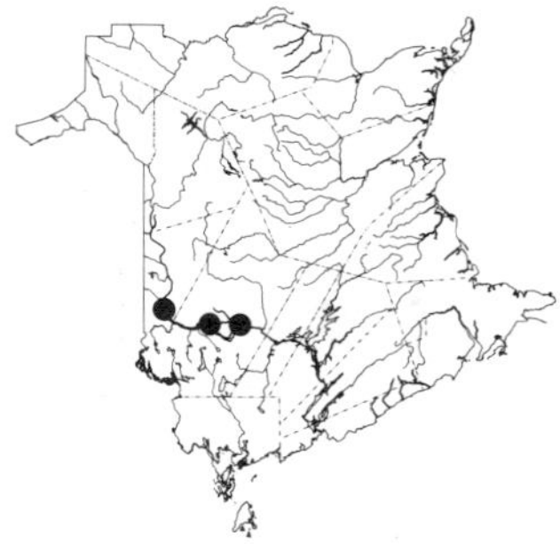

CELASTRUS SCANDENS

ACER NEGUNDO

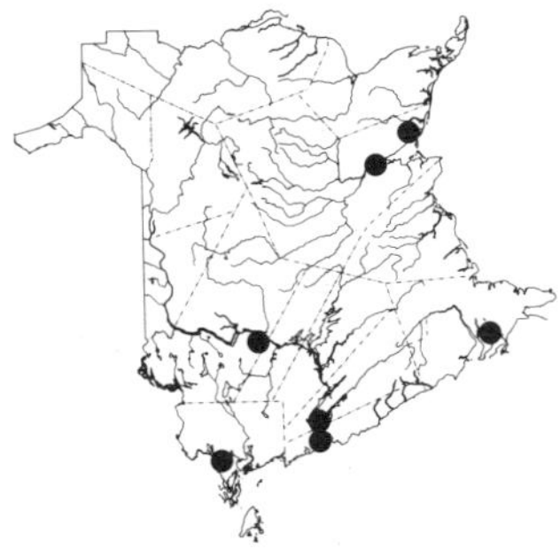

ACER PLATANOIDES

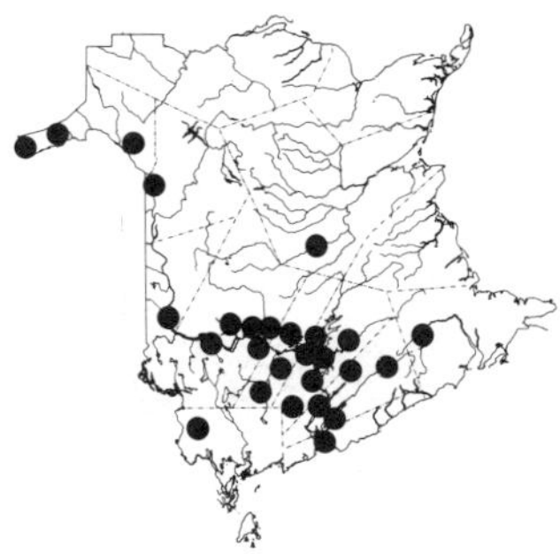

ACER SACCHARINUM

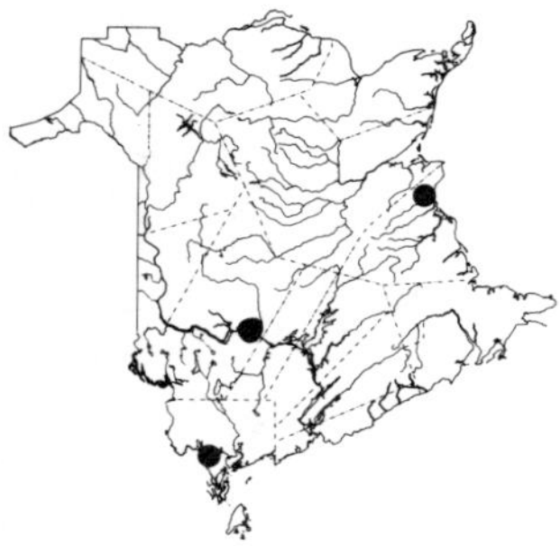

AESCULUS HIPPOCASTANUM

RHAMNUS ALNIFOLIA

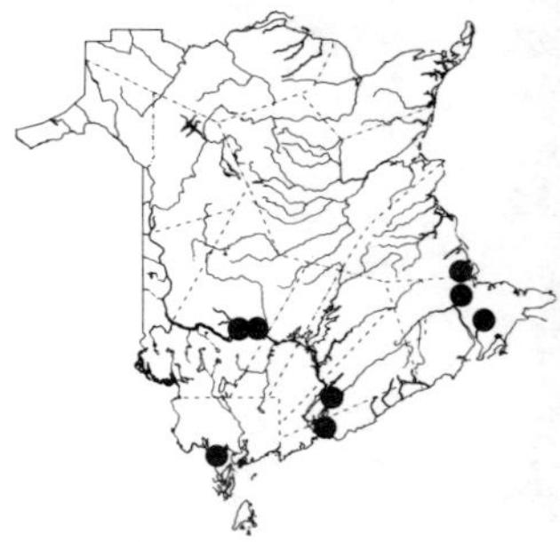

RHAMNUS CATHARTICA

RHAMNUS FRANGULA

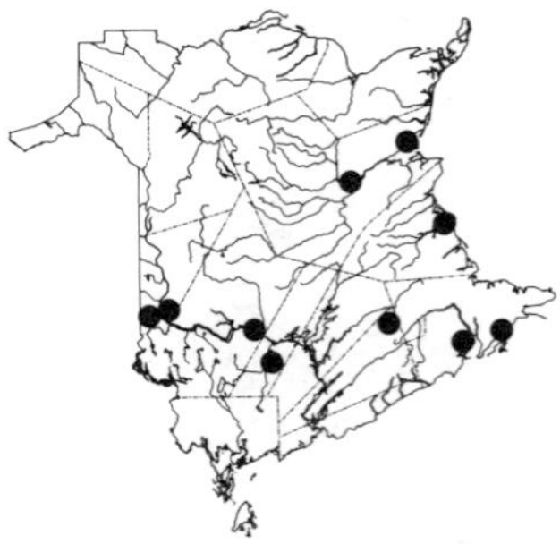

PARTHENOCISSUS INSERTA

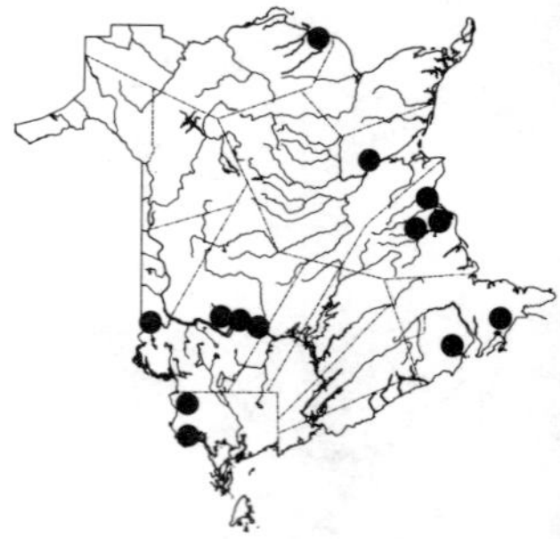

PARTHENOCISSUS QUINQUEFOLIA

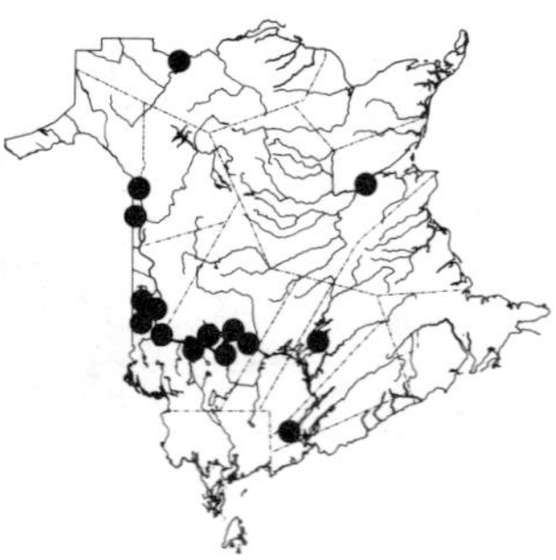

VITIS RIPARIA

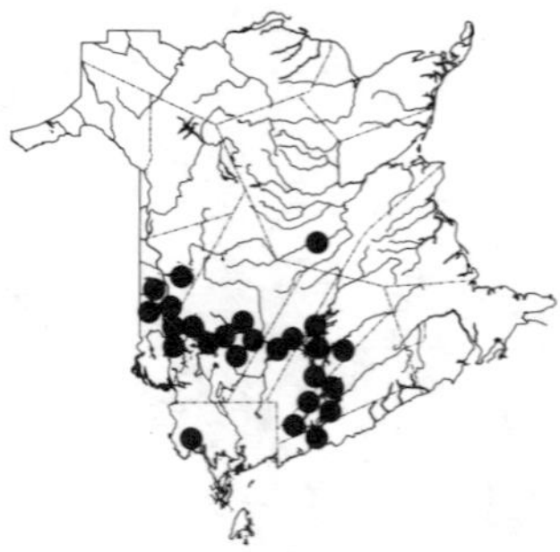

TILIA AMERICANA

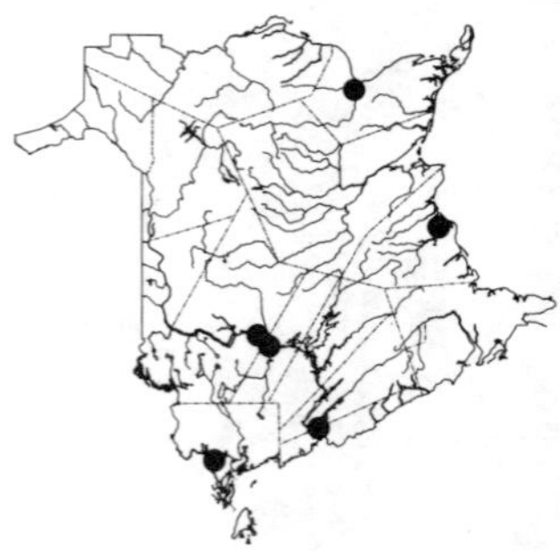

TILIA EUROPAEA

TILIA PLATYPHYLLOS

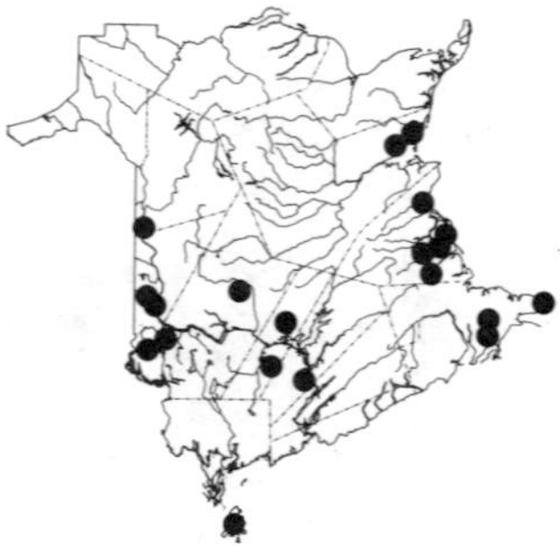

MALVA MOSCHATA

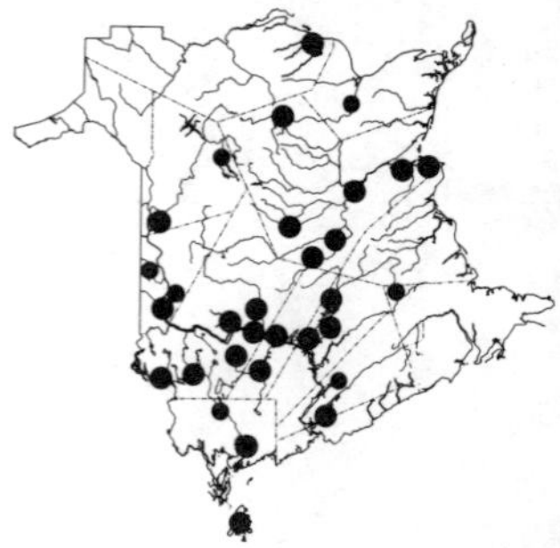

HYPERICUM BOREALE

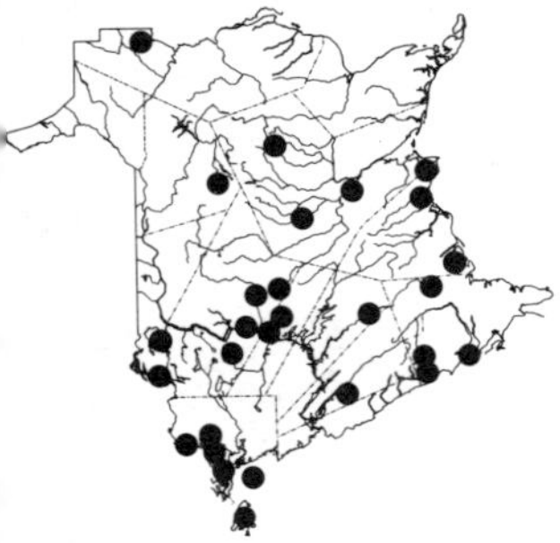

HYPERICUM CANADENSE

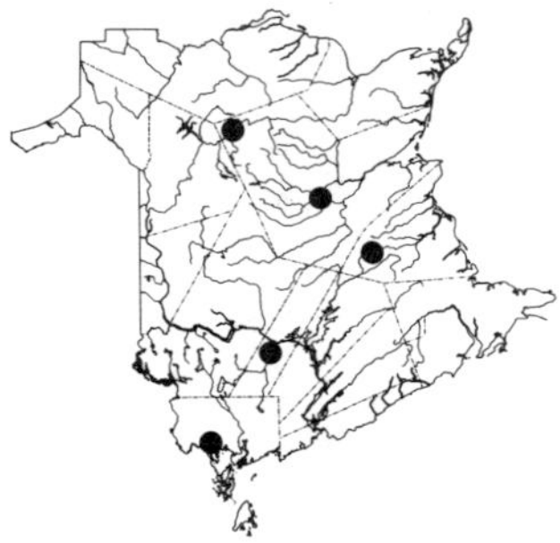

HYPERICUM DISSIMULATUM

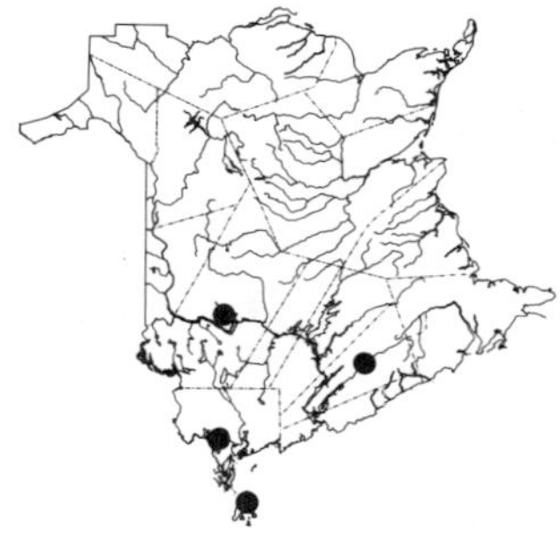

HYPERICUM MAJUS

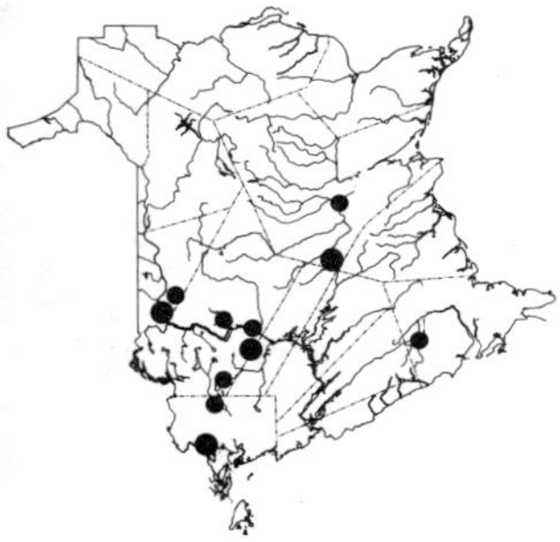

HYPERICUM MUTILUM

HYPERICUM VIRGINICUM

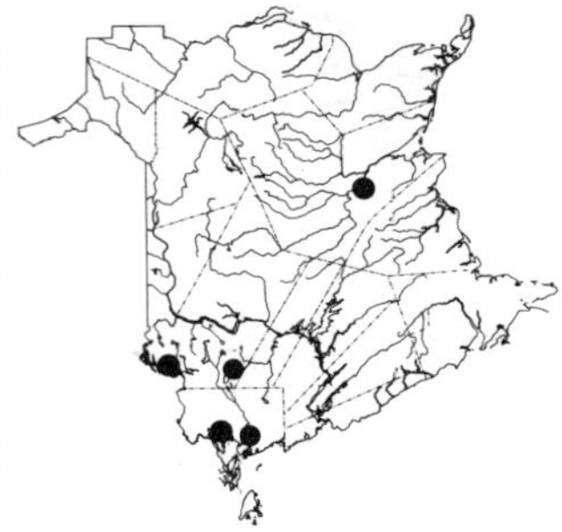

ELATINE MINIMA

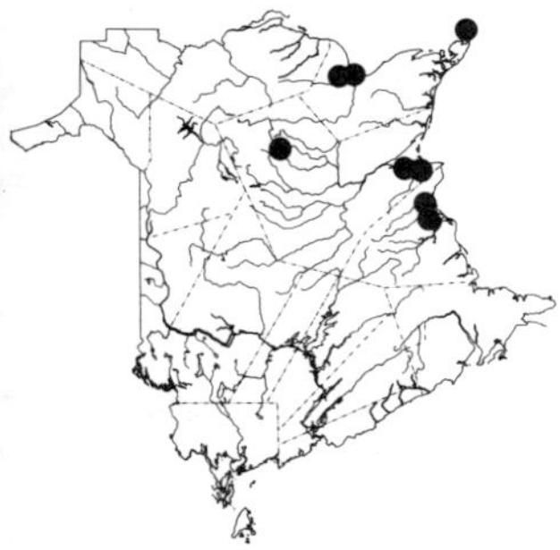

HUDSONIA TOMENTOSA

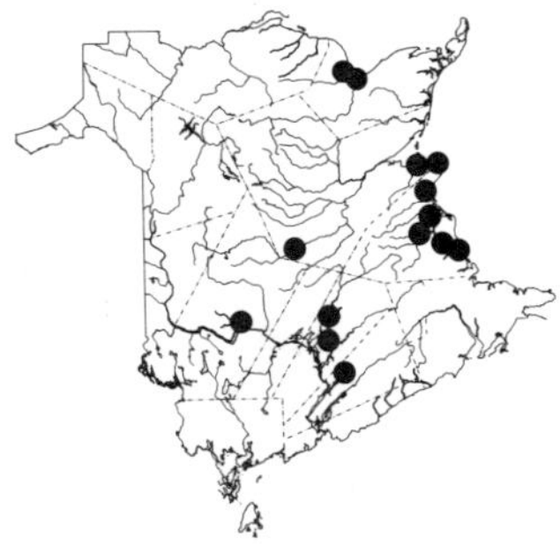

LECHEA INTERMEDIA

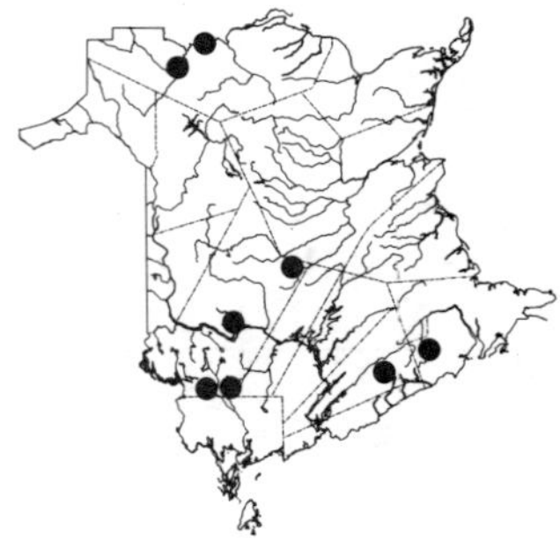

VIOLA ADUNCA

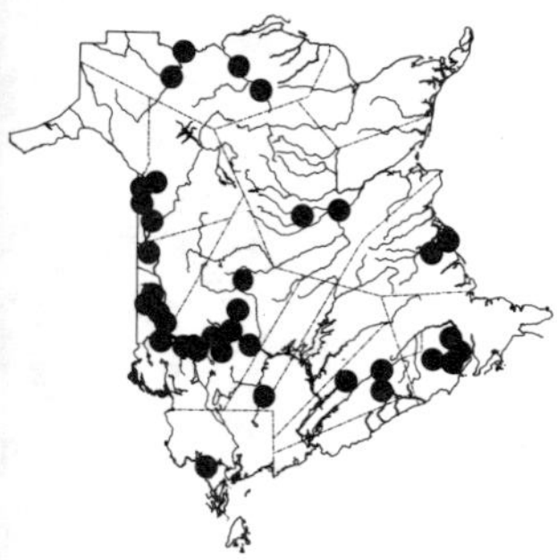

VIOLA CONSPERSA

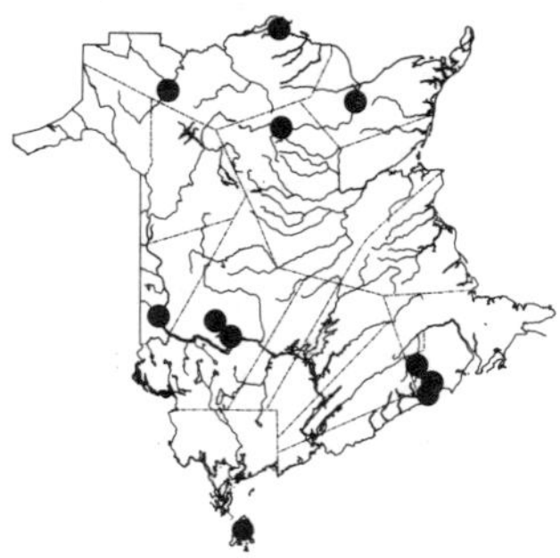

VIOLA LABRADORICA

VIOLA LANCEOLATA

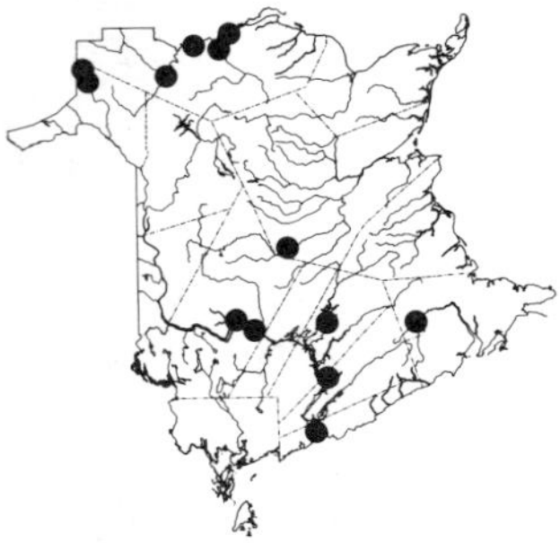

VIOLA NEPHROPHYLLA

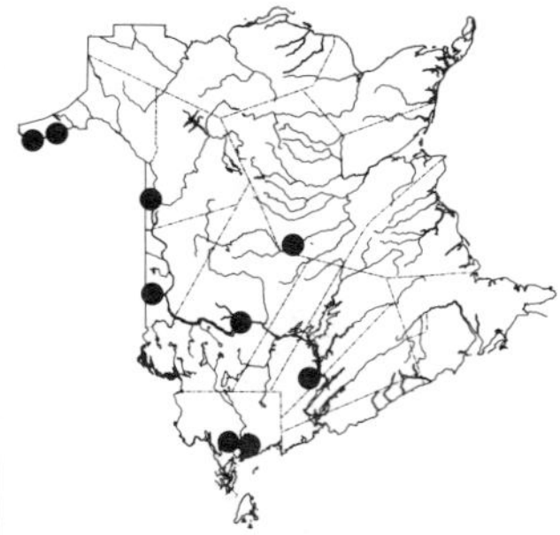

VIOLA NOVAE-ANGLIAE

VIOLA OBLIQUA X V. SEPTENTRIONALIS

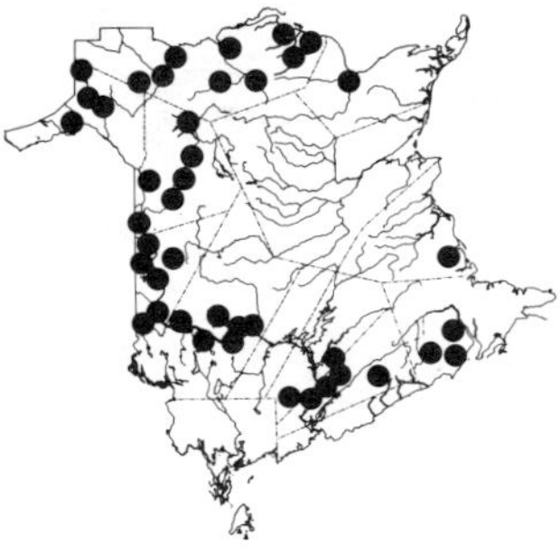

VIOLA PUBESCENS VAR. ERIOCARPA

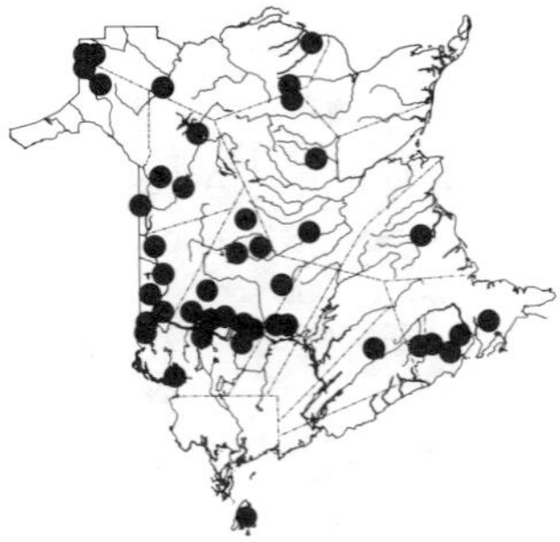

VIOLA RENIFOLIA

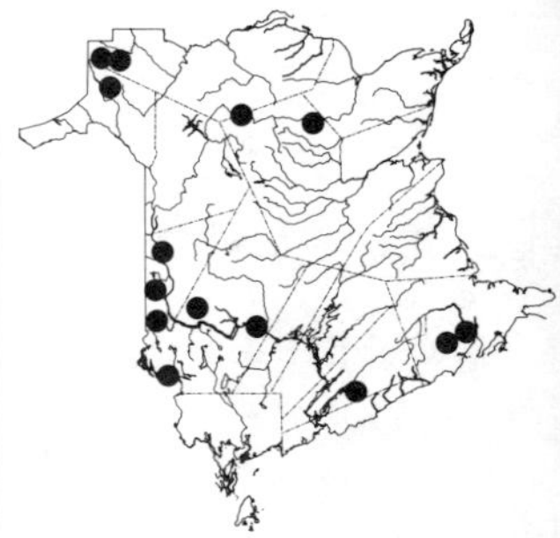

VIOLA SELKIRKII

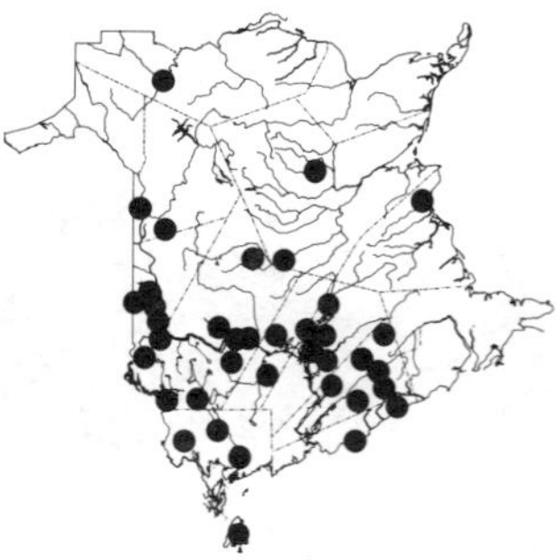

VIOLA SEPTENTRIONALIS

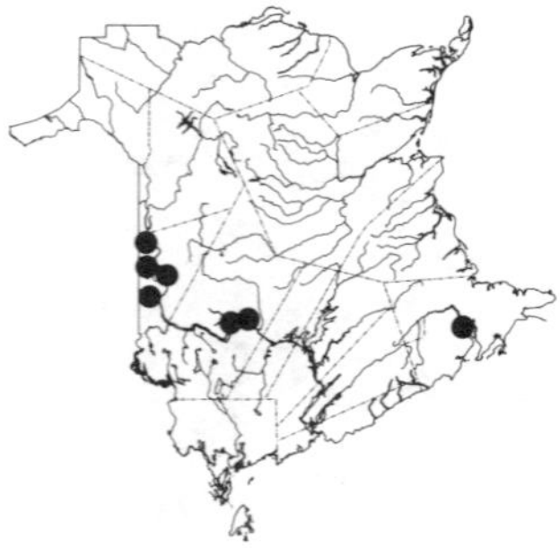

DIRCA PALUSTRIS

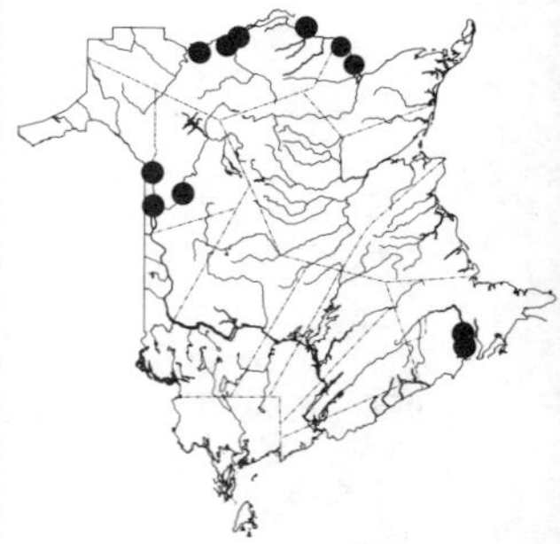

SHEPHERDIA CANADENSIS

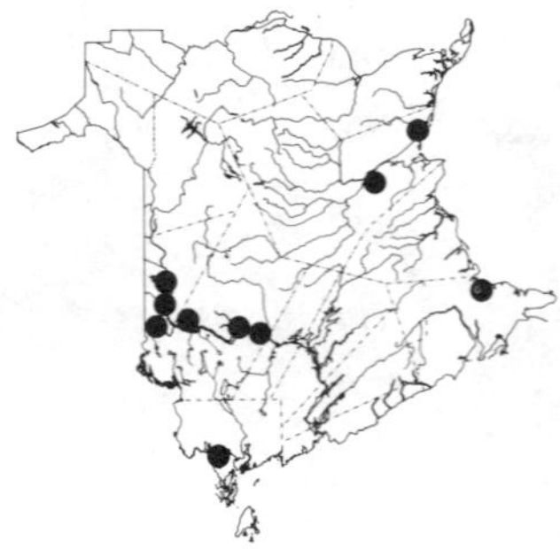

LYTHRUM SALICARIA

CIRCAEA ALPINA

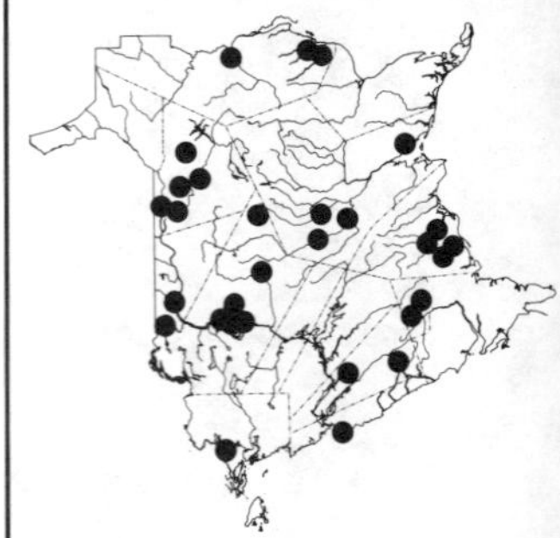

CIRCAEA X INTERMEDIA

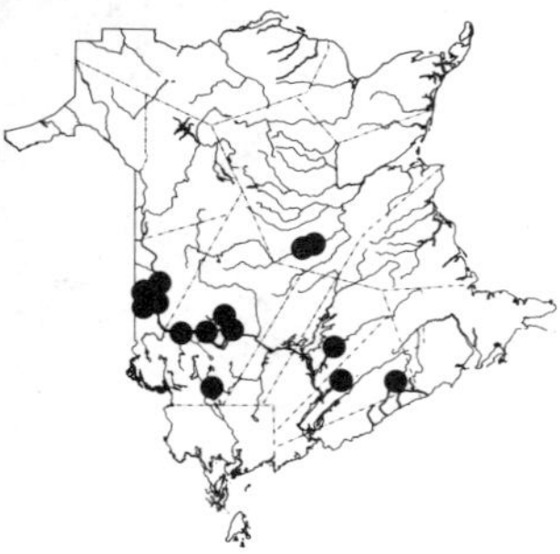

CIRCAEA LUTETIANA SSP. CANADENSIS

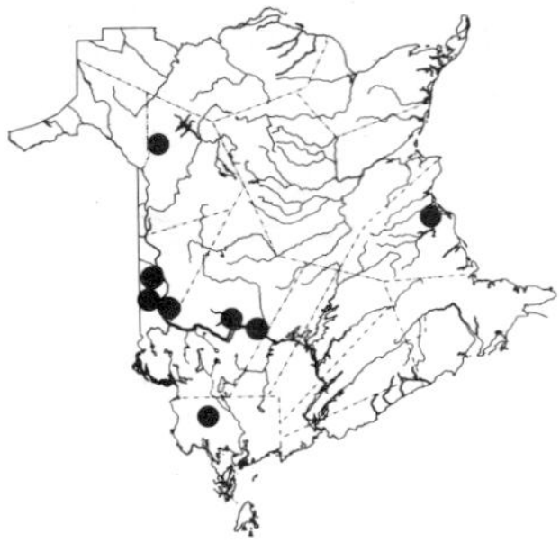

EPILOBIUM COLORATUM

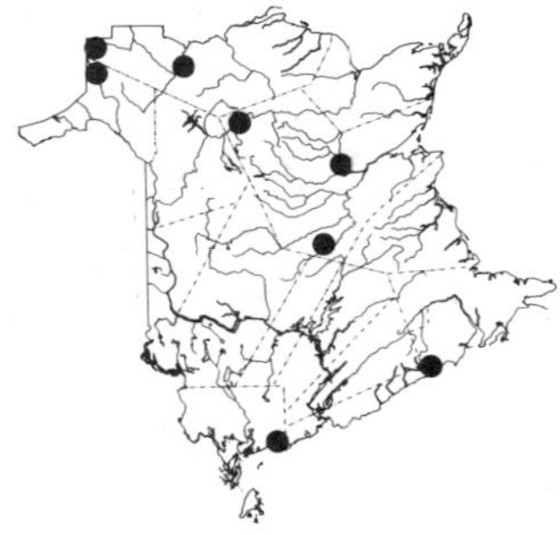

EPILOBIUM HORNMANNII

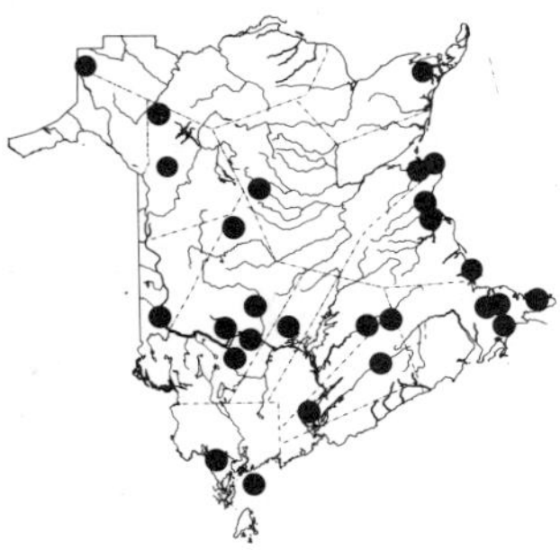

EPILOBIUM LEPTOPHYLLUM

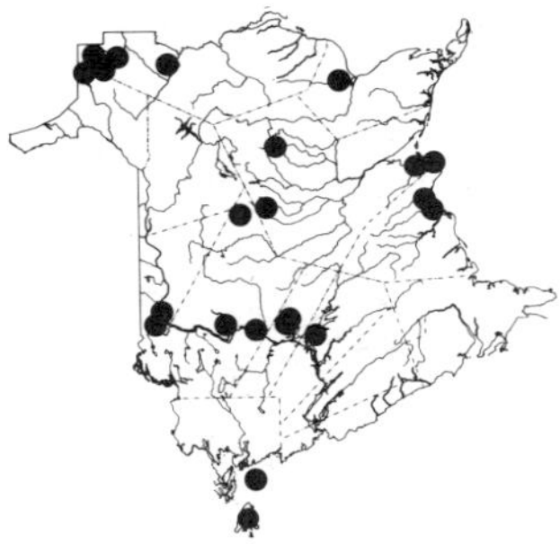

EPILOBIUM PALUSTRE

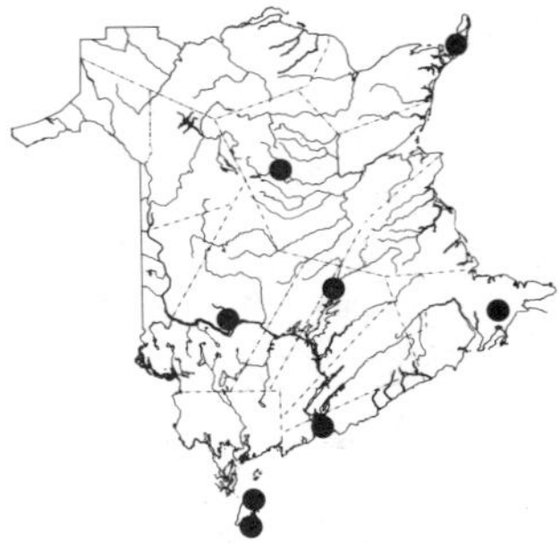

EPILOBIUM STRICTUM

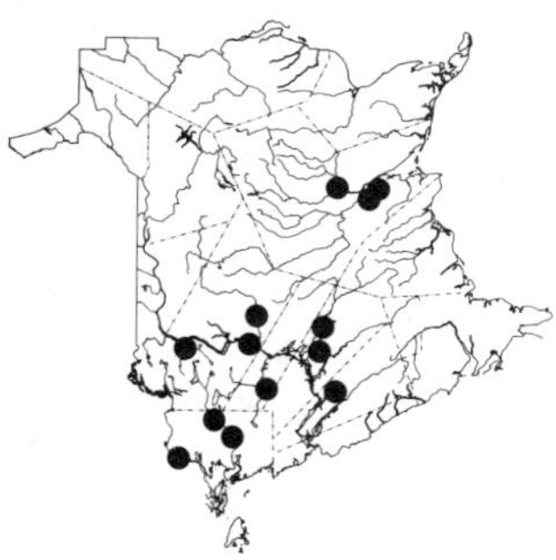

LUDWIGIA PALUSTRIS

OENOTHERA PARVIFLORA

OENOTHERA PERENNIS

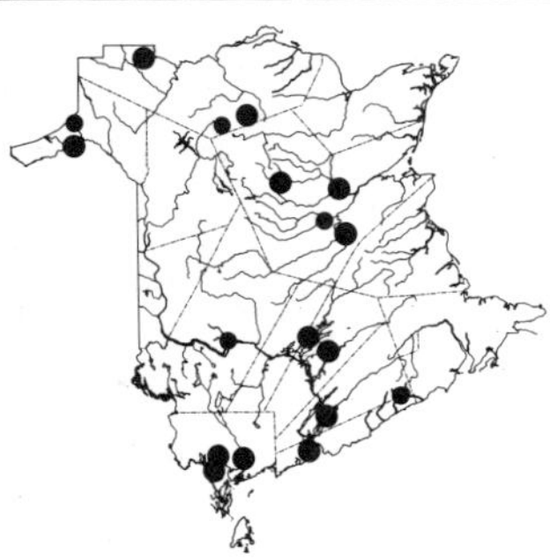

MYRIOPHYLLUM ALTERNIFLORUM

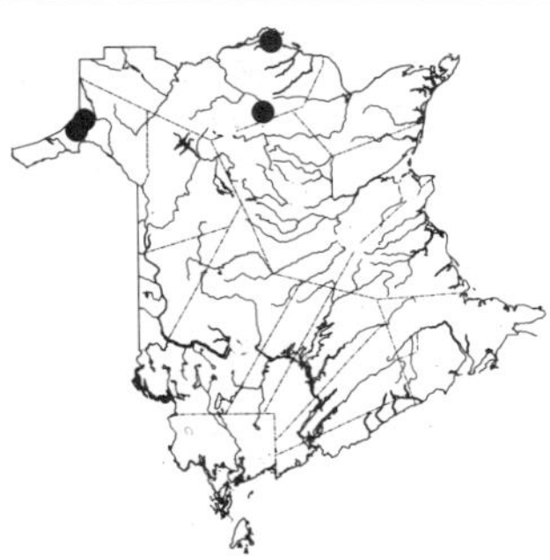

MYRIOPHYLLUM EXALBESCENS

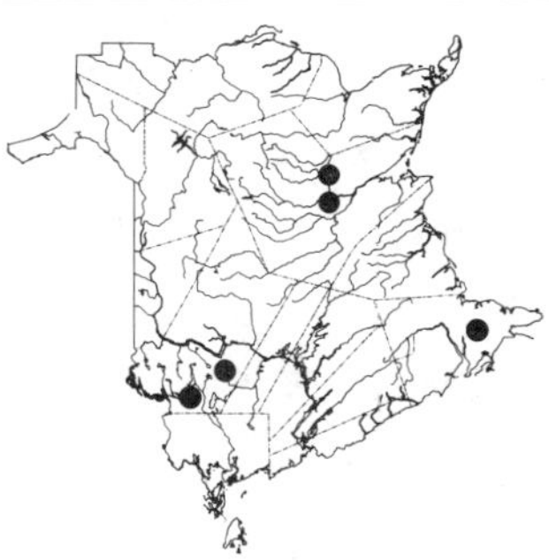

MYRIOPHYLLUM FARWELLII

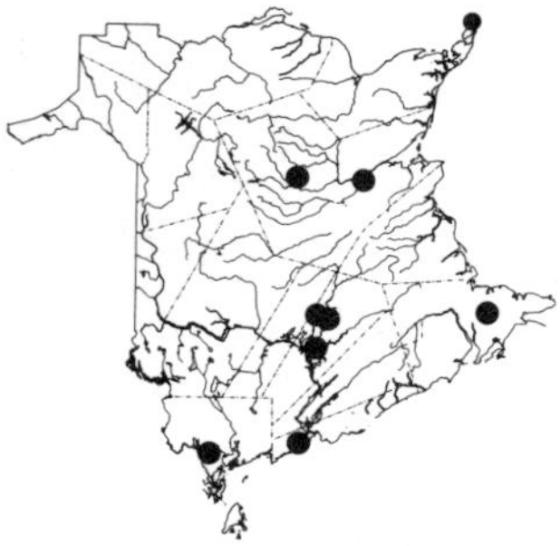

MYRIOPHYLLUM VERTICILLATUM

MYRIOPHYLLUM TENELLUM

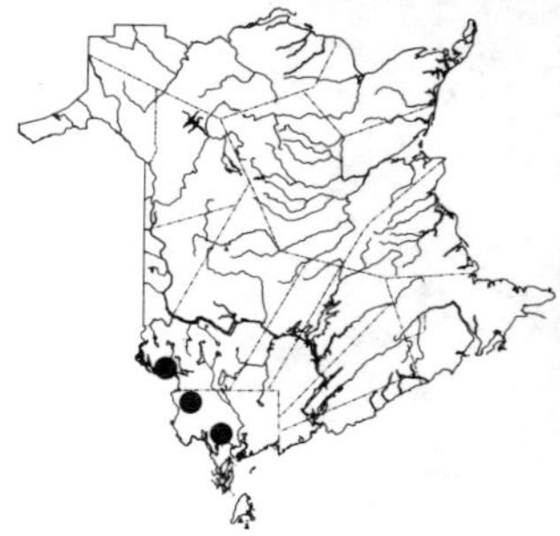

PROSERPINACA PALUSTRIS

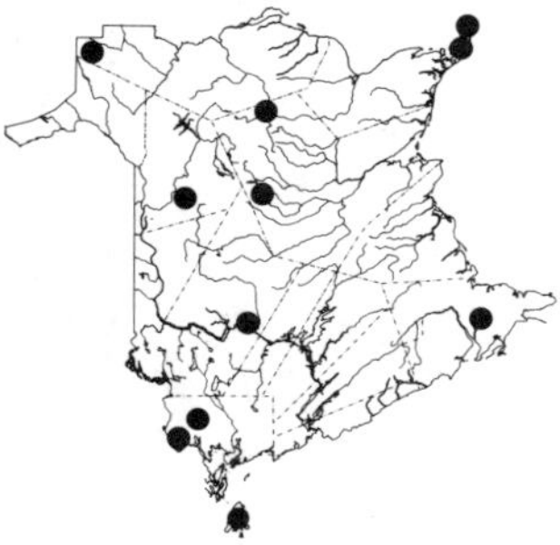

HIPPURIS VULGARIS

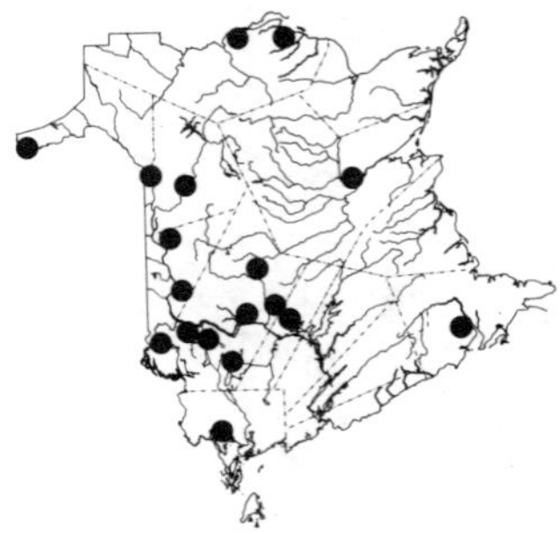

ARALIA RACEMOSA

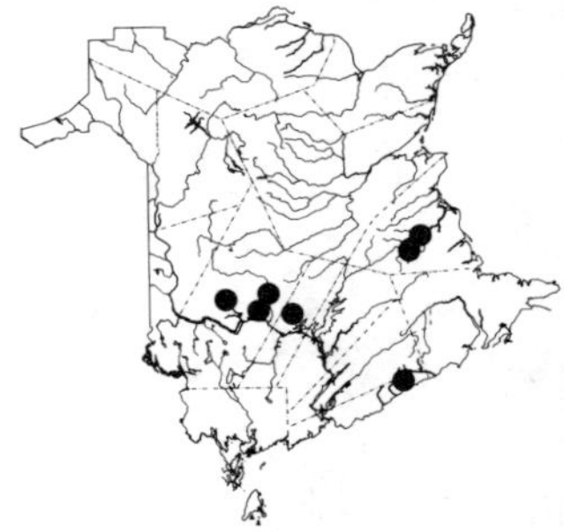

PANAX TRIFOLIUS

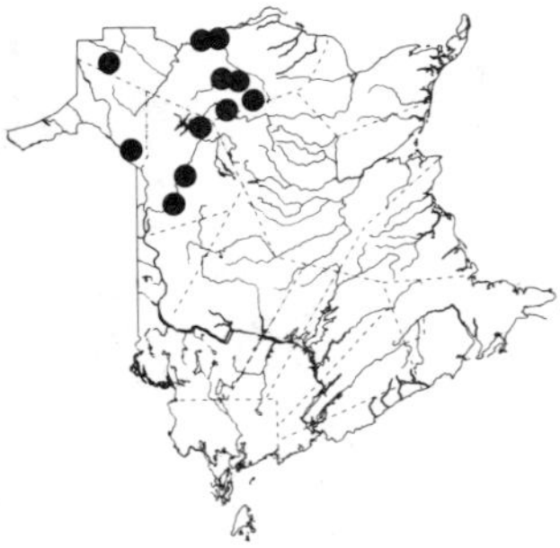

ANGELICA ATROPURPUREA

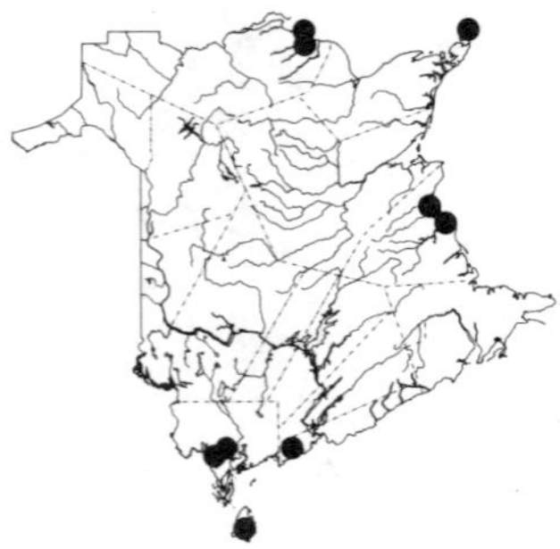

ANGELICA LUCIDA

CICUTA BULBIFERA

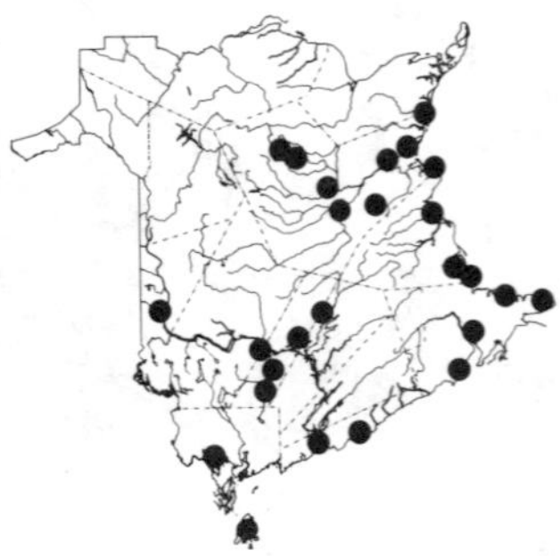

CICUTA MACULATA

CONIOSELINUM CHINENSE

HERACLEUM LANATUM

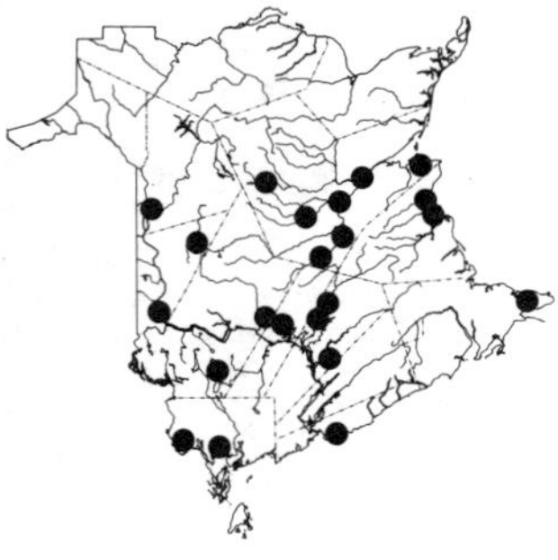

HYDROCOTYLE AMERICANA

LIGUSTICUM SCOTHICUM

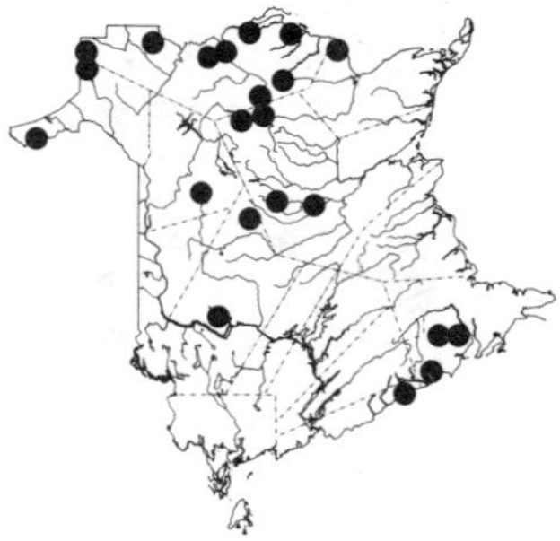

OSMORHIZA CHILENSIS

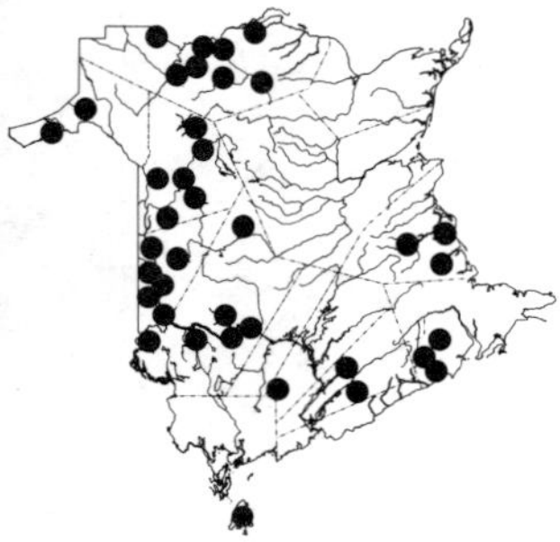

OSMORHIZA CLAYTONII

OSMORHIZA DEPAUPERATA

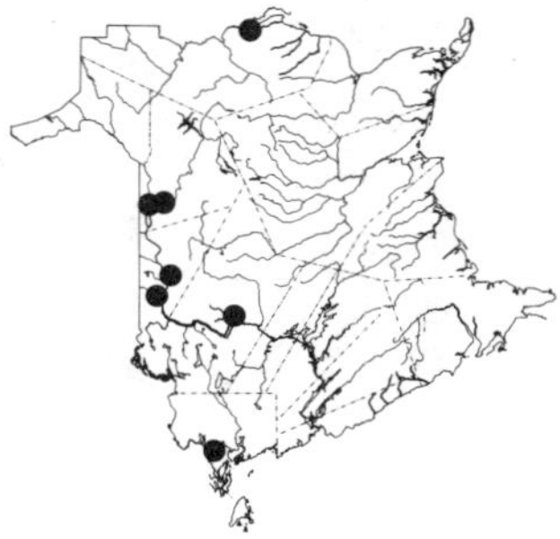

OSMORHIZA LONGISTYLIS

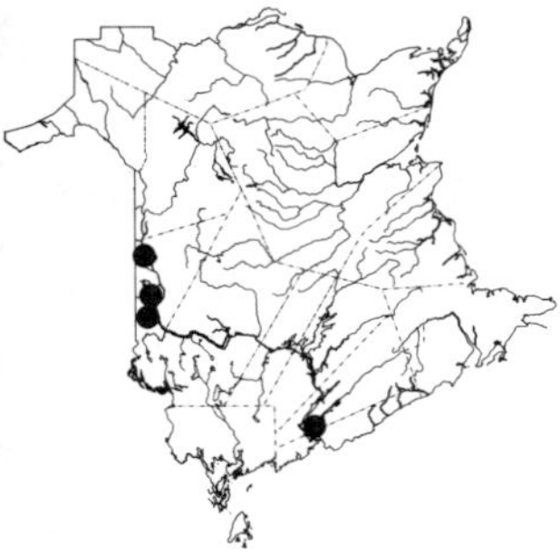

SANICULA GREGARIA

SANICULA MARILANDICA

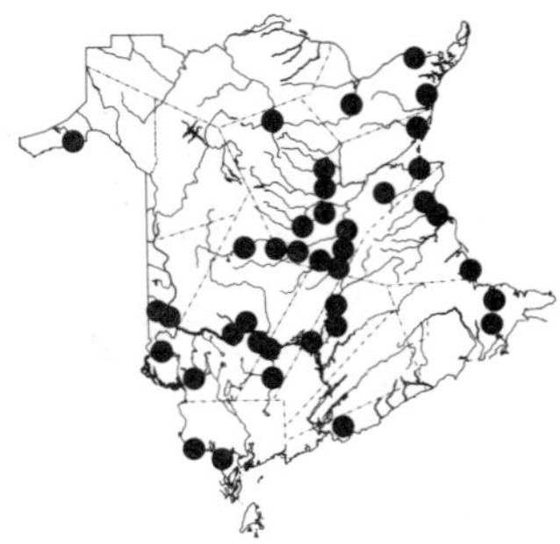

SIUM SUAVE

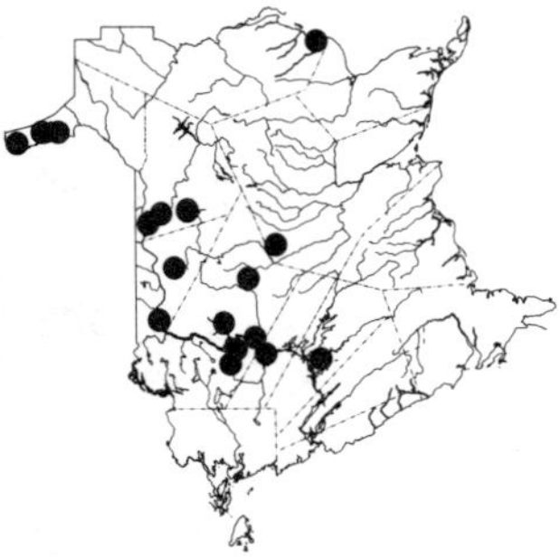

ZIZIA AUREA

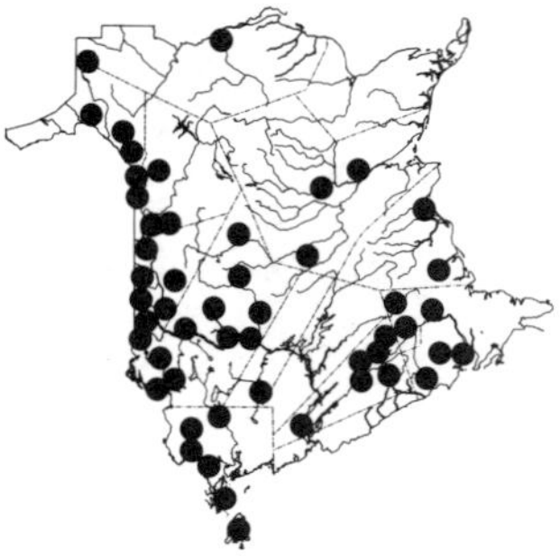

CORNUS ALTERNIFOLIA

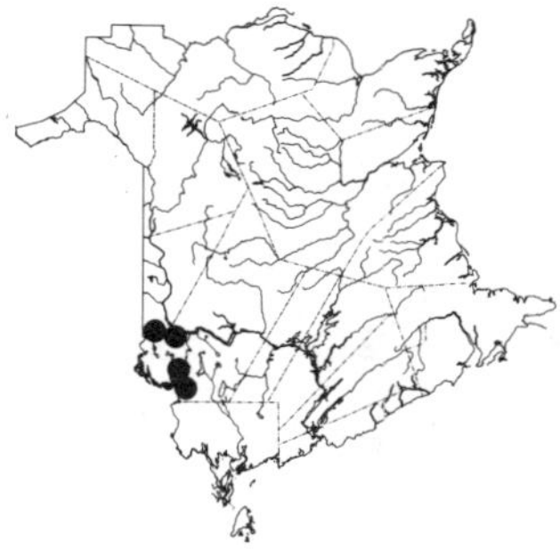

CORNUS AMOMUM

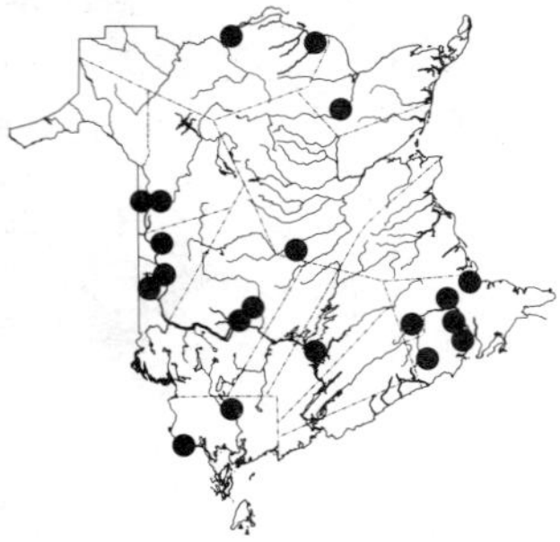

CORNUS RUGOSA

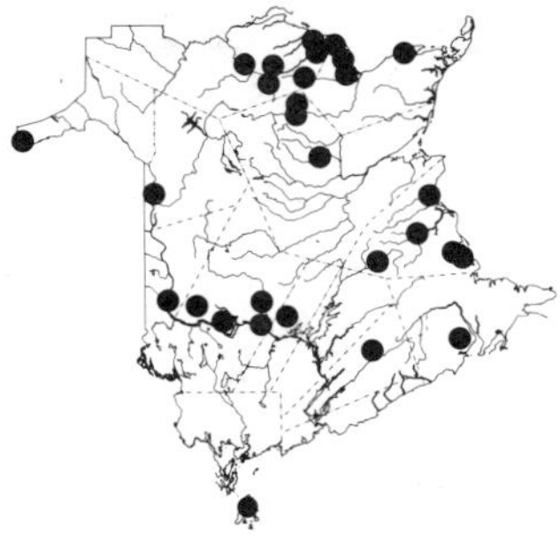

CHIMAPHILA UMBELLATA

MONESES UNIFLORA

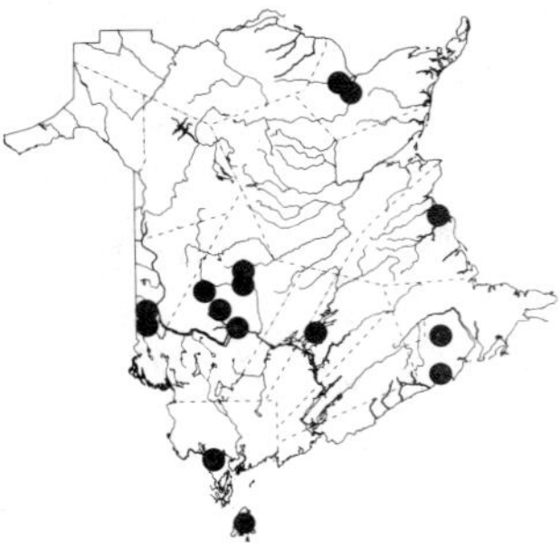

MONOTROPA HYPOPITYS

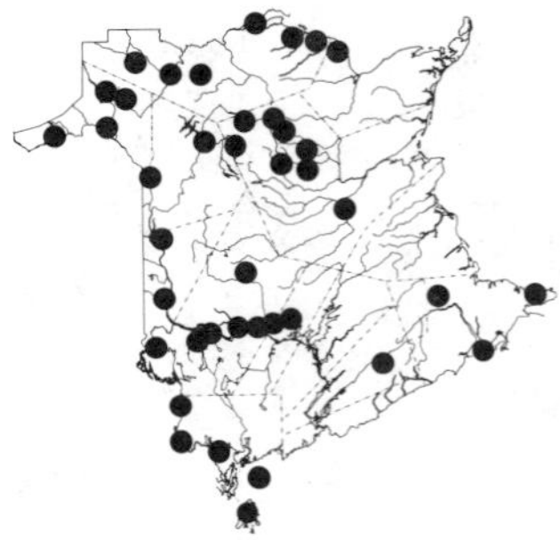

MONOTROPA UNIFLORA

ORTHILIA SECUNDA

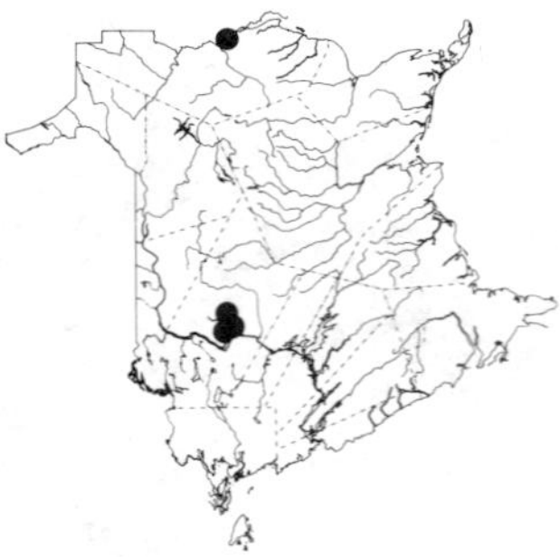

PTEROSPORA ANDROMEDEA

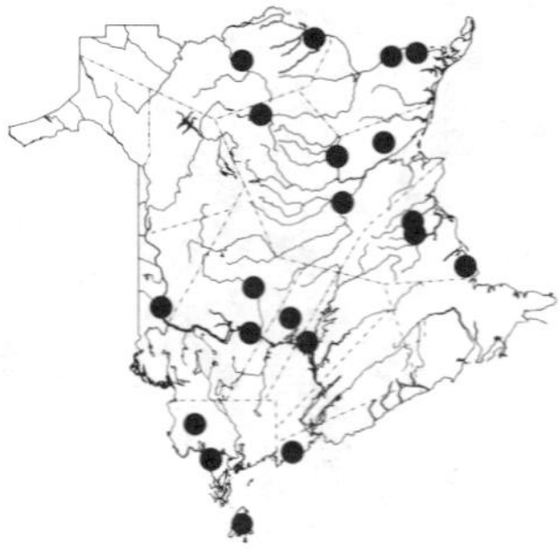

PYROLA AMERICANA

PYROLA ASARIFOLIA

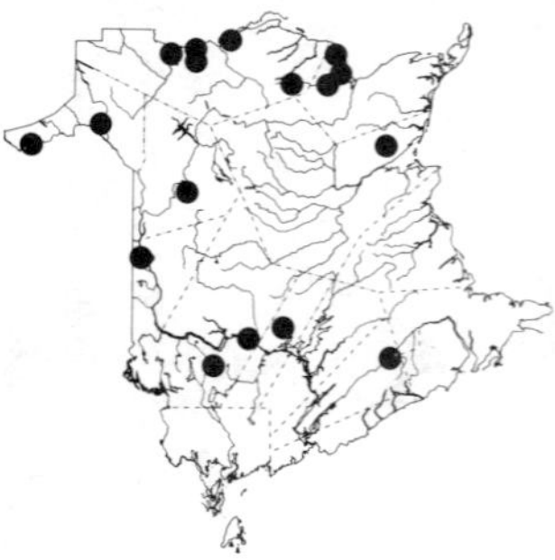

PYROLA CHLORANTHA

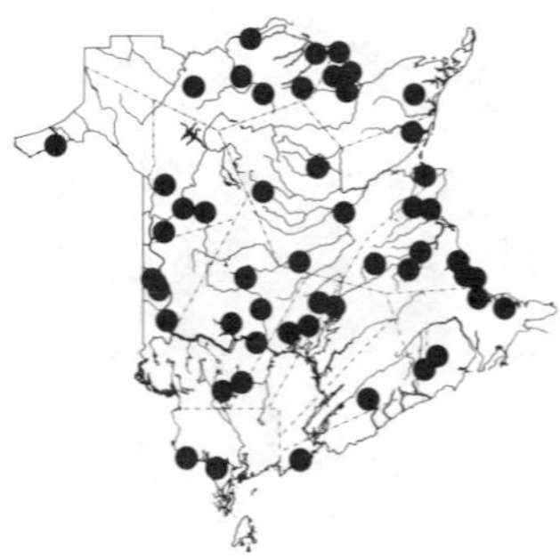

PYROLA ELLIPTICA

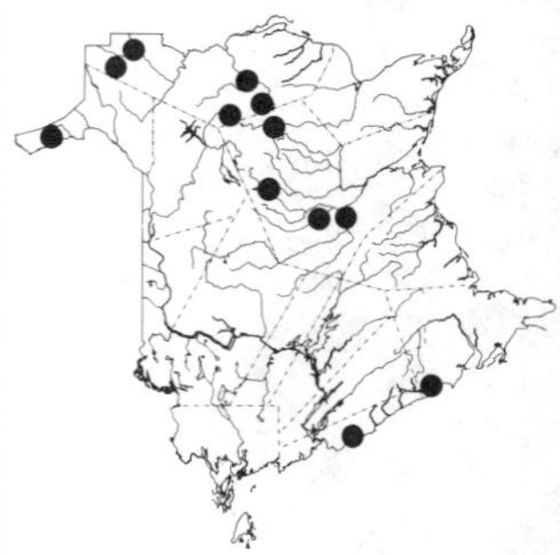

PYROLA MINOR

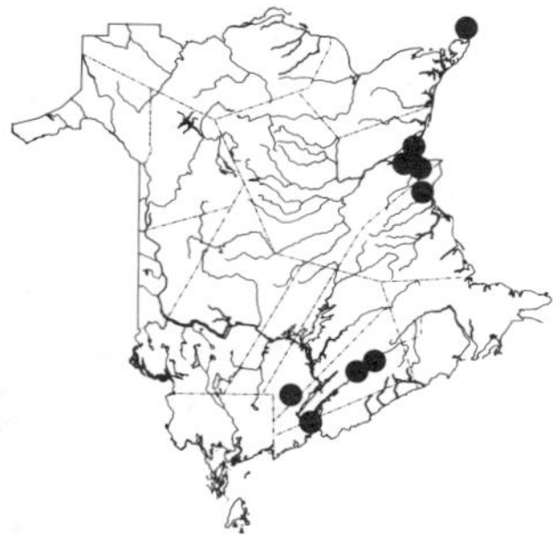
ARCTOSTAPHYLOS UVA-URSI

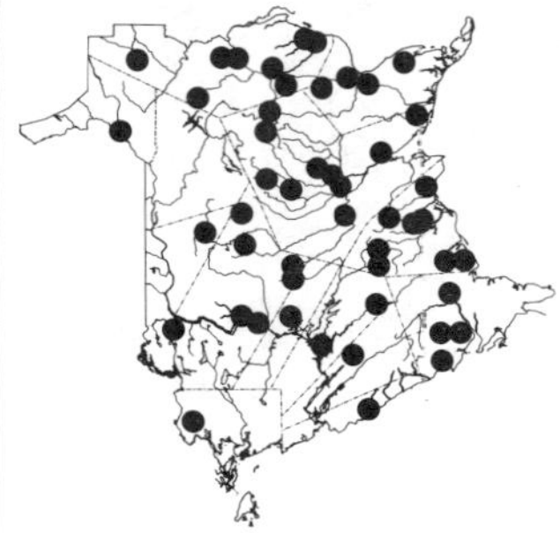
EPIGAEA REPENS

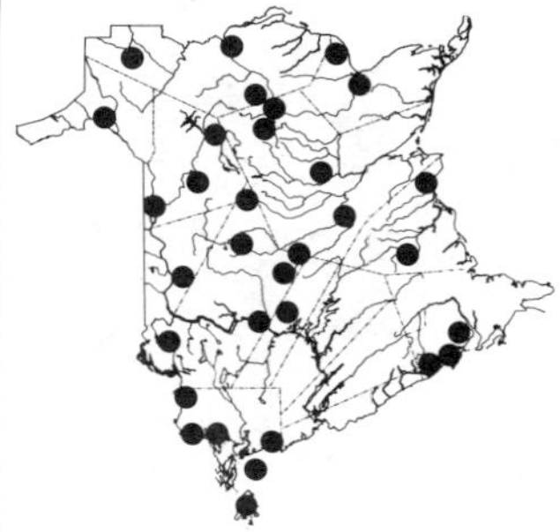
GAULTHERIA HISPIDULA

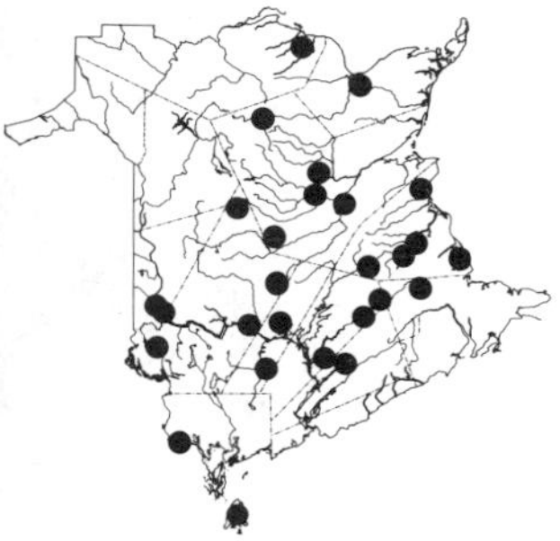
GAULTHERIA PROCUMBENS

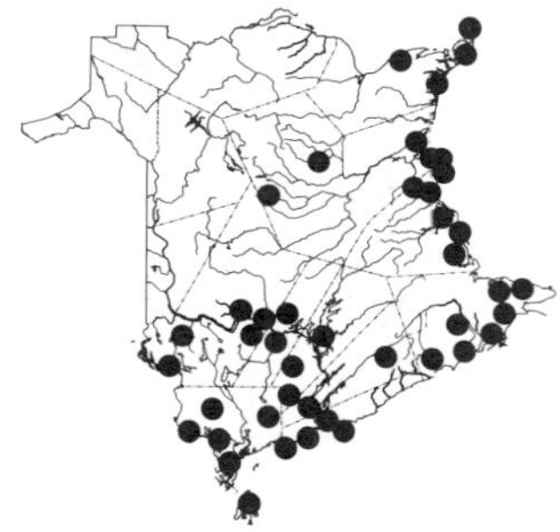
GAYLUSSACIA BACCATA

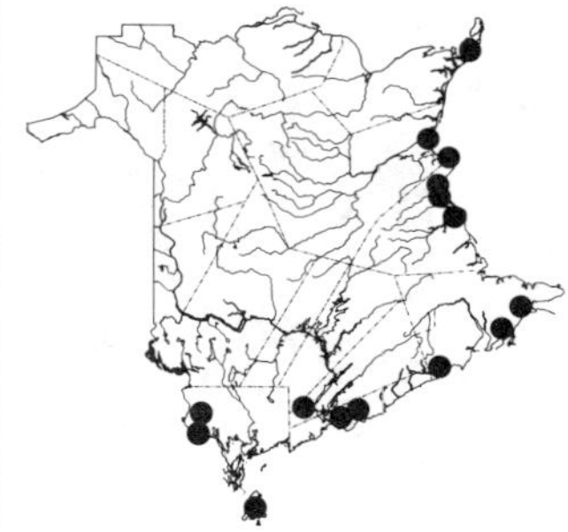
GAYLUSSACIA DUMOSA

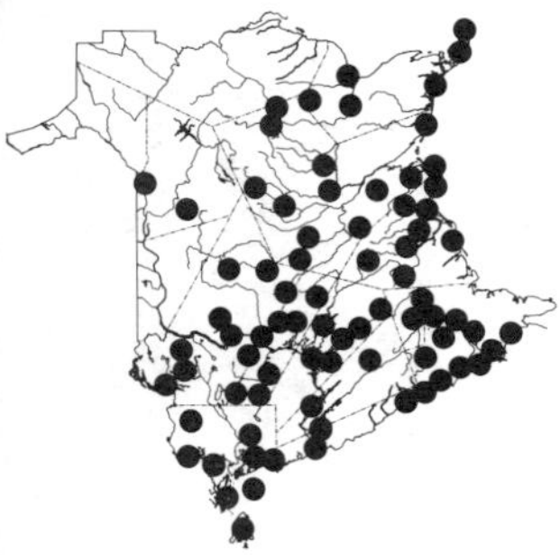
RHODODENDRON CANADENSE

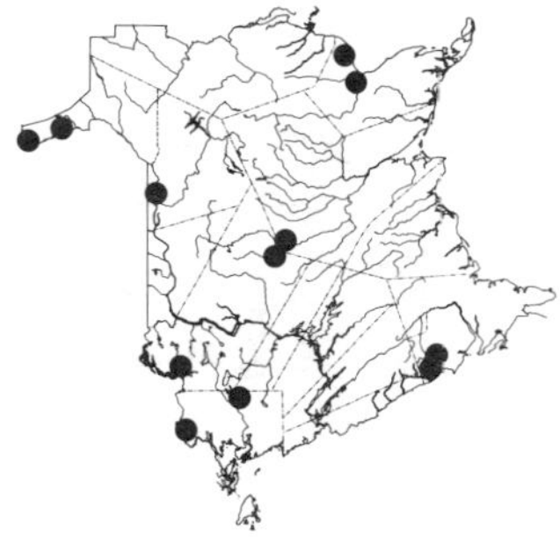
VACCINIUM CAESPITOSUM

VACCINIUM MACROCARPON

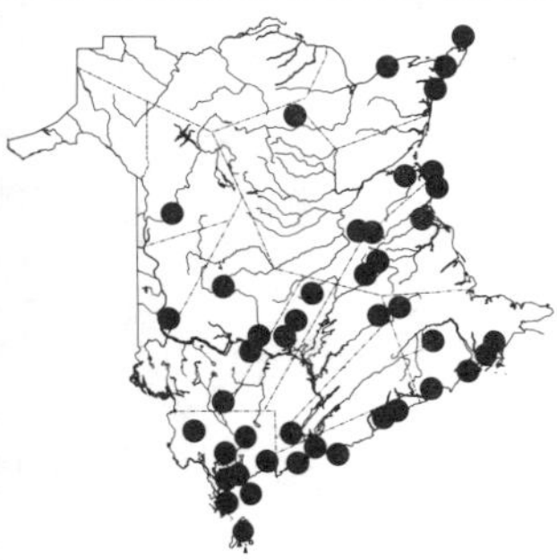
VACCINIUM OXYCOCCUS VAR. OVALIFOLIUM

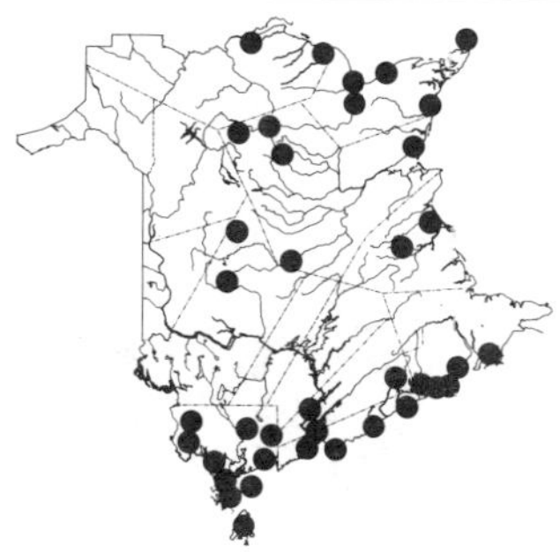
VACCINIUM VITIS-IDAEA VAR. MINUS

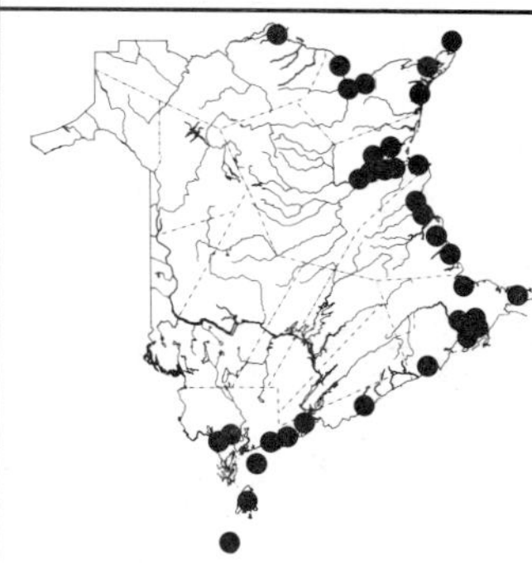
GLAUX MARITIMA

LYSIMACHIA CILIATA

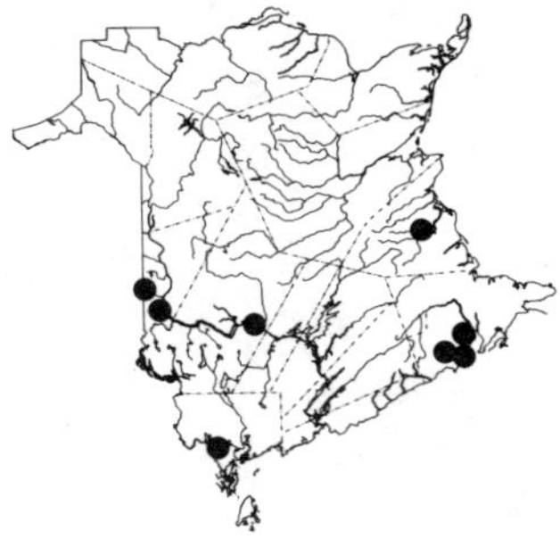

LYSIMACHIA NUMMULARIA

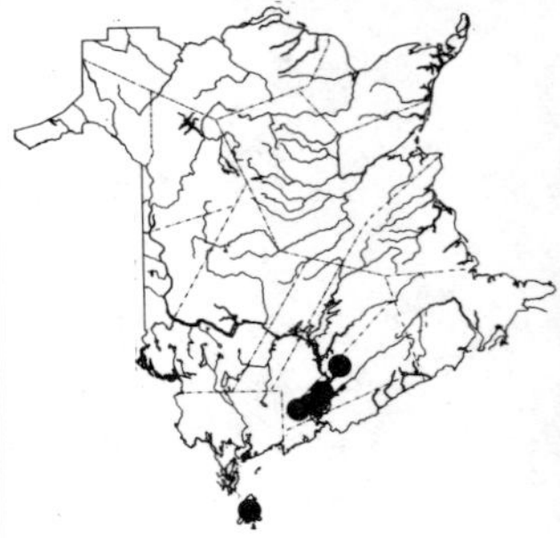

LYSIMACHIA QUADRIFOLIA

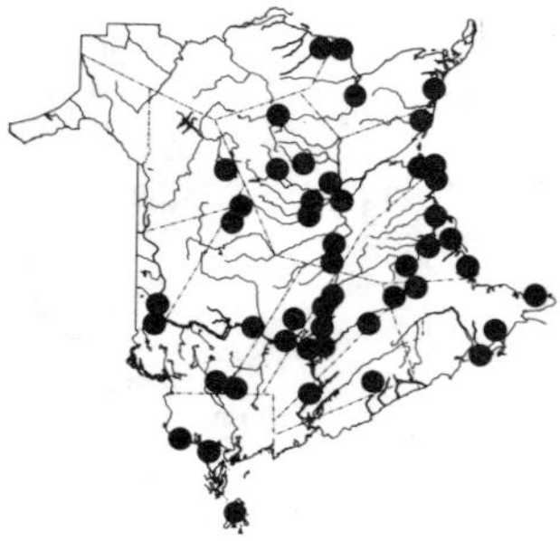

LYSIMACHIA TERRESTRIS

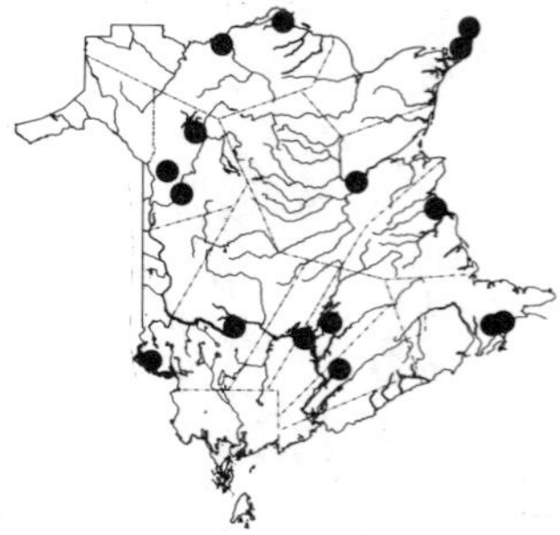

LYSIMACHIA THYRSIFLORA

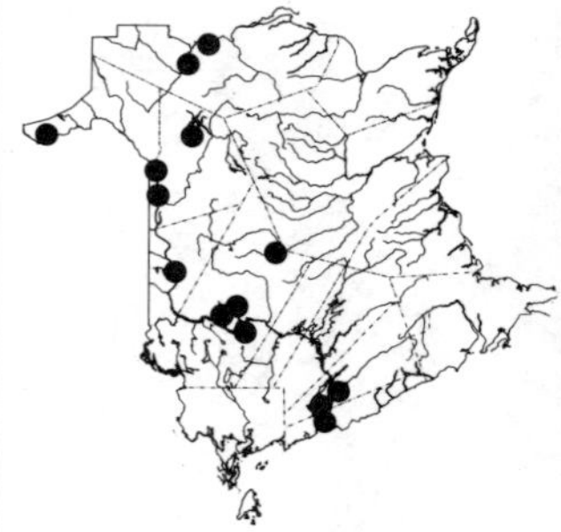

PRIMULA MISTASSINICA

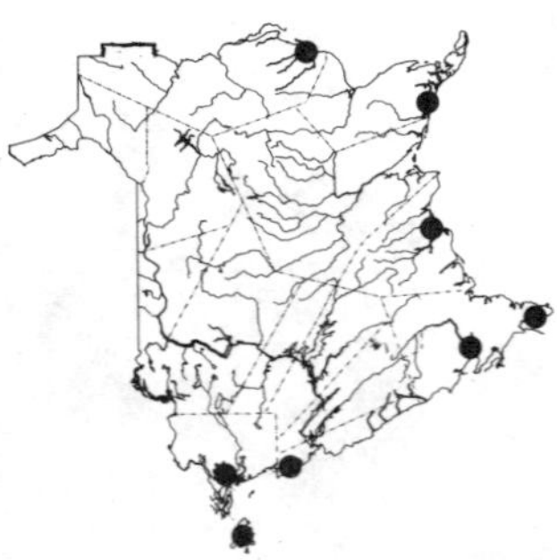

LIMONIUM CAROLINIANUM

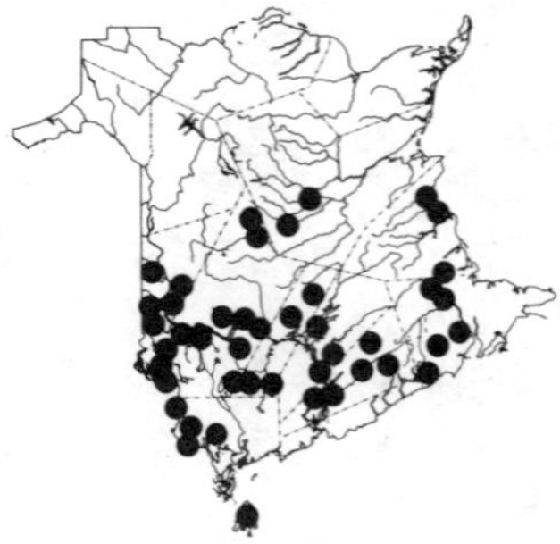

FRAXINUS AMERICANA

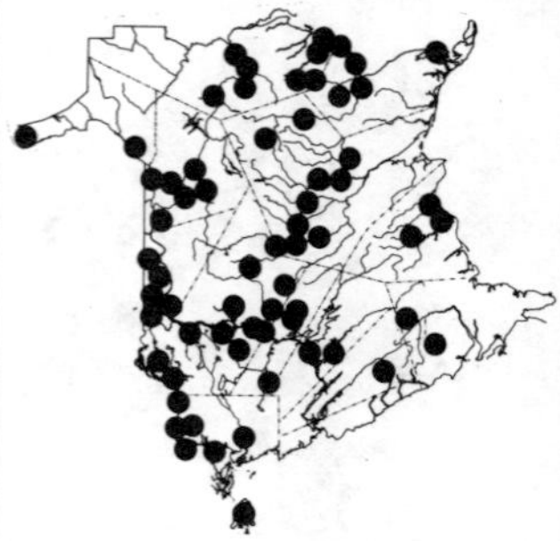

FRAXINUS NIGRA

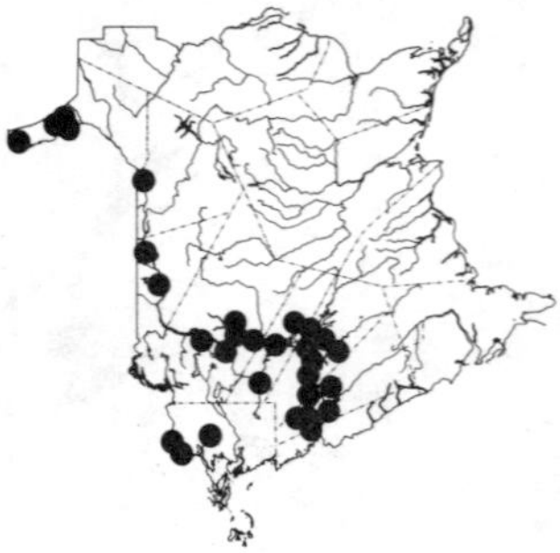

FRAXINUS PENNSYLVANICA

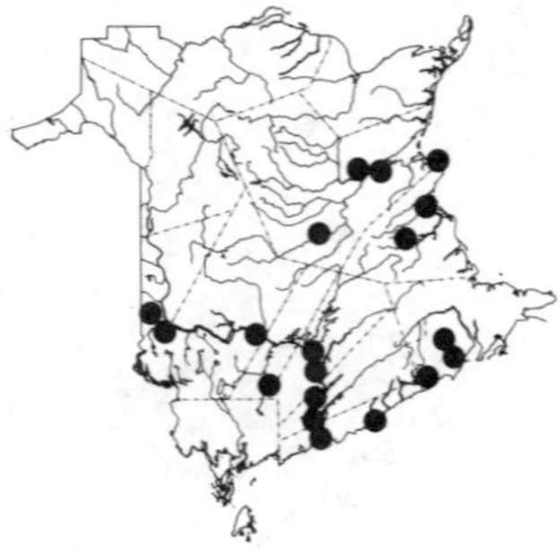

SYRINGA VULGARIS

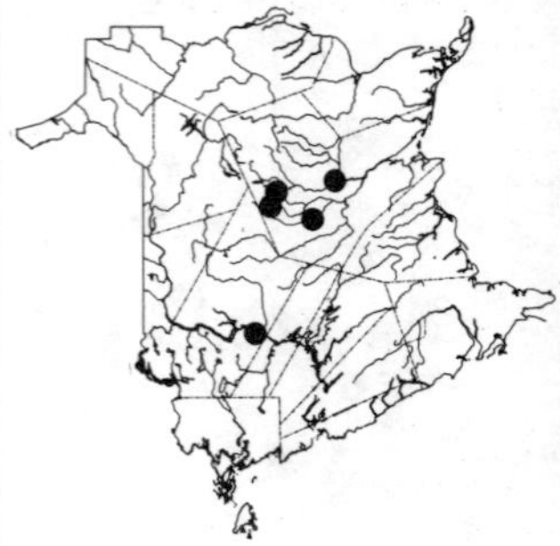

GENTIANA LINEARIS

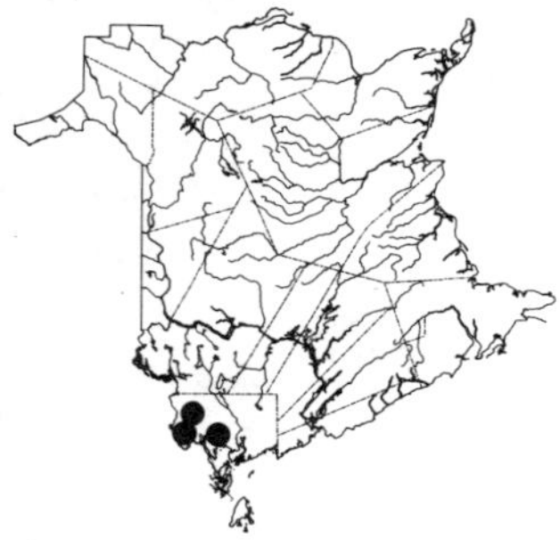

GENTIANA RUBRICAULIS

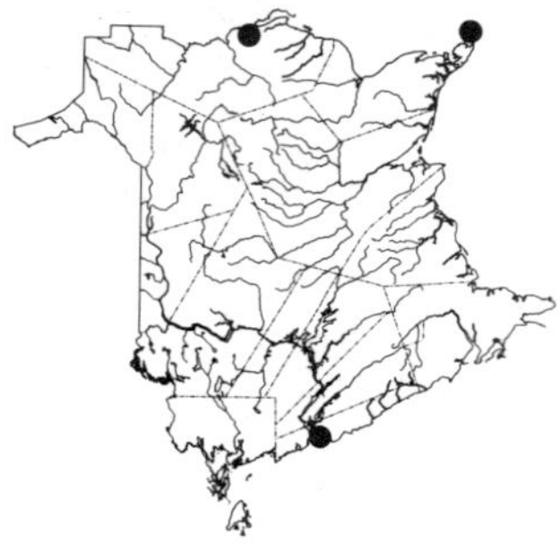

GENTIANELLA AMARELLA SSP. ACUTA

HALENIA DEFLEXA

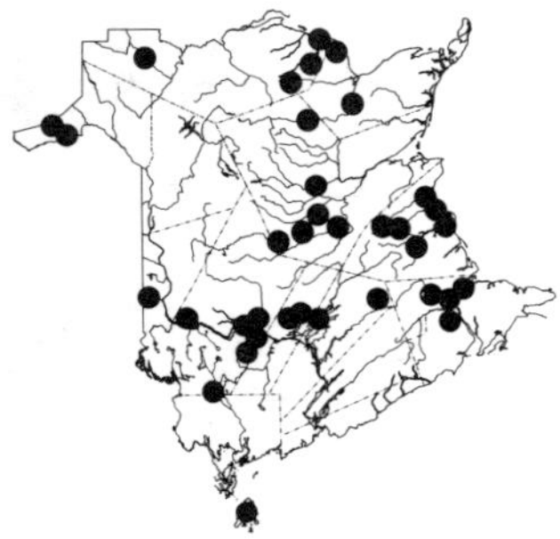

APOCYNUM ANDROSAEMIFOLIUM VAR. INCANUM

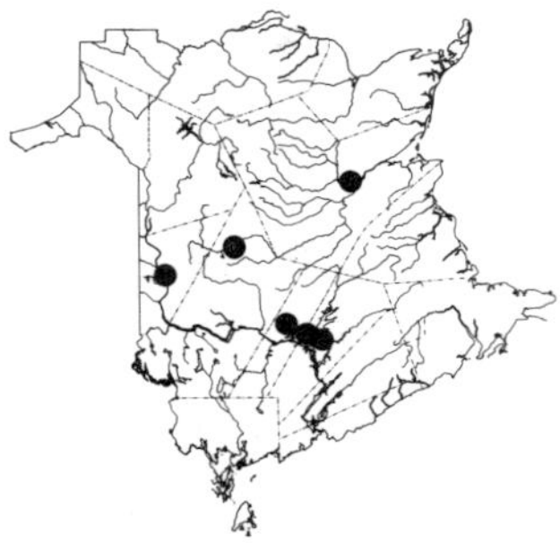

APOCYNUM X MEDIUM

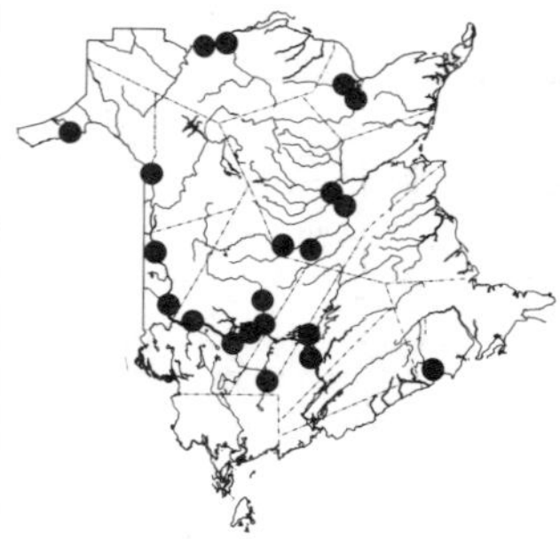

APOCYNUM SIBIRICUM

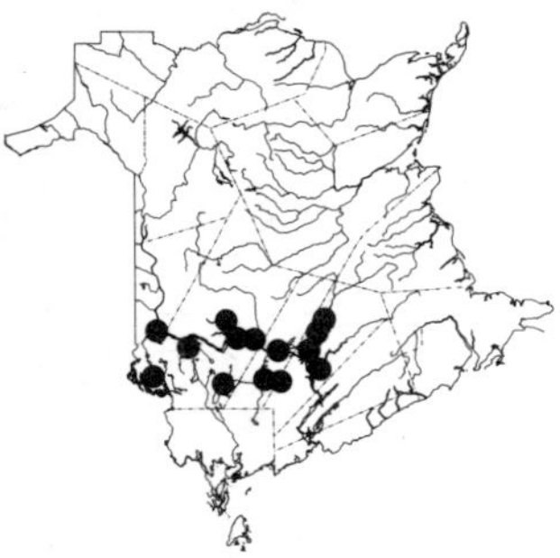

ASCLEPIAS INCARNATA

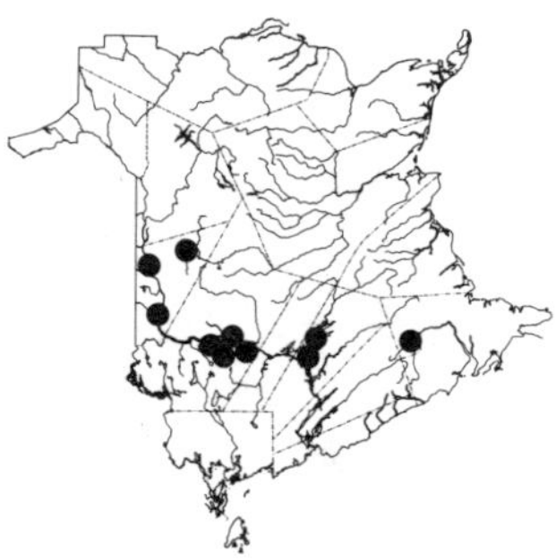

ASCLEPIAS SYRIACA

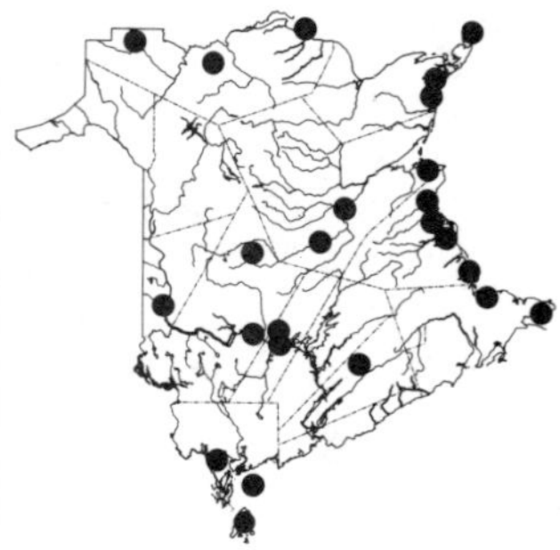

CONVOLVULUS SEPIUM

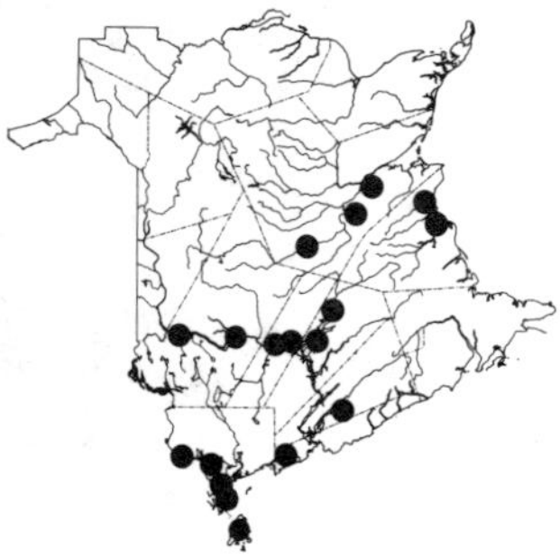

CUSCUTA GRONOVII

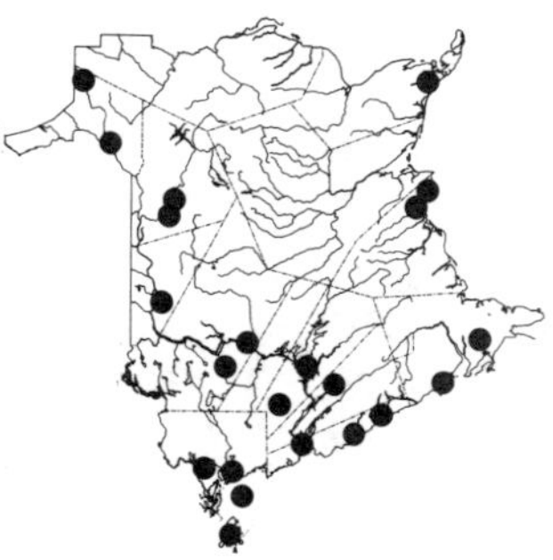

MENYANTHES TRIFOLIATA

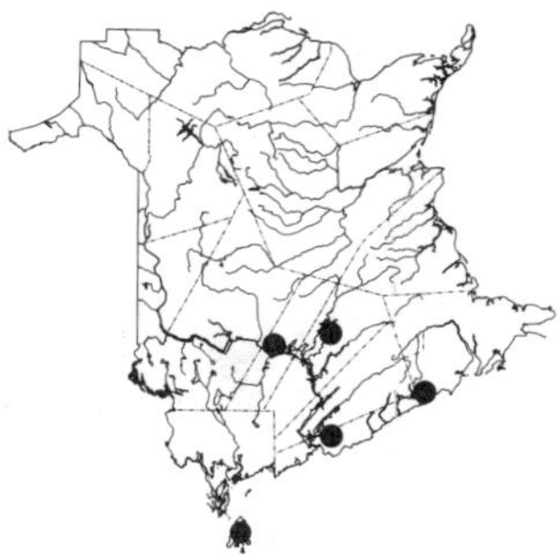

NYMPHOIDES CORDATA

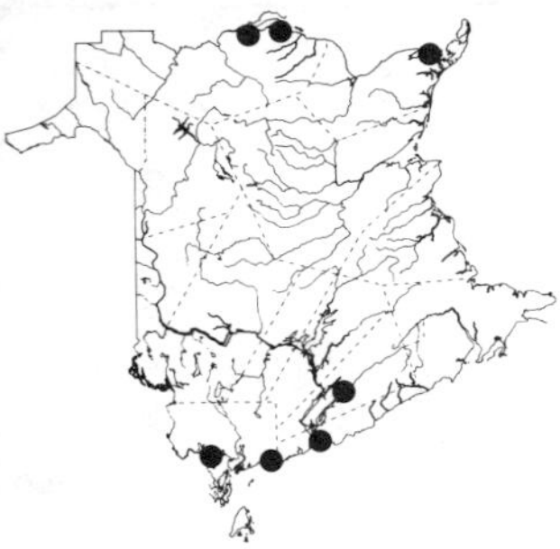

COLLOMIA LINEARIS

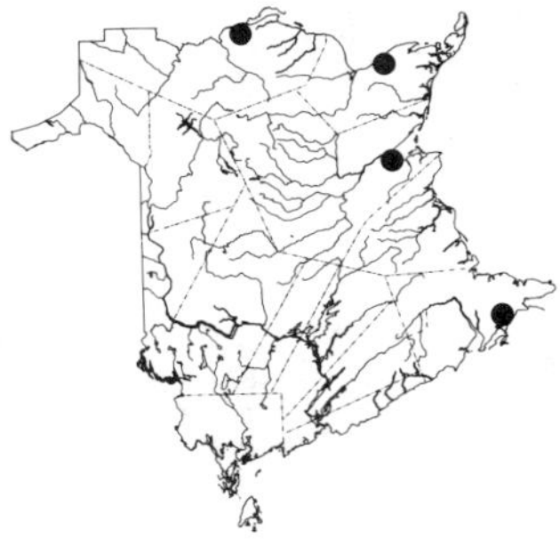

ANCHUSA ARVENSIS

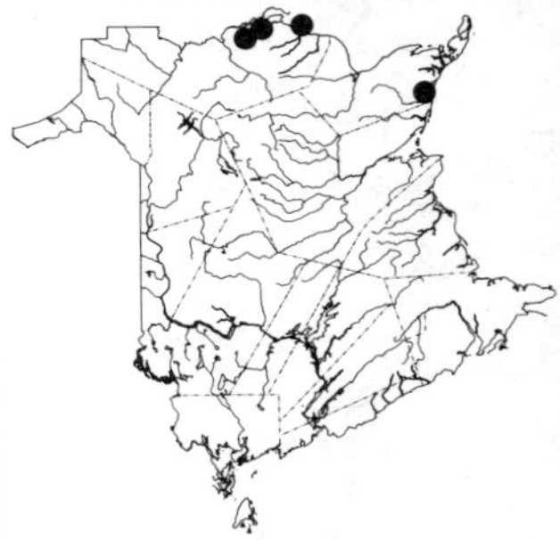

BUGLOSSOIDES OFFICINALE

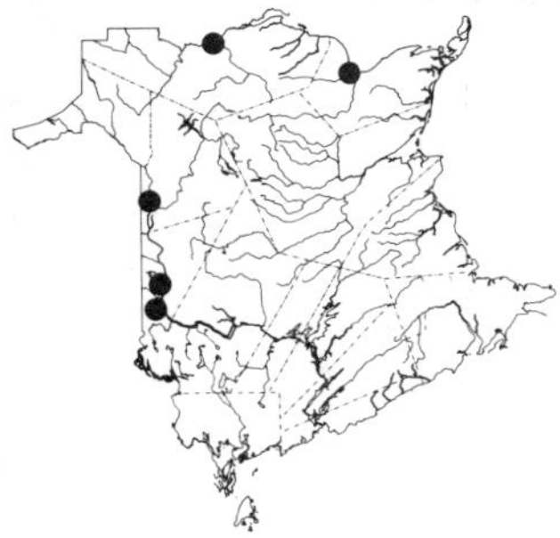

CYNOGLOSSUM BOREALE

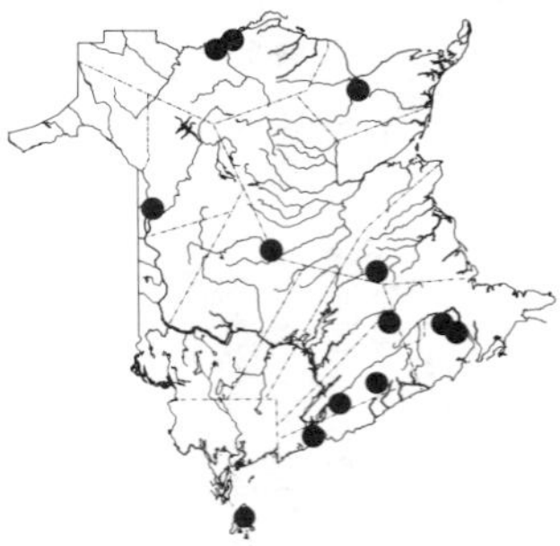

ECHIUM VULGARE

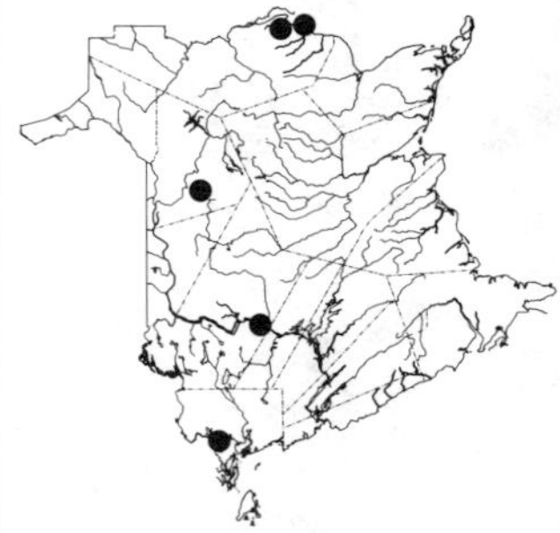

LAPPULA SQUARROSA

MERTENSIA MARITIMA

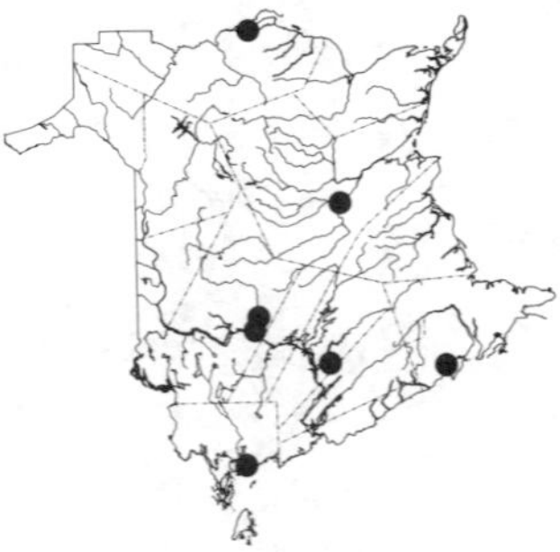

MYOSOTIS ARVENSIS

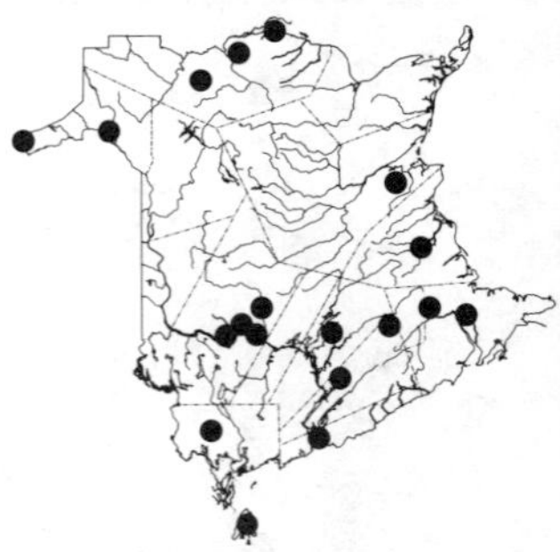

MYOSOTIS LAXA

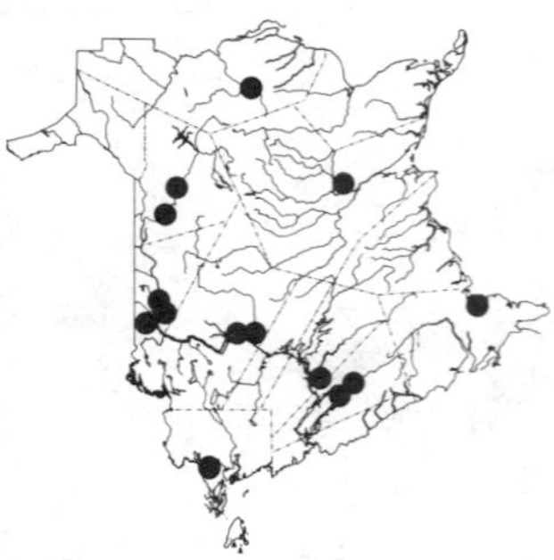

MYOSOTIS SCORPIOIDES

MYOSOTIS SYLVATICA

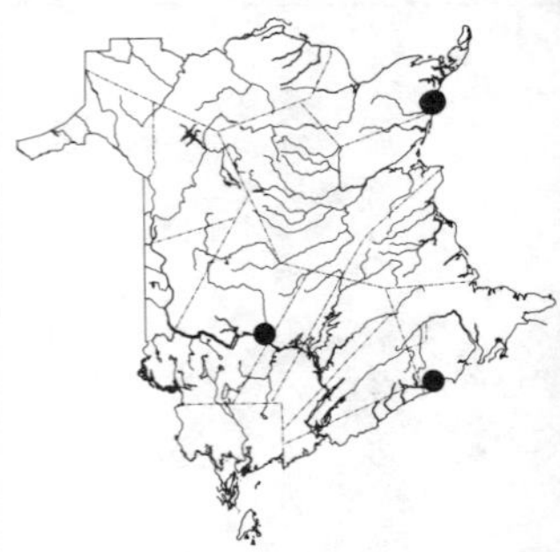

PLAGIOBOTHYS SCOULERI VAR. PENICILLATUS

BORAGINACEAE | VERBENACEAE | LAMIACEAE

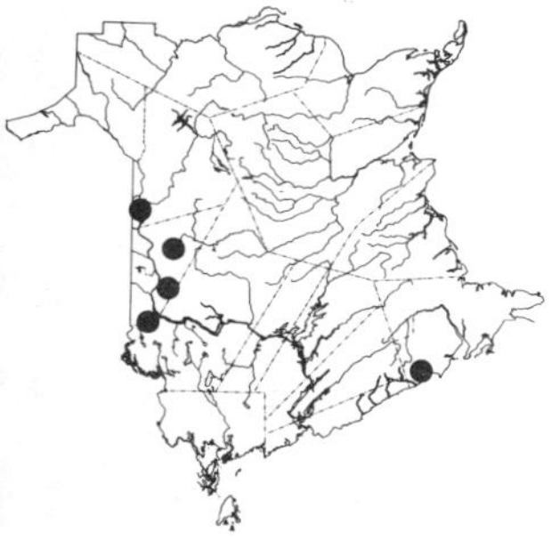

SYMPHYTUM ASPERUM

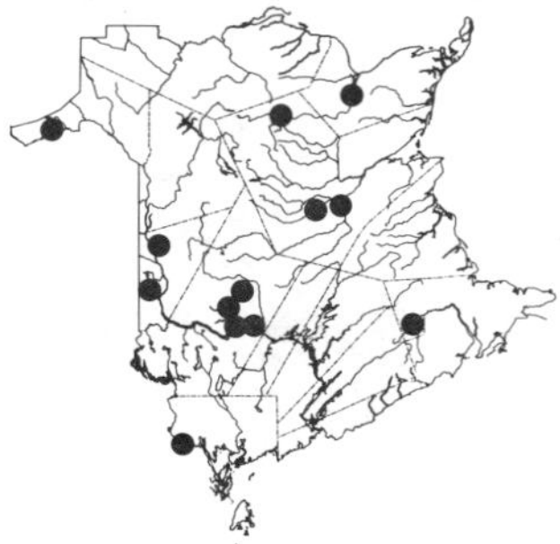

VERBENA HASTATA

VERBENA URTICIFOLIA

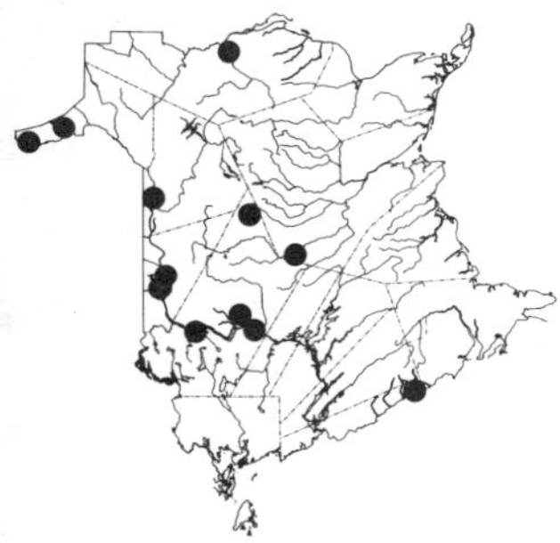

CLINOPODIUM VULGARE

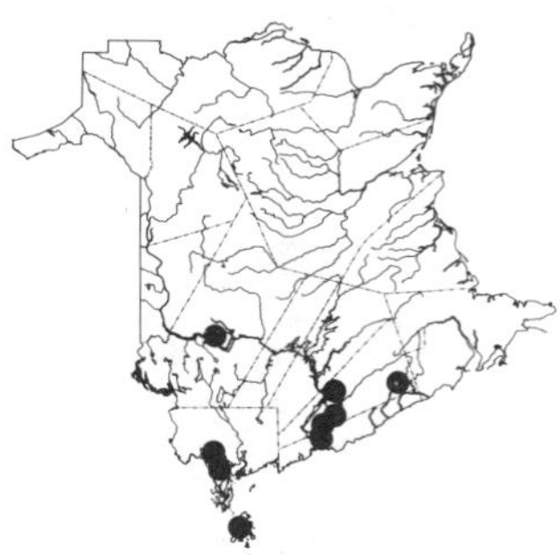

HEDEOMA PULEGIOIDES

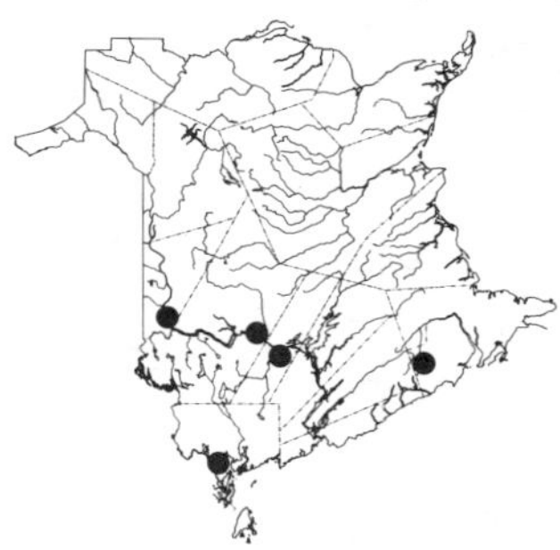

LEONURUS CARDIACA

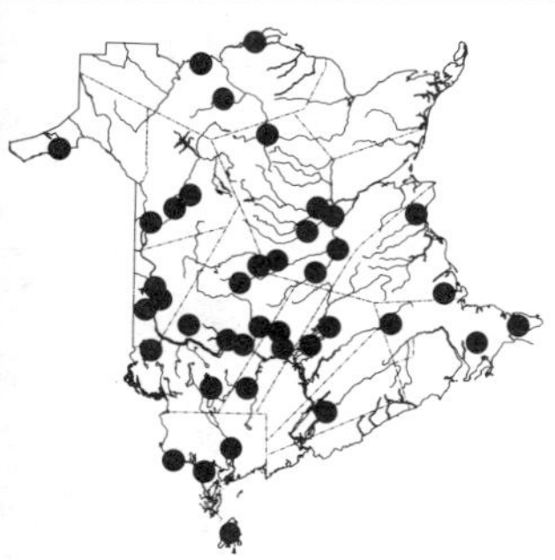

LYCOPUS AMERICANUS

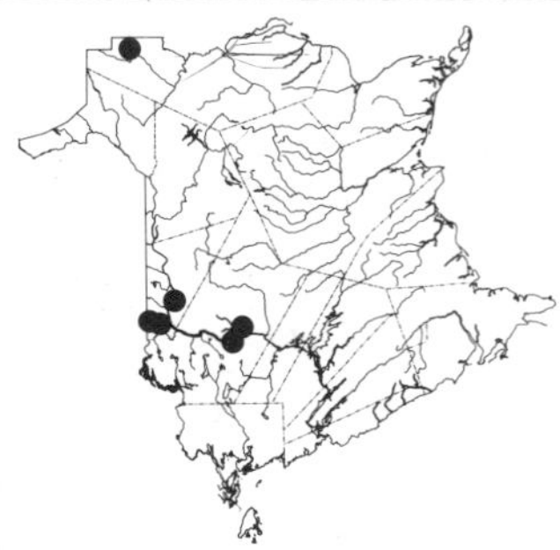

SCUTELLARIA X CHURCHILLIANA

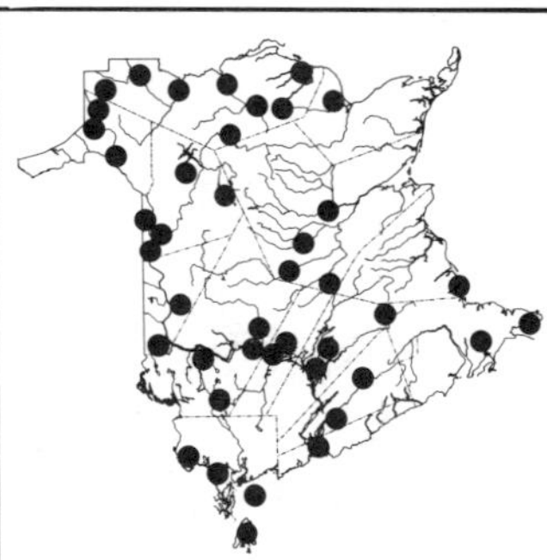

SCUTELLARIA GALERICULATA

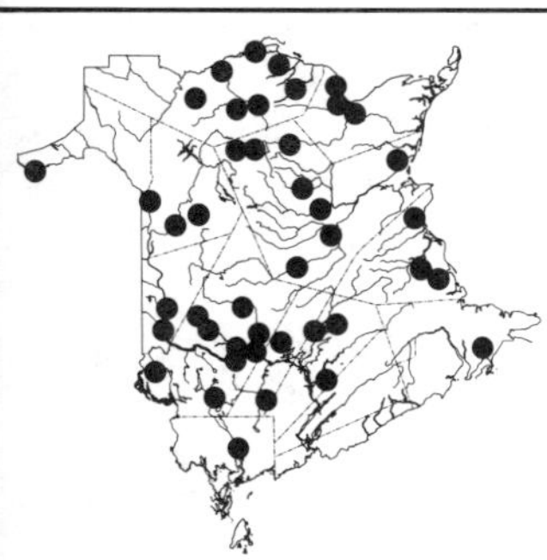

SCUTELLARIA LATERIFLORA

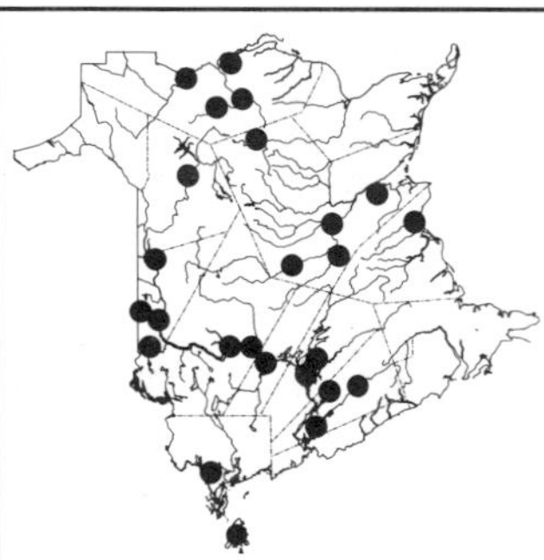

STACHYS PALUSTRIS

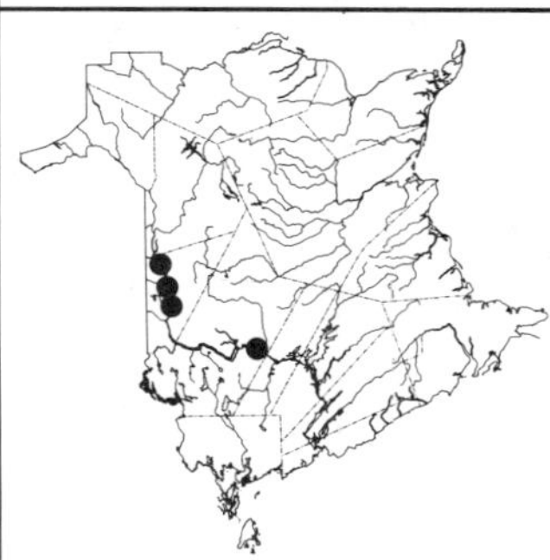

STACHYS TENUIFOLIA
VAR. PLATYPHYLLA

LAMIACEAE | SOLANACEAE | SCROPHULARIACEAE

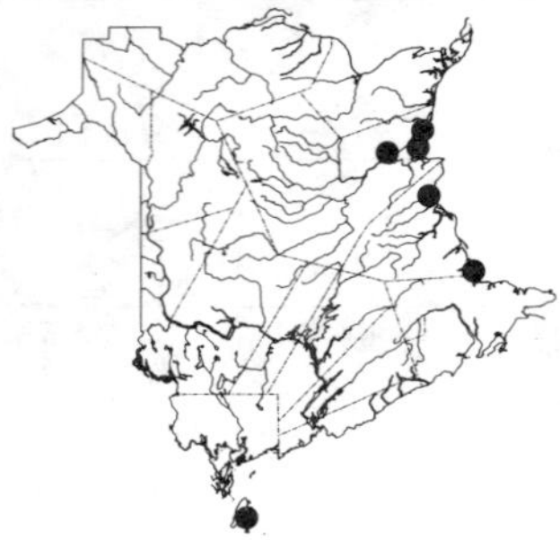

TEUCRIUM CANADENSE

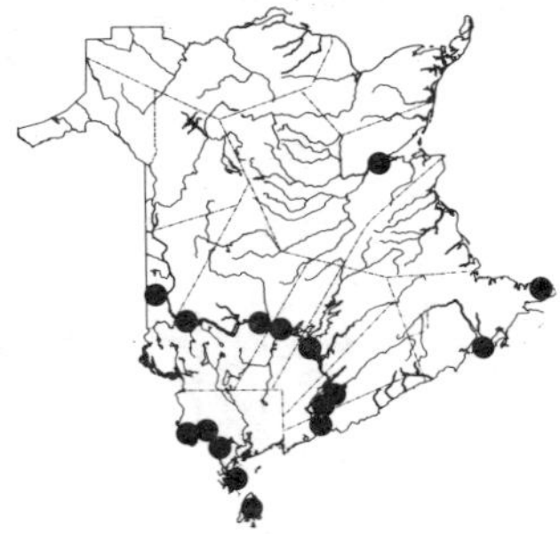

SOLANUM DULCAMARA

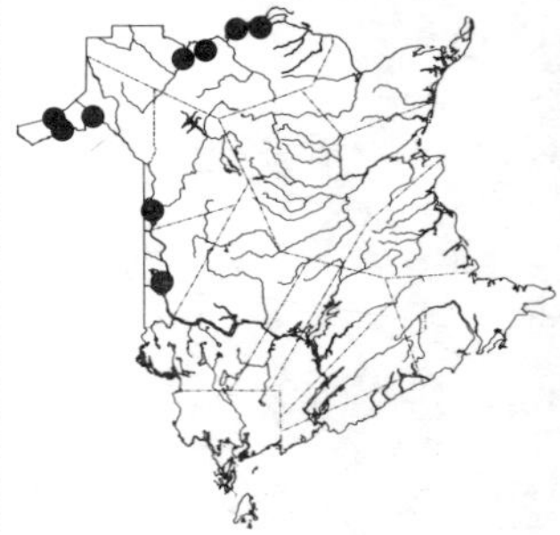

CASTILLEJA SEPTENTRIONALIS

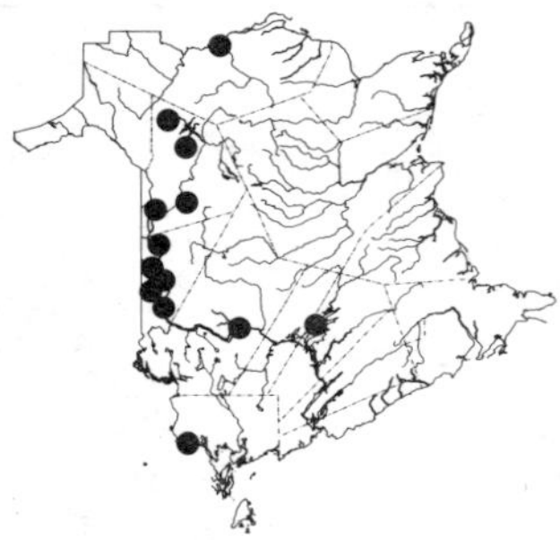

CHAENORRHINUM MINUS

CHELONE GLABRA

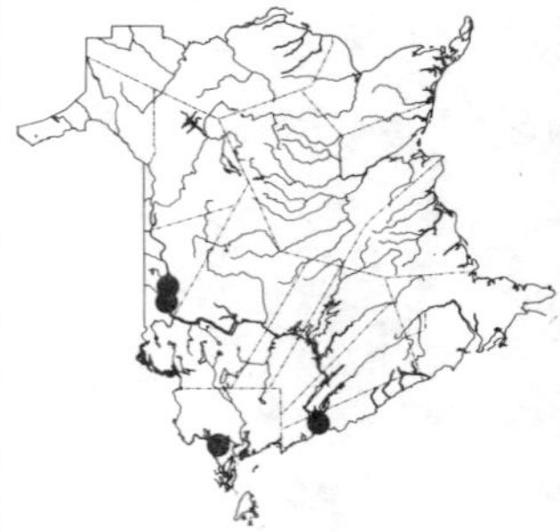

CYMBALARIA MURALIS

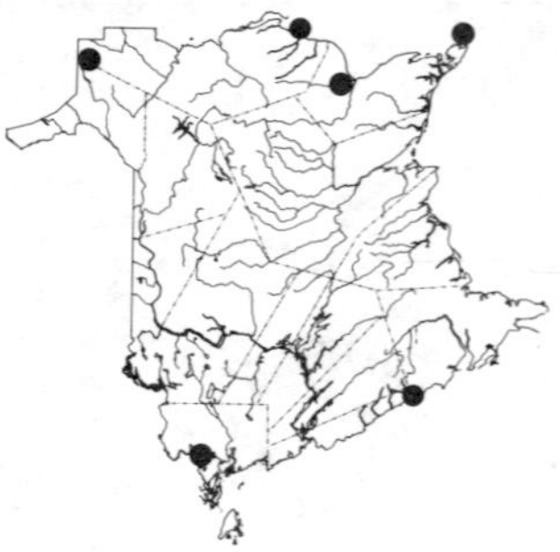

EUPHRASIA ARCTICA

EUPHRASIA NEMOROSA

EUPHRASIA RANDII

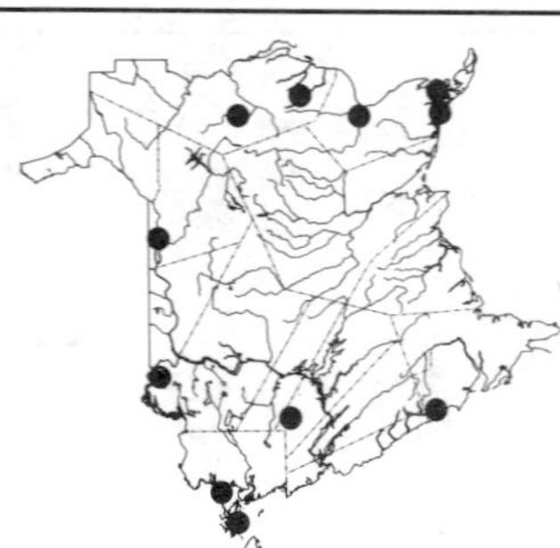

EUPHRASIA STRICTA

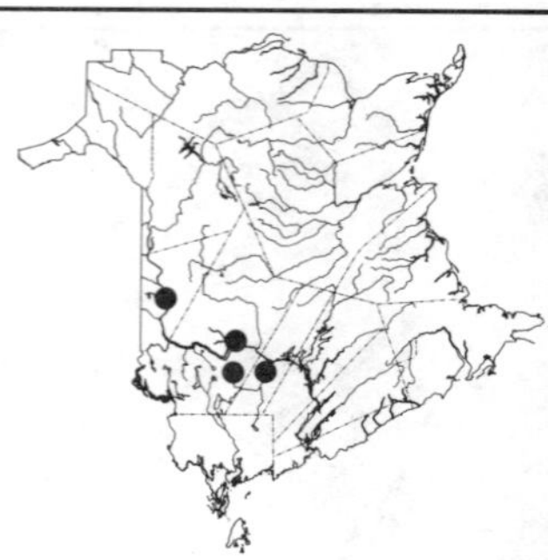

GRATIOLA NEGLECTA

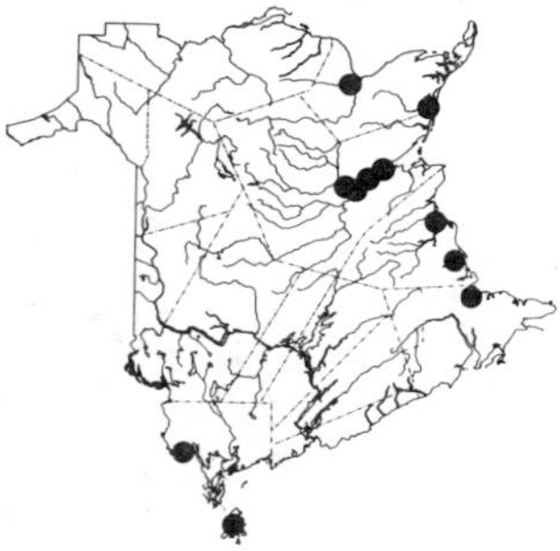

LIMOSELLA AUSTRALIS

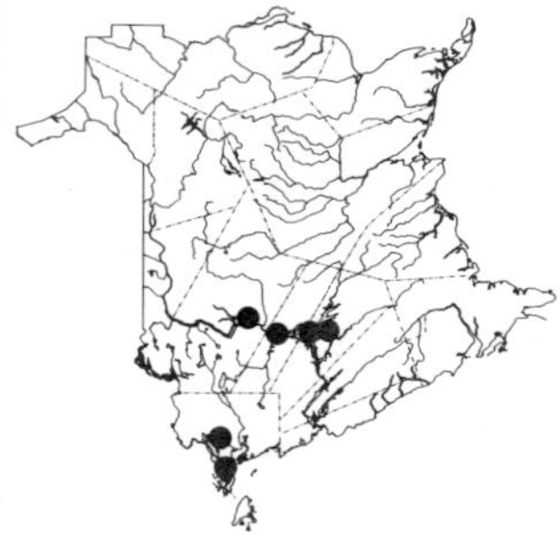

LINARIA CANADENSIS

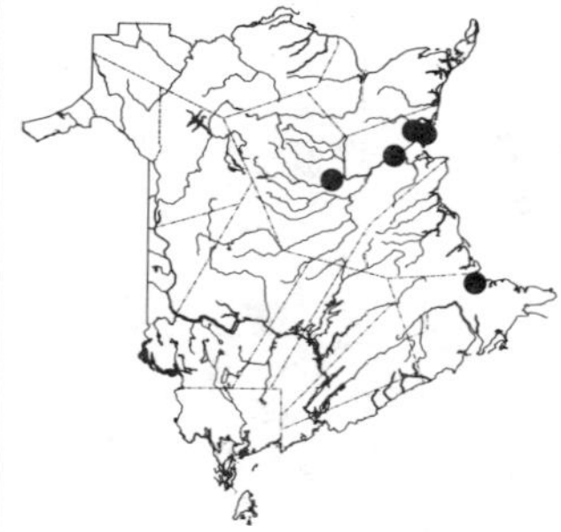

LINARIA REPENS

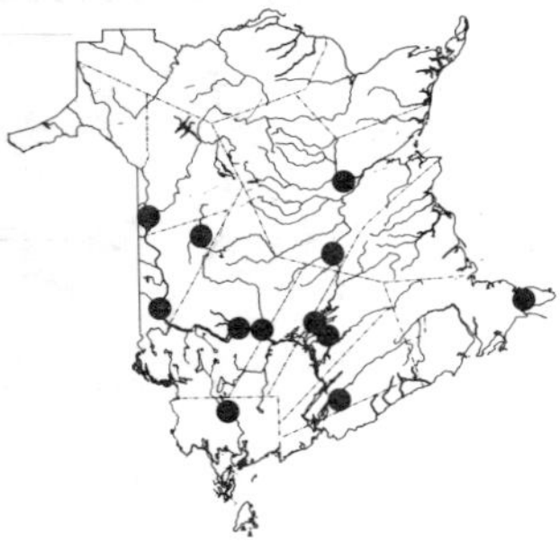

LINDERNIA DUBIA

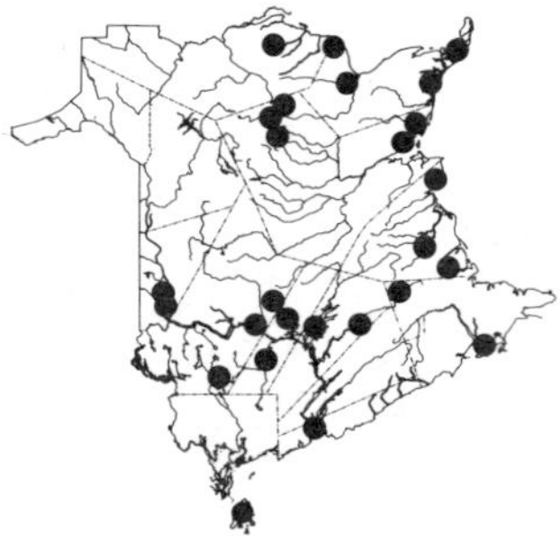

MELAMPYRUM LINEARE

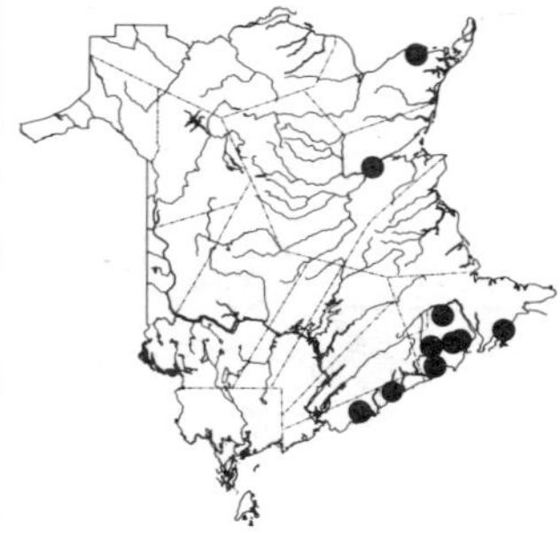

MIMULUS MOSCHATUS

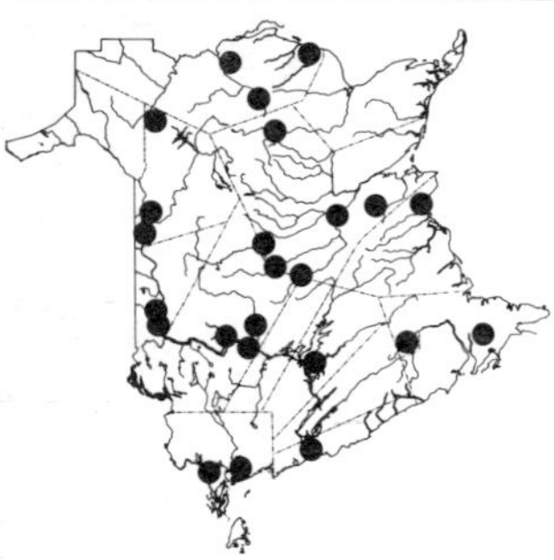

MIMULUS RINGENS

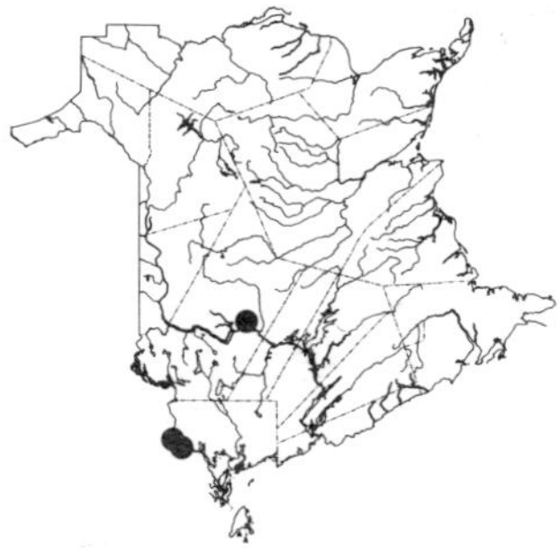

PEDICULARIS CANADENSIS

PEDICULARIS FURBISHIAE

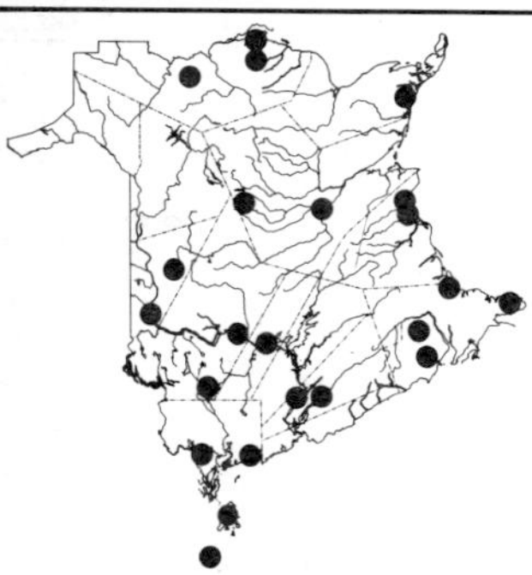

RHINANTHUS MINOR

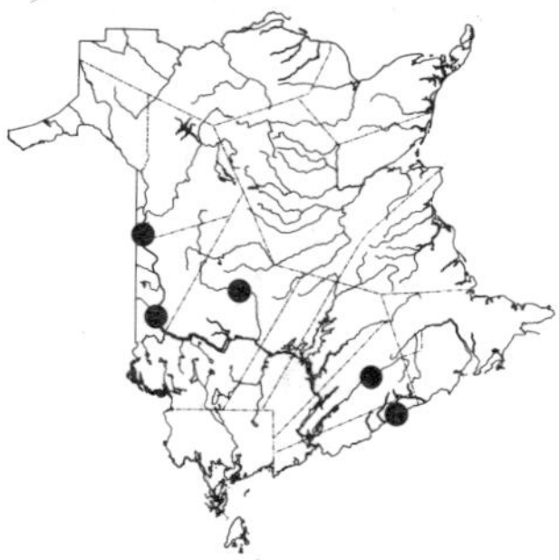

SCROPHULARIA LANCEOLATA

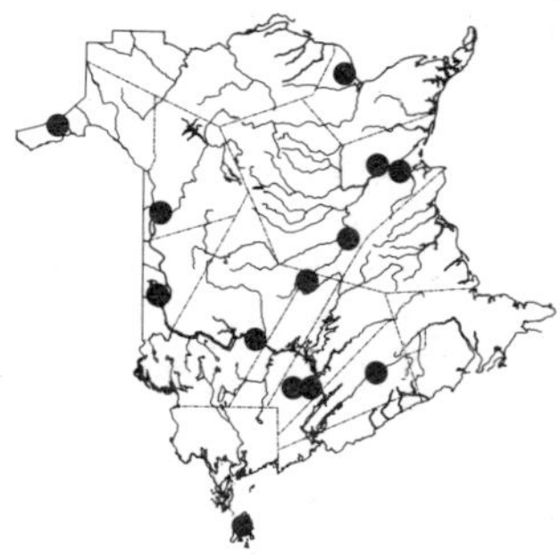

VERBASCUM THAPSUS

SCROPHULARIACEAE | OROBANCHACEAE | LENTIBULARIACEAE

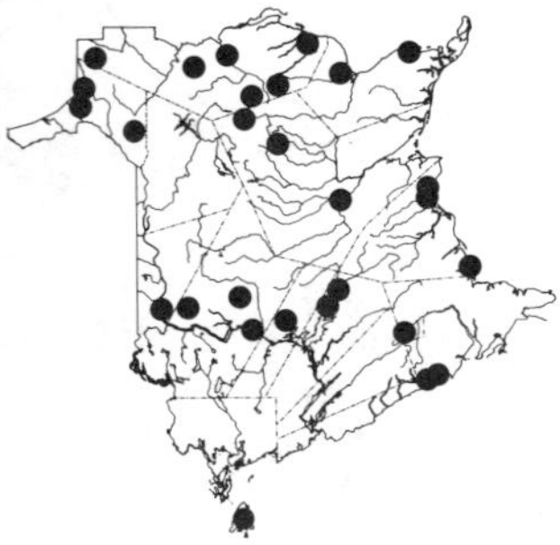

VERONICA AMERICANA

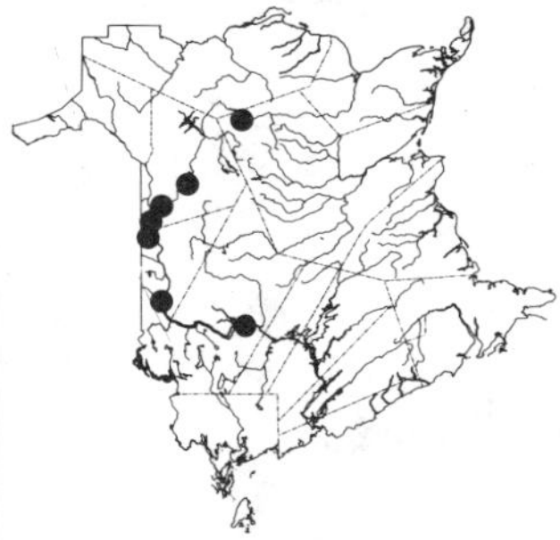

VERONICA CHAMAEDRYS

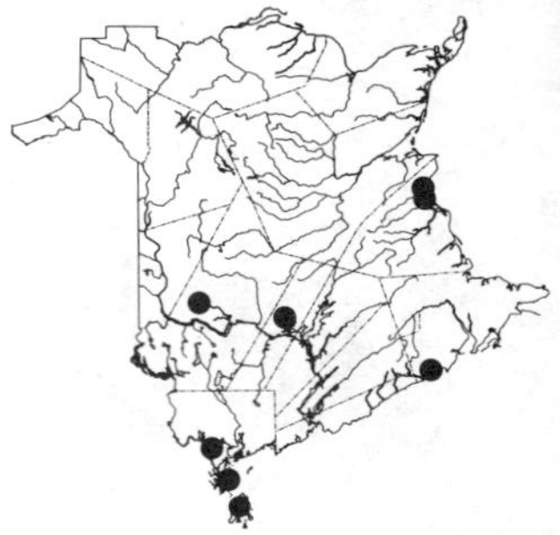

VERONICA LONGIFOLIA

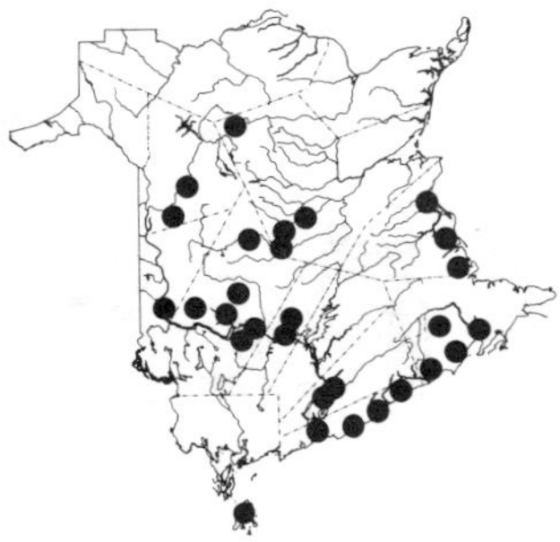

VERONICA OFFICINALIS

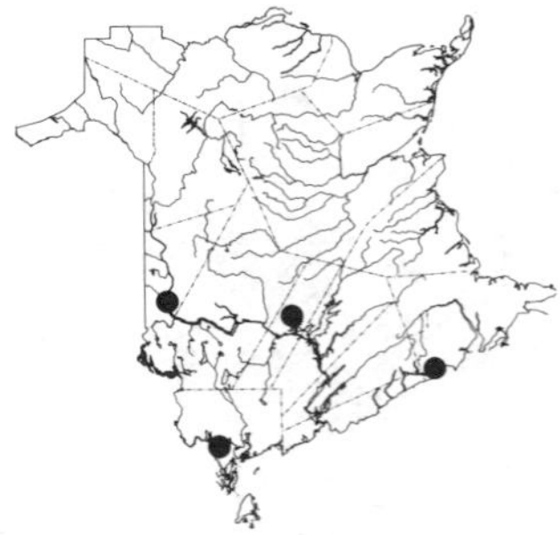

VERONICA PEREGRINA

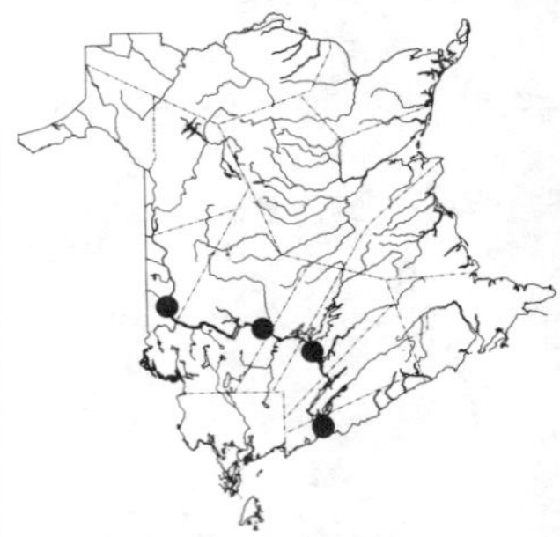

VERONICA PERSICA

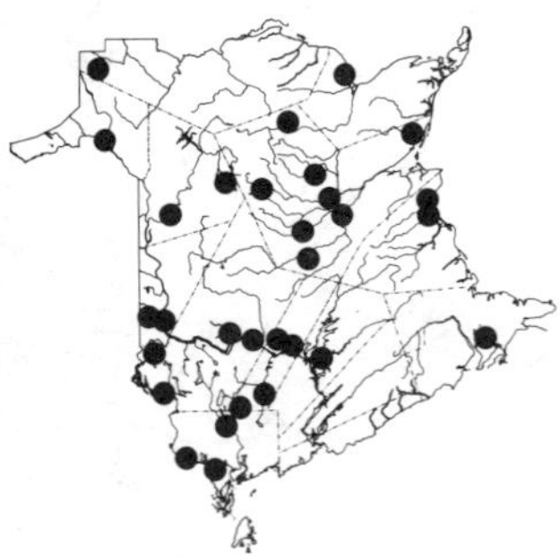

VERONICA SCUTELLATA

VERONICA SERPYLLIFOLIA

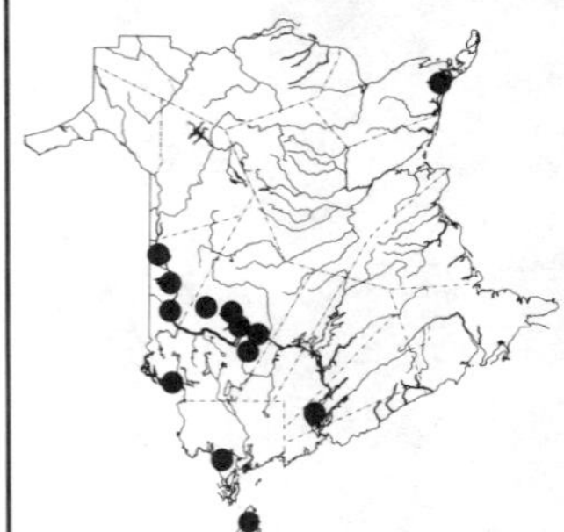

EPIFAGUS VIRGINIANA

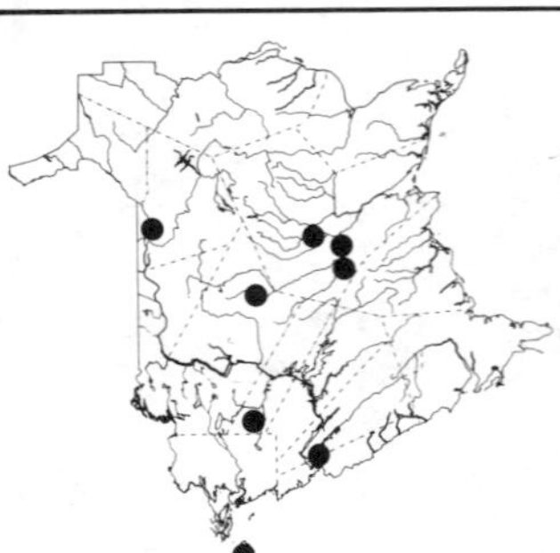

OROBANCHE UNIFLORA

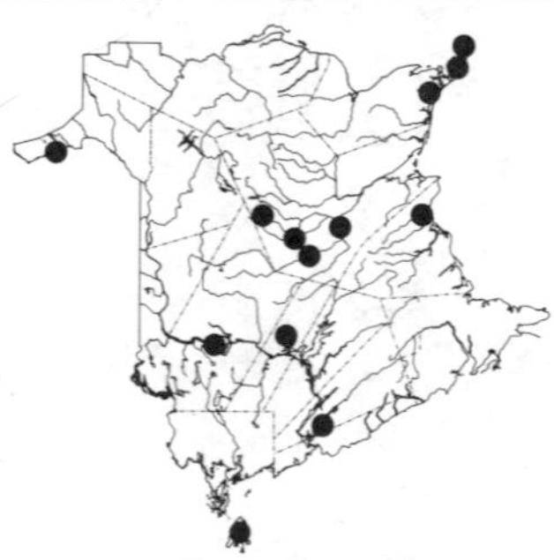

UTRICULARIA CORNUTA

UTRICULARIA GEMINISCAPA

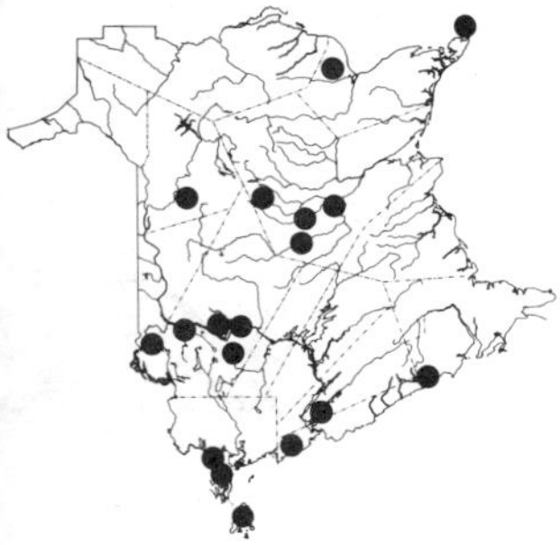

UTRICULARIA INTERMEDIA

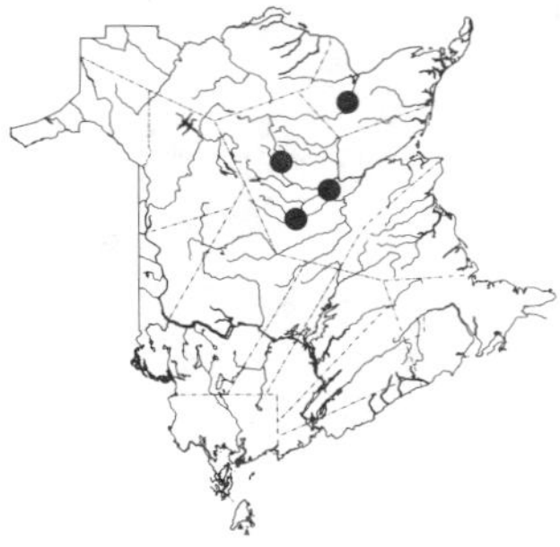

UTRICULARIA MINOR

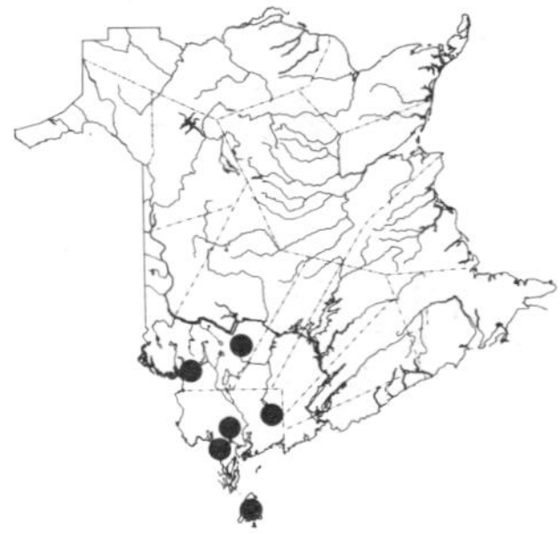

UTRICULARIA PURPUREA

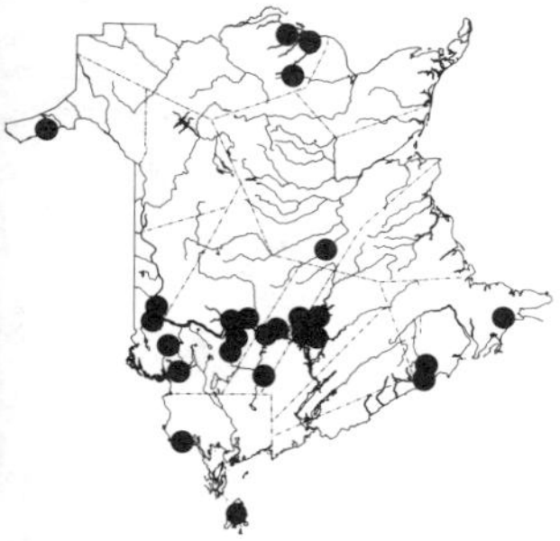

UTRICULARIA VULGARIS

PHRYMA LEPTOSTACHYA

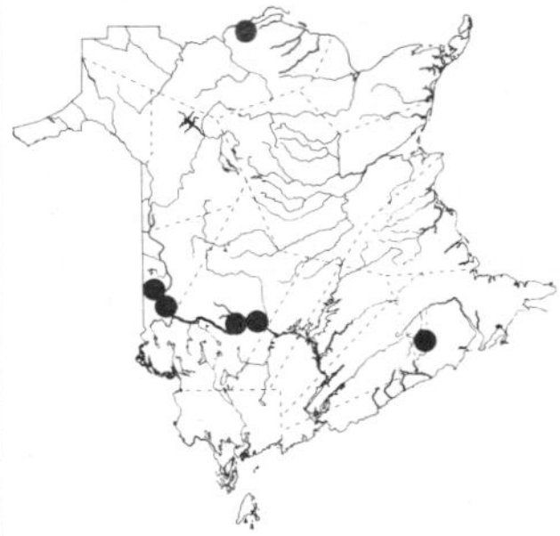

PLANTAGO LANCEOLATA

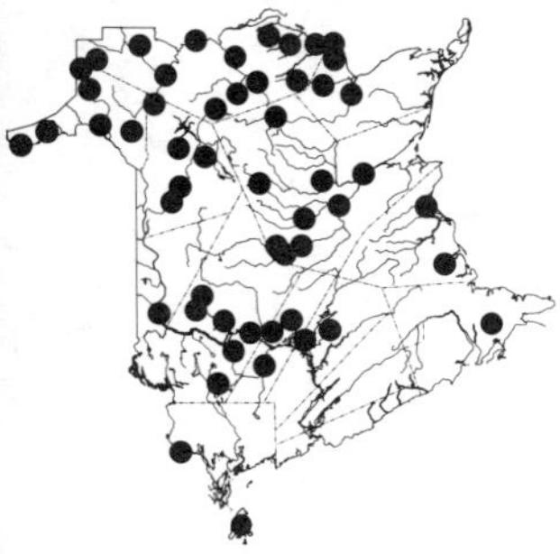

GALIUM ASPRELLUM

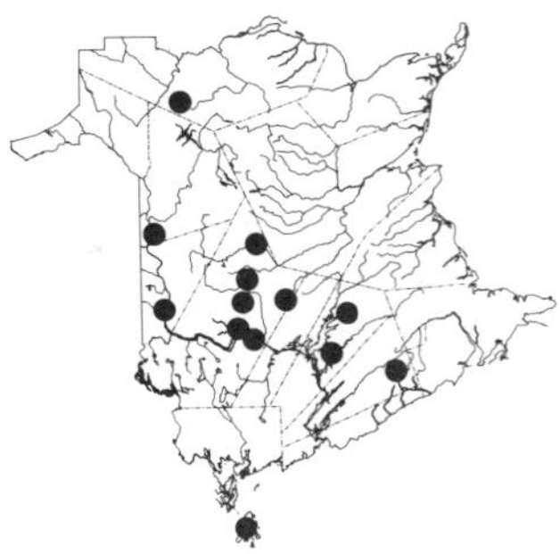

GALIUM BOREALE

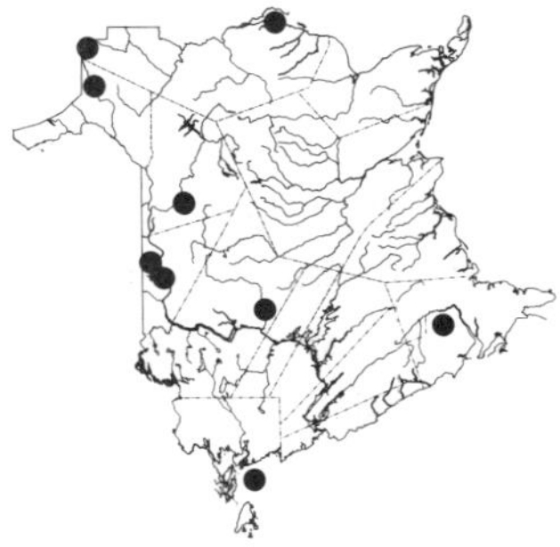

GALIUM LABRADORICUM

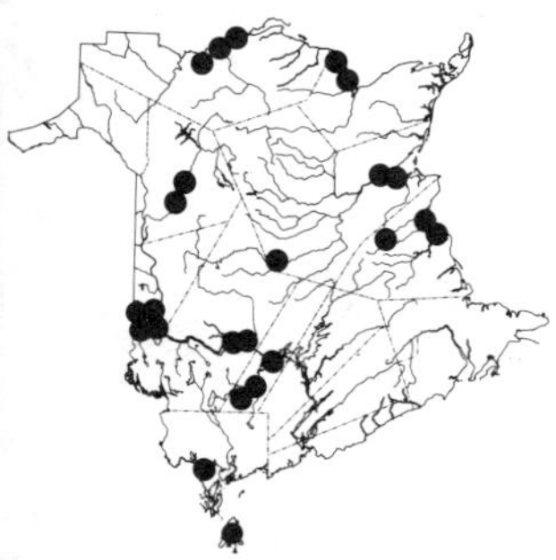

GALIUM MOLLUGO

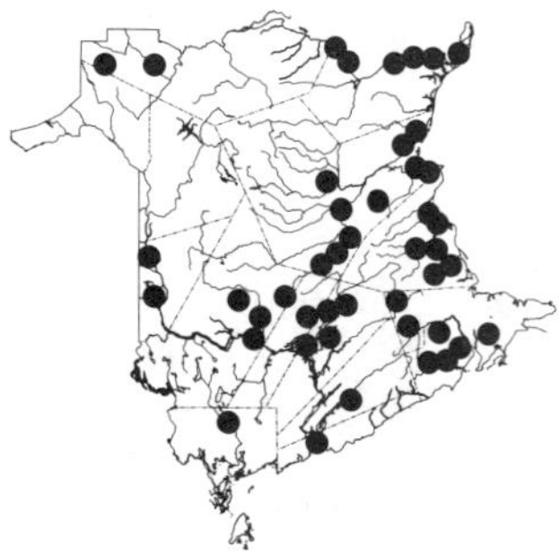

GALIUM OBTUSUM

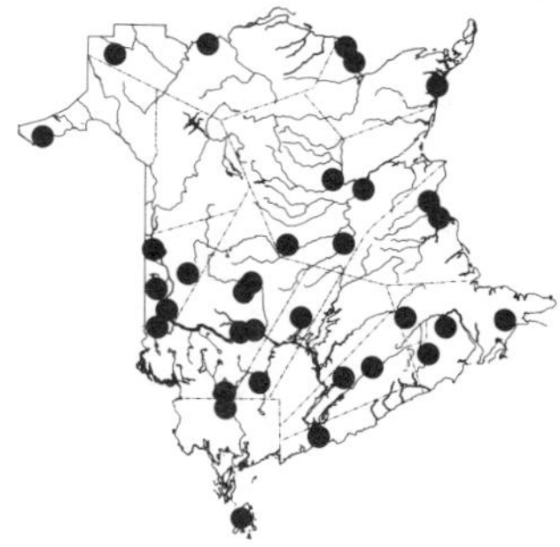

GALIUM PALUSTRE

RUBIACEAE

CAPRIFOLIACEAE

VALERIANACEAE

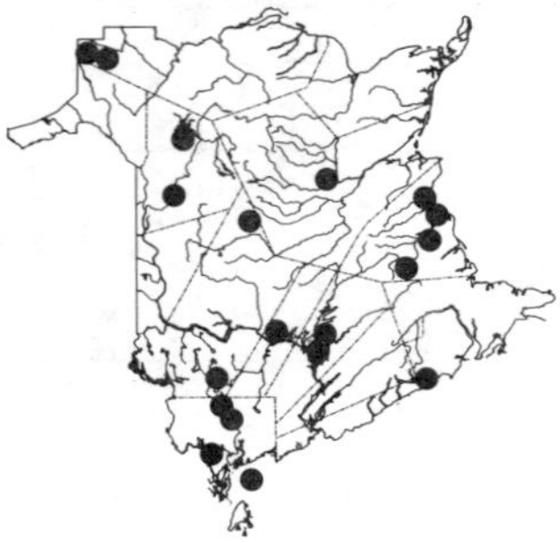

GALIUM TINCTORIUM

GALIUM TRIFIDUM

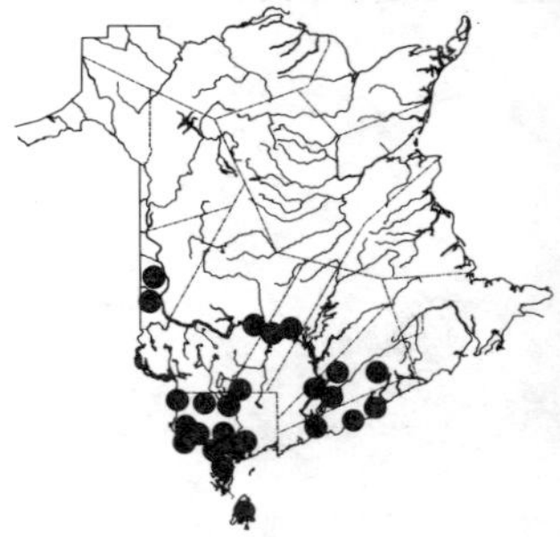

HEDYOTIS CAERULEA

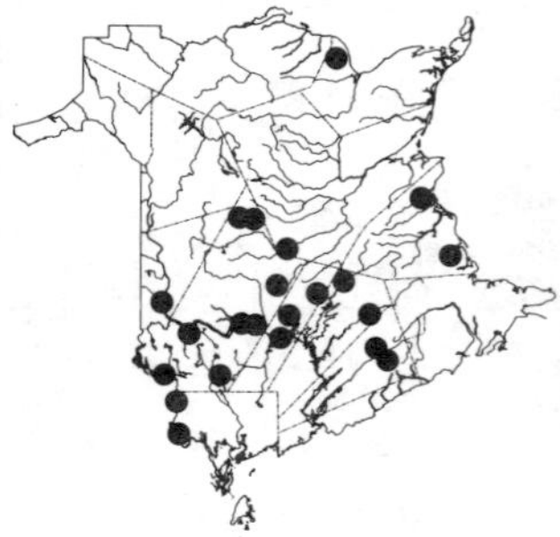

MITCHELLA REPENS

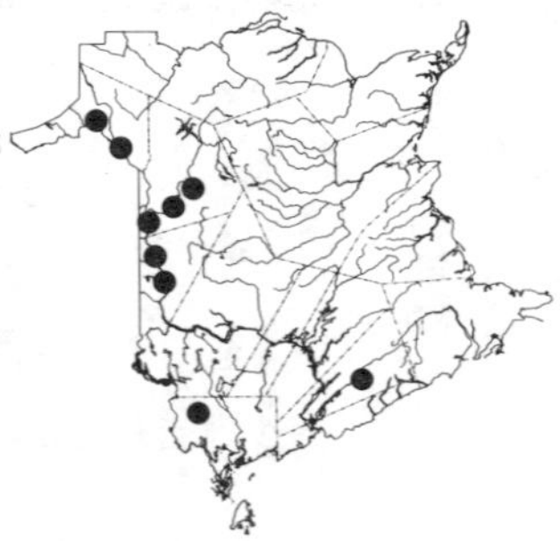

LONICERA OBLONGIFOLIA

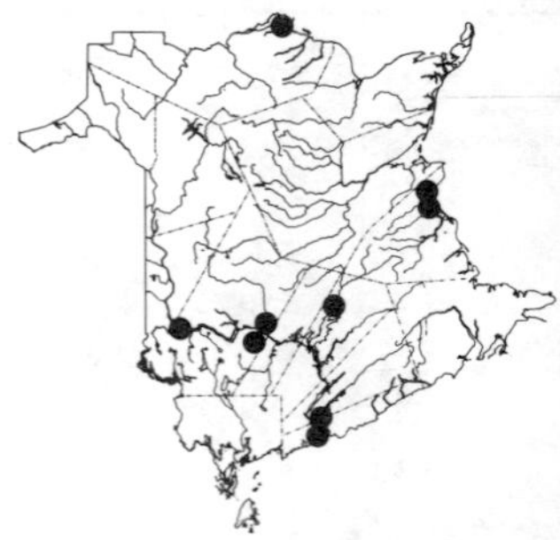

LONICERA TATARICA

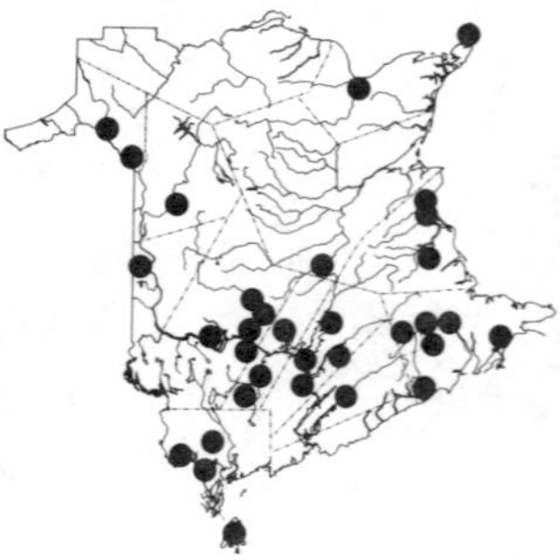

SAMBUCUS CANADENSIS

TRIOSTEUM AURANTIACUM

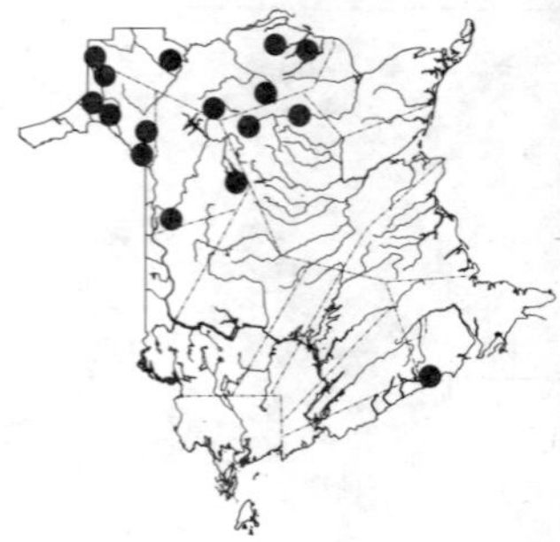

VIBURNUM EDULE

VIBURNUM LENTAGO

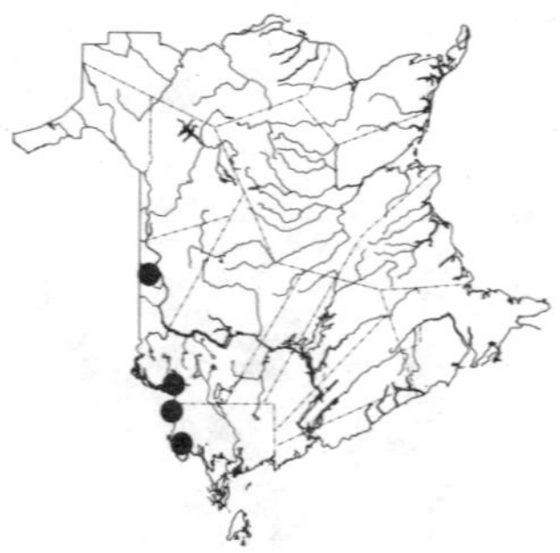

VIBURNUM RECOGNITUM

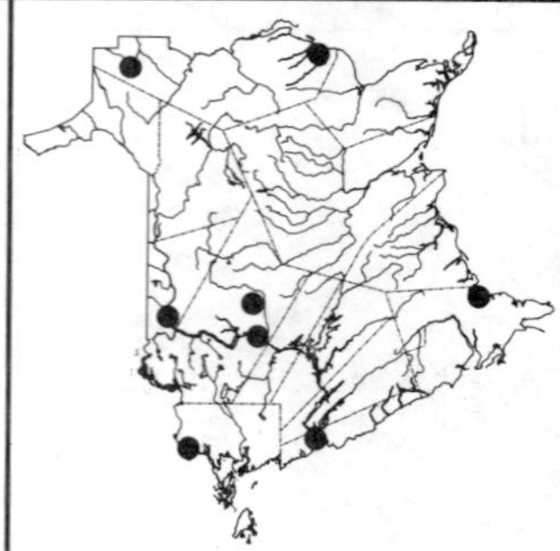

VALERIANA OFFICINALIS

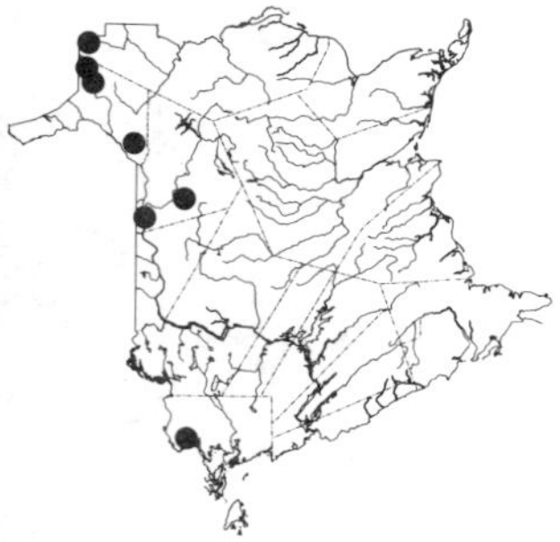

ALERIANA SITCHENSIS SSP. ULIGINOSA

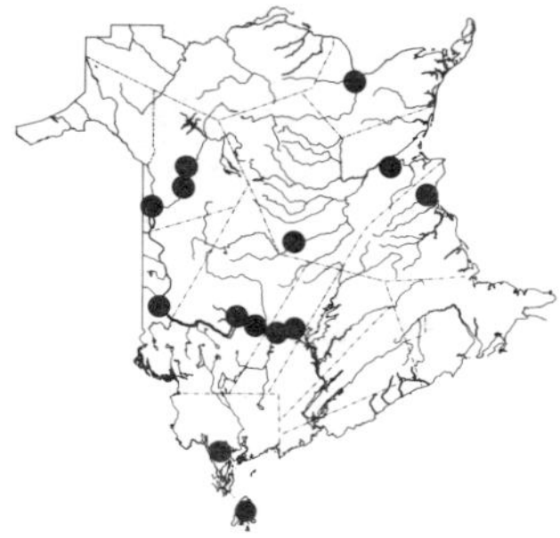

ECHINOCYSTIS LOBATA

CAMPANULA APARINOIDES

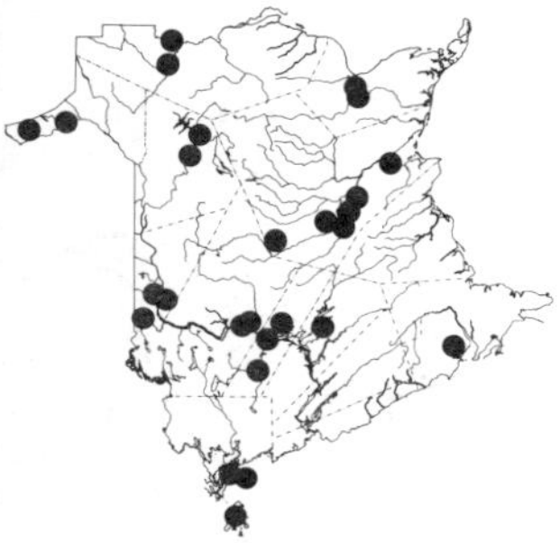

CAMPANULA ROTUNDIFOLIA

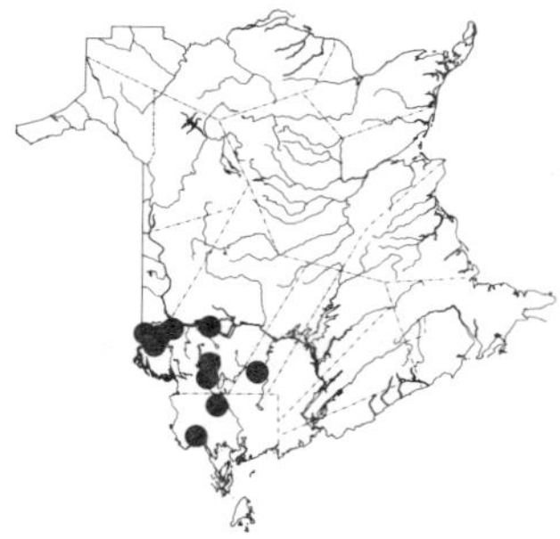

LOBELIA CARDINALIS

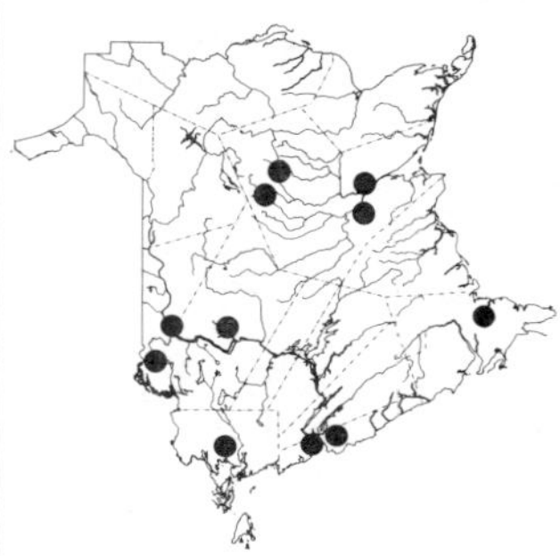

LOBELIA DORTMANNA

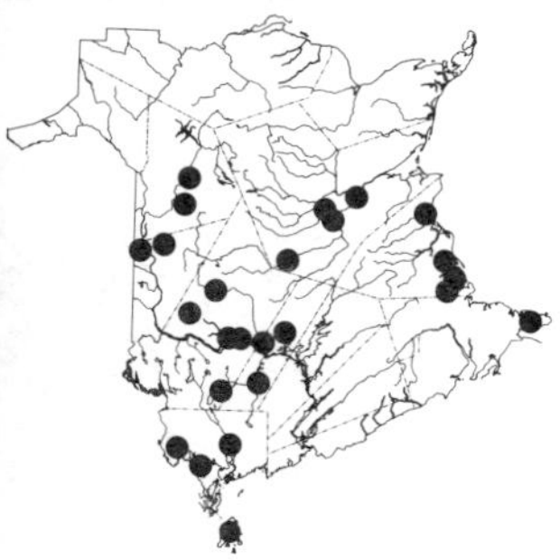

LOBELIA INFLATA

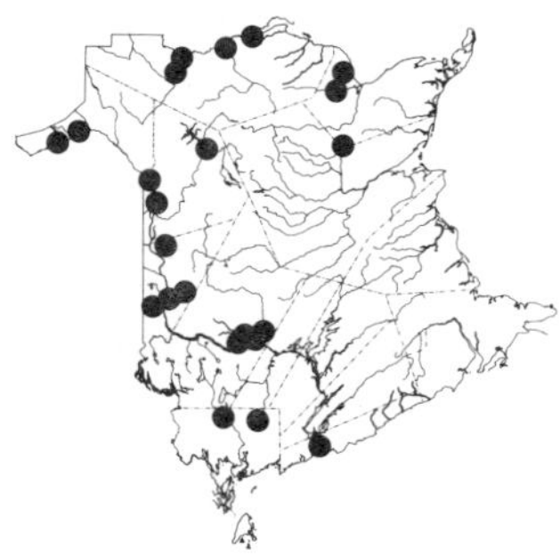

LOBELIA KALMII

ACHILLEA PTARMICA

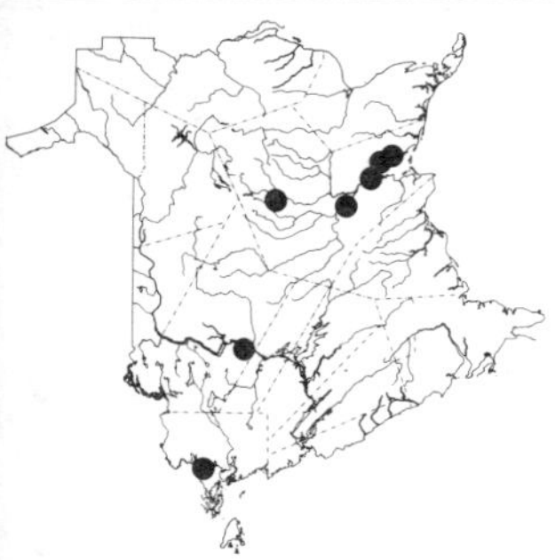

AMBROSIA TRIFIDA

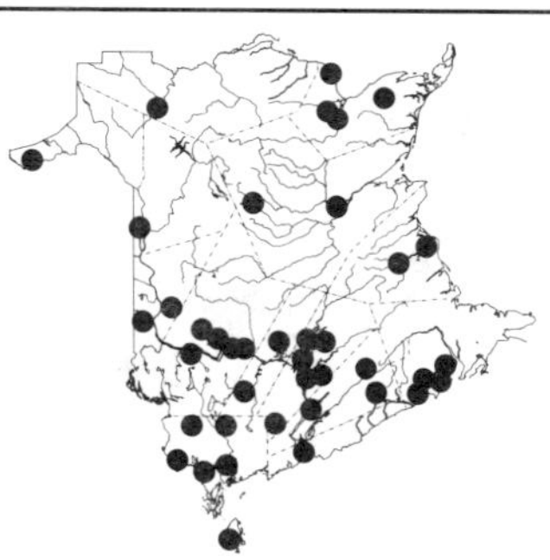

ANTENNARIA NEGLECTA

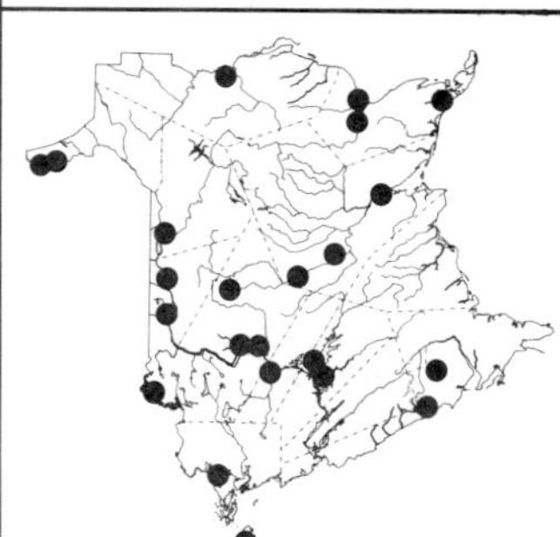

ANTENNARIA NEGLECTA VAR. ATTENUATA

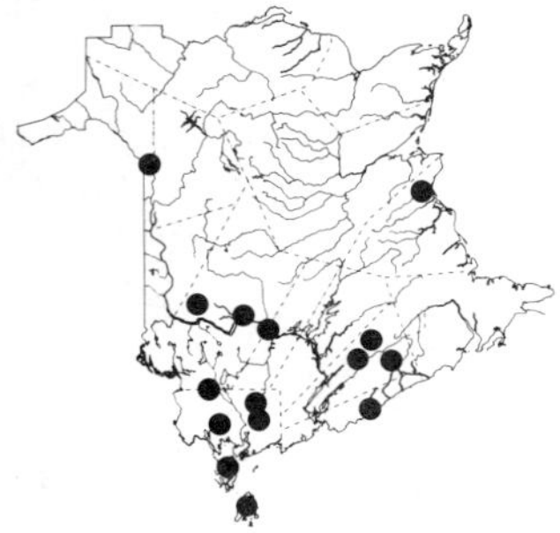

ANTENNARIA NEGLECTA VAR. RANDII

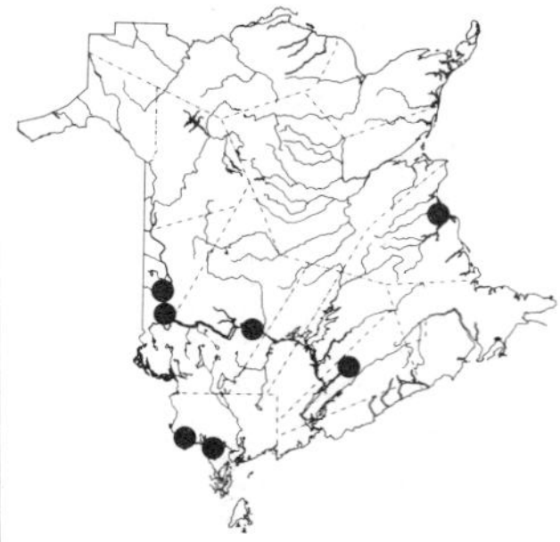

ANTHEMIS COTULA

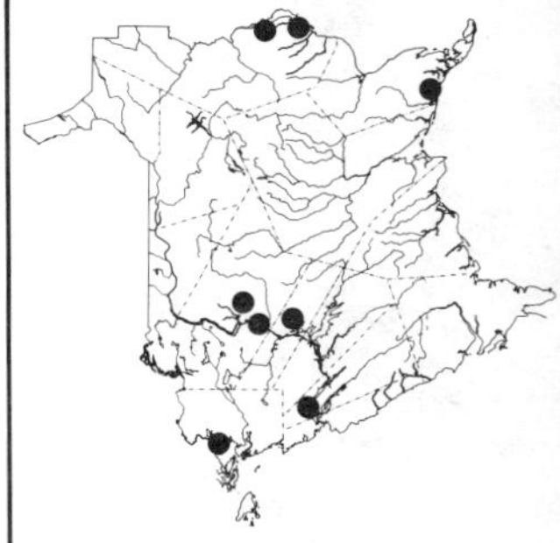

ANTHEMIS TINCTORIA

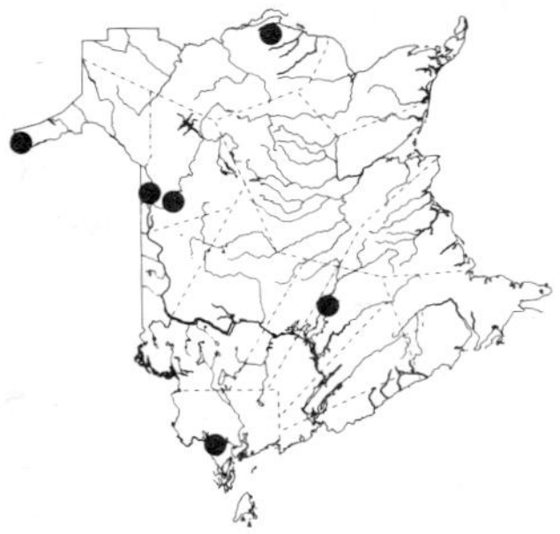

ARCTIUM LAPPA

ARCTIUM MINUS

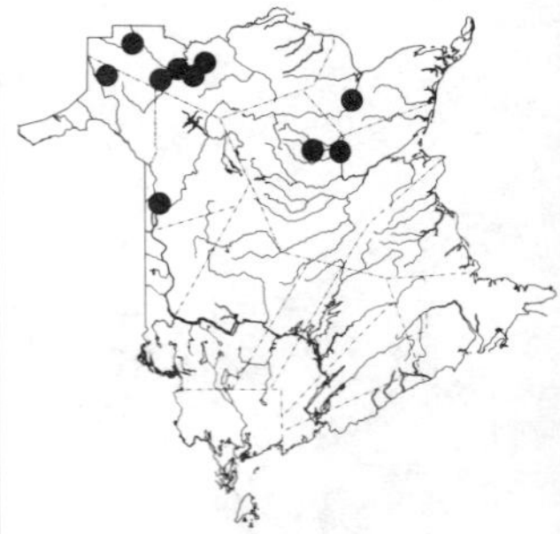

ARNICA MOLLIS

ARTEMISIA ABSINTHIUM

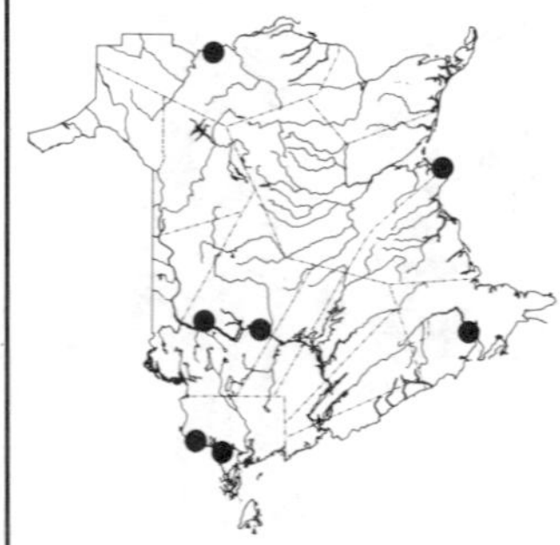

ARTEMISIA BIENNIS

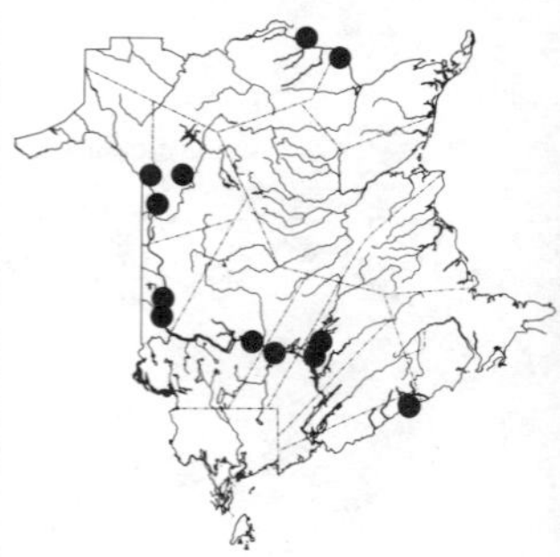

ARTEMISIA CAMPESTRIS SSP. CAUDAT

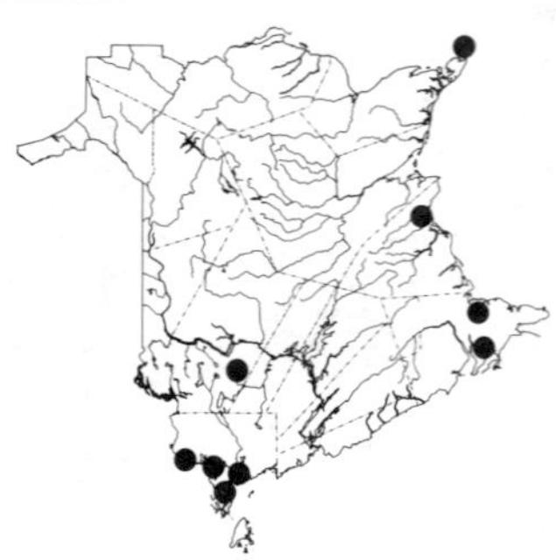

ASTER X BLAKEI

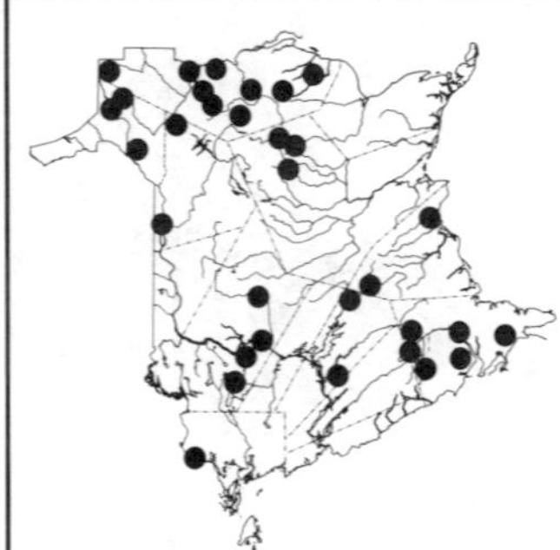

ASTER CILIOLATUS

ASTER CORDIFOLIUS

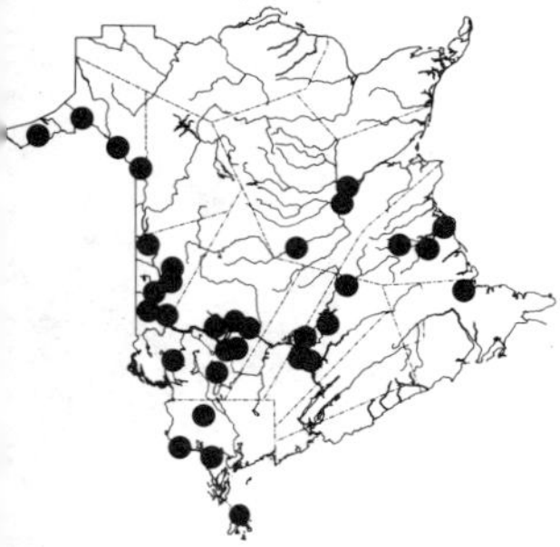

ASTER LANCEOLATUS

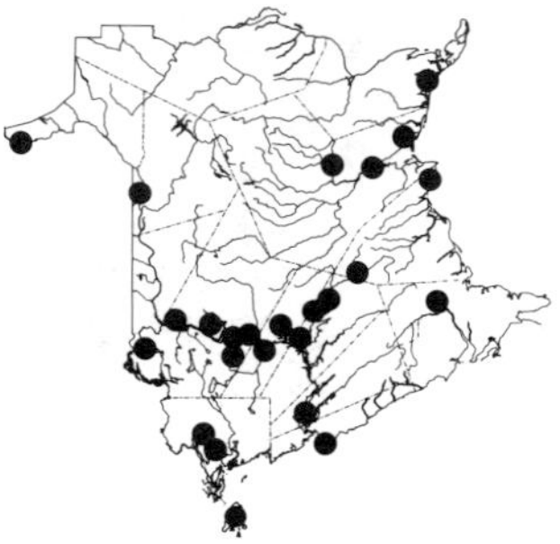

ASTER LATERIFLORUS

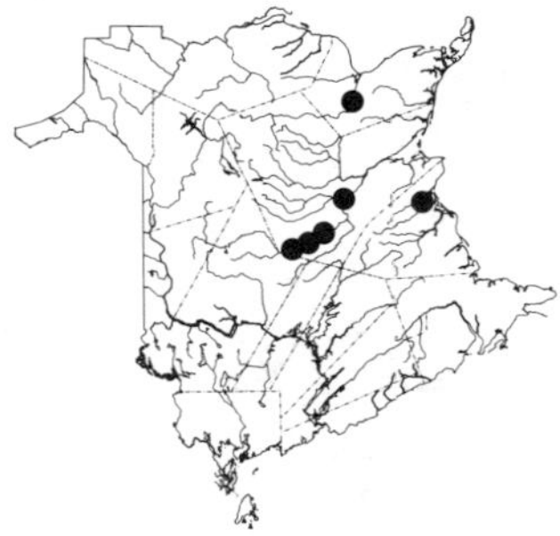

ASTER LINARIIFOLIUS

ASTER MACROPHYLLUS

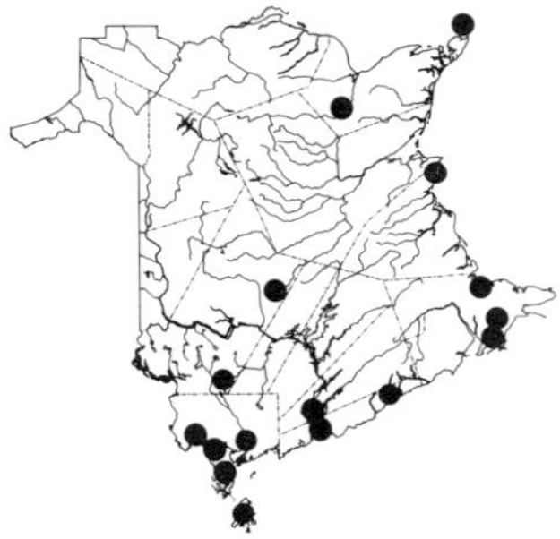

ASTER NEMORALIS

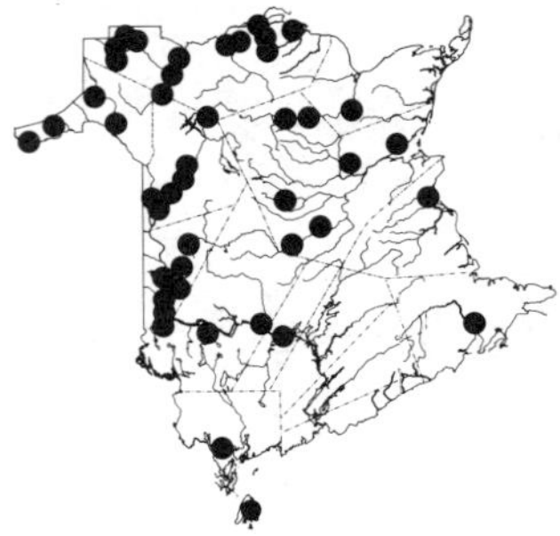

ASTER PUNICEUS

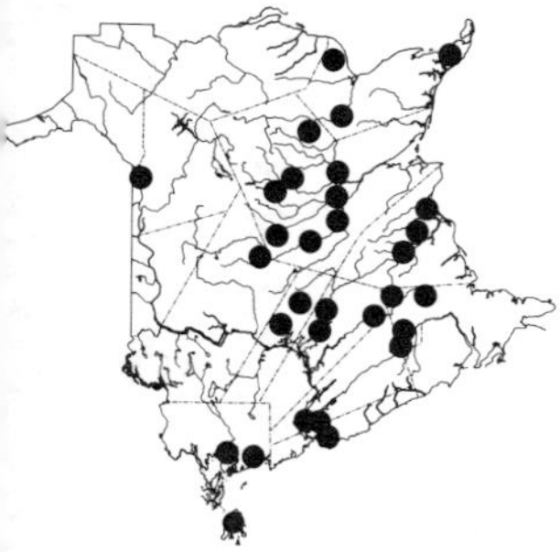

ASTER RADULA

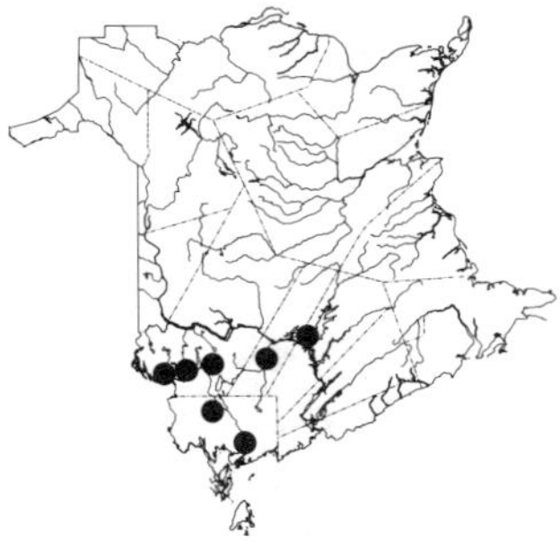

ASTER TRADESCANTII

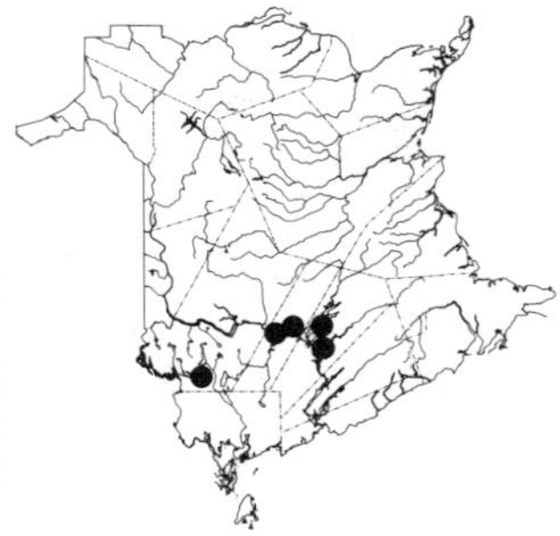

ASTER VIMINEUS VAR. SUBDUMOSUS

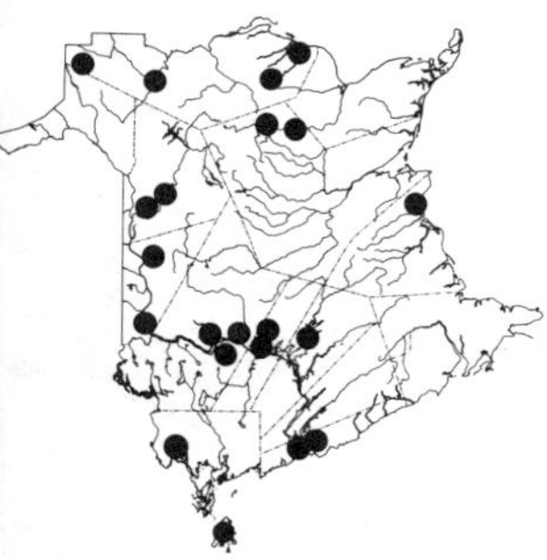

BIDENS CERNUA

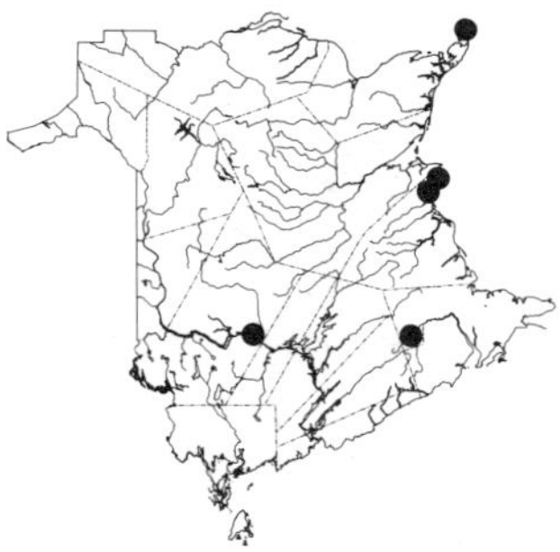

BIDENS CONNATA

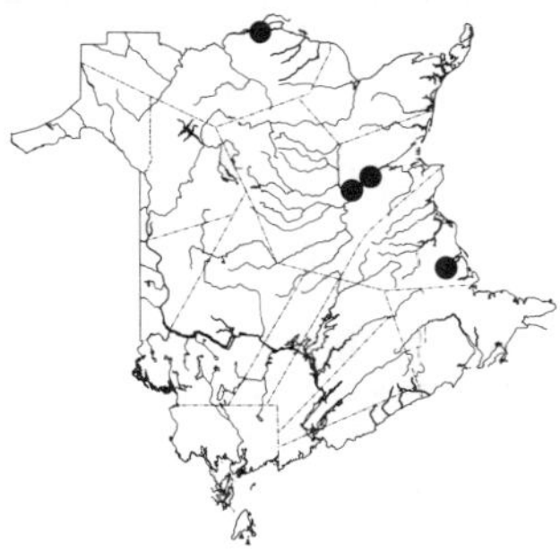

BIDENS HYPERBOREA VAR. COLPOPHILA

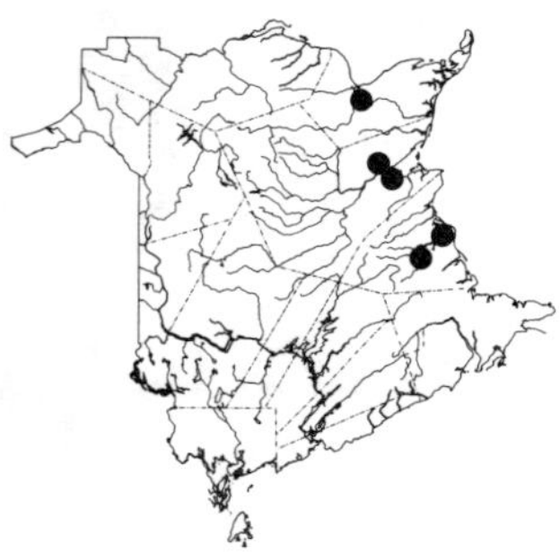

CARDUUS NUTANS

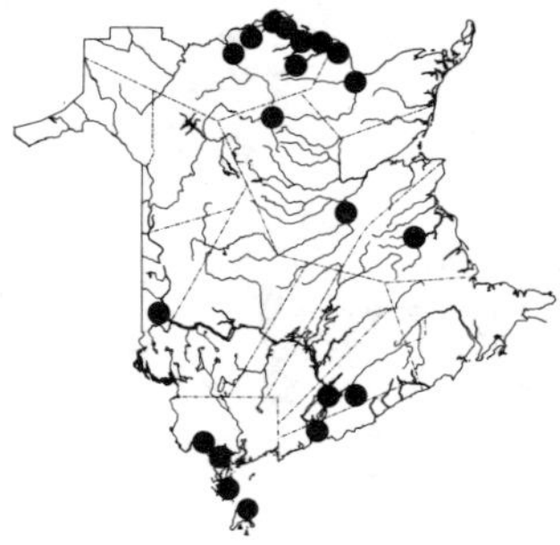

CENTAUREA NIGRA

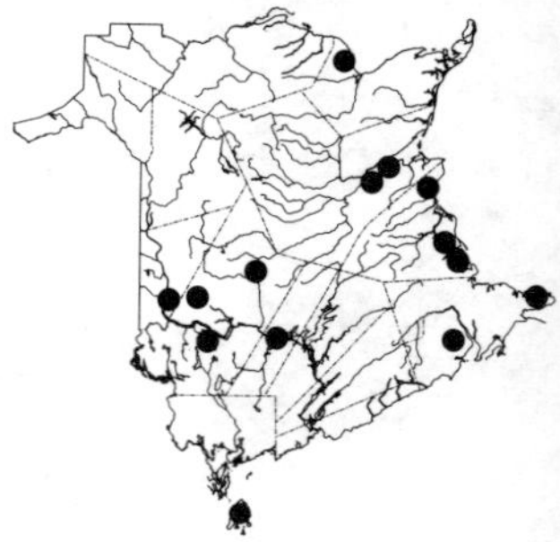

CICHORIUM INTYBUS

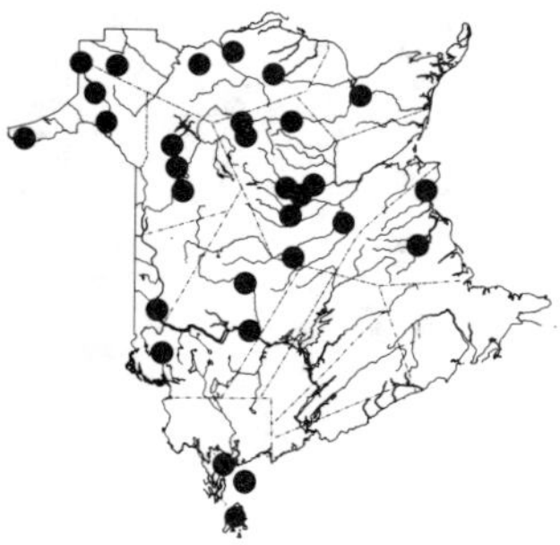

CIRSIUM MUTICUM

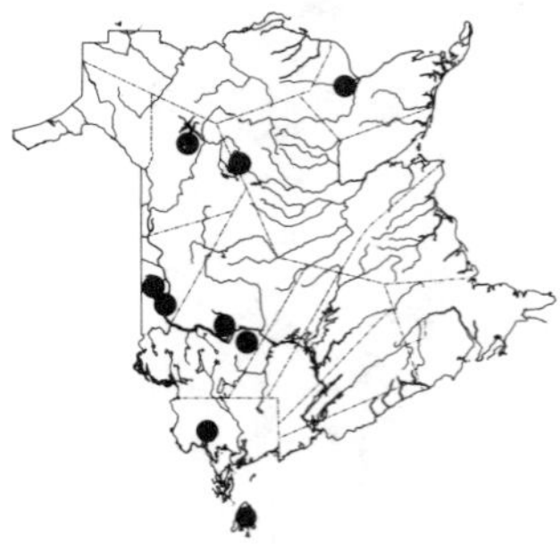

CONYZA CANADENSIS

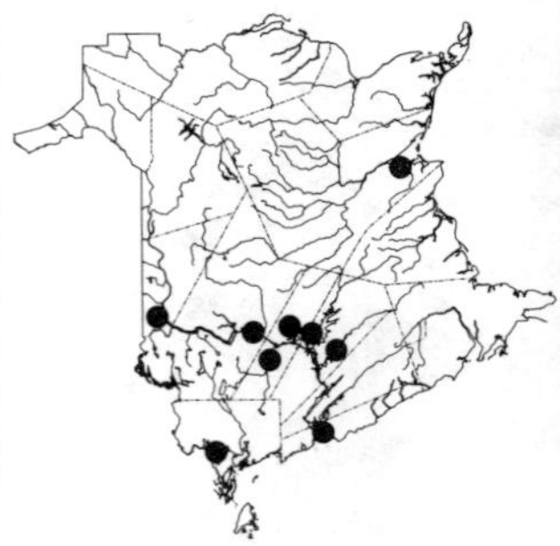

ERECHTITES HIERACIFOLIA

ERIGERON HYSSOPIFOLIUS

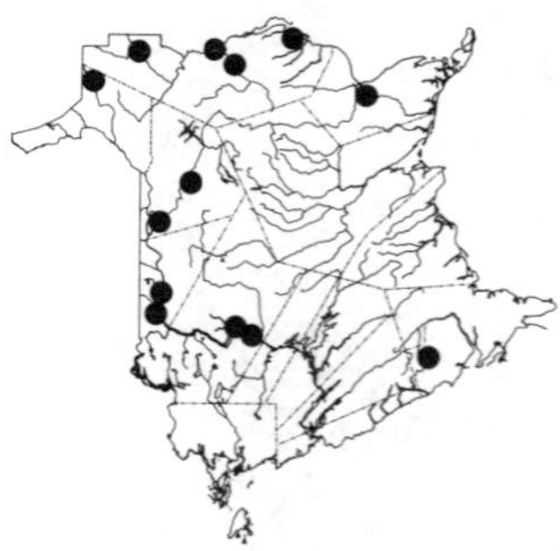

ERIGERON PHILADELPHICUS

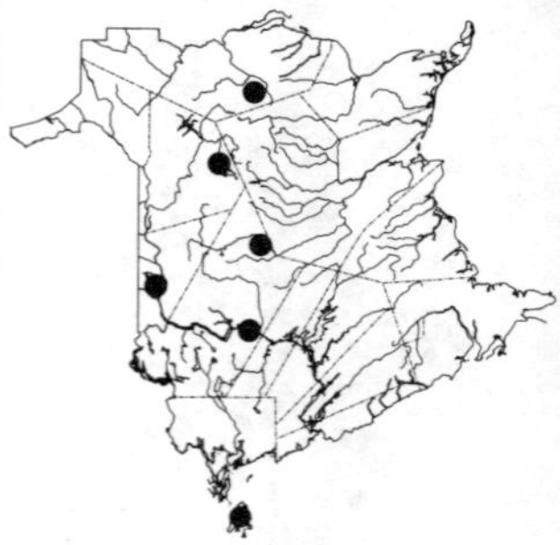

ERIGERON STRIGOSUS
VAR. SEPTENTRIONALIS

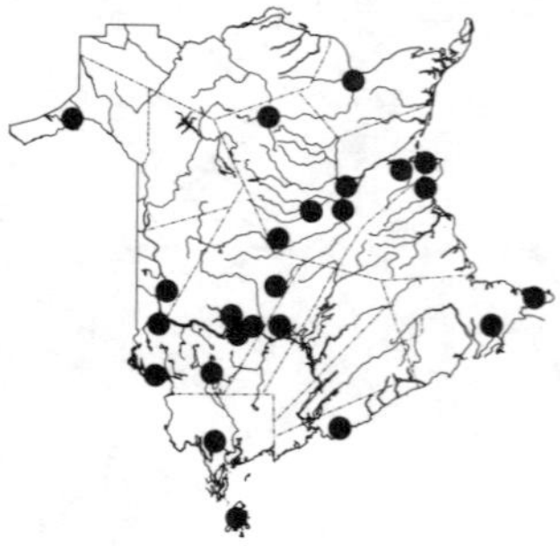

EUPATORIUM PERFOLIATUM

EUPATORIUM RUGOSUM

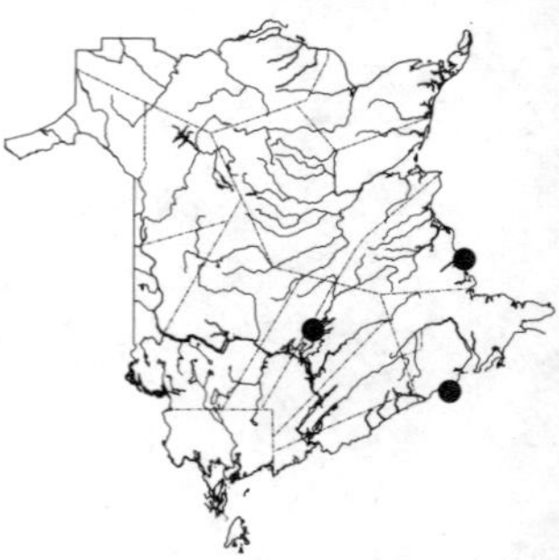

GNAPHALIUM OBTUSIFOLIUM

GNAPHALIUM SYLVATICUM

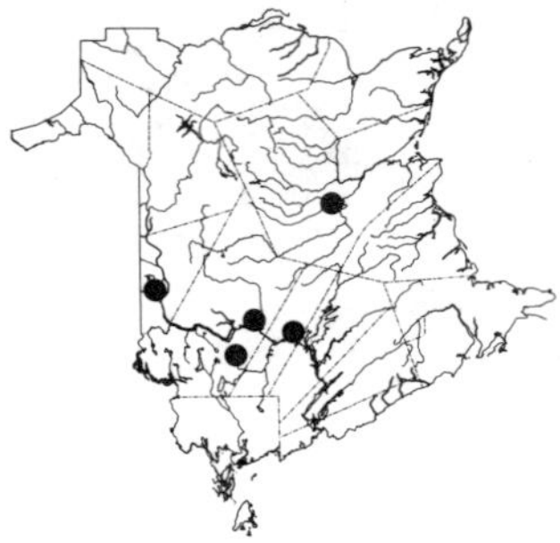

GNAPHALIUM VISCOSUM

HIERACIUM CANADENSE

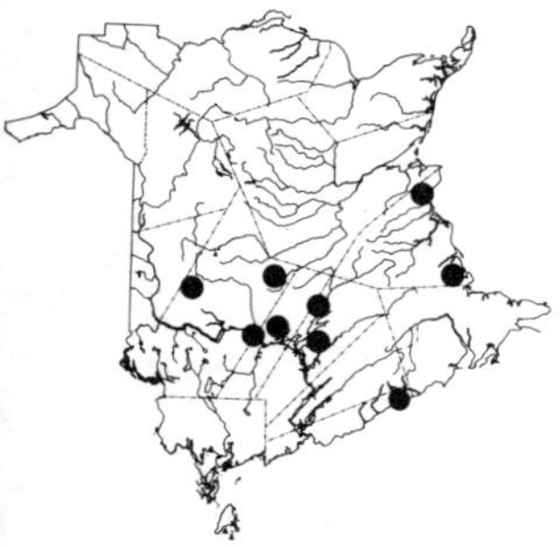

HIERACIUM X DUTILLYANUM

HIERACIUM KALMII

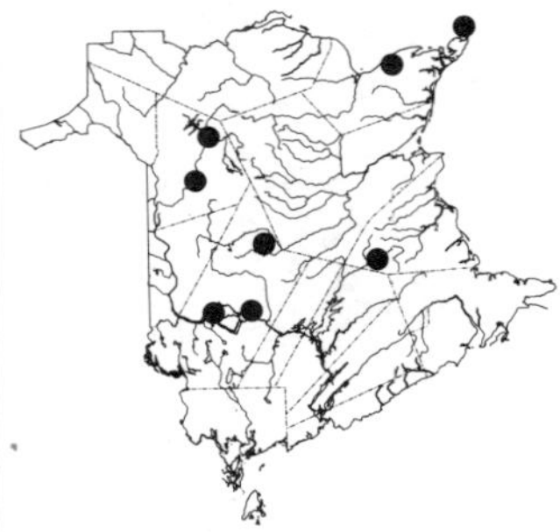

HIERACIUM PRAEALTUM
VAR. DECIPIENS

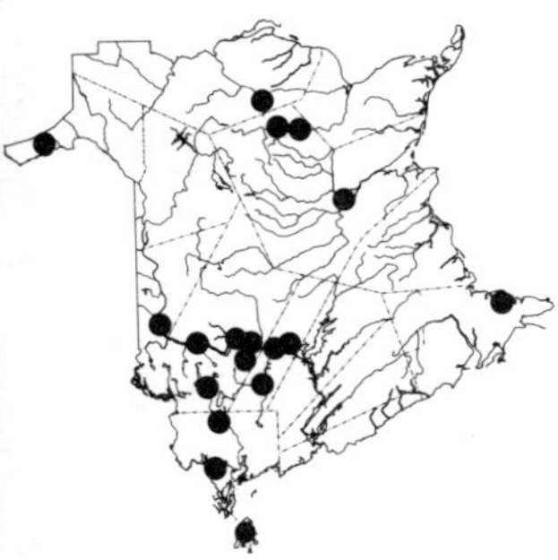

HIERACIUM SCABRUM

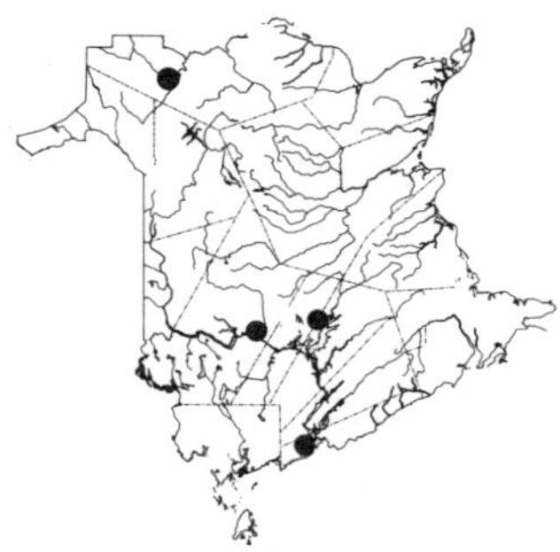

HIERACIUM UMBELLATUM
VAR. SCABRIUSCULUM

HIERACIUM VULGATUM

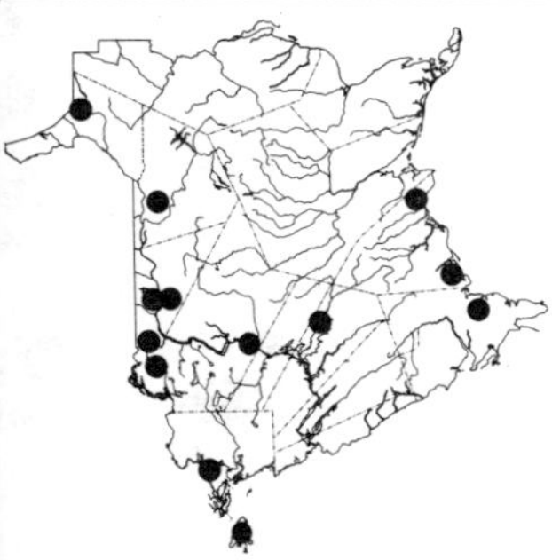

LACTUCA CANADENSIS

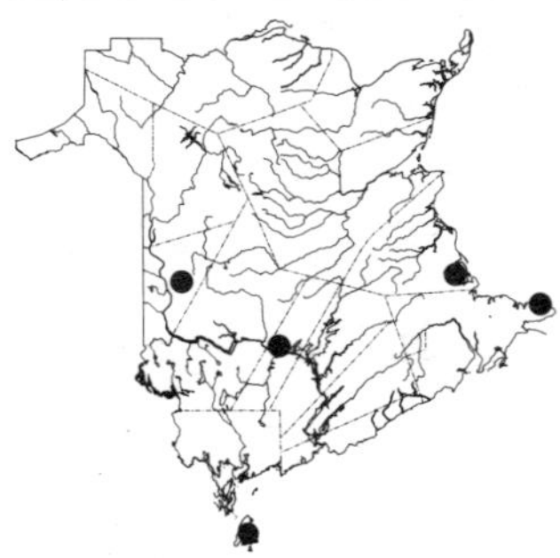

LACTUCA CANADENSIS
VAR. LONGIFOLIA

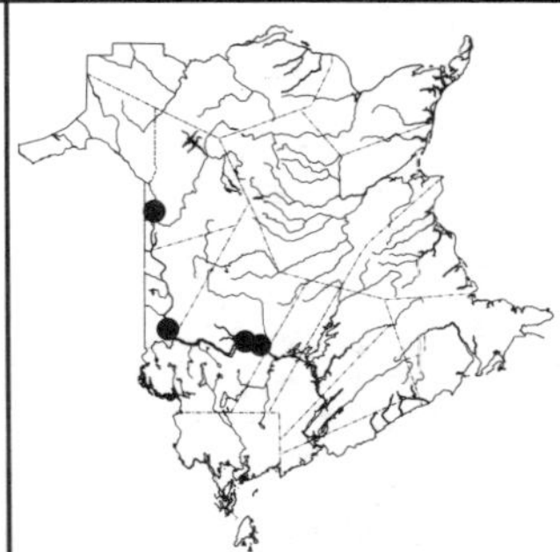

LAPSANA COMMUNIS

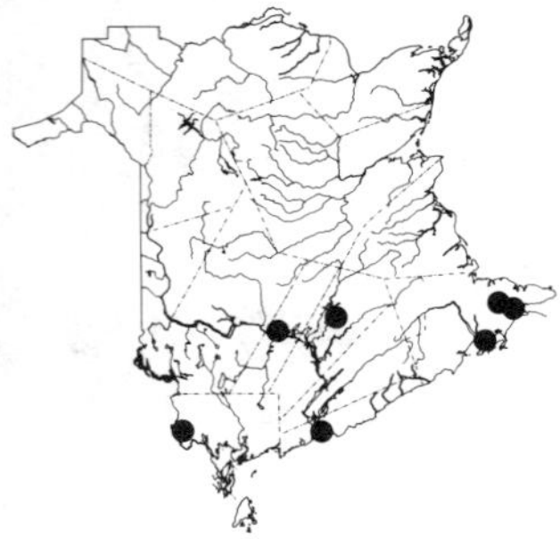

MEGALODONTA BECKII

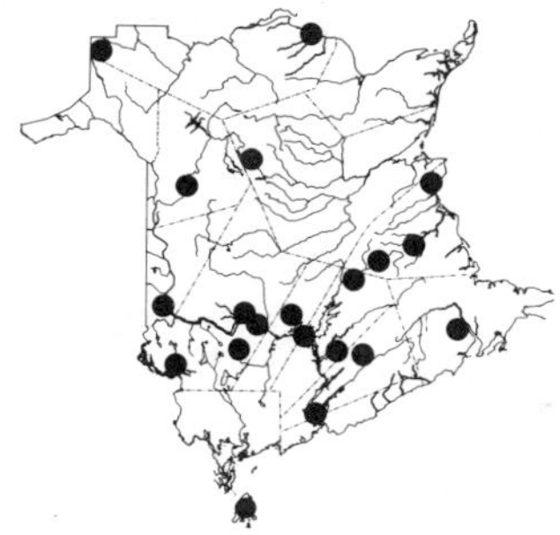

PETASITES FRIGIDUS VAR. PALMATUS

PRENANTHES ALTISSIMA

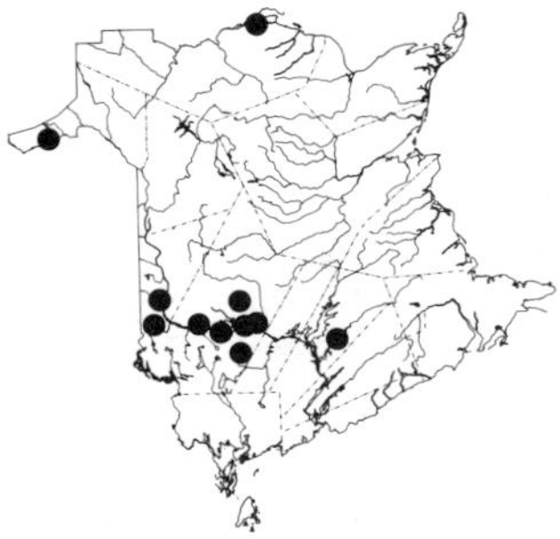

PRENANTHES RACEMOSA

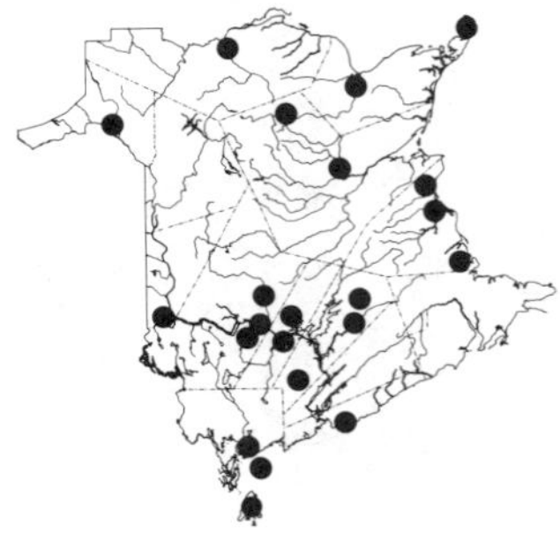

PRENANTHES TRIFOLIOLATA

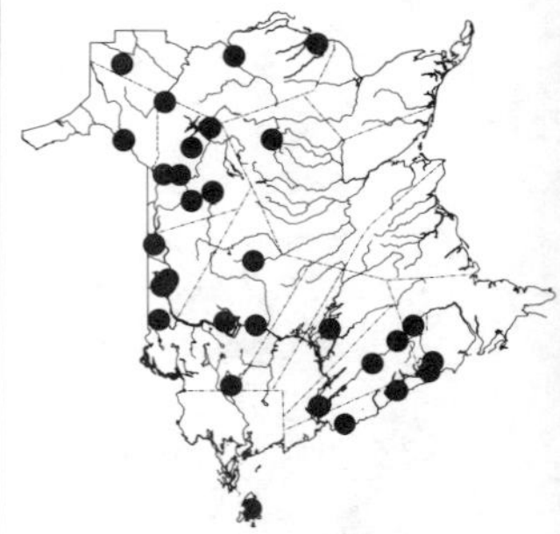

SENECIO AUREUS

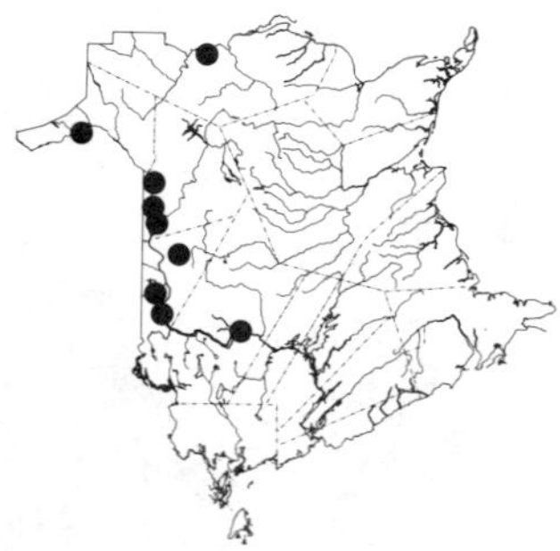

SENECIO PAUPERCULUS

SENECIO SCHWEINITZIANUS

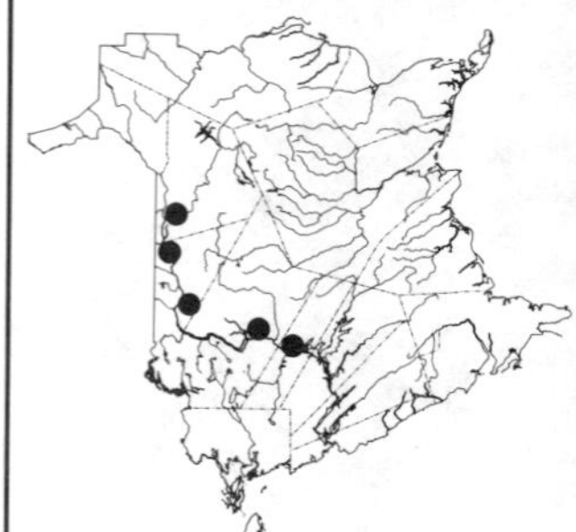

SOLIDAGO ALTISSIMA

SOLIDAGO BICOLOR

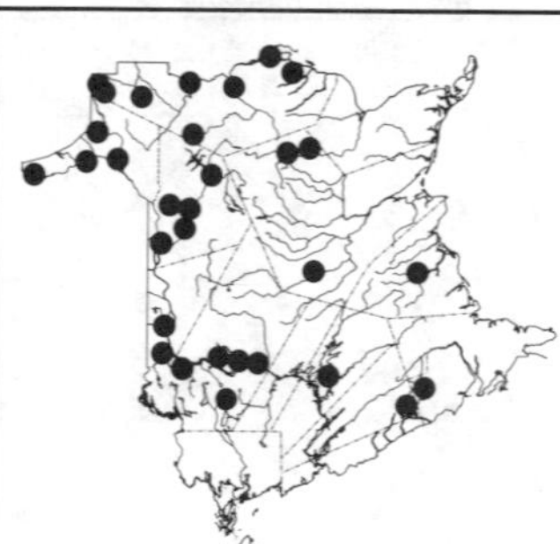

SOLIDAGO FLEXICAULIS

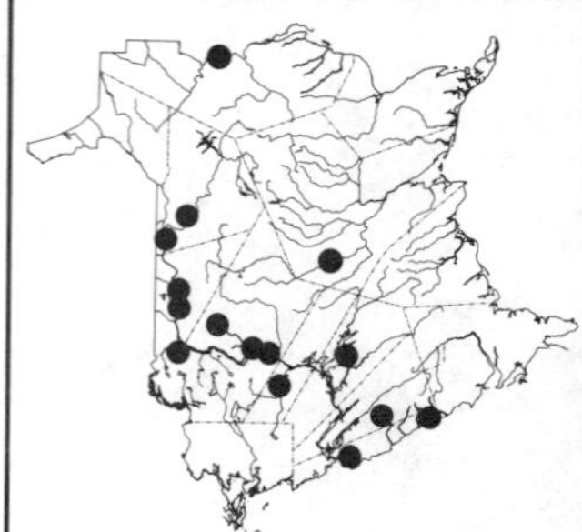

SOLIDAGO GIGANTEA

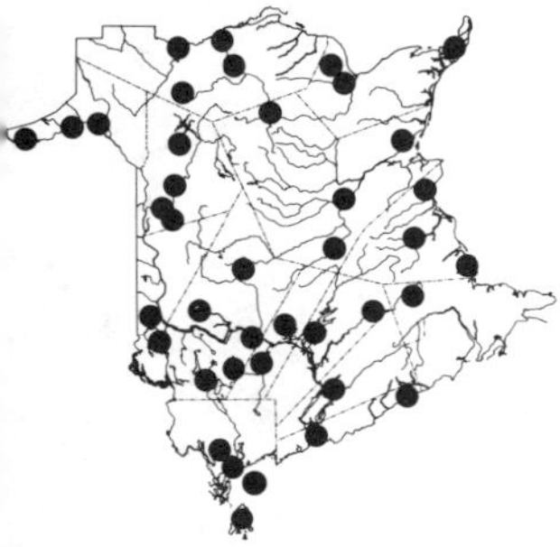

SOLIDAGO GRAMINIFOLIA

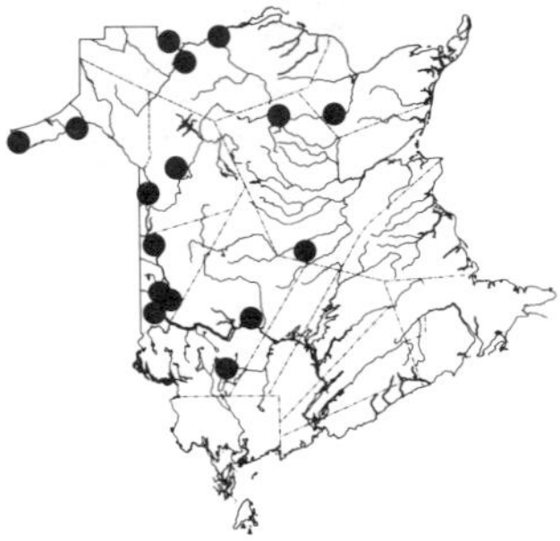

SOLIDAGO HISPIDA

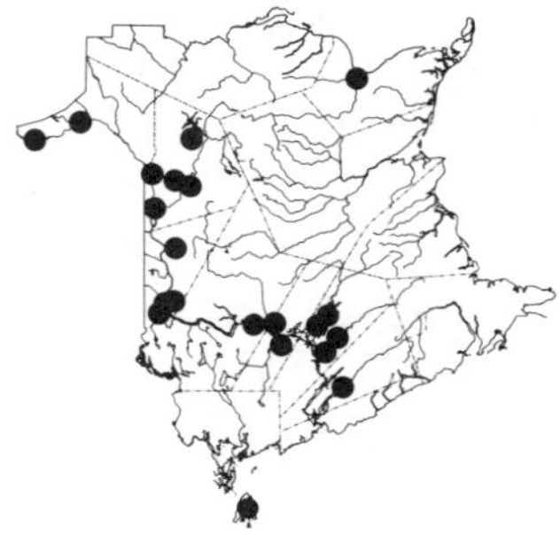

SOLIDAGO JUNCEA

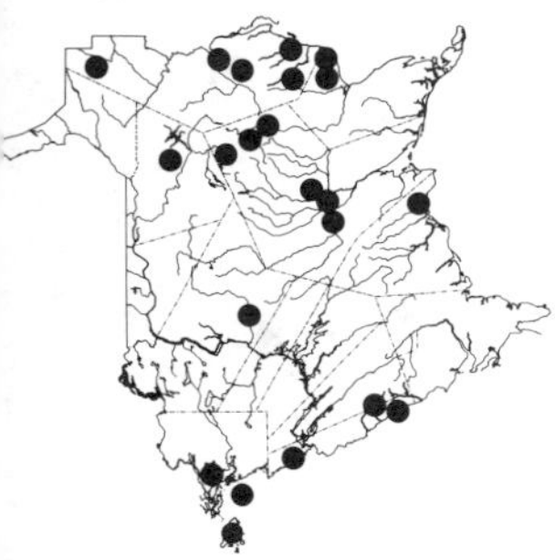

SOLIDAGO MACROPHYLLA

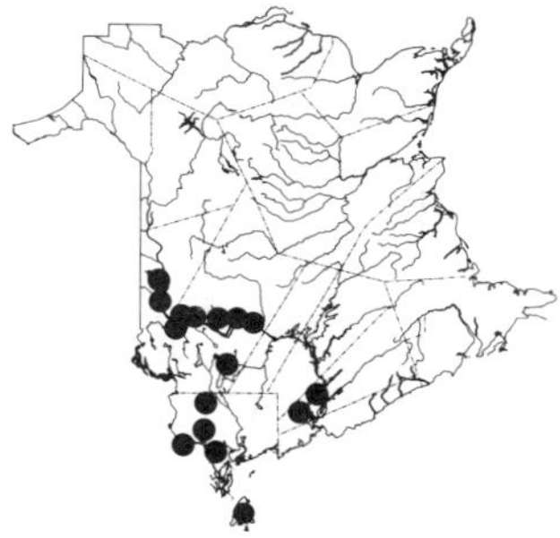

SOLIDAGO NEMORALIS

SOLIDAGO PUBERULA

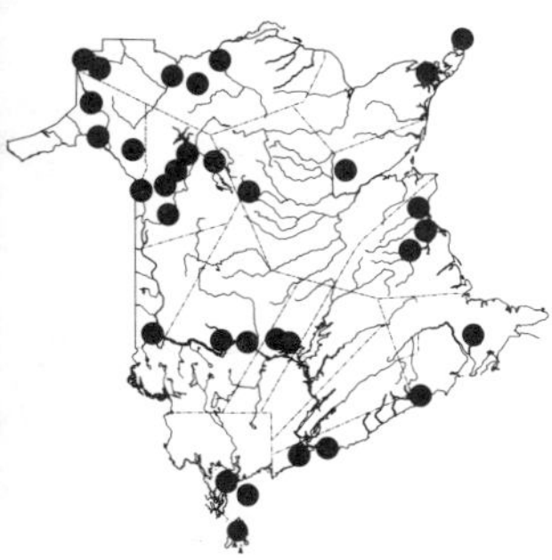

SOLIDAGO RUGOSA

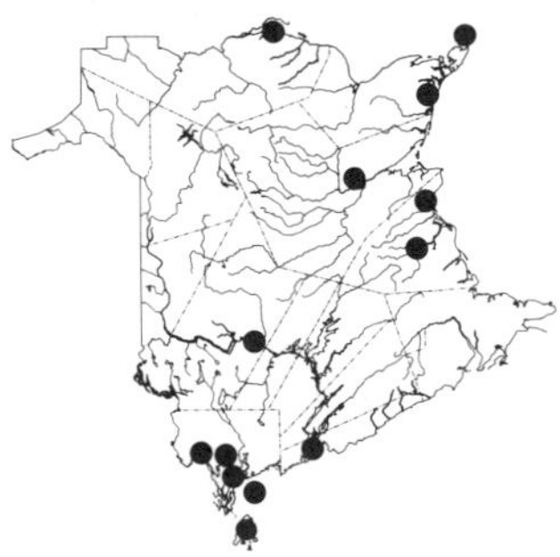

SOLIDAGO SEMPERVIRENS

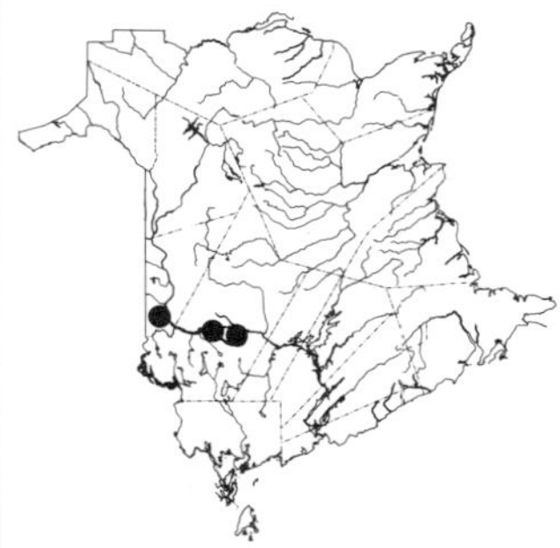

SOLIDAGO SPATHULATA VAR. GILLMANII

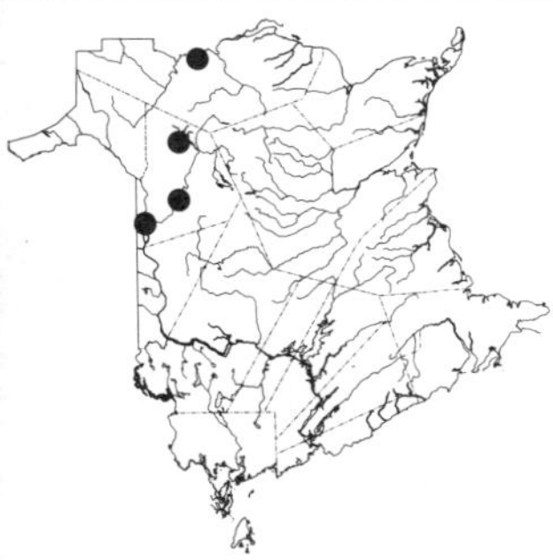

SOLIDAGO SPATHULATA SSP. RANDI VAR. RACEMOSA

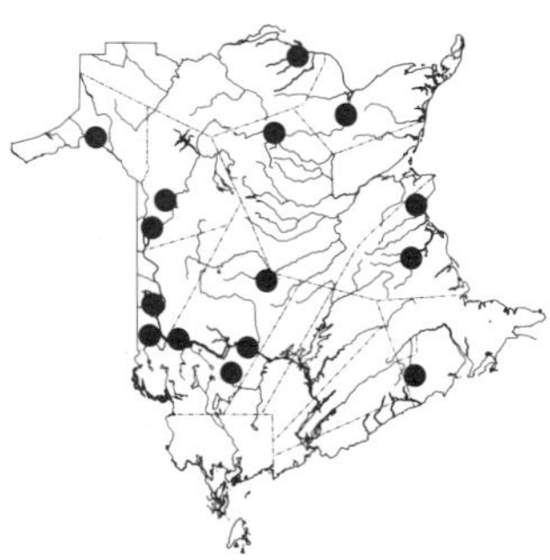

SOLIDAGO SQUARROSA

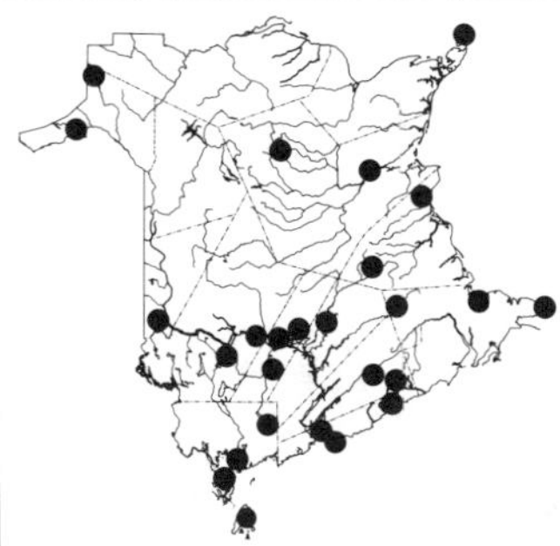

SOLIDAGO ULIGINOSA

TANACETUM HURONENSE VAR. JOHANNENSE

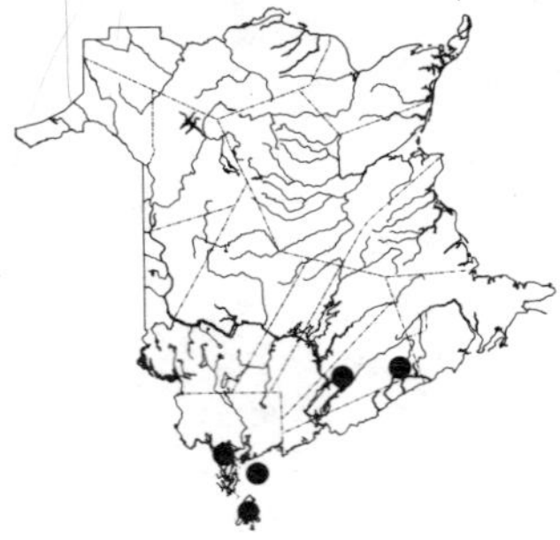

TARAXACUM LAEVIGATUM

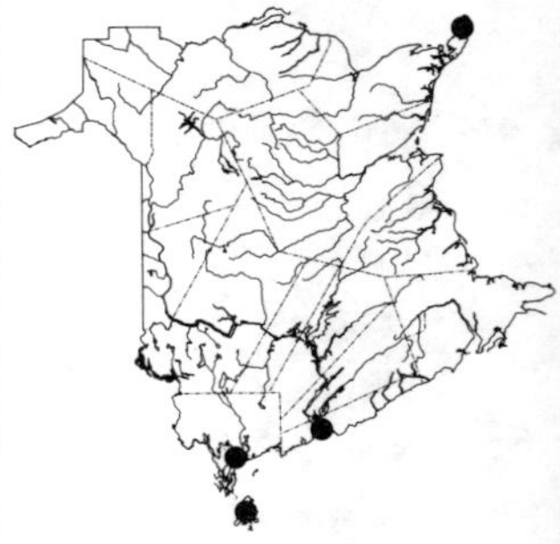

TARAXACUM LATILOBUM

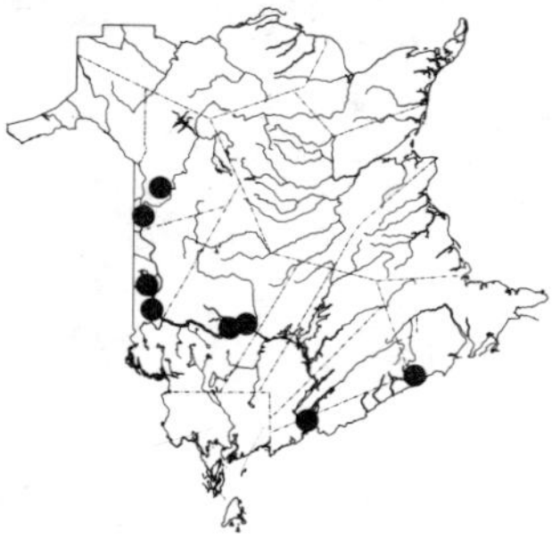

TRAGOPOGON PRATENSIS

TUSSILAGO FARFARA

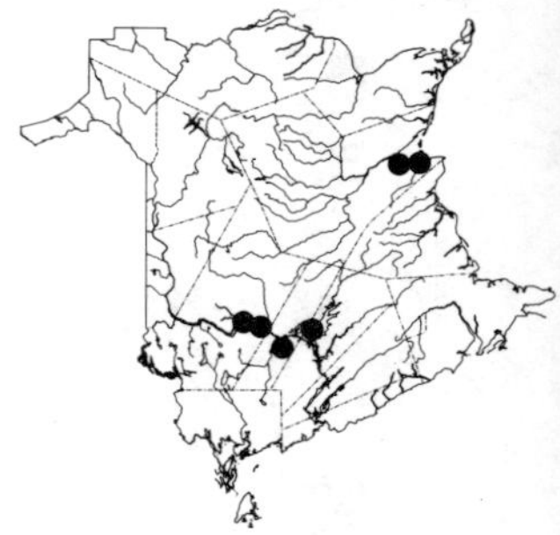

XANTHIUM STRUMARIUM